Topics in Applied Physics

Volume 144

Topics in Applied Physics is a well-established series of review books, each of which presents a comprehensive survey of a selected topic within the domain of applied physics. Since 1973 it has served a broad readership across academia and industry, providing both newcomers and seasoned scholars easy but comprehensive access to the state of the art of a number of diverse research topics.

Edited and written by leading international scientists, each volume contains high-quality review contributions, extending from an introduction to the subject right up to the frontiers of contemporary research.

Topics in Applied Physics strives to provide its readership with a diverse and interdisciplinary collection of some of the most current topics across the full spectrum of applied physics research, including but not limited to:

- Quantum computation and information
- Photonics, optoelectronics and device physics
- Nanoscale science and technology
- Ultrafast physics
- Microscopy and advanced imaging
- Biomaterials and biophysics
- Liquids and soft matter
- Materials for energy
- Geophysics
- Computational physics and numerical methods
- Interdisciplinary physics and engineering

We welcome any suggestions for topics coming from the community of applied physicists, no matter what the field, and encourage prospective book editors to approach us with ideas. Potential authors who wish to submit a book proposal should contact Zach Evenson, Publishing Editor:

zachary.evenson@springer.com

Topics in Applied Physics is included in Web of Science (2020 Impact Factor: 0.643), and is indexed by Scopus.

More information about this series at https://link.springer.com/bookseries/560

Hilmi Ünlü · Norman J. M. Horing
Editors

Progress in Nanoscale and Low-Dimensional Materials and Devices

Properties, Synthesis, Characterization, Modelling and Applications

Springer

Editors
Hilmi Ünlü
Department of Physics Engineering
Faculty of Science and Letters
Istanbul Technical University
Maslak, Istanbul, Turkey

Norman J. M. Horing
Department of Physics and Engineering Physics
Stevens Institute of Technology
Hoboken, NJ, USA

ISSN 0303-4216 ISSN 1437-0859 (electronic)
Topics in Applied Physics
ISBN 978-3-030-93459-0 ISBN 978-3-030-93460-6 (eBook)
https://doi.org/10.1007/978-3-030-93460-6

This Springer imprint is published by the registered company Springer Nature Switzerland AG
The registered company address is: Gewerbestrasse 11, 6330 Cham, Switzerland

My work in this book is dedicated to the memory of my mother Nergiz Ünlü and my father Yusuf Ünlü. I gratefully acknowledge the encouragement, assistance and forbearance provided by my wife, Yeliz Ünlü and my daughters Melek Cemile Ünlü, Merve Nuray Ünlü and İlknur Miray Ünlü, with love and devotion.

Hilmi Ünlü

My work in this book is dedicated to the memory of my mother, Rose Morgenstern Horing and the Joseph Morgenstern Family. I gratefully acknowledge the encouragement, assistance and forbearance provided by my wife, Jerilyn Gail Horing, with love and devotion.

Norman J. Morgenstern Horing

Preface

Semiconductors and nanoscale systems pervade modern life, in our cellphones, computers, radios, televisions, cars, boats, aircraft, even space exploration and medical equipment, wherever electronic devices have a function or play a role. Since the invention of the first transistor in 1949, there has been a continuous avalanche of growth and development in the science and technology of semiconductor materials and device applications. The thrust of the advances has involved the steady lowering of the dimensions in which confined charge carriers can move, ultimately reaching the nano-world. The requisite support basis of nanotechnology, laden with promising developments tantamount to a new industrial revolution, has been a qualitatively reliable understanding of the underlying physics of semiconducting materials accompanied by quantitatively precise predictions of device performance. This has led to new concepts and techniques of semiconductor fabrication that have facilitated the emergence of a generation of advanced devices with more complex functionality and much higher densities for electronic, computational and optical applications. Advances in the growth of semiconductor thin films of differing structural, electronic and optical properties, and the diminution of layer thickness approaching atomic dimensions, have provided new opportunities for fundamental scientific studies and technological applications of low dimensional semiconductor structures in new devices. Moreover, contemporary fabrication technologies have made it possible to reduce device dimensions to the point where size effects must be properly described quantum mechanically in order to reliably predict the potential and performance of low dimensional semiconductor systems for electronic and optical applications.

Modern crystal growth techniques including molecular beam epitaxy (MBE) and Metal Organic Chemical Vapor Deposition (MOCVD) have made it possible to control the alloy composition and doping in ternary and quaternary alloys based on group IV–IV, III–V and II–VI semiconductor compounds over atomic distances, as low as tens of angstroms. These artificially made microscale and nanoscale semiconductor heterostructures are easily grown as lattice mismatched substrates of epitaxial layers so thin that quantum mechanical effects are dominant. The resulting contemporary low dimensional semiconductor devices (e.g., Heterostructure Bipolar Transistors (HBTs) and Modulation Doped Field Effect transistors (MODFETs), nanowires, quantum dots, etc.) are known to operate much faster than conventional silicon devices (e.g., silicon Bipolar Junction Transistors (BJTs) and Metal Oxide Semiconductor Field Effect Transistors (MOSFETs)); this is crucial for the electronic and optical communication and computer industries.

Contemporary electronic communication and information technologies involving a high volume of information and the need for high speed capabilities and smaller size devices require development and production techniques that constantly evolve. To address this challenge and provide the needed improvements, the world's leading scientists and engineers have been carrying out intensive work in nanoscience and nanotechnology to create the future information and electronic communication technologies.

During the last two decades, large numbers of theoretical and experimental studies have been carried out on metals, semiconductors, ceramics, polymers and composites of low dimensions, which are intended for use in applications in electronic devices and telecommunications, biotechnology, energy conversion and storage, health sciences and medicine, including also construction and environmental issues. Nanotechnology now also impacts healthcare, including the drug industry and agriculture as well as electronics and communications, information processing and data storage, employing the multi-functionality of the new materials. Nanomaterials are very strong and lightweight, and they constitute very small components of computers as well as new sensors: They provide very rapid response nanoscale electronic and optoelectronic technologies of very small size and are also predicted to be important in high power defense systems.

The broad prevalence of semiconductors in modern life throughout the world suggests that the transition to nanodevices will have strong implications for marketing and national and international economies. With this in view, the United States of America, Russia, European Union, Japan, Korea, Taiwan and China (and also smaller nations) promote research in various fields of nanoscience and nanotechnology and have made huge investments. Rapidly increasing numbers of nanotechnology-based companies are being established at various scales worldwide, and state-supported university-industry cooperation is important. Especially, large industrial companies are making significant investments in nanotechnology for research and development.

Nanoscience and nanotechnology involve comprehensive interdisciplinary research between the sciences (physics, chemistry, biology, materials science) and engineering (electronics, computer, mechanical, chemical, construction, textiles, environment, etc.) fields. Nanoscience and nanotechnology, born as a concept put forward in the early 1960s, grew with intensive research and development gaining momentum in the last 20 years. The term "nano" refers to the size of a nanometer, one billionth of a meter (1nm = 1/1,000,000,000 m). At the nanoscale, new material properties emerge, both physical and chemical features that can influence the function and/or production of new materials that can be controlled.

This book describes some recent advances in the properties, synthesis, characterization, modeling and applications of low dimensional and nanoscale materials and devices in over 20 years of research. It begins with structural, electronic, optical and magnetic properties, methodology of synthesis, and characterization of quantum dots and nanowires. And special attention is focused here on "relativistic" Dirac materials, whose electrical conduction and sensing properties far exceed those of silicon-based materials, making them strong competitors to be the materials of choice for the next generation of electronic devices and computers. Recognition of the importance of such materials led to the award of the 2010 Nobel Prize to Geim and Novoselove for their groundbreaking work on Graphene. More complex issues associated with the environment and with energy production and storage follow in this book. Furthermore, important achievements in materials pertinent to the fields of biology and medicine are also reviewed, exhibiting an outstanding confluence of basic physical science with vital human endeavor.

We hope this book will be of interest to researchers in the semiconductor/device field in general, also graduate students and advanced undergraduates in physics, chemistry, biology and medicine. We have also endeavored to address subjects of interest to those in electrical, chemical, biological and mechanical engineering.

Hilmi Ünlü
İstanbul Technical University
Maslak, Istanbul, Turkey

Norman J. M. Horing
Stevens Institute of Technology
Hoboken, USA

Contents

Contributors

Meziani Abdelhakim Physics Department, Frères Mentouri University Constantine, Constantine, Algeria

Pinar Akkus Sut Genetics and Bioengineering Department, Yeditepe University, Faculty of Engineering and Architecture, Kayisdagi, Istanbul, Turkey

E. Akşahin Department of Physics, Yeditepe University, Atasehir, Istanbul, Turkey

Özden Akıncı Independent Researcher, Amstelveen, Netherlands

B. Alhalaili Department of Electrical and Computer Engineering, University of California at Davis, Davis, CA, USA;
Nanotechnology and Advanced Materials Program, Kuwait Institute for Scientific Research, Safat, Kuwait

Abdul-Rahman Allouche Institute of Light and Matter (ILM), UMR5306 University of Lyon1, CNRS, University of Lyon, Villeurbanne Cedex, France

Melda Patan Alper Department of Physics, Faculty of Arts and Sciences, Yeditepe University, Istanbul, Turkey

Hakan Aydın Faculty of Science and Letters, Department of Physics Engineering, İstanbul Technical University, Maslak, İstanbul, Turkey

R. A. W. Ayyubi Department of Physics, Quaid-i-Azam University, Islamabad, Pakistan

Telia Azzedine Electronic Department, Frères Mentouri University Constantine, Constantine, Algeria

Sina Bahrami Department of Physics, Stevens Institute of Technology, Hoboken, NJ, USA

Mesut Balaban Faculty of Science and Letters, Department of Physics, Yıldız Technical University, Esenler, İstanbul, Turkey

Antonios Balassis Department of Physics and Engineering Physics, Fordham University, Bronx, NY, USA

M. Batouche Laboratory of Quantum Physics of Matter and Mathematical Modeling, Mustapha Stambouli University, Mascara, Algeria

K. H. Bennemann Institute for Theoretical Physics, Freie Universität Berlin, Berlin, Germany

Damla Bulut ASELSAN Inc., Teknopark Istanbul, Istanbul, Turkey

Z. P. Çakar Department of Molecular Biology & Genetics, Faculty of Science and Letters, Istanbul Technical University, Maslak, Istanbul, Turkey;
Dr. Orhan Öcalgiray Molecular Biology, Biotechnology and Genetics Research Center (ITU-MOBGAM), Istanbul Technical University, Istanbul, Maslak, Turkey

Siwei Chen Mechanical Engineering Department, Stevens Institute of Technology, Hoboken, NJ, USA

Z. Faraei Department of Physics, Sharif University of Technology, Tehran, Iran; Department of Physics, Institute for Advanced Studies in Basic Sciences (IASBS), Zanjan, Iran

Vassilios Fessatidis Department of Physics and Engineering Physics, Fordham University, Bronx, NY, USA

Shichen Fu Mechanical Engineering Department, Stevens Institute of Technology, Hoboken, NJ, USA

M. L. Glasser Department of Physics, Clarkson University, Potsdam, NY, USA

Godfrey Gumbs Department of Physics and Astronomy, Hunter College of the City University of New York, New York, NY, USA

Hikmet Hakan Gürel Technology Faculty, Information Systems Engineering Department, Kocaeli University, Kocaeli, Turkey

B. Hacısalihoğlu Independent Researcher, Erzurum, Turkey

N. J. M. Horing Department of Physics, Stevens Institute of Technology, Hoboken, NJ, USA

Norman J. M. Horing Department of Physics, Stevens Institute of Technology, Hoboken, New Jersey, USA

S. L. Horton Department of Physics, Stevens Institute of Technology, Hoboken, NJ, USA

Rabia Ince Diamond Light Source, Harwell Science and Innovation Campus, Didcot OX11 ODE, United Kingdom

M. Saif Islam Department of Electrical and Computer Engineering, University of California at Davis, Davis, CA, USA

S. A. Jafari Department of Physics, Sharif University of Technology, Tehran, Iran

Kyungnam Kang Mechanical Engineering Department, Stevens Institute of Technology, Hoboken, NJ, USA

H. Karaağaç Department of Physics Engineering, Istanbul Technical University, Maslak, Istanbul, Turkey

Manolya Kukut Hatipoglu Materials Science and Nanotechnology Engineering Department, Yeditepe University, Faculty of Engineering and Architecture, Kayisdagi, Istanbul, Turkey;
The Food and Drug Administration, MO, USA

Harry Lenzing Department of Physics, Stevens Institute of Technology, Hoboken, NJ, USA

Svetlana A. Malinovskaya Department of Physics, Stevens University of Technology, Hoboken, NJ, USA

J. D. Mancini Department of Physical Science, Kingsborough Community College, CUNY, Brooklyn, NY, USA

Ryo Masuda Department of Mathematics and Physics, Hirosaki University, Hirosaki, Japan

Désiré Miessein Department of Physics and Engineering Physics, Fordham University, Bronx, NY, USA;
Department of Physics, Stevens Institute of Technology, Hoboken, NJ, USA

Takafumi Miyanaga Department of Mathematics and Physics, Hirosaki University, Hirosaki, Japan

Zaher M. Nassar Islamic University of Gaza, Gaza, Palestine

Mark Orman Department of Physics, Stevens Institute of Technology, Hoboken, New Jersey, USA

E. Peksu Department of Physics Engineering, Istanbul Technical University, Maslak, Istanbul, Turkey;
Department of Electrical and Computer Engineering, University of California at Davis, Davis, CA, USA

Chunlei Qu Department of Physics, Center for Quantum Science and Engineering, Stevens Institute of Technology, Hoboken, NJ, USA

Aneesh Ramaswamy Department of Physics, Stevens University of Technology, Hoboken, NJ, USA

Md. Rezaul Karim Faculty of Science and Letters, Department of Physics Engineering, İstanbul Technical University, Maslak, İstanbul, Turkey;
Department of Physics, Dhaka University of Engineering & Technology, Gazipur, Gazipur, Bangladesh

Oleksiy Roslyak Department of Physics and Engineering Physics, Fordham University, Bronx, NY, USA

K. Sabeeh Department of Physics, Quaid-i-Azam University, Islamabad, Pakistan

Bahadır Salmankurt Remote Education Center, Sakarya University of Applied Sciences, Sakarya, Turkey

T. Seddik Laboratory of Quantum Physics of Matter and Mathematical Modeling, Mustapha Stambouli University, Mascara, Algeria

Worasak Sukkabot Faculty of Science, Department of Physics, Ubon Ratchathani University, Ubon Ratchathani, Thailand

M. Tomak Department of Physics, Middle East Technical University, Ankara, Turkey

M. K. Torun Department of Physics, Faculty of Arts and Sciences, Yeditepe University, Istanbul, Turkey

Abdullah Zahid Turan TUBITAK Marmara Research Center Energy Institute, Kocaeli, Turkey

Mustafa Turan Department of Environmental Engineering, Istanbul Technical University, Istanbul, Turkey

V. Ustoglu Ünal Department of Physics, Yeditepe University, Atasehir, Istanbul, Turkey

Hilmi Ünlü Faculty of Science and Letters, Department of Physics Engineering, İstanbul Technical University, Maslak, Istanbul, Turkey

Aparajita Upali Borough of Manhattan Community College, CUNY, New York, NY, USA

Eui-Hyeok Yang Mechanical Engineering Department, Stevens Institute of Technology, Hoboken, NJ, USA

Ting Yu Department of Physics and Center for Quantum Science and Engineering, Stevens Institute of Technology, Hoboken, NJ, USA

Mehmet Hikmet Yukselici Faculty of Science and Letters, Department of Physics, Yıldız Technical University, Istanbul, Turkey

Xinyu Zhao Department of Physics and Center for Quantum Science and Engineering, Stevens Institute of Technology, Hoboken, NJ, USA

O. Zorlu Department of Physics, Yeditepe University, Atasehir, Istanbul, Turkey

Chapter 1
Modelling of Semiconductors for Low Dimensional Heterostructure Devices

Hikmet Hakan Gürel, Özden Akıncı, and Hilmi Ünlü

Abstract Advancement in the science and technology of low dimensional electronic and optical devices requires qualitatively reliable and quantitatively precise theoretical modelling of the structural, electronic and optical properties of semiconducting materials and their heterostructures to predict their potential profiles. In this chapter, we review the semiempirical tight binding and density functional theories of the modelling of electronic properties of III–V and II–VI binary/binary and ternary/binary compound semiconductor in low dimensional heterostructures. We also discuss the use of finite difference technique for modelling of electronic structure of two-dimensional quantum wells, one dimensional cylindrical nanowires and zero-dimensional spherical quantum dots. We focus on the semiempirical tight binding theory (with sp^3, sp^3s^* and $sp^3d^5s^*$ orbital sets) and density functional theory (DFT) based on modified Becke-Johnson exchange–correlation potential with a local density approximation (DFT-MBJLDA). We conclude that the NN sp^3d^5 TB model gives much more physical insight than the (2NN) sp^3s^* TB model, making use of the fictitious s^* state unnecessary in band structure calculations This is essential in the physically realistic and numerically accurate prediction of the device performance in technologically important bipolar and unipolar heterostructure devices that can proceed relatively independently of experiment. The semiempirical tight binding and density functional theories can be easily implemented in the charge transport in heterostructure devices and accurate design and simulation of low dimensional semiconductor devices for electronic and optical components in integrated circuits.

H. H. Gürel (✉)
Technology Faculty, Information Systems Engineering Department, Kocaeli University, 41001 Kocaeli, Turkey
e-mail: hhakan.gurel@kocaeli.edu.tr

Ö. Akıncı
Independent Researcher, Amstelveen, Netherlands

H. Ünlü
Faculty of Science and Letters, Department of Physics Engineering, İstanbul Technical University, 34469 Maslak, Istanbul, Turkey
e-mail: hunlu@itu.edu.tr

H. Ünlü and N. J. M. Horing (eds.), *Progress in Nanoscale and Low-Dimensional Materials and Devices*, Topics in Applied Physics 144,
https://doi.org/10.1007/978-3-030-93460-6_1

1.1 Introduction

Advances in the growth of semiconductor thin films of different structural, electronic and optical properties and with varying composition and layer thickness approaching nanometer has provided new opportunities and challenges in the fundamental science and technology of semiconductor devices for electronic and optoelectronic applications in integrated circuits [1]. The combination of contemporary growth and characterization of semiconducting materials and device fabrication technologies has resulted in the production of previously unknown semiconductor devices for fast signal processing and the discovery of some novel structures that are of special interest to solid state scientists and device engineers. Reducing device dimensions to nanometer scale require reliable description and accurate determination of quantum size effects in order to realistically describe the operation and predict the potential performance of low dimensional semiconductor devices.

Modern crystal growth techniques such as molecular beam epitaxy (MBE) and metal organic chemical vapor deposition (MOCVD) have made it possible to control the alloy composition and doping of semiconductor alloys based on group IV–IV, III–V and II–VI compounds over atomic distances. These artificially made microscale and nanoscale semiconductor heterostructures, are easily grown as lattice mismatched substrates [2] by the epitaxial layers are so thin that quantum mechanical effects are dominant. The resulting new low dimensional semiconductor devices (e.g., heterostructure bipolar transistors (HBTs) and modulation doped field effect transistors (MODFETs), nanowires, quantum dots, etc.) are known to operate much faster than conventional silicon devices (e.g., Si bipolar junction transistors (BJTs) and metal oxide semiconductor field effect transistors (MOSFETs)).

Consequently, large number of theoretical and experimental studies have been carried out on metals, semiconductors, ceramics, polymers and composites at low dimensions during the last three decades. Fabricated devices are to be used in electronic and telecommunications, biotechnology, energy conversion and storage, health sciences and medicine, construction and environmental issues. Nanomaterials are very strong and lightweight, and they constitute very small components of computers as well as new sensors: They provide very rapid response nanoscale electronic and optoelectronic technologies of very small size and are also predicted to be important in high power defense systems. Smaller size electronic and optoelectronic devices require reduction of the volume to increase the speed of these devices. This technological achievement is possible with increasing their production methods, which brings in many scientific and technological problems, and development of qualitatively reliable and quantitatively precise modelling and simulation techniques.

At microscale to nanoscale, the semiconducting materials may change with the size and new physical and chemical properties and their production can be controlled with the contemporary crystal growth and fabrication technologies. When the semiconductor composition changes abruptly across the interface between two constituents, the difference in their energy bandgaps is accommodated by the discontinuities in the conduction and valence bands, as shown in Fig. 1.1. The bandgap of

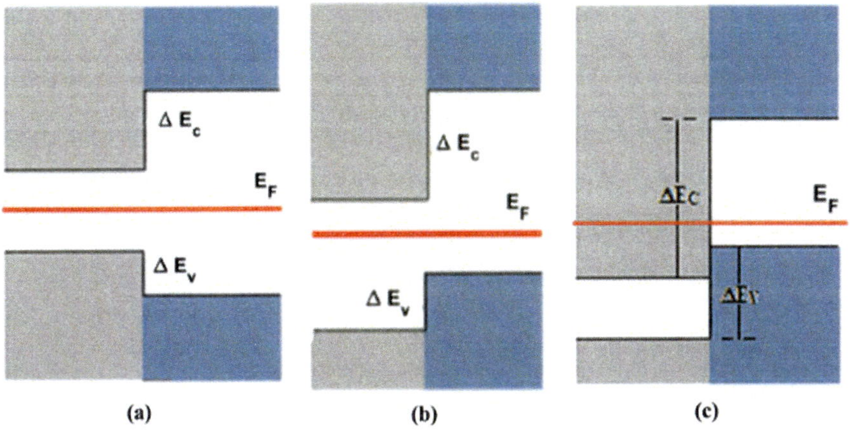

Fig. 1.1 Schematic band diagram of Type II (**a**) and Type III (**b**) heterostructures: the potential energy gradient tends to spatially separate electrons and holes on different sides of heterointerface

the barrier semiconductor overlaps (straddling lineup) that of the well and equilibrium Fermi level is near the middle of the bandgap on both sides, as shown in Fig. 1.1a. Both an electron and a hole tend to localize in the narrow-gap quantum well. Such heterostructure is useful in optoelectronic applications such as lasers, with both electrons and holes participate to device operation.

When the semiconductor composition changes abruptly across the interface between two constituents, the bandgap of the barrier semiconductor can also partially overlap (staggered lineup) that of the well and the equilibrium Fermi level is close to the conduction band (or valence band) on one side while it is near the middle of the bandgap on other side (Fig. 1.1b). The potential energy gradient tends to spatially separate electron and hole on different sides of heterointerface. Large potential barrier in conduction (or valence) band at heterointerface leads to a better electron (hole) confinement in field effect transistors with higher electron (hole) concentration. Position of equilibrium Fermi level determines the density of carrier confinement on either side of the heterointerface. When equilibrium Fermi level is close to conduction band of wide bandgap constituent (left) the density of electrons in its conduction band is greater but when it is closer to the valence band of narrow bandgap constituent (right) the number of holes is greater there.

If the bandgap of the barrier may not overlap at all (broken gap lineup), as shown in Fig. 1.1c, the equilibrium Fermi level is well above the conduction band minimum (or well below the valence band maximum) on one side while it is near the middle of the bandgap on the other side. Position of equilibrium Fermi level determines the carrier confinement on either side of the heterointerface. When equilibrium Fermi level is close to conduction band of wide bandgap constituent (left) the number of electrons in its conduction band is greater but when it is closer to the valence band of narrow bandgap constituent (right) the number of holes is greater there.

The impact of heterostructures on charge transport in three and two dimensional heterostructure devices can be summarized as follows. First of all, one can make bipolar transistors with high injection efficiency based on an heterostructure formed between two semiconductors (with different lattice and elastic constants). Under forward bias conditions, charged carriers flow from the wide bandgap semiconductor emitter to the narrow bandgap semiconductor base. Since charge injection efficiency is related exponentially to the change in the bandgap across the heterointerface, it can be made almost independent of doping levels [3]. Such transistor is known as heterojunction bipolar transistors (HBTs). In these devices, the electron (hole) injection factor is related to the conduction (valence) band offset and is also further enhanced by the electric field induced because of the charge transfer. Figure 1.2 shows schematic band diagram of forward biased emitter/base junction of Npn HBT. The potential barrier blocking electron emission from the emitter to the base is lowered and electrons injected from the emitter diffuse across the base and are collected in the collector. Holes injected from the base into the emitter are blocked by the valence band offset as $\Delta E_V/kT$.

Furthermore, is now possible to realize low dimensional structures such as quantum wells and superlattices, which were first proposed by Esaki and Tsu in 1970 [4]. Figure 1.3 shows a schematic band diagram of lattice matched abrupt AlGaAs/GaAs and AlGaAs/GaAs/AlGaAs multi-quantum wells. In these structures electrons and holes have dimensions of the order of the mean free path of charge carries.

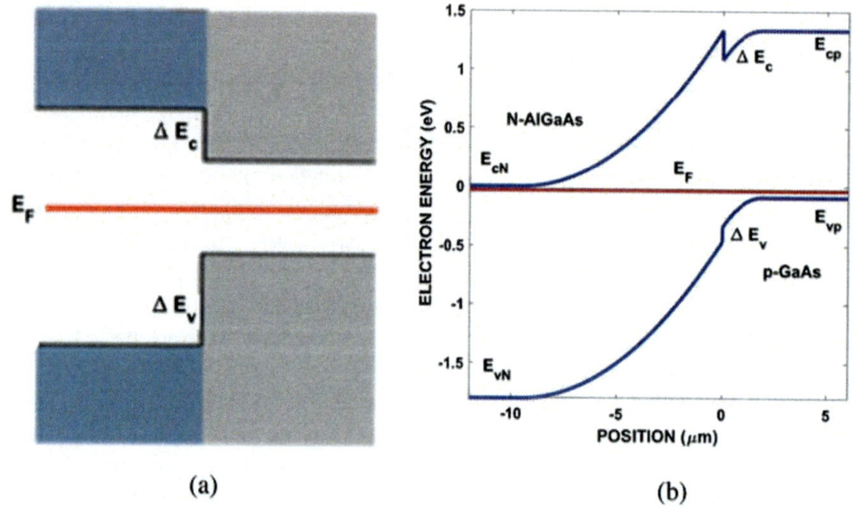

Fig. 1.2 The schematic equilibrium band diagram of intrinsic AlGaAs/GaAs emitter/base region of heterojunction bipolar transistor (HBT) when both sides are intrinsic (**a**) and doped (**b**). When N/p heteroemitter junction is forward biased, majority of electrons injected emitter electrons diffuse across the p-GaAs base. Some electrons recombine with holes at defect sites in base region

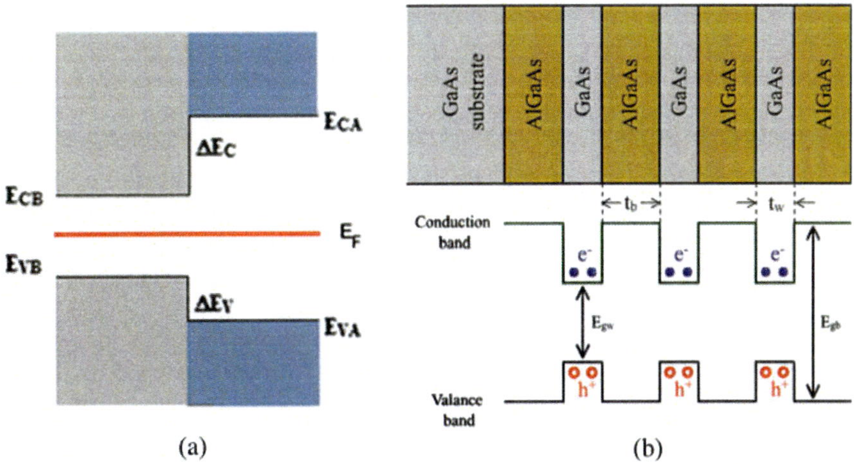

Fig. 1.3 Schematic view of band diagram of lattice matched abrupt AlGaAs/GaAs (**a**) and AlGaAs/GaAs/AlGaAs multi-quantum well (**b**). The difference in their energy bands is accommodated by the discontinuities in the conduction and valence bands. The spike ΔE_C in the conduction band and step ΔE_V in the valence band at heterointerface influences the carrier transport and influence the device performance

Semiconductor superlattices and quantum wells can now be grown by epitaxially depositing two semiconductors alternatively or introducing impurities into individual layers of a single semiconductor with thickness is smaller than the carrier mean free path. Periodic extension of thin-layer pairs imposes an artificial perturbation of the crystal structure and leads to novel electronic and optical properties. Second kind of superlattices are known as doping superlattices in which the dopant type alternates between n- and p-type. The motion of charge carriers parallel to the layers is not affected much by the potential barrier of the superlattice. Assuming parabolic bands and infinite barriers, the conduction band subband energy levels, relative to bottom of the conduction band, are described by

$$E_n = \frac{\hbar^2 n^2 \pi^2}{2m_n^* L_z}, \quad n = 1, 2, ... \tag{1.1}$$

where m_n^* is the effective mass of conduction electrons and L_z is the quantum well thickness.

The applications of heterostructures are not limited to HBTs and MODFETs. In some cases, a certain compound can be grown on top of the original material as ohmic (or Schottky) contact. For example, in GaAs/(Al, Ga). As laser diodes or modulation doped field effect transistors (MODFETs), a GaAs layer is always used on top of the (Al, Ga). A s layer in order to facilitate ohmic contacts since it is easier on smaller bandgap materials. In (In, Ga). As MESFETs, the introduction of a large bandgap GaAs or (In, Al). As as an overlayer before Schottky contact is found useful. In the case of modulation doped heterostructure field effect transistors (MODFETs), the wide-gap semiconductor is doped, from which charged carriers diffuse to the nominally undoped narrow gap semiconductor. Furthermore, it is also feasible to construct more or less undoped heterostructures in which the carriers are induced by the gate field in a way similar to inversion metal–oxide–semiconductor FETs (MOSFETs), in which the use of this inversion layer as FET channel allows one to obtain much improved threshold control.

Rapidly developed technologies and new electronic communication have intensified the urge for scientific and technological research and development that included structures with atoms and molecules as well as biological structures. Contemporary growth and fabrication technologies have further made it possible to reduce the device dimensions to nanometers where quantum size effects must be described in order to realistically describe the operation and reliably predict the potential and performance of low dimensional semiconductor devices for electronic and optical applications.

In a nanoscale semiconductor heterostructure devices, such as one-dimensional quantum wires and zero-dimensional quantum dots, the restricted movement of charged particles are investigated by using in quantum mechanical principles (e.g., *effective-mass approximation*) in literature [5–11]. The movement of discrete particles of (discrete) energies in three dimensions (quantized) for a spherical particle in a box can be obtained by solving Schrodinger equation with Hamiltonian defined as

$$H = \frac{\hbar^2}{2m_e^*}\nabla^2 + \frac{\hbar^2}{2m_h^*}\nabla^2 + V_e(\vec{r}_e) + V_h(\vec{r}_h) - \frac{e^2}{\varepsilon|\vec{r}_e - \vec{r}_h|} \quad (1.2)$$

First and second terms of the kinetic energy of electrons and space, the third and fourth terms of limiting the potential energy of the electrons and the gap between the last term is the Coulomb interaction energy. Interaction and exchange of energy here in the nanostructure and dielectric material between the interaction energy of the image (or polarization energy) has been neglected, such as energy terms. Experimental results are compared with the following quadratic expressions

$$E_{(n_e,l_e),(n_h,l_h)}(d) = E_g + \frac{2\hbar^2\pi^2}{d^2}\left[\frac{x^2_{n_e,l_e}}{m_e^*} + \frac{x^2_{n_h,l_h}}{m_h^*}\right] - \frac{3.572e^2}{\varepsilon d} - \frac{0.124e^4}{\hbar^2\varepsilon^2}\left[\frac{m_e^* m_h^*}{m_e^* + m_h^*}\right] \quad (1.3)$$

where $E_{\{(n_e,l_e),(n_h,l_h)\}}(d)$ is the measured core region energy-gap. $m_e^* = 0.13\text{m}_0$ ve $m_h^* = 0.43m_0$ and CdSe electron and hole effective masses with m_0 bare free electron mass, and $\varepsilon = K\varepsilon_0$ is the optical dielectric constant of core region. $x_{n,l}$ is $j_1(x_{n,l}) = 0$ first order Bessel function of the roots. Equation (1.3) can be used to calculate the valence band and conductive band transitions between (transition) (1s, 1s) (1p, 1p), (1d, 1d), (1f, 1f) and (2s, 2s) transition energies.

Advancement in the science and technology of low dimensional devices could not have been possible without a qualitatively reliable understanding of the basic physics of low dimensional semiconducting material structures and quantitatively precise potential predictions and performance of such devices. The electronic band structure of low dimensional bipolar and unipolar semiconductor devices are often calculated based on the following theoretical models: (i) First principle ab-initio methods, such as density functional theory (DFT) [12, 13], (ii) *methods*, such as local/empirical pseudopotential method [14], or (iii) *tight binding* method (also known as the linear combination of atomic orbitals (LCAO) method) [5–11, 15–24], and ***k.p*** *method* [25, 26]. The first principles ab-initio methods are computationally expensive and require heavy parallel computations. Since the ab-initio models are based on the calculations of ground state properties, they can only give limited physical insight about the energy band structure at high temperatures and pressures.

However, the semiempirical tight binding models are less expensive than the ab-initio methods and can easily be implemented in calculating electronic properties of low dimensional and heterostructures such as two-dimensional quantum wells, one dimensional nanowires and zero-dimensional quantum dots. We will discuss the semiempirical tight binding theory with first and second nearest neighbor (NN and 2NN) sp^3 and sp^3s^* atomic orbitals sets and first nearest neighbor (NN) sp^3d^5 orbitals set with spin–orbit coupling of cation (Al, Ga; In, Cd, Zn) and anion (P, As, Sb, S, Se, Te) atoms for calculating the electronic structure of III–V and II–VI compounds. In Sect. 1.5 we will compare the predictions of semiempirical TB with DFT that uses the modified Becke-Johnson exchange–correlation potential with the local density approximation (LDA), called MBJLDA functional, for calculating the band structure of group III–V and II–VI compounds and their alloys.

1.2 Strain in Low Dimensional Heterostructures

The composition variation in alloy constituent and lattice mismatch between two semiconductors will cause strain across the interface that will modify the electronic properties of both materials. As long as the thickness of epilayer is kept under a critical thickness, the lattice mismatch will be accommodated by uniform elastic strain, as shown in Fig. 1.4.

Strain will modify the structural and electronic properties of the constituent semiconductors in directions parallel and perpendicular to the growth direction. In order to emphasize the importance of heterostructures in low dimensional semiconductor systems it is essential to understand the interface formation and model band structure. Until the early 1980s, the studies of heterostructures had focused on semiconducting materials with close lattice matching with a substrate, such as an AlGaAs/GaAs heterostructure quantum well. The constraint on lattice matching in crystal growth is essential for relatively thick epitaxial layers since a large lattice mismatch can cause the occurrence of misfit dislocations with large densities. Such lattice-matched

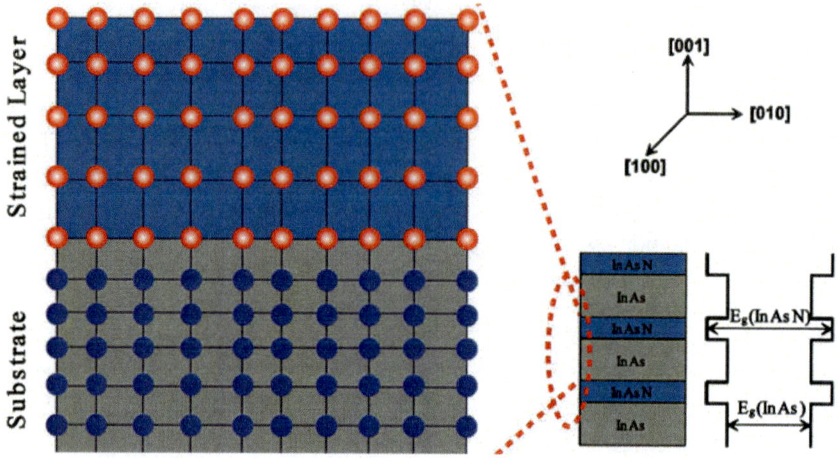

Fig. 1.4 Strained heterolayer epitaxy is a result of a growth of materials with dissimilar lattice constants. To accommodate the lattice mismatch the individual layers contract or expand in the plane of the epilayers and in the plane of the growth, respectively

growth will prevent the generation of misfit dislocations that would degrade charge transport and lower device performance.

The lattice constant of strained epilayer along the growth direction is equal to that of the substrate and is expanded or compressed in the direction parallel to the interface, shown in Fig. 1.4. Considering the (001) heterepitaxial growth, elastic theory allows decomposing the biaxial strain tensor into the sum of hydrostatic and uniaxial strains along the growth direction

$$\begin{aligned} \varepsilon_{xx} = \varepsilon_{yy} = \varepsilon_{f\|} &= \left(\frac{a_{f\|} - a_f}{a_f}\right), \\ \varepsilon_{zz} = \varepsilon_{f\perp} &= \left(\frac{a_{f\perp} - a_f}{a_f}\right) = 2\left(\frac{C_{12}}{C_{11}}\right)_f \varepsilon_{f\|} \end{aligned} \tag{1.4}$$

where $\varepsilon_{f\perp}$ and $\varepsilon_{f\|}$ are the strain components perpendicular and parallel to growth direction. The epilayer lattice constant will be equal to that of the substrate along the growth direction: $a_{f\|} = a_s$ and is expanded by the bulk value of its Poisson ratio perpendicular to the growth direction:

$$a_{f\perp} = a_f\left[1 - 2\left(\frac{C_{12}}{C_{11}}\right)_f \left(\frac{a_{f\|} - a_f}{a_f}\right)\right], \tag{1.5}$$

where,a_f is the bulk lattice constant and C_{11} and C_{12} are the bulk elastic stiffness constants of the strained layer. a_s is the bulk lattice constant of buffer /or substrate.

Equations (1.4) and (1.5), suggest that when a thin layer is deposited on a buffer (or substrate) with large mismatch, the epilayer will be under strain. The lattice constant of the epilayer along and perpendicular to the interface will be distorted to minimize its elastic energy. The interface strains due to lattice mismatch and lattice thermal expansion gradients over the crystal growth temperature can co-exist in heterolayers. The thermal expansion of lattice constants of epilayer and the substrate with the growth temperature cause the lattice constant of epilayer parallel and perpendicular to the interface to change with temperature

$$\begin{aligned} a_{f\parallel}(T) &= \alpha_{f\parallel}(T)T = \alpha_s(T)T, \\ a_{f\perp}(T) &= \alpha_{f\perp}(T)T = \alpha_f(T)T + 2\frac{C_{12f}}{C_{11f}}[\alpha_f(T) - \alpha_s(T)]T \end{aligned} \tag{1.6}$$

where $\alpha_s(T)$ and $\alpha_f(T)$ are the bulk linear thermal expansion coefficients of substrate and epilayer. Strain across the interface that will modify the electronic properties of both materials, including the band offsets. Compressive (or tensile) strain in epilayer results in an increase (or decrease) in conduction and valence band energy levels [27].

In the framework of so called statistical thermodynamic model, in which the conduction electrons and valence holes are treated as charged chemical particles, the conduction and valence band energy levels energy are expressed as a function of pressure at any temperature [28, 29]

$$\begin{aligned} E_c(T,P) &= E_c(0,P_0) + C^0_{cP}T(1 - \ln T) - \frac{a_c}{B}\left[P - \frac{P^2}{2B} - \frac{(1+B')}{6B^2}P^3\right], \\ E_v(T,P) &= E_v(0,P_0) + C^0_{vP}T(1 - \ln T) - \frac{a_v}{B}\left[P - \frac{P^2}{2B} - \frac{(1+B')}{6B^2}P^3\right], \end{aligned} \tag{1.7}$$

where P is the applied pressure, T is the temperature, $E_c(0, P_0)$ and $E_v(0, P_0)$ are the conduction or valence band edges at Γ, L and X high symmetry points with deformation potentials $a_{cl} = -B(\partial E_{cl}/\partial P)$ and $a_v = -B(\partial E_v/\partial P)$. B is the bulk modulus; $B' = \partial B/\partial P$ is its derivative. C^0_{iP} is the standard heat capacity of conduction electrons and valence holes. $C^0_{cP} = C^0_{nP} - C^0_{0P} = C^0_{pP} + \Delta C^0_P$ and $C^0_{vP} = C^0_{pP}$ are the standard heat capacities of conduction electrons and valence holes; $C^0_{nP} = C^0_{pP} = (5/2)k$, where k is the Boltzmann's constant. $\Delta C^0_P = C^0_{nP} + C^0_{pP} - C^0_{0P}$ is the heat capacity of reaction of free electron and hole formation obtained by fitting (1.6) to the experimental data [26, 27] and empirical pseudopotential bandgap energy [14] and at high symmetry points in the first Brillouin zone of semiconductors:

$$E_{gl}(T,P) = E_{gi} + \Delta C^0_P T(1 - \ln T) - \frac{a_{gl}}{B}\left[P - \frac{P^2}{2B} - \frac{(1+B')P^3}{6B^2}\right], \tag{1.8}$$

where $P = -2B_f C_f \varepsilon_{f//}$ and $P = -3B_s\varepsilon_{s||}$ for the epilayer and the substrate, respectively. In the case of (001) pseudomorphic growth, $\varepsilon = \varepsilon_{f||} = (a_{f||} - a_f)/a_f$ is the strain in the epilayer along the growth direction and $C_f = (C_{11} - C_{12})/C_{11}$. It should be noted that as the device dimension is reduced to nanoscale of the order of Bohr radius of charged carriers, the use of continuum elastic theory becomes questionable [2, 30]. In such a case one should use the atomic elasticity theory known as valence force field (VFF) approach [31, 32], especially as the lattice mismatch at heterointerface becomes large.

1.3 Composition Effects in Ternary/Binary Heterostructures

Semiconductor alloys based on IV–VI, III–V and II–VI compounds are important in fabricating low dimensional bipolar and unipolar heterostructure devices since their structural and electronic properties (e.g., lattice constants and bandgaps) can be tailored independently. Therefore, reliable and accurate determination of composition variation of lattice constant and bandgap energies are very important. In our recent studies [5–10, 15–22] the effects of composition and strain on the electronic properties of semiconductor alloy constituents in heterostructures are studied by using the semiempirical tight binding models and first principles DFT with MBJLDA functional in terms of host bond length and distorted bond length by the substitutional impurity without any adjustable parameter.

In determining the composition effects on tight binding parameters and in turn on ternary alloy semiconductor band structures, we employ the method of the modified virtual crystal approximation (MVCA) in which we formulate the composition dependence of the bond length of ternary material [5–10, 15–22]. The MVCA allows one to accurately take into account the effect of disorder-induced nonlinear variation of the bond length in the TB parameters used in calculating the band structure properties. The composition dependent bond length of ABC ternary is written as the sum of undistorted bond length ($d_{VCA} = (1-x)d^0_{AC} + xd^0_{BC}$) due to the virtual crystal approximation (VCA) [33] and the distorted bond length ($d_{relax}(x) = x(1-x)\delta_c(d_{BC}(x) - d_{AC}(x))$) due to cation–anion relaxation of binaries used to form ternary [5–10, 15–22]:

$$
\begin{aligned}
d(x) = d_{VGA}(x) + d_{relax}(x) &= (1-x)d^0_{AC} + xd^0_{BC} \\
&+ \delta_c(x)(1-x)(d_{BC}(x) - d_{AC}(x))
\end{aligned}
\tag{1.9}
$$

where $d_{AC}(x)$ and $d_{BC}(x)$ are the bond lengths of AC and BC binaries in an ABC ternary

$$d_{BC}(x) = (1-x)d^0_{AC} + xd^0_{BC} + (1-x)\xi_{BC:A}(d^0_{BC} - d^0_{AC}),$$
$$d_{AC}(x) = (1-x)d^0_{AC} + xd^0_{BC} + x\xi_{AC:B}(d^0_{AC} - d^0_{BC}), \tag{1.10a}$$

where d^0_{AC} and d^0_{BC} are the undistorted bond lengths of the host materials AC and BC and $\xi_{AC:B}$ and $\xi_{BC:A}$ are two dimensionless relaxation parameters [34],

$$\xi_{AC:B} = \frac{1}{1 + \frac{\alpha_{AC}}{6\alpha_{BC}}\left(1 + 10\frac{\beta_{AC}}{\alpha_{AC}}\right)}, \quad \xi_{BC:A} = \frac{1}{1 + \frac{\alpha_{BC}}{6\varepsilon_{AC}}\left(1 + 10\frac{\beta_{BC}}{\alpha_{BC}}\right)} \tag{1.10b}$$

where $\delta_c = \xi_{AC:B} - \xi_{BC:A}$ is the difference between dimensionless relaxation parameters.

Equation (1.9) can then be used to consider the composition effects on the band structures of the ternary constituent of heterostructures in the theoretical models such as first principles WIEN2K simulations package with the MBJLDA functional embedded in DFT and semiempirical sp^3, sp^3s^*, $sp^3d^5s^*$ and sp^3d^5 tight binding models. The diagonal or off-diagonal matrix elements in the NN sp^3d^5 and 2NN sp^3s^* TB Hamiltonian matrix elements for an ABC ternary are expressed as [5–10, 15–22]

$$E_{\alpha/\beta}(x) = (1-x)E_{\alpha/\beta}(AC) + xE_{\alpha/\beta}(BC) + x(1-x)\delta_c[E_{\alpha/\beta}(AC) - E_{\alpha/\beta}(BC)] \tag{1.11}$$

where $E_{\alpha/\beta}(AC)$ and $E_{\alpha/\beta}(BC)$ represent the fitted energies of the s, p and d states of anion and cation atoms forming the AC and BC binary compounds. This allows one to consider the composition variations of the diagonal and off-diagonal elements of the TB Hamiltonian matrix for a ternary ABC semiconductor can be taken as a nonlinear function of alloy composition.

1.4 Electronic Band Structure Modelling

In order to emphasize the importance of the electronic structure modelling of heterostructure in low dimensional semiconductor systems it is necessary to keep in mind that the motion of charge carriers in a semiconductor are subject to electric force or both electric and/or magnetic forces. If a constant force $\vec{F}$ is applied to an electron, change in its wavevector will be $d\vec{k}$. According to second law of Newton, the equation of motion of a charged particle can be written as

$$\frac{d\vec{k}}{dt} = \frac{1}{\hbar}\frac{d\vec{p}}{dt} = \vec{F} \tag{1.12}$$

where the force $\vec{F}$ can be due to an externally applied electric field ($\vec{E}$), magnetic field $\vec{B}$ or combination of both, known as Lorentz force $\vec{F} = q(\vec{E} + \vec{v} \times \vec{B})$. The velocity and acceleration of charge carriers such as electrons and holes can be calculated from the semiconductor band structure. In the framework of classical band structure model, the group velocity of the moving electrons and holes are given by the gradients of conduction and valence band energies in k-space.

$$\vec{v}_n = \frac{1}{\hbar}\frac{\partial E_n(k_n)}{\partial k_n} = \frac{1}{\hbar}\vec{\nabla}_k E_n(k_n), \quad \vec{v}_p = \frac{1}{\hbar}\frac{\partial E_p(k_p)}{\partial k_p} = \frac{1}{\hbar}\vec{\nabla}_k E_p(k_p) \tag{1.13}$$

The speed and acceleation of electronc and holes is affected by the electronic band structure of bulk or low dimensional heterostructure semiconductors. Furthermore, the effective masses of conduction electrons and valence holes are also described by the semconductor band structure. If the band structure is perfectly parabolic ($E(k) \propto k^2$), (1.12) reduces to the classical Newtonian expression for equation of motion. The accelerations and effective masses of of electrons and holes can be written as

$$\begin{aligned} \vec{a}_n &= \frac{d\vec{v}_n}{dt} = \left(\frac{1}{\hbar^2}\frac{\partial^2 E_n(k)}{\partial k_n^2}\right)\vec{F}_n \rightarrow \frac{1}{m_n^*} = \left(\frac{1}{\hbar^2}\frac{\partial^2 E_n(k)}{\partial k_n^2}\right), \\ \vec{a}_p &= \frac{d\vec{v}_p}{dt} = \left(\frac{1}{\hbar^2}\frac{\partial^2 E_p(k)}{\partial k_p^2}\right)\vec{F}_p \rightarrow \frac{1}{m_p^*} = \left(\frac{1}{\hbar^2}\frac{\partial^2 E_p(k)}{\partial k_p^2}\right) \end{aligned} \tag{1.14}$$

When the band structure is parabolic charged particles move faster and less scattered by the impurities. There are regions in the semiconductor band structure which are not parabolic. Therefore, a detailed study of semiconductor band structure is essential for a relaible modelling and accurate numerical simulation of charge transport in low dimensioal heterostructure devices.

Furthermore, in order to understand the charge transport in low dimensional semiconductor heterostructures, it is essential to reliably understand the formation and precisely determine the magnitude of the structural, electronic and optical properties across the interface between two semiconductors with dissimilar physical and chemical properties. When a widegap semiconductor is grown on a narrow bandgap semiconductor, the interface strain, caused by the lattice mismatch and thermal expansion gradient over the crystal growth temperature, will modify the structural and electronic properties of the constituents in directions parallel and perpendicular to the heterointerface [27, 35]. As with all semiconductor devices, the key issue to understand the impact of heterostructure on the performance of low dimensional semiconductor devices is the physically reliable and numerically precise determination of the electronic energy band structure across the heterointerface.

The theoretical calculations of structural and electronic properties of bulk semiconductors or low dimensional planar and core/shell semiconductor heterostructures are often carried out by:

(i) First principal ab-initio methods, such as density functional theory (DFT), which allows one to calculate the electronic structure from first principles in which there is no need for empirical fitting parameters,
(ii) Empirical methods, such as local/nonlocal empirical pseudopotential method (EPM), orthogonalized plane wave (OPW),
(iii) Semiempirical tight binding method, and
(iv) k.p approximation.

The ab-initio methods, which utilize variational method to calculate the ground state properties of many-body system, are computationally expensive and require heavy parallel computations. Therefore, ab-initio methods cannot be easily implemented for the study of nanoscale electronic and optoelectronic devices. In empirical methods, one performs calculations on systems with empirical parameters fit to experimental data at high symmetry points as derived from optical experiments. Since empirical methods involve solution of one-electron Schrödinger wave equation to calculate electronic band structure of semiconductors, they are less expensive than ab-initio methods and easily generates electronic band structure. In contrast, the *empirical* methods, such as semiempirical tight binding model (ETB) is known as not only simple but also reliable and easy implemented and has great advantages over first principles calculations in finding the electronic properties such as band structure, density of states, band gaps of the heterostructures. The semiempirical tight binding model is an atomistic approach and well suited to calculate the electronic band structure of low dimensional semiconductor heterostructures such as two-dimensional quantum wells, one dimensional quantum wires, and zero-dimensional quantum dots [36].

Quantitatively reliable and numerically precise modeling and simulation of electronic properties of compound semiconductors and their ternary alloys allows better prediction of their material properties [23, 37–40]. It is well known that the theoretical calculation of the electronic properties of heterostructures usually encounters serious computational and conceptual difficulties. Over the years various elaborate quantum mechanical models have been proposed over the years to determine magnitude of energy band gaps and band offsets (see [32] for a detailed review). Most of these quantum mechanical models are based on the zero temperature calculations of electronic band structure and therefore, can only give limited physical insight about the formation of the energy band structure across the heterointerface and of little use in development of nanoscale electronic and optical devices.

It is well known fact that the conventional density functional theory (DFT) allows one to calculate the electronic structure from first principles, in which there is no need for empirical fitting parameters. Although the DFT based calculations yield satisfactory results for the structural properties of semiconductors [27], they fail to produce satisfactory results for the electronic properties of the semiconductors when compared with experiment [41, 42]. It is thought that this is to the errors in formulation of the self-interaction effects and the discontinuity in exchange correlation potential energy in calculating the band gaps with the conventional DFT studies. Although the GW approximation or hybrid functional overcomes the band gap problem, they still result in band gap error on the order of 10–20% compared to experiment [43]. On the other hand, the semiempirical tight binding model, with various orbital bases (e.g., sp^3, sp^3s^*, sp^3d^5, and $sp^3s^*d^5$), although approximate, can realistically describe the electronic properties of semiconductors.

The semiempirical tight binding theory is an atomistic approach and because of its simplicity and ability to give better band structure predictions, it is well suited for the band structure studies of low dimensional structures (such as quantum wells and quantum dots [42]). There is, however, still argument about the use of the fictitious s* state in TB model (e.g., sp^3s^*, $sp^3s^*d^5$) in the electronic structure calculations. Since the inclusion of s^* excited state is done by modeling the average of p-d interactions, the sp^3s^* tight binding parameterization does not permit the inclusion of excited d-orbitals to the sp^3 basis and consequently, the actual contribution of the excited d-states in the band structure calculations is not reliably reflected. In this section, we shall give a brief discussion of the density functional theory (DFT) that uses the modified Becke-Johnson exchange potential with local density approximation (LDA) correlation potential (called MBJLDA) for obtaining band structure of group II–VI compounds and their ternary alloys, yielding more accurate fundamental band gaps.

In Sect. 2.2, we briefly describe the semiempirical second nearest neighbor (2NN) sp^3s^* and nearest neighbor (NN) sp^3d^5 tight binding parameterizations, with spin–orbit coupling of cation (Al,Ga, In, Cd, Zn) and VI anion (N, As, P, Sb, S, Se, Te) atoms, based on the parametrization presented in [44–46]. We will then present the results of our calculations for the electronic properties of CdX and ZnX (X = S, Se, Te) compounds and their heterostructures (CdZnS/CdS, CdSTe/CdTe and ZnSSe/ZnSe) and compare them with experimental data on band gap energies at high symmetry points in the Brillouin zone. In the following section we will give a brief discussion about the current progress in semiempirical tight binding theory which is based on nearest or second nearest neighbor sp^3, sp^3s^* [23], $sp^3d^5s^*$ [38, 39] and sp^3d^5 [40] orbitals sets to calculate the electronic structure of binary and ternary semiconductors and their binary/binary and ternary/binary heterostructures as a function of alloy composition, temperature, pressure and interface strain. We present the results of our 2NN tight binding calculations, with spin–orbit coupling [2, 35, 42] in calculating the composition, temperature, pressure and strain effects on the electronic properties (e.g., band structure, density of states, band gaps and band widths) of Ga and Al based III–V ternary and binaries, which are in good agreement with experiment [14, 37] for bandgap energies and electron effective mass.

1.5 Semiempirical Tight Binding Modelling

The elementary tight-binding method introduced in 1960s has become essentially first-principles theory and has been used in recent years to get approximate but very precise predictions of the bonding and electronic properties of semiconductors with both zinc-blende and wurtzite crystal structure. The linear combination of atomic-orbitals (LCAO) allows one to formulate the total energy in semiconductors which provides a clear way to think about all of the trends from system to system entirely in terms of the concepts. We will give a detailed discussion of the semiempirical tight binding theory which uses the second nearest neighbor (2NN) sp^3 and sp^3s^* atomic orbitals set and first nearest neighbor (NN) sp^3d^5 basis sets with spin–orbit coupling of cation (Al, Ga; In, Cd, Zn) and anion (P, As, Sb, S, Se, Te) atoms for calculating the electronic structure of III–V and II–VI compounds and their heterostructures. The results of DFT with MBJLDA functional and semiempirical NN sp^3d^5 tight binding theory are found to be in good agreement with measured band gaps of group III–V and II–VI compounds and their ternary/binary heterostructures.

1.5.1 Semiempirical sp^3 *Tight Binding Theory*

In the semiempirical tight binding theory of semiconductors one assumes that the valence electrons are tightly bound to their nuclei as in the free atom. Anion and cation atoms are brought together until their separations become comparable to the lattice constant of semiconductors, at which point their wave functions will overlap. In the Slater-Koster formalism [47], the crystal potential of a binary semiconductor is defined as a sum of the symmetrical spherical potentials around each atom. This allows the electronic wave function $\psi_{\alpha k}$ to be written as linear combination of atomic orbitals $\varphi_\alpha(k)$

$$|\psi_{\alpha k}\rangle = \sum_\alpha u_\alpha|\varphi_\alpha(k)\rangle = \sum_i \frac{e^{i.k.r_i}}{\sqrt{N}}|\varphi_\alpha(r - r_i)\rangle, \tag{1.15}$$

where **r** is the position of electron with respect to origin of coordinate system is real space, **k** is the wave vector, N is the number of atoms, and $\mathbf{r}_\mathrm{i}$ is the crystal lattice site. The energy state is given by the eigenvalue of the linear equation written in matrix form [44, 45, 47, 48];

$$\sum_\beta \left[H_{\alpha\beta}(k) - S_{\alpha\beta}(k)E\right]u_\beta = 0, \tag{1.16}$$

where α correspond to a cation s (p) orbital and β corresponds to an anion s (p) orbital and u_β is the wave function coefficient. The Bloch functions in (1.15) are taken to be orthogonal so that the overlap matrix $S_{\alpha\beta} = \langle\varphi_\alpha(k)|\varphi_\beta(k)\rangle$ is as an identity matrix. E is the eigenvalue and $H_{\alpha\beta} = \langle\varphi_\alpha(k)|H|\varphi_\beta(k)\rangle$ is the (2 × 2) Hamiltonian matrix. Equation (1.16) is rewritten as

$$\begin{pmatrix} H_{cc} & H_{ca} \\ H_{ac} & H_{aa} \end{pmatrix} - E\begin{pmatrix} 1 & S_{ca} \\ S_{ac} & 1 \end{pmatrix} = 0 \tag{1.17}$$

The diagonal ($E_{sa}, E_{sc}, E_{pa}, E_{pc}$) and off diagonal ($E_{ss}, E_{xx}, E_{s_a p_c}, E_{s_c p_a}, E_{xy}$) elements are

$$H_{aa} = \begin{bmatrix} E_s^a & 0 & 0 & 0 \\ 0 & E_p^a & 0 & 0 \\ 0 & 0 & E_p^a & 0 \\ 0 & 0 & 0 & E_p^a \end{bmatrix}, \quad H_{cc} = \begin{bmatrix} E_s^c & 0 & 0 & 0 \\ 0 & E_p^c & 0 & 0 \\ 0 & 0 & E_p^c & 0 \\ 0 & 0 & 0 & E_p^c \end{bmatrix} \tag{1.18}$$

$$H_{ac} = \begin{bmatrix} B_0 E_{ss} & B_1 E_{sp} & B_2 E_{ss} & B_3 E_{sp} \\ -B_0 E_{sp} & B_0 E_{xx} & B_3 E_{xy} & B_2 E_{xy} \\ -B_2 E_{sp} & B_3 E_{xy} & B_0 E_{xx} & B_1 E_{xy} \\ -B_3 E_{sp} & B_2 E_{xy} & B_1 E_{xy} & B_0 E_{xx} \end{bmatrix}, \quad H_{ca} = \begin{bmatrix} B_0^* E_{ss} & -B_1^* E_{sp} & B_2^* E_{ss} & B_3^* E_{sp} \\ B_1^* E_{sp} & B_0^* E_{xx} & B_3^* E_{xy} & B_2^* E_{xy} \\ B_2^* E_{sp} & B_3^* E_{xy} & B_0^* E_{xx} & B_1^* E_{xy} \\ B_3^* E_{sp} & B_2^* E_{xy} & B_1^* E_{xy} & B_0^* E_{xx} \end{bmatrix} \tag{1.19}$$

where diagonal elements and off-diagonal (interacting or hopping terms) elements are defined as

$$\begin{aligned}
&\langle p_x^c|H|p_x^a\rangle = \langle p_y^c|H|p_y^a\rangle = \langle p_z^c|H|p_z^a\rangle = E_{xx} B_0(k), \\
&\langle p_x^c|H|p_y^a\rangle = \langle p_y^c|H|p_x^a\rangle = E_{xy} B_3(k), \ \ \langle p_y^c|H|p_z^a\rangle = \langle p_z^c|H|p_y^a\rangle = E_{xy} B_1(k), \\
&\langle p_y^c|H|p_z^a\rangle = \langle p_z^c|H|p_y^a\rangle = E_{xy} B_1(k), \ \ \langle p_x^c|H|p_z^a\rangle = \langle p_z^c|H|p_x^a\rangle = E_{xy} B_2(k), \\
&\langle s^c|H|s^a\rangle = E_{ss} \sum_{n=1}^{4} e^{ik.r_n} = E_{ss} B_0(k), \ \ \langle s^c|H|p_x^a\rangle = E_{sp} \sum_{n=1}^{4} e^{ik.r_n} = E_{sp} B_1(k), \\
&\langle s^c|H|p_y^a\rangle = E_{sp} \sum_{n=1}^{4} e^{ik.r_n} = E_{sp} B_2(k), \ \ \langle s^c|H|p_z^a\rangle = E_{sp} \sum_{n=1}^{4} e^{ik.r_n} = E_{sp} B_3(k),
\end{aligned} \tag{1.20}$$

where B_i^* is the complex conjugate of B_i matrix elements that are expressed as

$$\begin{aligned}
B_0(k) &= 4Cos\left(\frac{k_x a}{2}\right)Cos\left(\frac{k_y a}{2}\right)Cos\left(\frac{k_z a}{2}\right) - 4i\,Sin\left(\frac{k_x a}{2}\right)Sin\left(\frac{k_y a}{2}\right)Sin\left(\frac{k_z a}{2}\right), \\
B_1(k) &= 4Cos\left(\frac{k_x a}{2}\right)Sin\left(\frac{k_y a}{2}\right)Sin\left(\frac{k_z a}{2}\right) + 4i\,Sin\left(\frac{k_x a}{2}\right)Cos\left(\frac{k_y a}{2}\right)Cos\left(\frac{k_z a}{2}\right), \\
B_2(k) &= -4Sin\left(\frac{k_x a}{2}\right)Cos\left(\frac{k_y a}{2}\right)Sin\left(\frac{k_z a}{2}\right) + 4i\,Sin\left(\frac{k_x a}{2}\right)Sin\left(\frac{k_y a}{2}\right)Cos\left(\frac{k_z a}{2}\right), \\
B_3(k) &= -4Sin\left(\frac{k_x a}{2}\right)Sin\left(\frac{k_y a}{2}\right)Cos\left(\frac{k_z a}{2}\right) - 4i\,Cos\left(\frac{k_x a}{2}\right)Cos\left(\frac{k_y a}{2}\right)Sin\left(\frac{k_z a}{2}\right),
\end{aligned} \tag{1.21}$$

where $i = \sqrt{-1}$ and $r_1 = (a/2)(1, 1, 1)$, $r_2 = (a/2)(1, -1, -1)$, $r_3 = (a/2)(-1, 1, -1)$ and $r_4 = (a/2)(-1, -1, 1)$ are displacement vectors of nearest

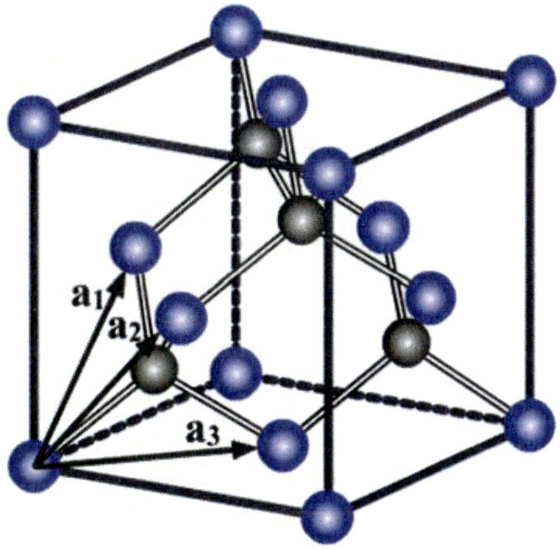

Fig. 1.5 Unit cell group III–V and II–VI compounds with zinc-blende crystal structure

neighbors. Figure 1.5 shows the schematic view of unit cells for tetrahedral semiconductors.

The on-site atomic energies E_{sa}, E_{sc}, E_{pa} and E_{pc} off-site atomic $E_{ss}, E_{xx}, E_{s_a p_c}, E_{s_c p_a}$ and E_{xy} dominate the acceptable features of electronic band structure of semiconductors. The two off-diagonal elements E_{ss} and E_{xx} are obtained from the bandgap at Γ high symmetry point ($k_x = k_y = k_z = 0$) in the first Brilloun zone for which B_1, B_2 and B_3 are all zero and B_0 is nonzero. The strain in a heterostructure will modify the off-site TB matrix elements with respect to their unstrained values and are often determined by assuming that they obey the Harrison Scaling Law [45]:

$$V_{ll'm}(\varepsilon) = V_{ll'm}(a/a_o)^{-\eta_{llm}} \tag{1.22}$$

where $V_{ll'm}(\varepsilon)$ and $V_{ll'm}$ are the strained and bulk value of interaction potential for anion and cation atoms. The exponents η_{llm} are determined to reproduce the strain dependent band structure of relevant semiconductors under hydrostatic pressure, namely the volume deformation potential $a_{gl} = -B(\partial E_{gl}/\partial P)$ of corresponding band gaps $E_{g\Gamma}$, E_{gL} and E_{gX} at high symmetry points.

In the framework of second nearest interaction (2NN) sp^3 TB model of Lohrer and Talwar [46, 49], with addition of spin–orbit interaction, considering a spherical symmetric potential to the Hamiltonian [25], greater improvement of electronic band structures of bulk binary semiconductors is obtained. Results are given in Fig. 1.6a and b, respectively, for the direct bandgap GaAs and indirect bandgap AlAs binary group III–V compound semiconductors, where the spin–orbit splitting of the valence bands are also shown.

Semiconductor alloys improve the performance of heterostructure devices because they allow the device designer to locally modify the band structure of the semiconductor (e.g., increasing direct bandgap) and in turn control the motion of the

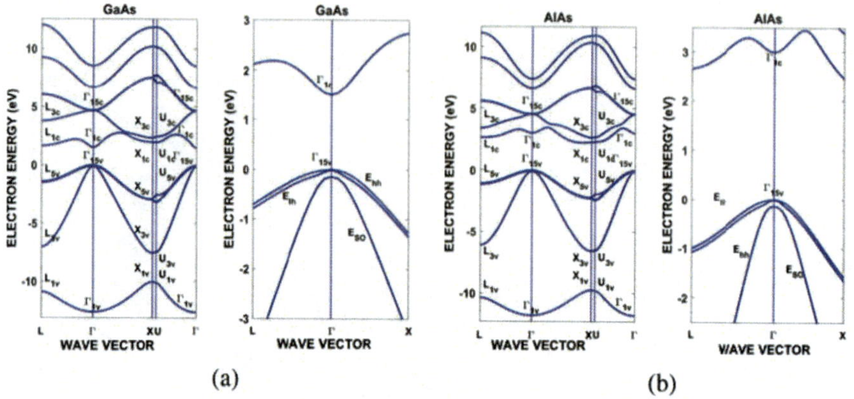

Fig. 1.6 Electronic band structure of direct bandgap GaAs **a** and indirect bandgap AlAs **b** calculated by using the 2NN sp^3 TB parametrization of Tran et al. [50]. Spin orbit splitting of the valence bands are also shown for both semiconductors

charge carriers. In order to understand how such local modification of band structure can affect the motion of positive and negative charge carriers, one needs to understand the composition variation of valence and conduction band energy levels of tetrahedral semiconductors as a function of **k** wave vector. Using the second nearest interaction (2NN) sp^3 TB parametrization of Lohrer and Talwar [46], with rotational spin interaction added to the Hamiltonian, calculated composition effects on band structure of AlGaAs/GaAs heterostructures are shown in Fig. 1.7 for 20 and 30% aluminum composition. Increasing Al composition from 20 to 30% results in increase in direct bandgap of AlGaAs ternary semiconductor.

The motion of negative and positive charged particles in conduction and valence bands are described by their group velocities and second law of Newton according to (1.13). In this concept, the perspective view of band structure of semiconductor components is important in order to accurately simulate the charge transport and performance of low dimensional heterostructure devices working under high strains and temperatures. Charge carriers move with a greater acceleration and much faster in parabolic regions of band structure, as described by (1.11)–(1.13). Probability of charge carriers being scattered by the local impurities will be lesser in parabolic regions than that are non-parabolic. Perspective plots of the energy band structures derived from 2NN sp^3 tight binding parametrization of Lohrer and Talwar [46] for AlGaAs of lattice matched AlGaAs/GaAs and from 2NN sp^3s^* tight binding model for InGaAs/GaAs heterostructures are plotted in Fig. 1.8. Using the perspective view we can easily visualize the effect of ternary alloy semiconductor band structure on the motion of negative and positive charge carriers in conduction and valence band energy levels, respectively.

Figure 1.9 compares the calculated composition effects on the direct and indirect bandgaps of lattice matched AlGaAs/GaAs heterostructures at high symmetry points of Brillouin zone for each constituent ternary semiconductor against experimental

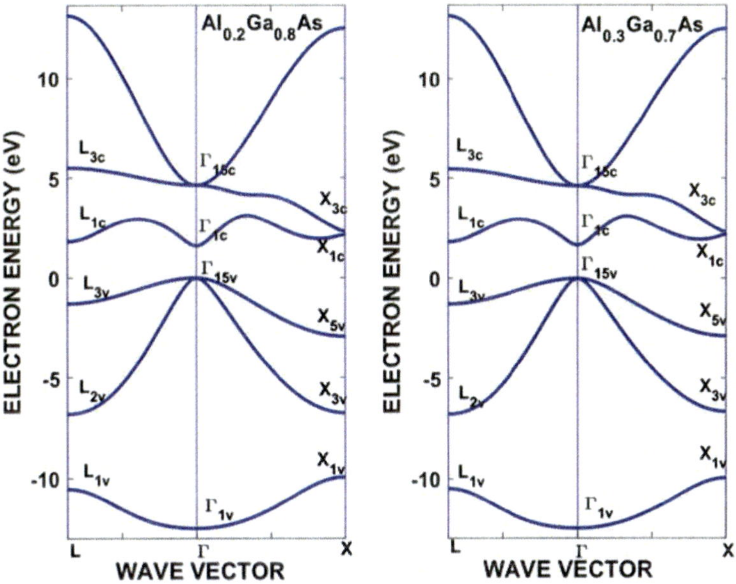

Fig. 1.7 Band structure of AlGaAs with x = 0.20 (left) and x = 0.30 (right) calculated with the 2NN sp^3 TB parameters of Loehr and Talwar [46]

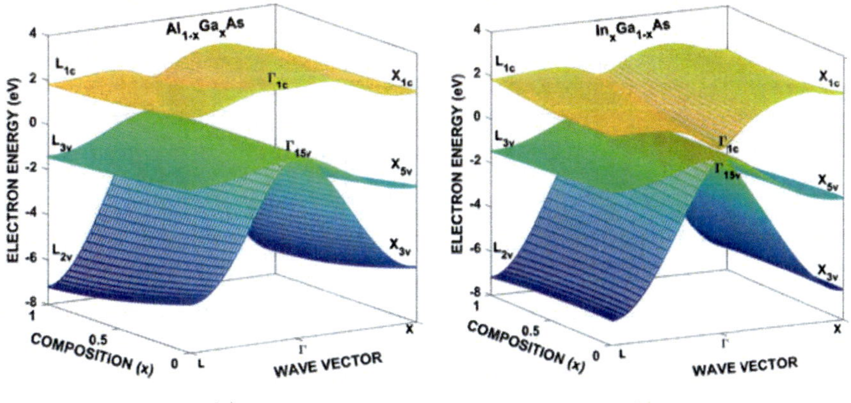

Fig. 1.8 Three-dimensional view of the composition effects on the band structure of AlGaAs in lattice matched AlGaAs/GaAs heterostructure (**a**) and InGaAs in coherently strained InGaAs/GaAs heterostructure (**b**) at room temperature

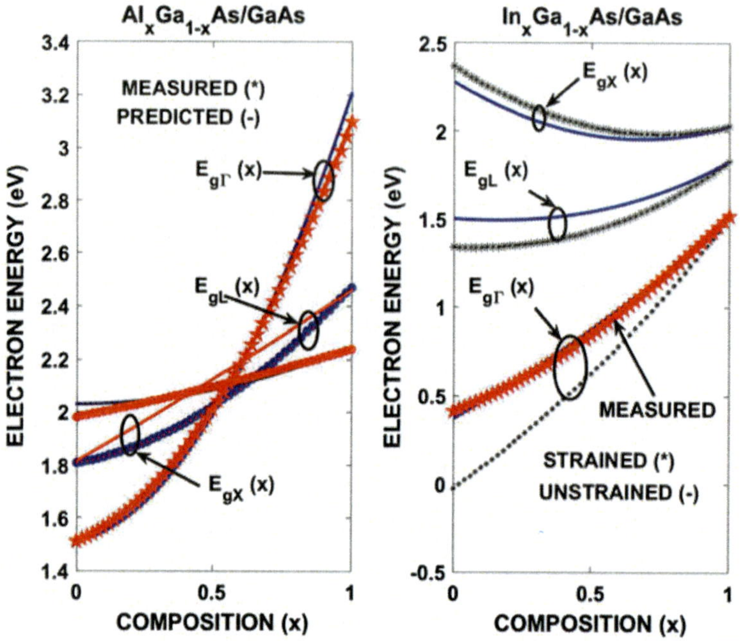

Fig. 1.9 Predicted strain effects on bandgaps of AlGaAs in lattice matched AlGaAs/GaAs (left) and of InGaAs in pseudomorphic InGaAs/GaAs (right) heterostructures

data. Since the lattice mismatch at the interface of AlGaAs/GaAs heterostructure is nearly zero and only thermal expansion due to linear expansion of lattice constants of AlGaAs and GaAs modify the band structure. Although the Slater-Koster type semiempirical 2NN sp^3s tight binding parametrization of electronic structures yields a good description of valence band dispersion curves, the conduction band dispersion curves are inaccurately given, especially in calculating the bandgap at the X symmetry point is not well reproduced. Since the predicted bandgaps depend on experimental data, in order to obtain reliable and accurate tight binding parameters which will produce reliable band structure, one must go through a fitting process that depends on the mapping of a large number of orbital coupling parameters on the set of observables; in many cases there are not many analytical expressions available. This is observed in determining the indirect band gaps at L and X high symmetry points.

One can overcome the difficulty associated with the 2NN sp^3s TB parametrization for L and X bandgaps by using a so-called statistical thermodynamic model [28, 29], which considers the conduction electrons and valence holes as distinct electrically chemical species, to study the interface strain and ternary composition effects on the electronic structure of heterostructures at symmetry points. In the framework of statistical thermodynamic model of semiconductors, one first calculates the valence band $E_v(x)$ of $A_xB_{1-x}C$ ternary alloy by writing the bond length and valence band energies according to the modified Vegard's rule described by (1.9) and (1.11), respectively. The bandgap energies at high symmetry points k = (2 π/a)(0; 0; 0) and k = (2 π/a)(1; 0; 0) are obtained by combining the Vegard's rule and Kane's *k.p approximation* used in determining the effective mass of electrons in conduction band at symmetry points [28, 29]:

$$\frac{1}{m_{n\Gamma}(x)}=\frac{x}{m_{nA\Gamma}(x)}+\frac{1-x}{m_{nB\Gamma}(x)}=1+\frac{P_{\Gamma}^{2}(x)}{3}\left(\frac{2}{E_{g\Gamma}(x)}+\frac{1}{E_{g\Gamma}(x)+\Delta(x)}\right) \tag{1.23}$$

$$\frac{1}{m_{ntL}(x)}=\frac{x}{m_{ntLA}(x)}+\frac{1-x}{m_{ntLB}(x)}=1+P_{L}^{2}(x)\left(\frac{2}{E_{gL}(x)}+\frac{1}{E_{gL}(x)+\Delta_{L}(x)}\right) \tag{1.24}$$

where $m_{nA\Gamma}(x)$ and $m_{nB\Gamma}(x)$ are the electron effective masses at Γ_{1C} conduction band valley and m_{ntLA} and m_{ntLB} are the transverse electron effective masses at L conduction valley of binaries AC and BC. $P_{\Gamma}(x)=xP_{A\Gamma}+(1-x)P_{B\Gamma}$ and $P_{L}(x)=xP_{LA}+(1-x)P_{LB}$ are the momentum matrix elements $\Delta(x)=x\Delta_{A}+(1-x)\Delta_{B}$ and $\Delta_{L}(x)=x\Delta_{LA}+(1-x)\Delta_{LB}$ are the spin–orbit energies. $P_{A\Gamma}$, $P_{b\Gamma}$, P_{AL} and P_{BL} of binaries AC and BC are obtained from (1.23) to (1.24) with measured $m_{n}(x)$, $m_{ntL}(x)$, $\Delta(x)$ and $\Delta_{L}(x)$ for composition of x = 0 and 1.

Inverting (1.23) and (1.24) gives quadratic equations for bandgaps $E_{g\Gamma}(x)$ and $E_{gL}(x)$. Solving the resultant quadratic equation, one obtains following expression for these bandgaps

$$E_{g\Gamma}(x)=\frac{3y_{\Gamma}^{-1}(x)-\Delta(x)}{2}+\frac{1}{2}\left[(\Delta(x)-3y_{\Gamma}^{-1}(x))^{2}+8\Delta(x)y_{\Gamma}^{-1}(x)\right]^{1/2} \tag{1.25}$$

$$E_{gL}(x)=\frac{2y_{\Gamma}^{-1}(x)-\Delta_{1}(x)}{2}+\frac{1}{2}\left[(\Delta_{1}(x)-2y_{\Gamma}^{-1}(x))^{2}+4\Delta_{1}(x)y_{\Gamma}^{-1}(x)\right]^{1/2} \tag{1.26}$$

$$y_{\Gamma}(x)=\frac{3(1-m_{n\Gamma}(x))}{m_{n\Gamma}(x)P_{\Gamma}^{2}(x)},\quad y_{L}(x)=\frac{(1-m_{nt\Gamma}(x))}{m_{nt\Gamma}(x)P_{L}^{2}(x)} \tag{1.27}$$

The indirect gap $E_{gX}(x)$ of $A_{x}B_{1-x}C$ ternary is determined from the following empirical expression

$$E_{gX}(x)=E_{g\Gamma}(x)+\left(E_{gXA}-E_{g\Gamma A}\right)x+\left(E_{gXB}-E_{g\Gamma B}\right)(1-x) \tag{1.28}$$

E_{gXA}, $E_{gXB,}E_{g\Gamma A}$, and $E_{g\Gamma B}$ are the indirect and direct bandgap energies at Γ and X high symmetry points of AC and BC binary semiconductors.

Figure 1.10 compares the measured and calculated alloy composition variation of the direct and indirect bandgaps of AlGaAs ternary of AlGaAs/GaAs and InGaAs ternary of InGaAs/GaAs heterostructures, respectively. There is excellent

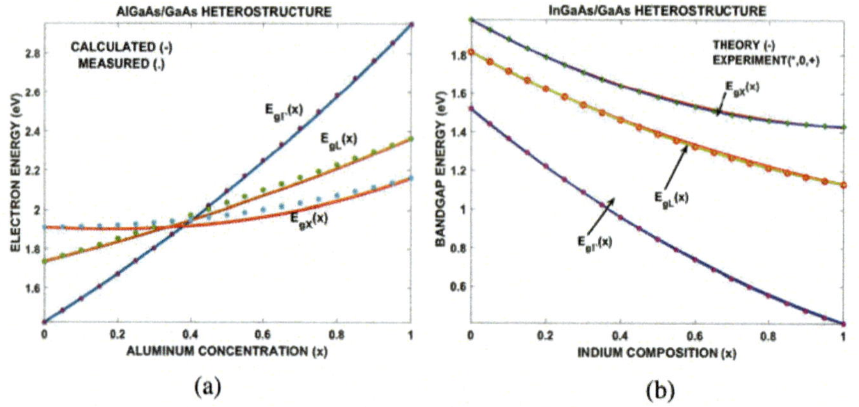

Fig. 1.10 Predicted strain effects on bandgaps of AlGaAs in lattice matched AlGaAs/GaAs (left)

agreement between the predictions of (1.25)–(1.27) with measured bandgaps for both AlGaAs/GaAs and InGaAs/GaAs heterostructures on (001) GaAs buffer. In the calculations following parameters were used: $m_{n\Gamma}/m_0 = 0.063$ and 0.08, $m_{nt}/m_0 = 0.0754$ and 0.0964, $\Delta_s/(eV) = 0.35$ and 0.29, $\Delta_1/(eV) = 0.23$ and 0.21, $E_{g\Gamma}/(eV) = 1.425$ and 2.95, $E_{gL}/(eV) = 1.734$ and 2.363, and $E_{gX}/(eV) = 1.911$ and 2.161 for GaAs and AlAs binary compounds, respectively and of InGaAs in pseudomorphic InGaAs/GaAs (right) heterostructures.

1.5.2 Semiempirical sp^3s^* *Tight Binding Theory*

In order to eliminate the inaccuracy of the 2NN sp^3 TB parametrization in predicting the conduction band dispersion curves, mainly in calculating the bandgap at the X symmetry point, Vogl et al. [51] added the higher lying excited d-states to the semiempirical nearest neighbour sp^3 TB parametrization. In this new (NN) sp^3s* tight binding theory there are 13 TB parameters, which are determined by comparing the predicted bandgaps with those produced by empirical pseudopotential model [14] at high symmetry points. As addition, the inclusion of 2NN interactions of cation and anion atoms in the sp^3s^* TB model yields better fit of conduction band dispersion curve at the L symmetry point. Adding the spin–orbit coupling in the 2NN sp^3s^* TB model increases the size of the TB Hamiltonian matrix (10 × 10) to (20 × 20), which is diagonalized for each **k** vector to obtain the band structure [51].

After the optimization of the diagonal matrix elements we then can have a realistic TB parametrization of the off-diagonal matrix elements of 2NN sp^3s* orbitals set, which are obtained by fitting the obtained band gaps at high symmetry points to those produced by empirical pseudopotential theory [14]. Table 1.1 lists the optimized TB parameters and Table 1.2 [5, 16, 17] lists the direct and indirect bandgaps at high symmery points for AlAs, GaAs, InAs, AlN, GaN and InN group III–V binary

Table 1.1 Optimized 2NN sp^3s^* TB parameters for AlAs, GaAs, InAs, AlN, GaN and InN obtained by fitting the band structure to experimental bandgaps at high symmetry points

(eV)	AlAs	GaAs	InAs	AlN	GaN	InN
$E_{s,a}$	−7.5253	−8.4399	−9.5381	−11.505	−12.915	−12.860
$E_{p,a}$	1.1627	0.9252	0.7733	4.3815	3.1697	1.9800
$E_{s,c}$	0.9833	−2.6569	−2.7219	0.5047	−1.5844	−0.3994
$E_{p,c}$	3.5867	3.5523	3.5834	10.2184	9.0302	8.0200
$E_{s*,a}$	7.4863	6.6235	7.2730	12.0400	12.2000	10.6300
$E_{s*,c}$	6.7267	7.4249	6.6095	13.7400	12.2000	13.0000
$4V_{s,s}$	−6.642	−6.4210	−5.6052	−9.8077	−8.8996	−4.2285
$4V_{x,x}$	1.8730	1.9850	1.8398	6.6900	5.3500	3.9800
$4V_{x,y}$	4.2919	4.9100	4.3977	8.9400	8.6200	7.4100
$4V_{sa,pc}$	5.1106	4.2390	3.0205	7.8500	6.4000	3.8100
$4V_{pa,sc}$	5.4965	5.15358	5.3894	7.6800	7.2400	6.1900
$4V_{s*a,pc}$	4.5216	3.80624	3.2191	8.0300	7.0600	6.8800
$4V_{pa,s*c}$	4.9950	4.7009	3.7234	2.4700	1.8200	3.3600
ε_{sx}	0.2459	0.2459	0.1441	−1.4000	0.9500	0.6150
ε_{xy}	−0.1050	−0.1050	0.0249	6.9000	1.0100	0.7100
λ_a	0.0072	0.0553	0.1385	0.0035	0.0035	0.0035
λ_c	0.1721	0.1338	0.1290	0.0070	0.0410	0.1100

Table 1.2 Bandgaps at symmetry points of AlAs, GaAs, InAs, AlN, GaN and InN compounds obtained using the 2NN sp^3s^* TB model [5, 16, 17] and experimental data [43] are given in parenthesis

(eV)	AlAs	GaAs	InAs	AlN	GaN	InN
$E_{g\Gamma}$	3.13 (3.13)	1.52 (1.52)	0.43 (0.42)	5.99 (6.00)	3.30 (3.27)	0.90 (0.90)
E_{gX}	2.23 (2.23)	1.98 (2.03)	2.28 (2.50)	4.90 (4.80)	4.70 (4.70)	2.83 (3.00)
E_{gL}	2.58 (2.54)	1.82 (1.85)	1.61 (1.43)	8.63	6.10	3.81

compound semiconductors which may be used to calculate their bulk and ternary electronic band structures.

Figure 1.11 describes the band structure of AlAs, GaAs, InAs, AlN, GaN, and InN binary compounds determined by using the optimized diagonal and off-diagonal matrix elements of 2NN sp^3s^* TB Hamiltonian listed in Table 1.1. The interaction parameters $4V_{s*,p}$ and $4V_{p,s*}$ are adjusted to fit to the X bands and 2NN interaction parameters ε_{sx} and ε_{xy} to get a good fit to the L bands in reproducing the empirical pseudopotential energy bands [14]. Adding the excited s* state to the sp^3 orbitals set on the cation and anion atoms with 2NN interactions and spin–orbit

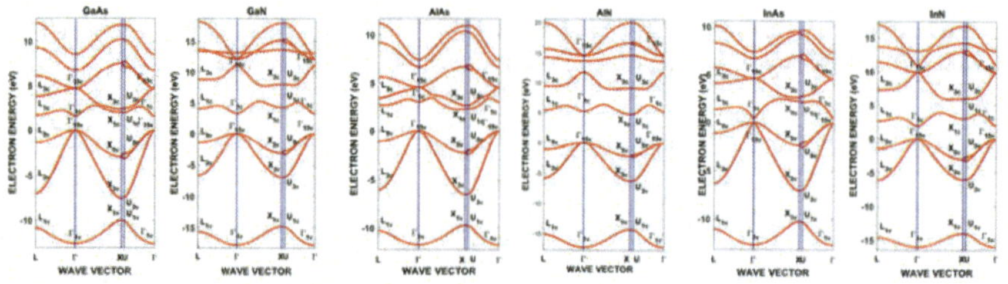

Fig. 1.11 Electronic band structure of GaAs, GaN, AlAs, AlN, InAs and InN III–V compound semiconductor calculated by using the *2NN* sp^3s^* TB parametrization

coupling of p-states, improves the simulation of the conduction band structure of III–V compounds, especially at the X symmetry point, accurately reproducing the empirical pseudopotential bands at symmetry points of conduction bands.

As an example of implementing the modified virtual crystal approximation (MVCA) for the calculation of composition effects in semiempirical NN or *2NN* sp^3s^* tight binding model, the tight binding parameters given in Table 1.1 is used in the calculations of the electronic structures of GaAsN, InasN, AlGaN and InGaN group III-nitride ternaries in k-space for various alloy compositions. By implementing the modified virtual crystal approximation (MVCA) [21, 22, 52] in the semiempirical NN or 2NN sp^3s^* tight binding model, one can then calculate the composition effects on tight binding parameters. Using bandgap energies in Table 1.2, tight binding parameters are optimized and used in the calculations of the electronic structures of AlGaN, InGaN and GaAsN group III-nitride ternary alloys in k-space for several compositions. Increasing the alloy concentration we can increase the direct bandgap region of the ternary semiconductors in ternary/binary heterostructures. Figure 1.12a–d show the expected trend in band structures of GaAsN, InAsN, AlGaN and InGaN ternaries of GaAsN/GaN, InAsN/InN, AlGaN/GaN and InGaN/GaN heterostructures.

As stated earlier, the perspective view of band structure of ternary alloy semiconductor is important in order to accurately simulate the charge transport and performance of low dimensional heterostructure devices working under high temperatures and strains. Charge carriers move with a greater acceleration and much faster in parabolic regions of band structure, as described by (1.11)–(1.15). Probability of being scattered by the local impurities will be lesser in parabolic regions than that are non-parabolic regions of the semiconductor band structure. Figure 1.13a and b show the perspective plots of the band structures of of GaAsN, InAsN, AlGaN and InGaN group III—ternaries of GaAsN/GaN, InAsN/InN, AlGaN/GaN and InGaN/GaN group III-nitride heterostructures, plotted as a function of wave vector for the entire composition range. These plots allow us to visualize the dynamics of charge carriers as a function of alloy composition and predict and simulate the performance of low dimensional ternary/binary heterostructure devices for electronic and optical applications.

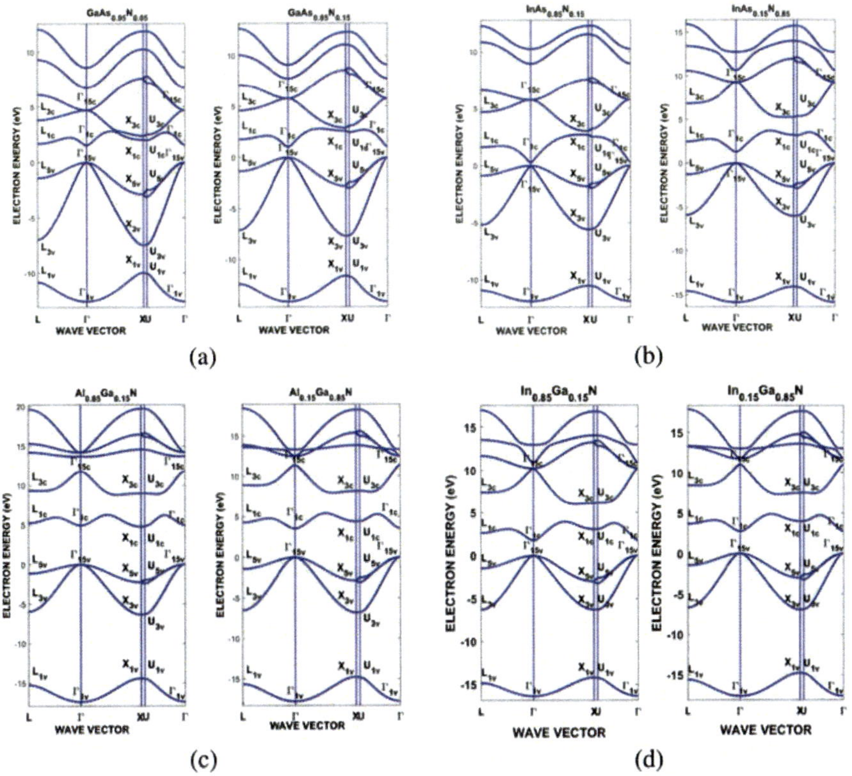

Fig. 1.12 Two-dimensional plots of composition effects on band structure of GaAsN, InAsN, AlGaN and InGaN ternaries of GaAsN/GaN (**a**), InAsN/InN (**b**), AlGaN/GaN (**c**) and InGaN/GaN (**d**) III-nitride ternary/binary heterostructures calculated by using the *2NN* sp^3s^* TB parameters

Furthermore, Fig. 1.14a and b show the alloy omposition effect on band structure of AlInN ternary of AlInN/InN heterostructure that are calculated by using the *2NN* sp^3s^* TB parameters.

The perspective plot in Fig. 1.14b allow us to better visualize the dynamics of charge carriers as a function of alloy composition and predict and simulate the performance of low dimensional ternary/binary heterostructure devices for electronic and optical applications.

The composition effects on the energy band properties of AlGaN/GaN, InGaN/GaN and GaAsN/GaAs ternary/binary group III-nitride heterostructures are summarized in Fig. 1.15 for AlGaN/GaN (left), InGaN/GaN (middle) and GaAsN/GaAs (right) heterostructures.

Predicted fundamental bandgaps of AlGaN, InGaN, and GaAsN ternaries at the Γ, L and X high symmetry points, especially at Γ symmetry point, are in excellent agreement with experiment [41]. The fundamental band gaps of AlGaN and InGaN nitrides increase with composition. However, the bandgap of GaAsN diluted nitride at the Γ

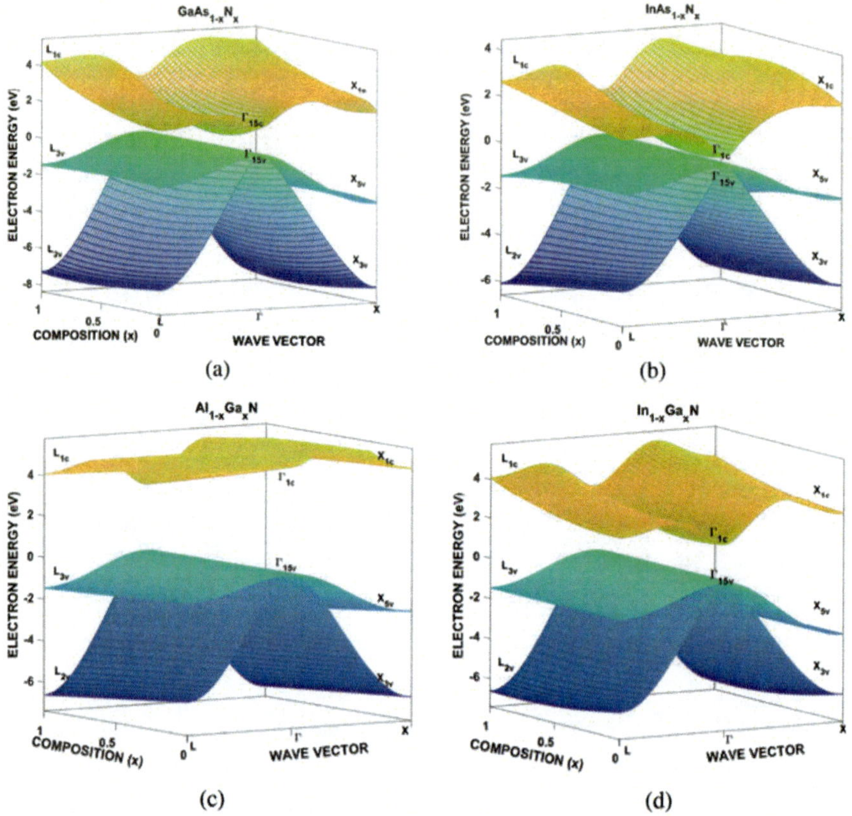

Fig. 1.13 Perspective views of band structure of GaAsN (**a**), InAsN (**b**), AlGaN (**c**) and InGaN (**d**) ternaries of GaAsN/GaN, InAsN/InN, AlGaN/GaN and InGaN/GaN III-nitride ternary/binary heterostructures calculated by using the *2NN* sp^3s^* TB parameters

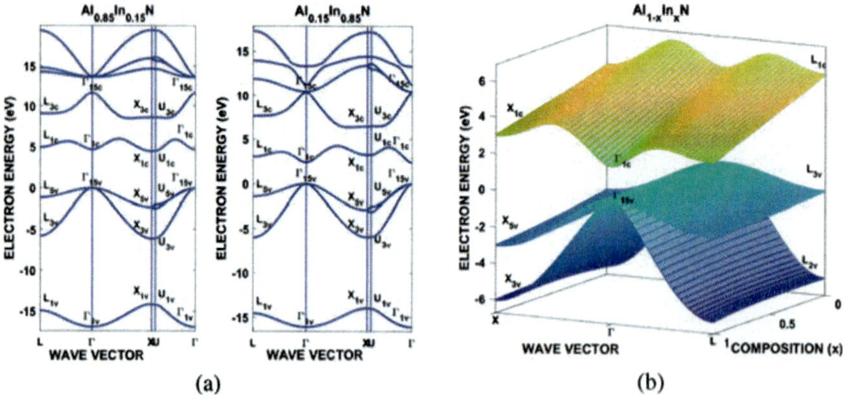

Fig. 1.14 Two-dimensional (**a**) and perspective (**b**) views of composition effects on band structure of InAsN ternary calculated by using the *2NN* sp^3s^* TB parameters

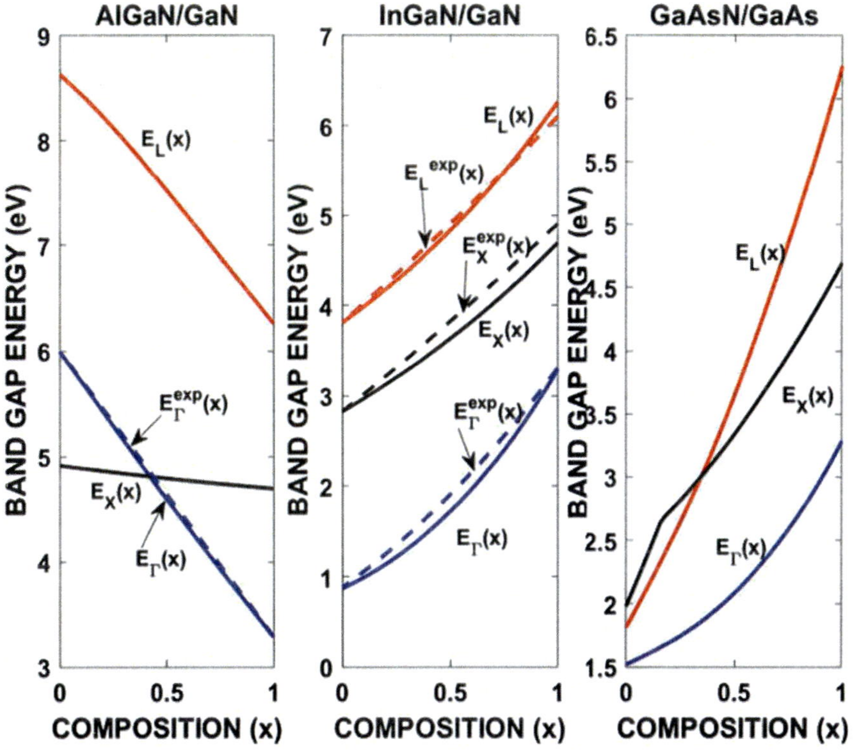

Fig. 1.15 Comparison of measured and predicted band gaps of AlGaN, InGaN and GaAsN ternaries in AlGaN/GaN (left), InGaN/GaN (middle) and GaAsN/GaAs (right) heterosructures calculated by using optimized *2NN* sp^3s^* TB parameter listed in Table 1.1

symmetry point decreases with alloy composition for $x < 0.25$, and then increases for $0.26 < x < 1$, in agreement with experiment. The lattice mismatch induced interface strain effects on the fundamental band gaps at Γ, L and X symmetry points can be quite large when the band gap (or conduction band) deformation potential is large. It is interesting to note that the optimized *2NN* sp^3s^* tight binding parameters in Table 1.1 gives excellent agreement with measured fundamental bandgaps of AlGaN, InGaN and GaAsN ternary semiconductors over the entire composition range without any empirical fitting or any adjustable parameter. These results suggest that by using the optimized 2NN sp^3s^* TB parameters for bulk GaAs, GaN, AlN and InN binary compounds, given in Table 1.1, the 2NN sp^3s^* TB model allows one to determine the nonlinear composition dependence of the principal band gaps of nitride-based ternary semiconductors without any empirical fitting nor any adjustable parameter.

1.5.3 *Semiempirical* sp^{3}d^5 *Tight Binding Theory*

In this section we discuss the calculation of the semiconductor electronic structure by using the semiempirical sp^3d^5 tight binding parametrization, with first nearest neighbor (NN) and spin–orbit coupling (SO) of p-states considered. The NN sp^3d^5 TB Hamiltonian matrix is written as

$$H_{\alpha\beta} = \langle\varphi_\alpha(k)|H\left|\varphi_\beta(k)\right\rangle = E_{\alpha\beta} + \sum_{i\neq 0} I_{\alpha\beta}(0,i)e^{ik.r_i} + H_{so} \qquad (1.29)$$

where $\varepsilon_{\alpha\beta}$ is the on-site energy for β orbital (s, p, d) at the atomic site α (cation and anion) and represents the intra-atomic integrals, which couple atomic orbitals located in the same cell. $I_{\alpha\beta}(0, i)$ represents the first nearest neighbor interaction integrals (hopping term) which couples atomic orbitals located in different cells. The nine atomic-like states (s; x, y, z; xy, yz, zx, $x^2 - y^2$, $3z^2 - r^2$) basis set describes each atom. There are 19 independent matrix elements: 9 on-site and 10 off-site elements. Adding the spin–orbit coupling to the sp^3d^5 orbitals basis set adds two extra tight binding parameters. The spin–orbit interaction is given by $\lambda_a =< x_a \uparrow |H_{so}|z_a \downarrow>$ for anion and $\lambda_c =< x_c \uparrow |H_{so}|z_c \downarrow>$ for cation atom. The role of λ_a is to reproduce the bulk zone center splitting between the split-off band and the light and heavy hole bands.

Group II–VI alloy semiconductors are considered as useful materials in fabricating heterojunction photovoltaic devices because their band gaps and lattice parameters can be tailored so that ternaries or quaternaries with different compositions can be grown on a suitable lattice matched substrate. The potential barrier in the conduction and the potential step in the valence bands at interface influences the carrier transport and device performance [31], making the reliable and accurate modelling of interface energy band structure highly crucial to understand the impact of group II–VI heterostructures in the design and optimization of semiconductor photovoltaic devices. We use the sp^3d^5 TB model with nearest neighbor interactions and spin–orbit coupling of p-states are considered to calculate the band structure of group II–VI ternary/binary structures. As example, we shall focus on band structure, band gaps ZnS, ZnSe, ZnTe, CdSe, CdS and CdTe II–VI compound semiconductors for which we were able to find some experimental data to compare. We will compare the predictions of nearest neighbor (NN) sp^3d^5 tight binding model, with spin–orbit coupling of p-states, to those of widely used second nearest neighbor (2NN) sp^3s^* model, with spin–orbit coupling of p-states [42]. Table 1.3 give the bandgap energies used to optimize tight binding parameters for the NN sp^3d^5 and 2NN sp^3s^* TB models that are needed to calculate the band structure of CdS, ZnS and CdSe binaries.

The matrix elements are determined by fitting the nonlocal pseudopotential energy band dispersion relations to those obtained from the tight binding model for bulk semiconductors. In doing so, one starts with calculating the electronic band structure by estimating the values of on-site matrix elements and off-site matrix elements and

Table 1.3 Bandgap energies of CdS, CdSe and ZnS bulk semiconductors, obtained from the fitting of the ETB model, with sp^3d^5 and sp^3s^* orbital basis sets

Bandgap (eV)	CdS		ZnS		CdSe	
	sp^3s^*	sp^3d^5	sp^3s^*	sp^3d^5	sp^3s^*	sp^3d^5
E_g^Γ	2.476	2.555	3.680	3.702	1.887	1.937
E_g^X	4.341	4.696	5.103	5.190	3.784	3.779
E_g^L	3.983	5.193	4.810	4.641	3.097	3.027

then carry out a least-squared error minimization fitting procedure at a number of high symmetry points in the valence and conduction band dispersion curves to fit band gap energies obtained from the empirical nonlocal pseudopotential method [42].

Table 1.4 gives the optimized tight binding parameters for 2NN sp^3s^* TB model used in the band structure calculations of group II–VI binary compounds such as CdS, ZnS and CdSe.

Table 1.5 give the optimized tight binding parameters for the NN sp^3d^5 TB model used in the band structure calculations of group II–VI binary compounds such as CdS, ZnS and CdSe.

Figure 1.16 compare the electronic band structure of CdS, ZnS and CdSe calculated using NN sp^3d^5 and 2NN sp^3s^* TB models, both reproducing the band gaps as well as the valence band and conduction band dispersion curves at Γ, X and

Table 1.4 The optimized *s*, *p* and *s** TB parameters for CdS, CdSe and ZnS compounds obtained from fitting 2*NN* sp^3s^* TB model [6] to pseudopotential bands at symmetry points [14]

Parameters (eV)	CdS	CdSe	ZnS
$E_{s,a}$	−11.5300	−11.610	−9.63
$E_{p,a}$	0.5300	1.4800	1.326
$E_{s,c}$	1.8300	1.1100	0.03
$E_{p,c}$	5.8700	6.5200	4.73
$E_{s^*,a}$	7.1300	8.0800	7.53
$E_{s^*,c}$	6.8700	8.0200	5.72
$4V_{s,s}$	−3.0700	−6.3000	−4.64
$4V_{x,x}$	1.7600	3.1100	2.64
$4V_{x,y}$	4.2300	5.0000	5.36
$4V_{s,p}$	2.1700	5.1600	4.57
$4V_{p,s}$	5.4800	5.1700	5.54
$4V_{s^*,p}$	1.9900	2.8900	3.05
$4V_{p,s^*}$	3.0600	1.7500	2.49
ε_{sx}	0.1000	0.2000	0.0
ε_{xy}	−0.0100	−0.1500	0.0
λ_a	0.0250 [8]	0.0250 [8]	0.1434
λ_c	0.0130 [14]	0.0270 [14]	0.0591

Table 1.5 Tight binding parameters for NN sp^3d^5 TB calculations of electronic band structure of CdS, ZnS, and CdSe II–VI compounds

(eV)	CdS	ZnS	CdSe
s_c	1.8300	1.3600	0.0300
p_c	5.8700	6.5200	4.7300
$d_c(t_2)$	−6.8300	−5.8200	−7.3100
$d_c(e)$	−7.4400	−6.2100	−7.8100
s_a	−11.5300	−14.6100	−9.6300
p_a	0.9300	1.7800	1.3260
$d_a(t_2)$	14.4300	15.5400	15.2600
$d_a(e)$	13.1500	13.6000	14.1000
$ss\sigma$	−0.8440	−1.0050	−1.2600
$sp\sigma$	2.2729	2.5387	2.3989
$pp\sigma$	1.68963	1.3343	1.9789
$pp\pi$	2.5500	4.3775	3.3400
$ps\sigma$	−0.10750	−0.3725	−0.6800
$ds\sigma$	0.0000	−2.5900	−1.2200
$dp\sigma$	−1.1400	−0.0500	−0.0100
$dp\pi$	−1.2900	−2.7800	−1.0900
$sd\sigma$	1.7500	1.3700	1.5200
$pd\sigma$	1.5580	2.1100	1.7800
$pd\pi$	−0.3500	−0.4500	−0.3200
λ_a	0.0250	0.0250	0.1434
λ_c	0.07600	0.0270	0.0591

L symmetry points, which cannot be done with NN sp^3 and 2NN sp^3 TB models. The NN sp^3d^5 TB parametrization, including the spin¯orbit coupling of p-states, is adequate to accurately reproduce the band gaps at Γ, X and L symmetry points and both the valence band and conduction band dispersion curves. When compared with the 2NN sp^3s^* TB model, the NN sp^3d^5 TB model better simulates both the valence band and conduction band dispersion curves. This is due to the fact that the inclusion of s^* excited state is done by modeling the average of p-d interactions and it does not permit the inclusion of excited d-orbitals to sp^3 basis. Therefore, the actual behavior and contribution of the excited d-states in the band structure calculations is not reliably reproduced, suggesting that the sp^3s^* TB model is of limited value for determination of optical properties of semiconductors involving high symmetry points. However, we should add that both of these TB models with NN sp^3d^5 and 2NN sp^3s^* orbitals sets are semiempirical and their basis lies in a good description of the electronic band structures of tetrahedral semiconductors.

Using the sp^3d^5 tight binding model parameters given in Table 1.5 we can also calculate the composition effects on the band structure of ternary/binary heterostructures. Figure 1.17 shows the comparison of the sp^3d^5 and sp^3s^* TB models prediction of the electronic band structure of $Cd_{1-x}Zn_xS$ ternary semiconductor along $\Gamma - L$

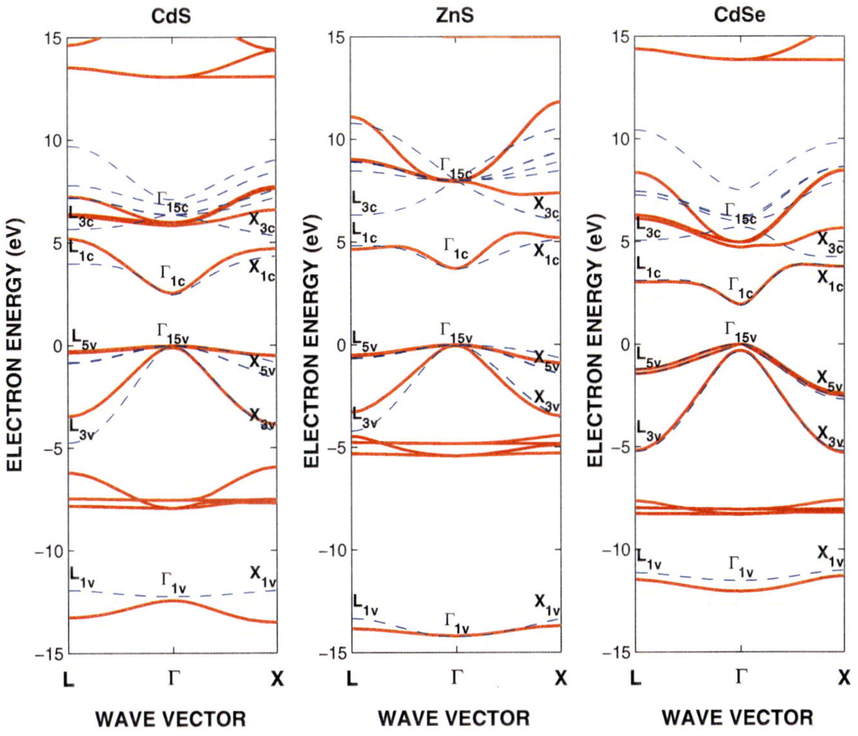

Fig. 1.16 Energy band structure of CdS, CdSe and ZnS bulk semiconductors along $\Gamma - L$ and $\Gamma - X$ directions, obtained by using the ETB model with NN sp^3d^5(solid line) and 2NN sp^3s^* (dashed line) orbital basis sets for the tight binding parameters given in Tables 1.1 and 1.5

and $\Gamma - X$ directions, with Zn composition x = 0.25 and 0.75, obtained by using given in Table 1.5. As noted in Sect. 1.2, there is a considerable lattice mismatch induced interface strain in II–VI ternary/binary heterostructures over most of the composition range. Interface strain causes changes in the bond length and tight binding parameters. Interface strain will modify the off-site tight binding matrix elements representing the nearest neighbor interactions, known as the hopping strength, with respect to their unstrained values and are often determined by using (1.22), obeying the universal Harrison scaling law [45]: $V^s_{ll'm} = V_{ll'm}(a/a_o)^{-\eta_{llm}}$. Here a and a_o are the strained and unstrained values of inter-atomic distance between nearest neighbors. η_{llm} is adjusted in order to fit the first pressure derivatives of band gaps $E_{g\Gamma}$, E_{gL} and E_{gX} at Γ, L and X symmetry points to their experimental values.

Semiempirical tight binding theory with sp^3d^5 orbital basis set is based on the representation of atomic interactions and need to be fitted to bulk material parameters. The fitting process is a difficult one since it depends on the mapping of large number of orbital coupling parameters on the set of observables and there are not many analytical expressions available. Therefore, accurate and reliable determination of the strain effects on the energy levels by fitting the off-site tight binding matrix

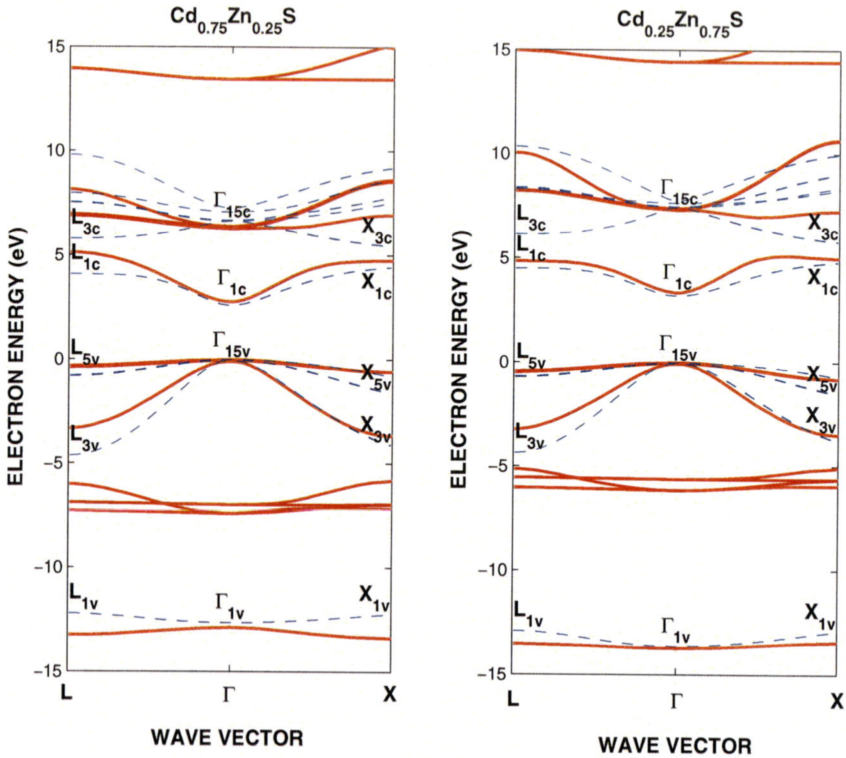

Fig. 1.17 Comparison of the sp^3d^5 and sp^3s^* TB models prediction of the electronic band structure of $Cd_{1-x}Zn_xS$ ternary semiconductor along $\Gamma - L$ and $\Gamma - X$ directions, with Zn composition x = 0.25 and 0.75, obtained by using given in Tables 1.4 and 1.5

elements to set of observables is difficult. We shall overcome this obstacle by using the statistical thermodynamic model [28] study the strain effects on energy levels.

Since the measured band gaps are near 0 K we ignored the logarithmic term, which accounts the electron–phonon interactions for temperature dependence. The predicted band gaps at high symmetry points, especially at Γ point, are in excellent agreement with experimental data [14] for these ternary semiconductors. Although the inclusion of s^* excited state in the 2NN sp^3 TB model yields accurate calculations of the conduction band structure features at the X high symmetry point, it does so by modelling the average of p-d interactions and is insufficient to determine energy levels above 6.0 eV. Jancu et al. [38, 39] suggested that from a solid-state physics point of view, the actual behaviour of excited d-states is not reliably addressed in the sp^3s^* tight binding model of valence band and conduction band dispersion curves involving the Γ, X and L high symmetry points of semiconductors which will be discussed in the next sub-section.

1.5.4 *Semiempirical* $sp^3d^5s^*$ *Tight Binding Theory*

Jancu et al. [38, 39] pointed out that the excited d-states contribute to both the valence band maximum at the Γ symmetry point and to the conduction band dispersion curves at the X and L symmetry points and is explicitly included in NN $sp^3d^5s^*$ orbitals set. The band gaps at high symmetry points in the Brillouin zone and optimized tight binding parameters given by Jancu et al. [38], The electronic band structure of AlAs, GaAs, AlSb, GaP and InP group III–V compounds are calculated. As can be seen from Figs. 1.18 and 1.19, both the valence band and conduction band dispersion curves of AlAs, GaAs, AlSbi GaP and InP obtained by using the NN $sp^3d^5s^*$ TB model overcomes most of the limitations of the earlier 2NN sp^3 and sp^3s^* TB models. Furthermore, accurate description of the second conduction band and the transverse effective masses at the X- and L-symmetry points that are in good agreement with experiment suggests that the NN $sp^3d^5s^*$ TB parametrization makes it possible to accurately calculate the electronic and optical properties involving symmetry points at the edge of the Brillouin zone of tetrahedral semiconductors.

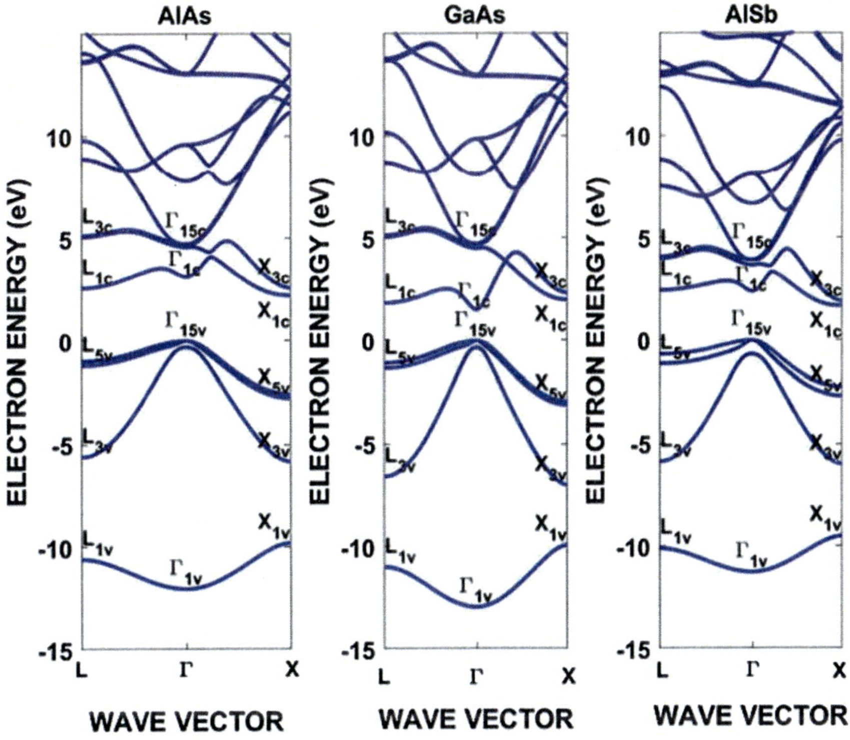

Fig. 1.18 Energy band structure of AlAs, GaAs and AlSb III–V binary compound semiconductors along $\Gamma - L$ and $\Gamma - X$ directions, obtained by using 2NN $sp^3d^5s^*$ orbital basis sets

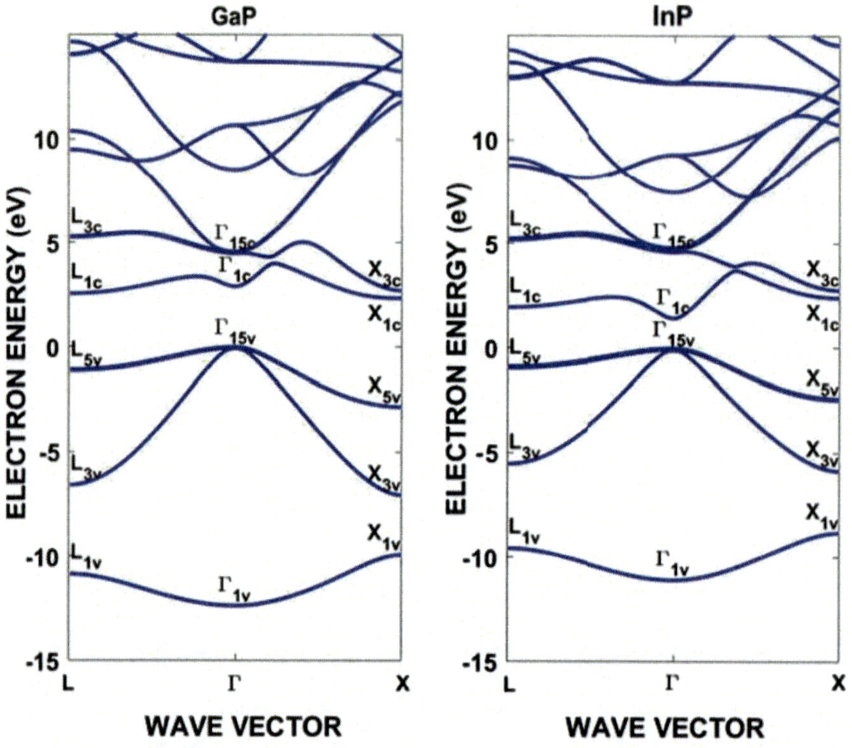

Fig. 1.19 Energy band structure of GaP and InP III–V binary compound semiconductors along $\Gamma - L$ and $\Gamma - X$ directions, obtained by using 2NN $sp^3d^5s^*$ orbital basis sets

As can be seen from Figs. 1.18 and 1.19. the tight binding model with NN $sp^3d^5s^*$ basis set eliminates the limitations of sp^3 and sp^3s^* basis sets. The NN $sp^3d^5s^*$ tight binding model is reliable for the calculation of dispersion curves for energies up to 6 eV above the valence band maximum, correctly reproducing the orbital character of band edges, and their behavior under strain, at Γ, X and L symmetry points of Brillouin zone of these semiconductors.

1.6 Density Functional Theory Modelling

The classical density functional theory (DFT) is based on the original work of Kohn and his collaborators [12, 13], which allows us to have a parameter-free description of structural and electronic properties of elementary and compound semiconductors. According to the Kohn-Hohenberg-Sham theory [12, 13], the wave functions of single particle states $\varphi_{\mathrm{i,k}}(\mathrm{r})$ are determined by the solution of the following set of equations similar to Schrödinger equation

$$H\varphi_{i,k} = \left(-\frac{\hbar^2}{2m}\nabla_i^2 + V_{ext} + V_{eff}\right)\varphi_{i,k} = E_i\varphi_{i,k} \tag{1.30}$$

$$V_{eff}(r) = V_H(r) + V_{xc}(r) = V_H(r) + \frac{\delta E_{xc}(n(r))}{\delta n(r)} \tag{1.31}$$

$$n(\vec{r}) = \sum_{i=1}^{N}\sum_{k} f_{i,k}|\varphi_{i,k}(r) > |^2 \tag{1.32}$$

where V_{ext} is the external potential of the nuclei, $V_{eff}(r)$ is the effective potential, $E_{xc}(n(r))$ is the exchange–correlation functional and $n(\vec{r})$ is the charge density which is expressed in terms of single-particle electronic orbitals. Equation (1.32) must be solved consistently with (1.30) and (1.31). Although the set of (1.30)–(1.32) yields an exact solution to the problem, there are approximations involved in the determination of the exchange–correlation potential $V_{xc}(r)$ for which Local Density Approximation (LDA) and Generalized Gradient Approximations (GGA) have proven to be effective for a large number of semiconductors.

Although classical DFT calculations based on (1.30)–(1.32) give satisfactory results for ground state properties (e.g., total energies, lattice constant, bulk modulus), their predictions are relatively poor for the electronic properties (e.g., band gaps and effective masses) [53]. The bandgap energies predicted by the DFT calculations are too small compared to experimental data [43] and predicted by empirical pseudopotential theory [14]. Even with GW and hybrid functionals, conventional DFT results in bandgap error on the order of 10–20% as compared with experimental data [43]. In order to overcome this obstacle, the exchange–correlation potential contribution to the DFT band gap is shifted by using a so-called "scissor operator", in accordance with the suggestion of Fiorentini and Balderschi [53]. The difference between the LDA and experimental band gaps (ΔE) scales with the optical dielectric constant ($\Delta E \cong 9.1/\varepsilon_\infty$) [8, 9].

Furthermore, Tran and Blaha [50] combined the Becke-Johnson exchange potential and the local density approximation (LDA) correlation potential in DFT band structure calculations. They called the modified Becke-Johnson density functional (mBJLDA), which is an exchange–correlation (XC) potential that is obtained as the functional derivative of the XC-energy functional E_{XC} with respect to the electron density $n(r)$ ($V_{xc}^{mBJ} = \delta E_{XC}[n(r)]/\delta n(r)$) taken from LDA. Although the computational cost of DFT with the mBJLDA functional is comparable with DFT-LDA and DFT-GGA, its predicted bandgaps are in good agreement with experimental data [34, 43] and empirical pseudopotential theory [14]. We used the WIEN2K simulations with various functionals [50, 54–57] embedded in DFT to calculate the electronic properties of group III–V and II–VI binary compound semiconductors as well as their ternary/binary heterostructures.

The band structures of GaAs, AlAs, AlSb, GaP and InP binary materials were investigated by using WIEN2K simulations embedded in DFT with various exchange–correlation potentials [54–57]. The best fit with pseudo potential theory calculations [38], experimental bandgaps [43] and $sp3s^*d^5$ tight binding calculations [38] are obtained. The obtained results [21] shown in Table 1.6.

The results are shown in Fig. 1.20 for GaAs, AlAs and AlSb group III–V binary compounds.

Table 1.6 The comparison of the results obtained for III–V compounds

Compound	GGA1 [54]	GGA2 [55]	GGA3 [56]	GGA4 [57]	LDSA [58]	Exp [43]	TB [38]	PP [38]
GaAs	1.348	1.223	1.273	1.331	1.116	1.519	1.519	1.220
AlAs	3.035	2.915	2.942	3.004	2.820	3.130	3.130	2.880
InAs	0.721	0.711	0.709	0.732	0.721	0.418	0.418	0.310
GaP	2.811	2.690	2.731	2.809	2.596	2.895	2.895	2.850
AlP	4.419	4.295	4.313	4.394	4.212	3.630	3.630	4.380
InP	1.602	1.479	1.517	1.594	1.397	1.424	1.424	1.440
AlSb	2.519	2.440	2.454	2.491	2.322	2.384	2.384	2.230

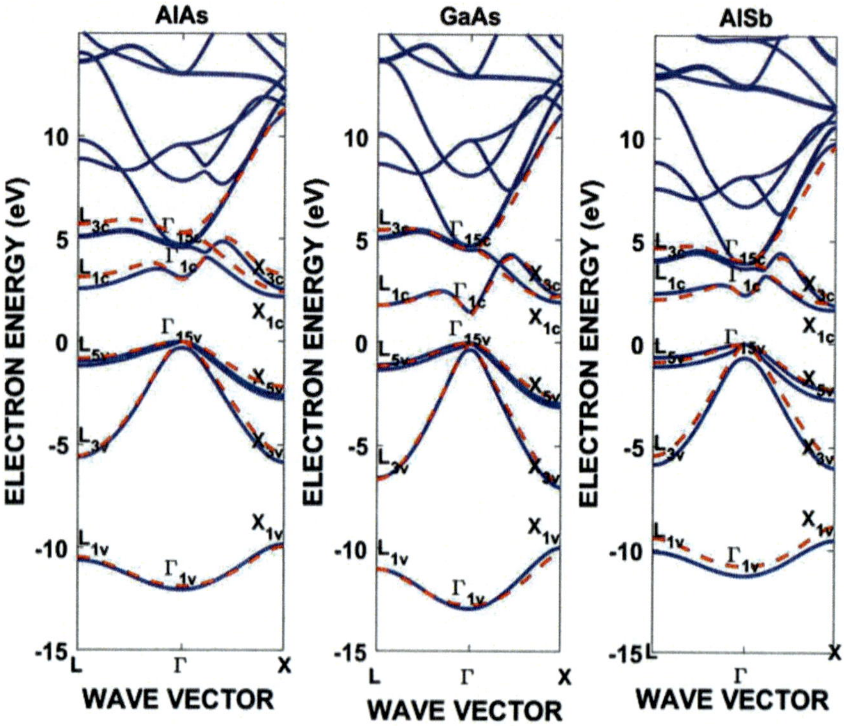

Fig. 1.20 Energy band structure of AlAs, GaAs and AlSb III–V binary compounds along $\Gamma - L$ and $\Gamma - X$ directions, obtained by using Wien2k (solid lines) and 2NN $sp^3d^5s^*$ (dashed lines) orbital basis sets for the tight binding parameters

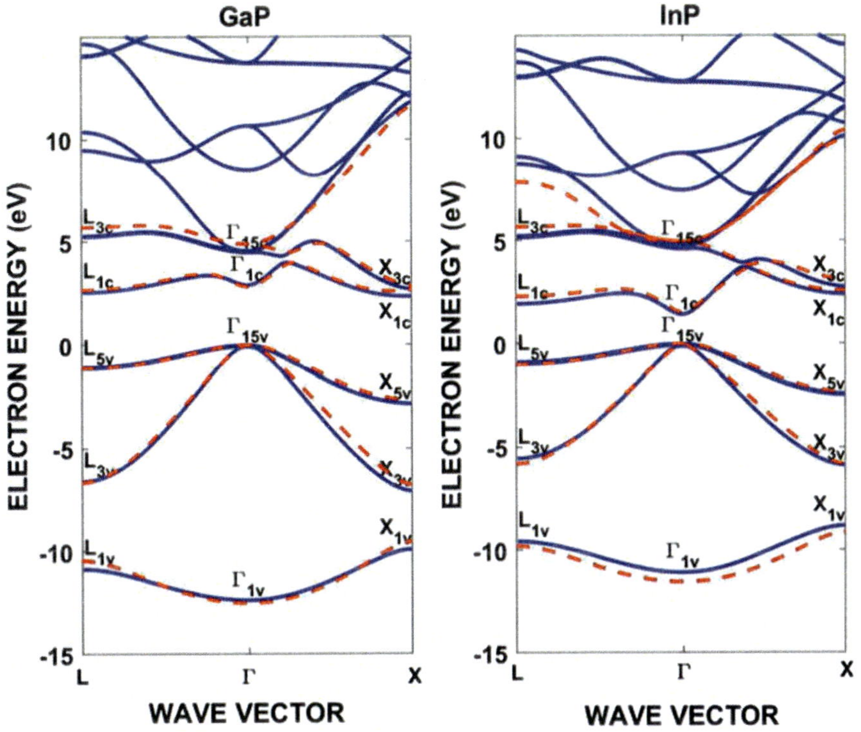

Fig. 1.21 Energy band structure of GaP and InP III–V binary compounds along $\Gamma - L$ and $\Gamma - X$ directions, obtained by using Wien2k (solid lines) and 2NN $sp^3d^5s^*$ (dashed lines) orbital basis sets for the tight binding parameters

The comparison of results for GaP and InP is also shown in Fig. 1.21 As can be seen from the figures, the DFT results show a perfect agreement with the $sp3s^*d^5$ TB results.

The interface strain effects on the fundamental bandgaps of III-nitrides based ternary/binary heterostructures are shown in Fig. 1.22 for AlGaN/GaN, InGaN/GaN and GaAsN/GaAs as functions of for the entire composition range ($0 \leq x \leq 1$). As shown in Fig. 1.22, the interface strain effect on the fundamental bandgaps of InGaN/GaN (middle) and GaAsN/GaAs (right) dilute nitrides based heterostructures is rather large because of the larger conduction band deformation potentials.

Group II–VI binary compounds and their ternary alloy semiconductors are considered as useful materials in fabricating heterojunction photovoltaic devices because their band gaps and lattice parameters can be tailored so that ternaries or quaternaries with different compositions can be grown on a suitable lattice matched substrate. The potential barrier in the conduction and the potential step in the valence bands at interface influences the carrier transport and device performance, making the reliable and accurate modelling of interface energy band structure highly

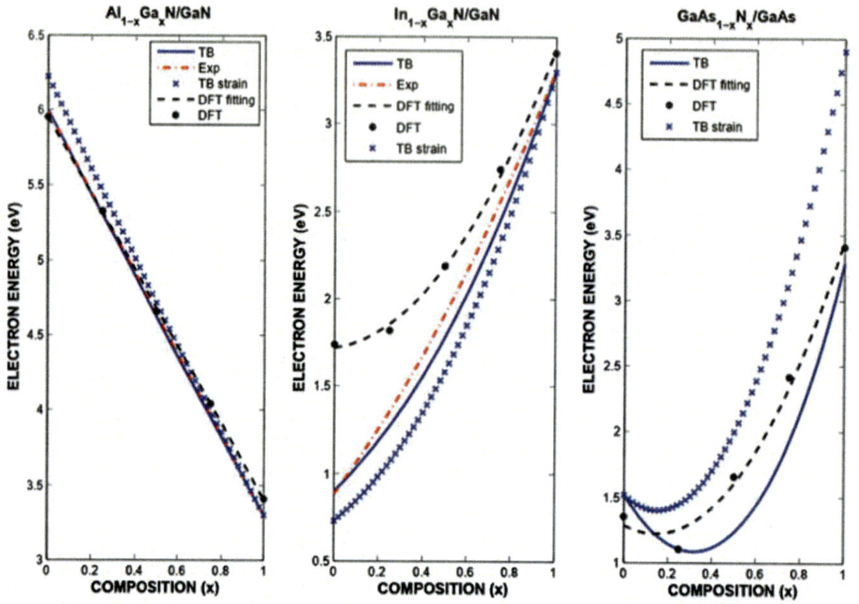

Fig. 1.22 Strain effects on the bandgaps of AlGaN/GaN, InGaN/GaN and GaAsN/GaAs heterostructures as a function of composition, obtained using the WIEN2K package with GGA [54] and corrected by scissor operator, and 2NN sp3s* TB model

crucial to understand the impact of group II–VI heterostructures in the design and optimization of photovoltaic devices.

Therefore, it is important to comparatively discuss the use of DFT with mBJLDA functional and the semiempirical NN sp^3d^5 and 2NN sp^3s^* TB models to calculate the electronic band structures of ZnSSe/ZnSe and CdSTe/CdTe heterostructures. Figure 1.23 shows the crystal structure of CdSe and ZnS compounds. CdX and ZnX (X = S, Se, Te) have a zinc-blende crystal structure (space group F-43 m No: 216) in which the Cd/Zn atoms are located at (0,0,0) and S (or Se) at (0.25,0.25,0.25). Converged results are obtained using 10,000 k points in the first Brillouin zone with $R_{MT}K_{max} = 8.50$, where R_{MT} represents the smallest muffin-thin radius and K_{max} is the maximum size of the reciprocal lattice vectors. LSDA, GGA and MBJLDA, as implemented in the WIEN2K simulations package, are used for exchange and correlation potentials in the calculations. SCF iterations are repeated until the total energy converges to a point $<10^{-4}$ Ryd. The WIEN2K simulations with the MBJLDA functional DFT yields a remarkable improvement over LDA and GGA in the calculation of the structural properties (e.g., lattice constants and bulk modulus) and electronic properties (e.g., bandgaps) of these compounds. However, we should point out that the MBJLDA underestimates the band gaps between −2.50 and −7% for CdTe, ZnS, ZnSe, ZnTe and overestimates up to ~4% for CdS and CdSe.

Fig. 1.23 Crystal structure of ZnS and CdSe compounds

Using WIEN2K simulations package with the mBJLDA functional based DFT, band structure of CdS, CdSe, CdTe, ZnS, ZnSe and ZnTe II–VI compounds are calculated and the results are compared with those obtained by using NN sp^3d^5 and 2NN sp^3s^* TB models. The comparison of the band structure dispersion curves obtained by using the WIEN2k simulation package with the MBJLDA functional embedded in DFT and the semiempirical tight binding theories with NN sp^3d^5 and 2NN sp^3s^* orbitals sets is shown in Fig. 1.24 for CdS and ZnS compound semiconductors.

Figure 1.25 shows the band structures of CdSe, CdTe and ZnSe obtained using WIEN2K with DFT-MBJLDA (dark-solid lines), NN sp^3d^5 TB (red-solid line) and 2NN sp^3s^* TB (dashed-blue line) models.

As demonstrated in Figs. 1.24 and 1.25 for CdX and ZnX (X = S, Se, Te) binaries of ZnSSe/ZnSe and CdSTe/CdTe heterostructures, the WIEN2K simulations with DFT-MBJLDA functional and semiempirical NN sp^3d^5 and 2NN sp^3s^* TB methods described in this work accurately reproduce the band gaps and band dispersion curves at high symmetry points of Brillouin zone when they are compared with empirical pseudopotential calculations for these binary and ternary semiconductors. the three different approaches almost equally well reproduce the band gaps at Γ, X and L high symmetry points and valence band and conduction band dispersion curves in the entire Brillouin zone of bulk compound semiconductors. When the two semiempirical NN sp^3d^5 and 2NN sp^3s^* tight binding parameterizations are compared with each other, they both give reasonable descriptions of the energy levels in the vicinity of the bottom of the conduction band and the top of the valence band of both binary and ternary II–VI compounds. However, the NN sp^3d^5 TB model does better than the 2NN sp^3s^* TB model in accurately reproducing band gaps as well as valence band and conduction band dispersion curves fitted to the empirical pseudopotential

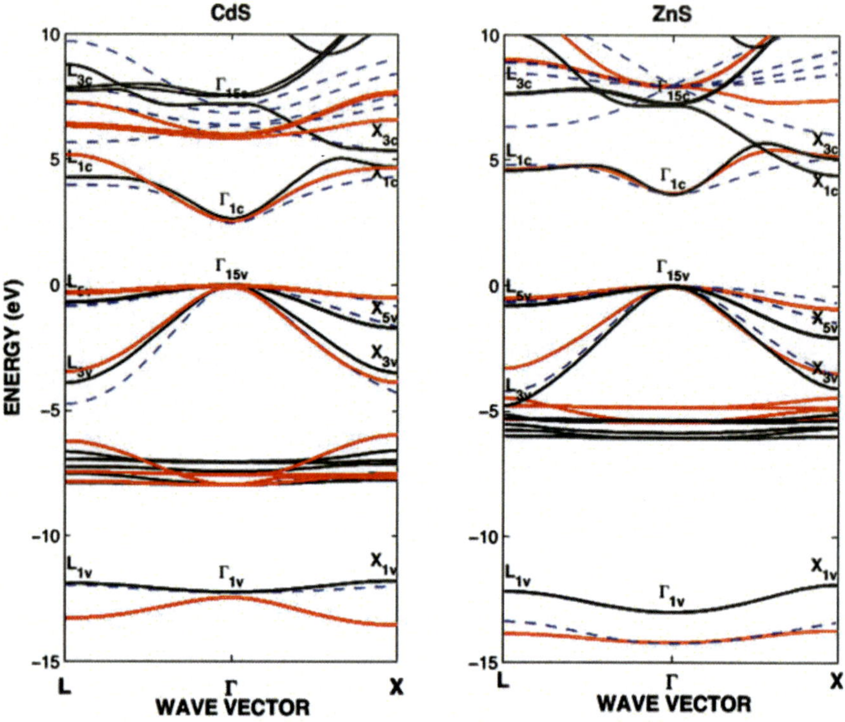

Fig. 1.24 Band structures of CdS and ZnS group II–VI compound semiconductors obtained using WIEN2K with DFT-MBJLDA (dark-solid lines), NN sp^3d^5 TB (red-solid line) and 2NN sp^3s^* TB (dashed-blue line) models

calculations at the L high symmetry point of these compounds, obviating the need for any fictitious excited s^* state in the semiempirical tight binding model calculations. Furthermore, since the inclusion of the fictitious s^* excited state is done by modelling the average of p-d interactions, the 2NN sp^3s^* TB parametrization does not permit the inclusion of excited d-orbitals to a sp^3 basis set. Consequently, the actual behavior and contribution of excited d-states in the band structure calculations is reliably and accurately reflected in the NN sp^3d^5 TB model. We should also point out that both the NN sp^3d^5 and 2NN sp^3s^* TB models are semiempirical and they rely on a good description of the band structures that are produced exactly by WIEN2K simulations package with MBLDA functional embedded in DFT [50, 59] and empirical pseudopotential theory [14].

We now discuss the use of DFT with mBJLDA functional and of the semiempirical NN sp^3d^5 and 2NN sp^3s^* TB models to calculate the composition variation of electronic band structures of ZnSSe/ZnSe and CdSTe/CdTe heterostructures. The predicted composition and interface strain effects on the fundamental band gaps of ZnSSe and CdSTe ternary alloy semiconductors in ZnSSe/ZnSe and CdSTe/CdTe

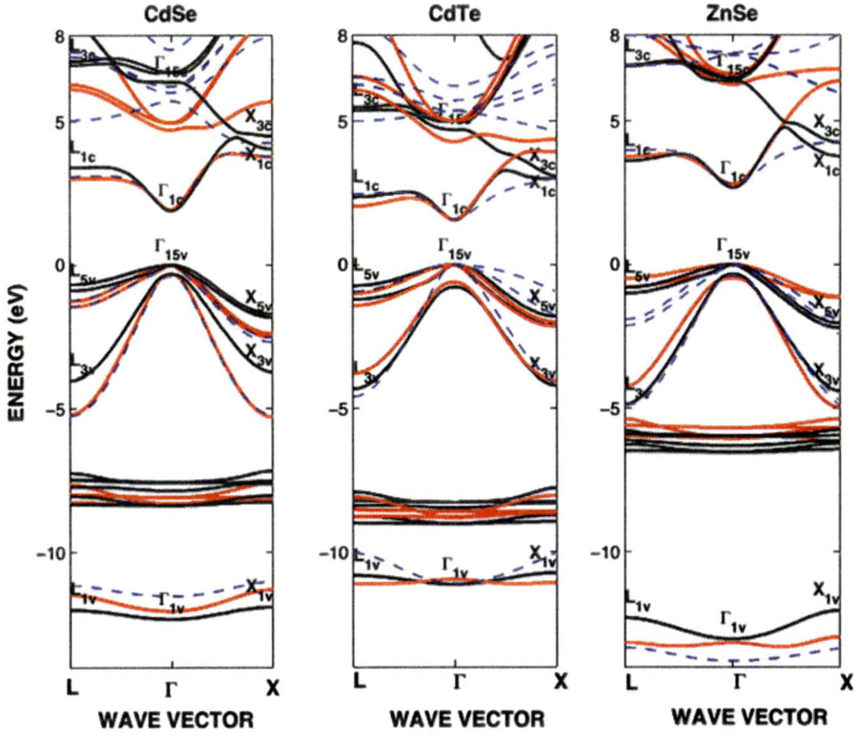

Fig. 1.25 Band structures of CdSe, CdTe and ZnSe obtained using WIEN2K with DFT-MBJLDA (dark-solid lines), NN sp^3d^5 TB (red-solid line) and 2NN sp^3s^* TB (dashed-blue line) models

heterostructures are shown in Fig. 1.26, respectively. The calculations are carried out using the WIEN2K simulations based on DFT with the MBJLDA functional (dark-solid line) and the semiempirical NN sp^3d^5 TB model (red-solid lines) and 2NN sp^3s^* TB models (dashed-blue lines) with tight binding parameters given in Tables 1.4 and 1.5, integrated with the statistical thermodynamic model [28, 29]. Since the measured band gaps are near 0 K, we ignored the logarithmic term, which accounts for the electron–phonon interactions for temperature dependence. We can state that strain effect on the fundamental bandgaps can be quite large when the interface strain increases for large deformation potential and high alloy composition. The predicted fundamental band gaps at the Γ high symmetry point, are in excellent agreement with experiment [43].

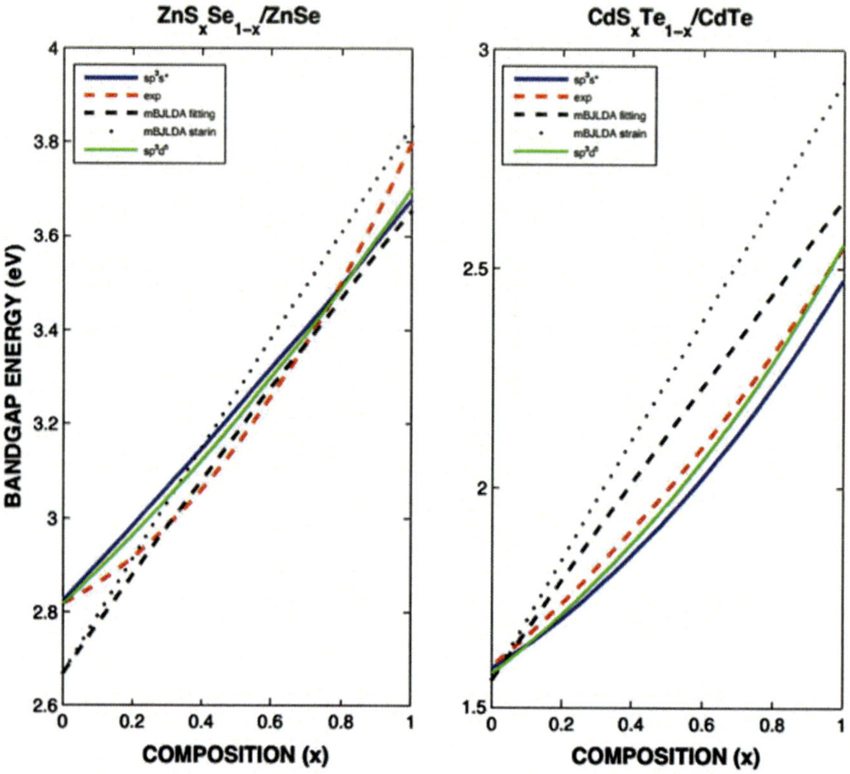

Fig. 1.26 Strain effects on the bandgaps of ZnSSe and CdSTe in ZnSSe/ZnSe and CdSTe/CdTe heterostructures as a function of composition, obtained using the WIEN2K simulation package with DFT-MBJLDA (dark-solid line), and the NN sp^3d^5 TB (red-solid line) and 2NN sp^3s^* TB (dashed-blue line) models

1.7 Tight Binding and DFT-MBJLDA Modelling of Band Offsets

One of the key issues in low dimensional semiconductor heterostructure device technology is the theoretical understanding the formation and determining the composition and strain variation of their conduction and valence band structure across the interface. Conduction and valence band offsets across the interface control the electronic properties and considerably influence carrier transport and performance of bipolar and unipolar heterojunction devices (e.g., HBTs and MODFETs).

Therefore, reliable and accurate modeling of interface band structure is crucial for the designing the functionality of group III–V and II–VI heterostructures and their optimization in low dimensional semiconductor electronic and optical devices. In this section, we will first focus on the effects of composition and interface strain on electronic band structure, band gaps and band offsets of ternary/binary group

III–V and II–VI compounds based heterostructures for which experimental data are available for comparison. In this context, we compare the predictions of the nearest neighbor (NN) sp^3d^5 tight binding model, with spin¯orbit coupling of p-states, to those of the widely used second nearest neighbor (2NN) sp^3s^* model, with spin–orbit coupling of p-states. Tables 1.1, 1.2, 1.3, 1.4 and 1.5 give the number of tight binding parameters for the 2NN sp^3s^* TB and NN sp^3d^5 tight TB models used in the calculations.

The valence band offset across an ABC/AC ternary/binary heterostructure can be obtained by taking the difference between the valence band energies of the constituents that are screened with the optical dielectric constants of the constituents [29]:

$$\Delta E_v = \left(\frac{E_v}{\varepsilon_\infty}\right)_{BC} - \left(\frac{E_v}{\varepsilon_\infty}\right)_{ABC}, \tag{1.33}$$

$E_v = E_v(\Gamma_{15})$ is top of the valence band at Γ high symmetry point and $\varepsilon_\infty(ABC)$ and $\varepsilon_\infty(BC)$ are the optical dielectric constants of the ABC ternary and BC binary constituents

$$\varepsilon_\infty(ABC) = x\varepsilon_\infty(AC) + (1-x)\varepsilon_\infty(BC) + -x(1-x)\delta_c(\varepsilon_\infty(BC) - \varepsilon_\infty(AC)), \tag{1.34}$$

The conduction band offset for a given Γ, L and X high symmetry point of the Brillouin zone is given as the difference between the respective band gap differences, written as [29]:

$$\Delta E_{ci} = E_{ciABC} - E_{ciBC} = \Delta E_{gi} - \Delta E_v, \tag{1.35}$$

where $E_{ci} = E_{\Gamma_{6c}}, E_{L_{6c}}$ and $E_{X_{6c}}$ are the minimums of conduction bands at Γ, L and X high symmetry points, determined with the proposed form of the 2NN sp^3 TB model by using the optimized tight binding parameters. $\Delta E_{gi} = E_{gi}(ABC) - E_{gi}(BC)$ is the difference between band gaps $E_{gi}(ABC)$ and $E_{gi}(BC)$ of ABC ternary and BC binary compound semiconductors.

Figure 1.27 shows the composition and strain variations of the band gap difference and valence band offsets of lattice matched AlGaAs/GaAs and coherently grown InGaAs/GaAs heterostructures. calculated by using the 2NN sp^3 TB parametrization of Lohrer and Talwar [46]. Results indicate that the valence band offsets vary almost linearly with alloy composition, but the conduction band offsets vary nonlinearly. This is due to fact that the valence band deformation potential is smaller than that of the conduction band at Γ, L and X high symmetry points. There is almost perfect agreement between the unstrained and strained energy levels in AlGaAs/GaAs heterostructure since strain is due to the difference between the linear thermal expansion coefficients. However, in the case of coherently grown InGaAs/GaAs heterostructure strain causes a nonlinear variation of the bandgaps and conduction

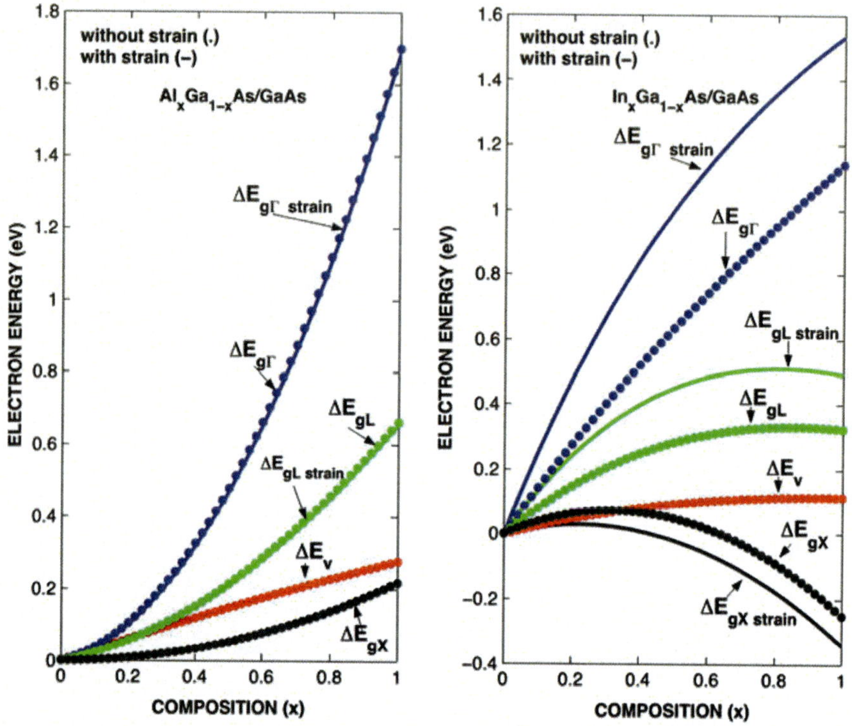

Fig. 1.27 Composition effects on bandgap differences and valence band offsets of lattice matched AlGaAs/GaAs (left) and strained InGaAs/GaAs (right) heterostructures

band offset since strain is due to the lattice mismatch and the difference between the thermal expansion coefficients of GaAs barrier and InGaAs quantum well, coupled with the larger conduction band deformation potentials.

Figure 1.28 shows the results of calculations done by using the statistical thermodynamic model of semiconductors, giving in good agreement with predictions of 2NN sp^3 TB parametrization.

In second step we compare the predictions of the nearest neighbor (NN) sp^3d^5 tight binding model. Table 1.1 give the number of tight binding parameters for the 2NN sp^3s^* TB and NN sp^3d^5 tight TB models used in the calculations. Figure 1.29 shows the composition and strain variations of the band gap difference and valence band offsets in AlGaAs/GaAs heterostructure.

The band offsets in AlGaN/GaN, InGaN/GaN and GaAsN/GaAs heterostructures are shown in Fig. 1.30 for functions of interface strain for the entire composition range ($0 \leq x \leq 1$). The strain effect on the valence band offsets in AlGaN/GaN (left), InGaN/GaN (middle) and GaAsN/GaAs (right) dilute nitrides based heterostructures is rather small because of the smaller valence band deformation potentials. The predicted valence band offsets are in good agreement with experimental findings [34, 43]. The strain effects on the conduction band offsets can be quite large since the

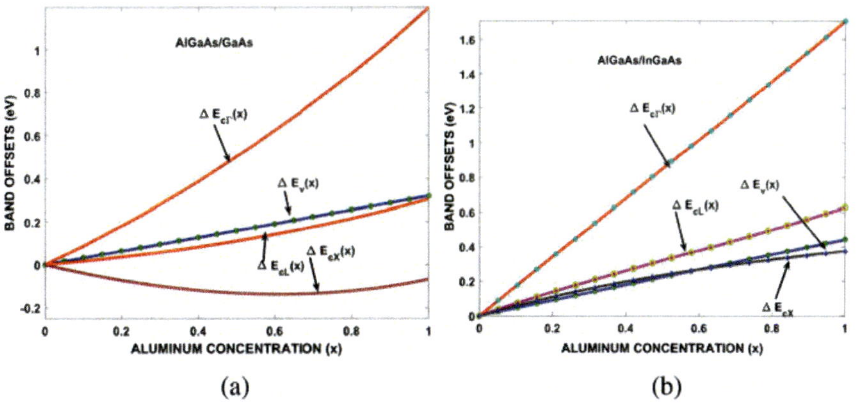

Fig. 1.28 Calculated composition effect in conduction and valence band offsets in AlGaAs/GaAs (**a**) and AlGaAs/InGaAs (**b**) heterostructures

Fig. 1.29 Composition effects on bandgap differences and valence band offsets of lattice matched AlGaAs/GaAs

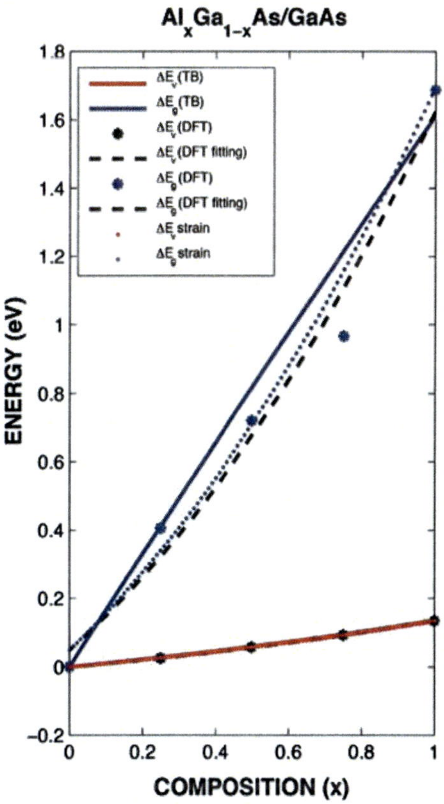

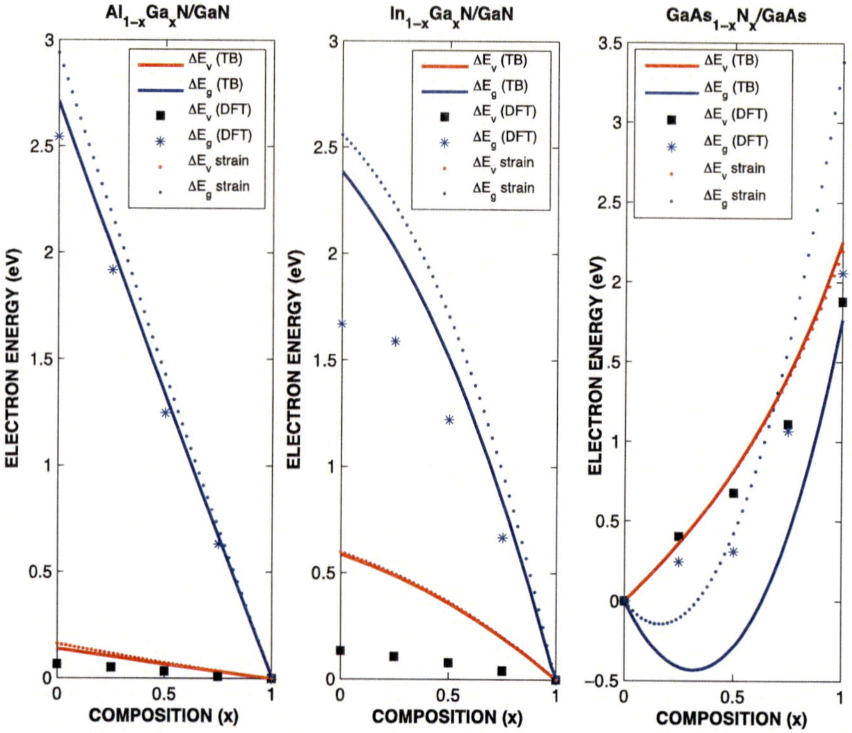

Fig. 1.30 Strain effects on the band offsets of AlGaN/GaN (left), InGaN/GaN (middle) and GaAsN/GaAs (right) heterostructures. obtained using the WIEN2K with DFT-MBJLDA (dark-solid line) compared with the NN sp^3d^5 (red-solid line) and 2NN sp^3s^*(dashed-blue line) TB models

lattice mismatch increases with an increase in the ternary mole fraction. As we pointed out earlier, there is a considerable lattice mismatch across many of the ternary/binary heterointerfaces. Interface strain due to lattice mismatch causes a shift in the lattice constant of the epilayer: $a = (1+\varepsilon)a_0$, where ε is the symmetric strain tensor. Therefore, the bond lengths and on-site and off-site TB matrix elements will be modified with strain. Consequently, the conduction and valence band offsets will be modified with respect to their unstrained values. As an example, Fig. 1.31 shows the composition and strain effects on conduction and valence band offsets of CdSTe/CdTe and ZnSSe/ZnSe heterostructures calculated by using the WIEN2K simulation package based on the DFT with MBJLDA functional [50] and semiempirical NN sp^3d^5 and 2NN sp^3s^* TB models for optimized tight binding parameters.

As can be seen in Fig. 1.31, the interface strain effects on the conduction band offsets of ZnSSe/ZnSe and CdSTe/CdTe heterostructures can be quite large when the interface strain increases for large the deformation potential and high alloy composition.

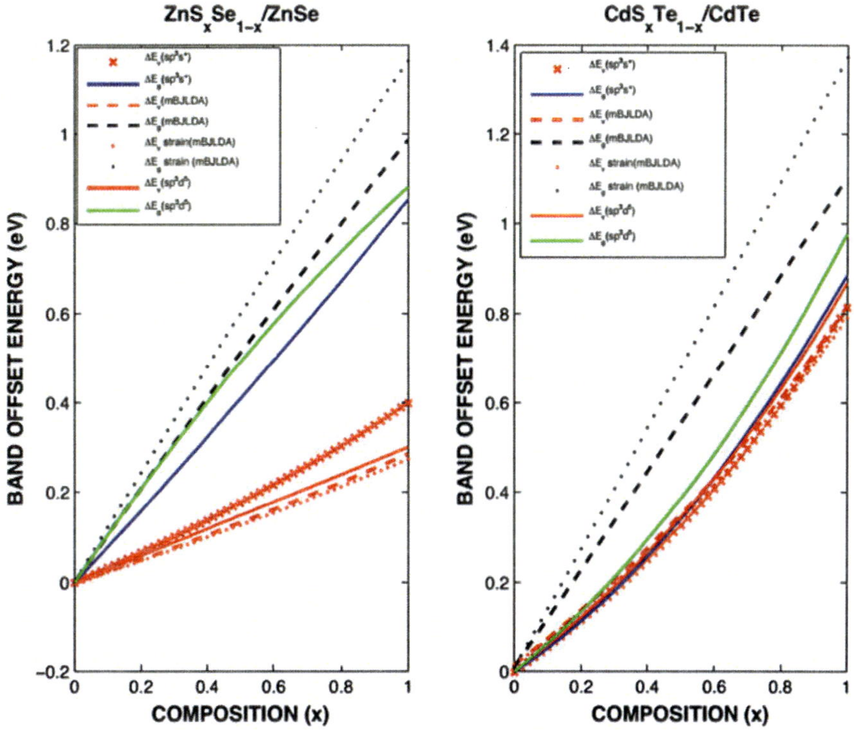

Fig. 1.31 Strain effects on the band offsets of ZnSSe/ZnSe and CdSTe/CdTe heterostructures, obtained using the WIEN2K with DFT-MBJLDA (dark-solid line) compared with the NN sp^3d^5 (red-solid line) and 2NN sp^3s^*(dashed-blue line) TB models

1.8 Pressure Effects on Structure and Electronic Properties

We now discuss the use of DFT as implemented in the package WIEN2k code [59] to calculate the structural and electronic properties of III-nitrides under high pressure. For As and N based compounds generalized gradient approximation (GGA) [54] are used for exchange and correlation potential for the calculations. The WIEN2K package is used in the adaptation of the partitioning the unit cell into non-overlapping atomic spheres (centered at atomic sites) and interstitial region. In the atomic spheres, a linear combination of radial functions times spherical harmonics are used and in the interstitial region a plane wave expansion is used. Each plane wave is augmented by an atomic-like function in every atomic sphere. We have adopted sphere radii values for GaAs, Ga and As, 2.1, 2.2 Bohr respectively, and in the cases of GaN, AlN and InN Ga, Al, In and N we use 1.89, 1.8, 2.1 and 1.6 Bohr, respectively. Self-consistent iteration is used for calculation and convergence criteria $R_{MT}K_{max}$ (where R_{MT} is the smallest atomic sphere radius in the unit cell and K_{max} is the interstitial plane wave cut-off) was set to 8.50. The iteration repeated until total energy converges to less than 10^{-4} Ryd.

1.8.1 Structural Parameters

The connection between lattice constant and pressure and volume are given by Murnaghan's equation of state [8]:

$$P = \left(\frac{B_0}{B_0'}\right)\left(\left(\frac{V_0}{V}\right)^{B_0'} - 1\right), \quad V(P) = V_0\left(1 + B_0'\left(\frac{P}{B_0}\right)\right)^{-1/B_0'} \tag{1.36}$$

where V_0 and V are the volume at ambient and P pressure, respectively. The lattice constant (a_0) was determined by fitting the total energy as a function of volume to Murnaghan's equation of state from which we get the volume or lattice constant (a_0), bulk modulus (B_0) and its pressure derivative (B_0'). The calculated lattice constant differs from experimental values just 2%. According to DFT it is well known issue and these values are in acceptable limitation. Also, calculated bulk modulus and its pressure derivative have similar trends with available experimental and other calculated data.

1.8.2 Electronic Properties

The first principles method with local density approximation (LDA) and related generalized approximation (GGA) is known to give reasonably accurate good result for the total energies and ground state structure properties of semiconductors and insulators, but it leads to a severe underestimate of the band gap by typically 30–50% [21]. This error is often corrected empirically by the so-called scissors operator [21, 53]. Fiorentini and Baldereschi [53] noted that the missing self-energy term was related to the screened Coulombic hole. They found that the difference between the LDA and experimental band gaps, ΔE, scales with the optical part of the dielectric constant ε_∞ as $\Delta E \approx 9.1/\varepsilon_\infty$. Calculated band gap energies shifted by using "*scissor operator*" and results are in better agreement with the experimental values. Band structure and density of states of GaAs are shown in Fig. 1.32 for ambient pressure, P = 10 GPa and P = 40 GPa, respectively.

The effects of composition and on electronic band structure of III–V nitrides based heterostructures are calculated by using the so called modified virtual crystal approximation (MVCA) [15, 16] as nonlinear function of composition. As an example, we calculated the band structure of dilute GaAsN ternary semiconductor at ambient and high pressures. The ternary band gap is shifted by $\Delta E \approx 9.1/\varepsilon_\infty(GaAs_{1-x}N_x)$, with 9.1 as fitting parameter [53] and dielectric constant ε_∞ given as

$$\begin{aligned}\varepsilon_\infty(GaAs_{1-x}N_x) &= (1-x)\varepsilon_\infty(GaAs) + x\varepsilon_\infty(GaN) \\ &\quad - x(1-x)\delta c(\varepsilon_\infty(GaAs) - \varepsilon_\infty(GaN))\end{aligned} \tag{1.37}$$

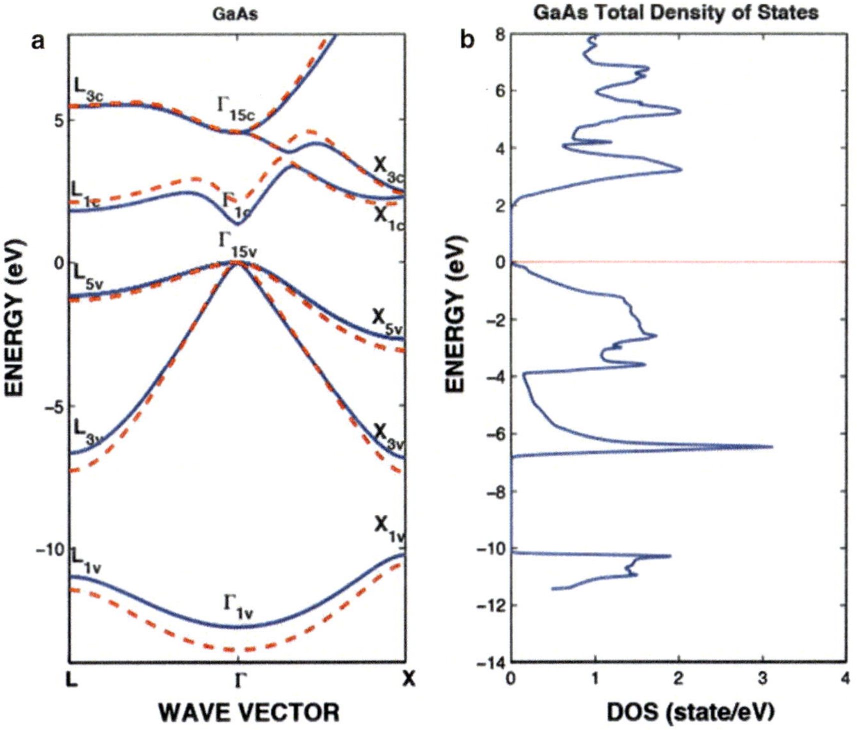

Fig. 1.32 Bulk band structure calculated by DFT with scissor operator at P = 0 GPa (solid) and P = 10 GPa for GaAs (**a**) (dash) and P = ambient and P = 40 GPa for GaN (**b**) (dash)

where δ_c is difference between dimensionless relaxation parameters [14]. Figure 1.33 shows the bulk band structure calculated by DFT with scissor operator at P = 0 GPa (solid) and P = 10 GPa for GaAs (**a**) (dash) and P = ambient and P = 40 GPa for GaN (**b**) (dash). Predictions of composition effects (star) calculated by DFT calculation for various compositon values (x = 0 0.25, 0.50, 0.75, 1) (points and squares) on the band gap energies of GaAsN for the P = ambient GPa (dash) and P = 15 GPa (dash-dot) Second order polinomial fitting due to DFT calculation (dash line, dash-dot) shown in (**c**).

1.9 Finite Difference Method for Low Dimensional Structures

Low-dimensional structures are classified as quantum wells, quantum wires, and quantum dots. Here the dimension specifies the number of directions in which the charge carrier (electron or hole) can move freely. Quantum wells are created by adding a different type of semiconductor layer between two semiconductor layers

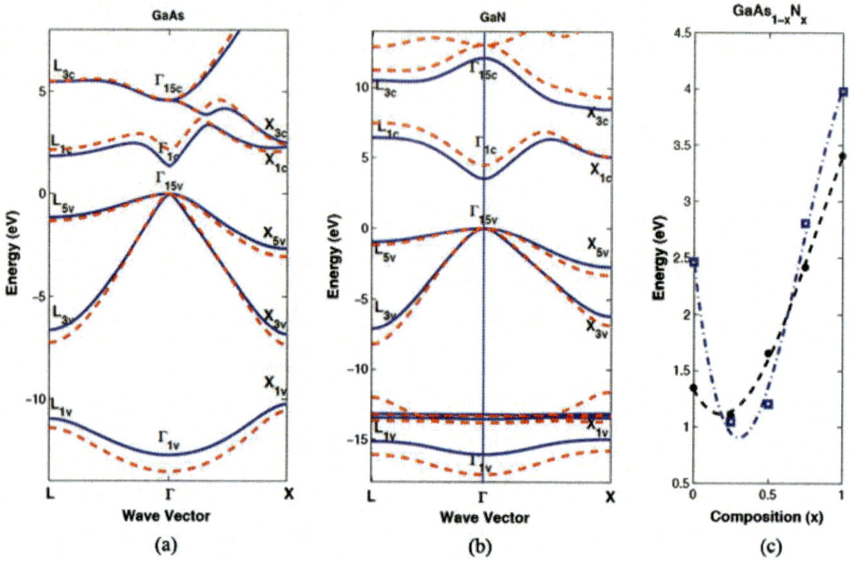

Fig. 1.33 Bulk band structure calculated by DFT with scissor operator at P = 0 GPa (solid) and P = 10 GPa for GaAs (**a**) (dash) and P = ambient and P = 40 GPa for GaN (**b**) (dash). Predictions of composition effects (star) calculated by DFT calculation for various compositon values (x = 0 0.25 0.50 0.75 1) (points and squares) on bands of GaAsN dilute ternary for the P = ambient GPa (dash) and P = 15 GPa (dash-dot) Second order polinomial fitting due to DFT calculation (dash line, dash-dot) shown in (**c**)

of the same type. In quantum wells, charge carriers can act as free particles in two dimensions, while their movement towards the different layer (in the direction of crystal growth) is limited in one dimension and their energies are quantized. In quantum wire structures, the motion of the carriers is quantized in two dimensions. In quantum dot structures, the motion of the carrier is quantized in all three dimensions.

In examining the properties of an electron trapped in low-dimensional structures, we can find the energy eigenvalues and waveforms of electron by solving the time-independent Schrödinger equation. The solution of the differential equations we come across constitutes our main subject. We do not always encounter analytically solvable differential equations. In most of the problems encountered in quantum mechanics, it is very difficult or impossible to solve the Schrödinger equation of the system analytically and determine the energy levels and wave functions. Thus, in cases where the exact solution of the Schrodinger equation is not done directly, numerical methods are applied. In this subject, one of the most used numerical methods is the Finite Difference Method. Here, we will discuss the use of finite difference method to study the electronic properties of AlGaAs/GaAs two-dimensional quantum wells, one dimensional quantum wires, and zero-dimensional quantum dots, respectively. Supposing the independent variable is measured at regular and equal intervals. Accordingly, we can create the following Table given in Fig. 1.34.

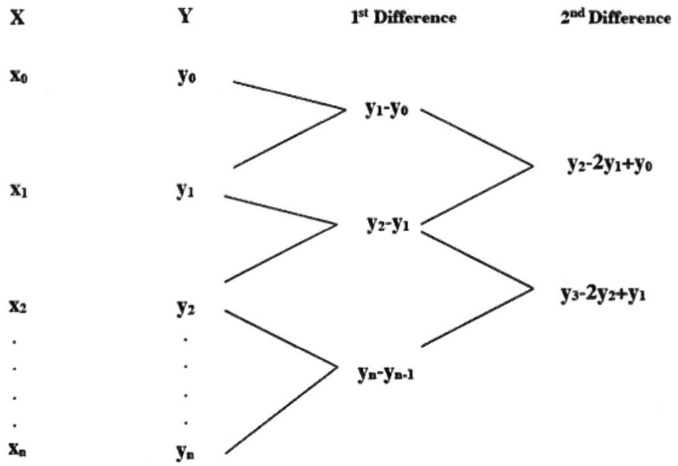

Fig. 1.34 Finite difference table

Table 1.7 The forward, backward (reverse) and central difference operators

Forward difference operator	Reverse difference operator	Central difference operator
$\Delta y(x) = y(x+h) - y(x)$ $\Delta y_0 = y_1 - y_0$ $\Delta y_1 = y_2 - y_1$	$\nabla y(x) = y(x) - y(x-h)$ $\nabla y_n = y_n - y_{n-1}$ $\Delta y_1 = y_1 - y_0$	$\delta y(x) = y(x+h/2) - y(x-h/2)$ $\delta(x+h/2) = y(x+h) - y(x)$ $\delta y_{1/2} = y_1 - y_0 = \Delta y_0 = \Delta y_1$ $\delta y_{3/2} = y_2 - y_1 = \Delta y_1 = \Delta y_2$

Difference Operators. Any differential equation can be solved numerically with the help of the forward, backward (reverse) and central difference operators listed in Table 1.7.

1.9.1 Application of Finite Difference Method to Quantum Wells

Semiconductor quantum wells are structures in which the motion of the electron is limited in only one dimension and can move freely in the other two dimensions. These structures are obtained by placing a semiconductor material with a small forbidden energy gap as a thin layer into the material with a larger forbidden energy gap, as seen in the figure. Quantum well is made of a very thin flat semiconductor layer is placed between the two layers. The difference between the conduction band energies of the two materials restricts the electrons to a thin layer.

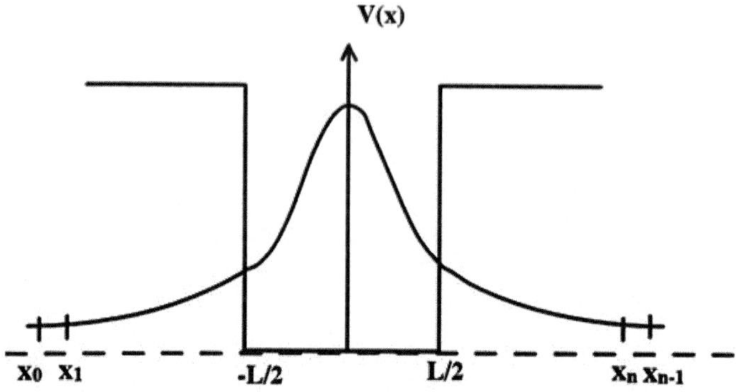

Fig. 1.35 Application of finite difference method to quantum well with finite barrier height

Depending on the height of the potential wall, a finite and infinite quantum well can be created wherein the electron is trapped. The potential height here can be controlled by the composition of the wide bandgap ternary semiconductor used as barrier. The infinite and finite potential barriers in a quantum well are defined as

$$V(x, y) = \begin{cases} 0 & -L/2 \le x \le L/2 \\ \infty & otherwise \end{cases} \tag{1.38a}$$

$$V(x, y) = \begin{cases} V_0 & -\infty \le x \le -L/2 \\ 0 & -L/2 \le x| \le L/2 \\ V_0 & L/2 \le x \le +\infty \end{cases} \tag{1.38b}$$

The solution of Schrodinger equation for the wave function and energy eigenvalues can be found in any classical textbook on quantum mechanics and will not be repeated here. Figure 1.35 shows the schematic diagram of a quantum well with finite barrier height to be studied by using the finite difference method.

If we apply the finite difference method to the solution of the quantum well, by taking different points, the first and second derivatives are written as follows

$$\frac{d\psi}{dx} = \frac{\Delta\psi}{\Delta x} = \frac{\psi_i - \psi_{i-1}}{x_i - x_{i-1}} \tag{1.39}$$

$$\frac{d^2\psi}{dx^2} = \frac{d}{dx}\left(\frac{d\psi}{dx}\right) = \frac{\Delta}{\Delta x}\left(\frac{d\psi}{dx}\right) = \frac{\psi_{i-1} - 2\psi_i + \psi_{i+1}}{dx^2} \tag{1.40}$$

For ith point, the Schrödinger wave equation is written as

$$\frac{\psi_{i-1} - 2\psi_i + \psi_{i+1}}{dx^2} + (V(x_i) - E)\psi_i = 0 \tag{1.41}$$

If we write equation for each point, we can then write N number of equations. The values of x_0 and ψ_0 are known from the initial conditions and we can set $\psi_0 = 0$ For point $i = 1, 2,..$ and so on, we write

$$\begin{aligned}
&-\frac{1}{dx^2}(\psi_0 - 2\psi_1 + \psi_2) + (V(x_1) - E)\psi_1 = 0,\\
&-\frac{1}{dx^2}(\psi_1 - 2\psi_2 + \psi_3) + (V(x_2) - E)\psi_2 = 0,\\
&\ldots
\end{aligned} \tag{1.42}$$

which can be rewritten as

$$\begin{aligned}
&-\frac{1}{dx^2}\left[\left(-2 - V(x_1)dx^2\right)\psi_1 + \psi_2\right] = E\psi_1,\\
&-\frac{1}{dx^2}\left[\psi_1 + \left(-2 - V(x_2)dx^2\right)\psi_2 + \psi_3\right] = E\psi_2,\\
&\ldots
\end{aligned} \tag{1.43}$$

If we write equation for each point, then we can write N number of equations. Representation of the wave function in a finite difference table for quantum well is given in Fig. 1.36. We can write these equations in matrix form, and it is possible to solve these set of linear equations with numerically.

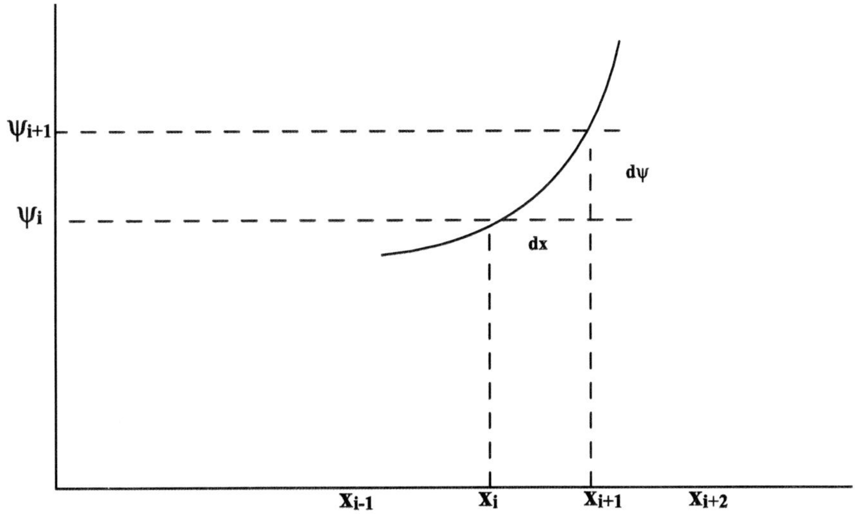

Fig. 1.36 Representation of the wave function in a finite difference table for quantum well

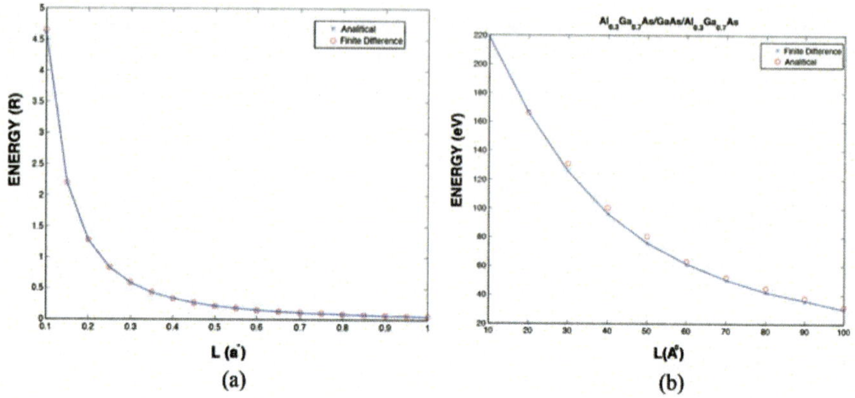

Fig. 1.37 Infinite and finite barrier GaAs quantum well. It is reproduced from [21]

$$-\frac{1}{dx^2}\begin{pmatrix} -2-V(x_1)dx^2 & 1 & 0 & 0 \ldots \\ 1 & -2-V(x_2)dx^2 & 1 & 0 \ldots \\ 0 & 1 & -2-V(x_3)dx^2 & 1 \ldots \\ . & . & . & . \ldots \\ . & . & . & . \ldots \end{pmatrix}\begin{pmatrix} \psi_1 \\ \psi_2 \\ \\ \\ \psi_N \end{pmatrix} = E\begin{pmatrix} \psi_1 \\ \psi_2 \\ \\ \\ \psi_N \end{pmatrix} \tag{1.44}$$

The solution of the GaAs quantum well with finite and infinite potential barriers, using finite difference method is given in Fig. 1.37. This figure also compares the numerical solution with the analytical solution, also expressed as a function of the width of the well. As can be seen in the figure, numerical and theoretical solutions are in harmony with each other. Numerical and theoretical solutions of the AlGAs/GaAs/AlGaAs, with $x = 0.30$, quantum wells are given. The desired potential height is obtained by varying the concentration of the AlGaAs material used as barrierr. In this example, $x = 0.3$ and the potential height is taken as approximately 228 meV. Numerical and theoretical solutions are in harmony with each other in the well problem with finite height potential barrier.

1.9.2 Application of Finite Difference Method to Quantum Wires

In quantum wires, electron motion is limited in two directions is limited in two directions. That is the electron is trapped by potential barriers in the x and y directions. We performed the same calculation that we did in quantum wells, in quantum wires. Figure 1.38 shows how the ground state energy of a GaAs quantum wire surrounded by an infinite potential barrier height.

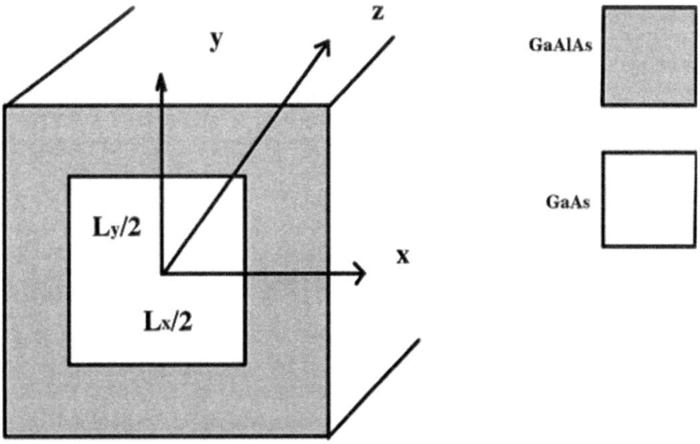

Fig. 1.38 Schematic view of one dimensional AlGaAs/GaAs quantum wire

In order to find the energy eigenvalues and wave functions of an electron trapped in an one dimensional infinitely long quantum wire, we need to solve the Schrödinger equation with potential looks like

$$V(x, y) = \begin{cases} 0 & |x| \le L_x/2 \ |y| \le -L_y/2 \\ \infty & |x| > L_x/2 \ |y| > -L_y/2 \end{cases} \quad (1.45)$$

We can write the single electron Schrodinger equation for the quantum wire of infinite length as

$$\left(-\frac{\partial^2}{\partial x^2} - \frac{\partial^2}{\partial y^2} + V(x, y)\right)\varphi(x, y) = \left(E_x + E_y\right)\varphi(x, y) \quad (1.46)$$

Since particle is not confined along the z-direction, it will behave as free particle in this direction and will be quantized in (*x*, *y*) directions. Therefore, solution of Schrodinger equation will give the following wave function and energy eigenvalues

$$\psi_0(x, y, z) = \psi_0(x, y)\psi_0(z) = A\cos\left(\frac{\pi}{L_x}x\right)\cos\left(\frac{\pi}{L_y}y\right)e^{ik_z z},$$
$$E_0 = -\frac{\hbar^2}{2m^*}\left(\frac{\pi^2}{L_x^2} + \frac{\pi^2}{L_y^2}\right) + \frac{\hbar^2 k_z^2}{2m^*} \quad (1.47)$$

In the case of one-dimensional finite length quantum wire, potential barrier is expressed as

$$V(x, y) = \begin{cases} 0 & |x| \le L_x/2, \ |y| \le -L_y/2 \\ V_0 & |x| > L_x/2, \ |y| > -L_y/2 \end{cases} \quad (1.48)$$

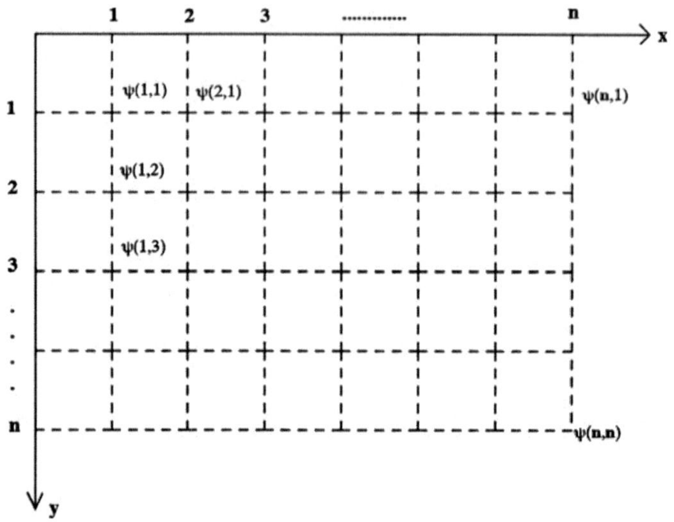

Fig. 1.39 Representation of the wave function in a finite difference table

and the Schrodinger equation is written as

$$-\frac{\hbar^2}{2m^*}\left(\frac{\partial^2}{\partial x^2}+\frac{\partial^2}{\partial y^2}+\frac{\partial^2}{\partial z^2}+V(x,y)\right)\psi(x,y,z)=E\psi(x,y,z) \tag{1.49}$$

which can be solved analytically. However, for some different potential profiles, the analytical solution may be very difficult or impossible. In such cases, numerical methods such as the finite difference method can be used. We take the following steps in applying the finite difference method to the quantum wires. We take the wave functions with equal steps on the x and y axes, shown in Fig. 1.39.

Using the definition of second derivative, we write the Schrödinger equation $\varphi(1,1)$ as

$$-\frac{\partial^2\psi(1,0)}{\partial x^2}-2\psi(1,1)+\psi(1,2)-\frac{\partial^2\psi(0,1)}{\partial y^2}-2\psi(1,1)+\psi(2,1)$$
$$=\big(E_{x1}+E_{y2}-V(1,1)\big)\psi(1,1) \tag{1.50}$$

Writing similar equations for $\psi(1,1)$, $\psi(1,2)$,......, $\psi(n,n)$, we then construct a matrix equation

$$\begin{pmatrix} \frac{4}{dx^2}+V(1,1) & -\frac{1}{dx^2} & 0 & 0\ 0\ 0\ \frac{1}{dy^2} & 0 & \cdots \\ -\frac{1}{dx^2} & \frac{4}{dx^2}+V(1,2) & -\frac{1}{dx^2} & 0\ 0\ 0\ 0 & \frac{1}{dy^2} & \cdots \\ 0 & -\frac{1}{dx^2} & \frac{4}{dx^2}+V(1,3) & -\frac{1}{dx^2}\ 0 \ldots \ldots & \ldots & \cdots \\ \cdot & & & & & \\ \cdot & & & & & \\ \cdot & & & & & \\ \cdot & & & & & \\ \cdot & & & & & \\ \ldots & 0 & 0 & \frac{1}{dy^2}\ 0\ 0\ 0 & -\frac{1}{dx^2} & \frac{4}{dx^2}+V(n,n) \end{pmatrix} \begin{pmatrix} \psi(1,1) \\ \psi(1,2) \\ \cdot \\ \cdot \\ \cdot \\ \cdot \\ \cdot \\ \psi(n,n) \end{pmatrix} = E \begin{pmatrix} \psi(1,1) \\ \psi(1,2) \\ \cdot \\ \cdot \\ \cdot \\ \cdot \\ \cdot \\ \psi(n,n) \end{pmatrix} \quad (1.51)$$

With a program that solves this matrix, we can find the E energy eigenvalues and $\psi(x, y)$ wave functions. In cases where the analytical solution is very difficult or impossible, we can quickly find both the energy eigenvalues and the wave functions by numerically solving the matrix we have constructed. We performed the same calculation that we did in quantum wells, in quantum wires. Figure 1.40a and b, respectively, show how the ground state energy of a GaAs quantum wire with its cross section when the wire surrounded by an infinite potential barrier (1.40) and surrounded by a varying height finite barrier (1.40). When the potential barrier decreases, the ground state energy also decreases at low dimensions.

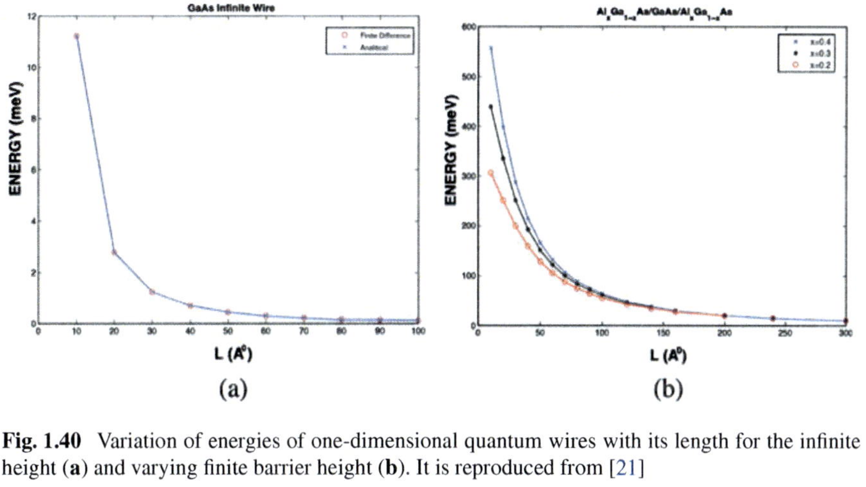

Fig. 1.40 Variation of energies of one-dimensional quantum wires with its length for the infinite height (**a**) and varying finite barrier height (**b**). It is reproduced from [21]

1.9.3 Finite Difference Method Applied to Quantum Dots

So far, semiconductor heterostructures in which the electron is confined in one or two directions have been considered. The energy spectrum of the electron in these structures became discrete in one or two dimensions. This changes the intensity of the situations drastically. However, in these structures, the particle can move freely in at least one direction. Structures in which the movement of electrons is restricted in three dimensions are called quantum dot structures.

We first write the three-dimensional Schrodinger equation for a cartesian quantum box, which is the generalization of the rectangular cross-section quantum wires and additional confinement along the x-axis. Considering the case of an infinite potential separating the interior and exterior of the quantum box, the three-dimensional Schrodinger equation within the box is written as

$$\left[\left(-\frac{\hbar^2}{2m^*}\frac{\partial^2}{\partial x^2}+V(x)\right)+\left(-\frac{\hbar^2}{2m^*}\frac{\partial^2}{\partial y^2}+V(y)\right)+\left(-\frac{\hbar^2}{2m^*}\frac{\partial^2}{\partial z^2}+V(z)\right)\right]\psi=E\psi \tag{1.52}$$

Using separation of variables (1.49) can be written in each direction as

$$\left(-\frac{\hbar^2}{2m^*}\frac{\partial^2}{\partial x^2}+V(x)\right)\psi(x)=E_x\psi(x) \tag{1.53a}$$

$$\left(-\frac{\hbar^2}{2m^*}\frac{\partial^2}{\partial y^2}+V(y)\right)\psi(y)=E_y\psi(y) \tag{1.53b}$$

$$\left(-\frac{\hbar^2}{2m^*}\frac{\partial^2}{\partial z^2}+V(z)\right)\psi(z)=E_z\psi(z) \tag{1.53c}$$

For an infinitely deep three-dimensional potential, $V(x)=0$, $V(y)=0$, $V(z)=0$, we then have the following wave functions and total energy of a single particle, respectively

$$\psi(x)=\sqrt{\frac{2}{L_x}}\sin\left(\frac{\pi n_x}{L_x}x\right),\quad \psi(y)=\sqrt{\frac{2}{L_y}}\sin\left(\frac{\pi n_y}{L_y}y\right),\quad \psi(z)=\sqrt{\frac{2}{L_z}}\sin\left(\frac{\pi n_z}{L_z}z\right) \tag{1.54a}$$

$$E=E_x+E_y+E_z=\frac{\hbar^2\pi^2}{2m^*}\left(\frac{n_x^2}{L_x^2}+\frac{n_y^2}{L_y^2}+\frac{n_z^2}{L_z^2}\right) \tag{1.54b}$$

In case of spherical quantum dots, the three-dimensional Schrodinger equation is written as

$$-\frac{\hbar^2}{2m^*}\left[\frac{1}{r^2}\frac{\partial}{\partial r}\left(r^2\frac{\partial}{\partial r}\right)+\frac{1}{r^2\sin\theta}\frac{\partial}{\partial\theta}\left(\sin\theta\frac{\partial}{\partial\theta}\right)+\frac{1}{r^2\sin\theta}\frac{\partial^2}{\partial\phi^2}\right]\psi(r,\theta,\phi)$$
$$=E\psi(r,\theta,\phi) \tag{1.55}$$

where m^* is the constant effective mass of a single particle and E is the energy associated with the confinement along the radiur r. Using separation of variables, the wave function is written as $\psi(r,\theta,\phi)=R(r)Y(\theta,\phi)$. After some steps, the radial form of (1.55) can be written as

$$\frac{d^2R(\rho)}{d\rho^2}+\frac{2}{\rho}\frac{dR(\rho)}{d\rho}+\left[1-\frac{l(l+1)}{\rho^2}\right]R(\rho)=0 \tag{1.56}$$

where $\rho=kr$. This equation is known as the Bessel differential equation and its general solution is

$$R(\rho)=A_{jl}(\rho)+B_{nl}(\rho) \tag{1.57}$$

Because Neumann functions are divergent at r = 0, setting B = 0 we obtain the wave function of the particle. Analytical solutions of the particle energy values can be obtained from the roots of the Bessel functions. For example, when l = 0 we get

$$j_0(ka)=\frac{\sin(ka)}{ka}=0\rightarrow E_{n,0}(ka)=\frac{\hbar^2\pi^2n^2}{2m^*a^2} \tag{1.58}$$

In applying the finite difference method to the core or core/shell spherical quantum dots we take the following steps. In step one, we take $\theta=\pi/2$ in (1.56) and rewrite the Schrödinger equation in the (x, y, 0) plane as

$$\left(-\frac{\partial^2}{\partial r^2}-\frac{1}{r^2}\frac{\partial^2}{\partial\phi^2}-\frac{2}{r^2}\frac{\partial}{\partial r}\right)\varphi(r,\phi)=\left(E_r+E_\phi\right)\varphi(r,\phi) \tag{1.59}$$

We take the wave functions with equal steps on the r and ϕ axes, shown in Fig. 1.41.

Substituting the 1st and 2nd derivatives of wave function in the (r,ϕ) Table of finite difference method shown in Fig. 1.41, we write

$$\begin{aligned}H(1,1)=&-\frac{1}{dr^2}[\varphi(1,1)-2\varphi(1,2)+\varphi(1,3)]\\&-\frac{1}{r^2d\phi^2}[\varphi(4,1)-2\varphi(1,1)+\varphi(1,2)]\\&-\frac{2}{rdr}[\varphi(1,2)-\varphi(1,1)+V(1,1)\varphi(1,1)]=\left(E_{\phi1}+E_{r1}\right)\varphi(1,1)\end{aligned} \tag{1.60}$$

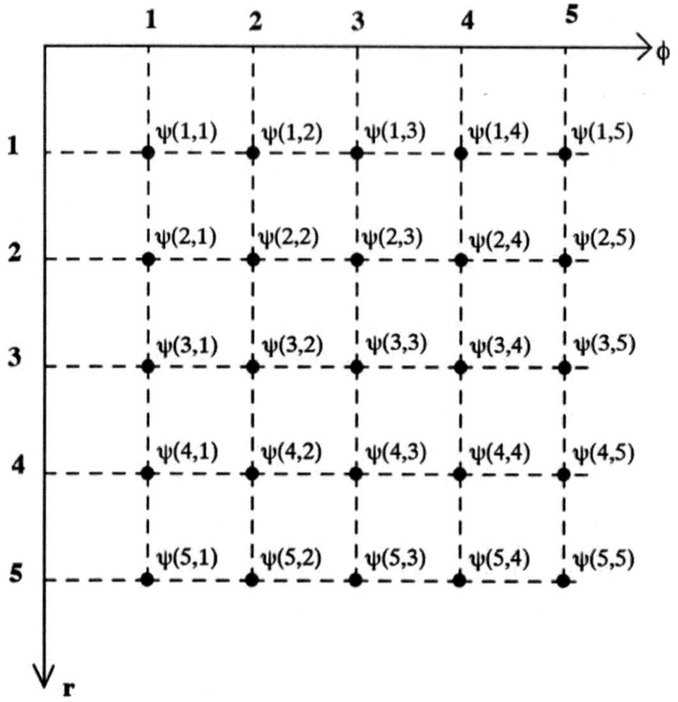

Fig. 1.41 Representation of the wave function in a table of finite difference method

$$\begin{aligned} H(2,4) = &-\frac{1}{dr^2}[\varphi(2,3) - 2\varphi(2,4) + \varphi(2,5)] \\ &- \frac{1}{r^2 d\phi^2}[\varphi(1,4) - 2\varphi(2,4) + \varphi(3,4)] \\ &- \frac{2}{rdr}[\varphi(2,5) - \varphi(2,4) + V(2,4)\varphi(2,4)] = \left(E_{\phi 2} + E_{r4}\right)\varphi(2,4) \end{aligned} \tag{1.61}$$

$$\begin{aligned} H(5,5) = &-\frac{1}{dr^2}[\varphi(5,5) - 2\varphi(5,4) + \varphi(5,3)] \\ &- \frac{1}{r^2 d\phi^2}[\varphi(2,5) - 2\varphi(2,4) + \varphi(3,4)] \\ &- \frac{2}{rdr}[\varphi(2,5) - \varphi(2,4) + V(2,4)\varphi(2,4)] = \left(E_{\phi 2} + E_{r4}\right)\varphi(2,4) \end{aligned} \tag{1.62}$$

As in the case of quantum wires, the variation of ground state energies of spherical quantum dots with different potential barriers with the diameter of the quantum dot is then calculated. Figure 1.42 shows the variation of the ground state energy

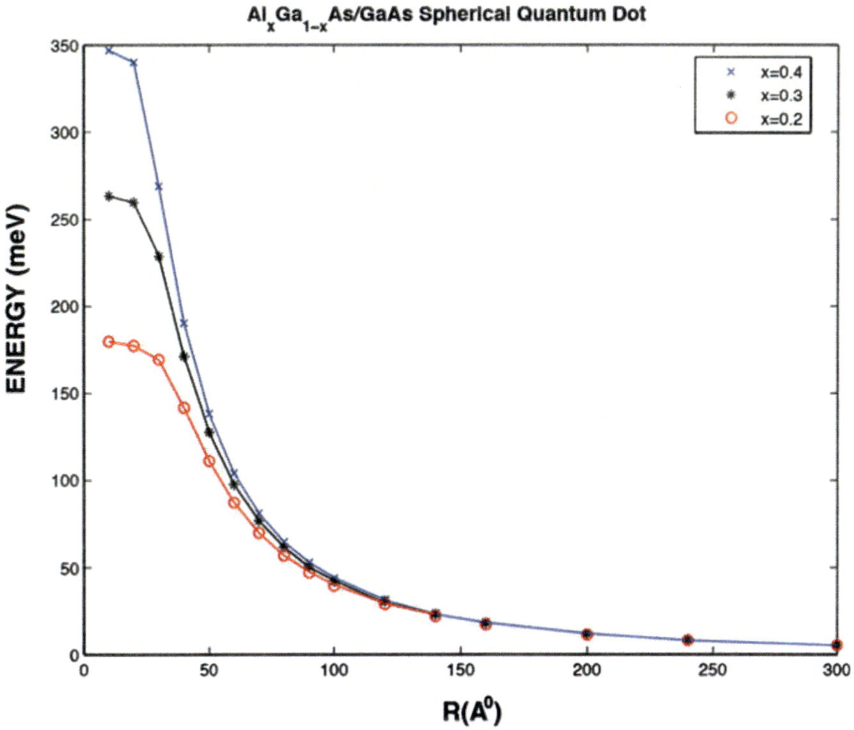

Fig. 1.42 Variation of the ground state energies with radius of spherical core/shell AlGaAs/GaAs QDs with different ternary alloy shell composition. It is reproduced from [21]

obtained for the lattice matched AlGaAs/GaAs spherical core/shell quantum dot. As with quantum wires, a decrease in ground state energy is also observed at small radius when the potential barrier decreases. As the radius increases, the height of the potential barrier loses its importance, and the ground state energy converges to a certain value.

1.10 Conclusion

Advances in the growth of semiconductor thin films with layer thickness approaching atomic dimensions and fabrication of low dimensional electronic and optical devices has provided new opportunities in fundamental science and technology of semiconductors and semiconductor devices. Such advancement could not have been possible without a qualitatively reliable understanding of the basic physics of semiconducting materials and quantitatively precise potential predictions and performance of devices, leading to new concepts in the semiconductor growth that allowed previously many unknown devices with more complex functionality and much higher densities for electronic and optical applications.

This review has presented comparative predictions of electronic band structures of group III–V and II–VI compounds and their heterostructures obtained using the semiempirical tight binding theory based on the 2NN sp^3, 2NN sp^3s^*, NN $sp^3d^5s^*$ and NN sp^3d^5 orbitals sets in comparison with the WIEN2K simulations package based on the density functional theory (DFT) with the modified modified Becke-Johnson exchange–correlation potential with the local density approximation (LDA), called MBJLDA functional. We conclude that DFT with MBJLDA functional and the semiempirical NN sp^3d^5 TB model are in good agreement with experimental data for band gaps and band offsets and have considerable potential in the design and optimization of group III–V and II–VI compounds and their ternaries used for the fabrication of electronic and optical devices. When compared to the 2NN sp^3s^* TB model, NN sp^3d^5 TB parametrization better simulates both the valence band and conduction band dispersion curves. The actual contribution of the excited d-states in the band structure calculations is reliably reproduced in the case of NN sp^3d^5 orbitals set tight binding parameterization. We also discuss the use of finite difference technique for modelling of electronic structure of two-dimensional quantum wells, one dimensional cylindrical nanowires and zero-dimensional spherical quantum dots. We conclude that DFT with MBJLDA and semiempirical NN sp^3d^5 TB model show good agreement with experiment for band gaps and band offsets. When these semiempirical tight binding theories and density functional theory with MBJLDA are integrated in the finite difference method they may have considerable impact in the design and optimization of group III–V and II–VI compounds and their ternaries used for the fabrication of low dimensional electronic and optical devices.

References

1. H. Ünlü, N.J.M. Horing (eds.), *Low Dimensional Semiconductor Structure: Characterization and Applications* (Springer, 2013)
2. H. Morkoç, H. Ünlü, H. Zabel, N. Otsuka, Solid State Technol. **31**, 71 (1988)
3. H. Kroemer, Quasi-electric and quasi-magnetic fields in non-uniform semiconductors. RCA Rev. **18**, 332–342 (1957)
4. L. Esaki, R. Tsu, IBM J. Res. Develop. **14**, 61 (1970)
5. Ö. Akıncı, H.H. Gürel, H. Ünlü, Physica Status Solidi (c) **5**(2), 478 (2008)
6. H.H. Gürel, Ö. Akıncı, H. Ünlü, Thin Solid Films **516**, 7098 (2008)
7. Ö. Akıncı, H.H. Gürel, H. Ünlü, J. Nanosci. Nanotechnol. **8**, 540 (2008)
8. Ö. Akıncı, H.H. Gürel, H. Ünlü, Thin Solid Films **517**, 2431 (2009)
9. H.H. Gürel, Ö. Akıncı, H. Ünlü, Superlattices Microstruct. **51**, 725 (2012)
10. H.H. Gürel, H. Ünlü, Mater. Sci. Semicond. Process. **16**, 1619 (2013)
11. J.L. Martins, A. Zunger, Phys. Rev. B **30**, 6217 (1984)
12. P. Hohenberg, W. Kohn, Phys. Rev. **136B**, 864 (1964)
13. W. Kohn, J.L. Sham, Phys. Rev. **140A**, 1133 (1965)
14. M.L. Cohen, J.R. Chelikowsky, Electronic Structure and Optical Properties of Semiconductors, 2nd edn. (Springer, 1989)
15. H.H. Gürel, Ö. Akıncı, H. Ünlü, Comput. Mater. Sci. **33**, 269 (2005)

16. Ö. Akıncı, H.H. Gürel, H. Ünlü, Thin Solid Films **511–512**, 684 (2006)
17. H.H. Gürel, Ö. Akıncı, H. Ünlü, Superlattices Microstruct. **40**(4–6), 588 (2006)
18. H.H. Gürel, Ö. Akıncı, H. Ünlü, Physica Status Solidi (c) **4**(2), 316 (2007)
19. H.H. Gürel, M.sc Thesis, Istanbul Technical University (2005)
20. Ö. Akıncı, M.sc Thesis, Istanbul Technical University (2005)
21. H.H. Gürel, Ph.D. Thesis, Istanbul Technical University (2011)
22. Ö. Akıncı, Ph.D. Thesis, Istanbul Technical University (2012)
23. T.B. Boykin, L.J. Gamble, G. Klimeck, R.C. Bowen, Phys. Rev. B **59**, 7301 (1999)
24. Y. Fu, K.A. Chao, Phys. Rev. B **43**, 4119 (1991)
25. A.D. Carlo, Semicond. Sci. Tech. **18**, R1 (2001)
26. E.P. O'Reilly, A. Lindsay, S. Tomic, M.K. Saadi, Semicond. Sci. Technol. **17**, 870 (2002)
27. H. Morkoç, H. Ünlü, G. Ji, *Principles and Technology of MODFETs*, vols. 1, 2 (Wiley, 1991)
28. H. Ünlü, Solid State Electron. **35**, 1343 (1992)
29. H. Ünlü, Phys. Status Solidi B **216**, 107(1999); B 223, 195 (2001); B 229 581 (2002); B 235, 248 (2003)
30. C. Pryor, J. Kim, L.M. Wang, A.J. Willamson, A. Zunger, J. Appl. Phys. **83**, 2548 (1998)
31. P. Keating, Phys. Rev. **145**, 637 (1966)
32. H. Ünlü, N.J.M. Horing, J. Dabrowski (eds.) *Low Dimensional and Nanostructured Materials and Devices* (Springer, 2015)
33. M. Rabah, B. Sahraui, B. Bouhafs, B. Abbar, H. Abid, Phys. Stat. Solidi B **238**, 156 (2003)
34. I. Vurgaftman, J.R. Meyer, L.R. Ram-Mohan, J. Appl. Phys. **89**, 5815 (2001)
35. H. Morkoç, H. Ünlü, in *Semiconductors and Semimetals*, **24** (R. Dingle, Editor), 135 (1987)
36. H. Kroemer, Proc. IRE **45**, 1535 (1957)
37. H. Kroemer, Rev. Modern Phys. **73**, 783 (2000)
38. M. Jancu, F.R.Scholz, F.Beltram, F. Bassani, Phys. Rev. B **57**, 6493 (1998)
39. J.M. Jancu, F. Bassani, F. Della Sala, R. Schols, Appl. Phys. Lett. **81**, 4838 (2002)
40. S.Sapra, N. Shanthi, D.D. Sarma, Phys. Rev. B **66**, 205202 (2002)
41. H. Ünlü, H. Morkoç, Solid State Technol. **31**, 83 (1988)
42. H. Ünlü, H. Morkoç, and S. Iyer, in *Gallium Arsenide Technology*, **2**(D. K. Ferry, ed., 1990), 231
43. O. Madelung (ed.), *Numerical Data and Functional Relationships in Science and Technology*, 17a (Springer, 1982); ibid 17d, Springer (1984)
44. D.J. Chadi, Phys. Rev. B **16**, 790 (1977)
45. W.A. Harrison, *Electronic Structure and the Properties of Solids* (Freeman, 1980)
46. J.P. Loehr, Talwar, Phys. Rev. B, **55**, 4353 (1997)
47. J.C. Slater, G.F. Koster, Phys. Rev. **94**, 1498 (1954)
48. D.J. Chadi, M.L. Cohen, Phys. Status Solidi **68**, 405 (1975)
49. D.N. Talwar, C.S. Ting, Phys. Rev. B **25**, 2660 (1982)
50. F. Tran, P. Blaha, Phys. Rev. Lett. **102**, 226401 (2009)
51. P.Vogl, H.P. Hjalmarson, J.D. Dow, J. Phys. Chem. Solids **44**, 365 (1983)
52. N. Lucas, H. Zabel, H. Morkoç, H. Ünlü, Appl. Phys. Lett. **52**, 2117 (1988)
53. V. Fiorentini, A. Baldereschi, Phys. Rev. B **51**, 17196 (1995)
54. J.P. Perdew, S. Burke, M. Ernzerhof, Phys. Rev. Let. **77**, 3865–3868 (1996)
55. Z. Wu, R. Cohen, Phys. Rev. B **73**, 235116–235123 (2006)
56. J.P. Perdew, Phys. Rev. Let. **100**, 136406–136410 (2008)
57. A. Mattsson, J. Chem. Phys. **128**, 084714–084725 (2008)
58. J.P. Perdew, Y. Wang, Phys. Rev. B **45**, 13244–13249 (1992)
59. P. Blaha, K. Schwarz, G.K.H. Madsen, D. Kvasnicka, J. Luitz, *WIEN2k: An Augmented Plane Wave Plus Local Orbitals Program For Calculating Crystal Properties*, Vienna University of Technology, Vienna, Austria (2001)

Chapter 2
Strain in Microscale and Nanoscale Semiconductor Heterostructures

Hilmi Ünlü

Abstract Reliable modelling and precise determination of the strain effects on the structural, electronic and optical properties of microscale planar and nanoscale core/shell semiconductor heterostructures is essential for the true prediction of their potential in fabricating electronic and optical devices operating at high temperatures. In this chapter, a thermoelastic model is described to determine strain effects on the structural, electronic and optical properties of microscale planar and nanoscale core/shell semiconductor heterostructures as a function dimensions, anisotropy of elastic constants, lattice mismatch, ternary ally composition and temperature. The model uses lattice mismatch induced *shrink fit* condition for strain across heterointerfaces to find contact pressure and then strain in order to consider the effect of the difference between lattice constants, linear expansion coefficients, anisotropy of elastic constants of constituent semiconductors. Qualitative discussion is given about the strain effects on anisotropy in linear expansion coefficient of GaAs epilayer in GaAs/Si (001) planar heteroepitaxy. Furthermore, details of the modelling and calculations of strain effects on core gaps and band offsets of core/shell type I and type II heterostructure spherical quantum dots and cylindrical nanowires are discussed by using parabolic two band **k.p** *effective mass approximation.* Calculations suggest a parabolic increase (decrease) in core bandgaps and conduction band offsets in GaAs/InAs (GaSb), CdSe/CdS, CdSe/ZnS, and ZnSe/ZnS quantum dots. Furthermore, a parabolic decrease in core bandgap and conduction band offset of GaSb/InAs(InSb) and ZnSe/CdS QDs as core (shell) diameter increases for a fixed shell (core) diameter. Excellent agreement between predicted and measured core bandgaps in CdSe and ZnSe based core/shell QDs suggests that proposed model can be a good design tool for the process simulation of microscale planar heteroepitaxial devices and nanoscale core/shell heterostructure nanowire and quantum dot based electronic and optical devices for high temperature applications.

H. Ünlü (✉)
Faculty of Science and Letters, Department of Physics Engineering, İstanbul Technical University, Maslak Istanbul 34469, Turkey
e-mail: hunlu@itu.edu.tr

H. Ünlü and N. J. M. Horing (eds.), *Progress in Nanoscale and Low-Dimensional Materials and Devices*, Topics in Applied Physics 144,
https://doi.org/10.1007/978-3-030-93460-6_2

2.1 Introduction

Since the invention of the first transistor in 1949, device scientists and engineers have witnessed an amazing theoretical and experimental interest and development in semiconductor science and technology [1]. Advances in the epitaxial growth of semiconductor thin films of different structural, electronic and optical properties and with layer thickness approaching atomic dimensions has provided new opportunities and challenges in fundamental science and technology of semiconductors for device applications [2–105]. Contemporary fabrication technologies have further made it possible to reduce the device dimensions to the point where quantum size effects must be considered in order to realistically describe the operation and reliably predict the potential and performance of low dimensional and nanoscale semiconductor devices for electronic and optical applications.

Semiconductor nanoscale core/shell heterostructures represent a new class of materials and a shift from conventional two-dimensional thin films to three-dimensional structures [8]. One can grow group IV elemental and III-V and II-VI compound semiconductor core/shell quantum dots (QDs) and nanowires (NWs) directly on inexpensive substrates such as Si, rather than lattice-matched but more expensive III-V substrates such as GaAs. This capability, along with other unique properties (quantum confinement and light trapping), makes core/shell QDs and NWs of great interest for the next generation electronic and optical devices with improved performance, new functionalities, and reduced production cost. The core/shell semiconductor nanowires and quantum dots can be synthesized as single or multiple heterostructures in order to isolate the active region from surface defects and traps, allowing one to facilitate better electronic properties (e.g., charge transport and mobility) and optical properties (e.g., luminescence efficiency and sharper linewidths, etc.) Isolating active core region from surface defects is a fundamental precondition for the use of nanocrystals in biological labeling and light-emitting devices, which rely on their emission properties.

When two semiconductors with different physical properties are brought in thermal contact to form heterostructure, the difference between conduction and valence band energy levels of constituent semiconductors are accommodated by discontinuities ΔE_C and ΔE_V at heterointerface, controlling charge transport and optical exciton in nanoscale electronic and optical devices, respectively [2]. Figure 2.1 shows the schematic band diagrams of Type I heterostructures, respectively, in which conduction and valence bands are aligned in straddling type ($\Delta E_c = \Delta E_g - \Delta E_v$) or staggered type ($\Delta E_c = \Delta E_g + \Delta E_v$), respectively. Here $\Delta E_g = E_{gB} - E_{gA}$ is the difference between bandgaps.

The conduction band of barrier semiconductor overlaps (straddling lineup) that of the well semiconductor and equilibrium Fermi level is near the middle of the bandgap on both sides of interface. Consequently, electrons and holes tend to localize in the narrow bandgap well in nearly equal number. Therefore, *Type I* band alignment in core/shell nanostructures results in an electron–hole pair excited near interface which tends to localize in core region and exciton energy is result of direct exciton transition. Such heterostructure is useful in fabricating optoelectronic devices such as lasers, with both electrons and holes participate to device operation.

Figure 2.2 shows the band structure of normal and inverted *Type II* heterostructures formed between two intrinsic semiconductors. The bandgap of the shell partially overlaps (staggered lineup) that of the core and equilibrium Fermi level is close to conduction (or valence band) on one side while it is near to middle of the bandgap on other side.

In nanoscale Type II core/shell heterostructures, conduction band edge of shell region is located in bandgap of core region leading to a local separation of electron and hole in these regions, respectively. Electrons (holes) are confined to core (shell) region as result of indirect exciton transitions. The local indirect bandgap of nanocrystal is equal to $E_g^{id} = E_{gA} - \Delta E_v$ in electron–hole confinement (Fig. 2.1a) and $E_g^{id} = E_{gB} - \Delta E_c$ in hole-electron confinement (Fig. 2.2b).

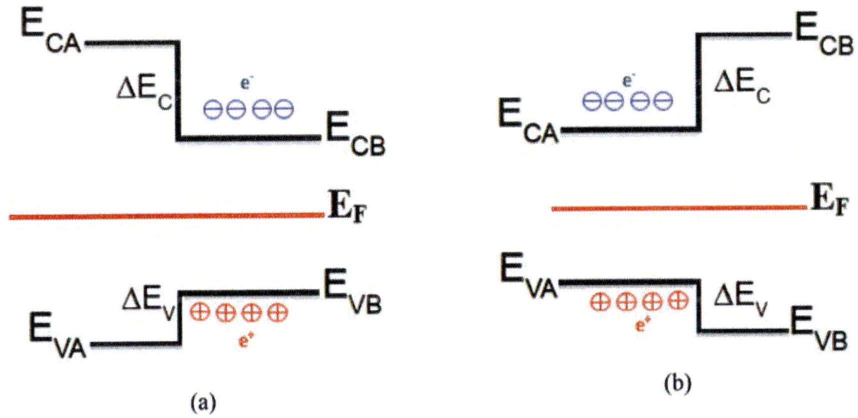

Fig. 2.1 Schematic band diagram of normal (**a**) and inverted (**b**) *Type I* heterostructures

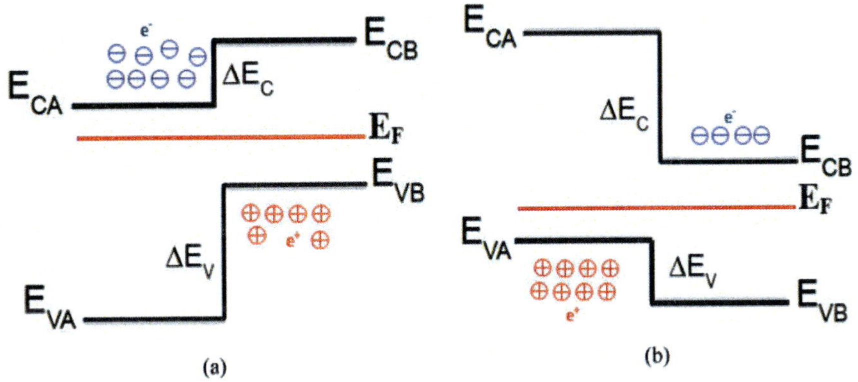

Fig. 2.2 Schematic energy band diagram of normal (**a**) and inverted (**b**) Type II heterostructure

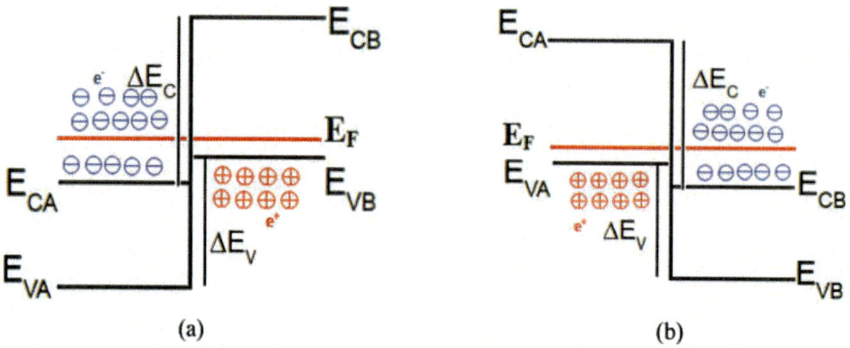

Fig. 2.3 Schematic band diagram of normal (**a**) and inverted (**b**) Type III heterostructures

The extreme case of Type II band alignment is known as *Type III* heterostructure, shown in Fig. 2.3. The bandgap of the barrier does not overlap at all (broken gap lineup). The equilibrium Fermi level is well above the conduction band minimum (or well below the valence band maximum) on one side while it is near the middle of the bandgap on the other side. Position of equilibrium Fermi level determines the carrier confinement on either side of the heterointerface. When equilibrium Fermi level is close to conduction band of wide bandgap constituent the number of electrons is greater but when it is closer to the valence band of narrow bandgap constituent the number of holes is greater there.

Contemporary crystal growth techniques including molecular beam epitaxy (MBE) and metal organic chemical vapor deposition (MOCVD) have made it possible to control the alloy composition and doping in semiconductor alloys based on group IV–IV elemental and III–V and II–VI compound semiconductor over atomic distances, as low as few nanometers. These artificially made semiconductor heterostructures are easily grown on lattice mismatched substrates. The epitaxial layers are so thin so that quantum mechanical effects dominate charge transport. The resulting contemporary low dimensional and nanoscale electronic and optoelectronic semiconductor devices (e.g., heterostructure bipolar transistors (HBTs), modulation doped field effect transistors (MODFETs), etc.) are known to operate much faster than conventional silicon devices (e.g., Si bipolar junction transistors (BJTs) and metal oxide semiconductor field effect transistors (MOSFETs)), which is crucial for the electronic and optical communication and computer industries.

As a result of intensive research and development activities in the growth and fabrication, the size of existing semiconductor devices in order of 5 to 10 nm can now be achieved. Advancement in the semiconductor technology could not have been possible without a qualitatively reliable physical understanding of the semiconducting materials and quantitatively precise potential predictions of devices, leading to new concepts in the semiconductor growth that allowed previously many unknown devices with more complex functionality and much higher densities.

In this chapter we will give a detailed discussion about the modelling of structural and elastic properties of semiconductors and how they can be implemented in theoretical models for the calculations of electronic and optical properties of microscale planar and nanoscale core/shell heterostructures. We will give details of the derivation of analytical expressions for the strain effects in these heterostructures by using continuum elastic theory of thermoelastic bodies. In Sect. 2, we give derivation of strain expressions in heterostructures with planar geometry such as quantum wells and superlattices. In Sects. 3 and 4, we give detailed derivations of strain expressions in nanoscale spherical and cylindrical core/shell heterostructures. In Sect. 5, detailed discussion of results will be given for calculating the strain effects on core band gap and band offset in GaAs/InAs(GaSb), GaSb/InAs(InSb), CdSe/CdS(ZnS) and ZnSe/CdS(ZnS) core/shell heterostructures.

2.2 Strain in Planar and Core/Shell Heterostructures

When two semiconductors with different physical properties are brought in thermal contact to form a coherently grown a microscale planar or nanoscale spherical or cylindrical core/shell heterostructure, strain across the interface can alter electronic and optical properties of constituent semiconductors relative to their bulk values which will influence the device operation and performance. Numerous models have been proposed over the years to calculate the strain effects on electronic properties of microscale planar and nanoscale core/shell heterostructures. In case of microscopic planar heteroepitaxy of thin films, the original works of Frank and van der Merve [3] and Matthews and Jesser [4] are used to model strain effects in microscale planar heteroepitaxial thin films and quantum wells [6]. In case of core/shell nanoscale core/shell heterostructures, the continuum elasticity theory of Eshelby [5] have been used in literature.

In a coherently grown microscale planar heterostructures, shown in Fig. 2.4, when lattice constants and linear thermal expansion coefficients of constituent semiconductors are different from each other, strain influences the energy band structure and charge transport across the interface, operation and performance of heterostructure devices. Strain effects in microscopic semiconductor heterostructures have been extensively studied. The in plane (along interface) and out of plane (along growth direction) strains in heterolayer are defined as [4]

$$\varepsilon_{xx}^{f} = \varepsilon_{xx}^{f} = \varepsilon_{\parallel}^{f} = \frac{a_{\parallel}^{f} - a_f}{a_f} = \varepsilon_m \tag{2.1}$$

$$\varepsilon_{zz}^{f} = \varepsilon_{\perp}^{f} = \frac{a_{\perp} - a_f}{a_f} = -\frac{2\nu_f}{1-\nu_f}\varepsilon_m \tag{2.2}$$

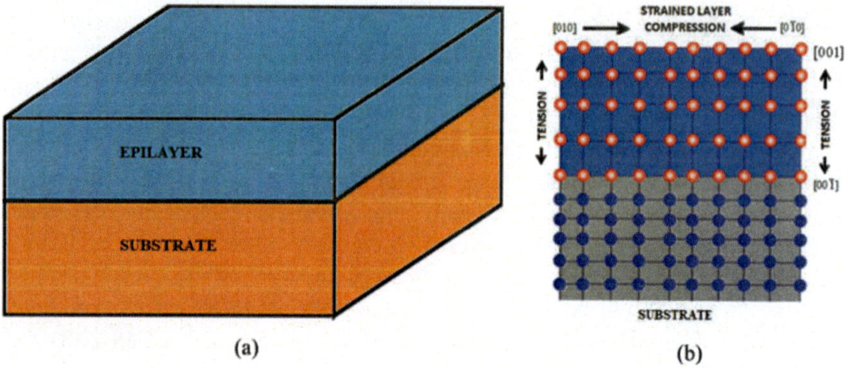

Fig. 2.4 Schematic cross-sectional view and interface strain effects in planar heterostructure

where $a_{\|}^{f}$ and $a_{\perp}^{f}$ are in plane (along interface) and out of plane (along growth direction) distorted lattice constants. a_f and a_s are relaxed lattice constant of epilayer in bulk form and thick substrate, respectively.$\varepsilon_m = (a_f - a_s)/a_s$ is lattice mismatch and ν_f is Poisson ratio of epilayer, defined as $\nu = C_{12}/(C_{11} + C_{12})$ with C_{11} and C_{12} elastic stiffness constants of bulk semiconductor.

Theoretical understanding the physics and determining the strain effects in one dimensional cylindrical and zero-dimensional spherical core/shell semiconductor heterostructures is still an obstacle. Figure 2.5 shows the schematic cross-sectional view of a spherical or cylindrical core/shell heterostructures. Interface imposes isotropic compression in core region and non-isotropic tension along radial and tangential directions in the shell region, respectively. In studying the strain effects in nanoscale structures, currently classical continuum elasticity and atomistic theories are used to study the strain effects in nanoscale structures. In a coherently grown nanoscale core/shell heterostructures, strain across the interface differently influences the energy band structure and charge transport, operation and performance of heterostructure devices. Qualitatively reliable and quantitatively precise modelling and simulation of strain effects on the morphology and electronic properties of semiconductor constituents is important for a better prediction of their potential in making nanoscale devices.

According to continuum elastic theory [5], strain on the core and shell sides of interface in spherical core/shell quantum dots (QDs) are described by

$$\varepsilon_{ir} = \varepsilon_{i\theta} = \varepsilon_{i\varphi} - \frac{2E_m(1 - 2\nu_i)\varepsilon_{im}}{(1 + \nu_m)E_i + 2(1 - 2\nu_i)E_m} \tag{2.3}$$

$$\varepsilon_{mr} = -\frac{2E_i(1 + \nu_m)\varepsilon_{im}}{(1 + \nu_m)E_i + 2(1 - 2\nu_i)E_m} \tag{2.4}$$

$$\varepsilon_{mr} = -\frac{2E_i(1 + \nu_m)\varepsilon_{im}}{(1 + \nu_m)E_i + 2(1 - 2\nu_i)E_m} \tag{2.5}$$

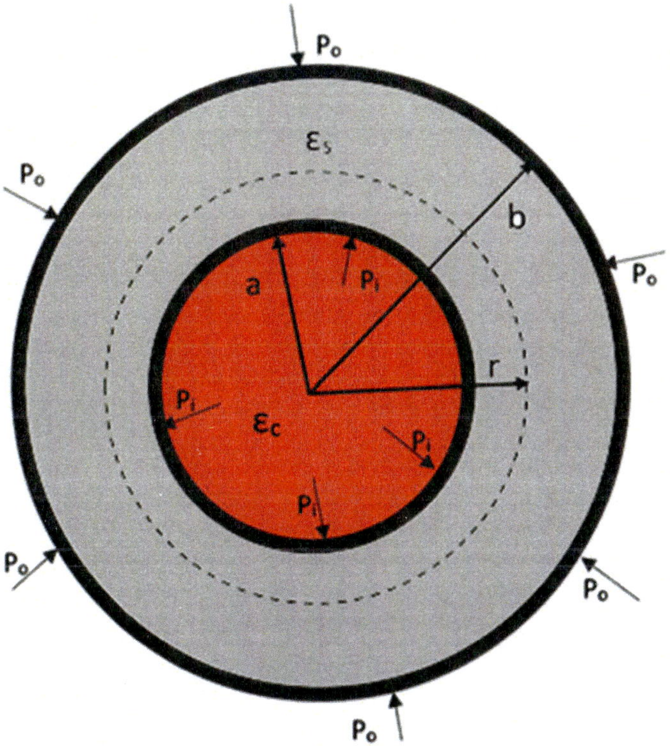

Fig. 2.5 Schematic cross-sectional view of core/shell heterostructure

where ε_{ir}, $\varepsilon_{i\theta}$, and $\varepsilon_{i\varphi}$ are radial and tangential strains in core region. ε_{mr} and $\varepsilon_{mt} = \varepsilon_{m\theta} = \varepsilon_{m\varphi}$ are radial and tangential strains in shell region. $\varepsilon_{im} = (a_i - a_m)/a_m$ is the lattice mismatch, where a_i and a_m are, respectively, the lattice constants of core and shell constituents. $E_i(E_m)$ and ν_i (ν_m) are the Young modulus and Poisson ratio of core and shell constituents. They are, respectively, defined as $E = (C_{11} - C_{12})(C_{11} + 2C_{12})/(C_{11} + C_{12})$ and $\nu = C_{12}/(C_{11} + C_{12})$ with C_{11} and C_{12} elastic stiffness constants of bulk semiconductors.

Likewise, in cylindrical core/shell heterostructure nanowires (NWs), the continuum elasticity theory [5] describes the strain on core and shell sides of interface according to the following equations

$$\varepsilon_i^* = \frac{(1 - \nu_i - 2\nu_i^2)E_m\varepsilon_{im}}{(1 + \nu_m)E_i + (1 - \nu_i - \nu_i^2)E_m} - \nu_i\varepsilon_{im}, \tag{2.6}$$

$$\varepsilon_{mr}^* = \frac{-E_i(1 + \nu_m)(1 - \nu_i)\varepsilon_{im}}{E_m(1 - \nu_i - 2\nu_i^2) + E_i(1 + \nu_m)}, \tag{2.7}$$

$$\varepsilon^*_{m\theta} = \frac{E_i(1+\nu_m)(1-\nu_i)\varepsilon_{im}}{E_m(1-\nu_i-2\nu_i^2)+E_i(1+\nu_m)} \tag{2.8}$$

where ε_i^* is the isotropic hydrostatic strain ($\varepsilon_i^* = \varepsilon_{ir}^* = \varepsilon_{i\theta}^* = \varepsilon_{i\varphi}^*$) in core region and ε^*_{mr} and $\varepsilon^*_{m\theta}$are radial and tangential components of hydrostatic strain in shell region.

When one of the constituents of microscale planar and nanoscale core/shell heterostructures is a semiconductor alloy (e.g., AlGaAs/InGaAs, GaAs/InGaAs, CdSe/CdZnS, ZnSe/ZnCdS), the composition variation of structural properties of alloy constituent will alter the sign and modify the magnitude interface strain. Therefore, in addition to mismatches between lattice constants and thermal expansion coefficients, it is important to include the composition contribution to the interface strain effects on electronic properties of microscale planar and nanoscale core/shell heterostructures prior to any device fabrication. Furthermore, the composition dependent strain effect on core lattice constant of cylindrical or spherical binary/ternary core/shell heterostructure will be in a way that is not seen in microscale planar geometry. Modified virtual crystal approximation [95, 96] can be implemented in the thermoelastic model [97, 98] to study the alloy effects on strain and in turn on the lattice constant, band gaps and band offsets in binary/ternary and ternary//binary core/shell heterostructures.

In the framework of modified virtual crystal approximation (MVCA) lattice constant of $A_{1-x}B_xC$ ternary alloy shell of binary/ternary heterostructure is defined as [95, 96]:

$$\begin{aligned} a_m(x) &= (1-x)a_{AC}(x) + (x)a_{BC}(x) \\ &= (1-x)a^0_{AC} + xa^0_{BC} - x(1-x)(\xi_{AC:B} - \xi_{BC:A})\left(a^0_{AC} - a^0_{BC}\right) \end{aligned} \tag{2.9}$$

where $a_{AC}(x)$ and $a_{BC}(x)$ are lattice constants of AC and BC bulk binary semiconductors forming $A_{1-x}B_xC$ ternary semiconductor shell, defined as

$$a_{AC}(x) = a^0_{AC} + x\xi_{BC:A}\left(a^0_{AC} - a^0_{BC}\right),\ a_{BC}(x) = a^0_{BC} + (1-x)\xi_{AC:B}\left(a^0_{BC} - a^0_{AC}\right) \tag{2.10}$$

where a^0_{AC} and a^0_{BC} are the undistorted lattice constants of host AC and BC binary semiconductors and $\xi_{AC:B}$ and $\xi_{BC:A}$ are dimensionless relaxation parameters [89]

$$\xi_{AC:B} = [1 + \frac{\alpha_{AC}}{6\alpha_{BC}}(1 + 10\frac{\beta_{AC}}{\alpha_{AC}})]^{-1},\ \xi_{BC:A} = [1 + \frac{\alpha_{BC}}{6\alpha_{AC}}(1 + 10\frac{\beta_{BC}}{\alpha_{BC}})]^{-1} \tag{2.11}$$

where α and β are force constants and are related to elastic stiffness constants as

$$C_{11} + 2C_{12} = (3\alpha + \beta)/a - 0.355s,\ C_{11} - C_{12} = 4\beta/a + 0.053s \tag{2.12}$$

where $s = e^2 Z *^2 / d^4 \varepsilon$. When disorder-induced nonlinear variation of the lattice constant is ignored, (2.9) reduces to classical virtual crystal approximation (VCA) of lattice constant of ternary alloy semiconductor constituent: $a_m^0(x) = (1-x)a_{AC}^0 + xa_{BC}^0$. One then defines lattice mismatch at spherical or cylindrical heterointerface of binary/ternary core/shell heterostructure as $\varepsilon_{im}(x) = (a_i - a_m(x))/a_m(x)$. The effects of composition on thermal properties (e.g., linear expansion coefficient), elastic properties, and valence and conduction band edges and their deformation potentials of $A_{1-x}B_xC$ ternary alloy component of core/shell QDs are calculated by using a nonlinear expression [95, 96]

$$P_{\alpha\beta}^{ABC}(x) = (1-x)P_{\alpha\beta}^{AC} + (x)P_{\alpha\beta}^{BC} - x(1-x)(\xi_{AC:B} - \xi_{BC:A})(P_{\alpha\beta}^{AC} - P_{\alpha\beta}^{BC}) \tag{2.13}$$

where $P_{\alpha\beta}^{AC}$ and $P_{\alpha\beta}^{AC}$ are the bulk parameters of AC and BC binaries.

Equations (2.1)–(2.13) suggest that the continuum elasticity theories do not fully consider the effect of dimensions, alloy composition, difference between linear thermal expansion coefficients, free thermal expansion of lattice constants and anisotropy of elastic constants of semiconductors forming microscale planar and nanoscale core/shell heterostructures. It is well known that group IV elemental and groups III–V and II–VI compound semiconductors tend to expand due to ΔT temperature rise during crystal growth and still retain their original crystal structure [91]. Therefore, one must consider the effects of lattice thermal expansion of constituent semiconductors and thermal strain across the heterointerface developed during crystal growth and must be added to interface strain due to local mechanical stress caused by the lattice mismatch at the interface.

Furthermore, most of the group IV elemental and groups III–V and II–VI compound semiconductors have elastic anisotropy [92]. Therefore, one must also take into account the anisotropy of elastic parameters (e.g., Young's modulus and Poisson's ratio) in the modelling of strain effects on electronic, and optical properties of nanoscale core/shell heterostructures. The anisotropy of elastic properties of semiconductor may have decisive role on the reliable modelling the operation and accurate prediction and simulation of the performance of microscale and nanoscale heterostructure devices. In quantifying the anisotropy of cubic crystals, we can use Zener anisotropy index, also known as *anisotropy ratio* [92], ratio of the maximum and minimum values of shear modulus of cubic crystal

$$A = 2C_{44}/(C_{11} - C_{12}) \tag{2.14}$$

where C_{11}, C_{12} and C_{44} are crystal elastic constants of a crystal with cubic symmetry. Zener's anisotropy factor A is equal to unity when the crystal is elastically isotropic. Deviation less than or greater than unity corresponds to the degree of elastic anisotropy possessed by the crystal. The extension of Zener's anisotropy index to crystals with hexagonal symmetry creates ambiguities. Ranganathan and Ostoja-Starzewski [93] proposed a universal elastic anisotropy index (AU) to quantify the elastic anisotropy of crystals of all classes, including cubic and hexagonal crystals. Average values of anisotropic elastic moduli ($\tilde{E}$ and $\tilde{\nu}$) of group IV elemental and groups III-V and II-VI compound semiconductors are calculated by [92]

$$\tilde{E} = \frac{\mu(3\lambda + 2\mu)}{\lambda + \mu}, \quad \tilde{\nu} = \frac{\lambda}{2(\lambda+\mu)} \tag{2.15}$$

$$\lambda = C_{12} - H/5, \quad \mu = C_{44} - H/5, \quad H = 2C_{44} + C_{12} - C_{11}$$

where λ and μ are Lame's parameter and shear modulus, respectively. H is called anisotropy factor of crystal with cubic structure [92]; A semiconductor with cubic crystal structure is elastically isotropic when $H = 0$ and anisotropic when $H > 0$. Deviation greater than unity corresponds to the degree of elastic anisotropy possessed by the crystal. The bulk modulus is then defined as $\tilde{B} = \tilde{E}/3(1 - \tilde{\nu})$.

In order to emphasize the importance of the structural and elastic properties of constituent semiconductors in strain modelling it is also necessary to incorporate the anisotropy of elastic properties in the stress–strain relations. Since thermal expansion of volume elements cannot proceed freely, the total strain can be thought to consist of algebraic sum of the thermal and the elastic strains produced by the resistance of the medium to thermal expansion [94]. The Hooke's law is modified to

$$\varepsilon_{ij} = \frac{1}{\tilde{E}}[(1 + \tilde{\nu})\sigma_{ij} - \tilde{\nu}\sigma_{kk}\delta_{ij}] + \alpha\Delta T\delta_{ij} \tag{2.16}$$

$$\sigma_{ij} = \frac{\tilde{E}}{(1 + \tilde{\nu})}[\varepsilon_{ij} + \frac{\tilde{\nu}}{(1 - 2\tilde{\nu})}\varepsilon_{kk}\delta_{ij}] - \frac{\tilde{E}}{(1 - 2\tilde{\nu})}\alpha\Delta T\delta_{ij} \tag{2.17}$$

where ε_{ij} and σ_{ij} are the strain and stress components, respectively. $\tilde{E}$ and $\tilde{\nu}$ are the anisotropic elastic moduli, respectively. The second term ($\alpha\Delta T$) in (2.16) is the thermal strain due to temperature change ΔT. While the second term in (2.17) $\tilde{E}\alpha\Delta T/(1 - 2\tilde{\nu})$ is the thermal stress due to the growth temperature. In system of rectangular cartesian coordinates, the subscripts x, y, z in (2.16) and (2.17) are substituted for *i* and *j*, respectively. In system of cylindrical and spherical coordinates, subscripts rr, $\theta\theta$, zz and rr, $\theta\theta$, $\varphi\varphi$ are substituted for *i* and *j*, respectively.

2.3 Strain in Microscale Planar Heterostructures

Thermal changes in a thermoelastic body are accompanied by the shifts in relative positions of particles composing the body. Since the thermal expansion of volume elements cannot proceed freely, the total strain can be thought to consist of sum of the thermal and elastic strains produced by the resistance of the medium to thermal expansion [90–92]. Therefore, thermal strains are added to elastic strains due to local mechanical stress due to lattice mismatch, so that Hooke's law is modified to the stress–strain relations given by (2.16) and (2.17). In system of rectangular coordinates, the components of displacement vector in the x, y and z directions are u_x,u_y,and u_z, respectively, and the strain–displacement relations are written as [94]

$$\varepsilon_{xx} = \frac{\partial u_x}{\partial x}, \quad \varepsilon_{yy} = \frac{\partial u_y}{\partial y}, \quad \varepsilon_{zz} = \frac{\partial u_z}{\partial z} \tag{2.18}$$

Since we are dealing with thermoelastic body, there are no shear strains ($\varepsilon_{xy} = \varepsilon_{yz} = \varepsilon_{zx} = 0$). The use of (2.16) and (2.17) for modelling of strain in microscale planar heterostructures (shown in Fig. 2.4) are subject to the following boundary conditions;

i. stress is zero across heterointerface along growth direction (out of plane),
ii. substrate is free of stress in any crystal directions, and
iii. in plane elastic strain has a lattice mismatch induced discontinuity at heterointerface, so called *shrink fit condition* defined as

$$\varepsilon_{\parallel}^{f} - \varepsilon_{\parallel}^{s} = \varepsilon_m = \frac{a_{\parallel}^{f} - a_f}{a_f} = \frac{a_s - a_f}{a_f} \tag{2.19}$$

where $\varepsilon_{\parallel}^{f}$ and $\varepsilon_{\parallel}^{s}$ are in plane strains in epilayer and thick (001) substrate, respectively, and, $\varepsilon_m = (a_s - a_f)/a_f$ is the lattice mismatch. Equations (2.1) and (2.2) are rewritten for the in- and out-of plane strains in heterolayer grown on a thick substrate

$$\varepsilon_{xx}^{f} = \varepsilon_{xx}^{f} = \varepsilon_{\parallel}^{f} = \frac{a_{\parallel}^{f} - a_f}{a_f} = \frac{1}{\tilde{E}_f}\left[(1 - \tilde{\nu}_f)\sigma_{\parallel}^{f} - \tilde{\nu}_f \sigma_{\perp}^{f}\right] + \alpha_f \Delta T \tag{2.20a}$$

$$\varepsilon_{zz}^{f} = \varepsilon_{\perp}^{f} = \frac{a_{\perp} - a_f}{a_f} = \frac{1}{\tilde{E}_f}\left[\sigma_{\perp}^{f} - 2\tilde{\nu}_f \sigma_{\parallel}^{f}\right] + \alpha_f \Delta T \tag{2.20b}$$

where $a_{\parallel}^{f}$ and $a_{\perp}^{f}$ are in- and out- of plane distorted lattice constants, and a_f is relaxed lattice constant of epilayer, respectively. $\Delta T = T - T_{\circ} = T$ is temperature change relative to $T_{\circ} = 0K$. Since stress is zero in all directions in a thick (001) substrate, we can write

$$\sigma_{xx}^{s} = \sigma_{yy}^{s} = \sigma_{zz}^{s} = \sigma_{\perp}^{s} = \sigma_{\parallel}^{s} = 0 \tag{2.21a}$$

$$\varepsilon_{xx}^{s} = \varepsilon_{yy}^{s} = \varepsilon_{zz}^{s} = \varepsilon_{\perp}^{s} = \varepsilon_{\parallel}^{s} = \alpha_s T \tag{2.21b}$$

where α_s is linear thermal expansion coefficient of substrate lattice constant.

Substituting (2.20a) and (2.20b) into (2.19), with continuous stress condition along growth direction ($\sigma_\perp^s = \sigma_\perp^f = 0$), the *shrink-fit* condition given by (2.19) is then written as

$$\frac{(1-\tilde{\nu}_f)}{\tilde{E}_f}\sigma_\parallel^f + \alpha_f T - \alpha_s T = \varepsilon_m \tag{2.22}$$

from which one can write the in-plane stress in epilayer given by the following expression

$$\sigma_\parallel^f = \frac{\tilde{E}_f}{(1-\tilde{\nu}_f)}[\varepsilon_m + (\alpha_s - \alpha_f)T] \tag{2.23}$$

On substituting (2.23) into (2.20a) with $\sigma_{zz}^f = \sigma_\perp^f = 0$ and into (2.20b) with $\sigma_\perp^s = \sigma_\perp^f = 0$, the in-plane and out of plane strains are expressed as

$$\varepsilon_{xx}^f = \varepsilon_{xx}^f = \varepsilon_\parallel^f = \frac{a_\parallel^f - a_f}{a_f} = \varepsilon_m + (\alpha_s - \alpha_f)T + \alpha_f T \tag{2.26a}$$

$$\varepsilon_{zz}^f = \varepsilon_\perp^f = \frac{a_\perp - a_f}{a_f} = -\frac{2\tilde{\nu}_f}{1-\tilde{\nu}_f}\varepsilon_m - \frac{2\tilde{\nu}_f}{1-\tilde{\nu}_f}(\alpha_f - \alpha_s)T + \alpha_f T \tag{2.24b}$$

The epilayer lattice constant distorted in plane and out of plane are then given by

$$a_\parallel^f = a_f(1+\varepsilon_\parallel) = a_f[1 + \varepsilon_m(T) + [\alpha_s(T) - \alpha_f(T)]T + \alpha_f(T)T] \tag{2.25a}$$

$$a_\perp^f = a_f(1+\varepsilon_\perp) = a_f[1 - \frac{2\tilde{\nu}_f}{1-\tilde{\nu}_f}[\varepsilon_m(T) + (\alpha_s(T) - \alpha_f(T))T] + \alpha_f(T)T] \tag{2.25b}$$

As an application, we discuss the anisotropy of thermal expansion of GaAs epilayer grown on Si (001) substrate. Equations (2.25a) and (2.25a) are used to calculate the in-and out- of plane strain distorted lattice constant ($a_\parallel^f$ and $a_\perp^f$) of GaAs epilayer on Si (001). Elastic stiffness parameters $C_{11} = 11.8$ and $C_{12} = 5.30$(in $10^{10} dyn/cm^2$) [103] for GaAs are used. The bulk lattice constants and linear expansion coefficients for GaAs and Si are fitted to following expressions

$$a(T) = a_0(1 + A + BT + CT^2 + DT^3) \tag{2.26a}$$

$$\alpha(T)/10^{-6}K^{-1} = \frac{1}{a}\frac{\partial a}{\partial T} = B + 2CT + 3DT^2 \tag{2.26b}$$

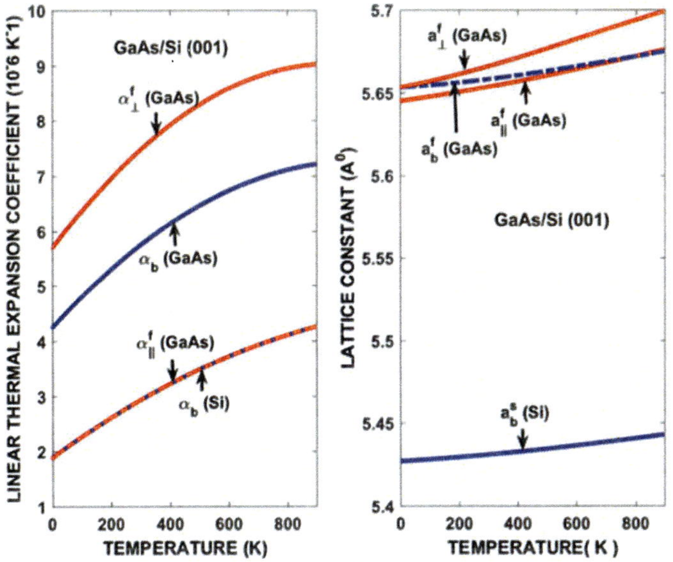

Fig. 2.6 Predicted temperature variation of in plane and out of plane linear thermal expansion coefficients (**a**) and lattice constants (**b**) of GaAs on Si (001) heterostructure

where $a_0 = 0.565325$ (0.543108) nm is bulk lattice constant of GaAs (Si) at 300 K. Constants A, B, C, and D are empirical fitting parameters [103]: $A/10^{-2} = -0.147$ (-0.071), $B/10^{-6}\ K^{-1} = 4.239$ (1.887), $C/10^{-9}\ K^{-2} = 2.916$ (1.934), and $D/10^{-12}\ K^{-3} = -0.936$ (–0.4544) for bulk GaAs (Si).

Figure 2.6 shows the composition effects on in plane and out of plane components of interface strain and lattice constant in InGaAs quantum well of AlGaAs/InGaAs planar heterostructure.

In plane and out of plane lattice constants as well as that of bulk GaAs all nearly converge at about 490 °C (average value of initial growth temperature.) Out of plane linear thermal expansion coefficient ($\alpha_\parallel^f(T)$) of GaAs epilayer is equivalent to that of Si (001) substrate ($\alpha_\parallel^s(T) = \alpha_s(T)$), but smaller than that of bulk GaAs ($\alpha_f(T)$) over the entire temperature range. However, out of plane hermal expansion coefficient $\alpha_\perp^f(T)$ of GaAs epilayer exceeds $\alpha_\parallel^f(T)$ by Poisson ratio. Similar observation is also true for in- and out- of plane distortions of GaAs lattice constant.

Results are in excellent agreement with high resolution x-ray scattering technique findings of Lucas et al [7], who measured the anisotropy in linear expansion coefficient of GaAs grown on Si (001) substrate. Table 2.1 gives the comparison of predictions with experimental data. X-ray diffraction measurements [7] has shown that epitaxial GaAs thin films on vicinal Si (001) substrate exhibit tetragonal distortion at 300 K. In plane thermal expansion of GaAs film follows the thermal expansion of Si (001) substrate. Out of plane thermal expansion of GaAs thin film exceeds bulk value.

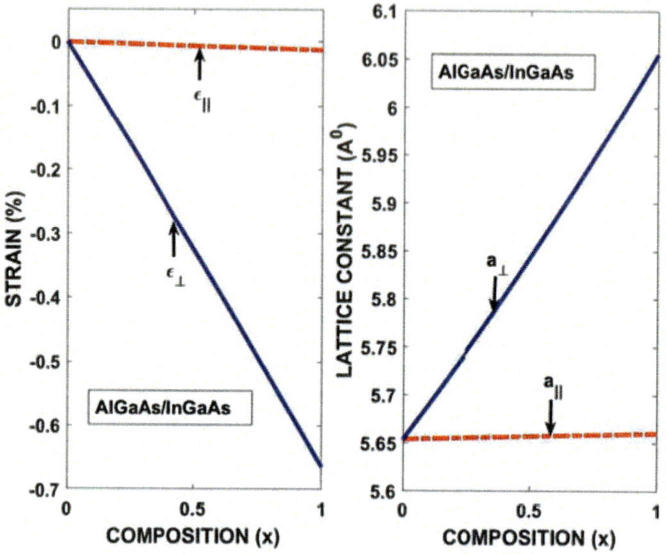

Fig. 2.7 Composition effects on in and out of plane components of interface strain and lattice constant of InGaAs quantum well in AlGaAs/InGaAs heterostructure

Table 2.1 Comparison of predicted and measured in plane and out of plane linear thermal expansion coefficients and lattice constants of GaAs/Si (001) heterostructure

Parameter	GaAs (predicted)	GaAs (measured)	Si (Predicted)	Si (measured)
$\alpha_{\perp}(10^{-6}K^{-1})$	7.9689	8.40	3.5095	3.51[7]
$\alpha_{\parallel}(10^{-6}K^{-1})$	3.5094	3.46	3.5095	3.46[7]
$a_{\perp}(nm)$	0.56667	0.56602	0.5431	–
$a_{\parallel}(nm)$	0.56592	0.56483	0.5431	–

Semiconductor alloys based on IV-VI, III-V and II-VI compounds are important in fabricating low dimensional bipolar and unipolar heterostructure devices since their structural and electronic properties (e.g., lattice constants and bandgaps) can be tailored independently. Therefore, reliable and accurate determination of composition variation of lattice constant and bandgap energies are very important. In the theoretical determination of composition effects on the structural properties such as lattice constants, a virtual crystal approximation (VCA) is often used in which the compositional disorder effect is neglected. Since in VCA the alloy potential is taken as the concentration weighted average of the constituent potentials the bandgap energy is linear function of alloy composition.

However, many experimental studies report that the bandgap energies of semiconductor alloys are nonlinear functions of composition. Furthermore, it is believed that the compositional disorder, which is related to the differences of electronegativity of atoms forming a ternary semiconductor, plays a major role in determining its bowing of the bandgap energy when lattice mismatch induced strain plays a crucial role in heterostructure electronic properties. In our recent studies [96–98] the effects of composition and strain on the electronic properties of semiconductor alloys in heterostructures are implemented in the semiempirical tight binding models and first principles density functional theory (DFT) in terms of host bond length and distorted bond length by the substitutional impurity without any adjustable parameter [94, 95]. In determining the composition effects on the band structure we employ the method of the modified virtual crystal approximation (MVCA) in which one formulates the composition effects on the bond length of ternary material. The MVCA allows one to accurately considers the effect of disorder-induced nonlinear variation of the lattice constant on the band structure parameters. As an example, Fig. 2.7 shows the composition variation of the lattice constants and interface strain in AlGaAs/InGaAs heterostructure. Compressive interface strain decreases the lattice constant of the InGaAs ternary as composition increases. Whereas tensile strain increases the lattice constant of InGaAs ternary as composition increases, such an increase or decrease in lattice constant due to interface strain will change the electronic properties.

2.4 Strain in Spherical Core/Shell Heterostructures

Consider a hollow sphere with inner and outer radius a and b (shown in Fig. 2.8) representing a nanoscale spherical core/shell heterostructure formed between two semiconductors with different physical properties. The inner region is core ($0 < r < a$) and outer region is shell ($a < r < b$), which are subject to inner and outer pressures and uniform temperature., the tangential displacements as well as the shear stresses and shear strains are zero ($\sigma_{r\theta} = \sigma_{r\varphi} = \sigma_{\varphi\theta} = 0$ and $\varepsilon_{r\theta} = \varepsilon_{r\varphi} = \varepsilon_{\varphi\theta} = 0$) because of spherical symmetry The strain is related to the radial displacement $u_r = u(r)$ by [94]

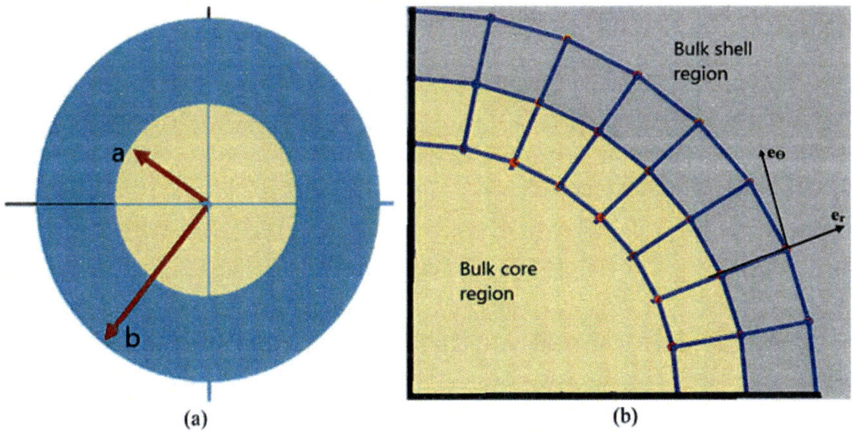

Fig. 2.8 Schematic view of a spherical core/shell heterostructure quantum dot (**a**) and radial and tangent components of strain across the core/shell interface (**b**), respectively

$$\varepsilon_r = \varepsilon_{rr} = \frac{\partial u_r}{\partial r}; \ \ \varepsilon_t = \varepsilon_{\theta\theta} = \varepsilon_{\varphi\varphi} = \frac{u_r}{r} \tag{2.27}$$

The modified stress–strain relations (Hooke's law) are written as

$$\varepsilon_{rr} = \frac{1}{\tilde{E}}[\sigma_{rr} - \tilde{\nu}(\sigma_{\theta\theta} + \sigma_{\varphi\varphi})] + \alpha\Delta T \tag{2.28a}$$

$$\varepsilon_{\theta\theta} = \frac{1}{\tilde{E}}[\sigma_{\theta\theta} - \tilde{\nu}(\sigma_{rr} + \sigma_{\varphi\varphi})] + \alpha\Delta T \tag{2.28b}$$

$$\varepsilon_{\varphi\varphi} = \frac{1}{\tilde{E}}[\sigma_{\varphi\varphi} - \tilde{\nu}(\sigma_{rr} + \sigma_{\theta\theta})] + \alpha\Delta T \tag{2.28c}$$

where $\varepsilon_{rr} = \varepsilon_r$ and $\varepsilon_{\theta\theta} = \varepsilon_{\varphi\varphi} = \varepsilon_t$ are radial and tangential strains with corresponding stresses $\sigma_{rr} = \sigma_r$ and $\sigma_{\theta\theta} = \sigma_{\varphi\varphi} = \sigma_t$. $\tilde{E}$ and $\tilde{\nu}$ are the anisotropic elastic moduli, respectively. In the absence of body forces, thc equilibrium equation for spherical core/shell structure is written as [94]

$$\frac{d\sigma_r}{dr} + \frac{2}{r}(\sigma_r - \sigma_t) = 0, \tag{2.29}$$

which can be solved by using the following boundary conditions:

(i) stress is continuous across interface ($\sigma_{ir}(a) = \sigma_{mr}(a) = -P_{im}$),
(ii))there is no stress outside core/shell structure ($\sigma_{mr}(b) = P_o = 0$), and
(iii) tangential strain has a lattice mismatch discontinuity at interface, so that we introduce a *shrink fit condition* [94], defined as

$$[r(\varepsilon_{m\theta} - \varepsilon_{i\theta})]_{r=a} = a\varepsilon_{im} = a(a_i - a_m)/a_m \tag{2.30}$$

where a is the radius of the core region. Solution of (2.29) in core region, subject to boundary conditions (i) (ii) and (iii), yields $\sigma_{ir} = \sigma_{i\theta} = \sigma_{i\varphi} = \sigma_i = -P_{im}$. Substituting these results into (2.28a), (2.28b), and (2.28c) one finds the radial and tangential strains in core region as

$$\varepsilon_{ir} = \varepsilon_{i\theta} = \varepsilon_{i\varphi} = \frac{1-2\tilde{\nu}_i}{\tilde{E}_i}\sigma_i + \alpha_i T = -\frac{1-2\tilde{\nu}_i}{\tilde{E}_i}P_{im} + \alpha_i T \tag{2.31}$$

Solving (2.29) in shell region, one finds the radial and tangential stresses

$$\sigma_{mr} = \frac{a^3b^3(P_o - P_{im})}{(b^3 - a^3)r^3} + \frac{a^3P_{im} - b^3P_o}{(b^3 - a^3)} \tag{2.32a}$$

$$\sigma_{m\theta} = \sigma_{m\varphi} = -\frac{a^3 b^3 (P_o - P_{im})}{2(b^3 - a^3) r^3} + \frac{a^3 P_{im} - b^3 P_o}{(b^3 - a^3)} \tag{2.32b}$$

which reduce to $\sigma_{mr} = -P_{im}(a/r)^3$ and $\sigma_{m\theta} = \sigma_{m\varphi} = P_{im}(a^3/2r^3)$ when $b \gg a$ and $P_o = 0$ [94]. Substituting (2.32a) and (2.32b) in (2.28a), (2.28b) and (2.28c) at $r = a$ with $P_o = 0$, radial and tangential strains on shell side of interface can be written as

$$\varepsilon_{mr} = \frac{P_{im}}{\tilde{E}_m(b^3 - a^3)}[(1 - 2\tilde{v}_m)a^3 - (1 + \tilde{v}_m)b^3] + \alpha_m T \tag{2.33a}$$

$$\varepsilon_{mt} = \varepsilon_{m\theta} = \varepsilon_{m\varphi} = \frac{P_{im}}{\tilde{E}_m(b^3 - a^3)}[(1 - 2\tilde{v}_m)a^3 - (1 + \tilde{v}_m)b^3] + \alpha_m T \tag{2.33b}$$

where $\varepsilon_{mt} = \varepsilon_{m\theta} = \varepsilon_{m\varphi}$ is the tangential component of interface strain in shell region. Combining (2.33a) and (2.33b) with (2.31), contact pressure at core/shell interface becomes

$$P_{im} = \frac{2\tilde{E}_i \tilde{E}_m [1 - (a/b)^3][\varepsilon_{im} + (\alpha_i - \alpha_m)T]}{[(1 + \tilde{v}_m)\tilde{E}_i + 2(1 - 2\tilde{v}_i)\tilde{E}_m] + 2[(1 - 2\tilde{v}_m)\tilde{E}_i - (1 - 2\tilde{v}_i)\tilde{E}_m](a/b)^3} \tag{2.34}$$

On substitution of (2.34) into (2.31), strain acting on core side becomes equal to

$$\varepsilon_i = -\frac{2(1 - 2v_i)E_m[1 - (a/b)^3][\varepsilon_{im} + (\alpha_i - \alpha_m)T]}{(1 + v_m)E_i + 2(1 - 2v_i)E_m + 2[(1 - 2v_m)E_i - (1 - 2v_i)E_m](a/b)^3} + \alpha_i T \tag{2.35}$$

On substitution of (2.34) into (2.33a) and (2.33b), strains acting on shell side become equal to

$$\varepsilon_{mr} = \frac{2E_i}{(b^3 - a^3)} \frac{[1 - (a/b)^3][(1 - 2v_m)a^3 - (1 + v_m)b^3][\varepsilon_{im} + (\alpha_i - \alpha_m)T]}{E_i(1 + v_m) + 2E_m(1 - 2v_i) + 2[(1 - 2v_m)E_i - (1 - 2v_i)E_m](a/b)^3} + \alpha_m T \tag{2.36a}$$

$$\varepsilon_{mt} = \frac{2E_i}{(b^3 - a^3)} \frac{[1 - (a/b)^3][(1 - 2v_m)a^3 + (1/2)(1 + v_m)b^3][\varepsilon_{im} + (\alpha_i - \alpha_m)T]}{E_i(1 + v_m) + 2E_m(1 - 2v_i) + 2[(1 - 2v_m)E_i - (1 - 2v_i)E_m](a/b)^3} + \alpha_m T \tag{2.36b}$$

In the absence thermal strain for $a \ll b$, (2.35), (2.36a) and (2.36b), respectively, reduce to (2.3), (2.4) and (2.5) given by Eshelby [5] for radial and tangential strains in core and shell regions.

In a coherently grown nanoscale core/shell heterostructures, strain across the interface differently influences the energy band structure and charge transport, operation and performance in heterostructure devices. In order to understand the effect of core

Table 2.2 Properties of some group III-V compounds used in the model calculations [103]

Parameter	GaAs	InAs	GaSb	InSb
a(nm)	0.56533	0.60584	0.581	0.541
E_g (eV)	1.519	0.41	0.809	0.235
$-E_v$ (eV)	13.31	12.3	11.6	11.7
a_g	– 8.80	– 6.30	– 8.30	– 7.0
b_v	– 0.85	1.00	0.79	1.00
$\varepsilon_\infty/\varepsilon_0$	10.86	11.6	14.2	15.3
C_{11} (10^{11} dyn/cm^2)	11.88	8.329	8.838	6.608
C_{12} (10^{11} dyn/cm^2)	5.38	4.526	4.027	3.531
C_{44} (10^{11} dyn/cm^2)	5.94	3.959	4.320	3.027
$\alpha_{th}(10^{-6}$ K$^{-1})$	6.03	5	6.35	5.04
$\alpha(10^{-4}$ eV)	5.5	3.07	5.3	2.7
β(K)	225	191	234	106

and shell diameters on the magnitude of interface strain acting on core and shell sides of GaAs and GaSb based heterostructure core/shell spherical quantum dots and cylindrical nanowires we used the material parameters listed in Table 2.2.

We will use these structural, electronic and elastic material properties of constituent semiconductors and core and shell dimensions to calculate the interface strain on the core and shell sides of some important group III-V compound semiconductors-based core/shell heterostructure spherical quantum dots and cylindrical nanowires. Specifically, we will focus on the strain variation with core and shell dimensions, lattice mismatch and thermal expansion difference, and anisotropy of elastic properties in GaAs/InAs(GaSb) and GaSb/InAs(InSb) heterostructure quantum dots and nanowires.

Group II-VI compounds semiconductors such CdSe, ZnSe, CdS and ZnS are also used as for the colloidal synthesis of core/shell heterostructure quantum dots and nanowires. The structural, electronic, optical and elastic material parameters of group II–VI compound semiconductors such as CdSe, ZnSe, CdS and ZnS compounds are given Table 2.3. We will use these material parameters the strain variation with core and shell dimensions, lattice mismatch and thermal expansion difference, and anisotropy of elastic properties in CdSe/ZnS, CdSe/CdS, ZnSe/ZnS, and ZnSe/CdS heterostructure core/shell spherical quantum dots and cylindrical nanowires. We will also discuss the strain effects on the bandgaps and band offsets in these core/shell heterostructures and compare the results of the calculations with experimental data to verify the thermoelastic strain model outlined above.

In order to quantify effect of the elastic anisotropy on the strain and electronic and optical properties of core/shell heterostructure QDs and NWs, in addition to lattice and thermal mismatches ($\varepsilon_{im} = (a_i - a_m)/a_m$ and $\delta\alpha_{th}(10^{-6}K^{-1})$), we introduce so called *elastic anisotropy mismatch* [98]

Table 2.3 Properties of some group II-VI compounds used in calculations [103]

Parameter	CdSe	ZnSe	CdS	ZnS
a(nm)	0.607	0.5668	0.581	0.541
E_g (0)/eV	1.766	2.807	2.445	3.865
$-E_v$ /eV	12.71	13.49	12.55	13.42
Δ/eV	0.410	0.424	0.070	0.092
a_g (eV)	– 2.89	– 5.1	– 2.9	– 5.2
a_v (eV)	0.90	1.23	0.40	0.83
$\varepsilon_\infty/\varepsilon_0$	5.80	5.56	5.24	5.20
C_{11} (10^{11} dyn/cm^2)	6.67	8.57	7.70	10.11
C_{12} (10^{11} dyn/cm^2)	4.63	5.07	5.39	6.46
C_{44} (10^{11} dyn/cm^2)	2.23	4.05	2.36	4.46
$\alpha_{th}(10^{-6}\ K^{-1})$	7.30	7.60	4.05	6.9
$\alpha(10^{-4}$ eV)	6.96	5.58	3.45	10
β(K)	281	187	208	600

$$\xi_{im} = (A_i - A_m)/A_m \qquad (2.37)$$

where A_i and A_m are *anisotropy ratios* defined by Zener for core and shell semiconductors with cubic symmetry. Furthermore, we will also introduce *elastic anisotropy mismatch*, by using the elastic H parameter and compare the results with those due to (2.37) by using the following expression $\eta_{im} = (H_i - H_m)/H_m$, where H_i and H_m are defined by Hirt [92] for core and shell semiconductors constituents with cubic crystal symmetry. Using the elastic constants given in Tables 2.2 and 2.3, we calculated the lattice mismatch and elastic anisotropy mismatches ξ_{im} and η_{im}. Table 2.4 lists the calculated the lattice, thermal mismatch, and elastic anisotropy mismatch for GaAs and GaSb III-V based and CdSe and ZnSe II-VI based ore/shell heterostructures.

Table 2.4 Lattice mismatch, thermal expansion mismatch, and elastic anisotropy mismatch for GaAs and GaSb based III-V and CdSe and ZnSe based II–VI core/shell heterostructures

Core/Shell	$\varepsilon_{im}(\%)$	$\delta\alpha_{th}(10^{-6}K^{-1})$	$\xi_{im}(\%))$	$\eta_{im}(\%)$
GaAs/InAs	–6.68	1.03	–12.21	30.74
GaAs/GaSb	–2.67	–0.32	1.77	40.50
GaSb/InAs	–4.10	1.35	–13.74	–6.95
GaSb/InSb	3.565	6.35	–8.72	28.61
CdSe/CdS	4.326	3.25	7.00	0.414
CdSe/ZnS	12.329	0.4	–10.50	–54.07
ZnSe/CdS	–2.712	3.55	13.30	90.87
ZnSe/ZnS	4.750	0.7	–5.30	–12.78

Because of the large anisotropy ratio predicted by η_{im} given by (2.37), we will limit our discussion to elastic anisotropy mismatch ξ_{im} given by (2.37) in calculating the elastic anisotropy effects on strain.

Calculations are carried out to investigate the effect of core diameter variation of interface strain in GaAs/InAs(GaSb), GaSb/InAs(InSb), CdSe/ZnS, (CdS) and ZnSe/ZnS(CdS) nanoscale spherical core/shell heterostructures and the results are shown in Fig. 2.9. Figure 2.9a and b, respectively, show the core diameter variation of strain acting on core and shell sides of interface suggest that strain is nearly constant on both sides. In the case of GaAs/InAs and GaSb/InSb heterostructures,ε_{im} and ξ_{im} are large and negative, resulting in a large interface strain. In GaAs/GaSb and GaSb/InSb heterostructures lattice mismatch is negative and large and elastic anisotropy ratio is also negative and large, but interface strain is relatively smaller than that in GaAs/InAs and GaSb/InSb heterostructures. In CdSe/ZnS and ZnSe/ZnS heterostructures for which both ε_{im} and ξ_{im} are large, anisotropic elastic moduli results in larger compressive strain than isotropic ones. In CdSe/CdS and ZnSe/CdS heterostructures (Fig. 2.9c and 2.9d),ε_{im} is large and positive and ξ_{im} is large and negative. Consequently, the interface strain is relatively smaller than that in CdSe/ZnS

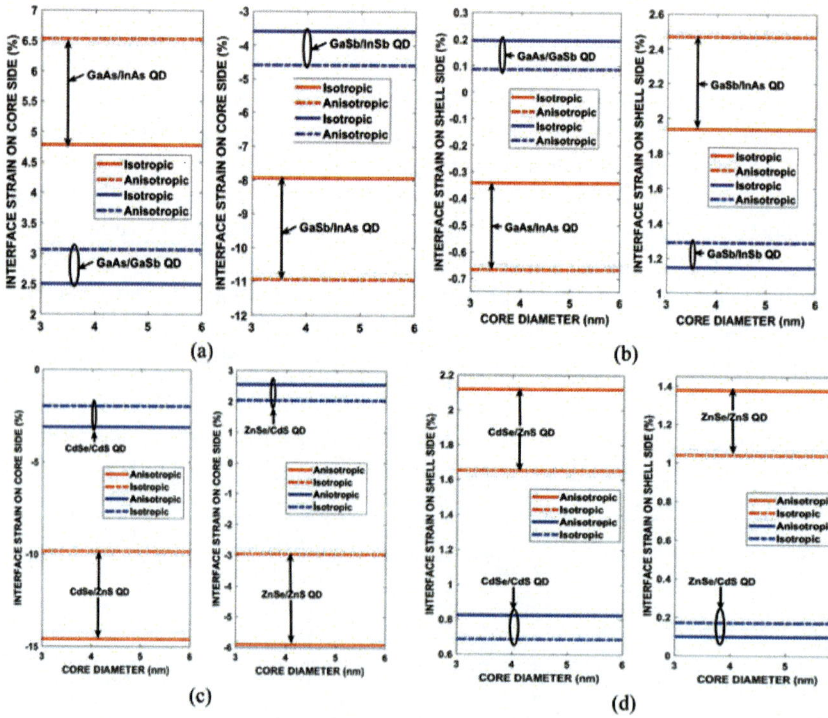

Fig. 2.9 Effect of core diameter on interface strain acting on core (**a**, **c**) and shell (**b**, **d**) sides of nanoscale GaAs/InAs(GaSb) and GaSb/InAs(InSb) III–V and CdSe/ZnS(CdS) and CdSe/CdS(ZnS) II-VI compounds based heterostructure core/shell QDs

and ZnSe/ZnS heterostructures. This has an important effect about the prediction of the shell diameter variation of quantum confinement in the design of core/shell heterostructures.

Figure 2.10 compare the results of the calculations for shell diameter effect on strain acting on core (a, c) and shell (b, d) sides of GaAs/InAs(GaSb) and GaSb/InAs(InSb) and CdSe/ZnS(CdS) and CdSe/CdS(ZnS) QDs with core diameter $d_i = 3.0$ nm at 300 K. The anisotropic elastic moduli result in larger strains than isotropic ones in GaAs/InAs, GaSb/InAs, CdSe/ZnS and ZnSe/ZnS for which both ε_{im} and ξ_{im} are large.

A similar magnitude of strain is predicted in GaAs/GaSb, GaSb/InSb, CdSe/CdS and ZnSe/CdS for which ε_{im} is small and ξ_{im} is large. Increasing shell diameter causes tensile and compressive strain on core side of interface in GaAs/InAs(GaSb) and GaSb/InAs(InSb) heterostructures, respectively, with increasing magnitude. Whereas, increasing shell diameter causes compressive and tensile and strain on shell side of interface in GaAs/InAs(GaSb) and GaSb/InAs(InSb), respectively, with decreasing magnitude.In CdSe/ZnS(CdS) and ZnSe/ZnS QDs, shell diameter causes

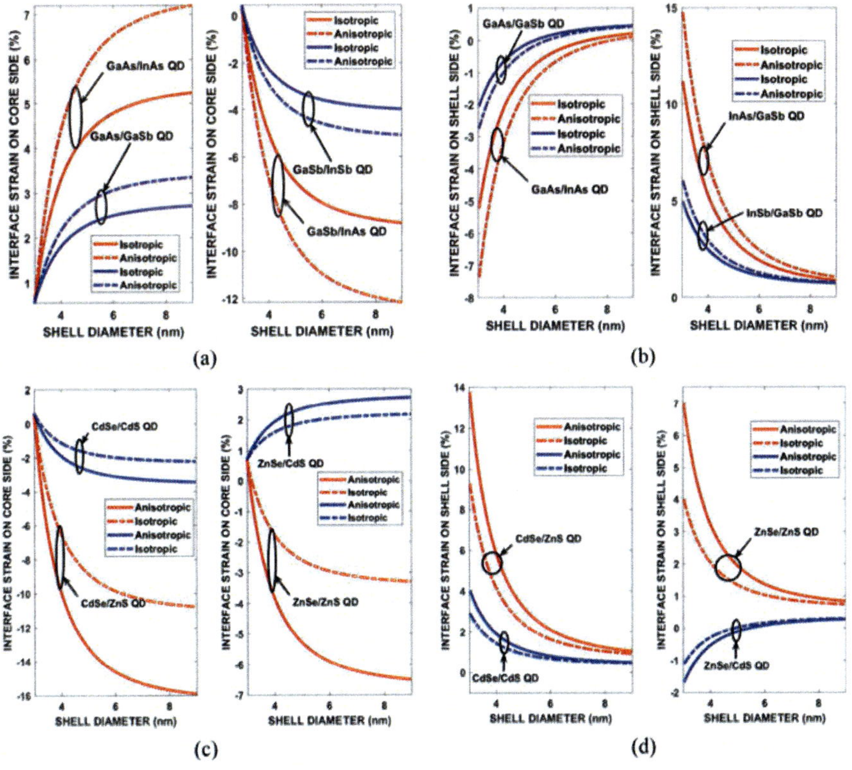

Fig. 2.10 Effect of shell diameter on interface strain acting on core (**a**, **c**) and shell (**b**, **d**) sides of nanoscale GaAs/InAs(GaSb) and GaSb/InAs(InSb) III-V and CdSe/ZnS(CdS) and CdSe/CdS(ZnS) II-VI compounds based heterostructure core/shell QDs for core diameter $d_i = 3.0$ nm at 300 K

compressive strain acting on the core side and becomes nearly constant above 8.0 nm. Whereas, increasing shell diameter causes tensile and strain on shell side of interface in CdSe/ZnS (CdS) and ZnSe/ZnS, respectively. Strain is tensile on core side and compressive on shell side of ZnSe/CdS interface. It becomes less important as shell diameter is increased above some critical diameter. This has an important consequence about the effect of shell diameter variation of quantum confinement in spherical core/shell heterostructure QDs. Drawing vertical and horizontal straight lines tangent to strain curve, intersection point will give the upper limit for elastic strain and shell diameter for which one can observe strong confinement regime.

The coherent growth of binary/ternary core/shell heterostructures is essential in order to increase degree of quantum confinement by varying core and shell direct bandgaps. Figure 2.11 shows the composition effects on strain and lattice constant in core side of ZnSe/CdZnS and CdSe/CdZnS core/shell heterostructure QDs for core diameter $d_i = 3.0$ nm at 300 K with isotropic and anisotropic elastic constant properties. As composition increases the difference between isotropic and anisotropic elastic properties tend to increase. Compressive strain acting on core causes its lattice constant to increase with increasing composition. This will increase direct bandgap energy and in turn the limit on quantum confinement, which will be discussed later on. It turns out that the anisotropy of elastic constants has negligible effects on the strain and lattice constants but tends to increase as the ternary alloy composition increases towards one.

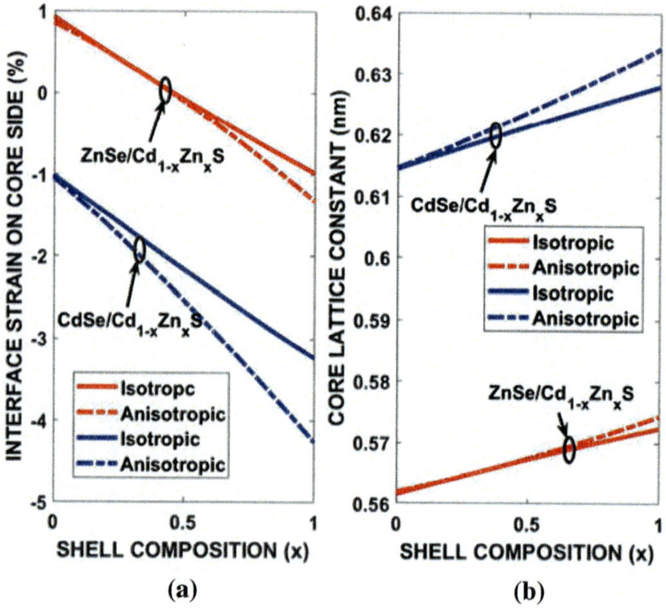

Fig. 2.11 Composition effects on strain (**a**) and lattice constant (**b**) in core region of ZnSe/CdZnS and and CdSe/CdZnS heterostructures for core diameter $d_i = 3.0$ nm at 300 K

2.5 Strain in Cylindrical Core/Shell Heterostructures

Consider an infinitely long concentric cylindrical core/shell semiconductor heterostructure with inner and outer radius a and b (a<b) and is subject to an inner and outer pressures Pi and Po, respectively, and uniform temperature distribution throughout nanostructure (Fig. 2.12). The core region is strained along z-axis due to lattice mismatch $\varepsilon_{iz} = \varepsilon_{im} = (a_i - a_m)/a_m$ but the shell region is unstrained ($\varepsilon_{mz} = \partial u_z/\partial z = 0$). Here a_i and a_m are lattice constants of core and shell regions, respectively.

Focusing on the cross-sectional view in Fig. 2.12b, because of cylindrical symmetry, the tangential displacements as well as the shear stresses and shear strains are all zero ($\sigma_{r\theta} = \sigma_{r\varphi} = \sigma_{\varphi\theta} = 0$ and $\varepsilon_{r\theta} = \varepsilon_{r\varphi} = \varepsilon_{\varphi\theta} = 0$). The components of the displacement vector in r, θ, and z directions are u_r,u_θ, and u_z. The strain–displacement relations are [91, 94]

$$\varepsilon^*_{rr} = \frac{\partial u_r}{\partial r} \; ; \varepsilon^*_{\theta\theta} = \frac{u_r}{r} + \frac{1}{r}\frac{\partial u_\theta}{\partial \theta} \; ; \varepsilon^*_{zz} = \frac{\partial u_z}{\partial z} \tag{2.38}$$

where $\varepsilon^*_{rr} = \varepsilon^*_r$ and $\varepsilon^*_{\theta\theta} = \varepsilon^*_t$ are radial and tangential strains. Since thermal expansion of volume elements cannot proceed freely, total strain is equal to sum of thermal strain and local strain (i.e., lattice mismatch) due to resistance of medium to thermal expansion. Applying (2.16) to system of cylindrical coordinates, the stress–strain relation for the cylindrical core region are written as [94]

$$\varepsilon^*_{ir} = \frac{1}{\tilde{E}_i}[\sigma^*_{ir} - \tilde{\nu}_i(\sigma^*_{i\theta} + \sigma^*_{iz})] + \alpha_i T \tag{2.39a}$$

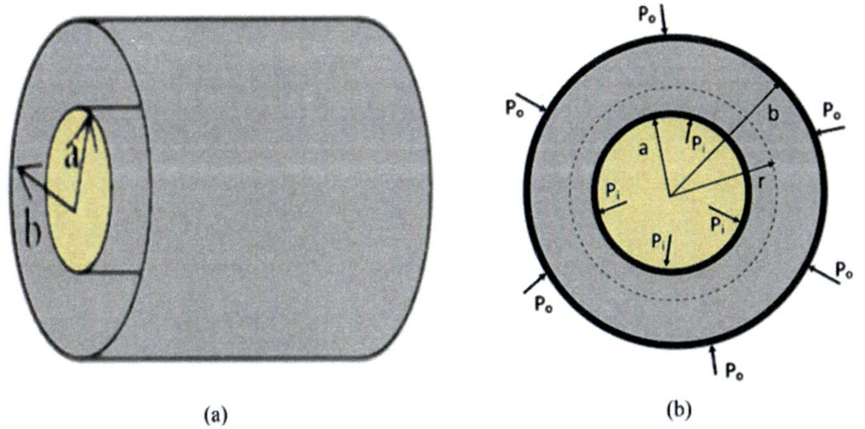

Fig. 2.12 Schematic cross-sectional view of cylindrical core/shell heterostructure

$$\varepsilon_{i\theta}^* = \frac{1}{\tilde{E}_i}[\sigma_{i\theta}^* - \tilde{\nu}_i(\sigma_{ir}^* + \sigma_{iz}^*)] + \alpha_i T \tag{2.39b}$$

$$\varepsilon_{iz}^* = \frac{1}{\tilde{E}_i}[\sigma_{iz}^* - \tilde{\nu}_i(\sigma_{ir}^* + \sigma_{i\theta}^*)] + \alpha_i T \tag{2.39c}$$

where ε_{ir}^*, $\varepsilon_{i\theta}^*$ and ε_{iz}^* are the radial and tangential strains and σ_{ir}^*, $\sigma_{i\theta}^*$ and σ_{iz}^* are the corresponding stresses. $\tilde{E}_i$ and $\tilde{\nu}_i$ are anisotropic Young's modulus and Poisson's ratio, respectively. In the shell region, there is no displacement along z-direction, $\varepsilon_{mz}^* = \partial u_z/\partial z = 0$, and stress–strain relations are given by the following equations

$$\varepsilon_{mr}^* = \frac{1}{\tilde{E}_m}[\sigma_{mr}^* - \tilde{\nu}_m(\sigma_{m\theta}^* + \sigma_{mz}^*)] + \alpha_m T \tag{2.40a}$$

$$\varepsilon_{m\theta}^* = \frac{1}{\tilde{E}_i}[\sigma_{m\theta}^* - \tilde{\nu}_m(\sigma_{mr}^* + \sigma_{mz}^*)] + \alpha_m T \tag{2.40b}$$

$$\sigma_{mz}^* = \tilde{\nu}_m(\sigma_{mr}^* + \sigma_{m\theta}^*) - \tilde{E}_m\alpha_m T \tag{2.40c}$$

where $\tilde{E}_m$, $\tilde{\nu}_m$, and α_m are, respectively, anisotropic Young's modulus, Poisson ratio, and linear thermal expansion coefficient of semiconductor forming shell region. Radial and tangential stresses satisfy following equilibrium equation for axial symmetry [94]

$$\frac{d\sigma_r}{dr} + \frac{1}{r}(\sigma_r - \sigma_\theta) = 0 \tag{2.41}$$

which can be solved by using the following boundary conditions.

(i) radial stress is zero at center ($\sigma_{ir}(0) = 0$) and continuous across interface ($\sigma_{ir}(a) = \sigma_{mr}(a) = -P_i$),
(ii) there is no stress outside core/shell nanowire structure ($\sigma_{mr}(b) = P_o = 0$),
(iii) Lattice mismatch at cylindrical core/shell interface is $\varepsilon_{im}^* = (a_i - a_m)/a_m$, and
(iv) tangential strain has a lattice mismatch discontinuity at interface and so called *shrink fit* condition is defined as [94]

$$[r(\varepsilon_{m\theta} - \varepsilon_{i\theta})]_{r=a} = a\varepsilon_{im} \tag{2.42}$$

where $\varepsilon_{i\theta}^*$ and $\varepsilon_{m\theta}^*$ are tangential strains on core and shell sides, given by (2.39a) and (2.40b).

In core region, one can set $\varepsilon_{ir}^* = \varepsilon_{i\theta}^*$ in (2.39a) and (2.39b) and find $\sigma_{ir}^* = \sigma_{i\theta}^*$ and substituting them into (2.40c), with $\varepsilon_{iz}^* = \varepsilon_{im} = (a_i - a_m)/a_m$, stress along z-axis

σ_{iz} is equal to

$$\sigma_{iz}^* = 2\nu_i\varepsilon_{ir}^* + E_i\varepsilon_{im} - E_i\alpha_i T \tag{2.43}$$

Substituting $\sigma_{ir}^* = \sigma_{i\theta}^* = \sigma_i^* = -P_{im}^*$ and σ_{iz}^* back into (2.39a) and (2.39b) one finds the radial and tangential strains in core region as

$$\varepsilon_i^* = \varepsilon_{ir}^* = \varepsilon_{i\theta}^* = -\frac{(1-\nu_i-2\tilde{\nu}_i^2)}{\tilde{E}_i}P_{im}^* - \nu_i\varepsilon_{im} + (1+\tilde{\nu}_i)\alpha_i T \tag{2.44}$$

where P_{im}^* is contact pressure at core/shell interface. Solving (2.41) in shell region one finds [94]

$$\sigma_{mr}^* = -P_{im}^*, \quad \sigma_{m\theta}^* = (a^2+b^2)P_{im}^*/(b^2-a^2) \tag{2.45}$$

for radial and tangential stresses at heterointerface ($r = a$), respectively. Radial and tangential strains given by (2.49) and (2.40b) at $r = a$ can then be written as

$$\varepsilon_{mr}^* = \frac{P_{im}^*}{E_m(b^2-a^2)}[(1-\nu_m-2\nu_m^2)a^2 - (1+\nu_m)b^2] + (1+\nu_m)\alpha_m T \tag{2.46a}$$

$$\varepsilon_{mt}^* = \frac{P_{im}^*}{E_m(b^2-a^2)}[(1-\nu_m-2\nu_m^2)a^2 + (1+\nu_m)b^2] + (1+\nu_m)\alpha_m T \tag{2.46b}$$

where $\varepsilon_{mt}^* = \varepsilon_{m\theta}^* = \varepsilon_{m\varphi}^*$ is the tangential strain in shell region. Combining (2.44) and (2.46b) with (2.42) for *shrink fit condition*, one finds interface contact pressure

$$P_{im}^* = \frac{\tilde{E}_i\tilde{E}_m(b^2-a^2)[(1-\tilde{\nu}_i)\varepsilon_{im} + \delta\varepsilon_{th}]}{\tilde{E}_m(b^2-a^2)(1-\tilde{\nu}_i-2\tilde{\nu}_i^2) + \tilde{E}_i[(1-\tilde{\nu}_m-2\tilde{\nu}_m^2)a^2 + (1+\tilde{\nu}_m)b^2]} \tag{2.47}$$

where $\delta\varepsilon_{th}(T) = (1+\tilde{\nu}_i)\alpha_i T - (1+\tilde{\nu}_m)\alpha_m T$ is thermal strain due to difference between linear expansion coefficients of core and shell constituents in bulk form. In the absence of thermal strain (2.44) and (2.46a), and (2.46b) reduce to (2.6) and (2.7) and (2.8) for $a << b$.

On substitution of (2.47) into (2.44), strain acting on core side is written as

$$\varepsilon_i = -\frac{E_m(1-\nu_i-2\nu_i^2)(b^2-a^2)[(1-\nu_i)\varepsilon_{im} + \delta\varepsilon_{th}]}{E_m(b^2-a^2)(1-\nu_i-2\nu_i^2) + E_i[(1-\nu_m-2\nu_m^2)a^2 + (1+\nu_m)b^2]} - \nu_i\varepsilon_{im} + (1+\nu_i)\alpha_i T \tag{2.48}$$

When core region radius is much smaller than that of shell region ($a << b$), (2.48) reduces to

$$\varepsilon_i = \frac{(1-v_i-2v_i^2)E_m[\varepsilon_{im}+\delta\varepsilon_{th}]}{[(1+v_m)E_i+(1-v_i-v_i^2)E_m]} - v_i\varepsilon_{im} + (1+v_i)\alpha_i T \quad (2.49)$$

Likewise, on substitution of (2.47) into (2.46a) and (2.46b), radial and tangential strains acting on shell side of the heterointerface can be written as

$$\varepsilon_{mr} = -\frac{E_i[(1+v_m)b^2-(1-v_m-2v_m^2)a^2][(1-v_i)\varepsilon_{im}+\delta\varepsilon_{th}]}{E_m(b^2-a^2)(1-v_i-2v_i^2)+E_i[(1-v_m-2v_m^2)a^2+(1+v_m)b^2]} + (1+v_m)\alpha_m T \quad (2.50a)$$

$$\varepsilon_{mt} = \frac{E_i[(1+v_m)b^2+(1-v_m-2v_m^2)a^2][(1-v_i)\varepsilon_{im}+\delta\varepsilon_{th}]}{E_m(b^2-a^2)(1-v_i-2v_i^2)+E_i[(1-v_m-2v_m^2)a^2+(1+v_m)b^2]} + (1+v_m)\alpha_m T \quad (2.50b)$$

When core region radius is much smaller than that of shell region ($a << b$), radial and tangential strains are maximum on shell side given by the following expressions

$$\varepsilon_{mr} = \frac{-E_i(1+v_m)[(1-v_i)\varepsilon_{im}+\delta\varepsilon_{th}]}{E_m(1-v_i-2v_i^2)+E_i(1+v_m)} + (1+v_m)\alpha_m T \quad (2.51a)$$

$$\varepsilon_{m\theta} = \frac{E_i(1+v_m)[(1-v_i)\varepsilon_{im}+\delta\varepsilon_{th}]}{E_m(1-v_i-2v_i^2)+E_i(1+v_m)} + (1+v_m)\alpha_m T \quad (2.51b)$$

Equations (2.49), (2.51a) and (2.51b) reduce to (2.6), (2.7), and (2.8) given by Eshelby [4] for strains in core and shell regions of a cylindrical core/shell nanowire.

In order to understand the effect of core and shell diameters on strain acting on core and shell sides of group III-V based cylindrical core/shell heterostructures, we use material parameters listed in Tables 2.3 and 2.4 to calculate the magnitude of strain acting on core and shell sides of GaAs/InAs(GaSb) and GaSb/InAs(InSb) cylindrical core/shell heterostructure as a function of core and shell diameters. Figure 2.13a and b show the results of calculations carried out about the core diameter variation of interface strain in GaAs/InAs(GaSb) and GaSb/InAs(InSb) heterostructures.

The interface strain is nearly constant on both core and shell sides of their heterointerfaces. In GaAs/InAs, GaAs/GaSb and GaSb/InAs heterostructures lattice mismatch $\varepsilon_{im} = (a_i - a_m)/a_m$ is large and negative which results in compressive lattice mismatch strain. In GaSb/InSb heterostructure ε_{im} is large and positive which results in tensile lattice mismatch strain. Elastic anisotropy mismatch ξ_{im} is large and negative for GaAs/InAs, GaSb/InAs and GaSb/InSb heterostuctures.

Figure 2.14a and b show the shell diameter variation of interface strain acting on the core and shell sides in GaAs/InAs(GaSb) and GaSb/InAs heterostructures.

Strain acting on core (shell) side of interface in GaAs/InAs(GaSb) and GaSb/InAs heterostructures is tensile (compressive) and parabolically increases in magnitude as shell diameter is increased and becomes nearly constant above 8.0 nm in all of these heterostructures. Whereas, in the case of GaSb/InSb heterostructure strain is tensile and increases parabolically in magnitude. However, increasing shell diameter causes compressive strain on shell side of interface in GaAs/InAs and GaSb/InSb, and tensile

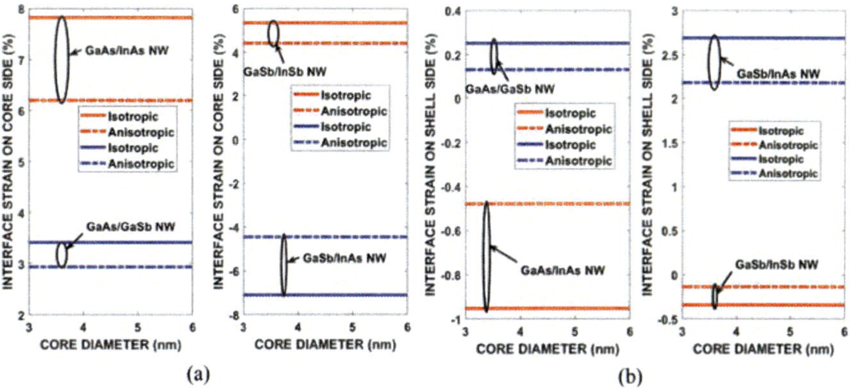

Fig. 2.13 Core diameter variation of interface strain on core (**a**) and shell (**b**) sides of GaAs/InAs, GaAs/GaSb, GaSb/InAs, and GaSb/InSb NWs with shell diameter $d_m = 2d_i$ at 300 K

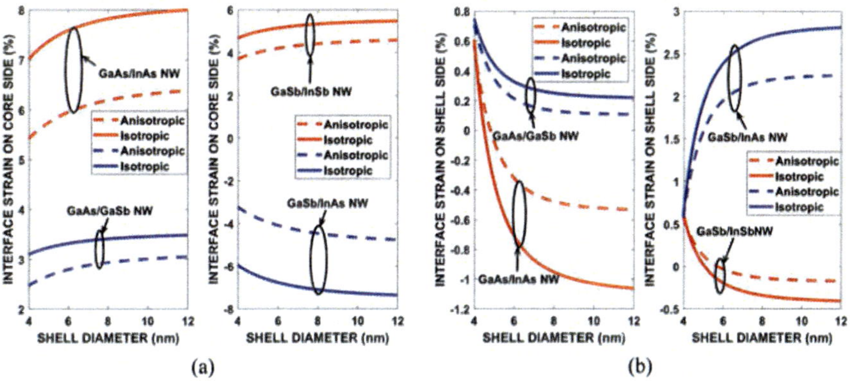

Fig. 2.14 Shell diameter effect on strain acting on core (**a**) and shell (**b**) sides of group III-V GaAs/InAs, GaAs/GaSb, GaSb/InAs, and GaSb/InSb core/shell NWs for $d_i = 3.0$ nm at 300 K

strain in GaAs/GaSb and GaSb/InAs heterostructures, respectively, with parabolic increase in magnitude. This result is contrary to the effect of core diameter on the interface strain. This has an important consequence about the effect of shell diameter variation of quantum confinement in cylindrical core/shell heterostructure NWs. It becomes less important as shell diameter is increased above some critical diameter. If one draws vertical and horizontal straight lines that are tangent to strain curve, intersection point will give the upper limit for elastic strain and shell diameter for which one can observe strong confinement regime.

2.6 Interface Strain and Morphology in Core/Shell QDs

When the lattice constant and linear thermal expansion coefficient of core region are larger than those of the shell region in core/shell heterostructure QDs and NWs, interface strain due to lattice and thermal mismatches causes the core region to be under isotropic compression and shell region to be under compression in radial direction and tension in tangential direction [100, 101]. Consequently, the interfacial strain energy stored on the core and shell sides of heterointerfaces in core/shell QDs depends on the diameters of constituents. In case of similar lattice parameter and elastic constants between constituent semiconductors forming core/shell heterostructure, uniform and coherent shell growth is possible only for small size core and thin shell region. However, when lattice and thermal expansion mismatches between core and shell constituents are significant, then thick shell growth can cause large interface elastic strain energy [100]. As it is evident from Table 2.4, there is significant lattice mismatch across interfaces of GaAs/InAs(GaSb), GaSb/InAs(InSb) CdSe/ZnS(CdS) and ZnSe/ZnS(CdS) core/shell heterostructures, which is expected to result in substantial elastic strain energy on both sides of interface, which can affect the morphology of core/shell nanostructures.

Continuum elastic theory [94] can be used to have a quantitative treatment of elastic. strain energy in determining the core and shell morphologies. The total elastic energy stored in spherical core/shell QDs is written as sum of elastic strain energies stored on core and shell side

$$E = \frac{1}{2}\int_0^a \sigma_{ij}\varepsilon_{ij}dV + \frac{1}{2}\int_a^b \sigma_{ij}\varepsilon_{ij}dV = \frac{1}{2}\int_0^a 3\sigma_i\varepsilon_i dV + \frac{1}{2}\int_a^b (\sigma_{mr}\varepsilon_{mr} + 2\sigma_{mt}\varepsilon_{mt})dV \tag{2.52}$$

where $\sigma_{ij} = \sigma_{ir} = \sigma_{i\theta} = \sigma_{i\varphi} = -P_{im}$ is stress and $\varepsilon_{ij} = \varepsilon_{ir} = \varepsilon_{i\theta} = \varepsilon_{i\varphi} = \varepsilon_i$ is corresponding strain in core region. σ_{mr} and $\sigma_{m\theta}$ are radial and tangential stresses and ε_{mr} and $\varepsilon_{m\theta}$ are corresponding strains, respectively, acting on shell side. Evaluating integral expression in (2.52) one finds

$$E_{im} = -2\pi a^3 P_{im}\varepsilon_i - \frac{2\pi a^3 b^3}{(b^3 - a^3)} P_{im}\left((\varepsilon_{mr} + \varepsilon_{mt})\ln(\frac{b}{a}) - (\varepsilon_{mr} + 2\varepsilon_{mt})\frac{(b^3 - a^3)}{3}\right) \tag{2.53}$$

where first and second terms are, respectively, the elastic strain energies stored on core and shell sides of spherical core/shell QD. Similar expression can be found for cylindrical core/shell nanowires.

Figure 2.15 shows the elastic strain energies stored on core and shell sides of of CdSe/ZnS (CdS), ZnSe/ZnS(CdS) spherical core/shell QDs as a function of core and shell dimensions. Figure 1.15a and b show the core and shell diameter dependence

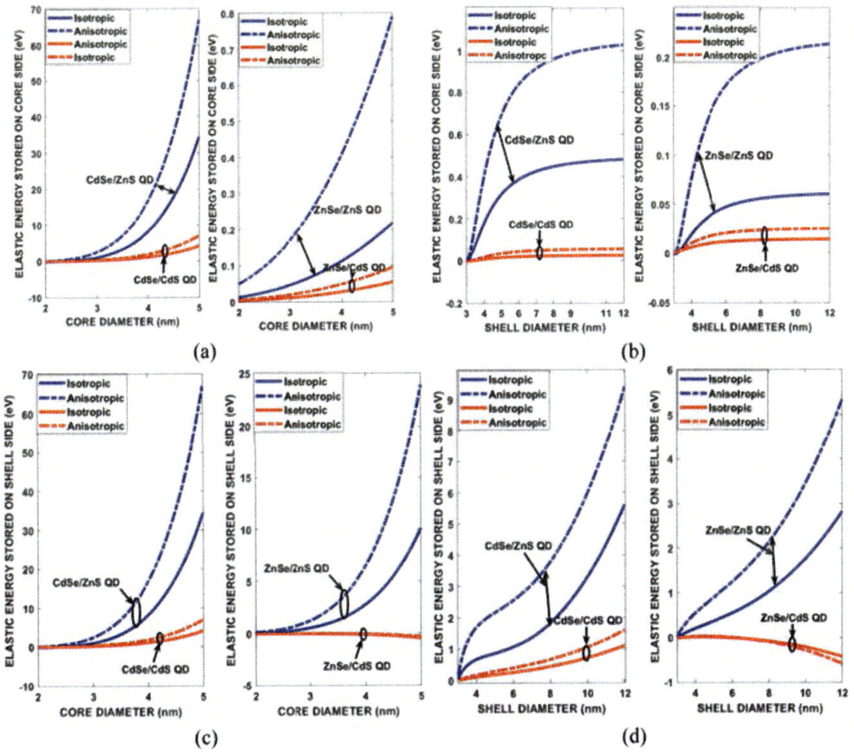

Fig. 2.15 Variation of elastic strain energy stored on core and shell sides of interfaces in CdSe/ZnS, CdSe/CdS, ZnSe/ZnS, and ZnSe/CdS core/shell QDs as a function of core diameter with $d_m = 2d_i$ and shell diameter with $d_i = 3.0$ nm, respectively, at 300 K

of interfacial elastic strain energies stored on core side and Fig. 1.15c and d show the core and shell diameter dependence of elastic strain energies stored on shell side of interfaces in CdSe/ZnS, CdSe/CdS, ZnSe/ZnS, ZnSe/CdS core/shell QDs for core and shell diameters of $d_i = 3.0$ nm and $d_m = 6.0$ nm, respectively.

Core diameter increase causes large parabolic increase in interfacial elastic energy stored on core side of CdSe/ZnS (Fig. 2.15a) and on shell side of CdSe/ZnS and ZnSe/ZnS QDs (Fig. 2.15c) for which lattice mismatch is large and positive and anisotropy mismatch is large and negative. Increase in shell diameter causes small increase (0.05 to 0.20 eV) in elastic strain energy stored on core side of CdSe/ZnS and moderate increase (0.05 to 0.90 eV) in that of ZnSe/ZnS QDs (Figs. 2.15b) and on shell sides (Figs. 2.15d). Using anisotropic elastic moduli in calculating elastic energy stored on both sides of interface yields larger values than using isotropic ones when lattice mismatch is large (e.g., CdSe/ZnS, ZnSe/ZnS core/shell QDs). However, using both isotropic and anisotropic elastic moduli yields similar and negligible increase in elastic energy when lattice mismatch is relatively small (e.g., CdSe/CdS and ZnSe/CdS core/shell QDs). The elastic energy calculations carried out here for abrupt core/shell heterointerfaces. Following Cai et al. [101], one can extend the use of (2.52) and (2.53) to spherical core/shell heterostructure QDs with compositionally graded core/shell interfaces.

2.7 Bandgaps and Band Offsts in Core/Shell Heterostructures

One of the critical key issues in microscale and nanoscale heterostructure semiconductor device technology is the qualitative understanding the formation and precise determination of the magnitude of band offsets across heterointerfaces. The band offsets dominate various device properties such as carrier injection efficiency in heterostructure bipolar transistors (HBTs), carrier confinement in modulation doped field effect transistors (MODFETs), quantum yield of photoluminescence in small scale and nanoscale electronic and optical devices, and have received considerable attention among device scientists and engineers over the years. Because potential barrier in the conduction band and the potential step in the valence band at the interface significantly influence carrier transport and device performance, reliable and accurate modeling of interface band structure is crucial for the designing the functionality of spherical core/shell heterostructure quantum dots and their performance optimization.

In order to understand the size effect on the core band gap of spherical bare core and core/shell QDs, Efros [37], Brus [38] and then Kayanuma [44, 45] used two band *effective band approximation.* Solving Schrödinger equation for a particle in a spherical box one obtains the first excited state (1 s-1 s) energy (core band gap) of quantum dots as [38]

$$E_g^{nc}(d) = E_g^b + \frac{2\hbar^2\pi^2}{m_{cv}^* d^2} - \frac{3.572e^2}{\varepsilon_\infty d} + 0.124\frac{m_{cv}^*}{\hbar^2\varepsilon_\infty^2} \tag{2.54}$$

where E_g^b is bulk bandgap of core semiconductor, $m_{cv}^* = m_e^* m_h^*/(m_e^* + m_h^*)$ is the reduced effective mass of electron–hole pair with effective masses of electrons and holes m_e^* and m_h^*, respectively, m_0 is the bare electron mass, and ε_∞ is the bulk value of optical dielectric constant of constituent semiconductor. Second term in (2.54) is the sum of the confinement energies of electrons and holes in the core region. Third term is the Coulomb attraction energy and the fourth term is the Rydberg electron–hole correlation energy, which is negligible when ε_∞ is considerably large.

Defining the strong confinement regime by the absence of the substantial electron–hole correlations, Kayanuma [44, 45] found that this is not limited to $d \leq 2a_B^*$, but remains valid up to about $d = 2a_B^*$. Here, $a_B^* = 2\varepsilon_\infty\hbar^2/m_{cv}^* e^2$ is the effective Bohr radius. For $d \gg 2a_B^*$, the exciton is envisioned as a quasiparticle moving around inside the core region with little energy increment due to confinement [79]. Although effective mass approximation is useful to understand variation of core band gap with diameter, it quantitatively fails to explain the results for QDs with small size (<5 nm). The reason was considered to be the use of bulk values for effective mass of electrons and holes [46].

In the framework of two band effective mass approximation, solving Schrödinger equation in cylindrical coordinates gives the following expression for the core band gap of nanowires [99]

$$E_g^{nw}(d) = E_g^b + \frac{2\hbar^2\pi^2}{m_{cv}^* d^2} + \delta E_s \tag{2.55}$$

where E_g^b is bandgap of core semiconductor in bulk form. $2\hbar^2\pi^2/m_{cv}^* d^2$ and δE_s are, respectively, the quantum confinement and surface the contributions to the core bandgap energy. For a constant core diameter (2a) the δE_s is negligible when $2a \ll l$. Coulomb interaction is not included in the calculation of nanowire band gap since correlation between electron and hole positions along the NW length are not easily considered with this kind of approximation, possibly leading to a small overestimation for the nanowire bandgap.

Since the first proposal of using two band *effective mass approximation* [37, 38] to calculate core bandgap, other theoretical models have been developed over the years. However, these models do not consider the effects of core and shell dimensions and difference between linear thermal expansion coefficients, elastic constants, and finally the effect of spin–orbit splitting of constituent semiconductors in the computation of elastic strain in QDs and NWs. Kane's parabolic two band *k.p approximation* [105] can be effectively used for a qualitative understanding of quantum confinement and strain effects on conduction and valence band edges of QDs and NWs. In the framework of Kane's *k.p approximation* the states are expanded in a finite set of Bloch states close to an extremum k_0 of band structure inside the Brillouin zone, with spin–orbit interaction effect is considered which leads to following expressions for conduction and valence band energy levels [106]

$$E_c(k) = -\frac{E_g}{2} + \frac{\hbar^2 k^2}{2m_0} + \frac{E_g}{2}\left(1 + \frac{\hbar^2 k^2 (E_g + \Delta)}{2m_{cv}^* E_g (E_g + 2\Delta/3)}\right)^{1/2}, \tag{2.56a}$$

$$E_v(k) = -\frac{E_g}{2} + \frac{\hbar^2 k^2}{2m_0} - \frac{E_g}{2}\left(1 + \frac{\hbar^2 k^2 (E_g + \Delta)}{2m_{cv}^* E_g (E_g + 2\Delta/3)}\right)^{1/2} \tag{2.56b}$$

where $E_g = E_c(0) - E_v(0)$ is the unperturbed band gap and Δ is the contribution of spin–orbit interaction to band gap and conduction and valence band energies. The quantum size effect in core region of spherical and cylindrical core/shell nanostructures is given by the dispersion relation with $|k| = 2\pi/d$, where d is the core diameter. One can expand square root term in (2.56a) and (2.56b) by using the Binomial approximation as $(1 + x)^{1/2} = 1 + x/2 - x^2/4 + ..$ and write the following equation for core bandgap

$$E_g^{nc} = E_g^{bi}(\varepsilon_i) + \frac{2\hbar^2\pi^2}{m_{cv}^* d^2}\frac{(E_g^{bi}(\varepsilon_i)+\Delta)}{(E_g^{bi}(\varepsilon_i)+2\Delta/3)} - \frac{1}{4}E_g\left(\frac{4\pi^2\hbar^2(E_g^{bi}(\varepsilon_i)+\Delta)}{m_{cv}^* d^2 E_g^{bi}(\varepsilon_i)(E_g^{bi}(\varepsilon_i)+2\Delta/3)}\right)^2$$
$$= E_g^{bi}(\varepsilon_i) + \frac{2\pi^2\hbar^2}{m_{cv}^* d^2}\delta_{sp}\left(1 - \frac{2\pi^2\hbar^2}{m_{cv}^* d^2}\delta_{sp}\right) \quad (2.57)$$

where $E_g^{bi}(\varepsilon_i)$ and $\delta_{sp} = \left(E_g^{bi}(\varepsilon_i) + \Delta_i\right)/\left(E_g^{bi}(\varepsilon_i) + 2\Delta_i/3\right)$ are, respectively, the hydrostatic strain dependent bulk bandgap and correction factor which considers the spin–orbit interaction contribution to quantum confinement. The correction factor δ_{sp} is close to unity for wide bandgap semiconductors such as GaAs and ZnSe ($E_g \gg \Delta$). In a recent work we have shown that its effect on quantum confinement is between 5 to 8% for wide gap III-V and II-VI compounds. However, correction factor δ_{sp} is larger than unity for narrow bandgap semiconductors for which bandgap is nearly equal to or smaller than spin–orbit energy ($E_g \ll \Delta$). δ_{sp} increases the quantum confinement about 20% for InAs and GaSb narrow gap compound semiconductors.

Therefore, it is essential to add the spin–orbit interaction contribution to quantum confinement in a realistic modelling of electronic properties of narrow gap/narrow gap (or wide gap). Core/shell QDs. Adding the Coulomb interaction and Rydberg correlation energies in (2.57), one obtains the core band gap of bare core and core/shell QD with *Type I* band alignment

$$E_g^{nc}(\varepsilon_i) = E_g^{bi}(\varepsilon_i) + \frac{2\pi^2\hbar^2}{m_{cv}^* d^2}\delta_{sp}\left(1 - \frac{2\pi^2\hbar^2}{m_{cv}^* d^2}\delta_{sp}\right) - \frac{3.572e^2}{\varepsilon_\infty d} + \frac{0.124e^4}{\hbar^2 m_{cv}^* \varepsilon_\infty^2} \quad (2.58)$$

Likewise, in a spherical core/shell QD with *Type II* band alignment core bandgap can be expressed

$$E_g^{ni}(\varepsilon_i) = E_g^{bi}(\varepsilon_i) - \Delta E_v(\varepsilon_i) + \frac{2\hbar^2\pi^2}{m_{cv}^* d^2}\delta_{sp}\left(1 - \frac{2\pi^2\hbar^2}{m_{cv}^* d^2}\delta_{sp}\right) - \frac{3.572e^2}{\varepsilon_\infty d} + \frac{0.124e^4}{\hbar^2 m_{cv}^* \varepsilon_\infty^2} \quad (2.59a)$$

$$E_g^{ni}(\varepsilon_i) = E_g^{bm}(\varepsilon_i) - \Delta E_c(\varepsilon_i) + \frac{2\hbar^2\pi^2}{m_{cv}^* d^2}\delta_{sp}\left(1 - \frac{2\pi^2\hbar^2}{m_{cv}^* d^2}\delta_{sp}\right) - \frac{3.572e^2}{\varepsilon_\infty d} + \frac{0.124e^4}{\hbar^2 m_{cv}^* \varepsilon_\infty^2} \quad (2.59b)$$

where $E_g^{bi}(\varepsilon_i) = E_g^{bi} + \delta E_g^{bi}(\varepsilon_i)$ and $E_g^{bm}(\varepsilon_m) = E_g^{bm} + \delta E_g^{bm}(\varepsilon_i)$ are strain dependent core and shell bandgaps with $\delta E_g^{bi}(\varepsilon_i)$ and $\delta E_g^{bm}(\varepsilon_i)$ strain shifts relative to their bulk values E_g^{bi} and E_g^{bm} at T = 0 K. $\Delta E_c(\varepsilon_i)$ and $\Delta E_v(\varepsilon_i)$ are the band offsets of spherical core/shell heterostructure.

In case of cylindrical core/shell nanowires with *Type I* band alignment core bandgap is found from the solution of Schrödinger equation in a cylindrical coordinate system in *effective mass approximation*, given by the following expression

$$E_g^{nw}(\varepsilon_i) = E_g^{bi}(\varepsilon_i) + \frac{2\pi^2\hbar^2}{m_{cv}^* d^2}\delta_{sp}\left(1 - \frac{2\pi^2\hbar^2}{m_{cv}^* d^2}\delta_{sp}\right) + \delta E_s \quad (2.60)$$

In a cylindrical core/shell nanowire with *Type II* band alignment core bandgap is expressed as

$$E_g^{nw}(\varepsilon_i) = E_g^{bi}(\varepsilon_i) - \Delta E_v(\varepsilon_i) + \frac{2\hbar^2\pi^2}{m_{cv}^* d^2}\delta_{sp}\left(1 - \frac{2\pi^2\hbar^2}{m_{cv}^* d^2}\delta_{sp}\right) + \delta E_s \tag{2.61a}$$

$$E_g^{ni}(\varepsilon_i) = E_g^{bm}(\varepsilon_i) - \Delta E_c(\varepsilon_i) + \frac{2\hbar^2\pi^2}{m_{cv}^* d^2}\delta_{sp}\left(1 - \frac{2\pi^2\hbar^2}{m_{cv}^* d^2}\delta_{sp}\right) + \delta E_s \tag{2.61b}$$

where δE_s is a contribution to NW core bandgap due to surface to volume ratio, defined in (2.55) and can be neglected in practical cases since capping core with shell isolates the active core from surface defects and traps, allowing one to facilitate better electronic and optical properties.

Valence band offsets are obtained by taking difference between the valence band widths of shell and core constituent semiconductors in bulk form, screened by their optical dielectric constants

$$\Delta E_v(\varepsilon) = \frac{1}{\varepsilon_{m\infty}} E_{mv}(\varepsilon_m) - \frac{1}{\varepsilon_{i\infty}} E_{iv}(\varepsilon_i) \tag{2.62}$$

where $E_{iv}(\varepsilon_i)$ ($E_{mv}(\varepsilon_m)$) and $\varepsilon_{i\infty}(\varepsilon_{m\infty})$ are valence band widths and optical dielectric constants of core and shell semiconductors, respectively. Valence band energies are obtained by either using the (a) empirical *sp3* or *sp3s** tight binding theories or (b) density functional theory. Strain effects on dielectric constants will be neglected for the sake of simplicity. Hydrostatic strain effects on conduction band offset is obtained by subtracting (adding) valence band offset from (to) band gap difference as.

$$\Delta E_{clI}(\varepsilon) = \Delta E_{gl}(\varepsilon) - \Delta E_v(\varepsilon),\ for\ Type\ I \text{interface band alignment} \tag{2.63a}$$

$$\Delta E_{clII}(\varepsilon) = \Delta E_{gl}(\varepsilon) + \Delta E_v(\varepsilon),\ for\ Type\ II \text{interface band alignment} \tag{2.63b}$$

where $E_{cl} = E_{\Gamma_{6c}}, E_{L_{6c}}$ and $E_{X_{6c}}$ conduction bands and $E_{gl} = E_{g\Gamma}, E_{gL}$ and E_{gX} are the lowest bandgaps at k $= 0(2\pi/a)(0;\ 0;\ 0)$, k $= (2\pi/a)(1;\ 0;\ 0)$, and k $= (2\pi/a)(1/2;\ 1/2;\ 1/2)$ in the Brillouin zone of the constituent semiconductors.$\Delta E_g(\varepsilon) = E_{gl}^m(\varepsilon_m) - E_{gl}^i(\varepsilon_i)$ is strain dependent band gap difference between shell and core constituent semiconductors at high symmetry points of the first Brillouin zone. The accuracy of the results is as good as the input parameters used in calculations.

Accuracy of valence and conduction band offsets prediction can be improved if one obtains them in terms of measured core band gaps, extracted from the first exciton peak energy of UV–Vis optical absorption spectra. In doing so, we use (2.59a) and (2.59b) to write

$$\Delta E_v(\varepsilon) = E_g^{ib} - E_g^{nc}(\exp) + \delta E_g^{bi}(T, \varepsilon_i) + \frac{2\hbar^2\pi^2}{m_{cv}^* d^2} - \frac{3.572e^2}{\varepsilon_\infty d} + \frac{0.124e^4}{\hbar^2 m_{cv}^* \varepsilon_\infty^2} \tag{2.64a}$$

$$\Delta E_c(\varepsilon) = E_g^{mb} - E_g^{nc}(\exp) + \delta E_g^{mb}(T, \varepsilon_m) + \frac{2\hbar^2\pi^2}{m_{cv}^* d^2} - \frac{3.572e^2}{\varepsilon_\infty d} + \frac{0.124e^4}{\hbar^2 m_{cv}^* \varepsilon_\infty^2} \tag{2.64b}$$

for spherical core/shell QDs. Here $E_g^{nc}(\exp)$ and d are measured core bandgap and diameter of core/shell QD, respectively. $E_g^{nc}(\exp)$ is obtained from the first exciton peak energy of the UV-absorption spectra. Core diameter d is obtained from the UV–Vis absorption spectra, X-ray diffraction (XRD), and high-resolution transmission electron microscopy (HRTEM) measurements, respectively [107, 108]. In case of cylindrical core/shell heterostructures we rewrite (2.61a) and (2.61b) as

$$\Delta E_v(\varepsilon_i) = E_g^{bi}(\varepsilon_i) - E_g^{nw}(\exp) + \frac{2\hbar^2\pi^2}{m_{cv}^* d^2}\delta_{sp}\left(1 - \frac{2\pi^2\hbar^2}{m_{cv}^* d^2}\delta_{sp}\right) + \delta E_s \tag{2.65a}$$

$$\Delta E_c(\varepsilon_i) = E_g^{bm}(\varepsilon_i) - E_g^{ni}(\exp) + \frac{2\hbar^2\pi^2}{m_{cv}^* d^2}\delta_{sp}\left(1 - \frac{2\pi^2\hbar^2}{m_{cv}^* d^2}\delta_{sp}\right) + \delta E_s \tag{2.65b}$$

where $E_g^{nc}(\exp)$ and d are measured core bandgap and diameter of core/shell NWs, respectively.

2.8 Strain Effects on Bandgaps and Band Offsets

Figure 2.16 shows the conduction minimum relative to valence band maximum at the Γ; high symmetry point of the first Brillouin zone of a direct bandgap semiconductor. In coherently grown microscale thin films, compressive (or tensile) strain in epilayer increase (or decrease) conduction and valence band energy levels of semiconductor band structure. Hydrostatic strain, which corresponds to the relative volume change of the strained epilayer, shifts the conduction band minimums relative to the average valence band maximum at Γ; point. The uniaxial component of biaxial strain tensor splits the heavy-hole, light-hole and split-off valence band edges relative to the average valence band edge. The heavy-hole, light-hole and split-off band energies relative to the average valence band edge E_v are [6]

$$E_{vh}(\varepsilon) = E_v(\varepsilon) + \frac{1}{3}\Delta - \frac{1}{2}\delta E, \tag{2.66a}$$

$$E_{vl}(\varepsilon) = E_v(\varepsilon) - \frac{1}{6}\Delta + \frac{1}{4}\delta E + \frac{1}{2}\sqrt{\Delta^2 + \Delta\delta E + \frac{9}{4}\delta E^2}, \tag{2.66b}$$

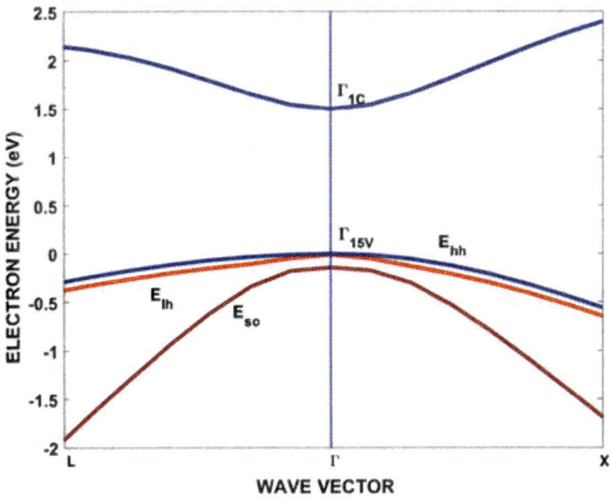

Fig. 2.16 Band structure of bulk direct bandgap GaAs, showing conduction band minimum, heavy hole, light hole and spin–orbit splitting of valence band maximum

$$E_{vs}(\varepsilon) = E_v(\varepsilon) - \frac{1}{6}\Delta + \frac{1}{4}\delta E - \frac{1}{2}\sqrt{\Delta^2 + \Delta\delta E + \frac{9}{4}\delta E^2} \tag{2.66c}$$

$$\delta E = 2b(\varepsilon_{zz} - \varepsilon_{xx}) = 2b(\varepsilon_{\perp} - \varepsilon_{||}) \tag{2.66d}$$

where b is the shear deformation potential which describes splitting of valence band energy due to the [001] uniaxial strain.$E_v(\varepsilon)$ is the average valence band maximum under hydrostatic strain.

Reliable modelling and precise determination of electronic properties (such as band gaps and band offsets) of spherical and cylindrical core/shell heterostructures is essential for the true and realistic prediction of their potential use in fabricating nanoscale electronic and optical devices operating at high temperatures. However, the determining the strain effects on electronic and optical properties of nanoscale core/shell one and zero-dimensional semiconductor heterostructures is still an obstacle. Qualitative understanding of strain shifts in core gaps of spherical core/shell type I and type II heterostructure QDs can be achieved by using the classical parabolic two band *effective mass approximation* [38]. We will discuss the effects of using isotropic and anisotropic elastic moduli in calculating hydrostatic strain effects on bandgaps and band offsets in wide bandgap GaAs/GaSb(InAs) and GaSb/InAs(InSb) group III-V and CdSe/ZnS(CdS) and ZnSe/ZnS(CdS) group II-VI compounds based heterostructure core/shell QDs as a function of core and shell diameters at any temperature. The material parameters are given Tables 2.1 and 2.2 for group III-V and II-VI compounds, respectively.

In calculating the hydrostatic strain effects on energy levels on both sides of heterointerface one can use so called the statistical thermodynamic model of semiconductors [102]. The energy bandgap at high symmetry points in the first Brillouin zone of semiconductors with zinc blende or hexagonal crystal structure is expressed as a function of temperature and pressure as

$$E_{gl}(T,P) = E_{gi} + \Delta C_P^0 T(1-\ln T) - \frac{a_{gl}}{B}[P - \frac{P^2}{2B} - \frac{(1+B')P^3}{6B^2}] \quad (2.67)$$

where E_{gl} is the band gap at standard temperature $T_0(K)$ and standard pressure $P_0(1atm).a_{gl} = -B(\partial E_{gl}/\partial P)$ is deformation potential and B is bulk modulus with its $B^{'} = \partial B/\partial P$. The logarithmic term represents the electron–phonon interactions contribution to shift in bandgap with temperature increase. Third term represents shifts in bandgap to free linear expansion of lattice constant of constituents with temperature increase and strain due to lattice and thermal mismatches.ΔC^0_{gP} is standard state heat capacity of reaction for formation of electron–hole pair, which is obtained by fitting bulk band gap $E_g(T,P)$ to experimental data [103], using the empirical expression of Varshni [104]

$$E_g(T) = E_g(0) + \frac{\alpha T^2}{(T+\beta)} \quad (2.68)$$

where α and β are fitting constants for bulk semiconductor.

Strain effects on electronic and optical properties of microscopic planar semiconductor heterostructures have been extensively studied and is reasonably well understood. In a planar heterostructure on (001) substrate, the hydrostatic pressure is expressed as $P = -2B_f C_f \varepsilon_{f//}$ and $P = -3B_s\varepsilon_{s||}$ for epilayer and substrate, respectively. Using (2.67) strain shifts in the bandgap energies of epilayer and substrate can be written as

$$\delta E_g^{bf}(T,\varepsilon) = \Delta C^0_{iP} T(1-\ln T) + 2C_f a_{gf}\left(1 + C_f\varepsilon_f - \frac{4C_f^2(1+B_i^{'})}{3}\varepsilon_f^2\right)\varepsilon_f$$
$$\delta E_g^{bs}(T,\varepsilon) = \Delta C^0_{sP} T(1-\ln T) + 3a_{gs}(1 + \frac{3}{2}\varepsilon - \frac{(1+B_m^{'})}{2}\varepsilon^2)\varepsilon \quad (2.69)$$

where $\varepsilon = \varepsilon_{f||} = (a_{f||} - a_f)/a_f$ is strain in epilayer along the growth direction and $C_f = (C_{11} - C_{12})/C_{11}$ is the elastic constants ratio for epilayer.

However, in a coherently grown nanoscale core/shell heterostructures, strain across the interface differently influences the energy band structure and charge transport, operation and performance in heterostructure devices. When the diameter of the nanoparticle becomes less or comparable to the Bohr diameter of the exciton, several size quantization effects occurs such as localization of electron and hole pair inside the quantum dot and widening of core band gap relative to bulk value. Strain shifts of core and shell bandgap energies $\delta E_g^{bi}(\varepsilon_i)$ and $\delta E_g^{bm}(\varepsilon_i)$ relative to their bulk values E_g^{bi} and E_g^{bm} at T = 0 K are still determined by using the statistical thermodynamic model of semiconductors [102]. The hydrostatic strain shifts in core and shell bandgaps are written as

$$\delta E_g^{bi}(T, \varepsilon_i) = \Delta C_{iP}^0 T(1 - \ln T) - \frac{a_{gi}}{B_i}(P_i - \frac{P_i^2}{2B_i} - \frac{(1 + B_i^{'})}{3B_i^2}P_i^3) \tag{2.70a}$$

$$\delta E_g^{bm}(T, \varepsilon_m) = \Delta C_{mP}^0 T(1 - \ln T) - \frac{a_{gm}}{B_m}(P_m - \frac{P_m^2}{2B_m} - \frac{(1 + B_m^{'})}{3B_m^2}P_m^3) \tag{2.70b}$$

where $P_i(\varepsilon_i, T) = -3B_i\varepsilon_i$ and $P_m(T, \varepsilon_m) = -B\varepsilon_m$ are hydrostatic pressures acting on core and shell sides of interface of spherical core/shell heterostructures, respectively. Here $\varepsilon_i = \varepsilon_{ir} = \varepsilon_{i\theta} = \varepsilon_{i\varphi}$ and $\varepsilon_m = \varepsilon_{mr} + \varepsilon_{m\theta} + \varepsilon_{m\varphi}$, are strains defined in Sect. 2.3. a_{gi} (a_{gm}). is core (shell) bandgap deformation potential, and $B_i(B_m)$ is bulk modulus of core (shell) region with $B_i^{'} = \partial B_i/\partial P$ and $B_m^{'} = \partial B/\partial P$. In cylindrical core/shell heterostructures, $P_i^*(\varepsilon_i^*, T) = -3B_i\varepsilon_i^*$ and $P_m^*(T, \varepsilon_m^*) = -B\varepsilon_m^*$ are hydrostatic pressures acting on core and shell sides of interfaces, respectively. Here $\varepsilon_i^* = \varepsilon_{ir}^* = \varepsilon_{i\theta}^* = \varepsilon_{i\varphi}^*$ and $\varepsilon_m^* = \varepsilon_{mr}^* + \varepsilon_{m\theta}^* + \varepsilon_{m\varphi}^*$, are strains defined in Sect. 2.4.

We first begin with the temperature variation of interface strain acting on core and shell sides of GaAs/InAs(GaSb) and GaSb/InAs(InSb) NWs respectively, shown in Fig. 2.17a and b for shell diameter $d_m = 2d_i$ with core diameter $d_i = 3.00$ nm. In all cases, interface strains monotonically increase with temperature. When lattice mismatch ε_{im} and anisotropy ratio mismatch ξ_{im} are both large, calculated strain difference between using isotropic and anisotropic elastic moduli is relatively large. Strain is tensile on core and shell sides of GaAs/InAs(GaSb) and GaSb/InAs(InSb) NWs but it is compressive on core side of GaAs/GaSb and GsSb/InSb NWs and on shell side of GaAs/InAs NWs.

Figure 2.18a and b show the temperature variation of interface strain acting on core and shell sides of CdSe/ZnS(CdS) and ZnSe/ZnS(CdS) heterostructure QDs with shell diameter $d_m = 2d_i$ ($d_i = 3.00$ nm). Strain is compressive on core side of CdSe/ZnS(CdS) and ZnSe/ZnS QDs and tensile in ZnSe/CdS QDs. Strain on core and

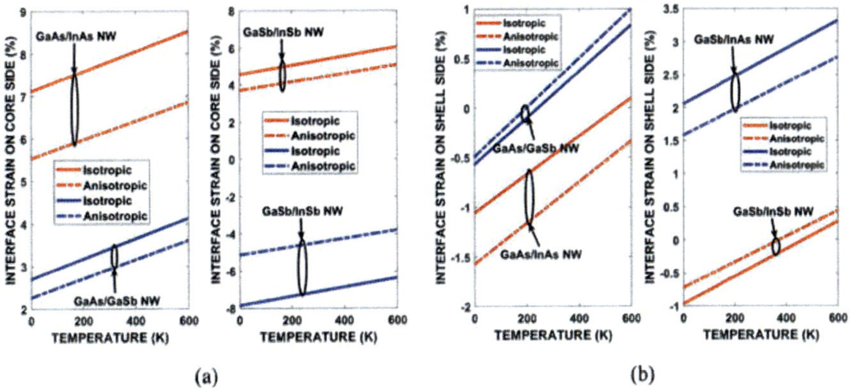

Fig. 2.17 Temperature variation of strain, calculated by using isotropic and anisotropic elastic moduli, on core side (**a**) and shell side (**b**) of interfaces in GaAs/InAs(GaSb) and GaSb/InAs(InSb) heterostructure core/shell NWs with shell diameter $d_m = 2d_i$ ($d_i = 3.00$ nm)

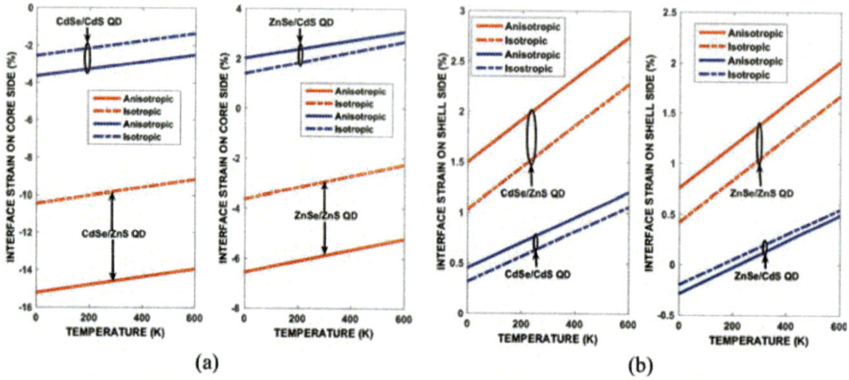

Fig. 2.18 Temperature variation of strain, calculated by using isotropic and anisotropic elastic moduli, on core side (**a**) and shell side (**b**) of interfaces in CdSe/ZnS(CdS) and ZnSe/ZnS(CdS) heterostructure QDs with shell diameter $d_m = 2d_i$ ($d_i = 3.00$ nm)

shell sides of ZnSe/CdS QD is tensile and compressive below 200 K (tensile between 200 and 600 K), respectively. Strain is compressive on core sides of CdSe/ZnS(CdS) and ZnSe/ZnS QDs and tensile in ZnSe/CdS QD. Strain on core and shell sides of ZnSe/CdS QD is tensile and compressive below 200 K, respectively.

Figure 2.19a and b compare the temperature variation of core band gaps of CdSe/ZnS, CdSe/CdS, ZnSe/ZnS and ZnSe/CdS QDs and valence and conduction band offsets of ZnSe/ZnS and ZnSe/CdS QDs, respectively. Anisotropic elastic moduli result in higher band gaps in CdSe/ZnS, and ZnSe/ZnS core/shell QDs for which both ε_{im} and ξ_{im} are large. Using anisotropic and isotropic elastic moduli result in a similar core band gaps in CdSe/CdS and ZnSe/CdS heterostructures.

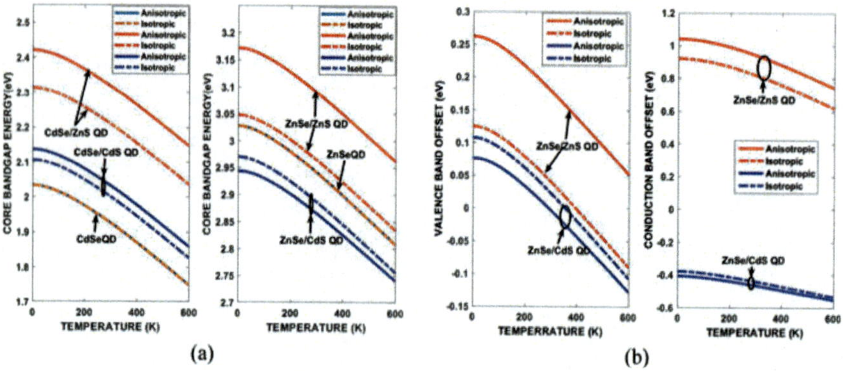

Fig. 2.19 Comparison of temperature variation of core bandgaps of CdSe/ZnS, CdSe/CdS, ZnSe/ZnS and ZnSe/CdS core/shell heterostructure QDs against those of bare CdSe and ZnSe core QDs for $d_i = 3.0\ nm$ and $d_m = 2d_i$, respectively (**a**) and valence and conduction band offsets in ZnSe/ZnS and ZnSe/CdS core/shell QDs for core (shell) diameter $d_i = 3.0\ nm$ ($d_m = 2d_i$)

Using anisotropic elastic moduli results in larger band offsets than isotropic ones in the case of ZnSe/ZnS heterostructure, but the shift is small in ZnSe/CdS. Since valence band deformation potentials of core and shell semiconductors are small, strain shifts in valence band offsets is small at any temperature. The deformation potentials of conduction band edges of core and shell components are larger than those of valence band edges in both ZnSe/ZnS and ZnSe/CdS heterostructures. Therefore, using anisotropic elastic moduli gives larger shifts in conduction band offsets than isotropic ones in these heterostructures. The conduction band offset is positive and large below 400 K, negative and small above 400 K, respectively. Valence band offsets are small and negative, which suggest a Type I interface band alignment below 400 K and Type II band alignment above 400 K in ZnSe/ZnS and ZnSe/CdS heterostructure, respectively.

Figure 2.20a and b, respectively, show the effect of increasing core (shell) diameter on core band gaps of CdSe/ZnS(CdS) and, ZnSe/ZnS(CdS) core/shell heterostructures with fixed shell (core) diameter at 300 K. In Fig. 1.20a, calculated bandgaps of bare CdSe and ZnSe core QDs are also given for comparison purpose.

There is a parabolic decrease in core bandgap with core diameter increase with fixed shell diameter $d_m = 2d_i$ (with $d_i = 3.00\ nm$). Increasing shell diameter with fixed core diameter (e.g., $d_i = 3.00\ nm$) causes parabolic increase in core band gaps of CdSe/ZnS, CdSe/CdS, and ZnSe/ZnS and decrease in that of ZnSe/CdS, respectively. Anisotropic elastic moduli causes about 0.10 - 0.20 eV higher bandgaps than isotropic ones in CdSe/ZnS and ZnSe/ZnS core/shell QD. However, they both yield nearly same core bandgaps for CdSe/CdS and ZnSe/CdS core/shell QDs.

Core bandgaps of CdSe/ZnS, CdSe/CdS, ZnSe/ZnS and ZnSe/CdS heterostructures become nearly constant above 7.00 nmcore diameter and above 6.00 nm shell diameter in all of these heterostructures, indicating that there is difference between strain effects on band gaps as a function of core and shell diameters. Therefore, increasing core and shell diameters further above 6.00 nm decreases the degree of

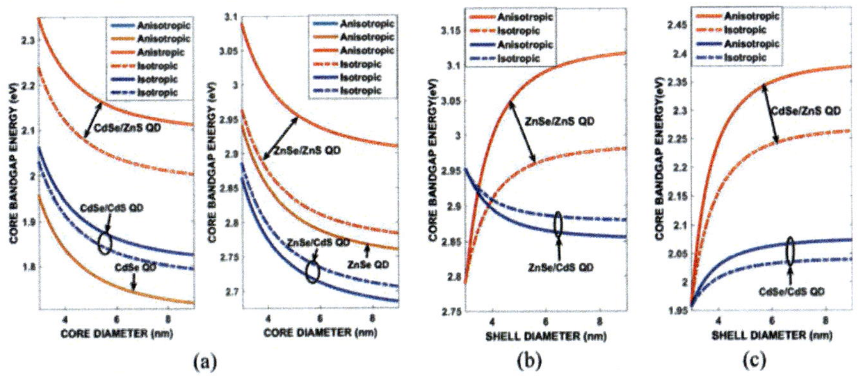

Fig. 2.20 Comparison of core and shell diameter effects on core bandgaps in CdSe/ZnS, CdSe/CdS, ZnSe/ZnS, and ZnSe/CdS core/shell QDs for shell diameter $d_m = 2d_i$, with $d_i = 3.0$ nm (**a**) and shell diameter for core diameter $d_i = 3.0$ nm (**b**) at 300 K

quantum confinement in these core/shell QDs. If one draws vertical and horizontal straight lines that are tangent to parabolic curves, intersection point will give the upper limit for strong confinement regime. Beyond the critical thickness that is greater than twice of Bohr diameter, the exciton is envisioned as a quasiparticle moving around inside the core region with little energy increase due to confinement. This agrees with the prediction of Kayanuma [44, 45], who proposed that strong confinement is not limited to $d \leq a_B^*$, but remains valid up to about $d = 2a_B^*$, provided it is defined by the absence of the substantial electron–hole correlations.

Figure 2.21a–d, respectively, show the effects increasing core (shell) diameter, with fixed shell (core) diameter, on the valence band offsets and conduction band offsets in spherical CdSe/ZnS, CdSe/CdS, ZnSe/ZnS, and ZnSe/CdS core/shell QDs. In CdSe/CdS and ZnSe/CdS heterostructures lattice mismatch and valence band deformation potentials of core and shell constituents are small and both isotropic and anisotropic elastic moduli result in similar and negligible strain shifts in valence band offsets for core and shell diameters. However, in CdSe/ZnS and ZnSe/ZnS

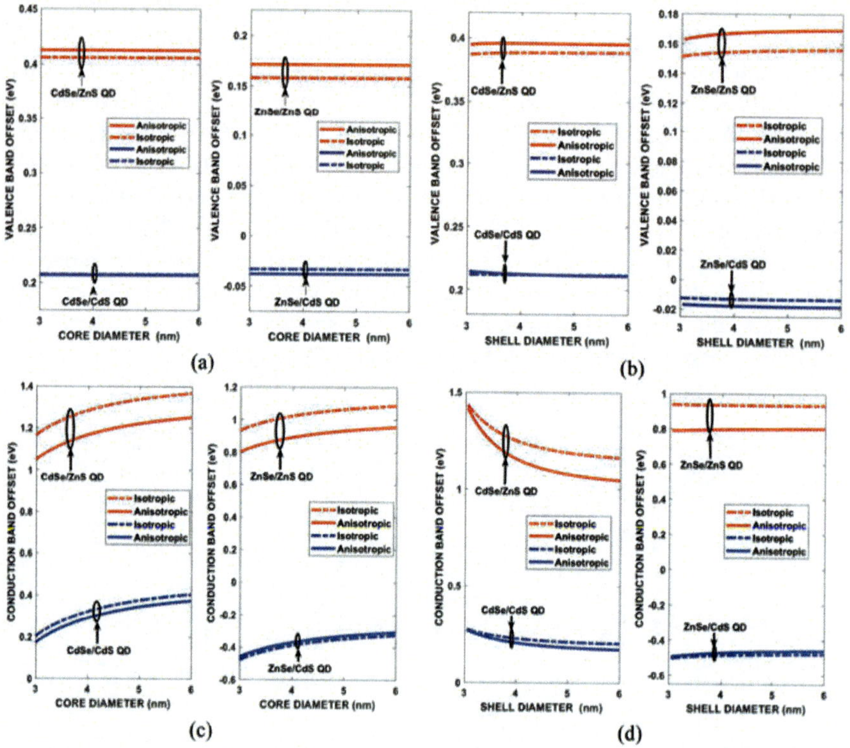

Fig. 2.21 Comparison of core and shell diameter effects on valence band offsets (**a, b**) and conduction band offsets (**c, d**) of CdSe/ZnS, CdSe/CdS, ZnSe/ZnS, and ZnSe/CdS core/shell QDs for shell diameter $d_m = 2d_i$ (**a, d**) and core diameter $d_i = 3\ nm$ (**b, c**) at 300 K, respectively

heterostructures, lattice mismatch is large and still valence band deformation potentials of core and shell are small using anisotropic elastic moduli result in valence band offsets of the order of 0.10–0.20 eV larger than those by using isotropic ones for core and shell diameter below 6.00 nm.Valence band offsets are small but not negative, which suggest that CdSe/ZnS and ZnSe/ZnS heterostructure may have Type I (Type II) band alignment if conduction band offset is positive (negative).

Increasing the core (shell) diameter with fixed shell (core) diameter causes large parabolic increase (decrease) in conduction band offset at interfaces of CdSe/ZnS, CdSe/CdS and ZnSe/ZnS heterostructures. However, increasing core (shell) diameter with fixed shell (core) diameter causes small parabolic increase in conduction band offset at interface of ZnSe/CdS heterostructure. Since conduction band deformation potentials of constituents are larger than those of valence bands, hydrostatic strain causes significant shifts in conduction band offsets than in valence band offsets. Using anisotropic elastic moduli results in about 0.10–0.20 eV higher strain shift in conduction band offsets than isotropic ones in case of CdSe/ZnS and ZnSe/ZnS heterostructure.

Semiconductor alloys based on the group IV–VI, III–V and II–VI compounds are important in fabricating low dimensional bipolar and unipolar heterostructure devices since their structural and electronic properties (e.g., lattice constants and bandgaps) can be tailored independently. Therefore, reliable and accurate determination of composition variation of lattice constant and bandgap energies are very important. In the theoretical determination of composition effects on the structural properties such as lattice constants, a virtual crystal approximation (VCA) is often used [16, 17] in which the compositional disorder effect is neglected. Since in VCA the alloy potential is taken as the concentration weighted average of the constituent potentials the bandgap energy is linear function of alloy composition. However, many experimental studies report that the bandgap energies of semiconductor alloys is nonlinear function of composition. Furthermore, it is believed that the compositional disorder, which is related to the differences of electronegativity of atoms forming a ternary semiconductor, plays a major role in determining its bowing of the bandgap energy when lattice mismatch induced strain plays a crucial role in heterostructure electronic properties.

In our recent studies [95, 96] the effects of composition and strain on the electronic properties of semiconductor alloy constituents in heterostructures are implemented in the semiempirical tight binding models and first principles DFT with MBJLDA functional in terms of host bond length and distorted bond length by the substitutional impurity without any adjustable parameter. The MVCA allows one to accurately considers the effect of disorder-induced nonlinear variation of the lattice constant on the structural parameters used in the semiempirical tight binding models and first principles DFT with MBJLDA functional calculating the band structure properties.

As an alternative method for calculating composition effects on the energy band structure is to use the thermodynamic approach [102]. One first calculates the valence band energy $E_v(x)$ of $A_x B_{1-x} C$ ternary alloy by writing the bond length and valence band energies according to the modified Vegard's rule described by (2.9) and (2.11), respectively. The bandgap energies at high symmetry points k = (2 π/a)(0; 0; 0) and k = (2 π/a)(1; 0; 0) can be obtained by combining the Vegard's rule and Kane's *three band k.p approximation* at symmetry points one writes [102]:

$$\frac{1}{m_{n\Gamma}(x)} = \frac{x}{m_{nA\Gamma}(x)} + \frac{1-x}{m_{nB\Gamma}(x)} = 1 + \frac{P_\Gamma^2(x)}{3}\left[\frac{2}{E_{g\Gamma}(x)} + \frac{1}{E_{g\Gamma}(x) + \Delta(x)}\right] \quad (2.71)$$

$$\frac{1}{m_{ntL}(x)} = \frac{x}{m_{ntLA}(x)} + \frac{1-x}{m_{ntLB}(x)} = 1 + P_L^2(x)\left[\frac{2}{E_{gL}(x)} + \frac{1}{E_{gL}(x) + \Delta_L(x)}\right] \quad (2.72)$$

where $m_{nA\Gamma}(x)$ and $m_{nB\Gamma}(x)$ are the electron effective masses at Γ_{1C} conduction band valley and m_{ntLA} and m_{ntLB} are the transverse electron effective masses at L conduction valley of binaries AC and BC. $P_\Gamma(x) = xP_{A\Gamma} + (1-x)P_{B\Gamma}$ and $P_L(x) = xP_{LA} + (1-x)P_{LB}$ are the momentum matrix elements $\Delta(x) = x\Delta_A + (1-x)\Delta_B$ and $\Delta_L(x) = x\Delta_{LA} + (1-x)\Delta_{LB}$ are the spin–orbit energies. $P_{A\Gamma}$, $P_{b\Gamma}$, P_{AL} and P_{BL} of binaries AC and BC are obtained from (2.71) and (2.72) with measured $m_n(x)$,$m_{ntL}(x)$,$\Delta(x)$ and $\Delta_L(x)$ for x = 0 and 1. Inverting (2.71) and (2.72) one obtains quadratic equations for bandgaps $E_{g\Gamma}(x)$ and $E_{gL}(x)$ and solving the resultant equations yields

$$E_{g\Gamma}(x) = \frac{3y_\Gamma^{-1}(x) - \Delta(x)}{2} + \frac{1}{2}\left[(\Delta(x) - 3y_\Gamma^{-1}(x))^2 + 8\Delta(x)y_\Gamma^{-1}(x)\right]^{1/2} \quad (2.73)$$

$$E_{gL}(x) = \frac{2y_\Gamma^{-1}(x) - \Delta_1(x)}{2} + \frac{1}{2}\left[(\Delta_1(x) - 2y_\Gamma^{-1}(x))^2 + 4\Delta_1(x)y_\Gamma^{-1}(x)\right]^{1/2} \quad (2.74)$$

$$y_\Gamma(x) = 3(1 - m_{n\Gamma}(x))/m_{n\Gamma}(x)P_\Gamma^2(x); \quad y_L(x) = (1 - m_{nt\Gamma}(x))/m_{nt\Gamma}(x)P_L^2(x)$$

The indirect gap $E_{gX}(x)$ of $A_xB_{1-x}C$ ternary is determined from the following expression

$$E_{gX}(x) = E_{g\Gamma}(x) + [E_{gXA} - E_{g\Gamma A}]x + [E_{gXB} - E_{g\Gamma B}](1-x) \quad (2.75)$$

E_{gXA}, E_{gXB}, $E_{g\Gamma A}$, and $E_{g\Gamma B}$ are the indirect bandgaps of AC and BC binary compound constituents.

If one of the constituents of spherical core/shell QD is a ternary semiconductor alloy (e.g., CdSe/CdZnS, ZnSe/ZnCdS), alloy composition of ternary shell can also influence interface strain and in turn core bandgap of spherical binary/ternary core/shell QDs and can be calculated in the framework of modified virtual crystal approximation [97, 98]. Figure 2.22 shows the predicted and measured [107, 108] composition variation of core bandgap energy and conduction and valence band offsets in ZnSe/CdZnS and CdSe/CdZnS core/shell heterostructures. As can be

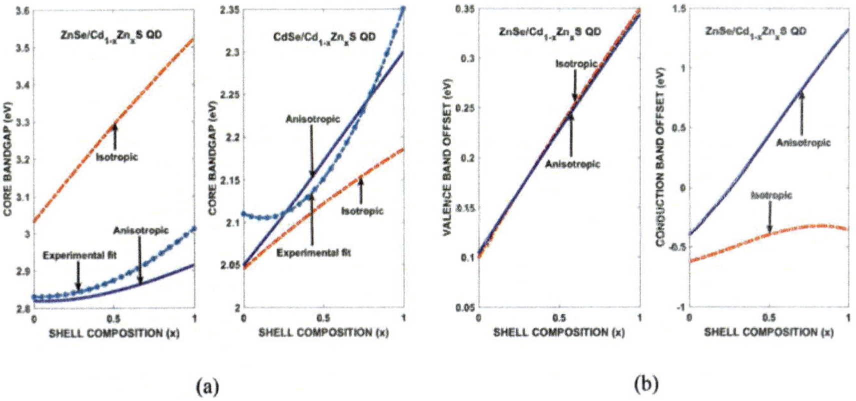

Fig. 2.22 Comparison of using isotropic and anisotropic elastic moduli in calculating the shell region composition effects on core bandgap core and on valence and conduction band offsets at interface of ZnSe/CdZnS and CdSe/CdZnS core/shell heterostructures

seen there is some difference between the use of isotropic and anisotropic elastic anisotropy ratio.

2.9 Comparison of Measured and Predicted Core Bandgaps

When the diameter of the nanoparticle becomes less or comparable to the Bohr diameter of the exciton, quantization effects such as localization of electron and hole pair inside the quantum dot and widening of core bandgap relative to bulk value, will take place. Changes in bandgaps of bare core and core/shell quantum dots are experimentally observed by the conventional UV–Vis optical absorption spectroscopy, in which a blue shift is observed with decreasing diameter, as first absorption peak energy.

$$E_g^{nc} = hc/\lambda_{\max}, \tag{2.76}$$

where h is Planck constant, c is speed of light, and $\lambda_{\max}$ is maximum wavelength of first exciton peak. Brus [38] has developed a theory to explain the spectral shift and calculate core bandgap with a parabolic two band effective band approximation. Figure 2.23a and b compare calculated and UV-absorption measured [109] core band of CdSe/ZnS core/shell QD, against calculated core band gap of bare CdSe QD, as a function of temperature for core diameters: 1.90 and 2.40 nm (Fig. 1.20a) and for 3.60 and 5.20 nm (Fig. 1.23b) with shell diameter $d_m = 2d_i$, respectively.

Since core bandgap shift due to electron–phonon interactions and carrier confinement effects are equal in both bare CdSe core and CdSe/ZnS core/shell QDs, the difference between calculated core band gaps of both QDs is due to hydrostatic strain

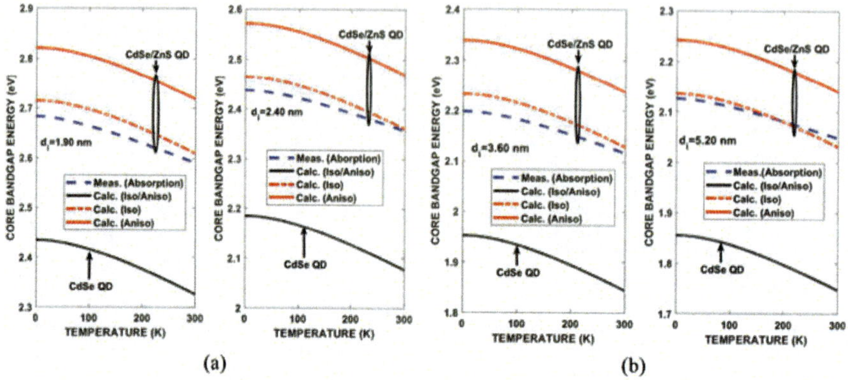

Fig. 2.23 Comparison of UV-absorption measured and calculated (with isotropic and anisotropic elastic moduli) core bandgap in CdSe/ZnS core/shell QD against calculated band gap of bare CdSe QD as a function of temperature for core diameters $d_i = 1.90\ nm$ and 2.40 nm (Fig. 10a) and $d_i = 3.60\ nm$ and 5.20 nm (Fig. 10b) with shell diameter $d_m = 2d_i$

in core region of latter. Since anisotropic elastic moduli results in larger compressive elastic strain than isotropic ones on core side, it results in about 0.10 eV higher core band gap when compared with experimental data.

Calculated and measured core band gaps are also compared in Table 2.5 against the calculated bandgap of bare CdSe core QD for core diameters $d_i\ (nm) = 2.40, 3.00, 3.40,$ and 4.40, respectively, at 300 K. In addition to Fig. 2.23a and b, Table 2.5 also demonstrates that one achieves better agreement with experimental and predicted core band gaps of core/shell QD if one uses isotropic elastic moduli than anisotropic ones in band gap calculations.

Excellent agreement between model predictions and experimental data shown in Fig. 2.23a and b and Table 2.5 for core band gaps of CdSe/ZnS QDs suggests that, despite the common view, the parabolic two band *effective mass approximation* can explain the common expected results such as core band gap and quantum confinement for QDs with very small sizes (i.e., less than 5.00 nm), provided interface strain effects are properly formulated. We conclude that provided strain effects are properly formulated, EMA is still a useful tool to quantitatively explain the quantum confinement and core bandgap variation with diameters less than 5.0 nm.

Table 2.5 Comparison of measured and calculated core band gap (eV) of CdSe/ZnS QD relative to that of bare CdSe core QD for $d_i = 2.40\ nm, 3.00, 3.40$ and $4.40\ nm$ with $d_m = 2d_i$ at 300 K

Core diameter	CdSe/ZnS QD Measured bandgap	CdSe/ZnS QD Isotropic	CdSe/ZnS QD Anisotropic	CdSe QD Isotropic/Anisotropic
$d_i = 2.40$	2.3664 [109]	2.3607	2.4696	2.0771
$d_i = 3.00$	2.2024 [109]	2.2099	2.3188	1.9263
$d_i = 3.40$	2.1232 [109]	2.1505	2.2595	1.8669
$d_i = 4.40$	2.0666 [109]	2.0664	2.1754	1.7828

In order to make a concluding remark about the reliability and accuracy of the proposed model for core band gaps of core/shell heterostrucure QDs with Type II band alignment (Fig. 2.2a and b), core bandgaps calculated from (2.63a) and (2.63b) are also compared against experimental data for CdSe/ZnS, CdSe/CdS, ZnSe/ZnS, and ZnSe/CdS core/shell QDs [107, 108]. In doing so, valence band offsets are first determined from (2.62) with bulk valence band widths that are calculated by using density functional theory based on modified Becke–Johnson exchange potential with local density approximation (mBJLDA) [95].

In second step, using these theoretical band offsets, core band gaps of core/shell heterostructure QDs are calculated from (2.64a) and (2.64b) by using isotropic and anisotropic elastic moduli. Table 2.6 compares calculated and measured core bandgaps of CdSe/ZnS, CdSe/CdS, ZnSe/ZnS, and ZnSe/CdS core/shell heterostructure QDs with shell diameters $d_m = 1.5d_i$ and $d_m = 2d_i$ for fixed core diameter $d_i = 3.00\ nm$ at 300 K. As can be seen from comparison in Table 2.6, there is excellent agreement between measured and predicted core bandgaps for almost all core/shell heterostructure QDs discussed in this work. The difference between core band gaps predicted by using isotropic and anisotropic elastic moduli can be explained with a close inspection of (2.59a) and (2.59b), which suggest that valence and conduction band offsets play decisive role in determining core band gaps of core/shell QDs with Type II interface band alignment (Table 2.7).

Since band offsets used in (2.59a) and (2.59b) to predict core band gaps of core/shell QDS with Type II heterointerface alignment are theoretically determined, they can be parameter dependent. For example, in calculating core bandgaps and band offsets, we assumed that strain does not have much influence on electron and hole effective masses and optical dielectric constant of core region. Therefore, we used their bulk values in band offset and bandgap calculations. However, there can be strain effects on electron and hole effective masses through band gap energy according to following expression due to eight band *k.p approximation* [85]

$$m_n^* = m_o / \left[(1+2F) + \frac{\left(E_g^{bi}(\varepsilon_i) + 2\Delta_i/3\right)E_P}{E_g^{bi}(\varepsilon_i)\left(E_g^{bi}(\varepsilon_i) + \Delta_i\right)} \right], \quad F = \frac{1}{m_0} \sum_r \frac{|\langle S|p_x|u_r\rangle}{E_c - E_r} \tag{2.77}$$

where $E_g^{bi}(\varepsilon_i)$ is strain dependent core bandgap, Δ_i is the spin–orbit coupling, E_P is Kane's matrix element, and F is the Kane parameter for the effects of higher conduction band levels. Detailed theoretical studies [85] suggest that the difference between eight band *k.p approximation* calculations and experimental values of electron mass is about 10–20%.

Table 2.6 Comparison of measured and predicted core band gaps of CdSe/ZnS, CdSe/CdS, ZnSe/ZnS, and ZnSe/CdS QDs at 300 K

Core/shell QD	Measured core band gap (eV)	Calculated core band gap (eV)			
		$d_m = 1.5d_i$ (isotropic)	$d_m = 1.5d_i$ (anisotropic)	$d_m = 2d_i$ (isotropic)	$d_m = 2d_i$ (anisotropic)
ZnSe/ZnS	3.080 [107]	2.9322	3.0359	2.9645	3.0904
ZnSe/CdS	2.850 [107]	2.8999	2.8837	2.8868	2.8656
CdSe/ZnS	2.255 [108]	2.1623	2.2598	2.2099	2.3188
CdSe/CdS	2.309 [108]	1.9872	2.0131	2.0015	2.0329

Table 2.7 Comparison of calculated and extracted valence band offsets in ZnSe/ZnS and ZnSe/CdS heterostructure core/shell QDs with Type II interface band alignment (with $d_m = 2d_i$)

Nanostructure	Calculated (Extracted) valence band offsets (eV)				
Type II Core/shell QD	Calculated (Extracted)	$d_i = 3.0$ nm (Isotropic)	$d_i = 3.0$ nm (Anisotropic)	$d_i = 4.0$ nm (Isotropic)	$d_i = 4.0$ nm (Anisotropic)
ZnSe/ZnS	mBJLDA (EMA)	0.1563 (0.0408)	0.1695 (0.1799)	0.1563 (–0.0479)	0.1695 (0.0911)
ZnSe/CdS	mBJLDA (EMA)	– 0.0133 (0.0235)	– 0.0181 (–0.0025)	– 0.0133 (–0.0653)	– 0.0181 (–0.0913)

2.9.1 Comparison of Predicted and Extracted Band Offsets

The qualitative understanding of elastic strain effects on core band gaps are described by (2.33), (2.54) and on band offsets are described by (2.57) and (2.58) in spherical core/shell heterostructure QDs with Type II interface band alignment, respectively. Using calculated valence band widths due to density functional theory based on mBJLDA [96] and other material parameters [25] given in Table 1.2 and lattice, thermal, and anisotropy mismatches given in Table 1.3, hydrostatic strain effects on band offsets of ZnSe/ZnS and ZnSe/CdS core/shell heterostructure QDs with Type II band alignment are calculated from (2.63a) and (2.63b) for core (shell) diameter $d_{i\,=}$ 3.00 ($d_m = 6.00$ nm) and 4.00 nm ($d_m = 8.00$ nm). The calculated and extracted valence band offsets are compared in Table 1.7 for 3.00 and 4.00 nmcore diameters. Comparison in Table 1.7 suggests that using isotropic and anisotropic elastic moduli give positive and negative values for valence band offsets in ZnSe/ZnS and ZnSe/CdS heterostructures, respectively. On the other hand, using isotropic elastic moduli, (2.65a) gives positive values for 3.00 nm and negative values for 4.00 nm, and they are small for these heterostructures. With anisotropic elastic moduli, (2.65a) gives similar results. Since valence band edge deformation potentials of constituents are small, difference between calculated and extracted valence band offsets is about few meV. Using isotropic and anisotropic elastic moduli do not have significant effect on strain shift in valence band offsets. Considering the uncertainties used as input parameters, there is good agreement between predicted and extracted valence band offsets in Type II core/shell heterostructures.

We should point out that in calculating core bandgaps and band offsets in the framework of parabolic two band *k.p approximation*, we assumed that elastic strain effect on electron and hole effective masses is negligible. However, more realistic calculation of strain effects on electronic properties such as core band gaps and band offsets of core/shell QDs with Type I and Type II heterointerface band alignment would require replacing the parabolic *two band effective mass approximation* with a more comprehensive *eight band k.p approximation* [85–88], in which higher conduction band energy levels and degeneracy of valence band energy levels must be considered.

2.9.2 Conclusions and Suggestions

Using the conventional stress–strain relations for thermo-elastic bodies, coupled with lattice mismatch induced discontinuity in elastic strain at core/shell heterointerface so called *shrink fit* condition, analytical expressions are derived for elastic strain acting on core and shell sides of heterointerfaces and its effects on core bandgaps and band offsets in type I and type II spherical core/shell heterostructure QDs as a function of core and shell diameters at any temperature. Since potential barrier in the conduction band and the potential step in the valence band at the interface significantly influence carrier transport and device performance, reliable and accurate modeling of interface strain effects on band structure across the nanoscale core/shell heterointerface is crucial for the designing the functionality of spherical core/shell heterostructure QDs and their optimization for maximum performance in nanoscale electronic and optical devices. In achieving that goal, we found that the strain modified parabolic two band effective mass approximation can be used to determine valence and conduction band offsets in Type II core/shell QDs provided core band gaps are experimentally known. Furthermore, we also found that the anisotropic elastic moduli can play important role in determining band offsets and core band gaps and core and shell morphologies in spherical core/shell QDs when lattice mismatch is large.

We can conclude that if both conduction and valence band offsets are positive, core/shell QD has a *Type I* interface band alignment. When valence and conduction band offsets are both negative, core/shell QD has inverted *Type I* interface band alignment. When both band offsets are negative and positive, core/shell QD has first kind of *Type II* interface band alignment Negative valence band offset and positive conduction band offset imply second kind of Type II interface band alignment. Furthermore, we show that core and shell diameters can play important role in determining core and shell morphologies of spherical core/shell QDs. Excellent agreement found between model predictions and measured core bandgaps and extracted band offsets suggests that proposed model can be a good predictive process design and simulation tool for spherical core/shell heterostructure QDs for high temperature electronic and optical device applications.

Acknowledgements The author greatly acknowledges the financial support by the Research Foundation of İstanbul Technical University (İTÜ BAP Project No: 39029).

References

1. H. Ünlü, N.J.M. Horing, J. Dabrowski, *Low Dimensional and Nanostructured Materials and Devices* (Springer, 2015)
2. H. H. Kroemer, Proc. IRE **45**, 1535 (1957)
3. F. Frank, J. van der Merwe, Proc. R. Soc. London Ser. A **198**, 205 (1949)
4. J. Matthews, W. Jesser, Philos. Mag. **15**, 1097 (1967)
5. J. Eshelby, Proc. R. Soc. London Ser. A **241**, 376 (1957)

6. G.C. Osbourne, Phys. Rev. B **27**, 5126 (1983)
7. N. Lucas, H. Zabel, H. Morkoç, H. Ünlü, Appl. Phys. Lett. **52**, 2117 (1988)
8. A.. R. Kortan, R. Hull, R. L. Opila, M. G. Bawendi, M. L. Steigerwald, P. J. Carroll, L.E. Brus, A. Am. Chem. Soc. **112**(4), 1327–1332 (1990)
9. Y. Wang, N. Herron, Phys. Rev. B **42**, 7253 (1990).
10. C.B. Murray, D.J. Norris, M.G. Bawendi, J. Am. Chem. Soc. **115**, 8706 (1993)
11. T. Vossmeyer, L. Katsiks, M. Giersig, I.G. Popovic, K. Diesner, A. Chemseddine, A. Eychmüller, H. Weller, J. Appl. Phys. **90**, 265 (2001)
12. A. van Dijcken, A.H. Janssen, M.H.P. Smitsmans, D. Vanmaekelbergh, A. Meijerink, Chem. Mater. **10**, 3513 (1998)
13. W.W. Yu, L. Qu, W. Guo, X. Peng, Chem. Mater. **15**, 2854 (2003)
14. V.N. Soloviev, A. Eichhöfer, D. Fenske, U. Banin, J. Am. Chem. Soc. **122**, 2673 (2000)
15. O.I. Mićić, C.J. Curtis, K.M. Jones, J.R. Sprague, A.J. Nozik, J. Phys. Chem. **98**, 4966 (1994)
16. A.A. Guzelian, J.E.B. Katari, A.V. Kadavanich, U. Banin, K. Hamad, E. Juban, A.P. Alivisatos, R.H. Wolters, C.C. Arnold, J.R. Heath, J. Phys. Chem. **100**, 7212 (1996)
17. O.I. Mićić, K.M. Jones, A. Cahill, A.J. Nozik, J. Phys. Chem. B **102**, 9791 (1998)
18. O.I. Mićić, S.P. Ahrenkiel, A.J. Nozik, Appl. Phys. Lett. **78**, 4022 (2001)
19. A.A. Guzelian, U. Banin, A.V. Kadavanich, X. Peng, A.P. Alivisatos, Appl. Phys. Lett. **69**, 1432 (1996)
20. U. Banin, Y.W. Cao, D. Katz, O. Millo, Nature (London) **400**, 542 (1999)
21. Y. Wang, A. Suna, W. Mahler, R. Kasowski, J. Chem. Phys. **87**, 7315 (1987)
22. N.F. Borrelli, D.W. Smith, J. Non-Cryst, Solids **180**, 25 (1994)
23. R. Thielsch, T. Böhme, R. Reiche, D. Schläfer, H.D. Bauer, H. Böttcher, Nanostruct. Mater. **10**, 131 (1998)
24. E.K. Lipovskii, V. Petrikov, I. Kang, A. Olkhovets, T. Krauss, M. Thomas, J. Silcox, F. Wise, Q. Shen et al., Appl. Phys. Lett. **71**, 3406 (1997)
25. A.D. Andreev, E.V. Kolobkova, A. Lipovskii, J. Appl. Phys. **88**, 750 (2000)
26. B.L. Wehremberg, C. Wang, P. Guyot-Sionnest, J. Phys. Chem. B **106**, 10634 (2002)
27. H. Yu, J. Li, R.A. Loomis, P.C. Gibbons, L.W. Wang, W.E. Buhro, J. Am. Chem. Soc. **125**, 16168 (2003)
28. Z. Tang, N.A. Kotov, M. Giersig, Science **297**, 237 (2002)
29. M.S. Gudiksen, J. Wang, C.M. Lieber, J. Phys. Chem. B **105**, 4062 (2001)
30. M.S. Gudiksen, J. Wang, C.M. Lieber, J. Phys. Chem. B **106**, 4036 (2002)
31. H. Yu, J. Li, R.A. Loomis, L.W. Wang, W.E. Buhro, Nat. Mater. **2**, 517 (2003)
32. X. Peng, L. Manna, W. Yang, J. Wickham, E. Scher, A. Kadavanich, A.P. Alivisatos, Nature (London) **404**, 59 (2000)
33. L.L. Li, J. Hu, W. Yang, A.P. Alivisatos, Nano Lett. **1**, 349 (2001)
34. D. Katz, T. Wizansky, O. Millo, E. Rothenberg, T. Mokari, U. Banin, Phys. Rev. Lett. **89**, 086801 (2002)
35. S.P. Ahrenkiel, O.I. Mićić, A. Miedaner, C.J. Curtis, J.M. Nedeljković, A.J. Nozik, Nano Lett. **3**, 833 (2003)
36. S. Kan, T. Mokari, E. Rothemberg, U. Banin, Nat. Mater. **2**, 155 (2003)
37. A.L. Efros, A.L. Efros, Sov. Phys. Semicond. **16**, 772 (1982)
38. L.E. Brus, J. Chem. Phys. **80**, 4403 (1984)
39. Y. Nosaka, J. Phys. Chem. **95**, 5054 (1991)
40. J.M. Ferreyra, C.R. Proetto, Phys. Rev. B **60**, 10672 (1999)
41. K.K. Nanda, F.E. Kruis, H. Fissan, Nano Lett. **1**, 605 (2001)
42. K.K. Nanda, F.E. Kruis, H. Fissan, J. Appl. Phys. **95**, 5035 (2004)
43. G. W. Bryant, Phys. Rev. B **37**, 8763 (1988)
44. Y. Kayanuma, Phys. Rev. B **38**, 9797 (1988)
45. Y. Kayanuma, H. Momiji, Phys. Rev. B **41**, 10261 (1990)
46. H. Fu, L.W. Wang, A. Zunger, Phys. Rev. B **57**, 9971 (1998)
47. G.E. Tudury, M.V. Marquezini, L.G. Ferreira, L.C. Barbosa, C.L. Cesar, Phys. Rev. B **62**, 7357 (2000)

48. J. Pèrez-Conde, A.K. Bhattacharjee, Solid State Commun. **110**, 259 (1999)
49. C. Delerue, G. Allan, M. Lannoo, Phys. Rev. B **48**, 11024 (1993)
50. M.V. Rama Krishna, R.A. Friesner, Phys. Rev. Lett. **67**, 629 (1991)
51. L.W. Wang, A. Zunger, J. Phys. Chem. **98**, 2158 (1994)
52. L.W. Wang, A. Zunger, Phys. Rev. B **53**, 9579 (1996)
53. E. Rabani, H. Balázs, J. Berne, L.E. Brus, J. Chem. Phys. **110**, 5355 (1999)
54. J. Li, L.W. Wang, Nano Lett. **3**, 101 (2003)
55. B. Zorman, R.A. Friesner, J. Chem. Phys. **118**, 5937 (2003)
56. H. Fu, A. Zunger, Phys. Rev. B **56**, 1496 (1997)
57. J. Li, L.W. Wang, Nano Lett. **4**, 29 (2004)
58. A.J. Williamson, A. Zunger, Phys. Rev. B **61**, 1978 (2000)
59. R.S. Kane, R.E. Cohen, R. Silbey, J. Phys. Chem. **100**, 7928 (1996)
60. A.J. Read, R.J. Needs, K.J. Nash, L.T. Canham, P.D.J. Calcott, A. Qteish, Phys. Rev. Lett. **69**, 1232 (1992)
61. A.J. Read, R.J. Needs, K.J. Nash, L.T. Canham, P.D.J. Calcott, A. Qteish, Phys. Rev. Lett. **70**, 2050 (1993)
62. R.J. Needs, S. Bhattacharjee, K.J. Nash, A. Qteish, A.J. Read, L.T. Canham, Phys. Rev. B **50**, 14223 (1994)
63. T. Trindade, P. O'Brien, N.L. Pickett, Chem. Mater **13**, 3843 (2001)
64. Y. Masumoto, K. Sonobe, Phys. Rev. B **56**, 9734 (1997)
65. K.V. Shah, V. Sudarsan, M. Goswami, A. Sarkar, S. Manikandan, R. Kumar, B.I. Sharma, V.K. Shrikhande, G.P. Kothiyal, Bull. Mater. Sci. **26**, 715 (2003)
66. S. Ninomiya, S. Adachi, J. Appl. Phys. **78**, 1182 (1995)
67. G.A.M. Alcalde, W.G., Semicond. Sci. Technol.15, 1082 (2000)
68. Z.C. Feng, L.S. Kim, S. Perkowitz, Y.P. Feng, H.C. Poon, K.P. Williams, G.D. Pitt, J. Cryst. Growth **138**, 239 (1994)
69. F. Vallée, Phys. Rev. B (1994)
70. W.Z. Shen, H.Z. Wu, P.J. McCann, J. Appl. Phys. **91**, 3621 (2002)
71. M. Ziesmann, D. Heitmann, L.L. Chang, Phys. Rev. B **35**, 4541 (1987)
72. M. Grundmann, O. Stier, D. Bimberg, Phys. Rev. B **52**, 11969 (1995)
73. D.A. Faux, J.R. Downes, E.P. O'Reilly, J. Appl. Phys. **82**, 3754 (1997)
74. A.J. Williamson, A. Zunger, Phys. Rev. B **58**, 6724 (1998)
75. X. Zhang, P. Sharma Phys. Rev. B **72** 195345 (2005)
76. T.O. Cheche, V. Barna, Y.-C. Chang, Superlattices Microstruct. **60**, 475 (2013)
77. C. Pryor, J. Kim, L.-W. Wang, A.J. Williamson, A. Zunger, J. Appl. Phys. **83**, 2548 (1998)
78. W. Sukkabot, Mater. Sci. Semicond. Process. **47**, 57 (2016)
79. G.T. Einovell, Phys. Rev. B **45**, 3410–3417 (1992)
80. P.E. Lippens, M. Lannoo, Phys. Rev. B **39**, 10935–10942 (1989)
81. H. Tan Pham, L. Van Duong, M.T. Nguyen, J. Phys. Chem. C **118**(41), 24181–24187 (2014)
82. A. Franceschetti, A. Zunger, Phys. Rev. Lett. **78**, 915–918 (1997)
83. . D. Valerini, A. Cretí, M. Lomascolo, L. Manna, R. Cingolani, M. Anni, Phys. Rev. B **71**, 235409 (2005)
84. P. Jing, Jinju Zheng, Micho Ikezawa, Xueyan Liu, Shaozhe Lv, Xianggui Kong, Jialong Zhao, and Yasuaki Masumoto. J. Phys. Chem. C **113**, 13545–13550 (2009)
85. C.E. Pryor, M.-E. Pistol, Phys. Rev. B **72**, 205311 (2005)
86. D.J. Norris, A.L. Efros, M. Rosen, M.G. Bawendi, Phys. Rev. B **53**, 1637 (1996)
87. D.J. Norris, M.G. Bawendi, Phys. Rev. B **53**, 1638 (1996)
88. . D.J. Norris, Chapter 2, in *Nanocrystal Quantum Dots*, ed. by V.I. Klimov, 2nd ed. (CRC Press, Boca Raton, 2010), pp. 63–96
89. R.M. Martin, Phys. Rev. B **1**, **4005**, 197
90. A.F. Bower, *Applied Mechanics of Solids* (CRC Press, 2009)
91. J.R. Barber, *Elasticity*, 3rd edn. (Springer, Dordrecht, 2010)
92. J.P. Hirt, J. Lothe, *Theory of Dislocations* (McGraw-Hill, 1968)

93. S.I. Ranganathan, M. Ostoja-Starzewski, Phys. Rev. Lett. **101**, 055505 (2008). The authors define so called universal anisotropy index: , where and are average shear modulus and and are called Vogit and Reuss average bulk modulus of bulk solid materials [17], with and . In case of cubic crystals, reduces to , which is equivalent of Zener's elastic anisotropy ratio A, given by (2.17). Using elastic constants in Table 1, one finds that yields the same values given by (2.17) for group III-V and II-VI compounds with cubic symmetry.
94. A.S. Saada, *Elasticity Theory and Applications* (Pergamon, New York, 1974)
95. H.H. Gürel, Ö. Akıncı, H. Ünlü, Comput. Mater. Sci. **33**, 269 (2005)
96. H. Gürel, H. Ünlü, Mater. Sci. Semicond. Process. **36**, 1619 (2013)
97. . H. Ünlü, AIP Conf. Proc. **1935**, 050003-1–050003-4 (2018)
98. H. Ünlü, Eur. Phys. J. Appl. Phys. **86**, 30401 (2019)
99. H. Ünlü, J. Mater. Electron. Devic. **1**, 24–33 (2019)
100. K. Gong, D.F. Kelley, J. Chem. Phys. **141**, 194704 (2014)
101. X. Cai, H. Mirafzal, K. Nguyen, V. Leppert, and Kelley. J. Phys. Chem. C **116**, 8118 (2012)
102. H. Ünlü, Solid State Electron. **35**, 1343 (1992)
103. O. Madelung (ed.), *Numerical Data and Functional Relationships in Science and Technology*, vol. 17/a (Springer, 1982) and Part d of vol. 17, (1984)
104. Y.P. Varshni, Physica **34**, 149 (1967)
105. E.O. Kane, J. Phys. Chem. Solids **1**, 249 (1957)
106. H.E. Ghassan, Al-Shabeeba, A.K. Arof, Eur. Phys. J. Plus **128**, 153 (2013)
107. N. Gheshlaghi, H.S. Pisheh, H. Ünlü, Superlattices Microstruct. **111**, 156 (2017)
108. H.S. Pisheh, N. Gheshlaghi, H. Ünlü, Physica E **85**, 334 (2017)
109. K. Abhishek Joshi, Y. Narsingi, M.O. Manasreh, E.A. Davis, B.D. Weaver, Appl. Phys. Lett. **89**, 131907 (2006)

Chapter 3
Synthesis, Characterization and Modelling of Colloidal Quantum Dots

Md. Rezaul Karim, Mesut Balaban, Hakan Aydın, Hilmi Ünlü, and M. Hikmet Yükselici

Abstract Colloidal semiconductor nanocrystals with their diameters range between 2–10 nm have received great theoretical and experimental interest for both optical and electronical applications such as solar cells, light emitting diodes (LEDs), lasers and fluorescence imaging over the last few decades due to their size dependent optical, physical and chemical properties. In this chapter, we present a review about the experimental and theoretical study about strain effects on core band gap and diameter of spherical bare CdSe core and CdSe/ZnS core/shell quantum dots (QDs) synthesized by using colloidal technique at varying temperatures. We will discuss the results of the structural, optical and dielectric characterizations. High resolution transmission electron microscopy (HRTEM) and x-ray diffraction (XRD) characterizations indicate that CdSe and CdSe/ZnS QDs have average particle sizes about 3.50 nm and 4.84 nm, respectively. Ultraviolet visible (UV–Vis) absorption and fluorescence emission spectroscopy measurements of first optical peak energies show that the compressive strain causes an increase (decrease) in the core band gap (diameter) of spherical CdSe/ZnS core/shell QDs at any temperature. Elastic strain modified effective mass approximation (EMA) predicts that there is a parabolic decrease (increase) in the core bandgap (diameter) of QDs with temperature. The diameter of spherical bare CdSe core and CdSe/ZnS core/shell QDs calculated by using strain modified EMA, with core bandgap extracted from absorption spectra are in excellent agreement with HRTEM data.

Md. Rezaul Karim (✉) · H. Aydın · H. Ünlü
Faculty of Science and Letters, Department of Physics Engineering, İstanbul Technical University, 34485 Maslak, İstanbul, Turkey

H. Ünlü
e-mail: hunlu@itu.edu.tr

Md. Rezaul Karim
Department of Physics, Dhaka University of Engineering & Technology, Gazipur, 1700 Gazipur, Bangladesh

M. Balaban · M. H. Yükselici
Faculty of Science and Letters, Department of Physics, Yıldız Technical University, 34220 Esenler, İstanbul, Turkey

H. Ünlü and N. J. M. Horing (eds.), *Progress in Nanoscale and Low-Dimensional Materials and Devices*, Topics in Applied Physics 144,
https://doi.org/10.1007/978-3-030-93460-6_3

3.1 Introduction

Colloidal semiconductor nanocrystals, also known as quantum dots (QDs), with their diameters range between 2–10 nm have gained remarkable interest for both optical and electronical applications such as solar cells [1–4], light emitting diodes (LEDs) [5–7], lasers [8, 9], fluorescence imaging and for fundamental studies [10–15] over the last few decades because of their size-dependent physical and chemical properties. These small-sized nanostructures have a large surface-to-volume ratio and constitute a class of materials intermediate between molecular and bulk forms of matter. QDs can be synthesized in core or core–shell forms due to interest and generally are made from III–V and II–VI compounds family of semiconductors in the periodic table. CdSe core and CdSe/ZnS(CdS) core/shell nanocrystals, with a direct band gap, are among the mostly investigated nanocrystals due to their bright luminescence in the visible range of optical spectrum with varying particle size.

Advances in the synthesis and fabrication of materials with dimensions approaching size of an atom have made it possible to control and manipulate the degrees of freedom of charge carriers to produce charge confinement in nanoscale structures such as one-dimensional nanowires (NWs) and zero-dimensional quantum dots (QDs). Because of the strong confinement imposed in all three spatial directions, QDs behave similarly to atoms. Due to similar diameters of NCs that in the scale of de Broglie wavelength or Bohr exciton radius, size quantization effect is observed. The conduction band or valence band energy levels are split into so called "sub-bands" or discrete energy levels depending on the dimensionality of the confined structure. As the size of nanocrystals gets smaller, the effective band gap and energy amount to be excited increase whereas the emission wavelength decreases.

When CdSe and ZnS are brought in contact to form a heterostructure CdSe/ZnS core/shell QD by using a crystal growth technique, conduction and valence band edges of core and shell regions are aligned in a way that an electron–hole pair excited near heterointerface (Type I structure) tend to localize in the core region. The exciton energy in Type I core/shell QDs is the result of direct exciton transition inside the core region (e.g., CdSe/ZnS and CdSe/CdS QDs). When electrons and holes are spatially separated between core and shell, it is said that the QD exhibits type-II confinement (e.g., CdSe/CdTe QDs). Figure 3.1 shows the schematic cross-sectional view and band diagram of a spherical heterostructure. The difference in band gaps of CdSe and ZnS is accommodated by the spike ΔE_c in conduction band and potential step ΔE_v in valence band at heterointerface, causing the electron and hole confinement in the CdSe core region.

Once an electron–hole pair is created by the absorption of a photon, they interact with each other by means of their opposite charges, forming a quasi-particle called "exciton". The total energy of an exciton indicates the color (wavelength) of the light emitted by the quantum systems at nanoscale. The dispersion or energy versus momentum (which is proportional to the wave vector k) curves are parabolic just as for classical free particles with some modifications.

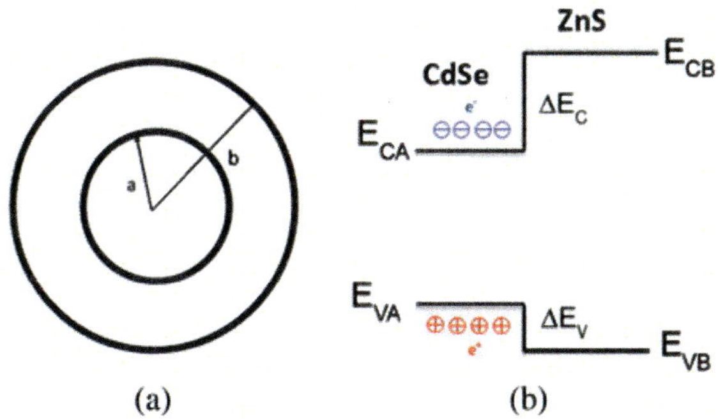

Fig. 3.1 Schematic cross-sectional view (**a**) and band diagram (**b**) of pseudomorphic CdSe/ZnS heterostructure core/shell quantum dot

As a nanoscale heterostructure core/shell QD is formed between two semiconductors with different lattice constants and thermal expansion coefficients, elastic strain develops across heterointerface. Elastic strain can modify the structural, electronic, optical and dielectric properties of constituent semiconductors in a way that is not seen in two-dimensional quantum wells and super lattices. If core lattice constant is greater than shell lattice constant, core (shell) region will be under compressive (tensile) elastic strain, resulting in increase (decrease) in core (shell) band gap energy and decrease (increase) in diameter. Reliable and precise determination of magnitude of interface strain effects on structural and electronic properties of core/shell QDs become important in determining their tunable properties such as core bandgap and diameter, which are essential for predicting their potential as electronic and optical devices.

In this chapter, it is aimed to discuss the structural and optical characterizations and parabolic two band *k.p effective mass approximation* modelling of the elastic strain effects on the bandgap and diameter of both CdSe core and CdSe/ZnS core/shell nanocrystals that are prepared at relatively low temperatures and characterizing them by altering the temperature, time and initial Cd:Se precursor ratios. The bare CdSe core and CdSe/ZnS core/shell QDs are synthesized by using colloidal technique at low temperatures and characterized by using x-ray diffraction (XRD), high resolution transmission electron microscopy (HRTEM), UV–Vis. absorption, and fluorescence emission techniques, respectively. Colloidal synthesis of CdSe core and CdSe/ZnS core/shell QDs is summarized in Sect. 3.2. The results of HRTEM and XRD characterization of size of QDs will be discussed in Sects. 3.3 and 3.4, respectively. The optical UV–Vis absorption and fluorescence emission spectral analysis and dielectric properties of QDs will be presented in Sects. 3.5 and 3.6, respectively. In Sects. 3.7, 3.8 and 3.9 we will discuss the effects of initial Cd:Se precursor ratios and temperature on the nanocrystal formation, emission quality and stability of nanocrystals. In Sect. 3.10 we will discuss the core diameter of QDs calculated by using strain modified two band effective mass approximation, with core bandgap extracted from UV–Vis absorption spectra, will be compared with results of HRTEM analysis.

3.2 Synthesis of CdSe Core and CdSe/ZnS Core/Shell QDs

Owing to the increasing demands for high quality (mono-dispersity, high crystallinity, narrow emission spectrum and high quantum yield) nanocrystals (NCs) it is important to develop low-cost, green and mass producible synthesis routes. Nanocrystals synthesized by a chemical route allow us to control their sizes and distribution. Altering the concentrations of reactants or changing the processing times at different temperatures of production may result in various properties of colloidal NCs [16].

3.2.1 Synthesis of CdSe Core QDs

In the preparation of synthesis of CdSe core NCs, the method developed by He and Gu [17] has been modified to use in our Nanostructure Semiconductor Research Laboratory (NANOSEMLAB). Layout of synthesis can be seen in Fig. 3.2.

The procedure summary is given in Fig. 3.3. At first, cadmium acetate and oleic acid have been dissolved with diphenyl ether in three neck flask. The reaction mixture has been heated 140 °C under stirring and continuous argon flow. When temperature raised to 140 °C, we waited for one hour. Then the mixture was cooled to 70 °C.

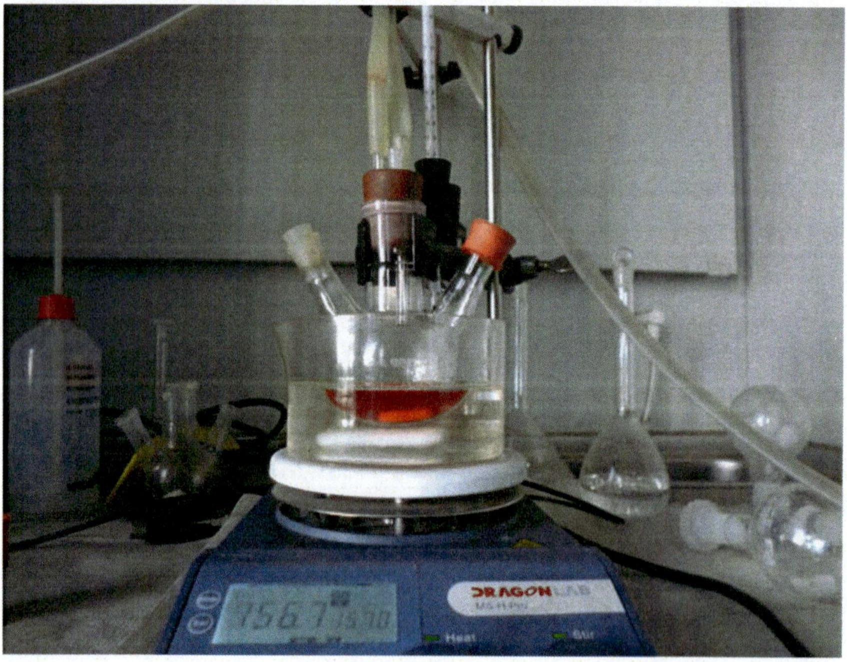

Fig. 3.2 Layout of synthesis of CdSe core and CdSe/ZnS core/shell nanocrystals

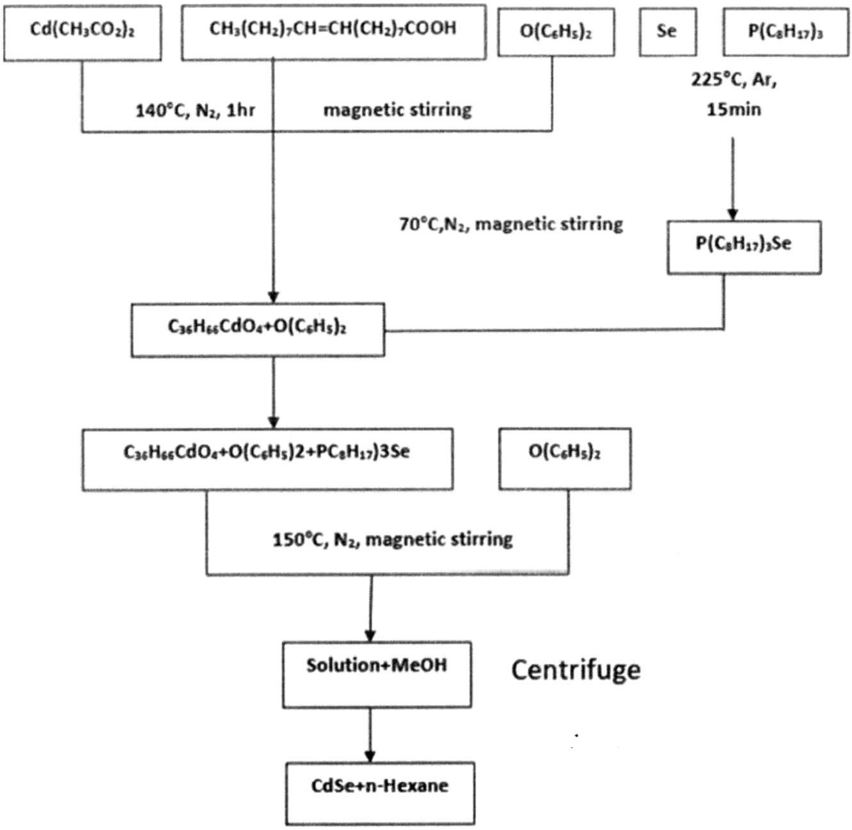

Fig. 3.3 Block diagram for the synthesis of CdSe nanocrystals

After then trioctylphosphine-selenide (TOPSe) solution having 200 °C temperature under the vacuum has been added to mixture and heated to 110–180 °C. Samples have been taken to tubes for 1, 5, 10, 15 and 20 min. After samples cooling to room temperature, the crude solution has washed with methanol and isolated by centrifugation to remove excess insoluble organics and salts that may have formed during the reaction. A precipitation using hexane as the solvents and methanol as the nonsolvent is repeated three times to narrow the size distribution and remove excess organics. After fine isolation of growth CdSe, the precipitation has been dissolved with different volume of hexane. The reaction was monitored with Shimadzu UV-3600 UV-VIS-NIR Spectrophotometer with aliquots taken at different time and temperature. The UV-3600 is equipped with three detectors: a photomultiplier tube (PMT) for the UV-Vis region, and InGaAs and PbS detectors for the NIR region. The three detectors ensure high sensitivity over the entire measurement range and help achieve the world's lowest noise level with <0.00003 Abs noise at 1500 nm. A high-performance double monochromator ensures ultra-low stray light (0.00005% or less at 340 nm) at

Fig. 3.4 CdSe core QDs synthesized at 160 °C with Cd:Se ratio of 1:10 are shown under ambient light

high resolution. Measurement range varies from 185 to 3300 nm, enabling analysis in a wide variety of applications.

In order to investigate the effect of temperature, time and initial ratio of the precursors on nucleation and growth kinetics; different combinations can be tried for synthesis of CdSe core QDs. Size quantization effect is evident in Fig. 3.4. Color transition from light yellow to red indicates the growth of NCs, thus resulting as decrease in the optical absorption band gap.

3.2.2 Growth of ZnS Shells on CdSe Core

The CdSe/ZnS nanocrystals have been synthesized by using modifications of Zhu et al. [18]. The summary of the procedure is given in Fig. 3.5.

Typically, the resulting core solution, Zn $(OAc)_2 \cdot 2H_2O$ (0.085 mmol) and S powder (0.085 mmol) were mixed in the reaction vessel. The reaction volume was adjusted to 15 mL by adding paraffin liquid. Next, with stirring, the mixture was degassed at 80 °C for 20 min. Afterward, temperature was set to 160–170 °C for the shell growth under N_2 atmosphere. To monitor the reaction, aliquots were taken at different times. The reaction mixture was cooled to room temperature after 50 min. To grow shell ZnS with different thicknesses around a CdSe core, a seeding-growth technique [11] was applied. After samples cooling to room temperature, crude solution has washed with methanol and isolated by centrifugation to remove excess insoluble organics and salts that may have formed during the reaction. The reaction was monitored with Shimadzu UV–Vis NIR absorption spectrometer with aliquots taken at different time and temperature.

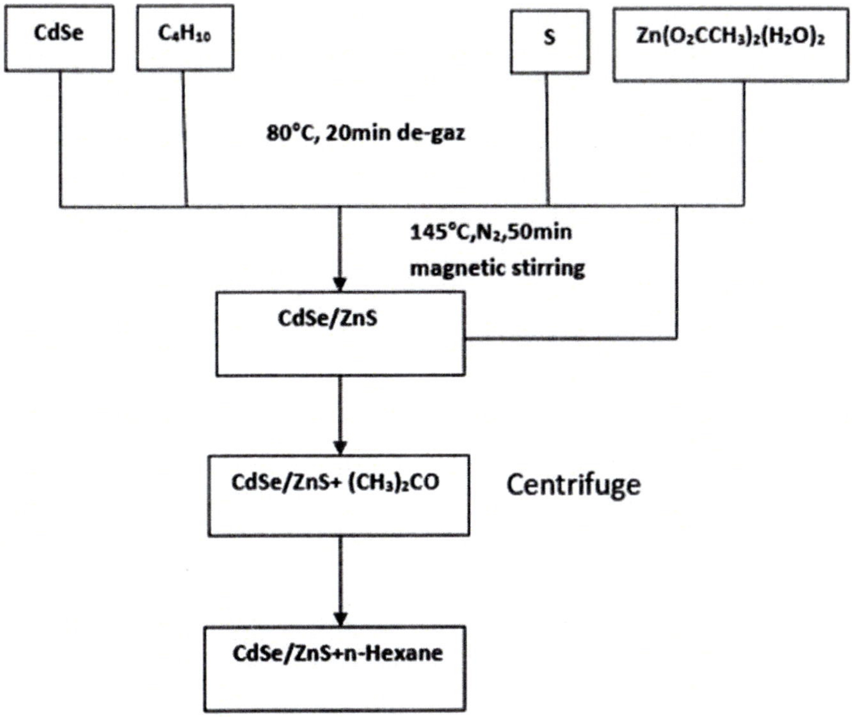

Fig. 3.5 Block diagram for the synthesis of CdSe/ZnS core/shell nanocrystals

3.3 HRTEM Characterization

The structural properties of the synthesized bare CdSe core and CdSe/ZnS core/shell nanocrystals were characterized by using high resolution transmission electron microscopy (HRTEM) with at an acceleration voltage of 200 kV. A drop of dispersed nanocrystals diluted in n-hexane dropped over an amorphous carbon substrate supported on a copper grid of 400 mesh for taking images. Figures 3.6a, b show the HRTEM images of bare CdSe core NCs synthesized at 160 °C for 1 and 20 min of reaction time, respectively. Figure 3.6c, d show the HRTEM images of CdSe/ZnS core/shell NCs synthesized at 160 °C and 170 °C for 20 min of reaction time, respectively. HRTEM images shown in Fig. 3.6 indicate that both the bare CdSe core and CdSe/ZnS core/Shell NCs are uniform in size and shape. The nanocrystal size becomes larger as reaction time is increased. We estimate the average diameters of bare CdSe core and CdSe/ZnS core/shell NCs are 3.50 nm and 4.84 nm, respectively.

One can estimate thickness of ZnS shell by subtracting size of bare CdSe core from that of CdSe/ZnS core/shell QD. However, since changes in core size during shell deposition is inevitable due to strain across core/shell interface, TEM images after the shell deposition may not be a reliable reference to estimate the exact core size.

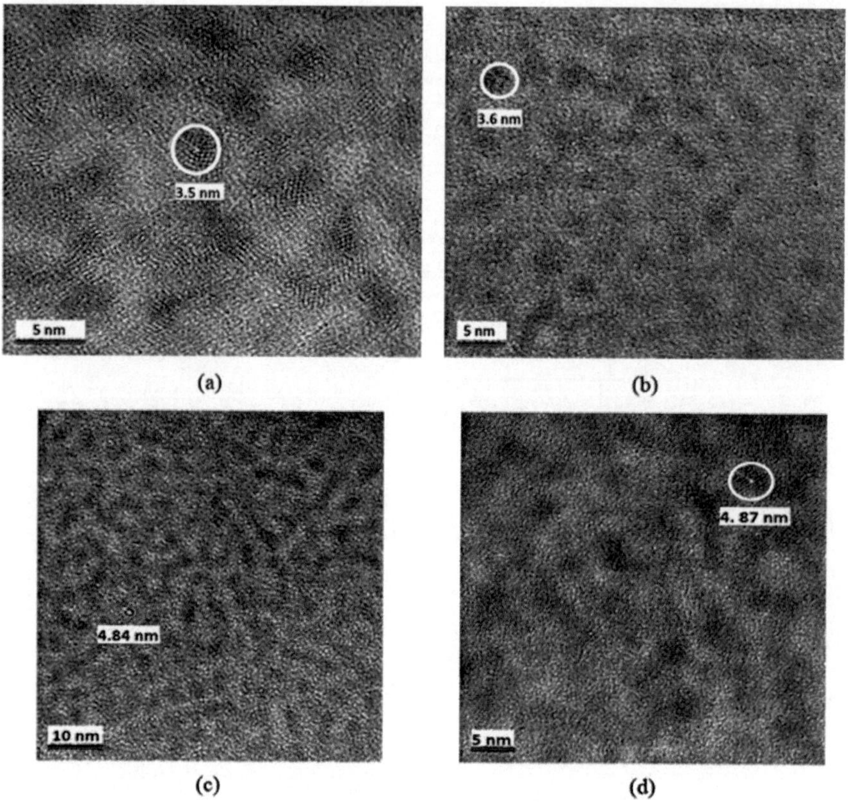

Fig. 3.6 HRTEM images of bare CdSe core QD synthesized at 160 °C for 1 min. (**a**) and 20 min. (**b**) reaction times, and of CdSe/ZnS core/shell QD synthesized at 160 °C (**c**) and 170 °C (**d**) for 20 min. reaction time, respectively

3.4 XRD Characterization

X-ray diffraction technique is commonly used for determining the mean size of single crystal nanoparticles. X'Pert[3] MRD (XL) X-ray diffractometer operating at 45 kV/40 mA using Copper Kα line (λ = 1.5406 Å) was used in the structural characterization of bare CdSe core and CdSe/ZnS core/shell nanocrystals. Purification of NCs in paraffin had to be ensured for a good quality diffraction pattern. As an amorphous phased material, paraffin drastically effects diffraction pattern of produced nanocrystals with size varying between 3.00 and 5.00 nm. Figure 3.7 shows the comparison of XRD patterns of bare CdSe core quantum dots prepared at 170 °C and 190 °C, respectively.

The XRD spectra for two samples (synthesized at 170 °C for 2 min and at 190 °C for 15 min were fitted by a Gausian profile for each peak and a quadratic function for the background. The strong peaks at around $2\theta_{hkl} = 25^0$ and at

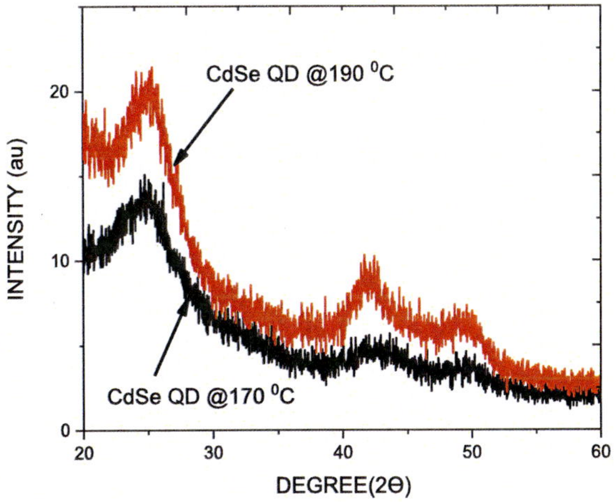

Fig. 3.7 The XRD patterns of bare CdSe core nanocrystals prepared at 170 °C for 2 min. and at 190 °C for 15 min, respectively

between 42–43° are consistent with (002) and (110) planes of CdSe hexagonal crystal structure, respectively, showing the characteristics features of hexagonal crystalline structure according to JCPDS (Joint Committee on Powder Diffraction Standards) data base (file No. 77–2100). The strain is calculated using the relation $\varepsilon = -(d_{hkl}^{nc} - d_{hkl}^{bulk})/d_{hkl}^{bulk}$ and Bragg law for hexagonal structure. We find $d_{002}^{bulk} = c/l = 7.0109/2 = 3.50545A$, $d_{002}^{nc} = 3.5556A$, $\varepsilon = -1.335 \times 10^{-2}$ for sample #2 prepared at 170 °C for 2 min and $d_{002}^{nc} = 3.05079$ with $\varepsilon = -6.39 \times 10^{-4}$ for sample #3 prepared at 190 °C for 15 min, respectively (Table 3.1).

Figure 3.8. compares XRD patterns of uncapped CdSe core and capped with ZnS shell (CdSe/ZnS core/shell), respectively. The XRD pattern of bare CdSe core NCs exhibits broad peaks at 2θ values of 25° related to (111), 42° to (220) and 48° to (311) crystalline plane for low temperature synthesized zinc-blende CdSe JCPDS data base (file No. 77–2100). There is a slight shift in the XRD pattern after capping CdSe core with ZnS shell to form CdSe/ZnS core/shell NC. The broad nature of the peaks suggests nanocrystalline particles.

Here, we point out that there is no drastic change in the diffraction patterns of CdSe/ZnS core/shell NCs relative to that of bare CdSe core NCs. This is because

Table 3.1 XRD plane assignments for core CdSe QDs

Sample	145 °C for 15 min		170 °C for 2 min		190 °C for 15 min	
Assignment	$2\theta_{hkl}$	$\Delta(2\theta_{hkl})_{size}$	$2\theta_{hkl}$	$\Delta(2\theta_{hkl})_{size}$	$2\theta_{hkl}$	$\Delta(2\theta_{hkl})_{size}$
(002)	–	–	25.0479	2.4618	25.3714	1.7702
(110)	–	–	42.6989	4.766	42.5080	2.3819

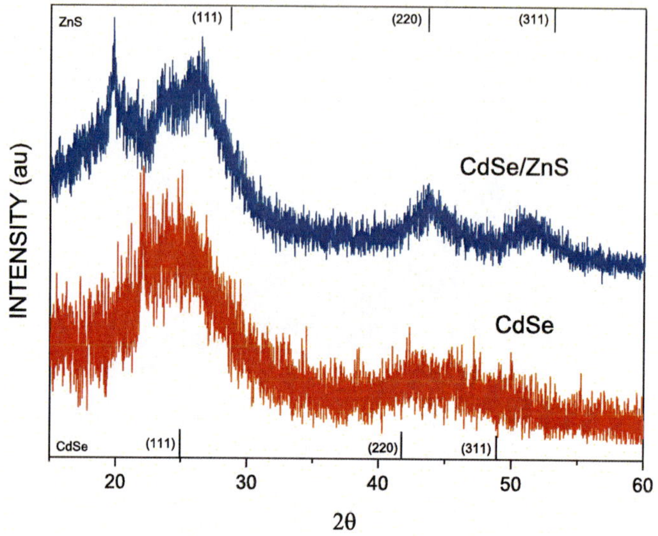

Fig. 3.8 The XRD pattern of bare CdSe core and CdSe/ZnS core/shell nanocrystals

the thickness of ZnS shell layer is very small (about one monolayer). Although during purification almost all of paraffin was extracted from aliquots, paraffin as an amorphous element suppresses peaks in the XRD pattern. The XRD patterns in Fig. 3.8 provide strong evidence that the nanocrystals prepared at considerably low temperatures have a zinc-blende (ZB) crystal structure. No reflection pattern from wurtzite lattice structure (102) at $2\theta \approx 35°$ and (103) at $2\theta \approx 46°$) is found in the XRD pattern of our nanocrystals, which is considered as a further evidence that both bare CdSe core and CdSe/ZnS core/shell nanocrystals have zinc-blende crystal structure. Table 3.2 gives the detailed analysis of the XRD spectra of CdSe/ZnS core/shell QDs and CdSe QD.

Diameter of uncapped (bare core) and capped (core/shell) nanocrystals and elastic strain can be estimated by using so called Williamson-Hall (W–H) analysis of XRD profile, which supposes that particle diameter (d) and strain (ε) contribute to the line broadening and defined as

Table 3.2 Details of W–H parameters for bare CdSe core and CdSe/ZnS core/shell QDs

QD	2θ	2θ (rad)	FWHM	FWHM (rad)	ßcosθ	sinθ
CdSe/ZnS	24.737	0.4315	3.500	0.061	0.059	0.214
	41.4811	0.723	3.418	0.059	0.055	0.353
	49.203	0.858	3.439	0.06	0.054	0.416
CdSe	24.261	0.423	8.353	0.145	0.132	−0.422

$$\frac{\beta \cos\theta}{K\lambda} = \frac{1}{d} + 4\varepsilon \frac{\sin\theta}{K\lambda} \tag{3.1}$$

where d is the coherent scattering length (nanocrystal size); K is a constant whose value is often taken as 0.9; and ε internal strain in percent (%). β is the integral width of the sample (in rad) calculated in the first step as full width at half-maximum (FWHM) by the Gaussian fitting. A plot is drawn with $\sin\theta$ along x-axis and $\beta \cos\theta$ along y-axis for as-prepared CdSe/ZnS core/shell nanocrystal, as shown in Fig. 3.9. From the linear fit to the data in Table 3.2, strain from the slope of the fit is about −2.56% and using (3.1) we find diameter of CdSe/ZnS core/shell QD to be about 4.67 nm.

The estimated nanocrystal size is about 4.30 nm. The magnitude of the estimated compressive strain in the CdSe core of CdSe/ZnS core/shell nanocrystal is about –2.56%. Furthermore, (3.1) suggests that CdSe core region of the CdSe/ZnS core/shell nanoscale heterostructure is under compressive strain with a magnitude of 3.95% due to the 12% lattice mismatch at the CdSe/ZnS interface, which is in good agreement with the strain estimated from the XRD measurement. Equation (3.1) can also be used to determine core size from XRD data provided interface strain is theoretically known. Rewriting (3.1) for core diameter one finds the following relation

$$d = \frac{k\lambda}{\beta \cos\theta - 4\varepsilon \sin\theta} \tag{3.2}$$

The magnitude of compressive strain in CdSe core of CdSe/ZnS core/shell quantum dots estimated from W–H analysis is about 2.56% although the lattice

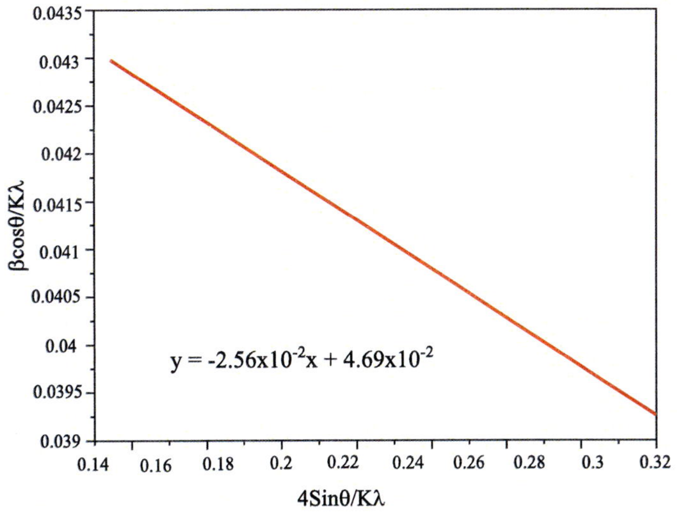

Fig. 3.9 WH plot of XRD pattern of CdSe/ZnS core/shell nanocrystal

mismatch between CdSe core and ZnS shell is about $\Delta a/a = (a_i - a_m)/a_m = 12.3\%$. This result is confirmed with the recent theoretical strain model [19] which considers total strain across core/shell heterointerface as sum of the thermal strain and the elastic strain produced by the resistance of the medium to thermal expansion, and modifies the Eshelby's inclusion strain in spherical heterostructure core/shell quantum dot as

$$\varepsilon_i = \frac{a_i(\varepsilon_i) - a_i}{a_i} = -\frac{2E_m(1-2\nu_i)[\varepsilon_{im}(T) + (\alpha_i - \alpha_m)T]}{E_i(1+\nu_m) + 2E_m(1-2\nu_i)} + \alpha_i(T)T \quad (3.3)$$

where $\varepsilon_{im} = (a_i(T) - a_m(T))/a_m(T)$ is the temperature dependent lattice mismatch with $a_i(T) = a_i(1+\alpha_i\Delta T)$ and $a_m(T) = a_m(1+\alpha_m\Delta T)$. Here a_i and a_m are lattice constants and α_i and α_m are the linear thermal expansion coefficients of core and shell at room temperature. $\varepsilon_{th} = (\alpha_i - \alpha_m)\Delta T$ is so called the thermal strain, due to the difference between α_i and α_m of core and shell. E_i (E_m) is Young's modulus and ν_i (υ_m) is Poisson's ratio of core (shell) semiconductor in bulk form, defined as $E = (C_{11} - C_{12})(C_{11} + 2C_{12})/(C_{11} + C_{12})$ and $\nu = C_{12}/(C_{11} + C_{12})$, where C_{11} and C_{12} are elastic stiffness constants.

Since linear thermal expansion coefficients of CdSe and ZnS are close to each other, and in colloidal synthesis difference between growth temperature and room temperature is not very high so that thermal strain contribution in (3.3) can be neglected. Using elastic constants $C_{11} = 10.2$, 6.67 GPa and $C_{12} = 6.46$, 4.63 GPa, linear thermal expansion coefficients $\alpha_i = 7.30 \times 10^{-6} K^{-1}$, $\alpha_m = 4.05 \times 10^{-6} K^{-1}$ and lattice constants 0.607 and 0.41 nm for ZnS and CdSe (3.3) gives $\varepsilon_i = -3.65\%$ interface strain at CdSe/ZnS heterointerface, which is opposite and smaller that lattice mismatch $\Delta a/a = (a_i - a_m)/a_m = 12.3\%$. Substituting $\varepsilon_i = -3.65\%$ in (3.1) with K = 0.25, one obtains d = 4.26 nm for core diameter of CdSe/ZnS heterostructure core/shell QD. This is in excellent agreement with core diameter obtained from HRTEM analysis discussed in Sect. 3.3.

3.5 Optical Absorption and Emission Characteristics

3.5.1 UV–Vis Characterization

Colloidal semiconductor QDs made from II-VI and III-V groups of the periodic table are found such a new class of fluorescent labels that is frequently employed in fluorescence imaging [20, 21]. Shimadzu UV-3600 UV–VIS-NIR Spectrophotometer is used to measure the UV–Vis absorption spectra of bare CdSe core and CdSe/ZnS heterostructure core/shell QDs diluted in *n*-hexane as a function of process time and growth temperature and are shown in Figs. 3.10 and 3.11, respectively. The first exciton transition energies in UV–Vis optical absorption spectra are known as the core band gaps of spherical bare CdSe core and CdSe/ZnS core/shell QDs which are determined from the following empirical expression [22]

$$E_g^{nc}(d) = \frac{hc}{\lambda_{\max}} \tag{3.4}$$

where $E_g^{nc}(d) = E_{cv} = E_c - E_v$ is the bandgap measured at wavelength $\lambda_{\max}$ at which the absorption of nanoparticles is maximum. c and h are respectively, speed of light and Planck's constant.

The values of $E_g^{nc}(d)$ obtained from (3.4) for the maximum wavelength for bare CdSe core and CdSe/ZnS core/shell QDs at several growth temperatures and process times. In Sect. 3.10, we will use these results to discuss the temperature and interface strain effects on core band gap and diameter in CdSe core and CdSe/ZnS core/shell QDs.

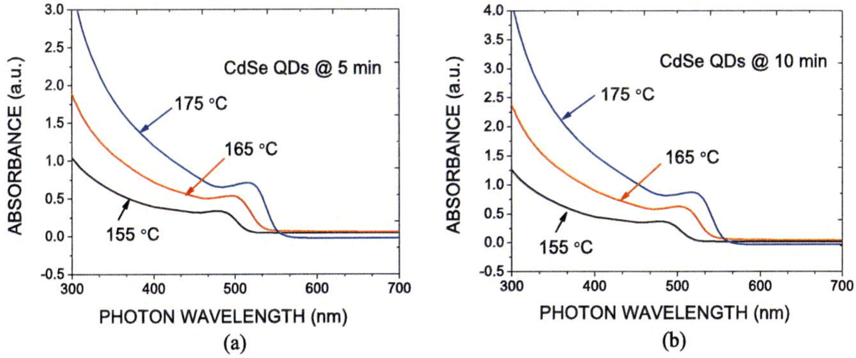

Fig. 3.10 Absorbance spectra versus wavelength of CdSe core QDs synthesized at 5 min. (**a**) and 10 min. (**b**) at various temperatures

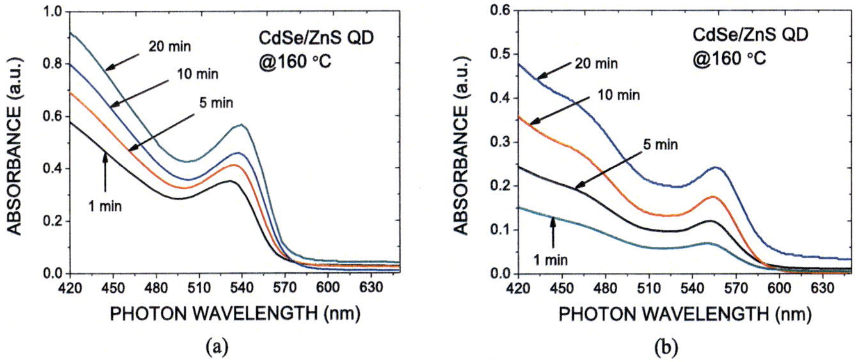

Fig. 3.11 Absorbance spectra versus wavelength of CdSe/ZnS heterostructure core/shell QDs synthesized at 160 °C (**a**) and170 °C (**b**) reaction temperature at different times (min)

3.5.2 Fluorescence Characterization

The photoluminescence (PL) brightness, measured by PL quantum yield (QY) and the stability of the emission of QDs strongly depend on synthesis route. Generation of the inorganic outer layer (shell) is the most critical step for producing highly emissive materials. This layer ensures that the materials possess the exceedingly high and stable QYs (>30%) which are pivotal for imaging and light emitting/absorbing applications. Importance of the shell layers was first identified by Hines et al. after coating CdSe NCs with ZnS [23]. Since that effort, all types of QDs have been prepared with appropriate shells and core–shell systems are now default structures for QDs. Analysis of PL features was done via Varian Cary Eclipse Fluorescence Spectrophotometer in our laboratory. The aliquots were diluted with *n*-hexane directly for characterization. The quantum yield of bare CdSe core and CdSe/ZnS core/shell QDs were calculated following the procedure by comparing with a standard (Rhodamine-101 in ethanol), with an assumption of its QYs as 95%, and using the data from the fluorescence and absorbance spectra of QDs, estimated using the following expression [24].

$$\varphi_x = \varphi_s \left(\frac{I_x}{I_s}\right)\left(\frac{A_s}{A_x}\right)\left(\frac{n_x^2}{n_s^2}\right) \tag{3.5}$$

where I_x (sample) and I_s (standard) are integrated emission peaks, upon 480 nm excitation; A_x (sample) and A_s (standard) are absorption areas at 480 nm; n_x (sample) and n_s (standard) are refractive indices of solvents; and Φ_x and Φ_s are FL QYs for measured and standard samples.

Fluorescence emission spectra of bare CdSe QD samples synthesized at several growth temperatures for 5 and 10 min of process times are shown in Fig. 3.12 as a function of photon wavelength. The fluorescence intensity and the emission peaks increase with increasing time and temperatures to the particle size. The emission

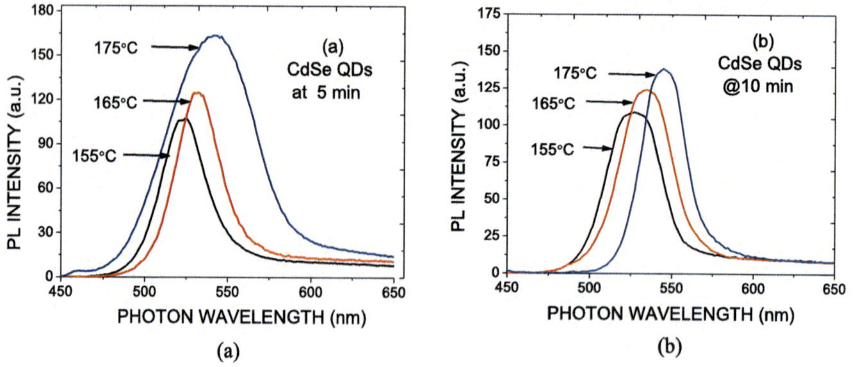

Fig. 3.12 Fluorescence emission spectra versus wavelength of CdSe core QDs at 5 min. (**a**) and 10 min. (**b**) synthesized at 155 °C, 165 °C, and 175 °C, respectively

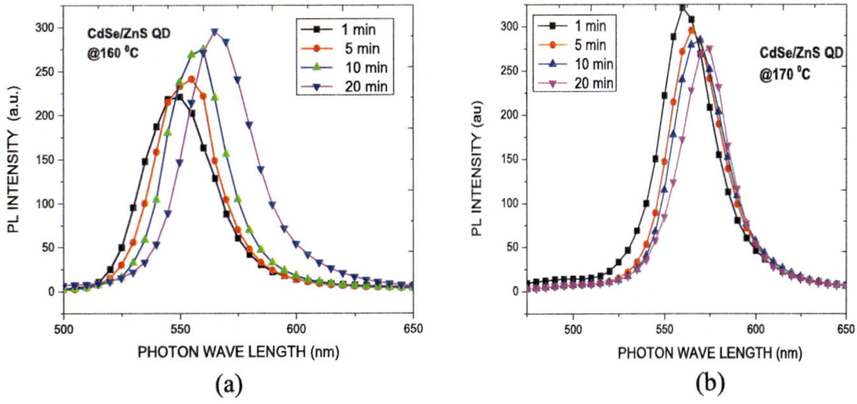

Fig. 3.13 Fluorescence emission spectra versus photon wavelength of CdSe/ZnS core/shell QDs synthesized at 1600° C (**a**) and 170° (**b**) for different reaction times (min)

peaks of CdSe/ZnS QDs range from 550 to 570 nm, and the corresponding full width at half-maximum (FWHM) of the band-edge luminescence was maintained between 32–36 nm.

From the emission spectra, it can be observed that the PL output wavelengths corresponding to the peaks are in the range between 507 to 542 nm at 170 °C for different times. The fluorescence emission peaks increase with increasing temperatures depending on particle size. As expected, the absorption maxima occur at higher energies than the emission maxima and the difference between absorption and emission wavelengths is the Stokes shift. The synthesized particle sizes have Stokes shifts of ranges between 29 to 72 nm for different time and temperatures. These results are given in Table 3.5 and are shown in Fig. 3.13 which indicate that the Stokes shift increases as the nanoparticle size decreases.

3.5.3 UV–Vis, PL and Stokes Shift

Figure 3.14a, b compare the growth temperature effects on the absorption and emission spectra of bare CdSe core QDs synthesized at 155, 165, and 175 C for 5 and 10 min of reaction times. The position of maximum wavelength at which absorption and emission coefficients of bare CdSe core NCs are maximum tends to slightly shift to higher wavelengths as temperature is increased. Emission peaks vary between 507 to 542 nm for different reaction times.

Furthermore, Figs. 3.15a, b compare the effect of growth temperature on the absorption and emission coefficients of CdSe/ZnS heterostructure core/shell QDs synthesized at 160, and 170 °C for 1, 5 10, and 20 min of reaction times. The emission peaks of core/shell NCs range from 550 to 570 nm and corresponding to the full width at half-maximum (FWHM) of band-edge luminescence between 32 and 36 nm. One can observe from the PL emission spectra that the PL output

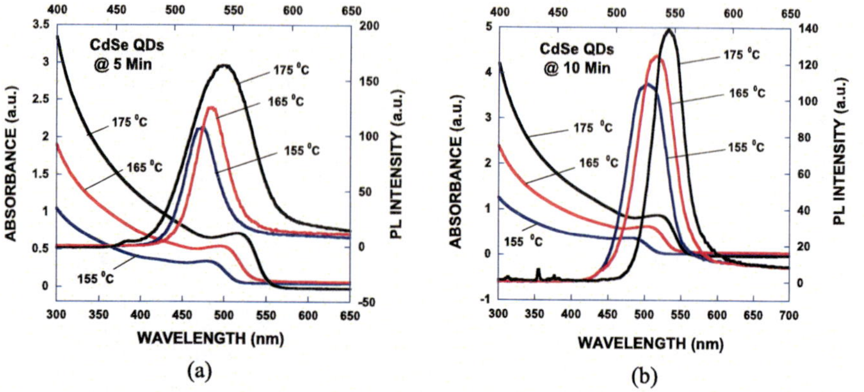

Fig. 3.14 Absorbance and emission spectra of bare CdSe core NCs synthesized at 155, 165 and 175 °C for 5 min. (**a**) and 10 min. (**b**), plotted as a function of photon wavelength

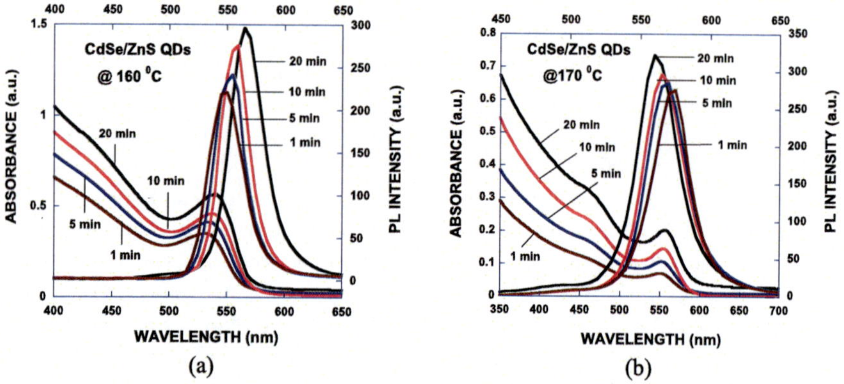

Fig. 3.15 Absorbance and emission spectra of CdSe/ZnS core/shell NCs synthesized at 160 °C (**a**) and170 °C (**b**) for different reaction times, plotted as a function of photon wavelength

wavelengths corresponding to the emission peaks of CdSe/ZnS core/shell NCs are in the range between 507 to 542 nm for different reaction times. The PL emission peaks increase with temperature.

The maximum wavelengths of UV–Vis absorption and PL emission and corresponding Stokes shifts and full width half maximum (FWHM) for bare CdSe QD synthesized from 150 to 175 °C are given in Table 3.3 for 5 and 10 minutes of reaction times.

The peak intensity values of UV–Vis absorption spectra and PL emission spectra, Stokes shift, full width half maximum (FWHM), and quantum yield (QY) of CdSe/ZnS core/shell QDs synthesized at 160 °C and 170 °C are listed in Table 3.4 for reaction times between 1 and 20 min., respectively.

Table 3.3 Maximum wavelengths of UV–Vis absorption and PL emission spectra and corresponding Stokes shift and full width half maximum (FWHM) of bare CdSe core QDs at different temperatures for 5 and 10 min of reaction times

Reaction time (min)	Growth Temp. (°C)	UV–Vis λmax (nm)	PL λmax (nm)	Stokes shift (nm)	FWHM (nm)
5	150	452	521	69	31
	155	476	525	49	33
	160	484	527	43	42
	165	496	532	36	65
	170	508	538	30	43
	175	512	541	29	48
10	150	453	525	72	30
	155	481	527	46	32
	160	488	530	42	33
	165	500	535	35	32
	170	512	542	30	44
	175	516	545	29	62

Table 3.4 Absorption and emission peak intensities, full width and half maximum (FWHM) and quantum yield (QY) of CdSe/ZnS NCs synthesized at 160 and 170 °C at various times

Temperature (°C)	Time (min)	Peak absorbance wavelength (nm)	Peak emission wavelength (nm)	Stokes shift (nm)	FWHM (nm)	QY (%)
160	1	532	550	18	32	27
	5	535	553	18	33	41
	10	538	554	16	32	45
	20	539	556	17	36	36
170	1	549	560	11	35	28
	5	552	563	11	34	40
	10	555	565	10	33	44
	20	557	570	13	33	42

The fluorescence quantum yield (QY) of CdSe/ZnS core/shell QDs synthesized at 160 and 170 °C increases monotonically from 27 to 45% with reaction time increase and decreases as reaction time is increased further. This means that one can optimize reaction time to get maximum quantum yield. The FWHM of CdSe/ZnS core/shell QDs synthesized at 170 °C (160 °C) decreases (increases) as reaction time is increased. The difference between the first exciton energy of absorption spectra and the PL peak energy in emission spectra at maximum wavelength is known as Stokes shift, given as $\Delta E = 2S\hbar\omega_p$, where $\hbar\omega_p$ is the energy of the photon coupled

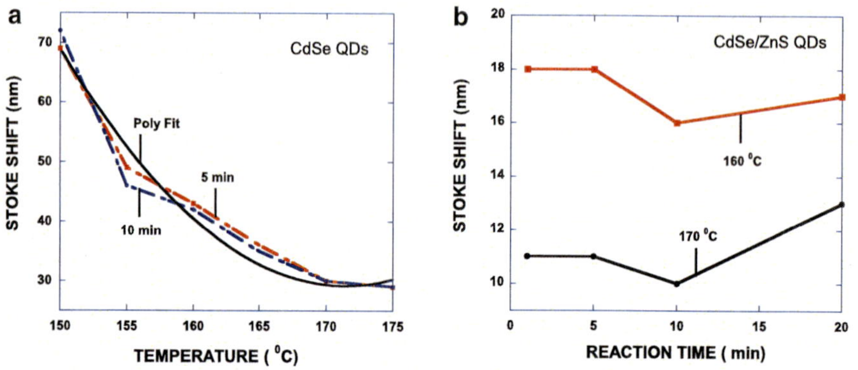

Fig. 3.16 Stokes shift in bare CdSe core QD (**a**) as a function of temperature and that in CdSe/ZnS core/shell QD (**b**) as a function of reaction time, respectively

to the electron (25 meV for CdSe) and S is called the Huang-Rhys factor [25] measure of the strength of electron–phonon coupling.

Figure 3.16a shows the temperature dependence of maximum wavelength of absorption and emission spectra and Stokes shift of bare CdSe core QD and CdSe/ZnS QD as a function of temperature for several reaction times. Maximum wavelength of emission and absorption spectral data given in Table 3.4 are used to study the effect of growth temperature and reaction time on the Stokes shifts in CdSe/ZnS core/shell QD. Figure 3.16b shows the reaction time dependence of Stokes shift of CdSe/ZnS QD synthesized at 160 and 170 °C.

Stokes shift in bare CdSe core NCs is observed to have a parabolic decrease with temperature increase and can be fitted to following expression

$$\Delta E(T) = -1.15 \times 10^{-2} T^2 + 4.75T - 436.44. \tag{3.6}$$

which indicates that increasing temperature decreases magnitude of Stokes shift and corresponding Huang-Rhys factor because of the increasing effect of electron–phonon interaction with temperature increase. Furthermore, reaction time dependence of Stokes shift in spherical CdSe/ZnS core/shell QD synthesized at 160 and 170 °C shown in Fig. 3.16b indicates that increasing reaction time has varying effect on magnitude of Stokes shift and corresponding Huang-Rhys factor.

3.6 Dielectric Spectroscopy Characterization

The optical properties of any semiconductor structure is generally described by the complex dielectric constant $\varepsilon(\omega) = \varepsilon^{'}(\omega) - i\varepsilon^{''}(\omega)$, where $\varepsilon^{'}(\omega)$ and $\varepsilon^{''}(\omega)$ are the real and imaginary parts of $\varepsilon(\omega)$, respectively. The real and imaginary parts of the complex dielectric constants, dielectric losses, and ac conductivity of bare CdSe core and CdSe/ZnS heterostructure core/shell QDs were investigated as a function of frequency at different temperatures. We used the Broadband System Novo-Control Concept-80, which includes the low–high frequency spectrometer BDS-80 (from 3–20 MHz) measurements for the materials investigation in combination with Quatro Cryosystem (−160 °C–400 °C) and a program complex for visualization and treatment of data. The measurements were carried out in the frequency range from 0.1 Hz to 10 MHz. The method for determining $\varepsilon^{'}(\omega)$ and $\varepsilon^{''}(\omega)$ is the comparison of capacity of air filled capacitor with that containing the dielectric material.

Figures 3.17a and 3.17b, respectively, compares the frequency dependency $\varepsilon^{'}(\omega)$ and $\varepsilon^{''}(\omega)$ of CdSe QDs synthesized at 130 °C, 140 °C, 150 °C, 160 °C, and 170 °C, respectively. $\varepsilon^{'}(\omega)$ and $\varepsilon^{''}(\omega)$ rapidly decrease with increasing frequency and then become nearly constant in high frequency range (over 100 Hz). However, the low frequency values of $\varepsilon^{'}(\omega)$ and $\varepsilon^{''}(\omega)$ at 130 °C and 140 °C is somewhat smaller than those synthesized at 150 °C, 160 °C, and 170 °C, respectively, but they approach to values of latter ones at frequencies over 100 Hz.

Figures 3.18a, b compare the frequency variations of $\varepsilon^{'}(\omega)$ and $\varepsilon^{''}(\omega)$) of bare CdSe core and CdSe/ZnS core/shell QDs (synthesized at 170 °C) after subtracting the hexane contribution according to a simple approximation: $\varepsilon^{QD}(\omega) = \varepsilon_{total}(\omega) - \varepsilon_{hexane}(\omega)$. The frequency variation of $\varepsilon^{'}(\omega)$ and $\varepsilon^{''}(\omega)$ for CdSe/ZnS core/shell QDs at low frequencies is slightly different from those of bare CdSe core QDs. The change in $\varepsilon^{'}(\omega)$ for CdSe/ZnS core/shell QD is nearly the same for that of bare CdSe core

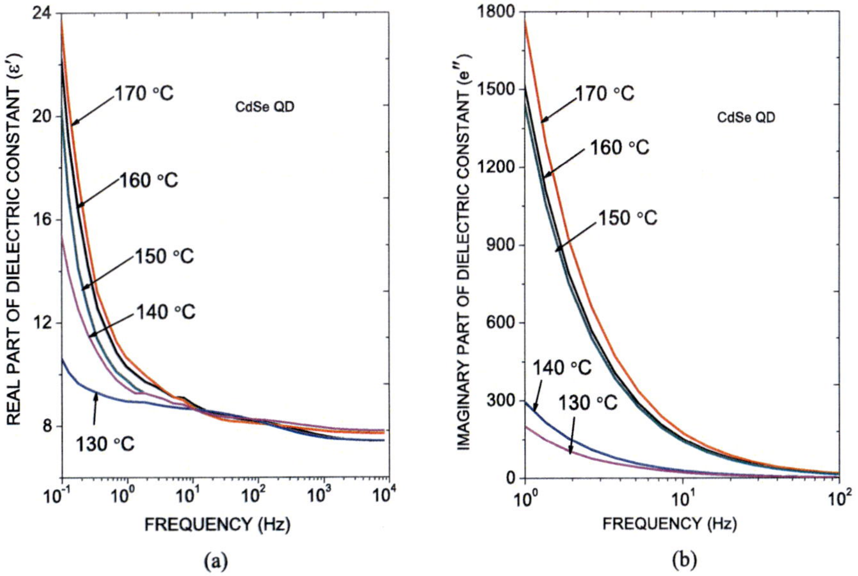

Fig. 3.17 Temperature effect on real part ε′ (left) and imaginary part (ε″) of complex dielectric constant (ε) of CdSe QD synthesized at 130 °C, 140 °C, 150 °C, 160°C, and 170 °C, respectively

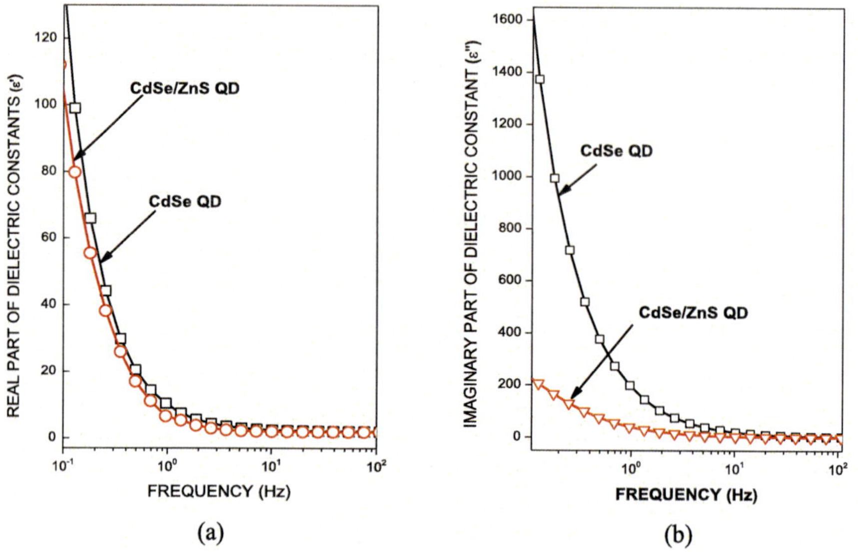

Fig. 3.18 Comparison of real (**a**) and imaginary (**b**) parts of dielectric constant of bare CdSe and CdSe/ZnS core/shell QDs synthesized at 170 °C

QD. However, the change in $\varepsilon^{''}(\omega)$ for CdSe/ZnS core/shell QDs is noticeable smaller than that of bare CdSe core QDs at low frequencies.

The dielectric loss ($\tan\delta = \varepsilon^{''}/\varepsilon^{'}$) and its measurement as a function of frequency for different temperatures makes it possible to determine the activation energy and relaxation time of dipolar orientation process in solid. The conduction behavior of nanocrystal can be understood from the conduction loss expression defined as $(\tan\delta)_c = \sigma/\varepsilon^{'}\omega$, where σ is nearly independent of frequency and is equal to the dc conductivity in practical cases. The ac conductivity of the bare CdSe and CdSe/ZnS core/shell QDs are calculated using the relation $\sigma_{ac} = \omega\varepsilon(\omega)\tan\delta = 2\pi f\varepsilon_r\varepsilon_0\tan\delta$ with $\varepsilon_r\varepsilon_0 = C/C_0$. Here C is the capacitance of the dielectric material and C_0 is the capacitance of reference material (air filled capacitor).

Figures 3.19a, b, respectively, compare the temperature effect on dielectric loss and ac conductivity of CdSe QDs synthesized at 130 °C, 140 °C, 150 °C, 160 °C, and 170 °C, respectively, as a function of frequency. Figure 3.19a suggests that, similar to the dielectric constant, the dielectric loss is strongly dependent on the frequency of the applied field. The dielectric loss decreases with increasing frequency at almost all temperatures but appears to achieve saturation in the higher frequency range of 1 kHz and above, at all the temperatures. Figure 3.19b suggests that electrical A.C. conductivity increases with an increase in temperatures and frequency. The results reveal that the A.C. conductivity varies almost linearly with the applied frequency in the high range and increases with different temperatures.

Figures 3.20a, b, respectively, show the low frequency behavior of dielectric loss

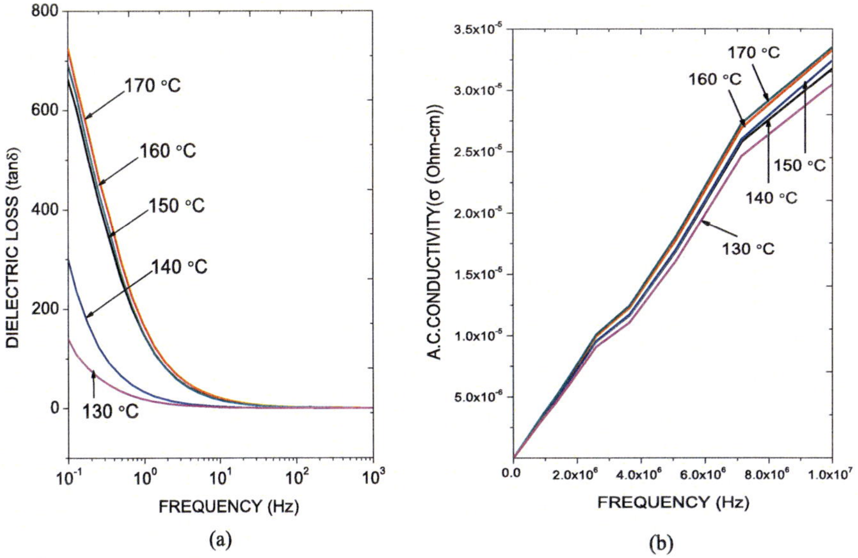

Fig. 3.19 Temperature effect on dielectric loss and ac conductibity of CdSe QD synthesized at 130 °C, 140 °C, 150 °C, 160 °C, and 170 °C, respectively

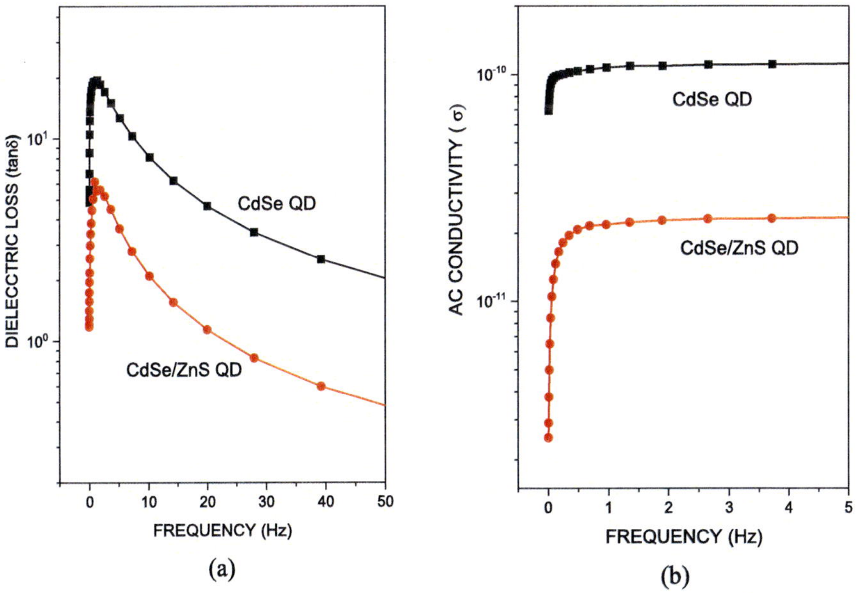

Fig. 3.20 Comparison of dielectric loss (**a**) and AC conductivity (**b**) of bare CdSe core and CdSe/ZnS core/shell QDs synthesized at 170 °C, with hexane contribution to real and imaginary parts of the dielectric constants for both nanocrystals is extracted

($\tan\delta = \varepsilon''/\varepsilon'$) and AC conductivity of bare CdSe core and CdSe/ZnS core/shell QDs synthesized at 170 °C with hexane contribution to the complex dielectric constant is removed. Figure 3.20a shows that dielectric loss decreases with increase in frequency but appears to achieve saturation in the higher frequency range. The dielectric loss of CdSe/ZnS QD is smaller than that of bare CdSe QD at low frequencies. However, it becomes nearly constant at high frequencies for both nanocrystals since $\varepsilon'(\omega)$ and $\varepsilon''(\omega)$ are almost independent of frequency in that regime. Figure 3.20b suggests that A.C. conductivity increases with an increase in temperatures and frequency. Furthermore, we observe that, just like dielectric loss, A.C. conductivity of CdSe/ZnS QD is smaller than that of bare CdSe core QD at low frequencies. However, just like the dielectric loss, A.C. conductivity becomes nearly constant at high frequencies for both nanocrystals since $\varepsilon'(\omega)$ and $\varepsilon''(\omega)$ are almost independent of frequency in that regime.

It is observed in Fig. 3.18 that the real and imaginary parts of complex dielectric constant of bare CdSe core and CdSe/ZnS heterostructure core/shell QDs decrease exponentially with increasing frequency and then attains almost a constant value in the high frequency region. Furthermore, Fig. 3.20 shows that dielectric loss and ac conductivity of bare CdSe core QD and CdSe/ZnS core/shell QD, respectively, increase at small frequencies and suddenly decay and saturate as frequency increases. We attribute these changes to the temperature dependent interface strain effect on dielectric constant of bare CdSe core and CdSe/ZnS heterostructure core/shell QDs. The magnitude of the complex dielectric constant $\varepsilon(\omega)$ of bare CdSe core and CdSe/ZnS core/shell quantum dots is nearly constant at high frequencies ($\omega \gg 1 \times 10^6 Hz$), plotted in Fig. 3.21. The expression $\varepsilon_\infty^{QD} = [\varepsilon_\infty'^2 + \varepsilon_\infty''^2]^{1/2}$ yields

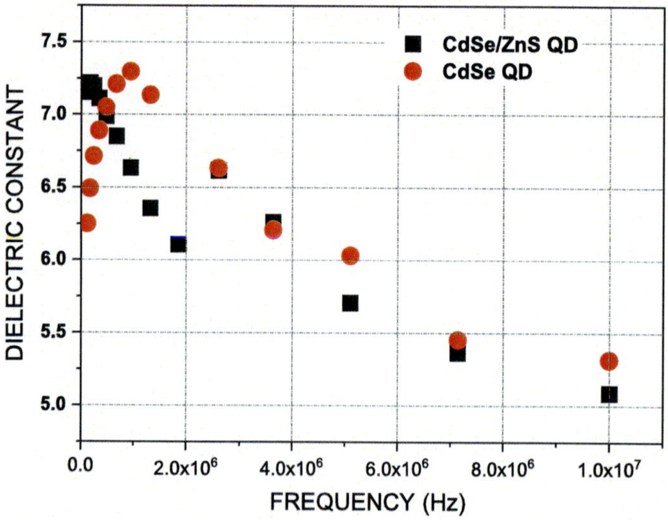

Fig. 3.21 Comparison of dielectric constant of bare CdSe core and CdSe/ZnS heterostructure core/shell QDs at high frequencies

$\varepsilon_{\infty}^{c} = 5.40\varepsilon_{\circ}$ and $\varepsilon_{\infty}^{cs} = 5.10\varepsilon_{\circ}$ at 1×10^{7} Hz for CdSe core and CdSe/ZnS core/shell QDs, respectively. Using the bulk dielectric constants of CdSe and ZnS we have $\varepsilon_{\infty}^{cs} = (\varepsilon_{\infty}^{CdSe} + \varepsilon_{\infty}^{ZnS})/2 = (5.8\varepsilon_{\circ} + 5.10\varepsilon_{\circ})/2 = 5.40\varepsilon_{\circ}$ for $\varepsilon_{\infty}^{cs}$ of CdSe/ZnS core/shell QDs. This is in good agreement with the dielectric measurements of dielectric constant at high frequencies.

Strain effects on dielectric constants of bare CdSe core and CdSe/ZnS heterostructure core/shell QDs can be qualitatively understood by using parabolic two band Penn model of dielectric constant of semiconductors [26]. We can write the following expressions for the dielectric constants of bare CdSe core and heterostructure core/shell CdSe/ZnS quantum dots

$$\varepsilon_{\infty}^{c} = 1 + (\hbar\omega_{p}^{c}/E_{g}^{c})^{2}; \ \ \varepsilon_{\infty}^{cs} = 1 + (\hbar\omega_{p}^{cs}/E_{g}^{cs})^{2} \tag{3.7}$$

where $\omega_{p}^{c}(\omega_{p}^{cs})$ and $E_{g}^{c}(E_{g}^{cs})$ are, respectively, the plasma frequency and band gap of bare CdSe core (CdSe/ZnS heterostructure core/shell) QDs. To a first order approximation, we can take plasma frequencies of core and core/shell QDs equal ($\hbar\omega_{p}^{c} \approx \hbar\omega_{p}^{cs}$) and relate the dielectric constant of heterostructure core/shell QD to that of core QD as

$$\varepsilon_{\infty}^{cs} = \varepsilon_{\infty}^{c} + \frac{1}{E_{gcs}^{2}} - \frac{1}{E_{gcs}^{2}} = \varepsilon_{\infty}^{c} + \frac{E_{gc}^{2} - E_{gcs}^{2}}{E_{gc}^{2}E_{gcs}^{2}} \tag{3.8}$$

where E_{gcs} and E_{gc} are obtained from UV–Vis absorption data, according to (3.4) or obtained from two band effective mass approximation, which will be discussed in Sect. 3.10. Equation (3.8) suggests that the dielectric constant of core/shell QD is shifted from that of core QD by factor of $(E_{gc}^{2} - E_{gcs}^{2})/E_{gc}^{2}E_{gcs}^{2}$. Since E_{gcs} and E_{gc} depends on the diameter of QD, (3.8) proves the fact that the real part of optical dielectric constant (ε') with increase in nanoparticle size, approaching the bulk value for larger particles. The dielectric constant of all synthesized samples is high at lower frequencies and decreases with increasing frequency. The dielectric loss tangent (*tanδ*) decreases with frequency and the sample with minimum particle size shows large value of loss tangent at a specific frequency.

3.7 Precursor Ratio Effect on Nanoparticle Growth

Since the growth kinetics of semiconductor nanocrystals are dependent on temperature, time and initial precursor ratio for any synthesis route, influence of changes in these variables on the final properties CdSe QDs were investigated. The most important parameter, Cd:Se molar ratio of the precursor is found having the ability of change the average particle size, number of density and size distribution by "focusing" and "defocusing" the particle growth in the solution [10].

Qu and Peng (2002), have investigated the influence of the initial ratio between Cd and Se precursors on the temporal evolution of the ensemble PL QY of CdSe NCs during their growth in a coordinating solvent [27]. The PL QY was observed to increase monotonically during growth to a maximum value (named as bright point) and then gradually decrease. The position and temporal width of the bright point, the highest QY, the growth kinetics and the sharpness of the PL peak were all reported to be strongly dependent on the initial Cd:Se ratio. The existence of the bright point was interpreted as a signature of an optimal surface structure of NCs grown under given conditions [27]. Talapin et al. investigated the distribution of properties within ensembles of collodially grown CdSe NCs by analyzing size-selected fractions. An excess of the metal cation precursor (Cd) was used and a huge difference was observed between the PL efficiencies of fractions size-selected from the same ensemble. This behavior was attributed to differences in surface disorder of the NCs as a consequence of the Ostwald ripening growth mechanism. The particles with the lowest growth rate within the ensemble were assumed to have the lowest degree of surface disorder and therefore the highest PL QY at any given reaction conditions [28].

Our investigations of the optical and electronical properties of synthesized CdSe core and CdSe/ZnS core–shell NCs indicate that depending on the diameter of CdSe NCs, the red shift was observed in the absorption spectra. In coherent with the situation, a decrease in the optical absorption band gap of NCs was recorded, thus proving the size quantization effect. The FWHM values of CdSe core and CdSe/ZnS core–shell NCs were found to vary between 25–31 nm without Ostwald ripening, indicating the luminescence quality of synthesized particles in narrow size distribution. For the synthesis' with altering the Cd:Se initial precursor ratio at a fixed temperature, it was observed that the samples with a high Cd:Se ratios are able to nucleate and grow faster. However, it was explored that the influence of operating temperature becomes dominant on the growth of NCs as the Cd:Se initial precursor ratio decreases. It was attributed to unfavorable temperatures for the growth of Cd rich samples. For a fixed temperature, it was investigated that there is an increase for FWHM values obtained from PL spectra by reducing the Cd:Se initial precursor ratio. The same situation is also valid for deposition of ZnS shells over CdSe core NCs. PL QY results were showed it is possible to increase the efficiency of CdSe core NCs with covering ZnS shells over them. Average particle diameters obtained from high resolution TEM images were suitable with the derived particle diameters from theoretical calculations by using effective mass approximation.

Figure 3.22 shows the diameter distribution of NCs grown with different Cd:Se molar ratios at given synthesis temperature. As the amount of Cd precursor increases, NCs grow into larger diameters very fast and growth is the predominant process over nucleation, which reduces the number of density and increases the average particle size under same experimental conditions, coherent with the study of Bhattacharjee et al. [29]. Increase in the NC diameter is also responsible for red shift in the emission and absorption spectra of CdSe QDs. On the other hand, the operating temperature influences the growth of NCs when it compared in identical chemical compounds and elapsed time.

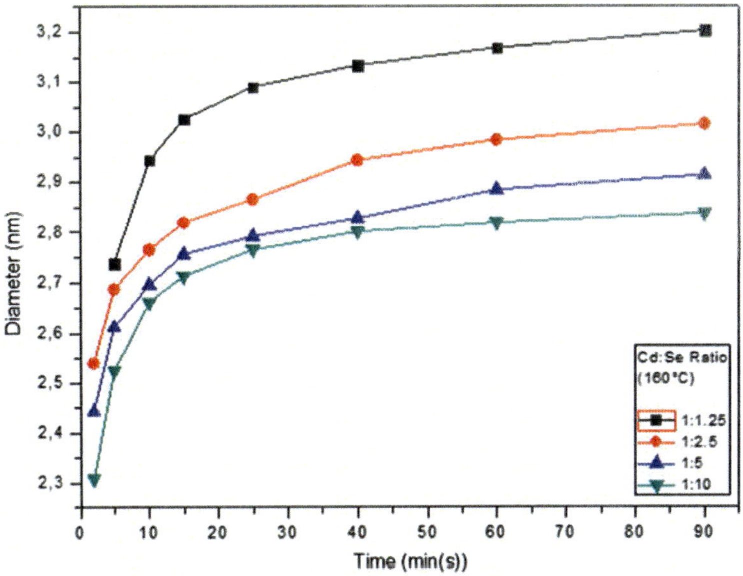

Fig. 3.22 Growth of CdSe NCs with different Cd:Se molar ratios at 160 °C

As shown in the Fig. 3.23a, b, difference in the diameter of grown NCs increases with the excessive amount of initial Se precursor, therefore indicating that the effect of operating temperature is evident for the Se rich samples. However, lower temperatures are not suitable for the ideal growth of Cd rich samples.

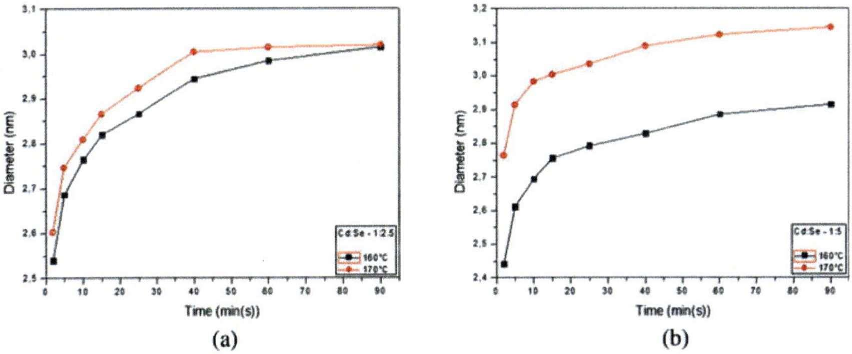

Fig. 3.23 Effect of temperature on CdSe NCs growth for Cd:Se ratio of 1:2.5 (**a**) and CdSe NCs growth for Cd:Se ratio of 1:5 (**b**) grown at 160 and 170 °C

3.8 Emission Quality and PL Yield

Photolumiscence analysis of the CdSe quantum dots dispersed in *n*-hexane was done via Varian Cary Eclipse Fluorescence Spectrophotometer. Excitation wavelentgh was chosen as 350 nm. Figure 3.24 shows the PL properties of the CdSe quantum dots growth for Cd:Se ratio of 1:1.25 at 160 and 170 °C. It is seen that the red shift is observed in the PL spectra with the duration of the synthesis.

First excitonic absorption peak energy and PL emission peak energy and corresponding Stokes shifts for the CdSe QDs synthesized at 160 and 170 °C are given in Table 3.5 from 5 to 90 min of reaction times.

Stokes shift of the CdSe QDs synthesized at 160 °C for 90 min of reaction is given in the Fig. 3.25a which is 147 meV. The relation of the synthesis duration and temperature on the Stokes shift can be seen in Fig. 3.25b.

FWHM of an emission peak is the measure of color purity of the emission where the smaller values indicate more pure emission and narrow size distributions without Ostwald ripcning. Tablc 3.6 shows the average FWHM values obtained from PL spectras with respect to various conditions. It is noticable that, the average FWHM value increase as the initial Cd:Se precursor ratio decreases.

Figure 3.26 also shows the normalized PL spectra of CdSe NCs and CdSe/ZnS NCs grown over them. The red shift that suitable with the one at absorption spectra indicates the growth of ZnS shells over CdSe core NCs. Table 3.7 and Fig. 3.26 shows the PL properties of CdSe core and CdSe/ZnS core-shell nanostructures for given conditions. A noticable decrease in the PL QY was recorded for second and third cycles of ZnS shell growth process, which is not expected in order to obtain good emission quality by passivating the core surface. In general, a low PL QY is considered as a result of the surface states located in the band gap of the NCs, which act as trapping states for the photogenerated charges. These surface trapping sites are originated from the dangling bonds of some of the surface atoms. The ligands on the surface of NCs may remove some or all of the surface trapping states and increase the PL QY of NCs. Theoretical studies indicate that the efficiency of the electronic

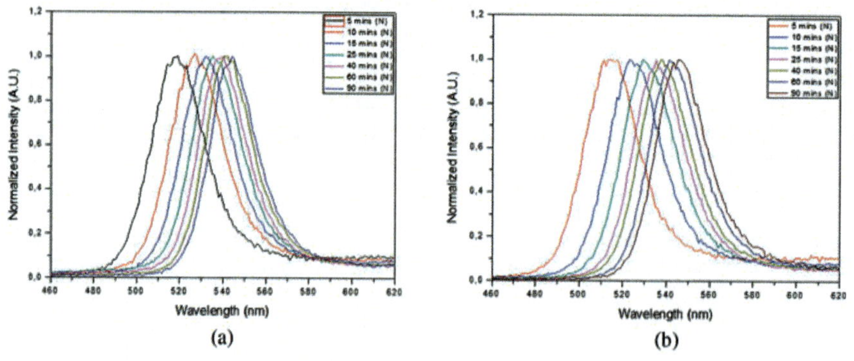

Fig. 3.24 Normalized PL spectra of CdSe quantum dots grown at 160 °C (**a**) and 170 °C (**b**)

Table 3.5 1st excitonic absorption peak energy and PL emission energy peaks and corresponding Stokes shift of CdSe core quantum dots synthesized at 160 and 170 °C for various durations

Temperature (C)	Synthesis duration (min.)	Peak absorbance energy (eV)	Peak emission energy (eV)	Stokes shift (meV)
170	5	2.510	2.412	98
	10	2.460	2.375	85
	15	2.431	2.344	87
	25	2.403	2.318	85
	40	2.385	2.305	80
	60	2.357	2.288	70
	90	2.340	2.271	69
160	5	2.600	2.394	206
	10	2.557	2.348	208
	15	2.525	2.331	195
	25	2.500	2.313	187
	40	2.460	2.296	164
	60	2.441	2.288	153
	90	2.427	2.279	147

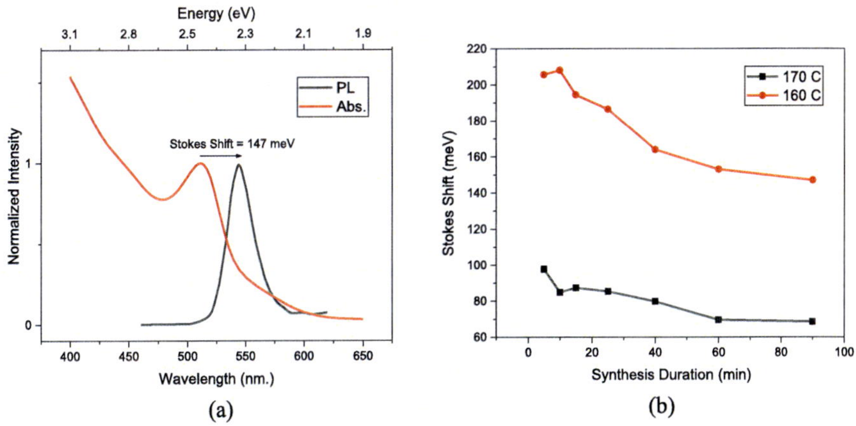

Fig. 3.25 Stokes shift of CdSe quantum dots grown at 160 °C for 90 min. (**a**). Stokes shift of as a function of reaction time for 160 and 170 °C synthesis temperature (**b**)

Table 3.6 Effect of Cd:Se initial precursor ratio on the average FWHM values of CdSe NCs

Sample	Molar ratio	Temperature (°C)	Average FWHM (nm)
CdSe	1:1.25	170	29.856
CdSe	1:5	170	30.156
CdSe	1:10	170	30.956

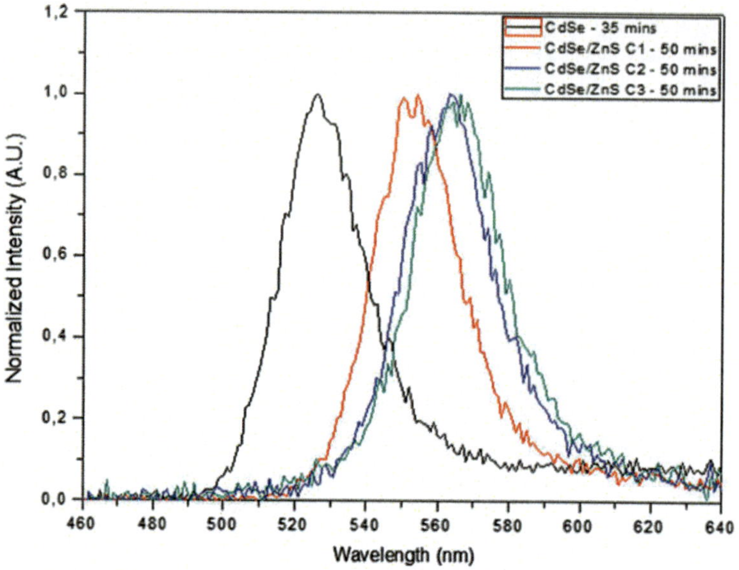

Fig. 3.26 Normalized PL spectra of CdSe and CdSe/ZnS NCs

Table 3.7 PL properties of CdSe QDs grown at 160 °C with Cd:Se ratio of 1:1.25 and CdSe/ZnS QDs deposited over with number of cycles

Sample	Cycle number	Reaction time (min)	FWHM (nm)	Peak site (nm)	QY (%)
CdSe	–	35	27	526	29.18
CdSe/ZnS	1	50	27.7	553	42.46
CdSe/ZnS	2	50	28.9	562	21.65
CdSe/ZnS	3	50	29.5	565	18.38

passivation provided by the surface ligands depend strongly on the surface structure and the nature of the surface states of the NCs themselves. If the surface ligands could provide a good electronic passivation for the surface states of the NCs, a high PL QY is expected.

However, decrease in the PL QY for stable conditions can be taken as ordinary, where the most of reports in the literature include the efficiencies of better samples. Chattopadhyay et al. have demonstrated that there is an ideal ratio between the diameter of core and shell thickness for different thicknesses of CdSe core QDs [30]. Normally, the resulting core solution of CdSe NCs was used for the shell growth in the originated article. In this study, the CdSe core QDs for the ZnS shell growth were choosen from the samples with highest PL intensity within each synthesis and aliquots were taken for once at constant time intervals to monitor the growth of ZnS shells.

Table 3.8 PL properties of CdSe QDs grown at 170 °C with Cd:Se ratio of 2:1 and CdSe/ZnS QDs deposited over within one cycle for given time intervals

Sample	Cycle number	Reaction time (min)	FWHM (nm)	Peak site (nm)	QY (%)
CdSe	–	45	28.3	535	14.49
CdSe/ZnS	1	10	28.7	542	40.87
CdSe/ZnS	2	20	29.1	546	24.97
CdSe/ZnS	3	30	30.3	520	22.31

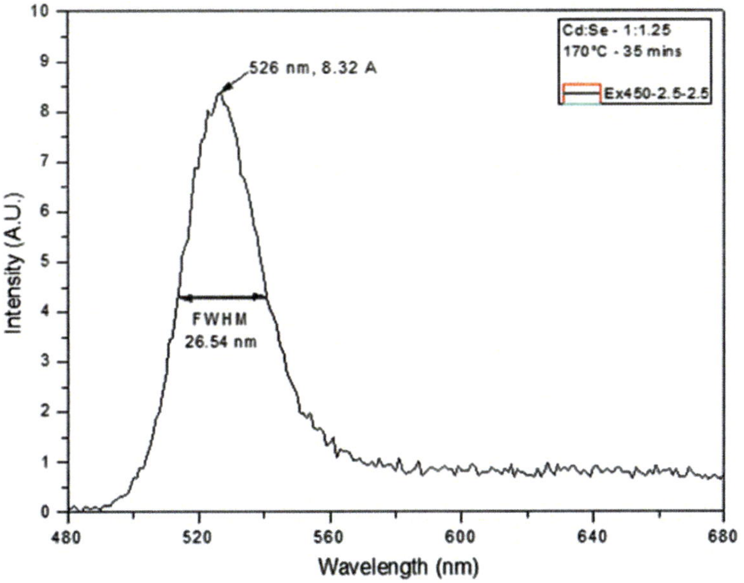

Fig. 3.27 PL spectra of CdSe QDs synthesized at 170 °C with Cd:Se ratio of 1:1.25

Table 3.8 shows the PL properties of CdSe core and CdSe/ZnS core-shell nanocrystals grown on throughout one cycle during the synthesis for given conditions. PL QY reaches its best in early stages of shell coverage process due to small diameter of choosen CdSe nanocrystals.

Fig. 3.27 shows the PL characteristics of choosen CdSe core nanocrystal with small FWHM value, proving narrow particle size distribution.

3.9 Stability of CdSe Quantum Dots

The synthesis of CdSe QDs are described earlier in this chapter. Firstly, cadmium stearate is prepared by heating the mixture of CdO (0.01 mol) and stearic acid (0.02 mol) at 170 °C for 15 min in order to use in further reactions. Then, 2 mL

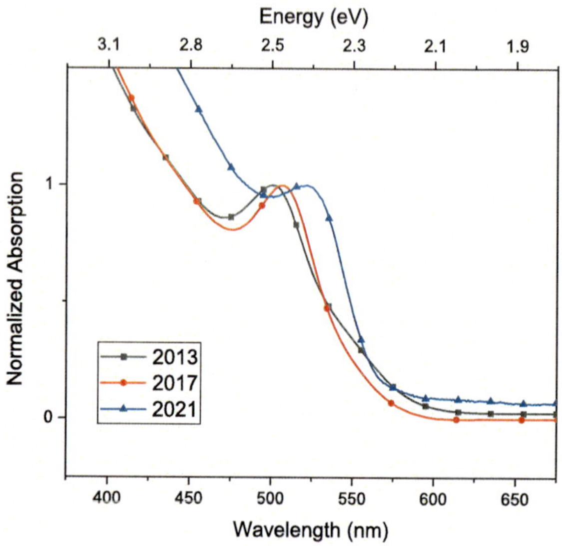

Fig. 3.28 Normalized absorption spectrum of as-synthesized CdSe QDs at 160 °C for 90 min. with Cd:Se ratio of 1:5. Each measurement was taken four years apart

TOP and Se powder (0.1 mmol) are added into a flask and are mixed in an ultrasonic cleaner at the room temperature for several seconds. For a typical synthesis of CdSe QDs, cadmium stearate (0.2 mmol) is added into a three-neck flask with 16 mL paraffin liquid. The cadmium stearate-paraffin mix is degassed at room temperature and then is heated to a chosen temperature (160 °C) in an oil-bath heater under nitrogen dry gas flow. When the heat is reaches to aimed temperature, TOP-Se solution is rapidly injected into the reaction flask for the growth of CdSe QDs. After a certain time (90 min.) of reaction the mixture is left to cool down to room temperature.

The absorption spectra of CdSe QDs diluted in *n*-hexane were recorded with Shimadzu UV-3600 UV–VIS-NIR Spectrophotometer. In order to monitor long life stability of the CdSe QDs, absorption measurements were done four years apart from each measurement (Fig. 3.28).

CdSe QDs were kept as synthesized without purification steps and at ambient temperature in a dark environment. Effect of the initial precursor ratio was also monitored for the aging properties of CdSe QDs (Fig. 3.29). Long term stability of CdSe QDs is investigated by absorption spectroscopy. 1st excitonic absorption peak shifts of CdSe QDs due to the aging is around 20 nm after eight years. As seen by Fig. 3.29 total peak shift is smaller for the Cd:Se ratio of 1:10, as expected since there is less Cd precursor left in the remaining solution.

3.10 Strain Effects on Size and Core Bandgap

The conventional two band effective mass approximation [31] gives a qualitative understanding of confinement effects on core bandgap and diameter of quantum dots. In this model, the solution of Schrödinger equation for a particle in a spherical

Fig. 3.29 First excitonic absorption peak shift of as-synthesized CdSe quantum dots due to aging. CdSe QDs were synthesized at 160 °C for 90 min. duration with different Cd:Se ratios

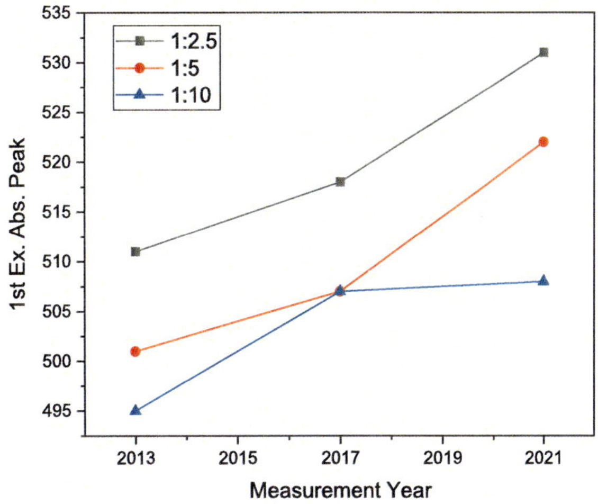

box gives the first excited state (1 s-1 s) energy (core band gap) of nanoparticle, given by following expression

$$E_g^{nc}(d) = E_g^b + \frac{2\hbar^2\pi^2}{m_{cv}^* d^2} - \frac{3.572e^2}{\varepsilon_\infty d} + \frac{0.124e^4}{\hbar^2 m_{cv}^* \varepsilon_\infty^2} \tag{3.9}$$

where E_g^b is bandgap, $m_{cv}^* = m_e^* m_h^*/(m_e^* + m_h^*)$ is the reduced effective mass of electron hole pair with effective masses of electrons and holes m_e^* and m_h^*, respectively, and ε_∞ is the optical dielectric constant of CdSe in bulk form. The second and third terms, respectively, represent the confinement energy and Coulomb interaction energy with a $1/d^2$ and $1/d$ dependence on QD diameter. Finally, the last term is the Rydberg correlation energy, which is negligibly small for when ε_∞ of semiconductor component is large.

Since spherical core/shell QDs are grown from two semiconductors with different lattice constants and thermal expansion coefficients at temperatures above 300 K, there will be elastic strain developed across heterointerface [16]. Elastic strain can modify the structural and electronic properties of core and shell constituents in a way that is not seen in two-dimensional quantum wells and superlattices. In order to understand the strain effects on structural and electronic properties of bare CdSe core and ZnSe//ZnS core/shell QDs, we will use the recent extension of the universal Eshelby's elastic strain model [32] to study strain effects in nanoscale spherical core/shell heterostructures at any temperature [19, 33, 34]. According to this model one assumes that core radius is much smaller than that of shell constituent (a < < b) of spherical core/shell structure and writes strain in core region as [35]

$$\varepsilon_i = \frac{a_i(\varepsilon_i) - a_i}{a_i} = \alpha_i(T)T - \frac{2E_m(1 - 2\nu_i)[\varepsilon_{im} + (\alpha_i - \alpha_m)T]}{E_i(1 + \nu_m) + 2E_m(1 - 2\nu_i)} \tag{3.10}$$

where $\varepsilon_{im} = (a_i - a_m)/a_m$ and $(\alpha_i - \alpha_m)T$ are lattice and thermal mismatches, respectively. Here,a_i and a_m are lattice constants and α_i and α_m are the linear expansion coefficients of core and shell at 300 K. E_i (E_m) is Young's modulus and ν_i (υ_m) is Poisson's ratio of core (shell) semiconductor in bulk form, defined as $E = (C_{11} - C_{12})(C_{11} + 2C_{12})/(C_{11} + C_{12})$ and $\nu = C_{12}/(C_{11} + C_{12})$, where C_{11} and C_{12} are elastic stiffness constants.

Using the elastic constants $C_{11} = 10.2$ GPa and 6.67 GPa and $C_{12} = 6.46$ GPa and 4.63 GPa, linear thermal expansion coefficients $\alpha_i = 7.30 \times 10^{-6} K^{-1}$, $\alpha_m = 6.9 \times 10^{-6} K^{-1}$, and lattice constants $a_i = 0.607$nm and $a_m = 0.41$nm for CdSe and ZnS [22], (3.10) gives $\varepsilon_i = -3.70\%$ compressive strain at interface of spherical CdSe/ZnS core/shell QD, which has positive lattice mismatch $\Delta a/a = (a_i - a_m)/a_m = 12.3\%$ at room temperature. Therefore, we can conclude that reliable modelling and precise determination of the magnitude of strain effects on electronic structure of nanoscale heterostructures is extremely important for the predicting their potential and simulation of the performance of nanoscale electronic and optical devices. In doing so, one must then modify (3.9) to following expression in order to take into account strain effects on core band gap for a reliable understanding of charge confinement in CdSe/ZnS core/shell QDs [33]

$$E_g^{nc}(\varepsilon_i) = E_g^{bc}(T, \varepsilon_i) + \frac{2\hbar^2\pi^2}{m_{cv}^* d^2} - \frac{3.572e^2}{\varepsilon_\infty d} + \frac{0.124e^4}{\hbar^2 m_{cv}^* \varepsilon_\infty^2} \tag{3.11}$$

where $E_g^{bc}(T, \varepsilon_i)$ is strain dependent bandgap of core semiconductor in bulk, which is obtained by using statistical thermodynamic model of semiconductors [34] as

$$E_g^{bc}(T, \varepsilon_i) = E_g^{bc}(0) + \Delta C_{iP}^0 T(1 - \ln T) - \frac{a_{gi}}{B_i}\left(P - \frac{P^2}{2B_i} - \frac{(1 + B_i')}{3B_i^2} P^3\right) \tag{3.12}$$

where P is the hydrostatic pressure acting on band structure of core region, given by

$$P = -3B_i\varepsilon_i(T) = -3B_i\alpha_i(T)\Delta T + 3B_i \frac{2E_m(1 - 2\nu_i)[\varepsilon_{im} + (\alpha_i - \alpha_m)T]}{E_i(1 + \nu_m) + 2E_m(1 - 2\nu_i)} \tag{3.13}$$

Logarithmic term $\Delta C_P^0 T(1 - \ln T)$ in (3.12) represents lattice vibration contribution to band gap change. ΔC_P^0 is heat capacity of reaction for formation of electron–hole pair obtained by fitting bulk band gap calculated from (3.11) to measured bandgap [17], fitted to Varshni equation [36]

$$E_g^b(T) = E_g^b(0) + \frac{\alpha T^2}{\beta + T} \tag{3.14}$$

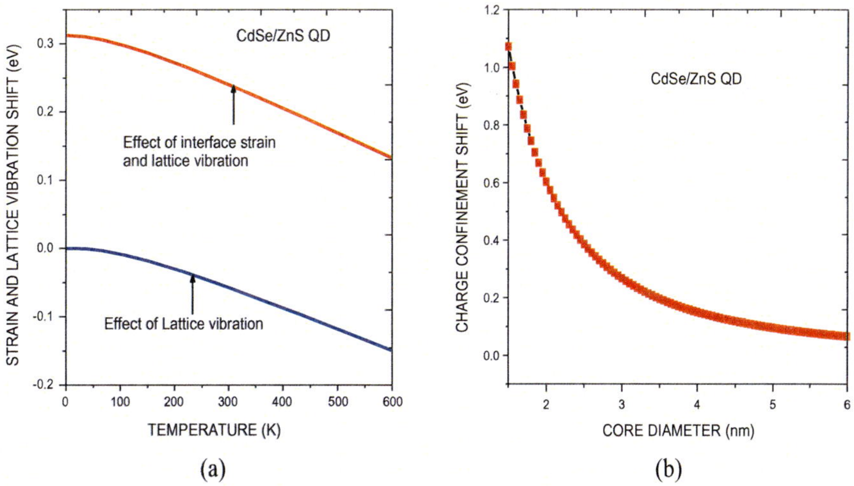

Fig. 3.30 Effects of interface strain and electron–phonon interactions (**a**) and charge confinement effects (**b**) on band gap decrease of CdSe/ZnS core/shell QDs, respectively

where $\alpha = 4.09 \times 10^{-4}$ and $\beta = 187$ are constants for bulk CdSe. We can now use (3.9) to get a qualitative understanding of the variation of core band gap with its diameter at any temperature.

Using the material parameters for CdSe and ZnS bulk semiconductors [22] we calculated the effects of interface strain, electron–phonon interactions, and charge confinement on the shift in core band gap of bare CdSe core QD and CdSe/ZnS core/shell QD as a function of temperature. Figures 3.30a, b compare the contribution of interface strain and electron–phonon interaction (Fig. 3.30a) and charge confinement (Fig. 3.30b) on band gap decrease in band gaps of CdSe/ZnS core/shell QD. Figure 3.30a indicates that electron–phonon interaction contribution to core bandgap decrease is always less than zero and interface strain and lattice vibration tends to decrease core bandgap with temperature increase. However, Fig. 3.30b indicates that charge confinement has parabolic dependence on core diameter and becomes nearly flat as core diameters is increased above 6 nm.

Once first exciton energy (core bad gap) is determined from UV–Vis absorption data using (3.4),core diameter quantum dot can also be calculated by converting (3.9) to a quadratic equation as d is variable: $Ad^2 + Bd + C = 0$, where A, B, and C are material parameter dependent coefficients. The positive root of such quadratic equation will give the diameter of QDs. This will then allow one to calculate the core diameter of bare core and core shell QDs from the first excited state energy extracted from UV–Vis absorption spectra. Figures 3.31a, b show the measured core bandgaps and calculated diameter of bare CdSe core QDs synthesized at temperatures between 155 °C and 180 °C and processed at 1, 5, 10, 15 and 20 min, respectively.

A polynomial fit to core bandgap of bare CdSe QDs, processed at 5 min, suggests a parabolic temperature of the core bandgap, given as

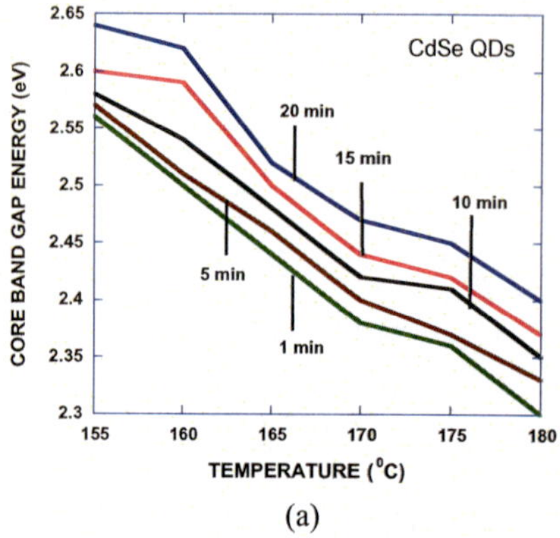

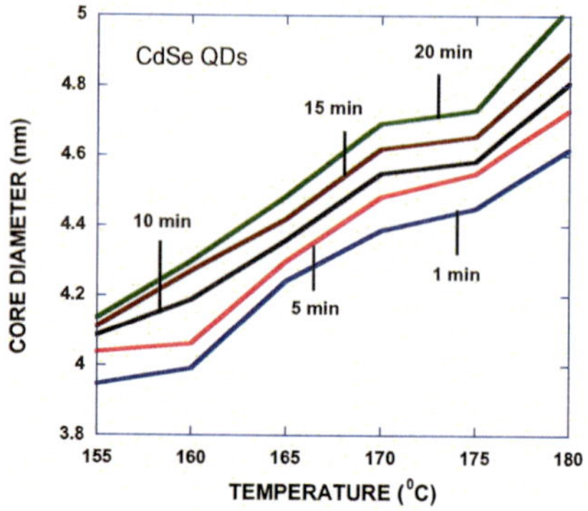

Fig. 3.31 Temperature variation of core band gap energy (**a**) and diameter (**b**) of bare CdSe core QDs processed at 1, 5, 10, and 20 min of reaction times, respectively

$$E_g^{nc}(T) = aT^2 + bT + c \tag{3.15}$$

where a, b and, c are coefficients. The parabolic fit in Fig. 3.31a yields $a = 4.43 \times 10^{-4}K^{-2}$, $b = -0.16K^{-1}$, and $c = 16.26$ for 10 min of reaction time. The parabolic decrease of core bandgap of bare CdSe QDs with temperature increase shown in Fig. 3.31a, b is consistent with the fact that bandgap of most of the groups III-V and II-VI compound semiconductors tend to decrease with increasing temperature, due to electron–phonon interaction and free expansion of lattice constant with increasing temperature. Similar polynomial fit to core diameter of is given by

$$d^{nc}(T) = a^*T^2 + b^*T + c^*, \tag{3.16}$$

where a^*, b^*, and c^* are coefficients: $a^* = -5.04 \times 10^{-4}/K^2$, $b^* = 0.18/K$, and $c^* = -12.96$ for 10 min of reaction time. Since there is only free thermal expansion of lattice constant in bare core QDs, shift in their band gap and diameter is mainly due to electron–phonon interaction with temperature increase.

Since the lattice constant and thermal expansion coefficient of CdSe core are greater than those of ZnS shell of CdSe/ZnS core/shell QDs, CdSe core region will be under compressive strain, resulting in increase (decrease) in its band gap energy (diameter) as temperature is increased. Table 3.9 gives list of UV–Vis maximum wavelength of absorption coefficient and corresponding first excited state energies (core bandgaps) of CdSe/ZnS QDs core/shell QDS synthesized at 160 and 170 °C and processed for 1, 5, 10, and 20 min, respectively.

Using the results given in Table 3.9, unstrained and strained values of core diameters are compared in Fig. 3.32 as a function of core bandgap of CdSe/ZnS core/shell QDs synthesized at 160 °C and 170 °C for various reaction times, respectively. Calculated unstrained and strained diameters show a gradual parabolic decrease (increase) as core bandgap increase (decreases) at both temperatures (Fig. 3.32).

Table 3.9 UV–Vis maximum wavelength and corresponding measured band gap and calculated unstrained and strained core diameter of CdSe/ZnS QDs synthesized at 160 °C and 170 °C for different reaction times (min)

Temperature (°C)	Time (min)	λ_{max} (nm)	E_g^{nc} (eV)	d (nm)	$d(\varepsilon_i)$ (nm)
160	1	532	2.34	3.89	4.35
	5	535	2.32	3.92	4.39
	10	538	2.31	3.95	4.43
	20	539	2.30	3.96	4.45
170	1	549	2.26	4.02	4.61
	5	552	2.25	4.05	4.66
	10	555	2.23	4.08	4.71
	20	557	2.23	4.11	4.74

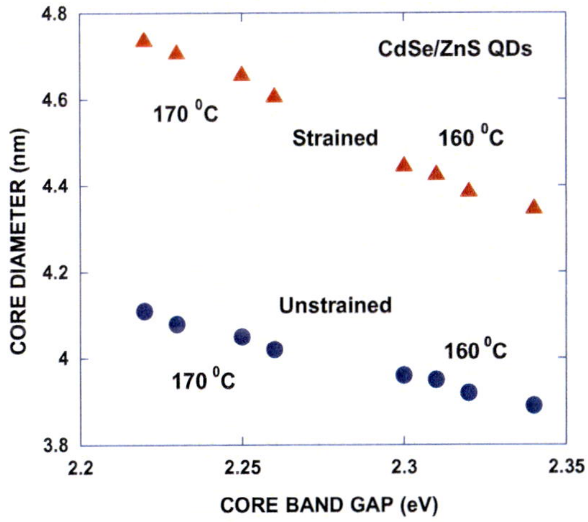

Fig. 3.32 Core diameter of CdSe/ZnS core/shell QD plotted and as a function of core bandgap with and without strain at 160 °C and 170 °C (b), respectively

The strained core diameters are in excellent agreement with HRTEM results, which indicates that the bare CdSe core and core/shell CdSe/ZnS QDs synthesized at 160 °C (a) and 170 °C (b) for 20 min of reaction time have average particle sizes about 3.50 nm and 4.84 nm, respectively.

3.11 Conclusion

We presented the results of a comprehensive study of elastic strain effects on core band gap and diameter of spherical bare CdSe core and CdSe/ZnS heteostructure core/shell quantum dots that are grown by a colloidal technique at various temperatures. The XRD analysis suggests that synthesized CdSe core and CdSe/ZnS core/shell QDs have zinc blende crystal structure. HRTEM measurements indicate that bare CdSe core and CdSe/ZnS core/shell QDs have average particle sizes about 3.50 nm and 4.84 nm, respectively. UV–Vis absorption spectra measurements show that the increase of growth temperature leads to a decrease (increase) in the bandgap (diameter) of CdSe/ZnS core/shell QDs. Furthermore, we also show that the effect of elastic strain to the band gap change is greater than that of lattice vibration in CdSe/ZnS core/shell QDs at any temperature. The compressive strain causes an increase (decrease) in the core band gap (diameter) of spherical CdSe/ZnS core/shell QDs. Elastic strain modified effective mass approximation predicts that there is a parabolic decrease (increase) in the core bandgap (diameter) of CdSe/ZnS core/shell QDs with temperature. The calculated core diameter of bare CdSe core and CdSe/ZnS core/shell QDs are in excellent agreement with HRTEM measurements.

Acknowledgements The authors greatly acknowledge the financial support by the Research Foundation of İstanbul Technical University (İTÜ BAP Project No: 34537).

References

1. J. Albero, P. Riente, J.N. Clifford, M.A. Pericàs, E. Palomares, J. Phys. Chem. C **117**, 13374 (2013)
2. A. Nadarajah, T. Smith, R. Könenkamp, Nanotechnology **23**, 485403 (2012)
3. A.B. Kashyout, H.M.A. Soliman, M. Fathy, E.A. Gomaa, A.A. Zidan, Int. J. Photoenergy **2012**, 1 (2012)
4. C. Li, L. Yang, J. Xiao, Y.-C. Wu, M. Søndergaard, Y. Luo, D. Li, Q. Meng, B.B. Iversen, Phys. Chem. Chem. Phys. **15**, 8710 (2013)
5. C.-Y. Huang, Y.-K. Su, T.-C. Wen, T.-F. Guo, M.-L. Tu, IEEE Photonics Technol. Lett. **20**, 282 (2008)
6. C. Shen, K. Li, Q. Hou, H. Feng, X. Dong, IEEE Photonics Technol. Lett. **22**, 884 (2010)
7. M.-J. Chen, J.-R. Yang, M. Shiojiri, Semicond. Sci. Technol. **27**, 074005 (2012)
8. L. Deng, L. Han, Y. Xi, X. Li, W.-P. Huang, IEEE Photonics J. **4**, 1600 (2012)
9. J.H. Yan, C.G. Wang, H. Zhang, C. Cheng, Laser Phys. Lett. **9**, 529 (2012)
10. X. Peng, J. Wickham, A.P. Alivisatos, J. Am. Chem. Soc. **120**, 5343 (1998)
11. W. Mi, J. Tian, J. Jia, W. Tian, J. Dai, X. Wang, J. Phys. D. Appl. Phys. **45**, 435303 (2012)
12. A.J. Peter, C.W. Lee, Chinese Phys. B **21**, 087302 (2012)
13. S. Buckley, K. Rivoire, J. Vučković, Reports Prog. Phys. **75**, 126503 (2012)
14. L. Hu, H. Wu, Z. Wan, C. Cai, T. Xu, T. Lou, B. Zhang, New J. Phys. **14**, 013059 (2012)
15. H. Ünlü, N.J.M. Horing, J. Dabrowski (eds.), *Low-Dimensional and Nanostructured Materials and Devices* (Springer International Publishing, Cham, 2016)
16. M.R. Karim, M. Balaban, H. Ünlü, Adv. Mater. Sci. Eng. **2019**, 1 (2019)
17. R. He, H. Gu, Colloids surfaces a physicochem. Eng. Asp. **272**, 111 (2006)

18. C.-Q. Zhu, P. Wang, X. Wang, Y. Li, Nanoscale Res. Lett. **3**, 213 (2008)
19. M.R. Karim, H. Ünlü, in *TURKISH Phys. Soc. 33RD Int. Phys. Congr.* (AIP Conference Proceedings, 2018), p. 050004
20. Y. Xing, Q. Chaudry, C. Shen, K.Y. Kong, H.E. Zhau, L.W. Chung, J.A. Petros, R.M. O'Regan, M.V. Yezhelyev, J.W. Simons, M.D. Wang, S. Nie, Nat. Protoc. **2**, 1152 (2007)
21. C. Yang, H. Xie, Y. Li, J.-K. Zhang, B.-L. Su, J. Colloid Interface Sci. **393**, 438 (2013)
22. O. Modelung, editor, *Numerical Data and Functional Relationships in Science and Technology*, vol. 17/a (Springer, 1982)
23. M.A. Hines, P. Guyot-Sionnest, J. Phys. Chem. **100**, 468 (1996)
24. M. Grabolle, M. Spieles, V. Lesnyak, N. Gaponik, A. Eychmüller, U. Resch-Genger, Anal. Chem. **81**, 6285 (2009)
25. A. Joshi, K.Y. Narsingi, M.O. Manasreh, E.A. Davis, B.D. Weaver, Appl. Phys. Lett. **89**, 131907 (2006)
26. D.R. Penn, P hys. Rev. 128. 2093 (1 9 6 2)
27. L. Qu, X. Peng, J. Am. Chem. Soc. **124**, 2049 (2002)
28. D.V. Talapin, A.L. Rogach, E.V. Shevchenko, A. Kornowski, M. Haase, H. Weller, J. Am. Chem. Soc. **124**, 5782 (2002)
29. B. Bhattacharjee, C.-H. Hsu, C.-H. Lu, W.H. Chang, Phys. E Low-Dimensional Syst. Nanostructures **33**, 388 (2006)
30. S. Chattopadhyay, P. Sen, J.T. Andrews, P.K. Sen, J. Phys. Conf. Ser. **365**, 012037 (2012)
31. L.E. Brus, J. Chem. Phys. **80**, 4403 (1984)
32. J.D. Eshelby, Proc. R. Soc. London. Ser. A. Math. Phys. Sci. **241**, 376 (1957)
33. H. Ünlü, in *Turkish Phys. Soc. 33RD Int. Phys. Congr.* (AIP Conference Proceedings, 2018), p. 050003
34. H. Ünlü, Eur. Phys. J. Appl. Phys. **86**, 30401 (2019)
35. H. Ünlü, Solid. State. Electron. **35**, 1343 (1992)
36. Y.P. Varshni, Physica **34**, 149 (1967)

Chapter 4
Synthesis of Transition Metal Dichalcogenides (TMDs)

Kyungnam Kang, Siwei Chen, Shichen Fu, and Eui-Hyeok Yang

Abstract Two-dimensional (2D) materials or van der Waals materials typically have strong in-plane covalent bonds and weak out-of-plane van der Waals forces. The van der Waals materials form stable atomically-thin structures. Graphene can be produced via mechanical exfoliation from highly ordered pyrolytic graphite (HOPG), from which many unique and superior properties have been revealed. The graphene research's success and the lack of semiconductor properties have led to the exploration of other inorganic 2D materials beyond graphene. These materials include transition metal dichalcogenides (TMDs), phosphorene, and MXene. TMDs have attracted considerable attention as core materials for next-generation semiconductor devices owing to their unique electrical, mechanical, chemical, and optical properties. This chapter discusses several methods to synthesize TMDs and to manipulate the properties of TMDs.

4.1 Introduction

TMDs are composed of three layers; top and bottom layers of chalcogen atoms and a middle layer of transition metal atoms. The TMDs are showing unique optical and electrical properties. The bandgaps of TMDs such as WS_2, MoS_2, WSe_2, and $MoSe_2$ change from indirect to direct bandgap when the materials are thinning from bulk to monolayer [1–3], which are in the range of 1–2 eV [4–7]. The type II heterojunction can be facilely fabricated with TMDs owing to their band alignments [8–10]. Similar to strong light-interaction materials, monolayer TMDs can absorb up to 5–10% of incident light [11]. Combining their unique properties, including direct bandgap, type II heterojunction, and strong light-interaction, facilitates efficient relaxation of interlayer excitons for photodetectors and solar cells [12–14]. The second-harmonic generation can be enhanced in TMD monolayers because of the broken inversion

K. Kang · S. Chen · S. Fu · E.-H. Yang (✉)
Mechanical Engineering Department, Stevens Institute of Technology, Hoboken, NJ, USA
e-mail: eyang@stevens.edu

H. Ünlü and N. J. M. Horing (eds.), *Progress in Nanoscale and Low-Dimensional Materials and Devices*, Topics in Applied Physics 144,
https://doi.org/10.1007/978-3-030-93460-6_4

symmetry [15]. The strong spin–orbit coupling and inequivalent valleys permit circularly polarized lights to excite electrons in K-valley and -K-valley due to the valley-dependent optical selection rules. The excited electrons in K-valley and -K-valley are in spin-up state and spin-down state, respectively, due to the spin-valley locking [16]. Moreover, the doping of TMDs changes their bandgaps, carrier mobilities, and magnetic properties [17] for various applications, including electronic, optical devices, energy harvest, sensors, and catalysts [18–22]. In this chapter, the synthesis of TMD monolayers with tailored material properties is introduced. Furthermore, the benefits and issues of different synthesis techniques are discussed.

4.2 Mechanical Exfoliation

The first graphite monolayer called graphene was isolated using the mechanical exfoliation method [23]. The mechanical exfoliation method uses mechanical forces to obtain van der Waals layers; high-quality monolayers can be isolated from natural bulk crystals [24], although the method is not scalable, as compared to CVD-growth methods [25, 26]. There are several types of mechanical exfoliation methods, such as scotch-tape, ball milling, roll milling, gel-assisted exfoliation, metal-assisted exfoliation, and layer-resolved splitting methods [27–32]. The scotch-tape, metal-assisted exfoliation, and LRS methods are introduced here since the scotch-tape method is the first-demonstrated mechanical exfoliation method, and LRS is a method that enables wafer-scale monolayer exfoliation. The metal-assisted exfoliation is similar to the scotch-tape method, while it is a polymer-free process and is facile to obtain large monolayers.

4.2.1 Scotch-Tape Method

Novoselov et al. used to isolate 2D monolayers from a bulk crystal for the first time. Figure 4.1 shows the process of the scotch-tape method [33]. A piece of graphite was located on the middle of an adhesive tape and exfoliated by repeated folding and unfolding the tape. The graphene flakes are transferred onto a SiO_2 substrate by applying a uniform pressure onto the graphite/tape layers to increase the adhesion between the SiO_2 substrate and graphene layers. Finally, the adhesive tape was gently removed to leave graphene on the SiO_2 substrate. Novoselov et al. also reported the exfoliation of TMD monolayers using the scotch-tape method for the first time [34]. They measured the mobility of the exfoliated MoS_2 monolayers was between 0.5 and 3 cm^2/Vs.

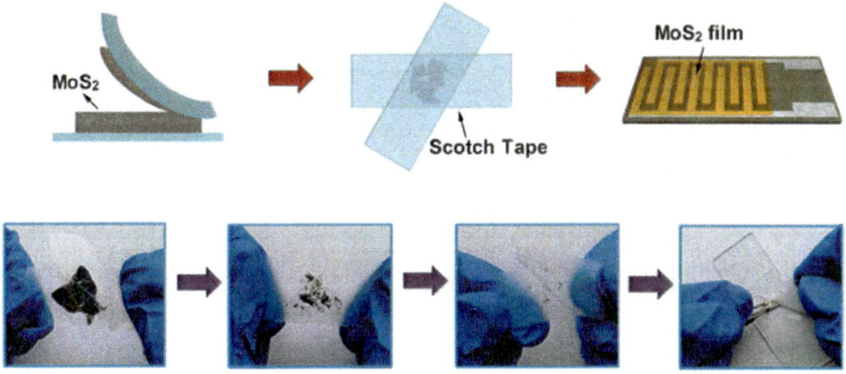

Fig. 4.1 Procedures of the Scotch-tape method to exfoliate monolayers of 2D materials

4.2.2 Metal-Assisted Method

Although the scotch-tape method is a very useful process to obtain monolayer TMDs, it does not allow the production of large-area monolayers. The metal-assisted method allows us to produce large-area flakes as compared to the scotch-tape method. For the metal-assisted method, the adhesion between metal and TMD is critical. Since gold has a strong semi-covalent interaction with chalcogen atoms, Desai et al. developed a gold-assisted exfoliation method that combines the deposited gold layer (100–150 nm) with a thermal release tape, as shown in Fig. 4.2. The exfoliated monolayer MoS_2 can reach about 500 μm^2 [32]. Velicky et al. performed a further study on the mechanism of the gold exfoliation of TMDs [35]. The STEM images showed that the distance between the gold atoms and the top sulfur atom of MoS_2 was 3.5 Å, which is larger than the covalent bond between gold and sulfur (2.2 Å), indicating the interaction between gold and sulfur atoms was not a chemical bond but a strong van der Waals bond. Missing gold atoms on the surface was studied by DFT simulation,

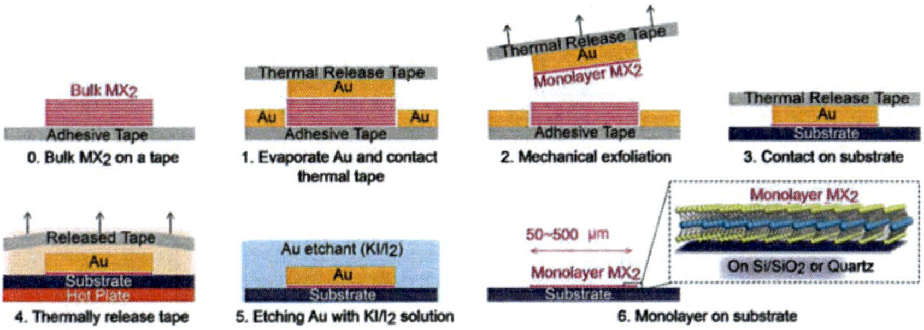

Fig. 4.2 Illustration of the gold-assisted exfoliation process

which showed that large vacancies in gold surfaces could reduce the overall binding force between the TMD and gold surfaces.

4.2.3 Layer-Resolved Splitting (LRS) Method

Since the Scotch-tape and metal-assisted exfoliation methods are not scalable, researchers put their efforts into improving the scalability. Shim et al. have developed a new method called the layer-resolved splitting technique [31]. This technique can isolate a 2-inch wafer size WS_2 monolayer. For this technique, CVD-grown multi-layer WS_2 on a sapphire substrate was used for exfoliation. The top surface of these multilayer TMDs can be nonuniform and discontinuous, but the bottom layer should be uniform and continuous. Ni thick film was deposited on top of the multilayer WS_2, followed by the delamination of the entire WS_2 layers from the sapphire substrate. The bottom side (i.e., continuous and uniform)faced up. Another thick Ni film was deposited to exfoliate WS_2 films layer-by-layer. The different interfacial toughness between 2D material and Ni, between layers in 2D material, and between 2D material and sapphire were 1.4 J/m^2, 0.45 J/m^2, and 0.26 J/m^2, respectively. The quality of the monolayers obtained using the LRS method was lower than that of the scotch-tape method because the LRS method used CVD-grown samples.

Here, it is worth introducing the thinning method, which removes layers to produce a thin film of TMDs. For the thinning process, few-layered TMD flakes are located on a substrate and use thermal energy or laser for the thinning process. Lu et al. applied thermal energy to sublimate TMD flakes from the upper layer [36]. In this experiment, a laser was produced monolayer TMDs by removing extra layers from exfoliated thick TMDs. Hu et al. used a 532 nm laser applied with 2.5mW for 0.2 s.

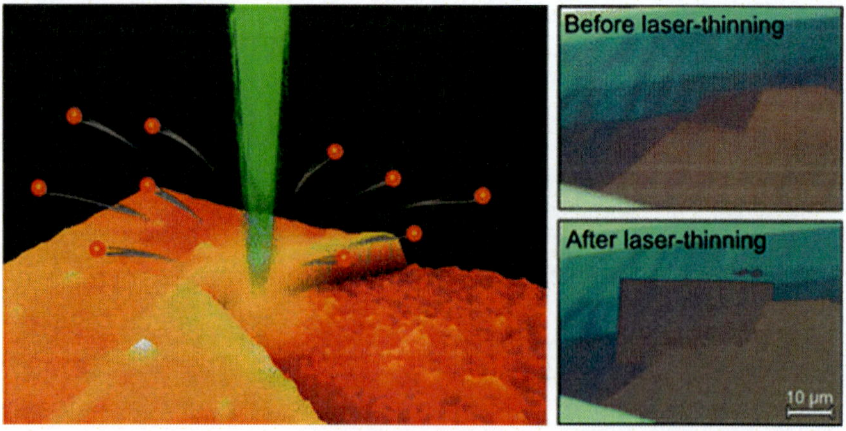

Fig. 4.3 Schematic illustration and photo-images of the laser-assisted thinning of TMDs (Left), before (top right), and after laser thinning (bottom right) of MoS_2

After 13 scans, the original 10-layer MoS_2 was made into monolayer MoS_2 with a size of around 10 microns [37]. The mechanism behind the laser thinning of the TMD multilayers was the thermal thinning when MoS_2 was heated to 603 K. A simulation suggests the surface can be around 669 K while under the laser exposure, and by controlling the exposure time and power, this method can make bulk MoS_2 into the desired thickness. Figure 4.3 left is the schematic illustration of the laser thinning method. As shown in Fig. 4.3 bottom right, this method enabled the fabrication of a square pattern of MoS_2 trilayer, bilayer, and monolayer from a bulk material [38]. The thickness of MoS_2 was confirmed by Raman measurements, as the frequency interval between $E^1{}_{2g}$ and A_{1g} Raman peaks is around 25.3 cm^{-1} for bulk and 19.4 cm^{-1} for monolayers [39]. Unlike other exfoliation and thinning methods, this method comes with control of the location and structure of monolayer TMDs. This method has potential applications in novel devices with a structural design.

4.3 Liquid-Phase Exfoliation

The liquid-phase exfoliation is a very useful method to achieve large-scale and mass production of 2D materials at low cost for various applications. It is useful for fabricating thin-film transistors, inkjet printed electronics, conductive electrodes, and nanocomposites [40–43]. However, the liquid phase exfoliation has clear drawbacks such as small grain size, high defect density, high possibility of contamination by chemical groups, and possible phase transition of exfoliated TMDs. Although several liquid-phase exfoliation methods have been reported, solvent-based and ion intercalation methods are the two most popular methods [44, 45].

4.3.1 Organic Solvent-Based Exfoliation Method

The procedure of the solvent-based exfoliation method consists of immersion, insertion, exfoliation, and stabilization. Figure 4.4 is the schematic diagram of the procedures of solvent-based exfoliation. The solvents must fully immerse the 2D material for providing efficient exfoliation during sonication. Furthermore, they need to exfoliate the material at high concentration and keep from the restacking of exfoliated 2D material. To satisfy the conditions, surface tension, Hildebrand, and Hansen solubility parameters must be taken into consideration in order to determine the proper solvent for a given 2D material. Based on the solvent requirements, IPA/water, acetone/water, and THF/water are the best well-known solvents. The optimal volumetric ratio of solvent to water depends on 2D material. Shen et al. suggested a 1:1 IPA/water ratio for graphene, hBN, WS_2, and $MoSe_2$ and 7:3 for MoS_2 [44]. The N-methyl pyrrolidone (NMP) is an effective solvent for graphene exfoliation due to the strong interaction between the graphene surface and NMP [46]. A pyrene containing ethanol solution has been used for the TMD exfoliation. The pyrene physically adsorbs on the

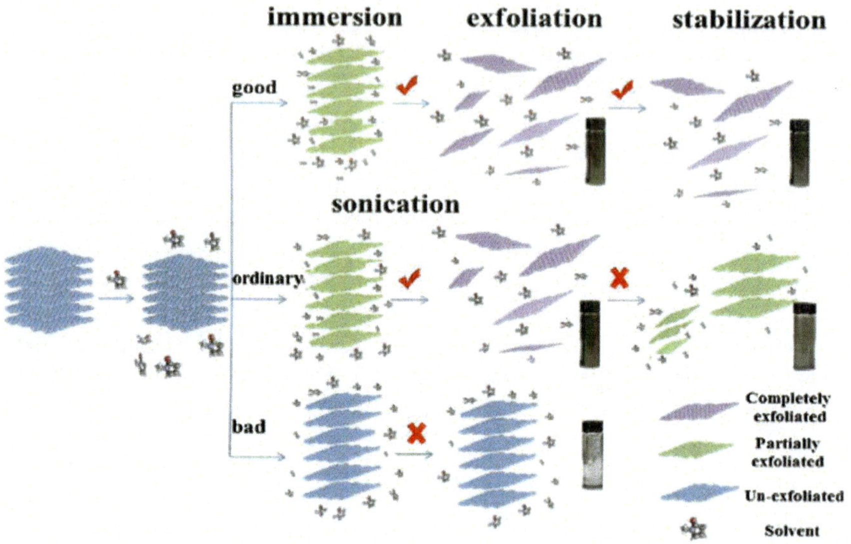

Fig. 4.4 Schematic illustration of the solvent-based exfoliation method

surface of TMD, which is emerged into the solution. During sonication, the pyrene adsorbed TMD surface layer slides to form TMD nanosheets. The pyrene immediately adheres to the freshly exposed TMD surface. The exfoliation and pyrene adsorption cycle can be repeated [47].

4.3.2 *Ion Intercalation Method*

The fundamental principle behind the ion intercalation method is the intercalation of impurities between layers of a bulk TMD crystal to increase the interlayer spacing. Then the increased interlayer spacing reduces van der Waals force for exfoliation. The intercalants include alkali metal, organometallic, polymers, and atomic species. Lithium (Li^+)-ion is a good material because of its high reduction potential and high mobility. For lithium intercalation, n-butyllithium (n-BuLi) solution in hexane has been widely used. The n-Bu^- transfers an electron to TMD layers and Li^+ ion intercalates for the charge balance. Ultrasonication or microwaves have been used to improve Li^+ intercalation efficiency. The Li^+-ion intercalated TMD bulk crystal was used to exfoliate thin layers using hydrolyzation and sonication. Figure 4.5 a is an illustration of a Li^+ intercalated exfoliation process [48]. Zeng et al. introduced the advanced Li^+-ion intercalation method using an electrochemical approach (Fig. 4.5b). The Li^+ intercalation of the electrochemical approach is a fast and controllable method [45]. For the Li^+ intercalation process, a voltage was applied between an anodic Li foil

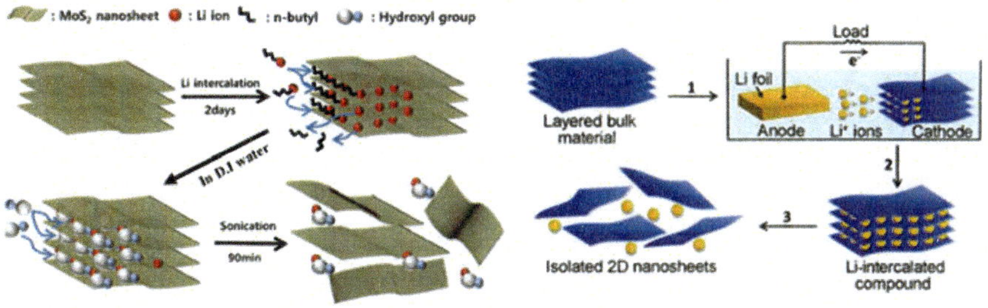

Fig. 4.5 Schematic diagram of the ion intercalation method. **a** lithium intercalation and exfoliation process, and **b** electrochemical lithium intercalation and exfoliation process for TMD materials

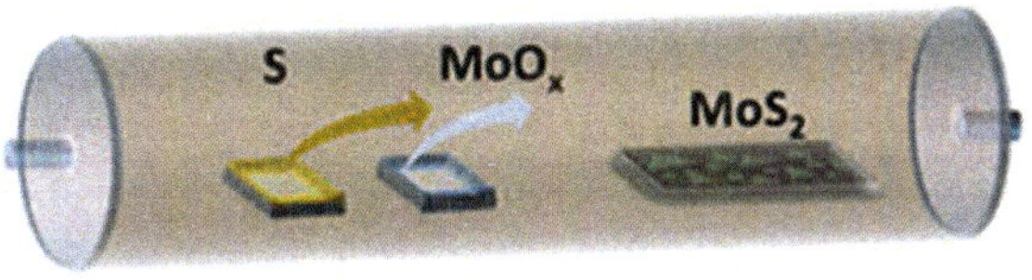

Fig. 4.6 Schematic of a typical thermal CVD setup for TMD growth

and a cathodic bulk TMD in an electrolyte. During the process, the Li^+ ions intercalated between the TMD layers. The Li^+ ion-intercalated TMD bulk crystal was then agitated to generate TMD nanosheets. This exfoliation process can lead to structural deformation of exfoliated materials. The Li^+ intercalation is associated with charge transfer from n-BuLi to the TMD crystal, and the charge transfer changes the structure of TMD from 2H to 1 T [49]. This phase change was more favorable when the Li^+ dosage increased. However, this intercalation-induced phase transformation was reversible by annealing process or exposure to infrared (IR) radiation [50, 51].

4.4 Chemical Vapor Deposition (CVD)

The CVD method for the growth of monolayer 2D materials was first reported in 2007 for graphene growth [52]. Since then, the CVD method has been implemented for material growth and synthesis owing to its cost-effective and scalable production. Now it can control the growth location, the number of layers, the grain size, and the doping impurities in addition to achieving the wafer-scale growth [53–61]. In this section, three different types of CVD methods are described for the TMD growth: thermal CVD, metal–organic CVD (MOCVD), and chemical vapor transport (CVT) methods. Those methods are distinguished by the precursors that are used. The melting point of precursors affects the vapor phase transport, which is crucial in the chemical reactions during the TMD growth. Ideally, one would use metal and chalcogen precursors with melting temperatures similar to each other to achieve steady and stoichiometrically vapor to achieve high-quality growth. In reality, the melting temperature of pure elements S, Se, and Te all have the melting temperatures lower than 500 °C, while Mo, W, Re, and Nb all have a high melting temperature above 2000 °C [62]. This difference in their melting temperatures means the growth of TMDs by using elemental precursors alone is difficult to achieve. To this end, so researchers started developing transitional metal oxide in thermal CVD growth and metal–organic oxide in MOCVD growth as well.

Transitional metal source	Melting temperature (C)	Chalcogen source	Melting temperature (C)
Mo	2633	S	115
MoO_3	802	$(C_2H_5)_2S$	−103.8
$MoBr_3$	500	H_2S	−82
$Mo(CO)_6$	150	Se	220
W	3414	$SeCl_4$	305
WO_3	1473	$(CH_3)_2Se$	−87.2
WCl_6	275	H_2Se	−65.7
$W(CO)_6$	150	Te	450
Nb	2477	$(C_2H_5)_2Te$	−10
$NbCl_5$	205	–	–
Re	3185	–	–
ReF_6	18	–	–

4.4.1 Thermal Chemical Vapor Deposition

Before discussing the thermal-CVD growth mechanism of TMD monolayers, it is worth noting the difference in the growth mechanism between graphene and TMDs. For the growth of graphene with the thermal-CVD method, the introduced hydrocarbon gases are decomposed on the surface of the metal substrate, and carbon atoms dissolve into the substrate. The carbon atoms segregate and form a graphene layer during the cooling down of the substrate due to the solubility difference according to the temperature. Thus thin metal foils have been used as a substrate because the carbon solubility of the substrate is key for graphene growth [63–65]. Unlike the formation process of graphene, the chemical reaction between precursors is the main route for the synthesis of TMDs. In general, powder forms of transition metal oxide and chalcogen are used as precursors. The precursors will evaporate at high temperatures and adsorb on a substrate where the chalcogenization of transition metal oxide occurs to form TMDs [66–68]. Imanishi et al. deposited MoS_2 film with a thermal-CVD method in 1992 [69], and Lee et al. grew MoS_2 monolayer on a SiO_2 substrate in 2012 [70]. They used MoO_3 powder and sulfur powder as precursors. Nitrogen was used as a carrier gas. A SiO_2 substrate was treated with reduced graphene oxide, perylene-3,4,9,10-tetracarboxylic acid tetrapotassium salt (PTAS),

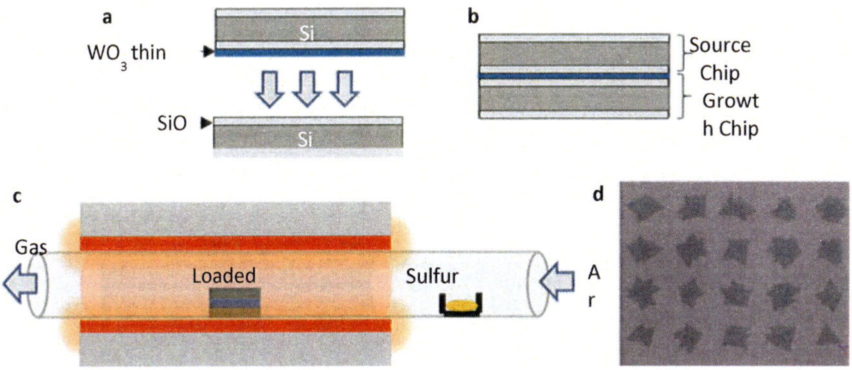

Fig. 4.7 Schematic diagram of the contact growth method. **a** transition metal oxide thin film deposited SiO_2, **b** face-to-face contacted chips, **c** set up for contact growth, and **d** localized growth of MoS_2 monolayers

or perylene-3,4,9,10-tetracarboxylic dianhydride (PTCDA) to promote MoS_2 monolayer growth. Figure 4.6 is a diagram of a typical thermal-CVD setup for TMD growth [71]. A contact growth method was suggested for location-specific growth [72]. This method utilized a thin film of transition metal oxide deposited onto a SiO_2 substrate as a transition metal source instead of using a transition metal oxide powder. The substrate was put onto a bare SiO_2 substrate face-to-face, where TMDs were grown on both substrates (Fig. 4.7a). This contacted substrate set (Fig. 4.7b) was located in the middle of the tube, and chalcogen powder was placed upstream (Fig. 4.7c). This method allowed the formation of well-aligned TMD monolayers (Fig. 4.7d) [53].

In general, the transition metal oxide and the chalcogen powder are placed at different temperature zones due to the difference in their sublimation temperatures. Therefore, the thermal-CVD furnace, which usually has two independent heating zones, is adequate to control the evaporation of the precursors. The gas-phase of transition metal oxide is adsorbed on the substrate, and chalcogen gas is delivered to the substrate surface by an inert carrier gas (Ar or N). The TMD monolayers are formed on the substrate surface after the chalcogenization of pre-adsorbed transition metal oxide. Hydrogen gas occasionally is introduced to improve the reduction of transition metal oxide resulting in better chalcogenization [73]. The typical growth temperatures are between 750 and 950 °C. Fundamental experiments show the relationship between the ratio of transition metal (M) to chalcogen (X), growth temperature, edge structure, and shape the evolution of TMDs. As shown in Fig. 4.8a, Wang et al. demonstrated that the crystal domain would have a triangle shape with M zigzag edges if the M:X ratio is greater than 1:2. If the M:X ratio is less than 1:2, the domain would have a triangle shape with X zigzag edges. When the M:X ratio is equal to 1:2, the domain would have a hexagonal or a truncated triangle shape with alternative M and X zigzag edges [74]. Yang et al. added one more factor, growth temperature, to a connection between the M:X ratio and the shape of the domain. As shown in Fig. 4.8b, they separated three noticeable growth conditions dictating

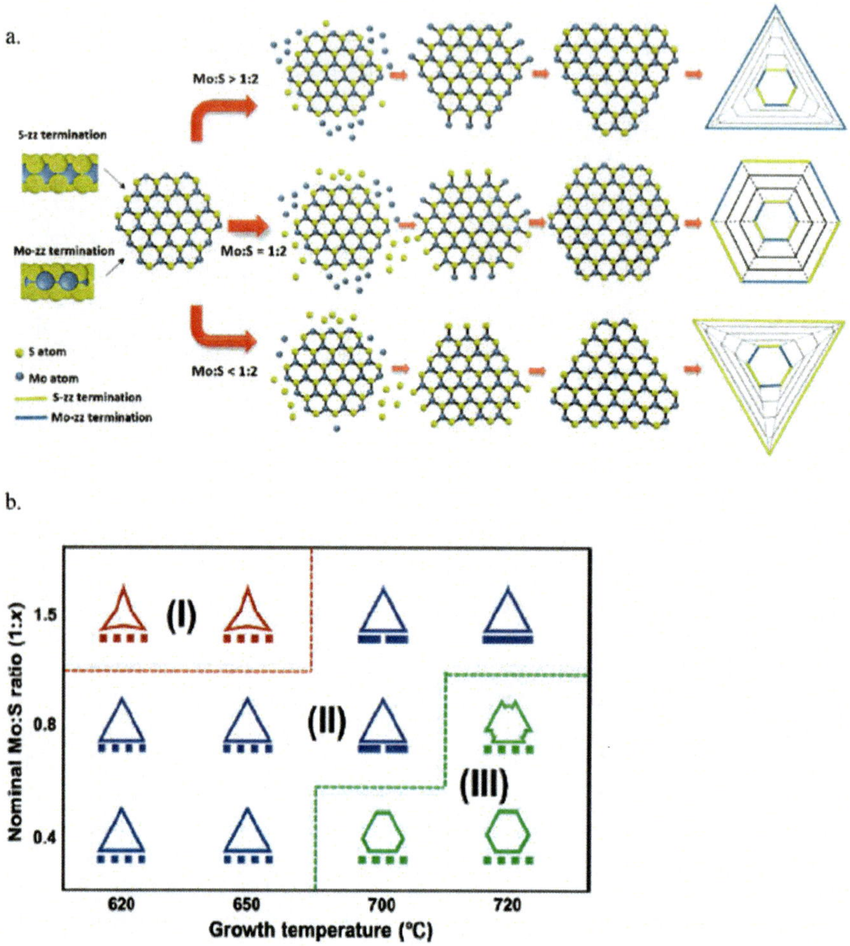

Fig. 4.8 Illustration of domain shape and growth parameters. **a** domain shape according to the M:X ratio, and **b** domain shape with respect to the nominal M:X ratio and growth temperature

domain shapes into a three-point star, triangle, and hexagonal flakes [75]. After the nucleation, the grain size of TMDs increases, and the adjacent grains start merging together. At this moment, the grains would like to stop the growth rather than overlapping and keeping to grow vertically [76]. Thus, large-area monolayers of TMDs can be grown with a thermal-CVD method. An experimental result recently published shows the growth of wafer-scale single-crystalline WS_2, and MoS_2 monolayers. Lee et al. used single-crystalline hexagonal boron nitride (SC-hBN) grown using self-collimation of B and N edges inherently. They used sodium tungstate dihydrate ($Na_2WO_4{\cdot}2H_2O$) dissolved in acetylacetone as a W precursor, sodium molybdate dihydrate ($Na_2MoO_4{\cdot}2H_2O$) in acetylacetone as a Mo precursor, liquid ammonium sulfide solution ($(NH_4)_2S$) as an S precursor, and carrier gases of H_2 and Ar. The triangle shape of WS_2 and MoS_2 monolayers were grown all over the substrate with an aligned direction [77].

In addition to the growth of Mo- and W-based TMDs, the thermal CVD method was used to grow Ta- and Te-based materials. Those materials come with unique properties owing to crystal structure and electronic band structure. TaS_2 has several phases due to different sulfur atom stacking location and can be tuned by gating [78]. 1 T'-WTe_2 is a semi-metallic material with unique large and non-saturating magnetoresistance at low temperatures. Wang et al. used $TaCl_5$ and sulfur as a precursor for 1 T-TaS_2 growth [79]. The growth occurred at 850 °C in a hydrogen and argon environment, which produces a 5 μm monolayer single crystal on an hBN substrate. Li et al. synthesized a few-layered WTe_2 with WCl_6 precursor at 500 °C, with the layer thickness controlled by adjusting the amount of WCl_6 and its distance from the growth substrate [80].

4.4.2 Metal–Organic Chemical Vapor Deposition (MOCVD)

The TMD growth method for scale-up production is critical for commercialization, and MOCVD is a good technique for it. Although MOCVD is a relatively recently developed method for TMD growth, it has been a well-known method for the deposition of thin-film semiconductors [81–83]. For the growth of TMDs using MOCVD, gases of organic molecules containing transition metal (Mo or W) and chalcogen (S or Se) are introduced over a substrate and decomposed by thermal energy to deposit TMD thin films on the substrate. Especially, the MOCVD method can precisely control the partial pressure of the precursors, which permits a uniform deposition of TMDs on large size of a substrate. Figure 4.9 a is the schematic diagram of a typical MOCVD setup [84]. Kang et al. deposited uniform monolayers and few layers of MoS_2 and WS_2 on a 4-inch SiO_2 substrate using molybdenum hexacarbonyl ($Mo(CO)_6$), tungsten hexacarbonyl ($W(CO)_6$), ethylene disulfide ($(C_2H_5)_2S$), Ar, and H_2 [85]. The Ar was used as a carrier gas, and H_2 improved the grain size and crystalline quality. Eichfeld et al. synthesized WSe_2 thin films on various substrates. They have shown the effect of temperature, pressure, the transition metal to chalcogen ratio, and the substrate for the morphology of films [86]. The TEM image shows the size of grain and well-stitched grain boundaries (Fig. 4.9b) [85]. The average grain size was around 1 μm, obtained under high pressure, growth temperature, and with an optimized Se:W ratio. This growth process needs 26 h to synthesize a 4-inch MoS_2 monolayer. The slow growth rate needs to be improved, and Kalanyan et al. reported a much-improved deposition rate. They deposited a few layers of MoS_2 films in 90 s using the pulsed MOCVD method with bis(tert-butylimido)-bis(dimethylamido)molybdenum and diethyl disulfide precursors [87]. The MOCVD method can control the number of layers and grow uniform film on a large area, but it requires toxic precursors.

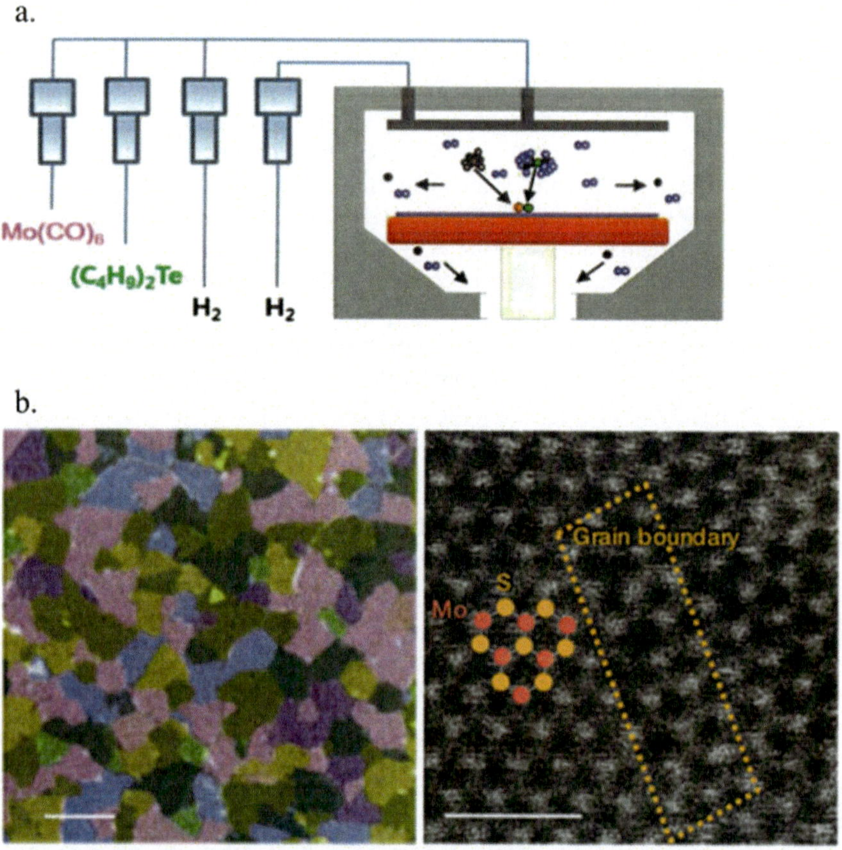

Fig. 4.9 Illustration of the MOCVD process **a** Schematic diagram of MOCVD growth, and **b** false-colored TEM image of MoS_2 monolayers (left), (scale bar 1 μm), and STEM image of stitched grain boundary (right)

4.4.3 Chemical Vapor Transport (CVT) Method

The concept of the CVT method was invented in the middle of the nineteenth century for the growth of single-crystal materials. Schafer conducted systematic research of CVT and elaborated on the migration process. Fischer et al. employed sealed ampoules for the first time. Figure 4.10a shows a setup of a typical CVT method and a grown crystal [88]. A powder form of precursor (AB(s)) is in the source zone (high temperature) with a gas form of transport agent (L(g)). The evaporated precursor is going to decompose and react with the transporting agent. Then, the gases move to the low-temperature area, which is called the sink or deposition zone. The reverse reaction occurs at the deposition zone resulting in the reformation of a single crystalline structure. As shown in Fig. 4.10b, there are two different routes to

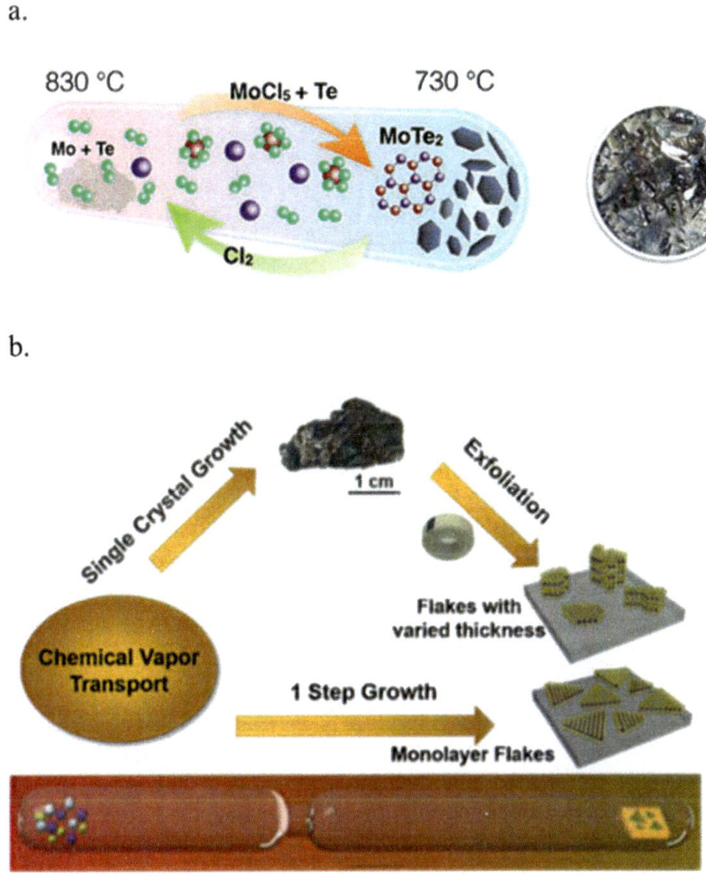

Fig. 4.10 Illustration of a CVT growth **a** typical CVT growth for $MoTe_2$ (left) and grown crystals (right), **b** schematic process for TMD monolayer from CVT grown crystals

obtain TMD monolayers with CVT [89]. One is the exfoliation from a bulk crystal of TMDs, which is grown with the CVT method, and the other is the growth of TMD monolayers on a substrate located at the deposition zone in the ampoule. Dave et al. synthesized MoS_2 and $MoSe_2$ bulk crystals with the CVT method. They used Mo, S, and S_2 as precursors and I_2 as a transporting agent [90]. Interestingly, both crystals were p-type. Ubaldini et al. demonstrated a chloride-driven CVT method to grow MoS_2, $MoSe_2$, and $MoTe_2$ bulk crystals [91]. They used Mo, S, Se, and Te with $MoCl_5$. They found that the ratio of transition metal to transition metal chloride is an important parameter and the ratio depends on the atomic number of chalcogen, such as 50 for sulfide and 15 for telluride. Hu et al. deposited MoS_2 monolayers on a mica substrate. MoO_3 and S were used as precursors, and I_2 was used as a transport agent [89]. This process allows avoiding the mechanical exfoliation process for obtaining monolayer MoS_2 after the bulk MoS_2 crystal growth with CVT. CVT provides high-quality TMD samples which are comparable to that of a mechanical exfoliated sample, but the experimental preparation is complicated and laborious compared to the CVD growths.

4.5 Molecular Beam Epitaxy (MBE)

The use of MBE in fabricating semiconductor devices can be traced back to the 1960s. The deposition requires an ultra-high vacuum (UHV), which typically goes below 10^{-8} to 10^{-12} Torr [92]. During the MBE epitaxial film growth, the molecular source is provided from the effusion cells by heating the solid material or providing the gas source. During the generation of molecular sources, no chemical reaction is involved; the chemical reaction is taking place on the target substrate (Fig. 4.11a) [71, 93–95]. By controlling the shutters of individual effusion cells, MBE can make the sharp atomically thin layers between epitaxy layers or doping the epitaxy layer precisely. Another advantage of using MBE is that the UHV environment can produce high purity of grown films. MBE is one of the first scalable methods for TMD monolayer fabrication. Starting from 1980s, Atsushi Koma synthesized monolayer $MoSe_2$ on $CaF_2(111)$ substrate [96]. Since the doping of TMDs can be achieved by introducing an extra molecular beam source, MBE has the potential for fabricating heterostructure with a doped layer. Fu et al. synthesized MoS_2 monolayers epitaxially on hBN [97]. Figure 4.11 b shows an AFM image of a seamless MoS_2 monolayer grown on an hBN/sapphire wafer. For MoS_2 to nucleate on hBN, the growth process was divided into two steps: In the first step, the growth substrate was kept at 750 °C with a high concentration of Mo source, in order to promote the nucleation of MoS_2. Three hours later, the substrate temperature was adjusted to 750 °C with a lower Mo flux, to initiate the growth of crystals. By doing so, large-scale growth of TMD monolayers was achieved, while it took approximately 10 h to grow a 2-inch wafer size of MoS_2 monolayers [94, 97].

4.6 Doping/Alloy of Transition Metal Dichalcogenides

Although TMDs have been widely studied for the applications due to the inherent optical and electrical properties, the tailoring of the material properties allows a better material selection for specific applications and improve the performance of devices. While several methods are available for the manipulation of TMD properties, doping is an essential method. The doping methods can be sub-categorized into substitutional, interstitial, and charge doping [58, 98–102]. In this chapter, we focus on substitutional doping. The substitutional doping represents the impurity atoms to replace host atoms to form covalent bonds; thus, the products are stable, and

Fig. 4.11 MBE deposition system **a** schematic diagram of MBE deposition, **b** AFM image of MBEgrown MoS_2. Scale bar is 1 μm

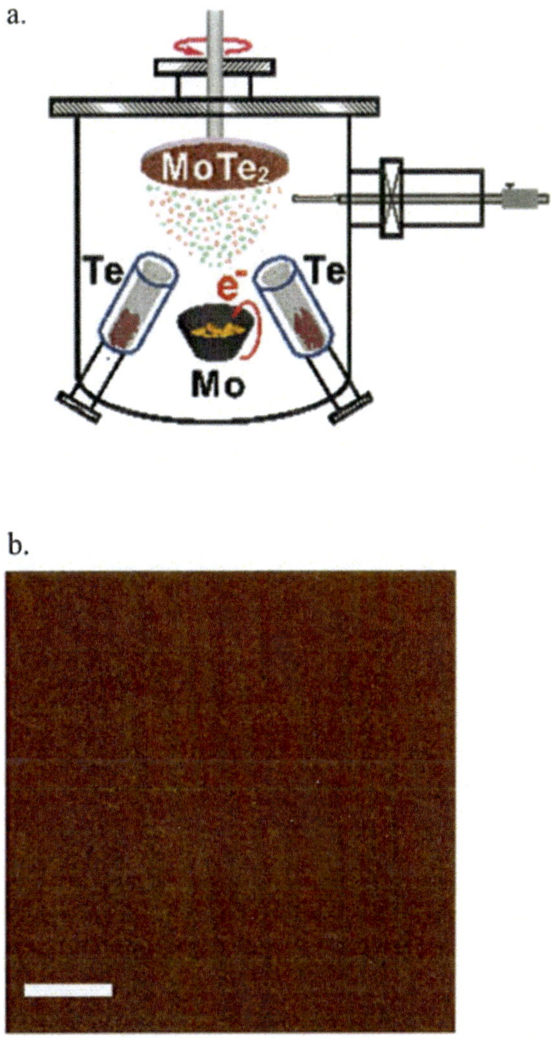

the property degradation is less likely than that of other doping types [103]. TMDs consist of cation elements of the transition metal (Mo and W) and anion elements of chalcogenide (S and Se). Thus, two different types of cation and anion substitutional dopings are discussed.

4.6.1 Substitution of Cation Elements in TMDs

The substitutional cation doping is a method where the host atoms of transition metal in TMD are replaced by the impurity atoms. The tungsten doping on Mo-based TMDs and vice versa are well-known cation substitution processes to manipulate the optical properties. The TMDs have their own peak value of photoluminescence (PL), and the peak position can be shifted by substitutional doping of transition metal atoms and chalcogen atoms. Figure 4.12a contains the graphs of PL intensities of WS_2, MoS_2, WSe_2, and $MoSe_2$ monolayers on SiO_2 substrate at room temperature [104]. The PL peaks are 2.03, 1.88, 1.67 and 1.57 eV for WS_2, MoS_2, WSe_2 and $MoSe_2$ respectively [105]. Tongay et al. controlled the concentration of W and Mo to fabricate $Mo_{1-x}W_xSe_2$ monolayer for tuning the PL peak in Fig. 4.12b [106]. Other metal atoms have also been used for the substitutional cation doping to tailor the property. Rhenium (Re) has been used for donors acting as an n-type dopant, and Niobium (Nb) has been used for acceptor acting as p-type dopants. Zhang et al. doped MoS_2 with Re and observed the shift of the Fermi level up by 0.5 eV resulted from the degenerate n-type doping [107]. The I-V curves of doped and undoped MoS_2 monolayers were measured using the conductive atomic force microscopy (CAFM) tip. Figure 4.12c illustrates a Schottky barrier junction between the CAFM tip and pristine MoS_2 [107]. However, the behavior of Re-doped MoS_2 is closer to metal than to a semiconductor. Suh et al. used Nb as a p-type dopant to transit from inherently n-type MoS_2 to extrinsic p-type doped MoS_2. The 0.5% Nb doping concentration gave an ohmic contact between the doped MoS_2 and the Ti electrode instead of the expected Schottky barrier. The p-n homojunction of vertically stacked Nb-doped and undoped MoS_2 showed the gate tunable current rectification, as shown in Fig. 4.12d [108]. The substitutional doping enabled 2D dilute magnetic semiconductors (DMS) at room temperature for the spintronics and valleytronics applications. The substitution of Fe atoms at Mo sites in MoS_2 monolayers facilitated the ferromagnetism of Fe:MoS_2 at room temperature [109]. The room temperature superconducting quantum interference device (SQUID) and optically detected magnetic resonance (ODMR) measurement showed unambiguous hysteresis (Fig. 4.12e) and frequency widening (Fig. 4.12f). Yun et al. showed vanadium-doped WSe_2 monolayers exhibiting ferromagnetism at room temperature as well [110]. The magnetic force microscopy (MFM) (Fig. 4.12g) showed a domain with phase-contrast separated by domain wall (i) and dendritic patterns in monolayer (ii) and multilayer (iii).

4.6.2 Substitution of Anion Elements in TMDs

The anion substitutional doping is a technique that replaces host atoms of chalcogenides in TMDs with non-metal dopants. Li et al. grew Se-doped MoS_2 monolayers ($MoS_{2x}Se_{2-2x}$) with different Se concentrations [111]. As shown in Fig. 4.13a, the PL

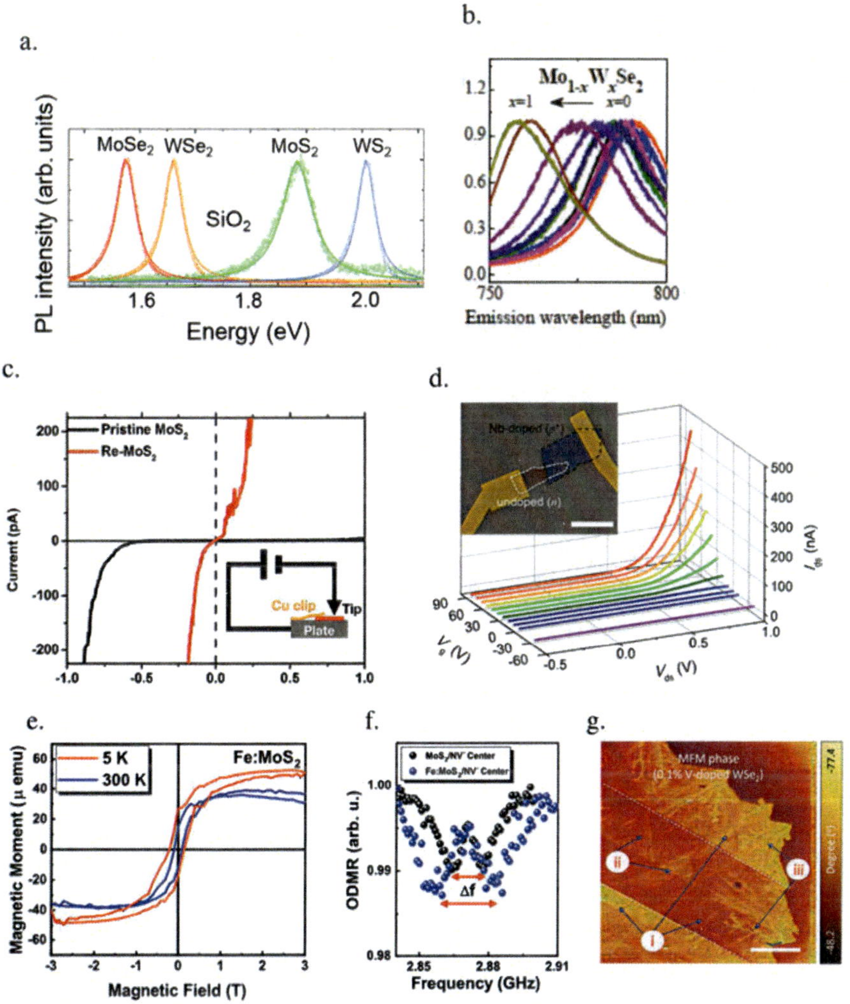

Fig. 4.12 Substitutionally doped TMDs **a** normalized photoluminescences of TMDs, **b** normalized photoluminescences of Mo1-xWxSe$_2$ alloy with different x ($0 \leq x \leq 1$) value, **c** I-V curve of pristine and Re-doped MoS_2 monolayers., **d** Nb-doped-MoS_2 monolayers are p-type. The fabricated p-n homojunction shows current rectification, **e** M-H loop of Fe: MoS_2 monolayers, **f** ODMR spectra of NV^- centers coated on MoS_2 and Fe: MoS_2 monolayers, **g** MFM image showing ferromagnetic domain strips (10 μm scale bar)

peak changes from 659 to 789 nm according to the Se concentrations. The transition from WS_2 to WSe_2 arising from the substitutional doping of Se on WS_2 monolayers changed not only the PL intensities but also the semiconductor type from n- to p-type. The WS_2, MoS_2, and $MoSe_2$ showed inherently n-type semiconductor properties, but WSe_2 showed a p-type property. Thus the transition from WS_2 to WSe_2 changed the PL peak positions as well as the semiconductor type from n to p-type. Duan et al.

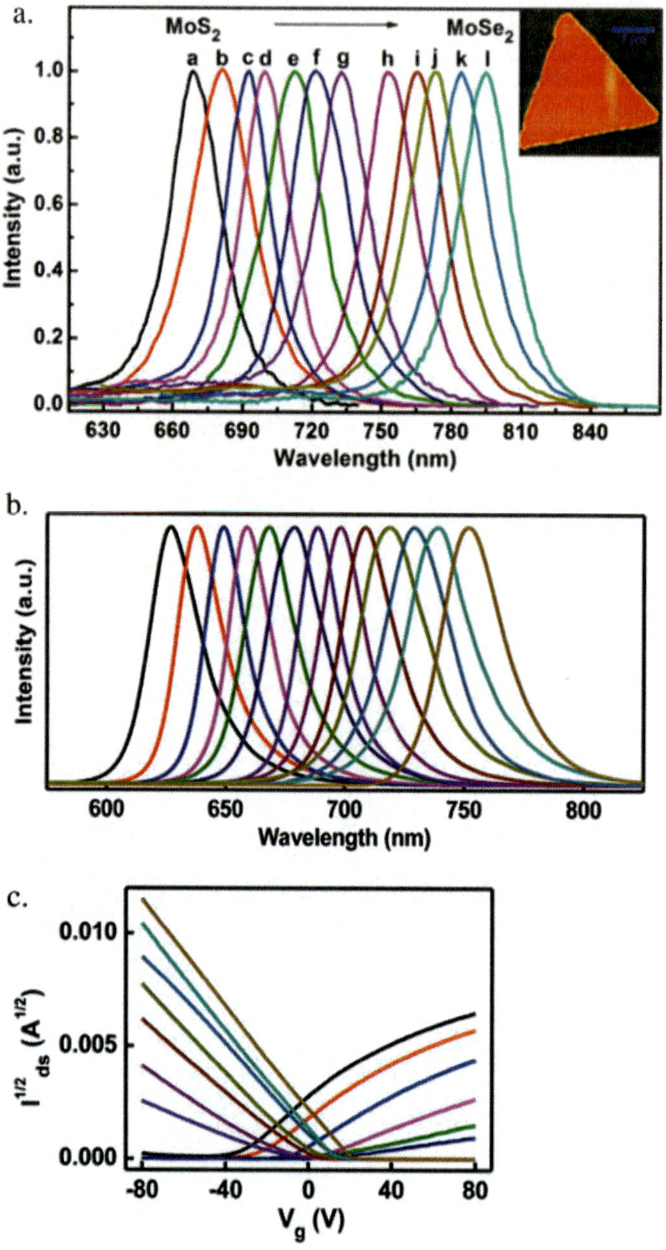

Fig. 4.13 TMD alloys via substitutional doping **a** normalized photoluminescences of $MoS_{2x}Se_{2-2x}$ alloy with different x ($0 \leq x \leq 1$) value, **b** normalized photoluminescences of $WS_{2x}Se_{2\text{-}2x}$ alloy with different x ($0 \leq x \leq 1$) value, **c** I-V curve of n-type semiconductor when it is WS_2 rich alloy and p-type when it is WSe_2 rich alloy, **d** chlorine doped MoS_2 and WS_2 have low contact resistance between Ni and doped material, **e** nitrogen-doped MoS_2 changes from n-type to p-type

Fig. 4.13 (continued)

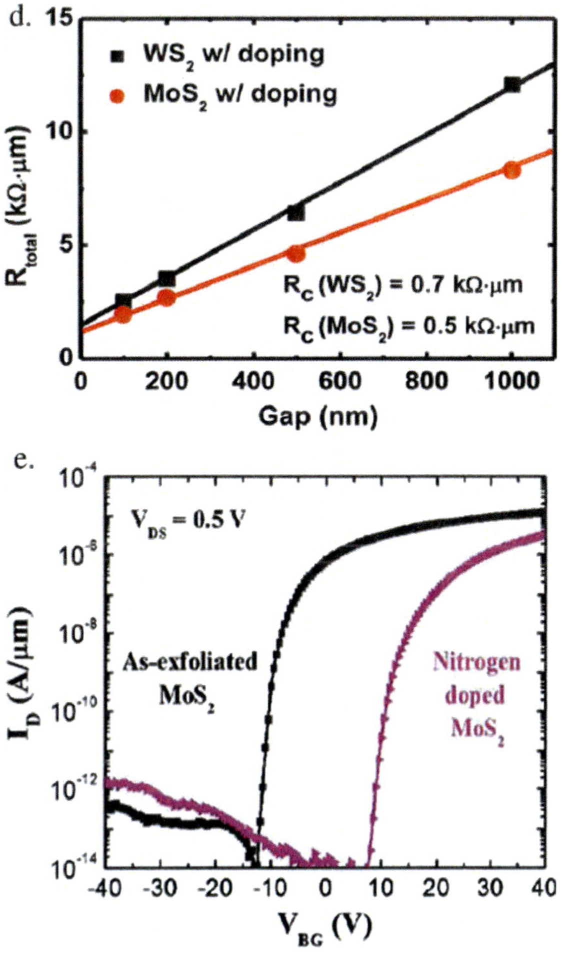

measured the PL intensities and the threshold voltages of $WS_{2x}Se_{2-2x}$ with different values of x. Figure 4.13b, c show the PL changes and p-type in WSe_2 and n-type in WS_2 [112]. Yang et al. doped a few-layered WS_2 and MoS_2 with chloride molecules (Cl) as a dopant [113]. As shown in Fig. 4.13d, the Cl-doped WS_2 and MoS_2 reduced contact resistance and Schottky barrier width. Azcatl et al. used nitrogen to replace sulfur atoms of MoS_2. The nitrogen-doped MoS_2 FET in Fig. 4.13e depicts a positive shifted threshold voltage (Vth), which represents nitrogen-doped MoS_2 is a p-type material [114]. Besides the substitutional doping, the conversion of the top-layer chalcogen atoms can also be achieved. The product is called Janus Monolayers, in which a layer of transitional metal atoms is sandwiched between two layers of chalcogen, halogen, or pnictogen atoms. Rawat et al. proposed a WSeTe/WSTe based hetero-bilayers for nano-piezotronics applications [115]. Lu et al. converted CVD

MoS_2 using an H_2 plasma stripping and thermal salinization [116]. The plasma stripping replaced top layer S atoms with H atoms, which later was replaced by Se atoms. However, during the plasma treatment, the previously grown TMD crystal was shown to be damaged, which limits the application of the Janus fabrication method.

The tailoring of TMD properties is important for several applications. The substitutional doping of ferromagnetic materials into TMDs generates a stable magnetic phase, and the 2D dilute magnetic semiconductors are thus drawing attention due to the application for spintronic and memory devices.

4.7 Summary

The TMDs are 2D semiconductor materials showing unique electrical, mechanical, and optical properties, which can complement graphene (metallic material) and hexagonal boron nitride (hBN, insulator) for the next-generation semiconductor applications. In this chapter, a variety of TMD growth methods have been introduced. Although the basic concepts of production methods can be shared with other van der Waals materials, unique approaches would be required for TMDs. Mechanical exfoliation methods use a bulk crystal of TMDs to produce high-quality samples quickly and easily. Remarkably, the layer-resolved splitting method gives wafer-size TMD monolayers using CVD-grown multilayer TMDs on a SiO_2 substrate. Liquid-phase exfoliations enable mass production of a few-layer TMDs. The CVD methods can permit the growth of wafer-scale TMD monolayers, the control the number of TMD layers and grain sizes. The doping of the TMDs for tailoring properties has also been introduced. The doping processes change the bandgap, type, mobility, and magnetic properties. The knowledge in the synthesis of TMDs and their alloys (via doping) is essential for the research and applications of TMDs.

References

1. Y. Zhang, T.-R. Chang, B. Zhou et al., Direct observation of the transition from indirect to direct bandgap in atomically thin epitaxial $MoSe_2$. Nat. Nanotechnol. **9**, 111–115 (2014)
2. K.F. Mak, C. Lee, J. Hone et al., Atomically thin MoS_2: a new direct-gap semiconductor. Phys Rev Lett **105**, 2–5 (2010)
3. I.G. Lezama, A. Arora, A. Ubaldini et al., Indirect-to-Direct Band Gap Crossover in Few-Layer $MoTe_2$. Nano Lett. **15**, 2336–2342 (2015)
4. H.J. Conley, B. Wang, J.I. Ziegler et al., Bandgap engineering of strained monolayer and bilayer MoS_2. Nano Lett. **13**, 3626–3630 (2013)
5. W. Zhao, Z. Ghorannevis, L. Chu et al., Evolution of electronic structure in atomically thin sheets of WS_2 and WSe_2. ACS Nano **7**, 791–797 (2013)
6. A. Arora, M. Koperski, K. Nogajewski et al., Excitonic resonances in thin films of WSe_2: from monolayer to bulk material. Nanoscale **7**, 10421–10429 (2015)
7. M.M. Ugeda, A.J. Bradley, S.-F. Shi et al., Giant bandgap renormalization and excitonic effects in a monolayer transition metal dichalcogenide semiconductor. Nat. Mater **13**, 1091–1095 (2014)

8. V.O. Özçelik, J.G. Azadani, C. Yang, et al. Band alignment of two-dimensional semiconductors for designing heterostructures with momentum space matching. Phys. Rev. B **94**, 035125 (2016)
9. C.-H. Lee, G.-H. Lee, A.M. van der Zande et al., Atomically thin p–n junctions with van der Waals heterointerfaces. Nat. Nanotechnol. **9**, 676–681 (2014)
10. J. Kang, S. Tongay, J. Zhou, et al., Band offsets and heterostructures of two-dimensional semiconductors. Appl. Phys. Lett. **102**, 012111 (2013)
11. X. Li, H. Zhu, Two-dimensional MoS_2: Properties, preparation, and applications. J. Mater. **1**, 33–44 (2015)
12. A. Dodda, A. Oberoi, A. Sebastian et al., Stochastic resonance in MoS_2 photodetector. Nat. Commun. **11**, 4406 (2020)
13. L.-Y. Gan, Q. Zhang, Y. Cheng et al., Photovoltaic heterojunctions of fullerenes with MoS_2 and WS_2 monolayers. J. Phys. Chem. Lett. **5**, 1445–1449 (2014)
14. M. Bernardi, M. Palummo, J.C. Grossman, Extraordinary sunlight absorption and one nanometer thick photovoltaics using two-dimensional monolayer materials. Nano Lett. **13**, 3664–3670 (2013)
15. H. Zeng, J. Dai, W. Yao et al., Valley polarization in MoS_2 monolayers by optical pumping. Nat Nanotechnol. **7**, 490–493 (2012)
16. D. Xiao, G.-B. Liu, W. Feng, et al., Coupled spin and valley physics in monolayers of MoS_2 and other group-VI dichalcogenides. Phys. Rev. Lett. **108**, 196802 (2012)
17. M. Zhong, C. Shen, L. Huang, et al., Electronic structure and exciton shifts in Sb-doped MoS_2 monolayer. npj 2D Mater Appl. **3**, 1 (2019)
18. H. Kwon, P.J. Jeon, J.S. Kim, et al., Large scale MoS_2 nanosheet logic circuits integrated by photolithography on glass. 2D Mater, **3**, 044001 (2016)
19. O. Lopez-Sanchez, D. Lembke, M. Kayci et al., Ultrasensitive photodetectors based on monolayer MoS_2. Nat Nanotechnol. **8**, 497–501 (2013)
20. C. Li, Q. Cao, F. Wang et al., Engineering graphene and TMDs based van der Waals heterostructures for photovoltaic and photoelectrochemical solar energy conversion. Chem. Soc. Rev. **47**, 4981–5037 (2018)
21. S. Barua, H.S. Dutta, S. Gogoi et al., Nanostructured MoS_2-based advanced biosensors: a review. ACS Appl. Nano Mater. **1**, 2–25 (2018)
22. L. Madauß, I. Zegkinoglou, H. Vázquez Muiños et al., Highly active single-layer MoS_2 catalysts synthesized by swift heavy ion irradiation. Nanoscale **10**, 22908–22916 (2018)
23. K.S. Novoselov, A.K. Geim, S.V. Morozov et al., Two-dimensional gas of massless Dirac fermions in graphene. Nature **438**, 197–200 (2005)
24. H. Li, J. Wu, Z. Yin et al., Preparation and applications of mechanically exfoliated single-layer and multilayer MoS_2 and WSe_2 nanosheets. Acc. Chem. Res. **47**, 1067–1075 (2014)
25. H. Yuan, D. Dubbink, R. Besselink et al., The rapid exfoliation and subsequent restacking of layered titanates driven by an acid-base reaction. Angew. Chemie **54**, 9239–9243 (2015)
26. S. Masubuchi, M. Morimoto, S. Morikawa et al., Autonomous robotic searching and assembly of two-dimensional crystals to build van der Waals superlattices. Nat. Commun. **9**, 1413 (2018)
27. H. Li, J. Wu, X. Huang et al., Rapid and reliable thickness identification of two-dimensional nanosheets using optical microscopy. ACS Nano **7**, 10344–10353 (2013)
28. A.E. Del Rio-Castillo, C. Merino, E. Díez-Barra et al., Selective suspension of single layer graphene mechanochemically exfoliated from carbon nanofibres. Nano Res. **7**, 963–972 (2014)
29. J. Chen, M. Duan, G. Chen, Continuous mechanical exfoliation of graphene sheets via three-roll mill. J. Mater. Chem. **22**, 19625 (2012)
30. P. Institution of Engineering and Technology, P.T. Baine, J.H. Montgomery, et al., Micro Nano Lett. Institution of Engineering and Technology (2006)
31. J. Shim, S.-H. Bae, W. Kong et al., Controlled crack propagation for atomic precision handling of wafer-scale two-dimensional materials. Science **362**, 665–670 (2018)
32. S.B. Desai, S.R. Madhvapathy, M. Amani et al., Gold-mediated exfoliation of ultralarge optoelectronically-perfect monolayers. Adv. Mater. **28**, 4053–4058 (2016)

33. W. Li, Y. Zhang, X. Long, et al., Gas sensors based on mechanically exfoliated MoS_2 nanosheets for room-temperature NO_2 detection. Sens. **19**, 2123 (2019)
34. K.S. Novoselov, D. Jiang, F. Schedin, et al., Two-dimensional atomic crystals. Proc. Natl. Acad. Sci. U S A **102**, 10451 LP–10453 (2005)
35. M. Velický, G.E. Donnelly, W.R. Hendren et al., Mechanism of gold-assisted exfoliation of centimeter-sized transition-metal dichalcogenide monolayers. ACS Nano **12**, 10463–10472 (2018)
36. X. Lu, M.I.B. Utama, J. Zhang et al., Layer-by-layer thinning of MoS_2 by thermal annealing. Nanoscale **5**, 8904–8908 (2013)
37. L. Hu, X. Shan, Y. Wu et al., Laser thinning and patterning of MoS_2 with layer-by-layer precision. Sci. Rep. **7**, 15538 (2017)
38. A. Castellanos-Gomez, M. Barkelid, A.M. Goossens et al., Laser-thinning of MoS_2: on demand generation of a single-layer semiconductor. Nano Lett. **12**, 3187–3192 (2012)
39. H. Li, Q. Zhang, C.C.R. Yap et al., From bulk to monolayer MoS_2: evolution of raman scattering. Adv. Funct. Mater. **22**, 1385–1390 (2012)
40. F. Torrisi, T. Hasan, W. Wu et al., Inkjet-printed graphene electronics. ACS Nano **6**, 2992–3006 (2012)
41. X. Zeng, H. Hirwa, S. Metel et al., Solution processed thin film transistor from liquid phase exfoliated MoS_2 flakes. Solid State Electron **141**, 58–64 (2018)
42. P. Blake, P.D. Brimicombe, R.R. Nair et al., Graphene-based liquid crystal device. Nano Lett. **8**, 1704–1708 (2008)
43. J. Xiao, D. Choi, L. Cosimbescu et al., Exfoliated MoS_2 nanocomposite as an anode material for lithium ion batteries. Chem. Mater. **22**, 4522–4524 (2010)
44. J. Shen, Y. He, J. Wu et al., Liquid phase exfoliation of two-dimensional materials by directly probing and matching surface tension components. Nano Lett. **15**, 5449–5454 (2015)
45. Z. Zeng, Z. Yin, X. Huang et al., Single-layer semiconducting nanosheets: high-yield preparation and device fabrication. Angew. Chemie **50**, 11093–11097 (2011)
46. S.P. Ogilvie, M.J. Large, G. Fratta et al., Considerations for spectroscopy of liquid-exfoliated 2D materials: emerging photoluminescence of N-methyl-2-pyrrolidone. Sci. Rep. **7**, 16706 (2017)
47. D. Wang, F. Wu, Y. Song et al., Large-scale production of defect-free MoS_2 nanosheets via pyrene-assisted liquid exfoliation. J. Alloys Compd. **728**, 1030–1036 (2017)
48. J.H. Lee, W.S. Jang, S.W. Han et al., Efficient hydrogen evolution by mechanically strained MoS $_2$ nanosheets. Langmuir **30**, 9866–9873 (2014)
49. S. Shi, Z. Sun, Y.H. Hu, Synthesis, stabilization and applications of 2-dimensional 1T metallic MoS_2. J. Mater. Chem. A **6**, 23932–23977 (2018)
50. X. Fan, P. Xu, D. Zhou et al., Fast and efficient preparation of exfoliated 2H MoS_2 nanosheets by sonication-assisted lithium intercalation and infrared laser-induced 1T to 2H phase reversion. Nano Lett. **15**, 5956–5960 (2015)
51. G. Eda, H. Yamaguchi, D. Voiry et al., Photoluminescence from chemically exfoliated MoS_2. Nano Lett. **11**, 5111–5116 (2011)
52. P. Blake, E.W. Hill, A.H. Castro Neto, et al., Making graphene visible. Appl. Phys. Lett. **91**, 063124 (2007)
53. X. Wang, K. Kang, S. Chen, et al., Location-specific growth and transfer of arrayed MoS_2 monolayers with controllable size. 2D Mater. **4**, 025093 (2017)
54. J. Jeon, S.K. Jang, S.M. Jeon et al., Layer-controlled CVD growth of large-area two-dimensional MoS_2 films. Nanoscale **7**, 1688–1695 (2015)
55. H. Xu, W. Zhou, X. Zheng et al., Control of the nucleation density of molybdenum disulfide in large-scale synthesis using chemical vapor deposition. Materials (Basel) **11**, 870 (2018)
56. E.Z. Xu, H.M. Liu, K. Park et al., p-Type transition-metal doping of large-area MoS_2 thin films grown by chemical vapor deposition. Nanoscale **9**, 3576–3584 (2017)
57. K. Zhang, S. Feng, J. Wang et al., Manganese doping of monolayer MoS_2: the substrate is critical. Nano Lett. **15**, 6586–6591 (2015)

58. T. Hallam, S. Monaghan, F. Gity, et al., Rhenium-doped MoS_2 films. Appl. Phys. Lett. **111**, 203101 (2017)
59. S. Das, M. Demarteau, A. Roelofs, Nb-doped single crystalline MoS_2 field effect transistor. Appl. Phys. Lett. **106**, 173506 (2015)
60. Y. Kim, H. Bark, G.H. Ryu, et al., Wafer-scale monolayer MoS_2 grown by chemical vapor deposition using a reaction of MoO_3 and H_2 S. J. Phys. Condens. Matter. **28**, 184002 (2016)
61. H. Yu, M. Liao, W. Zhao et al., Wafer-scale growth and transfer of highly-oriented monolayer MoS_2 continuous films. ACS Nano **11**, 12001–12007 (2017)
62. H. Li, Y. Li, A. Aljarb et al., Epitaxial growth of two-dimensional layered transition-metal dichalcogenides: growth mechanism, controllability, and scalability. Chem. Rev. **118**, 6134–6150 (2018)
63. Q. Yu, J. Lian, S. Siriponglert, et al., Graphene segregated on Ni surfaces and transferred to insulators. Appl. Phys. Lett. **93**, 113103 (2008)
64. A. Reina, S. Thiele, X. Jia et al., Growth of large-area single- and Bi-layer graphene by controlled carbon precipitation on polycrystalline Ni surfaces. Nano Res. **2**, 509–516 (2009)
65. K.F. McCarty, P.J. Feibelman, E. Loginova et al., Kinetics and thermodynamics of carbon segregation and graphene growth on Ru(0 0 0 1). Carbon N Y **47**, 1806–1813 (2009)
66. J. You, M.D. Hossain, Z. Luo, Synthesis of 2D transition metal dichalcogenides by chemical vapor deposition with controlled layer number and morphology. Nano Converg. **5**, 26 (2018)
67. J.D. Cain, F. Shi, J. Wu et al., Growth mechanism of transition metal dichalcogenide monolayers: the role of self-seeding fullerene nuclei. ACS Nano **10**, 5440–5445 (2016)
68. D. Zhou, H. Shu, C. Hu et al., Unveiling the growth mechanism of MoS_2 with chemical vapor deposition: from two-dimensional planar nucleation to self-seeding nucleation. Cryst. Growth Des. **18**, 1012–1019 (2018)
69. N. Imanishi, K. Kanamura, Z. Takehara, Synthesis of MoS[sub 2] thin film by chemical vapor deposition method and discharge characteristics as a cathode of the lithium secondary battery. J. Electrochem. Soc. **139**, 2082 (1992)
70. Y.-H. Lee, X.-Q. Zhang, W. Zhang et al., Synthesis of large-area MoS_2 atomic layers with chemical vapor deposition. Adv. Mater. **24**, 2320–2325 (2012)
71. T.H. Choudhury, X. Zhang, Z.Y. Al Balushi et al., Epitaxial growth of two-dimensional layered transition metal dichalcogenides. Annu. Rev. Mater. Res. **50**, 155–177 (2020)
72. K. Kang, K. Godin, Y.D. Kim et al., Graphene-assisted antioxidation of tungsten disulfide monolayers: substrate and electric-field effect. Adv. Mater. **29**, 1603898 (2017)
73. K. Kang, K. Godin, E.H. Yang, The growth scale and kinetics of WS_2 monolayers under varying H_2 concentration. Sci. Rep. **5**, 13205 (2015)
74. S. Wang, Y. Rong, Y. Fan et al., Shape evolution of monolayer MoS_2 crystals grown by chemical vapor deposition. Chem. Mater. **26**, 6371–6379 (2014)
75. S.Y. Yang, G.W. Shim, S.-B. Seo et al., Effective shape-controlled growth of monolayer MoS_2 flakes by powder-based chemical vapor deposition. Nano Res. **10**, 255–262 (2017)
76. D. Dumcenco, D. Ovchinnikov, K. Marinov et al., Large-area epitaxial monolayer MoS_2. ACS Nano **9**, 4611–4620 (2015)
77. J.S. Lee, S.H. Choi, S.J. Yun, et al., Wafer-scale single-crystal hexagonal boron nitride film via self-collimated grain formation. Science (80), **362**, 817–821 (2018)
78. Y. Yu, F. Yang, X.F. Lu et al., Gate-tunable phase transitions in thin flakes of 1T-TaS2. Nat Nanotechnol. **10**, 270–276 (2015)
79. X. Wang, H. Liu, J. Wu, et al., Chemical growth of 1T-TaS(2) monolayer and thin films: robust charge density wave transitions and high bolometric responsivity. Adv. Mater. **30**, e1800074 (2018)
80. J. Li, S. Cheng, Z. Liu et al., Centimeter-scale, large-area, few-layer 1T′-WTe_2 films by chemical vapor deposition and its long-term stability in ambient condition. J. Phys. Chem. C **122**, 7005–7012 (2018)
81. J. Nishizawa, T. Kurabayashi, Mechanism of gallium arsenide MOCVD. Vacuum **41**, 958–962 (1990)

82. J. Nishizawa, T. Kurabayashi, On the reaction mechanism of GaAs MOCVD. J. Electrochem. Soc. **130**, 413 (1983)
83. T. Institution of Electrical Engineers, S. Hattori, S. Sakai, et al., Electron. Lett. [Institution of Electrical Engineers] (1984)
84. T. Kim, H. Park, D. Joung et al., Wafer-scale epitaxial 1T′, 1T′–2H mixed, and 2H phases $MoTe_2$ thin films grown by metal–organic chemical vapor deposition. Adv. Mater. Interfaces **5**, 1800439 (2018)
85. K. Kang, S. Xie, L. Huang et al., High-mobility three-atom-thick semiconducting films with wafer-scale homogeneity. Nature **520**, 656–660 (2015)
86. S.M. Eichfeld, L. Hossain, Y.-C. Lin et al., Highly scalable, atomically thin WSe_2 grown *via* metal-organic chemical vapor deposition. ACS Nano **9**, 2080–2087 (2015)
87. B. Kalanyan, W.A. Kimes, R. Beams et al., Rapid wafer-scale growth of polycrystalline 2H-MoS_2 by pulsed metal-organic chemical vapor deposition. Chem. Mater. **29**, 6279–6288 (2017)
88. T. Goldstein, S.-Y. Chen, J. Tong et al., Raman scattering and anomalous Stokes–anti-Stokes ratio in $MoTe_2$ atomic layers. Sci. Rep. **6**, 28024 (2016)
89. D. Hu, G. Xu, L. Xing et al., Two-dimensional semiconductors grown by chemical vapor transport. Angew. Chemie **56**, 3611–3615 (2017)
90. M. Dave, R. Vaidya, S.G. Patel et al., High pressure effect on MoS_2 and $MoSe_2$ single crystals grown by CVT method. Bull. Mater. Sci. **27**, 213–216 (2004)
91. A. Ubaldini, J. Jacimovic, N. Ubrig et al., Chloride-driven chemical vapor transport method for crystal growth of transition metal dichalcogenides. Cryst. Growth Des. **13**, 4453–4459 (2013)
92. S. Tiefenbacher, H. Sehnert, C. Pettenkofer et al., Epitaxial films of WS_2 by metal organic van der Waals epitaxy (MO-VDWE). Surf. Sci. **318**, L1161–L1164 (1994)
93. Q. He, P. Li, Z. Wu, et al., Molecular beam epitaxy scalable growth of wafer-scale continuous semiconducting monolayer MoTe $_2$ on inert amorphous dielectrics. Adv. Mater., 1901578 (2019)
94. R.F.C. Farrow, Molecular beam epitaxy: applications to key materials. Noyes Publications (1995)
95. N. Briggs, S. Subramanian, Z. Lin, et al., A roadmap for electronic grade 2D materials. 2D Mater **6**, 22001 (2019)
96. A. Koma, K. Saiki, Y. Sato, Heteroepitaxy of a two-dimensional material on a three-dimensional material. Appl. Surf. Sci. **41–42**, 451–456 (1990)
97. D. Fu, X. Zhao, Y.-Y. Zhang et al., Molecular beam epitaxy of highly crystalline monolayer molybdenum disulfide on hexagonal boron nitride. J. Am. Chem. Soc. **139**, 9392–9400 (2017)
98. S. Sasaki, Y. Kobayashi, Z. Liu, et al., Growth and optical properties of Nb-doped WS_2 monolayers. Appl. Phys. Express **9**, 071201 (2016)
99. S.K. Pandey, H. Alsalman, J.G. Azadani et al., Controlled p-type substitutional doping in large-area monolayer WSe_2 crystals grown by chemical vapor deposition. Nanoscale **10**, 21374–21385 (2018)
100. M.R. Laskar, D.N. Nath, L. Ma, et al.. p-type doping of MoS $_2$ thin films using Nb. Appl. Phys. Lett. **104**, 092104 (2014)
101. S. Mouri, Y. Miyauchi, K. Matsuda, Tunable photoluminescence of monolayer MoS_2 via chemical doping. Nano Lett. **13**, 5944–5948 (2013)
102. W.H. Chae, J.D. Cain, E.D. Hanson, et al., Substrate-induced strain and charge doping in CVD-grown monolayer MoS_2. Appl. Phys. Lett. **111**, 143106 (2017)
103. Y. Kim, H. Bark, B. Kang et al., Wafer-scale substitutional doping of monolayer MoS_2 films for high-performance optoelectronic devices. ACS Appl. Mater. Interfaces **11**, 12613–12621 (2019)
104. F. Cadiz, E. Courtade, C. Robert, et al., Excitonic linewidth approaching the homogeneous limit in MoS_2-based van der waals heterostructures. Phys Rev X **7**, 021026 (2017)
105. J. Gusakova, X. Wang, L.L. Shiau et al., Electronic properties of bulk and monolayer TMDs: theoretical study within DFT framework (GVJ-2e method). Phys. Status Solidi **214**, 1700218 (2017)

106. S. Tongay, D.S. Narang, J. Kang, et al., Two-dimensional semiconductor alloys: monolayer Mo_{1-x} W_x Se_2. Appl. Phys. Lett. **104**, 012101 (2014)
107. K. Zhang, B.M. Bersch, J. Joshi et al., Tuning the electronic and photonic properties of monolayer MoS_2 via in situ rhenium substitutional doping. Adv. Funct. Mater. **28**, 1706950 (2018)
108. J. Suh, T.-E. Park, D.-Y. Lin et al., Doping against the native propensity of MoS_2: degenerate hole doping by cation substitution. Nano Lett. **14**, 6976–6982 (2014)
109. S. Fu, K. Kang, K. Shayan et al., Enabling room temperature ferromagnetism in monolayer MoS_2 via in situ iron-doping. Nat. Commun. **11**, 2034 (2020)
110. S.J. Yun, D.L. Duong, D.M. Ha, et al., Ferromagnetic order at room temperature in monolayer WSe(2) semiconductor via vanadium dopant. Adv Sci (Weinheim, Baden-Wurttemberg, Ger) **7**, 1903076 (2020)
111. H. Li, X. Duan, X. Wu et al., Growth of alloy MoS_{2x} Se 2(1–x) nanosheets with fully tunable chemical compositions and optical properties. J. Am. Chem. Soc. **136**, 3756–3759 (2014)
112. X. Duan, C. Wang, Z. Fan et al., Synthesis of WS_{2x} Se_{2-2x} alloy nanosheets with composition-tunable electronic properties. Nano Lett. **16**, 264–269 (2016)
113. L. Yang, K. Majumdar, H. Liu et al., Chloride molecular doping technique on 2D materials: WS_2 and MoS_2. Nano Lett. **14**, 6275–6280 (2014)
114. A. Azcatl, X. Qin, A. Prakash et al., Covalent nitrogen doping and compressive strain in MoS_2 by remote N_2 plasma exposure. Nano Lett. **16**, 5437–5443 (2016)
115. A. Rawat, R. Ahammed, et al., Solar energy harvesting in type II van der waals heterostructures of semiconducting group III monochalcogenide monolayers. J. Phys. Chem. C **123**, 12666–12675 (2019)
116. A.-Y. Lu, H. Zhu, J. Xiao et al., Janus monolayers of transition metal dichalcogenides. Nat Nanotechnol. **12**, 744–749 (2017)

Chapter 5
II-VI Semiconductor Quantum Dots: The Evolution of Color Purity with Structure

Mehmet Hikmet Yukselici, M. K. Torun, Asuman A. Bozkurt, Melda Patan Alper, Zaher M. Nassar, Damla Bulut, and Mesut Balaban

Abstract II-VI semiconductor quantum dots such as binary CdTe, CdSe, CdS, ZnO and ternary $CdS_{1-x}Se_x$, $Cd_xZn_{1-x}S$ embedded in glass or deposited on glass substrates as thin film or colloids in aqueous solution structures have potential for technological applications such as light emitting diodes (LED), gas, pressure and temperature sensing detectors and two- or three-dimensional imaging, and solar cells. In this chapter we present a comprehensive study on nanometer-sized crystals fabricated by various routes to exploit the defect structure, progression of stoichiometry, growth kinetics, strain due to lattice mismatch and quantum size effect by employing optical (transmission, photoluminescence and Raman spectroscopy) and structural (SEM and XRD) characterization techniques. We finally scrutinize the evolution of color or emission wavelength with crystallite size dependent band gap energy through colorimetry.

5.1 Introduction to II-VI Semiconductor Quantum Dots in Glass and Quantum Size Effect

In the properties and customization of nanodevices that are increasingly being incorporated into information technology, electrical, magnetic, mechanical, and biological systems, the sample composition and physics of quantum dots embedded in matrices

M. H. Yukselici (✉) · A. A. Bozkurt · M. Balaban
Faculty of Science and Letters, Department of Physics, Yildiz Technical University, 34220 Istanbul, Turkey
e-mail: yukseli@yildiz.edu.tr

D. Bulut
ASELSAN Inc., Teknopark Istanbul, 34906 Istanbul, Turkey

M. K. Torun · M. Patan Alper
Department of Physics, Faculty of Arts and Sciences, Yeditepe University, 34755 Istanbul, Turkey

Z. M. Nassar
Islamic University of Gaza, P.O. Box 108, Gaza, Palestine

H. Ünlü and N. J. M. Horing (eds.), *Progress in Nanoscale and Low-Dimensional Materials and Devices*, Topics in Applied Physics 144,
https://doi.org/10.1007/978-3-030-93460-6_5

are of fundamental significance [1–6]. Liquid nanoparticles are within a free medium through which they may quickly migrate and thus join to form larger colloids with neighboring nanoparticles; this causes a limit to their shelf-life. It is well known that quantum dots are mostly synthesized by wet chemistry which makes them sensitive to atmosphere [7]. The exciton-binding energy is increased even more by nanoparticles embedded in matrices with a smaller dielectric constant and a greater energy gap than in the standard dimensional systems. It activates improved optical effects. The significance of the solid matrix comes from the very long shelf-life of quantum dots embedded in glass since they are within a restricted medium. The crystal achieves new optical and electronic properties when the size of the bulk semiconductor is reduced to a nanometer scale. Energy levels are quantized as in atoms and the energy difference between successive levels widens with the inverse of the size squared [8–16]. The quantum dot radii are studied in strong containment limits with an effective mass model [17]. First of all, the band structure must be well examined to use the effective mass model. As a consequence of this analysis, the Urbach tail method [17, 18] discusses the defect levels below the bandgap. The Urbach tail is associated with amorphous materials and with localized states at the band edge.

5.2 Quantum Size Effect

A quantum dot is a semiconductor whose three-dimensional excitons are confined [19]. An exciton is a pair of bound electron-holes that are attracted by Coulomb force to each other. An electron–hole pair is bound inside a Bohr exciton radius in a bulk semiconductor. If the electrons and holes are squeezed below the Bohr radius, then the characteristics of the semiconductor can change and size effect sets in. Bohr Radius is the sum of electron Bohr Radius (a_e) and hole Bohr Radius (a_h)

$$a_r = a_e + a_h \tag{5.1}$$

If $R >> a_r$, we call it weak containment; if $R << a_r$, we call it strong confinement. a_r is around 7.5 (nm) for CdTe. Exciton activity is the most important aspect of the theory. Exciton behaves like an atom; the three-dimensional model of a particle in a box is a good approximation [20]. The Schrödinger Equation is the fundamental equation in quantum mechanics, and its solutions provide us with the wave description of particles. The energy E is the only unknown component.

$$-\frac{\hbar^2}{2m}\frac{d^2\psi_{II}}{dx^2} + 0 = E \tag{5.2}$$

$$\frac{d^2\Psi_{II}}{dx^2} = -\frac{2m}{\hbar^2}E \tag{5.3}$$

$$\frac{d^2\psi}{dx^2} = -k^2\psi_{II} \quad k^2 = \frac{2mE}{\hbar^2} > 0, \quad \Rightarrow k = \sqrt{\frac{2m}{\hbar^2}E} \tag{5.4}$$

Since the potential well is symmetric, we have three solutions, one even, two other odd solutions.

$$\psi_{II} = C\cos(kx)\ for\ even\ parity$$

$$\psi_{II} = D\sin(kx)\ for\ odd\ parity$$

At *a/2*, for even parity a, is the width of a quantum well,

$$C\cos\left(k\frac{a}{2}\right) = Ae^{-\kappa\frac{a}{2}} \tag{5.5}$$

$$-Ck\sin\left(\kappa\frac{a}{2}\right) = -A\kappa e^{-\kappa\frac{a}{2}} \tag{5.6}$$

These two equations are transcendental (5.5, 5.6), which means that they cannot be solved analytically, we must solve them numerically. Solving one-dimensional time-independent Schrödinger Equation numerically, these solutions are obtained for finite well [21–23]. Figure 5.1 shows the energy levels against radius size as 0.5 nm and Table 5.1 shows the calculations of energy levels for a particle in a potential well from 0.50 nm to 5.00 nm radius in size.

CdTe quantum dots have been grown through heat treatment processes in borosilicate glass by diffusion limited growth. In the production of quantum dot-based devices, size dispersion, average quantum dot radius, and the number of quantum

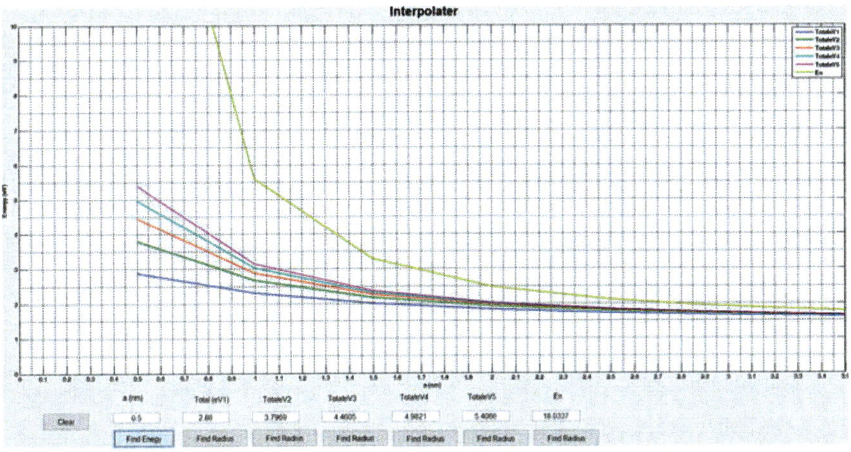

Fig. 5.1 Energy levels for a particle in a potential well. Energy plotted against (nm) for v = 0.5 (eV)

Table 5.1 Calculations of energy levels for a particle in a potential well. Energy listed against a (nm)

A (nm)	Total eV 1	Total eV 2	Total eV 3	Total eV 4	Total eV 5	En
0.50	2.88	3.80	4.46	4.98	5.41	18.03
1.00	2.31	2.67	2.88	3.03	3.14	5.59
1.50	2.01	2.17	2.26	2.32	2.37	3.28
2.00	1.83	1.92	1.97	2.00	2.02	2.48
2.50	1.72	1.78	1.81	1.83	1.84	2.10
2.75	1.69	1.73	1.75	1.77	1.78	1.99
3.00	1.65	1.69	1.71	1.72	1.73	1.90
3.25	1.64	1.66	1.68	1.69	1.69	1.83
3.50	1.61	1.64	1.65	1.66	1.66	1.78
3.75	1.60	1.61	1.62	1.63	1.63	1.73
4.00	1.58	1.60	1.60	1.61	1.61	1.70
4.35	1.56	1.58	1.58	1.59	1.59	1.66
4.50	1.56	1.57	1.57	1.58	1.58	1.64
4.75	1.55	1.56	1.56	1.56	1.57	1.62
5.00	1.54	1.55	1.55	1.55	1.55	1.61

dots per unit volume are major quantities to be regulated. During the quenching of the glass melt, and then through the diffusion of ions through the borosilicate glass, quantum dots in the glass first nucleate begin to develop at the temperature of heat treatment. When the relatively small quantum dots disappear and redeposit on larger quantum dots when the reactants (Cd, Te, Se, or Zn) have been precipitated, this condition is referred to as Ostwald ripening. These are the three distinct phases of precipitation: nucleation, natural growth, and ripening of the Ostwald [24]. The creation of the new phase at the first stage is based on the thermal shift of the medium, which takes the atoms to new positions relative to the untreated medium. Nucleation is known as this method. The nuclei's critical radius a^* is given by

$$a^* = \frac{2\sigma}{\Delta G_v} \tag{5.7}$$

The formation of nanocrystals, rate of nucleation R is given by

$$R = A \exp\left[-\left(\frac{\Delta G_c + \Delta G_a}{k_B T}\right)\right] \tag{5.8}$$

where $\Delta G_c = 16\pi\sigma^3/3\Delta G_v^2$ is the free energy to form a critical nucleus, ΔG_a is the activation energy for the atoms to jump across the nucleus matrix interface, ΔG_v is the free energy of bulk per unit volume, σ is the interface free energy per unit area. The distribution of the nuclei radius is given by *P(a)*.

$$P(a) = P_0 \exp\left(-\frac{4\pi\sigma(a - a^*)^2}{3K_B T} \right) \tag{5.9}$$

Supersaturation value controls the number of nuclei and the critical radius. The higher supersaturation value, the smaller is the critical radius and the smaller clusters can grow. From here, the standard deviation Δa of the nuclei radius of average size $a_{av} = a^*$ can be written as

$$\Delta a = \left(\frac{3K_B T}{8\pi\sigma} \right)^{\frac{1}{2}} \tag{5.10}$$

During the nucleation process, the average radius size and size distribution depends only on σ and T. Doped borosilicate glass with the properties of CdTe nanoparticles is found to be strongly size dependent [20]. Nanoscale systems are considered to have different properties than bulk systems. In this analysis, absorption spectroscopy is first used to detect the bandgap of CdTe nanoparticles, then the radii of the CdTe quantum dots doped into the borosilicate glass are calculated by using bandgap energy. Some mathematical and physical calculations and properties rely on the theoretical basis of estimating the radii of the quantum dots. Absorption spectroscopy is an analytical technique based on the measurement of radiation passing through the sample's volume of light absorption. Molecules can be excited by the absorption of light during this passage. The transmitted light intensity is lower than the incident light intensity. The following Beer-Lambert equation is used to measure the absorbance and optical density.

$$A = \log\left(\frac{I_0}{I} \right) \tag{5.11}$$

To find bandgap energies $(OD \times hv)^2$ versus hv curves are plotted in Fig. 5.2, the straight lines on the graph show the bandgap energies.

$$OD = \frac{A}{L} \tag{5.12}$$

where, OD means Optical Density, A absorbance, L thickness of the sample.

The fingerprints for crystal structures are the vibrational frequencies (or modes). Various crystals have various frequencies. In Table 5.2, some of the fundamental vibrational frequencies are mentioned [24–26] below.

The crystal under investigation is excited by laser light at frequency ω_1 and the frequency of scattered light is measured the difference between the two gives the vibrational frequency of the crystal. However, due to the nanometer size of the crystal (phonon confinement) and the strain between the nanocrystal and the host matrix, the vibrational frequencies we find in Raman measurements could be shifted. That is why the distinction between the shift due to phonon containment and strain

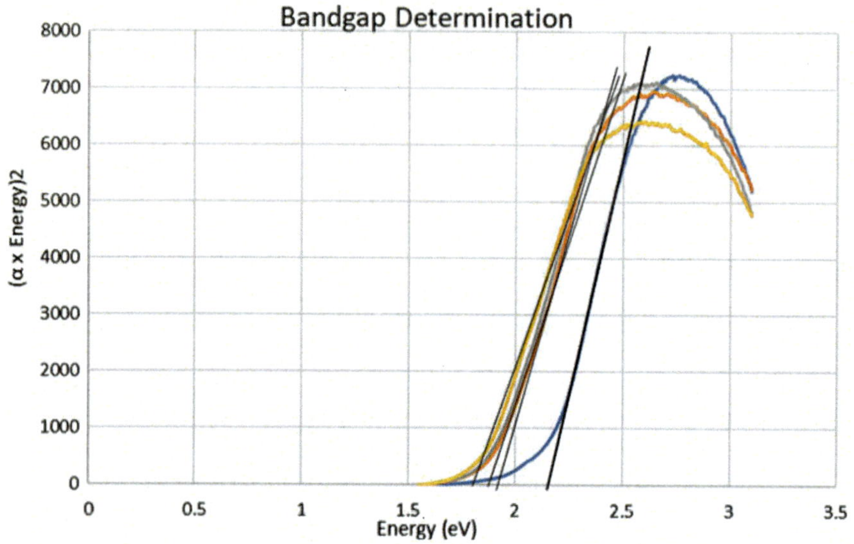

Fig. 5.2 $(OD \times h\nu)^2$ plotted versus photon energy $h\nu$ to find bandgap energy

Table 5.2 Fundamental vibrational frequencies

Crystals	Vibrational frequencies (cm^{-1})
CdTe	170.8
CdS	301.0
CdSe	210.3
ZnTe	217
ZnS	320

is an important task. The Grüneisen method [26] can be used to measure the order of magnitude. Grüneisen proposed that the ratio of percentage decrease in vibrational frequencies (ω) of the individual atoms to percentage increase in volume (V) is a constant which depends on the properties of the solid, that is,

$$\gamma = -\frac{\partial \ln \omega}{\partial \ln \mathrm{V}} = \frac{\frac{d\omega}{\omega}}{\frac{dV}{V}} \tag{5.13}$$

where γ is the Grüneisen parameter.

Table 5.3 shows the three samples for size confinement versus radius table where P_i is Raman node, γ is Grüneisen parameter and LO_B, is Bulk structure as 166.73 in Solving the equation for three different samples, as received, 12 and 24 h $\Delta\omega^c$. Correlation between $R_{(nm)}$ size confinement shift and radius can be seen in Table 5.4.

Quantum dot sensitized solar cells have been manufactured using cadmium-based quantum dot materials in the II-VI range, including CdTe, CdSe, and CdSe similar

Table 5.3 Three samples for size confinement versus radius table

	Radius (nm)	P_i	LO_B	Grüneisen parameter γ
As received	2.14	159.926	166.73	0.91
12 h	2.35	153.638	166.73	0.91
24 h	2.86	153.698	166.73	0.91

Table 5.4 Size confinement and radius

	Radius (nm)	Size confinement shift
As received	2.14	7.03
12 h	2.35	0.58
24 h	2.86	2.99

in scale. The size and structure of the synthesized QDs were determined using TEM and XRD analysis. The optical properties of the samples were characterized using absorption spectroscopy. The quantum dot scale was determined theoretically using the Effective Mass approximation. The performance of three solar cells sensitized with CdTe, CdSe, and CdS QDs is compared. On the basis of the differences in the effective masses of the charge carriers [27].

For technical applications in opto-electronics, nanocrystals embedded in the glass matrix have a robust and stable structure up to 250 °C. The findings of this work are focused on optical research, including measurements of optical transmission and Raman scattering. They should be backed by structural measurements such as High-Resolution Transmission Electron Microscope (HRTEM) and small angle X-ray scattering (SAXS) to assess relative weight of the results and conclusions drown.

5.3 Synthesis of Quantum Dots in Aqueous Solution

Colloidal aqueous synthesis of quantum dots gained significant increased research interest for the last 30–40 years because of their size dependent properties [28–34]. To prepare water dispersed QDs is relatively inexpensive, less toxic, environmentally friendly, effective and highly reproducible. In general, three components, organic solvents, precursor solution and surfactants were used to synthesize aqueous QDs. An aqueous synthesis process starts with the reaction between metal ions and anion predecessor in the presence of an appropriate capping reagent or stabilizer. The subsequent process of the synthesis, the nanoparticles' nuclei forms and then more particles deposit onto the formed nucleus called nucleation. After nucleation, particle growth stage take place, the growth rate of nanoparticles is slower than that of the first stage and they reach the desired size over time. In this type of growth, the average nanoparticle radius is proportional to the square root of the applied temperature. If the process is not quenched, this event is called Ostwald Ripening if the particles continue to grow where the average quantum dot size increases with the cube root of the annealing time [35–39]. The radius of the nanoparticle is controlled by the annealing time.

There are several methods utilized to synthesis quantum dots such as physical, chemical and biological and these synthesis techniques have the ability to control nanoparticle size, surface chemistry and composition. In this study, the chemical syntheses of CdTe nanocrystals are performed in aqueous medium. CdTe QDs are synthesized in aqueous phase where Mercaptopropionic Acid (MPA) acts as capping reagent which has some unique properties such as strongly emitting and stable colloids.

5.3.1 Aqueous Synthesis of CdTe Quantum Dots

For environmental and biological reasons, aqueous synthesis has been used in this study because of the fast, secure and stable formation of quantum dots. In general, highly luminescent CdTe nanoparticles were synthesized by mixing Te precursor and cadmium ion solution with capping ligand MPA in aqueous phase [40–44]. CdTe quantum dots are synthesized by a two-step method: Synthesis of NaHTe and preparation of Cadmium Ion Solution.

5.3.1.1 Synthesis of NaHTe (Precursor Solution)

HPLC (Sartorius Water Purification Systems) water used to clean the glass vial (2–3 times cleaning process), glass vial dried at 60 °C in the furnace. Tellurium and $NaBH_4$ precursors were mixed in a glass tube with a 1:4 ratio (1.5312 g $NaBH_4$ and 0.3828 g Te). Te powder color is black and $NaBH_4$ is white, their mixture which dissolved in 5 ml HPLC water is hyaline purple after one day in the refrigerator.

5.3.1.2 Preparation Method of Cadmium Ion Solution

1 mmol cadmium chloride is dissolved in 150 ml HPLC water and the mixture pH should be adjusted to 7.0 by adding sodium hydroxide into three neck flask. 0.2 ml of mercaptopropionic acid (or TGA or cysteine) is added into the above mixture. 3 ml of the Te precursor solution (0.3 mmol) is swiftly injected into the reaction mixture. The reaction mixture will become reddish brown in colour indicating the formation of CdTe nuclei in the solution. The mixture should then be aged (reflux) at 99 °C until the desired CdTe QD size (emission colour) is achieved. The CdTe QDs can be purified using centrifugation to remove excess surfactants and unreacted precursors in 6000 rpm. The mixture usually changes color in the first 10–15 minutes, with the onset of growth starting from golden yellow to dark red over time. With conditions relevant combinations, complete growth occurred between a few minutes

to several hours given in Fig. 5.4. Six solution samples were taken at different times and solutions in the vials contain various colour samples shown in Fig. 5.3 because each sample has different sized CdTe nanoparticles.

Developing simple, fast, cost-effective, high quantum yield methods for preparing QDs of controlled size and shape is always an important area of research. There are many state-of-the-art methods for synthesizing QDs such as electro deposition, organometallic way, aqueous route with small thiols as stabilizers or polymers and solution phase reduction methods [45–50]. The CdTe nanoparticles directly produced in aqueous phase with thiols capping ligand of MPA as a stabilizer. In this reproducible and clean process, Te-Cd molar ratio and pH value have great importance to fabricate ultrafast QDs. The advantages of QDs prepared by the aqueous method with thiols (MPA) are: they have high quantum yields, synthesis takes place low temperature and they are suitable for biological applications if they modified. The method studied in the research is practical, fast, stabile, accurate, low-cost and highly convenient and suitable for quality chemical synthesis. Other possible synthesis techniques are also evaluated.

The UV visible spectrum measurements were performed between 330–630 nm given in Fig. 5.4. All CdTe nanoparticles showed wide absorption and their absorption peaks lay between 430 and 530 nm, rely on the radius and structure of the CdTe nanoparticles. CdTe QDs exciton peaks have enormous blue shifted due to quantum confinement effect. With the long heat treatment time, the maximum absorption peak of QDs shifted to shorter wavelengths with decreasing size of the CdTe nanoparticles due to quantum confinement effect. The size of the CdTe NCs could be controlled by the temperature and heat treatment time and easily monitored by absorption spectrum. UV visible spectra revealed slightly red-shifted absorption bands as the particles produced in the 23-min synthesis went from 488 to 514 nm of the particles taken

Fig. 5.3 CdTe QDs in aqueous solution with different sizes

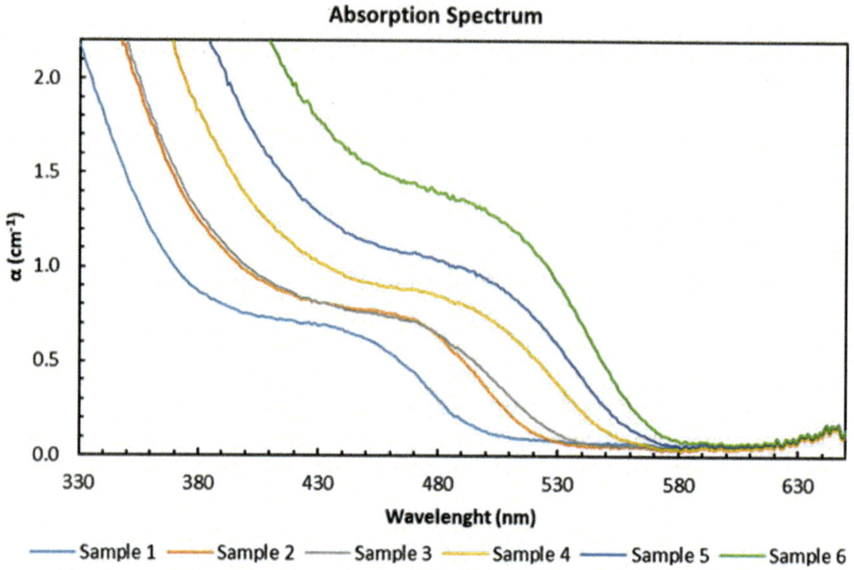

Fig. 5.4 The UV visible spectra of aqueous phase prepared CdTe QDs samples which were extracted at different heating time periods: Sample 1 (23 min), Sample 2 (74 min), Sample 3 (120 min), Sample 4 (200 min), Sample 5 (255 min), Sample 6 (310 min)

in the 310-min synthesis. This shift indicated that the nanoparticle size increased throughout the synthesis.

CdTe QDs with MPA capping agent were synthesized in aqueous medium were synthesized in aqueous medium successfully with chemical colloidal process. The synthesis process, which takes place at a low temperature of about 100 degrees, is a very simple, fast and inexpensive method. According to these absorption spectra results the absorption wavelength gradually increase with increasing the growth time (wavelength blue-shift). The size of QD-CdTe is well controlled by reaction time and growth temperature.

5.4 Investigation of Optical and Structural Properties of CdTe Thin Films

Because of its band gap energy which matches the solar radiation to get high-efficiency energy conversion, cadmium telluride (CdTe) is a promising semiconductor for many applications such as light-emitting diodes, photovoltaic devices, and fluorescent labels [51]. CdTe have a high optical absorption coefficient initiating from near-IR and rising to more than 10^4 cm^{-1}. The bandgap energy (E_g) of its bulk material is ~ 1.5 eV, this value matches the solar radiation spectrum to get best conversion efficiency [52]. Quantum size effect in CdTe thin films has been investigated in references [53, 54]. Reducing the grain size of the bulk material to nanoscale gives the material new conducting, optical, and structural properties; energy levels are quantized and the bandgap energy increases as size decreases [55]. We employ optical absorption (ABS) for evaluating the size of grains and Urbach energy (E_U). X-ray diffraction (XRD) technique is employed to study the structural characterization. Broadening the peaks of XRD is mostly a result of grain size and strain effects. Each effect of them is estimated by Williamson-Hall (W-H) method [55]. We also employ Raman spectroscopy to investigate the structural property; nano-scale grains and strain results in phonon frequency shift. The results of absorption, X-ray diffraction, and Raman spectroscopies are combined together to know the grain size effect on strain and structural disorder.

Five thin films of CdTe are prepared by depositing the compound molecules on glass substrates by using physical vapor deposition (PVD) method. The properties of the films are then optically and structurally investigated employing optical ABS, Raman, and XRD spectroscopies. Bandgap energy of the film (E_g^{film}) is evaluated and the grain size is calculated from E_g^{film} value. The bandgap energy is blueshifted by 0.71 eV when thickness (t) decreases from 500 to 100 nm. E_U related to the long-wavelength tail width increases from 0.585 to 0.827 eV when t decreases from 500 to 100 nm, respectively. Widening and shifting of X-ray diffraction lines are mostly produced from the strain and grain size effects. In Raman spectra, relative to the bulk phonon frequency, shifts in phonons frequency (LO-mode phonon is blueshifted, while LA-mode and TO-mode phonons are redshifted) are observed. The results of optical absorption, Raman, and X-ray diffraction, spectroscopies are combined for investigating grain size evolution effect, structural disorder, and grain-size-dependent strain.

5.4.1 Experimental Details

CdTe thin films are grown on glass slides at 150 °C and 2.2×10^{-5} torr. Powder of CdTe compound is thermally evaporated from tantalum crucible onto the glass slides in the "VAKSIS PVD-HANDY/25-TE" chamber. Film thickness was estimated using a quartz sensor fixed near the substrate.

The optical absorption experiment is conducted at room temperature. The measurements are obtained using silicon detector put at the output of a computer-controlled motorized 1/8 m Oriel monochromator. A tungsten light source powered by a dc power supply is employed for shining the samples. In Fig. 5.5 is the transmittance plot shown. Assuming transparency of the samples at wavelengths 850 through 900 nm, we compute the reflectance (R) from the relation $R = 1 - T_{max}$ where T_{max} is the maximum transmittance within that range. Also, we consider that the reflectance values do not vary considerably as a function of wavelength (λ) and compute the optical density (OD) according to (5.14):

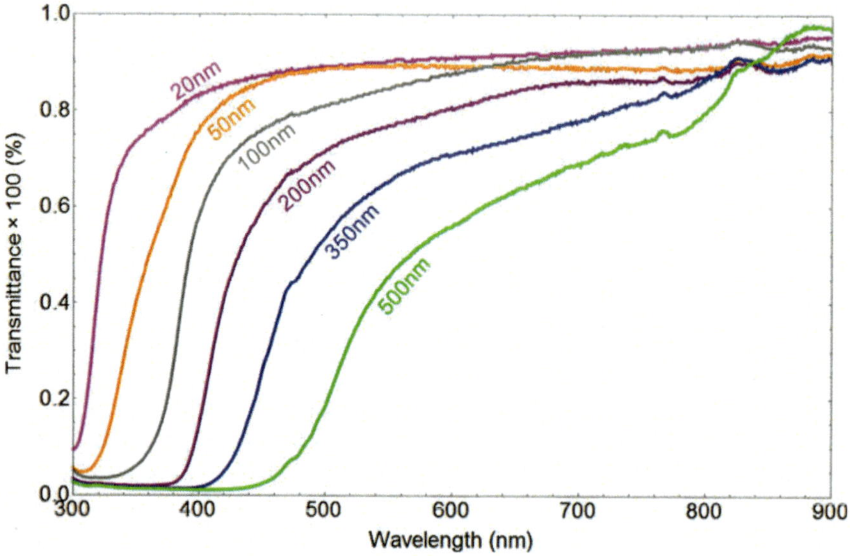

Fig. 5.5 Transmittance versus wavelength [55]. (Reproduced with permission. Copyright 2016, Wiley-VCH)

$$OD = \ln\left(\frac{I_0}{I}\right) - \ln\left(\frac{I_0}{I}\right)_{\min} \quad (5.14)$$

where the term $\ln\left(\frac{I_0}{I}\right)_{min}$ is due to the reflection effect at the interfaces.

X-ray diffraction patterns of CdTe thin films recorded on a panalytical multipurpose diffractometer (MPD) are shown in Fig. 5.6. The wavelength of the X-ray used is $CuKa_\alpha$ $(\lambda) = 1.5406$ A°. The instrumental error is less than 2 mrad.

A 250-mm-focal-length Renishaw spectrometer is used, at room temperature, for Raman spectroscopy. A Rayleigh line rejection filter is used in Raman measurements. The system can measure frequencies >100-cm^{-1}. A 532-nm-wavelength is used for exciting the CdTe films within a 1-mm-size spot. According to the estimation for the spectral region investigated in this study, the accuracy of the system is ±0.1 cm^{-1}. The Raman spectra, plotted according to the model discussed in Sect. 5.3.6 [56], are shown in Fig. 5.7. From the figure we notice that different frequencies were recorded for different phonon modes.

Figure 5.7a shows peak centers at wavenumbers around 123 cm^{-1} due to zone center LA phonons at the X critical point. The peak center at about 140 cm^{-1} in Fig. 5.7b is due to the Γ-critical-point transverse optical (TO) phonon, the peak of about 162 cm^{-1} in Fig. 5.7c results from X-critical-point LO phonons.

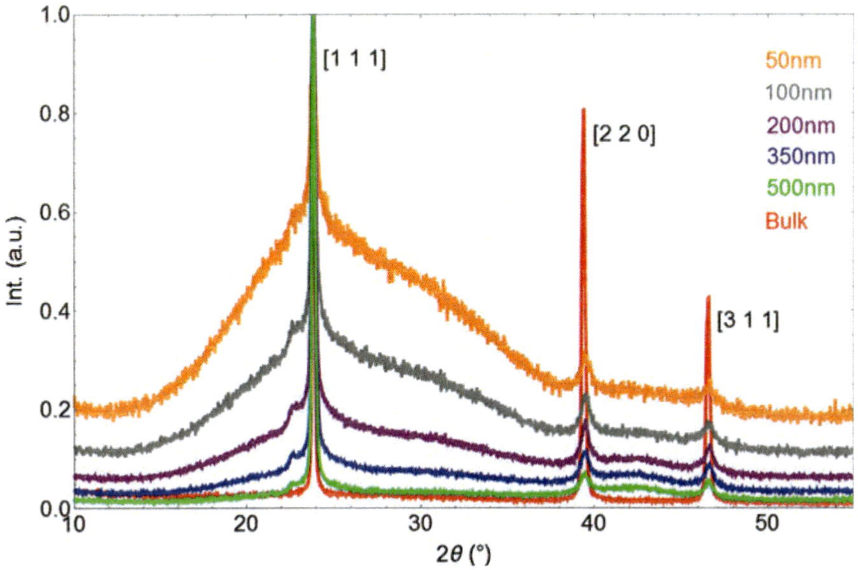

Fig. 5.6 Thin films and powder CdTe XRD patterns [55]. (Reproduced with permission. Copyright 2016, Wiley-VCH)

5.4.2 *Effect of Grain Size and Strain on Bandgap Energy*

Square of OD (OD^2) is plotted versus E of the incident light as shown in Fig. 5.8. The bandgap energy of the film (Table 5.5) is found by extending the straight part of the curve shown in the figure. It is noticed that E_g^{film} is blue shifted by 0.71 eV when R is reduced from 2.055 to 1.6 nm.

Strain, produced from dangling bonds at interfaces and surfaces, and quantum size effect, produced from the charge confinement in the nano-scale film, may both lead to this blueshift. The average grain size (R_{ave}) is computed from the measured value of E_g^{film} using the following equation [57]:

$$E_g^{film}(eV) = E_g^{bulk}(eV) - \frac{0.14}{R_{ave}(nm)} + \frac{0.376}{\mu[R_{ave}(nm)]^2} + \Delta E_{strain} \tag{5.15}$$

where μ represents the electron-hole pair reduced mass. The second term of (5.15) is the contribution of Coulombic effect, the third term is the contribution due to charge confinement, and the fourth one (ΔE_{strain}) is due to the strain effect and is calculated from the following relation [55, 58]:

$$\Delta E_{strain} = E_g^{film}|_{strain} - E_g^{bulk} = B\frac{dE_g}{dP}\left[1 - \left(\frac{a_{film}}{a_{bulk}}\right)^3\right] \tag{5.16}$$

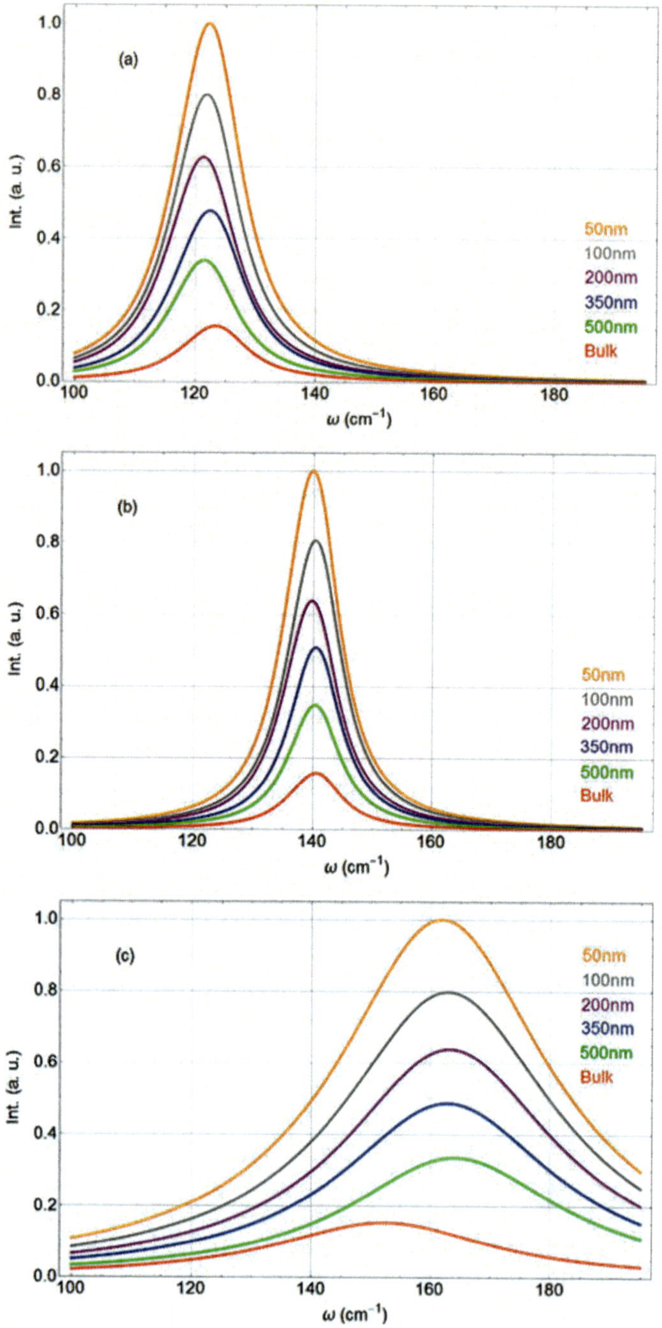

Fig. 5.7 Intensity versus Raman shift—a comparison between the different phonons frequencies: **a** X-critical-point LA phonon; **b** G-critical-point TO phonon; **c** X-critical-point LO phonon [55]. (Reproduced with permission. Copyright 2016, Wiley-VCH)

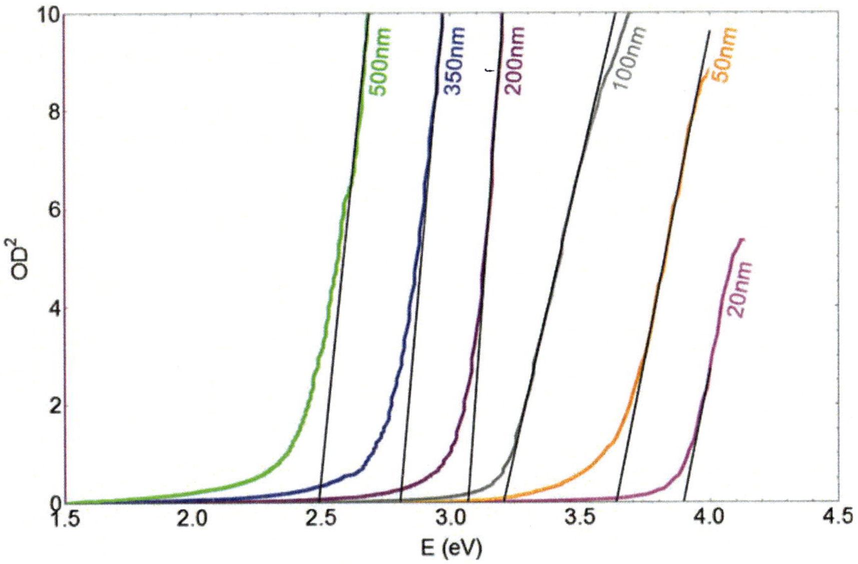

Fig. 5.8 Square value of optical density against energy for CdTe thin film [55]. (Reproduced with permission. Copyright 2016, Wiley-VCH)

Table 5.5 Bandgap energy and average grain size are evaluated for CdTe thin films [55]

Thickness (nm)	50	100	200	350	500
E_g^{film}(eV)	3.64	3.20	3.06	2.80	2.49
R_{ave}(nm)	1.432	1.600	1.665	1.818	2.055
E_U(eV)	0.894	0.827	0.762	0.657	0.585

where $B = 4.24 \times 10^{10} Pa$ is the bulk modulus for CdTe crystal, $\frac{dE_g}{dP} = 79 \pm \frac{2meV}{GPa}$ is the coefficient of pressure-dependence bandgap energy [59]. ΔE_{strain} was evaluated at about 2.2% of the 0.71 eV total blueshift that takes place in the bandgap energy with decreasing the thickness by 400 nm (from 500 to 100 nm) [55]. Substituting $E_g^{bulk} = 1.44$eV and $\mu = 0.08$ (in the unit of rest mass of electron (m_0)) in (5.2) gives R_{ave} values (Table 5.5). It is noticed that R_{ave} is directly proportional to the thickness. The E_g^{film}–R_{ave} data is fitted (Fig. 5.9) to an exponential equation given by

$$E_g^{film} = E_g^{bulk}\text{Exp}\left(\frac{A}{R^{1.5}}\right) \tag{5.17}$$

where the constant A is found by fitting ($A = 1.607$ in this study). As it is seen from Table 5.5, the film contains films consist of nano-size grains. Therefore, we propose

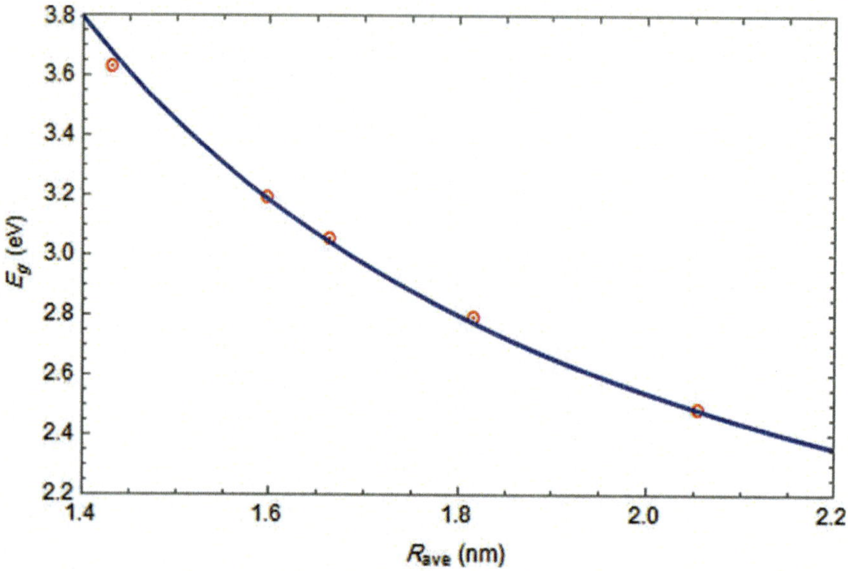

Fig. 5.9 Bandgap energy as a function of grain size (the data is fitted to an equation given by 5.17)

that grain size rather than film thickness determines the bandgap energy. The data analysis shows that reducing grain size results in a blueshift in the energy gap.

5.4.3 *Urbach Energy*

The exponential-shape tail appeared before the absorption edge (Fig. 5.8) may be due to the effect of structural and thermal disorder that can be determined by using Urbach rule. The relationship between OD and photon energy (E) is [60]

$$OD(E) = OD_0\,\mathrm{Exp}\left(-\frac{E_g - E}{E_U}\right) \tag{5.18}$$

where OD_0 is the optical density at $E=E_g$, and E_U is known as Urbach energy. Figure 5.9 is a plot for ln(OD) as a function of E according to (5.19):

$$\ln(OD) = \left(\ln(OD_0) - \frac{E_g}{E_U}\right) + \left(\frac{1}{E_U}\right) \tag{5.19}$$

The straight lines in Fig. 5.10 are linear fits to the straight part of the ln(*OD*)-*E* curve. The Urbach energy is calculated from the slopes of these lines, where $E_{U=}$ 1/slope. The calculated values of E_U are shown in Table 5.5. Urbach energy

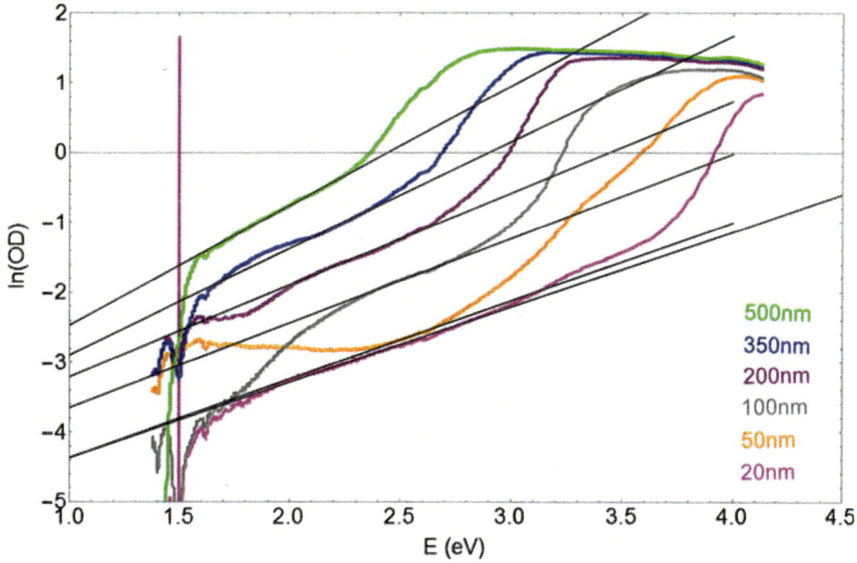

Fig. 5.10 Natural logarithm of OD is plotted versus energy. The straight lines are the linear fits to the straight part of the ln(OD)-E curve [55]. (Reproduced with permission. Copyright 2016, Wiley-VCH)

versus thickness (t) of the film is plotted in Fig. 5.11. The curve is fitted as a quadratic equation to the data given in Table 5.5:

$$E_U(eV) = 0.945 - 0.0011t + 7.7 \times 10^{-7}t^2, \quad tinnm. \tag{5.20}$$

It is seen from 5.20 and Fig. 5.11 that Urbach energy varies considerably and approximately linearly at small values of thicknesses (t), whereas it varies inconsiderably and quadratically at high values with varying film thickness. Accordingly, and referring to 5.18 and 5.20 we notice, at small values of t and a certain value of E_g, that decreasing film thickness leads to a significant increasing of E_U and hence a significant increasing of OD (see 5.18). This means that optical density varies considerably more at low values of thickness than at high values. Since E_U and structural disorder are related to each other, we conclude that structural disorder increases, more significantly at low values of thicknesses than at high values, as the film thickness decreases.

E_U values depend on both disorder due to structure and disorder due to temperature [61]. In this study, Urbach tail width is determined by only the structural disorder because the optical absorption spectra are recorded at constant temperature.

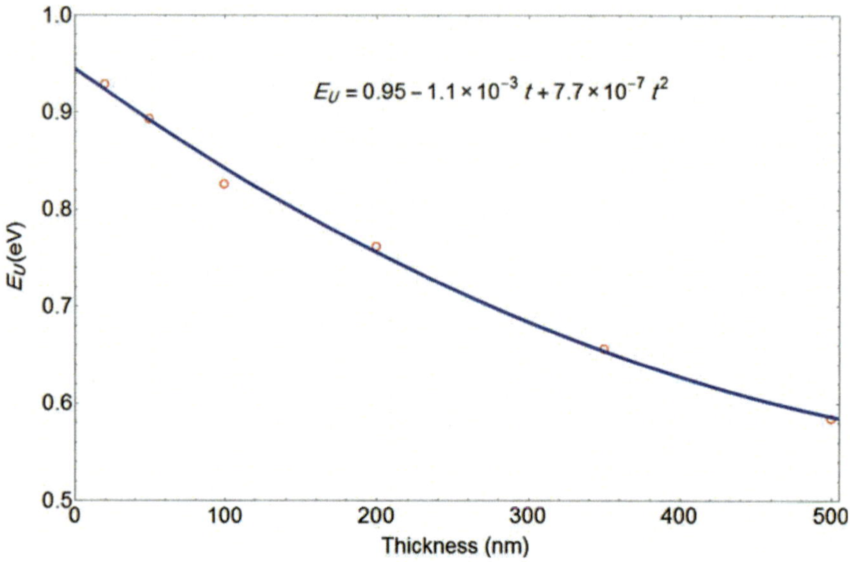

Fig. 5.11 Urbach energy against thickness [55]. (Reproduced with permission. Copyright 2016, Wiley-VCH)

5.4.4 XRD Spectra

The crystal growth in this work is cubic zinc-blende mode structure. Fitting the lines in the XRD spectra shown in Fig. 5.6 to a function given by adding Gaussian, Lorentzian, and a polynomial (due to the background effect) of a suitable order, the line center (θ_{hkl}) its associated wavevector (K), full width at half maximum [FWHM, i.e., $\Delta(2\theta_{hkl})$ or ΔK], lattice constant (a), and interplanar distance (d_{hkl}) are obtained. The obtained values of these unknowns are tabulated in Table 5.6 [55]. The position of most-intense line (at about $2\theta = 24°$), is due to the (1 1 1) plane which indicates that the crystal grows in the direction [111]. The wide background confirmed for the 50-nm-thickness sample is mostly due to the scattering from the substrate. The polynomial used in fitting is to estimate the intensity scattered from the substrates. The lattice parameter (a) in the [111] direction reduces by $\Delta a = \left|\frac{a_{500} - a_{50}}{a_{50}}\right| \times 1000 = 0.54‰$ with increasing thickness from 50 to 500 nm. For the strongest line, full width at half maximum, resulting from the (1 1 1) plane, reduces as thickness increases.

The lines of XRD broaden as a result of grain size, strain, and instrumental error according to the relation

$$[\Delta(2\theta_{hkl})]^n = \left[\Delta(2\theta_{hkl})_{size}\right]^n + \left[\Delta(2\theta_{hkl})_{strain}\right]^n + \left[\Delta(2\theta_{hkl})_{inst}\right]^n \tag{5.21}$$

where n is an integer and it depends on the profile by which one represents the line widening. For Lorentzian profile n = 1, and for Gaussian profile n = 2 [62, 63]. For

Table 5.6 Parameters obtained from fitting XRD patterns plotted in Fig. 5.6 [55]

t (nm)	hkl	K (nm^{-1})	ΔK (nm^{-1})	d_{hkl} (nm)	a (nm)	$\epsilon_{hkl} \times 10^{-3}$
50	111	2.68391	0.04688	0.372591	0.645346	0.3800
	220	4.38554	0.06087	0.228022	0.644944	2.1024
	311	5.13473	0.09718	0.194752	0.645920	1.0010
100	111	2.68626	0.03406	0.372265	0.644782	1.2545
	220	4.38686	0.06782	0.227953	0.644750	2.4026
	311	5.13203	0.08847	0.194855	0.646260	0.4754
200	111	2.68495	0.02907	0.372446	0.645096	0.7672
	220	4.38555	0.06250	0.228022	0.644942	2.1046
	311	5.1363	0.08650	0.194693	0.645723	1.3064
350	111	2.68695	0.02610	0.372169	0.644616	1.5110
	220	4.38446	0.06303	0.228078	0.645103	1.8566
	311	5.13594	0.06769	0.194706	0.645768	1.2364
500	111	2.6842	0.02444	0.372550	0.645276	0.4880
	220	4.38057	0.06034	0.228281	0.645676	0.9702
	311	5.13393	0.06201	0.194783	0.646021	0.8454
bulk	111	2.68289	0.01905	0.372732	0.645591	
	220	4.37632	0.02318	0.228502	0.646303	
	311	5.12959	0.02513	0.194947	0.646567	

size-dependent broadening, the Debye-Scherrer equation is [64]:

$$\Delta(2\theta_{hkl})_{size} = \frac{0.94\lambda}{2R_{ave}\cos\theta_{hkl}} \tag{5.22}$$

the value 0.94 refers to the shape factor [65]. The strain-dependent broadening and the coefficient of axial strain ϵ_{hkl} due to the pressure on hkl planes are given by 5.23 [66] and 5.24, respectively:

$$\Delta(2\theta_{hkl})_{strain} = \varepsilon \tan\theta_{hkl} \tag{5.23}$$

$$\varepsilon_{hkl} = \frac{d_{hkl} - d_{hkl}^{bulk}}{d_{hkl}^{bulk}} \tag{5.24}$$

The instrumental error $\Delta(2\theta_{hkl})_{inst}$ is neglected since it is less than 2 mrad. Substituting the values of d_{111} from Table 5.6 in 5.24, the strain coefficient (ϵ_{111}) can be calculated. The result shows that the strain increases by 61.1% when thickness decreases from 500 to 100 nm.

X-ray diffraction patterns are studied in reciprocal space employing Williamson-Hall (W-H) method [63]. XRD lines are fitted for all samples, and the parameters R and ϵ are calculated (Table 5.7). Williamson-Hall method is explained in Sect. 5.3.5.

Table 5.7 Average grain sizes and strains obtained by different ways [55]

t (nm)	$R_{\rm Abs}$ (nm)	$R_{\rm Ram}$ (nm)	$R_{\rm WH}$ (nm)	$\epsilon_{\rm Gr} \times 10^{-3}$	$\epsilon_{\rm WH} \times 10^{-3}$
50	1.432	1.380	10.705	10.551	−18.850
100	1.600	1.485	11.555	−9.333	−16.938
200	1.665	1.602	12.048	−8.951	−16.341
350	1.818	1.750	14.127	−8.196	−15.170
500	2.055	1.984	15.009	−7.319	−13.822

The negative sign for e means compressive strain

5.4.5 Williamson-Hall Analysis of X-Ray Diffraction

X-ray diffraction peaks are broadened presumably as a result of mainly strain and finite grain size. W-H method is an analysis used to separate the contributions of strain and size effects on line broadening. The following details show this method:

We ignore the instrumental effect $\Delta(2\theta_{hkl})_{inst}$ from 5.21 and substitute 5.22, 5.23 in 5.21 to obtain;

$$[\Delta(2\theta_{hkl})]^{\rm n} = \left[\frac{0.94\lambda}{2R_{ave}\cos\theta_{hkl}}\right]^{\rm n} + [\varepsilon\tan\theta_{hkl}]^{\rm n} \tag{5.25}$$

We use Bragg's law and multiply LHS side and RHS side of 5.25 by $\left(\frac{\cos\theta_{hkl}}{\lambda}\right)^n$ to get

$$\Delta K = \frac{0.94\lambda}{2R_{ave}} + \frac{\varepsilon}{2}K_0 \tag{5.26}$$

for fitting to Lorentzian function (n = 1), and

$$\Delta K = \sqrt{\left(\frac{0.94\lambda}{2R_{ave}}\right)^2 + \left(\frac{\varepsilon}{2}K_0\right)^2} \tag{5.27}$$

for fitting to Gaussian function (n = 2), where K0 is the wavevector at the peak center and ΔK is the FW@HM of the peak profile. We fit the intensity (I) versus the wavevector variable (K) to a function obtained by adding Lorentzian (L) and Gaussian (G) functions (Voigt function V [67]):

$$V = \frac{A}{2}(L+G) = \frac{A}{2}\left(\frac{1}{1+x^2} + e^{(-ln2)x^2}\right),\ x = \frac{K-K_0}{\Delta K/2} \tag{5.28}$$

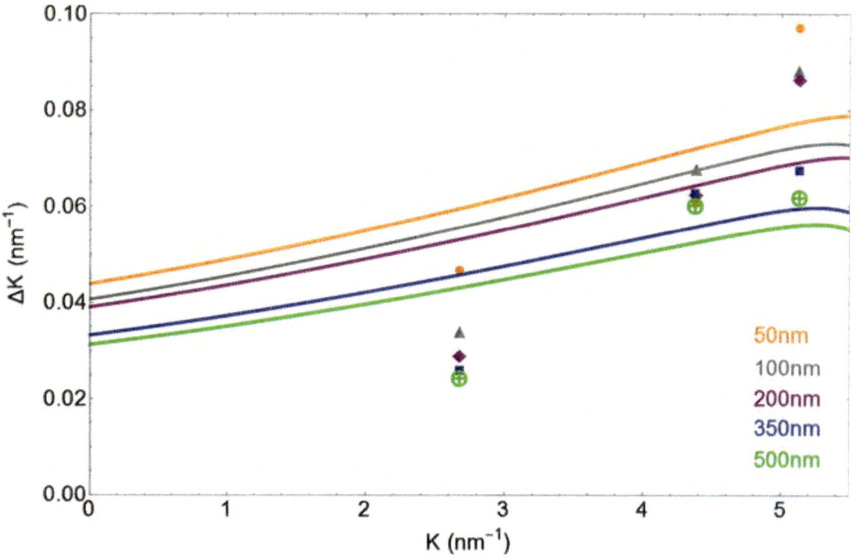

Fig. 5.12 W–H plot fitted to the Maclaurin series of 5.29 [55]. (Reproduced with permission. Copyright 2016, Wiley-VCH)

$$\Delta K = \frac{1}{2}\left[\frac{0.94\lambda}{2R_{ave}} + \frac{\varepsilon}{2}K_0 + \sqrt{\left(\frac{0.94\lambda}{2R_{ave}}\right)^2 + \left(\frac{\varepsilon}{2}K_0\right)^2}\right] \tag{5.29}$$

Plotting ΔK against K_0 results in a plot (Fig. 5.12) from which one can obtain R_{ave} and ϵ. The procedures followed for analyzing XRD patterns are as follows:

1. We convert the X-ray diffraction patterns, shown in Fig. 5.6, is converted from the real space (I versus θ_{hkl}) to the reciprocal space (I versus K) where I is the intensity.
2. Each line in X-ray diffraction spectrum is fitted to the Voigt function given by 5.28.
3. The FWHM (ΔK) of the X-ray diffraction lines is plotted versus the wavenumber at the line center (K_0)—Fig. 5.12.
4. The best-fit values of R_{ave} and ϵ are found by fitting the $K_0 - \Delta K$ graph to the Maclaurin series of the function given by 5.29, where this function is expanded up to $(k_0)^{20}$ [55].

5.4.6 *Raman Spectra*

Raman scattering is employed to investigate the structural properties. Phonon modes produced from Raman scattering are shifted as a result of small size and strain decreasing. In the following subsections are the phonon confinement model [56] and Grüneisen model [68] explained.

5.4.6.1 Phonon Confinement Model

Raman spectrum (intensity of light against Raman shift) of first order is given by the equation [56]

$$I(\omega) = \int \frac{|C(0,q)|^2}{[\omega - \omega(q)]^2 + (\Gamma_0/2)^2} d^3q \tag{5.30}$$

with

$$|C(0,q)|^2 = \mathrm{Exp}\left(\frac{-d^2q^2}{2\beta}\right) \tag{5.31}$$

where q is the phonon wavevector, $\omega(q)$ is the phonon dispersion curve, Γ_0 is the natural linewidth (FWHM) of the zone-center optical phonon in the infinite/bulk crystal, d is the diameter of the spherical grain, and β is a constant by which it is decided how fast the wavefunction decays as it approaches the grain boundary [56]. For simplicity of analyzing problem, we consider a spherical Brillouin zone (BZ) and we assume the optical phonon dispersion curve isotropic. These assumptions are assumed when only a small part of BZ, located at the center of Γ point (the center of BZ), contributes to Raman scattering. The dispersion curve of the optical phonon is then approximated to the analytical function [56]

$$\omega(q) = \omega_0 - \Delta\omega \sin^2\left(\frac{qa}{4}\right) \tag{5.32}$$

where ω_0 is the zone-center optical phonon frequency, $\Delta\omega$ is the phonon dispersion curve width, and a is the lattice constant. Fitting the Raman spectra measurements, for each sample, to a profile given by 5.30 results in the unknown parameters including grain size. The grain size obtained from Raman analysis is listed in Table 5.7. The phonon modes and their confinement-dependent shifts are shown in Table 5.8.

5.4.6.2 Grüneisen Model

The shift in phonon frequency taken place in Raman spectra is mainly due to two effects: (i) phonon confinement and (ii) strain. The total shift in phonon frequency (relative to the bulk) is given by

$$\omega_P - \omega_0 = \delta\omega_c - \delta\omega_s \tag{5.33}$$

Table 5.8 The observed phonon modes and their confinement-dependent shifts [55]

t (nm)	Phonon mode	$\omega_P^{meas.}$(cm^{-1})[a]	$\omega_P^{calc.}$(cm $^{-1}$)[b]	$\delta\omega_c$(cm^{-1})[c]	$\delta\omega_c$(cm^{-1})[d]
50	X-point LA	124.333	122.234	1.161	1.75943
	Γ-point TO	141.799	139.913	0.656	0.48126
	X-point LO	162.605	161.912	9.645	10.7863
100	X-point LA	123.934	121.839	1.556	1.57469
	Γ -point TO	142.116	140.324	0.245	0.43073
	X-point LO	165.209	162.906	10.639	10.7564
200	X-point LA	123.678	121.318	2.077	1.51321
	Γ -point TO	141.633	139.785	0.784	0.41391
	X-point LO	163.231	163.228	10.961	10.7457
350	X-point LA	124.87	122.440	0.955	1.385S6
	Γ -point TO	141.768	140.478	0.091	0.37908
	X-point LO	165.199	162.880	10.613	10.7221
500	X-point LA	123,341	121.495	1.900	1.22603
	Γ -point TO	141.639	140.377	0.192	0.33536
	X-point LO	164.735	164.119	11.852	10.6893
Bulk	X-point LA	125.364	123.395		
	Γ -point TO	142.169	140.569		
	X-point LO	155.649	152.267		

[a] $\omega_P^{meas.}$ is the line center obtained by fitting Raman spectra data

[b] $\omega_P^{calc.}$ is the line center evaluated using Fig. 5.7a–c (model in [56])

[c] Peak shift due to confinement evaluated from $\omega_P^{calc.}$

[d] Line shift, caused by confinement, evaluated from the equations shown in Figs. 5.13a–c

where ω_P is the measured Raman line center for the film, ω_0 is the line center for the bulk, $\delta\omega_c$ is the line shift caused by phonon confinement, and $\delta\omega_s$ is the line shift due to strain. The relation between $\frac{\delta\omega_s}{\omega_0}$ and the hydrostatic strain (ϵ) is given by [68]

$$\frac{\delta\omega_s}{\omega_0} = (1 + 3\varepsilon)^{-\gamma} - 1 \tag{5.34}$$

where γ is the Grüneisen parameter for the film material. Equation 5.34 can be rearranged as

$$\varepsilon = \frac{1}{3}\left[\left(\frac{\delta\omega_s + \omega_0}{\omega_0}\right)^{-\frac{1}{\gamma}} - 1\right] \tag{5.35}$$

Substituting $\delta\omega_s$ from 5.33 into 5.35, one gets

$$\varepsilon = \frac{1}{3}\left[\left(\frac{\omega_P - \delta\omega_c}{\omega_0}\right)^{-\frac{1}{\gamma}} - 1\right] \tag{5.36}$$

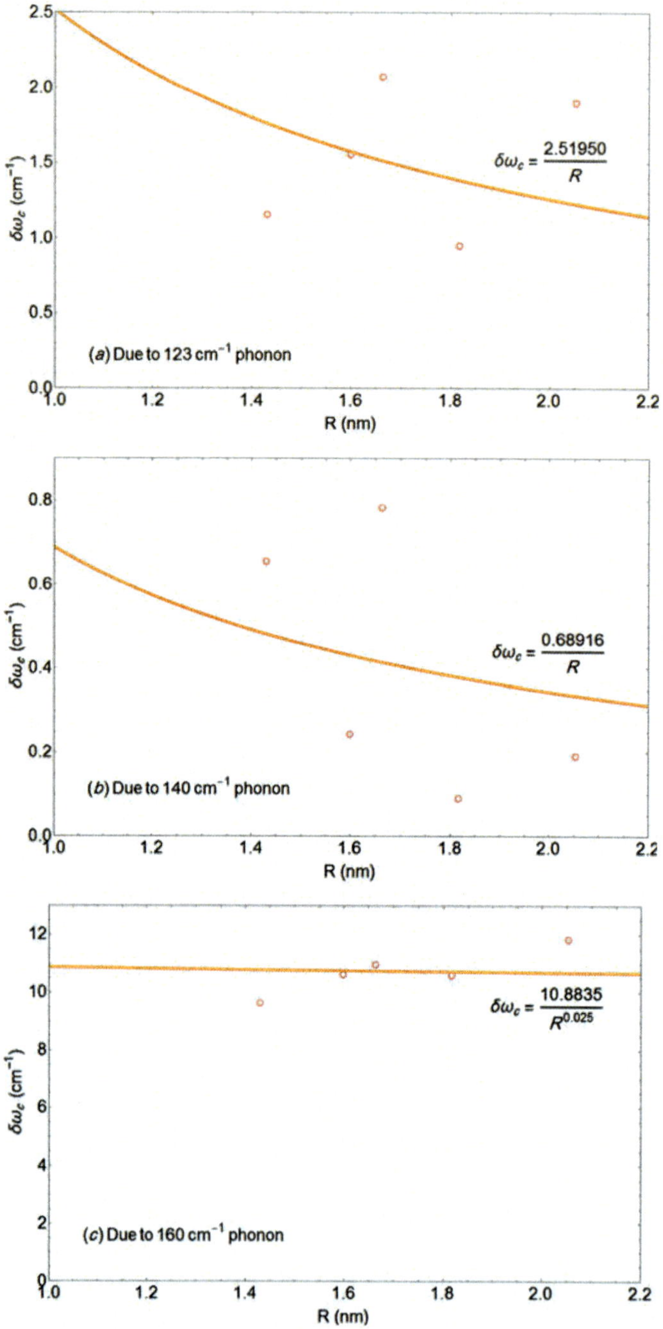

Fig. 5.13 The Raman frequency shift caused by the **a** 123 cm^{-1}, **b** 140 cm^{-1}, and **c** 160 cm^{-1} phonon confinements versus grain size [55]. (Reproduced with permission. Copyright 2016, Wiley-VCH)

which correlate the hydrostatic strain ϵ and the shift $\delta\omega_c$ due to phonon confinement. Grüneisen parameter (γ) is estimated from [69] ($\gamma = 0.5$ for CdTe at temperature 300 K), ω_P is the measured peak center for the thin film and ω_0 is the phonon frequency of bulk. The shift $\delta\omega_c$ is obtained in terms of R, as shown in Fig. 5.13a–c, according to the explanation in [56]. The strain determined by Grüneisen model (ϵ_{Gr}) is evaluated by, first, calculating the strain ϵ from 5.36 at all peaks of Raman spectrum for one sample and then the mean value of ϵ is calculated for the same sample in order to find the value of ϵ_{Gr} we obtained. Similarly, ϵ_{Gr} can be computed for the other samples (Table 5.7).

5.4.7 Conclusion

Optical ABS characterization shows that, when the grain size decreases from 2.055 to 1.6 nm, E_U increases from 0.585 to 0.827 eV, and E_g^{film} is blueshifted from 2.49 to 3.20 eV, i.e. ΔE_g^{film}=710 meV [55]. Equation 5.16 can be rearranged as [58]

$$E_g^{film}|_{strain} = E_g^{bulk} + B\frac{dE_g}{dP}\left[1 - \left(\frac{a_{film}}{a_{bulk}}\right)^3\right] \tag{5.37}$$

Substituting E_g^{bulk}, a_{bulk} and a_{100}, from Table 5.6, for the highest-intense line, E_g^{100} can be evaluated. Likewise, E_g^{500} can be calculated. Computing $|E_g^{500} - E_g^{100}|$, a blueshift of $7.8 \pm 0.2meV$ in the energy gap is obtained as a result of increasing the strain as the film thickness decreases from 500 to 100 nm. This conclusion is in agreement with the Grüneisen model given by 5.36; decreasing R results in increasing compressive strain (Fig. 5.15). Simultaneously, this leads to increasing the redshift relative to the finite material as a result of phonon confinement, which means that a redshift in phonon frequency produces as thickness decreases from 500 to 100 nm.

The most contribution of the 710-meV total blueshift in energy gap measured as the thickness decreases from 500 to 100 nm is due to size-dependent quantum confinement of charges; the contribution of strain is a redshift of just ~ 1.1% of this total shift, i.e. ~ 7.8 meV, the contribution of Coulomb effect is a redshift of ~19.4 meV; while the contribution of the charged particle confinement is a blueshift of ~723 meV. Accordingly, it is concluded that the size-dependent confinement of the charged particles (the 3rd term in the RHS of 5.15) is the main contribution to the total blueshift observed in the energy gap [55]. It is also noticed that the contribution in the shift in bandgap energy produced due to strain is less than the contribution due to Coulomb energy. Furthermore, from the second and seventh columns in Table 5.6 we notice that the wavevector K is affected by the compressive axial strain ϵ_{hkl}. We also extrapolate from Table 5.8 that the phonon frequencies (and hence the peak shift $\delta\omega$) are also sensitive to the compressive strain according to the Grüneisen model.

The center of the strongest Raman peak for the powder sample is 123.4 cm^{-1}, which is near 3.65 THz (= 121.7 cm^{-1}) for the X-critical-point LA-phonon mode

[22]. Relative to the powder, the phonon wavenumber at this peak is redshifted, for all thin film samples, as shown in Table 5.8 and Fig. 5.13a. The Γ-critical-point TO-phonon wavenumber of bulk is 4.20 THz (= 140.0 cm^{-1}) [22]. The TO-phonon wavenumber of bulk in this study is 140.6 cm^{-1}. It is redshifted for all films as shown in Table 5.8 and Fig. 5.13b. Referring to [69], the X-critical-point LO-phonon wavenumbers of bulk are 4.03; 4.86, and 4.92 THz (134.3; 162.0, and 164.0 cm^{-1}, respectively). The resultant of the 134.3 and 164.0 cm^{-1} wavenumbers results in a single peak centered at a wavenumber of $\omega_0 = \sqrt{\frac{(\omega_1^2+\omega_3^2)}{2}} \approx 150cm^{-1}$, which plays the role of Fröhlich mode [70]. The phonon mode of powder in this study is centered at around 152.3 cm^{-1}, which is near the frequency of Fröhlich mode, as shown in Table 5.8 and Fig. 5.13c. Compared to the wavenumber of the powder, the wavenumber of phonons, for all film samples, is blue shifted to around 163.0 cm^{-1}, which also has the value of the X critical point LO-phonon mode. The decrease observed in LA and TO phonon frequency relative to phonon frequency for the bulk may be due to two contributions: (i) phonon confinement and (ii) compressive strain [56].

Being the exciton Bohr radius of CdTe ($a_B = 7.3nm$ [71]) is relatively high compared to other compounds of Group II-VI, this makes the quantum size effect in CdTe possible at greater sizes since the charge confinement is assorted by comparing the grain radius with Bohr radius; for $R > a_B$ the confinement is weak, for $R < a_B$ the confinement is strong, and for $R = a_B$ it is intermediate. As noticed from Table 5.5, since R_{ave} is less than a_B of the bulk, the confinement is strong. Reducing R_{ave} leads to stronger phonon confinement, which causes a redshift in the phonon frequency [56].

The values of R_{ave} and ϵ found by the different techniques are tabulated in Table 5.7 for comparison. The grain size R_{ave}, is plotted against thickness in Fig. 5.14. The curves in this figure are fitted to quadratic equations given by [55]

$$R_{Abs} = 0.66 + 0.16\sqrt{t} - 7.5 \times 10^{-3}t + 6.19 \times 10^{-6}t^2 \tag{5.38}$$

$$R_{WH} = 2.84 - 6.52 \times 10^{-2}\sqrt{t} - 6.59 \times 10^{-3}t - 3.66 \times 10^{-6}t^2 \tag{5.39}$$

$$R_{Ram} = 0.90 + 0.90 \times 10^{-2}\sqrt{t} - 3.83 \times 10^{-3}t - 3.66 \times 10^{-6}t^2 \tag{5.40}$$

We notice that the particle size increases as film thickness increases. It is also noticed from Table 5.7 and Fig. 5.14 that the grain size evaluated by XRD overestimates that obtained by ABS and Raman spectroscopies.

The strain ϵ is plotted as a function of the grain size R_{Abs} in Fig. 5.15. The curve is fitted to the equation [68],

$$\varepsilon(R) = \varepsilon_T + \varepsilon_\sigma = \varepsilon_T - \frac{2}{3}k\left(\frac{b}{2R^2} + \frac{\sigma_\infty}{R}\right) \tag{5.41}$$

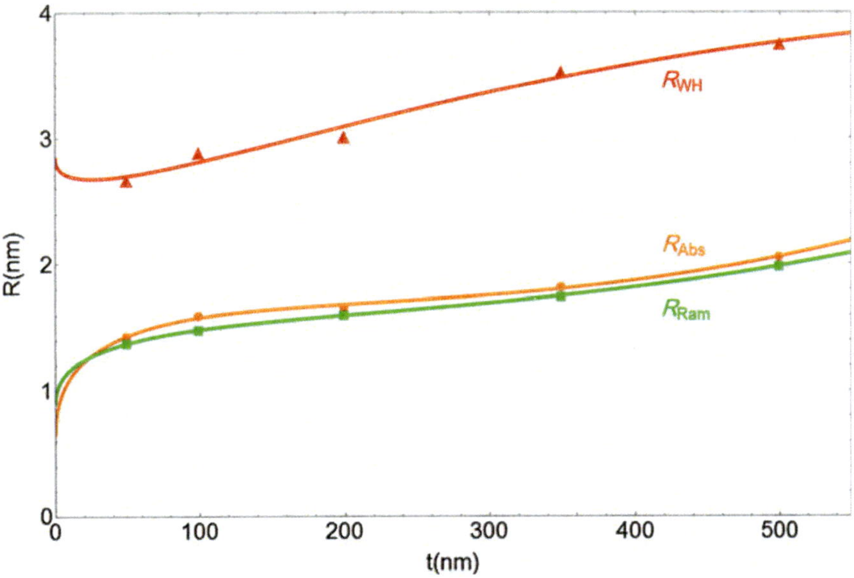

Fig. 5.14 Grain radius R_{WH} (red, triangles), R_{Abs} (orange, circles), and R_{Ram} (green, squares) against thickness t (R_{WH} axis is multiplied by 4) [55]. (Reproduced with permission. Copyright 2016, Wiley-VCH)

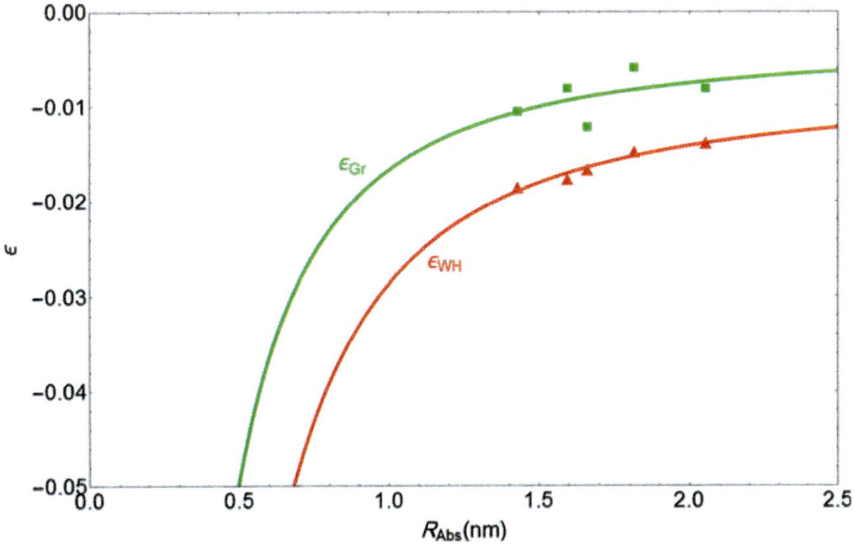

Fig. 5.15 The strains ϵ_{WH} and ϵ_{Gr} against the grain size R_{Abs} [55]. (Reproduced with permission. Copyright 2016, Wiley-VCH)

where ϵ_T (it is obtained by the fitting) is the strain produced from the difference of thermal expansion coefficient between the film and the glass slide, ϵ_σ is the strain produced from the surface-free energy, $\kappa = 7.86\times10^{-12}m^2/N$ [59] is the CdTe compressibility factor, b is an unknown constant to be found, and $\sigma_\infty = 0.725\text{N/m}$ [71] is the surface tension σ for the flat-slab CdTe bulk.

Fitting ϵ against R (Fig. 5.15) results in the following equations:

$$\varepsilon_{WH}(\text{R}) = -7.99 \times 10^{-3} - 5.24 \times 10^{-12}\left(\frac{3.21 \times 10^9}{R^2} + \frac{7.25 \times 10^8}{R}\right) \quad (5.42)$$

$$\varepsilon_{Gr}(\text{R}) = -3.18 \times 10^{-3} - 5.24 \times 10^{-12}\left(\frac{1.847 \times 10^9}{R^2} + \frac{7.25 \times 10^8}{R}\right) \quad (5.43)$$

It is seen that the strain obtained by Grüneisen model has the same order as the strain obtained by W–H method (Table 5.7). It is also noticed from Table 5.8 that the phonon frequencies of Raman lines (and hence the line shift $\delta\omega$) is sensitive to the compressive strain according to the Grüneisen model.

5.5 Difficulties in the Thin Film Growth of ZnO and Defect Structure

The experimental data, results and discussions given in this section are mainly cited from and/or based on [72]. ZnO is a significant member of II-VI semiconductor family and has been scrutinized for many years due to its applications as photodetectors, transparent thin film transistors, LEDs and laser diodes that function in UV region [73–76]. It has a wide direct band gap energy (3.3 eV) in the near-UV region and a large free-exciton binding energy (60 meV) at room temperature [77, 78].

There are a number of thin film growthing processes for ZnO, including PVD and non-PVD techniques. The preparation techniques used in published reports can be classified into (i) chemical vapor deposition [79], spray pyrolysis [80], sol–gel [81] (non-PVD); (ii) vacuum deposition [82, 83], sputter deposition [84], pulsed laser deposition and molecular beam epitaxy [85] (PVD). The structural properties of ZnO thin films such as crystallite orientation, particle size, layer resistivity, carrier mobility or optical transparency are affected by growthing method [79]. Most of the ZnO thin film deposition/coating methods mentioned above have some disadvantages in terms of containing some toxic chemicals, long reaction times, low purity that may occur in the films formed, and inadequacy of large surface area deposition. Despite its low cost and low toxicity, thermal evaporation of ZnO in vacuum attracted less attention by the research groups [86–88].

One of the methods to obtain ZnO thin films by thermal evaporation under vacuum is depositing Zn films from a Zn target and oxidizing these films by controlling a flow of Ar, Oxygen and Hydrogen gas mixture in same chamber simultaneously [82]. Fouad et al., who prepared ZnO thin films by thermal evaporation deposition under vacuum, studied with a vacuum pressure of 5×10^{-3} Torr and a temperature (for deposition and oxidation) range from 350 to 650 °C and a deposition time of 10–30 min. The other method to obtain thermally evaporated ZnO thin film is first to deposit a Zn thin film in vacuum and then to expose these Zn films to heat treatment in open atmosphere [83]. Rusu et al. prepared ZnO thin films by thermal oxidation in air of as-deposited Zn thin films deposited in vacuum. The oxidation temperature was in a range of 600–700 K, while the oxidation time changed in between 20–30 min.

In this part of the study, we consider the growth of ZnO on glass slides by using annealing process of Zn deposited films. Firstly, Zn thin films were deposited by thermal evaporation technique under vacuum (T-PVD) and subsequently heat-treated in air in between 1–3 h and 450–600 °C. Zn powder had a purity of 99.995% (Sigma Aldrich-324930). The value of vacuum pressure was on the order of 10^{-5} Torr in vacuum chamber unit (PVD-HANDY/25-TE/VAKSIS) while the current passing through the tungsten boat was ultimately 70 A [72]. A schematic view of vacuum chamber unit is represented in Fig. 5.16.

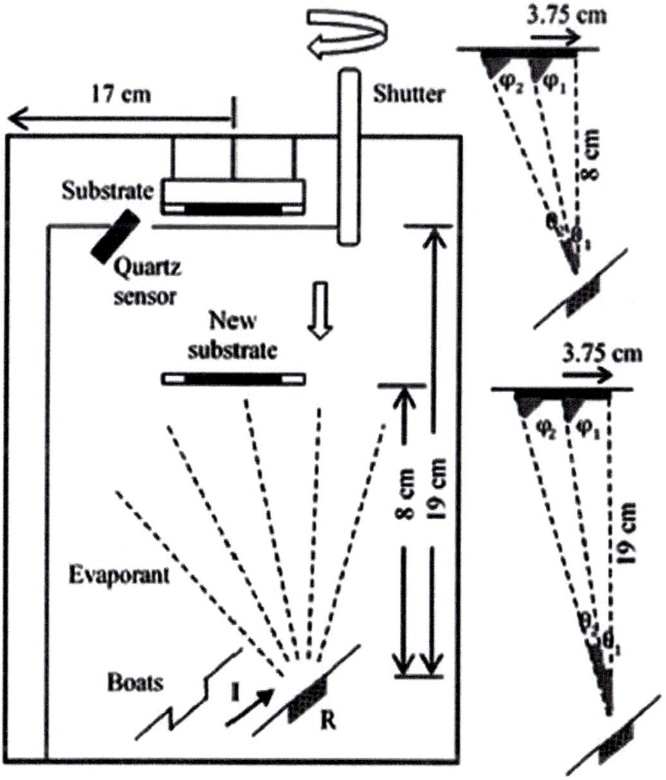

Fig. 5.16 Deposition geometry 1 (lower right corner) and 2 (upper right most corner). The angles for geometries 1 and 2 are, in degree; $\theta_1 = 11.14, \theta_2 = 10.36, \varphi_1 = 78.86, \varphi_2 = 68.50 and \theta_1 = 25.07, \theta_2 = 18.06, \varphi_1 = 64.93, \varphi_2 = 71.94$ respectively [72]

We reveal that deposition of Zn powder on glass slide significantly depended on the distance between the source (the boat) and the substrate. The deposition rate which depends on the vaporization rate is affected by the geometry of the components in the vacuum chamber unit. According to the following Hertz-Knudsen equation,

$$\frac{dN}{dT} \propto (p - p*)/T^{-1/2} \tag{5.44}$$

vaporization rate is directly proportional to the difference between the vapor pressure of the evaporant (p) and the pressure of the vacuum chamber (p*) at a certain temperature [89].

In addition to this, it is revealed that Zinc vapour is quite volatile. Hence we expect a high equilibrium vapour pressure of 10–2 Torr of Zn at its melting point of around 415 °C while the pressure decreases ($p \to 0$) [90]. For the distances above 12 cm, we could not obtain Zn deposition on the glass slide despite the inner walls of the whole vacuum chamber were deposited by the source [72].

Cosine deposition distribution points out a relationship between mass per unit area (dm/dA) and angular distribution of the deposition flux (see θ and φ in Fig. 5.16) and the source-substrate distance (r) [89]:

$$\frac{dm}{dA} \propto \frac{\cos\theta\cos\varphi}{r^2} \tag{5.45}$$

The ratio for mass per unit area of the first set of angles of Geometry 1 to the first set of angles for Geometry 2 is 4.4, while the mentioned ratio is 11.4 for the second set of angles. Therefore, in order to maximize deposition of Zn powder on glass slide it is obvious to use Geometry 2 instead of Geometry 1.

As mentioned above, Zn deposited samples were heat treated in air at temperatures in between 450 and 600 °C for durations from 1 to 3 h to grow ZnO.

Raman measurements were performed on Renishaw 250 mm focal length in Via Reflex Spectrometer system which has a Rayleigh line rejection filter allowing Raman spectrum to 100 cm^{-1} from a cw laser line at 532 nm at room temperature. Each vibrational mode was fitted by a Lorentzian function with a polynomial background and the best fit parameters were obtained for each peak position.

Raman scattering results were represented in Fig. 5.17 for the following samples: (a) as-deposited Zn thin film, (b) ZnO thin film heat-treated at 500 °C for 2 hours and (c) ZnO thin film heat-treated at 600 °C for 3 hours [72].

Raman measurements are sensitive to the composition; hence we can provide information about defect structure by identifying the Raman modes. The most intense peak at about 439 cm^{-1} in the spectra is denoted as E_2^{high} mode and this mode involves only the vibration of oxygen atoms at high frequency in ZnO while the E_2^{low} mode is related to the vibration of the heavy Zn sublattice (not shown) [91–94]. The peak at 331.94 cm^{-1} is due to the difference mode between E_2^{high} and E_2^{low} modes. The peaks at 382.06 and 582.091 cm^{-1} are previously presumed to be A_1(TO) and A_1(LO) Raman modes [95].

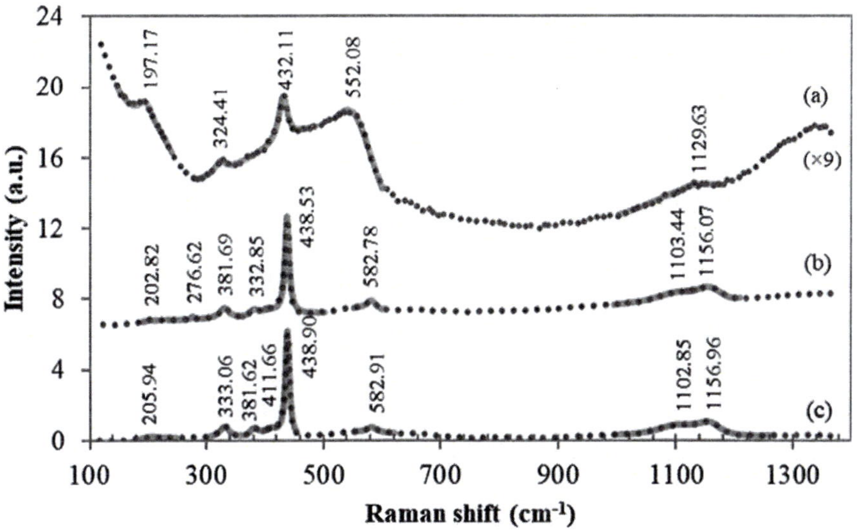

Fig. 5.17 Raman spectra of as-deposited Zn thin film (**a**), heat treated at 500 °C for 2 **h** (**b**) and 600 °C for 3 **h** (**c**) [72]

Nikitento et al. who studied with ZnO powder and Wang et al. who identified the line at ~1100 cm^{-1} as the acoustic combination of A_1 and E_2 symmetries also reported the broad peaks in between 1000 and 1200 cm^{-1} previously as indicated in Fig. 5.17b [96, 97]. The line at 1156.95 cm^{-1} is assigned to the third optical overtone of A_1 symmetry mode [97]. We presume that these two lines overlap as shown in Fig. 5.17a for as deposited Zn thin film while same lines are resolved for the ZnO thin films obtained by heat treatment at 600 °C for 3 h as indicated in Fig. 5.17c.

The peaks observed at 205.94, 331.94 and 439 cm^{-1} as seen in Fig. 5.17b are red-shifted by 7 to 9 cm^{-1} with respect to the peaks in Fig. 5.17b. This red-shift can arise from the possible oxygen vacancies in Zn rich ZnO thin film [93]. It can also be employed that the presence of the peak at about 582 cm^{-1} for heat treated samples, that has no correspondence in Fig. 5.17a might be originated from oxygen deficiency or zinc interstitials [98].

In this work we consider the difficulties in the thin film growth of ZnO deposited by thermal evaporation technique in vacuum followed by heat-treatment in air and present the results of Raman spectra of these thin films in order to provide information about defect structure. The red shift of some vibrational modes in the Raman spectra for as-deposited thin film compared to the heat-treated films and the presence of a line at around 582 cm^{-1} point out that ZnO thin films have defects such as oxygen vacancies or Zn interstitials. The most intense Raman mode attributed to high frequency oxygen vibration for as-deposited Zn film is red-shifted by about 7 cm^{-1} compared to the Raman mode for the heat-treated ZnO thin film which is presumed to originate from oxygen vacancies in Zn rich ZnO thin film. We also reveal that the deposition of Zinc on glass slide highly depends on the distance between the source and the target in consistent with the Hertz-Knudsen vaporization rate.

5.6 Colorimetric Evaluation of Group II-VI Quantum Dots in Glass Matrix

Colorimetry is the science and technology used to physically determine and define human color perception [99]. Color in an any medium is typically achieved by applying a colorant (dye or pigment) to the medium. Quantum dots have been investigated for various applications for many years [7]. Considering their unique luminescent properties of quantum dots, they can be used as a colorant for different mediums. Semiconductor quantum dots embedded glass matrixes (QDEGs) can provide chemical stability, thermal stability, and mechanical stability for numerous high endurance applications [100, 101]. In this part, we evaluate colorimetric properties of QDEGs within Commission Internationale de l'Elcairage (CIE) advices scope.

5.6.1 Materials and Methods

Commercial Cd, S, Se doped RG695 Schott filter glass was chosen to prepare our samples. Typical composition percentage of semiconductor doped Schott color filters is given as 52% SiO_2, 20% K_2O, 20% ZnO, 5% B_2O_3 and 3% of doping oxides which are CdO, SO_3, SeO_3, TeO_3 [102]. To dissolve the doping oxides, glass layers subjected to melting process at 1000 °C on a platinum substrate in an alumina boat for 15 min. After melting process, to avoid reformation of the doping oxides inside the glass layers while cooling down, glass layers quenched rapidly to room temperature by cold air flow. Color of the glass layers transformed from dark reddish to clear. Then the glass layers annealed at 450 °C for 5 h to reduce stress in glass samples and to start nucleation of nanocrystals. Subsequently, the glass samples were sanded down to width of 0.3–0.6 mm to distribute heat evenly inside the glass layers for the second annealing. Finally, glass layers annealed at different temperatures for various periods of time to develop nanocrystals in the glass layers. As shown in shown in Table 5.9 QDEG samples labeled according to estimated average quantum dot radius.

Optical transmission measurements were done by a system having a 1/8 m monochromator with a 1200 line/mm grating and a silicon photodetector. A 50 W tungsten lamp which is driven by a constant current source was used as the light source. Optical bandgap and average quantum dot radius calculations details can be found in previous publications [103]. Basically, bandgap and radius of the quantum dots are given by equations;

Table 5.9 Quantum dot embedded glass layers labeled by estimated average quantum dot radius. Second and third column show secondary annealing temperature and duration. Fourth column is the thickness of the glass samples. Fifth column shows estimated average particle radius doped in the glass layers

Sample	2nd Annealing Temp. (°C)	Duration (hour)	Thickness (mm)	Estimated Av. Particle Radius (nm)
S1	600	0.5	0.386	–
S2	625	1	0.415	2.1
S3	625	2	0.376	2.2
S4	625	4	0.39	2.5
S5	625	8	0.498	2.7
S6	650	2	0.401	3.1
S7	650	4	0.402	3.4
S8	650	8	0.342	3.5
S9	675	2	0.39	4.5
S10	675	4	0.345	4.7
S11	As received	As received	0.55	Bulk

$$E_{ex}(R) = E_g \frac{\hbar\pi^2}{\mu R^2} \tag{5.46}$$

$$R = \frac{\hbar\pi}{2\sqrt{\mu(E_{ex} - E_g)}} \tag{5.47}$$

where E_{ex} is first exciton peak energy value of the sample, $E_g = 1.83$ eV is the bulk band gap of $CdS_{0.08}Se_{0.92}$ crystal [104] and the second part is confinement energy for an exciton. Effective mass of an exciton is $\mu = 0.106\ m_0$ electron rest mass [105].

CIE L* a* b* space is a three-dimensional space contrary to two dimensional CIE 1931 color space. CIE L* a* b* space encompasses the complete range of human color perception where L* axis denotes lightness, a* axis denotes green and red color shades, b* axis denotes blue and yellow color shades. CIE L* a* b* values calculated with the advices of the CIE Technical Report [106]. Basically;

$$\begin{aligned} L^* &= 116f(G_y) - 16 \\ a^* &= 500\left[f(G_x) - f(G_y)\right] \\ b^* &= 200\left[f(G_y) - f(G_z)\right] \end{aligned} \tag{5.48}$$

where;

$$G_x = (X/X_n);\ G_y = (Y/Y_n);\ G_z = (Z/Z_n) \tag{5.49}$$

where;

$$f(G_i) = G_i^{1/3}\ if\ G_i > (6/29)^3 \tag{5.50}$$

else;

$$f(G_i) = (841/108)G_i + 4/29\ if\ G_i \leq (6/29)^3 \tag{5.51}$$

where indice i changes as X, Y and Z which are CIE tristimulus values. For our colorimetric calculations, illumination has selected as CIE standard D65 illuminant. D65 is common selection for various standards and applications.

As illustrated in Fig. 5.18, the CIE definition of dominant wavelength (λ_d) is "Wavelength of the monochromatic stimulus that, when additively mixed in suitable proportions with the specified achromatic stimulus, matches the color stimulus considered". Also, the definition of complimentary wavelength (λ_c) is given by CIE as "Wavelength of the monochromatic stimulus that, when additively mixed in suitable proportions with the color stimulus considered, matches the specified achromatic stimulus" [106]. Excitation purity (p_e) is defined as;

$$p_e = \frac{a}{a+b} \tag{5.52}$$

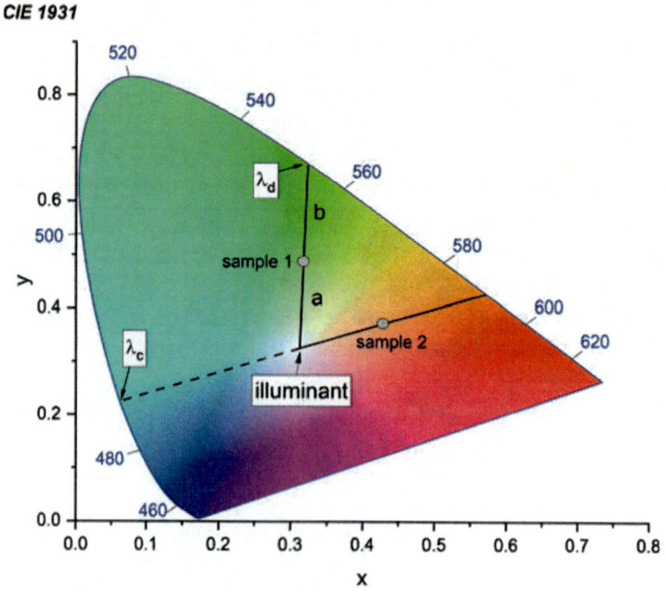

Fig. 5.18 Dominant wavelength (λ_d), complimentary wavelength (λ_c) and excitation purity (p_e) are displayed on color chromaticity diagram

the proportion of the distance from the sample's chromaticity coordinates to the illuminant coordinates, to the distance of the dominant wavelength coordinates of the sample to the illuminant coordinates (5.52).

5.6.2 Results and Discussions

We present optical transmittance behavior of QDEG samples in Fig. 5.19. One can see that increase of temperature and duration of thermal annealing red shifts absorption band edge of QDEGs and approach to RG695 Schott filter glass which has bulk $CdS_{0.08}Se_{0.92}$ nanocrystals [107]. By regulating the heat treatment conditions, the bandgap of doping quantum dots can be modified over the entire visible range [105].

The CIE XYZ model is the foundation of all colorimetric methods. On the other hand, these coordinates are not very intuitive, since X, Y and Z are mathematical parameters that do not give much detailed information about the measured color [99]. Therefore, color coordinates of QDEGs are given in Fig. 5.19 has shown in CIE 1931 color space to aid visualization.

Taking into consideration of Figs. 5.19 and 5.20, it is easy to realize that color of the glass matrix caused by change in the size of $CdS_{0.08}Se_{0.92}$ QDs as an effect of thermal annealing. It is observed from CIE chromaticity diagram that the transparent sample

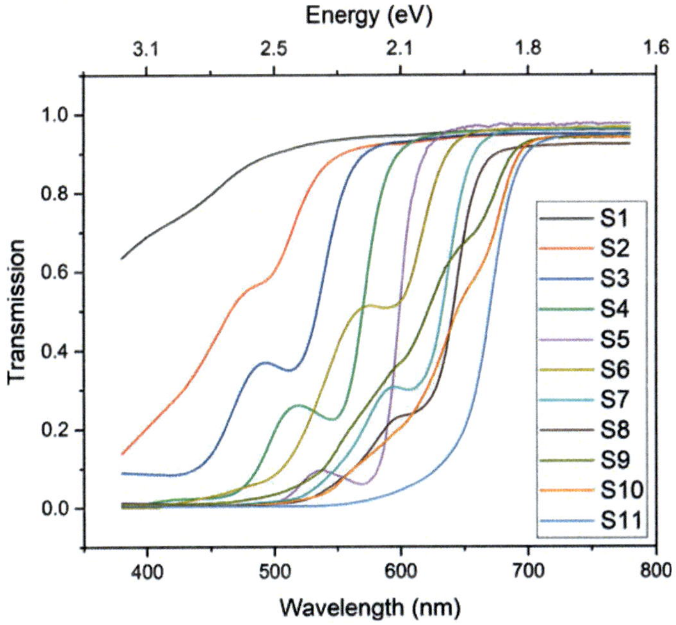

Fig. 5.19 Optical transmission characteristics of $CdS_{0.08}Se_{0.92}$ quantum dots in glass matrix for various temperature and durations of thermal annealing

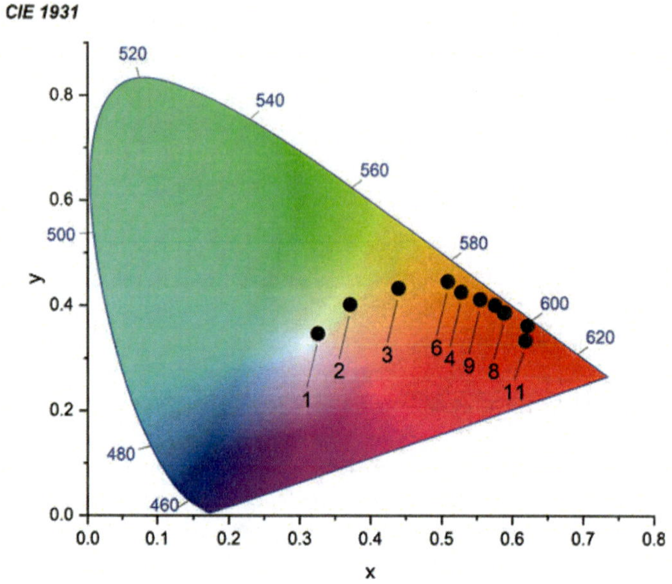

Fig. 5.20 CIE xy chromaticity diagram of QDEGs. Higher temperature of annealing and time subjected samples approaches to bulk crystal color coordinates

locate near the white point of the diagram. Moreover, from the CIE 1931 gamut, glass matrix which are subjected higher temperature of annealing and durations shows shades of yellow and red colors. Simply by changing the thermal annealing conditions the color of the glass matrix can be altered due to QDs acting as a colorant.

Even though CIE 1931 color space aids to visualize coloring effect of $CdS_{0.08}Se_{0.92}$ quantum dots, CIE L* a* b* space can give clearer picture in order to understand colorimetric performance. Each dot in Fig. 5.21 represents QDEG samples. Sample number 1 shows higher lightness L* value since it is transparent to all visible range and sample number 11 gives darkish red color due to the low visible region transmission. Color shades gets darker with the increasing radius of the quantum dots in the glass matrix.

Excitation purity and estimated average quantum dot radius has an asymptotic relation as shown in Fig. 5.22. Radius of the $CdS_{0.08}Se_{0.92}$ QDs in RG695 is considered around 6 nm [108]. Considering the 5.52 and the definition of the excitation purity, estimated average particle radius of the $CdS_{0.08}Se_{0.92}$ QDs in glass matrix affects location of the QDEG chromaticity coordinates. When $CdS_{0.08}Se_{0.92}$ QDs' estimated average particle radius is smaller than 3 nm, color purity is relatively lower than when the particle radius is bigger than 3 nm. This might be due to the QDEGs' optical transmission behavior since color coordinates are calculated with the transmission spectra of QDEGs and D65 stimulus. Taking into consideration of optical transmission characteristics of $CdS_{0.08}Se_{0.92}$ quantum dots in glass matrix (Fig. 5.19) one can see that QDEGs have wide visible transmission characteristics also have

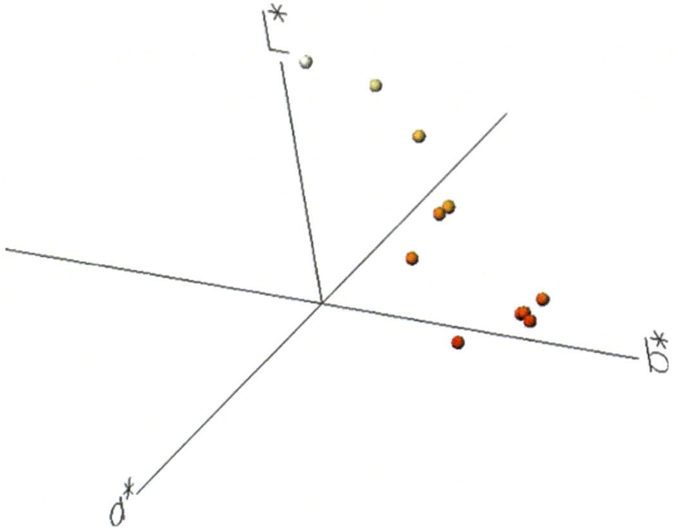

Fig. 5.21 CIE L* a* b* visualization of QDEG samples

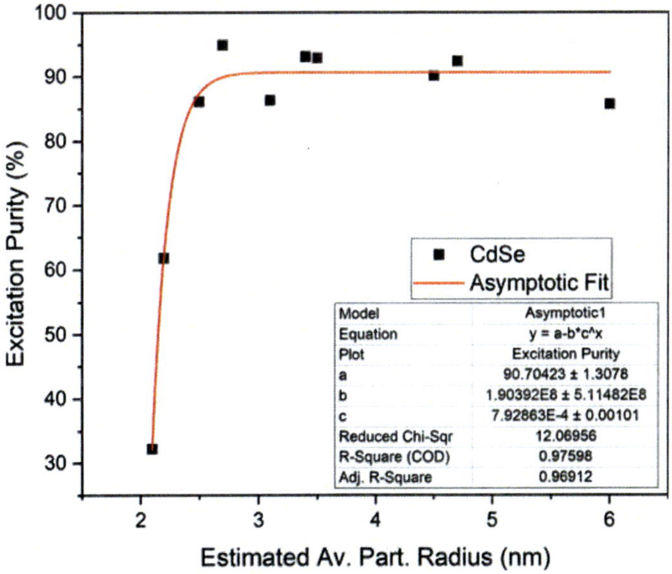

Fig. 5.22 Color purity of the QDEG samples with respect to estimated average particle radius shows asymptotic behavior

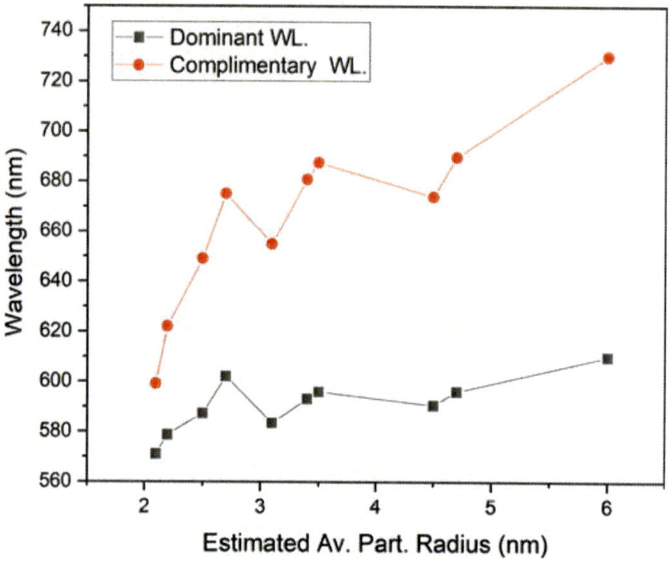

Fig. 5.23 Dominant and complimentary wavelength relation with respect to average particle size

relatively small particle radius QDs. Therefore, those QDEGs allows to pass wider portion of the D65 stimulus and that causes to move QDEGs' color coordinates to closer location of D65 stimulus. Consequently, wider band gap QDs embedded glass layers have lower color purity due to quantum confinement effect.

In Fig. 5.23 the relation between QD radius and dominant and complimentary wavelength relation is given. In order to aid visualization each dot connected for both dominant and complimentary wavelength regarding average QD radius.

Colorimetric properties of QDEGs evaluated by their CIE chromaticity diagram and CIE L* a* b* space since CIE XYZ model is not visual friendly. 11 samples of QDEGs were prepared by melt quenching technique and subjected to various thermal annealing processes. These processes helped to see the effect of quantum confinement of semiconductor $CdS_{0.08}Se_{0.92}$ QDs as a colorant for glass matrixes. Coloring behavior is related to particle size.

References

1. A.L. Efros, A.L. Efros, Sov. Phys. Semicond **16**, 772 (1982)
2. A.I. Ekimov, A.A. Onushchenko, Sov. Phys. Semicond. **16**, 775 (1982)
3. J. Warnock, D.D. Awschalom, Phys. Rev. B **32**, 5529 (1985)
4. B.G. Potter, J.H. Simmons, Phys. Rev. B **37**, 10838 (1988)
5. M.H. Yükselici, A. Aşıkoğlu Bozkurt, Ç. Allahverdi, Z. Nassar, D. Bulut, B. Can Ömür, M.K. Torun, A.T. İnce, in *Low-Dimensional Nanostructured Mater. Devices* (2016), pp. 327–350
6. M.K. Torun, *Growth of Group II-VI Quantum Dots (CdTeSe) in Glass Studied Through Optical Absorption and Raman Spectroscopies*. Yeditepe University (2017)

7. J. Xue, X. Wang, J.H. Jeong, X. Yan, Chem. Eng. J. **383**, 123082 (2020)
8. A.I. Ekimov, A.A. Onushchenko, ZhETF Pisma Redaktsiiu **34**, 363 (1981)
9. R.K. Jain, R.C. Lind, J. Opt. Soc. Am. **73**, 647 (1983)
10. A. Nozik, Phys. E Low-Dimensional Syst. Nanostructures **14**, 115 (2002)
11. A.J. Sutherland, Curr. Opin. Solid State Mater. Sci. **6**, 365 (2002)
12. M.G. Bawendi, P.J. Carroll, W.L. Wilson, L.E. Brus, J. Chem. Phys. **96**, 946 (1992)
13. F. Hache, M.C. Klein, D. Ricard, C. Flytzanis, J. Opt. Soc. Am. B **8**, 1802 (1991)
14. O.R. Ochoa, C. Colajacomo, E.J. Witkowski, J.H. Simmons, B.G. Potter, Solid State Commun. **98**, 717 (1996)
15. C. Flytzanis, F. Hache, M.C. Klein, D. Ricard, in *Optical Complex Systems* (International Society for Optics and Photonics, 1990), pp. 75–80
16. M.K. Torun, A.T. İnce, M.H. Yükselici, in *10th Nanoscience and Nano Technology Conference* (Istanbul, 2014)
17. J.D. Dow, D. Redfield, Phys. Rev. B **5**, 594 (1972)
18. F. Urbach, Phys. Rev. **92**, 1324 (1953)
19. L. Brus, Appl. Phys. A Solids Surfaces **53**, 465 (1991)
20. D.A.B. Miller, in: *Quantum Dynamic Simple Systems*, ed. by G.-L. Oppo, S.M. Barnett, E. Riis, M. Wilkinson (Institute Physics, London, 1996), p. 226
21. S. Chandra, *Shooting Method in Solving Quantum Structure Related Problems*. Jadavpur University (2013)
22. J. Killingbeck, J. Phys. A. Math. Gen. **20**, 1411 (1987)
23. H.G. Hugdal, P. Berg, Eur. J. Phys. **36**, 045013 (2015)
24. I.M. Lifshitz, V.V. Slyozov, J. Phys. Chem. Solids **19**, 35 (1961)
25. N. Romčević, M. Romčević, R. Kostić, D. Stojanović, G. Karczewski, R. Galazka, Microelectron. J. **40**, 830 (2009)
26. E.C.F. da Silva, D. Strauch, *Landolt-Börnstein Semiconductors Subvolume E* (Springer-Verlag, Berlin Heidelberg, 2012)
27. T.K. Nideep, M. Ramya, M. Kailasnath, Superlattices Microstruct. **141**, 106477 (2020)
28. N. Gaponik, A.L. Rogach, Phys. Chem. Chem. Phys. **12**, 8685 (2010)
29. X. Lou, Q. Zeng, Y. Zhang, Z. Wan, J. Qin, Z. Li, J. Mater. Chem. **22**, 5581 (2012)
30. Q. Ma, E. Ha, F. Yang, X. Su, Anal. Chim. Acta **701**, 60 (2011)
31. Y. Liu, F. Dai, R. Zhao, X. Huai, J. Han, L. Wang, J. Mater. Sci. **54**, 8571 (2019)
32. A. Aboulaich, D. Billaud, M. Abyan, L. Balan, J.-J. Gaumet, G. Medjadhi, J. Ghanbaja, R. Schneider, A.C.S. Appl. Mater. Interfaces **4**, 2561 (2012)
33. L.M. Liz-Marzán, P. Mulvaney, J. Phys. Chem. B **107**, 7312 (2003)
34. W.E. Mahmoud, S.J. Yaghmour, Opt. Mater. (Amst.) **35**, 652 (2013)
35. G. O'brien, *Nanoscale Electronic Properties of Conjugated Polymers and Nanocrystals*. University College Cork (2006)
36. M.H. Yükselici, Ç. Allahverdi, J. Lumin. **128**, 537 (2008)
37. M. Zanella, A.Z. Abbasi, A.K. Schaper, W.J. Parak, J. Phys. Chem. C **114**, 6205 (2010)
38. L. Jing, S.V. Kershaw, Y. Li, X. Huang, Y. Li, A.L. Rogach, M. Gao, Chem. Rev. **116**, 10623 (2016)
39. D. Pan, Q. Wang, L. An, J. Mater. Chem. **19**, 1063 (2009)
40. R. Moradian, M. Elahi, A. Hadizadeh, M. Roshani, A. Taghizadeh, R. Sahraei, Int. Nano Lett. **3**, 56 (2013)
41. L. Feng, H. Kuang, X. Yuan, H. Huang, S. Yi, T. Wang, K. Deng, C. Tang, Y. Zeng, Spectrochim. Acta Part A Mol. Biomol. Spectrosc. **123**, 298 (2014)
42. K.-T. Yong, W.-C. Law, I. Roy, Z. Jing, H. Huang, M.T. Swihart, P.N. Prasad, J. Biophotonics **4**, 9 (2011)
43. S.Y. Choi, J.P. Shim, D.S. Kim, T. Kim, K.S. Suh, J. Nanomater. **2012**, 1 (2012)
44. W. Wang, L. Cheng, W. Liu, Sci. China Chem. **57**, 522 (2014)
45. Y. Shi, Z. Ma, N. Cui, Y. Liu, X. Hou, W. Du, L. Liu, T. Gangsheng, Nanoscale Res. Lett. **9**, 121 (2014)
46. J. Lim, M. Park, W.K. Bae, D. Lee, S. Lee, C. Lee, K. Char, ACS Nano **7**, 9019 (2013)

47. D. Zhou, M. Lin, Z. Chen, H. Sun, H. Zhang, H. Sun, B. Yang, Chem. Mater. **23**, 4857 (2011)
48. Y. Shi, P. He, X. Zhu, Mater. Res. Bull. **43**, 2626 (2008)
49. L. Jing, K. Ding, S. Kalytchuk, Y. Wang, R. Qiao, S.V. Kershaw, A.L. Rogach, M. Gao, J. Phys. Chem. C **117**, 18752 (2013)
50. Y. Shi, J. Wang, S. Li, Z. Wang, X. Zang, X. Zu, X. Zhang, F. Guo, G. Tong, J. Mater. Res. **28**, 1940 (2013)
51. V. Dzhagan, I. Lokteva, C. Himcinschi, X. Jin, J. Kolny-Olesiak, D.R.T. Zahn, Nanoscale Res. Lett. **6**, 1 (2011)
52. J.J. Loferski, J. Appl. Phys. **27**, 777 (1956)
53. S.N. Sharma, S. Kohli, A.C. Rastogi, Phys. E Low-Dimensional Syst. Nanostructures **25**, 554 (2005)
54. A.C. Rastogi, S.N. Sharma, S. Kohli, Semicond. Sci. Technol. **15**, 1011 (2000)
55. Z.M. Nassar, M.H. Yükselici, A.A. Bozkurt, Phys. Status Solidi **253**, 1104 (2016)
56. A.K. Arora, M. Rajalakshmi, T.R. Ravindran, Encycl. Nanosci. Nanotechnol. 499 (2004)
57. F. Henneberger, J. Puls, C. Spiegelberg, A. Schulzgen, H. Rossman, V. Jungnickel, A.I. Ekimov, Semicond. Sci. Technol. **6**, A41 (1991)
58. A. Aşıkoğlu, M.H. Yükselici, Semicond. Sci. Technol. **26**, 055012 (2011)
59. S. Adachi, *Handbook on Physical Properties of Semiconductors* (Kluwer Academic Publishers, Dordrecht, 2004)
60. M.V. Kurik, Phys. Status Solidi **8**, 9 (1971)
61. G.D. Cody, T. Tiedje, B. Abeles, B. Brooks, Y. Goldstein, Phys. Rev. Lett. **47**, 1480 (1981)
62. B.E. Warren, *X-Ray Diffraction* (Dover Pub. Inc., New York, 1990)
63. G. Williamson, W. Hall, Acta Metall. **1**, 22 (1953)
64. G. Gordillo, J.M. Flórez, L.C. Hernández, Sol. Energy Mater. Sol. Cells **37**, 273 (1995)
65. V. Mote, Y. Purushotham, B. Dole, J. Theor. Appl. Phys. **6**, 6 (2012)
66. C. Suryanarayana, M.G. Norton, *X-Ray Diffraction: A Practical Approach*, 1st edn. (Plenum Press, New York, 1998)
67. J.F. Kielkopf, J. Opt. Soc. Am. **63**, 987 (1973)
68. V. Spagnolo, G. Scamarcio, M. Lugará, G.C. Righini, Superlattices Microstruct. **16**, 51 (1994)
69. H.-M. Kagaya, T. Soma, Phys. Status Solidi **124**, 37 (1984)
70. A.G. Rolo, M.I. Vasilevskiy, N.P. Gaponik, A.L. Rogach, M.J.M. Gomes, Phys. Status Solidi **229**, 433 (2002)
71. F. Wu, J.W. Lewis, D.S. Kliger, J.Z. Zhang, J. Chem. Phys. **118**, 12 (2003)
72. D. Bulut, M.H. Yükselici, Mater. Res. Express **6**, 056408 (2019)
73. A. Janotti and C.G. Van de Walle, Rep. Prog. Phys. **72**, 126501 (2009)
74. C. Soci, A. Zhang, B. Xiang, S.A. Dayeh, D.P.R. Aplin, J. Park, X.Y. Bao, Y.H. Lo, D. Wang, Nano Lett. **7**, 1003 (2007)
75. R.L. Hoffman, B.J. Norris, J.F. Wager, Appl. Phys. Lett. **82**, 733 (2003)
76. M. Willander, O. Nur, Q. X. Zhao, L.L. Yang, M. Lorenz, B. Q. Cao, J. Zúñiga Pérez, C. Czekalla, G. Zimmermann, M. Grundmann, A. Bakin, A. Behrends, M. Al-Suleiman, A. El-Shaer, A. Che Mofor, B. Postels, A. Waag, N. Boukos, A. Travlos, H.S. Kwack, J. Guinard, D. Le Si Dang, Nanotechnology **20**, 332001 (2009)
77. D.M. Bagnall, Y.F. Chen, M.Y. Shen, Z. Zhu, T. Goto, T. Yao, J. Cryst. Growth **184–185**, 605 (1998)
78. W.I. Park, Y.H. Jun, S.W. Jung, G.-C. Yi, Appl. Phys. Lett. **82**, 964 (2003)
79. M. Purica, Thin Solid Films **403–404**, 485 (2002)
80. N. Lehraki, M.S. Aida, S. Abed, N. Attaf, A. Attaf, M. Poulain, Curr. Appl. Phys. **12**, 1283 (2012)
81. C.-Y. Tsay, K.-S. Fan, Y.-W. Wang, C.-J. Chang, Y.-K. Tseng, C.-K. Lin, Ceram. Int. **36**, 1791 (2010)
82. O. Fouad, A. Ismail, Z. Zaki, R. Mohamed, Appl. Catal. B Environ. **62**, 144 (2006)
83. D.I. Rusu, G.G. Rusu, D. Luca, Acta Phys. Pol. A. **119**, (2011)
84. W. Yang, Z. Liu, D.-L. Peng, F. Zhang, H. Huang, Y. Xie, Z. Wu, Appl. Surf. Sci. **255**, 5669 (2009)

85. R. Triboulet, J. Perrière, Prog. Cryst. Growth Charact. Mater. **47**, 65 (2003)
86. S.J. Chen, Y.C. Liu, J.G. Ma, D.X. Zhao, Z.Z. Zhi, Y.M. Lu, J.Y. Zhang, D.Z. Shen, X.W. Fan, J. Cryst. Growth **240**, 467 (2002)
87. R.K. Gupta, N. Shridhar, M. Katiyar, Mater. Sci. Semicond. Process. **5**, 11 (2002)
88. Y.G. Wang, S.P. Lau, H.W. Lee, S.F. Yu, B.K. Tay, X.H. Zhang, H.H. Hng, J. Appl. Phys. **94**, 354 (2003)
89. D.M. Mattox, *Handbook of Physical Vapor Deposition (PVD) Processing*, 2nd edn. (Elsevier, 2007)
90. C. Yaws, *Handbook of Vapor Pressure*, vol. 4 (Gulf Pub. Co., USA, 1995)
91. H. Morkoç, Ü. Özgür, *Zinc Oxide: Fundamentals, Materials and Device Technology* (Wiley, 2009)
92. R. Zhang, P.-G. Yin, N. Wang, L. Guo, Solid State Sci. **11**, 865 (2009)
93. M. Šćepanović, M. Grujić-Brojčin, K. Vojisavljević, S. Bernik, T. Srećković, J. Raman Spectrosc. **41**, 914 (2010)
94. R. Cuscó, E. Alarcón-Lladó, J. Ibáñez, L. Artús, J. Jiménez, B. Wang, and M. J. Callahan, Phys. Rev. B **75**, 165202 (2007)
95. C. Kumari, A. Pandey, A. Dixit, J. Alloys Compd. **735**, 2318 (2018)
96. V.A. Nikitenko, V.G. Plekhanov, S.V. Mukhin, M.V. Tkachev, J. Appl. Spectrosc. **63**, 290 (1996)
97. R.P. Wang, G. Xu, P. Jin, Phys. Rev. B **69**, 113303 (2004)
98. R. Jothilakshmi, V. Ramakrishnan, R. Thangavel, J. Kumar, A. Sarua, M. Kuball, J. Raman Spectrosc. **40**, 556 (2009)
99. N. Ohta, A.R. Robertson, *Colorimetry Fundamentals and Applications* (John Wiley & Sons, Ltd., 2005)
100. M. Xia, J. Luo, C. Chen, H. Liu, J. Tang, Adv. Opt. Mater. **1900851**, 1900851 (2019)
101. Y. Wei, H. Ebendorff-Heidepriem, J. Zhao, Adv. Opt. Mater. **7**, 1 (2019)
102. G. Banfi, V. Degiorgio, B. Speit, J. Appl. Phys. **74**, 6925 (1993)
103. H. Ünlü, M.R. Karim, H.H. Gürel, Ö. Akıncı, in ed. by H. Ünlü, N.J.M. Horing (Springer, Berlin, 2013), pp. 1–17
104. O. Madelung (ed.), *Semiconductors: Properties of II–VI Compounds* (Springer, Berlin Heidelberg, 1982)
105. B. Can Ömür, A. AşIkoğlu, Ç. Allahverdi, M. H. Yükselici, J. Mater. Sci. **45**, 112 (2010)
106. CIE, *Colorimetry*, 3rd edn. (2004)
107. M. Gandais, M. Allais, Y. Zheng, and M. Chamarro, J. Phys. IV JP **4**, 47 (1994)
108. P. D. Persans, A. Tu, Y.-J. Wu, and M. Lewis, J. Opt. Soc. Am. B **6**, 818 (1989)

Chapter 6
Recent Progress in Magnetic Nanostructures Studied by Synchrotron Radiation

Takafumi Miyanaga and Ryo Masuda

Abstract The recently developed synchrotron radiation source provides high performance X-ray beam and opens new technique to study for the nanostructured science and devices. In this chapter we discuss two topics: (1) XMCD and XAFS studies for cluster-layered Fe/Cr film and (2) Mössbauer study for Fe surface magnetism. The cluster-layered Al_2O_3/Cr(70Å) /[Fe(1.2Å) /Cr(10.5Å)]$_{30}$/Cr(12Å) film shows the minimum point in the resistivity. In this film, the magnetic moment of Fe atoms decreases as the temperature increases and disappears around the critical temperature. And further, the local structure around Fe atoms shows rather disorder, which may be related the fluctuation of magnetic structure obtained from the mean field theory. The surface magnetism of Fe (001) was studied in an atomic layer-by-layer fashion by using the in-situ ^{57}Fe probe layer method with a synchrotron Mössbauer source. The observed magnetic hyperfine field exhibits a remarkable decrease at the surface and an oscillatory behavior with increasing its depth in the upper four layers below the surface. These results provide a first experimental evidence for the magnetic Friedel oscillations, which penetrate several layers from the Fe (001) surface. From these two topics, we present high utility of the applications of synchrotron radiation to magnetic nanostructure studies.

Abbreviations and Symbols

CEMS	Conversion Electron Mössbauer Spectroscopy
EXAFS	Extended X-ray Absorption Fine Structure
GMR	Giant magnetoresistance
IXS	Inelastic X-ray Scattering
NMC	Nuclear Monochromator Crystal
RI	Radioactive Isotope
SMS	Synchrotron Mössbauer Source

T. Miyanaga (✉) · R. Masuda
Department of Mathematics and Physics, Hirosaki University, Hirosaki, Japan
e-mail: takaf@hirosaki-u.ac.jp

H. Ünlü and N. J. M. Horing (eds.), *Progress in Nanoscale and Low-Dimensional Materials and Devices*, Topics in Applied Physics 144,
https://doi.org/10.1007/978-3-030-93460-6_6

SR	Synchrotron Radiation
XAS	X-ray Absorption Spectroscopy
XAFS	X-ray Absorption Fine Structure
XANES	X-ray Absorption Near Edge Structure
XMCD	X-ray Magnetic Circular Dichroism
XPS	X-ray Photoelectron Spectroscopy

6.1 Introduction

Magnetic nanostructures have many considerable interests due to their unique magnetic properties and potential of IT applications as sensing materials, spintronics, data storage, and optoelectronic technologies [1]. The important factors to govern the magnetic properties of nanostructures are the chemical composition of the element, shape and size of materials, surface morphology, anisotropy of atomic arrange, layer thickness, and molecular interaction for the system. Drastic change in magnetic properties is observed when the particle size reduces to a critical magnetic scale, such as the exchange length, for example, GMR. In the basic physics, the broken symmetry at surfaces and interfaces is key concept which may increase or decrease magnetic moment per atoms in the vicinity of surface and interface. To understand the magnetic nanostructure, the behaviors in surface and interface are quite important, where the atomic arrangement is well controlled on surface and in multilayers.

In recent years, highly intense (or brilliant), high energy and energy-tunable X-ray from the SR is applicable, which opens new field on the materials science, semiconductors, magnetic materials, catalysis, battery materials, chemically functional materials, even in biological systems. Especially, for the study of magnetic materials, XMCD using the circular polarization of X-ray provides information about the local magnetic environment around atoms with good S/N ratio [1, 2]. On the other hand, Mössbauer spectroscopy is a method to study the local electronic and vibrational states of functional materials, which includes the valence, structural change, magnetic ordering, and Debye temperature. These techniques are also progress by using the SR source with high brilliance[1] and energy-tunability.

In this chapter, first we present the application of XMCD and XAFS to study for thin films: cluster-layered Fe/Cr film which shows interesting Kondo-like behavior in the resistivity. Second, we discuss about the Mössbauer spectroscopy from the basic introduction to recent applications: the surface magnetism of Fe(001) was studied in an atomic layer-by-layer fashion by using the in situ ^{57}Fe probe layer method with a SR source.

[1] In this chapter, "brilliance" is defined in the manner of X-ray optics as follows: Brilliance $= \frac{\text{Photons/Time}}{\text{Vertical and horizontal divergence of X-ray source}\cdot\text{Vertical and horizontal size of the source}\cdot 0.1\%\text{bandwidth of the X-rays}}$. The typical unit is $\frac{\text{Photons/seconds}}{\text{mrad}^2\cdot\text{mm}^2\cdot 0.1\%\text{bandwidth}}$ or $\frac{\text{Photons/seconds}}{\text{arcsecond}^2\cdot\text{mm}^2\cdot 0.1\%\text{bandwidth}}$. The energy width used in this definition is "0.1%bandwidth" and thus depends on the X-ray energy. For example, 0.1%bandwidth of 14.413 keV X-rays is 14.413 eV.

6.2 XMCD and XAFS Study for Thin Film

In recent years, XMCD becomes powerful technique for the study of magnetic materials. When the difference in X-ray absorption coefficient of left- and right-hand circular polarized X-rays is measured for magnetic materials, an XMCD spectra can be obtained. The difference of the magnetic dichroism deduce the atomic magnetic moment in the ferromagnetic materials. XMCD has several attractive features in comparison with other techniques. (1) XMCD gives information about the spin- and orbital-magnetization individually, while most of other methods are sensitive only to total magnetization. (2) This technique is element specific, by measuring the dichroism around the specific absorption edges. (3) XMCD can be applied to determine extremely small magnetic moments and to study small quantities of atoms in thin films and on surfaces because of its high sensitivity. This makes XMCD to study nano-scale structures, which are connected to modern electric data storage device etc.

6.2.1 Methodology

We present here brief introduction to XMCD method with support from ref [3]. In a circular polarized electric field, an X-ray is described as a helical wave around the propagation direction **k**. The circular polarized electromagnetic wave can be quantum mechanically described and in this case the photon with z-direction of the propagation is in an eigenstate of the angular momentum operator, J_z. For right-circular polarized, RCP (or Left-, LCP), photons, the eigenvalue of J_z is $+\hbar$(or $-\hbar$). Linearly polarized photons have the expectation value, $\langle J_z \rangle = 0$, since they are equally superposed by RCP and LCP photons. A circular polarized photon in eigenstate of J_z can be governed by the selection rules for the conservation of angular momentum in electronic transition, that makes the interpretation of XMCD signal be simple.

Let us understand the origin of the XMCD signal by simple atomic model and consider an electronic transition from a core state to a bound state at higher energy as a starting point. The schematical energy level described above discussion in appear in Fig. 6.1. It is well known in the atomic physics that the probability of electric transitions is controlled by selection rules for the change in the quantum numbers describing the initial and final states. A main mechanism is the interaction between the electric field of X-ray and the electric dipole moment operator. The parity of the initial and final states is given by their orbital quantum numbers ℓ and non-zero matrix element is obtained for the change of ℓ, $\Delta\ell$, only if $\Delta\ell = \pm 1$, which is so called "dipole transition selection rule". Since the X-ray photon is annihilated in the absorption process, the angular momentum J_z must be transferred to the sample. For the circular polarized photon, the change in magnetic quantum number, Δm, is permitted if $\Delta m = +1$ for RCP photons or $\Delta m = -1$ for LCP photons.

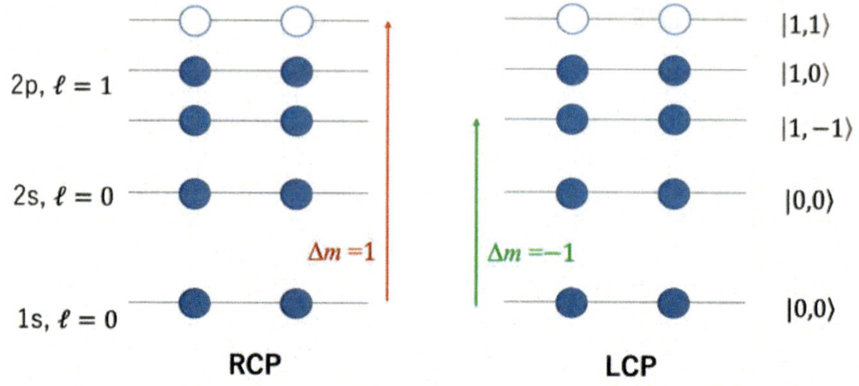

Fig. 6.1 The simplified energy level diagram to describe XMCD origin

As general XMCD geometry, a magnetic field is applied parallel to the direction of photon propagation z. Through the Zeeman effect, the magnetic field breaks the orbital degeneracy of the $2p$ state into the separate states $|\ell, m\rangle = |1, -1\rangle, |1, 0\rangle$, and$|1, -1\rangle$l. To simplify, only the orbital quantum numbers are shown in Fig. 6.1 but the spin–orbit interaction is neglected. The Pauli exclusion principle give a limit for the occupation in each state to two electrons. Possible transitions are restricted by the dipole selection rule, and the selection rule $\Delta m = +1$ for RCP photons and $\Delta m = -1$ for LCP photons. For RCP photons, the transition from 1 s state $|0, 0\rangle$ to unoccupied $2p$ state $|1, 1\rangle$ is allowed, but the transition from 1 s state $|0, 0\rangle$ to occupied $2p$ state $|1, -1\rangle$ is forbidden. In the latter case the transition cannot proceed because the $\Delta m = -1$ state is already fully occupied. In this case a large XMCD signal can be expected. The sensitivity of an XMCD is greatly enhanced when the photon energy is in the vicinity of an absorption edge. This is particularly true in the case that the core electron is excited to a strongly magnetically polarized final state. For the first series of transition metals, this means that the photon excited a core electron to a $3d$ state, while for the rare earths it is a $4f$ state. The edges which interacts to these states for a dipole transition are L_{II} ($2p_{1/2} \rightarrow 3d$) and L_{III} ($2p_{3/2} \rightarrow 3d$) in the case of the transition metals, and are M_{IV} ($3d_{3/2} \rightarrow 4f$) and M_{V} ($3d_{5/2} \rightarrow 4f$) in the case of rare earths.

One of the reasons that XMCD has become such a popular technique is that the orbital, m_{orb}, and spin, m_{spin}, moments can be obtained separately from the use of *sum rules*. We write the sun rule in the case of the $3d$ metals. The m_{orb}, and m_{spin} are obtained from the following equations:

$$\begin{cases} m_{orb,\theta} = -\dfrac{4}{3\alpha\cos\beta} \cdot \dfrac{A_\theta + B_\theta}{C} n_h \mu_B & (6.1) \\ m_{spin} = -\dfrac{1}{\alpha\cos\beta} \dfrac{2A_\theta - 4B_\theta}{C} n_h \mu_B & (6.2) \end{cases}$$

where the A_θ and B_θ are the integrations of L_{III}-edge and L_{II}-edge of XMCD spectrum, and.

$$\begin{cases} A_\theta = \left\{\int_{L3} (\mu_+ - \mu_-) d\omega\right\}_\theta & (6.3) \\ B_\theta = \left\{\int_{L2} (\mu_+ - \mu_-) d\omega\right\}_\theta & (6.4) \\ C = \int_{L2+L3} (\mu_+ + \mu_-) d\omega & (6.5) \end{cases}$$

$$\begin{cases} n_h & \text{: number of holes in d state} \\ \mu_B & \text{:Bohr magneton} \\ \mu_+ & \text{:absorption intensity from LCP photon} \\ \mu_- & \text{:absorption intensity from RCP photon} \\ \omega & \text{:frequency(energy)of incident X} - \text{ray} \end{cases}$$

θ is the angle between normal direction of the sample and the direction of photon propagation z. α is polarizability and β is the angle between direction of the magnetization and of the photon propagation z.

XMCD measurements are performed for several different way depending on the X-ray source and energy of the edge. The circular polarized X-ray can be obtained from a bending magnet out of the orbital plane from the plane center; depending on whether the source is viewed above or below the orbital plane the electrons will be seen to rotate in a clockwise or anti-clockwise direction. This circular rotation of the electron beams makes an angular momentum to the photons, which can be then used for XMCD measurements. On the other hand, helical undulators can produce the quite intense circular polarized X-ray beams. For high energy X-ray, a diamond phase retarder can be used by producing the circular polarized photons. In either case, the magnetically dichroic signal can be obtained by reversing the helicity of photons or by reversing the direction of magnetization; these two situations are completely equivalent. At hard X-ray anergy region XMCD experiments are usually performed in a transmission mode, while for soft X-rays the XMCD signals are recording either fluorescent X-ray, with e.g. semiconducting detector, or by measuring the photoelectron yield. The typica XMCD measurement set up by transmission mode is shown in Fig. 6.2.

In the latter part of this section, we discuss about the EXAFS results. The detailed description of EXAFS technique has already appeared in the previous book chapter [4].

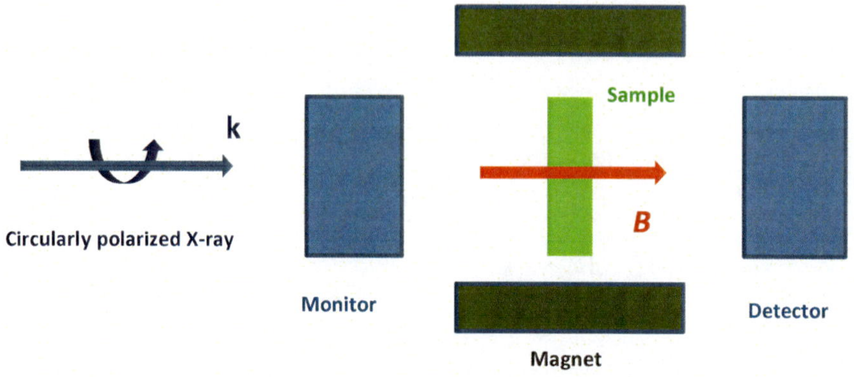

Fig. 6.2 Typical set up of XMCD measurement by transmission mode

6.2.2 *XMCD and XAFS for Cluster-Layered Fe/Cr Films*

One of the prominent features of Fe/Cr multilayer is GMR which was found in 1988 [5, 6]. This effect has opened the wide application of this materials to magnetic storages and sensors. On the other hand, the correlation between magnetic and transport properties has been long studied in such a multilayer structures and granular thin films with ferromagnetic nanoparticles [7–9]. In 2006, Kondo-like behavior, which shows the minimum point in temperature dependence of the resistivity, has been observed in cluster-layered Fe/Cr nanostructures [10]. In such ultrathin film, the small amount of Fe shows superparamagnetic feature, but no GMR behavior appears.

In the MgO based thin-films as MgO/Cr(80Å)/[Fe(t_{Fe}Å)/Cr(10Å)]$_n$; (t_{Fe} is the average thickness of Fe layer), Fe atoms are not continuously arranged but exist as clusters in the range of $t_{Fe} < 5$Å as shown schematically in Fig. 6.3. The system shows superparamagnetic behavior for $t_{Fe} < 2$Å, and its resistivity shows minimum point (Kondo-like behavior) for $t_{Fe} = 0.3 \sim 1.2$Å [10]. To reveal the origin and mechanism of such an interesting behavior in such Fe multilayer films, the magnetic and atomic structure are investigated from the local point of view.

In this section, XMCD and EXAFS to study the local magnetic structures and the mechanism of the characteristic magnetic feature of the Al_2O_3 based cluster-layered Fe/Cr nanostructures are discussed [11].

The samples discussed in this section are Al_2O_3 based cluster-layered Fe/Cr films:

(1) Al_2O_3/Cr(100Å)/[Fe(8Å)/Cr(10.5Å)]$_2$/Cr(20Å): (G1),
(2) Al_2O_3/Cr(70Å)/[Fe(1.2Å)/Cr(10.5Å)]$_{30}$/Cr(12Å): (K1)
(3) Al_2O_3/Cr(63Å)/$Fe_{0.17}Cr_{0.83}$/Cr(24Å)): (A1)

which are all grown by molecular beam epitaxy (MBE) [10]. G1 sample shows GMR and K1 sample shows Kondo-like behavior. In K1 sample the minimum point is appear in the resistivity at ~170 K as shown in Fig. 6.4.

First, the XMCD result for G1 and K1 samples was discussed. XMCD was measured at BL7A in KEK-PF, Tsukuba Japan. The applied magnetic field was 1.0 T

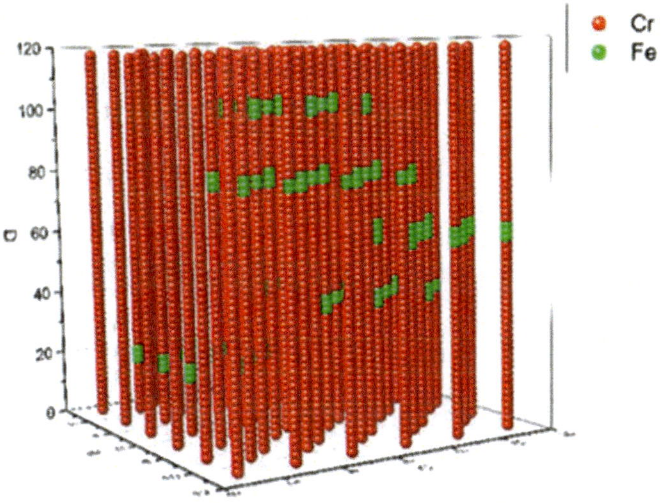

Fig. 6.3 Schematic model for cluster-layered Fe/Cr film (red: Cr, green: Fe)

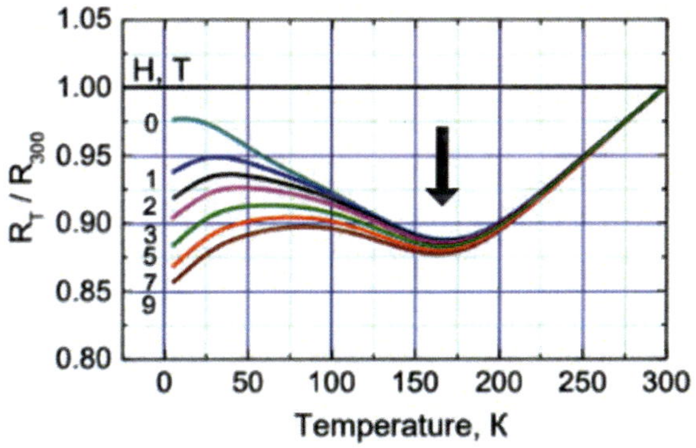

Fig. 6.4 The temperature dependence of the resistivity ratio with that for 300 K of Al_2O_3/Cr(70Å)/[Fe(1.2Å)/Cr(10.5Å)]$_{30}$/Cr(12Å) (K1 sample) under several magnetic field (0 ~ 9 T)

and circular polarizability of incident X-ray was 0.7. Fe L_{II} and L_{III} edge XMCD were measured by fluorescence mode using SDD (silicon drift detector). Angular dependent ($\theta = 30°$, 45°, 55°, 70°, where θ is the angle between normal direction of the sample film and X-ray beam direction). XMCD spectra were measured at temperature range of 77 K ~ 300 K.

The G1 sample which shows GMR presents ferromagnetic feature thorough the present temperature ranges (77 ~ 300 K) and shows clear Fe $L_{II,III}$-edge XMCD which does not depend on temperature. On the other hand, A1 sample does not present any XMCD signal in this temperature range (77 ~ 300 K).

The typical X-ray Absorption Spectra (XAS) of Fe L_{II} and L_{III} edge for K1 sample measured at 158 K with different circular polarizations are shown in Fig. 6.5. The XMCD in Fe $L_{II, III}$-edge spectra for K1 sample at $\theta = 70°$ are shown in Fig. 6..6. Clear XMCD peaks were observed at 77 K but the intensity decreases as temperature

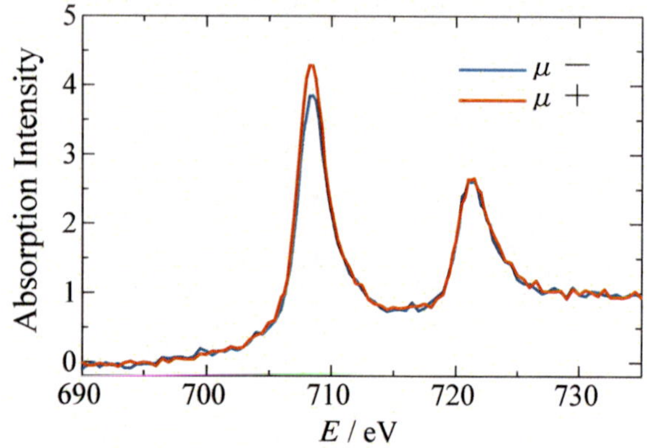

Fig. 6.5 Fe L_{II} and L_{III} edge XAS spectra for K1 sample at 158 K. $\theta = 70^{\circ}$. This figure is reproduced from ref [11]

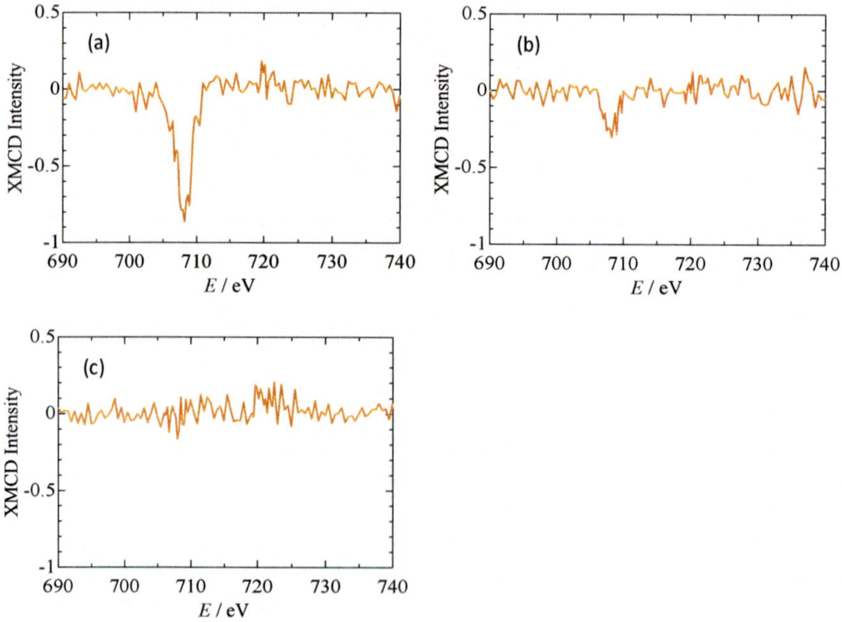

Fig. 6.6 Fe L_{II}- and L_{III}-edge XMCD spectra for K1 sample at **a** 77 K, **b** 170 K, **c** 223 K. $\theta = 70^{\circ}$. These figures are reproduced from ref [11]

increases (170 K and 223 K). Similar clear XMCD were observed for another angle of $\theta = 30°, 45°, 55°$.

Using the sum rule for XMCD [12], the contribution from spin and orbital magnetic moments can be derived from the ratio of intensities of L_{III} and L_{II} XMCD signal individually. Figure 6.7 shows temperature dependence of the magnetic moments from spin- and orbital contribution for K1 sample measured at (a) $\theta = 30°$ and (b) 70°. Each magnetic moment is decreases with increasing temperature and the XMCD peaks disappear higher than around 200 K.

Now we discuss the total magnetic moment obtained by sum of spin- and orbital-moment. Temperature dependence of the total magnetic moment are shown for K1 sample measured at $\theta = 30°$ and 70°. The magnetic moment is decreases as temperature increases and the XMCD peaks disappear more than ~200 K (Fig. 6.8).

The temperature dependence of the total magnetic moments obtained from XMCD was analyzed by mean field theory. The applied function is

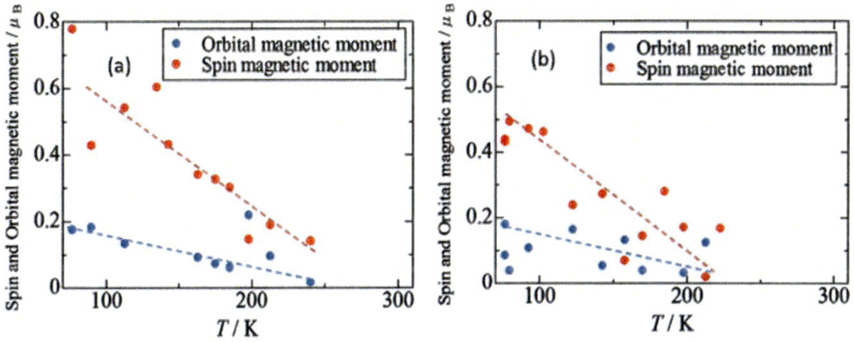

Fig. 6.7 Temperature dependence of spin-(red) and orbital-(blue) magnetic moment derived from XMCD using sum rule for **a** $\theta = 30°$ and **b** $\theta = 70^{\circ}$

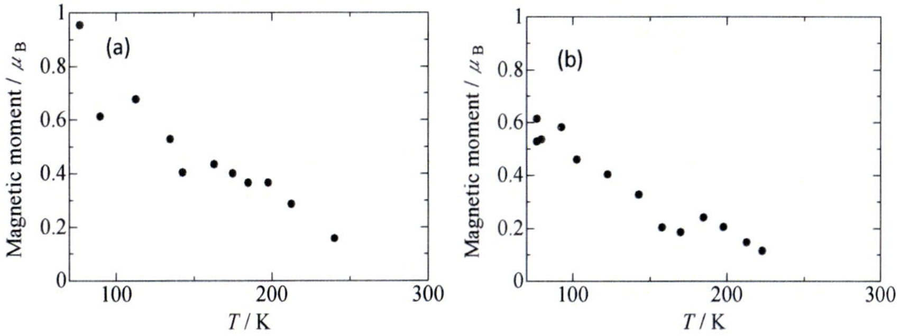

Fig. 6.8 Temperature dependence of total magnetic moment derived from XMCD for **a** $\theta = 30^{\circ}$ and **b** $\theta = 70^{\circ}$. These figures are reproduced from ref [11]

$$m_j(T; T_C\Delta) = \frac{m_j^0}{T_C^{3/2}} \int_T^{\infty} L\left(T_C' - T_C, \Delta\right)\sqrt{T_C'^{3/2} - T^{3/2} dT_C'}$$
$$= m_j^0 \mathrm{Re}\sqrt{\left(\frac{T_C - i\Delta/2}{T_C}\right)^{3/2} - \left(\frac{T}{T_C}\right)^{3/2}} \tag{6.6}$$

where the convolution function $L(T, \Delta)$ is expressed as

$$L(T, \Delta) = \frac{\Delta/(2\pi)}{T^2 + (\Delta/2)^2}. \tag{6.7}$$

Figure 6.9 shows the fits of the temperature dependence of the magnetic moment with the mean field theory. The result that a diffuse phase transition from the ferromagnetic to the paramagnetic state with the Curie temperature distributed in the interval $\Delta = 56$ K around the average value $T_c = 129$ K was obtained. In this temperature range, strong fluctuations occur for thc system, and they should induce an increase in electrical resistance.

The Curie temperature is reduced by several times with the thick films. in the two-dimensional magnetic tri-layers with spin fluctuations, such phenomenon has been also observed [13].

Second, we discuss the local structures for these samples from the EXAFS. Fe K-edge EXAFS were measured at BL9C in KEK-PF, Tsukuba Japan in Fluorescence mode using SSD (Ge-solid state detector) and Lytle detector measured at 25 ~ 300 K and angle $\theta = 45^\circ$.

The local structure around Fe atom for G1, K1 and A1samples were studied. Figure 6.10 shows (a) Fe K-edge XAFS $k^2\chi(k)$ and (b) Fourier transform for K1 sample at various temperature as an example. Same quality of EXAFS data was

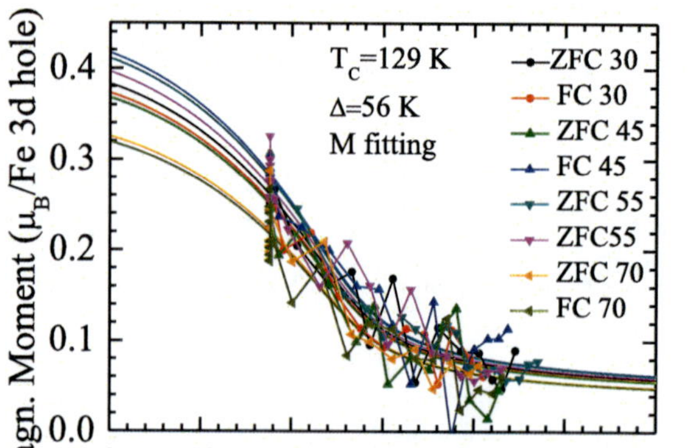

Fig. 6.9 The fits of the temperature dependence of the magnetic moment with mean field theory for zero field cooling (ZFC) and field cooling (FC) at several angle θ

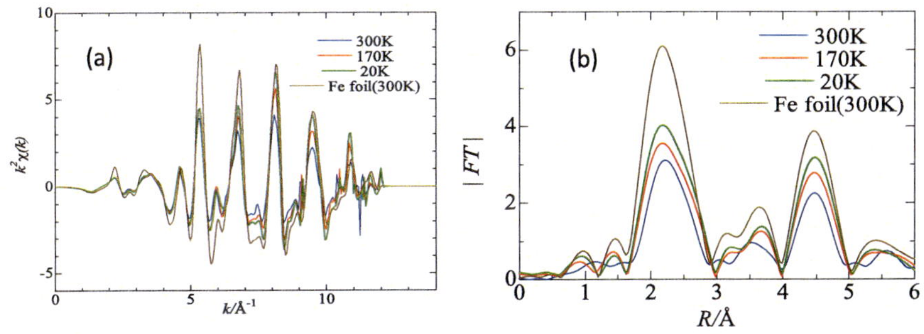

Fig. 6.10 **a** Fe K-edge EXAFS $k^2\kappa(k)$ and **b** Fourier transform for K1 sample at various temperature with Fe foil as standard sample. These figures are reproduced from ref [11]

obtained for both G1 and A1 samples. The data was analyzed at the range of $k = 2 \sim 12$Å^{-1} by use of Athena & Artemis code [14].

The temperature dependence of (a) interatomic distance and (b) Debye–Waller factor (or mean square relative displacement) of 1st nearest neighbor (1NN) Fe–Fe atomic pair are shown in Fig. 6.11. The interatomic distance of 1NN Fe–Fe for K1 sample is similar to that for A1 sample, and longer than that for G1 at lower temperature. These results mean that the local structure around Fe of K1 sample is similar to that of A1 sample. Debye–Waller factor generally indicates the disorder or fluctuation of the relative atomic positions from the X-ray absorbing atom.

The Debye–Waller factor of K1 sample is almost same as that of A1 sample and larger than that of G1 sample; this result presents that the local structure around Fe atoms for K1 sample is disordered like as alloy sample (A1).

Electron spin density fluctuations induced by local structural disorder associated with the transition cause temperature anomalies of electrical resistivity and other transport properties due to carrier scattering on fluctuations.

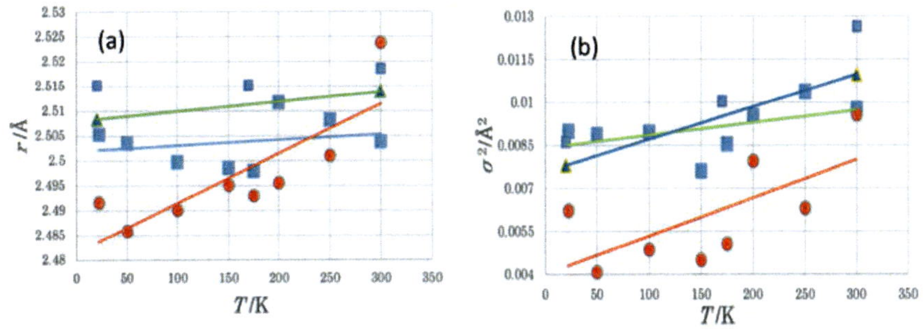

Fig. 6.11 Temperature dependence of **a** interatomic distance and **b** Debye–Waller factor of 1NN Fe–Fe. K1; blue square, G1; red circle, A1; green triangle. These figures are reproduced from ref [11]

In this subsection, existence of local magnetic and atomic structures of cluster-layered Fe/Cr thin films are shown by XMCD and EXAFS. The magnetic moment of K1 sample is around 0.8 μ_B at 77 K for K1 sample and decreases as temperature increases and finally disappear near the minimum point of electric resistivity. Analysis of the temperature dependence of the magnetic moment suggests that the strong spin fluctuations occur in the system, and they may induce an increase in electrical resistance. The EXAFS result shows that the local structure disorder of cluster-layered film is close to the FeCr alloy film. The connection between the magnetic fluctuation and the local structure disorder can be expected.

6.2.3 *Other Applications*

Here we shortly introduce the other applications of circular polarized X-ray to magnetic materials. The magnetic EXAFS has advantages in both of EXAFS and XMCD. In the magnetic EXAFS, the amplitude is proportional not only to the coordination number but also to the magnetic moment of X-ray absorbing and photoelectron scattering atoms because of spin–orbit coupling.

The magnetic EXAFS was applied to study Ni_3Mn magnetic alloy, where a coupling between the structure disorder and magnetic properties is prominent. The magnetic EXAFS for Mn using circular polarized X-ray was discussed about the effect of atomic arrangement in order–disorder couplings to magnetic properties [15, 16]. Figure 6.12 is the Fourier transforms for Mn *K*-edge of (a) conventional EXAFS and (b) magnetic EXAFS for $Ni_{0.75}Mn_{0.25}$ alloy which was heat-treated at 693 K in 500 h to order the atomic arrangement ($S = 0.78$) [15]. In the ordered Ni_3Mn alloy, the Mn atom has a large magnetic moment (about 3.15 β_B) and is located at the 2nd nearest neighbor (NN) and 4th NN from the X-ray absorbing Mn atom. The FT modulus peak intensity of 2nd and 4th NN in Fig. 6.12b becomes larger than that in Fig. 6.12a because of the large magnetic moment of Mn. From

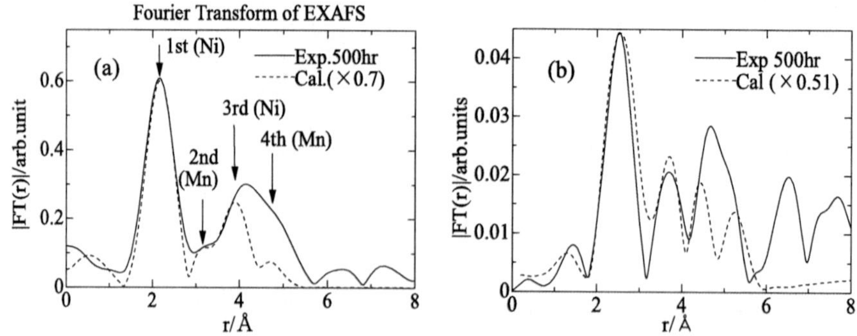

Fig. 6.12 Mn K-edge Fourier transforms of **a** non-magnetic and **b** magnetic EXAFS for ordered (500 h annealing) $Ni_{0.75}Mn_{0.25}$ alloy. Dotted lines are the theoretical calculation spectra

the ratio of the intensities of the 1st (Ni) and 2nd (Mn) peaks, the ratio of magnetic moment, Mn/Ni, can be evaluated. From the band calculation, the theoretical value of the magnetic moment ratio, Mn/Ni, is 5.2 which is close to the magnetic EXAFS result for 500 h annealing sample (Mn/Ni = 5.9) [15]. Experimental result shows that Mn/Ni magnetic moment increases for longer annealing time or more proceeded to order. This result indicates that the magnetic moment of Mn atoms in the highly ordered Ni_3Mn phase is larger than that in the less ordered phase.

6.3 Mössbauer Spectroscopy for Thin Films Using Synchrotron Radiation

6.3.1 Mössbauer Spectroscopy for Thin Films

Mössbauer spectroscopy is a method to study the local electronic and vibrational states in a sample; that is, valence, structural change, magnetic ordering, thermal Debye characteristic temperature, and so on. This method is used in wide scientific areas: e. g. physics, chemistry, biology, planetary science. It is of course applied to study the nanostructured materials. In fact, it has a unique property, which might be especially effective for the study of thin films—"isotope-specificity". What is isotope-specificity? You may know various element-specific methods, such as XAS, XAFS, XMCD, XPS, and resonant IXS. The key to select the elements focused on in these methods is the following two: the elemental absorption edges by the quantum electronic energy levels observed in the absorption-type measurement using X-rays and the atomic elementally characteristic X-rays observed by the scattering after the excitation of the absorption edges. Similarly, the isotope specificity is achieved by the nuclear energy levels. The nuclear energy levels are mainly constructed by the strong interactions in the nuclei, as the atomic energy levels are by the electromagnetic interactions in the atoms. In fact, the nuclear energy levels of different isotopes differ, as the atomic energy levels of different elements do. Some atomic energy levels and nuclear energy levels of iron and nickel isotopes are shown in Table 6.1.

The difference of the nuclear energy levels allows us to distinguish the isotopes among elements through the nuclear resonance with photons. This is the isotope-specificity. Therefore, if we can arrange the isotope atoms to be measured in the position under investigation, such as surfaces, interfaces, and functional positions, we can obtain the information from the position. Recently, we can arrange atoms in the spatial resolution of one atomic monolayer and thus we can obtain the information in the one-monolayer spatial resolution using this isotope-specificity, as shown in Fig. 6.13. Here, we discuss the Mössbauer spectroscopy for thin films, especially focusing on the isotope-specificity.

Now the next point is "What information we can obtain from the Mössbauer active isotopes?". The simplest answer is "the nuclear energy levels"—not the electronic levels. Is it really informative for the study on the condensed matter physics? The

Table 6.1 Some atomic and nuclear energy levels of stable isotopes of iron and nickel

Isotopes	Atomic energy levels—K absorption edge [17]	Nuclear energy levels—first excited state [18]
^{54}Fe	7.110747(20) keV	1408.19(19) keV
^{56}Fe		846.7778(19) keV
^{57}Fe		14.4129(6) keV
^{58}Fe		810.7662(20) keV
^{58}Ni	8.331486(20) keV	1454.21(9) keV
^{60}Ni		1332.514(4) keV
^{61}Ni		67.414 (7) keV
^{62}Ni		1172.98 (10) keV
^{64}Ni		1345.75(5) keV

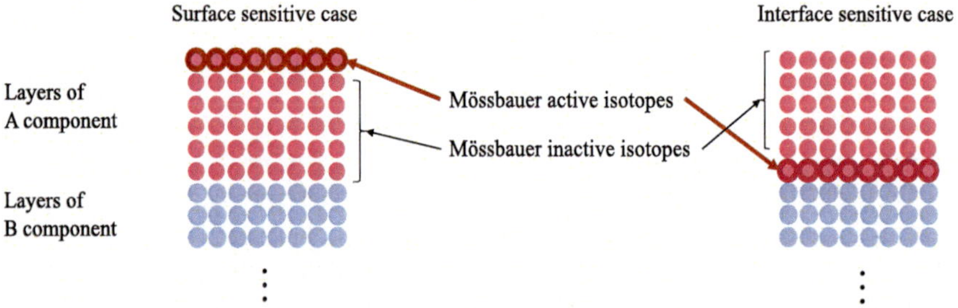

Fig. 6.13 The schematic figure of an usage of the isotope-specificity in the study of a thin film. The Mössbauer "active" isotopes are arrenged in the position to be studied and the Mössbauer "inactive" isotopes are arranged in the other positions. We can obtain the information from the active isotope, although all of the surface, interface and the inner layers in the layer A consist of the same elemental components

answer is "Yes", because the nuclear levels are affected by the surrounding electrons. The nuclear levels are mainly constructed by the strong interaction between the nucleons and a little modulated by the electromagnetic interactions with the surrounding electrons. This interaction between a nucleus and the surrounding electrons are called "hyperfine interactions"; the energy scale of the hyperfine interactions is by far smaller than that of the nuclear levels, since the former is below the order of μeV and the latter is typically the order of keV or MeV. Despite, the interactions can be detected in principle because of the very narrow energy width of the nuclear levels, typically from neV to μeV (Such nuclear levels are usually selected in Mössbauer spectroscopy, although there are many nuclear levels whose energy width is wider than them, e. g. even MeV!). The main contributions of the hyperfine interactions are the following three, as shown in Fig. 6.14: the electric monopole interaction, electric quadrupole interaction, and magnetic dipole interaction. The effect by the electric monopole interaction is usually called "isomer shift" and an indicator for the oxidation state, electric negativity of ligands, character of

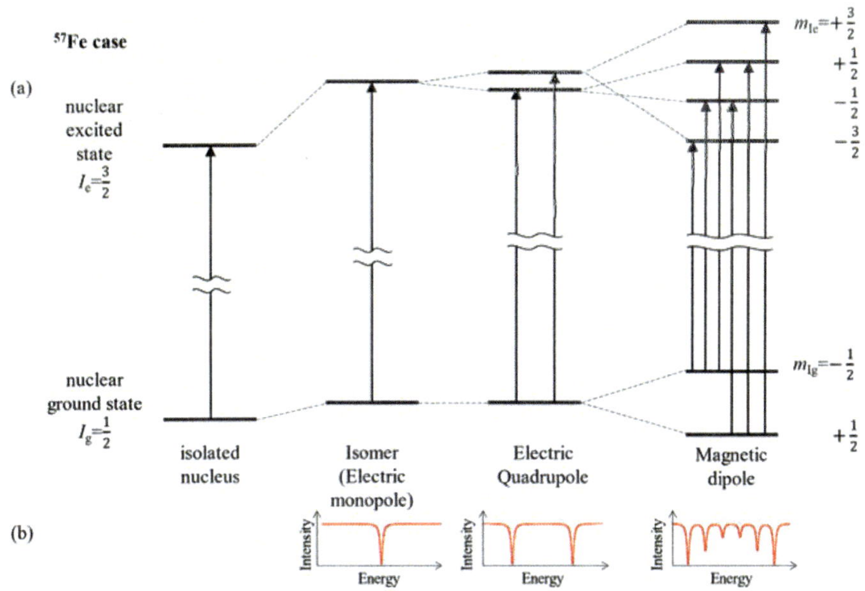

Fig. 6.14 The schematic figure for a typical hyperfine splitting in a ^{57}Fe nucleus. (**a**) ^{57}Fe nuclear structures in the case of each main contribution of hyperfine interaction. The arrows show the allowed nuclear transitions in this M1 transition case. The most left one is the original structure in the isolated nuclei without surrounding electrons (i. e. without hyperfine interactions). The dashed lines show the relation of degeneration of nuclear energy levels. Note that the right two figures also includes the effect of the isomer shift because it is always included in energy domain Mössbauer spectroscopy. (**b**) Typical Mössbauer absorption spectra observed in each case. Note the magnitude of energy of the relating phenomena: the energy of the first nuclear excited state is 14.4129(6) keV (energy difference shown in the leftmost figure in (**a**)), the energy shift or splitting due to the hyperfine interactions are typically below the order of μeV, and the energy width (full width of half maximum) of the first excited state is 4.64 (1) neV

bonds including the distance to the neighboring atoms, and spin states. The relation between the isomer shift values and the oxidation states is shown in literature [for example, refs. 19–21]. The effect by the electric quadrupole interaction is often called "quadrupole splitting" or "quadrupole shift": the latter phrase is usually used in the case of the combination of the electric quadrupole interaction and magnetic dipole interaction, especially the case that the electric quadrupole interaction is perturbation to the magnetic dipole interaction. The electric quadrupole interaction is an indicator for the symmetry of surrounded electrons and atoms, including molecular symmetry, ligand symmetry, symmetry of crystal structure. It is also an indicator for oxidation state, character of bonds, and spin states. The magnetic dipole interaction reflects the local magnetic field at the nucleus and thus the value is often called "magnetic hyperfine field" or "internal magnetic field". This is an indicator for the spin states and magnetic orderings including the ferromagnetism, ferrimagnetism, anti-ferromagnetism. These values are simply calculated from the absorption energies from the Mössbauer spectra, typically shown in Fig. 6.14b [22]. Some computer software packages are very useful to analyze them [for example, 23 and 24]. For the interpretation of these values, the theoretical calculation and/or previous experimental literature is very useful. For the theoretical calculation, the calculation of the inner-core electrons with relativistic effect is essential because the Mössbauer nucleus is surrounded by the inner-core electrons with relativistic momentum [25, 26]. As for the experimental literature, various substances have been measured by many researchers and summarized in many textbooks and databases, especially for those for ^{57}Fe [For example, 19–21, 27]. When there is some chemical species or different sites due to the breaking of some symmetry such as low crystallographic symmetry, the Mössbauer spectra are represented by the superposition of the Mössbauer spectra of each species and/or sites. This site-analysis capability is essentially important to the non-ideal surface and/or interfaces, such as surface of an alloy with partial oxidization.

To measure the Mössbauer spectra reflecting hyperfine interactions, we should resolve the energy difference of neV. Is it really possible? In principle, "Yes" by "Mössbauer effect". In 1957, R. L. Mössbauer discovered the recoilless, i.e. elastic, nuclear resonant absorption/scattering process [28]. This effect guarantees the precise determination of energy changes even in the range of neV. The next point is how we actually resolve it. It is impossible to resolve such small energy difference by commonly-used energy-resolving detectors, such as an solid-state detector. Therefore, we use the Doppler effect of light. Consider that photons are incident on a nucleus and the energy of the photons E is quite similar to that of a nuclear excited level of the nucleus. We move the nucleus for the Doppler effect by the velocity of the nucleus v. Then, the energy shift ΔE is

$$\Delta E = \frac{v}{c}E, \tag{6.8}$$

where c is the speed of light. In the case of ^{57}Fe, 1 mm/s corresponds to 48.076(2) neV. The actual measurement system of conventional Mössbauer spectroscopy using radioactive isotopes (RI) as an incident photon source (that is, the established system without SR) is shown in Fig. 6.15. The velocity of the RI source is controlled by a velocity controlling system and thus the energy of the incident photons are modulated by the velocity. This incident photons show a simple monochromatic

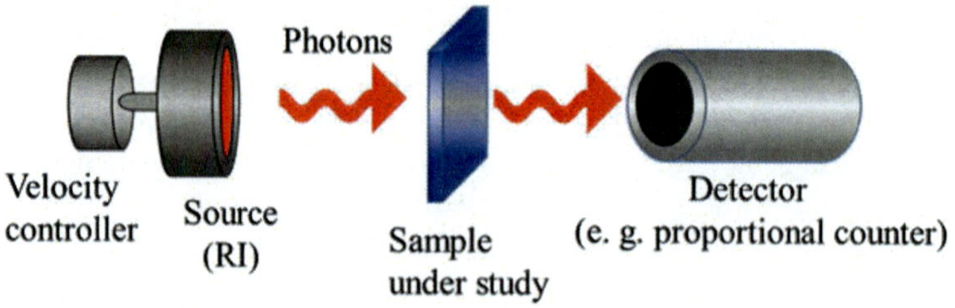

Fig. 6.15 The schematic figure for a typical Measurement system of Mössbauer spectra

energy profile with the energy width of the nuclear resonance. In the case of ^{57}Fe, it is 4.64 neV in the ideal case. The incident photons penetrate the sample under study and then they are detected by the detector. We measured the photon intensity depending on the velocity; when the incident photon energy coincides with one of the allowed nuclear transitions of ^{57}Fe nuclei in the sample, the transmission intensity decreases due to the nuclear resonant absorption. Thus, we obtain the absorption type energy spectra where the x-axis is velocity, i. e. Mössbauer spectra of the sample. This measurement system is non-contact and non-destructive using hard X-rays and thus we can easily combine the vacuum chamber, thermal furnace, and other environmental cells. This property is advantageous for the surface science.

Mössbauer spectroscopy using RI source have been also applied to thin films for the study of the surface and interface science. However, many thin films are synthesized on substrates, which are often quite thick for the incident photons to penetrate. This problem can be solved by measuring the scattering spectra, similarly to the fluorescent XAFS compared with the transmission XAFS. In fact, after the nuclear resonant absorption by the nuclear level, photons and electrons are scattered in its de-excitation processes. A well-known method of this scattering-type measurement is conversion electron Mössbauer spectroscopy (CEMS) [29], where the internal conversion electrons in the nuclear internal conversion processes are detected. As shown in Fig. 6.16, the thin film samples are arranged in a proportional counter with the ionization gas and the scattered electrons are efficiently detected by the gas counter. Using the isotope-specificity, even the Mössbauer spectra of interfaces with one monolayer spatial resolution is measured, although it takes tens of hours for one spectrum [30]. In addition, the thin film sample is arranged in the gas counter at CEMS system and thus some complex instrumentation and treatment is required for the the study of special environmental conditions: low temperature, vacuum, and electric current application.

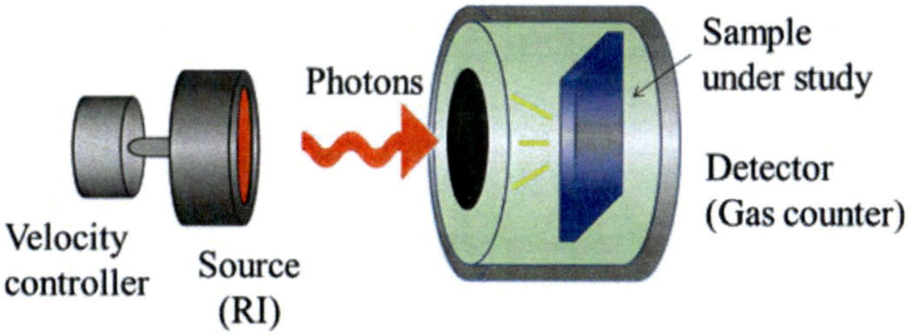

Fig. 6.16 The schematic figure for a typical CEMS system

6.3.2 Synchrotron Mössbauer Source

To overcome this problem, SR is used as an incident photon source. SR is a source with high brilliance: the brilliance of SR is more than 10^5 times higher than that of the commonly-used RI source. Because of its high brilliance, SR is also easily combined with various advanced X-ray technique. In the study of thin films, the combination of total reflection technique is very effective. We can apply transmission-type measurement system with a modulation in the sample arrangement: instead of the transmission through the sample, the photons are totally reflected by the thin film or the substrate under the film. The high brilliance will also allow us short measurement time. We see an example of SR using method applied to iron thin film later in this chapter.

There are some ways to combine Mössbauer spectroscopy with SR. This is because the property of the SR is different from that of photons from RI and we cannot simply use SR instead of the photons from RI. One clear difference between the photons from RI and SR is its energy width. The energy width of photons from RI is around that of nuclear resonance and that of SR hard X-rays are usually around a few eV, even after the standard Si crystal monochromator. Even when we use the high-resolution monochromator using a higher index of Si crystal monochromator, the energy width is around a few meV or sub-meV.

One solution to this difference is to monochromate the SR in the energy width of the nuclear resonance. This method is called synchrotron Mössbauer source (SMS) method [31, 32]. The schematic drawing of the SMS method in the case of thin film study is shown in Fig. 6.17. The key component is the nuclear monochromator crystal (NMC), which is developed by Prof. Smirnov [33]. This is nearly perfect single crystal of $FeBO_3$ and the Bragg diffraction by the 'forbidden' index such as (111) and (333) is used. Because the indices satisfy forbidden rule, there is no electronic scattering in principle. However, the nuclear scattering has another symmetry because the nuclear

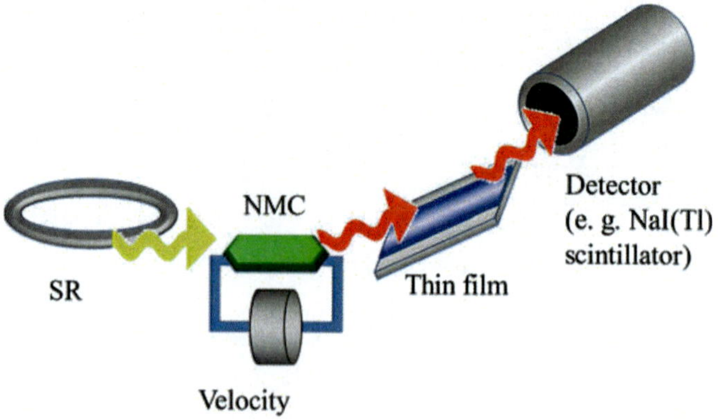

Fig. 6.17 The schematic figure for a typical SMS system for thin film. The total reflection technique at thin film sample is combined with SMS method for the efficient measurement

scattering is also affected by the magnetic order, which is anti-ferromagnetic around room temperature. In this case, the forbidden rules for nuclear resonant scattering differs from those for electronic scattering and the nuclear resonant scattering from $FeBO_3$ (111) and (333) is allowed. In that case, Bragg scattering occurs by purely nuclear resonant scattering process without electronic scattering process. Thus, only pure nuclear Bragg scattering is allowed and X-rays with the energy width of the nuclear resonance and energy profile reflecting the nuclear hyperfine structure are filtered from the NMC. With the combination of this pure nuclear Bragg scattering and some other conditions, i. e. the selection rules for nuclear resonant scattering using the polarization of SR and the heating up of the NMC crystal just below its Néel temperature to manage the hyperfine structure due to the magnetic hyperfine field, the incident SR after the NMC shows both the simple monochromatic energy profile and very narrow energy width similar to the photons from RI, as well as extremely high brilliance compared to those from RI. These photons filtered from SR through the NMC is often called SR Mössbauer radiation. Furthermore, the combination of SR and NMC looks a source of incident photons for Mössbauer spectroscopy using RI with extremely high brilliance, this system is called SMS. This method is first developed in European Synchrotron Radiation Facility (ESRF) in 1990's [31] and then began to be widely used in SPring-8 in 2000's [34]. In fact, after the success of the control of SR Mössbauer radiation by the velocity of the NMC through the Doppler effect of light, we can arrange various environmental cells and the applicability of this method becomes very wide [35, 36]. Now we can simply use the SMS as a source for Mössbauer spectroscopy instead of the RI source. It is noted that the SR Mössbauer radiation is non-destructive to the samples in spite of its high brilliance because its radiation power is by far smaller than the SR used in the other method, such as XRD and XAFS. The reason is that the intensity of SR Mössbauer radiation is very low owing to its very narrow energy width. In fact, the energy width of the incident SR for the other method is typically a few eVs, although that of ^{57}Fe Mössbauer radiation is typically ten neVs. This means that the radiation power of SR Mössbauer radiation is as weak as 10^{-8} times that of the SR for the other methods. In many cases, not all the incident photons are absorbed by the sample and the radiation power is less than this estimation. In fact, the SR Mössbauer radiation at a sample is typically tens of thousands counts and thus the total radiation power of the incident radiation is sufficiently below 1 nW. It is also noted that the NMC described here is only about that for ^{57}Fe Mössbauer spectroscopy. Those for the other Mössbauer active isotopes can be synthesized in principle but have not been succeeded until 2021.

For the Mössbauer spectra of thin films, the total reflection technique is often combined. The SR Mössbauer radiation is incident on the thin film in a grazing angle and its direction is modified by the total reflection by the thin film and/or the substrate. Therefore, the SR does not penetrate the substrate and efficiently penetrate the thin film for the reasonable absorption due to the thickness of the resonant nuclei in the photon path. Then the totally reflected SR is detected by an detector, usually an NaI(Tl) scintillator, to obtain the photon intensity depending on the velocity of

NMC. When we arrange the grazing incidence angle small[37],[2] the obtained spectra are absorption type and can be analyzed using software packages for conventional Mössbauer spectroscopy using RI [22, 23].

6.3.3 *Mössbauer Spectroscopy with Monoatomic Layer Spatial Resolution*

By the combination of the isotope specificity of Mössbauer spectroscopy and the SMS method, the magnetic Friedel oscillation near the surface of iron thin film is studied [38]. Here some experimental details of this study are introduced.

The magnetic Friedel oscillation is related to the magnetic property at the surface. It is theoretically predicted that the magnetic moment M_{Fe} is 30% enhanced at the surface of Fe(001) and it shows damping-oscillatory behavior with increasing the depth from the surface [39, 40]. Furthermore, it is also predicted that the magnetic hyperfine field H_{int} at the surface is 30% reduced relative to those at bulk [41].[3] This strong modulation of H_{int} was experimentally studied by an advanced in-situ CEMS using RI source [42]. However, the H_{int} was not so modulated at the surface in the study. The authors discussed the contamination to the sample surface from residual gas in spite of the ultra-high vacuum environment of $<2.7 \times 10^{-8}$ Pa, because even such tiny residual gas might change the sample top surface due to the long measurement time of several weeks in total. Until 2020, the magnetic Friedel oscillation has not been experimentally observed in the viewpoint of magnetic hyperfine field.

However, using the SMS method, the measurement time should be drastically reduced. Therefore, the magnetic Friedel oscillation was recently tried to be confirmed experimentally using SMS. Five Fe(001) thin film samples were synthesized on the MgO(001) substrate with the size $10 \times 10 \times 0.5$ mm^3 under the vacuum of typically 10^{-8} Pa. Here, the thickness of the Fe films was 5 nm and almost all of each film is made from ^{56}Fe isotope atoms except the monoatomic layer to be studied. The monoatomic layer to be studied consisted of ^{57}Fe atoms arranged with 0.6 Å thickness and the arranged layer differs each other for the five samples: the top surface (first layer), second layer, third layer, fourth layer, and seventh layer. A schematic drawing of the samples is shown in Fig. 6.18a. The Mössbauer spectra of these samples were measured by the SMS measurement system. In this case, the $FeBO_3$(111) reflection was used for the NMC and 14.4 keV SR Mössbauer radiation for ^{57}Fe with 15.4 neV energy width was filtered from the incident SR. The Mössbauer radiation was vertically focused by an elliptical mirror. The beam size

[2] The incident angle is sufficiently smaller than the critical angle of the electronic scattering. In this condition, the dominant process in the total reflection is electronic scattering and absorption-type spectra are obtained.

[3] The magnetic Friedel oscillation is essentially the fluctuation of electronic spin-polarization at the surface. This causes both modulation of *d* electrons resulting in the fluctuation of atomic magnetic moments and modulation of *s* electrons resulting in the fluctuation of magnetic hyperfine fields.

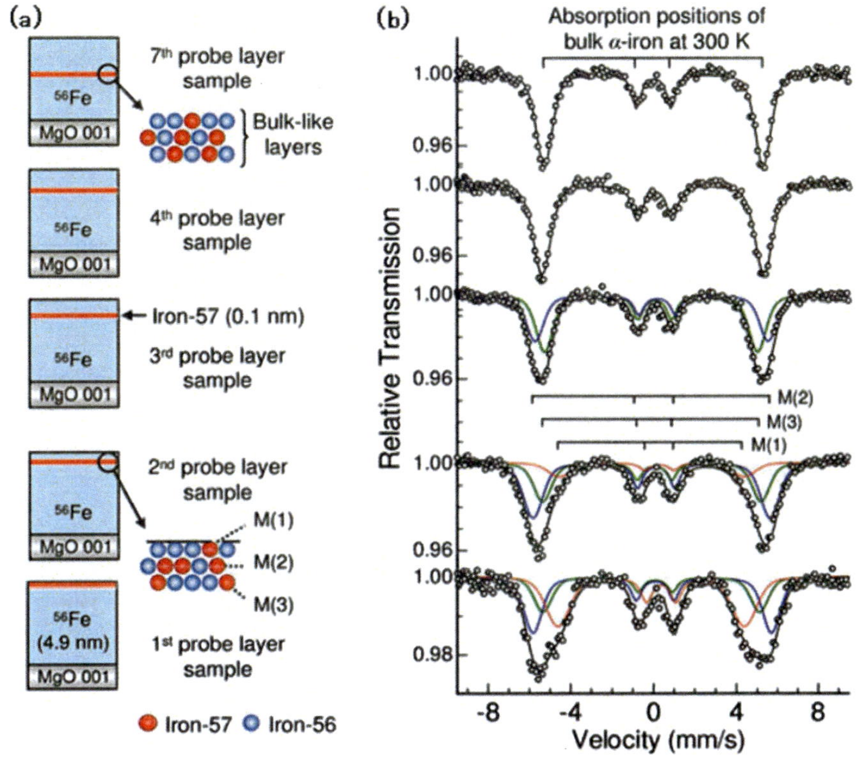

Fig. 6.18 The sample thin films and corresponding Mössbauer spectra. (**a**) Schematic drawings of the five sample thin films for the Mössbauer spectrosocpy with monolayer spatial resolution. (**b**) Mössbauer spectra of the sample drawn in the left of this figure. Black solid lines represent the fitted curves. Red, blue, and green lines represent three different magnetic components. Note that the ^{57}Fe isotope atoms oozed into neighboring layers. These figures are reproduced from [38]

was 15 μm (vertical) × 1.6 mm (horizontal) and the beam flux was about 2.9 × 10^4 photons/sec. This focused Mössbauer radiation was introduced to the sample thin films. Each thin film was in a vacuum chamber, where the film was transferred without exposure to lower vacuum after their synthesis. The Mössbauer radiation was totally reflected from the sample at an incident angle of 0.1° with a reflectivity of about 80% and was detected by an NaI(Tl) scintillator.

The obtained Mössbauer spectra are shown in Fig. 6.18b in the right side of each sample figures in Fig. 6.18a. Here, each spectrum was obtained within a few hours from sample preparation and thus the contamination from the residual gas is expected to be significantly less than that in [38]. The spectra of the samples where ^{57}Fe atoms were arranged in the first, second, and third layer exhibited complex profiles composed of two or three different magnetic components. This reflects the situation that the ideal ^{57}Fe layer in the samples was surrounded by finely distributed unexpected ^{57}Fe atoms, which stemmed from the random deposition and surface diffusion of iron atoms during the growth process. Therefore, the main component, i. e. the component with the largest absorption area in each spectrum, should show the information of the target layer. The reason why similar three components were not observed in the spectra of the samples with the deeper ^{57}Fe layer is that the neighboring layers also showed the same Mössbauer spectra. Based on this consideration, the experimental magnetic hyperfine field are summarized in Fig. 6.19 with the results of theoretical calculation; the theoretical calculation is performed by the local density approximation combined with the optimized effective potential method using the exact exchange-potential for core states with the consideration to the relativistic effect. Both theoretically calculated magnetic hyperfine fields and experimentally evaluated ones clearly show the oscillatory behavior with increasing the depth from the surface. In addition, the experimentally evaluated ones quite agree with those by the theoretical calculation except those of the top surface layer. At the top surface,

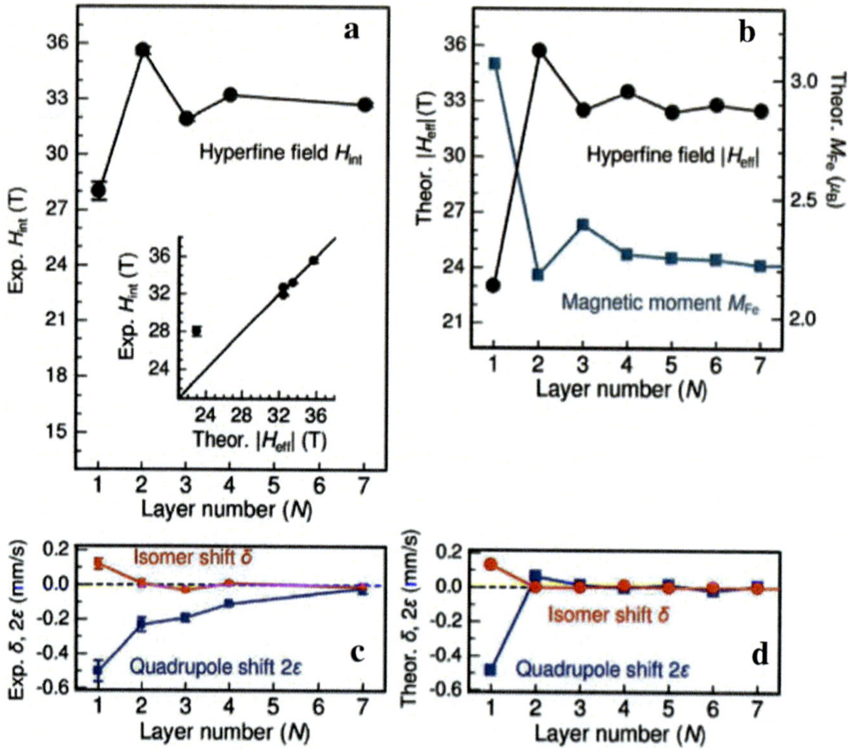

Fig. 6.19 The experimentally evaluated and theoretically calculated parameters of hyperfine interactions of ^{57}Fe atoms in each depth layer from the Fe(001) surface. **a** Experimentally evaluated magnetic hyperfine fields. The inner graph showed the relation between the experimental and theoretical ones. **b** The theoretically calculated mangetic hyperfine fields and the Fe atomic magnetic moments. **c** Experimeltally evaluated isomer shifts and quadrupole shifts. **d** Theoretically calculated isomer shifts and quadrupole shifts. All lines among data points are eye guides. These figures are reproduced from [38]

although the 30% reduction of the magnetic hyperfine field compared to the bulk one is predicted, only 20% reduction is experimentally evaluated. This might be because the calculations assume an ideal Fe(001) surface, although it is not strange that there is the surface defects such as edges, steps, kinks, and vacancies, which might cause the difference of coordination number around the surface atoms of actual samples from that of the theoretical calculation. Furthermore, the isomer shifts and quadruple shifts are also compared in Fig. 6.18. The isomer shifts also showed that the surface charge density was modulated but it was already bulk-like at one atomic layer below the surface. The theoretical quadrupole shifts also showed the same results. The experimental quadrupole shifts showed some deviation from theoretical ones, but this agrees with the not-ideally-flat surface of the samples. Thus, although there is some effect of the surface defects, the magnetic Friedel oscillation predicted by theoretical calculation was first clearly confirmed by the SMS experiments with one atomic monolayer spatial resolution using its isotope-specific property.

6.3.4 Other Applications

As seen in the example above, the SMS method is very effective to study thin films. Furthermore, this method can be also applied to thin films without such special isotope treatment. In fact, when we measure all of a thin film and do not require the special spatial information, we can use Mössbauer spectroscopy as an element specific method: natural iron material includes 2.2% ^{57}Fe isotope atoms. For example, with such a natural Fe containing sample, i. e., the sample where the isotope abundance of Fe was not controlled, the direction of magnetic ordering in anti-ferromagnetic hematite Fe_2O_3 thin film with 20 nm thickness have been studied by SMS [43].

Here, another excellent property of SR is noted: the nearly perfect linear polarization. This property can be used to select the nuclear transition to be measured; the nuclear transition probability depends on the direction of the hyperfine fields (i. e. magnetic hyperfine field and the electronic field gradient at the resonant nucleus) with respect to the polarization and the propagation vector of the incoming incident photons. This property was already used in the study of magnetic Friedel oscillation as the four-absorption line spectra in Fig. 6.17, while the generally allowed transition is six, as shown in Fig. 6.13. This selection property was used for the convenience in analysis in this case. Moreover, this polarization of SR is also used to the selection of the magnetization direction of samples in Mössbauer spectroscopy. We can produce circularly polarized X-rays from linearly polarized ones using an X-ray polarizing element such as phase retarder[44]. The spin-direction selective Mössbauer spectroscopy is also available using the circularly polarized SR [45], similarly to the XMCD.

As described, SMS is practically a method for ^{57}Fe until 2021. How about the other isotopes and/or other elements? Indeed, the broad energy spectra is further another characteristic property of SR and photons in any energy in the broad energy range can be used. The third generation SR facility can produce hard X-ray photons below 100 keV with quite high intensity and we can use SR as an incident source for Mössbauer spectroscopy using various isotopes; for example, we can use 67.4 keV SR photons, whose energy is the nuclear resonance energy of ^{61}Ni. We can use other SR photons whose energy is that of other isotopes. For the Mössbauer spectroscopy with various isotopes, some other methods have been developed, such as nuclear resonant forward scattering [46] and SR-based Mössbauer absorption spectroscopy [47]. For example, ^{119}Sn SR-based Mössbauer absorption spectra of Sn-including thin film was measured in the literature [48].

Summary

In this chapter, we introduce the recent application of SR source to the study of magnetic nanostructures. First example is XMCD and XAFS and second one is Mössbauer spectroscopy. These methods are already widespread and applied several magnetic nanostructures. On the other hand, nanostructure sciences and technology are now in progress. The importance of these techniques will be increasing in future.

Acknowledgements For the study of cluster-layered Fe/Cr films, the SR experiments of XMCD and XAFS were conducted at the Photon Factory in KEK under Proposals 2017G558. This work was supported by JPSJ KAKENHI Grant Number JP17K05011 and JP20K05295., and also supported by the state assignment of Minobrnauki of Russia (theme "Quantum" No. AAAA-A18-118020190095-4).

For the description of Mössbauer spectroscopy, authors (especially R. M.) gratefully acknowledge Dr. T. Mitsui and Dr. M. Saito for critical and instructive advice. The experiments for the magnetic Friedel oscillation were conducted at SPring-8 under Proposals Nos. 2017A3551, 2017B3551, 2018A3552, 2018B3551, 2018B3552, 2019A3551 and partially supported by JSPS KAKENHI Grants Numbers 16H03875, 17H07376, 17K18373, and 18K13985.

References

1. J. Stöhr, H.C. Siegmann, *Magnetism: From Fundamentals to Nanoscale Dynamics*, (Springer, Berlin-Hiderberg, 2006)
2. S.W. Lovesey, S.P. Collins, *X-ray Scattering and Absorption by Magnetic Materials* (Oxford University Press, New York, 1996)
3. J. Als-Nielsen, D. McMorrow, *Element of Modern X-ray Physics* (Wiley, Chichester, 2011)
4. T. Miyanaga, T. Azuhata, Recent Progress in XAFS study for semiconducting thin films, In *Low-Dimensional and Nanostructured Materials and Devices*, H. Ünlü, N.J.M Horning, J. Dabowski, Eds. pp.149 (Springer, Switzerland, 2016).
5. M.N. Baibich, J.M. Broto, A. Fert, F. Nguyen Van Dau, F. Petroff, P. Eitenne, G. Creuzet, A. Friederich, J. Chazelas, Phys. Rev. Lett. **61**, 2472 (1988)
6. S.S.P. Parkin, N. More, K.P. Roche, Phys. Rev. Lett. **64**, 2304 (1990)
7. O.Z. Sefrioui, J.L. Mendez, E. Navarro, A. Cebollada, F. Briones, P. Crespo, A. Hernando, Phys. Rev. B **64**, 224431 (2001)
8. A.K. Majumdar, A.F. Hebard, A. Singh, D. Temple, Phys. Rev. B 054408 (2002)
9. B.G. Almeida, J.B. Sousa, R. Shad, V.V. Moshchalkov, Y. Bruynseraede, J. Magn. Magn. Mater., **157/158**, 730 (1996)

10. V.V. Ustinov, L.N. Romashev, M.A. Milayev, A.V. Korolev, T.P. Krinitsina, A.M. Burkhanov, J. Magn. Magn. Mater. **300**, 148 (2006)
11. T. Miyanaga, Y. Ikeda, Y. Hasunuma, D. Ponomarev, V. Grebennikov, Y.A. Babanov, J. Phys. Conf. Ser., **1389**, 012146 (2019).https://doi.org/10.1088/1742-6596/1389/1/012146
12. B.T. Thole, P. Carra, F. Setle, G. van der Laan, Phys. Rev. Lett., **68**, 1943 (1992)
13. H. Wende, C. Sorg, M. Bernien, A. Scherz, P.J. Jensen, N. Ponpandian, K. Beberschke, Phys. Stat. Sol. (b), **243**, 165 (2006)
14. B. Ravel, M. Newville, J. Synchrotron Radiat. **12**(4), 537 (2005)
15. T. Miyanaga, T. Okazaki, R. Maruko, K. Takegahara, S. Nagamatsu, T. Fujikawa, H. Kon, Y. Sakisaka, J. Synchrotron Rad., **10,** 113 (2003). https://doi.org/10.1107/S0909049502022549
16. T. Miyanaga, T. Ogasawara, T. Okazaki, Y. Sakisaka, K. Okamoto, S. Nagamatsu, T. Fujikawa, J. Magn. Magn. Mater. **310**, e601 (2007).https://doi.org/10.1016/j.jmmm.2006.10.703
17. R.D. Deslattes, E.G. Kessler Jr., P. Indelicato, L. de Billy, E. Lindroth, J. Anton, J.S. Coursey, D.J. Schwab, C. Chang, R. Sukumar, K. Olsen, R.A. Dragoset, X-ray Transition Energies (version 1.2) (2005). https://www.nist.gov/pml/x-ray-transition-energies-database [accessed 30 Jun. 2021]. National Institute of Standards and Technology, Gaithersburg, MD. Originally published as: R.D. Deslattes, E.G. Kessler Jr., P. Indelicato, L. de Billy, E. Lindroth, J. Anton, Rev. Mod. Phys. **75**, 35 (2003); E.A. McCutchan, Evaluated Nuclear Structure Data File Search and Retrieval, Last updated 2021–06–29
18. Natinal Nuclear Data Center, Brookheaven National Laboratory, New York. https://www.nndc.bnl.gov/ensdf/. Accessed 30 Jun. 2021
19. G.K. Shenoy, F.E. Wagner (eds.), *Mössbauer Isomer Shifts* (North-Holland, Amsterdam, 1978)
20. N.N. Greenwood, T.C. Gibb, *Mössbauer Spectroscopy* (Chapman and Hall, London, 1971)
21. P. Gütlich, E. Bill, A.X. Trautwein, *Mössbauer spectroscopy and transition metal chemistry: fundamentals and applications*, (Springer, Berlin, 2011).https://doi.org/10.1007/978-3-540-88428-6
22. T. Zemčík, Czech. J. Phys. B **18**, 551 (1968). https://doi.org/10.1007/BF01698220
23. Z. Klencsár, Hyperfine Interact. **217**, 117 (2013). https://doi.org/10.1007/s10751-012-0732-2
24. C. Prescher, C. McCammon, L. Dubrovinsky, A. Moss, J. Appl. Crystallogr. **45**, 329 (2012). https://doi.org/10.1107/S0021889812004979
25. H. Akai, M. Akai, S. Blügel, B. Drittler, H. Ebert, K. Terakura, R. Zeller, P.H. Dederichs, Prog. Theor. Phys. Suppl. **101**, 11 (1990). https://doi.org/10.1143/PTP.101.11
26. H. Akai, T. Kotani, **3**, 120–121 (1999). https://doi.org/10.1023/A:1017057408403
27. Mössbauer Effect Reference and Data Journal. https://medc.dicp.ac.cn/Journal.php
28. R.L. Mössbauer, Z. Phys. **151**, 124 (1958). https://doi.org/10.1007/BF01344210
29. K.R. Swanson, J.J. Spijkerman, J. Appl. Phys. **41**, 3155 (1970). https://doi.org/10.1063/1.1659379
30. J. Korecki, U. Gradmann, Phys. Rev. Lett. **55**, 2491 (1985). https://doi.org/10.1103/physrevlett.55.2491
31. A.I. Chumakov, M.V. Zelepukhin, G.V. Smirnov, U. van Bürck, R. Rüffer, R. Hollatz, H.D. Rüter, E. Gerdau, Phys. Rev. B **41**, 9545 (1990). https://doi.org/10.1103/physrevb.41.9545
32. G.V. Smirnov, U. van Bürck, A.I. Chumakov, A.Q.R. Baron, R. Rüffer, Phys. Rev. B **55**, 5811 (1997). https://doi.org/10.1103/PhysRevB.55.5811
33. G.V. Smirnov, M.V. Zelepukhin, W. van Burk, JETP Lett. **43**, 352 (1986) (Translated from Russian.) http://jetpletters.ru/ps/1404/article_21315.shtml
34. T. Mitsui, M. Seto, S. Kikuta, N. Hirao, Y. Ohishi, H. Takei, Y. Kobayashi, S. Kitao, S. Higashitaniguchi, R. Masuda, Jpn. J. Appl. Phys. **46**, 821 (2007). https://doi.org/10.1143/JJAP.46.821
35. T. Mitsui, M. Seto, R. Masuda, Jpn. J. Appl. Phys. **46**, L930 (2007). https://doi.org/10.1143/JJAP.46.L930
36. T. Mitsui, N. Hirao, Y. Ohishi, R. Masuda, Y. Nakamura, H. Enoki, K. Sakaki, M. Seto, J. Synchrotron Radiat. **16**, 723 (2009). https://doi.org/10.1107/s0909049509033615
37. T. Mitsui, R. Masuda, M. Seto, E. Suharyadi, K. Mibu, J. Synchrtron Radiat. **19**, 198 (2012). https://doi.org/10.1107/s0909049511049958

38. T. Mitsui, S. Sakai, S. Li, T. Ueno, T. Watanuki, Y. Kobayashi, R. Masuda, M. Seto, and H. Akai, Phys. Rev. Lett. **125**, 236806 (2020)https://doi.org/10.1103/PhysRevLett.125.236806
39. C.S. Wang, A.J. Freeman, Phys. Rev. B **24**, 4364 (1981). https://doi.org/10.1103/PhysRevB.24.4364
40. A.J. Freeman, H. Krakauer, S. Ohnishi, D.S. Wang, M. Weinert, E. Wimmer, J. Phys. Colloq. **43**, C7-167 (1982). https://doi.org/10.1051/jphyscol:1982725
41. S. Ohnishi, A.J. Freeman, M. Weinert, Phys. Rev. B **28**, 6741 (1983). https://doi.org/10.1103/PhysRevB.28.6741
42. W. Kiauka, K. Debusmann, W. Keune, R.A. Brand, N. Hosoito, D. Liljequist, Solid State Commun. **58**, 641 (1986). https://doi.org/10.1016/0038-1098(86)90237-1
43. T. Mitsui, K. Mibu, M. Seto, M. Kurokuzu, S.P. Pati, T. Nozaki, M. Sahashi, J. Phys. Soc. Jpn. **85**, 063061 (2016).https://doi.org/10.7566/JPSJ.85.063601
44. K. Hirano, K. Izumi, T. Ishikawa, S. Annaka, S. Kikuta, Jpn. J. Appl. Phys. **30**, L407 (1990). https://doi.org/10.1143/jjap.30.l407
45. T. Mitsui, Y. Imai, R. Masuda, M. Seto, K. Mibu, J. Synchrotron Rad. **22**, 427 (2015). https://doi.org/10.1107/S1600577514028306
46. J.B. Hastings, D.P. Siddons, U. van Bürck, R. Hollatz, U. Bergmann, Phys. Rev. Lett. **66**, 770 (1991). https://doi.org/10.1103/PhysRevLett.66.770
47. M. Seto, R. Masuda, S. Higashitaniguchi, S. Kitao, Y. Kobayashi, C. Inaba, T. Mitsui, Y. Yoda, Phys. Rev. Lett. **102**, 217602 (2009).https://doi.org/10.1103/PhysRevLett.102.217602
48. K. Mibu, M. Seto, T. Mitsui, Y. Yoda, R. Masuda, S. Kitao, Y. Kobayashi, E. Suharyadi, M. Tanaka, M. Tsunoda, H. Yanagihara, E. Kita, Hyperfine Interact. **217**, 127 (2013). https://doi.org/10.1007/s10751-012-0734-0

Chapter 7
Quantum Dynamics and Statistical Thermodynamics of Nanostructured Dirac-Like Materials in a Magnetic Field

Norman J. M. Horing

Abstract This chapter is focused on the Landau-quantized dynamics and statistical thermodynamics of nanostructured Dirac-like materials in a normal magnetic field, as exemplified by Graphene and the Group VI Dichalcogenides. The chapter is based on notes on lectures presented at the UNESCO International Center for Theoretical Physics in Trieste, Italy in May 2019. They are composed of slides presented at the lectures, which were carefully prepared to provide clear explanation of the material to audience members whose native language may not have been English, using the standard terms and mathematical symbols of theoretical physics. Sufficient detail is provided to facilitate understanding the derivations and results presented and relevant references are listed.

7.1 Introduction

The Landau quantized retarded Green's function is discussed in detail for the Group VI Dichalcogenides in position-time representation and also in position-frequency representation, and the energy spectrum is exhibited. This Green's function is derived in a closed-form integral representation in terms of elementary functions (also, as a Whittaker function), and it is expanded in a series of Laguerre polynomials. As a (2×2) matrix, both diagonal and off-diagonal elements of the Green's function are exhibited. The thermodynamic Green's function and spectral weight are discussed as well as the retarded function. In regard to nanostructures, we employ a model Graphene quantum dot to examine its role in the dynamical Green's function and in the energy spectrum; we also discuss a model Dichalcogenide quantum wire, again to determine its role in the Green's function and spectrum. A model Dichalcogenide quantum anti-dot lattice is employed to explore its role in the Landau quantized Green's function dynamics and spectrum, explicitly exhibiting Landau mini-band formation in the spectrum.

N. J. M. Horing (✉)
Department of Physics, Stevens Institute of Technology, Hoboken, NJ 07030, USA
e-mail: nhoring@stevens.edu

H. Ünlü and N. J. M. Horing (eds.), *Progress in Nanoscale and Low-Dimensional Materials and Devices*, Topics in Applied Physics 144,
https://doi.org/10.1007/978-3-030-93460-6_7

The statistical thermodynamics of the Landau-quantized Group VI Dichalcogenides is discussed using the Sondheimer-Wilson formulation of the Grand Potential in terms of the canonical partition function, which we obtain from the associated retarded Green's function, leading to the Helmholtz Free Energy, Magnetic Moment, Entropy and Specific Heat at constant volume, all of which are discussed and explicitly exhibited here. Typical de Haas-van Alphen oscillatory phenomenology is seen in the degenerate statistical regime and low temperature corrections to it are exhibited with their quantum magnetic field effects. The material of this chapter is based on the references at the end of the chapter, which may be consulted for further detailed information [1–7].

7.2 Dirac "Relativistic" Materials

- $Hamiltonian \sim H \sim \gamma p \cdot (Pauli\ Matrix)$
  - Graphene ($\gamma \cong 10^6$ m/s)
  - Group VI Dichalcogenides
  - Silicene
  - Topological Insulators
  - "Diced" Lattice

Dichalcogenides	γ (m/s) (lattice speed parameter)
Mo S_2	5.96×10^5
W S_2	7.43×10^5
Mo Se_2	5.28×10^5
W Se_2	6.68×10^5

Ref: D. Xiao, G. B. Liu, W. Feng, X. Xu, W. Yao, Phys. Rev. Lett. **108**, 196802 (2012)

7.3 Calculations A: Graphene and Dichalcogenides

- Quantum Dynamics and Spectra in Magnetic Field
- Green's Functions/Propagators; Landau Quantization
  - Retarded; Thermodynamic; Spectral Weight Function
  - Control Wave Packet Propagation by Pseudo-spin Initial State Preparation
- Model Nanostructures in Graphene and Dichalcogenides
  - Q-Dots; Q-Wires; Q-Antidot Lattice: Propagators; Spectrum
  - Landau Minibands in Dichalcogenides and Graphene Q-Antidot Lattice
- Twisted Graphene Double Layer in Normal Magnetic Field

7.4 Calculations B

Statistical Thermodynamics of Dirac Materials

- Grand Potential; Helmholtz Free Energy; Partition Fns.; Entropy in Quantizing Magnetic Field
  - Magnetic Moments—Landau Quantization
- de Haas—van Alphen Oscillations (degenerate regime)
  - Temperature Dependence (approach to $\boldsymbol{T} = \mathbf{0}$)
- Specific Heat (standard characterization technique).

7.5 Diced Lattice Calculations

- Zero Field
  - Green's Function Derived; Propagator and Spectrum
  - All Thermodynamic Functions Evaluated
- Magnetic Field—Landau Quantization
  - Green's Function and Spectrum Evaluated
  - Thermodynamic Functions Evaluated.

7.6 Work in Progress and Planned

- Dichalcogenide Polarizability, Landau Quantized
  - Magneto-Plasmon Spectrum (pristine and in device configuration with substrate)
  - Damping and Static Screening in Magnetic Field
  - van der Waals interaction
- Magneto-Conductivity Tensor, Landau Quantized
  - Magneto-Polariton Spectrum; Dichalcogenides
- Quantum Dots, Wires in "*DICED*" Lattice (with and without magnetic field)
- Quantum Anti-Dot Lattice in "*DICED*" Material
  - Zero Field
  - Magnetic Field with Landau Minibands
- Landau Quantization Effects on Twisted Graphene Layers.

7.7 Hamiltonian: $H \propto \vec{p}$; $\vec{\pi} = \vec{p} + \frac{e\vec{A}}{c}$

- Group VI Dichalcogenides

The low-energy Dichalcogenides pseudo-spin-½ $\vec{k} \cdot \vec{p}$ model Hamiltonian is block diagonal, with four 2 × 2 blocks (each block for a specific choice of spin and valley), and the individual 2 × 2 blocks are like that of graphene (except for an additional spin term and an energy shift):

$$H = \gamma\sigma_x\pi_x + \gamma_\nu\sigma_y\pi_y - g\sigma_z - E_{s_z},$$

where $\vec{\pi} = (\pi_x, \pi_y) = \vec{p} + e\vec{A}$; ($\hbar = \mathrm{c} = 1$); $\vec{\sigma}$ = Pauli spin matrices; $\gamma_\nu = \gamma\,\mathrm{sign}(\nu)$ and $\nu = \pm 1$ as valley index; $s_z = \pm 1$ = spin index; $E_{s_z} = s_z\nu\lambda/2$; $g = -E_{s_z} + \Delta/2$; λ = spin splitting and Δ = energy gap; γ is a characteristic lattice speed determined by the tight binding hopping parameter and lattice spacing. (For the Hamiltonian of Graphene: $Limit\,\Delta \to 0$, $\lambda \to 0$, $E_{s_z} \to 0$; $I_{2D} = 2D\,unit\,matrix$.)

7.8 Green's Function Equa. and Magnetic Field Gauge

- Retarded Green's Function (subscript + denotes time ordering of Q-field operators ψ, ψ^+):

$$G_r\left(\vec{r}, \vec{r}'; t, t'\right) = -i\epsilon\left\langle\left(\psi\left(\vec{r}, t\right)\psi^+\left(\vec{r}'t'\right)\right)_+\right\rangle_{\text{vacuum}},$$

- subscript "vacuum" denotes vacuum expectation value.

- Green's Fn. Equation (Vector potential: $\vec{A} = \vec{B} \times \vec{r}/2$)

$$\left(i\frac{\partial}{\partial t} - H\left(\vec{\nabla}, \vec{A}, \vec{B}\right)\right)G_r\left(\vec{r}, \vec{r}'; t, t'\right) = I_{2D}\delta^{(2)}\left(\vec{r} - \vec{r}'\right)\delta(t - t').$$

- Gauge Transform: $\vec{A}(r) \to \vec{B} \times \left(\vec{r} - \vec{r}'\right)/2 = \vec{B} \times \vec{R}/2$

$$i\frac{\partial}{\partial T} - H\left(\vec{\nabla}_R, \vec{B} \times \vec{R}/2, \vec{B}\right)G_r'\left(\vec{R}, T\right) = I_{2D}\delta^{(2)}(\vec{R})\delta(T),$$

where ($\vec{R} = \vec{r} - \vec{r}' = [x - x', y - y']$; $T = t - t'$)

$$G_r\left(\vec{r}, \vec{r}'; t, t'\right) = C\left(\vec{r}, \vec{r}'\right)G_r'(\vec{R}; T),$$

with $G_r'(\vec{R}, T)$ translationally invariant and

$$C\left(\vec{r}, \vec{r}'\right) = \exp\left[\frac{ie}{2}\vec{r} \cdot \vec{B} \times \vec{r}' - \phi(\vec{r}) + \phi\left(\vec{r}'\right)\right].$$

and $\phi(\vec{r})$ is an arbitrary gauge function, which we take to vanish.

- Matrix elements of $G_r \to G_{ij}$

7.9 Retarded Green's Function Equation

- Canonical Momentum

$$\Pi_X \to \Pi_{XY} \equiv -i\partial/\partial X + eBY/2,$$

$$\Pi_Y \to \Pi_{YX} \equiv -i\partial/\partial Y - eBX/2.$$

- Position/Frequency representation (retardation: $\omega \to \omega + i0^+$)

$$\left(\omega I_{2D} - \gamma\sigma_x\Pi_{XY} - \gamma_v\sigma_y\Pi_{YX} - g\sigma_z - E_{s_z}\right)G_r'\left(\vec{R},\omega\right) = I_{2D}\delta^{(2)}\left(\vec{R}\right);$$

$$\omega_- G_{11}' - \left[\gamma\Pi_{XY} - i\gamma_v\Pi_{YX}\right]G_{21}' = \delta^{(2)}\left(\vec{R}\right),$$

$$\omega_+ G_{21}' = \left[\gamma\Pi_{XY} + i\gamma_v\Pi_{YX}\right]G_{11}',$$

$$\omega_+ G_{22}' - \left[\gamma\Pi_{XY} - i\gamma_v\Pi_{YX}\right]G_{12}' = \delta^{(2)}\left(\vec{R}\right),$$

$$\omega_- G_{12}' = \left[\gamma\Pi_{XY} + i\gamma_v\Pi_{YX}\right]G_{22}',$$

where $\omega_\pm = \omega - E_{s_z} \pm g$.

7.10 Diagonal Green's Function Analysis

$$\left[\omega_\mp - \omega_\pm^{-1}\{\gamma^2(\Pi_{XY}^2 + \Pi_{YX}^2) + i\gamma\gamma_v(\Pi_{XY}\Pi_{YX} - \Pi_{YX}\Pi_{XY})\}\right]G_{\substack{11\\22}}'\left(\vec{R};\omega\right) = \delta^{(2)}\left(\vec{R}\right).$$

Identity: $\Pi_{XY}\Pi_{YX} - \Pi_{YX}\Pi_{XY} = ieB$;

$G_{\substack{11\\22}}' = G_{11}'$ or G_{22}' correspond to the upper or lower of $\pm, \mp$ sign:

$$\left[\omega_\mp + eB\gamma\gamma_v/\omega_\pm\right]G_{\substack{11\\22}}'\left(\vec{R};\omega\right) + \frac{\gamma^2}{\omega_\pm}\left\{\nabla_R^2 - (eBR/2)^2 - ieB\left(X\frac{\partial}{\partial Y} - Y\frac{\partial}{\partial X}\right)\right\}G_{\substack{11\\22}}'\left(\vec{R};\omega\right) = \delta^{(2)}\left(\vec{R}\right).$$

7.11 Conservation of Angular Momentum

Definition: $L_Z = \frac{1}{i}\left(X\frac{\partial}{\partial Y} - Y\frac{\partial}{\partial X}\right) = L_z + L_{z'}$ vanishes in application to the Green's function,

$$L_Z G_r\left(\vec{x}, t; \vec{x}', t'\right) = (L_Z + L_{Z'})G_r\left(\vec{x}, t; \vec{x}', t'\right) = 0,$$

due to conservation of the z-component of angular momentum. Creation of a particle at $\left(\vec{x}', t'\right)$ provides angular momentum which is subsequently eliminated with its annihilation at $\left(\vec{x}, t\right)$. Verify by expansion of $G_r\left(\vec{x}, t; \vec{x}', t'\right)$ in angular momentum eigenfunctions: With arbitrary gauge function $\phi \equiv 0$, L_Z commutes with $C\left(\vec{x}, \vec{x}'\right)$, so

$$L_Z G_r'\left(\vec{R}; \omega\right) = 0.$$

7.12 Diagonal Green's Function Solution

Definitions: $\Omega_{\mp} = \omega_{\mp} \pm eB\gamma\gamma_\nu/\omega_{\mp}$; $M_{\pm} = \omega_{\pm}/2\gamma^2$.

- 2D Oscillator: $\left[\Omega_{\mp} + \left\{\nabla_R^2 - (eBR/2)^2\right\}/2M_{\pm}\right]G'_{\substack{11\\22}}\left(\vec{R}; \omega\right) = \delta^{(2)}\left(\vec{R}\right)$.

- Solution: $G'_{\substack{11\\22}}\left(\vec{R}; \omega\right) = -\frac{eB}{4\pi}\int_0^\infty d\tau \frac{e^{i\Omega_{\mp}\tau}}{\sin(\gamma^2 eB\tau/\omega_{\pm})}\exp\left\{\frac{ieBR^2}{4\tan(\gamma^2 eB\tau/\omega_{\pm})}\right\}$;

  – Expand in Laguerre polynomials L_n (notation: $1_\nu = \{+1, -1\}$ for $\{K, K'\}$ points)

- $$G'_{\substack{11\\22}}\left(\vec{R}; \omega\right) = \frac{eB\omega_{\pm}}{2\pi}\exp\left(\frac{-eBR^2}{4}\right)\sum_{n=0}^{\infty}\frac{L_n\left(\frac{eBR^2}{2}\right)}{\omega_{\pm}^2 \mp 2g\omega_{\pm} - (2n + 1 \mp 1_\nu)\gamma^2 eB}.$$

(Off-diagonal $G_{\substack{21\\12}}$ are obtained from these diagonal elements $G_{\substack{11\\22}}$.)

7.13 Dichalcogenide Energy Spectrum

- Landau Quantization

  Without the energy shifts/spin-splitting under consideration, Dirac-like Landau-quantized energies for graphene are given by $\epsilon_{n\pm}^2 = (2n + 1 \mp 1_\nu)\gamma^2 eB$, so Dichalcogenide denominators are

$$D_{n\pm} = \omega_\pm^2 \mp 2g\omega_\pm - \epsilon_{n\pm}^2$$
$$= \left(\left[\omega_\pm \mp g\right] \pm' \sqrt{g^2 + \epsilon_{n\pm}^2}\right)\left(\left[\omega_\pm \mp g\right] \mp' \sqrt{g^2 + \epsilon_{n\pm}^2}\right).$$

  Correspondingly, the Dichalcogenide energy spectrum is given by $D_{n\pm} = 0$,

- $\omega_\pm = \omega - E_{s_z} \pm g = \pm g \mp' \sqrt{g^2 + \epsilon_{n\pm}^2}; \, and \quad \omega_n = E_{s_z} \pm' \sqrt{g^2 + \epsilon_{n\pm}^2}.$

7.14 Off-Diagonal Elements

Notation: $\hat{\varrho} \equiv \sqrt{g^2 + \epsilon_n^2}$ with $\epsilon_n^2 \equiv (2n + 2)\gamma^2 eB$.

$$\omega_\pm G'_{\substack{21\\12}}\left(\vec{R}, \omega\right) = \left[\gamma \Pi_{XY} \pm i\gamma_\nu \Pi_{YX}\right] G'_{\substack{11\\22}}\left(\vec{R}, \omega\right).$$

The results for K, K' separately are given by: $\left(L_n^\alpha(x) = \text{Laguerre Polynomials.}\right)$

- $K(1_\nu = +1) : G'_{\substack{21\\12}}\left(\vec{R}, \omega\right) = \dfrac{\gamma(eB)^2}{2\pi}(iX \mp Y)e^{-eBR^2/4} \displaystyle\sum_{n=0}^{\infty} L_n^1\left(\frac{eBR^2}{2}\right)/D_n,$

- $K'(1_\nu = -1) : G'_{\substack{21\\12}}\left(\vec{R}, \omega\right) = \dfrac{\gamma(eB)^2}{2\pi}(iX \pm Y)e^{-eBR^2/4} \displaystyle\sum_{n=0}^{\infty} L_n^1\left(\frac{eBR^2}{2}\right)/D_n,$

where $D_n \equiv \left(\omega - E_{s_z} + \hat{\varrho}\right)\left(\omega - E_{s_z} - \hat{\varrho}\right)$.

Note: Off-diagonal elements are interchanged with interchange of K, K'.

7.15 Other Representations (Notation: $\rho = \sqrt{g^2 + \epsilon_{n\pm}^2}$)

- Momentum/Frequency Representation—Dichalcogenides

$$G'_{\substack{11\\22}}\left(\vec{p};\omega\right) = -i\int_0^\infty d\tau \frac{e^{i\Omega_\mp\tau}}{\cos\left(\gamma^2 eB\tau/\omega_\pm\right)}\exp\left\{\frac{-i\tan(\gamma^2 eB\tau/\omega_\pm)}{eB}p^2\right\},$$

- Position/Time Representation—Dichalcogenides

$$G'_{\substack{11\\22}}\left(\vec{R};T\right) = -i\eta_+(T)(eB/2\pi)e^{-iE_{s_z}T}\exp(-eBR^2/4)$$
$$\times\sum_{n=0}^{\infty}L_n\left(eBR^2/2\right)\{\cos(\varrho T)\mp(ig/\varrho)\sin(\varrho T)\},$$

- Position/Frequency Whittaker Function—Graphene

$$G'_{\substack{11\\22}}\left(\vec{R};\omega\right) = (M_\pm/\pi)Z_2(R;\omega),$$

$Z_2(R)$ is the second Frobenius solution of the Bessel Wave Equation.

7.16 Thermodynamic Green's Function and Spectral Weight Matrix A

- Thermal Avg. Weight $e^{-\beta(H-\mu N)}$, instead of vacuum average.
- Thermodynamic Green's Function: ($\epsilon = \pm 1$ for $t \gtrless t'$ and subscript "+" denotes time order of field operators; η_+(T) is the Heaviside unit step function.)

$$G(\vec{r},t;\vec{r}',t') = -i\epsilon\frac{Tr[e^{-\beta(H-\mu N)}(\psi(\vec{r},t)\psi^+(\vec{r}',t'))_+]}{Tr[e^{-\beta(H-\mu N)}]}.$$

- $G\left(\vec{r},\vec{r}';T\right) = \eta_+(T)G^>\left(\vec{r},\vec{r}';T\right) + \eta_+(-T)G^<\left(\vec{r},\vec{r}';T\right).$
- Define Spectral Weight Matrix $\boldsymbol{A}(\omega)$ in terms of $G^>, G^<$:

$$G^>\left(\vec{R};T\right) - G^<\left(\vec{R};T\right) \equiv -i\boldsymbol{A}\left(\vec{R};T\right) \equiv -i\int\frac{d\omega}{2\pi}e^{-i\omega T}\boldsymbol{A}\left(\vec{R};\omega\right).$$

- Gauge: $[G, G^>, G^<, \boldsymbol{A}] = C\left(\vec{r},\vec{r}'\right)\left[G', G'^>, G'^<, \boldsymbol{A}'\right](\vec{R}).$

- Spectral weight: $A\left(\vec{R},\omega\right) = C\left(\vec{r},\vec{r}'\right)A'\left(\vec{R},\omega\right)$ satisfies the *homogenous* Green's function equation ($G^>$, $G^<$ satisfy the same homogeneous equation) and the sum rule

$$\int (d\omega/2\pi)A\left(\vec{r},\vec{r}';\omega\right) = \delta^{(2)}\left(\vec{R}\right)I_{2D} = A'\left(\vec{R},T=0\right).$$

- Thermal Green's Function: $iG^{\left\{\substack{<\\>}\right\}}\left(\vec{r},\vec{r}';\omega\right) = \left\{\substack{1\pm f(\omega)\\ \pm f(\omega)}\right\}A\left(\vec{r},\vec{r}';\omega\right)$, where $f(\omega)$ is Fermi–Dirac statistical distribution function.

7.17 Spectral Weight Matrix (Matrix Elements of $A \to A_{ij}$)

$$\text{Notation: } \varrho \equiv \sqrt{g^2+\epsilon_{\text{n}\pm}^2};\quad \Omega_{c\pm} \equiv 2\gamma^2 eB/\omega_\pm;\ \text{M}_\pm = \Omega_\pm/2\gamma^2,$$

- Time Rep: $\left[i\partial/\partial\tau - p^2/2M_\pm + M_\pm\Omega_{c_\pm}^2\nabla_p^2/8\right]A'_{\substack{11\\22}}\left(\vec{p},\tau\right) = 0.$
- Solution: $A'_{\substack{11\\22}}\left(\vec{p},\tau\right) = \sec\left(\Omega_{c_\pm}\tau/2\right)exp\left\{-ip^2\tan\left(\Omega_{c_\pm}\tau/2\right)/M_\pm\Omega_{c_\pm}\right\}.$
- Spectral weight from *Retarded* Green's fn. G_r :
$A\left(\vec{r},\vec{r}',\omega\right) = -2C\left(\vec{r},\vec{r}'\right)\text{Im}G'_r\left(\vec{R},\omega\right)$; $\left[\omega\to\omega+i0^+ \text{ in } G'_r\right]$;

$$A_{\substack{11\\22}}\left(\vec{r},\vec{r}',\omega\right) = C\left(\vec{r},\vec{r}'\right)\frac{eB}{2}\exp\left(\frac{-eBR^2}{4}\right)\left(1\pm\frac{g}{\omega-E_{s_z}}\right)\sum_{n=0}^{\infty}L_n\left(\frac{eBR^2}{2}\right)\sum_{\pm'}\delta(\omega-E_{s_z}\pm'\rho),$$

$$A'_{\substack{11\\22}}\left(\vec{p},\omega\right) = 2\pi\exp\left(-\frac{p^2}{eB}\right)\left(1\pm\frac{g}{\omega-E_{s_z}}\right)\sum_{n=0}^{\infty}(-1)^n L_n\left(\frac{2p^2}{eB}\right)\sum_{\pm'}\delta(\omega-E_{s_z}\pm'\varrho).$$

- Density of States: $D(\omega) = (1/2\pi)TrA(\vec{R},\omega) = (A'_{11}(\vec{R}=0) + A'_{22}(\vec{R}=0))\frac{area}{2\pi}$.

7.18 Model Function Dot Green's fn. G_{dot}—Graphene

- Model Dot Potential: $U(\vec{r}) = \alpha\delta^{(2)}(\vec{r})$; $\alpha = \int d^2\vec{r}\; U(\vec{r}) < 0$
  (Dot radius is smallest length parameter; $U(\vec{r})$ is dot potential depth.)
- Dyson-Type Integral Equation For G_{dot}: (G_r is 2d retarded G-function without dot.)

$$G_{dot}\left(\vec{r_1}, \vec{r_2}; \omega\right) = G_r\left(\vec{r_1}, \vec{r_2}; \omega\right) + \int d^2\vec{r_3}\, G_r\left(\vec{r_1}, \vec{r_3}; \omega\right) U\left(\vec{r_3}\right) G_{dot}\left(\vec{r_3}, \vec{r_2}; \omega\right):$$

- Integrate, solve algebraically,

$$G_{dot}\left(\vec{r_1}, \vec{r_2}; \omega\right) = G_r\left(\vec{r_1}, \vec{r_2}; \omega\right) + \alpha G_r\left(\vec{r_1}, 0; \omega\right) \frac{1}{I - \alpha G_r(0, 0; \omega)} G_r\left(0, \vec{r_2}; \omega\right),$$

- Inversion: (Note: $G_r(0, 0; \omega) = G_r'(0, 0; \omega)$ since $C(0, 0) = 1$)

$$[I - \alpha G_r(0, 0; \omega)]^{-1} = \Delta^{-1} \begin{pmatrix} 1 - \alpha G_{22}' & \alpha G_{12}' \\ \alpha G_{21}' & 1 - \alpha G_{11}' \end{pmatrix}.$$

7.19 Landau Quantized Energy Spectrum: Graphene-Dot

- Dispersion Relation ($\hbar = c = 1$)

$$\Delta = \left(1 - \alpha G_{11}'(0, 0)\right)\left(1 - \alpha G_{22}'(0, 0)\right) - \alpha^2 G_{12}'(0, 0) G_{21}'(0, 0).$$

- Divergence at $\vec{r} \equiv 0$ is due to artificial confinement of Q-well to zero radius.
  For a *small* integration radius "a" we expect an average "smear":
  $G'(0, 0; \omega) \Rightarrow G'(a; \omega)$: Note $G_{12;21}' \propto (\gamma eBa/\omega) G_{11;22}' \ll G_{11;22}'$.
- Disp. Rel.: $1 \cong \alpha \frac{eB}{2\pi} \omega e^{\left(\frac{-eBa^2}{4}\right)} \sum_{n=0}^{\infty} L_n\left(\frac{eBa^2}{2}\right) / \left(\omega^2 - 2\gamma^2 eB(n + [1 \mp 1_\nu]/2)\right)$.
- Parameters: B=500gauss; $a \sim 10^{-9}$m; $\gamma = 10^6 \frac{\text{m}}{\text{s}}$; $\overline{\omega} \equiv \left(\frac{2\gamma^2 eB}{\hbar}\right)^{\frac{1}{2}} \sim \frac{10^{13}}{s}$;
  $\frac{\gamma eBa}{\hbar\overline{\omega}} \sim 10^{-3}$; $\frac{\alpha eB}{2\pi\overline{\omega}\hbar^2} \sim 10^{-3}$.
- Approximate solution of dispersion relation for $\alpha \sim a^2 \times$ potential well depth very small:

$$\omega_{n\pm} = \frac{\alpha eB}{4\pi} e^{-eBa^2/4} L_n\left(\frac{eBa^2}{2}\right) \pm \frac{1}{2}\left(\left(\frac{\alpha eB}{2\pi}\right)^2 e^{-eBa^2/2} L_n^2\left(\frac{eBa^2}{2}\right) + 8\gamma^2 eB\left(n + \frac{[1 \mp 1_\nu]}{2}\right)\right)^{1/2}.$$

7.20 Model Q-Wire Green's Function G^W—Dichalcogenide

- Model Wire Potential: $V(\vec{r}) = \beta\delta(x)$; $\beta = \int dx V(x) < 0$.
  (Wire width is smallest length parameter.)

- Dyson-Type Integral Equation for G^W: (G_r is 2d retarded G_r-function without wire.)

$$G^W(\vec{r}_1, \vec{r}_2; \omega) = G_r(\vec{r}_1, \vec{r}_2; \omega) + \int d^{(2)}\vec{r}_3\, G_r(\vec{r}_1, \vec{r}_3; \omega) V(x_3) G^W(\vec{r}_3, \vec{r}_2; \omega).$$

- Integrate:

$$G^W(x_1, y_1; x_2, y_2) = G_r(x_1, y_1; x_2, y_2) + \beta \int dy_3 G_r(x_1, y_1; x_3 = 0, y_3) G^W(x_3 = 0, y_3; x_2, y_2).$$

- Gauge: $G^W\left(\vec{r}_1, \vec{r}_2; \omega\right) = C\left(\vec{r}_1, \vec{r}_2\right) G'^W\left(\vec{r}_1, \vec{r}_2; \omega\right)$.

- Restrict Propagation to Wire Alone ($x_1 \equiv x_2 \equiv 0$); $C(y_1, y_2) \equiv 1$;
- Fourier Transform: $G^W(0, y_1; 0, y_2) = G^W(y_1 - y_2) \Rightarrow G^W\left(p_y\right)$,

$$G^W(0, y_1; 0, y_2) = G_r(0, y_1; 0, y_2) + \beta \int dy_3 G_r(0, y_1; 0, y_3) G^W(0, y_3; 0, y_2).$$

- Solve algebraically for p_y-transform

$$G^W(p_y) = G_r(p_y) + \beta G_r(p_y)\left[1 - \beta G_r(p_y)\right]^{-1} G_r(p_y).$$

7.21 Q-Wire Green's Fn. Elements (G_r review)

- Diagonal:

$$G'_{\substack{11\\22}}(p_y) = \frac{eB}{2\pi} \sum_{n=0}^{\infty} \frac{\omega_\pm}{D_{n\pm}} \int_{-\infty}^{\infty} dY \cos\left(p_y Y\right) L_n\left(\frac{eB}{2} Y^2\right) \exp\left(-\frac{eB}{4} Y^2\right)$$

and $G'_{\substack{11\\22}}(p_y) = \sqrt{\frac{eB}{\pi}} \sum_{n=0}^{\infty} \frac{\omega_{\pm}}{D_{n\pm}} \frac{(-1)^n}{n!} f_n(p_y)$, ($H_n$ are Hermite polynomials)

where $f_n(p_y) \equiv \exp\left(\frac{-p_y^2}{eB}\right) \sum_{k=0}^{n} (-1)^k \binom{n}{k} \frac{(2n-2k)!}{(n-k)!} H_{2k}\left(\sqrt{\frac{2}{eB}} p_y\right)$.

- Off Diagonal: $G'_{\substack{21\\12}}(p_y) = \sqrt{\frac{eB}{\pi} \sum_{n=0}^{\infty} \frac{(-1)^n}{D_{n\pm} n!} \Lambda_{\pm}^{op}(p_y) f_n(p_y)}$,

  where $\Lambda_{\pm}^{op}(p_y) = i\gamma_v \frac{eB}{2} \frac{\partial}{\partial p_y} \pm \gamma p_y$.

7.22 Model Q-Wire Eigenenergy Dispersion Relation

$$\beta = \int dx V(x) \sim wV_0 \sim small$$

- Dispersion Relation: $\det\left(I_{2D} - \beta G'_r(p_y)\right) = 0 = 1 - \beta\left[G'_{11} + G'_{22}\right] + \beta^2\left[G'_{11}G'_{22} - G'_{21}G'_{12}\right]$;

$$0 = 1 - \beta\sqrt{\frac{eB}{\pi}} \sum_{n=0}^{\infty} \frac{(-1)^n}{n!}\left[\frac{\omega_+}{D_{n+}} + \frac{\omega_-}{D_{n-}}\right] f_n(p_y)$$
$$+ \beta^2 \frac{eB}{\pi} \sum_{n=0}^{\infty}\sum_{m=0}^{\infty} \frac{(-1)^n(-1)^m}{n!m!} \frac{\left[\omega_+\omega_- f_n(p_y) f_m(p_y) - (\Lambda_{+}^{\mathrm{op}}(p_y) f_n(p_y))(\Lambda_{-}^{\mathrm{op}}(p_y) f_m(p_y))\right]}{D_{n+}D_{m-}}.$$

- For a very narrow wire, $w \sim \beta$ is the *smallest* length: A root can occur only if $D_{n\pm} \cong O(\beta)$, in which case the other denominator, $D_{\mp} \cong O(\beta^0) = O(1)$. Therefore, most $\beta^2 \Sigma_n \Sigma_m$—terms on the right $\cong O\left(\frac{\beta^2}{\beta}\right) = O(\beta)$ compared to the preceding term $\cong O\left(\frac{\beta}{\beta}\right) = O(1)$. However, the term $m = n - 1_\nu$ has $D_{m-} = D_{n-1_{\nu,-}} = D_{n+} = O(\beta)$, so that term of $\Sigma_n \Sigma_{m=n-1_\nu}$ contributes competitively with the first term in producing the nth-root. Dropping all terms of higher order in β, the nth root is determined by (choose $\pm \to +$; similar results obtain for $\pm \to -$):

$$0 = 1 - \frac{\beta}{D_{n+}}\sqrt{\frac{eB}{\pi}}\left[\frac{(-1)^n}{n!}\omega_+ f_n(p_y) + \frac{(-1)^{n-1_\nu}}{(n-1_\nu)!} f_{n-1_\nu}(p_y)\right]$$
$$- \frac{\beta^2}{D_{n+}^2} \frac{eB/\pi}{n!(n-1_\nu)!}\left[\omega_+\omega_- f_n(p_y) f_{n-1_\nu}(p_y) - \left(\Lambda_{+}^{\mathrm{op}}(p_y) f_n(p_y)\right)\left(\Lambda_{-}^{\mathrm{op}}(p_y) f_{n-1_\nu}(p_y)\right)\right].$$

7.23 Landau Quantized Dichalcogenide Q-Wire Energy Spectrum

- Dispersion Relation: (Notation: $\rho_{n\pm} \equiv \sqrt{g^2 + \epsilon_{n\pm}^2}$.)

$$0 = 1 + \frac{\beta}{D_{n+}}\Phi_{n+} + \frac{\beta^2}{D_{n+}^2}\psi_n,$$

where we have defined

$$\Phi_{n+} = -\sqrt{\frac{eB}{\pi}}\left[\frac{(-1)^n}{n!}\omega_+ f_n(p_y) + \frac{(-1)^{n-1_\nu}}{(n-1_\nu)!}f_{n-1_\nu}(p_y)\right],$$

$$\psi_n = -\frac{eB/\pi}{n!(n-1_\nu)!}\big[\omega_+\omega_- f_n(p_y) f_{n-1_\nu}(p_y) - \big(\Lambda_+^{\text{op}}(p_y) f_n(p_y)\big)\big(\Lambda_-^{\text{op}}(p_y) f_{n-1_\nu}(p_y)\big)\big].$$

- Quadratic Solution: $D_{n+} = -\frac{\beta}{2}\Phi_{n+} \pm'' \frac{\beta}{2}\sqrt{\Phi_{n+}^2 - 4\psi_n}$.
  Note: $D_{n+} = (\omega - E_{s_z} - \rho_{n+})(\omega - E_{s_z} + \rho_{n+}) \cong (\omega - E_{s_z} - \rho_{n+})(2\rho_{n+} + O(\beta))$.

- Q-Wire Mode Energy to Order β: (nth root)

$$\omega = E_{S_Z} + \rho_{n+} - \frac{\beta}{4\rho_{n+}}\left(\Phi_{n+} \mp'' \sqrt{\Phi_{n+}^2 - 4\psi_n}\right).$$

7.24 Model Q-Anti-dot Lattice Dichalcogenide Landau Minibands

- Dyson-Type Integral Equa. for G^L of Lattice (G_r is a 2D- retarded G-Fn without antidot lattice.)

$$G^L(x_1, x_2; y_1, y_2; \omega) = G_r(x_1, x_2; y_1, y_2; \omega) + \int dx_3 \int dy_3 G_r(x_1, x_3; y_1, y_3; \omega) U(x_3, y_3) G^L(x_3, x_2; y_3, y_2; \omega).$$

- Krönig-Penney Model Potential for Q-Anti-Dot Lattice in magnetic field

$$U(\vec{r}) = U(x, y) = \alpha \sum_{n=-\infty}^{\infty} \delta(x - nd)\delta(y);\ d = \text{Lattice Period}$$

($\alpha = U_0 a^2 > 0$ is a product of antidot potential barrier height with its area),

$$G^L(x_1, x_2; y_1, y_2; \omega) = G_r(x_1, x_2; y_1, y_2; \omega) + \alpha \sum_{n=-\infty}^{\infty} G_r(x_1, nd; y_1, 0) G^L(nd, x_2; 0, y_2).$$

- To solve, set $x_1 = md$:

$$G^L(md, x_2; 0, y_2) = G_r(md, x_2; 0, y_2) + \alpha \sum_{n=-\infty}^{\infty} G_r(md, nd; 0, 0) G^L(nd, x_2; 0, y_2).$$

- Lack of spatial translational invariance is a problem, except for propagation directly on the lattice: $(y_1 \equiv y_2 \equiv 0)$; $C\left(\vec{r}, \vec{r}'\right) \to 1$;

$$G_r\left(\vec{r}, \vec{r}'; t, t'\right) = C\left(\vec{r}, \vec{r}'\right) G'_r\left(\vec{r} - \vec{r}', t - t'\right) = G'_r\left(\vec{r} - \vec{r}', t - t'\right).$$

7.25 Lattice G^L-Fn. In Magnetic Field: Analysis

- Propagation Directly on Q-Anti-dot Lattice: $(y_1 \equiv y_2 \equiv 0)$.
- Define $\dot{G}_r$: $G_r(md, nd; 0, 0) \equiv \dot{G}_r([m - n]d, 0, 0)$.
- Lattice Periodicity: $\tilde{G}^L(p) = \sum_{r=-\infty}^{\infty} e^{ipdr} G^L(rd)$.
- Fourier Coefficient: $G^L(md) = \frac{d}{2\pi} \int_{-\pi/d}^{\pi/d} dp e^{-ipdm} \tilde{G}^L(p)$,

  and

$$\dot{G}^L([m - n]d) = \frac{d}{2\pi} \int_{-\pi/d}^{\pi/d} dp e^{-ipd[m-n]} \dot{\tilde{G}}^L(p).$$

- Fourier Series to Crystal Momentum Representation:

$$\begin{aligned} \tilde{G}^L(p) &= \tilde{G}_r(p) + \alpha \dot{\tilde{G}}_r(p) \sum_{n=-\infty}^{\infty} e^{ipdn} G^L(nd) \\ &= \tilde{G}_r(p) + (\alpha d/2\pi) \dot{\tilde{G}}_r(p) \int_{-\pi/d}^{\pi/d} dq \left(\Sigma_{n=-\infty}^{\infty} e^{in[p-q]d}\right) \tilde{G}^L(q). \end{aligned}$$

7.26 Solution for Lattice G^L-Function

- Poisson Sum Formula:

$$\tilde{G}^L(p) = \tilde{G}_r(p) + \alpha \sum_{m=-\infty}^{\infty} \int_{-\pi/d}^{\pi/d} dq\, \delta(p - q - 2\pi m/d) \dot{\tilde{G}}_r(p) \tilde{G}^L(q);$$

- Integrate over first Brillouin Zone:

$$\tilde{G}^L(p) = \tilde{G}_r(p) + \alpha \dot{\tilde{G}}_r(p) \tilde{G}^L(p),$$

- whence $\tilde{G}^L(p; x_2; 0, 0; \omega) = \left[1 - \alpha \dot{\tilde{G}}_r(p; 0, 0; \omega)\right]^{-1} \tilde{G}_r(p; x_2; 0, 0; \omega)$.
- Lattice Green's Function: 2D Krönig-Penney model in magnetic field

$$G^L(x_1, x_2; 0, 0; \omega) = G_r(x_1, x_2; 0, 0; \omega) + \alpha \sum_{n=-\infty}^{\infty} G_r(x_1, nd; 0, 0; \omega)$$
$$\times \frac{d}{2\pi} \int_{-\pi/d}^{\pi/d} dp e^{-ipdn} \left[1 - \alpha \dot{\tilde{G}}_r(p; 0, 0; \omega)\right]^{-1} \tilde{G}_r(p; x_2; 0, 0; \omega).$$

7.27 Q-Anti-dot Lattice Energy Spectrum: Landau Minibands

- Dispersion Relation: $\det\left[1 - \alpha \dot{\tilde{G}}_r(p, 0, 0; \omega)\right] = 0$,

$$1 = \alpha\left[\dot{\tilde{G}}_{11}(p) + \dot{\tilde{G}}_{22}(p)\right] - \alpha^2\left[\dot{\tilde{G}}_{11}(p)\dot{\tilde{G}}_{22}(p) - \dot{\tilde{G}}_{21}(p)\dot{\tilde{G}}_{12}(p)\right];$$

- Diagonal Elements → Landau Quantized Dichalcogenides G^L-Fn:

$$\dot{\tilde{G}}'^{Diag}_{\substack{11\\22}}(p) = \sum_{n=0}^{\infty} (\omega \pm N_n)/D_{n,s_z,\nu,\pm}; \quad N_n \equiv \frac{eB}{2\pi} \sum_{r=-\infty}^{\infty} e^{ipdr} e^{-eBd^2r^2/4} L_n(eBd^2r^2/2);$$

- Off-Diagonal Elements: $\dot{\tilde{G}}_{\substack{21\\12}}(p) = \sum_{n=0}^{\infty} M_n/D_{n,s_z,\nu,\pm}$;

$$M_n \equiv \frac{i\gamma(eB)^2 d}{2\pi} \sum_{r=-\infty}^{\infty} e^{ipdr} r e^{\frac{-eBd^2r^2}{4}} \left\{\frac{[1 \mp 1_\nu]}{2} L_n\left(\frac{eBd^2r^2}{2}\right) + L^1_{n-1}\left(\frac{eBd^2r^2}{2}\right)\right\}.$$

- Dispersion Relation: (Notation $\rho_{n\pm} = \sqrt{g^2 + \epsilon_{n\pm}^2}$)

$$0 = 1 - \alpha \sum_{n=0}^{\infty} N_n \omega_+ / D_{n+} - \alpha \sum_{n=0}^{\infty} N_n \omega_- / D_{n-} + \alpha^2 \sum_{n=0}^{\infty} \sum_{m=0}^{\infty} \frac{\omega_+ \omega_- N_n N_m - M_n M_m}{D_{n+} D_{m-}},$$

where $D_{n\pm} = \left(\omega - E_{s_z} \mp' \rho_{n\pm}\right)\left(\omega - E_{s_z} \pm' \rho_{n\pm}\right)$.

7.28 Dispersion Relation Analysis for Small Anti-dot Area

- Parameters: $\alpha \sim a^2 U_0$; a^2 = Dot Area = Small; (Note: $D_{n,+} = D_{n-1_{\nu},-}$).
- For a root $D_{n,+} = 0(\alpha) = D_{n-1_{\nu},-}$, so neglecting smaller terms from other poles,

$$0 = 1 - \frac{\alpha}{D_{n,+}} \left(N_n \omega_+ + N_{n-1_\nu} \omega_-\right) + \frac{\alpha^2}{D_{n,+}^2} \left(\omega_+ \omega_- N_n N_{n-1_\nu} - M_n M_{n-1_\nu}\right).$$

Note that the "n"-series in M_n and N_n must be terminated when the spin-split/shifted magnetic energy $\rho_{n\pm} = \sqrt{g^2 + \epsilon_{n\pm}^2}$ reaches the limit of validity of the Dirac linear-momentum low-energy Hamiltonian approximation (due to band bending). Solve quadratic eqn. for $D_{n\pm}$:

$$D_{n\pm} = \frac{\alpha}{2}\left(N_n \omega_+ + N_{n-1_\nu} \omega_-\right) \pm'' \frac{\alpha}{2} \sqrt{\left(N_n \omega_+ + N_{n-1_\nu} \omega_-\right)^2 - 4\left(\omega_+ \omega_- N_n N_{n-1_\nu} - M_n M_{n-1_\nu}\right)}.$$

Recall $D_{n\pm} = \left(\omega - E_{s_z} \mp' \rho_{n\pm}\right)\left(\omega - E_{s_z} \pm' \rho_{n\pm}\right) \Rightarrow \left(\omega - E_{s_z} \mp' \rho_{n\pm}\right)\left(\pm' 2\rho_{n\pm}\right)$ to order α for root near $\omega = E_{s_z} +' \rho_+ + 0(\alpha)$: (Choose $\pm = +$.)

$$\omega = E_{s_z} +' \rho_{n+} +' D_{n+} / \left(\pm' 2\rho_{n+}\right).$$

Since $D_{n+} \sim 0(\alpha)$ explicitly, its further dependence on α implicitly through ω_+, ω_- may be neglected to order α.

7.29 Landau Minibands

$$N_n = \frac{eB}{2\pi}\Big[1 + 2\Sigma_{r=1}^{\infty} \cos(pdr) e^{-eBd^2r^2/4} L_n(eBd^2r^2/2)\Big],$$

$$M_n = \frac{-\gamma(eB)^2 d}{\pi}\left[\Sigma_{r=1}^{\infty} r \sin(pdr) e^{-eBd^2r^2/4}\left\{\frac{[1 \mp 1_\nu]}{2} L_n(eBd^2r^2/2) + L_{n-1}^1(eBd^2r^2/2)\right\}\right].$$

- The n'th eigenenergy mode is given by

$$\omega_n = E_{s_z} \pm' \rho_{n\pm} \pm' \frac{\alpha}{4\rho_{n\pm}}\big(N_n\omega_+ + N_{n-1_\nu}\omega_-\big)$$
$$\pm'' \left(\frac{\pm''\alpha}{4\rho_{n\pm}}\right)\sqrt{\big(N_n\omega_+ + N_{n-1_\nu}\omega_-\big)^2 - 4\big(\omega_+\omega_- N_n N_{n-1_\nu} - M_n M_{n-1_\nu}\big)}.$$

- Since $\cos(pdr)$, $\sin(pdr)$ vary over a restricted range $-1 \le \cos(pdr), \sin(pdr) < 1$, the discrete Landau level has spread into a band of finite width, ie: a Dichalcogenide Landau miniband of width proportional to α has been formed by the antidot lattice.

7.30 Statistical Thermodynamics of Group VI Dichalcogenides in Magnetic Field

$$k_B = \text{Boltzmann Constant}$$

- Grand Potential: Ω; $\beta = \big(k_B T'\big)^{-1}$; T' is Kelvin Temperature
- Helmholtz Free Energy F; μ = Chemical Potential
- Grand Partition Function: $Z = \prod_{E_\gamma}\big(1 + e^{-\beta(E_\gamma-\mu)}\big)$–FermiDirac
- Ordinary Partition Function: $\hat{Z} = \Sigma_{E_\gamma} e^{-\beta E_\gamma} = \mathrm{Tr} e^{-\beta H}$
- Entropy: $S = -\big(\partial F/\partial T'\big)_{N,V\,\text{constant}}$; $\big(dF = -pdV - SdT' + \mu dN\big)$
- Magnetic Moment: $M = -(\partial F/\partial B)_{T,V,N}$ (B is Magnetic Field)
  - de Haas-van Alphen(dHvA) Oscillations
- Specific Heat: $C_V = T'\big(\partial S/\partial T'\big)_V$.

7.31 Thermodynamic Functions: Relations

- Grand Potential Ω:

$$\Omega = \mathrm{F} - \mu N = -k_B T' \ln Z = -k_B T' \Sigma_{E_\gamma} \ln\left(1 + e^{-\beta(E_\gamma - \mu)}\right),$$

and writing the E_γ-summand as an inverse Laplace transform (contour c)

$$B(E) \equiv -k_B T' \ln\left(1 + e^{-\beta(E-\mu)}\right) = \int_c \frac{ds}{2\pi i} e^{sE} p(s),$$

with $p(s)$ as the Laplace transform of $B(E)$,

$$p(s) = \int_0^\infty dE e^{-sE} B(E),$$

- we have $\Omega = \mathrm{F} - \mu N = \int_c \frac{ds}{2\pi i} p(s) \sum_{E_\gamma} e^{sE_\gamma} = \int_c \frac{ds}{2\pi i} p(s) \hat{Z}(\beta \to -s)$.

7.32 Wilson's Evaluation in Terms of Ordinary Partition Function

- A.H. Wilson: $\Omega = \mathrm{F} - \mu N = \int_c \frac{ds}{2\pi i} \left(\frac{\hat{Z}(\beta \to -s)}{s^2} \right) \left(s^2 p(s)\right),$

- Laplace Transform Convolution Theorem: (*f(E)* is the Fermi-Dirac distribution.)

$$\Omega = \mathrm{F} - \mu N = \int_0^\infty dE \int_c \frac{ds}{2\pi i} e^{Es} \frac{\hat{Z}(s)}{s^2} \int_c \frac{ds'}{2\pi i} e^{Es'} s'^2 p(s'),$$

$$\int_c \frac{ds'}{2\pi i} e^{Es'} s'^2 p(s') = \frac{\partial^2}{\partial E^2} \int_c \frac{ds'}{2\pi i} e^{Es'} p(s') = \frac{\partial^2 B(E)}{\partial E^2} = \frac{\partial f(E)}{\partial E},$$

$$\text{whence } \Omega = \mathrm{F} - \mu N = \int_c \frac{ds}{2\pi i} \frac{\hat{Z}(s)}{s^2} \int_0^\infty dE e^{Es} \frac{\partial f(E)}{\partial E},$$

- and $\Omega = \frac{-\beta}{4} \int_c \frac{ds}{2\pi i} \frac{\hat{Z}(s)}{s^2} \int_0^\infty dE e^{Es} \mathrm{sech}^2\left(\frac{[E-\mu]\beta}{2}\right).$

7.33 Retarded Green's Fn. and Ordinary Partition Function

- Time Translation Operator $= e^{-iHT}$; $\hat{Z}(\beta) = \mathrm{Tr}\, e^{-\beta H}$,
  $\hat{Z}(\beta) = \int d\vec{x}\, \mathrm{Tr}\Big(iG_{r,T>0}(\vec{x}, \vec{x}; T \to -i\beta)\Big)$.

- Dichalcogenides: Landau Quantized Green's Function (Retarded; Diagonal Elements).

$$iG_{\substack{11\\22}}\left(\vec{x}, \vec{x}'; T\right) = \eta_+(T)\exp\left(\frac{ie}{2}\left[\vec{x} \cdot \vec{B} \times \vec{x}'\right]\right)\frac{eB}{2\pi}e^{-iE_{s_z}T}\exp\left(-\frac{eBR^2}{4}\right)$$
$$\times \sum_{n=0}^{\infty} L_n\left(\frac{eBR^2}{2}\right)\left\{\cos\left(\sqrt{g^2+\epsilon_{n\pm}^2}\,T\right) \mp \frac{ig}{\sqrt{g^2+\epsilon_{n\pm}^2}}\sin\left(\sqrt{g^2+\epsilon_{n\pm}^2}\,T\right)\right\},$$

with $\epsilon_{n\pm}^2 = (2n + 1 \mp 1_\nu)\gamma^2 eB$ (definition:$1_\nu = \mathrm{sign}(\nu) = \pm 1$). Also,

- $$\int d\vec{x}\, \mathrm{Tr}\Big(iG_{r,T>0}\left(\vec{x}, \vec{x}; T\right)\Big) = (\mathrm{area})\frac{eB}{4\pi}e^{-iE_{s_z}T}$$
$$\times \sum_{s_z=\pm 1}\sum_{\nu=\pm 1}\sum_{n=0}^{\infty}\sum_{\pm}\sum_{\pm'}\left(1 \pm (\pm')\frac{g}{\sqrt{g^2+\epsilon_{n\pm}^2}}\right)e^{\pm' i\sqrt{g^2+\epsilon_{n\pm}^2}\,T}.$$

7.34 Thermodynamic Green's Function and Spectral Weight A

- Grand Canonical Ensemble Average (subscript "+" denotes time ordering of field operators $\psi(\vec{x}, t)$, $\psi^+(\vec{x}', t')$ and $\epsilon = \pm 1$ for $t \gtrless t'$; $\eta_+(T)$ is the Heaviside unit step function);

$$G(\vec{x}, t; \vec{x}'t') = \frac{-\left[i\epsilon \mathrm{Tr}\left(e^{-\beta(H-\mu N)}\left(\psi\left(\vec{x}, t\right)\psi^+(\vec{x}', t')\right)_+\right)\right]}{\mathrm{Tr}\left(e^{-\beta(H-\mu N)}\right)},$$

$$G\left(\vec{x}, \vec{x}'; T\right) = \eta_+(T)G^>\left(\vec{x}, \vec{x}'; T\right) + \eta_+(-T)G^<\left(\vec{x}, \vec{x}'; T\right).$$

- Spectral Weight Matrix $\mathbf{A}(\vec{x}, \vec{x}'; \omega)$ relates to $G^>$, $G^<$ as

$$\left\{\begin{matrix} G^> \\ G^< \end{matrix}\right\}\left(\vec{x}, \vec{x}'; \omega\right) = -i\left\{\begin{matrix} 1 - f_0(\omega) \\ -f_0(\omega) \end{matrix}\right\}\mathbf{A}\left(\vec{x}, \vec{x}'; \omega\right) \text{ for Fermions,}$$

and

$$G^{>}\left(\vec{x},\vec{x}';T\right)-G^{<}\left(\vec{x},\vec{x}';T\right)=-i\mathbf{A}\left(\vec{x},\vec{x}';T\right)=-i\int\frac{d\omega}{2\pi}e^{-i\omega T}\mathbf{A}\left(\vec{x},\vec{x}';\omega\right).$$

- $G^{>}, G^{<}, \mathbf{A}$ all satisfy *homogeneous* G-fn. equation; Sum rule for $\mathbf{A}\left(\vec{x},\vec{x}';\omega\right)$:

$$\int_{-\infty}^{\infty}(d\omega/2\pi)\mathbf{A}\left(\vec{x},\vec{x}';\omega\right)=\delta\left(\vec{x}-\vec{x}'\right)I_{2D};\ \int_{-\infty}^{\infty}(d\omega/2\pi)\mathbf{A}\left(\vec{p},\omega\right)=I_{2D}.$$

- Relation of Spectral Weight to Retarded G-fn. (G_r) and Density of States $D(\omega)$.

$$\left(\vec{R}=\vec{x}-\vec{x}';\left\{G,G^{>},G^{<},\mathbf{A}\right\}=C\left(\vec{x},\vec{x}'\right)\left\{G',G'^{>},G'^{<},\mathbf{A}'\right\}(\vec{R}).\right)$$

$$\text{Retarded Green's Fn}:\mathbf{A}(\omega)=-2ImG_r(\omega).$$

$$\text{Density of States}:D(\omega)=-\frac{1}{\pi}ImTrG_r\left(\vec{x},\vec{x}';\omega\right)=(\frac{1}{2\pi})Tr\mathbf{A}(\omega).$$

7.35 Landau Quantized Dichalcogenide Spectral Weight

- Diagonal Elements: (Matrix elements of $\mathbf{A} \to A_{ij}$)

$$A'_{\substack{11\\22}}\left(\vec{x},\vec{x}';\omega\right) = \frac{eB}{2}\exp\left(\frac{-eBR^2}{4}\right)\left(1 \pm \frac{g}{\omega - E_{s_z}}\right)$$
$$\times \sum_{n=0}^{\infty}\sum_{\pm'} L_n\left(\frac{eBR^2}{2}\right)\delta\left(\omega - E_{s_z} \pm' \sqrt{g^2+\epsilon_{n\pm}^2}\right).$$

- $G^<$-Trace:

$$-iTrG^<\left(\vec{x},\vec{x}';T\right) = \exp\left(\frac{ie}{2}[\vec{x}\cdot\vec{B}\times\vec{x}']\right)\frac{eB}{4\pi}e^{-\frac{eBR^2}{4}}\sum_{s_z=\pm1}\sum_{\nu=\pm1}\sum_{n=0}^{\infty}\sum_{\pm}\sum_{\pm'}$$
$$\times f_0\left(E_{s_z} \mp' \sqrt{g^2+\epsilon_{n\pm}^2}\right)L_n\left(\frac{eBR^2}{2}\right)\left(1 \pm \frac{(\mp'1)g}{\sqrt{g^2+\epsilon_{n\pm}^2}}\right)e^{-i\left(E_{s_z}\mp'\sqrt{g^2+\epsilon_{n\pm}^2}T\right)}.$$

- Density:

$$n = -iTrG'^<\left(\vec{R}=0;T=0\right) = \frac{eB}{4\pi}\sum_{s_z=\pm1}\sum_{\nu=\pm1}\sum_{n=0}^{\infty}\sum_{\pm}\sum_{\pm'}$$
$$\times\left(1 \pm \frac{(\mp'1)g}{\sqrt{g^2+\epsilon_{n\pm}^2}}\right)f_0\left(E_{s_z} \mp' \sqrt{g^2+\epsilon_{n\pm}^2}\right).$$

7.36 Dichalcogenide Grand Potential: Degenerate Regime

- Degenerate Regime: $\mu\beta \to \infty$
- Wilson: $\Omega = \mathrm{F} - \mu N = -\frac{\pi}{\beta}\int_c \frac{ds}{2\pi i} e^{s\mu} \frac{\hat{Z}(s)}{s\,\sin(\pi s/\beta)}$ (per unit area),

$$\Omega = \mathrm{F} - \mu n = -\frac{eB}{4\beta}\sum_{s_z=\pm 1}\sum_{\nu=\pm 1}\sum_{\pm}\sum_{\pm'}\sum_{n=0}^{\infty}\left(1 \mp (\pm' 1)\frac{g}{\sqrt{g^2+\epsilon_{n\pm}^2}}\right)\int_c \frac{ds'}{2\pi i}\frac{\exp\left[s'\beta\left(\mu - E_{s_z} \pm' \sqrt{g^2+\epsilon_{n\pm}^2}\right)\right]}{s'\sin(\pi s')}.$$

- To evaluate the s'-integral, we exponentiate the integrand factor as $1/s' = \beta\int_0^\infty dx e^{-s'\beta x}$ so that $\left(z = \pi s'\right)$.

$$\int_c \frac{ds'}{2\pi i}\frac{e^{s'\beta E_\gamma}}{s'\sin(\pi s')} = \beta\int_0^\infty dx \int_c \frac{dz}{2\pi i}\frac{e^{z\beta[E_\gamma/\pi - x]}}{\sin(z)}.$$

Noting that the contour of z-integration along c is a straight line from $z = -i\infty + 0^+$ to $z = +i\infty + 0^+$, we consider closing the contour with a parallel line (c') from $i\infty - \pi^+$ to $-i\infty - \pi^+$ on which $dz_{c'} = -dz_c$ and $\sin\left(z'\right) = -\sin(z)$.

7.37 Contour Integral for Ω: Degenerate Regime

The closed contour integrand $\oint = \int_c + \int_{c'}$ has the residue "1" at $z = 0$, so that

$$\oint \frac{dz}{2\pi i}\cdots = \int_c \frac{dz_c}{2\pi i}\cdots + \int_{c'}\frac{dz_{c'}}{2\pi i} = \left(1 + e^{-\pi\beta[E_\gamma/\pi - x]}\right)\int_c \frac{dz}{2\pi i}\frac{e^{z\beta[E_\gamma/\pi - x]}}{\sin(z)} = 1.$$

Consequently, the x-integration of 7.36 is given by

$$\int_0^\infty dx \frac{1}{1 + e^{-\pi\beta[E_\gamma/\pi - x]}} = \frac{1}{\pi\beta}\ln\left(1 + e^{\beta E_\gamma}\right),$$

and, for the degenerate regime under consideration, we obtain the Grand Potential Ω as

- $$\Omega = F - \mu n = \frac{-eB}{4\pi\beta}\sum_{s_z=\pm 1}\sum_{\nu=\pm 1}\sum_{n=0}^{\infty}\sum_{\pm}\sum_{\pm'}\left(1 \mp (\pm')\frac{g}{\sqrt{g^2+\epsilon_{n\pm}^2}}\right) \times \ln\left(1 + exp\left[\beta\left(\mu - E_{s_z} \pm' \sqrt{g^2+\epsilon_{n\pm}^2}\right)\right]\right).$$

7.38 Grand Potential in the Degenerate Regime: Further Comments

This indicates the behavior at very low temperature to be given approximately by ($\eta_+(x) = 1$ for $x > 0$, 0 for $x < 0$ and $\eta_-(x) = 1$ for $x < 0, 0$ for $x > 0$; $\eta_+(x) + \eta_-(x) = 1$ for all x):

- $$\Omega = F - \mu n = \frac{-eB}{4\pi} \sum_{s_z=\pm 1} \sum_{\nu=\pm 1} \sum_{n=0}^{\infty} \sum_{\pm} \sum_{\pm'} \left(1 \mp (\pm') \frac{g}{\sqrt{g^2 + \epsilon_{n\pm}^2}} \right)$$
$$\times \left\{ \eta_+ \left(\mu - E_{s_z} \pm' \sqrt{g^2 + \epsilon_{n\pm}^2} \right) \left[\mu - E_{s_z} \pm' \sqrt{g^2 + \epsilon_{n\pm}^2} \right] \right.$$
$$\left. + \beta^{-1} exp \left(-\beta \left| \mu - E_{s_z} \pm' \sqrt{g^2 + \epsilon_{n\pm}^2} \right| \right) \right\},$$

on a per-unit-area basis.

The first term is the zero temperature limit, and the second exhibits finite temperature corrections (small). The first term is oscillatory in the de Haas-van Alphen sense due to abrupt population changes as the magnetic field varies, causing successive vanishings of the argument of $\eta_+ \left(\mu - E_{s_z} \pm' \sqrt{g^2 + \epsilon_{n\pm}^2} \right)$ as Landau levels cross the Fermi energy.

7.39 Magnetic Moment of Landau Quantized Dichalcogenides

$$M = -\left(\frac{\partial F}{\partial B} \right)_{T,V,N} = -\left(\frac{\partial \Omega}{\partial \beta} \right)_{T,V,N}.$$

$$M_{\text{Deg}} = \Delta M_{\text{Deg}} (\text{T}' = 0) + \Delta M_{\text{Deg}} (\text{T}' > 0).$$

- Degenerate Regime :

- $$\Delta M_{\text{Deg}} (\text{T}' = 0) = \frac{\text{e}}{4\pi} \sum_{s_z=\pm 1} \sum_{\nu=\pm 1} \sum_{n=0}^{\infty} \sum_{\pm} \sum_{\pm'} \eta_+ \left(\mu - E_{s_z} \pm' \sqrt{g^2 + \epsilon_{n\pm}^2} \right)$$
$$\times \left\{ \left(\mu - E_{s_z} \pm' \sqrt{g^2 + \epsilon_{n\pm}^2} \right) \left(1 \mp \frac{(\pm' 1) g}{\sqrt{g^2 + \epsilon_{n\pm}^2}} \pm \frac{(\pm' 1) g \epsilon_{n\pm}^2}{2 \left(g^2 + \epsilon_{n\pm}^2 \right)^{\frac{3}{2}}} \right) \right.$$
$$\left. + \left(1 \mp \frac{(\pm' 1) g}{\sqrt{g^2 + \epsilon_{n\pm}^2}} \right) \left(\frac{(\pm' 1) \epsilon_{n\pm}^2}{2 \sqrt{g^2 + \epsilon_{n\pm}^2}} \right) \right\},$$

- $$\Delta M_{\mathrm{Deg}}(\mathrm{T}' > 0) = \frac{\mathrm{e}}{4\pi\beta} \sum_{s_z=\pm1} \sum_{\nu=\pm1} \sum_{n=0}^{\infty} \sum_{\pm} \sum_{\pm'} e^{-\beta\left|\mu - E_{s_z} \pm' \sqrt{g^2+\epsilon_{n+}^2}\right|} \left\{ \left(1 \mp (\pm') \frac{g}{\sqrt{g^2+\epsilon_{n\pm}^2}}\right) \times \left(1 - \frac{\beta\epsilon_{n\pm}^2}{2\sqrt{g^2+\epsilon_{n\pm}^2}}\right) \pm \frac{(\pm')g\epsilon_{n\pm}^2}{2\left(g^2+\epsilon_{n\pm}^2\right)^{3/2}} \right\}.$$

Note: de Haas-van Alphen Oscillations in $\Delta M_{\mathrm{Deg}}(\mathrm{T}' = 0)$ are *Not Simply Periodic* in Dirac materials.

7.40 Entropy of Landau Quantized Dichalcogenides; Specific Heat

- $$S = -\left(\frac{\partial F}{\partial T'}\right)_{N,V} = -\frac{\partial\beta}{\partial T'}\frac{\partial(F-\mu n)}{\partial\beta} = k_B\beta^2\frac{\partial(F-\mu n)}{\partial\beta} = k_B\beta^2\left(\frac{\partial\Omega}{\partial\beta}\right)_{N,V}.$$

  Despite unbounded negative energy branch of spectrum, entropy S vanishes as $T' \to 0$. (Notation: $E_\gamma \equiv E_{s_z} \mp' \sqrt{g^2+\epsilon_{n\pm}^2}$.)
- $S = \frac{\partial\beta}{\partial T'}\frac{\partial}{\partial\beta}\left(\beta^{-1}\sum_{E_\gamma}\ln\left(1+e^{-\beta(E_\gamma-\mu)}\right)\right) \Rightarrow 0$ as $T' \to 0$.
- Degenerate Regime:

$$S_{Deg} = k_B\frac{eB}{4\pi}\sum_{s_z=\pm1}\sum_{\nu=\pm1}\sum_{n=0}^{\infty}\sum_{\pm}\sum_{\pm'}\left(1 \mp (\pm'1)\frac{g}{\sqrt{g^2+\epsilon_{n\pm}^2}}\right)\left\{\ln(1+\exp[\beta(\mu-E_\gamma)]) - \frac{\beta[\mu-E_\gamma]}{1+\exp[-\beta(\mu-E_\gamma)]}\right\}.$$

- Simplifying for large β as $T' \sim 0$, the curly bracket becomes

$$S_{Deg} = \left\{\exp(-\beta|\mu-E_\gamma|)[1+\beta|\mu-E_\gamma|]\right\}.$$

- Specific Heat at Constant Volume: $C_v = T'\left(\frac{\partial S}{\partial T'}\right)_V = -\beta\left(\frac{\partial S}{\partial\beta}\right)_V$;

$$C_v = k_B\beta^2\frac{eB}{4\pi}\sum_{s_z=\pm1}\sum_{\nu=\pm1}\sum_{n=0}^{\infty}\sum_{\pm}\sum_{\pm'}\left(1 \mp (\pm'1)\frac{g}{\sqrt{g^2+\epsilon_{n\pm}^2}}\right)\left\{\exp(-\beta|\mu-E_\gamma|)|\mu-E_\gamma|^2\right\}.$$

Note: Both the entropy and the specific heat are devoid of dHvA oscillations.

References

1. N.J.M. Horing, Aspects of the theory of graphene, Trans. Royal Soc. A **368**, 5525–5556 (2010)
2. N.J.M. Horing, S.Y. Liu, Energy spectrum and density of states for a graphene quantum dot in a magnetic field, J. Phys. Condens. Matter **22**, 025502 (2010)
3. N.J.M. Horing, Landau quantized dynamics and spectra for group vi dichalcogenides, including a model quantum wire, AIP Adv. **7**, 065316 (2017)
4. N.J.M. Horing, J.D. Mancini, Thermal and magnetic properties of Landau quantized group vi dichalcogenide carriers in the approach to the degenerate limit, J. Phys. Commun. 4, 095006 (2020): Also Corrigendum: J. Phys. Commun. 5, 089501 (2021)
5. N.J.M. Horing, Dichalcogenide Landau miniband dynamics and spectrum in an antidot superlattice, AIP Adv. **10**(3) (2020)
6. Chapter 9 of this book.
7. Chapter 16 of this book.

Chapter 8
T-3 "DICED" LATTICE Quantum Dynamics and Statistical Thermodynamics (a) Zero Magnetic Field and (b) Landau Quantized

Norman J. M. Horing

Abstract This chapter addresses the pseudospin-1 T-3 "Diced" Lattice in regards to its quantum dynamics, spectrum and statistical thermodynamics, both with and without an impressed normal magnetic field. As in the preceding chapter, it is based on lecture notes composed of slides that have been carefully prepared to provide clear explanations of the material to audience members whose native language may not have been English, using the standard terms and mathematical symbols of theoretical physics. Considerable detail is presented to enhance understanding of the derivations and results described on the slides and pertinent references are provided.

8.1 Introduction

The (3×3) pseudospin-1 matrix retarded Green's functions for the "Diced" lattice are discussed in detail, both with and without the presence of a normal magnetic field, describing its quantum dynamics and energy spectra, with discrete Landau quantized energy levels in the presence of a magnetic field. These Green's functions are used in the determination of the corresponding canonical partition functions, which are, in turn, employed to evaluate the statistical thermodynamic functions for the T-3 "Diced" lattice, both with and without a magnetic field. These functions include the Grand Potential, Helmholtz Free Energy, the Grand Partition Function, Entropy, Specific Heat, and Magnetic Moment in the case with a magnetic field. In these evaluations, the Sondheimer-Wilson formulation is employed and the evaluations of all the statistical thermodynamic functions are carried out in the degenerate statistical regime (including the determination of temperature dependence in the approach to the zero temperature limit). The material of this chapter is based on the references at the end of the chapter, which may be consulted for further detailed information [1–5].

N. J. M. Horing (✉)
Department of Physics, Stevens Institute of Technology, Hoboken, NJ 07030, USA
e-mail: nhoring@stevens.edu

H. Ünlü and N. J. M. Horing (eds.), *Progress in Nanoscale and Low-Dimensional Materials and Devices*, Topics in Applied Physics 144,
https://doi.org/10.1007/978-3-030-93460-6_8

8.2 Dynamics and Statistical Thermodynamics of the T-3 Diced Lattice

- Zero Field
  - Green's Function Derived; Propagation and Spectrum
  - All Thermodynamic Functions Evaluated
- Magnetic Field—Landau Quantization
  - Green's Function and Spectrum Evaluated
  - Thermodynamic Functions Evaluated
  - Degenerate Statistical Regime.

8.3 "Diced" Lattice: Retarded Green's Fn. G^{ret} at Zero Field

- Pseudospin 1
- Hamiltonian: $\vec{K}$-Momentum/Frequency Rep. ($\alpha = \hbar v/\sqrt{2}$; v is characteristic lattice speed):

$$H = \alpha \begin{bmatrix} 0 & K_- & 0 \\ K_+ & 0 & K_- \\ 0 & K_+ & 0 \end{bmatrix} \text{ where, } K_\pm = K_x \pm i K_y,$$

- Green's Fn: $(I\omega - H)G^{ret} = I$ ($\omega \to \omega + i0^+$ for retardation)

$$G^{\text{ret}}\left(\vec{K}, \omega\right) = \begin{bmatrix} G_{11} & G_{12} & G_{13} \\ G_{12}^* & G_{22} & G_{23} \\ G_{13}^* & G_{23}^* & G_{33} \end{bmatrix}; \text{ Define } D \equiv \left(\omega^2 - 2\alpha^2 K^2\right);$$

$$\begin{aligned} G_{11} = G_{33} &= (\omega D)^{-1}\left[\omega^2 - \alpha^2 K^2\right]; & G_{22} &= \omega D^{-1} \\ G_{12} &= \alpha K_- D^{-1}; & G_{12}^* = G_{21} &= \alpha K_+ D^{-1} \\ G_{23} &= \alpha K_- D^{-1}; & G_{23}^* = G_{32} &= \alpha K_+ D^{-1} \\ G_{13} &= \alpha^2 K_-^2 (\omega D)^{-1}; & G_{13}^* = G_{31} &= \alpha^2 K_+^2 (\omega D)^{-1} \end{aligned}$$

- Energy Spectrum: $\omega = \pm\sqrt{2}\,\alpha\, K$.

8.4 Statistical Thermodynamic Functions: Diced Lattice

- Ordinary Partition Fn. $\hat{Z}$: Cutoff K_m at linear K-limit of validity of Dirac linear $\vec{K}$-approximation $\left(\beta = \frac{1}{\kappa_B T'}\right)$

$$\hat{Z}(\beta) = \text{trace}\left(e^{-\beta H}\right) = \int d^2x \text{Tr}\left(\hat{i} G^{\text{ret}}_{T>0}(\vec{x}, \vec{x}; T \to -i\beta)\right),$$
$$= (\text{area}) i \int \frac{d^2K}{(2\pi)^2} \left[\int \frac{d\omega}{2\pi} e^{-i\omega T} \text{Tr} G^{\text{ret}}\left(\vec{K}, \omega\right)\right]_{T \to i\beta},$$

$$\text{Tr}\, G^{\text{ret}}\left(\vec{K}, \omega\right) = \frac{1}{\omega} + \sum_{\pm} \frac{1}{\omega \pm \sqrt{2\alpha^2 K^2}};$$

$$\hat{Z}(\beta) = \frac{1}{2\pi} \int_0^{K_m} dK \cdot K \left(1 + \sum_{\pm} \exp\left(\pm\sqrt{2}\alpha\beta K\right)\right),$$

- $\hat{Z}(\beta) = \dfrac{K_m^2}{4\pi} - \dfrac{\cosh\left(\sqrt{2}\alpha\beta K_m\right)}{2\pi\alpha^2\beta^2} + \dfrac{K_m \sinh\left(\sqrt{2}\alpha\beta K_m\right)}{\sqrt{2}\pi\alpha\beta} + \dfrac{1}{2\pi\alpha^2\beta^2}.$

8.5 Grand Potential Ω

- Sondheimer-Wilson Formulation:

$$\Omega = \mathrm{F} - \mu N = \int_c \frac{ds}{2\pi i} \frac{\hat{Z}(s)}{s^2} \int_0^\infty dE e^{Es} \frac{\partial f_0(E)}{\partial E},$$

where F is Helmholtz Free Energy, N number, μ chemical potential, i.e. "c" denotes the inverse Laplace transform integration contour. $f_0(E)$ is the Fermi–Dirac distribution and $\beta = 1/\kappa_B T'$ is inverse thermal energy.

$$\frac{\partial f_0(E)}{\partial E} = -\frac{\beta}{4} \text{sech}^2\left(\frac{[E - \mu]\beta}{2}\right),$$

and introduce the variable $z = [E - \mu]\beta/2$, so that

$$\int_0^\infty dE\, e^{Es} \frac{\partial f_0(E)}{\partial E} = -\frac{1}{2} e^{s\mu} \int_{-\mu\beta/2}^\infty dz\, e^{2sz/\beta} \text{sech}^2(z).$$

- Degenerate Regime: In the degenerate regime $\mu\beta \to \infty$, so the lower limit of the z-integral may be taken as $-\mu\beta/2 \to -\infty$, with the result

$$\int_0^\infty dE\, e^{Es} \frac{\partial f_0(E)}{\partial E} = -\frac{\pi}{\beta} \frac{s e^{s\mu}}{\sin\left(\frac{\pi s}{\beta}\right)}.$$

8.6 Degenerate Regime Continued: Ω Calculation

Putting $s' = s/\beta$,

$$\Omega = \mathrm{F} - \mu N = -\frac{\pi}{\beta} \int_z \frac{ds'}{2\pi i} \frac{e^{s'\beta\mu} \hat{Z}(\beta s')}{s' \sin(\pi s')},$$

and using 8.4 $\left(v \equiv \sqrt{2}\alpha;\, \hbar \to 1\right)$

$$\hat{Z}(\beta s') = \frac{K_m^2}{4\pi} + \sum_{\pm} \int_0^{K_m} \frac{dK}{2\pi} K e^{\pm v\beta K s'}.$$

In the s'-integral of Ω, we exponentiate the integrand denominator factor $1/s' = \beta \int_0^\infty dx e^{-s'\beta x}$, so that a particular term with energy E_γ contributes as

$$\int_c \frac{ds'}{2\pi i} \frac{e^{s'\beta E_\gamma}}{s' \sin(\pi s')} = \beta \int_0^\infty dx \int_c \frac{dz}{2\pi i} \frac{e^{z\beta[E_\gamma/\pi - x]}}{\sin(z)}.$$

8.7 Contour Integration for Ω

Since the contour of z-integration along c is a straight line from $z = -i\infty + 0^+$ to $+i\infty + 0^+$, we consider closing the contour with a parallel line (c') from $i\infty - \pi^+$ to $-i\infty - \pi^+$ on which $dz_{c'} = -dz_c$ and $\sin(z_c') = -\sin(z_c)$. Moreover, the closed contour integrand $\oint = \int_c + \int_{c'}$ has the residue "1" at $z = 0$, so that

$$\oint \frac{dz}{2\pi i} \cdots = \int_c \frac{dz_c}{2\pi i} \cdots + \int_{c'} \frac{dz_{c'}}{2\pi i} \cdots = \left(1 + e^{-\pi\beta[E_\gamma/\pi - x]}\right) \times \int_c \frac{dz_c}{2\pi i} \frac{e^{z\beta[E_\gamma/\pi - x]}}{\sin(z)} = 1.$$

Consequently, the x-integration of Ω is given by

$$\int_0^\infty dx \frac{1}{1 + e^{-\pi\beta[E_\gamma/\pi - x]}} = \frac{1}{\pi\beta} \ln\left(1 + e^{\beta E_\gamma}\right).$$

8.8 Ω In the Degenerate Regime

- $\Omega = \dfrac{-K_m^2}{4\pi\beta} \ln\left(1 + e^{\beta\mu}\right) - \dfrac{1}{2\pi\beta} \sum_\pm I_\pm,$

  where $I_\pm = \int_0^{K_m} dK\, K \ln\left(1 + e^{\beta[\mu \pm vK]}\right)$.
  Since $K_m \gg \mu/v \gg k_B T/v$ in the degenerate regime, $I_+ \equiv \frac{\beta\mu K_m^2}{2} + \frac{\beta v K_m^3}{3}$, and I_- is given by a parts-integration as

$$I_- = \frac{K_m^2}{2} \ln\left(1 + e^{\beta(\mu - vK_m)}\right) + \frac{\beta v}{2} \int_0^{K_m} dK\, K^2 f_0(vK).$$

- Grand Potential in the Degenerate Regime

$$\Omega = -\frac{K_{\rm m}^2}{4\pi\beta} [\ln(1 + e^{\beta[\mu - vK_m]}) + \ln(1 + e^{\beta\mu})] - \frac{v}{12\pi} K_m^3 f_0(vK_m) - \frac{vK_m^3}{6\pi} - \frac{\mu K_m^2}{4\pi} - \frac{\mu^3}{6\pi v^2} - \frac{\pi\mu\left(\kappa_B T'\right)^2}{12v^2}.$$

- Neglecting exponentially small terms, Ω is given by

$$\Omega \equiv -\frac{vK_m^3}{6\pi} - \frac{\mu K_m^2}{2\pi} - \frac{\mu^3}{6\pi v^2} - \frac{\pi\mu\left(\kappa_B T'\right)^2}{12v^2}.$$

8.9 Entropy and Specific Heat: Degenerate Regime

The entropy may be determined using the thermodynamic relation

$$dF = -PdV - SdT' + \mu dN$$

by variation holding both area (volume) and number fixed, with the result

$$S = -\left(\frac{\partial F}{\partial T'}\right)_{N,V,\mu} = -\left(\frac{\partial(F - \mu n)}{\partial T'}\right)_{N,V,\mu} = -\left(\frac{\partial \Omega}{\partial T'}\right)_{N,V,\mu}.$$

- Entropy in Degenerate Regime:

$$S_{Deg} = \frac{\pi \mu \kappa_B^2 T'}{6v^2}.$$

- Specific Heat at Constant Volume in Degenerate Regime:

$$C_v = T'\left(\frac{\partial S}{\partial T'}\right)_V = \frac{\pi \mu \kappa_B^2 T'}{6v^2}.$$

8.10 T-3 "Diced" Lattice in Quantizing Magnetic Field B

- Hamiltonian ($\alpha = \hbar v/\sqrt{2}$; v is characteristic lattice speed; $\hbar \to 1$)

$$H = \begin{bmatrix} 0 & \alpha\pi_- & 0 \\ \alpha\pi_+ & 0 & \alpha\pi_- \\ 0 & \alpha\pi_+ & 0 \end{bmatrix},$$

  where $\pi_x = \frac{1}{i}\frac{\partial}{\partial x} + \frac{eB}{2}y$; $\pi_y = \frac{1}{i}\frac{\partial}{\partial y} - \frac{eB}{2}x$ and $\pi_+ = \pi_x + i\pi_y$; $\pi_- = \pi_x - i\pi_y$.

- Green's Function: (I is 3×3 unit matrix); matrix elements of G are $G_{ij}(i, j = 1, 2, 3)$

$$(\omega I - H)G(\vec{r}, \vec{r}'; \omega) = I\delta^{(2)}(\vec{r} - \vec{r}'), \left[\omega \to \omega + i0^+ \text{ for the retarded function}\right],$$

  where $\vec{r} = (x, y)$; $\vec{r}' = (x', y')$; and $\vec{R} = \vec{r} - \vec{r}'$, $X = x - x'$, $Y = y - y'$;

- Gauge Considerations: Gauge and Translational Invariance

$$G(\vec{r}, \vec{r}'; \omega) = C(\vec{r}, \vec{r}')G'(\vec{R}; \omega),$$

where $C\left(\vec{r},\vec{r}'\right) = \exp\left(\frac{ie}{2}\vec{r}\cdot\vec{B}\times\vec{r}'\right)$. The resulting translationally invariant equation for $G'(R,\omega)$ is given by

$$\left(\omega I - H(\vec{R},\omega)\right)G'(\vec{R},\omega) = I\delta^{(2)}(\vec{R}).$$

8.11 Green's Function Equations (9 Elements G_{ij})

- $G_{ij}\left(\vec{R},\omega\right) = G^*_{ji}\left(-\vec{R},\omega\right)$.
- Symmetry $G_{33}\left(\vec{R},\omega\right) = G_{11}\left(\vec{R},\omega\right)$.
- Green's Function (Elements) Equations:

$$\begin{aligned} \omega G'_{11} - \alpha\pi_- G'_{21} &= \delta^2(\vec{R}), \\ -\alpha\pi_+ G'_{11} + \omega G'_{21} - \alpha\pi_- G'_{31} &= 0, \\ -\alpha\pi_+ G'_{21} + \omega G'_{31} &= 0, \end{aligned}$$

$$\begin{aligned} \omega G'_{12} - \alpha\pi_- G'_{22} &= 0, \\ -\alpha\pi_+ G'_{12} + \omega G'_{22} - \alpha\pi_- G'_{32} &= \delta^2(\vec{R}), \\ -\alpha\pi_+ G'_{22} + \omega G'_{32} &= 0, \end{aligned}$$

$$\begin{aligned} \omega G'_{13} - \alpha\pi_- G'_{23} &= 0, \\ -\alpha\pi_+ G'_{13} + \omega G'_{23} - \alpha\pi_- G'_{33} &= 0, \\ -\alpha\pi_+ G'_{23} + \omega G'_{33} &= \delta^2(\vec{R}). \end{aligned}$$

- Detailed Equations:

$$\left\{\omega + 2\frac{\alpha^2}{\omega}\left[\frac{\partial^2}{\partial X^2} + \frac{\partial^2}{\partial Y^2} - \left(\frac{eB}{2}\right)^2 (X^2+Y^2)\right]\right\} G'_{22}\left(\vec{R};\omega\right) = \delta^2\left(\vec{R}\right),$$

$$\left\{\omega + 2\frac{\alpha^2}{\omega}\left[\frac{\partial^2}{\partial X^2} + \frac{\partial^2}{\partial Y^2} - \left(\frac{eB}{2}\right)^2 (X^2+Y^2)\right]\right\} G'_{21}\left(\vec{R};\omega\right) = \frac{\alpha}{\omega}\pi_+\delta^2\left(\vec{R}\right);\ G'_{12} = G'^*_{21};$$

$$\left\{\omega + 2\frac{\alpha^2}{\omega}\left[\frac{\partial^2}{\partial X^2} + \frac{\partial^2}{\partial Y^2} - \left(\frac{eB}{2}\right)^2 (X^2+Y^2)\right]\right\} G'_{23}\left(\vec{R};\omega\right) = \frac{\alpha}{\omega}\pi_-\delta^2\left(\vec{R}\right);\ G'_{32} = G'^*_{23};$$

$$G'_{11}(\vec{R};\omega) = G'_{33}(R;\omega) = \frac{1}{2}G'_{22}(\vec{R},\omega) + \frac{1}{2\omega}\delta^2(\vec{R});$$

$$G'_{13} = \frac{\alpha}{\omega}\pi_- G'_{23};\ G'_{31} = \frac{\alpha}{\omega}\pi_+ G'_{21} = G'^*_{13}.$$

8.12 $G'_{ij}(\vec{R}, \omega)$ Solutions; Energy Spectrum

- Landau Levels: $\epsilon_n \equiv \omega_n = \pm\sqrt{2(2n+1)\alpha^2 eB}$ $(n = 0 \ldots \infty)$.

$$G'_{22}(\vec{R}; \omega) = -\frac{eB}{4\pi}\int_0^\infty d\tau \frac{e^{i\omega\tau}}{\sin\left(\frac{2eB\alpha^2\tau}{\omega}\right)} \exp\left(\frac{ieB(X^2+Y^2)}{4\tan\left(\frac{2eB\alpha^2\tau}{\omega}\right)}\right);$$

$$G'_{22}(R; \omega) = \frac{eB}{2\pi}\omega\epsilon^{-eB\frac{R^2}{4}} \sum_{n=0}^\infty L_n\left(\frac{eBR^2}{2}\right)\frac{1}{\omega^2 - 2(2n+1)\alpha^2 eB}.$$

$$G'_{23}(\vec{R}; \omega) = \frac{\alpha(eB)^2}{2\pi}e^{-eB\frac{R^2}{4}}[iX+Y]\sum_{n=1}^\infty \frac{L^1_{n-1}\left(eB\frac{R^2}{2}\right)}{\omega^2 - 2(2n+1)\alpha^2 eB};$$

$$G'_{32}(\vec{R}, \omega) = G'^*_{23}(-\vec{R}, \omega).$$

$$G'_{31}(\vec{R}; \omega) = \frac{\alpha^2(eB)^3}{2\pi\omega}e^{-eB\frac{R^2}{4}}[X+iY]^2\sum_{n=1}^\infty \frac{L^2_{n-2}\left(eB\frac{R^2}{2}\right)}{\omega^2 - 2(2n+1)\alpha^2\epsilon B};$$

$$G'_{13}(\vec{R}; \omega) = G'^*_{31}(-\vec{R}; \omega).$$

$$G'_{11}(\vec{R}; \omega) = G'_{33}(\vec{R}; \omega) = \frac{1}{2}G'_{22}(\vec{R}; \omega) + \frac{1}{2\omega}\delta^2(\vec{R});$$

$$G'_{13}(\vec{R}; \omega) = \frac{\alpha}{\omega}\pi_- G'_{23}(\vec{R}; \omega); \quad G'_{31}(\vec{R}; \omega) = G'^*_{13}(-\vec{R}; \omega).$$

8.13 Grand Potential Ω for Diced Lattice In Magnetic Field

- Sondheimer-Wilson Formulation.
  $\hat{Z}(\beta)$ is the ordinary Partition Function

$$\beta = 1/\kappa_B T'; \ \kappa_B = \text{Boltzmann Constant}$$

$$T' = \text{Kelvin Temperature}$$

$$\mu = \text{Chemical potential}$$

$$\Omega = F - \mu N = -\frac{\beta}{4}\int_c \frac{ds}{2\pi i}\frac{\hat{Z}(s)}{s^2}\int_0^\infty dE e^{Es}\text{sech}^2\left(\frac{[E-\mu]\beta}{2}\right).$$

- Partition Function in terms of Retarded Green's Function (per unit area)

$$\hat{Z}(\beta) = \int d^2x \, \mathrm{Tr}\big(iG^{ret}_{T>0}(x, x; T \to -i\beta)\big) = \frac{eB}{2\pi} \sum_{n=0}^{n_{\max}} \sum_{\pm} e^{\pm\epsilon_n\beta} = \frac{eB}{\pi} \sum_{n=0}^{n_{\max}} \cosh(\epsilon_n\beta),$$

The n-series over Landau levels (on right above) terminates at $n_{\max}$, which labels the highest Landau eigen-energy for which the linear momentum approximation of the Dirac Diced Hamiltonian is valid (due to band bending).

8.14 Ω for Landau Quantized Diced Lattice: Degenerate Regime: $\mu\beta \to \infty$

$$\Omega = F - \mu N = -\frac{\pi}{\beta} \int\limits_{c} \frac{ds}{2\pi i} \frac{e^{s\mu} \hat{Z}(s)}{s \sin\left(\frac{\pi s}{\beta}\right)} = -\frac{e\beta}{2} \sum_{n}^{n_{\max}} \sum_{\pm} \int_0^{\infty} dx \int\limits_{c} \frac{dz}{2\pi i} \frac{e^{z\beta[(\mu\pm\epsilon_n)/\pi - x]}}{\sin(z)},$$

or, on a per-unit-area basis ($n = N/area$; $z = \pi s/\beta$)

$$\Omega = F - \mu n = -\frac{eB}{2\pi} \sum_{n=0}^{n_{\max}} \sum_{\pm} \frac{1}{\beta} \ln\big(1 + e^{|\beta(\mu\pm\epsilon_n)|}\big).$$

- For low temperature such that $\beta(\mu \pm \epsilon_n) \gg 1$, we have

$$\Omega = -\frac{eB}{2\pi} \sum_{n=0}^{n_{\max}} \sum_{\pm} \left\{ \eta_+(\mu \pm \epsilon_n)[\mu \pm \epsilon_n] + \frac{1}{\beta} e^{-\beta|\mu\pm\epsilon_n|} \right\},$$

where $\eta_+(x) = 1$ for $x > 0$; 0 for $x < 0$.

8.15 Magnetic Moment M of Diced Lattice: Degenerate Regime $(T' \to 0)$

- Degenerate: $(\beta(\mu \pm \epsilon_n) \gg 1; \mu > 0)$.
- Magnetic Moment/Area:

$$M = -\left(\frac{\partial F}{\partial B}\right)_{T,V,N} = -\left(\frac{\partial \Omega}{\partial B}\right)_{T,V,N},$$

$$M = \frac{e}{2\pi}\sum_{n=0}^{n_{max}}\sum_{\pm}\left\{\eta_+(\mu \pm \epsilon_n)\left(\left[\mu \pm \frac{\epsilon_n}{2}\right] + \frac{\epsilon_n}{2}\right) + \frac{e}{2\pi}\sum_{n=0}^{n_{max}}\sum_{\pm}\frac{1}{\beta}e^{-\beta|\mu\pm\epsilon_n|}\left(1 - \frac{\beta\epsilon_n}{2}\right)\right\}.$$

- Density n ($f_0(x)$ is the Fermi–Dirac distribution):

$$n = \frac{eB}{2\pi}\sum_{n=0}^{n_{\max}}\sum_{\pm} f_0(\mp\epsilon_n).$$

- Magnetic Moment: Degenerate Regime ($T' = 0$):

$$M_{T'=0} = \frac{e}{2\pi}\sum_{n=0}^{n_{\max}}\sum_{\pm}\eta_+(\mu \pm \epsilon_n)\left(\left[\mu \pm \frac{\epsilon_n}{2}\right] + \frac{\epsilon_n}{2}\right) > 0.$$

This shows that the T-3 Diced lattice is paramagnetic, $M_{T'=0} > 0$, in the degenerate regime. As usual, the singular function $\eta_{+'}(\mu \pm \epsilon_n)$ signals the presence of de Haas-van Alphen oscillatory phenomenology as variation of the magnetic field gives rise to successive vanishings of $\mu - \epsilon_n \to 0$, as successive Landau levels ϵ_n cross the Fermi level μ.

8.16 Magnetic Moment M of Diced Lattice: Temperature Corrections ΔM in the Approach to $T' = 0$

$$\Delta M = \frac{e}{2\pi}\sum\nolimits_{n=0}^{n_{\max}}\sum\nolimits_{\pm}\frac{1}{\beta}e^{-\beta|\mu\pm\epsilon_n|}\left(1 - \frac{\beta\epsilon_n}{2}\right).$$

This is the temperature correction to ΔM to the total magnetization in the degenerate statistical regime in the approach to zero temperature. The total magnetic moment (per unit area) is given by

$$M \to M + \Delta M.$$

Note that in the zero temperature limit the (weak) temperature-dependent part, ΔM, is devoid of de Haas-van Alphen oscillations and vanishes exponentially as $\kappa_B T' \exp\left(-\frac{|\mu\pm\epsilon_n|}{\kappa_B T'}\right)$ as $T' \to 0$.

8.17 Entropy and Specific Heat of Landau-Quantized Diced Lattice

(All per unit area.)

- Entropy S:

$$S = -\left(\frac{\partial F}{\partial T'}\right)_{N,V,\mu} = -\left(\frac{\partial \Omega}{\partial T'}\right)_{N,V,\mu} = \kappa_B \beta^2 \left(\frac{\partial \Omega}{\partial \beta}\right)_{N,V,\mu},$$

- Degenerate Regime:

$$S = \kappa_B \frac{eB}{2\pi} \sum_{n=0}^{n_{\max}} \sum_{\pm} e^{-\beta|\mu \pm \epsilon_n|}(1 + \beta|\mu \pm \epsilon_n|).$$

- Specific Heat at Constant Volume C_v:

$$C_v = T' \frac{\partial S}{\partial T'} = -\beta \frac{\partial S}{\partial \beta} = \kappa_B \beta^2 \frac{eB}{2\pi} \sum_{n=0}^{n_{\max}} \sum_{\pm} [\mu \pm \epsilon_n]^2 e^{-\beta|\mu \pm \epsilon_n|}.$$

Note absence of dHvA oscillatory behavior in both the entropy, S, and specific heat, C_v.

8.18 Summary: T-3 Diced Lattice—Zero Field Statistical Thermodynamic Degenerate Regime

(All on a per-unit-area basis.)

- Grand Potential:

$$\Omega = -\frac{v K_m^3}{6\pi} - \frac{\mu K_m^2}{2\pi} - \frac{\mu^3}{6\pi v^2} - \frac{\pi \mu (\kappa_B T')^2}{12 v^2}.$$

- Entropy $S_{Deg} = \frac{\pi \mu \kappa_b^2 T'}{6v^2}$.
- Specific Heat at Constant Volume $C_v = \frac{\pi \mu \kappa_B^2 T'}{6v^2}$.

8.19 Summary: T-3 Diced Lattice—Magnetic Field Statistical Thermodynamics (A) Degenerate Regime

- Grand Potential:

$$\Omega = -\frac{\mathrm{eB}}{2\pi}\sum_{n=0}^{n_{\max}}\sum_{\pm}\frac{1}{\beta}\ln\left(1+e^{|\beta(\mu\pm\epsilon_n)|}\right);$$

$$\Omega_{\text{Low Temp}} \to -\frac{\mathrm{eB}}{2\pi}\sum_{n=0}^{n_{\max}}\sum_{\pm}\left\{\eta_+(\mu\pm\epsilon_n)[\mu\pm\epsilon_n]+\frac{1}{\beta}e^{-\beta|\mu\pm\epsilon_n|}\right\}.$$

- Magnetic Moment:

$$M = \frac{e}{2\pi}\left(\sum_{n=0}^{n_{\max}}\sum_{\pm}\eta_+(\mu\pm\epsilon_n)\left(\left[\mu\pm\frac{\epsilon_n}{2}\right]+\frac{\epsilon_n}{2}\right)+\frac{1}{\beta}e^{-\beta|\mu\pm\epsilon_n|}\left(1-\frac{\beta\epsilon_n}{2}\right)\right).$$

Note dHvA oscillatory behavior in the zero temperature limit of M. The finite temperature correction is devoid of dHvA oscillations.

8.20 Summary: T-3 Diced Lattice—Magnetic Field Statistical Thermodynamics (B) Degenerate Regime

- Entropy:

$$S = \kappa_B\frac{eB}{\pi}\sum_{n=0}^{n_{\max}}\sum_{\pm}e^{-\beta|\mu\pm\epsilon_n|}(1+\beta|\mu\pm\epsilon_n|).$$

- Specific Heat at Constant Volume:

$$C_v = \kappa_B\beta^2\frac{eB}{2\pi}\sum_{n=0}^{n_{\max}}\sum_{\pm}|\mu\pm\epsilon_n|^2e^{-\beta|\mu\pm\epsilon_n|}.$$

Note absence of dHvA oscillatory behavior in both entropy and specific heat.

References

1. N.J.M. Horing, M.L. Glasser, J.D. Mancini, Green's function analysis of "diced lattice" statistical thermodynamics, Phys. Scripta **96**(1), 015806 (2021)
2. M.L. Glasser, N.J.M. Horing, Exact temperature and density dependencies of the statistical thermodynamic functions of the pseudospin-1 diced lattice carriers. Following chapter in this volume
3. N.J.M. Horing, Landau quantized dynamics and spectrum of the diced lattice, J. Phys. Condens. Matter 33, 015302 (2020)
4. N.J.M. Horing, M.L. Glasser, Statistical thermodynamics and magnetic moment of the Landau quantized T-3 diced lattice in the degenerate regime, submitted for publication
5. Chapter 14 of this book

Chapter 9
Exact Temperature and Density Dependencies of the Statistical Thermodynamic Functions of the Pseudospin-1 Diced Lattice Carriers

M. L. Glasser and Norman J. M. Horing

Abstract In this work we analyze the exact temperature and density dependencies of the thermodynamic properties of the two dimensional pseudospin-1 T-3 Diced Lattice. Starting from the Gibbs canonical partition function, we determine the exact grand canonical ensemble thermodynamic potential, entropy and specific heat at arbitrary temperature and density, and confirm results in the degenerate and non-degenerate statistical regimes.

PAC 73.20-r · 73.43-r · 05.70 Np · 05.70 Ce

9.1 Introduction

This work addresses the thermodynamic properties of the Fermion pseudospin-1 system whose Hamiltonian is

$$\mathcal{H} = (\hbar v/\sqrt{2}) \begin{bmatrix} 0 & K_- & 0 \\ K_+ & 0 & K_- \\ 0 & K_+ & 0 \end{bmatrix}, \tag{9.1}$$

which was introduced by Bercioux, et al. [1] in 2009, where v is an effective lattice speed, $K_\pm = K_x \pm i K_y$ and $\vec{K}$ is the 2D- crystal momentum. This system has attracted attention recently, particularly by Malcom and Nicol [2] who studied its electronic polarizability and related properties. This system falls into the class of Dirac materials, which includes Group VI Dichalcogenides [3], Topological Insulators [4], Silicene [5] and, most notably, Graphene [6–12].

M. L. Glasser
Department of Physics, Clarkson University, Potsdam, NY 13699, USA

N. J. M. Horing (✉)
Department of Physics, Stevens Institute of Technology, Hoboken, NJ 07030, USA
e-mail: nhoring@stevens.edu

H. Ünlü and N. J. M. Horing (eds.), *Progress in Nanoscale and Low-Dimensional Materials and Devices*, Topics in Applied Physics 144,
https://doi.org/10.1007/978-3-030-93460-6_9

In Sect. 9.2, we derive the exact temperature/density dependencies of the Grand Potential, Entropy, Chemical Potential and Specific Heat, which are all expressed in closed form, and in Sect. 9.3 we confirm the behavior of these quantities in the degenerate and non-degenerate statistical regimes which may be relevant to possible experimental conditions. The results are summarized and discussed in Sect. 5.

9.2 Calculations

In the body of this work we set $\hbar = 1$ and introduce the dimensionless momentum $\vec{k} = \vec{K}/K_m$, where K_m is a cut-off introduced to restrict the band structure to the relativistic region where the dispersion is linear in momentum. We also introduce the characteristic energy $a = \hbar v K_m$ corresponding to the highest energy in the linear portion of the spectrum. The partition function, discussed in Appendix A, is

$$\hat{Z}(s) = \frac{K_m^2}{\pi}[\frac{1}{4} + \frac{1}{a^2s^2} - \frac{\cosh(as)}{a^2s^2} + \frac{\sinh(as)}{as}]. \tag{9.2}$$

To proceed, we note that the Wilson-Sondheimer formula [13, 14] giving the Grand Thermodynamic Potential (Appendix B) may be written as

$$\Omega = -\kappa_B T \sum_{\{\alpha\}} \ln[1 + e^{-\beta(E_\alpha - \mu}] = -\frac{\beta}{4}\int_0^\infty dt \frac{z(t)}{\cosh^2[\frac{\beta}{2}(t-\mu)]}, \tag{9.3}$$

(κ_B being Boltzmann's constant) where the $\alpha-$ sum is over all the energy levels, E_α, T is absolute temperature, $\beta = 1/\kappa_B T$, μ is the chemical potential and (Appendix C)

$$z(t) = \int_{c-i\infty}^{c+i\infty} \frac{ds}{2\pi i s^2} e^{st}\hat{Z}(-s) = \frac{K_m^2}{12\pi a^2}\left[t^3 + 6a^2t + 2a^3 - \left(t^3 - 3a^2t + 2a^3\right)\theta(t-a)\right]. \tag{9.4}$$

Here, $\theta(z)$ denotes the Heaviside unit step function and, in conformance with the prefactor K_m^2, Ω and subsequent extensive quantities refer to unit area. We point out that, apart from the system parameters K_m and a, Ω is a function of the state variables T (or β) and μ (or density n) since the volume (i.e. area) is fixed. Once the density is specified, as in dealing with a specific sample, μ is itself a function of T. Details of the derivation of (9.4) are provided in Appendix C. The function $z(t)$ is basically a cubic polynomial in t and behaves nearly linearly over its range. We find that the integral resulting from inserting it into (9.3) can be evaluated exactly, with the result (Appendix D)

$$\Omega = -\frac{K_m^2}{12\pi\beta}\{\beta(3\mu - a) + a\beta\tanh(\beta\mu/2) + 9\ln(1 + e^{\beta\mu})$$
$$-\frac{6}{(a\beta)^2}[a\beta Li_2(-e^{\beta(a-\mu)}) - Li_3(-e^{\beta(a-\mu)}) + Li_3(-e^{-\beta\mu})]\}, \tag{9.5}$$

where

$$Li_n(z) = \sum_{k=1}^{\infty}\frac{z^k}{k^n} \tag{9.6}$$

is the polylogarithm [15] (Appendix E).

The chemical potential is the Lagrange coefficient associated with the constraint that the system contain N particles and is closely related to the energy necessary to add or remove one particle. Since in most situations the areal density n is fixed, it is useful to know how μ is related to $n = -\partial\Omega/\partial\mu$. From (9.5) we obtain

$$n = \frac{K_m^2}{12\pi}\left\{12 - \frac{9}{1 + e^{\beta\mu}} + \frac{a\beta}{\cosh(\beta\mu) + 1} - \frac{6}{a\beta}\ln[1 + e^{\beta(a-\mu)}]\right.$$
$$\left. - \frac{6}{a^2\beta^2}\left[Li_2\left(-e^{\beta(a-\mu)}\right) - Li_2\left(-e^{-\beta\mu}\right)\right]\right\}. \tag{9.7}$$

Next, the entropy $S = \kappa_b\beta^2\partial\Omega/\partial\beta$, is given by

$$S = \frac{\kappa_B K_m^2}{24\pi}\left[\frac{1}{a^2\beta^2}(36Li_3(-e^{\beta(a-\mu)})\right.$$
$$+ 12\beta(\mu - 3a)Li_2(-e^{\beta(a-\mu)}) - 12\beta\mu Li_2(-e^{-\beta\mu}) - 36Li_3(-e^{-\beta\mu}) - a^3\beta^4\mu sech^2(\beta\mu/2))$$
$$\left. + 18ln(e^{\beta\mu} + 1) + 6(\frac{2\mu}{a} - 2)ln(e^{\beta(a-\mu)} + 1) - \frac{18\beta\mu}{1 + e^{-\beta\mu}}\right]. \tag{9.8}$$

9.3 Degenerate Limit

As $T \to 0$ the system Fermions condense into the lowest states up to the Fermi energy, which in the limit coincides with μ. In this case it is convenient to write (9.3) as

$$\Omega = \int_0^{\infty} dt f_0'(t)z(t) = -\frac{z(0)}{1 + e^{\beta\mu}} - \int_0^{\infty} dt\ f_0(t)z'(t), \tag{9.9}$$

where ($f_0(t)$ is the Fermi-Dirac statistical distribution function)

$$f_0(t) = \frac{1}{1 + e^{\beta(t-\mu)}}. \quad (9.10)$$

The last integral is (after the change of variable $\beta(t - \mu) \to t$)

$$\frac{1}{\beta}\int_{-\beta\mu}^{\infty} dt\, \frac{z'(\mu + t/\beta)}{e^t + 1} = \frac{1}{\beta}\left[\int_0^{\beta\mu} dt\, z'(\mu - t/\beta)\left(1 - \frac{1}{e^t + 1}\right)\right.$$
$$\left. + \int_0^{\infty} dt\, \frac{z'(\mu + t/\beta)}{e^t + 1}\right]. \quad (9.11)$$

In the degenerate limit, $\beta\mu \to \infty$, so

$$\Omega = -\frac{z(0)}{1 + e^{\beta\mu}} - z(0) + z(\mu) - \frac{1}{\beta}\int_0^{\infty} \frac{z'(\mu - t/\beta) - z'(\mu + t/\beta)}{e^t + 1} dt. \quad (9.12)$$

Now,

$$z'(\mu \pm t/\beta) = z'(\mu) \pm \frac{t}{\beta} z''(\mu) + \frac{t^2}{2\beta^2} z'''(\mu) + \cdots \quad (9.13)$$

so

$$\Omega_{deg} = \left(\frac{2 + e^{\beta\mu}}{1 - e^{\beta\mu}}\right) z(0) - z(\mu) + \frac{z''(\mu)}{2\beta^2}\int_0^{\infty} \frac{t}{1 + e^t} dt + \cdots. \quad (9.14)$$

Since we can ignore $e^{-\beta\mu}$ for any Fermi system in the degenerate regime,

$$\Omega_{deg} = Const. - z(\mu) - \frac{\pi^2}{6} z''(\mu)(k_b T)^2 + O((k_b T)^4). \quad (9.15)$$

Since the Fermi temperature is on the order of kiloKelvins, it is this regime which applies to most experimental situations. Also, since a is roughly the maximum energy within the linear Dirac regime, we can assume $\mu < a$, in which case one has from (9.4) and (9.15)

$$\Omega_{deg} = -\frac{K_m^2}{12\pi a^2}\left[\mu^3 + 6a^2\mu + 2a^3\right]. \quad (9.16)$$

More details are provided in reference [16].

9.4 Non-degenerate Limit

To investigate the high temperature, low density behavior, we rewrite (9.3) in the form

$$\Omega = \int_{c-i\infty}^{c+i\infty} \frac{ds}{2\pi i} \frac{\hat{Z}(-s)}{s^2} \int_0^\infty dt\, e^{st} f_0'(t). \tag{9.17}$$

In the nondegenerate case it is appropriate to write

$$f_0'(t) \approx -\beta e^{\beta\mu} e^{-\beta t}, \tag{9.18}$$

and, therefore, the right hand side of (9.17) is simply the inverse Laplace transform of a Laplace transform, yielding

$$\Omega_{nd} = -\frac{1}{\beta} e^{\beta\mu} \hat{Z}(\beta). \tag{9.19}$$

where $\hat{Z}(\beta)$ is given in (9.2). Consequently, Ω_{nd} and density n are related by

$$n = -\frac{\partial \Omega_{nd}}{\partial \mu} = e^{\beta\mu} \hat{Z}(\beta) = -\beta\Omega_{nd} \quad or \quad \Omega_{nd} = -\beta^{-1} n. \tag{9.20}$$

Note that this relationship is valid for any Fermi system in the nondegenerate statistical regime.

Further discussion is available in [16].

9.5 Discussion

The thermodynamic quantities obtained above are all expressed in terms of the chemical potential, μ, so it is important to determine how μ depends on the electron density n. Except in the degenerate regime, where this has been carried out in [16], one must solve (9.7) for μ, a formidable task. Therefore, we proceed numerically by specifying n, a and looking at null contours of the n-β-μ plots. In addition, in order to avoid the complication of an additional parameter, we introduce the scaled density $\nu = 12\pi n / K_m^2$ and solve (9.7)

$$\nu - \left\{ 12 - \frac{9}{1 + e^{\beta\mu}} + \frac{a\beta}{\cosh(\beta\mu) + 1} \right.$$
$$\left. - \frac{6}{a\beta} \ln[1 + e^{\beta(a-\mu)}] - \frac{6}{a^2\beta^2} [Li_2\left(-e^{\beta(a-\mu)}\right) - Li_2\left(-e^{-\beta\mu}\right) \right\} = 0 \tag{9.21}$$

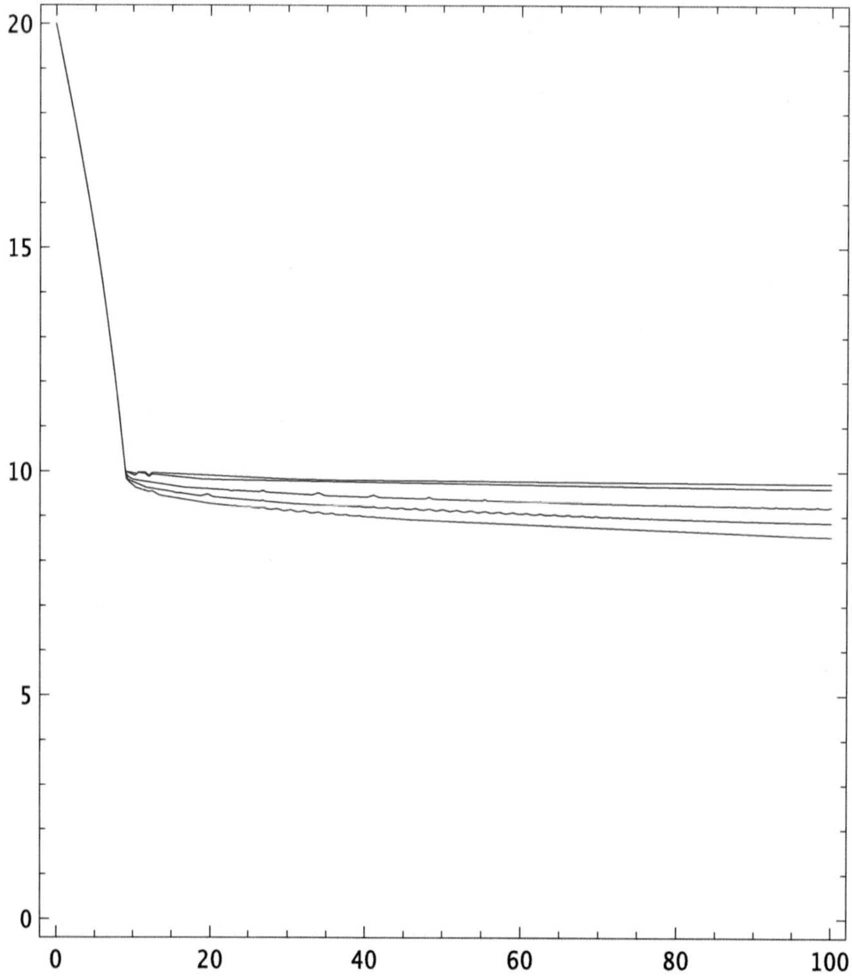

Fig. 9.1 μ versus ν for $a = 10$ at $T = 4$, 10, 100 and 300K

for μ vs ν. Because the behavior is not very sensitive to a, we consider only the two values $a = 10$ and $a = 100$ atomic units ($\hbar = 1$). A reasonable temperature range is $4K < T < 400K$ corresponding roughly to $3 \cdot 10^3 < \beta < 4 \cdot 10^5$. We assume $0 < \nu < 1000$. From the results shown in Figs. 9.1 and 9.2 it appears that the chemical potential is virtually independent of temperature over most of the range, but it rises steeply at low values of ν, indicating a density-induced sharp phase transition near $\nu = 10$, remaining nearly constant at $\mu = a$ at all higher densities.

Next, we examine the grand thermodynamic potential (9.5). Introducing the scaled version $\bar{\Omega} = -12\pi\Omega/K_m^2$, one has

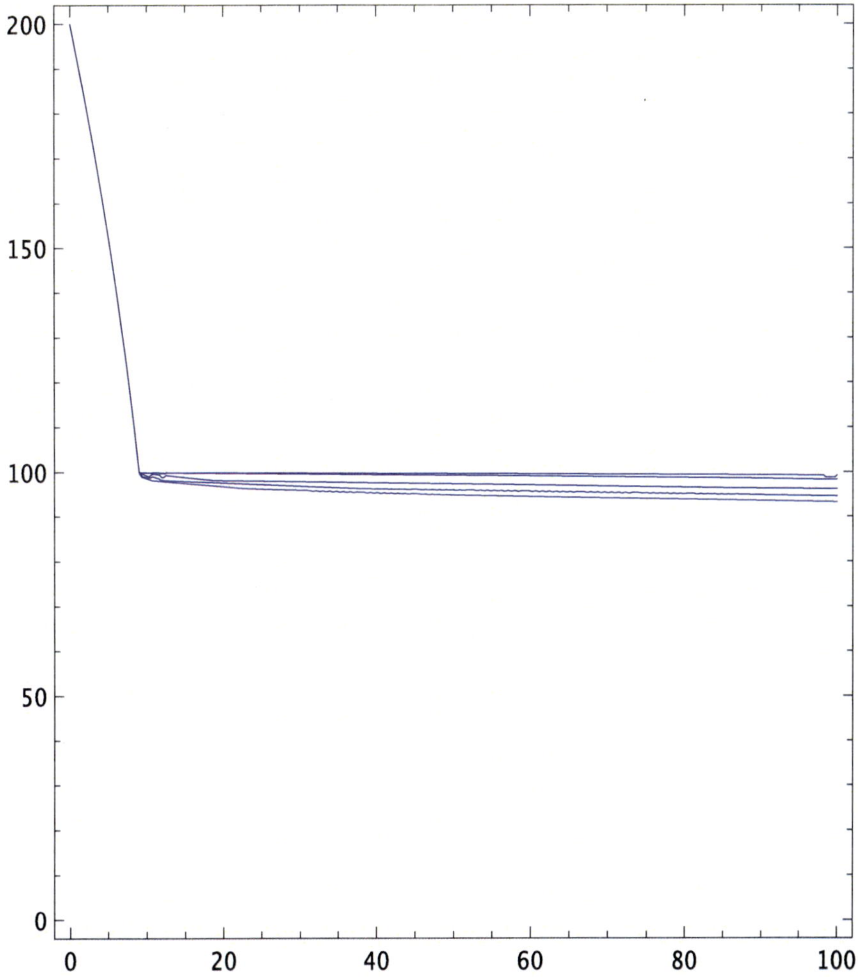

Fig. 9.2 μ versus ν for $a = 100$ at $T = 4$, 10, 100 and 400K

$$\bar{\Omega} = A(a, \beta, \mu) + B(a, \beta, \mu), \tag{9.22}$$

where

$$A(a, \beta, \mu) = 3\mu - a(1 - \tanh(\beta\mu/2)) + \frac{9}{\beta}\ln(1 + e^{\beta\mu}) \tag{9.23}$$

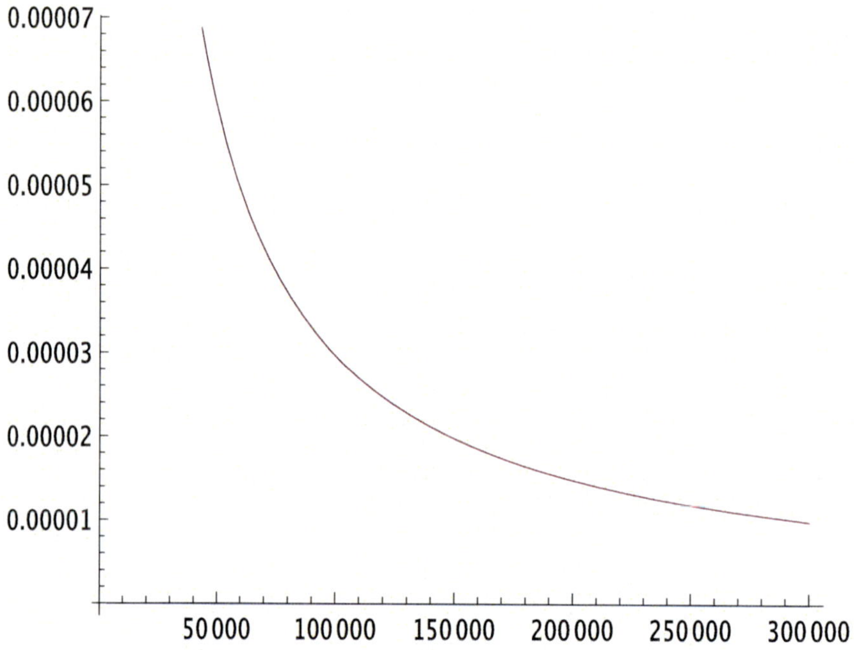

Fig. 9.3 $\bar{\Sigma}$ versus inverse temperature β for $400K > T > 4K$. Here $a = 10$, $\nu = 20$

$$B(a, \beta, \mu) = -\frac{6}{(a\beta)^2}[aLi_2(-e^{\beta(a-\mu)}) - \frac{1}{\beta}\{Li_3(-e^{\beta(a-\mu)}) - Li_3(-e^{\beta\mu})\}]. \tag{9.24}$$

Numerically, we find that A and B, and so $\bar{\Omega}$, are nearly temperature independent over the range $4K < T < 400K$ and that A is nearly independent of a over this range. Consequently, quantities related to temperature derivatives of the thermodynamic potential, such as the entropy and specific heat, will be small, as seen for the dimensionless entropy of Fig. 9.3. Of course, at higher temperatures where states no longer described by (9.1) are occupied, the situation will be different.

The scaled entropy defined as

$$\bar{\Sigma} \equiv \frac{12\pi S}{k_B K_m^2} \tag{9.25}$$

is sketched in Fig. 9.3.

The Specific Heat is given by

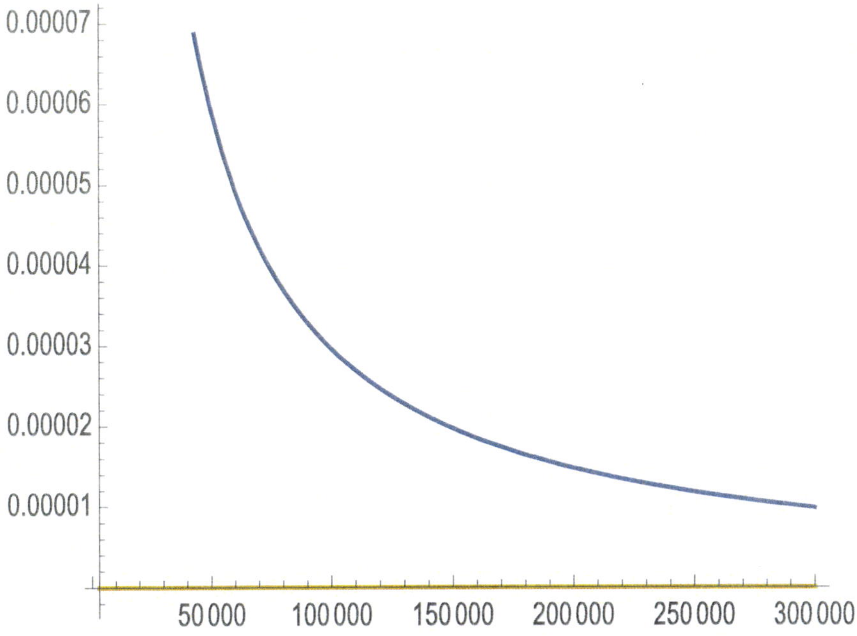

Fig. 9.4 C_V versus inverse temperature β for $400K > T > 4K$. Here $a = 10$, $v = 10$

$$C_V = -\frac{\kappa_B K_m^2}{12\pi}\beta\left\{\frac{1}{a^2\beta^3}\left[\beta^2\left(\frac{a^3\beta^2\mu(2-\beta\mu\tanh(\frac{\beta\mu}{2}))}{\cosh(\beta\mu)+1} - 6(3a^2-4a\mu+\mu^2)\log(e^{\beta(a-\mu)}+1)\right.\right.\right.$$
$$\left. + 6\mu^2\log(e^{-\beta\mu}+1)\right) + 12\beta(2\mu-3a)Li_2(-e^{\beta(a-\mu)}) + 36Li_3(-e^{\beta(a-\mu)}) - 24\beta\mu Li_2(-e^{-\beta\mu})$$
$$\left.\left. - 36Li_3(-e^{-\beta\mu})\right] + \frac{6e^{a\beta}(a-\mu)^2}{a(e^{a\beta}+e^{\beta\mu})} + \frac{9\beta\mu^2}{e^{\beta\mu}+1} - \frac{9\beta\mu^2}{(e^{\beta\mu}+1)^2}\right\} \quad (9.26)$$

and is shown in Fig. 9.4 for $a = 10$, $100K > T > 4K$.

Acknowledgement It is a pleasure to acknowledge the assistance of Prof. Dr. J. D. Mancini in checking equations in this paper.

Appendix

A: Partition Function

Let $\vec{K} = K_m\vec{k}$ and $a = \hbar v K_m$:

$$H = \frac{a}{\sqrt{2}} \begin{bmatrix} 0 & k_- & 0 \\ k_+ & 0 & k_- \\ 0 & k_+ & 0 \end{bmatrix} = \mathcal{U}^{-1} \begin{pmatrix} 0 & 0 & 0 \\ 0 & ak & 0 \\ 0 & 0 & -ak \end{pmatrix} \mathcal{U}, \tag{9.27}$$

where the unitary matrix $\mathcal{U}$ consists of the normalized eigenvectors of H, with eigenvalues $\omega = 0, \ \pm ak$. The associated partition function is

$$\hat{Z}(s) = \sum_{\{\omega\}} e^{-s\omega} = \frac{K_m^2}{2\pi} \int_0^1 dk\, k[1 + 2\cosh(ask)]$$

$$= \frac{K_m^2}{\pi} \left[\frac{1}{4} + \frac{1}{a^2 s^2} (1 + as \sinh as - \cosh as) \right], \tag{9.28}$$

since $\int dx\, x \cosh(ax) = [(ax)\sinh(ax) - \cosh(ax)]/a^2$.

B: Derivation of (9.3)

Since any function is the inverse Laplace transform of its Laplace transform,

$$f(E) = k_B T \ln[1 + e^{\beta(E-\mu)}] = \int_{c-i\infty}^{c+i\infty} \frac{ds}{2\pi i} e^{Es} \int_0^\infty dt e^{-st} f(t). \tag{9.29}$$

From (9.3),

$$\Omega = \int_{c-i\infty}^{c+i\infty} \frac{ds}{2\pi i} \hat{Z}(-s)\varphi(s), \tag{9.30}$$

where $\varphi(s)$ is the second integral in (9.29). By multiplying and dividing the integrand of (9.30) by s^2 and noting that $s^2\varphi(s)$ is the Laplace transform of $f''(t) = -(\beta/4)\text{sech}^2\left(\left[\frac{\beta}{2}(t-\mu)\right]\right)$, and with $z(t)$ defined in (9.4), one has the Sondheimer-Wilson formula.

C: Derivation of (9.4)

$$\hat{Z}(-s) = \hat{Z}(s) = \frac{K_m^2}{\pi} \left[\frac{1}{4} + \frac{1}{p^2} + \frac{\sinh(p)}{p} - \frac{\cosh(p)}{p^2} \right] \tag{9.31}$$

where $p = as$. Hence,

$$\frac{\hat{Z}(s)}{s^2} = \frac{K_m^2 a^2}{\pi} \left[\frac{1}{4p^2} + \frac{1}{p^4} + \frac{e^p}{2} \left(\frac{1}{p^3} - \frac{1}{p^4} \right) - \frac{e^{-p}}{2} \left(\frac{1}{p^2} + \frac{1}{p^4} \right) \right]. \tag{9.32}$$

Denoting the inverse Laplace transform with respect to x by L_x^{-1}, we write

$$z(t) = \frac{K_m^2 a^2}{\pi}\left[L^{-1}_{t/a}[\frac{1}{4p^2}+\frac{1}{p^4}]+\frac{1}{2}L^{-1}_{(t+a)/a}[\frac{1}{p^3}-\frac{1}{p^4}]-\frac{1}{2}L^{-1}_{(t-a)/a}[\frac{1}{p^3}+\frac{1}{p^4}]\right], \tag{9.33}$$

since an exponential factor just shifts the argument of the inverse Laplace transform, which is t/a after changing the variable of integration from s to p. Finally, using

$$L_x^{-1}[p^{-n}] = \frac{1}{(n-1)!}x^{n-1}\theta(x) \tag{9.34}$$

we obtain

$$z(t) = \frac{K_m^2}{12\pi a^2}\left[t^3 + 6a^2 t + 2a^3 - \left(t^3 - 3a^2 t + 2a^3\right)\theta(t-a)\right]. \tag{9.35}$$

D: Calculation of Ω

By inserting (9.4) into (9.3) one has a linear combination of the four elementary integrals

$$\begin{aligned}
\int_0^\infty \mathrm{sech}^2(\frac{1}{2}\beta(\mathrm{t}-\mu))\,\mathrm{dt} &= \frac{4\mathrm{e}^{\beta\mu}}{\beta(\mathrm{e}^{\beta\mu}+1)}\\
\int_0^\infty t\,\mathrm{sech}^2(\frac{1}{2}\beta(\mathrm{t}-\mu))\,\mathrm{dt} &= \frac{4\ln(\mathrm{e}^{\beta\mu}+1)}{\beta^2}\\
\int_0^\infty t^2\,\mathrm{sech}^2(\frac{1}{2}\beta(\mathrm{t}-\mu))\,\mathrm{dt} &= -\frac{8\mathrm{Li}_2(-\mathrm{e}^{\beta\mu})}{\beta^3}\\
\int_0^\infty t^3\,\mathrm{sech}^2(\frac{1}{2}\beta(\mathrm{t}-\mu))\,\mathrm{dt} &= \frac{4(\beta^3\mu^3 - 6\mathrm{Li}_3(-\mathrm{e}^{-\beta\mu}) + \pi^2\beta\mu)}{\beta^4},
\end{aligned} \tag{9.36}$$

which follow from the elementary integrals

n	$\int \frac{x^n}{\cosh^2(ax)}dx =$
0	$\frac{\tanh(ax)}{a}$
1	$\frac{ax\tanh(ax) - \log(\cosh(ax))}{a^2}$
2	$\frac{Li_2\left(-e^{-2ax}\right) + ax\left(-ax + a\tanh(ax)x - 2\log\left(1+e^{-2ax}\right)\right)}{a^3}$
3	$\frac{2a^2\left[-ax + a\tanh(ax)x - 3\log\left(1+e^{-2ax}\right)\right]x^2 + 6aLi_2\left(-e^{-2ax}\right)x + 3Li_3\left(-e^{-2ax}\right)}{2a^4}.$

(9.37)

E: The Polylogarithm for small and large arguments

For z small,

$$Li_2[-e^{-z}) \approx -\frac{\pi^2}{12} + \ln(2)z - \frac{z^2}{4}, \tag{9.38}$$

$$Li_3(-e^{-z}) \approx -\frac{3}{4}\zeta(3) + \frac{\pi^2}{12}z - \frac{1}{2}\ln(2)z^2. \tag{9.39}$$

For large z

$$Li_2(-e^{-z}) \approx -e^{-z} + \frac{1}{4}e^{-2z}, \tag{9.40}$$

$$Li_3(-e^{-z}) \approx -e^{-z} + \frac{1}{8}e^{-2z}. \tag{9.41}$$

References

1. D. Bercioux, D.F. Urban, H. Grabert, W. Haüsler, Phys. Rev. **A80**, 063683 (2009)
2. J.D. Malcolm, E.J. Nicol, Phys. Rev. B **93**, 165433 (2016)
3. *Applications of Dichalcogenides S, Se, Te*, Ed. G.K. Ahluwalia, Springer (2017)
4. S.Q. Shen, *Topological Insulators* (Springer, 2012)
5. M.J.S. Spencer, *Silicene* (Springer, 2016)
6. M. Katsnelson, *Graphene*: Carbon in Two-Dimensions (Cambridge UP, 2012)
7. H. Aoki, M.S. Dresselhaus, *Physics of Graphene* (Springer, 2013)
8. E.L. Wolf, *Graphene*: A New Paradigm in Condensed Matter and Device Physics (Oxford UP, 2013)
9. T.O. Wehling, A.M. Black-Shaffer, A.V. Balatsky, ArXiv 1405.5774v1 (2014)
10. J. Weng, S. Deng, Z. Liu, Z. Liu, The rare 2D materials with Dirac cones. Nat. Sci. Rev. **2**, 22–39 (2015)
11. N.J.M. Horing, Trans. Roy. Soc **A368**, 5525 (2010)
12. N.J.M. Horing, *Quantum Statistical Field Theory* (Oxford UP, 2017)
13. E.H. Sondheimer, A.H. Wilson, Proc. Roy. Soc. London **210**, 273 (1951)
14. A.H. Wilson *Theory of Metals*, Cambridge UP, Sec. 6.6 (1965)
15. NIST on-line Handbook of Mathematical Functions https://dlmf.nist.gov/25.12#ii
16. N.J.M. Horing, M.L. Glasser, J.D. Mancini, *Statistical Thermodynamics of the Diced Lattice* Physica Scripta 96 #1, 015806 (January 20, 2021)

Chapter 10
Non-Markovian Fermionic Quantum State Diffusion Approach

Xinyu Zhao and Ting Yu

Abstract In this chapter the non-Markovian fermionic quantum state difusion approach is discussed. Within the open system plus environment frame work, the non-Markovian fermionc quantum diffusion theory is established based on non-commutative Grassmann Gaussian processes. We show that this ferminionic quantum stochastic approach is useful in handling quantum processes involving multi-time scales such as decoherence, dissipation and non-equilibrium quantum transport. Particularly, our fermionic quantum state diffusive theory can provide more detailed information about the system's Markov-non-Markvoian crossover features.

10.1 Introduction

The rapid development of experimental quantum technology and its application to the fabrication and manipulation of quantum devices (e.g., photonic devices, quantum dots, nanomechanical oscillators, etc.), has caused a resurgent interest in the theory of open quantum systems. One problem, in particular, is the fact that no system can be completely isolated from its environment (bath, reservoir etc.), and thus the dynamics of a system of interest will be noticeably affected by any couplings to its environments [1, 2]. When an open system is coupled to a Markov environment, the Lindblad master equation may be used as a critical tool to study the open system dynamics [3]. Additionally, the temporal behaviors of quantum open systems are essential for understanding many fundamental issues such as quantum dissipation, decoherence and quantum transport processes in different time scales, to name a few [4–6]. If the Born-Markov approximation is unable to describe the evolution of a quantum system, as is the case if the coupling between system and environment is not weak or if the environment cannot be approximated by a broadband bath, one must extend the standard Markov theory to a more general non-Markovian environ-

X. Zhao · T. Yu (✉)
Department of Physics and Center for Quantum Science and Engineering,
Stevens Institute of Technology, Hoboken, NJ 07030, USA
e-mail: tyu1@stevens.edu

H. Ünlü and N. J. M. Horing (eds.), *Progress in Nanoscale and Low-Dimensional Materials and Devices*, Topics in Applied Physics 144,
https://doi.org/10.1007/978-3-030-93460-6_10

ment. Several attempts have been proposed to derive the evolution equation of open quantum systems beyond the Markov approximation [2, 7–12]. Notably, the non-Markovian quantum state diffusion (NMQSD) approach has shown the potential to solve large systems (multi-qubit or multi-cavity) [13–18], and as a computing tool, the approach may be used in several domains such as high-precision measurement [19], entanglement dynamics [20], quantum control [21] and coherence dynamics of large molecules in biophysics [22] and more. Therefore, it is worth extending the NMQSD approach for the bosonic baths to the fermionic cases where the non-Markovian features have played an increasingly important role [23–27].

Here, we show how to derive a fermionic stochastic Schrödinger equation for an open quantum system embedded in a fermionic bath called the fermionic NMQSD equation [28, 29]. To illustrate our approach, we solve several models as examples by using this new technique, including a one-qubit dissipative model, a two-qubit dissipative model, the quantum Brownian motion in a fermionic bath, and a multiple-particle model. In the first example, we give the explicit analytical solution without any approximation in a special case. In the second example, we show how to construct the crucial functional derivative Q operator contained in the fermionic NMQSD equation. In the third example, we consider a continuous variable model where a Brownian particle is immersed in a bath of fermionic particles. The last example involves a genuine multi-particle system that has been solved exactly by our NMQSD approach. Finally, the differences between the bosonic bath and the fermionic bath are discussed.

With the fermionic NMQSD approach, we are able to extend our theory to a situation where a quantum system is coupled to a hybrid bath containing both fermions and bosons [30]. By introducing Gaussian noise and Grassmann noise simultaneously, we manage to obtain a hybrid NMQSD for the hybrid bath model. The applications for these results include the example of a quantum dot model where the quantum dots may interact with two fermionic reservoirs (source and drain) and other agents such as a phonon bath. In this context, the dynamics of the quantum dot system are determined by both the fermionic reservoirs and the bosonic bath. The hybrid NMQSD approach is capable of taking into account of the environmental effects arising from both types of environments. We also use several examples to illustrate the hybrid NMQSD approach and identify the different impacts of fermionic bath and bosonic bath on the system dynamics.

In summary, a non-Markovian treatment towards solving quantum systems that are coupled to a fermionic or hybrid bath is developed. The environmental influences on the system are encoded into a set of random variables forming a stochastic process. Therefore, taking the partial trace over the environment is simplified to taking the statistical average over the stochastic variables, thus allowing us to obtain a master equation governing the evolution of the reduced density matrix. It is our hope that the fermionic/hybrid NMQSD approach could pave a new way to solve the dynamics of open quantum systems coupled to fermionic/hybrid baths, particularly to investigate temporal behaviors beyond Markov dynamics. The approaches described in this chapter is expected to provide a new method to deal with the non-equilibrium quantum transport processes.

10.2 The NMQSD Theory for Quantum System Coupled to Fermionic Baths

10.2.1 The General Stochastic Schrödinger Equation and the Corresponding Master Equation

Typically, a quantum system coupled to a fermionic bath can be described by the following Hamiltonian

$$H_{tot} = H_S + H_B + H_{int}, \tag{10.1}$$

where H_S is the Hamiltonian of the system of interest, $H_B = \omega_i c_i^\dagger c_i$ is the Hamiltonian of the fermionic bath which is composed of a set of fermions, and $H_{int} = \sum_i g_i(c_i^\dagger L + H.c.)$ describes the interaction between the system and the fermionic bath (or reservoir). The annihilation and creation operators c_i and $c_i^\dagger$ satisfy the fermionic anti-commutation relation $\{c_i, c_j^\dagger\} = \delta_{ij}$. Here, the operators H_S and L are not specified, so that this Hamiltonian may represent various types of systems and interactions. In the interaction picture, we can rewrite the Hamiltonian as

$$H_{tot}^I(t) = H_S + \sum_i g_i(c_i^\dagger L e^{i\omega_i t} + H.c.). \tag{10.2}$$

Before deriving the fundamental dynamical equation governing the quantum system coupled to a fermionic bath, it is necessary to specify the commutation relations between the system and its environment. For a fermionic bath, the operators living in the system Hilbert space can be either commutative or anti-commutative with the operators living in the Hilbert space of the bath. Consider a system that is composed of fermions, for example, a single electron in the quantum dot, the system Hamiltonian can be written as $H_S = \omega_d d^\dagger d$, where d is the annihilation operator of the electron in the system. Then the commutation relation between d and c_i, the annihilation operator of the bath mode "i", is anti-commutative $\{d, c_i\} = 0$. However, if the system is composed of a set of bosons or spins, then the commutation relation between the system and bath operators can be commutative. Therefore, to be more precise, we should discuss these two cases separately. It will be shown that different commutation between the system and bath can result in different NMQSD equations and the corresponding master equations.

10.2.1.1 Case 1: System of Interest Commutes with Bath Operators

This case can naturally arise when the Hamiltonian (10.1) describes an effective fermionic bath. For example, if we consider a spin-chain bath, the spin bath can be transformed into an effective fermionic bath by performing the Jordan-Wigner transformation and the Fourier transformation [28, 31]. In the case of an effective fermionic bath, the creation and annihilation operators $c_i^\dagger$ and c_i can commute with any operators living in the Hilbert space of the system H_S. The following example clearly shows that a spin-chain bath can be transformed into an effective fermionic bath. Consider a quantum system interacting with a *XX* spin chain with the following Hamiltonian:

$$H_{tot} = H_S + H_B + H_{int}, \tag{10.3}$$

$$H_B = \sum_i (\sigma_i^+ \sigma_{i+1}^- + \sigma_{i+1}^+ \sigma_i^-), \tag{10.4}$$

$$H_{int} = L^\dagger \sigma_1^- + \sigma_1^+ L. \tag{10.5}$$

After performing the Jordan-Wigner transformation,

$$\sigma_j^- = \exp(-i\pi \sum_{k=1}^{j-1} c_k^\dagger c_k) c_j, \tag{10.6}$$

and then the Fourier transformation [31],

$$c_j = \frac{1}{\sqrt{N}} \sum_{p=-N/2}^{N/2} \exp(-ij\phi_p) a_p, \tag{10.7}$$

the original Hamiltonian (10.3–10.5) becomes

$$H_{tot} = H_S + H_B + H_{int}, \tag{10.8}$$

$$H_B = \sum_{p=-N/2}^{N/2} 2\cos\phi_p a_p^\dagger a_p, \tag{10.9}$$

$$H_{int} = \frac{1}{\sqrt{N}} \sum_{p=-N/2}^{N/2} [L^\dagger \exp(-i\phi_p) a_p + \exp(i\phi_p) a_p^\dagger L]. \tag{10.10}$$

where the operators a_p and $a_p^\dagger$ satisfy the fermionic commutation relation $\{a_p, a_q^\dagger\} = \delta_{pq}$. This effective Hamiltonian obtained from the above transformations takes the standard form given by (10.1).

In order to derive the fundamental dynamical equation governing the dynamics of a system described by the Hamiltonian (10.1), we introduce the fermionic coherent states defined as,

$$|\xi_i\rangle = (1 - \xi_i c_i^\dagger)|0\rangle, \tag{10.11}$$

where ξ_i are Grassmann variables satisfying the following properties $\{\xi_i, \xi_j\} = 0$, $\{\xi_i, \xi_j^*\} = 0$. Generally, the coherent state can be expanded in terms of Fock states as $|\xi_i\rangle = |0\rangle - \xi_i c_i^\dagger|0\rangle$. For the multi-mode environment, one can expand the coherent state as $|\xi\rangle = |\xi_1\rangle \otimes |\xi_2\rangle \otimes |\xi_3\rangle \otimes \ldots$. Similar to the bosonic NMQSD approach [13], one can also define a stochastic state vector as

$$|\psi(t, \xi^*)\rangle = \langle\xi|\psi_{tot}(t)\rangle, \tag{10.12}$$

where $|\psi_{tot}(t)\rangle$ is the total state vector for the system and environment, and $\langle\xi|$ is a coherent state representation for the environment. It should be noted that the fermionic state $|\psi(t, \xi^*)\rangle$ defined above is physically different from that defined in the bosonic case. The fermionic state $|\psi(t, \xi^*)\rangle$ should be regarded as a state vector in the "super-space" containing non-commutative stochastic variables. With the coherent state representation, we can derive the dynamical equation for the stochastic state vector as [28]

$$\begin{aligned}\frac{\partial}{\partial t}|\psi(t, \xi^*)\rangle &= \langle\xi|\frac{\partial}{\partial t}|\psi_{tot}(t)\rangle \\ &= \langle\xi| - iH_{tot}^I(t)|\psi_{tot}(t)\rangle \\ &= [-iH_S + L\xi_t^* - L^\dagger \int_0^t ds K(t, s)\frac{\delta}{\delta\xi_s^*}]|\psi(t, \xi^*)\rangle,\end{aligned} \tag{10.13}$$

where $\xi_t^* = -i\sum_i g_i^* e^{i\omega_i t}\xi_i^*$ is the Grassmann Gaussian noise and $K(t, s) = \sum_i |g_i|^2 e^{-i\omega_i(t-s)}$ is the correlation function of the fermionic bath. The Grassmann noise ξ_t^* satisfies

$$M\{\xi_t^*\} = \int \prod_i d\xi_i^* d\xi_i e^{-\xi_i^*\xi_i}\xi_t^* = 0, \tag{10.14}$$

$$M\{\xi_t\xi_s^*\} = \int \prod_i d\xi_i^* d\xi_i e^{-\xi_i^*\xi_i}\xi_t\xi_s^* = K(t, s), \tag{10.15}$$

where $M\{\cdot\} \equiv \int \prod_i d\xi_i^* d\xi_i e^{-\xi_i^*\xi_i}[\cdot]$ denotes the statistical mean over the Grassmann noise. In the NMQSD equation, we use $\frac{\delta}{\delta\xi_s^*}\psi_t(\xi^*)$ to denote the left-functional-derivative with respect to the Grassmann variables. Throughout this chapter, we will always use the $\frac{\delta}{\delta\xi_s^*}$ to represent the left-functional-derivative unless specifically indicated otherwise. Similar to the formal bosonic NMQSD equation [13, 14], the fermionic NMQSD equation contains a time-nonlocal Grassmann functional derivative. In order to find a time-local NMQSD equation, one can introduce a time-dependent Q operator defined as

$$\frac{\delta|\psi(t, \xi^*)\rangle}{\delta\xi_s^*} = Q(t, s, \xi^*)|\psi(t, \xi^*)\rangle. \tag{10.16}$$

With this Q operator, the exact stochastic NMQSD equation can be written as

$$\frac{\partial}{\partial t}|\psi(t,\xi^*)\rangle = [-iH_S + L\xi_t^* - L^\dagger \bar{Q}]|\psi(t,\xi^*)\rangle \tag{10.17}$$

where $\bar{Q}(t,\xi^*) = \int_0^t ds K(t,s) Q(t,s,\xi^*)$. This fermionic NMQSD equation is directly derived from the microscopic Hamiltonian without any approximation, it is the fundamental equation governing the dynamics of the open quantum system. In the fermionic NMQSD equation, we have introduced a new type of stochastic process ξ_t^*. The solution of our NMQSD equation $|\psi(t,\xi^*)\rangle$ for a single realization of ξ_t^* is called a Grassmann quantum trajectory. By construction, the reduced density matrix of the open system can be recovered by the statistical mean over the Grassmann noise $\rho = M[|\psi(t,\xi^*)\rangle\langle\psi(t,\xi)|]$. Although the fermionic NMQSD equation looks formally similar to the bosonic case, the dynamic behaviors of the system governed by the two types of equations can be quite different due to distinct differences between the bosonic and fermionic particles. Mathematically, the most striking difference between the bosonic and fermionic NMQSD equations is that the former contains a complex Gaussian noise while the latter is driven by a non-commutative Grassmann Gaussian noise. The detailed difference between the bosonic bath and fermionic bath will be investigated in the following sections.

In order to solve the fermionic NMQSD equation, one need to first derive a dynamic equation for the Q operator. Consider the consistency condition (CC),

$$\frac{\delta}{\delta\xi_s^*}\frac{\partial}{\partial t}|\psi(t,\xi^*)\rangle = \frac{\partial}{\partial t}\frac{\delta}{\delta\xi_s^*}|\psi(t,\xi^*)\rangle. \tag{10.18}$$

Applying the NMQSD (10.17) to CC, the left-hand side (LHS) is

$$LHS = [-iH_S Q - L\xi_t^* Q - L^\dagger(\frac{\delta}{\delta\xi_s^*}\bar{Q}) - L^\dagger\bar{Q}(-\xi^*)Q]|\psi(t,\xi^*)\rangle. \tag{10.19}$$

On the other hand, the right-hand side (RHS) becomes

$$RHS = \frac{\partial}{\partial t}(Q)|\psi(t,\xi^*)\rangle + [-iQH_S + QL\xi_t^* - QL^\dagger\bar{Q}]|\psi(t,\xi^*)\rangle. \tag{10.20}$$

Equating LHS and RHS, the equation of Q operator can be obtained:

$$\frac{\partial}{\partial t}Q = -i[H_S, Q] - \{L\xi_t^*, Q\} - L^\dagger\bar{Q}(-\xi^*)Q + QL^\dagger\bar{Q} - L^\dagger\frac{\delta}{\delta\xi_s^*}\bar{Q}. \tag{10.21}$$

In the derivation above, we have used the following commutation relations:

$$\frac{\delta_l}{\delta\xi_s^*}[\xi_t^*\psi_t(\xi^*)] = -\xi_t^*\frac{\delta_l}{\delta\xi_s^*}\psi_t(\xi^*), \tag{10.22}$$

and

$$\frac{\delta_l}{\delta \xi_s^*}[\bar{Q}\psi_t(\xi^*)] = \bar{Q}(-\xi^*)\hat{Q}\psi_t(\xi^*) + (\frac{\delta_l}{\delta \xi_s^*}\bar{Q})\psi_t(\xi^*). \tag{10.23}$$

where the sign of $\bar{Q}(-\xi^*)$ depends on the functional form of the noise contained in $\bar{Q}$. If Q only contains odd order noise terms, $\bar{Q}(-\xi^*) = -\bar{Q}(\xi^*)$. With the differential equation for Q operator and the initial condition

$$Q(t, s = t, \xi^*) = L, \tag{10.24}$$

the exact Q operator may be fully determined. Besides the exact Q operator determined in (10.21), there is also a systematic perturbative solution for Q operator. For most practical problems, the perturbation approach is more useful. Similar to the bosonic case, the operator Q can be expanded as [14, 32, 33]

$$\begin{aligned} Q(t, s, \xi^*) = Q^{(0)}(t, s) &+ \int_0^t Q^{(1)}(t, s, s_1)\xi_{s_1}^* ds_1 \\ &+ \int_0^t \int_0^t Q^{(2)}(t, s, s_1, s_2)\xi_{s_1}^* \xi_{s_2}^* ds_1 ds_2 + ... \\ &+ \int_0^t ... \int_0^t Q^{(n)}(t, s, s_1, ...s_n)\xi_{s_1}^* ...\xi_{s_n}^* ds_1 ...ds_n \\ &+ .... \end{aligned} \tag{10.25}$$

Substituting this expansion into (10.21), one can derive the dynamic equations of the coefficients for each order $Q^{(n)}$. Particularly, the zeroth-order term $Q^{(0)}(t, s)$ will satisfy the following equation (neglect all the noise terms)

$$\frac{\partial}{\partial t} Q^{(0)}(t, s) = -i[H_S, Q^{(0)}(t, s)] - [L^\dagger \bar{Q}^{(0)}(t), Q^{(0)}(t, s)], \tag{10.26}$$

where $\bar{Q}^{(0)}(t) = \int_0^t Q^{(0)}(t, s)K(t, s)ds$, and the initial condition is

$$Q^{(0)}(t, s = t) = L. \tag{10.27}$$

The fundamental equations governing the dynamics of the open quantum system coupled to a fermionic bath have been derived. However, the solution of the NMQSD equation $|\psi(t, \xi^*)\rangle$ may not fully describe the dynamic evolution of the system. Actually, it only gives a formal realization called fermionic quantum trajectory generated by a specific realization of the Grassmann noise ξ_t^*. Physically, in order to get the full picture of the evolution of the system of interest, one need to reproduce the reduced density matrix from the Grassmann stochastic state vector $|\psi(t, \xi^*)\rangle$:

$$\begin{aligned}\rho(t) &= \sum_n \langle n|\psi_{tot}\rangle\langle\psi_{tot}|n\rangle \\ &= \int \prod_i d\xi_i^* d\xi_i \exp(-\sum_j \xi_j^*\xi_j)\sum_n \langle n|\xi\rangle\langle\xi|\psi_{tot}\rangle\langle\psi_{tot}|n\rangle \\ &= \int \prod_i d\xi_i^* d\xi_i \exp(-\sum_j \xi_j^*\xi_j)\sum_n \langle\xi|\psi_{tot}\rangle\langle\psi_{tot}|n\rangle\langle n|-\xi\rangle \\ &= \int \prod_i d\xi_i^* d\xi_i \exp(-\sum_j \xi_j^*\xi_j) P_t \\ &= M\{P_t\}, \end{aligned} \tag{10.28}$$

where $P_t \equiv |\psi(t,\xi^*)\rangle\langle\psi(t,-\xi)|$ is the stochastic density operator. Given the relation (10.28), the physical meaning of the NMQSD equation become clear. By choosing a random realization of the noise ξ_t^* (reflecting the states of the environment), the evolution of the reduced density operator is decomposed into many pure-state quantum trajectories $|\psi(t,\xi^*)\rangle$. However, taking the statistical average over all of these trajectories, the reduced density matrix is reproduced. Therefore, the complicated properties of the environment are all encoded into the noise function ξ_t^*, so that tracing out the environment is equivalent to taking average over all the realizations of the noises. Based on relation (10.28), the master equation can be derived as

$$\begin{aligned}\frac{d}{dt}\rho &= \frac{d}{dt}M\{P_t\} \\ &= M\{(-iH_S + L\xi_t^* - L^\dagger\bar{Q})P_t\} + M\{P_t(iH_S - \xi_t L^\dagger - \bar{Q}^\dagger(-\xi)L)\} \\ &= -i[H_S,\rho] + LM\{\xi_t^* P_t\} - M\{P_t\xi_t\}L^\dagger - L^\dagger M\{\bar{Q}P_t\} - M\{P_t\bar{Q}^\dagger(-\xi)\}L. \end{aligned} \tag{10.29}$$

In order to establish an explicit master equation, one needs to handle the terms like $M\{P_t\xi_t\}$ etc. These terms can be simplified by a Novikov-type theorem for the Grassmann noise:

$$M\{P_t\xi_t\} = -M\{\bar{Q}P_t\}, \tag{10.30}$$

$$M\{\xi_t^* P_t\} = M\{P_t\bar{Q}^\dagger(-\xi)\}. \tag{10.31}$$

In fact, the above theorem can be proved as follows

$$\begin{aligned}&M\{P_t\xi_t\} \\ &= \int \prod_i d\xi_i^* d\xi_i \exp(-\sum_k \xi_k^*\xi_k)\, |\psi(\xi^*)\rangle\,\langle\psi(-\xi)|\,(i\sum_j g_j e^{-i\omega_j t}\xi_j) \\ &= -i\sum_j g_j e^{-i\omega_j t}\int \prod_i d\xi_i^* d\xi_i [\,|\psi(\xi^*)\rangle\,\langle\psi(-\xi)|\,\frac{\partial_l}{\partial\xi_j^*}\exp(-\sum_k \xi_k^*\xi_k)] \end{aligned}$$

$$
\begin{aligned}
&= i\sum_j g_j e^{-i\omega_j t}\int\prod_i d\xi_i^* d\xi_i \Big(\frac{\partial_l}{\partial(-\xi_j^*)}\,|\psi(\xi^*)\rangle\langle\psi(-\xi)|\Big)\exp(-\sum_k \xi_k^*\xi_k)\\
&= -i\sum_j g_j e^{-i\omega_j t}\int\prod_i d\xi_i^* d\xi_i [\exp(-\sum_k \xi_k^*\xi_k)\Big(\int ds\frac{\partial\xi_s^*}{\partial\xi_j^*}\frac{\delta_l}{\delta\xi_s^*}\Big)P_t]\\
&= -\int ds\sum_j |g_j|^2 e^{-i\omega_j(t-s)}\int\prod_i d\xi_i^* d\xi_i [\exp(-\sum_k \xi_k^*\xi_k)\frac{\delta_l}{\delta\xi_s^*}P_t]\\
&= -\int\prod_i d\xi_i^* d\xi_i \exp(-\sum_k \xi_k^*\xi_k)\int ds K(t,s)Q(t,s,\xi^*)P_t]\\
&= -M\{\bar{Q}P_t\}.
\end{aligned}
\tag{10.32}
$$

It should be noted that one needs to distinguish the left-derivative (left-functional-derivative) from the right-derivative (right-functional-derivative) when the derivatives are performed. Similarly, one can prove $M\{\xi_t^* P_t\} = M\{P_t\bar{Q}^\dagger(-\xi)\}$. With the help of the Novikov-type theorem for the Grassmann noise, the exact master equation can be formally written as

$$
\frac{\partial}{\partial t}\rho = -i[H_S,\rho] + [L, M\{P_t\bar{Q}^\dagger(-\xi)\}] + [M\{\bar{Q}P_t\}, L^\dagger]. \tag{10.33}
$$

For example, if the operator Q is independent of the Grassmann noise, then the exact master equation is immediately obtained as,

$$
\frac{\partial}{\partial t}\rho = -i[H_S,\rho] + [L, \rho\bar{Q}^\dagger] + [\bar{Q}\rho, L^\dagger]. \tag{10.34}
$$

Moreover, in the Markov limit, $\bar{Q} = \gamma_f L$, this master equation reduces to the standard Lindblad master equation:

$$
\frac{\partial}{\partial t}\rho = -i[H_S,\rho] + \gamma_f[L, \rho L^\dagger] + \gamma_f[L\rho, L^\dagger]. \tag{10.35}
$$

In this part we have shown how to establish a fermion stochastic differential equation and the derivation of the corresponding master equation.

10.2.1.2 Case 2: System of Interest Anti-commutes with the Bath Operators

After discussing the case that system commutes with the fermionic bath, we will consider the case that system anti-commutes with the fermionic bath which often describes electronic system such like quantum dots. Following a similar procedure, one can derive an anti-commutative fermionic NMQSD equation as [29, 34]

$$\frac{\partial}{\partial t}|\psi(t,\xi^*)\rangle = [-iH_S - L\xi_t^* - L^\dagger \int_0^t dsK(t,s)\frac{\delta}{\delta\xi_s^*}]|\psi(t,\xi^*)\rangle \tag{10.36}$$

Since the operator L typically anti-commutes with the fermionic bath, the derivation is slightly different from the commutative case [28, 29, 34], there is a minor difference between (10.13) and (10.36). Similarly, one can also define the fermionic Q operator. However, it satisfies a different differential equation in the anti-commutative case as

$$\frac{\partial}{\partial t}Q = [-iH_S - L\xi_t^* - L^\dagger\bar{Q}] + L^\dagger\frac{\delta}{\delta\xi_s^*}\bar{Q}, \tag{10.37}$$

with the initial condition

$$Q(t, s = t, \xi^*) = L, \tag{10.38}$$

where $\bar{Q}(t,\xi^*) = \int_0^t dsK(t,s)Q(t,s,\xi^*)$. Finally the NMQSD equation can be also written in a compact form as

$$\frac{\partial}{\partial t}|\psi(t,\xi^*)\rangle = [-iH_S - L\xi_t^* - L^\dagger\bar{Q}]|\psi(t,\xi^*)\rangle. \tag{10.39}$$

and the corresponding master equation can be derived as

$$\frac{d}{dt}\rho = -i[H_S,\rho] + [L, M\{P_t\bar{Q}^\dagger(-\xi)\}] + [M\{\bar{Q}P_t\}, L^\dagger] \tag{10.40}$$

10.2.2 Examples of Solving Fermionic Bath with Fermionic NMQSD Equation

In the last section, we have built up the framework of the fermionic NMQSD theory. Particularly, the fermionic NMQSD equation and the corresponding master equation are derived. In this section, we will illustrate the power of the newly developed fermionic NMQSD theory by solving some examples. First, we will solve a single qubit (two-level system) coupled to a fermionic bath. Through this simple model, we show how to apply the general fermionic NMQSD approach to a specific model. Particularly, in a limiting case where the environment only contains one fermion, this model can be easily solved by ordinary quantum mechanics. It is shown that the results obtained by the fermionic NMQSD approach are identical to that obtained by solving the Schrödinger equation of the whole system. Second, we solve a two-qubit system coupled to a dissipative fermionic bath. In this case, we show how to construct the fermionic Q operator with noise dependent terms. Third, we use the example of a quantum Brownian particle in fermionic bath to illustrate a case representing a continuous variable system. Finally, we solve an N-fermion system by the fermionic

NMQSD approach. Besides, with this example we compare the difference between fermionic bath and bosonic bath.

10.2.2.1 Effective Fermionic Bath: One-Qubit Coupled to a Spin Chain

In the first example, we consider an effective fermionic bath transformed from a spin chain, and a single qubit is coupled to the spin chain bath. According to the discussion in Sect. 10.2.1.1, the Hamiltonian can be finally transformed into a case where the single-qubit is coupled to an effective fermionic bath as

$$H_{tot}^{I}(t) = H_S + \sum_i \omega_i c_i^\dagger c_i + \sum_i g_i (c_i^\dagger L + H.c.), \tag{10.41}$$

where $H_S = \frac{\omega}{2}\sigma_z$ and $L = \sigma_-$. Since the effective fermionic bath is transformed from a spin chain, the system may commute with the bath, so we can use the theory developed in Sect. 10.2.1.1 to solve this problem. According to (10.21), the solution for Q can be obtained as

$$Q(t,s) = x_1(t,s)\sigma_-, \tag{10.42}$$

with the initial condition

$$Q(t, s=t) = L = \sigma_-, \tag{10.43}$$

and the coefficient $x_1(t,s)$ is shown to satisfy

$$\frac{\partial}{\partial t} x_1(t,s) = [i\omega + X_1(t)]x_1(t,s), \tag{10.44}$$

where $X_1(t) = \int_0^t x_1(t,s)K(t,s)ds$, and $K(t,s)$ is the correlation function, and the initial condition is given by $x_1(t,s) = 1$. Thus, the exact Q operator can be fully determined. Finally, the explicit NMQSD equation for this model is

$$\frac{\partial}{\partial t}|\psi(t,\xi^*)\rangle = [-i\frac{\omega}{2}\sigma_z + \sigma_-\xi_t^* - X_1(t)\sigma_+\sigma_-]|\psi(t,\xi^*)\rangle, \tag{10.45}$$

and the exact master equation is

$$\begin{aligned} \frac{d}{dt}\rho &= -i[H_S,\rho] + [L,\rho Q^\dagger] + [Q\rho, L^\dagger] \\ &= -i\frac{\omega}{2}(\sigma_z\rho - \rho\sigma_z) + X_1^*(t)(\sigma_-\rho\sigma_+ - \rho\sigma_+\sigma_-) \\ &+ X_1(t)(\sigma_-\rho\sigma_+ - \sigma_+\sigma_-\rho). \end{aligned} \tag{10.46}$$

With this exact master equation, the dynamics of this model can be fully determined.

It is worth noting that this Q operator has the same form as the bosonic case [14]. Besides, the master equation is also formally identical to that obtained for the model with a single qubit coupled to a bosonic bath. The reason that fermionic bath and bosonic bath have the same influence on the single qubit system is because the initial state of the environment (bosonic or fermionic) is assumed to be a vacuum state. Therefore, there exists only one excitation in the total system. In this sense, the bosonic bath and the fermionic bath are identical.

Now, we consider a very special case for the one-qubit model where the "environment" contains only one fermion. By analytically solving this model, we show explicitly that the results obtained from the fermionic NMQSD approach are identical to those predicted by the ordinary quantum mechanics. The model is described by the following Hamiltonian,

$$H_{tot} = \frac{\omega}{2}\sigma_z + \omega_b c^\dagger c + g(\sigma_- c^\dagger + \sigma_+ c) \tag{10.47}$$

and the zero-temperature correlation function becomes

$$K(t,s) = |g|^2 e^{-i\omega_b(t-s)}. \tag{10.48}$$

Substituting the correlation function into the expression of $X_1(t) = \int_0^t x_1(t,s) K(t,s)ds$, the differential equation for $X_1(t)$ is written as

$$\frac{\partial}{\partial t}X_1(t) = |g|^2 - i\omega_b X_1(t) + i\omega X_1(t) + X_1(t)^2. \tag{10.49}$$

For simplicity, we consider the resonance case $\omega_b = \omega$, then the solution $X_1(t)$ is given by

$$X_1(t) = |g| \tan(|g|\, t). \tag{10.50}$$

From the master equation (10.46), one can calculate the time evolution for the off-diagonal elements in the density matrix.

$$\frac{d}{dt}\rho_{21} = \frac{d}{dt}\langle\sigma_+\rangle = \mathrm{tr}(\frac{d}{dt}\rho\sigma_+) = i\omega\rho_{21} - X_1^*(t)\rho_{21}. \tag{10.51}$$

Finally, the time evolution for the matrix element ρ_{21} can be obtained as

$$\rho_{21}(t) = \rho_{21}(0)e^{i\omega t}\cos\left[|g|\, t\right]. \tag{10.52}$$

This result shows that the coherence (off-diagonal elements in density matrix) will decrease and increase periodically.

On the other hand, we can easily solve this simple case using elementary quantum mechanics. Since this is only a two-body problem, one can solve the evolution for the whole system in a straightforward manner. One can check that elementary quantum mechanics gives rise to the identical results obtained by the fermionic NMQSD approach in (10.52).

10.2.2.2 Example 2: Coupled Two-Qubit Dissipative Model

In this section, we consider a system consisting of a pair of coupled two-level systems (spins or some other effective two-level models) interacting with a common fermionic bath. We will show how to construct the exact and approximate Q operator in this example. The total Hamiltonian of this model can be written in a standard form as (10.1) with

$$H_S = \omega_A \sigma_z^A + \omega_B \sigma_z^B + J_{xy}(\sigma_+^A \sigma_-^B + \sigma_-^A \sigma_+^B) + J_z \sigma_z^A \sigma_z^B, \tag{10.53}$$

$$L = \kappa_A \sigma_-^A + \kappa_B \sigma_-^B, \tag{10.54}$$

where κ_A and κ_B are constants describing different coupling strengths for the two qubits.

The exact Q operator of this model can be determined as

$$Q(t, s, \xi^*) = \sum_{i=1}^{4} f_i(t, s) Q_i + \int_0^t ds' f_5(t, s, s') \xi_{s'}^* Q_5, \tag{10.55}$$

where the basis operator are given by $Q_1 = \sigma_-^A$, $Q_2 = \sigma_-^B$, $Q_3 = \sigma_z^A \sigma_-^B$, $Q_4 = \sigma_z^B \sigma_-^A$, and $Q_5 = \sigma_-^A \sigma_-^B$. Substituting (10.55) into (10.21), one can derive the equations for the time-dependent coefficients f_i $(i = 1, 2, 3, 4, 5)$. The details can be found in [28]. As we can see in [28], the exact Q operator is noise dependent. However, it is possible to neglect the last term in the exact Q operator and the Q operator is reduced to the zeroth-order Q operator as

$$Q^{(0)}(t, s) = \sum_{i=1}^{4} f_i(t, s) Q_i. \tag{10.56}$$

From (10.26), we can derive the differential equation for the coefficients as

$$\begin{aligned} \frac{\partial}{\partial t} f_1(t, s) &= +2i\omega_A f_1 - i J_{xy} f_3 + 2i J_z f_4 + \kappa_A F_1 f_1 - \kappa_B F_1 f_3 \\ &\quad + \kappa_B F_3 f_1 + \kappa_B F_3 f_4 + \kappa_B F_4 f_3 + \kappa_A F_4 f_4, \end{aligned} \tag{10.57}$$

$$\begin{aligned} \frac{\partial}{\partial t} f_2(t, s) &= +2i\omega_B f_2 - i J_{xy} f_4 + 2i J_z f_3 + \kappa_B F_2 f_2 - \kappa_A F_2 f_4 \\ &\quad + \kappa_B F_3 f_3 + \kappa_A F_3 f_4 + \kappa_A F_4 f_2 + \kappa_A F_4 f_3, \end{aligned} \tag{10.58}$$

$$\frac{\partial}{\partial t} f_3(t,s) = +2i\omega_B f_3 - iJ_{xy} f_1 + 2iJ_z f_2 - \kappa_A F_2 f_1 + \kappa_B F_2 f_3 + \kappa_A F_3 f_1 + \kappa_A F_4 f_2 + \kappa_B F_3 f_2 + \kappa_A F_4 f_3, \tag{10.59}$$

$$\frac{\partial}{\partial t} f_4(t,s) = +2i\omega_A f_4 - iJ_{xy} f_2 + 2iJ_z f_1 - \kappa_B F_1 f_2 + \kappa_A F_1 f_4 + \kappa_B F_3 f_1 + \kappa_B F_3 f_4 + \kappa_A F_4 f_1 + \kappa_B F_4 f_2, \tag{10.60}$$

where $F_i(t) = \int_0^t ds K(t,s) f_i(t,s)$ $(i = 1, 2, 3, 4)$, and the initial conditions are

$$f_1(t, s = t) = \kappa_A, \tag{10.61}$$

$$f_2(t, s = t) = \kappa_B, \tag{10.62}$$

$$f_3(t, s = t) = 0, \tag{10.63}$$

$$f_4(t, s = t) = 0, \tag{10.64}$$

With the zeroth-order Q operator, the master equation can be explicitly written in the following form

$$\frac{d}{dt}\rho = -i[H_S\rho - \rho H_S] + \{\sum_{i=1}^{4} F_i^*[L\rho\bar{Q}_i^\dagger - \rho\bar{Q}_i^\dagger L] + H.c.\}. \tag{10.65}$$

In the following numerical simulation, we consider a simple case, in which all the parameters are symmetric for the two qubits, i.e., $\omega_A = \omega_B = \omega$, $\kappa_A = \kappa_B = 1$. Then, we can derive the following master equation

$$\begin{aligned}
\frac{d}{dt}\rho = &-i\omega[(\sigma_z^A + \sigma_z^B)\rho - \rho(\sigma_z^A + \sigma_z^B)] - iJ_z[\sigma_z^A\sigma_z^B\rho - \rho\sigma_z^A\sigma_z^B] \\
&- iJ_{xy}[(\sigma_+^A\sigma_-^B + \sigma_+^B\sigma_-^A)\rho - \rho(\sigma_+^A\sigma_-^B + \sigma_+^B\sigma_-^A)] \\
&+ \{F_1^*[(\sigma_-^A + \sigma_-^B)\rho\sigma_+^A - \rho\sigma_+^A(\sigma_-^A + \sigma_-^B)] \\
&+ F_2^*[(\sigma_-^A + \sigma_-^B)\rho\sigma_+^B - \rho\sigma_+^B(\sigma_-^A + \sigma_-^B)] \\
&+ F_3^*[(\sigma_-^A + \sigma_-^B)\rho\sigma_z^A\sigma_+^B - \rho\sigma_z^A\sigma_+^B(\sigma_-^A + \sigma_-^B)] \\
&+ F_4^*[(\sigma_-^A + \sigma_-^B)\rho\sigma_z^B\sigma_+^A - \rho\sigma_z^B\sigma_+^A(\sigma_-^A + \sigma_-^B)] + H.c.\}.
\end{aligned} \tag{10.66}$$

Although the master equation derived above is valid for an arbitrary correlation function, for numerical simulations, it is convenient to consider a specific example of the correlation function. It is known that the spectral density of a fermionic bath may be given by [10],

$$J(\omega) = \frac{\Gamma\gamma^2/2\pi}{(\omega - \Omega)^2 + \gamma^2}, \tag{10.67}$$

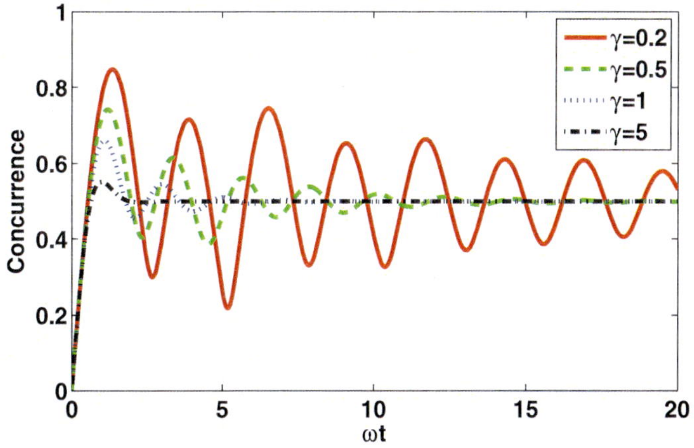

Fig. 10.1 Time evolution of concurrence for different γ. The other parameters are $\omega_A = \omega_B = \frac{\omega}{2}$, $J_{xy} = 0.5$, $J_z = 0$, $\Gamma = 1$, $\Omega = \pi/4$

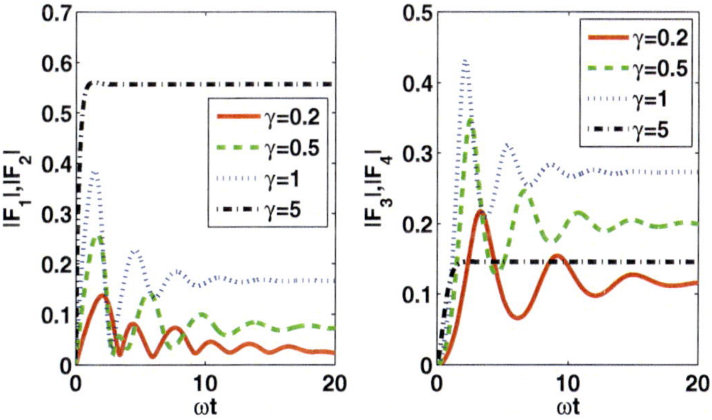

Fig. 10.2 Time evolution of $|F_1(t)|$, $|F_2(t)|$, $|F_3(t)|$, and $|F_4(t)|$. In the symmetric case, $|F_1(t)| = |F_2(t)|$, $|F_3(t)| = |F_4(t)|$. The other parameters are $\omega_A = \omega_B = \frac{\omega}{2}$, $J_{xy} = 0.5$, $J_z = 0$, $\Gamma = 1$,$\Omega = \pi/4$

which is called the Lorentzian spectral density. Therefore, the correlation function is

$$K(t,s) = \frac{\Gamma\gamma}{2}\exp[-(\gamma + i\Omega)|t-s|], \tag{10.68}$$

where γ indicates the correlation time of the bath.

In Fig. 10.1, we plot the time evolution of the concurrence of the two qubits. Given different non-Markovian correlation times, the dynamic evolution of entanglement shows different properties. In Fig. 10.2, we plot the evolution of the time-dependent coefficients $|F_i(t)|$. In the Markov limit, all of those coefficients are time-independent constants, however, in the non-Markovian regime with relatively small γ, they become time-dependent. In the long time limit, they converge to constants, but the behavior at the early stage of the evolution is related to the non-Markovian properties of the environment.

10.2.2.3 Example 3: Quantum Brownian Particle in a Fermionic Bath

In this subsection, we consider a continuous-variable model consisting of a Brownian particle interacting with a fermionic bath (a similar model is discussed in [24]). The Hamiltonian of the Brownian particle is given by,

$$H_S = \omega_m(p^2 + q^2). \tag{10.69}$$

The coupling between the system and the fermionic bath is described by the coupling operator $L = q$, as a result, the interaction Hamiltonian can be given by

$$H_{int} = q \sum_i g_i(c_i^\dagger + c_i). \tag{10.70}$$

Applying our NMQSD approach to this model, it can be easily shown that Q operator is a polynomial with arbitrary high orders of p, q:

$$Q = x_1(t,s)q + x_2(t,s)p + x_3(t,s,\xi^*)pq + x_4(t,s,\xi^*)p^2 + x_5(t,s,\xi^*)q^2 + ... \tag{10.71}$$

A useful approximation is to neglect all the noise-dependent terms, then we obtain the so-called zeroth-order approximate Q as

$$Q^{(0)} = x_1(t,s)q + x_2(t,s)p. \tag{10.72}$$

Substituting this approximate $Q^{(0)}$ operator into (10.26), we can derive the differential equations for the coefficients $x_1(t,s)$ and $x_2(t,s)$ as

$$\frac{\partial}{\partial t}x_1(t,s) = 2\omega_m x_2(t,s) + iX_2(t)x_1(t,s) - 2iX_1(t)x_2(t,s), \tag{10.73}$$

$$\frac{\partial}{\partial t}x_2(t,s) = -2\omega_m x_1(t,s) - iX_2(t)x_2(t,s). \tag{10.74}$$

where $X_i(t) = \int_0^t ds K(t,s)x_i(t,s)$ $(i = 1, 2)$. The initial conditions for coefficients $x_1(t,s)$ and $x_2(t,s)$ are

$$x_1(t, s = t) = 1, \tag{10.75}$$

$$x_2(t, s = t) = 0. \tag{10.76}$$

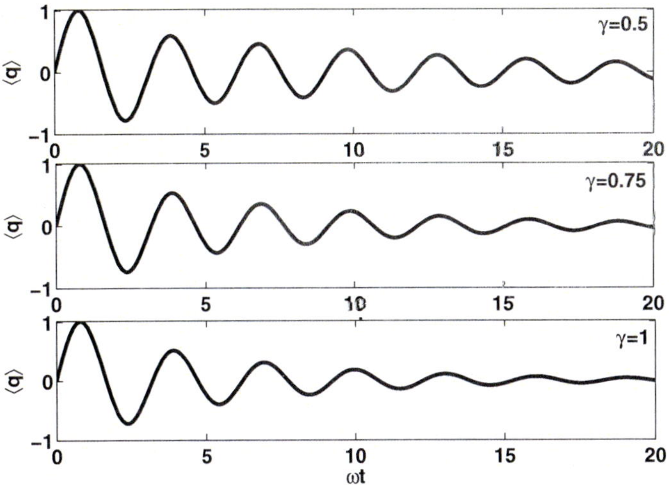

Fig. 10.3 Time evolution of the mean values of operator q in different environments. The parameter γ describs the memory effect. The other parameters are $\omega_m = \omega = 1$, $\Omega = \pi/2$

Using this approximate $Q^{(0)}$ operator, the master equation can be written as

$$\frac{d}{dt}\rho = -i\omega_m[(p^2+q^2)\rho - \rho(p^2+q^2)] \\ + \{X_1^*[q\rho q - \rho q q] + X_2^*[q\rho p - \rho p q] + H.c.\}. \tag{10.77}$$

With the approximate Q operator, the master equation derived here is still in the non-Markovian regime. It should be noted that the X_2^* (including its complex conjugation X_2) does not exist in the Markov limit. The approximation used here is also different from the weak-coupling approximation since the approximate Q still contains the higher-order terms of the coupling constant. From the master equation we can derive the evolution equations for all the mean values of operators q and p,

$$\frac{d}{dt}\langle q\rangle = 2\omega_m\langle p\rangle, \tag{10.78}$$

$$\frac{d}{dt}\langle p\rangle = -2\omega_m\langle q\rangle - iX_1^*\langle q\rangle - iX_2^*\langle p\rangle + iX_1\langle q\rangle + iX_2\langle p\rangle. \tag{10.79}$$

In Fig. 10.3, we plot the time evolution of $\langle q\rangle$ for different values of γ. In order to show the transition from non-Markovian to Markov regimes, the Ornstein-Uhlenbeck noise $K(t,s) = \frac{\gamma}{2}e^{-(\gamma+i\Omega)|t-s|}$ is chosen in our numerical simulations. When γ is very large, the correlation function $K(t,s)$ is close to a delta function and the dynamics reduces to the Markovian case. Figure 10.3 shows how the evolution of $\langle q\rangle$ is affected by γ. This is a unique phenomenon in the non-Markovian case.

10.2.2.4 Example 4. N-fermion System Coupled To a Fermionic Bath

In this example, we will establish the exact time-local fermionic NMQSD equation and the corresponding master equation for a genuine multipartite system coupled to a fermionic bath. We show that, using the fermionic NMQSD approach, the exact Q operator of the N-qubit model can be easily determined.

More specifically, The Hamiltonian we considered can be written in the standard form as (10.1) with

$$H_S = \sum_{i=1}^{N_S} A_i a_i^\dagger a_i. \tag{10.80}$$

Different from the previous examples, the system here is also composed of a set of fermions where $a_i^\dagger$ and a_i are fermionic creation and annihilation operators. The coupling operator is given as

$$L = \sum_{i=1}^{N_S} a_i. \tag{10.81}$$

This Hamiltonian could be an effective Hamiltonian transformed from a set of spins. For example, suppose that we have a long spin chain with N sites, if the first N_S ($N_S < N$) sites are treated as the system and the other N_B sites are treated as the bath ($N_S + N_B = N$), then performing the Jordan-Wigner transformations for both the system and the bath, we may obtain this type of effective Hamiltonian [28, 31].

We can show that the exact Q operator of this model takes the following form

$$Q = \sum_{i=1}^{N_S} x_i(t, s) a_i, \tag{10.82}$$

and the coefficients in Q operator are governed by the following differential equations

$$\frac{\partial}{\partial t} x_j(t, s) = i A_j x_j(t, s) + \sum_{i=1}^{N_s} X_j(t) x_i(t, s), \tag{10.83}$$

where $X_j(t) = \int_0^t K(t, s) x_j(t, s) ds$. So, $\bar{Q}(t) = \sum_{i=1}^{N_S} X_i(t) a_i$. The exact master equation of this model is

$$\frac{\partial}{\partial t}\rho = -i[\sum_{i=1}^{N_S} A_i a_i^\dagger a_i, \rho] + \{[\sum_{j=1}^{N_S} a_j, \rho \sum_{i=1}^{N_S} X_i^*(t) a_i^\dagger] + H.c.\} \tag{10.84}$$

10.2.2.5 Fermionic Versus Bosonic Baths

In this example, we show the difference between bosonic bath and fermionic bath in a specific example, the N-fermion model discussed above. It is instructive to consider a simple case with two fermions in the system ($N_S = 2$). The Hamiltonian is then given by

$$H_S = \omega_1 a_1^\dagger a_1 + \omega_2 a_2^\dagger a_2, \tag{10.85}$$

$$L = a_1 + a_2, \tag{10.86}$$

it is easy to show that the exact $\bar{Q}$ operator is

$$\bar{Q} = X_1(t)a_1 + X_2(t)a_2 \tag{10.87}$$

where $X_1(t)$ and $X_2(t)$ can be determined in (10.83) as $N_S = 2$ case. Then, the explicit master equation can be written as

$$\begin{aligned}\frac{d}{dt}\rho = &-i\omega_1(a_1^\dagger a_1\rho - \rho a_1^\dagger a_1) - i\omega_2(a_2^\dagger a_2\rho - \rho a_2^\dagger a_2)\\ &+ \{X_1^*(t)(a_1\rho a_1^\dagger - \rho a_1^\dagger a_1) + X_1^*(t)(a_2\rho a_1^\dagger - \rho a_1^\dagger a_2)\\ &+ X_2^*(t)(a_1\rho a_2^\dagger - \rho a_2^\dagger a_1) + X_2^*(t)(a_2\rho a_2^\dagger - \rho a_2^\dagger a_2) + H.c.\}.\end{aligned} \tag{10.88}$$

On the other hand, we can also solve this model exactly if the two effective fermions (transformed from spins) are coupled to a bosonic bath. The Hamiltonian takes the same form as (10.85–10.86). The Hamiltonian of the bath and the interaction Hamiltonian are given by $H_B = \sum_i \omega_i c_i^\dagger c_i$ and $H_{int} = \sum_i (g_i c_i^\dagger L + H.c.)$ respectively. The difference between bosonic bath and fermionic bath arise from the definition of the operators c_i and $c_i^\dagger$. In the bosonic bath case, c_i and $c_i^\dagger$ represent a set of bosonic annihilation and creation operators satisfying $[c_i, c_j^\dagger] = \delta_{ij}$. In the fermionic bath case, the operators c_i and $c_i^\dagger$ represent fermionic annihilation and creation operators satisfying $\{c_i, c_j^\dagger\} = \delta_{ij}$. Using the non-Markovian NMQSD approach for bosonic bath [13, 14], the bosonic NMQSD equation can be derived as

$$\frac{\partial}{\partial t}|\psi(t, s, z^*)\rangle = [-iH_S + L^\dagger z_t^* - L^\dagger \bar{O}]|\psi(t, s, z^*)\rangle, \tag{10.89}$$

where $\bar{O}(t, z^*) = \int_0^t ds K(t, s)O(t, s, z^*)$. In the bosonic NMQSD equation, the noise $z_t^* = -i\sum_i g_i^* e^{i\omega_i t} z_i^*$ is the complex (not Grassmann) Gaussian noise. The exact $\bar{O}$ operator is determined as follows [17]

$$\begin{aligned}\bar{O}(t, z^*) = &X_1(t)a_1 + X_2(t)a_2 + X_3(t)a_1^\dagger a_1 a_2\\ &+ X_4(t)a_2^\dagger a_1 a_2 + i\int_0^t ds' X_5(t, s')z_{s'}^* a_1 a_2.\end{aligned} \tag{10.90}$$

The coefficients in (10.90) satisfy the following differential equations

$$\frac{\partial}{\partial t}x_1(t,s) = i\omega_a x_1 + x_1 X_1 + x_2 X_1, \tag{10.91}$$

$$\frac{\partial}{\partial t}x_2(t,s) = i\omega_b x_2 + x_1 X_2 + x_2 X_2, \tag{10.92}$$

$$\begin{aligned}\frac{\partial}{\partial t}x_3(t,s) = i\omega_b x_3 - x_4 X_2 + x_3 X_2 + x_2 X_3 \\ + x_3 X_3 - x_3 X_4 - x_2 X_4 - iX_5,\end{aligned} \tag{10.93}$$

$$\begin{aligned}\frac{\partial}{\partial t}x_4(t,s) = i\omega_a x_4 + x_4 X_1 + x_1 X_4 - x_1 X_3 \\ - x_3 X_1 + x_4 X_3 - x_4 X_4 - iX_5,\end{aligned} \tag{10.94}$$

$$\begin{aligned}\frac{\partial}{\partial t}x_5(t,s,s') = i\omega_a x_5 + i\omega_b x_5 + x_5 X_1 + x_5 X_2 \\ + x_5 X_3 - x_5 X_4 + x_1 X_5 + x_2 X_5,\end{aligned} \tag{10.95}$$

with the initial conditions

$$x_1(t, s=t) = 1, \tag{10.96}$$

$$x_2(t, s=t) = 1, \tag{10.97}$$

$$x_3(t, s=t) = 0, \tag{10.98}$$

$$x_4(t, s=t) = 0, \tag{10.99}$$

$$x_5(t, s=t, s') = 0, \tag{10.100}$$

$$ix_5(t, s, s'=t) = 2(x_2 - x_1) + x_3 + x_4, \tag{10.101}$$

and

$$X_j(t) = \int_0^t K(t,s)x_j(t,s)ds \ (j = 1 \text{ to } 4), \tag{10.102}$$

$$X_5(t,s') = \int_0^t K(t,s)x_5(t,s,s')ds. \tag{10.103}$$

We use this particular example to illustrate some differences between the bosonic and fermionic baths. As shown above, we can find the exact Q (O) operators for both the fermionic bath and the bosonic counterpart. Since the exact dynamic evolution of the system will be fully determined by Q (O) operators, so we may compare the difference between the two operators given in (10.87) and (10.90), respectively. The

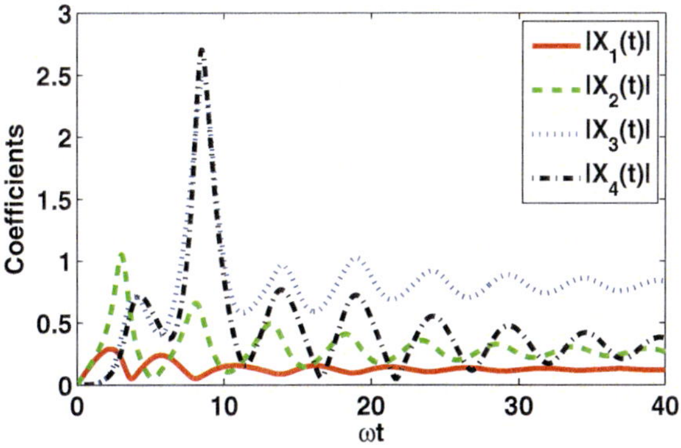

Fig. 10.4 Dynamic evolution of the coefficients in $\bar{Q}$ (for fermionic bath) and $\bar{O}$ (for bosonic bath) operators. The parameters are $\omega_1 = 2$, $\omega_2 = \omega = 1$, $\gamma = 0.4$, $\Omega = \pi/4$

first two terms are the same for both the Q and O operators, and the difference arises from other terms. When $X_1(t)$ and $X_2(t)$ are dominant, there is no big difference between the fermionic and bosonic baths. For example, when $\omega_1 = \omega_2$, two operators Q and O are exactly the same. However, we found that the extra terms $X_3(t)$ and $X_4(t)$ occurred in O may become important under certain conditions. As shown in Fig. 10.4, the fermionic and bosonic baths may result in very different dynamics.

10.2.3 Summary

In this section, we have developed a novel technique called the fermionic state diffusion approach that is a useful tool for studying an open quantum system coupled to a fermionic bath. By using the Grassmann coherent states, the exact fermionic NMQSD equation and the corresponding master equation are derived. It is worth to note that (10.33) and (10.40) are directly derived from the microscopic Hamiltonian without using any approximations (especially without the Markov approximation). Therefore, these equations can successfully capture the important dynamical features beyond the Markov evolution. To illustrate the fermionic NMQSD approach, we have solved several interesting models and discussed the non-Markovian features of the dynamics including a comparison between the bosonic bath and the fermionic bath.

10.3 NMQSD Theory for a Quantum System Coupled to a Hybrid Bath

In the last section, we have developed a systematic approach to solving open quantum systems coupled to a fermionic bath. However, in a more practical case, an environment can be very complex, and it may contain both bosons and fermions, so a unified description of a hybrid bath is needed [35–37]. From a more fundamental point of view, the dynamics of open quantum systems embedded in one or more environments has attracted a wide-spread interest in the area of quantum foundation such as decoherence, dissipation, quantum measurement etc. [2, 4–6, 17, 21, 38–43].

Zhang et al. [44] derived a master equation applicable to either bosonic bath or fermionic bath. A further step is to develop a dynamic equation for an open quantum system coulped to a hybrid environment containing both bosonic bath and fermionic bath simultaneously [45, 46]. For example, quantum dots can interact with two fermionic reservoirs (source and drain) and a phonon (bosonic) environment. In this case, the combined environment is a hybrid one. In this example, both the fermionic baths and the bosonic bath may significantly affect the dynamics of the quantum dots of interest. A dynamical equation incorporating two kinds of baths can provide us with a powerful tool to study many realistic open system problems. In this part, we will achieve this goal by developing a hybrid NMQSD approach. The bosonic NMQSD approach and fermionic NMQSD have been developed respectively [13, 28, 29, 34]. In the NMQSD approach for either bosons or fermions, the central idea is to encode the influence of the environment into a stochastic process z_t^* (bosonic) or ξ_t^* (fermionic). Taking the statistic average over the stochastic variables is equivalent to taking the partial trace over the environment yielding the reduced density matrix. The difference between the bosonic and the fermionic environments is that z_t^* is a complex Gaussian noise arising from the use of the bosonic coherent states, while ξ_t^* is a Grassmann stochastic process due to the fermionic coherent state representation. Therefore, for a hybrid bath, we may introduce two types of noises simultaneously and derive a stochastic equation with both the bosonic Gaussian noise and the fermionic Grassmann noise.

10.3.1 Hybrid Baths: Commutative and Anti-commutative Cases

The hybrid open system we considered is illustrated in Fig. 10.5, where a quantum system interacts with a combined bath consisting of both bosons and fermions. The model can be described by the following Hamiltonian [30]

$$H_{tot} = H_S + H_{FB} + H_{BB} + H_{FI} + H_{BI}, \tag{10.104}$$

where H_S describes the Hamiltonian of the system, $H_{BB} = \sum_r \Omega_r b_r^\dagger b_r$ and $H_{FB} = \sum_k \epsilon_k c_k^\dagger c_k$ represent the bosonic bath and the fermionic bath, respectively. Here "b_r" and "c_k" are the annihilation operators for a single mode of the bosonic bath and the fermionic bath respectively. The interaction between the system and the combined bath is given by $H_{BI} = \sum_r \lambda_r b_r^\dagger L_b + \text{H.c.}$ and $H_{FI} = \sum_k \mu_k c_k^\dagger L_f + \text{H.c.}$, where L_b, and L_f are the bosonic and fermionic coupling operators, respectively. Typically, the bosonic bath commute with both the fermionic bath and the system no matter

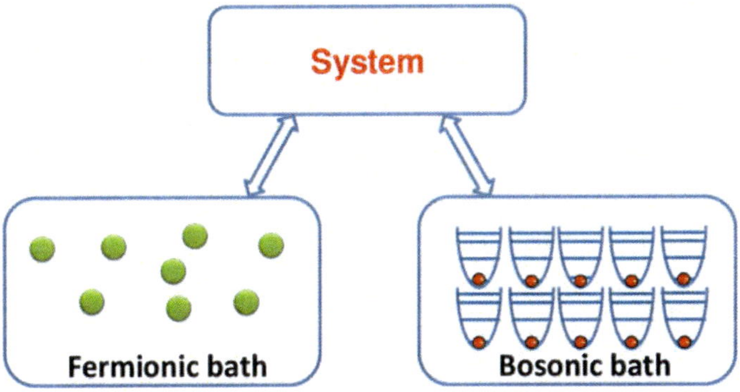

Fig. 10.5 An example of commutative hybrid bath. The original spin-chain bath on the left side can be transformed into an effective fermionic bath [28]. In this case, the effective fermionic bath commutes with the system since the original spin-chain commutes with the system

whether the system is composed of fermions or bosons. However, the commutation relation between system and fermionic bath indicates that we should treat the hybrid case in two different categories.

Case 1: System of interest commutes with the fermionic bath
In this case, the Hamiltonian (10.104) typically describes an effective fermionic bath. For example, if we consider a spin-chain bath, then the spin bath may be transformed into an effective fermionic bath by performing the Jordan-Wigner transformation and the Fourier transformation [28, 31]. In the case of an effective fermionic bath, the creation and annihilation operators $c_k^\dagger$ and c_k will commute with any operators living in the Hilbert space of the system.

Case 2: System of interest anti-commutes with the fermionic bath
In case 2, the fermionic bath is composed of many electrons and the system of interest is also composed of electrons. Since the electrons in the fermionic bath and system are indistinguishable, the fermionic bath will anti-commute with the system. One typical example in this case is the quantum dot model where the Hamiltonian of the system is $H_S = \omega_d d^\dagger d$, where the operator "$d$" defined in the quantum dot system and the operators "c_k" defined the fermionic bath satisfy the anti-commutation relations $\{d, c_k\} = 0$.

For the two categories described above, we will develop two different schemes in the following sections illustrated with a few concrete examples.

10.3.2 Commutative Hybrid Bath

10.3.2.1 The General Stochastic SchrÖdinger Equation

First, we consider the case where the system commutes with an effective fermionic bath. In this case, the total Hamiltonian may be transformed into the interaction picture:

$$H_{tot}^{int} = H_S + (\sum_k \mu_k c_k^\dagger L_f e^{i\epsilon_k t} + \sum_r \lambda_r b_r^\dagger L_b e^{i\Omega_r t} + \text{H.c.}). \tag{10.105}$$

By introducing the bosonic coherent states and fermionic coherent states (See Appendix)

$$|z\rangle = \prod_r \exp\{z_r b_r^\dagger\}|0\rangle, \tag{10.106}$$

$$|\xi\rangle = \prod_k (1 - \xi_k c_k^\dagger)|0\rangle, \tag{10.107}$$

A generalized stochastic state vector can be defined as

$$|\psi_t(z^*, \xi^*)\rangle = \langle z^*, \xi^*|\psi_{tot}(t)\rangle. \tag{10.108}$$

Throughout this chapter, we use the short-notation $|\psi_t\rangle$ to denote the stochastic pure state if no confusion arises. It should be noted that the two types of noise variables introduced here are different. For the bosonic coherent states, z_r is a complex variable, while for the fermionic coherent states, ξ_k is a Grassmann variable satisfying anti-commutation relations $\{\xi_i, \xi_j\} = 0$. Applying the Schrödinger equation to the total system, the dynamic equation for the stochastic state vector may be derived as

$$\begin{aligned}\frac{\partial}{\partial t}|\psi_t\rangle &= -i\langle z^*, \xi^*|H_{tot}^{int}(t)|\psi_{tot}(t)\rangle \\ &= [-iH_S + L_f \xi_t^* - L_f^\dagger \int ds K_f(t,s)\frac{\delta_l}{\delta \xi_s^*} \\ &\quad + L_b z_t^* - L_b^\dagger \int ds K_b(t,s)\frac{\delta}{\delta z_s^*}]\,|\psi_t\rangle, \end{aligned} \tag{10.109}$$

where $K_b(t,s) = \sum_r \lambda_r^2 e^{-i\Omega_r(t-s)}$ and $K_f(t,s) = \sum_k \mu_k^2 e^{-i\epsilon_k(t-s)}$ are two correlation functions for the bosonic and fermionic baths respectively. Equation (10.109) is a dynamical equation governing the generalized stochastic state vector $|\psi_t\rangle$. Note that this equation contains two different stochastic processes:

$$z_t^* = -i \sum_r z_r^* e^{i\Omega_r t}, \tag{10.110}$$

$$\xi_t^* = -i \sum_k \xi_k^* e^{i\epsilon_k t}, \tag{10.111}$$

where z_t^* is a complex Gaussian process, and ξ_t^* is a Grassmann Gaussian process. They satisfy the following statistical relations

$$M_b\{z_t\} = M_b\{z_t^*\} = 0, \;\; M_b\{z_t z_s^*\} = K_b(t,s), \tag{10.112}$$

$$M_f\{\xi_t\} = M_f\{\xi_t^*\} = 0, \;\; M_f\{\xi_t \xi_s^*\} = K_f(t,s). \tag{10.113}$$

The statistical averages over the complex noise and the Grassmann noise are defined as $M_b\{\cdot\} = \int \prod_r \frac{1}{\pi} e^{-|z_r|^2} dz_r^2 [\cdot]$ and $M_f\{\cdot\} = \int \prod_k d\xi_k^* d\xi_k e^{-\xi_k^* \xi_k} [\cdot]$, respectively.

The difficulty of solving (10.109) arises from the functional derivatives. However, similar to the technique used in Refs. [14, 28, 29, 34], we can replace these functional derivatives by two time dependent operators denoted as O and Q, respectively:

$$\frac{\delta}{\delta z_s^*} |\psi_t\rangle = O(t, s, z^*, \xi^*) |\psi_t\rangle, \tag{10.114}$$

$$\frac{\delta}{\delta \xi_s^*} |\psi_t\rangle = Q(t, s, \xi^*, z^*) |\psi_t\rangle, \tag{10.115}$$

then, the NMQSD equation can be written in a compact form as

$$\frac{\partial}{\partial t} |\psi_t\rangle = [-iH_s + L_f \xi_t^* - L_f^\dagger \bar{Q} + L_b z_t^* - L_b^\dagger \bar{O}] |\psi_t\rangle, \tag{10.116}$$

where $\bar{O}(t, z^*, \xi^*) = \int_0^t K_b(t,s) O(t, s, z^*, \xi^*) ds$, $\bar{Q}(t, z^*, \xi^*) = \int_0^t K_f(t,s) Q(t, s, z^*, \xi^*) ds$. The applications of the NMQSD equation are dependent on whether these time-dependent operators O and Q can be determined. Using the consistency condition,

$$\frac{\partial}{\partial t} \frac{\delta}{\delta z_s^*} |\psi_t\rangle = \frac{\delta}{\delta z_s^*} \frac{\partial}{\partial t} |\psi_t\rangle, \tag{10.117}$$

$$\frac{\partial}{\partial t} \frac{\delta}{\delta \xi_s^*} |\psi_t\rangle = \frac{\delta}{\delta \xi_s^*} \frac{\partial}{\partial t} |\psi_t\rangle, \tag{10.118}$$

we have

$$\frac{\partial}{\partial t}O = [-iH_s + L_f\xi_t^* - L_f^\dagger\bar{Q} + L_b z_t^* - L_b^\dagger\bar{O}, O] - L_b^\dagger\frac{\delta}{\delta z_s^*}\bar{O} - L_f^\dagger\frac{\delta}{\delta z_s^*}\bar{Q}, \tag{10.119}$$

$$\frac{\partial}{\partial t}Q = [-iH_s, Q] - \{L_f\xi_t^*, Q\} + [L_b z_t^*, Q] - L_f^\dagger\bar{Q}(-\xi^*)Q + QL_f^\dagger\bar{Q}$$
$$-L_b^\dagger\bar{O}(-\xi^*)Q + QL_b^\dagger\bar{O} - L_b^\dagger\frac{\delta_l}{\delta\xi_s^*}\bar{O} - L_f^\dagger\frac{\delta_l}{\delta\xi_s^*}\bar{Q}, \tag{10.120}$$

with the initial conditions

$$O(t, t, z^*, \xi^*) = L_b, \tag{10.121}$$

$$Q(t, t, z^*, \xi^*) = L_f. \tag{10.122}$$

Given these conditions, the O and Q operators may be fully determined for many physically interesting cases. As a result, the NMQSD equation (10.116) can be solved or applied to deriving the corresponding master equations. We can reproduce the reduced density matrix from the stochastic state vector $|\psi_t\rangle$ as

$$\rho(t) = M_b\{M_f\{P_t\}\} \tag{10.123}$$

where $P_t \equiv |\psi_t(z^*, \xi^*)\rangle\langle\psi_t(z^*, -\xi^*)|$ is the stochastic density operator. Given the relation (10.123), the physical meaning of the NMQSD equation become clear. By choosing a random realization of the noises z_t^* and ξ_t^* (reflecting the states of the environment), the evolution of the reduced density matrix is decomposed into many pure-states commonly called quantum trajectories $|\psi_t\rangle$. However, taking the statistical average over all of these trajectories, the reduced density matrix may be reproduced. Therefore, the complicated properties of the environment are all encoded into the noise functions z_t^* and ξ_t^*. Therefore, tracing out the environment is equivalent to taking average over all the realizations of the stochastic processes. Based on (10.123), the master equation can be derived as

$$\frac{d}{dt}\rho = -i[H_S, \rho] + [L_f, M_b\{M_f\{P_t\bar{Q}^\dagger(-\xi)\}\}] + [M_b\{M_f\{\bar{Q}P_t\}\}, L_f^\dagger]$$
$$+[L_b, M_b\{M_f\{P_t\bar{O}^\dagger\}\}] + [M_b\{M_f\{\bar{O}P_t\}\}, L_b^\dagger] \tag{10.124}$$

where the Novikov theorem for fermionic case [28] and the bosonic case [14] are used in the derivation. Although taking the statistical average $M_b\{\cdot\}$ and $M_f\{\cdot\}$ is by no means a simple process in general. When the operators O and Q are noise-independent, we may easily find the exact master equation,

$$\frac{d}{dt}\rho = -i[H_S, \rho] + \{[\bar{Q}\rho, L_f^\dagger] + [\bar{O}\rho, L_b^\dagger] + H.c.\}, \tag{10.125}$$

Actually, this case may apply to many interesting models [14, 28, 42] in which the exact O or Q operators do not contain noises. Moreover, in general cases, we can still expand O and Q into functional series and only take the first term (the zeroth order in terms of the noise variables) of the expansions as $O(t, s, z^*, \xi^*) \approx O^{(0)}(t, s)$ and $Q(t, s, z^*, \xi^*) \approx Q^{(0)}(t, s)$. This approximation is called the zeroth-order approximation. The validity and accuracy of this approximation are analyzed in [47].

10.3.2.2 Example 1: Two Qubits in a Hybrid Bath

In order to illustrate the NMQSD approach for the hybrid bath we have discussed above, we will solve several interesting examples in details. In the first example, we will consider two qubits interacting with two dissipative baths, one is a bosonic bath and the other the fermionic bath. In the context of the general model described by Fig. 10.5 and (10.104), this particular example is defined as,

$$H_S = \frac{\omega}{2}(\sigma_z^A + \sigma_z^B), \tag{10.126}$$

$$L_b = L_f = \sigma_-^A + \kappa_B \sigma_-^B, \tag{10.127}$$

where κ_B is a parameter describing the coupling strength of the second qubit. First, we will investigate the case $\kappa_B = 1$, namely the second qubit is involved in the system evolution. A special case where $\kappa_B = 0$ will be considered later, which means the second qubit evolves independently from the other parts and the model reduces to a single qubit case. Given this specific model, the NMQSD equation can be written as

$$\frac{\partial}{\partial t}|\psi_t\rangle = [-i\frac{\omega}{2}(\sigma_z^A + \sigma_z^B) + (\sigma_-^A + \sigma_-^B)(\xi_t^* + z_t^*) - (\sigma_+^A + \sigma_+^B)(\bar{Q} + \bar{O})]|\psi_t\rangle, \tag{10.128}$$

where the exact O and Q operators can be determined as

$$O = f_1(t,s)O_1 + f_2(t,s)O_2 + i\int_0^t ds' f_3(t,s,s')z_{s'}^* O_3 + i\int_0^t ds' f_4(t,s,s')\xi_{s'}^* O_4, \tag{10.129}$$

$$Q = g_1(t,s)Q_1 + g_2(t,s)Q_2 + i\int_0^t ds' g_3(t,s,s')z_{s'}^* Q_3 + i\int_0^t ds' g_4(t,s,s')\xi_{s'}^* Q_4. \tag{10.130}$$

For this case, the basis operators are $O_1 = \sigma_-^A + \sigma_-^B$, $O_2 = (\sigma_z^A + \sigma_z^B)(\sigma_-^A + \sigma_-^B)$, $O_3 = O_4 = \sigma_-^A\sigma_-^B$, $Q_1 = \sigma_-^A + \sigma_-^B$, $Q_2 = (\sigma_z^A + \sigma_z^B)(\sigma_-^A + \sigma_-^B)$, $Q_3 = Q_4 = \sigma_-^A\sigma_-^B$. The time-dependent coefficients satisfy the following relations

$$\frac{\partial}{\partial t} f_1(t,s) = i\omega f_1 + 4f_1F_2 + 4f_1G_2 + iF_3 + iG_3, \tag{10.131}$$

$$\frac{\partial}{\partial t} f_2(t,s) = i\omega f_2 + f_1(4F_2 + 4G_2 - F_1 - G_1) - \frac{i}{2}F_3 \\ + f_2(2F_1 + 2G_1 - 4F_2 - 4G_2) - \frac{i}{2}G_3, \tag{10.132}$$

$$\frac{\partial}{\partial t} f_3(t,s,s') = 2i\omega f_3 + 2f_1F_3 + 2f_1G_3 - 4f_2F_3 - 4f_2G_3 + 2f_3F_1 + 2f_3G_1, \tag{10.133}$$

$$\frac{\partial}{\partial t} f_4(t,s,s') = 2i\omega f_4 + 2f_1F_4 + 2f_1G_4 - 4f_2F_4 - 4f_2G_4 + 2f_4F_1 + 2f_4G_1, \tag{10.134}$$

$$\frac{\partial}{\partial t} g_1(t,s) = i\omega g_1 + 4g_1F_2 + 4g_1G_2 + iF_3 + iG_3, \tag{10.135}$$

$$\frac{\partial}{\partial t} g_2(t,s) = i\omega g_2 + g_1(4F_2 + 4G_2 - F_1 - G_1) - \frac{i}{2}F_3 \\ + g_2(2F_1 + 2G_1 - 4F_2 - 4G_2) - \frac{i}{2}G_3, \tag{10.136}$$

$$\frac{\partial}{\partial t} g_3(t,s,s') = 2i\omega g_3 + 2g_1F_3 + 2g_1G_3 - 4g_2F_3 - 4g_2G_3 + 2g_3F_1 + 2g_3G_1, \tag{10.137}$$

$$\frac{\partial}{\partial t} g_4(t,s,s') = 2i\omega g_4 - 2g_1F_4 - 2g_1G_4 + 4g_2F_4 + 4g_2G_4 + 2g_4F_1 + 2g_4G_1, \tag{10.138}$$

with the initial conditions

$$f_1(t,t) = g_1(t,t) = 1, \quad f_2(t,t) = g_2(t,t) = 0, \tag{10.139}$$

$$f_3(t,t,s') = f_4(t,t,s') = g_3(t,t,s') = g_4(t,t,s') = 0, \tag{10.140}$$

$$f_3(t,s,t) = f_4(t,s,t) = -4if_2(t,s), \tag{10.141}$$

$$g_3(t,s,t) = -4if_2(t,s), \; g_4(t,s,t) = -4if_1(t,s) + 4if_2(t,s), \tag{10.142}$$

where

$$F_i(t) = \int_0^t K_b(t,s) f_i(t,s) ds, \quad (i = 1, 2) \tag{10.143}$$

$$G_i(t) = \int_0^t K_f(t,s) g_i(t,s) ds, \quad (i = 1, 2) \tag{10.144}$$

$$F_i(t,s') = \int_0^t K_b(t,s) f_i(t,s,s') ds, \quad (i = 3, 4) \tag{10.145}$$

$$G_i(t,s') = \int_0^t K_f(t,s) g_i(t,s,s') ds. \quad (i = 3, 4) \tag{10.146}$$

In this example, the bosonic O operator contains the fermionic noise ξ^*, while the fermionic Q operator also contains the bosonic noise z^*. Notably, the O and Q operators in the hybrid bath case may not be just a simple combination of the operators appearing in a single bath case. In fact, there are many cross terms and they are mutually coupled. This reflects the fact that the effect of the hybrid bath can not be simply treated as the summation of individual baths either being a fermionic bath or a bosonic bath. Through coupling to the system, the cross talk between the two uncoupled baths can be generated.

10.3.2.3 Single Qubit Case: Consistency with the Standard Quantum Mechanics

The O and Q operators in the two-qubit example are rather complicated. It is still possible to find a simple form for some simple cases. Now we consider a special case, the single qubit case, namely $\kappa_B = 0$. In this case, the second qubit evolves independently from all the other parts, so it can be removed in the interaction picture. Therefore, the NMQSD equation reduces to

$$\frac{\partial}{\partial t}|\psi_t\rangle = [-i\frac{\omega}{2}\sigma_z^A + \sigma_-^A(\xi_t^* + z_t^*) - \sigma_+^A(\bar{Q} + \bar{O})]|\psi_t\rangle, \tag{10.147}$$

where the exact O and Q operators can be determined as

$$O = Q = f(t,s)\sigma_-^A, \tag{10.148}$$

The time-dependent coefficient $f(t,s)$ satisfies the equation

$$\frac{\partial}{\partial t}f(t,s) = [i\omega + F(t)]f(t,s), \tag{10.149}$$

where $F(t) = \int_0^t [K_b(t,s) + K_f(t,s)]f(t,s)ds$. Finally, the exact master equation for this simple model can be derived as

$$\frac{d}{dt}\rho = -i[H_S, \rho] + \{F(t)[\sigma_-\rho, \sigma_+] + H.c.\}, \tag{10.150}$$

In general, the bosonic bath and fermionic bath can be very complicated, so the correlation functions $K_b(t,s)$ and $K_f(t,s)$ may be very complicated. However, for the special case to be considered here, we can show the result derived from NMQSD approach is completely consistent with that obtained from the standard quantum mechanics. Consider the special case that there are only one boson and one fermion in the bosonic bath and the fermionic bath respectively, i.e., $H_{FB} = \epsilon c^\dagger c$, $H_{BB} = \Omega b^\dagger b$. Therefore, for this special case, the correlation functions reduce to

$$K_b(t,s) = \lambda^2 e^{-i\Omega(t-s)}, \tag{10.151}$$

$$K_f(t,s) = \mu^2 e^{-i\epsilon(t-s)}. \tag{10.152}$$

In the resonance case, $\omega = \epsilon = \Omega$, $\lambda = \mu$, the equation for $F(t)$ is

$$\frac{\partial}{\partial t} F(t) = 2\lambda^2 + F^2(t). \tag{10.153}$$

The solution is

$$F(t) = \sqrt{2}\lambda \tan(\sqrt{2}\lambda t). \tag{10.154}$$

From the master equation (10.150), the evolution of the off-diagonal element ρ_{21} is

$$\frac{d}{dt}\rho_{21}(t) = i\omega\rho_{21} - F^*(t)\rho_{21}. \tag{10.155}$$

Finally, the solution of $\rho_{21}(t)$ is

$$\rho_{21}(t) = \rho_{21}(0)e^{i\omega t}\cos(\sqrt{2}\lambda t). \tag{10.156}$$

On the other hand, this simple example can be easily solved by solving the whole system with Schrödinger equation since the total system only contains three particles. It is easy to check, for this simple example, solving the whole system will give the identical result to the result obtained by using the NMQSD approach.

10.3.2.4 Example 2: Single Qubit Coupled to a Dephasing Bosonic Bath and a Dissipative Fermionic Bath

In the second example, we will investigate a case that will involve an open system coupled to the bosonic bath and the fermionic bath in two different ways. The model we considered is

$$H_S = \frac{\omega}{2}\sigma_z, \tag{10.157}$$

$$L_b = \sigma_z, \quad L_f = \sigma_-. \tag{10.158}$$

According to the general discussion in Sect. 10.3.2, the NMQSD equation for this model can be formally written as

$$\frac{\partial}{\partial t}|\psi_t\rangle = [-i\frac{\omega}{2}\sigma_z + \sigma_-\xi_t^* - \sigma_+\bar{Q} + \sigma_z z_t^* - \sigma_z\bar{O}]|\psi_t\rangle, \tag{10.159}$$

In this example, the O and the Q operators can contain an infinite number of the noise terms, therefore, for simplicity, we will use the zeroth-order functional expansion to yield the approximate zeroth order operators $O^{(0)}$ and $Q^{(0)}$. In the zeroth-order approximation, we have

$$O^{(0)} = \sigma_z, \tag{10.160}$$

$$Q^{(0)} = g(t,s)\sigma_-, \tag{10.161}$$

where the function $g(t,s)$ satisfies

$$\frac{\partial}{\partial t} g(t,s) = [i\omega + G(t)]g(t,s), \tag{10.162}$$

where $G(t) = \int_0^t g(t,s)K_f(t,s)ds$. Finally, the corresponding approximate master equation is

$$\frac{d}{dt}\rho = -i[\frac{\omega}{2}\sigma_z, \rho] + \{G(t)[\sigma_-\rho, \sigma_+] + F(t)[\sigma_z\rho, \sigma_z] + H.c.\}, \tag{10.163}$$

where $F(t) = \int_0^t K_b(t,s)ds$. Different from the first example where the model can be solved exactly, we show how to use the zeroth-order approximation to derive an approximate master equation in the second example. In the real application of the NMQSD approach, a systematic approximation method is certainly more important since the exact O and Q are quite difficult to find in many realistic models. With this approximation approach, we can still solve these problems with a good accuracy.

10.3.3 Anti-commutative Hybrid Bath

10.3.3.1 The General Stochastic Schrödinger Equation

After discussing the commutative hybrid bath, we now turn to the case where the system of interest anti-commutes with the fermionic bath, which tyically describes an electronic system (e.g., quantum dots). Following a similar procedure, one can also derive a NMQSD equation for the anti-commutive hybrid bath as

$$\begin{aligned}\frac{\partial}{\partial t}|\psi_t\rangle &= -i\langle z^*, \xi^*|H_{tot}^{int}(t)|\psi_{tot}(t)\rangle \\ &= [-iH_S - L_f\xi_t^* - L_f^\dagger \int ds K_f(t,s)\frac{\delta_l}{\delta\xi_s^*} \\ &\quad + L_b z_t^* - L_b^\dagger \int ds K_b(t,s)\frac{\delta}{\delta z_s^*}]|\psi_t\rangle.\end{aligned} \tag{10.164}$$

Since the operators L_f typically anti-commutes with the fermionic bath, as to be seen below, the functional derivation is slightly different from that defined in the commutative case [28, 29, 34] and we may notice certain difference between (10.109) and (10.164). The bosonic noise z_t^*, the fermionic noise ξ_t^* and the corresponding correlation functions $K_b(t,s)$ and $K_f(t,s)$ are all defined in a similar way. We can also define the bosonic O operator and the fermionic Q operator, however, they satisfy different differential equations in the anti-commutative case as

$$\frac{\partial}{\partial t}O = [-iH_s - L_f\xi_t^* - L_f^\dagger\bar{Q} + L_b z_t^* - L_b^\dagger\bar{O}, O] - L_b^\dagger\frac{\delta}{\delta z_s^*}\bar{O} - L_f^\dagger\frac{\delta}{\delta z_s^*}\bar{Q},$$

$$\begin{aligned}\frac{\partial}{\partial t}Q &= [-iH_s, Q] + [L_f\xi_t^*, Q] + [L_b z_t^*, Q] - L_f^\dagger\bar{Q}Q + QL_f^\dagger\bar{Q}\\ &\quad -L_b^\dagger\bar{O}Q + QL_f^\dagger\bar{O} - L_b^\dagger\frac{\delta_l}{\delta\xi_s^*}\bar{O} + L_f^\dagger\frac{\delta_l}{\delta\xi_s^*}\bar{Q},\end{aligned} \tag{10.165}$$

with the initial conditions

$$O(t,t,z^*,\xi^*) = L_b, \tag{10.166}$$

$$Q(t,t,z^*\xi^*) = L_f. \tag{10.167}$$

Then, the density matrix of the open quantum system can be reproduced as

$$\rho(t) = M_b\{M_f\{P_t\}\} \tag{10.168}$$

where the average is performed for both the complex noise and the fermionic noise and the master equation can be derived as

$$\begin{aligned}\frac{d}{dt}\rho &= -i[H_S,\rho] + [L_b, M_b\{M_f\{P_t\bar{O}^\dagger\}\}] + [M_b\{M_f\{\bar{Q}P_t\}\}, L_b^\dagger]\\ &\quad +[L_f, M_b\{M_f\{P_t\bar{Q}^\dagger(-\xi)\}\}] + [M_b\{M_f\{\bar{Q}P_t\}\}, L_f^\dagger]\end{aligned} \tag{10.169}$$

It should be noted that, in the derivation of the master equation, a new version of the Novikov theorem [29, 34] has been used. This new version is different from either the Novikov theorem for fermionic bath [28] or the one for the bosonic bath [14]. Similarly, when O and Q are noise-independent, the master equation is reduced to

$$\frac{d}{dt}\rho = -i[H_S,\rho] + \{[\bar{Q}\rho, L_f^\dagger] + [\bar{O}\rho, L_b^\dagger] + H.c.\}. \tag{10.170}$$

10.3.3.2 Example 3: Quantum Dot Coupled to a Hybrid Bath

In order to show the details of solving an anti-commutative hybrid bath problem, we consider a realistic example that is the Anderson model in a bosonic environment (see Refs. [45, 46] for example). In this particular example the general Hamiltonian (10.104) becomes

$$H_S = \varepsilon d^\dagger d, \tag{10.171}$$

which describes a quantum dot, and

$$H_{FB} = \sum_{k,i=L,R} [\epsilon(k) - \mu_i] c_{ki}^\dagger c_{ki}, \quad H_{BB} = \sum_r \omega_r b_r^\dagger b_r, \tag{10.172}$$

represent two fermionic baths ("L" and "R") and one phonon bath.

$$H_{FI} = \sum_{k,i=L,R} t_{k,i} c_{ki}^\dagger d + H.c., \quad H_{BI} = \sum_r \lambda_r (d^\dagger d - \frac{1}{2})(b_r + b_r^\dagger), \tag{10.173}$$

are the interaction Hamiltonians describing the transport process and the dissipation process.

In a finite temperature case [48], we need to introduce two fictitious baths a_L and a_R with negative eigen-frequencies as:

$$\begin{aligned} H = H_S &+ \sum_{k,i=L,R} [\epsilon(k) - \mu_i] c_{ki}^\dagger c_{ki} + \{t_{ki} c_{ki}^\dagger d + \text{H.c.}\} \\ &+ \sum_r \lambda_r (d^\dagger d - \frac{1}{2})(b_r + b_r^\dagger) + \sum_r \Omega_r b_r^\dagger b_r \\ &+ \sum_{k,i=L,R} -[\epsilon(k) - \mu_i] a_{ki}^\dagger a_{ki}. \end{aligned} \tag{10.174}$$

Then, performing the Bogoliubov transformation gives

$$c_{ki} = \sqrt{1 - \bar{n}_{ki}} c'_{ki} + \sqrt{\bar{n}_{ki}} a'^\dagger_{ki} \quad (i = L, R), \tag{10.175}$$

$$a_{ki} = \sqrt{1 - \bar{n}_{ki}} a'_{ki} - \sqrt{\bar{n}_{ki}} c'^\dagger_{ki} \quad (i = L, R), \tag{10.176}$$

the total Hamiltonian become

$$\begin{aligned} H = H_S &+ \sum_{k,i=L,R} [\epsilon(k) - \mu_i] c'^\dagger_{ki} c'_{ki} \\ &+ \{t_{ki} (\sqrt{1 - \bar{n}_{ki}} c'^\dagger_{ki} + \sqrt{\bar{n}_{ki}} a'_{ki}) d + \text{H.c.}\} \\ &+ \sum_r \lambda_r (d^\dagger d - \frac{1}{2})(b_r + b_r^\dagger) + \sum_r \Omega_r b_r^\dagger b_r \\ &+ \sum_{k,i=L,R} -[\epsilon(k) - \mu_i] a'^\dagger_{ki} a'_{ki}. \end{aligned} \tag{10.177}$$

Redefining $\omega_{ki} = \epsilon(k) - \mu_i$, $g_{ki} = t_{ki}\sqrt{1-\bar{n}_{ki}}$, $f_{ki} = t_{ki}\sqrt{\bar{n}_{ki}}$, and using the interaction picture, the Hamiltonian can be written as

$$H_{int}(t) = H_S + \sum_r \lambda_r(d^\dagger d - \frac{1}{2})(b_r e^{-i\Omega_r t} + b_r^\dagger e^{i\Omega_r t})$$
$$+\{\sum_{k,i=L,R} g_{ki}e^{i\omega_{ki}t}c_{ki}'^\dagger d + f_{ki}e^{i\omega_{ki}t}a_{ki}' d + \text{H.c.}\} \quad (10.178)$$

By introducing the bosonic coherent state and the fermionic coherent states as follow:

$$|z\rangle = \prod_r \exp\{z_r b_r^\dagger\}|0\rangle, \quad (10.179)$$

$$|\xi_{ia}\rangle = \prod_k (1 - \xi_{kia}a_{ki}'^\dagger)|0\rangle \ (i = L, R), \quad (10.180)$$

$$|\xi_{ic}\rangle = \prod_k (1 - \xi_{kic}c_{ki}'^\dagger)|0\rangle \ (i = L, R). \quad (10.181)$$

The stochastic pure state vector can be defined as

$$|\psi_t(z^*, \xi_{La}^*, \xi_{Ra}^*, \xi_{Lc}^*, \xi_{Rc}^*)\rangle = \langle z^*, \xi_{La}^*, \xi_{Ra}^*, \xi_{Lc}^*, \xi_{Rc}^*|\psi_{tot}(t)\rangle. \quad (10.182)$$

Following the general approach discussed in the last section, the NMQSD equation governing the stochastic state vector for the quantum open system is derived as

$$\frac{\partial}{\partial t}|\psi_t\rangle = -i\langle z^*, \xi_{La}^*, \xi_{Ra}^*, \xi_{Lc}^*, \xi_{Rc}^*|H_{int}(t)|\psi_{tot}(t)\rangle$$
$$= H_{eff}|\psi_t\rangle,$$

where

$$H_{eff} = -iH_S + d^\dagger \int_0^t ds K_{La}(t,s)\frac{\delta}{\delta\xi_{La,s}^*} + d\xi_{La,t}^*$$
$$+d^\dagger \int_0^t ds K_{Ra}(t,s)\frac{\delta}{\delta\xi_{Ra,s}^*} + d\xi_{Ra,t}^* - d\int_0^t ds K_{Lc}(t,s)\frac{\delta}{\delta\xi_{Lc,s}^*} - d^\dagger\xi_{Lc,t}^*$$
$$-d\int_0^t ds K_{Rc}(t,s)\frac{\delta}{\delta\xi_{Rc,s}^*} - d^\dagger\xi_{Rc,t}^* - b^\dagger\int_0^t ds\alpha(t,s)\frac{\delta}{\delta z_s^*} + dz_t^*. \quad (10.183)$$

In this equation, we have introduced the following noises,

$$z_t^* = -i\sum_r z_r^* e^{i\Omega_r t}, \quad (10.184)$$

$$\xi_{ia,t}^{*} = -i\sum_{k}\xi_{kia}^{*}e^{-i\omega_{ki}t},\ (i = L, R), \tag{10.185}$$

$$\xi_{ic,t}^{*} = -i\sum_{k}\xi_{kic}^{*}e^{i\omega_{ki}t},\ (i = L, R), \tag{10.186}$$

and the corresponding correlation functions are

$$\alpha(t,s) = \sum_{r}\lambda_{r}^{2}e^{-i\Omega_{r}(t-s)}, \tag{10.187}$$

$$K_{ia}(t,s) = \sum_{k}g_{ki}^{2}e^{i\omega_{ki}(t-s)}\quad (i = L, R), \tag{10.188}$$

$$K_{ic}(t,s) = \sum_{k}f_{ki}^{2}e^{-i\omega_{ki}(t-s)}\quad (i = L, R), \tag{10.189}$$

Among the noises defined above, z_t^* is a complex Gaussian noise, and $\xi_{ia,t}^*$ and $\xi_{ic,t}^*$ are Grassmann Gaussian noises. They satisfy the following statistical relations

$$M_b\{z_t\} = M_b\{z_t^*\} = 0, \tag{10.190}$$

$$M_b\{z_t^* z_s\} = \alpha(t,s), \tag{10.191}$$

$$M_f\{\xi_{ia,t}^*\} = M_f\{\xi_{ia,t}\} = M_f\{\xi_{ic,t}^*\} = M_f\{\xi_{ic,t}\} = 0, \tag{10.192}$$

$$M_f\{\xi_{ia,t}^*\xi_{ia,s}\} = K_{ia}(t,s),\ M_f\{\xi_{ic,t}^*\xi_{ic,s}\} = K_{ic}(t,s). \tag{10.193}$$

Using the technique discussed in Sects. 10.3.2 and 10.3.3, the time-dependent operators O and Q are defined as

$$\frac{\delta}{\delta z_s^*}|\psi_t\rangle = O(t,s,z^*)|\psi_t\rangle, \tag{10.194}$$

$$\frac{\delta}{\delta\xi_{ia,s}^*}|\psi_t\rangle = Q_{ia}(t,s,\xi_{ia}^*)|\psi_t\rangle\quad (i = L, R), \tag{10.195}$$

$$\frac{\delta}{\delta\xi_{ic,s}^*}|\psi_t\rangle = Q_{ic}(t,s,\xi_{ic}^*)|\psi_t\rangle\quad (i = L, R). \tag{10.196}$$

and the zeroth-order approximation gives the solution of these operators as

$$O \approx f_1(t,s)d^{\dagger}d, \tag{10.197}$$

$$Q_{ic} \approx f_{ic}(t,s)d \ (i = L, R), \tag{10.198}$$

$$Q_{ia} \approx f_{ia}(t,s)d^{\dagger} \ (i = L, R), \tag{10.199}$$

while the coefficients satisfy

$$\frac{\partial}{\partial t} f_1(t,s) = 0 \tag{10.200}$$

$$\frac{\partial}{\partial t} f_{Lc}(t,s) = (i\varepsilon + F_1 + F_{La} + F_{Ra} + F_{Lc} + F_{Rc}) f_{Lc} \tag{10.201}$$

$$\frac{\partial}{\partial t} f_{Rc}(t,s) = (i\varepsilon + F_1 + F_{La} + F_{Ra} + F_{Lc} + F_{Rc}) f_{Rc} \tag{10.202}$$

$$\frac{\partial}{\partial t} f_{La}(t,s) = (-i\varepsilon - F_1 - F_{La} - F_{Ra} - F_{Lc} - F_{Rc}) f_{La} \tag{10.203}$$

$$\frac{\partial}{\partial t} f_{Ra}(t,s) = (-i\varepsilon - F_1 - F_{La} - F_{Ra} - F_{Lc} - F_{Rc}) f_{Ra} \tag{10.204}$$

where $F_1 = \int_0^t \alpha(t,s) f_1(t,s) ds$, $F_{Lc} = \int_0^t K_{Lc}(t,s) f_{Lc}(t,s) ds$, $F_{Rc} = \int_0^t K_{Rc}(t,s) f_{Rc}(t,s) ds$, $F_{La} = \int_0^t K_{La}(t,s) f_{La}(t,s) ds$, $F_{Ra} = \int_0^t K_{Ra}(t,s) f_{Ra}(t,s) ds$.

Finally, the master equation is derived as

$$\begin{aligned} \frac{\partial}{\partial t}\rho = & -i\varepsilon[d^{\dagger}d, \rho] + \{(F_{Lc} + F_{Rc})[d\rho, d^{\dagger}] \\ & +(F_{La} + F_{Ra})[d, d^{\dagger}\rho] \\ & +F_1[d^{\dagger}d\rho, d^{\dagger}d] + H.c.\} \end{aligned} \tag{10.205}$$

In the third example, the hybrid NMQSD approach is applied to the Anderson model in a bosonic dephasing environment. First, we show how to map a finite temperature problem into a zero temperature problem to apply the hybrid NMQSD approach in a finite temperature case. More importantly, we show the hybrid NMQSD approach provide us with a powerful tool to investigate the dynamics of a quantum system in a non-Markovian regime. Typically, in Markov case, all of the coefficients in the master equation, F_{Lc}, F_{Rc}, F_{La}, F_{Ra}, and F_1 are all constants. However, in (10.205), those coefficients are time-dependent, which reflects the non-Markovian behavior even if we only consider the zeroth order O and Q operators.

10.3.3.3 Fermionic Bath Versus Bosonic Bath

Based on the master equation derived in (10.205), we will compare the fermionic bath and bosonic bath. In order to show the impact of two different types of baths, we introduce two parameters in the original Hamiltonian to describe the coupling strength to the fermionic bath and bosonic bath. As it is shown in (10.173), t_{ki} determines the strength of the interaction between the system and fermionic bath, and λ_r determine the strength of the coupling to the bosonic bath. In the numerical simulation, we will introduce c_f and c_b to control the global coupling strength for the fermionic bath and the bosonic bath respectively. Namely, we replace t_{ki} by $\sqrt{c_f}t_{ki}$ and λ_r by $\sqrt{c_b}\lambda_r$. These two parameters will reflect the global coupling strength. If we take $c_b = 0$, then the bosonic bath is shut off, and we can see the evolution without the impact of the bosonic bath. In the numerical simulation, we use four Ornstein-Uhlenbeck noises $K_{mn}(t, s) = \frac{\Gamma_{mn}}{2} \exp[(-\gamma_{mn} + i\phi_{mn})|t - s|]$ $(m = L, R;\ n = a, c)$ to simulate the correlation functions K_{La}, K_{Lc}, K_{Ra}, K_{Rc}. The parameters are chosen as $\Gamma_{Lc} = 0.017$, $\gamma_{Lc} = 0.3$, $\phi_{Lc} = 1.1$, $\Gamma_{Rc} = 0.034$, $\gamma_{Rc} = 0.5$, $\phi_{Rc} = 1.65$, $\Gamma_{La} = 0.012$, $\gamma_{La} = 0.4$, $\phi_{La} = 0.75$, $\Gamma_{Ra} = 0.044$, $\gamma_{Ra} = 0.45$, $\phi_{Ra} = 1.2$.

Figures 10.6 and 10.7 show clearly the different effects of the bosonic bath and the fermionic bath. Generally, the fermionic bath contributes most to the transport process, while the bosonic bath most to the dephasing process. From Fig. 10.6, we can see the dephasing process (off-diagonal elements) is almost the same while changing the fermionic coupling strength will significantly change the transport process. From Fig. 10.7, we can see the transport process is barely affected by changing the bosonic coupling strength, as a comparison, but the dephasing rate is affected. This result can be also predicted by analyzing the master equation or the Hamiltonian. Since the coupling form of the bosonic bath is a dephasing type, definitely it will affect the dephasing process. And since the coupling form of the fermionic bath is an "energy exchange" type, it is not surprising to find that it mainly contributes to the transport process.

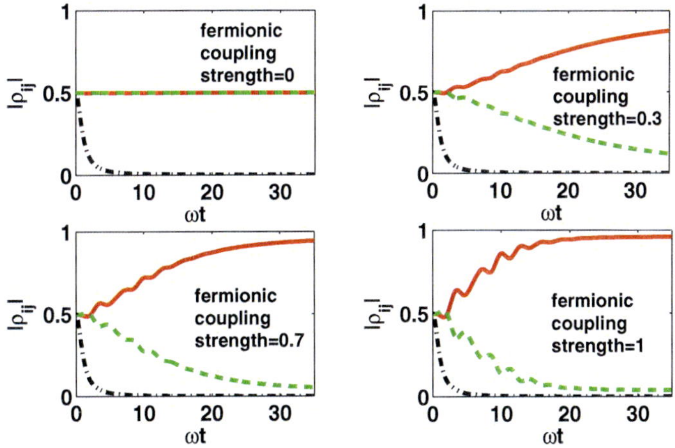

Fig. 10.6 Time evolution for different coupling strength of the fermionic bath. The coupling strength of the bosonic bath is fixed as 1. The red (solid), green (dashed), and black (dash-dotted) curves are the elements of density matrix ρ_{11}, ρ_{22}, ρ_{12} respectively

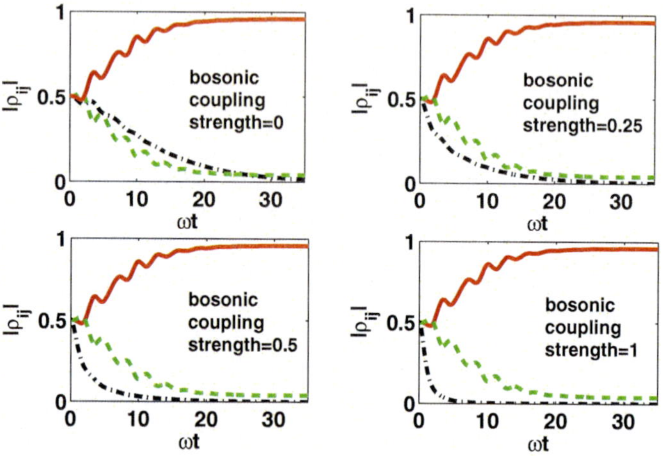

Fig. 10.7 Time evolution for different coupling strength of bosonic bath. The coupling strength of fermionic bath is fixed as 1. The red (solid), green (dashed), and black (dash-dotted) curves are the elements of density matrix ρ_{11}, ρ_{22}, ρ_{12} respectively

10.3.4 Summary

In this section, we have extended the NMQSD approach to an open system interacting with a hybrid environment which contains both bosons and fermions. By combining the bosonic and fermionic NMQSD approach, we have introduced two types of noises simultaneously to derive the NMQSD equation of the hybrid bath case. Based on the hybrid NMQSD equation, we also derive a master equation for the hybrid bath case. We have discussed the hybrid model with two separate cases (commutative and anti-commutative) together with a few specific examples. With these examples, we illustrated the consistency between the NMQSD approach and the standard quantum mechanics. In these cases, we have discussed both the exact and approximate solutions of O and Q operators.

10.4 Conclusion

In this chapter, the NMQSD approach was used to solve interesting open system problems where a quantum system interacts with an environment with fermions or with both bosons and fermions. In particular, we have discussed the derivations of exact master equations beyond the Markov approximation in two separate cases, namely whether the environment was commutative or anti-commutative with the system. In each case, we use several specific examples to show that the NMQSD approach can serve as a powerful theoretical tool in the study of the dynamic evolution of the fermionic/hybrid open systems. Moreover, it should be noted that the NMQSD is useful in investigating some early stage evolution caused by the memory effect of its environments since our approach is systematically derived from the microscopic model which goes beyond the Born-Markov approximation.

10.5 Appendix: Grassmann Algebra and Fermionic Coherent State

For convenience, we collect some useful facts about the Grassmann algebra and fermionic coherent state. The Grassmann numbers ξ_i are anti-commuting numbers. These numbers anti-commute with each other but commute with real or complex numbers x. The Grassmann numbers satisfy

$$\xi_i \xi_j = -\xi_j \xi_i, \quad \xi_i x = x \xi_i. \tag{A1}$$

The Grassmann numbers are also used in the fermionic path integral. A function with Grassmann variables ξ_i is called a Grassmann function $F(\xi_i)$. The derivative and integral of the Grassmann numbers are defined as

$$\frac{d}{d\xi_i} 1 = 0, \; \frac{d}{d\xi_i} \xi_i = 1, \tag{A2}$$

$$\int 1 d\xi_i = 0, \; \int \xi_i d\xi_i = 1. \tag{A3}$$

For Grassmann numbers, we have $\xi_i \xi_i = 0$, therefore, a polynomial expansion of $F(\xi_i)$ will only contain two terms as

$$F(\xi_i) = a_0 + a_1 \xi_i, \tag{A4}$$

where a_0 and a_1 are complex numbers. Then, we can derive a very important relation, the commutation relation of Grassmann functions. Consider the product of two Grassmann functions (multivariable)

$$\begin{aligned} F_1 & (\xi_1 ... \xi_n) F_2(\xi_1 ... \xi_n) \\ = & (a_0 + \textstyle\sum_i a_{1i}\xi_i + \sum_{i.j} a_{2ij}\xi_i\xi_j + \sum_{i.j,k} a_{3ij}\xi_i\xi_j\xi_k + ...) \\ & (b_0 + \textstyle\sum_i b_{1i}\xi_i + \sum_{i.j} b_{2ij}\xi_i\xi_j + \sum_{i.j,k} b_{3ij}\xi_i\xi_j\xi_k + ...). \end{aligned} \tag{A5}$$

Note that $\xi_i \xi_i = 0$ as stated above. Let us first consider a case where F_2 only contains terms of a certain order. For example, only first order coefficients are non-zero, $b_{1m} \neq 0$ ($F_2(\xi_1 ... \xi_n) = \sum_m b_{1m}\xi_m$). In this case, the following equalities can be established by using the commutation relation term by term:

$$
\begin{aligned}
a_0 \sum_m b_{1m}\xi_m &= a_0 \sum_m b_{1m}\xi_m, \\
\sum_i a_{1i}\xi_i \sum_m b_{1m}\xi_m &= \sum_m b_{1m}\xi_m \sum_i a_{1i}(-\xi_i), \\
\sum_{i.j} a_{2ij}\xi_i\xi_j \sum_m b_{1m}\xi_m &= \sum_m b_{1m}\xi_m \sum_{i.j} a_{2ij}(-\xi_i)(-\xi_j), \\
\sum_{i.j,k} a_{3ij}\xi_i\xi_j\xi_k \sum_m b_{1m}\xi_m &= \sum_m b_{1m}\xi_m \sum_{i.j,k} a_{3ij}(-\xi_i)(-\xi_j)(-\xi_k).
\end{aligned}
\tag{A6}
$$

More generally, we have,

$$
F_1(\xi_1 ... \xi_n)F_2(\xi_1 ... \xi_n) = F_2(\xi_1 ... \xi_n)F_1(-\xi_1, -\xi_2 ... - \xi_n), \tag{A7}
$$

With a short-hand notation the above equality may be simply written as

$$
F_1(\xi)F_2(\xi) = F_2(\xi)F_1(-\xi) \text{ (where } F_2 \text{ is first order homogeneous polynomial).} \tag{A8}
$$

Similarly, we also have this result when F_2 is of the third order, the fifth order etc., which means that the result holds if F_2 is an odd-order homogeneous polynomial, i.e.,

$$
F_1(\xi)F_2(\xi) = F_2(\xi)F_1(-\xi) \text{ (where } F_2 \text{ is an \textit{odd} order homogeneous polynomial).} \tag{A9}
$$

If F_2 is an even order homogeneous polynomial (for example the second order), we have

$$
\begin{aligned}
a_0 F_2 = a_0 F_2 &= F_2 a_0, \\
(\sum_i a_{1i}\xi_i)F_2 &= F_2 \sum_i a_{1i}\xi_i, \\
(\sum_{i.j} a_{2ij}\xi_i\xi_j)F_2 &= F_2 \sum_{i.j} a_{2ij}\xi_i\xi_j, \\
(\sum_{i.j,k} a_{3ij}\xi_i\xi_j\xi_k)F_2 &= F_2 \sum_{i.j,k} a_{3ij}\xi_i\xi_j\xi_k.
\end{aligned}
\tag{A10}
$$

More generally, we get

$$
F_1(\xi)F_2(\xi) = F_2(\xi)F_1(\xi) \text{ (where } F_2 \text{ is an \textit{even} order homogeneous polynomial)} \tag{A11}
$$

Note that here ξ is a general notation representing both $\xi_1, \xi_2, ..., \xi_n$ and ξ_1^*, $\xi_2^*, ..., \xi_n^*$.

Now we can define the so-called fermionic coherent states (we start with a single mode "i") as [49, 50]

$$
|\xi_i\rangle = (1 - \xi_i c_i^\dagger)|0\rangle \tag{A12}
$$

where $c_i^\dagger$ is the fermionic creation operator and ξ_i is a Grassmann variable which satisfies the following properties

$$\{\xi_i, \xi_j\} = 0, \tag{A13}$$
$$\xi_i \xi_i = 0, \tag{A14}$$
$$\{\xi_i, c_i\} = 0, \tag{A15}$$
$$\{\xi_i, c_i^\dagger\} = 0, \tag{A16}$$
$$\xi_i |0\rangle = |0\rangle \xi_i, \tag{A17}$$
$$\xi_i \langle 0| = \langle 0| \xi_i, \tag{A18}$$
$$\xi_i |1\rangle = \xi_i c_i^\dagger |0\rangle = -|1\rangle \xi_i, \tag{A19}$$
$$\xi_i \langle 1| = -\langle 1| \xi_i. \tag{A20}$$

Then, it is easy to check that the fermionic coherent state defined above has a property that is similar to a bosonic coherent state,

$$\begin{aligned} c_i |\xi_i\rangle &= c_i(|0\rangle - \xi_i |1\rangle) \\ &= 0 + \xi_i |0\rangle \\ &= \xi_i(|0\rangle - \xi_i |1\rangle) \\ &= \xi_i |\xi_i\rangle. \end{aligned} \tag{A21}$$

We have shown that we can use the fermionic coherent states to represent a fermionic bath. In this way, a non-commutative stochastic process can be introduced, such that the fundamental fermionic NMQSD equation and the corresponding non-Markovian master equation can be established.

References

1. C.W. Gardiner, P. Zoller, *Quantum Noise* (Springer, Berlin, 2004)
2. H.-P. Breuer, F. Petruccione, *The Theory of Open Quantum Systems* (Oxford University Press, 2007)
3. G. Lindblad, Commun. Math. Phys. **65**, 281 (1979)
4. W.H. Zurek, Rev. Mod. Phys. **75**, 715 (2003), https://doi.org/10.1103/RevModPhys.75.715
5. M. Schlosshauer, Rev. Mod. Phys. **76**, 1267 (2005a), https://doi.org/10.1103/RevModPhys.76.1267
6. W.H. Zurek, S. Habib, J.P. Paz, Phys. Rev. Lett. **70**, 1187 (1993), https://doi.org/10.1103/PhysRevLett.70.1187
7. B.L. Hu, J.P. Paz, Y. Zhang, Phys. Rev. D **45**, 2843 (1992), https://doi.org/10.1103/PhysRevD.45.2843
8. J. Halliwell, T. Yu, Phys. Rev. D **53**, 2012 (1996)
9. J.-H. An, W.-M. Zhang, Phys. Rev. A **76**, 042127 (2007), https://doi.org/10.1103/PhysRevA.76.042127
10. M.W.Y. Tu, W.-M. Zhang, Phys. Rev. B **78**, 235311 (2008), https://doi.org/10.1103/PhysRevB.78.235311

11. S. Maniscalco, F. Petruccione, Phys. Rev. A **73**, 012111 (2006), https://doi.org/10.1103/PhysRevA.73.012111
12. R. Feynman, F. Vernon, Ann. Phys. **281**, 547 (2000)
13. L. Diósi, N. Gisin, W.T. Strunz, Phys. Rev. A **58**, 1699 (1998), https://doi.org/10.1103/PhysRevA.58.1699
14. T. Yu, L. Diósi, N. Gisin, W.T. Strunz, Phys. Rev. A **60**, 91 (1999), https://doi.org/10.1103/PhysRevA.60.91
15. W.T. Strunz, T. Yu, Phys. Rev. A **69**, 052115 (2004), https://doi.org/10.1103/PhysRevA.69.052115
16. J. Jing, T. Yu, Phys. Rev. Lett. **105**, 240403 (2010), https://doi.org/10.1103/PhysRevLett.105.240403
17. X. Zhao, J. Jing, B. Corn, T. Yu, Phys. Rev. A **84**, 032101 (2011), https://doi.org/10.1103/PhysRevA.84.032101
18. C.J. Broadbent, J. Jing, T. Yu, J.H. Eberly, Ann. Phys. **327**, 1962 (2012)
19. H. Yang, H. Miao, Y. Chen, Phys. Rev. A **85**, 040101 (2012), https://doi.org/10.1103/PhysRevA.85.040101
20. J. Jing, T. Yu, Europhys. Lett. **96**, 44001 (2011)
21. E.L. Hahn, Phys. Rev. **80**, 580 (1950), https://doi.org/10.1103/PhysRev.80.580
22. J. Roden, A. Eisfeld, W. Wolff, W.T. Strunz, Phys. Rev. Lett. **103**, 058301 (2009), https://doi.org/10.1103/PhysRevLett.103.058301
23. C.P. Search, S. Pötting, W. Zhang, P. Meystre, Phys. Rev. A **66**, 043616 (2002), https://doi.org/10.1103/PhysRevA.66.043616
24. S.S. Sinha, D. Mondal, B.C. Bag, D.S. Ray, Phys. Rev. E **82**, 051125 (2010), https://doi.org/10.1103/PhysRevE.82.051125
25. K. Vladár, G.T. Zimányi, A. Zawadowski, Phys. Rev. Lett. **56**, 286 (1986)
26. L.-D. Chang, S. Chakravarty, Phys. Rev. B **31**, 154 (1985)
27. E.S. Hernández, C.O. Dorso, Phys. Rev. C **29**, 1510 (1984)
28. X. Zhao, W. Shi, L.-A. Wu, T. Yu, Phys. Rev. A **86**, 032116 (2012), https://doi.org/10.1103/PhysRevA.86.032116
29. W. Shi, X. Zhao, T. Yu, Phys. Rev. A **87**, 052127 (2013), https://doi.org/10.1103/PhysRevA.87.052127
30. X. Zhao, W. Shi, J. You, T. Yu, Ann. Phys. **381**, 121 (2017), ISSN 0003-4916, http://www.sciencedirect.com/science/article/pii/S0003491617301070
31. E. Barouch, B.M. McCoy, M. Dresden, Phys. Rev. A **2**, 1075 (1970)
32. D. Suess, A. Eisfeld, W.T. Strunz, Phys. Rev. Lett. **113**, 150403 (2014), https://doi.org/10.1103/PhysRevLett.113.150403
33. Z.-Z. Li, C.-T. Yip, H.-Y. Deng, M. Chen, T. Yu, J. Q. You, C.-H. Lam, Phys. Rev. A **90**, 022122 (2014), https://doi.org/10.1103/PhysRevA.90.022122
34. M. Chen, J.Q. You, Phys. Rev. A **87**, 052108 (2013), https://doi.org/10.1103/PhysRevA.87.052108
35. X. Hu, R. de Sousa, S.D. Sarma, *Foundations of Quantum Mechanics in the Light of New Technology* (World Scientific, 2002)
36. G. Ritschel, J. Roden, W.T. Strunz, A. Aspuru-Guzik, A. Eisfeld, J. Phys. Chem. Lett. **2**, 2912 (2011)
37. N. Lambert, F. Nori, Phys. Rev. B **78**, 214302 (2008), https://doi.org/10.1103/PhysRevB.78.214302
38. W.G. Unruh, Phys. Rev. A **51**, 992 (1995), https://doi.org/10.1103/PhysRevA.51.992
39. J. Jing, X. Zhao, J.Q. You, T. Yu, Phys. Rev. A **85**, 042106 (2012), https://doi.org/10.1103/PhysRevA.85.042106
40. T. Yu, J.H. Eberly, Phys. Rev. Lett. **93**, 140404 (2004), https://doi.org/10.1103/PhysRevLett.93.140404
41. J. Jing, X. Zhao, J.Q. You, W.T. Strunz, T. Yu, Phys. Rev. A **88**, 052122 (2013), https://doi.org/10.1103/PhysRevA.88.052122
42. X. Zhao, J. Jing, J.Q. You, T. Yu, Quantum Inform. Comp. **14**, 741 (2014)

43. H.M. Wiseman, G.J. Milburn, Phys. Rev. Lett. **70**, 548 (1993), https://doi.org/10.1103/PhysRevLett.70.548
44. W.-M. Zhang, P.-Y. Lo, H.-N. Xiong, M.W.-Y. Tu, F. Nori, Phys. Rev. Lett. **109**, 170402 (2012), https://doi.org/10.1103/PhysRevLett.109.170402
45. C.-H. Chung, K. Le Hur, M. Vojta, P. Wölfle, Phys. Rev. Lett. **102**, 216803 (2009), https://doi.org/10.1103/PhysRevLett.102.216803
46. C.-H. Chung, Phys. Rev. B **83**, 115308 (2011), https://doi.org/10.1103/PhysRevB.83.115308
47. J. Xu, X. Zhao, J. Jing, L.-A. Wu, T. Yu, J. Phys. A: Math. Theor. **47**, 435301 (2014), http://stacks.iop.org/1751-8121/47/i=43/a=435301
48. Y. Ting, Phys. Rev. A **69**, 062107 (2004), https://doi.org/10.1103/PhysRevA.69.062107
49. W.-M. Zhang, D.H. Feng, R. Gilmore, Rev. Mod. Phys. **62**, 867 (1990), https://doi.org/10.1103/RevModPhys.62.867
50. K.E. Cahill, R.J. Glauber, Phys. Rev. A **59**, 1538 (1999)

Chapter 11
Synthetic Spin-Orbit-Coupling in Ultracold Atomic Gases and Topological Superfluids

Chunlei Qu

Abstract Inspired by the discovery of topological insulating and topological superconducting states in electronic solid-state materials, there has been a significant research progress in the last few years towards the observation of various topological phenomena in other controllable quantum systems. Ultracold atomic gases hold great promise for the exploration of topological condensed matter physics due to their unprecedented controllability and the absence of disorder. This chapter reviews some aspects of the recent developments of topological physics in ultracold atomic gases. Specifically, we will discuss the experimental realization of one-dimensional synthetic spin-orbit-coupling in ultracold neutral atoms, the ground state and low-energy collective excitations of the spin-orbit-coupled Bose-Einstein condensate, and the appearance of Majorana quasi-particles in a spin-orbit-coupled Fermi gas.

11.1 Introduction

Since the observation of integer quantum Hall effect in a two-dimensional electron gas under a strong magnetic field [1], the field of topological condensed matter physics has been developing very rapidly. There is a surge of research interest in topological condensed matter physics in the last two decades due to the discovery of topological insulators and topological superconductors [2, 3]. These topological systems exhibit robust edge states that are immune against external perturbation and therefore they could be used for new functional devices and topological quantum computing.

Searching for topological phases in other synthetic materials such as photonic crystals [4], ultracold atomic gases [5], acoustic structures [6], etc., has been an active research area in recent years. Ultracold atomic gases provide a unique platform for the study of various interesting quantum physics because of the high tunability of

C. Qu (✉)
Department of Physics, Center for Quantum Science and Engineering, Stevens Institute of Technology, Hoboken, NJ, USA
e-mail: cqu5@stevens.edu

H. Ünlü and N. J. M. Horing (eds.), *Progress in Nanoscale and Low-Dimensional Materials and Devices*, Topics in Applied Physics 144,
https://doi.org/10.1007/978-3-030-93460-6_11

system parameters and the absence of disorder. Soon after the first experimental realization of Bose-Einstein condensate in 1995, people have observed many interesting macroscopic quantum phenomena, such as BEC-BCS crossover, superfluid to Mott-insulator phase transition, etc. [7].

The major challenge for the observation of topological phenomena with ultracold atomic gases is that atoms are charge neutral. Gauge field and spin-orbit-coupling, which are crucial ingredients for the emergence of topological physics, are absent for ultracold neutral atoms. To explore topological physics with atoms, we need to engineer synthetic gauge field and synthetic spin-orbit-coupling by exploiting the light and matter interaction. In this chapter, we will introduce the experimental realization of 1D synthetic spin-orbit-coupling with two-photon Raman transitions and then focus on the discussion of the hydrodynamics of the spin-orbit-coupled Bose-Einstein condensate. Then we will discuss topological superfluids of a spin-orbit-coupled Fermi gas. For a complete survey of this exciting research field, one can refer the recent review articles [8–10].

11.2 Spin-Orbit-Coupled Bose-Einstein Condensate

11.2.1 Synthetic Spin-Orbit-Coupling

Synthetic spin-orbit coupling was realized for the first time with ultracold bosonic atoms in the pioneering experiments by Ian Spielman's group [11, 12]. As illustrated in Fig. 11.1, a pair of counter-propagating laser beams couple two internal hyperfine states of the ^{87}Rb atoms. These two relevant energy levels can be defined as pseudospins and thus an effective spin-1/2 system is engineered. An atom in spin up state transits to spin down by absorbing a photon from one beam and emitting a photon into the other beam. During this two-photon Raman transition process, the atom acquires a $+2k_L\hat{e}_x$ recoil momentum along the horizontal direction. Conversely, an atom in a spin down state transits to spin up while acquiring a $-2k_L\hat{e}_x$ recoil momentum along the opposite direction.

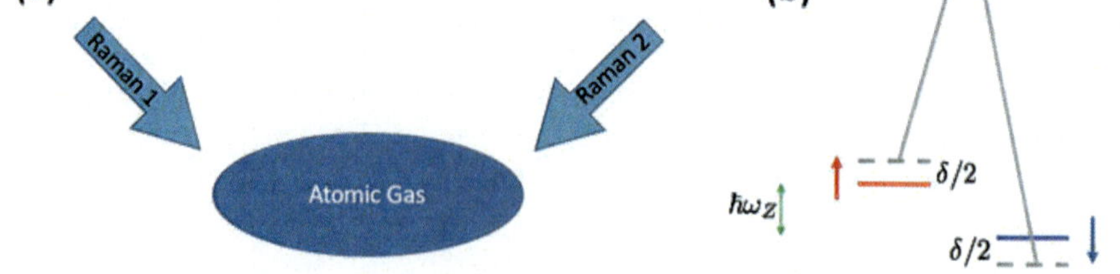

Fig. 11.1 Illustration of the two-photon Raman-transition process. **a** Configuration of laser setup. **b** Energy levels of the atoms and the Raman beam induced two-photon transition process

After a time-dependent unitary transformation and by considering the rotating wave approximation, the single-particle Hamiltonian becomes

$$H_0 = \begin{pmatrix} \frac{\hbar^2}{2m}(\mathbf{k} + k_L \hat{e}_x)^2 + \frac{\hbar\delta}{2} & \frac{\hbar\Omega}{2} \\ \frac{\hbar\Omega}{2} & \frac{\hbar^2}{2m}(\mathbf{k} - k_L \hat{e}_x)^2 - \frac{\hbar\delta}{2} \end{pmatrix} \tag{11.1}$$

where $\hbar\mathbf{k}$ is the momentum of the atom, $\hbar\Omega$ is the Raman coupling strength, and $\hbar\delta$ is the detuning. In terms of Pauli matrices, the single-particle Hamiltonian can be written in a compact form

$$H_0 = \frac{\hbar^2}{2m}(\mathbf{k} + k_L \sigma_z \hat{e}_x)^2 + \frac{\hbar\Omega}{2}\sigma_x + \frac{\hbar\delta}{2}\sigma_z \tag{11.2}$$

$$= \frac{\hbar^2\mathbf{k}^2}{2m} + \frac{\hbar^2 k_L}{m} k_x \sigma_z + \frac{\hbar\Omega}{2}\sigma_x + \frac{\hbar\delta}{2}\sigma_z + E_L \tag{11.3}$$

where $E_L = \hbar^2 k_L^2/2m$ is the recoil energy. The second term indicates that a one-dimensional spin-orbit-coupling (of the form $\sim k_x\sigma_z$) has been synthesized. The strength of the spin-orbit-coupling is set by the experimental setup. However, as demonstrated in a recent experiment [13], the strength of the synthetic spin-orbit-coupling can be tuned by a periodically modulated Raman coupling. The third Raman coupling term opens an energy gap at around $\mathbf{k} \sim 0$ and it mixes the two spin states, giving rise to two dressed state dispersions. The detuning term breaks the inversion symmetry of the dispersion. Direct diagonalization gives the explicit form of the two eigenenergies

$$E_\pm = \frac{\hbar^2 k^2}{2m} \pm \sqrt{\left(\frac{\hbar^2 k_L}{m} k_x + \frac{\hbar\delta}{2}\right)^2 + \left(\frac{\hbar\Omega}{2}\right)^2} \tag{11.4}$$

where $k^2 = k_x^2 + k_y^2 + k_z^2$.

As shown in Fig. 11.2a, for zero-detuning $\hbar\delta = 0$, the single-particle dispersion is symmetric with respect to $k_x = 0$ and the lower branch exhibits two local minima $\pm k_{min}$ if the Raman coupling $\hbar\Omega < 4E_L$. When the Raman coupling is larger than the critical value $\hbar\Omega_c = 4E_L$, the two minima merge into a single minimum at $k_x = 0$. Note that the spin composition is different at each momentum k_x as the Raman coupling mixes the two spin states into dressed states. For nonzero detuning, the symmetry between the two minima is broken (see Fig. 11.2b). However, it is still possible to have two local minima in the lower branch if the Raman coupling and detuning are small.

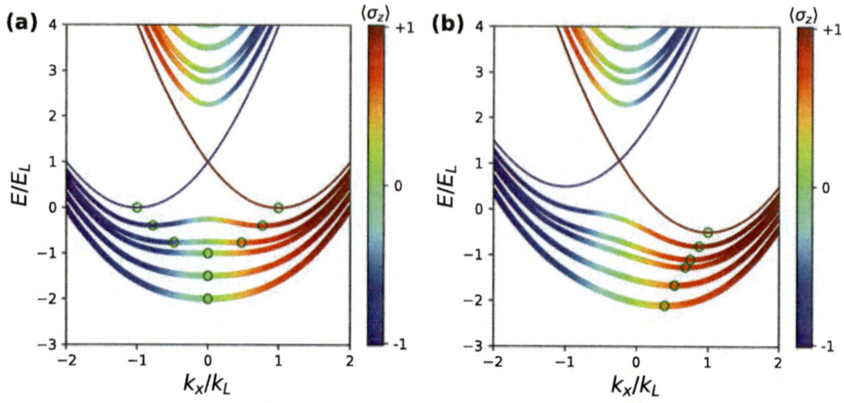

Fig. 11.2 Single-particle dispersion of the spin-orbit-coupling system. **a** $\hbar\delta = 0$, **b** $\hbar\delta = 1E_L$. In each panel, multiple dispersions are shown with the Raman coupling $\hbar\Omega = 0, 1, 2, 3, 4, 5E_L$ (from thin to thick lines). The green circles at the global minima of the lower band indicate the approximate position where the bosonic atoms can condense. The color at each point of the dispersion marks the spin polarization, i.e., the relative population imbalance between the two spin states

11.2.2 Mean-Field Description

At the mean-field level, the dynamics of the spin-orbit-coupled Bose-Einstein condensate is governed by the couple Gross-Pitaevskii (GP) equation [14]

$$i\hbar\frac{\partial}{\partial t}\Psi_1 = \left(\frac{(\mathbf{p}+\hbar k_L\hat{e}_x)^2}{2m}+\frac{\hbar\delta}{2}\right)\Psi_1 + V_{ext}\Psi_1 + \left(g_{11}|\Psi_1|^2+g_{12}|\Psi_2|^2\right)\Psi_1+\frac{\hbar\Omega}{2}\Psi_2 \tag{11.5}$$

$$i\hbar\frac{\partial}{\partial t}\Psi_2 = \left(\frac{(\mathbf{p}-\hbar k_L\hat{e}_x)^2}{2m}-\frac{\hbar\delta}{2}\right)\Psi_2 + V_{ext}\Psi_2 + \left(g_{12}|\Psi_1|^2+g_{22}|\Psi_2|^2\right)\Psi_2+\frac{\hbar\Omega}{2}\Psi_1 \tag{11.6}$$

where

$$V_{ext}(x,y,z) = \frac{1}{2}m\omega_x^2x^2+\frac{1}{2}m\omega_y^2y^2+\frac{1}{2}m\omega_z^2z^2 \tag{11.7}$$

is the external harmonic potential with trapping frequency $(\omega_x, \omega_y, \omega_z)$, and $g_{ij} = 4\pi\hbar a_{ij}/m$ $(i, j = 1, 2)$ are the interaction constants between the two spin states i and j. The scattering lengths are denoted by a_{ij}. The order parameters of the two spin states are normalized according to

$$\int\left(|\Psi_1|^2+|\Psi_2|^2\right)d^3\mathbf{r} = N \tag{11.8}$$

with N the total number of condensed atoms.

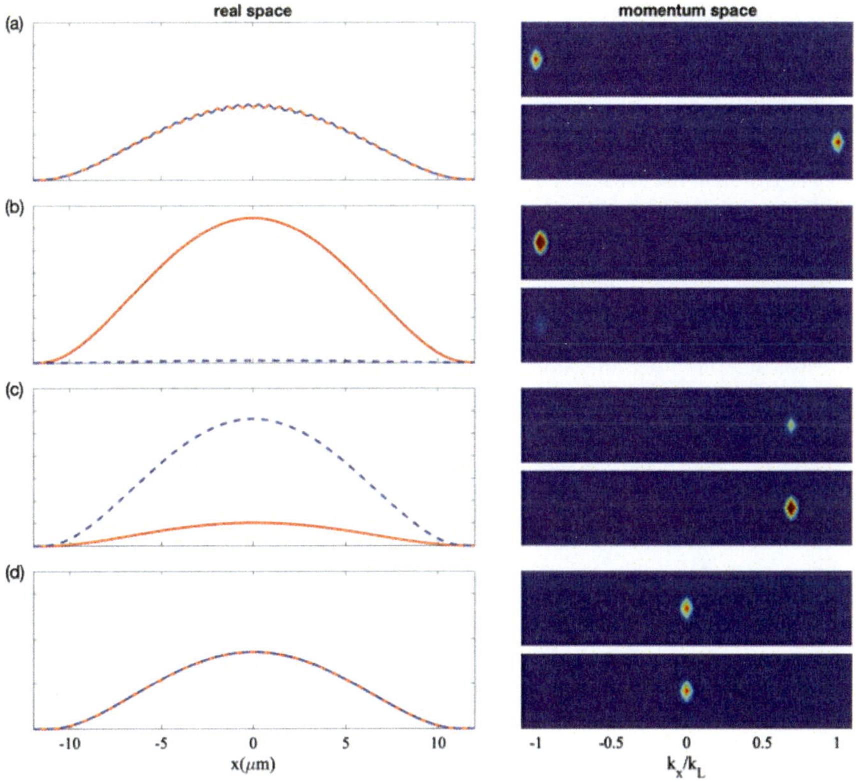

Fig. 11.3 Typical density profiles of the spin-orbit-coupled Bose-Einstein condensates at various Raman coupling strengths **a** $\Omega = 0.1E_L$, **b** $\Omega = 1E_L$, **c** $\Omega = 3E_L$, **d** $\Omega = 5E_L$. The detuning $\delta = 0$ for all panels. The density distributions have been shown in both real and momentum spaces. The condensate exhibits a density modulation in the strip phase in (**a**) due to the interference of the two dressed states at $k_x \sim \pm k_L$. **b, c** are the density distributions in the plane-wave phase with a single dressed state locating at either k_{min} or $-k_{min}$. Panel (**d**) corresponds to the case of single-minimum phase where the condensate locates at $k_x = 0$

The ground state and non-equilibrium dynamics of the spin-orbit coupled Bose-Einstein condensate can be obtained by numerically solving the coupled GP equations. In the presence of harmonic trapping potential, the momentum of the condensate is no longer a good quantum number. However, since the trapping potential is usually very weak, the atoms occupy a very narrow range of the momentum states around the global minimum of the lower branch of the single-particle dispersion. Mean-field interaction plays a key role in determining the phase diagram of true ground state.

As shown in Fig. 11.3, for zero detuning ($\hbar\delta = 0$), as the Raman coupling $\hbar\Omega$ increases, the system evolves into the following interesting phases [15]:

- *stripe phase*—It occurs when $0 < \hbar\Omega < \hbar\Omega_{c,1}$ where $\hbar\Omega_{c,1}$ is determined by the ratio between interspin coupling and intraspin coupling. For typical experimental parameters, $\hbar\Omega_{c,1} \approx 0.2E_L$. In the stripe phase, there is an equal superposition of two dressed states at the two degenerate minima and their interference induces a density modulation in real space, see Fig. 11.3a. Note that the spin compositions of the two dressed states are opposite to each other, due to the spin-momentum locking.
- *plane-wave phase*—It occurs when $\hbar\Omega_{c,1} < \hbar\Omega < \hbar\Omega_c$ where $\hbar\Omega_c \approx 4E_L$, i.e., the critical Raman coupling strength at which the two minima of the lower band merge into one. The bosonic atoms spontaneously choose one of the two minima as the ground state. Since this state corresponds to the occupation of a single dressed state at a definite momentum, it is called plane-wave phase. As Raman coupling increases, the spin polarization of the dressed state decreases and the system becomes spin balanced at $\hbar\Omega_c$, see Fig. 11.3b, c.
- *single-minimum phase*—When $\hbar\Omega > \hbar\Omega_c$, the two minima merge into a single minimum at $k_x = 0$. Further increasing the Raman coupling does not change the spin polarization, see Fig. 11.3d.

Around the global minimum, the lower band dispersion can be approximated by a quadratic function

$$H_{eff} \approx \frac{\hbar^2}{2m^*}(\mathbf{k} - k_{\min}\hat{e}_x)^2 \tag{11.9}$$

which is analogous to the Hamiltonian of a charged particle in an external magnetic field. Hence, $k_{min}(\Omega, \delta)\hat{e}_x$ can be regarded as a *synthetic gauge field* $\mathbf{A}_{syn}$ for the neutral atoms. It has only an x-component as it originates from the one-dimensional spin-orbit-coupling. The corresponding *synthetic magnetic field* is defined as the curl of the synthetic gauge field

$$\mathbf{B}_{syn} = \nabla \times \mathbf{A}_{syn} \tag{11.10}$$

which is nonzero only if $\mathbf{A}_{syn}$ depends on the transverse coordinates y or z. Since k_{min} is a function of Raman coupling and detuning, we can engineer a nonzero synthetic magnetic field by applying a y-position-dependent Raman coupling or detuning. When the gradient of the position dependent detuning is larger than a critical value, vortices appear as a signature of the rotating velocity field or the synthetic magnetic field [11]. An example of the generated vortices is shown in Fig. 11.4.

11.2.3 Hydrodynamic Theory

The low-energy collective excitations of a Bose-Einstein condensate can be accurately described by the hydrodynamic theory [16]. In this section, we discuss the

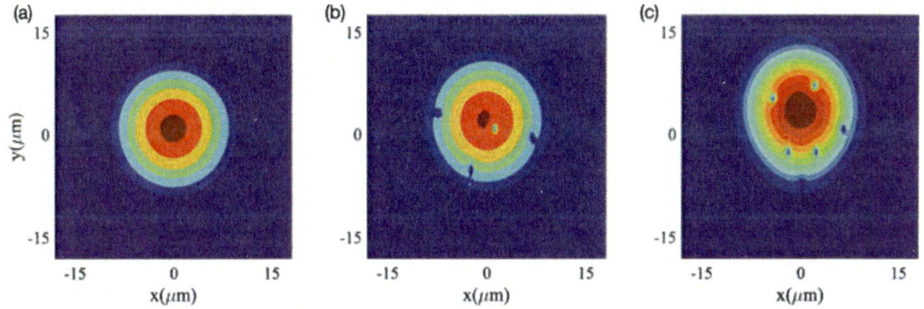

Fig. 11.4 Generation of vortices in a spin-orbit-coupled Bose-Einstein condensate by increasing the strength of the synthetic magnetic field or the gradient of the Raman detuning. The gradient of the applied external Raman detuning is **a** 0.1 kHz/μm, **b** 0.2 kHz/μm, **c** 0.3 kHz/μm, respectively

hydrodynamic theory for the spin-orbit-coupled Bose-Einstein condensate in the presence of a position-dependent detuning.

Without loss of generalities, we shall consider the single minimum phase, i.e., the Raman coupling $\hbar\Omega > \hbar\Omega_c$. The detuning is zero at the trap center and it varies linearly as a function of position y

$$\hbar\delta(y) = \eta k_L y \tag{11.11}$$

We will consider the small η regime so that there are no vortices generated in the condensate. Under the Thomas-Fermi approximation, the total density of the condensate can be written as

$$n_0 = \frac{\mu - V_{ext}}{g} \tag{11.12}$$

where μ is the chemical potential and $g \equiv g_{ij}$ are the nonlinear coupling constants (For simplicity, we have assumed that the inter- and intra-spin couplings are the same).

To derive the hydrodynamic equations, we start from the Lagrangian density of the spin-orbit-coupled Bose-Einstein condensate

$$\mathcal{L} = \sum_{j=1}^{2} \frac{i\hbar}{2}\left(\Psi_j^* \frac{\partial}{\partial t}\Psi_j - \Psi_j \frac{\partial}{\partial t}\Psi_j^*\right) - \mathcal{E} \tag{11.13}$$

where the energy density is given by

$$\mathcal{E} = \sum_j \left[\frac{\hbar^2}{2m} |\nabla \Psi_j|^2 + V_{ext} |\Psi_j|^2 \right] + \frac{1}{2} \sum_{i,j} g_{ij} |\Psi_i|^2 |\Psi_j|^2 + \frac{\Omega}{2} (\Psi_1^* \Psi_2 + \Psi_2^* \Psi_1)$$
$$+ \eta k_L y (|\Psi_1|^2 - |\Psi_2|^2) + \frac{i\hbar^2 k_L}{m} \left(\Psi_1^* \nabla_x \Psi_1 - \Psi_2^* \nabla_x \Psi_2 \right) \quad (11.14)$$

The order parameter for the two components of the condensate can be parameterized as

$$\begin{pmatrix} \Psi_1 \\ \Psi_2 \end{pmatrix} = \begin{pmatrix} \sqrt{\frac{n+s}{2}} e^{\phi + \phi_R/2} \\ \sqrt{\frac{n-s}{2}} e^{\phi - \phi_R/2} \end{pmatrix} \quad (11.15)$$

where $n = |\Psi_1|^2 + |\Psi_2|^2$ and $s = |\Psi_1|^2 - |\Psi_2|^2$ represent the total density and spin density, respectively. ϕ and ϕ_R represent the total and relative phases of the two components. Substituting 11.15 into the Lagrangian density, after ignoring the quantum pressure term, we find

$$\mathcal{L} \approx -n\hbar \frac{\partial \phi}{\partial t} - \frac{1}{2} s\hbar \frac{\partial \phi_R}{\partial t} - \frac{n\hbar^2}{8m} (\nabla \phi_R)^2 - \frac{n\hbar^2}{2m} (\nabla \phi)^2 - \frac{s\hbar^2}{2m} \nabla \phi \cdot \nabla \phi_R + \frac{\hbar^2 k_L}{m} s \nabla_x \phi$$
$$+ \frac{\hbar^2 k_L}{2m} n \nabla_x \phi_R + \frac{\hbar \Omega}{2} \sqrt{n^2 - s^2} \cos \phi_R - \frac{1}{2} n^2 g - V_{ext} n + \eta k_L y s \quad (11.16)$$

The equations of motion for the generalized coordinates can be obtained by the variation of the Lagrangian density with respect to the four variables. For ground state and low-energy excitations, only the lower band is relevant and thus the relative phase $\phi_R = 0$ as the two spin states are locked. Furthermore, in the single minimum phase, the spin density is much smaller than the total density, i.e., $s \ll n$. These considerations allow us to simplify the four differential equations into the following two

$$\frac{\partial n}{\partial t} + \frac{\hbar}{m^*} \nabla_x (n \nabla_x \phi) + \frac{\hbar}{m} \nabla_y (n \nabla_y \phi) - \frac{1}{\hbar} \frac{\Omega_c}{\Omega} \eta y \nabla_x n = 0 \quad (11.17)$$
$$\hbar \frac{\partial \phi}{\partial t} + \frac{\hbar^2}{2m^*} (\nabla_x \phi)^2 + \frac{\hbar^2}{2m} (\nabla_y \phi)^2 - \frac{\hbar \Omega}{2} + ng + V_{ext} - \frac{\Omega_c}{\Omega} \eta y \nabla_x \phi = 0 \quad (11.18)$$

where $m^*/m = (1 - \Omega_c/\Omega)^{-1}$ is the effective mass which is due to the modification of the dispersion along the k_x direction by the Raman transition induced synthetic spin-orbit-coupling.

To study low-energy collective excitations, we expand the density and phase around their equilibrium values

$$n = n_0 + \delta n, \qquad \phi = \phi_0 + \delta \phi \quad (11.19)$$

where the equilibrium phase ϕ_0 can be found by substituting the ansatz $\phi_0 = \alpha xy$, $s = 2\beta y n_0$ into the above equations. One finds

$$\alpha = 2\eta \frac{k_L^2}{\hbar\Omega} \frac{\omega_x^2}{\omega_{sc}^2}, \qquad \beta = \eta \frac{k_L}{\hbar\Omega} \frac{\omega_x^2 + \omega_y^2}{\omega_{sc}^2} \tag{11.20}$$

where $\omega_{sc} = \sqrt{\omega_x^2(m/m^*) + \omega_y^2}$ is the scissors mode frequency in the absence of detuning gradient ($\eta = 0$).

After substituting 11.19 into the hydrodynamic equations for the density and phase, the linearized differential equations become

$$\frac{\partial \delta n}{\partial t} + \frac{\hbar}{m^*}\nabla_x[n_0\nabla_x(\delta\phi)] + \frac{\hbar}{m}\nabla_y[n_0\nabla_y(\delta\phi)] - \frac{1}{\hbar}\frac{\Omega_c}{\Omega}\eta y\nabla_x(\delta n) + \alpha\left[\frac{\hbar}{m^*}y\nabla_x(\delta n) + \frac{\hbar}{m}x\nabla_y(\delta n)\right] = 0$$

$$\hbar\frac{\partial \delta\phi}{\partial t} + \alpha\left[\frac{\hbar^2}{m^*}y\nabla_x(\delta\phi) + \frac{\hbar^2}{m}x\nabla_y(\delta\phi)\right] + g\delta n - \frac{\Omega_c}{\Omega}\eta y\nabla_x\delta\phi = 0$$

These two equations allow us to explore the collective dynamics of various physical quantities for the spin-orbit-coupled Bose-Einstein condensate [17, 18].

11.2.4 Low-Energy Collective Modes

Due the presence of a synthetic magnetic field controlled by the position dependent detuning, some low energy collective modes may be coupled together. With the help of the hydrodynamic equations derived in the previous section, we shall study the dipole mode and scissors mode, respectively.

Dipole mode is the center-of-mass motion of the condensate. For regular Bose-Einstein condensate, the dipole modes along the three spatial directions are independent from each other. We can excite the dipole modes by shifting the harmonic trap by a small distance. The oscillation frequency of the excited dipole mode is equal to the corresponding trapping frequencies of that direction. In the presence of the synthetic magnetic field, the two dipole modes in the plane perpendicular to the direction of the synthetic magnetic field will be coupled together.

The center-of-mass position and the superfluid velocity of the condensate can be obtained straightforwardly from the hydrodynamic equations. The closed set of governing equations are found to be

$$\frac{d}{dt}\langle x\rangle - \frac{m}{m^*}\langle v_x\rangle + \omega_{eff}\langle y\rangle = 0 \tag{11.21}$$

$$\frac{d}{dt}\langle y\rangle - \langle v_y\rangle - \omega'_{eff}\langle x\rangle = 0 \tag{11.22}$$

$$\frac{d}{dt}\langle v_x\rangle + \omega_x^2\langle x\rangle + \omega'_{eff}\langle v_y\rangle = 0 \tag{11.23}$$

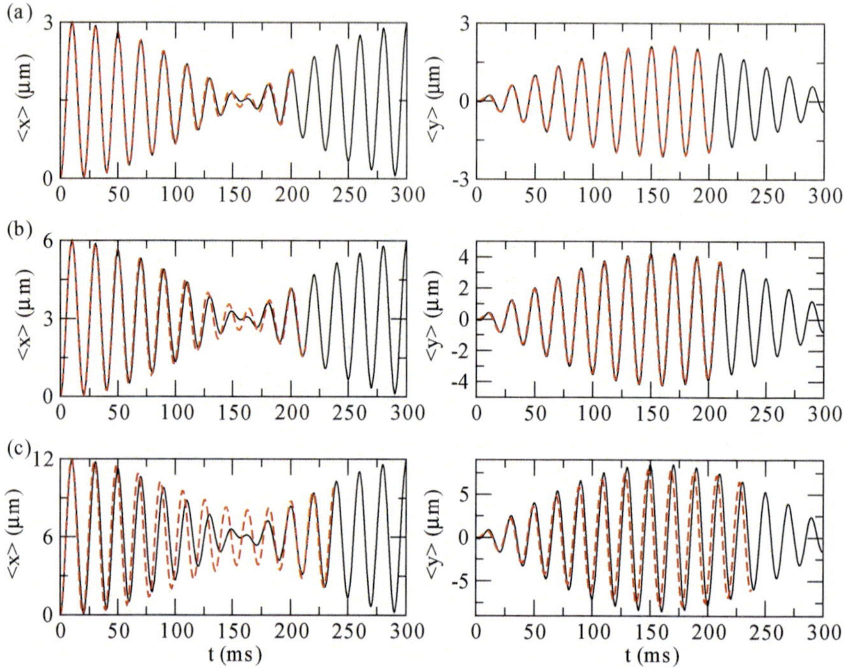

Fig. 11.5 Time dependence of the dipole mode $\langle x \rangle$ (first column) and $\langle y \rangle$ (second column) of a Bose-Einstein condensate in the presence of a synthetic magnetic field. The dynamics is excited by shifting the harmonic trap along x-direction by an initial displacement **a** $x_0 = 1.5\mu$m, **b** $x_0 = 3\mu$m, **c** $x_0 = 6\mu$m. The solid black lines correspond to the results obtained from hydrodynamic theory and the dashed red lines correspond to the results obtained from the GP simulation. Other parameters are: $\hbar\Omega = 8E_L, \eta = 0.0025E_L, \omega_x = 2\pi 50\sqrt{2}$Hz, $\omega_y = 50$Hz

$$\frac{d}{dt}\langle v_y \rangle + \omega_y^2 \langle y \rangle - \omega_{eff} \langle v_x \rangle = 0 \tag{11.24}$$

where we have introduced

$$\omega_{eff} = \frac{\eta}{\hbar}\frac{\Omega_c}{\Omega} - \alpha\frac{\hbar}{m^*}, \qquad \omega'_{eff} = \alpha\frac{\hbar}{m} \tag{11.25}$$

The above coupled differential equations indicate that the position dependent detuning or the synthetic magnetic field causes the precession of the dipole mode, in analogy with the Foucault precession exhibited by a classical pendulum in the rotating earth [18].

In Fig. 11.5, we compare the numerical results for the dipole oscillation obtained from the hydrodynamic theory and the GP simulation. The dynamics is excited by shifting the harmonic trap along x-direction by a displacement x_0. The condensate starts from a dipole oscillation along x-direction. After about 150ms, it is transformed

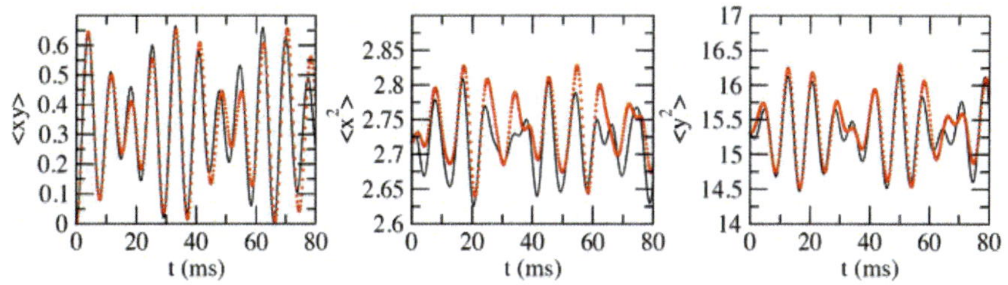

Fig. 11.6 Collective modes $\langle xy \rangle$, $\langle x^2 \rangle$, and $\langle y^2 \rangle$ excited by rotating the trap a small angle. The black solid lines correspond to the results obtained from the hydrodynamic theory and the red dots correspond to the results obtained from GP simulation

to a dipole oscillation along y-direction. The beating pattern indicates a precession of the dipole mode in the $x - y$ plane.

For x_0 much smaller than the Thomas-Fermi radius, we see very good agreement between the two results as hydrodynamic theory works very well in the linear regime. As the initial displacement x_0 increases, the discrepancy between the hydrodynamic theory and the GP simulation becomes more obvious. Nonetheless, the precession of the dipole mode persists in a wide parameter range for x_0.

Similar analysis can be directly generalized to the case of other collective modes, such as the scissors mode. For a Bose-Einstein condensate in the Thomas-Fermi approximation, it is easy to show that the angular rotation of the condensate is proportional to $\langle xy \rangle$ which can be calculated with our hydrodynamic equations. Straightforward calculation shows that the scissors mode $\langle xy \rangle$ is coupled to the widths of the condensate $\langle x^2 \rangle$, $\langle y^2 \rangle$. The closed set of differential equations are found to be

$$\frac{d}{dt}\langle xy \rangle - \frac{m}{m^*}\langle y v_x \rangle - \langle x v_y \rangle + \omega_{eff}\langle y^2 \rangle - \omega'_{eff}\langle x^2 \rangle = 0 \tag{11.26}$$

$$\frac{d}{dt}\langle x^2 \rangle - 2\frac{m}{m^*}\langle x v_x \rangle + 2\omega_{eff}\langle xy \rangle = 0 \tag{11.27}$$

$$\frac{d}{dt}\langle y^2 \rangle - 2\langle y v_y \rangle - 2\omega'_{eff}\langle xy \rangle = 0 \tag{11.28}$$

$$\frac{d}{dt}\langle x v_x \rangle + \omega'_{eff}\langle x v_y \rangle + \frac{3}{2}\omega_x^2\langle x^2 \rangle + \frac{1}{2}\omega_y^2\langle y^2 \rangle = 0 \tag{11.29}$$

$$\frac{d}{dt}\langle y v_x \rangle + \omega'_{eff}\langle y v_y \rangle + \omega_x^2\langle xy \rangle = 0 \tag{11.30}$$

$$\frac{d}{dt}\langle x v_y \rangle - \omega_{eff}\langle x v_x \rangle + \omega_y^2\langle xy \rangle = 0 \tag{11.31}$$

$$\frac{d}{dt}\langle y v_y \rangle - \omega_{eff}\langle y v_x \rangle + \frac{3}{2}\omega_y^2\langle y^2 \rangle + \frac{1}{2}\omega_x^2\langle x^2 \rangle = 0 \tag{11.32}$$

The scissors mode can be excited by rotating the harmonic trap a small angle. As shown in Fig. 11.6, the scissors mode $\langle xy \rangle$ is coupled to the widths of the condensate characterized by the two modes $\langle x^2 \rangle$ and $\langle y^2 \rangle$. We again find a good agreement between the results from the hydrodynamic theory and the GP simulation.

It is worth mentioning that the above derivations and analysis can be extended to the plane-wave phase or the single-minimum phase with an additional finite detuning.

11.3 Spin-Orbit-Coupled Fermi Gas and Topological Superfluid

One-dimensional synthetic spin-orbit-coupling has also been realized in an ultracold Fermi gas with the same Raman transition scheme. The interaction between fermionic atoms can be tuned with Feshbach resonance. When the interaction between atoms is attractive, Cooper pairs may be formed. We shall consider the s-wave pairing between the two spin states. The superfluid order parameter of the Fermi gas needs to be determined self-consistently. Similar to the nanowire-semiconductor system which may be driven to a topological superconducting phase with zero-energy Majorana quasi-particles appearing at the two ends of the nanowire [19], the ultracold femrionic superfluid may be driven to a topological superfluid phase. This allows us to search for Majorana quasiparticles with ultracold atoms and use the spin-orbit-coupled Fermi gas for topological quantum computation [20, 21].

11.3.1 Tight Binding Model and Bogliubov-De Gennes Equation

We assume that N fermionic atoms are confined in a one-dimensional deep optical lattice. The system can be described by a Fermi-Hubbard tight-binding model which is of the following form [22]

$$H = H_0 + H_Z + H_{so} + V_{int}, \tag{11.33}$$

where

$$H_0 = -t \sum_{\langle i,j \rangle,\sigma} c_{i\sigma}^\dagger c_{j\sigma} - \mu \sum_{i\sigma} n_{i\sigma} \tag{11.34}$$

describes the spin-independent nearest-neighbour hopping and the chemical potential. $\hat{n}_{i\sigma} = c_{i\sigma}^\dagger c_{i\sigma}$ is the particle number operator for atoms at lattice site i and spin σ. Hereafter, we use $t = 1$ as the basic energy unit.

$$H_Z = -h_z \sum_i (n_{i\uparrow} - n_{i\downarrow}) \tag{11.35}$$

is the out-of-plane Zeeman energy term. The spin-orbit-coupling term is described by

$$H_{\text{so}} = -\frac{\alpha}{2}\sum_i (c^\dagger_{i-1,\downarrow} c_{i\uparrow} - c^\dagger_{i+1,\downarrow} c_{i\uparrow} + \text{h.c.}) \tag{11.36}$$

The interaction between fermionic atoms is given by

$$V_{\text{int}} = -U \sum_i n_{i\uparrow} n_{i\downarrow} \tag{11.37}$$

where $U > 0$ indicates that the interaction between atoms is attractive. Under the mean-field approximation, the interaction term can be decomposed to the following form

$$V_{\text{int}} = \sum_i \Delta_i^* c_{i\downarrow} c_{i\uparrow} + \Delta_i c^\dagger_{i\uparrow} c^\dagger_{i\downarrow} - |\Delta_i|^2/U \tag{11.38}$$

where $\Delta_i = -U\langle c_{i\downarrow} c_{i\uparrow}\rangle$ denotes the superfluid order parameter. Note that for simplicity we have ignored the Hartree shift term in the decomposition. The particle number $n_{i\sigma} = \langle \hat{n}_{i\sigma}\rangle = \langle c^\dagger_{i\sigma} c_{i\sigma}\rangle$ and superfluid order parameter $\Delta_i = -U\langle c_{i\downarrow} c_{i\uparrow}\rangle$ can be determined self-consistently for a fixed chemical potential μ.

Using the Bogoliubov transformation, we obtain the real space Bogoliubov-de Gennes (BdG) equation

$$\sum_j \begin{pmatrix} H_{ij\uparrow} & \alpha_{ij} & 0 & \Delta_{ij} \\ -\alpha_{ij} & H_{ij\downarrow} & -\Delta_{ij} & 0 \\ 0 & -\Delta^*_{ij} & -H_{ij\uparrow} & -\alpha_{ij} \\ \Delta^*_{ij} & 0 & \alpha_{ij} & -H_{ij\downarrow} \end{pmatrix} \begin{pmatrix} u^n_{j\uparrow} \\ u^n_{j\downarrow} \\ -v^n_{j\uparrow} \\ v^n_{j\downarrow} \end{pmatrix} = E_n \begin{pmatrix} u^n_{j\uparrow} \\ u^n_{j\downarrow} \\ -v^n_{j\uparrow} \\ v^n_{j\downarrow} \end{pmatrix}, \tag{11.39}$$

where

$$H_{ij\uparrow} = -t\delta_{i\pm 1,j} - (\mu + h_z)\delta_{ij} \tag{11.40}$$

$$H_{ij\downarrow} = -t\delta_{i\pm 1,j} - (\mu - h_z)\delta_{ij} \tag{11.41}$$

$$\alpha_{ij} = \frac{1}{2}(j - i)\alpha\delta_{i\pm 1,j} \tag{11.42}$$

$$\langle \hat{n}_{i\sigma}\rangle = \sum_{n=1}^{2N} [|u_{i\sigma}|^2 f(E_n) + |v_{i\sigma}|^2 f(-E_n)] \tag{11.43}$$

$$\Delta_{ij} = -U\delta_{ij} \sum_{n=1}^{2N} [u^n_{i\uparrow} v^{n*}_{i\downarrow} f(E_n) - u^n_{i\downarrow} v^{n*}_{i\uparrow} f(-E_n)] \tag{11.44}$$

with $f(E) = 1/\left(1 + e^{E/T}\right)$. The BdG spectrum can be obtained from direct diagonalization of the real space BdG matrix. Due to the particle-hole redundancy, the eigenvalues of the BdG matrix always appear in pairs $\pm E_n$.

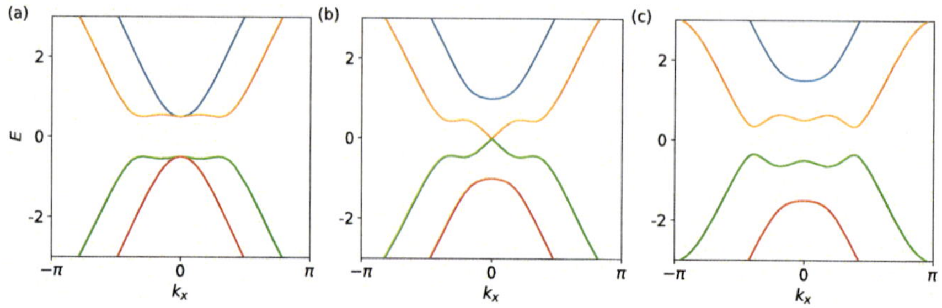

Fig. 11.7 BdG quasi-particle excitation spectrum as function of k_x by taking $k_y = 0$. The out-of-plane Zeeman fields for the three panels are **a** $h_z = 0$, **b** $h_z = 0.5$, **c** $h_z = 1.0$. The other parameters are $t = 1$, $\alpha = 2$, $h_x = 0$. The pairing order parameter has been chosen as $\Delta = 0.5$

11.3.2 Topological Phase Transition

Before presenting the self-consistent calculation results, we will first consider the BdG spectrum of the spin-orbit-coupled Fermi gas with a uniform superfluid order parameter Δ. In this situation, we can transform the tight-binding Hamiltonian to the momentum space. In the Nambu basis chosen as $(c_{\mathbf{k},\uparrow}, c_{\mathbf{k},\downarrow}, c^\dagger_{-\mathbf{k},\downarrow}, -c^\dagger_{-\mathbf{k},\uparrow})$, it is found to be

$$H_{BdG}(\mathbf{k}) = \begin{pmatrix} H_0(\mathbf{k}) & \Delta \\ \Delta & -\sigma_y H_0^*(\mathbf{k})\sigma_y \end{pmatrix} \tag{11.45}$$

where

$$H_0(\mathbf{k}) = -2t\cos(k_x) - \mu + \alpha\sin(k_x)\sigma_x - h_z\sigma_z \tag{11.46}$$

is the single particle Hamiltonian. The BdG quasi-particle excitation spectrum can be obtained by a direct diagonalization of the BdG matrix in k-space.

As shown in Fig. 11.7, the band gap between the particle and hole branches decreases as we increase the out-of-plane Zeeman field. At a critical out-of-plane Zeeman field, $h_z = h_{z,c}$, the energy gap closes, signaling a topological phase transition. For $h_z > h_{z,c}$ the gap reopens and the system evolves into a topological superconductor or topological superfluids. In the presence of boundaries, there exist localized bound states at the two ends of the one-dimensional lattice which can be shown to be Majorana quasi-particles, i.e., they are their own anti-particles [19].

The phase transition from a topologically trivial superfluid to a topologically nontrivial superfluid can be quantitatively determined by calculating the Pfaffian of BdG matrix. One finds that the system is in a topological superfluid phase if

$$\sqrt{\Delta^2 + (|\mu| - 2t)^2} < h_z < \sqrt{\Delta^2 + (|\mu| + 2t)^2} \tag{11.47}$$

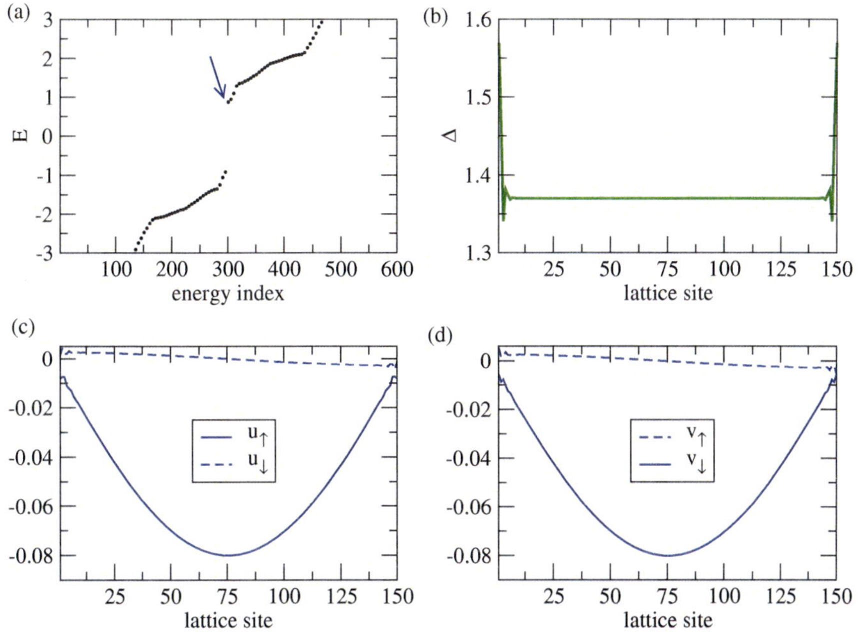

Fig. 11.8 **a** BdG spectrum of the spin-orbit-coupled Fermi gas at $h_z = 0.5t$. **b** The self-consistent superfluid order parameter Δ as a function of position. **c, d** The wavefunction of the lowest energy quasi-particle excitation (indicated by the blue arrow in panel **a**). Other system parameters: $U = 5t$, $N = 150$, $\alpha = 1.5t$, $\mu = -2t$

11.3.3 Topological Superfluids and Majorana Edge Modes

We self-consistently solve the real-space BdG equation for a 1D spin-orbit-coupled Fermi gas confined in a deep optical lattice and calculate the superfluid order parameter in each iteration until it converges. In the following numerical results, we set the system parameters as $N = 150$, $U = 5t$, $\alpha = 1.5t$, $\mu = -2t$ and focus on the effect of the out-of-plane Zeeman field h_z.

The numerical results for a smaller out-of-plane Zeeman field is shown in Fig. 11.8. Panel (a) shows the BdG spectrum. As we mentioned in the previous section, due to particle-hole symmetry of the BdG matrix, the eigenvalues of the system alway appear in paris $\pm E_n$. An energy gap exists around $E = 0$ with the lowest quasi-particle excitation energy marked by the blue arrow. The energy gap is a consequence of the competition between the superfluid gap Δ and the out-of-plane Zeeman field h_z. As shown in panel (b), the superfluid fluid gap is uniform in the bulk $\Delta \approx 1.38$ and it becomes rapidly oscillating near the two ends due to the edge effects. In panels (c, d), we plot the wavefunction $(u_\uparrow, u_\downarrow, v_\uparrow, v_\downarrow)$ for the first quasi-particle excitation (marked by the blue arrow in panel (a)). It is obvious that the wavefunction distributes mainly in the bulk of the system. These observations indicate that the system is in a non-topological phase.

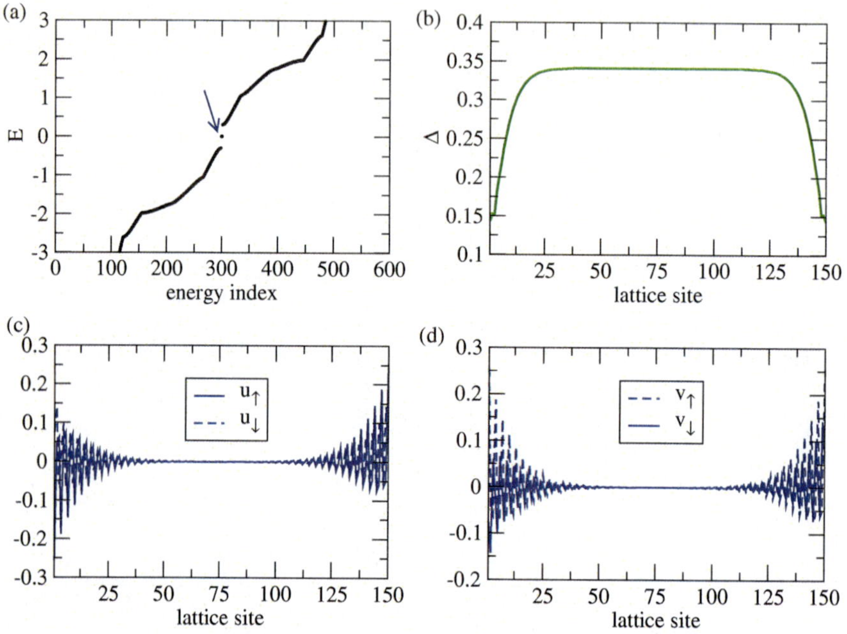

Fig. 11.9 **a**BdG spectrum of the spin-orbit-coupled Fermi gas at $h_z = 1.4t$. **b** The self-consistent superfluid order parameter Δ as a function of position. **c, d** The wavefunction of the lowest energy quasi-particle excitation (indicated by the blue arrow in panel **a**). Other system parameters: $U = 5t$, $N = 150$, $\alpha = 1.5t$, $\mu = -2t$

The same results for a larger out-of-plane Zeeman field ($h_z = 1.4t$) is shown in Fig. 11.9. A major difference is the appearance of two degenerate zero-energy Majorana quasi-particle modes marked by the blue arrow (Fig. 11.9a). The superfluid order parameter becomes smaller $\Delta \approx 0.34$ in the bulk because of the large Zeeman field. The wavefunction for the Majorana modes distributes mainly on the two ends of the system. The system is in a topological superfluid phase. It is worth mentioning that the two degenerate Majorana modes could be weakly coupled in a finite sized system. However, the splitting of the two eigenenergies is exponentially small for a large system size.

To identify the precise phase transition from nontopological phase to topological phase, we scan the out-of-plane Zeeman field from $h_z = 0$ to $h_z = 2t$. The results are presented in Fig. 11.10. Here, three quantities are presented: the blue-line with diamonds corresponds to the eigen-energy of the lowest quasi-particle excitation, the red solid line corresponds to the second eigenenergy of the BdG spectrum, and the green line with squares corresponds to the superfluid order parameter average over all the lattice sites. A sharp change appears at around $h_z \approx t$. Above this critical point, E_1 becomes identically zero, the order parameter dramatically changes to a different value and there is a large energy gap between the two lowest eigenenergies. These observations indicate that the system becomes a topological superfluid for $h_z > t$.

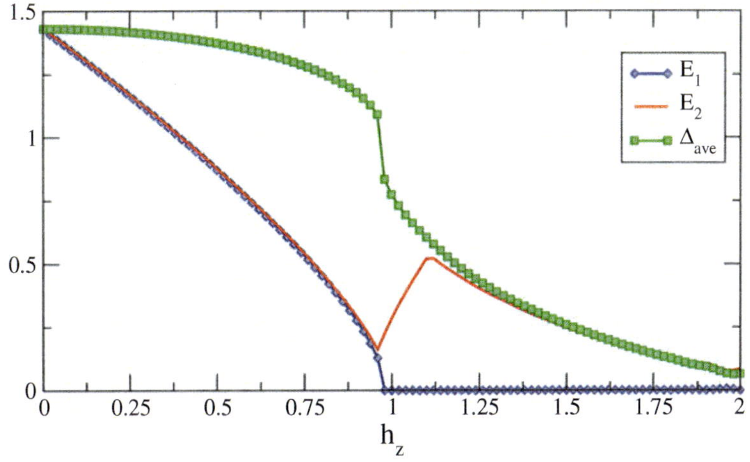

Fig. 11.10 The eigenenergies of the two lowest quasi-particle excitations E_1, E_2, and the average superfluid order parameter Δ_{ave} as a function of out-of-plane Zeeman field h_z. Other system parameters: $U = 5t$, $N = 150$, $\alpha = 1.5t$, $\mu = -2t$

Of course, as h_z further increases, the superfluid order parameter will be completely destroyed. The system will be in a non-superfluid phase.

Another interesting physics that will not be discussed here is the appearance of a new type of topological superfluid: the topological Fulder-Ferrel superfluid phase where the order parameter in the bulk becomes oscillatory $\Delta(x) \sim \Delta_0 e^{iQx}$. This can be realized by adding an in-plane Zeeman field which breaks the inversion symmetry of the BdG band structure.

Acknowledgements We would like to thank Chuanwei Zhang, Sandro Stringari, and Ming Gong for their close collaboration in this research field.

References

1. K.V. Klitzing, G. Dorda, M. Pepper, New method for high-accuracy determination of the fine-structure constant based on quantized hall resistance. Phys. Rev. Lett. **45**, 494 (1980)
2. M.Z. Hasan, C.L. Kane, Colloquium: topological insulators. Rev. Mod. Phys. **82**, 3045 (2010)
3. X.-L. Qi, S.-C. Zhang, Topological insulators and superconductors. Rev. Mod. Phys. **83**, 1057 (2011)
4. T. Ozawa et al., Topological photonics. Rev. Mod. Phys. **91**, 015006 (2019)
5. N. Goldman, J.C. Budich, P. Zoller, Topological quantum matter with ultracold gases in optical lattices. Nat. Phys. **12**, 639–645 (2016)
6. Z. Yang, F. Gao, X. Shi, X. Lin, Z. Gao, Y. Chong, B. Zhang, Topological acoustics. Phys. Rev. Lett. **144**, 114301 (2015)
7. I. Bloch, J. Dalibard, W. Zwerger, Many-body physics with ultracold gases. Rev. Mod. Phys. **80**, 885 (2008)

8. H. Zhai, Angular momentum of a Bose-Einstein condensate in a synthetic rotational field. Rep. Prog. Phys. **78**, 026001 (2015)
9. V. Galitski, I.B. Spielman, Spin-orbit-coupling in quantum gases. Nature **494**, 49–54 (2013)
10. N.R. Cooper, J. Dalibard, I.B. Spielman, Topological bands for ultracold atoms. Rev. Mod. Phys. **91**, 015005 (2019)
11. Y.-J. Lin, R. Compton, K. Jimenez-Garcia, J.V. Porto, I.B. Spielman, Synthetic magnetic fields for ultracold neutral atoms. Nature **462**, 628 (2009)
12. Y.-J. Lin, K. Jimenez-Garcia, I.B. Spielman, Spin-orbit-coupled Bose-Einstein condensates. Nature **471**, 83 (2011)
13. K. Jimenez-Garcia, L.J. LeBlanc, R.A. Williams, M.C. Beeler, C. Qu, M. Gong, C. Zhang, I.B. Spieman, Tunable spin-orbit coupling via strong driving in ultracold-atom systems. Phys. Rev. Lett. **114**, 125301 (2015)
14. L.P. Pitaevskii, S. Stringari, *Bose-Einstein Condensation and Superfluidity*, Oxford Science Publications (2016)
15. Y. Li, L.P. Pitaevskii, S. Stringari, Quantum tricriticality and phase transitions in spin-orbit coupled Bose-Einstein condensates. Phys. Rev. Lett. **108**, 225301 (2012)
16. S. Stringari, Collective excitations of a trapped Bose-Condensed gas. Phys. Rev. Lett. **77**, 2360 (1996)
17. G.I. Martone, Y. Li, L.P. Pitaevskii, S. Stringari, Anisotropic dynamics of a spin-orbit-coupled Bose-Einstein condensate. Phys. Rev. A **86**, 063621 (2012)
18. C. Qu, S. Stringari, Angular momentum of a Bose-Einstein condensate in a synthetic rotational field. Phys. Rev. Lett. **120**, 183202 (2018)
19. A. Yu Kitaev, Unpaired Majorana fermions in quantum wires. Phys.-Usp, **44**, 131 (2001)
20. C. Zhang, S. Tewari, R. M. Lutchyn, S. Das Sarma, $p_x + ip_y$ Superfluid from s-wave interactions of fermionic cold atoms. Phys. Rev. Lett. **101**, 160401 (2008)
21. M. Sato, Y. Takahashi, S. Fujimoto, Non-abelian topological order in s-wave superfluids of ultracold fermionic atoms. Phys. Rev. Lett. **103**, 020401 (2009)
22. C. Qu, M. Gong, C. Zhang, Fulde-Ferrell-Larkin-Ovchinnikov or Majorana superfluids: the fate of fermionic cold atoms in spin-orbit-coupled optical lattices. Phys. Rev. A **89**, 053618 (2014)

Chapter 12
Control with EIT: High Energy Charged Particle Detection

Aneesh Ramaswamy and Svetlana A. Malinovskaya

Abstract The strong non-linear optical response of atomic systems in electromagnetically induced transparency (EIT) states is considered as a means to detect the presence of small perturbations to steady states. For the 3 level system, expressions for the group velocity and group velocity dispersion (GVD) were derived and a quantum control protocol was established to account for the change in the chirp spectrum of a probe pulse when the steady state was perturbed. This was applied to the propagation of slow Cherenkov radiation in the medium due to the passage of a train of high-energy charged particles. The choice of the initial steady state, with focus on the slow light condition and strong narrowly confined dispersion, equated to the continuous trapping of Cherenkov polaritons in the medium along a narrow group cone, allowing for non-trivial fields to accumulate. Sweeping of the control field and detuning parameters in the field-atom parameter space showed the presence of optimal regions to maximize the first order perturbation in the coherences, creating changes in the optical responses that modify the chirp spectra of probe pulses.

12.1 Introduction

12.1.1 A Brief Review on EIT

The use of coherent interactions between light and matter yield phenomena in which there is drastic change in the optical response function. EIT is a phenomenon in which a narrow transparency window with strong non-linear optical effects is achieved due to interference between quantum excitation pathways. Of prime interest is the greatly reduced absorption and drastically increased dispersion effects. Whilst its effects has analogues in CPT (Coherent population trapping), which occurs in optically thin media, and ATS (Autler-Townes Splitting); EIT is a phenomenon that involves

A. Ramaswamy · S. A. Malinovskaya (✉)
Department of Physics, Stevens University of Technology, Hoboken, NJ 07030, USA
e-mail: smailnov@stevens.edu

H. Ünlü and N. J. M. Horing (eds.), *Progress in Nanoscale and Low-Dimensional Materials and Devices*, Topics in Applied Physics 144,
https://doi.org/10.1007/978-3-030-93460-6_12

modification of both the optical and material states of the coupled light-matter system, and hence occurs in optically thick systems. Of particular interest is the study of quantum fields in EIT systems as dark state polaritons, the pseudo-particle arising from the entanglement of the propagating light with the dipole transitions.

Theoretical studies have opened up several applications including generation of non-classical atomic ensembles, high-resolution spectroscopy and reversible quantum memories in optical systems. A proposed technique by Fleischhauer and Mewes to greatly increase storage time for use in quantum memory is in adiabatically reducing the control field's strength to bring the light to a halt, effectively mapping the light's quantum state to the atomic spin ensemble and then achieve reconstruction through restoring the control field's strength [1, 2]. An experiment realization of this protocol in a laser-cooled ensemble of a cloud of Rb-87 atoms in a magneto-optical trap achieved a favourable result of 0.036 storage efficiency for a transport distance of 1.26 mm (1/e size of the cloud) for a lifetime of 2.6 ms [3]. The approach and results show a high promise for EIT in developing robust mechanisms for quantum optical storage with increased transport distance, though reducing decoherence due to natural lifetime and dephasing is still a problem of study. EIT also has great promise in few photon systems and creating schemes for resonant non-linear interactions that create large single-photon phase shifts, which is of great interest in quantum information, such as in the development of high fidelity quantum gates. There have been a number of theoretical and experimental approaches to achieve high fidelity after quantum operations whilst reducing the characteristic residual absorption of EIT [4, 5].

Further uses of EIT include transmission with negligible dissipation, non-linear optics in the weak field regime, and slow light. EIT does not require the highly controlled experimental setups (e.g. ultra-low temperatures) for these phenomena to occur. Whilst EIT was initially discovered in atomic/molecular systems (chief of which is the three-level Λ system), it has been studied in optomechanical systems, plasmonics, coupled microresonators, solid-state physics and photonic crystals [6]. Especially considering the rise of interest in quantum metrology using atomic systems, EIT is a powerful candidate for investigation in increased performance and precision in atomic clocks, in high-resolution atomic interferometry and in magnetoptic measurements. Traditional weaknesses of EIT including low signal-to-noise ratio and fidelity, have been compensated for by the great versatility in control protocols that can be developed to target phenomena of interest. For example, an investigation into the frequency stability of atomic clocks found higher stability when tuning EIT towards detecting magneto-optically induced light polarization shifts rather than controlling the intensity of EIT fields [7]. This has naturally led to interest in EIT, for quantum control theory, towards developing efficient practical protocols for optimizing performance and precision.

Our study will be on the three-level Λ system, a simple starting point to gain a strong understanding in the features and challenges with EIT.

12.2 Developing a Picture of EIT with Three-Level Λ Systems

We consider the 3-level Λ system with stable energy levels $|1\rangle$, $|2\rangle$ and excited state $|3\rangle$. Only the 1–3,2–3 transitions are allowed. We will be considering the dynamics with the 1st Born approximation and the Markovian approximation. Most of our work will be based on a Lindbladian master equation.

An archetypal example of such a system is for the Rubidium D_2 transition fine splittings of the $5S_{1/2}$, $F = 1$ and $5P_{1/2}$, $F = 2$ energy levels (Fig. 12.1). We take our ground state, $|1\rangle$, as $[5S_{1/2}, F = 1, m_f = 0]$, the excited state, $|3\rangle$, as $[5P_{1/2}, F = 1, m_f = 1]$ and the Raman ground state, $|2\rangle$, as $[5S_{1/2}, F = 2, m_f = 0]$. The 1–3 transition frequency is 377 THz (not including the contributions from the fine splitting differences) and the Stokes shift is 6.384 GHz. The natural lifetime of the transition τ is 27.7 ns. The hyperfine transition dipole matrix elements for our chosen transitions is 2.54 Debye [8].

We first develop our model's interaction Hamiltonian.

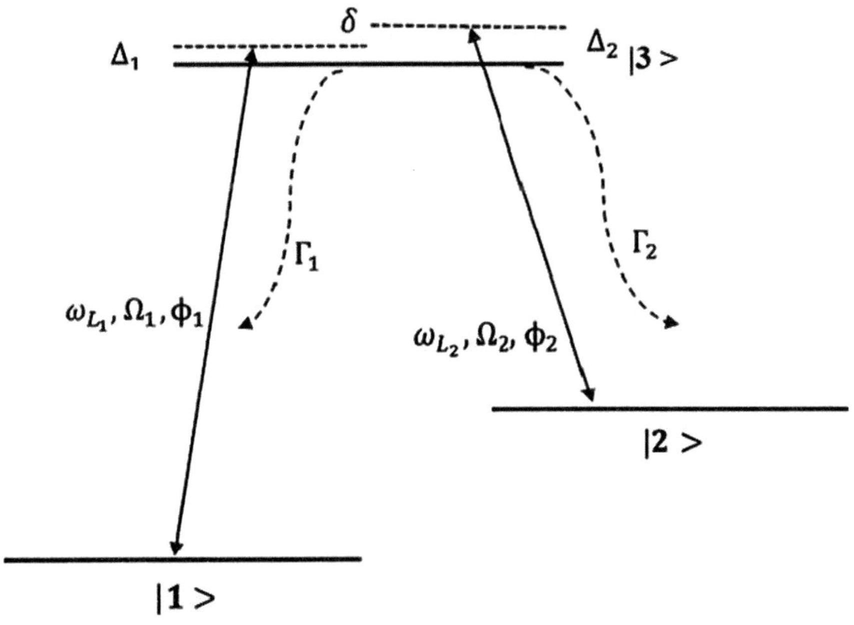

Fig. 12.1 Three level Λ system with laser frequency ω_L, Rabi frequency Ω, laser phase modulation ϕ, one-photon detuning Δ, two-photon detuning $\delta = \Delta_2 - \Delta_1$ and decay rate Γ

12.2.1 Hamiltonian in the Field Interaction Picture

We consider $|3\rangle$ to be the zero energy reference and we define the pump E_1 and Stokes E_2 electric fields as below

$$E_j(t) = E_j^0(t)\cos(\omega_{L_j}(t - t_c) + \phi_j(t)) \tag{12.1}$$

The Rabi frequency is given by:

$$\Omega_j(t) = \frac{|p_{j3}|E_j^0(t)}{\hbar} \tag{12.2}$$

The dipole moment can be complex, but we introduce the phase function into the oscillatory part of the field. We have:

$$p_{j3} = |p_{j3}|e^{i\theta_j}$$
$$\tilde{\phi}_j(t) = \phi_j(t) + \theta_j$$

We now define the interaction Hamiltonian H_{int} where:

$$H_{int} = \hat{U} H_{int} \hat{U}^{\dagger} \tag{12.3}$$

We use the unitary operator to go to the interaction picture and apply the rotating wave approximation (RWA) where:

$$H_{int}^{j3} = \frac{\hbar}{2}\Omega_1(t)(e^{i((2\omega_{L_j}+\Delta_j)(t-t_c)+\tilde{\phi}_j(t)} + e^{-i(\Delta_j(t-t_c)+\tilde{\phi}_j(t)}) \tag{12.4}$$

$$H_{int}^{3j} = (H_{int}^{j3})^*$$

And all other terms are zero. We ignore the fast oscillating term and transform back to the Schrodinger picture, and now consider the field interaction picture using the unitary operator:

$$\hat{U}(t) = e^{-i\omega_{L_j}t/\hbar} \tag{12.5}$$

We arrive at the field interaction Hamiltonian

$$H_{FE} = \frac{\hbar}{2}\begin{pmatrix} 2\Delta_1 & 0 & \Omega_1(t)e^{i\tilde{\phi}_1(t)} \\ 0 & 2\Delta_2 & \Omega_2(t)e^{i\tilde{\phi}_2(t)} \\ \Omega_1(t)e^{-i\tilde{\phi}_1(t)} & \Omega_2(t)e^{-i\tilde{\phi}_2(t)} & 0 \end{pmatrix}. \tag{12.6}$$

12.2.2 Lindblad Formalism: Finding EIT Steady States

Now we consider the case for open dynamics. We use $\Delta_2 = \Delta_1 - \delta$ and define the master equation with the superoperator $\mathcal{L}$:

$$\dot{\rho} = -\frac{i}{\hbar}[H_{FE}, \rho] + \mathcal{L}_{\mathcal{J}}\rho \tag{12.7}$$

$$\mathcal{L}_{\mathcal{J}}\rho = \sum_{j \in \mathcal{J}} \lambda_j \mathcal{F}_j(\rho) \tag{12.8}$$

$$\mathcal{F}_j(\rho) = F_j \rho F_j^\dagger - \frac{1}{2}\left\{F_j^\dagger F_j, \rho\right\}$$

$$\{F_j\} = \{|1\rangle\langle 3|, |2\rangle\langle 3|, |1\rangle\langle 1|, |2\rangle\langle 2|, |3\rangle\langle 3|\} \tag{12.9}$$

$$\{\lambda_j\} = \left\{\frac{\Gamma_1}{2}, \frac{\Gamma_2}{2}, \frac{\gamma_{12} + \gamma_2 - \gamma_3}{2}, \frac{\gamma_{12} - \gamma_2 + \gamma_3}{2}, \frac{-\gamma_{12} + \gamma_2 + \gamma_3}{2}\right\}$$

$$\mathcal{L}\rho = \begin{pmatrix} \Gamma_1\rho_{33} & -\gamma_{12}\rho_{12} & -(\frac{\Gamma_1 + \Gamma_2}{2} + \gamma_1)\rho_{13} \\ -\gamma_{12}\rho_{21} & \Gamma_2\rho_{33} & -(\frac{\Gamma_1 + \Gamma_2}{2} + \gamma_2)\rho_{23} \\ -(\frac{\Gamma_1 + \Gamma_2}{2} + \gamma_1)\rho_{31} & -(\frac{\Gamma_1 + \Gamma_2}{2} + \gamma_2)\rho_{32} & -(\Gamma_1 + \Gamma_2)\rho_{33} \end{pmatrix}. \tag{12.10}$$

We define Γ_j to be the natural decay rate from the excited state to the jth state and γ are dephasing parameters to account for thermal broadening and other losses of coherence. We then define the dephasing parameters:

$$\gamma_{31} = \frac{\Gamma_1 + \Gamma_2}{2} + \gamma_1 \tag{12.11}$$

$$\gamma_{32} = \frac{\Gamma_1 + \Gamma_2}{2} + \gamma_2$$

$$\gamma = \gamma_{12}.$$

We find stationary solutions for this open system that corresponds to an EIT state, using the Gell-Mann matrices as our basis and solve for the Bloch vector:

$$\langle \vec{\sigma} \rangle = \mathrm{Tr}\{\rho \vec{\sigma}\} \tag{12.12}$$
$$\langle \vec{\sigma} \rangle = (\langle \sigma_1 \rangle, ..., \langle \sigma_9 \rangle)$$

We solve for the density matrix terms in the Bloch equations using the below initial condition:

$$\left\langle \dot{\vec{\sigma}}(0) \right\rangle = 0 \tag{12.13}$$

We find the coherence terms, reducing equations to depend on one unknown ρ_{12}, assuming the populations for determined parameters are known and fixed:

$$\rho_{13} = \frac{i\gamma_{31} + \Delta_1}{2(\gamma_{31}^2 + \Delta_1^2)} \left(\Omega_1(t) e^{i\tilde{\phi}_1(t)} (\rho_{11} - \rho_{33}) + \Omega_2(t) e^{i\tilde{\phi}_2(t)} \rho_{12} \right) \tag{12.14}$$
$$\rho_{23} = \frac{i\gamma_{32} + \Delta_2}{2(\gamma_{32}^2 + \Delta_2^2)} \left(\Omega_1(t) e^{i\tilde{\phi}_1(t)} \rho_{21} + \Omega_2(t) e^{i\tilde{\phi}_2(t)} (\rho_{22} - \rho_{33}) \right).$$

The solution to the differential equation for ρ_{12} are obtained by substituting the above expressions for ρ_{13}, ρ_{23}:

$$\dot{\rho}_{12} = (-\Gamma' + i\omega')\rho_{12} + K. \tag{12.15}$$

$$\Gamma' = \gamma + \frac{\gamma_{31}}{4} \left(\frac{\gamma_{31}}{\gamma_{32}} \frac{\left(\frac{\Omega_1}{\gamma_{31}}\right)^2}{1 + \left(\frac{\Delta_2}{\gamma_{31}}\right)^2 \left(\frac{\gamma_{31}}{\gamma_{32}}\right)^2} + \frac{\left(\frac{\Omega_2}{\gamma_{31}}\right)^2}{1 + \left(\frac{\Delta_1}{\gamma_{31}}\right)^2} \right) \tag{12.16}$$

$$\omega' = \delta + \frac{\gamma_{31}}{4} \left(-\frac{\Delta_2}{\gamma_{31}} \left(\frac{\gamma_{31}}{\gamma_{32}}\right)^2 \frac{\left(\frac{\Omega_1}{\gamma_{31}}\right)^2}{1 + \left(\frac{\Delta_2}{\gamma_{31}}\right)^2 \left(\frac{\gamma_{31}}{\gamma_{32}}\right)^2} + \frac{\Delta_1}{\gamma_{31}} \frac{\left(\frac{\Omega_2}{\gamma_{31}}\right)^2}{1 + \left(\frac{\Delta_1}{\gamma_{31}}\right)^2} \right).$$

$$K = \frac{\gamma_{31}}{4} \left(-\left(\frac{\gamma_{31}}{\gamma_{32}}\right)^2 \frac{\Omega_1 \Omega_2}{\gamma_{31}^2} \frac{(\rho_{11} - \rho_{33}) e^{i(\tilde{\phi}_1(t) - \tilde{\phi}_2(t))}}{1 + \left(\frac{\Delta_2}{\gamma_{31}}\right)^2 \left(\frac{\gamma_{31}}{\gamma_{32}}\right)^2} \left(1 + i\frac{\Delta_2}{\gamma_{31}} \frac{\gamma_{31}}{\gamma_{32}}\right) \right)$$
$$+ \frac{\gamma_{31}}{4} \left(\frac{\Omega_1 \Omega_2}{\gamma_{31}^2} \frac{(\rho_{22} - \rho_{33}) e^{i(\tilde{\phi}_1(t) - \tilde{\phi}_2(t))}}{1 + \left(\frac{\Delta_1}{\gamma_{31}}\right)^2} \left(-1 + i\frac{\Delta_1}{\gamma_{31}}\right) \right)$$

In the weak field approximation, we generally see a significant oscillatory behaviour only for the case of non-zero detuning. The scattering parameter also depends mainly on dephasing between the two ground states when γ_{3j} is large. Now we solve for the initial value of ρ_{12} at the stationary point:

$$\rho_{12}^s = \frac{K}{\Gamma' - i\omega'}.$$

Next we find the expressions for the steady-state coherences. The expression of ρ_{13}^s and ρ_{23}^s is our main interest as it is a measure of the dispersion and the absorption of radiation near the resonance of the $(|1\rangle \to |3\rangle)$ transition and will be used to determine the optical response functions

$$\rho_{13}^s = \frac{\frac{\Delta_1}{\gamma_{31}} + i}{1 + \left(\frac{\Delta_1}{\gamma_{31}}\right)^2} \left(\frac{\Omega_1}{\gamma_{31}} e^{i\tilde{\phi}_1(t)} (\rho_{11} - \rho_{33}) + \frac{\frac{\Omega_2}{\gamma_{31}} \frac{K}{\gamma_{31}} \left(\frac{\Gamma'}{\gamma_{31}} + i \frac{\omega'}{\gamma_{31}} \right) e^{i\tilde{\phi}_2(t)}}{\left(\frac{\Gamma'}{\gamma_{31}}\right)^2 + \left(\frac{\omega'}{\gamma_{31}}\right)^2} \right) \tag{12.17}$$

$$\rho_{32}^s = \frac{\frac{\Delta_2}{\gamma_{32}} + i}{1 + \left(\frac{\Delta_2}{\gamma_{32}}\right)^2} \left(\frac{\Omega_2}{\gamma_{32}} e^{-i\tilde{\phi}_2(t)} (\rho_{22} - \rho_{33}) - \frac{\frac{\Omega_1}{\gamma_{32}} \frac{K}{\gamma_{32}} \left(\frac{\Gamma'}{\gamma_{32}} + i \frac{\omega'}{\gamma_{32}} \right) e^{-i\tilde{\phi}_1(t)}}{\left(\frac{\Gamma'}{\gamma_{32}}\right)^2 + \left(\frac{\omega'}{\gamma_{32}}\right)^2} \right) \tag{12.18}$$

We note that the system has a natural timescale dependent on the dephasing parameters γ_{3j}. For the rest of this chapter, all quantities will be expressed in terms of γ_{31}.

The above is the general results for the coherences for the steady-state solution. In the weak-field approximation, the fields being smaller than the natural decay rate implies that the steady state is unique as all the other eigenvalues of the Lindbladian have negative real parts. With a model of the microscopic dynamics understood, we now build the optical response functions.

12.2.3 Optical Response Functions

We find the expectation for the dipole density operator. First, we rotate from the field interaction picture back to the Schrödinger picture using (12.5):

$$\tilde{\rho}(t) = U(t)^{\dagger}\rho(t)U(t). \tag{12.19}$$

When we rotate back to the ground reference frame, the coherence terms oscillate with the frequencies of the fields driving the transitions:

$$\tilde{\rho}_{3j}(t) = \rho_{3j}(t)e^{-i\omega_{L_j}t}. \tag{12.20}$$

We compare the terms for the polarization field in terms of the electric field and susceptibility tensor, and the density matrix description. Here $\chi(\omega)$ is a generally non-linear function and depends on the frequency and the amplitude of the electric field. Then the polarization reads:

$$\mathbf{P}(\mathbf{r}, \omega) = \chi(\omega)\mathbf{E}(\mathbf{r}, \omega). \tag{12.21}$$

If we use a control protocol where we vary the field properties over time or if the envelope is time dependent, it's not obvious what the form of the polarization will be.

In the case where we have a superposition of monochromatic plane waves, the expression simplifies to:

$$\mathbf{P}(\mathbf{r}, t) = \sum_{k=1}^{n}(E_0/2(\chi_k(\omega, t)e^{-i\omega_k t} + \chi_k^{\dagger}(\omega, t)e^{i\omega_k t})) \tag{12.22}$$

For the case where we have a general field of the form $E(\mathbf{r}, t) = E_0(\mathbf{r}, t)cos(\omega_l t + \phi)$, where we take ϕ to be constant, we get the following:

$$\mathbf{P}(\mathbf{r}, t) = (\chi(\omega, t)E_0(\mathbf{r}, \omega))e^{-i\omega_l t} + (\chi^{\dagger}(\omega, t)E_0(\mathbf{r}, \omega))e^{i\omega_l t}. \tag{12.23}$$

For the rest of the chapter, we deal with CW light and monochromatic plane waves.

$$\begin{aligned} P(\omega) = \epsilon_0\chi(\omega, E)E &= \frac{\epsilon_0\hbar\Omega_1}{2|p_{13}|}(\chi_1^*(\omega)e^{i(\omega_{L_1}t+\phi_1)} + \chi_1(\omega)e^{-i(\omega_{L_1}t+\phi_1)}) \\ &+ \frac{\epsilon_0\hbar\Omega_2}{2|p_{23}|}(\chi_2^*(\omega, E)(e^{i(\omega_{L_2}t+\phi_2)} + \chi_2 e^{-i(\omega_{L_2}t+\phi_2)}) \end{aligned} \tag{12.24}$$

$$\begin{aligned} \langle p\rangle = \mathrm{Tr}\{\rho p\} = \zeta_1(\omega, t) + \zeta_2(\omega, t) &\equiv |p_{13}|/2(\rho_{13}e^{i(\omega_{L_1}t+\phi_1+\theta_1)} + \rho_{31}e^{-i(\omega_{L_1}t+\phi_1+\theta_1)}) \\ &+ |p_{23}|/2(\rho_{23}e^{i(\omega_{L_2}t+\phi_2+\theta_2)} + \rho_{32}e^{-i(\omega_{L_2}t+\phi_2+\theta_2)}). \end{aligned} \tag{12.25}$$

We have n as the number density of atoms/molecules:

$$\langle P\rangle = n\,\langle p\rangle\,.$$

We use a simple model of a homogeneous medium of a dilute weakly interacting gas. Comparing (12.24) and (12.25), we see that we have to match oscillating factors to get the electric susceptibility functions. Then it follows that:

$$\chi_j(\omega, E) = \frac{2n}{\hbar\epsilon_0}\frac{|p_{3j}|^2}{\Omega_j}\rho_{3j}e^{-i\theta_j}. \qquad (12.26)$$

At this point, we have enough information to find other optical response functions such as the complex refractive index, the phase and the group velocities and the Group Velocity Dispersion (GVD). Now we focus our attention on the region in the parameter space in which we develop our model for the EIT optical response functions.

12.3 Control Scheme

With the above results, now we can develop a scheme where we explore the slow light phenomenon and the strong dispersive effects of EIT. Our goal is to develop a control scheme for a given atomic system, which yields high transmission of the probe pulses with the group profiles containing frequencies belonging to the resonance window. The parameters of the system+field are chosen to satisfy this condition as well as to optimise changes in dispersion properties, such as the group velocity, GVD, and the transmitted pulse spectrum, so that small differences in the frequency spectrum of the probe pulses lead to significant deviations in group profiles of pulses with differing spectra propagating in the atomic medium.

12.3.1 The Concept

We choose a control Rabi frequency relatively large compared to the pump Rabi frequency, which gives a peak near resonance with sharp dispersion and low absorptivity, and we choose a value for the two-photon detuning to shift the transparency window relative to the resonance. Choosing non-zero detunings does contribute to some population of the other states, though our conditions are still valid for small detunings. We use units of γ_{31} to make relevant parameters dimensionless (time is in $(\gamma_{31}/2\pi)^{-1}$). We choose the following conditions:

$$\begin{aligned}\gamma_{31} &= 1\\ \Omega_j, \Delta_j &\ll 1\\ \Omega_1 &\ll \Omega_2\\ \rho_{11} &\approx 1\\ \omega' &\approx \Delta_2\nu_2 - \Delta_1\nu_1\\ \alpha &= \gamma_{32}/\gamma_{31}.\end{aligned} \tag{12.27}$$

Our next step is to refine the choice of parameters to optimise the optical response of the system to a weak probe pulse of form $E(t) = A(t)e^{i(\omega_{L_1}+x)t}$. The polarization $P(t)$ gains a chirp in the atomic medium and its spectrum is given by $P(\omega) = A_\omega(\omega - (\omega_{L_1} + x))\chi(\omega)$. The phases of the complex quantities A_ω, χ_ω are given by ϕ_A, ϕ_χ respectively such that:

$$P(\omega) = |A_\omega(\omega - (\omega_{L_1} + x))||\chi(\omega)|e^{i(\phi_A(\omega-(\omega_{L_1}+x))+\phi_\chi(\omega))}. \tag{12.28}$$

Each frequency component gains a phase delay and a group delay, $\tau_p(\omega)$, $\tau_g(\omega)$ with respect to the component in the vacuum:

$$\begin{aligned}\tau_p(\omega) &= -\frac{\phi_\chi(\omega)}{\omega}\\ \tau_g(\omega) &= -\frac{d\phi_\chi}{d\omega}.\end{aligned} \tag{12.29}$$

The above delays allow us to construct the chirp spectra for probe pulses that can yield phenomena such as anomalous dispersion, group velocity direction changes and pulse lengthening. We consider control of group velocity, GVD and the chirp spectrum in the general case and for transform-limited pulses.

12.3.1.1 Control Using Group Velocity, Group Velocity Dispersion and the Chirp Spectrum

If we wish to be in the neighbourhood of the minimum group velocities, we choose parameters such that $\left.\frac{d^2k'}{d\omega^2}\right|_{\omega=\omega_{L_1}+x} = 0$ (the local extremum of $\frac{dk'}{d\omega}$ in the resonance window). Here, the group velocity differences are small but we can focus on controlling the dispersion and extinction coefficients at low velocities.

To maximize the delay between the probe and pump pulses, we intend to maximize $\Delta T \propto \left(\frac{1}{v_g(\omega_{L1}+x)} - \frac{1}{v_g(\omega_{L1})}\right)$. As a general approach, we choose parameters where the group velocities are small in the interval $[\omega_{L1}, \omega_{L1} + x]$ but the difference in v_g is still significant.

If the second-order Taylor series for $n'(\omega)$ is a good approximation in the interval, the GVD can be used to estimate the time delay/chirp. We choose parameters such that $\frac{dn'}{d\omega}$ resembles a linear function near our chosen frequency, say ω_{L1}, giving us local extrema for the GVD. We have $\Delta T \propto GVD(\omega_{L1})x$.

For a probe pulse with a Gaussian envelope, the duration of the pulse (related to the standard deviation σ) is extended after traveling a distance d in the medium. We define $\sigma_{ex} = \sqrt{\sigma^2 + (0.5d \times GVD(\omega)/\sigma)^2}$ as the extended pulse standard deviation.

In general, we can always design our optical response function design the transmitted pulse spectrum of probe pulses for our purposes. Whilst not discussed in this work, the use of chirped pulses for the driving and probe fields can subdue undesired non-adiabatic effects during pulse propagation in a medium and add more control to shaping of the optical response functions and the spectra of post-medium pulses [9]. For sustaining EIT states in general high dimensional systems, spectral modulation by methods such as linear and sinuosidal spectral chirp can improve mitigation of decoherence effects and adiabatically transfer populations [10, 11].

In the next section, we develop approximate forms for the optical response functions and its derivatives in the control regime.

12.3.2 Optical Response Functions in the Control Regime

We consider the approximate form of the susceptibility $\chi = \chi' + i\chi''$ near resonance. We use $\Delta\omega = \omega - \omega_{31}$ as the argument of the functions.

$$\chi'(\Delta\omega) \approx \frac{2n|p_{31}|^2}{\hbar\epsilon_0}\left((\rho_{11}-\rho_{33})\Delta\omega + \frac{Q_1\Gamma'(\Delta_2\nu_2 - \Delta\omega\nu_1)}{(\Gamma'^2 + (\Delta_2\nu_2 - \Delta\omega\nu_1)^2)} + \frac{Q_2\Gamma'\Delta_2}{(\Gamma'^2 + (\Delta_2\nu_2 - \Delta\omega\nu_1)^2)}\right) \tag{12.30}$$

$$\chi''(\Delta\omega) \approx \frac{2n|p_{3j}|^2}{\hbar\epsilon_0}\left((\rho_{11}-\rho_{33}) - \frac{Q_3\Gamma'}{(\Gamma'^2 + (\Delta_2\nu_2 - \Delta\omega\nu_1)^2)}\right) \tag{12.31}$$

$$\begin{aligned}
Q_1 &= \frac{\Omega_2^2}{4\Gamma'}(\alpha^2(\rho_{11}-\rho_{33}) + (\rho_{22}-\rho_{33})) \\
&\quad + \frac{\alpha}{\nu_1}(\rho_{11}-\rho_{33}) + (\rho_{22}-\rho_{33}) \\
Q_2 &= \alpha(\rho_{11}-\rho_{33}) - \frac{\alpha}{\nu_1}(\rho_{11}-\rho_{33}) + (\rho_{22}-\rho_{33}) \\
Q_3 &= \frac{\Omega_2^2}{4}(\alpha^2(\rho_{11}-\rho_{33}) + (\rho_{22}-\rho_{33})).
\end{aligned}$$

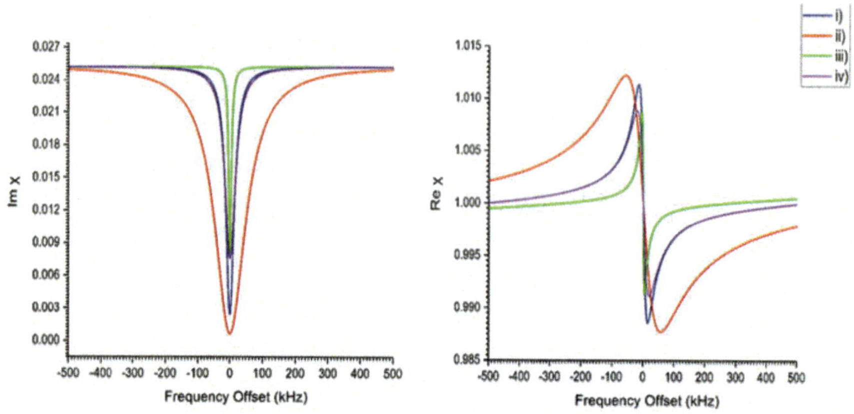

Fig. 12.2 Plot of the imaginary (left) and the real(right) part of the susceptibility of the system for parameters (Ω_2, γ_{12}) for the numbered configurations: (i) $(1/35, 4.3E-5)$, (ii) $(2/35, 4.3E-5)$ (iii) $(1/70, 4.3E-5)$ (iv) $(1/30, 1.7E-4)$ and $\Omega_1 = 0.1\Omega_2$

We see that the real susceptibility is the sum of a linear term that depends on the population in the 1st ground state plus a distribution (similar to a Fano resonance) with maximum value $Q_1/(2\Gamma')$ and $2\Gamma'$ being the full width between the local extrema. The second term is an overlapping Cauchy distribution with FWHM Γ' and amplitude $Q_2\Delta_2(\Gamma')^{-1}$. Similarly, the imaginary part of the susceptibility has a constant term plus a Cauchy distribution with FWHM $2\Gamma'$ (Fig. 12.2).

Using our scheme, we assume the following approximations for the parameters:

$$\Gamma' \approx \gamma + \gamma_{31}\left(\frac{\Omega_2^2}{4} + \frac{\alpha\Omega_1^2}{4}\right) \tag{12.32}$$

$$\nu_1 = (1 - \gamma_{31}\Omega_2^2/4) > 0 \qquad \nu_2 = (1 - \gamma_{31}(\alpha\Omega_1)^2/4) > 0.$$

Below, we use the approximations for the dispersion and extinction coefficients:

$$n'(\omega) \approx 1 + 0.5\chi'(\omega) \qquad n''(\omega) \approx \frac{\chi''(\omega)}{2}.$$

We get the expression for the group velocity. For brevity, we define $f(\Delta\omega) = (\Gamma')^2 + (\Delta_2\nu_2 - \Delta_1\nu_1)^2)$, then:

$$
\begin{aligned}
v_g(\Delta\omega)/c \approx{} & \frac{(f(\Delta\omega))^2}{\chi_0 Q_1 \Gamma' \nu_1} \\
& \times \left(\frac{(f(\Delta\omega))^2}{\chi_0 Q_1 \Gamma' \nu_1} (1 + 0.5\chi_0(\rho_{11} - \rho_{33})\Delta\omega) \right. \\
& + \frac{f(\Delta\omega)}{2 Q_1 \Gamma' \nu_1} (Q_1 \Gamma' (\Delta_2 \nu_2 - \Delta\omega \nu_1) + Q_2 \Gamma' \Delta_2) \\
& \left. + 0.5(\omega_{31} + \Delta\omega) \left(\frac{(f(\Delta\omega))^2}{Q_1 \Gamma' \nu_1} (\rho_{11} - \rho_{33}) - f(\Delta\omega) + \frac{2}{Q_1 \Gamma'} (\Delta_2 \nu_2 - \Delta\omega \nu_1) \right) \right)^{-1} .
\end{aligned}
\tag{12.33}
$$

The above expression implies that we can get infinite and negative group velocities if χ' has a negative gradient region with a large enough steepness. We note the group velocity about the saddle point near 0 of the real susceptibility reads:

$$
\begin{aligned}
\min_{|v_g|} v_{g,e}/c &\approx \beta_{\min} \left(1 + \mathcal{O}\left(\frac{\Gamma'}{\omega_{31}} \right) \right)^{-1} \qquad (12.34) \\
&= \frac{\Gamma'}{2\chi_0 Q_1 \nu_1 \omega_{31}} \left(1 + \frac{\Gamma'}{Q_1 \nu_1 \omega_{31}} \left(2\Gamma' + \Delta_2 \left(Q_2 + (\rho_{11} - \rho_{33}) \Gamma' \frac{\nu_2}{\nu_1} \right) \right) \right)^{-1}
\end{aligned}
$$

This is pretty close to the minimum (unsigned) group velocity, and we note that the second term $\ll 1$. We see the strong dependence of the slow light condition on Γ' and ω_{31}. Whilst we can use weaker fields to reduce the line broadening effect, the dephasing between the ground states (and therefore γ_{31}) sets a hard limit on the lowest value of Γ'.

With this, we can calculate the group velocity dispersion. In our scheme, our light fields are at least of THz frequencies, therefore the dominant terms in the GVD are those which include ω. We also use the fact that for a dilute medium $\chi', \chi'' \ll 1$. Then:

$$
\begin{aligned}
GVD(\Delta\omega)c \approx{} & -A\nu_1(\Delta_2\nu_2 - \Delta\omega\nu_1) f(\Delta\omega)^{-2} \\
& \times (1 + (\Delta_2\nu_2 - \Delta\omega\nu_1) f(\Delta\omega)^{-1})
\end{aligned}
\tag{12.35}
$$

$$
\begin{aligned}
A = {} & 2\omega\chi_0 \frac{\Omega_2^2}{4} (\nu_1(\alpha^2(\rho_{11} - \rho_{33}) + (\rho_{22} - \rho_{33})) \\
& + \Gamma'(\alpha(\rho_{11} - \rho_{33}) + (\rho_{22} - \rho_{33}))).
\end{aligned}
$$

For a light pulse, the polarized field $P(\omega)$ picks up a phase $\Delta\phi(\omega) = k(\omega)d$ after the pulse passes through a dispersive planar medium of length d. The pulse exiting the medium will be the Inverse Fourier Transform (IFT) of the post-medium $E(\omega)$ and

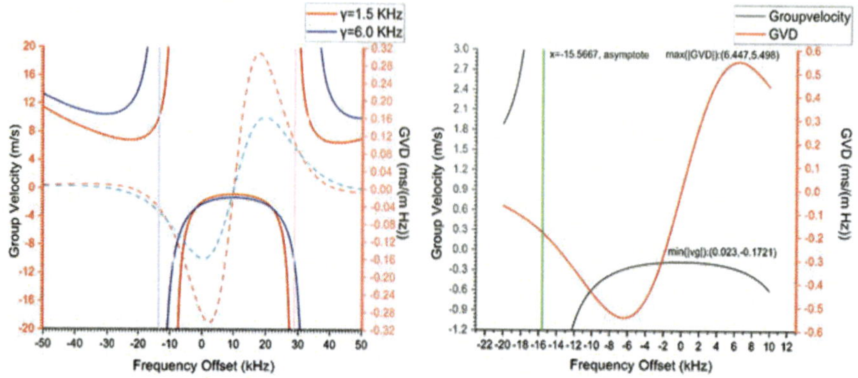

Fig. 12.3 (Left) Plot of the group velocity (solid line) and the GVD (dashed line) for the configuration in Sec 2 for two values of γ_{12} 1.5 kHZ, 6.0 kHZ and $\Omega_2 = 1/35$, $\Omega_1 = 0.1\Omega_2$. (Right) Plot of group velocity and the GVD showing anomalous dispersion and local minima

the time difference between peaks of two simultaneous probe pulses with different central frequencies ω_1, ω_2 is given by:

$$T = d\left(\frac{1}{v_g(\omega_1)} - \frac{1}{v_g(\omega_2)}\right). \tag{12.36}$$

Since the dispersion in the transparency window is strong in our scheme, a transform-limited pulse will have its spectral components accumulate different phases with the time-domain pulse broadening. The GVD is part of the estimate (GVD(ω_0)(bandwidth)d) of the chirp introduced to the pulse (valid for relatively small pulse bandwidths). Therefore, the group velocity, GVD and the chirp spectrum are effective tools to monitor responses to perturbations in the EIT medium by the way of determining how the properties of the polarization response to the probe pulses changes relative to the baseline case (Figs. 12.3 and 12.4).

An example we will consider is the passage of high-energy charged particles that induces Cherenkov radiation production in the medium. The probability of a photon being emitted during an individual high energy particle event is rather small. Tuning the medium to have Cherenkov radiation about frequencies which have slow group velocities will effectively ensure the buildup of the radiation over time as more events pass. The effect will be to introduce a growing electric potential that will affect the EIT state, which will manifest in probe pulses having a different chirp spectrum upon detection. We will explicitly consider the Cherenkov radiation produced in the EIT medium and give an analytic form to the field in our parameter regime.

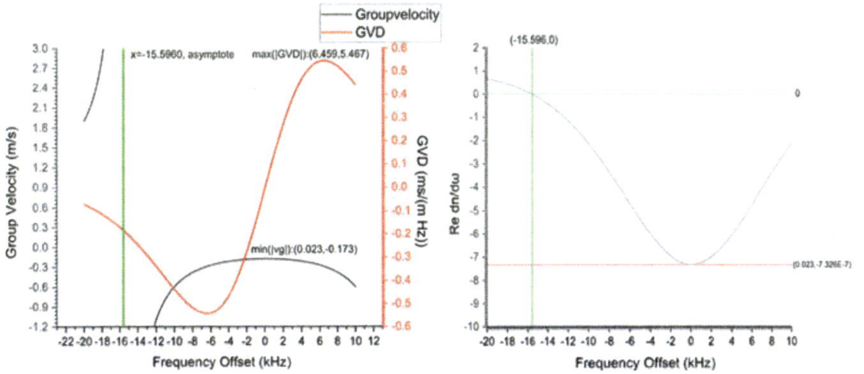

Fig. 12.4 (Left) Plot of the group velocity and the GVD showing anomalous dispersion and local minima with Ω_1 scaled by factor of $1.1\times$. Green line is the asymptote where anomalous dispersion occurs and the group velocity becomes infinite. We see how small perturbations in the pump Rabi frequency cause non-trivial changes in the dispersion of a pulse propagating through the medium. (Right) Plot showing the first derivative of the dispersion. The zero points and the local minima/maxima in the derivative correspond respectively to anomalous dispersion asymptotes and the local maxima/minima of the dispersion

12.4 Application: Detecting High-Energy Charged Particles Using EIT

Traditional detection of high energy charged particles have been handled in experiments using Cherenkov gas counters where ultra-sensitive detectors measure the small yield of light (essentially 1 photon per 10,000 events). The small yield of Cherenkov radiation, which can be calculated by the Frank-Tamm formula, makes the use of such ultra-sensitive detectors a necessity. But there has been a recent surge of interest in using techniques in atomic interferometry to measure low-energy physical phenomena; and using EIT with atomic spectroscopic techniques to measure changes in atomic media had been proposed in the 2000s [12].

We develop on these techniques and propose a semiclassical model which gives an approximate analytic solution for the Cherenkov radiation using the optical response functions calculated using the Lindbladian. The strong non-linear dispersion and high transmission properties of EIT can be used to create Cherenkov radiation with slow group velocities for a train of relativistic charged particles. At a classical level, the slow light condition ensures that the group profile of additive Cherenkov contributions will not smear and constructively increase in intensity. The Cherenkov radiation develops coherence with the transition dipoles after emission by the medium, essentially travelling as slow-moving dark polaritons confined to the medium for a duration proportional to the inverse of the group velocity. The model is developed for the case of a dilute homogeneous gaseous medium, but the results can easily be extended and optimised towards more complex systems such as EIT diffraction gratings and crystals.

12.4.1 A Model Using Fourier Transformed Maxwell's Equations

We consider a charged high energy particle traveling with relativistic velocity $\mathbf{v} = \beta c\hat{z}$ and having charge q in the lab frame. We use the treatment of a classical point particle, assuming that the width of the relativistic particle's wavefunction is much smaller than the wavelengths of emitted radiation. The below derivations are in the far-field regime where we assume the particle is sufficiently far away to neglect near-field radiation. The current density and its Fourier transform in the space and time domain is given by:

$$\begin{aligned} \mathbf{J} = \rho(x, y, z, t)\mathbf{v} &= q\delta(z - vt)\delta(x)\delta(y)\mathbf{v} \\ \mathbf{J}_{\omega,k} &= q\tilde{\delta}(\omega - \mathbf{k}\cdot\mathbf{v})\mathbf{v}. \end{aligned} \tag{12.37}$$

(where $\tilde{\delta}(z)$ is the generalized delta function).

We solve for the fields produced as a result of this current when placed in a homogeneous, anisotropic, non-magnetic medium driven to a EIT steady state. We use the derived optical response function ($\epsilon(\omega) = 1 + \chi_{31}(\omega - \omega_{31})$) with the Fourier transformed Maxwell equations to calculate the fields in the far-field limit. We start with the Fourier transformed Maxwell's equations for an anisotropic, non-magnetic medium:

$$\begin{aligned} \omega\mathbf{k}\cdot(\epsilon_0\check{\epsilon}\cdot\mathbf{E}_{\omega,k}) &= \rho_{\omega,\mathbf{k}} \\ i\omega\mathbf{k}\cdot\mathbf{B}_{\omega,k} &= 0 \\ \mathbf{k}\times\mathbf{E}_{\omega,k} &= -\omega\mathbf{B}_{\omega,k} \\ \mathbf{k}\times\mathbf{B}_{\omega,k} &= -i\mu\cdot\mathbf{J}_{\omega,k} + \omega\mu\epsilon_0\check{\epsilon}\cdot\mathbf{E}_{\omega,k}. \end{aligned} \tag{12.38}$$

And then introduce the scalar and vector potentials:

$$\mathbf{E}_{\omega,k} = i\mathbf{k}V_{\omega,\mathbf{k}} + i\omega\mathbf{A}_{\omega,k}$$

$$\omega\mathbf{k}\cdot(\check{\epsilon}\cdot\mathbf{A}_{\omega,k}) + (\mathbf{k}\cdot(\check{\epsilon}\cdot\mathbf{k}))V_{\omega,\mathbf{k}} = \frac{\rho_{\omega,\mathbf{k}}}{\epsilon_0}$$

$$\mathbf{k}\times(\check{\mu}^{-1}\cdot(\mathbf{k}\times\mathbf{A}_{\omega,k})) + \frac{\omega^2}{c^2}\check{\epsilon}\cdot(\mathbf{A}_{\omega,k} + \frac{c}{\omega}\mathbf{k}V_{\omega,\mathbf{k}}) = -\mu_0\mathbf{J}_{\omega,k}.$$

Here $\check{\epsilon}$ is the 3×3 rank 2 dielectric tensor and is a function of frequency. We choose to take our coordinate system to be along the dielectric axes (e_+, e_-, e_z) where the tensor is diagonal. We consider the case where only one transverse mode is active and the medium is nearly homogeneous. We use the Lorentz gauge condition

$$\mathbf{k}\cdot(\check{\epsilon}\cdot\mathbf{A}_{\omega,k}) + \frac{\omega|\check{\epsilon}\cdot\check{\mu}\cdot\check{\epsilon}|}{c^2}V_{\omega,\mathbf{k}} = 0$$

to arrive at the Helmholtz equations for the potentials [13]:

$$\mathbf{A}_{\omega,k} = \mu_0 q \left(k^2 - \frac{\omega^2}{c^2}\hat{\epsilon}\cdot\right)^{-1} \tilde{\delta}(\omega - \mathbf{k}\cdot\mathbf{v})\mathbf{v} \tag{12.39}$$

$$V_{\omega,\mathbf{k}} = \frac{q}{\epsilon\epsilon_0}\left(k^2 - \frac{\omega^2}{c^2}\hat{\epsilon}\cdot\right)^{-1} \tilde{\delta}(\omega - \mathbf{k}\cdot\mathbf{v}). \tag{12.40}$$

For lossy mediums, $\mathbf{k}^* \cdot \mathbf{E} \neq 0$ in general. We're interested in only the transverse mode in the far-field limit.

We take the inverse spatial Fourier transform of $\mathbf{A}_{\omega,k}$:

$$\mathbf{A}_\omega \cdot \hat{v} = \frac{1}{(2\pi)^3}\int_{\Gamma_k} \mu_0 q v \left(k^2 - \frac{\omega^2}{c^2}\check{\epsilon}\cdot\right)^{-1} \tilde{\delta}(\omega - \mathbf{k}\cdot\mathbf{v})e^{i\mathbf{k}\cdot\mathbf{r}} d^2k_\perp dk_z. \tag{12.41}$$

Expressing it in cylindrical coordinates and carrying out integration over k_z, we get:

$$\mathbf{A}_\omega \cdot \hat{v} = \mu_0 q e^{i\omega(z/v)} \frac{1}{(2\pi)^3}\int_0^\infty \left(k^2 - \frac{\omega^2}{c^2}\check{\epsilon}\cdot\right)^{-1}\left(\int_0^{2\pi} e^{ik_\perp r_\perp \cos(\theta)} d\theta\right) k_\perp dk_\perp \tag{12.42}$$

The angular integral is a representation of the zeroth order Bessel function.

The above integral, for the $k_\perp$ integrand, has 2 poles $\pm ik_r = \pm\omega c^{-1}\sqrt{\beta^{-2} - \epsilon(\omega)}$, where $\epsilon(\omega)$ is the scalar dielectric constant for the active transverse mode, in the complex domain corresponding to forward and backward radially propagating modes. Since the medium is dissipative, we consider only the physical forward propagating pole and use a semi-circular contour with bumps excluding the poles. $k_z = \omega v^{-1}$ fixes the possible values for the radial wavevector. For the case of no loss, the Cherenkov condition is $\mathrm{Re}\{\epsilon\} > \beta^{-2}$. We will use this later to determine the contributions for radiative modes and evanescent modes:

$$\mathbf{A}_\omega \cdot \hat{v} = \mu_0 q e^{i\omega(z/v)} \frac{1}{2\pi}\int_0^\infty \frac{k_\perp J_0(k_\perp r_\perp)}{k_\perp^2 + \frac{\omega^2}{c^2}\left(\beta^{-2} - \epsilon(\omega)\right)} dk_\perp. \tag{12.43}$$

The solution of the integral is the modified Bessel function K_0 evaluated at ik_r. Considering the treatment in [14] in extending the lossless case to the lossy case when we transform to the time domain, the frequency integration can be segmented into two regions $\mathrm{Re}\,(\epsilon(\omega) - \beta^{-2}) \gtrless 0$, which correspond respectively to the radiative and non-radiative regions. We define the radiative region: $\Gamma_1 = [\omega_{31} + a, \omega_{31} + b]$. We hence solve the final integral for $\mathbf{A}_\omega$, and similarly define V_ω:

$$\mathbf{A}_\omega(\mathbf{r}, \omega) \cdot \hat{v} = \begin{cases} \frac{\mu_0 q}{4\pi} e^{i\omega(z/v)} K_0(ik_r r_\perp) & \omega \in [\omega_{31} + a, \omega_{31} + b]^c \\ \\ \frac{\mu_0 q}{4\pi} e^{i\omega(z/v)} \frac{i\pi}{2} H_0^{(1)}(k_r r_\perp) & \omega \in [\omega_{31} + a, \omega_{31} + b] \end{cases} \tag{12.44}$$

$$V_\omega(\mathbf{r}, \omega) = \begin{cases} \frac{q}{4\pi\epsilon_0\epsilon v} e^{i\omega(z/v)} K_0(ik_r r_\perp) & \omega \in [\omega_{31} + a, \omega_{31} + b]^c \\ \\ \frac{q}{4\pi\epsilon_0\epsilon v} e^{i\omega(z/v)} \frac{i\pi}{2} H_0^{(1)}(k_r r_\perp) & \omega \in [\omega_{31} + a, \omega_{31} + b]. \end{cases} \tag{12.45}$$

In the far-field limit, $kr_\perp \gg 1$, asymptotic expansion of the Hankel function gives us $i\pi H_0^{(1)}(k_r r_\perp) \sim \sqrt{i\pi} e^{ik_r r} / \sqrt{k_r r}$. We can define the steepest descent contours by using the phase term with complex frequency $\tilde{\omega}$:

$$\frac{r_\perp}{c} \theta(\tilde{\omega}, r, z, t) = \frac{r_\perp}{c} i \left(\frac{\tilde{\omega} c}{r} \left(\frac{z}{\tilde{v}} - t \right) + \tilde{\omega} n'_\perp + i\tilde{\omega} n''_\perp \right) \tag{12.46}$$

The steepest descent contour will keep the imaginary part of the phase constant whilst the real part will have an additional term $-p$. We define the contour $\gamma_{\tilde{\omega}}(p)$, parameterized by real p by:

$$\theta(\gamma_{\tilde{\omega}}(p), r, z, t) = \theta(\tilde{\omega}, r, z, t) - p \tag{12.47}$$

About a point ω_i, if we can define a local analytic expansion of $\theta(\tilde{\omega}, r, z, t)$ extended to the complex plane, we can define an inverse function for θ and get $\gamma_{\tilde{\omega}}(p) = \theta^{-1}(\theta(\omega_i, r, z, t) + ip)$.

The method of steepest descent can be used to find any stationary points in the case when χ'' is very small. These group modes represent points of constructive interference of wavefronts on a group cone with vertex at $(0, vt)$. This involves finding the critical points of (12.46). Analyticity of the refractive index ensures that the critical points will be saddle points and there'll be no local extrema in the complex plane. The critical frequencies $\tilde{\omega}_i$ obey the below condition:

$$t - \frac{z}{v} - r_\perp \frac{dk_r}{d\tilde{\omega}} = 0. \tag{12.48}$$

We deform the integration contour on the real interval to the complex plane such that we pass through all critical points via their steepest descent contours and all other contributions are on segments, which are very small. For the time-domain fields in the radiative domain, we define the set of critical points $\mathcal{S}_c$ with each critical point defined on contour $\gamma_{\tilde{\omega}_i}(p)$. A larger upper limit P will give a better approximation of the integral

$$V(\mathbf{r},t) \approx \sum_{w_i \in \mathcal{S}_c} \frac{iq}{4\pi\epsilon_0 v} e^{ir_\perp c^{-1}\theta(\tilde{\omega}_i,r,z,t)} \frac{\sqrt{i\pi}}{\sqrt{2r_\perp}} \int_0^P \frac{e^{-r_\perp c^{-1}p}}{\epsilon\sqrt{k_\perp(\gamma_{\tilde{\omega}_i}(p))}} \frac{d\gamma_{\tilde{\omega}_i}}{dp}. \tag{12.49}$$

In the case where the critical points are non-degenerate and represent local maxima of the real part of the phase, we can use Morse's Lemma for complex functions to get a nicer form of the integral. Since the functions are holomorphic, there exists a neighbourhood near non-degenerate saddle points where phase $\theta(\tilde{\omega})$ is approximately quadratic. Accordingly, we can define the asymptotic expansion of (12.49)

$$V(\mathbf{r},t) \approx \frac{-iq}{4\pi\epsilon_0 r_\perp v} \sum_{\tilde{\omega}_i \in S_{t,z,r_\perp}} \frac{e^{r_\perp c^{-1} i\theta(\tilde{\omega}_i,t,z,r_\perp)}}{(-\mathrm{GVD}_r(\tilde{\omega}_i))^{1/2}} \left(\frac{1}{\sqrt{k_\perp(\tilde{\omega}_i)}} + \mathcal{O}\left(\frac{c}{r}\right)\right). \tag{12.50}$$

The critical points will be complex for a dissipative system and will be located near neighbourhoods of critical points for the non-dissipative system if the dissipation is weak. In the case where no critical points exist or the integration is difficult, we can calculate the Gauss-Laguerre quadrature of the integral [15]. However, when the dissipation is significant, the validity of the Cherenkov condition and the simplicity of the fields is not generally true.

An explicit evaluation of the integral in terms of complex error functions can be done in the far-field limit by approximating $k_\perp$ by a piecewise function composed of parabolas. As shown in Fig. 12.5, we consider the local quadratic best fit in the region and expand it about a central frequency, $\omega_{31}' = \omega_{31} + x$, in region $[a_i, b_i]$, $k_r \sim k_0 + \eta\Delta\omega + \sigma'(\Delta\omega)^2$. We evaluate the radial component of the E-field which has a Hankel function of first type with order 1. We use the asymptotic far-field limit and we get the below integral:

$$\mathbf{E}(\mathbf{r},t) \approx \frac{iq}{8\pi\epsilon_0 r_\perp v}\sqrt{ir_\perp\pi}\sum_{j\in I} e^{i\omega_{31}'(z/v-t)+ik_0 r_\perp} \times \int_{a_j}^{b_j} dw\, e^{i(w(z/v-t)+r_\perp(\eta w+\sigma' w^2))} \frac{\sqrt{k_r}\hat{r} + \frac{\omega}{v\sqrt{k_r}}(1-\beta^{-2}\epsilon)\hat{z}}{\epsilon(w)} \tag{12.51}$$

$$\mathbf{B}(\mathbf{r},t) \approx \frac{iq}{8\pi\epsilon_0 r_\perp c^2}\sqrt{ir_\perp\pi}\sum_{j\in I} e^{i\omega_{31}'(z/v-t)+ik_0 r_\perp} \times \int_{a_j}^{b_j} dw\, e^{i(w(z/v-t)+r_\perp(\eta w+\sigma' w^2))} \frac{\sqrt{k_r}}{\epsilon(w)}\hat{\phi}. \tag{12.52}$$

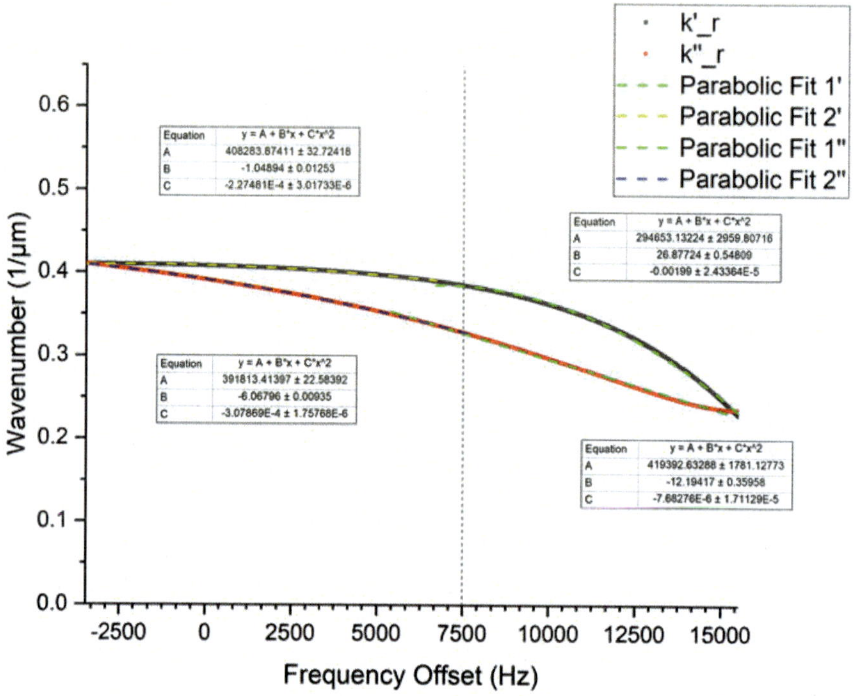

Fig. 12.5 Plots of real and imaginary parts of k_r, when the radiative modes are in the anomalous dispersion region, with piecewise parabolic approximations as functions of the frequency offset. For our particular system+field choice, we divide the radiative region into 2 regions with common endpoint 7500 Hz. Resonant points where Re $\epsilon = \beta^{-2}$ are at the intersections of real and imaginary parts, bounding the radiative region

We approximate the part of the integrand outside the exponential by a Taylor series in w with coefficients c_j. The solution to (12.51) is a sum of complex error functions. The integral of the exponential term is given below:

$$
\begin{aligned}
G_0(a,b) = i\tilde{\sigma}^{-1} \Bigg[& (e^{\tilde{\sigma}^2 b^2 r_\perp - it'b} - e^{\tilde{\sigma}^2 a^2 r_\perp - it'a}) e^{\Phi^2 t'^2/(2\sigma^2 r_\perp)} \\
& - e^{\tilde{\sigma}^2 b^2 r_\perp - it'b} \operatorname{erfcx}\left(\Phi\left(\frac{t'}{\sqrt{2r_\perp}\sigma} + i\frac{b\tilde{\sigma}^2\sqrt{r_\perp}}{\sqrt{2}\sigma}\right)\right) \\
& + e^{\tilde{\sigma}^2 a^2 r_\perp - it'a} \operatorname{erfcx}\left(\Phi\left(\frac{t'}{\sqrt{2r_\perp}\sigma} + i\frac{a\tilde{\sigma}^2\sqrt{r_\perp}}{\sqrt{2}\sigma}\right)\right)\Bigg]
\end{aligned}
\tag{12.53}
$$

$$
\sigma = |\tilde{\sigma}| \equiv |\sqrt{2i\sigma'}| \qquad \Phi = e^{-i\arg(\tilde{\sigma})/2} \qquad t' = (t - z/v) - \eta r_\perp.
$$

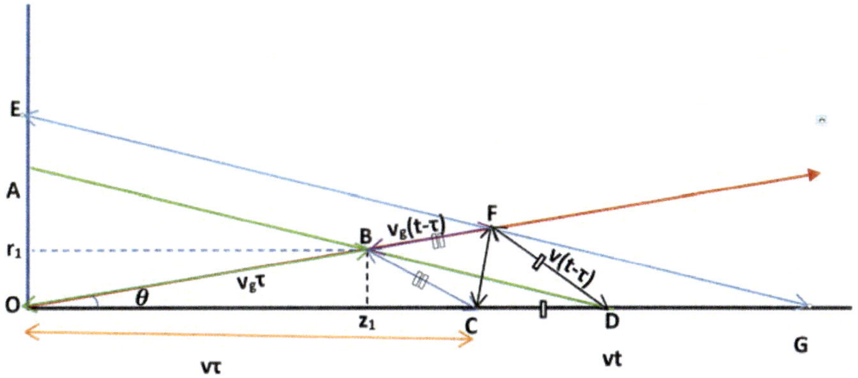

Fig. 12.6 Formation of the group cone for coherent Cherenkov radiation, showing the propagation of the group mode with angle $\theta = \theta_g$ and the radiation propagation distances over transit time τ and time of Cherenkov cone vertex t

The general expansion for any of the field terms with expressions of form $A \sum_{j\in I} c_j \int_{a_j}^{b_j} dw\, e^{i(w(z/v-t)+r_\perp(\eta w+\sigma' w^2))} f(w)$ is given by:

$$F(t,z,r) = A \sum_{j\in I} c_j \sum_{k=0}^{N} i^n \left.\frac{d^k f}{dx^k}\right|_{x=\omega'_{31}} \frac{d^k}{d(z/v-t)^k} G_0(a_j, b_j). \tag{12.54}$$

Expression (12.53) provides more information about the relations between the parameters in the Cherenkov field and system parameters. It's clear that η_j and σ_j are both related to derivatives of the radial wavenumber k_r about the central frequency x for each frequency interval $[a_j, b_j]$. We can estimate these values by looking at the susceptibility and seeing how altering parameters like Γ', Δ_2, γ_{12} affect the curvature and first derivative of k_r. We see that decreasing the nonlinearity in the susceptibility about region I_j reduces the inhomogeneous broadening of spectral components over propagation distance, suggesting that the linear dispersion region of EIT is ideal for decreasing dissipation and for controlling group velocities.

We finish this section by expanding on the geometry of Cherenkov radiation and checking that the causality conditions are satisfied. For the case where we have a single stationary point for a mode with group velocity v_g, we can find the critical frequency $\tilde{\omega}$ that solves (12.48). The phase speed at this frequency matches the charge velocity v and results in the phase matching condition where path difference $CF = 2BF\sin(\phi/2)$ where $\sin(\phi) = \dfrac{\sin(\theta)}{1 - \dfrac{v_g^2 - 2\vec{v}_g\cdot\vec{v}}{v^2}}$. θ_k is the angle of the group velocity vector with respect to the z-axis and θ_g is half the aperture of the group cone. We can define the transit time τ of radiation reaching (r, z) on the group cone generated by the vertex at time t (as in Fig. 12.6), where $\tilde{v}_r$ is the complex radial velocity or group velocity for wavenumber $k_\perp$:

$$\tilde{\tau} = t - \frac{\sqrt{(z - v\tilde{\tau})^2 + r^2}}{\tilde{v}_g(\tilde{\omega})} \tag{12.55}$$

In the case of weak dissipation, the real group velocity will be a relevant physical quantity and the below causality condition must be satisfied:

$$t - \frac{z}{v} - r\sqrt{\frac{1}{v_g(\omega)^2} - \frac{1}{v^2}} \geq 0. \tag{12.56}$$

Defining the group cone (12.48) and the time-retarded current density ensures that this condition is automatically satisfied. To satisfy non-degeneracy, we require that the complex GVD not be identically zero.

Care must be taken here, as the real group velocity defined by $v_g = \left(\mathrm{Re}\,\frac{dk}{d\omega}\right)^{-1}$ is not always a reliable measure of the signal velocity, especially in cases where we have significant dissipation and superluminal group velocities. For our purposes, we assume we're working with timescales where the group is still coherent (no smearing of the cone). In the next section, we'll consider the fields in the case of very small dissipation when we use the steepest descent method and the case of weak dissipation where we numerically solve for the fields.

12.4.2 Cherenkov Radiation in the EIT Regime

In general, the introduction of a nonzero loss term breaks the Cherenkov cone-like condition and dampens the field amplitude. For a large enough loss, the highly anisotropic character of the radiation and angular sensitivity to the charged particle's speed is weakened. Ideally we would want to get contributions close to the center of the transparency window, reduce the gaseous medium's particle density and introduce a background refractive index enhancement to reduce Im χ. For the case of EIT with Rubidium D1 line spectra using linearly polarized fields, we can select a transition between levels with magnetic number $m = 0$ and introduce a static magnetic field to cause large enough energy shifts for levels with the same nonzero m. For large enough detuning, these transitions will introduce a frequency-independent contribution to the susceptibility ϵ_∞.

We use the control scheme outlined in Sect. 12.3. The frequency range where Cherenkov radiation is transmitted is $[\omega_{31} + a, \omega_{31} + b]$, which contains all frequencies that contribute non-trivially to the radiation. We develop a scheme to evolve the fields corresponding to a mode ω_i after some time Δt.

To define the time-evolving field, we first state the group velocity vector for Cherenkov radiation:

$$\mathbf{v}_{g,Ch}(\Delta\omega) = \frac{\beta^{-1}\hat{z} + n_{\perp}\hat{r}}{(1+\chi') + 0.5(\omega_{31} + \Delta\omega)\dfrac{d\chi'}{d\Delta\omega}} \tag{12.57}$$

Suppose the wave propagates for small time Δt, such that the group profile remains coherent, from its initial point at $\{t, z, r_{\perp}\}$. Since the peak propagates with velocity $\mathbf{v}_{g,Ch}$, the new positions are given by $r' = r + (\mathbf{v}_{g,Ch}(\Delta\omega_i)\cdot\hat{r})\Delta t$, $z' = z + (\mathbf{v}_{g,Ch}(\Delta\omega_i)\cdot\hat{z})\Delta t$. We can express $v\Delta t + z' - z = \alpha(vt - z)$ and $r' - r = \beta r$ and they satisfy the cone condition if $\alpha = \beta$. This condition gives us the relation $\dfrac{v - \mathbf{v}_{g,Ch}(\Delta\omega_i)\cdot\hat{z}}{v\mathbf{v}_{g,Ch}(\Delta\omega_i)\cdot\hat{r}} = v_{g,r_{\perp}}^{-1}$, which can be obtained from (12.54). The cone condition, and the condition for non-degeneracy, is then automatically satisfied and we can define the one-cycle averaged Cherenkov Poynting vector, and the radiant power per unit length at a fixed radius in the axial direction $\zeta_z = 2\pi r_{\perp}\langle S\rangle$, for modes $\tilde{\omega}_i \in \mathcal{S}_{t,z,r_{\perp}}$:

$$\begin{aligned}\zeta_z(\mathbf{r} + \mathbf{v}_{g,Ch}(\Delta\tilde{\omega}_i)\Delta t, t + \Delta t) &= \sum_{\tilde{\omega}_i \in \mathcal{S}_{t,z,r_{\perp}}} \pi(r + (\mathbf{v}_{g,Ch}(\Delta\omega_i)\cdot\hat{r})\Delta t)\\ &\times \mu^{-1}\mathrm{Re}\,\{E_{r,\tilde{\omega}_i}(\mathbf{r},t)B^{*}_{\tilde{\omega}_i}(\mathbf{r},t)\}e^{-2\Delta t\,\mathrm{Im}\,\{\mathbf{k}\cdot\mathbf{v}_{g,Ch}(\Delta\tilde{\omega}_i)+\Delta\tilde{\omega}\}}.\end{aligned} \tag{12.58}$$

The peak intensity of the radiation is confined to a thin cylinder about the z-axis because the dissipation and the spatial coherence depend only on the bandwidth and the radial wavevector. For small r, the spatial phase $k_r'(\omega)r$ for different frequencies is small enough that the spectral components add coherently. The dissipation part $k_r''(\omega)r$'s shaping for the spectral profile is negligible for these distances. For large r, the phase differences are large enough that the phase is rapidly changing with frequency and the contributions to the field come from the critical points satisfying (12.48) or from nodes for the Gauss-Laguerre quadrature. The coherent addition of spectral components also depends on the retarded time through the relative spectral phase $\Delta\omega t'$, and hence on the radial components of group velocity and GVD. It follows that minimizing bandwidth and the range of k_r, for frequencies that contribute non-trivially to the radiant power, increases the coherence distance and time. The background susceptibility ϵ_{∞} and the system+field parameters can be used to control the radiation spectrum and the radial refractive index through manipulation of the susceptibilities (12.30, 12.31), as seen in Figs. 12.7 and 12.8. EIT and the slow light condition allow for strong correlations between spectral components such that the fields due to multiple charges can stay in this coherence regime. With a significant enough accumulation of Cherenkov radiation in the medium, we can expect a change

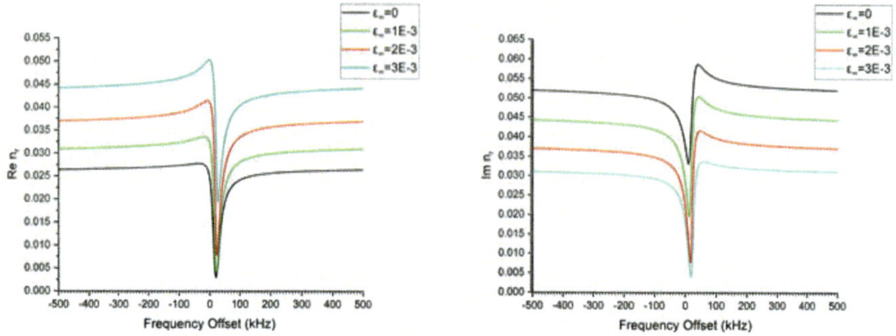

(a) $\mathrm{Re}\, n_r$ (b) $\mathrm{Im}\, n_r$

Fig. 12.7 Plots showing the radial refractive index n_r for various values of ϵ_∞. The decrease of $\mathrm{Im}\, n_r$ for large ϵ_∞ proportionally decreases the dissipation and is accompanied by increased and sharper dispersion

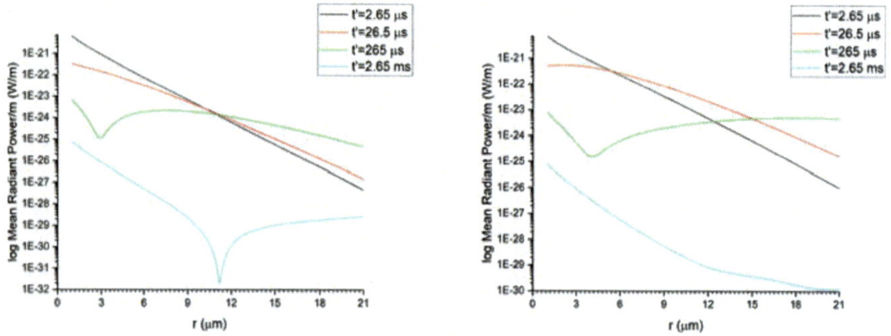

(a) $\Omega_2 = 1/35\gamma_{31}$ (b) $\Omega_2 = 2/35\gamma_{31}$

Fig. 12.8 Plots for $\zeta_z(\mathbf{r}, t)$ with varying r for different times

in the optical response functions. In the next section, we consider the perturbative effects to the EIT state due to small fields.

12.4.3 Perturbation to Steady States: Developing a Control Method to Detect High Energy Particles

We can consider a couple of approaches in detecting the presence of Cherenkov radiation. Use of ring Cherenkov particle detectors is the conventional approach, but we can also use the methods of atomic spectroscopy and interferometry by tuning the optical properties of the medium and its sensitivity to small perturbations.

Consider a setup of atoms in an EIT steady state exposed to Cherenkov radiation described by the Lindbladian $\mathcal{L}_{\mathcal{J}}$. We assume that $\mathcal{L}_{\mathcal{J}}$ has a unique steady state, with projection operator on the steady state subspace given by $\mathcal{P}_\infty = \sum_{\lambda_k} |\psi_{\lambda_k}\rangle\langle\psi_{\lambda_k}|$, and introduce the perturbation operator $g(t)\mathcal{O}$ (corresponding to $-i[V,\cdot]$ in the field interaction picture) to describe the Cherenkov radiation-atom interaction. The time evolution from an initial state is given by $\mathcal{T}e^{\int_{-\infty}^{t} d\tau(\mathcal{L}_\mathcal{J}+g(t)\mathcal{O})}|\rho_s\rangle$. We assume that the perturbation due to Cherenkov radiation is switched on at $t=0$ and $g(t)$ is slowly time-varying. We use the Kubo formula to solve for the first-order perturbation [16]:

$$\mathcal{T}_t^{(1)}|\rho_s\rangle = \int_{-\infty}^{t} d\tau g(\tau) e^{(t-\tau)\mathcal{L}_\mathcal{J}} \mathcal{O}|\rho_s\rangle \tag{12.59}$$

In the case where $g(t) = G\Theta(t)$, $\mathcal{T}_t^{(1)}|\rho_s\rangle = G\mathcal{L}_\mathcal{J}^{-1}\mathcal{O}|\rho_s\rangle$. The above equation assumes we start from the steady state of the unperturbed system. We then apply the perturbation to the system and calculate the first order perturbation to the steady state using the Kubo formula. The first order perturbation on a steady state describes a leakage from the asymptotic state subspace (which contains the steady states) to the subspace of decaying states, [16]. For the choice of parameters $\Omega_1 = 0.1$ MHz, $\Omega_2 = 1.0$ MHz, $\Delta_1 = \Delta_2 = 0$, the Lindbladian is diagonalizable with 1 physical state and 8 unphysical traceless states with negative real part of eigenvalues λ_k, $\left(\frac{\Omega_1^2+\Omega_2^2}{\gamma_{31}} \le |\lambda_k| \le \gamma_{31}\right)$. The dissipative energy gap, Δ_{dg}, is given by the minimum of the absolute value of real parts of the eigenvalues and represents the time-scale for which the perturbed wavefunction will have a contribution from states in the decaying subspace. For the weak-field case, it roughly follows the curve of $\frac{\Omega_1^2+\Omega_2^2}{\gamma_{31}}$. Increasing the dissipation gap to large values will decrease the leakage from the asymptotic subspace. We observe that small fields lead to almost no change in the coherence with application of the probe field as contributions due to dephasing negates any population in the other states. For large Rabi frequencies, we observe that the perturbation magnitude decreases, as the field strength increases Δ_{dg} and Γ'. However, as Fig. 12.9 shows, this effect is somewhat offset by choosing large $|\Delta_2|$. The perturbative effect seems to be strongest with detuning in the 20–50 kHZ range with control Rabi frequency in the 1–2 MHz range. There are trade-offs in choosing smaller Rabi frequencies in having to deal with larger absorptivity, while larger Rabi frequencies will broaden the optical response lineshapes.

Figure 12.9 shows regions in parameter spaces in which we can get significant change to ρ_{31}, whilst retaining choice for the real and imaginary parts, for small perturbations. In determining the full perturbation to the steady state, we can configure the initial system state of the detection system towards a desired change in the optical response functions n', n''. For example, we could be in the region with maximized GVD and the perturbed state will modify the chirp spectrum of the Cherenkov radiation or of another probe field and cause a delay in comparison to the pulses propagating in the unperturbed system. Small perturbations can also be very useful if the peak frequencies of the Cherenkov field or of another probe pulse are near

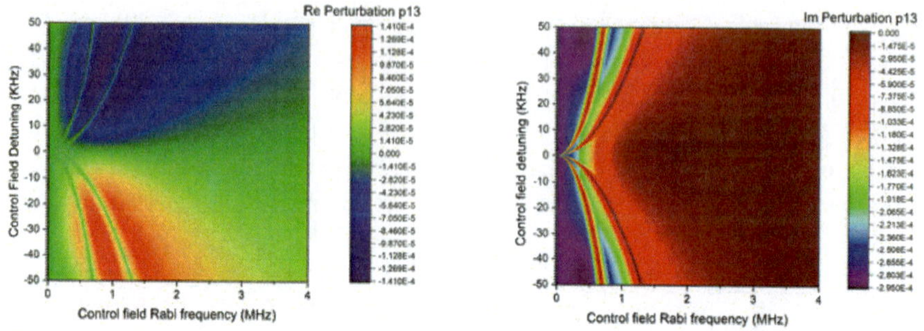

(a) $\mathrm{Re}\,\mathcal{T}_t^{(1)}\rho_{13,s}$ (b) $\mathrm{Im}\,\mathcal{T}_t^{(1)}\rho_{13,s}$

Fig. 12.9 Plots showing the real and imaginary parts of the perturbation $\mathcal{T}_t^{(1)}\rho_{13,s}$, due to perturbation $\Omega_{pert}\mathcal{O}$ after 1 μs (color axis). $\Omega_1 = 0.1\Omega_2$, $\gamma_{12} = 4.3E - 5$, $\Delta_2 = -5.7E - 4$, $\Omega_{pert} = 2.7E - 4$

the anomalous dispersion region, such that the perturbative response will lead to phenomena such as superluminal group velocities. Alternatively we could measure a different propagation angle for Cherenkov radiation due to change in η. This method allow us to observe the change in group profile/chirp spectrum of the Cherenkov radiation for a sequence of probe pulses.

12.4.3.1 Control Method Summary

We explicitly state our control protocol for detection of high energy particles. Our detection scheme is developed for two control objectives. The first is for the accumulation of slow Cherenkov radiation, it prioritises high transmission, nonlinear modes and slow group velocity to increase the yield of radiation. The second is for optimizing the detection of the Cherenkov radiation, involving choosing atomic and field parameters such that the optical response functions in the medium will develop a desired change in properties such as the chirp spectrum of a probe pulse after propagation through the medium. We start from the same base conditions as those listed in Sect. 12.3.1.

For the first control objective, we note that the properties of the Cherenkov radiation are determined by the parameters ω_{31}, $\mathbf{r}_\perp$, σ, η, ϕ, c_0 and the bandwidth $[a, b]$ (from Sect. 12.4.2). We set the cutoff speed to be equal to β, choose a Cherenkov parameter set to give us a desired form for the radiation, choose the system+field parameters to shape the radial wavenumber function $k_r(\omega)$ and determine the frequency interval $[\omega_{L1} + a, \omega_{L1} + b]$ (with a, b being the solutions to $1 - \beta^{-2} + \mathrm{Re}\{\chi\} = 0$ in the window centered at the peak frequency). The bandwidth for the coherent group mode is limited by the EIT linewidth Γ' and adjusting the control Rabi frequency Ω_2 is one way to broaden the linewidth, decrease dissipation and increase coherence times. The axial and radial group velocities are determined using (12.57) and (12.33) which can then be used to find any critical points satisfying (12.48) and then get the radiant power density (12.58). We can then optimise the power density for a train of high energy particles associated with some distribution of arrival times.

For the second control objective, we optimise the detection process by choosing the set of system+field parameters such that the initial state is perturbed by the Cherenkov field to a target state, which has different optical responses. We determine the regions in the plot of $\mathcal{T}^{(1)}(\rho_s)$ in the system+field parameter space which correspond to perturbed states we target. For example, we can choose to have the initial state to cause high GVD for a probe pulse and thus a significant enough perturbation that we get a measurable chirp and pulse delay for propagation in the perturbed medium.

This concludes our treatment for detecting high energy particles using EIT in atomic media. In the final section, we review our results and consider their extension to more complex systems.

12.5 Summary and Vision for Nanophotonics

We derived the optical response functions for the three-level Λ system driven to a EIT state in an open weakly coupled system using the Lindblad formalism to consider the effects of dephasing and decoherence after determining the field interaction Hamiltonian. We used the Bloch vector formalism to transition from the microscopic to the macroscopic picture where we discussed the optical response functions (dispersion and absorptivity), and showed how an EIT state with a choice of parameters would lead to a narrow transparency window with strong non-linear dispersion effects that created phenomena such as slow light and anomalous dispersion.

This led to our main focus which is on developing quantum control schemes, using the optical response and sensitivity features of atomic media in EIT states as a spectroscopic tools, to detect the passage of high-energy particles. Our model was developed in the regime of weak fields with ultranarrow resonances that capitalize on the properties of slow light and dissipation-free transmission. The group velocity dispersion (GVD) and features of the group velocity (such as in the presence of anomalous dispersion) (Sec 3) were considered in this regard.

We discussed an application of our approach to the detection of high-energy charged particles through the accumulative stimulation of Cherenkov radiation in a medium undergoing EIT (Sect. 12.4), and derived a weak-field approximation analytical model that allows us to calculate the group profile and its time-dependence. We then considered how small perturbations would affect the steady state configuration, determined the leakage from the steady-state subspace, the first order perturbation and the dissipation gap. In tandem with the known optical response functions, we showed that we can realize a control scheme that optimises the sensitivity of the optical response functions to the perturbation.

We note that whilst we spend most of our attention on the atomic and field properties, the high energy particles can also be controlled in a similar manner to the atomic transitions. Work in the field of PINEM (Photon induced near-field electron microscopy) [17] shows that high energy charged particle wavefunctions can be modulated over nanostructures with fringing evanescent fields that enhance probing of a train of randomly arriving high energy particles, introduce correlations between the particles, and increase the perturbation to a bound electron state [18]. We can control the Rabi phase for a random arrival process and increase coherent population build up in comparison to a train of uncorrelated wavefunctions. Further approaches include reverse engineering the high energy particle wavefunction from the photon spectral density function as well as through optically probing the atomic coherences [19].

Whilst this chapter focused on a particular atomic system, the results presented here would have similar analogues in more complex systems, such as photonic crystals and nanostructures, providing a strong relevance to quantum information theory and optical storage [20]. Coupling of the EM field to material modes in the system leads to formations of quasiparticles similar to dark state atomic EIT polaritons. Examples already researched include surface plasmon polaritons, in the case of introducing a substrate below the dielectric, or quasiparticles that form in bulk such as exciton-polaritons.

Surface plasmon polaritons (SPPs) show powerful applications in the production of tunable power-enhanced Cherenkov radiation. In [21], a dielectric substrate with a buffer layer deposited with multiple layers of graphene was used to enhance the production of Cherenkov radiation due to an electron beam by coupling the radiation to the SPP modes, resulted in increasing the intensity of the radiation by more than 2 orders of magnitude. The properties of the buffer, substrate and SPP modes can be used to control the intensity increase and the radiation frequency. Of particular interest is the analogous phenomenon of plasmon-induced transparency (PIT) realized in metamaterials consisting of two plasmonic resonators coupled to a waveguide [22]. One scheme utilizes coupling between a dark and bright resonator to achieve a transparency window and optical response functions similar to that of EIT in atomic media. Graphene metamaterials have seen wide applications in this field due to its low-loss, shiftable Fermi energies and electronic properties. A number of different geometrical approaches have theoretically showed PIT [23, 24] in the THz regime and this phenomena's applicability to quantum control. Due to the geometry of the resonators, spacings in the metamaterial, and properties of the graphene layer, there is high controllability for the electromagnetic response. The features of enhanced slow light and trapping as well as the amplification of signals in the presence of PIT suggests a possible extension for this chapter's results.

Exciton interactions are another area where this research can be extended. The strong interaction of excitons with matter and light, the capability to induce EIT, and the formation of strongly non-linear exciton-polariton states [25] suggest a future direction for sensing high energy particles with EIT in more complex media. One particular application is for systems with Rydberg excitonic states, which yield significantly higher nonlinearities in the optical response and few photon strongly correlated states that result from the Rydberg blockade effect [25]. With the introduction of the blockade effect, EIT is disrupted and the nonlinearities are transferred to the few photon states. Benefits of systems with Rydberg excitonic states include greatly increased interaction strengths and applications of ultralow field intensities, which reduces decoherence and line broadening effects. With the traditional features of EIT including the transparency window and slow light along with the strongly correlated few photon states and quantum control aspect for the Rydberg exciton system, there is good scope in detecting high energy particles through the production and control of slow moving Cherenkov polariton-excitons.

Acknowledgements Authors gratefully acknowledge the support from the Office of Naval Research, the Department of Energy and the Jefferson Laboratory.

References

1. M. Fleischhauer, M.D. Lukin, Dark-state polaritons in electromagnetically induced transparency. Phys. Rev. Lett. **84**, 5094–5097 (2000)
2. M. Fleischhauer, “Stopping” of light and quantum memories for photons, in *Coherence and Quantum Optics VIII*, ed. by N.P. Bigelow, J.H. Eberly, C.R. Stroud, I.A. Walmsley (Springer US, Boston, MA, 2003), pp. 127–127
3. W. Li, P. Islam, P. Windpassinger, Controlled transport of stored light. Phys. Rev. Lett. **125**, 150501 (2020)
4. P. Bienias, H.P. Büchler, Two photon conditional phase gate based on rydberg slow light polaritons. J. Phys. B At. Mol. Opt. Phys. **53**, 054003 (2020)
5. R. Beausoleil, S. Barrett, P. Kok, K. Nemoto, W. Munro, T. Spiller, Classical and quantum information processing with eit, in *Proc SPIE*, vol. 5735, (04, 2005) pp. 143–154
6. B. Peng, S. Ozdemir, W. Chen, F. Nori, L. Yang, What is and what is not electromagnetically induced transparency in whispering-gallery microcavities. Nat. Commun. **5**, 5082 (04, 2014)
7. M.A. Guidry, E. Kuchina, I. Novikova, E.E. Mikhailov, Characterization of frequency stability in eit-based atomic clocks using a differential detection scheme (2017)
8. D. Steck, Rubidium 87 d line data (01, 2003)
9. N.V. Vitanov, A.A. Rangelov, B.W. Shore, K. Bergmann, Stimulated raman adiabatic passage in physics, chemistry, and beyond. Rev. Mod. Phys. **89**, 015006 (2017)
10. S. Malinovskaya, Design of many-body spin states of rydberg atoms excited to highly tunable magnetic sublevels. Opt. Lett. **42**, 314 (01, 2017)
11. S.A. Malinovskaya, G. Liu, Harmonic spectral modulation of an optical frequency comb to control the ultracold molecules formation. Chem. Phys. Lett. **664**, 1–4 (2016)
12. I. Carusotto, M. Artoni, G.C. La Rocca, F. Bassani, Slow group velocity and cherenkov radiation. Phys. Rev. Lett. **87** (Jul 2001)
13. W.C. Chew, Vector potential electromagnetic theory with generalized gauge for inhomogeneous anisotropic media (2014)
14. M.H. Saffouri, Treatment of Cherenkov radiation from electron and magnetic charges in dispersive and dissipative media. Nuovo Cim. D **3**, 589 (1984)
15. S. Chandler-Wilde, I. Graham, S. Langdon, E. Spence, Numerical-asymptotic boundary integral methods in high-frequency acoustic scattering*. Acta Numer. **21**, 89–305 (2012)
16. V.V. Albert, Lindbladians with multiple steady states: theory and applications (2018)
17. S.T. Park, M. Lin, A.H. Zewail, Photon-induced near-field electron microscopy (PINEM): theoretical and experimental. New J. Phys. **12**, 123028 (2010)

18. A. Gover, A. Yariv, Free-electron-bound-electron resonant interaction. Phys. Rev. Lett. **124**, 064801 (2020)
19. A. Karnieli, N. Rivera, A. Arie, I. Kaminer, The coherence of light is fundamentally tied to the quantum coherence of the emitting particle. Sci. Adv. **7**(18) (2021)
20. D.N. Basov, A. Asenjo-Garcia, P.J. Schuck, X. Zhu, A. Rubio, Polariton panorama. Nanophotonics **10**(1), 549–577 (01 Jan2021)
21. T. Zhao, M. Hu, R. Zhong, G. Sen, C. Zhang, S. Liu, Cherenkov terahertz radiation from graphene surface plasmon polaritons excited by an electron beam. Appl. Phys. Lett. **110**, 231102 (06, 2017)
22. H. Yan, T. Low, F. Guinea, F. Xia, P. Avouris, Tunable phonon-induced transparency in bilayer graphene nanoribbons. Nano Lett. **14**(8), 4581–4586 (2014). (PMID: 25019702)
23. G. Wang, X. Zhang, X. Wei, K. Yang, Tunable plasmon-induced transparency through bright mode resonator in a metal-graphene terahertz metamaterial. Appl. Sci. **10**, 08 (2020)
24. A. Noual, M. Amrani, E.B. El Houssaine, Y. Pennec, B. Djafari-Rouhani, Terahertz plasmon-induced transparency and absorption in compact graphene-based coupled nanoribbons. Appl. Phys. A **125**, 02 (2019)
25. V. Walther, R. Johne, T. Pohl, Giant optical nonlinearities from rydberg excitons in semiconductor microcavities. Nat. Commun. **9** (Apr 2018)

Chapter 13
Probing Plasmons by EELS in Chiral Array of Hyperbolic Metasurfaces. The Role of Plasmon Canalization

Oleksiy Roslyak, Vassilios Fessatidis, Antonios Balassis, Godfrey Gumbs, and Aparajita Upali

Abstract We have derived the energy loss spectrum for a beam of energetic electrons moving parallel to a set of two-dimensional (2D)-metasurfaces capable of carrying hyperbolic plasmons due to their anisotropic conductivity. The formalism is presented for naturally occurring metasurfaces such as transition metal dichalcogenides (TMDCs), as well as for an engineered array of graphene or phosphorene nanoribbons. Our formalism is applicable in the long-wavelength limit where one neglects (graphene) or emphasizes the cross-polarization of the nanoribbons (phosphorene). When the nanoribbons are aligned, the plasmon propagation along oblique directions is not permitted. This allows for plasmon probing at much lower electron energies in contrast with isotropic graphene and offers drastic modification of the energy loss probability. Additionally, the structure is akin to a thin uni-axial crystal supporting extraordinary refracted beams with negative group velocity. We suggest that electron energy loss spectroscopy (EELS) is capable of identifying two resonances separately once the beam and the direction of the nanoribbons are mismatched. For a twisted double layer configuration, one may observe a peculiar interference effect between ordinary and extraordinary plasmon modes.

O. Roslyak · V. Fessatidis (✉) · A. Balassis
Department of Physics and Engineering Physics, Fordham University, Bronx, NY 10458, USA
e-mail: fessatidis@fordham.edu

O. Roslyak
e-mail: oroslyak@fordham.edu

A. Balassis
e-mail: balassis@fordham.edu

G. Gumbs
Department of Physics and Astronomy, Hunter College of the City University of New York, New York, NY 10065, USA
e-mail: ggumbs@hunter.cuny.edu

A. Upali
Borough of Manhattan Community College, CUNY, New York, NY 10007, USA
e-mail: uaparajita@bmcc.cuny.edu

H. Ünlü and N. J. M. Horing (eds.), *Progress in Nanoscale and Low-Dimensional Materials and Devices*, Topics in Applied Physics 144,
https://doi.org/10.1007/978-3-030-93460-6_13

13.1 Introduction

Two-dimensional (2D) metamaterials, also referred to as metasurfaces, exhibit a plethora of peculiar properties, such as beam shaping and steering via anomalous reflection and refraction, controllable surface refractive index, and nonlinear response. Conventional metasurfaces are made of subwavelength metallic nanoantennas arranged in a periodic pattern [1, 2]. This allows for a nontrivial coupling between localized surface plasmons (SP) of the individual nanoparticles and lattice surface plasmon resonances (LSPR). The LSPRs, in contrast to plasmons in three-dimensional (3D) metamaterials, do not exhibit volumetric losses. Recent advances in metasurfaces fabrication allow for excellent coupling with planar optical devices and waveguides. Some properties of metasurfaces can be replicated by anisotropic 2D materials. Whenever the surface allows for inductive and capacitive responses along two transverse directions, a new class of metamaterials exhibits itself: hyperbolic metasurfaces as well as extremely anisotropic 2D structures borrowing the anisotropy from an underlying dielectric substrate. Such structures are capable of both, strong confinement of one type of electromagnetic field, known as plasmons, while effectively radiating the other modes (polaritonic effect). This may lead to effective negative refraction and focusing of surface plasmons yielding extremely high local density of states (Purcell factor) and polarization anisotropy. The plasmon dispersion, due to the anisotropy, becomes a set of equal frequency contours which morph from elliptical to hyperbolic upon tunable topological transitions. The critical frequency range for such transitions depends on both the topology of the elements and morphology of the underlying chemical structure.

There are several plasmonic probing techniques. Plasmon-polaritons can be optically excited by normal incidence illumination. Several diffraction orders arising from collective modes are readily achievable. Only modes with effective dipole moments laying on the surface plane are usually of interest. Oblique incidence as well as localized near field probes may be used to analyze the response of the out-of-plane collective modes. The latter is usually associated with scanning near-field optical microscopy (SNOM) [3, 4]. This technique maps optical properties of individual plasmonic nanoparticles and metasurfaces with a spatial resolution of about 10 nm. The near-field mapping can occur with a large variety of excitation sources and is limited only by the laser tunability range. The accuracy of SNOM imaging relies on the dimensions of the optical probe. They must be small enough to obtain high spatial resolution. It must also be efficient at converting the optical near-field into a measurable far-field signal. Usually, this requires placing a probe in close proximity with the sample, typically within a few nm away from the surface, in order to convert the evanescent waves into propagating ones. Several aperture (microfiber) and apertureless (metallic nanotip) techniques are available. However, the probe itself may disturb the optical near-field, thereby producing images that are hard to interpret. The mechanical interaction between the probe and the surface creates additional artifacts.

There is an alternative way to excite resonances out from the plane. This alleviates the aforementioned SNOM issues. When an energetic relativistic electron (of a few keV) passes close to a metasurface (or approaches the surface), the time dependent current associated with the electron motion excites plasmons and interacts with the self induced plasmonic field. The overall interaction is frictional in nature. The electromagnetic field induced by a moving charge is accompanied by an evanescent wake field along the surface which allows for the exploration of the dispersion relation of elementary electronic excitations on the specimen surface outside the light cone. The gradient of the wake potential will then act back on the electron making it lose energy [5–7]. Out-of-plane dipole moments play an important role in this interaction. This EELS technique has been proven to also be an important tool in quantum plasmonics. It involves the analysis of electron scattering caused by the evanescent field of the sample. This interaction often involves a few atomic layers. Researchers are often interested in the dissipative component of the force acting back on the moving electron passing near a solid surface [8, 9]. Besides launching surface plasmons and losing energy in the process, impinging electrons may undergo energy gain, Smith-Purcell or Cherenkov radiation and ultimately cathodoluminescence [10].

Instead of investigating scattering events, we suggest an alternative approach. We assume that the electron's trajectory is not modified by underlying interactions with the surface thereby ignoring quantum recoil effects. In this approximation, the external charge acts as an infinite source of energy and momentum with a classical charge distribution $\rho_{ext} = -e\delta\left(\mathbf{r} - \mathbf{v}t\right)$. In reciprocal space, it is given by $\sim \delta\left(\mathbf{q} - \omega\mathbf{v}\right)$. The problem is to find the dissipative component of the induced force. The negative of this component, can be interpreted as a space-dependent stopping power, $S = (-Ze/v)\left[\partial\phi_{ind}(\mathbf{r})/\partial t\right]_{\mathbf{r}=\mathbf{v}t}$ [11]. The main feature of an EELS spectrum is the plasmon resonance [12–14]. In a classical description of an extended system of charged particles, a plasmon is a collective excitation of the surface electrons oscillating against a background of positively charged ions. The interpretation of an EELS spectrum is not a trivial task [10]. Besides collective excitations, there are features in the spectrum due to uncorrelated electron-hole interband transitions (Landau damping). Developing simplified theoretical models is of paramount importance for interpreting experimental results . Nevertheless, in the low energy transfer domain (low $\mathbf{q}$), the plasmonic response is still sharp and Landau damping may be neglected. Measurements of the angular distribution of inelastically scattered fast electrons give information regarding the imaginary part of the inverse dielectric function of the solid [12, 13], at the frequency and momentum of the energy transfer to the electrons in the solid. It also probes the surface response function.

A thorough understanding of the electronic structure is necessary for the EELS signal interpretation. Starting from density functional theory (DFT), linear response time-dependent DFT in frequency-reciprocal space within the random phase approximation (TDDFT-RPA) is used to describe the loss function $\mathrm{Im}\left[-1/\varepsilon(\mathbf{q}, \omega)\right]$ for isolated monolayer structures [15]. The next section is devoted to the detailed derivation of the EELS signal for stratified media where different strata are separated by generic metasurfaces described via an anisotropic conductivity tensor. The result is given in terms of the poles (plasmon resonances) for Green's tensor components aligned with the electronic beam propagation. The Green's tensor is derived via the transfer matrix approach accounting for anisotropic layers separating the strata. The formalism can be readily extended to the anisotropy of the underlying dielectrics as well. The main result is given in terms of the probability of the electron scattering into an energy interval for a layer configuration. This can be described via the set of angles that the main optical axes are making with respect to the beam's direction.

We applied the above described formalism to several realistic structures made of either naturally occurring or artificially constructed layers. Due to their specific anisotropy, those structures are capable of carrying hyperbolic rather than more conventional elliptical plasmon excitations of two types: transverse electric (TE) and magnetic (TM) waves. Naturally occurring hyperbolic materials are represented by transition metal dichalcogenide monolayers [15]. The artificial metasurfaces are represented by, for instance, patterned graphene layers with negligible coupling between highly eccentric graphene nanodots. We shall also look at structures capable of extreme plasmon canalization and squeezing, such as dense graphene nanoribbons or ribbons comprised of milled out naturally anisotropic materials such as phosphorene [16] or hexagonal boron nitride [17]. The former allows for plasmon canalization along the ribbons and the latter sets plasmons propagating across the ribbon array by utilizing the capacitive coupling between the ribbons. Both structures can be fabricated using standard e-beam lithography. On one hand, their plasmonic response is narrowly selected via their topology (the pitch and size). On the other hand, it can be finely tuned via electrical doping or mechanical stresses. That is, all ribbons can be simultaneously biased with a single metallic contact, thus enabling topological transitions in their plasmonic response, on-off switching of canalization and collimation for specific frequencies, narrow-band tunable filters and sensitive detectors.

We also present results for a twisted double-layer configuration where adjacent anisotropic layers are trying to canalize plasmons in different directions. The competitive canalization directions result in a peculiar plasmon dispersion. In some cases, it fully mimics one-dimensional (1D) plasmons. Numerical simulations and discussion of the results are the subject of the last section.

13.2 Formalism of the Electron Energy Loss for Stratified Media with Hyperbolic Metasurfaces

We now consider the EELS configuration shown in Fig. 13.1. An electron beam is traveling in the plane $z = b$ above the stratified media in the region $z \leq 0$. The beam is directed along the x–direction with velocity $\mathbf{v}$. The strata are formed by arranging 2D anisotropic conducting materials (gray areas in Fig. 13.1) and stacking them on top of each other at $z = z_n$. We assume that the layers have the same composition but, in general, their primary axes run at an arbitrary x_n–direction. The region just above the n–th layer is characterized by dielectric and magnetic constants, $\epsilon_n\epsilon_0$ and $\mu_n\mu_0$ respectively, where $\epsilon_0 = 1/4\pi$ is the electric permittivity of vacuum (CGS units), and $\mu_0 = \epsilon_0^{-1}c^{-2}$ being the magnetic permeability. Here, we restrict our attention to isotropic dielectric materials separating the layers.

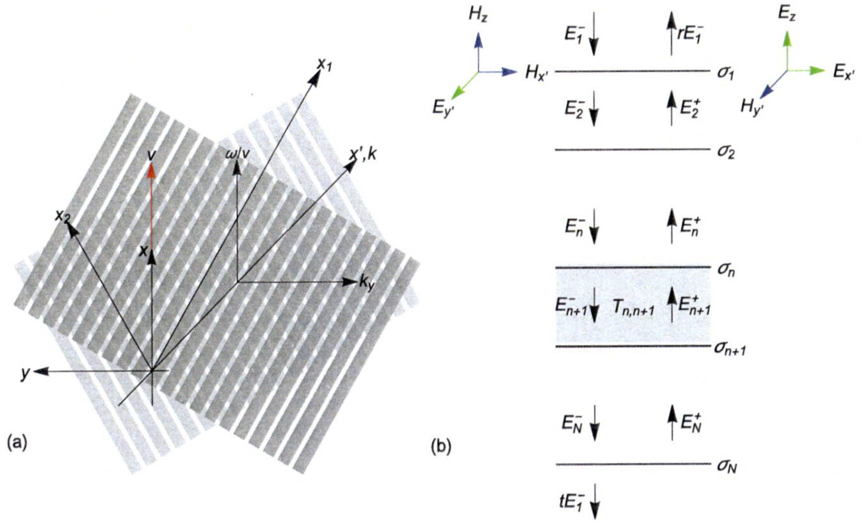

Fig. 13.1 Panel **a** schematic illustration of a beam of relativistic electrons traveling with velocity **v** ∥ **x** above an array of generic metasurfaces. The primary directions of each layer are indicated by the gray areas, we have $\sigma_{\parallel} = \sigma_{xx}$. Panel **b** a side view of the stratified structure with the layers being separated by the metasurfaces. TE/TM decomposition of the EM field is shown on the left/right of the panel. Backward/forward propagating components of the electric field $E_n^{\pm}$ for those waves are indicated by up/down arrows. The setup leads to results for the reflection r and transmission t coefficients via the transfer matrix method. The poles of the surface response functions $\sim \text{Im}\, r$ determine self-sustained plasmon-polariton oscillations propagating in the $\pm\mathbf{x}'$ direction. Each layer conductivity tensor must be written in this coordinate system. Plasmons propagating in the $\mathbf{x}' \parallel \mathbf{x}$ direction produce a self induced frictional force on the electron

The energy loss of an electron (with charge e) in the beam moving along the trajectory $\mathbf{r}_e(t) = (x = vt, 0, b)$ is given by [18]:

$$\Delta E = ev \int dt\, E_x^{ind}(\mathbf{r}_e(t), t) = \hbar \int_0^{\infty} d\omega\, \delta E(\omega), \tag{13.1}$$

where $\delta E = \Gamma_{\parallel}(\omega)\omega$ is the portion of energy given to electronic oscillations in the stratified media and the loss function $\Gamma_{\parallel}$ describes scattering probability (per trajectory segment L) off the otherwise preset linear path

$$\Gamma_{\parallel}(\omega) = \frac{ev}{\pi\hbar\omega} \int dt\, \text{Re}\left[e^{-i\omega t} E_x^{ind}(\mathbf{r}_e(t), \omega)\right] = \frac{e}{\pi\hbar\omega} \int dx\, \text{Re}\left[e^{-i\omega x/v} E_x^{ind}(\mathbf{r}, \omega)\right]. \tag{13.2}$$

Here, we have used the frequency representation of the induced field

$$
\begin{aligned}
E_x^{ind}\left(\mathbf{r}_e(t),t\right) &= \frac{1}{2\pi}\int\limits_{-\infty}^{\infty} E_x^{ind}\left(\mathbf{r}_e(t),\omega\right)e^{-i\omega t}d\omega \\
&= \frac{1}{2\pi}\int\limits_{0}^{\infty}\left[E_x^{ind}\left(\mathbf{r}_e(t),\omega\right)e^{-i\omega t} + E_x^{ind}\left(\mathbf{r}_e(t),-\omega\right)e^{i\omega t}\right]d\omega \\
&= \frac{1}{2\pi}\int\limits_{0}^{\infty}\left[E_x^{ind}\left(\mathbf{r}_e(t),\omega\right)e^{-i\omega t} + \left(E_x^{ind}\left(\mathbf{r}_e(t),\omega\right)\right)^{\star}e^{i\omega t}\right]d\omega \\
&= \frac{1}{\pi}\,\mathrm{Re}\int\limits_{0}^{\infty} E_x^{ind}\left(\mathbf{r}_e(t),\omega\right)e^{-i\omega t}d\omega .
\end{aligned}
$$

On the other hand, the induced field may be expressed via the respective component of the dyadic Green's function tensor $\bar{\bar{G}}$ (see Appendix 13.7 for details) as

$$
E_x^{ind}\left(\mathbf{r},\omega\right) = -4\pi i\omega\int d^3\mathbf{r}' G_{xx}\left(\mathbf{r},\mathbf{r}',\omega\right) j_x\left(\mathbf{r}',\omega\right), \tag{13.3}
$$

where the source current j_x in the frequency representation is given by

$$
\begin{aligned}
j_x\left(\mathbf{r}',\omega\right) &= -ev\int dt e^{i\omega t}\delta(x'-vt)\delta(y')\delta(z'-b) \\
&= -e\int dt e^{i\omega t}\delta(x'/v-t)\delta(y')\delta(z'-b) \\
&= -ee^{i\omega x'/v}\delta(y')\delta(z'-b).
\end{aligned} \tag{13.4}
$$

Substituting (13.4) into (13.3) and the resulting equation into (13.2), we obtain

$$
\Gamma_{\parallel}(\omega) = \frac{4e^2}{\hbar}\,\mathrm{Im}\int\int dxdx' e^{i(\omega/v)(x'-x)} G_{xx}\left((x,0,b),(x',0,b),\omega\right). \tag{13.5}
$$

Now, let us use the reciprocity relation for the Green's function

$$
\begin{aligned}
&\int\int dxdx' e^{i(\omega/v)(x'-x)} G_{xx}\left((x,0,b),(x',0,b),\omega\right) \\
&= \int\int dxdx' e^{i(\omega/v)(x-x')} G_{xx}\left((x',0,b),(x,0,b),\omega\right).
\end{aligned} \tag{13.6}
$$

Combination of (13.5) and (13.6) results in

$$\Gamma_{\parallel}(\omega) = \frac{4e^2}{\hbar} \int\int dx dx' \cos\left[\frac{\omega}{v}(x'-x)\right] \text{Im}\, G_{xx}\left((x,0,b),(x',0,b),\omega\right). \tag{13.7}$$

The spatial double integral in (13.7) exists only on the interval $-L/2 \leq x \leq L/2$, and assumes the form

$$\begin{aligned}
&\int\limits_{-L/2}^{L/2} dx \int\limits_{-\infty}^{\infty} dx' \cos\left[\frac{\omega}{v}(x'-x)\right] \exp\left[iq_x(x-x')\right] \\
&= \frac{1}{2}\sum_{s=\pm 1} \int\limits_{-L/2}^{L/2} dx \int\limits_{-\infty}^{\infty} dx' \exp\left[i(q_x + s\omega/v)(x-x')\right] \\
&= \frac{1}{2}\sum_{s'=\pm 1} \int\limits_{-L/2}^{L/2} dx \int\limits_{-\infty}^{\infty} dx' \exp\left[i(q_x + s'\omega/v)x'\right] \\
&= \pi L \sum_{s'=\pm 1} \delta(q_x + s'\omega/v).
\end{aligned} \tag{13.8}$$

The explicit form of Green's function in terms of Fresnel's reflection coefficients R (valid for $z > 0$), is given in [19] as

$$\begin{aligned}
G_{xx}\left((x,0,b),(x',0,b),\omega\right) &= \frac{i}{8\pi^2} \int\int \frac{dq_x dq_y}{q_{\parallel}^2} \exp\left[i(q_x(x-x') + 2q_{z1}b)\right] \\
&\times \left[\frac{R_{TE}(q_x,q_y)q_y^2}{q_{z1}c^2} - \frac{R_{TM}(q_x,q_y)q_x^2 q_{z1}}{\epsilon_1 \omega^2}\right].
\end{aligned} \tag{13.9}$$

Here, $q_{\parallel} = \sqrt{\omega^2/v^2 + q_y^2}$ is the magnitude of the in-plane momentum, and

$$q_{zn} = \sqrt{k^2\varepsilon_n - q_{\parallel}^2} = i\sqrt{q_{\parallel}^2 - \varepsilon_n \omega^2/c^2}, \tag{13.10}$$

is the out-of-plane momentum in the n-th layer. Using the identity $\int_{-\infty}^{\infty} dq_y f(q_y) = \sum_{s=\pm} \int_0^{\infty} dq_y f(sq_y)$ and inserting (13.8) and (13.9) into (13.7), we obtain

$$\Gamma_{\parallel}(\omega) = \frac{e^2 L}{2\pi\hbar v^2} \int_0^{\infty} \frac{dq_y}{q_{\parallel}^2} \text{Re}\left\{q_{z1} e^{2iq_{z1}b}\left[\left(\frac{q_y v}{q_{z1}c}\right)^2 R_{TE} - \frac{1}{\varepsilon_1} R_{TM}\right]\right\}. \tag{13.11}$$

In the non-retarded limit we can ignore polaritonic effects so the first term in the integrand, involving the term R_{TE}, can be omitted. The out of plane momentum can also be simplified by writing

$$\Gamma_{\parallel}(\omega) = -\frac{e^2 L}{2\pi\hbar v^2}\int_0^{\infty}\frac{dq_y}{q_{\parallel}^2}\,\mathrm{Re}\left\{q_{z1}e^{-2q_{\parallel}b}\frac{1}{\varepsilon_1}R_{TM}\right\}. \tag{13.12}$$

The Fresnel reflection coefficients are given in terms of the reflection matrix elements $\bar{\bar{r}}$

$$\begin{aligned} R_{TE} &= \sum_{s=\pm,\,s'=\pm} r_{TE,TE}(s\frac{\omega}{v}, s'q_y) + r_{TE,TM}(s\frac{\omega}{v}, s'q_y), \\ R_{TM} &= \sum_{s=\pm,\,s'=\pm} r_{TM,TM}(s\frac{\omega}{v}, s'q_y) + r_{TM,TE}(s\frac{\omega}{v}, s'q_y). \end{aligned} \tag{13.13}$$

Explicit forms of these matrix elements are calculated in the next section via the transfer matrix formalism.

13.3 Reflection and Transfer Matrices for Anisotropic Stratified Media

For $z < b$, we have no free charges and therefore the in-plane electromagnetic fields in the n-th strata must satisfy the following set of Maxwell's equations

$$\begin{aligned} \nabla^{\pm}\times\mathbf{E}_n &= i\omega\mu\mu_0\mathbf{H}_n, \\ \nabla^{\pm}\times\mathbf{H}_n &= -i\omega\varepsilon\epsilon_0\mathbf{E}_n, \\ \nabla^{\pm}\cdot\mathbf{E}_n &= \nabla^{\pm}\cdot\mathbf{H}_n = 0, \end{aligned}$$

with $\nabla^{\pm} = \pm iq_{zn}\mathbf{z} + iq_{x'}\mathbf{x}' + iq_{y'}\mathbf{y}'$, and the dispersion for the component $q_{zn}(\omega)$ is given by (13.10). Note that here we limit ourselves to the highly localized, non radiative surface bound modes for q_{z1} which are purely imaginary. In the TM, TE basis, the solutions of the above system of equations assume the following form

$$\begin{aligned} E_{n,TE,y'} &= E^+_{n,TE}e^{iq_{zn}z} + E^-_{n,TE}e^{-iq_{zn}z}, \\ H_{n,TE,x'} &= \frac{1}{Z_{TE,n}}\left(-E^+_{n,TE}e^{iq_{zn}z} + E^-_{n,TE}e^{-iq_{zn}z}\right), \\ E_{n,TM,x'} &= E^+_{n,TM}e^{iq_{zn}z} + E^-_{n,TM}e^{-iq_{zn}z}, \\ H_{n,TM,y'} &= \frac{1}{Z_{TM,n}}\left(E^+_{n,TM}e^{iq_{zn}z} - E^-_{n,TM}e^{-iq_{zn}z}\right), \end{aligned} \tag{13.14}$$

where implicit dependence on $\exp(i\mathbf{q}_{\parallel}\cdot\mathbf{r}')$ is assumed, and we have introduced the impedances

$$Z_{TE,n} = \frac{\omega\mu_n\mu_0}{q_{zn}} = \frac{\omega\mu_n}{cq_{zn}}Z_0, \quad Z_{TM,n} = \frac{q_{zn}}{\omega\varepsilon_n\epsilon_0} = \frac{q_{zn}c}{\omega\varepsilon_n}Z_0,$$

with $Z_0 = 4\pi/c = 1/\epsilon_0 c$ being the free space impedance. The unknown amplitudes can be presented in vector form as

$$\boldsymbol{\Psi}_n = \begin{pmatrix} E^+_{n,TE} \\ E^-_{n,TE} \\ E^+_{n,TM} \\ E^-_{n,TM} \end{pmatrix}.$$

Boundary conditions require continuity of the electric field, $\mathbf{E}^{\parallel}_n - \mathbf{E}^{\parallel}_{n+1} = 0$, and discontinuity of the magnetic field via induced currents on the interface, $\mathbf{H}^{\parallel}_n - \mathbf{H}^{\parallel}_{n+1} = \mathbf{z} \times \left(\bar{\bar{\sigma}}\mathbf{E}\right)$, with $\bar{\bar{\sigma}}$ being a 2D–conductivity tensor specific for the interface material [20]. At the $z = z_n$ interface, these boundary conditions applied to (13.14) yield the following equations (for brevity we replace q_{zn} by q_n)

$$\begin{aligned}
E^+_{n+1,TE}e^{iq_{n+1}z_n} + E^-_{n+1,TE}e^{-iq_{n+1}z_n} &= E^+_{n,TE}e^{iq_n z_n} + E^-_{n,TE}e^{-iq_n z_n}, \\
\sigma_{TE,n+1}\left(-E^+_{n+1,TE}e^{iq_{n+1}z_n} + E^-_{n+1,TE}e^{-iq_{n+1}z_n}\right) &= \sigma_{TE,n}\left(-E^+_{n,TE}e^{iq_n z_n} + E^-_{n,TE}e^{-iq_n z_n}\right) \\
&\quad + \sigma_{yx}\left(E^+_{n,TM}e^{iq_n z_n} + E^-_{n,TM}e^{-iq_n z_n}\right) \\
&\quad + \sigma_{yy}\left(E^+_{n,TE}e^{iq_n z_n} + E^-_{n,TE}e^{-iq_n z_n}\right), \\
E^+_{n+1,TM}e^{iq_{n+1}z_n} + E^-_{n+1,TM}e^{-iq_{n+1}z_n} &= E^+_{n,TE}e^{iq_n z_n} + E^-_{n,TE}e^{-iq_n z_n}, \\
\sigma_{TM,n+1}\left(E^+_{n+1,TM}e^{iq_{n+1}z_n} - E^-_{n+1,TM}e^{-iq_{n+1}z_n}\right) &= \sigma_{TM,n}\left(E^+_{n,TM}e^{iq_n z_n} - E^-_{n,TM}e^{-iq_n z_n}\right) \\
&\quad - \sigma_{xx}\left(E^+_{n,TM}e^{iq_n z_n} + E^-_{n,TM}e^{-iq_n z_n}\right) \\
&\quad - \sigma_{xy}\left(E^+_{n,TE}e^{iq_n z_n} + E^-_{n,TE}e^{-iq_n z_n}\right),
\end{aligned}$$

where the wave admittances are defined through the equations $\sigma_{TE(TM)} = 1/Z_{TE(TM)}$. The above equations can be recast in the following compact matrix form

$$\mathbf{D}_{n+1}\boldsymbol{\Phi}_{n+1}(z_n)\boldsymbol{\Psi}_{n+1} = \mathbf{D}_n\boldsymbol{\Phi}_n(z_n)\boldsymbol{\Psi}_n, \tag{13.15}$$

$$\mathbf{D}_{n+1} = \begin{pmatrix} 1 & 1 & 0 & 0 \\ -\sigma_{TE,n+1} & \sigma_{TE,n+1} & 0 & 0 \\ 0 & 0 & 1 & 1 \\ 0 & 0 & \sigma_{TM,n+1} & -\sigma_{TM,n+1} \end{pmatrix},$$

$$\mathbf{D}_n = \begin{pmatrix} 1 & 1 & 0 & 0 \\ -\sigma_{TE,n} + \sigma_{yy} & \sigma_{TE,n} + \sigma_{yy} & \sigma_{yx} & \sigma_{yx} \\ 0 & 0 & 1 & 1 \\ -\sigma_{xy} & -\sigma_{xy} & \sigma_{TM,n} - \sigma_{xx} & -\sigma_{TM,n} - \sigma_{xx} \end{pmatrix},$$

$$\boldsymbol{\Phi}_{n+1}(z_n) = \begin{pmatrix} e^{iq_{n+1}z_n} & 0 & 0 & 0 \\ 0 & e^{-iq_{n+1}z_n} & 0 & 0 \\ 0 & 0 & e^{iq_{n+1}z_n} & 0 \\ 0 & 0 & 0 & e^{-iq_{n+1}z_n} \end{pmatrix}.$$

Introducing the propagation matrix $\mathbf{P}_n = \mathbf{\Phi}_n(z_{n-1})\,\mathbf{\Phi}_n^{-1}(z_n)$ and cascading (13.15), we obtain the transfer matrix $\mathbf{T}_{1,N+1}$ via

$$\mathbf{\Phi}_1(z_1)\,\mathbf{\Psi}_1 = \mathbf{T}_{1,N}\mathbf{\Phi}_{N+1}(z_N)\,\mathbf{\Psi}_{N+1},$$

where

$$\mathbf{T}_{1,N} = \mathbf{D}_1^{-1}\left[\prod_{n=2}^{N}\mathbf{D}_n\mathbf{P}_n\mathbf{D}_n^{-1}\right]\mathbf{D}_{N+1} = \prod_{n=2}^{N}\mathbf{T}_{n-1,n}\mathbf{P}_n\mathbf{T}_{n,n+1}. \tag{13.16}$$

For a single interface we have

$$\mathbf{T}_{1,2} = \begin{pmatrix} \frac{\sigma_{TE,1}+\sigma_{TE,2}+\sigma_{yy}}{2\sigma_{TE,1}} & \frac{\sigma_{TE,1}-\sigma_{TE,2}+\sigma_{yy}}{2\sigma_{TE,1}} & \frac{\sigma_{yx}}{2\sigma_{TE,1}} & \frac{\sigma_{yx}}{2\sigma_{TE,1}} \\ \frac{\sigma_{TE,1}-\sigma_{TE,2}-\sigma_{yy}}{2\sigma_{TE,1}} & \frac{\sigma_{TE,1}+\sigma_{TE,2}-\sigma_{yy}}{2\sigma_{TE,1}} & -\frac{\sigma_{yx}}{2\sigma_{TE,1}} & -\frac{\sigma_{yx}}{2\sigma_{TE,1}} \\ \frac{\sigma_{xy}}{2\sigma_{TM,1}} & \frac{\sigma_{xy}}{2\sigma_{TM,1}} & \frac{\sigma_{TM,1}+\sigma_{TM,2}+\sigma_{xx}}{2\sigma_{TM,1}} & \frac{\sigma_{TM,1}-\sigma_{TM,2}+\sigma_{xx}}{2\sigma_{TM,1}} \\ -\frac{\sigma_{xy}}{2\sigma_{TM,1}} & -\frac{\sigma_{xy}}{2\sigma_{TM,1}} & \frac{\sigma_{TM,1}-\sigma_{TM,2}-\sigma_{xx}}{2\sigma_{TM,1}} & \frac{\sigma_{TM,1}+\sigma_{TM,2}-\sigma_{xx}}{2\sigma_{TM,1}} \end{pmatrix}.$$

Let us assume that the conducting layers do not allow electron transfer but are optically close to each other, i.e., $q(z_{n-1} - z_n) \ll 1$. Then the propagator matrix becomes identity and therefore $\mathbf{T}_{1,N+1} = \mathbf{T}_{1,2}(\sum_{n=1}^{N}\bar{\bar{\sigma}}_n)$. The reflection coefficients are given in terms of the elements of the transfer matrix $\mathbf{T}$ as follows

$$\begin{aligned} r_{TE,TE} &= \left(\frac{E_{1,TE}^{+}}{E_{1,TE}^{-}}\right)_{E_{1,TM}^{-}=0} = \frac{T_{14}T_{42}-T_{12}T_{44}}{T_{24}T_{42}-T_{22}T_{44}}, \\ r_{TE,TM} &= \left(\frac{E_{1,TM}^{+}}{E_{1,TE}^{-}}\right)_{E_{1,TM}^{-}=0} = \frac{T_{34}T_{42}-T_{32}T_{44}}{T_{24}T_{42}-T_{22}T_{44}}, \\ r_{TM,TM} &= \left(\frac{E_{1,TM}^{+}}{E_{1,TM}^{-}}\right)_{E_{1,TE}^{-}=0} = \frac{T_{24}T_{32}-T_{22}T_{34}}{T_{24}T_{42}-T_{22}T_{44}}, \\ r_{TM,TE} &= \left(\frac{E_{1,TE}^{+}}{E_{1,TM}^{-}}\right)_{E_{1,TE}^{-}=0} = \frac{T_{12}T_{24}-T_{14}T_{22}}{T_{24}T_{42}-T_{22}T_{44}}. \end{aligned} \tag{13.17}$$

13.4 EELS in Terms of Conductivity Tensor Elements

Each layer is characterized by an anisotropic conductivity tensor which in the (x_n, y_n) primary coordinates has the following generic form

$$\bar{\bar{\sigma}} = \begin{pmatrix} \sigma_{\parallel} & 0 \\ 0 & \sigma_{\perp} \end{pmatrix}. \tag{13.18}$$

We now introduce a set of angles θ_n which rotates the primary direction of each layer to the coordinate system of the electrons, $x_n \to x$, yielding a rotation matrix

$$\bar{\bar{R}}_n = \begin{pmatrix} \cos\theta_n & -\sin\theta_n \\ \sin\theta_n & \cos\theta_n \end{pmatrix} = \frac{1}{q_{\parallel}} \begin{pmatrix} q_{x_n} & -q_{y_n} \\ q_{y_n} & q_{x_n} \end{pmatrix}. \tag{13.19}$$

For a single interface, the conductivity tensor (in the frame of reference of a moving electron) is

$$\bar{\bar{\sigma}}_1 = \bar{\bar{R}}_1^T \bar{\bar{\sigma}} \bar{\bar{R}}_1 = \begin{pmatrix} \sigma_{\parallel}\cos^2\theta_1 + \sigma_{\perp}\sin^2\theta_1 & (\sigma_{\perp} - \sigma_{\parallel})\sin(2\theta_1)/2 \\ (\sigma_{\perp} - \sigma_{\parallel})\sin(2\theta_1)/2 & \sigma_{\parallel}\sin^2\theta_1 + \sigma_{\perp}\cos^2\theta_1 \end{pmatrix}. \tag{13.20}$$

The angle θ_1 denotes the angle between the major axis of the conductivity tensor (parallel) and the propagating electron velocity. For a twisted double layer we have

$$\sum \bar{\bar{\sigma}}_n = \begin{pmatrix} \sigma_{\parallel} + \sigma_{\perp} + (\sigma_{\parallel} - \sigma_{\perp})\cos(\delta\theta_{1,2})\cos(2\theta_{1,2}) & (\sigma_{\perp} - \sigma_{\parallel})\cos(\delta\theta_{1,2})\sin(2\theta_{1,2}) \\ (\sigma_{\perp} - \sigma_{\parallel})\cos(\delta\theta_{1,2})\sin(2\theta_{1,2}) & \sigma_{\parallel} + \sigma_{\perp} - (\sigma_{\parallel} - \sigma_{\perp})\cos(\delta\theta_{1,2})\cos(2\theta_{1,2}) \end{pmatrix}, \tag{13.21}$$

where $\theta_{1,2} = (\theta_1 + \theta_2)/2$ is the average twist of both layers with respect to the electron velocity, and $\delta\theta_{1,2} = \theta_1 - \theta_2$ is the relative twist angle between x_1 and x_2.

The reflection matrix elements in (13.17) can now be written as

$$\begin{aligned} r_{TE,TE} &= \frac{1}{\mathcal{D}}\left[\sigma_{xy}\sigma_{yx} + (\sigma_{TM1} + \sigma_{TM2} - \sigma_{xx})\left(\sigma_{TE1} - \sigma_{TE2} + \sigma_{yy}\right)\right], \\ r_{TE,TM} &= \frac{1}{\mathcal{D}}\left[2\sigma_{TE1}\sigma_{xy}\right], \\ r_{TM,TE} &= \frac{1}{\mathcal{D}}\left[2\sigma_{TM1}\sigma_{xy}\right], \\ r_{TM,TM} &= \frac{1}{\mathcal{D}}\left[\sigma_{xy}\sigma_{yx} + (\sigma_{TM1} - \sigma_{TM2} + \sigma_{xx})\left(\sigma_{TE1} + \sigma_{TE2} - \sigma_{yy}\right)\right], \end{aligned} \tag{13.22}$$

where $\mathcal{D}$, the denominator of the transfer matrix, is given by

$$\mathcal{D} = -\sigma_{xy}\sigma_{yx} + (\sigma_{TM1} + \sigma_{TM2} - \sigma_{xx})\left(\sigma_{TE1} + \sigma_{TE2} - \sigma_{yy}\right).$$

Substituting (13.22) into (13.13) and then making use of the resulting equation in (13.12) yields

$$\Gamma_{\parallel}(\omega) = \frac{e^2 L}{2\pi\hbar\varepsilon_1 v^2}\int_{\omega/v}^{\infty}\frac{dq_{\parallel}e^{-2q_{\parallel}b}}{\sqrt{q_{\parallel}^2 - \omega^2/v^2}}\,\mathrm{Im}\left[R_{TM}\left(q_{\parallel},\omega;\theta_1,\ldots,\theta_n\right) + R_{TM}\left(q_{\parallel},\omega;-\theta_1,\ldots,-\theta_n\right)\right], \tag{13.23}$$

where we have performed a change of variables.

The pair of symmetrized terms in (13.23) comes from the summation over the s, s' indices, as well as from a change of variables in Eq. (13.11):

$$q_{\|} = \sqrt{\omega^2/v^2 + q_y^2} \Rightarrow q_y = \sqrt{q_{\|}^2 - \omega^2/v^2} \Rightarrow dq_y = \frac{q_{\|}}{\sqrt{q_{\|}^2 - \omega^2/v^2}} dq_{\|}.$$

Noticing that $r_{TM,TE}$ is an odd function of θ_n, the final expression for the normalized energy loss can be simplified as

$$\begin{aligned}\Gamma &= \frac{\hbar\varepsilon_1 v^2}{e^2 L}\Gamma_{\|} = \frac{1}{2\pi}\int_{\omega/v}^{\infty} \frac{dq_{\|}}{\sqrt{q_{\|}^2 - \omega^2/v^2}} e^{-2q_{\|}b}\,\mathrm{Im}\left[r_{TM,TM}\right] \\ &= \sum_{i=\mathrm{TM,\,TE}} \frac{\theta\left(q_i - \omega/v\right)}{\sqrt{q_i^2 - \omega^2/v^2}} e^{-2q_i b} \\ &\times \mathrm{Re}\left[\sigma_{xy}\sigma_{yx} + \left(\sigma_{TM1} - \sigma_{TM2} + \sigma_{xx}\right)\left(\sigma_{TE1} + \sigma_{TE2} - \sigma_{yy}\right)\right]_{q_{\|}=q_i}.\end{aligned} \tag{13.24}$$

In the above expression, we have assumed a lossless conductivity tensor. The denominator of the reflection matrix has been factorized in terms of its poles as $\mathcal{D} = (q - q_{TM})(q - q_{TE})$. The two plasmonic branches are given by

$$q_{TM,TE} = \frac{i\omega\left[4 - \sigma_{xy}\sigma_{yx} + \sigma_{xx}\sigma_{yy} \pm \sqrt{(4 - \sigma_{xy}\sigma_{yx} + \sigma_{xx}\sigma_{yy})^2 - 16\sigma_{xx}\sigma_{yy}}\right]}{4\sigma_{xx}}, \tag{13.25}$$

where the TM (TE) subscripts stand for + (−) signs respectively.

It is worth presenting the transverse incidence of the electron case $\mathbf{r}_e(t) = (0, 0, z = vt)$. Using the G_{zz} component of the Green's tensor and denoting $\tan\theta = q_y/q_x$, we obtained the transverse energy loss function

$$\Gamma_{\perp} = \frac{2e^2}{\pi^2\hbar v^2}\int_0^{\infty} \frac{q_{\|}^2 dq_{\|}}{\left(q_{\|}^2 + \omega^2/v^2\right)^2}\int_0^{2\pi} d\theta\,\mathrm{Im}\left[R_{TM}(q_{\|}, \theta)\right] + \Gamma_{\mathtt{bulk}}. \tag{13.26}$$

The transverse case is not only obscured by the bulk contribution to the energy loss (part of the electron trajectory is inside the substrate) but also carries no information about a single layer orientation with respect to the electron velocity. Therefore, it makes sense to probe multi layered structures with this technique. In the next sections we shall focus our attention on (13.24) and its implications.

13.5 Energy Loss at Hyperbolic Metasurfaces

In the last decade several types of metasurfaces emerged as an alternative to conventional bulk metamaterials [21]. Several stacks of metasurfaces with weak interaction between layers also have been reported [22–24]. These devices are designed to collimate and guide plasmons in the desired directions via elliptical to hyperbolic topological transitions in the plasmon dispersion curve [25–27]. These topological properties are fully defined by the conductivity tensor (13.18). Some materials are naturally occurring, such as transition-metal dichalcogenides (TMDC). There are several techniques determining the conductivity of layers made of these materials. One is based on first-principles calculations of the bulk dielectric function $\bar{\bar{\epsilon}}\,(\omega)$ which can be done, for instance, with the Vienna *ab-initio* package. The electronic properties of exfoliated TMDC films are usually calculated with the HSE06 hybrid functional based on PAW pseudopotentials [28]. The layer conductivities become $\bar{\bar{\sigma}}\,(\omega) = Z_0^{-1}\,(-i\omega d/c)\left[\bar{\bar{\epsilon}}\,(\omega) - 1\right]$, with the effective thickness being $d_{\mathtt{WS}_2} \approx 6.15$ Å. Within the energy range $1.5\text{ eV} \leq \hbar\omega \leq 3.0\text{ eV}$ it yields Lorentzian components of the conductivity tensor (see Fig. 13.3 in [15])

$$\sigma_{\parallel,\perp} = \frac{1}{Z_0}\left(\sigma^{\infty}_{\parallel,\perp} + \frac{i\omega^0_{\parallel,\perp}\omega}{\omega^2 - \Omega^2_{\parallel,\perp} + i\omega\gamma_{\parallel,\perp}}\right). \tag{13.27}$$

Here $\sigma^{\infty} = i0.2$, $\omega^0_{\parallel,\perp} = 0.2$, $\gamma = 0.02$, $\Omega_{\parallel} = 1$, $\Omega_{\perp} = 1.2$, in units of $\Omega_{\parallel} \approx$ 2eV, for $\mathtt{WS}_2$ materials. There are several other methods, for the efficient calculation of spectral quantities of such 2D materials, based on the Chebyshev expansion of the material Green's functions known as Kernel Polynomial Methods [29].

One can construct the desired conductivity tensor of a more traditional metasurface by patterning graphene into nano-flakes of characteristic size D arranged into a lattice of area S. It was shown that the effective conductivity of such layer is proportional to the polarizability of the unit cell [30, 31]. For this case, (13.27) accepts vanishing $\sigma^{\infty}_{\parallel,\perp}$, Lorentzian frequency $\Omega^2_{\parallel,\perp} = \mathcal{L}_{\parallel,\perp}\omega_F/DZ_0$ and weight $\omega^0_{\parallel,\perp} = 4\pi D^2\mathcal{A}_{\parallel,\perp}\omega_F/S$. Here we have used the graphene isotropic conductivity $\sigma_{\parallel,\perp} = \sigma_G = i\omega_F/\omega$, with the Fermi frequency $\omega_F = 4E_F/137\hbar$. The geometry dependent unitless coefficients $\mathcal{L}$, $\mathcal{A}$ can be found using the polarizability formula for a perfectly conducting ellipsoid [32]. We have also neglected the lattice factor by assuming it to be dense in comparison to the plasmon wavelength $\sqrt{S} \ll 1/q_{\mathrm{SP}}$. The anisotropic Lorentzian conductivity in (13.27), depicted in Fig. 13.2, supports two classes of surface plasmons. Inductive plasmons are characterized by $\operatorname{Im}\sigma_{\parallel}\operatorname{Im}\sigma_{\perp} > 0$, with conventional elliptical dispersion $\omega_{\mathrm{SP}}\left(k_{x1}, k_{y1}\right) = \mathtt{constant}$. For our model this regime corresponds to $\omega_{\mathrm{SP}} < \Omega_{\parallel}$, given by the poles of the single layer transfer matrix $\mathcal{D}$. The plasmon ellipticity is broken once we enter the regime, $\Omega_{\parallel} < \omega_{\mathrm{SP}} < \Omega_{\perp}$. Here the parallel conductivity assumes capacitive character, $\operatorname{Im}\sigma_{\parallel}\operatorname{Im}\sigma_{\perp} < 0$. The plasmon dispersion, $\omega_{\mathrm{SP}}\left(k_{x1}, k_{y1}\right) = \mathtt{constant}$, splits into two complementary hyperbolas. The capacitive region does not support plasmons.

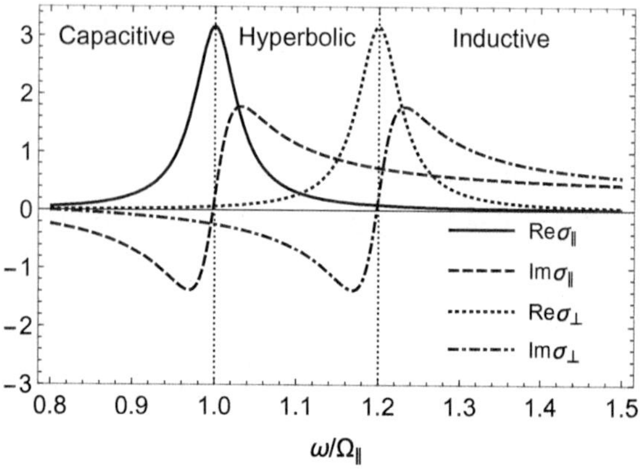

Fig. 13.2 The elements of the Lorentzian conductivity tensor in (13.27) as functions of the normalized frequency $\omega/\Omega_{\parallel}$

Isotropic surfaces, such as graphene, generate purely TM plasmons due to the inductive nature of their conductivities. For graphene the Fermi energy E_F is the only parameter. We introduce new variables - the normalized frequency $\omega \to \omega/\omega_F$, and the normalized wave number $q = cq_{\parallel}/\omega_F$. The latter is known as the squeezing factor indicating that when compared with the wavelength in free space c/ω, the polaritonic wavelength is squeezed by this factor. Note that for hyperbolic metasurfaces the canalization effect boosts the squeezing factor by two orders of magnitude. Therefore, even a low energy electron beam can couple to such a plasmon-polariton effectively, provided the right mutual orientation between the beam and the electron.

Now let us define the conductivity tensor components $\sigma_{\parallel} = \sigma_{\perp} = i\omega_F/\omega$. The plasmon dispersion given by (13.25) reduces to $q_{TM} = 2\omega^2$. There is no TE plasmon in the long wavelength approximation. Formally it can be expressed as $q_{TE} = -1$. Equation (13.24) assumes the simple form

$$\Gamma = \frac{\omega' \Theta\left(\omega' - 1\right)}{\sqrt{\omega'^2 - 1}} e^{-2q_{TM}b},$$

with $\omega' = \omega(v/c)$. For the multilayer response of graphene to EELS as well as nonlocal effects in graphene's conductivity see [33].

For metasurfaces the surface waves of pure TE or TM polarization can propagate only along the principal axes of the conductivity tensor $\theta_1 = 0,\ \pi/2$ since the off-diagonal components of the conductivity tensor disappear. According to (13.24) the TE wave does not contribute to the EELS. Corresponding poles are given by crossing of the plasmon dispersion in (13.25) with the electron line $q = \omega/v$ as shown in Fig. 13.3. At the crossing point EELS response spikes up and then recedes as $1/\sqrt{q - \omega^2/v^2}$. Withing the hyperbolic region provided $\frac{\pi}{4} \leq \theta_1 \leq \frac{3\pi}{4}$ both types of waves gets mixed and become quasi-TM, TE modes and both contribute to EELS as in Fig. 13.3b.

Although only the $r_{TM,TM}$ reflection coefficient contributes to EELS, it is worth to analyze individual contributions of the mixed $r_{TM,TE}$ reflection, as shown in Fig. 13.4.

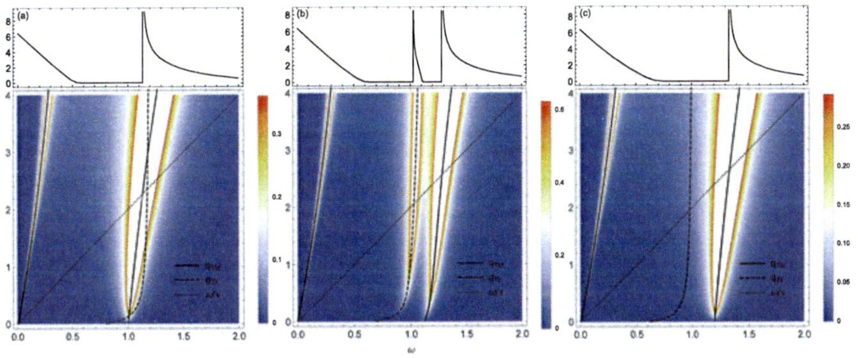

Fig. 13.3 Single layer EELS contribution from an electron running at $v/c = 0.5$ above ($b\Omega_{\parallel}/c = 0.1$) a stand-alone $\mathtt{WS_2}$ metasurface. The surface primary direction (x_1 denoted as $\parallel$ in Fig. 13.1) is at angles: **a** $\theta_1 = \pi/2$, **b** $\theta_1 = \pi/3$, **c** $\theta_1 = 0$ with respect to the propagating electron. The plasmon dispersion, as well as, the imaginary part of the $r_{TM,TM}$ (color density plot) are depicted at the bottom of each panel. The normalized frequency is given in units of $\Omega_{\parallel}$ and the propagation vector in units of $c/\Omega_{\parallel}$

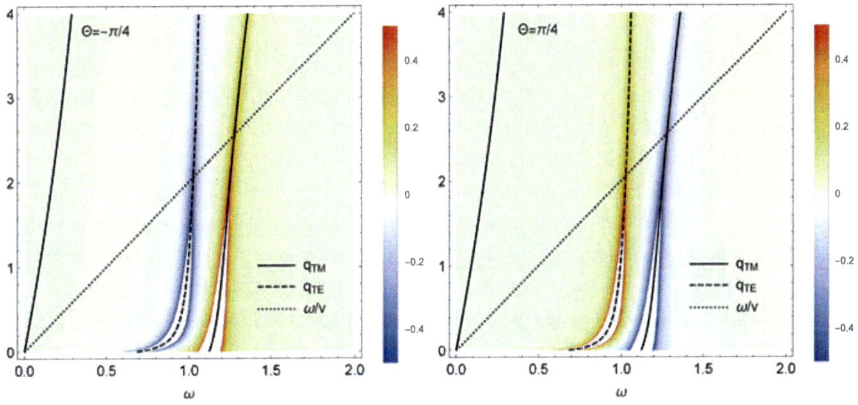

Fig. 13.4 Single layer extraordinary surface response function $\mathrm{Im}\left[r_{TM,TE}\right]$

The $\mathtt{TE}$ response to the $\mathtt{TM}$ component of the electron induced field is rather peculiar. For $\theta_1 > 0$ the $\mathtt{TE}$ wave gains its energy from the $\mathtt{TM}$ wave giving nonphysical negative contribution to EELS. However, their role changes upon $\theta_1 \to -\theta$ and the $\mathtt{TE}$ wave gives the same energy back to the $\mathtt{TM}$ mode. The overall effect on EELS vanishes.

It is informative to consider the crossing points $q_{TM,TE}(\omega) = \omega/v$ as a function of ω, θ_1 as shown in Fig. 13.5 and compare it with corresponding EELS. Low energy electrons lose their energy to plasmons in the inductive region as depicted in Fig. 13.5b. There is no considerable dependence on the metasurface orientation. The higher velocity can probe hyperbolic plasmons of quasi $\mathtt{TM}$ origin. In this region we can clearly observe plasmon canalization along $\parallel$ primary direction, thus only small

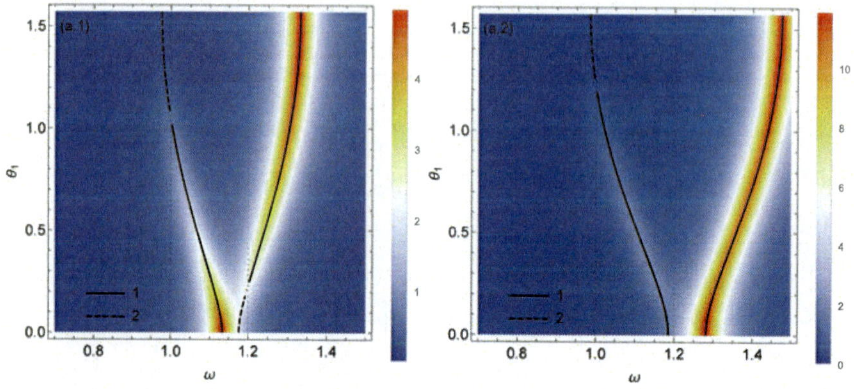

Fig. 13.5 Maximum EELS response given by (13.25) as well as $q = \omega/v$ for a single anisotropic metasurface with panels (a.1, a.2) corresponding to $\frac{v}{c} = 0.5$ and $\frac{v}{c} = 0.3$ respectively. Pole contributions from quasi TM, TE plasmons are labeled as curves 1, 2. Color theme depicts the overall ELLS for various angles (between the electron direction and the primary axis of the metasurfaces) and frequencies

θ_1 contribute. It was previously shown that the Poynting vector of the plasmonic eigenvalues is getting directed into the small angle [23]. In this regime hybridization between TE, TM plasmons asserts the group velocity and the Poynting vector running in four prescribed directions transverse to the hyperbolic asymptotic. An electron running along those asymptotes would not experience back action via plasmonic field. For individual metasurfaces it occurs at a small frequency range. But the effect can be magnified by assembling those into a twisted (by angle $\delta\theta_{12}$) double layer as highlighted in Fig. 13.1. We shall focus on the twisting angle effect and preset the double-layer into a symmetric configuration so that the electron runs in the middle of the twist angle $\theta_1 = -\theta_2$ thus making their average $\theta_{12} = 0$. The $\delta\theta_{12} = 0$ case, shown in Fig. 13.6c, just renormalizes the value $\omega^0 \rightarrow 2\omega^0$ in (13.27) thus blue-shifting the EELS maximum. For $\delta\theta_{12} = \pi/2$ we restore the isotropy of the system but the plasmonic branches are split in two. Quasi TM are the only branches contributing to the EELS. For $0 \leq \delta\theta_{12} \leq \pi/2$ plasmonic hybridization occurs.

According to our simulations at the crossing point $q_{TM} = q_{TE}$, which occurs at the frequencies given by $(4 - \sigma_{xy}\sigma_{yx} + \sigma_{xx}\sigma_{yy})^2 = 16\sigma_{xx}\sigma_{yy}$, the overall EELS contribution vanishes. In other words, it happens at the asymptotes of the hyperbolas where the plasmon momentum runs up to infinity. The corresponding twist angle can be revealed by EELS since the angle modifies hybridization in the hyperbolic region (see Fig. 13.7). For a twist above the critical (a.k.a. "magic") angle canalized plasmons can run in the direction of the propagating electron. Given their enhanced propagation length $\sim 1/\operatorname{Im} q$, the EELS acquires a significant boost in the hyperbolic regime $\Omega_{\parallel} \leq \omega \leq \Omega_{\perp}$. Now let us turn our attention to the ultimate canalization mechanism known as near zero σ.

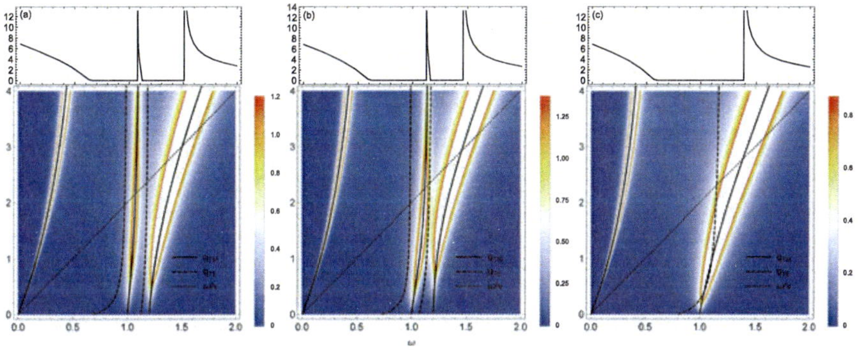

Fig. 13.6 Plasmon dispersion and EELS contribution in an anisotropic symmetric double-layer configuration $\theta_1 + \theta_2 = 0$, with the same notations as in Fig. 13.3. The three panels correspond to various twist angles: **a** $\delta\theta_{12} = \pi/2$, **b** $\delta\theta_{12} = \pi/3$, **c** $\delta\theta_{12} = 0$

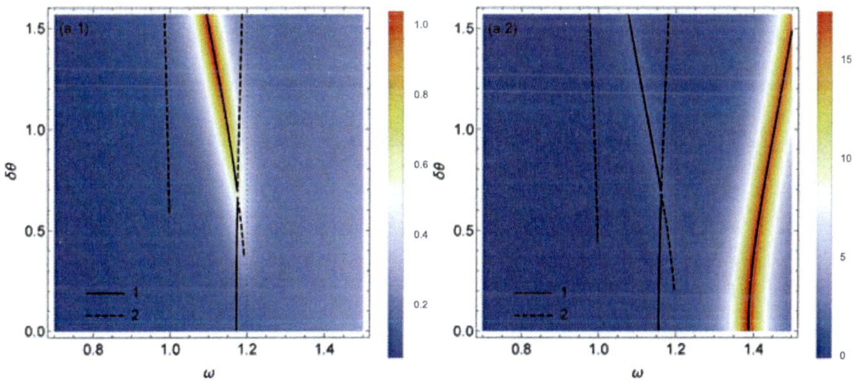

Fig. 13.7 Maximum EELS in a symmetric $\theta_{12} = 0$ double layer configuration for (a.1) $\frac{v}{c} = 0.3$, (a.2) $\frac{v}{c} = 0.5$, as a function of the twist angle. The overall notation is the same as in Fig. 13.5

Let us consider a set of graphene nanoribbons (shown as gray areas in Fig. 13.1) of width W arranged into a lattice with period P. The ribbons are cut along the armchair direction. The conductivity tensor component along that direction can be written as $\sigma_\parallel = \frac{W}{P}\sigma_{G,AC}$. The transverse component of the tensor is given by $\sigma_\perp = \frac{W\sigma_C\sigma_{G,ZZ}}{P\sigma_C + W\sigma_{G,ZZ}}$. Here $\sigma_{G,ZZ}$ is the conductivity along the zigzag direction. $\sigma_C = -iC\omega$ is associated with the near-field coupling between adjacent nanoribbons. It defines capacitive coupling across the lattice with the normalized capacitance given by $C = \frac{1}{\omega_C} = \frac{2P}{\pi c}\ln\left(\sin^{-1}\left(\frac{\pi}{2}\left(1 - \frac{W}{P}\right)\right)\right)$. The conductivity tensor can be written in Lorentzian form as in (13.27) with the following set of parameters: $\sigma_\infty = 0$, $\omega_\parallel^0 = \omega_\perp^0 = \frac{W}{P}\omega_F$, $\Omega_\perp^2 = \frac{W}{P}\omega_F\omega_C$. Along the primary direction it retains the Drude form $\Omega_\parallel = 0$. This effective-medium tensor model is valid when the pitch is much smaller than the plasmon wavelength $Pq \ll 1$. For typical graphene, with doping $E_F = 0.1$ eV, pitch

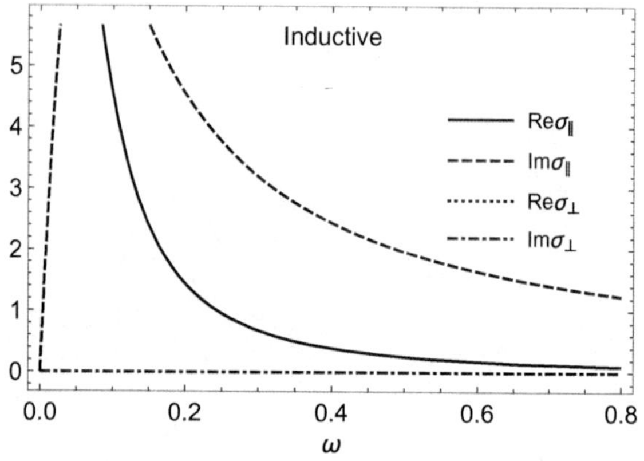

Fig. 13.8 A dense graphene nanoribbon array conductivity. The frequency is given in units of the Fermi frequency ω_F which depends on the doping level. The intrinsic losses are neglected

$P = 30$ nm and width $W = 20$ nm, the Lorentzian is centered at $\Omega_\perp \approx 15$ THz. Note that for graphene we can neglect the difference between zigzag and armchair conductivity $\sigma_{G,ZZ} = \sigma_{G,AC} = \sigma_G$. In order to achieve the "near sigma zero" regime we restrict ourselves to a dense array $\frac{W}{P} \approx 1$. Therefore, we can neglect the capacitive contribution $\sigma_C \ll \sigma_G$ which ensues vanishing $\sigma_\perp = 0$ and purely inductive $\sigma_\parallel = \sigma_G$ as shown in Fig. 13.8. In the case when Im $\sigma_\parallel \gg$ Im $\sigma_\perp$ one observes perfect plasmon canalization in the parallel direction. In other words, the plasmonic Poynting vector is fully oriented along that direction.

For a single layer of dense graphene nanoribbons the plasmon dispersion becomes directional

$$q_{TM,TE} = \frac{\omega}{\cos^2\theta_1}\left(\omega \pm \sqrt{\omega^2 + \sin^2(2\theta_1)/4}\right), \tag{13.28}$$

where we use the same conventions as for graphene. The TE−plasmonic branch disappears since $q_{TE} < 0$. The remaining branch can be probed by EELS as shown in Fig. 13.9.

The plasmon dispersion we obtained is reminiscent of the 1D-plasmon configuration. The vanishing EELS in the transverse direction (see Fig. 13.9) was confirmed experimentally for 1D systems in atomic scale metal wires arranged on Si(557) substrate [34]. Therefore, it is informative to derive the plasmonic dispersion of the graphene nanoribbon array without the effective media approximation. The 1D-electron density can be found by redefining $\omega_F = e^2 E_F Z_0 \pi \hbar^2 \frac{W}{P} = \frac{e^2 N_{2D} Z_0}{m^*}$, with $N_{2D} = N_{1D}/P = k_F/P$. Here we have used the fact that $N_{1D} = \sqrt{n_{2D}/W^2}$ is defined via the number of electrons n_{2D} within that circle. The effective mass, $m^* = \frac{\pi\hbar}{v_F W}$, depends on the width W and the 1D-Fermi energy $E_F = \hbar v_F N_{1D}\frac{W}{P}$. The procedure is equivalent to carving out ribbons from a uniform graphene sheet and scaling the conductivity by the filling factor W/P along the x_1 direction. This naive effective model has its grounds in the quantum description of the plasmons in graphene nanoribbons. Indeed, the 1D-model of a single nanoribbon supports plasmons of the form

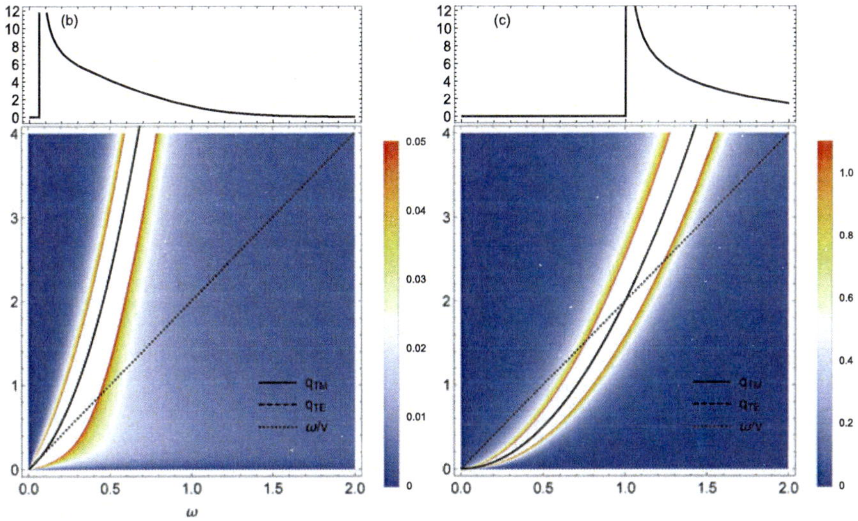

Fig. 13.9 A dense graphene nanoribbon array $\sigma_C \ll \sigma_G$, $\omega/\omega_F \to \omega$ **a** $\theta_1 = \pi/2$ no energy loss (not shown); **b** $\theta_1 = \pi/3$; **c** $\theta_1 = 0$, $v = 0.5c$. The top panels depict EELS and the bottom panels show the dispersion as well as the surface response function

$$\omega_{1D} = \sqrt{-\frac{2e^2 N_{1D}}{m^*} |q_{x1} W|^2 \ln(|q_{x1}| W)}.$$

A uniform collection of these nanoribbons supports anisotropic two dimensional plasmons [35]

$$\omega_{2D} = \sqrt{\frac{2e^2 N_{1D}}{m^*} |q_{x1} W|^2 \left[K_0(|q_{x1}| W) + 2\sum_{m=1}^{\infty} K_0(m|q_{x1}| P) \cos(m q_{y1} P) \right]}.$$

Here K_0 is the modified Bessel function of second kind. In the sparse limit $|q_{x1}| P \gg 1$ we recover $\omega_{2D} = \omega_{1D}$. In the limit of a dense nanoribbon array $q_\| P \ll 1$ we can use [36] identity

$$\begin{aligned}\sum_{m=1}^{\infty} K_0(mu)\cos(muw) &= \frac{1}{2}\left(C + \ln\frac{u}{4\pi}\right) + \frac{\pi}{2u\sqrt{1+w^2}} + \\ &+ \frac{\pi}{2}\sum_{m=1}^{\infty}\sum_{s=\pm}\left\{\frac{1}{\sqrt{u^2 + (2\pi m + swu)^2}} - \frac{1}{2\pi m}\right\} \\ &\simeq \frac{1}{2}\left(C + \ln\frac{u}{4\pi}\right) + \frac{\pi}{2u\sqrt{1+w^2}} \simeq \frac{\pi}{2u\sqrt{1+w^2}} \gg K_0(uW/d),\end{aligned}$$

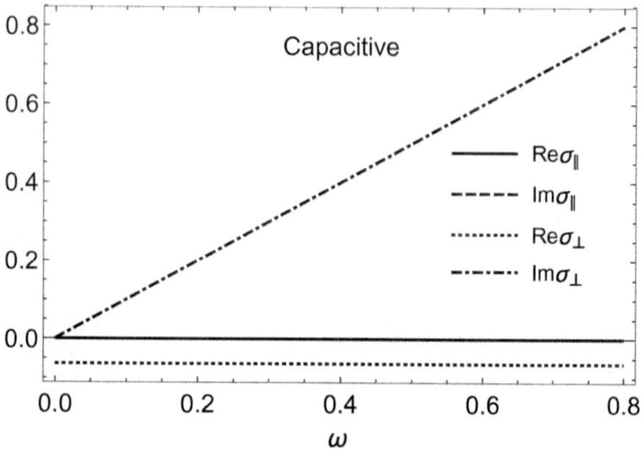

Fig. 13.10 The black phosphorous nanoribbon array conductivity. The frequency is given in units of the capacitive frequency ω_C which depends on the given ribbons topology and not sensitive to biasing effects. The intrinsic losses are neglected

where $u = |q_{x1}|\, P$ and $w = q_{y1}/\,|q_{x1}|$. Combining the above two equations we obtain

$$\omega_{2D} = \sqrt{\frac{2e^2 N_{1D}}{m^*} \left|q_{x1} W\right|^2 \frac{\pi}{u\sqrt{1+w^2}}} = \sqrt{\frac{2\pi e^2 n_{2D} W}{m^* P} \left|q_\parallel\right| \cos^2\theta_1}. \qquad (13.29)$$

For small θ_1 the above plasmon dispersion coincides with that given in (13.28). The $\sin^2(2\theta_1)$ term is a consequence of the effective media approximation. For a double layer with average $\theta_{12} = 0$ only one branch remains, $q_{TM} = \omega^2 \cos^{-2}(\delta\theta_{12}/2)$. Therefore, it will behave as an 1D-plasmon with the obvious replacement $\theta_1 \rightarrow \delta_{12}/2$.

Now let us turn our attention to combining the naturally occurring anisotropy of a material such as black phosphorous (BP) with the nanoribbon configuration into metasurface layers. The photonic properties of monolayer BP can be described by employing a simple semi-classical Drude model which comes from DFT simulations of 1 nm thin films as discussed in [37]. Proper choice of the electronic density $E_F =$ 0.4 eV biases the ribbon into $\sigma_{AC} \ll \sigma_C \ll \sigma_{ZZ}$. This, in turn, creates sigma-zero configuration $\sigma_\parallel = 0$ and $\sigma_\perp \approx \sigma_C$. The conductivity tensor is depicted in Fig. 13.10. Measuring the energy in units of ω_C and the wave number in units of c/ω_c in (13.25) yields the plasmonic branches

$$q_{TM,TE} = \frac{1}{\cos^2\theta_1}\left(1 \pm \sqrt{1 + \omega^2 \sin^2(2\theta_1)/4}\right). \qquad (13.30)$$

The TM plasmon is characterized by negative wavevector and only the TE plasmonic branch remains. For pitch $P = 40$ nm and ribbon width $W = 30$ nm, the typical resonant frequency is around 50 THz. The plasmon energy highly directional propagation occurs in the transverse direction which can be probed by EELS as shown in Fig. 13.11. Given the flat band dispersion in the transverse direction $\theta_1 = \pi/2$, we observe that the available propagation length stretches to infinity and is only limited by the intrinsic losses $\hbar\gamma$ which we have neglected in our simulations. It was shown in [16] that the Poynting vector in this direction is proportional to the

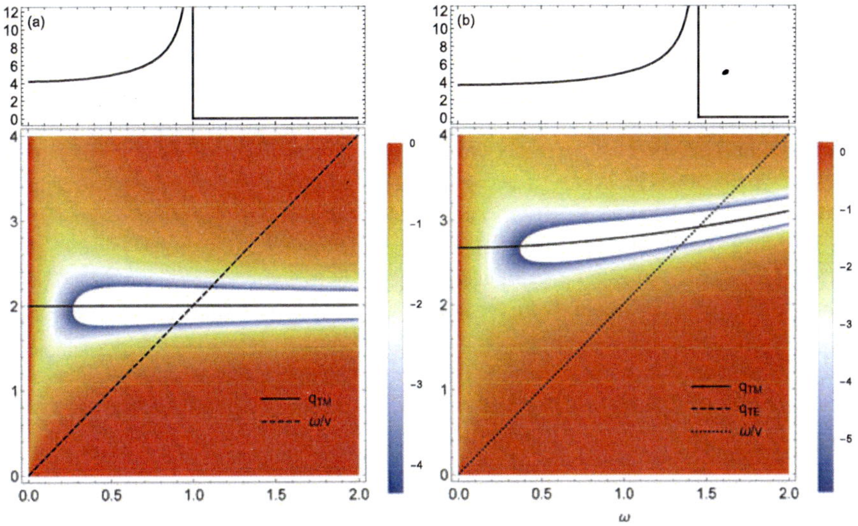

Fig. 13.11 EELS for $v = 0.5c$ electron induced by a black phosphorous nanoribbon array defined via the following set of parameters: conductivity $\sigma_{ZZ} \ll \sigma_C \ll \sigma_{AC}$, normalized frequency $\omega/\omega_c \to \omega$ and electron propagation angle **a** $\theta_1 = \pi/2$; **b** $\theta_1 = \pi/3$; **c** $\theta_1 = 0$ no energy loss (not shown). The colored density plot indicates the surface loss function given by the imaginary part of the reflection coefficient

real part of q_{TE}. For the symmetrical double layer configuration $\theta_1 = -\theta_2$ we obtain the plasmon dispersion $q_{TE} = \sin^{-1}(\delta\theta_{12})$, which has no analogy for standalone 1D plasmonics. It is also important to recognize that without the chosen limitation on the conductivity along zigzag and armchair directions, a BP metasurface can exhibit all properties of a Lorentzian metasurface: TE, TM wave hybridization and topological transitions between elliptical and hyperbolic regimes. Partial plasmon canalization along the hyperbolic branches is possible in such a system as was verified experimentally in [17].

13.6 Summary

In this chapter, we presented a theoretical framework which may be employed for probing plasmon excitations in stratified media with electron energy loss spectroscopy. The layers in the media are assumed separated with various types of metasurfaces thereby allowing for topological transitions from elliptical to hyperbolic types of plasmons. For a Lorentzian type of anisotropic conductivity, in naturally occurring TMDC or artificially designed patterned graphene metasurfaces, the role played by TE, TM hybridization in the hyperbolic region was investigated.

The main subject of this study is the interplay between the mutual orientation of the metasurface primary directions and the parallel beam propagation direction. We have indicated several conditions upon which EELS is maximized or minimized. The role of the twist angle in double layer configurations was elucidated. We have shown that in the hyperbolic region there is a critical "magic" angle below which the EELS vanishes since the Poynting vector associated with the hybridized modes runs in the direction transverse to the propagating electron above the metasurface. We have also investigated the effect of ultimate plasmon canalization in near sigma zero metasurfaces. Dense graphene and phosphorene nanoribbon arrays were chosen as model systems. Fundamental differences between such systems have their origin in the role of the capacitive coupling between the ribbons.

A symmetric double layer of graphene nanoribbons demonstrates plasmonic response identical to a set of 1D-plasmons propagating in the middle of the opening angle. EELS in the direction transverse to such plasmons disappears. The phosphorene configuration favors plasmon canalization in the direction transverse to the ribbons. The corresponding double layer offers no analogy to conventional single layer configurations.

Acknowledgements G.G. would like to acknowledge the support from the Air Force Research Laboratory (AFRL) through Grant No. FA9453-21-1-0046.

13.7 Appendix A. Green's Function Tensor for Anisotropic Stratified Media

The imaginary part of the single point reflected Green's tensor describes the enhancement in spontaneous emission rate of a stationary dipole $\boldsymbol{\mu}$ known as Purcell factor

$$1+\frac{6\pi c}{|\boldsymbol{\mu}|\,\omega}\,\boldsymbol{\mu}\cdot\mathrm{Im}\left[\bar{\bar{G}}\left(\vec{r}_0,\vec{r}_0,\omega\right)\right]\cdot\boldsymbol{\mu}, \tag{13.31}$$

$$\begin{aligned}\bar{\bar{G}}\left(\vec{r}_0,\vec{r}_0,\omega\right)=&\frac{i}{8\pi^2}\int_{-\infty}^{\infty}\int_{-\infty}^{\infty}dq_x dq_y e^{i2q_z b}\\&\times\left(r_{TE,TE}\bar{\bar{M}}_{TE,TE}+r_{TE,TM}\bar{\bar{M}}_{TE,TM}+r_{TM,TE}\bar{\bar{M}}_{TM,TE}+r_{TM,TM}\bar{\bar{M}}_{TM,TM}\right),\end{aligned} \tag{13.32}$$

$$\bar{\bar{M}}_{TE,TE}=\frac{1}{q_z^2q_\parallel^2}\begin{pmatrix}q_y^2 & -q_xq_y & 0\\ -q_xq_y & q_x^2 & 0\\ 0 & 0 & 0\end{pmatrix},\quad \bar{\bar{M}}_{TM,TM}=\frac{c^2q_z}{\omega^2q_\parallel^2}\begin{pmatrix}-q_x^2 & -q_xq_y & -q_xq_\parallel^2/q_z\\ -q_xq_y & -q_y^2 & -q_yq_\parallel^2/q_z\\ q_xq_\parallel^2/q_z & q_yq_\parallel^2/q_z & q_\parallel^4/q_z^2\end{pmatrix},$$

$$\bar{\bar{M}}_{TE,TM}=\frac{c}{\omega q_\parallel^2}\begin{pmatrix}-q_xq_y & -q_y^2 & -q_yq_\parallel^2/q_z\\ q_x^2 & q_xq_y & q_xq_\parallel^2/q_z\\ 0 & 0 & 0\end{pmatrix},\quad \bar{\bar{M}}_{TM,TE}=\frac{c}{\omega q_\parallel^2}\begin{pmatrix}q_xq_y & -q_x^2 & 0\\ q_y^2 & -q_xq_y & 0\\ -q_yq_\parallel^2/q_z & q_xq_\parallel^2/q_z & 0\end{pmatrix}.$$

References

1. N. Meinzer, W.L. Barnes, I.R. Hooper, Nat. Photonics **8**(12), 889 (2014)
2. S.M. Choudhury, D. Wang, K. Chaudhuri, C. DeVault, A.V. Kildishev, A. Boltasseva, V.M. Shalaev, Nanophotonics **7**(6), 959 (2018). https://doi.org/10.1515/nanoph-2017-0130
3. B.J. Bohn, M. Schnell, M.A. Kats, F. Aieta, R. Hillenbrand, F. Capasso, Nano Lett. **15**(6), 3851 (2015)
4. J. Babocky, A. Krizova, L. Strbkova, L. Kejik, F. Ligmajer, M. Hrtoň, P. Dvorak, M. Tyc, J. Collakova, V. Krapek et al., ACS Photonics **4**(6), 1389 (2017)
5. J. Lindhard, A. Winther, Dan. Vidensk. Selsk **4**, 34 (1964)
6. R. Ritchie, Phys. Rev. **114**(3), 644 (1959)
7. W. Brandt, M. Kitagawa, Phys. Rev. B **25**(9), 5631 (1982)
8. R. Nunez, P. Echenique, R. Ritchie, J. Phys. C Solid State Phys. **13**(22), 4229 (1980)
9. N. Horing, H. Tso, G. Gumbs, Phys. Rev. B **36**(3), 1588 (1987)
10. F.G. De Abajo, Rev. Mod. Phys. **82**(1), 209 (2010)
11. F. Flores, F. Garcia-Moliner, J. Phys. C Solid State Phys. **12**(5), 907 (1979)
12. H. Raether, *Excitation of Plasmons and Interband Transitions by Electrons*, vol. 88 (Springer, 2006)
13. W. Schülke, J. Schmitz, H. Schulte-Schrepping, A. Kaprolat, Phys. Rev. B **52**(16), 11721 (1995)
14. F. Wooten, *Optical Properties of Solids* (Academic Press, 2013)
15. Y. Li, A. Chernikov, X. Zhang, A. Rigosi, H.M. Hill, A.M. Van Der Zande, D.A. Chenet, E.M. Shih, J. Hone, T.F. Heinz, Phys. Rev. B **90**(20), 205422 (2014)
16. D. Correas-Serrano, A. Alù, J.S. Gomez-Diaz, Phys. Rev. B **96**(7), 075436 (2017)
17. P. Li, G. Hu, I. Dolado, M. Tymchenko, C.W. Qiu, F.J. Alfaro-Mozaz, F. Casanova, L.E. Hueso, S. Liu, J.H. Edgar et al., Nat. Commun. **11**(1), 1 (2020)
18. R.H. Ritchie, Phys. Rev. **106**(5), 874 (1957)
19. L. Novotny, B. Hecht, *Principles of Nano-Optics* (Cambridge University Press, 2012)
20. G. Gumbs, D. Huang, *Properties of Interacting Low-Dimensional Systems* (John Wiley & Sons, 2013)
21. S. Carr, D. Massatt, S. Fang, P. Cazeaux, M. Luskin, E. Kaxiras, Phys. Rev. B **95**(7), 075420 (2017)
22. Y. Zhao, J. Shi, L. Sun, X. Li, A. Alù, Adv. Mater. **26**(9), 1439 (2014)
23. G. Hu, A. Krasnok, Y. Mazor, C.W. Qiu, A. Alù, Nano Lett. **20**(5), 3217 (2020)
24. Y. Zhao, M. Belkin, A. Alù, Nat. Commun. **3**(1), 1 (2012)
25. H.N. Krishnamoorthy, Z. Jacob, E. Narimanov, I. Kretzschmar, V.M. Menon, Science **336**(6078), 205 (2012)
26. A.A. Sokolik, O.V. Kotov, Y.E. Lozovik (2020). arXiv:2011.13480
27. O. Yermakov, A. Ovcharenko, M. Song, A. Bogdanov, I. Iorsh, Y.S. Kivshar, Phys. Rev. B **91**(23), 235423 (2015)
28. H. Gu, B. Song, M. Fang, Y. Hong, X. Chen, H. Jiang, W. Ren, S. Liu, Nanoscale **11**(47), 22762 (2019)
29. A. Weiße, G. Wellein, A. Alvermann, H. Fehske, Rev. Mod. Phys. **78**(1), 275 (2006)
30. Y. Yermakov, D.V. Permyakov, F.V. Porubaev, P.A. Dmitriev, A.K. Samusev, I.V. Iorsh, R. Malureanu, A.V. Lavrinenko, A.A. Bogdanov, Sci. Rep. **8**(1), 1 (2018)
31. E. Mobini, A. Rahimzadegan, R. Alaee, C. Rockstuhl, Opt. Lett. **42**(6), 1039 (2017)
32. C.F. Bohren, D.R. Huffman, *Absorption and Scattering of Light by Small Particles* (John Wiley & Sons, 2008)
33. O. Roslyak, G. Gumbs, D. Huang, Physica E Low-dimensional Syst. Nanostructures **44**(9), 1874 (2012)
34. T. Nagao, S. Yaginuma, T. Inaoka, T. Sakurai, Phys. Rev. Lett. **97**(11), 116802 (2006)
35. V. Popov, T.Y. Bagaeva, T. Otsuji, V. Ryzhii, Phys. Rev. B **81**(7), 073404 (2010)
36. I.S. Gradshteyn, I.M. Ryzhik, *Table of Integrals, Series, and Products* (Academic Press, 2014)
37. Z. Liu, K. Aydin, Nano Lett. **16**(6), 3457 (2016)

Chapter 14
Landau Quantized Dynamics and Energy Spectra of Asymmetric Double-Quantum-Dot Systems: (a) Nonrelativistic Electrons; (b) Dirac T-3 "Diced" Lattice Carriers

N. J. M. Horing, J. D. Mancini, and S. L. Horton

Abstract This chapter is concerned with charge carrier propagation and the associated energy spectrum in asymmetric quantum double-dot systems embedded in a two dimensional host sheet subject to Landau quantization. The two dots are modeled by two Dirac delta function potential terms of differing depths (each of which would support just one subband state if the other were absent, if there were no magnetic field). The Dyson-like integral equation for the Green's function of such a double-dot system is solved exactly in closed form in terms of the infinite sheet Green's function for two dimensional charge carriers subject to Landau quantization in the absence of the quantum dots. The dispersion relation for the coupled double-quantum-dot subband energies is formulated and examined by analyzing the frequency poles of the Green's function, with Landau-quantization-like splintering of both of the dot levels by the magnetic field, modified by the proximity of the two dots. The effects of the asymmetry of the quantum dots in regard to their differing potential well depths are analyzed as functions of the well depth difference and dot separation. These studies are carried out here for *two* 2D systems: (A) nonrelativistic Landau quantized charge carriers, and (B) T-3 "Diced" lattice Dirac-like pseudospin 1 Landau quantized charge carriers.

N. J. M. Horing (✉) · S. L. Horton
Department of Physics, Stevens Institute of Technology, Hoboken, NJ 07030, USA
e-mail: nhoring@stevens.edu

J. D. Mancini
Department of Physical Science, Kingsborough Community College, CUNY, Brooklyn, NY 11235, USA

H. Ünlü and N. J. M. Horing (eds.), *Progress in Nanoscale and Low-Dimensional Materials and Devices*, Topics in Applied Physics 144,
https://doi.org/10.1007/978-3-030-93460-6_14

14.1 Introduction: Asymmetric Double-Quantum-Dot Green's Function

Newly conceived quantum-dot-based electronic devices for transport in nanostructured systems have often assumed that each constituent dot [1] involved in the device would support just one single energetically accessible energy subband level in the absence of a magnetic field, and that the role of the magnetic field can be represented merely in terms of a Peierls phase factor. However, this ignores another important effect of the magnetic field in that it induces a "splintering" of the single subband energy level into a proliferation of many Landau-quantized states. The latter states may be energetically accessible making the situation much more complicated, possibly interfering with the intended device function.

This chapter is concerned with charge carrier propagation and the associated energy spectrum in a two-dimensional asymmetric quantum double-dot system embedded in a two-dimensional host sheet subject to Landau quantization in a normal magnetic field. The two dots are represented by two model Dirac delta function potential terms (at spatially separated positions), each of which would support just one subband state if the other were absent, if there were no magnetic field (consonant with the assumption of just one single energetically accessible subband level per dot). The quantum dynamics of charge carriers in this double-dot system is formulated in terms of a Dyson-like integral equation for its Green's function, which is solved exactly in closed form in terms of the infinite sheet Green's function for two dimensional electrons subject to Landau quantization with no quantum dots. The dispersion relation for the asymmetric double dot subband energies is analyzed by examining the frequency poles of the Green's function, with Landau-quantization-like splintering of both of the dot levels by the magnetic field, modified by the proximity of the two quantum dots. Two 2D double quantum dot systems are examined in these studies: (A) nonrelativistic Landau quantized charge carriers, and (B) T-3 "Diced" lattice Dirac-like pseudospin 1 Landau quantized charge carriers.

We consider a planar double-quantum-dot system on a two dimensional sheet in a perpendicular magnetic field, represented by a double-well potential as ($\alpha_\pm < 0$)

$$U(\vec{r}) = \alpha_+ \delta^{(2)}(\vec{r} - \vec{r}_+) + \alpha_- \delta^{(2)}(\vec{r} - \vec{r}_-) = \sum_\pm \alpha_\pm \delta^{(2)}(\vec{r} - \vec{r}_\pm), \qquad (14.1)$$

where $\vec{r}$ = x$\hat{i}$+y$\hat{j}$ and the two potential wells defining the double dot are located at $\vec{r}_\pm = \pm\hat{i}d/2$ with strengths $\alpha_\pm = -U_{0\pm}A_\pm < 0$ representing the product of the well depths $U_{0\pm}$ at $\vec{r}_\pm$ times their common area $A_\pm \equiv a^2$. The retarded Green's function for charge carrier propagation on the sheet, $G(\vec{r}_1, \vec{r}_2, \omega)$, including the dot regions, obeys the Dyson-like integral equation (position/frequency representation; $\hbar \to 1$)

$$G(\vec{r}_1, \vec{r}_2; \omega) = G^B_{2D}(\vec{r}_1, \vec{r}_2; \omega) + \int d^2\vec{r}_3 G^B_{2D}(\vec{r}_1, \vec{r}_3; \omega)\, U(\vec{r}_3) G(\vec{r}_3, \vec{r}_2; \omega), \qquad (14.2)$$

or

$$G(\vec{r}_1, \vec{r}_2; \omega) = G_{2D}^B(\vec{r}_1, \vec{r}_2; \omega) + \sum_{\pm} \alpha_{\pm} \int d^2\vec{r}_3 G_{2D}^B(\vec{r}_1, \vec{r}_3; \omega)\delta^{(2)}(\vec{r}_3 - \vec{r}_{\pm})G(\vec{r}_3, \vec{r}_2; \omega), \quad (14.3)$$

where G_{2D}^B is the infinite sheet two dimensional Green's function for electron propagation on the sheet in a magnetic field in the absence of the quantum dot potential wells. Integration with respect to $\vec{r}_3$ yields

$$G(\vec{r}_1, \vec{r}_2; \omega) = G_{2D}^B(\vec{r}_1, \vec{r}_2; \omega) + \sum_{\pm} \alpha_{\pm} G_{2D}^B(\vec{r}_1, \vec{r}_{\pm}; \omega)G(\vec{r}_{\pm}, \vec{r}_2; \omega). \quad (14.4)$$

This algebraic equation is readily solved by setting $\vec{r}_1 = \vec{r}_{\pm'}$,

$$G(\vec{r}_{\pm'}, \vec{r}_2; \omega) = G_{2D}^B(\vec{r}_{\pm'}, \vec{r}_2; \omega) + \sum_{\pm} \alpha_{\pm} G_{2D}^B(\vec{r}_{\pm'}, \vec{r}_{\pm}; \omega)G(\vec{r}_{\pm}, \vec{r}_2; \omega), \quad (14.5)$$

whence

$$G_{2D}^B(\vec{r}_{\pm'}, \vec{r}_2; \omega) = \sum_{\pm} \left[\delta_{\pm',\pm} - \alpha_{\pm} G_{2D}^B(\vec{r}_{\pm'}, \vec{r}_{\pm}; \omega)\right] G(\vec{r}_{\pm}, \vec{r}_2; \omega). \quad (14.6)$$

The inversion of this 2×2 position space matrix equation yields (assuming that all elements of the 2×2 matrix $G_{2D}^B(\vec{r}_{\pm'}, \vec{r}_{\pm}; \omega)$ commute, which is assured if we are dealing with spinless, nonrelativistic "$p^2/2m$" electrons, but is *not* assured if those elements of the 2×2 matrix $G_{2D}^B(\vec{r}_{\pm'}, \vec{r}_{\pm}; \omega)$ are themselves 3×3 pseudospin 1 matrices of the "Diced" lattice carriers to be treated later in this chapter)

$$G\left(\vec{r}_{\pm}, \vec{r}_2; \omega\right) = \frac{D_{\mp,\mp} G_{2D}^B(\vec{r}_{\pm}, \vec{r}_2; \omega) + D_{\mp,\pm} G_{2D}^B(\vec{r}_{\mp}, \vec{r}_2; \omega)}{D_{\mp,\mp} D_{\pm,\pm} - D_{\pm,\mp} D_{\mp,\pm}}, \quad (14.7)$$

and $D_{\pm,\pm}$ is defined as

$$D_{\pm,\pm} = \delta_{\pm,\pm} - \alpha_{\pm} G_{2D}^B(\vec{r}_{\pm}, \vec{r}_{\pm}; \omega) \quad (14.8)$$

and also

$$D_{\pm,\mp} = \alpha_{\pm} G_{2D}^B(\vec{r}_{\mp}, \vec{r}_{\pm}; \omega). \quad (14.9)$$

The final Green's function is given by (14.4) and (14.6) as

$$G(\vec{r}_1, \vec{r}_2; \omega) = G^B_{2D}(\vec{r}_1, \vec{r}_2; \omega) + \sum_{\pm} \alpha_{\pm} G^B_{2D}(\vec{r}_1, \vec{r}_{\pm}; \omega) \\ \times \frac{D_{\mp,\mp} G^B_{2D}(\vec{r}_{\pm}, \vec{r}_2; \omega) + D_{\mp,\pm} G^B_{2D}(\vec{r}_{\mp}, \vec{r}_2; \omega)}{D_{\mp,\mp} D_{\pm,\pm} - D_{\pm,\mp} D_{\mp,\pm}}. \tag{14.10}$$

The coupled subband energy eigenstates associated with the double quantum dot arise from its frequency poles, as given by the dispersion relation:

$$D_{\mp,\mp} D_{\pm,\pm} - D_{\pm,\mp} D_{\mp,\pm} = 0. \tag{14.11}$$

14.2 Infinite Sheet Nonrelativistic 2D "$p^2/2m$" Magnetic Field Green's Function G^B_{2D}

The retarded nonrelativistic "$p^2/2m$" infinite sheet Green's function for a spinless 2D charge carrier (absent dots) subject to Landau quantization is well known as [2] (the magnetic field $\vec{B}$ is taken in the z-direction; $\hbar \to 1$; $c \to 1$)

$$G^B_{2D}(\vec{r}_1, \vec{r}_2; T) = -\eta_+(T) C(\vec{r}_1, \vec{r}_2) \frac{m\omega_c}{4\pi} \frac{\exp\left[\frac{im\omega_c(X^2+Y^2)}{4\tan(\omega_c T/2)}\right]}{\sin(\omega_c T/2)}, \tag{14.12}$$

in position-time representation. Alternatively, expanding the exponential on the right as a generator of Laguerre polynomials [3], we have ($\omega \to \omega + i0^+$)

$$G^B_{2D}(\vec{r}_1, \vec{r}_2; \omega) = C(\vec{r}_1, \vec{r}_2) \frac{m\omega_c}{2\pi} e^{-m\omega_c R^2/4} \sum_{n=0}^{\infty} L_n\left(\frac{m\omega_c R^2}{2}\right) \frac{1}{\omega - (n + \frac{1}{2})\omega_c}, \tag{14.13}$$

in position-frequency representation (above, $T = t_1 - t_2$, $X = x_1 - x_2$, $Y = y_1 - y_2$, $R^2 = X^2 + Y^2$, ω_c is the cyclotron frequency and $\eta_+(T)$ is the Heaviside unit step function). Furthermore, $C(\vec{r}_1, \vec{r}_2)$, which is in the nature of a Peierls phase factor, is given by ($\vec{A}(\vec{x})$ is the vector potential for the constant, uniform magnetic field; $\hbar = c = 1$)

$$C(\vec{r}_1, \vec{r}_2) = \exp\left[\frac{ie}{\hbar c} \int_{\vec{r}_2}^{\vec{r}_1} d\vec{x} \cdot \vec{A}(x)\right] = \exp\left[i(e/2)\vec{r}_1 \cdot \vec{B} \times \vec{r}_2\right], \tag{14.14}$$

and the $\vec{x}$-integral depends on the path in accordance with the Aharonov-Bohm effect (but it is taken on the right-hand-side of (14.14) as a straight line). Considering (14.12) in frequency representation, and setting $\vec{r}_1, \vec{r}_2 = \vec{r}_{\pm}$ as needed in (14.7), ($X = 0$, $Y = 0$) with $C(\vec{r}_1, \vec{r}_2) \to C(\vec{r}_{\pm}, \vec{r}_{\pm}) = 1$, we have

$$G^B_{2D}(0,0;\omega) = -\frac{m\omega_c}{4\pi}\int_0^\infty \mathrm{d}T\,\frac{e^{i\omega T}}{\sin(\omega_c T/2)}. \tag{14.15}$$

Expanding the integrand,

$$\frac{1}{\sin(\omega_c T/2)} = 2i\sum_{n=0}^{\infty} e^{-i\omega_c(n+1/2)T}, \tag{14.16}$$

the T-integral may be evaluated as

$$G^B_{2D}(0,0;\omega) = \frac{m\omega_c}{2\pi}\sum_{n=0}^{\infty}\frac{1}{\omega-(n+1/2)\omega_c}. \tag{14.17}$$

This series diverges as $n\to\infty$, as does the T-integral of (14.15) for $X = Y = 0$. The divergence is an artifact of assuming a $\delta^{(2)}(\vec{r})$ potential, which confines the dot to a single point. It may be removed by noting that the original integral equation involves integration over a small, but finite radius "a", so we should install a "smear" radius "a" putting $X^2 + Y^2 \to a^2$ (instead of "0") in (14.11), (14.12), leading to

$$G^B_{2D}(0,0;\omega) \Rightarrow G^B_{2D}(a;\omega) = -\frac{m\omega_c}{4\pi}\int_0^\infty \frac{\mathrm{d}T\,e^{i\omega T}}{\sin(\omega_c T/2)}\exp\left[\frac{im\omega_c a^2}{4\tan(\omega_c T/2)}\right], \tag{14.18}$$

and, installing the "smear" radius, we have,

$$G^B_{2D}(0,0;\omega) \Rightarrow G^B_{2D}(a;\omega) \Rightarrow \frac{m\omega_c}{2\pi i}\exp\left[-m\omega_c a^2/4\right]\sum_{n=0}^{\infty} L_n\left(\frac{m\omega_c a^2}{2}\right) \int_0^\infty \mathrm{d}T\exp[i([\omega-(n+1/2)\omega_c])T], \tag{14.19}$$

where we have again expanded the last exponential on the right hand side of the T-integrand of (14.18) in Laguerre polynomials [3], $L_n(x)$, that embody the Landau eigenstate matrix elements on the plane in closed form. In this context the T-integral of (14.19) can be carried out with the result

$$G^B_{2D}(0,0;\omega) \Rightarrow G^B_{2D}(a;\omega) = \frac{m\omega_c}{2\pi}e^{-m\omega_c a^2/4}\sum_{n=0}^{\infty} L_n\left(\frac{m\omega_c a^2}{2}\right)\frac{1}{\omega-[n+\left(\frac{1}{2}\right)]\omega_c}. \tag{14.20}$$

The n-series of (14.20) is slowly convergent [3]. As $n \to \infty$, $L_n(x) \sim n^{-1/4}$, so the summand goes like $1/n^{1.25}$, which ensures convergence [4]. Notwithstanding this convergence, there is reason to consider truncating the n-series of (14.20) in materials with band structure that bends away from the "$p^2/2m$" form assumed here at an energy

$$E_{max} \cong (n_{max} + 1/2)\omega_c, \tag{14.21}$$

which identifies the maximum Landau eigen-energy index n_{max} acceptable for inclusion in the series. Considered jointly with dot size "a" understood to be the smallest length parameter, so that $m\omega_c a^2/2 \ll 1$ and $L_n(m\omega_c a^2/2) \to 1$, we recover (14.17) cut off at n_{max}.

It should be noted that the dispersion relation of (14.6), (14.11) also involves Peierls phase factors $C(\vec{r}_\pm, \vec{r}_\mp)$ as $D_{\pm,\mp}D_{\mp,\pm} \sim C(\vec{r}_\pm \vec{r}_{\mp,})C(\vec{r}_{\mp,}\vec{r}_\pm) \sim \exp[\frac{ie}{2}(\vec{r}_\pm \cdot \vec{B} \times \vec{r}_\mp + \vec{r}_\mp \cdot \vec{B} \times \vec{r}_\pm] = 1$ so the Peierls phase factors cancel in the dispersion relation, but not in $G(\vec{r}_1, \vec{r}_2; \omega)$.

14.3 Dispersion Relation for a Nonrelativistic "$p^2/2m$" Asymmetric Double-Quantum-Dot in a Normal Magnetic Field

Finally, the results above yield the dispersion relation as (recall that $|\vec{r}_+ - \vec{r}_-| = d$)

$$1 = (\alpha_- + \alpha_+) \sum_{n=0}^{\infty} \Lambda_n \frac{1}{\omega - [n + (\frac{1}{2})]\omega_c} - \alpha_-\alpha_+ \left(\sum_{n=0}^{\infty} \Lambda_n \frac{1}{\omega - [n + (\frac{1}{2})]\omega_c} \right)^2$$
$$+ \alpha_-\alpha_+ \left(\sum_{n=0}^{\infty} \Gamma_n \frac{1}{\omega - [n + (\frac{1}{2})]\omega_c} \right)^2, \tag{14.22}$$

where (restore $\hbar$)

$$\Gamma_n = \frac{m\omega_c}{2\pi\hbar^2} \exp\left(\frac{-m\omega_c d^2}{4\hbar} \right) L_n \left(\frac{m\omega_c d^2}{2\hbar} \right)$$

and

$$\Lambda_n = \frac{m\omega_c}{2\pi\hbar^2} \exp\left(\frac{-m\omega_c a^2}{4\hbar} \right) L_n \left(\frac{m\omega_c a^2}{2\hbar} \right),$$

for an asymmetric quantum-double-dot in a nonrelativistic "$p^2/2m$" semiconductor (not a Dirac material) in a perpendicular magnetic field. One may expect that there are infinitely many roots of (14.22) corresponding to all the integer values of n, but since the equation involves second order (quadratic) terms in $(\omega - [n + 1/2]\omega_c)^{-1}$,

Table 14.1 First 5 calculated "upper" energies for differing well depths

n	$U_{0-}, U_{0+} =$ (25, 75) meV	$U_{0-}, U_{0+} =$ (100, 200) meV	$U_{0-}, U_{0+} =$ (100, 300) meV	$U_{0-}, U_{0+} =$ (200, 300) meV	$U_{0-}, U_{0+} =$ (200, 400) meV
n = 0	0.51230	0.53456	0.54515	0.55478	0.56435
n = 1	1.51054	1.52870	1.53999	1.54504	1.55517
n = 2	2.50977	2.52544	2.53724	2.53776	2.54893
n = 3	3.50970	3.52548	3.51230	3.53798	3.54880
n = 4	4.50984	4.52662	4.53790	4.54096	4.55119

there are actually two such coupled subband energy eigenvalues for each value of n, an "upper" (larger) value and a "lower" (smaller) value, corresponding to the Landau quantization of each of the two distinct quantum dot spectra, with modification due to the proximity of the two dots.

From this dispersion relation one can see that as d, the distance separating the two quantum dots, goes to infinity, $\Lambda_n \to 0$, we are left with the two *decoupled* quantum dots contributing separately to the dispersion relation as:

$$\left(1-\alpha_-\frac{m\omega_c}{2\pi\hbar^2}\exp\left(\frac{-m\omega_c a^2}{4\hbar}\right)\sum_{n=0}^{\infty}\frac{1}{\omega-[n+(\frac{1}{2})]\omega_c}L_n\left(\frac{m\omega_c a^2}{2\hbar}\right)\right)$$
$$\times\left(1-\alpha_+\frac{m\omega_c}{2\pi\hbar^2}\exp\left(\frac{-m\omega_c a^2}{4\hbar}\right)\sum_{n=0}^{\infty}\frac{1}{\omega-[n+(\frac{1}{2})]\omega_c}L_n\left(\frac{m\omega_c a^2}{2\hbar}\right)\right)=0. \tag{14.23}$$

On the other hand, when the dots approach each other, $d \to a$, the first and third terms on the right side of (14.22) cancel, yielding

$$1=(\alpha_-+\alpha_+)\sum_{n=0}^{\infty}\left[\frac{m\omega_c}{2\pi\hbar^2}e^{\left(\frac{-m\omega_c a^2}{4\hbar}\right)}L_n\left(\frac{m\omega_c a^2}{2\hbar}\right)\frac{1}{\omega-[n+(\frac{1}{2})]\omega_c}\right], \tag{14.24}$$

in which we have the dispersion relation of a single dot having the combined strength of the two coupled dots, $(\alpha_- + \alpha_+)$, as one should expect.

Our numerical analysis of the asymmetric double quantum dot spectrum based on (14.22) for $a = 5$ nm, $d = 5a = 25$ nm, $\hbar\omega_c = 2$ meV, $m = 0.067\, m_e$, carrying 5000 terms, yields the first 5 calculated "upper" and "lower" energies for each different U_- and U_+ pair as exhibited below in Tables 14.1 and 14.2, respectively.

We also consider an alternative analytic scheme for solving the dispersion relation of (14.22) when $\alpha\Lambda \ll 1$ and $\alpha\Gamma \ll 1$. This occurs for nanoscale parameters such as we have employed in the text above. In this situation the magnitudes of the prefactors $(\alpha_\pm\Lambda_n, \alpha_\pm\Gamma_n)$ on the right of (14.22) are very small due to the smallness of the dot area, and to achieve equality with the unit "1" on the left of (14.22), the frequency

Table 14.2 First 5 calculated "lower" energies for differing well depths

n	$U_{0-}, U_{0+} =$ (25, 75) meV	$U_{0-}, U_{0+} =$ (100, 200) meV	$U_{0-}, U_{0+} =$ (100, 300) meV	$U_{0-}, U_{0+} =$ (200, 300) meV	$U_{0-}, U_{0+} =$ (200, 400) meV
n = 0	0.48695	0.45966	0.44273	0.43053	0.39846
n = 1	1.48909	1.46897	1.45398	1.45048	1.43134
n = 2	2.48992	2.47241	2.45774	2.45772	2.44226
n = 3	3.49001	3.47139	3.45834	3.45582	3.43795
n = 4	4.48992	4.47139	4.45837	4.45367	4.43479

denominators must be close to zero, $\omega_n - [n + \frac{1}{2}]\omega_c \cong |\alpha_\pm \Lambda_n|$ or $|\alpha_\pm \Gamma_n|$. This is to say that the frequency roots are essentially determined by the nearest pole alone. In such a "single pole" approximation (14.22) becomes ($\hbar \to 1$)

$$1 = (\alpha_+ + \alpha_-) \frac{\Lambda_n}{\omega_n - [n + 1/2]\,\omega_c} + \alpha_+\alpha_- \frac{\Gamma_n^2 - \Lambda_n^2}{(\omega_n - [n + 1/2]\,\omega_c)^2}, \tag{14.25}$$

where we have neglected terms from the squares of series $\left(\sum_n \ldots\right)^2 = \sum_n \sum_m \ldots$ involving $n \neq m$: Denoting the small parameters collectively by $\lambda \equiv |\alpha_\pm \Lambda_n|$ or $|\alpha_\pm \Gamma_n|$, such terms are of the form ($n \neq m$) (assume $\omega - [n + \frac{1}{2}]\omega_c \cong \lambda$)

$$\begin{aligned} \frac{\lambda^2}{(\omega - [n + \frac{1}{2}]\omega_c)(\omega - [m + \frac{1}{2}]\omega_c)} &= \frac{\lambda^2}{\lambda(\omega - [m + \frac{1}{2}]\omega_c)} \cong \frac{\lambda^2}{\lambda[n - m]\omega_c} \\ &\cong \frac{\lambda}{[n - m]\omega_c} \ll 1, \end{aligned} \tag{14.26}$$

and as they are of order $O(\lambda)$, they are negligible compared to the second order pole terms having $n = m$ that are order $O\left(\frac{\lambda^2}{\lambda^2}\right) = O(1)$. Solving (14.25), we obtain

$$\omega_n - (n + 1/2)\,\omega_c = \left(\frac{(\alpha_- + \alpha_+)\,\Lambda_n \pm \sqrt{(|\alpha_+ - \alpha_-|)^2\,\Lambda_n^2 + 4\alpha_-\alpha_+\Gamma_n^2}}{2} \right). \tag{14.27}$$

Again, if $d \to \infty$, $\Gamma_n \to 0$ and two decoupled dot modes emerge as

$$\omega_n - (n + 1/2)\,\omega_c \to \left\{ \begin{matrix} \alpha_+\Lambda_n \\ \text{and} \\ \alpha_-\Lambda_n \end{matrix} \right\}, \tag{14.28}$$

while if $d \to a$, $\Gamma_n \to \Lambda_n$ and a single mode emerges having the combined strength of two dots as

$$\omega_n - (n + 1/2)\,\omega_c \to (\alpha_+ + \alpha_-)\,\Lambda_n. \tag{14.29}$$

In the Appendix, Tables 14.3 and 14.4 show the results for the first five "upper" and "lower" energies respectively, using the "single pole" evaluation approximation with the same parameters employed in Tables 14.1 and 14.2.

14.4 Infinite Sheet Landau Quantized Greens Function for the T-3 "Diced" Lattice

The Green's function $G(\vec{r}_1, \vec{r}_2; \omega)$ for the T-3 "Diced" lattice in a magnetic field is a 3×3 pseudospin 1 matrix with elements $(i,\ j = 1, 2, 3)$; again, we write $G^B_{2D} \rightarrow G$, dropping subscript "$2D$" and superscript "B" to make room for matrix element subscripts

$$G_{ij}\left(\vec{r}_1, \vec{r}_2; \omega\right) = C\left(\vec{r}_1, \vec{r}_2\right) G'_{ij}\left(\vec{R}; \omega\right), \tag{14.30}$$

in position/frequency representation. $\left(\text{As above, } C(\vec{r}_1, \vec{r}_2) = \exp\left(\frac{ie}{2}\vec{r}_1 \cdot \vec{B} \times \vec{r}_2\right).\right)$ Since G and G' (and $C(\vec{r}_1, \vec{r}_2)$) are hermitian for real ω, $G'_{ij} = G'^{*}_{ji}$, and the 9 elements have been determined as follows [5]:

$$G'_{11}\left(\vec{R}; \omega\right) = G'_{33}\left(\vec{R}; \omega\right) = \tfrac{1}{2} G'_{22}\left(\vec{R}; \omega\right) + \left(\tfrac{1}{2\omega}\right)\delta^2\left(\vec{R}\right), \tag{14.31}$$

and $(\hbar = 1)$

$$G'_{22}\left(\vec{R}; \omega\right) = \frac{eB}{2\pi}\omega e^{-eBR^2/4} \sum_{n=0}^{\infty} L_n\left(\frac{eBR^2}{2}\right) \frac{1}{\omega^2 - 2(2n+1)\alpha^2 eB}. \tag{14.32}$$

The frequency poles identify the Landau energy eigenvalues ϵ_n of the T-3 "Diced" lattice carriers as ($\hbar \rightarrow 1$; $\alpha = v/\sqrt{2}$ is a characteristic speed parameter of the T-3 "Diced" lattice)

$$\epsilon_n \equiv \omega_n = \pm\sqrt{2(2n+1)\alpha^2 eB} \qquad (n = 0 \ldots \infty). \tag{14.33}$$

Furthermore, the off-diagonal element $G'_{23}\left(\vec{R}; \omega\right)$ is given by

$$G'_{23}\left(\vec{R}; \omega\right) = \frac{\alpha (eB)^2}{2\pi} e^{-eB\frac{R^2}{4}} \left[iX + Y\right] \sum_{n=1}^{\infty} \frac{L^1_{n-1}\left(eB\frac{R^2}{2}\right)}{\omega^2 - 2(2n+1)\alpha^2 eB}. \tag{14.34}$$

Moreover, $G'_{31}\left(\vec{R};\omega\right)$ is given by

$$G'_{31}\left(\vec{R},\omega\right) = \frac{\alpha^2(eB)^3}{2\pi\omega} e^{-eB\frac{R^2}{4}}\left[X+iY\right]^2 \sum_{n=2}^{\infty} \frac{L^2_{n-2}\left(eB\frac{R^2}{2}\right)}{\omega^2 - 2(2n+1)\alpha^2 eB}. \tag{14.35}$$

These results for G'_{22}, G'_{23}, G'_{31} suffice to determine all 9 elements of G' since we also have the relation $G'_{11} = G'_{33} = (\frac{1}{2})G'_{22} + (1/2\omega)\delta^2\left(\vec{R}\right)$; and also $G'_{21} = G'^{*}_{23}$; as well as $G'_{32} = G'^{*}_{23}$, $G'_{13} = G'^{*}_{31}$, $G'_{23} = G'^{*}_{32}$, $G'_{12} = G'^{*}_{21}$ so $G'_{12} = G'_{32}$. In connection with these relations it should be noted that the diagonal elements G'_{22}, G'_{11}, and G'_{33} are real for real ω, while the off-diagonals are complex conjugates across the diagonal.

The solution of (14.6) for $G(\vec{r}_\pm, \vec{r}_2;\omega)$ (with $|r_{\pm'} - r_\pm| \to a$ as in (14.18); $|r_{\pm'} - r_\mp| \to d$), which is essential for solution of $G(\vec{r}_1, \vec{r}_2)$, involves the inversion of the matrix M defined as (in the interest of clarity, we briefly restore superscript "B" and subscript "$2D$" on G^B_{2D} to distinguish it from the double dot Green's function G),

$$M = \begin{pmatrix} 1-\alpha_+ G^B_{2D}(a;\omega) & -\alpha_- G^B_{2D}(d;\omega) \\ -\alpha_+ G^B_{2D}(d;\omega) & 1-\alpha_- G^B_{2D}(a;\omega) \end{pmatrix}, \tag{14.36}$$

in the context of $G^B_{2D}(\vec{r}_{\pm'}, \vec{r}_2;\omega) = MG(\vec{r}_\pm, \vec{r}_2;\omega)$ as the matrix product of (14.6). Each of the elements of M exhibited here is a 3×3 matrix in pseudospin 1 space. This inversion is complicated by the possibility that the 3D pseudospin matrices $G^B_{2D}(a;\omega)$ and $G^B_{2D}(d;\omega)$ do not generally commute, and we will not address this issue here. Nevertheless, considerable progress can be made with the determinant Det$M$ to deal with the determination of the eigenvalue spectrum of the T-3 "Diced" lattice asymmetric double quantum dot system in a magnetic field. (Henceforth, we again delete the superscript "B" and subscript "$2D$" on $G^B_{2D} \to G$ to make room for matrix element subscripts.) As noted below (14.21), the Peierls phase factor $C(\vec{r}_1, \vec{r}_2)$ cancels in the determinental dispersion relation for the eigenvalue spectrum, which therefore involves G' in place of G.

14.5 Eigenenergy Dispersion Relation for a T-3 "Diced" Lattice Asymmetric Double-Quantum Dot in a Normal Magnetic Field

The determinental energy eigenvalue dispersion relation for a Landau quantized asymmetric double-quantum-dot on a T-3 "Diced" lattice may be identified from (14.6) by noting that a frequency pole of its Green's function G requires the vanishing of Det$M$, so that

$$\mathrm{Det}M = \mathrm{Det}\left\{\left(I - \alpha_+ G'(a;\omega)\right)\left(I - \alpha_- G'(a;\omega)\right) - \alpha_+\alpha_- G'^2(d;\omega)\right\} = 0, \tag{14.37}$$

or, alternatively,

$$\mathrm{Det}\left\{I - [\alpha_+ + \alpha_-]G'(a;\omega) + \alpha_+\alpha_-[G'^2(a;\omega) - G'^2(d;\omega)]\right\} = 0. \tag{14.38}$$

Here, I represents the 3×3 pseudospin 1 unit matrix, just as G' is a 3×3 pseudospin 1 matrix, and Det is the 3×3 determinant. Again, there are two relatively simple limiting cases:

(a) if the dot-dot separation is very large, $G'(d;\omega) \to 0$ and (14.37) yields two separated, distinct single dot dispersion relations,

$$\mathrm{Det}\left\{I - \alpha_+ G'(a;\omega)\right\} = 0, \tag{14.39}$$

$$\mathrm{Det}\left\{I - \alpha_- G'(a;\omega)\right\} = 0; \tag{14.40}$$

(b) if the dot-dot separation is very small, $d \to a$, (14.38) takes the form

$$\mathrm{Det}\left\{1 - [\alpha_+ + \alpha_-]G'(a;\omega)\right\} = 0, \tag{14.41}$$

which is appropriate for a single dot having the combined strength $[\alpha_+ + \alpha_-]$. These 3 single dot limiting-case dispersion relations can be solved together as a generic single dot relation,

$$\mathrm{Det}\left\{1 - \lambda G'(a;\omega)\right\} = 0, \tag{14.42}$$

with $\lambda = \alpha_+, \alpha_-, \alpha_+ + \alpha_-$.

Considering $\lambda G'(a;\omega)$ of the generic single dot dispersion relation, (14.42), it is clear that the smallness of dot size a^2 leads to a relatively small value of λ (we take $U_0 \cong 100\,\mathrm{meV} = 1.6 \times 10^{-20}\mathrm{J}$),

$$\lambda = U_0 a^2 \cong 1.6 \times 10^{-38}\ \mathrm{Jm}^2, \tag{14.43}$$

where we have taken $a = 1\,\mathrm{nm}$. Estimating with parameter values $B = 1\mathrm{Tesla}$, $\alpha = 10^6\,\mathrm{m\,s}^{-1}$ and $\omega \cong \alpha\sqrt{eB/\hbar} = 4 \times 10^{13}\,\mathrm{s}^{-1}$ as a typical, non-resonant frequency value for the T-3 "Diced" lattice (ie: not an eigenenergy of the quantum dot), we have (restore $\hbar$)

$$G'(a;\omega) \to G'_{22}(a;\omega) \cong \frac{eB}{2\pi\omega\hbar^2} \cong 0.7 \times 10^{35}\ /\mathrm{Jm}^2, \tag{14.44}$$

so that for the typical value of ω,

$$\left|\lambda G'(a;\omega)\right| \to \left|\lambda G'_{22}(a;\omega)\right| \cong 10^{-3} \ll 1, \tag{14.45}$$

which is too small for a resonant solution of the dispersion relation, since the latter requires that such a resonant frequency/eigenenergy root $\bar{\omega}_n$ solution of the dot dispersion relation (14.42) must have $\left|\lambda G'(a;\bar{\omega}_n)\right| \rightarrow 1$. This can only occur if

$$\bar{\omega}_n^2 - \epsilon_n^2 \cong O(\lambda) \qquad (\text{recall } \epsilon_n = \sqrt{2(2n+1)\alpha^2 eB/\hbar}\ \text{s}^{-1}). \tag{14.46}$$

In the analysis of such roots for every positive integer value of n, the parameter values cited above yield off-diagonal elements $G'_{i\neq j}$ which are considerably smaller than the diagonal ones, for example,

$$\frac{G'_{12}(a;\omega)}{G'_{22}(a;\omega)} \cong \frac{\alpha eBa}{\hbar\omega} \cong 0.04 < 1, \tag{14.47}$$

and

$$\frac{G'_{31}(a;\omega)}{G'_{22}(a;\omega)} \cong \frac{\alpha^2(eB)^2a^2}{\hbar^2\omega^2} \cong 1.6 \times 10^{-3} \ll 1. \tag{14.48}$$

On this basis, the off-diagonal $G'_{i\neq j}(a;\omega)$ elements may be ignored, yielding the generic single dot dispersion relation in diagonal form:

$$0 = \text{Det}(1-\lambda G'(a;\omega)) = \text{Det}\begin{vmatrix} 1-\lambda G'_{11}(a;\omega) & 0 & 0 \\ 0 & 1-\lambda G'_{22}(a;\omega) & 0 \\ 0 & 0 & 1-\lambda G'_{33}(a;\omega) \end{vmatrix}, \tag{14.49}$$

or ($i = 1, 2, 3$)

$$1-\lambda G'_{ii}(a;\omega) \equiv 1 - \sum_{n=0}^{\infty} \frac{\lambda g_{n(ii)}(a;\omega)}{\omega^2-\epsilon_n^2}, \tag{14.50}$$

where

$$g_{n(11)}(a;\omega) = g_{n(33)}(a;\omega) = \frac{1}{2} g_{n(22)}(a;\omega) \tag{14.51}$$

and

$$g_{n(22)}(a;\omega) = \frac{eB\omega}{2\pi\hbar^2} e^{-eBa^2/4\hbar}\, L_n\left(\frac{eBa^2}{2\hbar}\right). \tag{14.52}$$

Bearing in mind the discussion of (14.43)–(14.46), and recalling that the smallness of $|\lambda G'(a;\omega)|$ for a typical frequency that is not resonant means that in the vicinity of the nth-resonance root $\overline{\omega}_n^2 - \epsilon_n^2 \cong O(\lambda)$, all other $n' \neq n$ terms have relatively large denominators,

$$\omega^2 - \epsilon_{n'}^2 \rightarrow \bar{\omega}_n^2 - \epsilon_{n'}^2 \cong \epsilon_n^2 - \epsilon_{n'}^2 \cong \alpha^2 eB/\hbar \cong O(\lambda^0) = O(1), \tag{14.53}$$

which sharply reduces the impact of those terms on the nth root in question, rendering them ($n \neq n'$) negligible in the context of the nth resonant frequency/eigenvalue root

$\bar{\omega}_n$ of the quantum dot. Thus, the nth root $\bar{\omega}_n$ is determined by its nearest pole alone (to the exclusion of all other poles) as

$$\bar{\omega}^2_{n(i,i)} = \epsilon_n^2 + \lambda g_{ii}(a; \epsilon_n), \tag{14.54}$$

in this "single pole" approximation.

Alternatively expressed, the generic single dot frequency/eigenvalue roots for the T-3 "Diced" lattice quantum dot subject to Landau quantization are given by (in inverse seconds)

$$\bar{\omega}_{n(22)} = \pm\epsilon_n \left(1 + \lambda \frac{eB}{2\pi\hbar^2} e^{-eBa^2/4\hbar}\, L_n\left(\frac{eBa^2}{2\hbar}\right) \Big/ 2\epsilon_n\right) \tag{14.55}$$

and the degenerate $\bar{\omega}_{n(11)}$ and $\bar{\omega}_{n(33)}$ frequency roots are

$$\bar{\omega}_{n(11)} = \bar{\omega}_{n(33)} = \pm\epsilon_n \left(1 + \frac{\lambda eB}{2\pi\hbar^2} e^{-eBa^2/4\hbar}\, L_n\left(\frac{eBa^2}{2\hbar}\right) \Big/ 4\epsilon_n\right), \tag{14.56}$$

where $\lambda = U_0 a^2 \cong 1.6 \times 10^{-3}\,\mathrm{Jm}^2$ and $\epsilon_n \cong \alpha\sqrt{eB/\hbar}$ in inverse seconds, for the parameters cited above.

Obtaining the 6 dot frequency/eigenvalue roots when the off-diagonal elements $G'_{i\neq j}$ need to be taken into consideration involves solving a 6th order determinental equation, and in the context of the "single pole" approximation it takes the form

$$\begin{aligned} 0 &= \mathrm{Det}\left|1 - \lambda G'(a;\omega)\right| \\ &= \mathrm{Det}\begin{vmatrix} \omega^2 - \epsilon_n^2 - \lambda g_{n(11)} & -\lambda g_{n(12)} & -\lambda g_{n(13)} \\ -\lambda g_{n(21)} & \omega^2 - \epsilon_n^2 - \lambda g_{n(22)} & -\lambda g_{n(23)} \\ -\lambda g_{n(31)} & -\lambda g_{n(32)} & \omega^2 - \epsilon_n^2 - \lambda g_{n(33)} \end{vmatrix}, \end{aligned} \tag{14.57}$$

where $g_{n(ij)} \equiv g_{n(ij)}(a;\omega)$ are given in (14.51), (14.52); also $g_{n(ij)} = g^*_{n(ji)}$ and $g_{n(23)} = g^*_{n(21)}$, where

$$g_{n(23)}(a;\omega) = \frac{i\alpha(eB)^2 a}{2\pi\hbar^3} e^{-eBa^2/4\hbar} L^1_{n-1}(eBa^2/2\hbar) \tag{14.58}$$

and

$$g_{n(31)}(a;\omega) = \frac{\alpha^2(eB)^3 a^2}{2\pi\omega\hbar^4} e^{-eBa^2/4\hbar} L^2_{n-2}(eBa^2/2\hbar). \tag{14.59}$$

Alternately, we have the single dot dispersion relation as

$$\begin{aligned} &\left(\omega^2 - \epsilon_n^2 - g_{11}\right)^2 \left(\omega^2 - \epsilon_n^2 - g_{22}\right) \\ &\quad - \left(\omega^2 - \epsilon_n^2 - g_{22}\right)|g_{13}|^2 - 2\left(\omega^2 - \epsilon_n^2 - g_{11}\right)|g_{12}|^2 - 2g_{13}|g_{12}|^2 = 0, \end{aligned} \tag{14.60}$$

which includes the off-diagonal terms.

In regard to the double dot dispersion relation, we proceed using (14.38) noting the necessary presence of terms of the form $G'^2(a \text{ or } d; \omega)$. In conformance with the discussion above, we again ignore the relatively small off-diagonal terms of G' here. Such $G'^2(a \text{ or } d; \omega)$–terms involve the product of two series

$$\alpha_+\alpha_- G'^2_{ii}(a \text{ or } d; \omega) = \alpha_+\alpha_- \sum_{n=0}^{\infty}\sum_{m=0}^{\infty} \frac{g_{n(ii)}g_{m(ii)}}{\left(\omega^2 - \epsilon_n^2\right)\left(\omega^2 - \epsilon_m^2\right)}. \tag{14.61}$$

Since both α_+ and α_- are proportional to λ, $\alpha_+\alpha_- \approx \lambda^2$, so both denominator factors must be proportional to λ to contribute significantly to the dispersion relation solution. This is satisfied for the subset of terms with $n = m$, which thus constitute a second order pole. However, terms having $n \neq m$ may involve one denominator factor $\omega^2 - \epsilon_n^2 \sim O(\lambda)$, but the other denominator factor is then $\omega^2 - \epsilon_m^2 \sim \epsilon_n^2 - \epsilon_m^2 \sim O(\lambda^0) = O(1)$, which is relatively large, sharply reducing the magnitude of the $n \neq m$ terms relative to the $n = m$ terms, rendering them negligible. In this context, the $G'^2_{ii}(a \text{ or } d)$ terms are approximately given by

$$\begin{aligned}\alpha_+\alpha_- \left(G'^2_{ii}(a;\omega) - G'^2_{ii}(d;\omega)\right) &= \alpha_+\alpha_- \sum_{n=0}^{\infty} \frac{g^2_{n(ii)}(a;\omega) - g^2_{n(ii)}(d;\omega)}{(\omega^2 - \epsilon_n^2)^2} \\ &\cong O\left(\frac{\lambda^2}{\lambda^2}\right) = O(1),\end{aligned} \tag{14.62}$$

and the double-dot dispersion relation for its eigenenergies is given by the single-pole approximation as ($i = 1, 2, 3$)

$$0 = 1 - (\alpha_+ + \alpha_-)\frac{g_{n(ii)}}{\omega^2 - \epsilon_n^2} + \alpha_+\alpha_- \frac{g^2_{n(ii)}(a;\epsilon_n) - g^2_{n(ii)}(d;\epsilon_n)}{(\omega^2 - \epsilon_n^2)^2}, \tag{14.63}$$

so the double dot eigenenergy solutions $\bar{\omega}_n$ are given by

$$\begin{aligned}\bar{\omega}^2_{n(ii)} - \epsilon_n^2 = \frac{1}{2}\Big\{&(\alpha_+ + \alpha_-)g_{n(ii)}(a;\epsilon_n) \\ &\pm\sqrt{(\alpha_+ + \alpha_-)^2 g^2_{n(ii)}(a;\epsilon_n) - 4\alpha_+\alpha_-(g^2_{n(ii)}(a;\epsilon_n) - g^2_{n(ii)}(d;\epsilon_n))}\Big\}.\end{aligned} \tag{14.64}$$

14.6 Conclusions

In summary, we have explicitly determined the key dynamical propagator of a spinless nonrelativistic "$p^2/2m$" electron on a planar double quantum dot system subject to Landau quantization in a magnetic field as the retarded Green's function given by ($\omega \to \omega + i0^+; \hbar \to 1$)

$$G(\vec{r}_1, \vec{r}_2; \omega) = G_{2D}^B(\vec{r}_1, \vec{r}_2; \omega) + \sum_{\pm} \alpha_{\pm} G_{2D}^B(\vec{r}_1, \vec{r}_{\pm}; \omega) \times \frac{D_{\mp,\mp} G_{2D}^B(\vec{r}_{\pm}, \vec{r}_2; \omega) + D_{\mp,\pm} G_{2D}^B(\vec{r}_{\mp}, \vec{r}_2; \omega)}{D_{\mp,\mp} D_{\pm,\pm} - D_{\pm,\mp} D_{\mp,\pm}}, \quad (14.65)$$

with the associated double dot eigenenergy dispersion relation as

$$D_{\mp,\mp} D_{\pm,\pm} - D_{\pm,\mp} D_{\mp,\pm} = 0, \quad (14.66)$$

where

$$D_{\pm,\pm} = \delta_{\pm,\pm} - \alpha_{\pm} G_{2D}^B(\vec{r}_{\pm}, \vec{r}_{\pm}; \omega) \quad (14.67)$$

and

$$D_{\pm,\mp} = \alpha_{\pm} G_{2D}^B(\vec{r}_{\mp}, \vec{r}_{\pm}; \omega). \quad (14.68)$$

While our calculations have focused on the smallness of dot size as the dominant consideration, with the attendant employment of the "single pole" approximation carefully discussed in the text, the fundamental double dot Green's function of (14.10) and dispersion relation (14.11) are valid without any such restriction.

Furthermore, the infinite sheet 2D Landau quantized Green's functions involved for the nonrelativistic "$p^2/2m$" system and for the relativistic T-3 "Diced" lattice are discussed in Sects. 14.2 and 14.4, respectively.

Considering the "Diced" lattice Green's function in general, (14.6) may be written as

$$\begin{pmatrix} G_{2D}^B(\vec{r}_+, \vec{r}_2; \omega) \\ G_{2D}^B(\vec{r}_-, \vec{r}_2; \omega) \end{pmatrix} = \mathcal{M} \begin{pmatrix} G(\vec{r}_+, \vec{r}_2; \omega) \\ G(\vec{r}_-, \vec{r}_2; \omega) \end{pmatrix}, \quad (14.69)$$

where

$$\mathcal{M} = \begin{pmatrix} 1 - \alpha_+ G_{2D}^B(\vec{r}_+, \vec{r}_+; \omega) & -\alpha_- G_{2D}^B(\vec{r}_+, \vec{r}_-; \omega) \\ -\alpha_+ G_{2D}^B(\vec{r}_-, \vec{r}_+; \omega) & 1 - \alpha_- G_{2D}^B(\vec{r}_-, \vec{r}_-; \omega) \end{pmatrix}. \quad (14.70)$$

Inversion of the matrix $\mathcal{M}$ to obtain $\mathcal{M}^{-1}$ must be done carefully since the off-diagonal matrix elements are themselves 3×3 pseudospin 1 matrices, and the off-diagonal elements do not commute with the diagonal elements in general. However, the determination of $\mathcal{M}^{-1}$ yields the "Diced" lattice Green's function of (14.4) as

$$G(\vec{r}_1, \vec{r}_2; \omega) = G_{2D}^B(\vec{r}_1, \vec{r}_2; \omega) + \sum_{\pm} \alpha_{\pm} G_{2D}^B(\vec{r}_1, \vec{r}_{\pm}; \omega) G(\vec{r}_{\pm}, \vec{r}_2; \omega), \quad (14.71)$$

where

$$G(\vec{r}_\pm, \vec{r}_2; \omega) = \begin{pmatrix} G(\vec{r}_+, \vec{r}_2; \omega) \\ G(\vec{r}_-, \vec{r}_2; \omega) \end{pmatrix} = \mathcal{M}^{-1} \begin{pmatrix} G^B_{2D}(\vec{r}_+, \vec{r}_2; \omega) \\ G^B_{2D}(\vec{r}_-, \vec{r}_2; \omega) \end{pmatrix}. \tag{14.72}$$

A further simplification can be achieved if $|\vec{r}_1|, |\vec{r}_2| \gg a, d$, when there is interest in propagation far from the region of the two dots, since one can then approximately write $G^B_{2D}(\vec{r}_1, \vec{r}_\pm; \omega) \cong G^B_{2D}(\vec{r}_1, 0; \omega)$ and $G^B_{2D}(\vec{r}_\pm, \vec{r}_2; \omega) \cong G^B_{2D}(0, \vec{r}_2; \omega)$, independent of $r_\pm$ because it is relatively small. In this case, we have the approximation

$$G(\vec{r}_1, \vec{r}_2; \omega) = G^B_{2D}(\vec{r}_1, \vec{r}_2; \omega) + G^B_{2D}(\vec{r}_1, 0; \omega)\widehat{\mathcal{M}}^{-1} G^B_{2D}(0, \vec{r}_2; \omega), \tag{14.73}$$

where

$$\widehat{\mathcal{M}}^{-1} = (1\ 1)\left(\mathcal{M}^{-1}\right)\begin{pmatrix}1\\1\end{pmatrix} = \left(\mathcal{M}^{-1}\right)_{11} + \left(\mathcal{M}^{-1}\right)_{12} + \left(\mathcal{M}^{-1}\right)_{21} + \left(\mathcal{M}^{-1}\right)_{22}, \tag{14.74}$$

which is the sum of all 3×3 pseudospin 1 elements of $\mathcal{M}^{-1}$. These results may be useful in the description of transport over relatively large distances. Over smaller distances this approximation is not valid, and (14.71), (14.72) would have to be employed: The same is true for the statistical thermodynamic properties of the "Diced" double-dot system, which involves evaluations at $\vec{r}_1 = \vec{r}_2$ in the region of the dots, as well as elsewhere.

In conclusion, the proliferation of states due to Landau quantization, which may be energetically accessible, needs to be taken into account in considerations of the quantum dynamics of transport through nanostructured dot systems in the presence of a normal magnetic field. In general, magnetic considerations restricted to the neglect of Landau quantization, are, in fact, insufficient. Our analysis of the eigen-energy dispersion relation for a spinless nonrelativistic "$p^2/2m$" 2D Landau-quantized double quantum dot, with the results shown in Tables 14.1 and 14.2 in the text above and in Tables 14.3, 14.4, and 14.5 in the Appendix, clearly exhibits the splintering of each zero-field dot subband level into a multitude of Landau levels, modified by the proximity of the two quantum dots. These same features are evident in our results for the eigenenergy spectrum of a Landau-quantized double quantum dot in a T-3 "Diced" lattice, as seen in (14.55), (14.56), and (14.64).

14.7 Appendix

Tables 14.5 and 14.6 exhibit the percentage difference between the first five calculated "upper" and "lower" energies and the corresponding results using the single pole approximation.

Table 14.3 First 5 "upper" energies for differing well depths using the single pole approximation

n	$U_{0-}, U_{0+} =$ (25, 75) meV	$U_{0-}, U_{0+} =$ (100, 200) meV	$U_{0-}, U_{0+} =$ (100, 300) meV	$U_{0-}, U_{0+} =$ (200, 300) meV	$U_{0-}, U_{0+} =$ (200, 400) meV
n = 0	0.49884	0.49580	0.49537	0.49236	0.49160
n = 1	1.49588	1.48472	1.48353	1.47180	1.46944
n = 2	2.49200	2.46952	2.46798	2.44268	2.43905
n = 3	3.48787	3.49765	3.51230	3.54091	3.40547
n = 4	4.48397	4.43646	4.53790	4.52662	4.37292

Table 14.4 First 5 "lower" energies for differing well depths using the single pole approximation

n	$U_{0-}, U_{0+} =$ (25, 75) meV	$U_{0-}, U_{0+} =$ (100, 200) meV	$U_{0-}, U_{0+} =$ (100, 300) meV	$U_{0-}, U_{0+} =$ (200, 300) meV	$U_{0-}, U_{0+} =$ (200, 400) meV
n = 0	0.48734	0.46273	0.45101	0.43853	0.42547
n = 1	1.47678	1.43326	1.40711	1.39150	1.36652
n = 2	2.46745	2.40880	2.40880	2.35453	2.31760
n = 3	3.48683	3.38683	3.33459	3.32355	3.27365
n = 4	4.44993	4.36522	4.29970	4.29446	4.23043

Table 14.5 Percent difference between the first 5 calculated "upper" energies and the single pole approximations

n	$U_{0-}, U_{0+} =$ (25, 75) meV (%)	$U_{0-}, U_{0+} =$ (100, 200) meV (%)	$U_{0-}, U_{0+} =$ (100, 300) meV (%)	$U_{0-}, U_{0+} =$ (200, 300) meV (%)	$U_{0-}, U_{0+} =$ (200, 400) meV (%)
n = 0	2.63	7.25	9.13	11.25	12.89
n = 1	0.97	2.87	3.67	4.74	5.51
n = 2	0.71	2.21	2.73	3.75	4.31
n = 3	0.62	0.79	1.73	2.94	4.04
n = 4	0.57	1.99	2.25	3.35	3.92

Table 14.6 Percent difference between the first 5 calculated "lower" energies and the single pole approximations

n	$U_{0-}, U_{0+} =$ (25, 75) meV (%)	$U_{0-}, U_{0+} =$ (100, 200) meV (%)	$U_{0-}, U_{0+} =$ (100, 300) meV (%)	$U_{0-}, U_{0+} =$ (200, 300) meV (%)	$U_{0-}, U_{0+} =$ (200, 400) meV (%)
n = 0	0.08	0.67	1.87	1.86	6.78
n = 1	0.82	2.43	3.22	4.07	4.53
n = 2	0.90	2.57	2.09	4.20	5.10
n = 3	0.06	2.44	3.58	3.83	4.80
n = 4	0.89	2.37	3.56	3.57	4.61

Generally, the results differ by a few percent, in a few cases about ten percent. Overall, the single pole approximation provides a reasonable estimate of the order of magnitude, with minimal numerical work.

References

1. P. Harrison (ed.), *Quantum Wells, Wires and Dots*, 2 edn. (Wiley-Interscience, 2005)
2. N.J.M. Horing, *Quantum Statistical Field Theory* (Oxford University Press, 2017), p. 121. Also, N.J.M. Horing, M. Yildiz, Ann. Phys. (NY) **97**, 216, Section 2 (1976)
3. A. Erdelyi, et al., *Bateman Manuscript Project; Higher Transcendental Functions*, vol. 2 (McGraw Hill, 1953), p. 189, #10.12.17 and p. 199
4. D. Widder, *Advanced Calculus* (Prentice-Hall Inc., 1947), p. 240
5. N.J.M. Horing, J. Phys. Cond. Matter **33**, 015302 (2021)

Chapter 15
Two Dimensional Magnetopolaritons and the Associated Landau Quantized Magnetoconductivity Tensor

Mark Orman and Norman J. M. Horing

Abstract We address the magnetopolariton spectrum and magnetoconductivity tensor of a two dimensional plasma of nonrelativistic charge carriers subject to Landau quantization in a normal magnetic field. The analysis is carried out in the random phase approximation for linear electromagnetic response. Various regimes of magnetic field strength and wave number are considered. An exact integral representation of the magnetoconductivity tensor is derived and several expansions in terms of modified Bessel functions and Laguerre polynomials are obtained, as well as a low wave number power expansion. These results encompass the nondegenerate and degenerate statistical regimes as well as intermediate field and quantum strong field strengths. The two-dimensional magnetopolariton dispersion relation is formulated and solved for the electromagnetic normal modes, and local and nonlocal magnetopolaritons/plasmons are discussed, including nonlocal Bernstein modes, all subject to Landau quantization. The leading nonlocal corrections to the local modes are shown to exhibit de Haas-van Alphen oscillatory quantum structure in the degenerate statistical regime.

Preface

This chapter is focused on the magnetopolariton normal mode spectrum and the associated linear magnetoconductivity tensor of a Landau quantized two-dimensional sheet of nonrelativistic electrons. The nonequilibrium Green's function techniques discussed can be fruitfully employed more generally, including to the determination of the magnetoconductivity tensors of 2D Dirac materials, for studies of their magnetopolariton spectra.

In Sect. 15.1 the electromagnetic Green's function propagator for a Landau quantized 2D plasma in 3D space is discussed and its frequency poles provide the dispersion relation for the 2D magnetopolariton spectrum in terms of the 2D magneto-

M. Orman (✉) · N. J. M. Horing
Department of Physics, Stevens Institute of Technology, Hoboken, New Jersey 07030, USA
e-mail: drmarkorman@gmail.com

H. Ünlü and N. J. M. Horing (eds.), *Progress in Nanoscale and Low-Dimensional Materials and Devices*, Topics in Applied Physics 144,
https://doi.org/10.1007/978-3-030-93460-6_15

conductivity tensor of the mobile charge carriers of the material sheet. Section 15.2 addresses the determination of the linear 2D magnetoconductivity tensor $\underset{\approx}{\sigma}_2$ in terms of an associated tensor $\underset{\approx}{L}(\vec{k}, \Omega) = (i\Omega/c)\, \underset{\approx}{\sigma}_2(\vec{k}, \Omega)$. (In this, we ignore the negligible Pauli spin current, but discuss it and provide some pertinent references.) The determination of the linear current response tensor $\underset{\approx}{L}(\vec{k}, \Omega)$ is formulated in terms of a grand canonical ensemble averaged current, but is then evaluated in terms of a related imaginary-time-periodic averaged nonequilibrium Green's function, whose Matsubara Fourier series coefficient provides the frequency (Ω) dependent linear current response tensor $\underset{\approx}{L}(\vec{k}, \Omega)$ and linear conductivity $\underset{\approx}{\sigma}_2(\vec{k}, \Omega)$ in momentum ($\vec{k}$) representation: The latter is finally evaluated by an analytic continuation in an imaginary time integration in terms of a product of equilibrium Green's functions—the RPA "ring" diagram—by linearization of the nonequilibrium Green's function equation of motion. Employing the Landau-quantized equilibrium Green's function involved, its non-translationally-invariant Peierls phase factor is separated and the remainder of the Green's function is spatially translationally invariant, leading to the conductivity tensor $\underset{\approx}{\sigma}_2(\vec{k}, \Omega)$ and $\underset{\approx}{L}(\vec{k}, \Omega)$ as functions of a single momentum/wavenumber variable $\vec{k}$ and frequency Ω.

The resulting linear current response tensor $\underset{\approx}{L}(\vec{k}, \Omega)$ of nonrelativistic electrons is presented in terms of a 3-fold integral representation that involves only elementary functions and the Fermi-Dirac distribution, incorporating the exact role of the magnetic field in its discretization of the 2D spectrum due to Landau quantization. The tensor elements $L^i_j(\vec{k}, \Omega)$ are also presented in a modified Bessel function expansion exhibiting resonant frequency pole structure at integer multiples of the cyclotron frequency, with coefficients involving a Fermi integration and an accompanying inverse Laplace transform. Both integrations are evaluated exactly in the nondegenerate statistical regime, and the resulting nondegenerate linear magnetoconductivity tensor elements L^i_j are exhibited explicitly. The zero temperature degenerate limit is also examined with exact execution of the Fermi integration, so the coefficients involve only an inverse Laplace transform replete with de Haas-van Alphen oscillatory structure in the L^i_j tensor elements. A further expansion of L^i_j at arbitrary temperature in terms of Laguerre polynomials expedites the execution of all three integrations, leaving the Fermi distribution intact at arbitrary temperature (but evaluated at multiples of $\hbar\omega_c$), and the tensor elements L^i_j are also presented in this form. This form is particularly useful for determining L^i_j in the degenerate quantum strong field limit, in which the magnetic field is sufficiently high that only the lowest Landau eigenstate is populated: This most extreme quantum manifestation of the magnetic field is also exhibited in the L^i_j tensor elements explicitly. Finally, we also exhibit a low wavenumber power expansion of the L^i_j tensor elements for arbitrary temperature and arbitrary field strength taking account of the leading nonlocality effects of the L^i_j tensor elements along with its local structure.

In Sect. 15.3 the 2D magnetopolariton spectrum is examined using the magnetoconductivity tensor of Sect. 15.2 in the low wavenumber approximation to determine the local modes and their wavenumber corrections, including the $n = 2$ quantum generalization of the Bernstein mode at arbitrary temperatures (statistical regime) and magnetic field strength, incorporating de Haas-van Alphen phenomenology. The evanescent wave dispersion relation is formulated exactly and simple approximate analytic solutions for the 2D magnetopolariton normal modes are obtained in the cases having ω_c much smaller and much larger than the other frequency parameters; also, the electrostatic limit $ck_{\parallel} \to \infty$ is studied. Quantum magnetic field effects, including de Haas-van Alphen oscillations, are fully incorporated in both the local modes and their nonlocal corrections, including the $n = 2$ Bernstein mode generalization. Furthermore the polarizations of all modes studied are determined.

15.1 Electromagnetic Propagator for a Landau Quantized 2D Plasma

This report is concerned with the magnetopolaritons and associated magnetoconductivity tensor of a Landau quantized 2D plasma sheet in a normal magnetic field. As such, it is appropriate to discuss the associated electromagnetic propagator, whose frequency poles define the normal electromagnetic polariton modes of the 2D system. In this context, we briefly review studies [1, 2] of the propagator and its dispersion relation.

Considering a uniform 3D background medium of dielectric constant $\varepsilon_b^{(3D)} \to \varepsilon_b$ and magnetic permeability $\mu = 1$, the Helmholtz equation for the dynamic electric field $\vec{E}(\vec{r}, \Omega)$ may be written in position/frequency representation as (c is the speed of light in vacuum; $\widehat{I} = \widehat{x}\widehat{x} + \widehat{y}\widehat{y} + \widehat{e}_z\widehat{e}_z$ is the 3D unit dyadic/tensor)

$$\left[\widehat{I}\left(\vec{\nabla}^2 + \frac{\Omega^2}{c^2}\varepsilon_b^{(3D)}\right) - \vec{\nabla}\vec{\nabla}\right]\vec{E}(\vec{r}, \Omega) + \frac{4\pi i\Omega}{c^2}\int d^3\vec{r}'\widehat{\sigma}_{fs}^{(2D)}(\vec{r}, \vec{r}'; \Omega)\vec{E}(\vec{r}'; \Omega) = -\frac{4\pi i\Omega}{c^2}\vec{J}_{ext}(\vec{r}; \Omega). \qquad (15.1.1)$$

Here, we have written the linear induced current $\vec{J}_{ind}(\vec{r}, \Omega)$ in terms of the non-local, spatially inhomogeneous 3×3 matrix dyadic conductivity tensor $\widehat{\sigma}(\vec{r}, \vec{r}'; \Omega)$ as

$$\vec{J}_{ind}(\vec{r}, \Omega) = \int d^3\vec{r}'\widehat{\sigma}_{fs}^{(2D)}(\vec{r}, \vec{r}'; \Omega)\vec{E}(\vec{r}'; \Omega), \qquad (15.1.2)$$

and transposed it to the left side of (15.1.1). The associated Green's function propagator $\widehat{\mathscr{G}}_{fs}(\vec{r}, \vec{r}'; \Omega)$ is defined to satisfy the equation

$$\vec{E}(\vec{r}, \Omega) = \frac{4\pi i\Omega}{c^2}\int d^3\vec{r}'\widehat{\mathscr{G}}_{fs}(\vec{r}, \vec{r}'; \Omega) \cdot \vec{J}_{ext}(\vec{r}'; \Omega), \qquad (15.1.3)$$

and it is therefore determined by the relation

$$\left[\widehat{I}\left(\vec{\nabla}^2+\frac{\Omega^2}{c^2}\varepsilon_b^{(3D)}\right)-\vec{\nabla}\vec{\nabla}\right]\widehat{\mathscr{G}}_{fs}(\vec{r},\vec{r}';\Omega)$$
$$+\frac{4\pi i\Omega}{c^2}\int d^3\vec{r}''\widehat{\sigma}_{fs}^{(2D)}(\vec{r},\vec{r}'';\Omega)\widehat{\mathscr{G}}_{fs}(\vec{r}'',\vec{r}';\Omega)=-\widehat{I}\delta^{3D}(\vec{r}-\vec{r}'). \tag{15.1.4}$$

In the absence of the 2D plasma, the bulk 3D Green's function $\widehat{\mathscr{G}}_{3D}(\vec{r},\vec{r}';\Omega)$ is given by

$$\left[\widehat{I}\left(\vec{\nabla}^2+\frac{\Omega^2}{c^2}\varepsilon_b^{(3D)}\right)-\vec{\nabla}\vec{\nabla}\right]\widehat{\mathscr{G}}_{3D}(\vec{r},\vec{r}';\Omega)=-\widehat{I}\delta^{(3D)}(\vec{r}-\vec{r}') \tag{15.1.5}$$

and it may be used to convert the differential relation (15.1.4) to an integral equation as follows ($\vec{k}_\parallel=k_x\widehat{x}+k_y\widehat{y}$):

$$\widehat{\mathscr{G}}_{fs}(\vec{r},\vec{r}';\Omega)=\widehat{\mathscr{G}}_{3D}(\vec{r},\vec{r}';\Omega)+\frac{4\pi i\Omega}{c^2}\int d^3\vec{r}''\int d^3\vec{r}'''\widehat{\mathscr{G}}_{3D}(\vec{r},\vec{r}'';\Omega)$$
$$\times\,\widehat{\sigma}_{fs}^{(2D)}(\vec{r}'',\vec{r}''';\Omega)\widehat{\mathscr{G}}_{fs}(\vec{r}''',\vec{r}',\Omega). \tag{15.1.6}$$

Due to spatial translational invariance in the plane of the plasmonic layer, the dyadic Green's function may be written in terms of a Fourier transform in the plasma plane $[\vec{r}_\parallel-\vec{r}'_\parallel\rightarrowtail\vec{k}_\parallel]$, so that (15.1.6) takes the form

$$\widehat{\mathscr{G}}_{fs}(\vec{k}_\parallel;z,z',\Omega)=\widehat{\mathscr{G}}_{3D}(\vec{k}_\parallel;z,z';\Omega)$$
$$+\frac{4\pi i\Omega}{c^2}\int dz''\int dz'''\widehat{\mathscr{G}}_{3D}(\vec{k}_\parallel;z,z'';\Omega)$$
$$\times\,\widehat{\sigma}_{fs}^{(2D)}(\vec{k}_\parallel;z'',z''';\Omega)\widehat{\mathscr{G}}_{fs}(\vec{k}_\parallel;z''',z';\Omega). \tag{15.1.7}$$

Considering the conductivity tensor to be confined to the plane $z=0$ of the 2D plasma, and to be homogeneous on the plasma plane, but nonlocal, we have

$$\widehat{\sigma}_{fs}^{(2D)}(\vec{k}_\parallel;z'',z''';\Omega)=\widehat{I}\widehat{\sigma}_{fs}^{(2D)}(\vec{k}_\parallel;\Omega)\delta(z'')\delta(z'''), \tag{15.1.8}$$

where $\widehat{\sigma}_{fs}^{(2D)}$ is the 2D surface current conductivity tensor. Application of the $\delta(z'')\delta(z''')$ functions in (15.1.7) reduces that integral equation to an algebraic matrix (dyadic) equation, whose solution is [1, 2]

$$\widehat{\mathscr{G}}_{fs}(\vec{k}_\parallel;z,z',\Omega)=\widehat{\mathscr{G}}_{3D}(\vec{k}_\parallel;z,z';\Omega)+\widehat{\mathscr{G}}_{3D}(\vec{k}_\parallel;z,0;\Omega)\gamma$$
$$\times\left[\widehat{I}-\widehat{\mathscr{G}}_{3D}(\vec{k}_\parallel;0,0;\Omega)\gamma\right]^{-1}\widehat{\mathscr{G}}_{3D}(\vec{k}_\parallel;0,z';\Omega), \tag{15.1.9}$$

where

$$\gamma = \frac{4\pi i\Omega}{c^2}\widehat{\sigma}^{(2D)}_{fs}(\vec{k}_\parallel;\Omega). \tag{15.1.10}$$

This Green's function has already been applied to the light pulse excitation of radiating polaritons in a 2D excitonic layer [2] and EM wave transmission-diffraction through a subwavelength nano-hole in a 2D plasmonic layer [1]. The collective electromagnetic magnetopolariton mode dispersion relation is given by the frequency poles of $\widehat{\mathscr{G}}_{fs}$, and this means that

$$\det\left[\widehat{I} - \widehat{\mathscr{G}}_{3D}(\vec{k}_\parallel;0,0;\Omega)\gamma\right] = \det\left[\widehat{I} - \frac{4\pi i\Omega}{c^2}\widehat{\mathscr{G}}_{3D}(\vec{k}_\parallel;0,0;\Omega)\widehat{\sigma}^{(2D)}_{fs}(\vec{k}_\parallel;\Omega)\right] = 0, \tag{15.1.11}$$

provides the requisite roots, Ω. The solution for $\widehat{\mathscr{G}}_{3D}(\vec{k}_\parallel;0,0;\Omega)$ of (15.1.5) is readily obtained by Fourier transforming $\vec{r}_\parallel - \vec{r}'_\parallel \to \vec{k}_\parallel$, with the result (bear in mind that $\widehat{e}_z = \widehat{z}$ is the unit vector in the perpendicular direction, which is parallel to the normal magnetic field; $k_\perp \equiv k_z \equiv \sqrt{k^2 - k_\parallel^2}$)

$$\widehat{\mathscr{G}}_{3D}(\vec{k}_\parallel;z,z';\Omega) = -\frac{e^{ik_\perp|z-z'|}}{2ik_\perp} \times\left\{\widehat{I} - \frac{1}{q_\Omega^2}[\vec{k}_\parallel\vec{k}_\parallel + k_\perp \mathrm{sgn}(z-z')(\vec{k}_\parallel\widehat{e}_z + \widehat{e}_z\vec{k}_\parallel) + \widehat{e}_z\widehat{e}_z\left(k_\perp^2 - 2ik_\perp\delta(z-z')\right)]\right\}, \tag{15.1.12}$$

with

$$q_\Omega \equiv k \equiv |\vec{k}| = \sqrt{k_\parallel^2 + k_\perp^2} \equiv (\Omega/c)\sqrt{\varepsilon_b^{(3D)}}, \tag{15.1.13}$$

in mixed $(\vec{k}_\parallel;z,z')$ Fourier representation.

As (15.1.9) calls for an evaluation of $\widehat{\mathscr{G}}_{3D}$ at $z = z' = 0$, so that $\mathrm{sgn}(0) = 0$ and $\delta(0)$ may be understood as representing the inverse plasma layer thickness $1/d$, we have

$$\widehat{\mathscr{G}}_{3D}(\vec{k}_\parallel;0,0;\Omega) = -\frac{1}{2ik_\perp}\left\{\widehat{I} - \frac{1}{q_\Omega^2}\left[\vec{k}_\parallel\vec{k}_\parallel + \widehat{e}_z\widehat{e}_z\left(k_\perp^2 - \frac{2ik_\perp}{d}\right)\right]\right\}. \tag{15.1.14}$$

Thus, the dispersion relation is given by

$$\det\left(\widehat{I} + \frac{2\pi\Omega}{c^2k_\perp}\left\{\widehat{I} - \frac{c^2}{\Omega^2\varepsilon_b^{(3D)}}[\vec{k}_\parallel\vec{k}_\parallel + \widehat{e}_z\widehat{e}_z(k_\perp^2 - 2ik_\perp/d)]\right\}\widehat{\sigma}^{(2D)}_{fs}(\vec{k}_\parallel,\Omega)\right) = 0. \tag{15.1.15}$$

This 3×3 matrix determinant is block diagonal and its 2×2 sub-block pertaining to the planar fields and currents of the plasma layer yields (note that $\widehat{I} = \underset{\sim}{I} + \widehat{e}_z\widehat{e}_z$ with $\underset{\sim}{I} = \widehat{x}\widehat{x} + \widehat{y}\widehat{y}$ as the 2D identity dyadic, and recall that $k^2 = \Omega^2 \varepsilon_b^{(3D)}/c^2$)

$$\det\left(\underset{\sim}{I} + \frac{2\pi\Omega}{c^2 k_\perp}\left\{\underset{\sim}{I} - \frac{[\vec{k}_\parallel \vec{k}_\parallel]}{k^2}\right\} \underset{\sim}{\sigma}(\vec{k}_\parallel, \Omega)\right) = 0. \quad (15.1.16)$$

Here, $\underset{\sim}{\sigma}$ is the 2×2 sub-block of $\widehat{\sigma}_{fs}^{(2D)}$ that refers to the lineal in-plane plasma layer conductivity dyadic/tensor.

The polarization of the in-plane electric field of the normal self-sustaining magnetopolariton modes of the plasma layer may be obtained from the homogeneous counterpart of (15.1.1) in the absence of any driving current, $\vec{J}_{ext}(\vec{r}; \Omega) \equiv 0$. Expressing this homogeneous counterpart of (15.1.1) as it pertains to the in-plane electric field on the plasma sheet in terms of $\mathscr{G}_{3D}(\vec{r}, \vec{r}'; \Omega)$ as shown in (15.1.14), we have

$$\left(\underset{\sim}{I} + \frac{2\pi\Omega}{c^2 k_\perp}\left\{\underset{\sim}{I} - \frac{[\vec{k}_\parallel \vec{k}_\parallel]}{k^2}\right\} \underset{\sim}{\sigma}(\vec{k}_\parallel, \Omega)\right) \cdot \left(E_x \widehat{x}\widehat{x} + E_y \widehat{y}\widehat{y}\right) = 0, \quad (15.1.17)$$

which determines the polarization of the modes in terms of E_x/E_y (but not the absolute magnitude of $\vec{E}$ as it is a homogeneous equation).

It might be noted that an alternative (albeit less general) derivation for the 2D magnetoplasma dispersion relation can be obtained using Gauss and Stokes Laws boundary conditions, and this is presented in Appendix 15.4.

In subsequent sections, it will be convenient to use a coordinate system in which $\vec{k}_\parallel \equiv \vec{k}_y$, $\vec{k}_x \equiv 0$, $\vec{k}_\perp \equiv \vec{k}_z$. With this coordinate system, the 2D magnetoplasma dispersion relation (15.1.16), reduces to the simple form

$$\det\left(\underset{\sim}{1} + \frac{2\pi k_\perp}{\epsilon_b^{(3D)}\Omega}\begin{pmatrix} k^2/k_\perp^2 & 0 \\ 0 & 1 \end{pmatrix} \cdot \underset{\sim}{\sigma}\right) = 0 \quad (15.1.18)$$

where $k_\perp \equiv i\left(k_\parallel^2 - \frac{\epsilon_b^{(3D)}\Omega^2}{c^2}\right)^{\frac{1}{2}}$ and $k_\parallel^2 > \frac{\epsilon_b^{(3D)}\Omega^2}{c^2}$ for evanescent waves.

15.2 2D Linear Conductivity Tensor

The linear charge transport properties of a Landau quantized 2D plasma may be determined from the nonlocal conductivity tensor defined as (notation: $1 = x_1, y_1, t_1$, etc.),

$$\underset{\sim}{\sigma}_2(1-2) = \left[\frac{\delta \vec{J}^s(1)}{\delta \vec{E}(2)}\right]_{\vec{E}\to 0}. \quad (15.2.1)$$

Here, $\vec{E}$ is the electric field along the plasma surface and $\vec{J}^s$ is the induced surface current density. It will be convenient to carry out our analysis in terms of a related 2D linear response tensor $\underset{\sim}{L}(1-2)$,

$$\underset{\sim}{L}(1-2) = \left[\frac{\delta \vec{J}^s(1)}{\delta \vec{A}(2)}\right]_{\vec{A}\to 0}, \tag{15.2.2}$$

where $\vec{E} = -\frac{1}{c}\frac{\partial \vec{A}}{\partial t} - \vec{\nabla}\phi \to \frac{i\Omega}{c}\vec{A} - i\vec{k}\phi$ in Fourier representation $(\vec{k}, \Omega)$, and $\vec{A}$ is the vector potential in the plane of the plasma, with ϕ as the scalar potential. For a translationally invariant medium the Fourier components of $\underset{\sim}{\sigma_2}$ and $\underset{\sim}{L}$ are related by

$$L(\vec{k}, \Omega) = \left(\frac{\partial \vec{J}_s}{\partial \vec{A}}\right)_{\vec{A}\to 0} = \left(\frac{\partial \vec{J}_s}{\partial \vec{E}} \cdot \frac{\partial \vec{E}}{\partial \vec{A}}\right)_{\vec{A}\to 0} = \frac{i\Omega}{c}\underset{\sim}{\sigma_2},$$

$$\text{or} \qquad \underset{\sim}{\sigma_2}(\vec{k}, \Omega) = \frac{c}{i\Omega}\underset{\sim}{L}(\vec{k}, \Omega). \tag{15.2.3}$$

To evaluate $\underset{\sim}{\sigma_2}$ or $\underset{\sim}{L}$, it is necessary to specify the average current density in the presence of perturbing fields. For a non-relativistic quantum plasma, the field theoretic expressions for the surface charge density and current density operators are given by $(\hbar \to 1)$

$$\rho^s = e\psi^\dagger\psi, \tag{15.2.4a}$$

and

$$\vec{\mathbf{J}}^s = \frac{e}{2m}\left(\frac{-\vec{\nabla}}{i}\psi^\dagger \cdot \psi + \psi^\dagger \cdot \frac{\vec{\nabla}}{i}\psi\right) - \frac{e^2}{mc}(\vec{A} + \vec{A}_m)\psi^\dagger\psi, \tag{15.2.4b}$$

where ψ is the second quantized fermion field operator for 2D electrons and $\vec{A}_m$ is the vector potential along the plasma surface due to a normal, constant, uniform, external magnetic field. We assume here that the electromagnetic fields may be treated as classical (c-number) quantities, whereas the particle fields are treated as second quantized operators.

It should be noted that the present nonrelativistic formulation of current and conductivity neglects the role of magnetic field spin energy associated with the $\vec{A}$-field as it pertains to the "$\mu_0\vec{H}_m \cdot \vec{\sigma}$" Hamiltonian term of (15.2.21). Consideration of it leads to the replacement

$$\mu_0\vec{H}_m \cdot \vec{\sigma} \to \mu_0\left(\vec{H}_m + \vec{\nabla} \times \vec{A}\right) \cdot \vec{\sigma}, \tag{15.2.5}$$

and following the usual procedures this leads to an additional current density operator term of the form $(\hbar \to 1)$

$$\vec{\mathbf{J}}^s = \frac{e}{2m}\vec{\nabla} \times \left(\psi^\dagger \vec{\sigma}\psi\right). \tag{15.2.6}$$

This "spin current" operator has been interpreted [3–5] in terms of the spin magnetic moment operator $e\vec{\sigma}/2m$, which, on a "per-unit-volume" basis yields the spin magnetization operator as

$$\vec{M} = \frac{e}{2m}\psi^\dagger \vec{\sigma}\psi, \tag{15.2.7}$$

whose curl is the corresponding spin current density operator

$$\vec{\mathbf{J}}^s = \vec{\nabla} \times \vec{M} = \frac{e}{2m}\vec{\nabla} \times \left(\psi^\dagger \vec{\sigma}\psi\right), \tag{15.2.8}$$

in accordance with electromagnetic interpretation [4, 5].

Spin current has recently received considerable attention, particularly in the presence of strong spin-orbit coupling that is crucial for creating and manipulating it. Such spin-orbit coupling is an automatic consequence of relativistic Dirac theory, and spin currents in Graphene and the Transition Metal Dichalcogenides have been under examination [6–13]. Thermodynamic, thermoelectric, ferromagnetic and superconducting aspects of spin currents have also been discussed [14–17], including speculation about its practical application in quantum computing and spintronics. As a charge transport current, it is unusual in being divergence-less (by vector identity),

$$\vec{\nabla} \cdot \vec{\mathbf{J}}^s = \frac{e}{2m}\vec{\nabla} \cdot \vec{\nabla} \times \left(\psi^\dagger \vec{\sigma}\psi\right) \equiv 0, \tag{15.2.9}$$

so it cannot affect the equation of continuity, $\partial\rho/\partial t = -\vec{\nabla} \cdot \vec{\mathbf{J}}^s$, which is unusual for a charge transport current. It has been interpreted as a "swirl"/circulatory-type current [3].

The present chapter does not treat the spin current as it has not been traditionally significant in nonrelativistic quantum transport theory or in the determination of nonrelativistic electromagnetic polariton spectra, upon which we focus with careful consideration of an impressed steady magnetic field $\vec{H}_m$ that dwarfs the magnetic wave field contribution of $\vec{A}$ in its direct comparison with the spin energy contribution of $\vec{\nabla} \times \vec{A}$. (The wave magnetic field strength is of the order of 10^{-6} Tesla, which is many orders of magnitude smaller than steady laboratory fields commonly of order 10–50 Tesla.) Nevertheless, there was an examination of the role of spin current effects on the nonrelativistic magnetoconductivity tensor by Bardos and Frankel [18]: In their words, they found that spin contributions do "not significantly affect the plasma dispersion in the plasma layer", and we do not deal further with spin currents in this chapter. However, the interested reader will find another timely chapter on spin currents in this book by K. H. Bennemann, under the title "Spin Dependent Thermoelectric Currents of Tunnel Junctions and Other Nanostructures: Onsager Response Theory".

If the 2D plasma is perturbed from thermodynamic equilibrium by an external field, the expectation value of the resulting nonequilibrium current operator $\vec{\mathbf{J}}^s$ is given by the grand canonical ensemble average,

$$< \vec{\mathbf{J}}^s >^{\beta} = \frac{\sum\limits_{N,E} \left[e^{-\beta(E-\zeta N)} < N, E|\vec{\mathbf{J}}^s|N, E > \right]}{\sum\limits_{N,E} e^{-\beta(E-\zeta N)}}, \tag{15.2.10}$$

where this average involves states of the initial equilibrium system; i.e. just prior to the perturbation. Here, ζ is the chemical potential for 2D electrons, $\beta = \frac{1}{KT}$ ($K \equiv$ Boltzmann's constant, T = absolute temperature), and $|N, E>$ are the energy/number eigenstates of the unperturbed 2D Hamiltonian. The preceding expression may also be written as

$$< \vec{\mathbf{J}}^s >^{\beta} = \frac{Tr[e^{-\beta(H_0-\zeta N)}\vec{\mathbf{J}}^s]}{Tr[e^{-\beta(H_0-\zeta N)}]}, \tag{15.2.11}$$

with H_0 and N representing the Hamiltonian and number density operators for the initial, unperturbed 2D plasma system and Tr indicating the trace operation (including a spin trace involved in the sum over energies). By virtue of (15.2.2),

$$\mathcal{L}_c^{\beta}(1-2) = \left[\frac{\delta < \vec{\mathbf{J}}^s(1) >^{\beta}}{\delta \vec{A}(2)} \right]_{\vec{A}\to 0}. \tag{15.2.12}$$

The angle bracket with superscript "β" indicates thermal averaging of the current density over the initial equilibrium ensemble, and the subscript "c" indicates that the response is causal in nature; i.e $\mathcal{L}_c^{\beta}(1-2) \equiv 0$ for $t_1 < t_2$.

Equation (15.2.12) specifies the actual physical linear current response of a perturbed 2D plasma. Although this quantity has a simple representation as the ensemble average of a current-current commutator, its actual evaluation can be challenging and we approach it from another point of view [19–23]. Martin and Schwinger [19–21] developed a technique that allows the indirect evaluation of $\mathcal{L}_c^{\beta}$ in terms of a closely related "periodic" response function. This latter quantity may be represented in terms of nonequilibrium Green's functions (involving time ordered products of two current operators) which provide a powerful mathematical framework for analyzing nonequilibrium response properties.

To be specific, one may define the following nonequilibrium ensemble averaged current density:

$$< \vec{\mathbf{J}}^s >^{i\tau} = \frac{Tr\Big[\Big(exp[-i\int\limits_0^{\tau}(H(t') - \zeta N)dt')]\Big)_+ \vec{\mathbf{J}}_s(t)\Big]}{Tr\Big[\Big(exp[-i\int\limits_0^{\tau}(H(t') - \zeta N)dt']\Big)_+\Big]}, \tag{15.2.13}$$

where $H(t')$ is the time dependent Hamiltonian in the presence of external, time dependent perturbations that drive the system out of equilibrium. The + subscripts in (15.2.13) denote a time ordering process that imparts an imaginary time periodicity to $< \vec{\mathbf{J}}^s >^{i\tau}$ above, as discussed in Appendix 15.5. The corresponding "periodic" 2-D linear response tensor is defined as

$$\mathcal{L}_p^{i\tau}(1-2) = \left[\frac{\delta < \vec{\mathbf{J}}^s(1) >^{i\tau}}{\delta \vec{A}(2)} \right]_{\vec{A}\to 0} . \tag{15.2.14}$$

Whereas $\mathcal{L}_c^{\beta}$ represents the actual causal physical response of the system driven from thermodynamic equilibrium, the tensor $\mathcal{L}_p^{i\tau}$ is a closely related mathematical quantity that facilitates the evaluation of $\mathcal{L}_c^{\beta}$. Its utility arises from the fact that it satisfies time-periodicity conditions in imaginary time $\tau = -i\beta$ (in consequence of the time translation character of the operator $[exp(-i\int_0^{\tau} H(t')dt')]_+$ under the trace average in (15.2.13)), so that $\mathcal{L}_p^{i\tau}$ may be expanded in a Fourer series over the imaginary time interval $0 < t_1, t_2 < \tau$. It can be shown [19–23] that the resulting Fourer series coefficients $\mathcal{L}_p^{i\tau}\left(\frac{2\pi\nu}{\tau}\right)$ $(\nu = 0, 1, 2, etc)$ are directly related to the frequency spectrum of $\mathcal{L}_c^{\beta}$ via the analytic continuation (ϵ is a positive infinitesimal)

$$\mathcal{L}_c^{\beta}(\Omega) = \mathcal{L}_p^{i\tau}\left(\frac{2\pi\nu}{\tau} \to \Omega + i\epsilon, i\tau \to \beta\right), \tag{15.2.15}$$

(This reflects the fact that the fundamental time interval of periodicity is imaginary since $\tau \to -i\beta$). By virtue of this relationship, one can analyze the linear response properties of a plasma in terms of imaginary-time-periodic functions, and then use the resulting solutions to determine the physical (causal) response. The advantage of this approach is that the linear physical response functions can be directly expressed in terms of periodic/antiperiodic equilibrium Green's functions, and these may be evaluated by employing various Green's function techniques.

In the present case, the 2D nonequilibrium one particle Green's function is defined as

$$\bar{G}^{i\tau}(1, 1') = -i <\in (\psi(1)\psi^{\dagger}(1'))_+ >^{i\tau} e^{i\zeta(t_1 - t_1')}. \tag{15.2.16}$$

Here, $\in \equiv \pm 1$ for Fermions (for Bosons it is always $+1$) depending on whether $t_1 \gtrless t_1'$, and the $+$ subscript denotes the time ordering of the field operators $\psi(1)$, $\psi^{\dagger}(1')$; $1 = t_1, \vec{r}_1 = (x_1, y_1)$, etc. The averaging procedure here is identical to that of (15.2.13). Defining the "velocity" operators ($\hbar \to 1$)

$$\vec{V} = \frac{1}{m}\left(\frac{\vec{\nabla}}{i} - \frac{e\vec{A}_m}{c}\right), \tag{15.2.17}$$

$$\vec{V}^{\dagger} = \frac{1}{m}\left(\frac{-\vec{\nabla}}{i} - \frac{e\vec{A}_m}{c}\right),$$

and employing (15.2.4b), (15.2.13) and (15.2.16), one can easily show that (neglecting spin current)

$$< \vec{\mathbf{J}}^s(1) >^{i\tau} = \frac{e}{i} \lim_{1' \to 1} \left\{ \frac{\vec{\nabla}^\dagger(1') + \vec{\nabla}(1)}{2} - \frac{e\vec{A}(1)}{mc} \right\} \bar{G}^{i\tau}(1, 1'^+), \tag{15.2.18}$$

with the $+$ superscript on t_1' in the Green's function indicating that t_1' is infinitesimally larger that t_1. Inserting this into (15.2.14), we have

$$\underset{\sim}{L}_p^{i\tau}(1-2) = \frac{-e^2}{mc} n^s \underset{\sim}{1} \delta(1-2) + \frac{e}{i} \lim_{1' \to 1} \left[\frac{\vec{\nabla}^\dagger(1') + \vec{\nabla}(1)}{2} \right] \left[\frac{\delta \bar{G}^{i\tau}(1, 1'^+)}{\delta \vec{A}(2)} \right]_{\vec{A} \to 0}. \tag{15.2.19}$$

Here, $\underset{\sim}{1}$ is the unit tensor in two dimensions, $\delta(1-2)$ is the Dirac delta function $\delta(1-2) = \delta(t_1 - t_2)\delta(x_1 - x_2)\delta(y_1 - y_2)$ (with z as the coordinate perpendicular to the plasma surface), and n^s is the equilibrium surface plasma density,

$$n^s = -i \lim_{1' \to 1} \bar{G}^{i\tau}(1, 1'^+)]_{\substack{\vec{A} \to 0 \\ i\tau \to \beta}} = \frac{Tr[e^{-\beta(H_0 - \zeta N)} \psi^\dagger(1)\psi(1)]}{Tr[e^{-\beta(H_0 - \zeta N)}]}. \tag{15.2.20}$$

To obtain the zero field limit for the Green's function derivative in (15.2.19), one can employ the differential equation for $\bar{G}^{i\tau}$ (in which we neglect collisional and phonon scattering interactions):

$$\left\{ i\frac{\partial}{\partial t_1} + \zeta - \mu_o \vec{\mathscr{H}}_m \cdot \vec{\sigma} - \frac{\left[\frac{\vec{\nabla}_1}{i} - \frac{e\vec{A}_m(1)}{c} - \frac{e\vec{A}(1)}{c} \right]^2}{2m} - e\Phi \right\} \bar{G}^{i\tau}(1, 1') = \delta(1 - 1'), \tag{15.2.21}$$

where ζ is the chemical potential, μ_o is the Bohr magneton, $\vec{\sigma} = (\sigma_1, \sigma_2, \sigma_3)$ are the Pauli spin matrices ($\mu_o \vec{\mathscr{H}}_m \cdot \vec{\sigma}$ is the Zeeman energy term due to electron spin coupling with the uniform external magnetic field), and Φ is the classical (c-number) scalar potential for the 2D plasma field. This differential equation can be derived in the usual manner [19–23] from the equation of motion for the 2D field operators $\psi(1)$ and $\psi^\dagger(1')$. Converting (15.2.21) into an integral equation,

$$\bar{G}^{i\tau}(1,1') = \bar{G}_o^{i\tau}(1,1') - \int dx_3 dy_3 dt_3 \bar{G}_o^{i\tau}(1,3)\Bigg\{\frac{e}{2mc}\left(\frac{\vec{\nabla}_3}{i} - \frac{e\vec{A}_m(3)}{c}\right)\cdot \vec{A}(3) + \frac{e}{2mc}\vec{A}(3)\cdot\left(\frac{\vec{\nabla}_3}{i} - \frac{e\vec{A}_m(3)}{c}\right) - \frac{e^2}{2mc}\vec{A}^2(3) - e\Phi(3)\Bigg\}\bar{G}^{i\tau}(3,1'), \tag{15.2.22}$$

where $\bar{G}_o^{i\tau}(1,1') = [\bar{G}^{i\tau}(1,1')]_{\vec{A}=0,\Phi=0}$ is the equilibrium Green's function solution to (15.2.21). Thus, we have

$$\left[\frac{\delta\bar{G}^{i\tau}(1,1')}{\delta\vec{A}(2)}\right]_{\vec{A}\to 0} = \frac{-e}{c}\lim_{2'\to 2}\left(\left[\frac{\vec{V}^{\dagger}(2') + \vec{V}(2)}{2}\right]\bar{G}_o^{i\tau}(1,2')\bar{G}_o^{i\tau}(2,1')\right). \tag{15.2.23}$$

Substituting into (15.2.19), one obtains

$$\underset{\sim}{L}_p^{i\tau}(1-2) = \frac{-e^2}{mc}n^s\underset{\sim}{1}\delta(1-2) + \frac{ie^2}{c}\lim_{\substack{1'\to 1\\ 2'\to 2}}\left(\left[\frac{\vec{V}^{\dagger}(1') + \vec{V}(1)}{2}\cdot\frac{\vec{V}^{\dagger}(2') + \vec{V}(2)}{2}\right]\cdot\bar{G}_o^{i\tau}(1,2')\bar{G}_o^{i\tau}(2,1')\right). \tag{15.2.24}$$

The preceding equation provides a concise formulation for the 2D plasma linear response in the presence of a uniform magnetic field. It should be noted that evaluation of the linear response in the limit of zero perturbing electromagentic field is tantamount to the Random Phase Approximation [21, 22].

One can readily show [21–23] that the equilibrium Green's function $\bar{G}_o^{i\tau}(1,2)$ in a constant, uniform magnetic field can be factored into the product form

$$\bar{G}_o^{i\tau}(1,2) = C(1,2)\bar{G}_o^{i\tau}(1-2). \tag{15.2.25}$$

Here, $\bar{G}_o^{i\tau}(1-2)$ is the gauge-independent translationally invariant portion of the Green's function, and the Peierls phase factor is given by

$$C(1,2) = exp\left(i\left\{\frac{e}{2c}\vec{r}_1\cdot(\vec{\mathscr{H}}_m\times\vec{r}_2) - \frac{e}{c}\phi(\vec{r}_1) + \frac{e}{c}\phi(\vec{r}_2)\right\}\right), \tag{15.2.26}$$

where $\vec{r}_1$ and $\vec{r}_2$ are position vectors along the plasma surface, $\vec{\mathscr{H}}_m = \mathscr{H}_m\hat{z}$ is a constant uniform magnetic field perpendicular to the plasma surface, and $\phi(\vec{r}_1)$ is an arbitrary gauge function for the vector potential $\vec{A}_m$ $\left(\vec{A}_m(1) = \frac{1}{2}\vec{\mathscr{H}}_m\times\vec{r}_1 - \vec{\nabla}_1\phi(\vec{r}_1)\right)$. Substituting (15.2.25) and (15.2.26) into (15.2.24), one obtains the matrix elements of $\underset{\sim}{L}_p^{i\tau}(1-2)$ as follows (take $\phi(\vec{r})\equiv 0$);

$$
\begin{aligned}
[L_p^{i\tau}(1-2)]_j^i &= \frac{-e^2}{mc}\delta_j^i\delta(1-2)n^s \qquad (15.2.27)\\
&+ \frac{ie^2}{4m^2c}\lim_{\substack{1'\to 1\\2'\to 2}}\left\{\left[\frac{\vec{\nabla}_{1'}-\vec{\nabla}_1}{i}\right]^i\left[\frac{\vec{\nabla}_{2'}-\vec{\nabla}_2}{i}\right]^j \bar{G}_o^{i\tau}(1'-2)\bar{G}_o^{i\tau}(2'-1)\right\}\\
&+ \frac{ie^2}{4m^2c}\lim_{\substack{1'\to 1\\2'\to 2}}\left\{\left[\frac{e}{c}\vec{\mathscr{H}}_m\times(\vec{r}_2-\vec{r}_1)\right]^i\left[\frac{\vec{\nabla}_{2'}-\vec{\nabla}_2}{i}\right]^j \bar{G}_o^{i\tau}(1'-2)\bar{G}_o^{i\tau}(2'-1)\right\}\\
&+ \frac{ie^2}{4m^2c}\lim_{\substack{1'\to 1\\2'\to 2}}\left\{\left[\frac{e}{c}\vec{\mathscr{H}}_m\times(\vec{r}_1-\vec{r}_2)\right]^j\left[\frac{\vec{\nabla}_{1'}-\vec{\nabla}_1}{i}\right]^i \bar{G}_o^{i\tau}(1'-2)\bar{G}_o^{i\tau}(2'-1)\right\}\\
&+ \frac{ie^2}{4m^2c}\lim_{\substack{1'\to 1\\2'\to 2}}\left\{\left[\frac{e}{c}\vec{\mathscr{H}}_m\times(\vec{r}_2-\vec{r}_1)\right]^i\left[\frac{e}{c}\vec{\mathscr{H}}_m\times(\vec{r}_1-\vec{r}_2)\right]^j\right.\\
&\qquad\left.\times\ \bar{G}_o^{i\tau}(1'-2)\bar{G}_o^{i\tau}(2'-1)\right\}.
\end{aligned}
$$

The thermodynamic equilibrium Green's function $\bar{G}_o^{i\tau}(1-2)$ satisfies the differential equation [21–23]

$$
\left\{i\frac{\partial}{\partial t_1}+\zeta-\mu_o\vec{\mathscr{H}}_m\cdot\vec{\sigma}-\frac{\left(\frac{\vec{\nabla}_1}{i}-\frac{e}{c}\frac{\vec{\mathscr{H}}_m\times(\vec{r}_2-\vec{r}_1)}{2}\right)^2}{2m}\right\}\bar{G}_o^{i\tau}(1-2) = \delta(1-2), \qquad (15.2.28)
$$

which has the solution [21, 23] below written in terms of the symbols $\bar{G}_{o\{\gtrless\}}^{i\tau=\beta}$ corresponding to the time orders $t_1\,\{\gtrless\}\,t_2$ (ie: $\bar{G}_o^{i\tau=\beta}(1-2) = \eta_+(t_1-t_2)\bar{G}_{o>}^{i\tau=\beta}(1-2)+\eta_+(t_2-t_1)\bar{G}_{o<}^{i\tau=\beta}(1-2)$):

$$
\begin{aligned}
\bar{G}_{o\{\gtrless\}}^{i\tau=\beta}(1-2) &= e^{i\zeta(t_1-t_2)}\int\frac{d^2\bar{p}}{(2\pi)^2}e^{i\bar{p}\cdot(\bar{1}-\bar{2})}\int\frac{d\omega}{2\pi}\left\{\begin{matrix}-i(1-f_o(\omega))\\ if_o(\omega)\end{matrix}\right\}e^{-i\omega(t_1-t_2)}\\
&\times\int_{-\infty}^{\infty}dT'e^{i\omega T'}sec\left(\frac{\omega_cT'}{2}\right)e^{-i\mu_o\mathscr{H}_m\cdot\sigma_zT'}\ exp\left[-i\left(\frac{\bar{p}^2}{m\omega_c}tan\left(\frac{\omega_cT'}{2}\right)\right)\right],
\end{aligned}
\qquad (15.2.29)
$$

where $\bar{p}\equiv(p_x,p_y)$, ω_c = cyclotron frequency = $\frac{e\mathscr{H}_m}{mc}$ and $f_o(\omega) = [1+e^{(\omega-\zeta)\beta}]^{-1}$ is the Fermi distribution. For $t_1 > t_2$, the term $-i(1-f_o(\omega))$ in the curly brackets applies for $\bar{G}_{o>}^{i\tau=\beta}$, whereas for $t_1 < t_2$, the term $if_o(\omega)$ applies for $\bar{G}_{o<}^{i\tau=\beta}$.

To obtain the wavenumber, frequency dependent linear response tensor, the preceding Green's function is substituted into (15.2.27) and the resulting expression is Fourier transformed in space and time using the Fourier series in imaginary time.

The imaginary time Fourier series coefficient $L_p^{i\tau}(\frac{2\pi\nu}{\tau})$ is then evaluated using an analytic continuation involving $\bar{G}^{i\tau=\beta}_{o\{\gtrless\}}$ in the complex time plane with deformation of the time integration contour, as explained in [21], Sects. 9.8 and 10.7. Thus obtaining $L_p^{i\tau}(\Omega)$, (15.2.15) provides the result for $L_c^{\beta}(\Omega)$. The mathematical details of this procedure are straightforward though tedious, and we omit them from the present chapter for the sake of brevity (the interested reader is referred to [21, 22] for further details). After performing the spin trace over the $\bar{G}_o\bar{G}_o$ products of (15.2.27) using (15.2.29), restoring factors of $\hbar$ (Planck's constant) and Fourier transforming from position representation to momentum representation, and (without loss of generality) restricting the propagation vector along the plasma surface to the "y" direction so that $k_x \equiv 0$, $k_y \equiv k_\parallel$, the final results are (tensor elements are denoted as L^i_j with $i, j = 1, 2$ for the two directions on the plane):

$$\begin{aligned}
L^i_j(k_\parallel, \Omega) &= \frac{-e^2}{mc} n^s \delta^i_j \qquad (15.2.30)\\
&+ \Bigg\{ \int d\omega \frac{f_o(\omega)}{\hbar^2} \int_{-i\infty+\delta}^{i\infty+\delta} \frac{ds}{2\pi i} \frac{e^{s\omega}}{2\pi} \frac{m\hbar\omega_c}{tanh\left(\frac{\hbar\omega_c s}{2}\right)} \left(\frac{-1}{\hbar}\right)\\
&\times \int_0^{-\infty} dT e^{-i(\Omega+i\epsilon)T} exp\left[\frac{\hbar k_\parallel^2}{m\omega_c}\left(\frac{cos\left[\frac{\omega_c}{2}(2T - is\hbar)\right] - cosh\left(\frac{\hbar\omega_c s}{2}\right)}{2sinh\left(\frac{\hbar\omega_c s}{2}\right)}\right)\right]\\
&\times \mathscr{L}^i_j(y = 2T - is\hbar) + (\hbar \to -\hbar)\Bigg\},
\end{aligned}$$

where (note that $y = 2T - is\hbar$ does *not* represent a spatial coordinate here)

$$\mathscr{L}^1_1 = \frac{ie^2}{mc}\frac{\hbar\omega_c}{2}\left(\frac{cos\left(\frac{\omega_c y}{2}\right)}{sinh\left(\frac{\hbar\omega_c s}{2}\right)}\right) + \frac{ie^2\hbar^2}{m^2c}\left(\frac{cos\left(\frac{\omega_c}{2}y\right) - cosh\left(\frac{\hbar\omega_c s}{2}\right)}{2sinh\left(\frac{\hbar\omega_c s}{2}\right)}\right)^2 k_\parallel^2, \qquad (15.2.31a)$$

$$\mathscr{L}^2_2 = \frac{ie^2}{mc}\frac{\hbar\omega_c}{2}\left(\frac{cos\left(\frac{\omega_c y}{2}\right)}{sinh\left(\frac{\hbar\omega_c s}{2}\right)}\right) - \frac{ie^2\hbar^2}{4m^2c}\left(\frac{sin\left(\frac{\omega_c y}{2}\right)}{sinh\left(\frac{\hbar\omega_c s}{2}\right)}\right)^2 k_\parallel^2, \qquad (15.2.31b)$$

$$\begin{aligned}
\mathscr{L}^1_2 = -\mathscr{L}^2_1 &= -\frac{ie^2}{mc}\frac{\hbar\omega_c}{2}\left(\frac{sin\left(\frac{\omega_c y}{2}\right)}{sinh\left(\frac{\hbar\omega_c s}{2}\right)}\right)\\
&+ \frac{ie^2\hbar^2}{2m^2c}\left(\frac{sin\left(\frac{\omega_c y}{2}\right)}{sinh\left(\frac{\hbar\omega_c s}{2}\right)}\right)\left(\frac{cos\left(\frac{\omega_c y}{2}\right) - cosh\left(\frac{\hbar\omega_c s}{2}\right)}{2sinh\left(\frac{\hbar\omega_c s}{2}\right)}\right) k_\parallel^2, \qquad (15.2.31c)
\end{aligned}$$

(δ^i_j is the Kroenecker delta and the parameters δ and $\in$ are infinitesimal positive quantities here).

It can readily be shown that the L^i_j terms of (15.2.30), (15.2.31a)–(15.2.31c) are conveniently expressed in terms of the quantity $L^0_0 = [\delta\langle c\rho^s\rangle^\beta/\delta\Phi]_{\Phi\to 0}$ (with ρ^s as the charge density operator and Φ as the scalar potential) as follows:

$$L^0_0 = \frac{ie^2c}{\hbar}\int_{-\infty}^{\infty} d\omega \frac{f_0(\omega)}{\hbar^2}\int_{-i\infty+\delta}^{i\infty+\delta}\frac{ds}{2\pi i}\frac{e^{s\omega}}{2\pi}\frac{m\hbar\omega_c}{\tanh(\hbar\omega_c s/2)}$$
$$\times\int_0^{-\infty} dT e^{-i(\Omega+i\epsilon)T} exp\left[\frac{\hbar k_\parallel^2}{m\omega_c}\left(\frac{\cos\left[\frac{\hbar\omega_c}{2}(2T-is\hbar)\right]-\cosh\left[\frac{\hbar\omega_c s}{2}\right]}{2\sinh\left(\frac{\hbar\omega_c s}{2}\right)}\right)\right]+(\hbar\to-\hbar); \tag{15.2.32}$$

so that

$$L^1_1 = -\frac{e^2n^s}{mc}-\frac{\omega_c^2}{c^2}\left(\frac{\partial}{\partial(k_\parallel^2)}+k_\parallel^2\frac{\partial^2}{\partial^2(k_\parallel^2)}\right)L^0_0$$
$$-\frac{\hbar\omega_c}{2mc^2}L^0_0\left\{\frac{1}{\left(\tanh\left(\frac{\hbar\omega_c s}{2}\right)\right)}\to\frac{1}{\left(\tanh\left(\frac{\hbar\omega_c s}{2}\right)\right)^2}\right\}, \tag{15.2.33a}$$

$$L^2_2 = -\frac{\Omega^2}{c^2k_\parallel^2}L^0_0, \tag{15.2.33b}$$

$$L^1_2 = -L^2_1 = -\frac{i\Omega\omega_c}{c^2}\frac{\partial L^0_0}{\partial(k_\parallel^2)}, \tag{15.2.33c}$$

where

$$n^s = \int_{-\infty}^{\infty} d\omega\frac{f_0(\omega)}{\hbar^2}\int_{-i\infty+\delta}^{i\infty+\delta}\frac{ds}{2\pi i}\frac{e^{s\omega}}{2\pi}\frac{m\hbar\omega_c}{\tanh\left(\frac{\hbar\omega_c s}{2}\right)}. \tag{15.2.34}$$

In the expression for L^1_1 above, the last term (in curly brackets) indicates that in the "s" contour integrand for L^0_0 (as per (15.2.32)), we replace the denominator factor $(\tanh\left(\frac{\hbar\omega_c s}{2}\right))$ by $(\tanh\left(\frac{\hbar\omega_c s}{2}\right))^2$. These results can be obtained by direct inspection of (15.2.30)–(15.2.32) and the use of the 2D gauge invariance requirement $L^\mu_\nu k^\nu_\parallel = 0$, where (μ,ν) have values $(0,1,2)$, $k^\nu_\parallel = (\frac{\Omega}{c},0,k_\parallel)$, and repeated indices imply summation over these indices (Einstein convention).

The preceding closed form integral representations for the frequency-wavevector dependent magnetoconductivity tensor elements $\sigma^i_j = \frac{c}{i\Omega}L^i_j$ are valid for all field strengths and temperatures and statistical regimes, viz degenerate, non-degenerate. They are particularly useful for obtaining evaluations at low and intermediate magnetic field strengths, and for generating low wavenumber power expansions.

A fuller appreciation of the effects of Landau quantization at arbitrarily strong magnetic field strengths (particularly in the degenerate limit) may be obtained by expanding the wavenumber dependent exponentials in the integrals of L^i_j in a modified Bessel function series and then employing the Hille-Hardy identity. The T-, s-, and ω- integrals in (15.2.32)–(15.2.33c) can then be readily evaluated. Specifically, we use the identity [24]

$$e^{[y\cos\phi]} = \sum_{n=-\infty}^{\infty} e^{in\phi} I_n(y); \quad I_n = I_{-n} \text{ are the modified Bessel functions.} \tag{15.2.35}$$

Setting, $y \equiv \frac{\hbar k_\parallel^2}{m\omega_c} \cdot \frac{1}{2\sinh\left(\frac{\hbar\omega_c s}{2}\right)}$, and $\phi \equiv \omega_c \left(T - \frac{is\hbar}{2}\right)$, we have

$$\exp\left[\frac{\hbar k_\parallel^2}{2m\omega_c} \cdot \left(\frac{\cos\left[\omega_c\left(T - \frac{is\hbar}{2}\right)\right] - \cosh\left(\frac{\hbar\omega_c s}{2}\right)}{\sinh\left(\frac{\hbar\omega_c s}{2}\right)}\right)\right] \tag{15.2.36}$$

$$= e^{-\frac{\hbar k_\parallel^2}{2m\omega_c}\coth\left(\frac{\hbar\omega_c s}{2}\right)} \sum_{n=-\infty}^{\infty} e^{in\omega_c T} e^{\left(\frac{n\hbar\omega_c s}{2}\right)} I_n\left(\frac{\hbar k_\parallel^2}{2m\omega_c\sinh\left(\frac{\hbar\omega_c s}{2}\right)}\right).$$

The resulting T-integrals in L_j^i (15.2.32)–(15.2.33c) can be evaluated immediately:

$$\int_0^{-\infty} dT e^{-i(\Omega+i\epsilon-n\omega_c)T} = \frac{-i}{\Omega_+ - n\omega_c}, \quad \Omega_+ \equiv \Omega + i\epsilon,$$

and then (15.2.32) yields L_0^0 as ($\Omega_+ \equiv \Omega + i\varepsilon$):

$$L_0^0 = e^2 c \int_{-\infty}^{\infty} \frac{d\omega f_0(\omega)}{\hbar^2} \int_{-i\infty+\delta}^{i\infty+\delta} \frac{ds}{2\pi i}\frac{e^{s\omega}}{2\pi}\frac{m\hbar\omega_c}{\tanh\left(\frac{\hbar\omega_c s}{2}\right)} \sum_{n=1}^{\infty} \frac{2n\omega_c}{\hbar(\Omega_+^2 - n^2\omega_c^2)}$$
$$\times\ 2\sinh(n\hbar\omega_c s/2) e^{-\left[\frac{\hbar k_\parallel^2}{2m\omega_c}\coth\left(\frac{\hbar\omega_c s}{2}\right)\right]} I_n\left(\frac{\hbar k_\parallel^2}{2m\omega_c}\frac{1}{\sinh\left(\frac{\hbar\omega_c s}{2}\right)}\right). \tag{15.2.37}$$

Furthermore, using (15.2.33a)–(15.2.33c), we have

$$(a)\ \ L_1^1 = \frac{-e^2}{mc} n^s + e^2 c \int_{-\infty}^{\infty} d\omega \frac{f_0(\omega)}{\hbar^2} \int_{-i\infty+\delta}^{i\infty+\delta} \frac{ds}{2\pi i}\frac{e^{s\omega}}{2\pi}\frac{m\hbar\omega_c}{\tanh\left(\frac{\hbar\omega_c s}{2}\right)} \sum_{n=1}^{\infty} 2\sinh\left(\frac{n\hbar\omega_c s}{2}\right)$$
$$\times \frac{2n\omega_c}{\hbar\left(\Omega_+^2 - n^2\omega_c^2\right)} e^{-\left[\frac{\hbar k_\parallel^2}{2m\omega_c}\coth\left(\frac{\hbar\omega_c s}{2}\right)\right]} \left\{\ldots\right\} I_n\left(\frac{\hbar k_\parallel^2}{2m\omega_c\sinh\left(\frac{\hbar\omega_c s}{2}\right)}\right), \tag{15.2.38a}$$

where $\left\{\ldots\right\} \equiv \frac{-\hbar^2 k_\parallel^2}{4m^2c^2}\coth^2\left(\frac{\hbar\omega_c s}{2}\right) + \left(\frac{\hbar\omega_c k_\parallel^2}{mc^2}\coth\left(\frac{\hbar\omega_c s}{2}\right) - \frac{\omega_c^2}{c^2}\right)\frac{\partial}{\partial k_\parallel^2} - \frac{\omega_c^2}{c^2}k_\parallel^2\frac{\partial^2}{\partial^2(k_\parallel^2)}$; and

$$(b)\quad L_2^2 = \frac{-e^2\Omega^2}{ck_\parallel^2}\int_{-\infty}^{\infty} d\omega\frac{f_0(\omega)}{\hbar^2}\int_{-i\infty+\delta}^{i\infty+\delta}\frac{ds}{2\pi i}\frac{e^{s\omega}}{2\pi}\frac{m\hbar\omega_c}{\tanh\left(\frac{\hbar\omega_c s}{2}\right)}\sum_{n=1}^{\infty}\frac{2n\omega_c}{\hbar\left(\Omega_+^2 - n^2\omega_c^2\right)}$$
$$\times\, 2\sinh\left(\frac{n\hbar\omega_c s}{2}\right)e^{-\left[\left(\frac{\hbar k_\parallel^2}{2m\omega_c}\right)\coth\frac{\hbar\omega_c s}{2}\right]}I_n\left(\frac{\hbar k_\parallel^2}{2m\omega_c}\frac{1}{\sinh\left(\frac{\hbar\omega_c s}{2}\right)}\right); \tag{15.2.38b}$$

and

$$(c)\quad L_2^1 = -L_1^2$$
$$= -\frac{ie^2\omega_c\Omega}{c}\int_{-\infty}^{\infty} d\omega\frac{f_0(\omega)}{\hbar^2}\int_{-i\infty+\delta}^{i\infty+\delta}\frac{ds}{2\pi i}\frac{e^{s\omega}}{2\pi}\frac{m\hbar\omega_c}{\tanh\left(\frac{\hbar\omega_c s}{2}\right)}\sum_{n=1}^{\infty}\frac{2n\omega_c}{\hbar\left(\Omega_+^2 - n^2\omega_c^2\right)}$$
$$\times\, 2\sinh\left(\frac{n\hbar\omega_c s}{2}\right)\exp\left[-\frac{\hbar k_\parallel^2}{2m\omega_c}\coth\left(\frac{\hbar\omega_c s}{2}\right)\right]$$
$$\left(\frac{-\hbar}{2m\omega_c}\coth\left(\frac{\hbar\omega_c s}{2}\right) + \frac{\partial}{\partial k_\parallel^2}\right)$$
$$\times\, I_n\left(\frac{\hbar k_\parallel^2}{2m\omega_c}\frac{1}{\sinh\left(\frac{\hbar\omega_c s}{2}\right)}\right). \tag{15.2.38c}$$

It is useful to recognize that two important statistical regimes are readily identified in the above results: First, the nondegenerate regime of relatively low density and high temperature is obtained as the leading term in the fugacity expansion of $f_0(\omega)$;

$$f_0(\omega) = \frac{1}{1 + e^{\beta(\omega-\zeta)}} \to e^{\zeta\beta}e^{-\omega\beta}, \tag{15.2.39}$$

whereupon the $\omega-$ and $s-$ integrals are just Laplace transform and inverse so the evaluation immediately yields the common integrand with the replacement $s \to \beta$. Thus, (15.2.37) yields L_0^0 in the nondegenerate regime as

$$L_0^0 = \frac{e^2 c}{2\pi\hbar^2}e^{\zeta\beta}\frac{m\hbar\omega_c}{\tanh(\hbar\omega_c\beta/2)}\sum_{n=1}^{\infty}\frac{2n\omega_c}{\hbar(\Omega_+^2 - n^2\omega_c^2)}2\sinh(n\hbar\omega_c\beta/2)$$
$$\times\exp\left[-\frac{\hbar k_\parallel^2}{2m\omega_c}\coth(\hbar\omega_c\beta/2)\right]I_n\left(\frac{\hbar k_\parallel^2}{2m\omega_c}\frac{1}{\sinh(\hbar\omega_c\beta/2)}\right), \tag{15.2.40}$$

and L_1^1, L_2^2, $L_2^1 = -L_1^2$ can then be obtained using (15.2.38a)–(15.2.38c) with the nondegenerate density expression

$$n^s = \frac{e^{\zeta\beta}}{2\pi\hbar^2} \frac{m\hbar\omega_c}{\tanh(\hbar\omega_c\beta/2)}, \tag{15.2.41}$$

which may be used to eliminate $e^{\zeta\beta}$ in L_0^0 and L_j^i in favor of the density n^s in the nondegenerate regime.

The other important regime to address is the degenerate regime of zero temperature, in which $f_0(\omega) = \eta_+(\zeta - \omega)$, so integrals of the form

$$\begin{aligned} J &= \int_{-\infty}^{\infty} d\omega\, f_0(\omega) \int_{-i\infty+\delta}^{i\infty+\delta} \frac{ds}{2\pi i} e^{s\omega} j(s) \\ &= \int_{-\infty}^{\infty} d\omega \eta_+(\zeta - \omega) \int_{-i\infty+\delta}^{i\infty+\delta} \frac{ds}{2\pi i} \frac{de^{s\omega}}{d\omega} \frac{j(s)}{s} \end{aligned} \tag{15.2.42}$$

can be evaluated by a parts integration as

$$J = -\int_{-\infty}^{\infty} d\omega \frac{d\eta_+(\zeta - \omega)}{d\omega} \int_{-i\infty+\delta}^{i\infty+\delta} \frac{ds}{2\pi i} e^{s\omega} \frac{j(s)}{s} = \int_{-i\infty+\delta}^{i\infty+\delta} \frac{ds}{2\pi i} e^{s\zeta} \frac{j(s)}{s} \tag{15.2.43}$$

since $d\eta_+(\zeta - \omega)/d\omega = -\delta(\omega - \zeta)$. This enables us to identify the principal physical feature of the degenerate regime, de Haas-van Alphen oscillations, which arise from the discrete isolated singularities of the s-integrand of L_0^0 and J at $s_n = \pm i2\pi n/\hbar\omega_c$ in the structure of $j(s)$, yielding characteristic oscillatory terms of the form $e^{s_n\zeta} = e^{\pm\frac{i2\pi n\zeta}{\hbar\omega_c}}$, as a result of Landau quantization of the spectrum. With this in view, L_0^0 can be written in the degenerate regime as

$$\begin{aligned} L_0^0 = \frac{e^2 c}{\hbar^2} \int_{-i\infty+\delta}^{i\infty+\delta} \frac{ds}{2\pi i} \frac{e^{s\zeta}}{2\pi s} \frac{m\hbar\omega_c}{\tanh\left(\frac{\hbar\omega_c s}{2}\right)} \sum_{n=1}^{\infty} \frac{2n\omega_c}{\hbar(\Omega_+^2 - n^2\omega_c^2)} \cdot 2\sinh\left(n\hbar\omega_c s/2\right) \\ \times e^{-\left[\frac{\hbar k_\parallel^2}{2m\omega_c} \coth\frac{\hbar\omega_c s}{2}\right]} I_n\left(\frac{\hbar k_\parallel^2}{2m\omega_c} \frac{1}{\sinh\left(\hbar\omega_c s/2\right)}\right). \end{aligned} \tag{15.2.44}$$

Inspection of the poles of the s-integrand of (15.2.44) shows that they produce de Haas-van Alphen oscillatory terms (except for the pole at $s = 0$) of the form ($r =$ integers $\neq 0$) $\cos(2\pi r\zeta/\hbar\omega_c)$, which have their maximum amplitude at $T = 0$. If they are neglected (Appendix 15.6), the remaining pole at $s = 0$ yields a "semiclassical" result as

$$L_0^0 = \frac{2e^2mc}{\pi\hbar^2}\sum_{n=1}^{\infty}\frac{n^2\omega_c^2}{\Omega_+^2 - n^2\omega_c^2}\int_{-i\infty+\delta}^{i\infty+\delta}\frac{ds}{2\pi i}e^{s\zeta}\frac{e^{-\frac{k_\parallel^2}{m\omega_c^2 s}}}{s}I_n\left(\frac{k_\parallel^2}{m\omega_c^2 s}\right),$$

where the s-integral is evaluated as ($\alpha \equiv k_\parallel^2/m\omega_c^2$)

$$\int_{-i\infty+\delta}^{i\infty+\delta}\frac{ds}{2\pi i}\frac{e^{s\zeta}}{s}e^{-\alpha/s}I_n\left(\frac{\alpha}{s}\right) = \left[J_0(\sqrt{2\alpha\zeta})\right]^2 = \left[J_0\left(\frac{k_\parallel v_F}{\omega_c}\right)\right]^2,$$

with $v_F \equiv \sqrt{2\zeta/m}$ (see "Tables of Integral Transforms", Erderlyi, et al., McGraw Hill, pg. 280 # 5.16.15 (1954)). Again, L_1^1, L_2^2 and $L_2^1 = -L_1^2$ can be obtained from L_0^0 using (15.2.38a)–(15.2.38c). Further discussion of this semiclassical model, which is devoid of Landau quantization effects that are of principal interest in this chapter, will be taken up in Appendix 15.6.

The degenerate expression for density at zero temperature is given by

$$n^s = \int_{-i\infty+\delta}^{i\infty+\delta}\frac{ds}{2\pi i}\frac{e^{s\zeta}}{2\pi\hbar^2}\frac{m\hbar\omega_c}{s\tanh(\hbar\omega_c s/2)}, \qquad (15.2.45)$$

which exhibits de Haas-van Alphen oscillations in the relation between the density and chemical potential (Fermi energy). More discussion of such integrals can be found in [23], including evaluation at intermediate magnetic field strengths, as well as low and high fields. In the semiclassical model, with its neglect of dHvA oscillations, the density is given by

$$n^s = \int_{-i\infty+\delta}^{i\infty+\delta}\frac{ds}{2\pi i}\frac{e^{s\zeta}}{2\pi\hbar^2}\frac{2m}{s^2} = \frac{m\zeta}{\pi\hbar^2}.$$

To develop yet another representation of L_j^i that spans all statistical regimes while focusing on the discrete states of the spectrum for arbitrary field strength and arbitrary wavenumber, we express the linear response tensor in terms of generalized Laguerre polynomials. This can be accomplished by applying the Hille-Hardy transformation to the modified Bessel function expansions of (15.2.37)–(15.2.38c). The resulting expressions for L_j^i may then be integrated exactly (in terms of Dirac delta functions) to provide a series solution whose contributing terms correspond to occupied Landau levels with the Fermi distribution function maintained intact, without approximation. Since few such levels are populated in the high field regime ($\hbar\omega_c \sim \zeta$), the generalized Laguerre polynomial series for L_j^i converges rapidly.

The Hille-Hardy transformation is given by [25]

$$e^{-\left(\frac{\hbar k_\parallel^2}{2m\omega_c}\coth\frac{\hbar\omega_c s}{2}\right)} I_n\left(\frac{\hbar k_\parallel^2}{2m\omega_c}\cdot\frac{1}{\sinh\left(\frac{\hbar\omega_c s}{2}\right)}\right) = \sum_{r=0}^{\infty}\frac{2r!}{(n+r)!}\sinh\left(\frac{\hbar\omega_c s}{2}\right)e^{-\left(\frac{\hbar\omega_c s}{2}(n+2r+1)\right)}$$

$$\times e^{-\left(\frac{\hbar k_\parallel^2}{2m\omega_c}\right)}\left(\frac{\hbar k_\parallel^2}{2m\omega_c}\right)^n\left[L_r^n\left(\frac{\hbar k_\parallel^2}{2m\omega_c}\right)\right]^2, \tag{15.2.46}$$

with L_r^n as the Laguerre polynomials. The "s" and "ω" integrals in (15.2.37), (15.2.38a)–(15.2.38c) may now be performed, yielding [22]

$$L_0^0 = e^2 c\left(\frac{m\omega_c}{2\pi\hbar^2}\right)\sum_{n=1}^{\infty}\sum_{r=0}^{\infty}\sum_{\pm}\sum_{\pm'}(\pm')\frac{r!}{(n+r)!}\frac{2n\omega_c}{(\Omega_+^2 - n^2\omega_c^2)}$$
$$\times e^{-\left(\frac{\hbar k_\parallel^2}{2m\omega_c}\right)}\left(\frac{\hbar k_\parallel^2}{2m\omega_c}\right)^n\left[L_r^n\left(\frac{\hbar k_\parallel^2}{2m\omega_c}\right)\right]^2 f_o(\omega = a_{rn}(\pm,\pm')), \tag{15.2.47}$$

and

$$L_1^1 = \frac{-e^2}{mc}n^s - \frac{\omega_c^2}{c^2}\left(\frac{\partial}{\partial k_\parallel^2} + \frac{k_\parallel^2\partial^2}{\partial^2(k_\parallel^2)}\right)L_0^0 - \frac{e^2\omega_c^2}{2\pi c\hbar}\left\{\sum_{n=1}^{\infty}\sum_{r=0}^{\infty}\sum_{p=0}^{\infty}{}'\right.$$
$$\times\sum_{\pm}\sum_{\pm'}(\pm')\frac{r!}{(n+r)!}\frac{2n\omega_c}{(\Omega_+^2 - n^2\omega_c^2)}f_o(\omega = p\hbar\omega_c + a_{rn}(\pm,\pm'))$$
$$\left.\times e^{-\left(\frac{\hbar k_\parallel^2}{2m\omega_c}\right)}\left(\frac{\hbar k_\parallel^2}{2m\omega_c}\right)^n\left[L_r^n\left(\frac{\hbar k_\parallel^2}{2m\omega_c}\right)\right]^2\right\}, \tag{15.2.48a}$$

$$L_2^2 = -\left(\frac{\Omega^2}{c^2 k_\parallel^2}\right)L_0^0, \tag{15.2.48b}$$

$$L_2^1 = -L_1^2 = -\left(\frac{i\omega_c\Omega}{c^2}\right)\frac{\partial L_0^0}{\partial k_\parallel^2}, \tag{15.2.48c}$$

where

$$n^s = \frac{m\omega_c}{2\pi\hbar}\sum_{\pm}\sum_{n=0}^{\infty} f_o\left(\left[n + \frac{1}{2} \pm \frac{1}{2}\right]\hbar\omega_c\right), \tag{15.2.49}$$

and

$$a_{rn}(\pm,\pm') = [(n + 2r + 1) - (\pm 1) - (\pm' n)]\frac{\hbar\omega_c}{2}. \tag{15.2.50}$$

(The prime superscript on the p-sum in (15.2.48a) indicates that the $p = 0$ term is to be divided by 2. It should be noted that we are taking spin splitting to be the same as Landau level separation).

The summation over the index "n" in (15.2.47) represents contributions to the 2D plasma response from various levels of cyclotron excitation at the frequencies $\Omega = \omega_c, 2\omega_c, 3\omega_c, etc.$ The summation over the index variable "r" corresponds to different Landau level contributions. The $(\pm)$ sum over $a_{rn}(\pm\pm')$ arises from the two electron spin states; the $(\pm')$ sum is a consequence of folding the cyclotron frequency summation $\sum_{n=-\infty}^{\infty} \to \sum_{n=1}^{\infty} \sum_{\pm'}$ when using (15.2.36) to obtain (15.2.47)–(15.2.48c).

It is of interest to note that the degenerate de Haas-van Alphen oscillations discussed above as complex exponentials, $e^{\pm i2\pi n\zeta/\hbar\omega_c}$, are manifested in a different form in (15.2.47)–(15.2.48c). Here, they arise from the sharply cutoff degenerate Fermi distributions at zero temperature,

$$f_0(\omega = a_{rn}(\pm, \pm')) = -\eta_+(\zeta - a_{rn}(\pm, \pm')), \tag{15.2.51}$$

which causes sharp changes of $f_0(\omega)$ as variation of the magnetic field in $a_{rn}(\pm, \pm')$ induces successive vanishings of the argument of $\eta_+(\zeta - a_{rn}(\pm, \pm'))$ for successive integers that label the discrete Landau quantized energy levels. Notwithstanding the difference of mathematical forms, these two representations of de Haas-van Alphen oscillations are equivalent.

In the degenerate limit of very strong magnetic fields, the 2D plasma electrons are restricted essentially to the ground state Landau level with the spins aligned anti-parallel to $\vec{\mathscr{H}}_m$. The dominant contribution to the 2D linear response tensor then comes from the lowest order Laguerre polynomial, $L_0^n(x) \equiv 1$. This can be shown by noting that for high magnetic fields such that $\hbar\omega_c \tilde{>} \zeta >> \beta^{-1}$, the definition of $a_{rn}(\pm\pm')$ in the degenerate Fermi distribution of (15.2.51) yields $(\zeta\beta \to \infty)$

$$\begin{aligned} f_o(\omega = a_{rn}(\pm, \pm')) &\cong \delta_{r,o}\delta(\pm, +)\delta(\pm', +') \\ f_o(\omega = p\hbar\omega_c + a_{rn}(\pm, \pm')) &\cong \delta_{r,o}\delta_{p,o}\delta(\pm, +)\delta(\pm', +'). \end{aligned} \tag{15.2.52}$$

Hence, in this degenerate quantum strong field limit, the Fermi distribution cuts off all the Laguerre series contributions beyond the $r = 0$ term, corresponding to the ground state Landau contribution. Under these conditions, (15.2.47)–(15.2.49) reduce to:

$$L_0^0 = e^2 cn^s e^{\left(\frac{-\hbar k_\parallel^2}{2m\omega_c}\right)} \sum_{n=1}^{\infty} \frac{1}{n!} \left(\frac{\hbar k_\parallel^2}{2m\omega_c}\right)^n \frac{2n\hbar\omega_c}{\hbar^2\Omega^2 - n^2\hbar^2\omega_c^2}, \tag{15.2.53}$$

$$L_1^1 = \frac{-e^2}{mc} n^s - \frac{e^2}{mc} n^s \left(\frac{\hbar\omega_c}{2}\right) e^{-\left(\frac{\hbar k_\|^2}{2m\omega_c}\right)} \left\{ \sum_{n=1}^{\infty} \frac{1}{n!} \left(\frac{\hbar k_\|^2}{2m\omega_c}\right)^{n+1} \right.$$

$$\left. \times \frac{2n\hbar\omega_c}{\hbar^2\Omega^2 - n^2\hbar^2\omega_c^2} \left(\frac{n}{\left(\frac{\hbar k_\|^2}{2m\omega_c}\right)} - 1 \right)^2 \right\}, \qquad (15.2.54a)$$

$$L_2^2 = -\left(\frac{\Omega^2}{c^2 k_\|^2}\right) L_0^0$$

$$= -e^2 c n^s \left(\frac{\Omega^2}{c^2 k_\|^2}\right) e^{-\left(\frac{\hbar k_\|^2}{2m\omega_c}\right)} \sum_{n=1}^{\infty} \frac{1}{n!} \left(\frac{\hbar k_\|^2}{2m\omega_c}\right)^n \cdot \frac{2n\hbar\omega_c}{\hbar^2\Omega_+^2 - \hbar^2 n^2\omega_c^2}, \qquad (15.2.54b)$$

$$L_2^1 = -L_1^2 = -\left(\frac{i\omega_c\Omega}{c^2}\right) \frac{\partial L_0^0}{\partial k_\|^2} = -e^2 c n^s \left(\frac{i\omega_c\Omega}{c^2}\right) e^{-\left(\frac{\hbar k_\|^2}{2m\omega_c}\right)} \sum_{n=1}^{\infty} \frac{1}{n!} \left(\frac{\hbar k_\|^2}{2m\omega_c}\right)^n \cdot \frac{\hbar}{2m\omega_c}$$

$$\times \frac{2n\hbar\omega_c}{\hbar^2\Omega_+^2 - \hbar^2 n^2\omega_c^2} \left(\frac{n}{\left(\frac{\hbar k_\|^2}{2m\omega_c}\right)} - 1 \right), \qquad (15.2.54c)$$

where

$$n^s = \frac{m\omega_c}{2\pi\hbar} \text{ (quantum strong field limit).} \qquad (15.2.55)$$

Another useful expansion for the 2D linear response tensor may be obtained in the low wave number regime. Referring to the general solution for L_j^i in (15.2.30), we note that at low wavenumbers, the wave number dependent exponential can be aproximated by the leading terms in an expansion in powers of $k_\|$ as follows:

$$\exp\left[\frac{\hbar k_\|^2}{m\omega_c} \left(\frac{cos\left[\omega_c(T - is\frac{\hbar}{2}]\right) - cosh\left(\hbar\omega_c\frac{s}{2}\right)}{2sinh\left(\hbar\omega_c\frac{s}{2}\right)} \right) \right]$$

$$= 1 + \frac{\hbar k_\|^2}{2m\omega_c} \left(\frac{cos\left[\omega_c(T - is\frac{\hbar}{2})\right] - cosh\left(\frac{\hbar\omega_c s}{2}\right)}{sinh\left(\hbar\omega_c\frac{s}{2}\right)} \right)$$

$$+ \frac{1}{2}\left(\frac{\hbar k_\|^2}{2m\omega_c}\right)^2 \left(\frac{\cos\left[\omega_c(T - is\frac{\hbar}{2})\right] - \cosh\left(\frac{\hbar\omega_c s}{2}\right)}{\sinh\left(\hbar\omega_c s/2\right)} \right)^2 + - - -. \qquad (15.2.56)$$

Employing this series in (15.2.30)–(15.2.31c) and integrating over T, one readily obtains

$$L_1^1 = \frac{-e^2 n^s}{mc} \frac{\Omega^2}{\Omega^2 - \omega_c^2} - \frac{e^2}{mc} \cdot \frac{\sigma^s}{mc^2} \frac{c^2 k_\parallel^2}{\omega_c^2} \left(\frac{4\omega_c^2}{\Omega^2 - (2\omega_c)^2} - \frac{3\omega_c^2}{\Omega^2 - \omega_c^2} \right) + \cdots \tag{15.2.57a}$$

$$L_2^2 = \frac{-e^2 n^s}{mc} \frac{\Omega^2}{\Omega^2 - \omega_c^2} - \frac{e^2}{mc} \cdot \frac{\sigma^s}{mc^2} \frac{c^2 k_\parallel^2}{\omega_c^2} \left(\frac{4\omega_c^2}{\Omega^2 - (2\omega_c)^2} - \frac{\omega_c^2}{\Omega^2 - \omega_c^2} \right) + \cdots \tag{15.2.57b}$$

$$\begin{aligned} L_2^1 &= -L_1^2 \\ &= -\frac{e^2}{mc} n^s \left(\frac{i\omega_c}{\Omega} \right) \frac{\Omega^2}{\Omega^2 - \omega_c^2} + \frac{2e^2}{mc} \frac{\sigma^s}{mc^2} \frac{c^2 k_\parallel^2}{\omega_c^2} \left(\frac{i\omega_c}{\Omega} \right) \\ &\quad \times \left(\frac{4\omega_c^2}{\Omega^2 - (2\omega_c)^2} - \frac{\omega_c^2}{\Omega^2 - \omega_c^2} \right) + \cdots\cdots, \end{aligned} \tag{15.2.57c}$$

where n^s, the surface charge density, is given by

$$n^s = \int_{-\infty}^{\infty} \frac{d\omega f_o(\omega)}{\hbar^2} \int_{-i\infty+\delta}^{i\infty+\delta} \frac{ds}{2\pi i} \frac{e^{s\omega}}{2\pi} \frac{m\hbar\omega_c}{tanh\left(\frac{\hbar\omega_c s}{2}\right)} = \int_{-\infty}^{\infty} d\omega f_o(\omega) D(\omega), \tag{15.2.58}$$

and σ^s, the surface energy density, is given by

$$\sigma^s = \int_{-\infty}^{\infty} \frac{d\omega f_o(\omega)}{\hbar^2} \int_{-i\infty+\delta}^{i\infty+\delta} \frac{ds}{2\pi i} \frac{e^{s\omega}}{2\pi} \frac{m\hbar\omega_c}{tanh\left(\frac{\hbar\omega_c s}{2}\right)} \frac{\hbar\frac{\omega_c}{2}}{tanh\left(\frac{\hbar\omega_c s}{2}\right)}. \tag{15.2.59}$$

Here, $D(\omega) = \frac{1}{\hbar^2} \int_{-i\infty+\delta}^{i\infty+\delta} \frac{ds}{2\pi i} \frac{e^{s\omega}}{2\pi} \frac{m\hbar\omega_c}{tanh(\frac{\hbar\omega_c s}{2})}$ is the density of states (per unit energy, per unit area) for a 2D Landau quantized plasma. This quantity may be readily evaluated by contour integration to yield [22]

$$D(\omega) = \frac{m}{\pi\hbar} \omega_c \eta_+(\omega) \sum_{n=-\infty}^{\infty} \delta(\omega - n\hbar\omega_c), \tag{15.2.60}$$

where $\eta_+(\omega) = \left\{ \begin{smallmatrix} 1 & \omega>0 \\ 0 & \omega<0 \end{smallmatrix} \right\}$ and $\eta_+(0) = \frac{1}{2}$.

Hence, from (15.2.58),

$$n^s = \frac{m\omega_c}{\pi\hbar} \left(\frac{f_o(0)}{2} + \sum_{n=1}^{\infty} f_o(n\hbar\omega_c) \right). \tag{15.2.61}$$

Similarly, contour integration of (15.2.59) yields [22]

$$\sigma^s = \frac{m\omega_c}{\pi\hbar}\sum_{n=1}^{\infty} n\hbar\omega_c f_o(n\hbar\omega_c), \tag{15.2.62}$$

where $f_o(\omega) = [1 + e^{(\omega-\zeta)\beta}]^{-1}$. These expressions show that the plasma surface charge density is the sum of thermally weighted contributions from each of the Landau eigenstates. (The presence of a sum rather than an integral is a consequence of the discreteness of these 2D Landau quantized states). It is clear that σ^s represents an average surface energy density, since by (15.2.62), it is the sum of thermally weighted energy contributions from each of the Landau eigenstates. The zero energy contribution from the ground state (and also the $\frac{1}{2}$ weighting factor in the ground state contribution to n^s in (15.2.61)) is due to our implicit assumption that the electrons are spin $\frac{1}{2}$ particles with a "g" factor equal to 2. It may be noted that the local limit ($k_\parallel \to 0$) of the 2D plasma conductivity is purely classical as a function of electron density n^s alone (but n^s itself is oscillatory in the de Haas-van Alphen sense as a function of $\zeta/\hbar\omega_c$). By contrast, the nonlocal contributions to plasma conductivity incorporate Landau quantization effects, e.g. through σ^s in (15.2.57), with σ^s specified by (15.2.62).

15.3 2D Magnetopolaritons in a Quantizing Magnetic Field

We now discuss the dispersion relation for evanescent electromagnetic normal modes propagating on the surface of a Landau-quantized 2D plasma. Our considerations will be restricted to low wave numbers $\left(\frac{\zeta k_\parallel^2}{m\omega_c^2} \text{ or } \frac{\hbar k_\parallel^2}{m\omega_c} \text{ or } \frac{k_\parallel^2}{m\omega_c^2\beta} \ll 1\right)$ where the plasma spectrum consists of local modes and a nonlocal "Bernstein"-type mode in the vicinity of $\Omega \approx 2\omega_c$. The dispersion relation for the local modes will be formulated exactly, with simple approximate solutions obtained in the cases where ω_c is very much smaller or very much larger than the other frequency parameters, and also in the electrostatic limit $ck_\parallel \to \infty$. In each of these three cases, we find that there is only one local solution to the plasma dispersion equation consistent with an evanescent wave of real frequency. The leading nonlocal quantum corrections to the local mode dispersion relations for these various cases is provided. The dispersion relation for the nonlocal "Bernstein"-type mode is also explored, with special consideration given to the electrostatic limit, low magnetic-field strength, and high surface-charge density.

(a) Local Limit

Summarizing the formulation of Sect. 15.1 for the electric field associated with an evanescent 2D plasma wave, we write its equation as (recall notation: $k_\perp = k_z$, $k_\parallel = k_y$, $k_x \equiv 0$; $k^2 = k_\perp^2 + k_\parallel^2 = \varepsilon_m\Omega^2/c^2$; $|k_\perp| = -ik_\perp = -ik_z$ and $e^{ik_\perp z} = e^{-|k_z|z}$ for evanescent waves; $\epsilon_m \equiv \epsilon_b^{(3D)}$ of Sect. 15.1 notation)

$$\left\{\mathbb{1}+\frac{2\pi i|k_\perp|}{\epsilon_m\Omega}\begin{pmatrix}\frac{-k^2}{|k_\perp|^2} & 0\\ 0 & 1\end{pmatrix}\begin{pmatrix}\sigma_1^1 & \sigma_2^1\\ \sigma_1^2 & \sigma_2^2\end{pmatrix}\right\}\begin{pmatrix}E_s^1\\ E_s^2\end{pmatrix}=0. \tag{15.3.1}$$

The plasma normal modes are thus determined by the condition (also see (15.1.18))

$$\det\left|\mathbb{1}+\frac{2\pi i|k_\perp|}{\epsilon_m\Omega}\begin{pmatrix}\frac{-k^2}{|k_\perp|^2} & 0\\ 0 & 1\end{pmatrix}\begin{pmatrix}\sigma_1^1 & \sigma_2^1\\ \sigma_1^2 & \sigma_2^2\end{pmatrix}\right|=0, \tag{15.3.2}$$

and the polarization of the modes is given by

$$\left(\frac{E_s^1}{E_s^2}\right)=\frac{1+2\pi i|k_\perp|\sigma_2^2/\epsilon_m\Omega}{2\pi i|k_\perp|\sigma_2^1/\epsilon_m\Omega}. \tag{15.3.3}$$

The values of Ω and $|k_\perp|=(k_\parallel^2-\epsilon_m\Omega^2/c^2)^{\frac{1}{2}}$ utilized in (15.3.3) are those determined by the dispersion relation (15.3.2).

Within the framework of a low wave-number approximation, the magnetoconductivity tensor may be expressed as (15.2.3), (15.2.57):

$$\sigma_1^1=\frac{ie^2\Omega}{m}\left[\frac{n^s}{\Omega_+^2-\omega_c^2}+\frac{\sigma^s k_\parallel^2}{m\Omega^2}\left(\frac{4}{\Omega_+^2-(2\omega_c)^2}-\frac{3}{\Omega_+^2-\omega_c^2}\right)+\cdots\right], \tag{15.3.4a}$$

$$\sigma_2^2=\frac{ie^2\Omega}{m}\left[\frac{n^s}{\Omega_+^2-\omega_c^2}+\frac{\sigma^s k_\parallel^2}{m\Omega^2}\left(\frac{4}{\Omega_+^2-(2\omega_c)^2}-\frac{1}{\Omega_+^2-\omega_c^2}\right)+\cdots\right], \tag{15.3.4b}$$

$$\sigma_2^1=-\sigma_1^2=-\frac{e^2\omega_c}{m}\left[\frac{n^s}{\Omega_+^2-\omega_c^2}+\frac{2\sigma^s k_\parallel^2}{m\Omega^2}\left(\frac{4}{\Omega_+^2-(2\omega_c)^2}-\frac{1}{\Omega_+^2-\omega_c^2}\right)+\cdots\right], \tag{15.3.4c}$$

where

$$n^s=\frac{m\omega_c}{\pi\hbar}\left(\frac{f_0(0)}{2}+\sum_{n=1}^{\infty}f_0(n\hbar\omega_c)\right), \tag{15.3.5a}$$

$$\sigma^s=\frac{m\omega_c}{\pi\hbar}\sum_{n=1}^{\infty}n\hbar\omega_c f_0(n\hbar\omega_c), \tag{15.3.5b}$$

and

$$f_0(\omega)\equiv\left[1+e^{\beta(\omega-\zeta)}\right]^{-1}. \tag{15.3.5c}$$

In the local limit, these further reduce to

$$\sigma_1^1\cong\sigma_2^2\cong\frac{ie^2\Omega}{m}\frac{n^s}{\Omega_+^2-\omega_c^2}, \tag{15.3.6a}$$

$$\sigma_2^1 \cong -\sigma_1^2 \cong -\frac{e^2\omega_c n^s}{m\left(\Omega_+^2 - \omega_c^2\right)}. \tag{15.3.6b}$$

Substitution of (15.3.6) into (15.3.2) results in the local-mode dispersion equation

$$\det \left| \mathbb{1} - \frac{\alpha c|k_\perp|}{\epsilon_m\left(\Omega_+^2 - \omega_c^2\right)} \begin{pmatrix} \frac{-k^2}{|k_\perp|^2} & 0 \\ 0 & 1 \end{pmatrix} \begin{pmatrix} 1 & \frac{+i\omega_c}{\Omega} \\ \frac{-i\omega_c}{\Omega} & 1 \end{pmatrix} \right| = 0, \tag{15.3.7}$$

where

$$\alpha \equiv \frac{2\pi e^2 n^s}{mc}. \tag{15.3.8}$$

With some algebraic manipulation, this may be re-expressed as

$$\left(\epsilon_m\Omega^2 - \alpha c|k_\perp|\right)(c|k_\perp| + \alpha) = \epsilon_m\omega_c^2 c|k_\perp|. \tag{15.3.9}$$

The polarization of the local modes—as determined from (15.3.3), (15.3.6), (15.3.8) and (15.3.9)—can be written in the convenient form

$$\left(\frac{E_s^1}{E_s^2}\right) = -\frac{i\epsilon_m\omega_c\Omega}{c|k_\perp|\,(\alpha + c|k_\perp|)}. \tag{15.3.10}$$

In the limit of zero magnetic field, the only evanescent wave permitted by (15.3.9) is the longitudinal Stern mode [26],

$$\Omega_0^2 = \frac{\alpha c|k_\perp|}{\epsilon_m} = \frac{\alpha c}{\epsilon_m}\left(k_\parallel^2 - \epsilon_m\Omega_0^2/c^2\right)^{\frac{1}{2}} = \left(\frac{\alpha^2}{2\epsilon_m}\right)\cdot\left(-1 + \left[1 + 4c^2k_\parallel^2/\alpha^2\right]^{\frac{1}{2}}\right). \tag{15.3.11}$$

(There are no transverse evanescent modes in the local zero magnetic-field limit.) For nonzero magnetic field ($\omega_c \neq 0$), one can obtain an exact solution of the local dispersion equation by eliminating Ω^2 in favor of $|k_\perp|$ in (15.3.9) (using $|k_\perp| = (k_\parallel^2 - \epsilon_m\Omega^2/c^2)^{\frac{1}{2}}$), and then determining the roots of the resultant cubic equation for $|k_\perp|$. Employing this procedure, we find the cubic dispersion relation

$$(c|k_\perp|)^3 + p\,(c|k_\perp|)^2 + q\,(c|k_\perp|) + r = 0, \tag{15.3.12}$$

where

$$p \equiv 2\alpha; \quad q \equiv \alpha^2 + \epsilon_m\omega_c^2 - c^2k_\parallel^2; \quad r \equiv -\alpha c^2 k_\parallel^2. \tag{15.3.13}$$

With the definitions

$$\mathrm{a} \equiv \left(q - \frac{p^2}{3}\right) = -\left(c^2k_\parallel^2 - \epsilon_m\omega_c^2 + \frac{\alpha^2}{3}\right), \tag{15.3.14a}$$

$$\mathrm{b} \equiv \left(r - \frac{pq}{3} + \frac{2p^3}{27}\right) = -\frac{\alpha}{3}\left(c^2k_\parallel^2 + 2\epsilon_m\omega_c^2 + \frac{2\alpha^2}{9}\right), \tag{15.3.14b}$$

$$D^2 = \left(\frac{\mathrm{b}}{2}\right)^2 + \left(\frac{\mathrm{a}}{3}\right)^3 = \left(\frac{\epsilon_m\omega_c^2}{3}\right)^3\left(1 + \frac{\alpha^2}{\epsilon_m\omega_c^2}\right)^2 - \frac{\epsilon_m\omega_c^2}{27}c^2k_\parallel^2\left(3\epsilon_m\omega_c^2 - 5\alpha^2\right) + \frac{c^4k_\parallel^4}{9}\left(\epsilon_m\omega_c^2 - \frac{\alpha^2}{12}\right) - \frac{c^6k_\parallel^6}{27}, \tag{15.3.14c}$$

$$A \equiv \left(-\frac{\mathrm{b}}{2} + D\right)^{\frac{1}{3}}, \tag{15.3.14d}$$

$$B \equiv \left(-\frac{\mathrm{b}}{2} - D\right)^{\frac{1}{3}}, \tag{15.3.14e}$$

the exact solutions for the roots of the cubic equation (15.3.12) are given by [24]:

$$c|k_\perp| = -\frac{2\mathrm{a}}{3} + (A + B), \tag{15.3.15a}$$

$$c|k_\perp| = -\frac{2\mathrm{a}}{3} - \frac{(A + B)}{2} + \frac{i(A - B)}{2}\sqrt{3}, \tag{15.3.15b}$$

$$c|k_\perp| = -\frac{2\mathrm{a}}{3} - \frac{(A + B)}{2} - \frac{i(A - B)}{2}\sqrt{3}, \tag{15.3.15c}$$

where the only acceptable roots have $c|k_\perp|$ real and positive, corresponding to evanescent wave solutions. The frequencies of these normal modes are obtained from

$$\epsilon_m\Omega^2 = c^2k_\parallel^2 - c^2|k_\perp|^2, \tag{15.3.16}$$

with $\Omega^2 > 0$ for real frequency waves. When the quantity D^2 in (15.3.14c) is greater than zero, the only real root for $c|k_\perp|$ is (15.3.15a). By contrast, when $D^2 \leq 0$ all three roots are real. For $D^2 < 0$, a convenient trigonometric solution is [24]

$$c|k_\perp| = -\frac{2\mathrm{a}}{3} + 2\sqrt{\frac{-\mathrm{a}}{3}}\cos\left(\frac{\phi}{3} + 120°n\right), \quad n = 0, 1, 2 \tag{15.3.17}$$

where

$$\cos\phi = +\sqrt{\frac{\mathrm{b}^2}{4}\bigg/\frac{(-\mathrm{a})^3}{27}}. \tag{15.3.18}$$

In any case, the only acceptable modes are those for which $|k_\perp| = -ik_z$ is both real *and* positive and $\Omega^2 > 0$.

It is apparent from the preceding expressions that the exact solutions of the cubic dispersion relation have a rather complicated dependence on the three frequency parameters $ck_{\|}$, ω_c and α, so that in general, these solutions must be evaluated by numerical means. However, there are a number of limiting cases in which the local dispersion relation can be treated analytically by the use of approximate methods. In the discussion that follows, we will explicitly consider the weak field limit $\omega_c \ll \alpha$ and $ck_{\|}$, the strong field limit $\omega_c \gg \alpha$ and $ck_{\|}$, and the electrostatic limit $c \to \infty$ (corresponding to $ck_{\|} \gg \omega_c$ and α).

When ω_c is very much smaller than both $ck_{\|}$ and α, one may obtain an approximate solution of the local dispersion equation by employing a simple iterative technique. Specifically, we multiply both sides of (15.3.9) by factor $\frac{\Omega^2+\alpha c|k_\perp|/\epsilon_m}{\epsilon_m(\alpha+c|k_\perp|)}$ to obtain

$$\left(\Omega^2\right)^2 - \left(\frac{\alpha c}{\epsilon_m}\right)^2 |k_\perp|^2 = \omega_c^2 c|k_\perp| \left(\frac{\Omega^2 + \alpha c|k_\perp|/\epsilon_m}{\alpha + c|k_\perp|}\right). \tag{15.3.19}$$

The left-hand side of (15.3.19) is expressible as

$$\left(\Omega^2\right)^2 + \frac{\alpha^2}{\epsilon_m}\Omega^2 - \left(\frac{\alpha c}{\epsilon_m}\right)^2 k_{\|}^2 = \left(\Omega^2 - \Omega_{0+}^2\right)\left(\Omega^2 - \Omega_{0-}^2\right), \tag{15.3.20}$$

where $\Omega_{0\pm}^2$ are the roots of the quadratic form and are hence *the zero magnetic field local solutions*, given by (15.3.11)

$$\Omega_{0\pm}^2 = \frac{\alpha^2}{2\epsilon_m}\left(-1 \pm \sqrt{1 + 4c^2k_{\|}^2/\alpha^2}\right). \tag{15.3.21}$$

(Note that $\Omega_{0+}^2 > 0$ but $\Omega_{0-}^2 < 0$, so the latter root does not correspond to a real wave.) Thus,

$$\left(\Omega^2 - \Omega_{0+}^2\right)\left(\Omega^2 - \Omega_{0-}^2\right) = \omega_c^2 c|k_\perp| \left(\frac{\Omega^2 + \alpha c|k_\perp|/\epsilon_m}{\alpha + c|k_\perp|}\right),$$

or

$$\left(\Omega^2 - \Omega_{0+}^2\right) = \omega_c^2 c|k_\perp| \left(\frac{\Omega^2 + \alpha c|k_\perp|/\epsilon_m}{\alpha + c|k_\perp|}\right)\frac{1}{\Omega^2 - \Omega_{0-}^2}. \tag{15.3.22}$$

The leading magnetic-field correction to Ω_{0+} (of order ω_c^2) is determined by putting $\Omega \to \Omega_{0+}$ and $|k_\perp| \to |k_\perp|_0$ on the right-hand side of (15.3.22). Since

$$\Omega_{0+}^2 - \Omega_{0-}^2 = \frac{\alpha^2}{\epsilon_m}\sqrt{1 + \frac{4c^2k_{\|}^2}{\alpha^2}},$$

we obtain

$$\Omega^2 - \Omega_{0+}^2 = \frac{\omega_c^2 c|k_\perp|_0\left(\Omega_{0+}^2 + \alpha c|k_\perp|_0/\epsilon_m\right)}{\frac{\alpha^2}{\epsilon_m}\left(\alpha + c|k_\perp|_0\right)\left(1 + 4c^2k_\parallel^2/\alpha^2\right)^{\frac{1}{2}}}, \tag{15.3.23}$$

where Ω_{0+}^2 is given in (15.3.21) and

$$|k_\perp|_0 = \left(k_\parallel^2 - \frac{\epsilon_m\Omega_{0+}^2}{c^2}\right)^{\frac{1}{2}} = \left[k_\parallel^2 - \frac{\alpha^2}{2c^2}\left(-1 + \sqrt{1 + 4c^2k_\parallel^2/\alpha^2}\right)\right]^{\frac{1}{2}}. \tag{15.3.24}$$

(In regard to the leading magnetic field correction to the Ω_{0-} root, that results in $\Omega^2 < 0$, which is unacceptable as a normal mode.)

If $\frac{ck_\parallel}{\alpha} \gg 1$, then $\Omega_{0+}^2 \approx \alpha ck_\parallel/\epsilon_m$, $|k_\perp|_0 \approx k_\parallel(1 - \frac{\alpha}{ck_\parallel})^{\frac{1}{2}}$, and (15.3.23) reduces to

$$\Omega^2 \cong \frac{\alpha ck_\parallel}{\epsilon_m} + \omega_c^2. \tag{15.3.25}$$

(Bardos and Frankel [18] suggest that the preceding local low magnetic field correction to Stern's result in (15.3.25) is not valid because it is inconsistent with a low wavenumber approximation for the plasma conductivity, such as shown in (15.2.56). We disagree. While it is correct that one cannot let $\omega_c \to 0$ for a low wavenumber expansion, there is a range of values $\omega_c << ck_\parallel$ such that the low wavenumber expansion is still valid, and over this range (15.3.25) holds true.) Alternatively, if $\frac{ck_\parallel}{\alpha} \ll 1$, then $\Omega_{0+}^2 \approx \frac{c^2k_\parallel^2}{\epsilon_m}\left(1 - \frac{c^2k_\parallel^2}{\alpha^2}\right)$, $|k_\perp|_0 \approx \frac{ck_\parallel^2}{\alpha}$, and

$$\Omega^2 \approx \frac{c^2k_\parallel^2}{\epsilon_m}\left(1 - \frac{c^2k_\parallel^2}{\alpha^2}\right) + 2\omega_c^2\left(\frac{ck_\parallel}{\alpha}\right)^4. \tag{15.3.26}$$

Hence, the presence of a weak magnetic field induces an ω_c^2 correction to the zero magnetic field dispersion results. It should be noted that the polarization of the weak field mode of (15.3.23) is essentially longitudinal since the numerator of (15.3.10) goes to zero as $\omega_c \to 0$, whereas the denominator tends to the zero field value $c|k_\perp|_0\left(\alpha + c|k_\perp|_0\right) > 0$.

When ω_c is very much larger than $ck_\parallel$ and α, one may employ another iterative technique to obtain an approximate solution of the local dispersion relation. Recalling that

$$\left(\epsilon_m\Omega^2 - \alpha c|k_\perp|\right)\left(\alpha + c|k_\perp|\right) = \epsilon_m\omega_c^2 c|k_\perp| \tag{15.3.27}$$

we can obtain the high magnetic field limit by noting that $|k_\perp| \to 0$ as $\omega_c \to \infty$ in order that the right-hand side be finite. (The left-hand side of (15.3.27) is bounded because evanescent waves require $|k_\perp| = \left(k_\parallel^2 - \epsilon_m\Omega^2/c^2\right)^{\frac{1}{2}} > 0$, or $\Omega^2 < c^2k_\parallel^2/\epsilon_m$). This suggests that we put the dispersion relation into the form

$$|k_\perp| = \frac{1}{\omega_c^2 c\epsilon_m}\left(\epsilon_m \Omega^2 - \alpha c|k_\perp|\right)(\alpha + c|k_\perp|)\,, \tag{15.3.28}$$

in order to obtain an iterative series for $|k_\perp|$ in powers of $\frac{1}{\omega_c}$. Replacing $|k_\perp|$ and Ω on the right-hand side of (15.3.28) by their values at large field strength $\left(|k_\perp| \to 0;\ \Omega^2 = \frac{c^2}{\epsilon_m}\left(k_\parallel^2 - |k_\perp|^2\right) \to \frac{c^2 k_\parallel^2}{\epsilon_m}\right)$, one obtains:

$$|k_\perp| \approx \alpha c k_\parallel^2/\epsilon_m \omega_c^2, \qquad \left(\omega_c \gg \alpha,\, ck_\parallel\right) \tag{15.3.29}$$

whence

$$\Omega^2 \approx \frac{c^2 k_\parallel^2}{\epsilon_m} - \frac{\alpha^2 c^4 k_\parallel^4}{\epsilon_m^3 \omega_c^4}, \qquad \left(\omega_c \gg \alpha,\, ck_\parallel\right) \tag{15.3.30}$$

to leading order in $\left(\frac{1}{\omega_c^2}\right)$. Substitution of (15.3.29) and (15.3.30) into (15.3.10) further shows that the high magnetic field mode is essentially transverse polarized $\left(E_s^1/E_s^2 \gg 1\right)$.

Another simple limiting case is obtained if we set $c \to \infty$ corresponding to the electrostatic plasma limit. In this regime,

$$|k_\perp| = \sqrt{k_\parallel^2 - \epsilon_m \frac{\Omega^2}{c^2}} \to k_\parallel, \tag{15.3.31}$$

and the dispersion relation

$$\left(\epsilon_m \Omega^2 - \alpha c|k_\perp|\right)(\alpha + c|k_\perp|) = \epsilon_m \omega_c^2 c|k_\perp|$$

results in

$$\left(\epsilon_m \Omega^2 - \alpha c k_\parallel\right) \cong \epsilon_m \omega_c^2. \tag{15.3.32}$$

Hence,

$$\Omega^2 \cong \frac{\alpha c k_\parallel}{\epsilon_m} + \omega_c^2, \qquad (c \to \infty) \tag{15.3.33}$$

which is in agreement with known results [27] (It should be noted that $\alpha c = \frac{2\pi e^2 n^s}{m}$ is independent of c.) The polarization of the electrostatic mode is purely longitudinal, since by (15.3.10), $E_s^1/E_s^2 \to 0$ as $c \to \infty$.

Thus, in the weak magnetic field, strong magnetic field, and electrostatic limits, there exists only one evanescent mode solution of the local dispersion relation.

It is worth noting that there are no purely "radiative" modes (Ω and $k_\perp$ both real) in the local limit of the 2D dispersion relation. For by (15.3.9) we have

$$\left(\epsilon_m \Omega^2 + i\alpha c k_\perp\right)(\alpha - ick_\perp) = -i\epsilon_m \omega_c^2 c k_\perp,$$

and if $k_\perp$ is real, this dispersion relation does not possess a purely real solution for Ω. (Complex values of Ω correspond to waves that are either unstable or damped.) Also, magnetic field effects on the local 2D plasma spectrum are essentially classical in nature (except for de Haas-van Alphen oscillations of the density as a function of $\zeta/\hbar\omega_c$). However, as shown immediately below, such quantization effects are manifest when nonlocal terms are included.

(b) **Nonlocal and Quantum Corrections to the Local Modes**

The low wave-number approximation for the magnetoconductivity tensor in (15.3.4) can conveniently be rewritten as

$$\sigma_1^1 = \left(\frac{ie^2n^s\Omega}{m(\Omega^2-\omega_c^2)}\right)\left(1+S_{1,NL}^1\right) \tag{15.3.34a}$$

$$\sigma_2^2 = \left(\frac{ie^2n^s\Omega}{m(\Omega^2-\omega_c^2)}\right)\left(1+S_{2,NL}^2\right) \tag{15.3.34b}$$

$$\sigma_2^1 = -\sigma_1^2 = -\left(\frac{e^2n^s\omega_c}{m(\Omega^2-\omega_c^2)}\right)\left(1+S_{2,NL}^1\right), \tag{15.3.34c}$$

$$\text{where } S_{1,NL}^1 \equiv \left(\frac{\sigma^s}{n^s}\right)\left(\frac{k_\parallel^2}{m\Omega^2}\right)\left(\frac{4(\Omega^2-\omega_c^2)}{\Omega^2-(2\omega_c)^2}-3\right) \tag{15.3.35a}$$

$$S_{2,NL}^2 \equiv \left(\frac{\sigma^s}{n^s}\right)\left(\frac{k_\parallel^2}{m\Omega^2}\right)\left(\frac{4(\Omega^2-\omega_c^2)}{\Omega^2-(2\omega_c)^2}-1\right) \tag{15.3.35b}$$

$$S_{2,NL}^1 \equiv \left(\frac{2\sigma^s}{n^s}\right)\left(\frac{k_\parallel^2}{m\Omega^2}\right)\left(\frac{4(\Omega^2-\omega_c^2)}{\Omega^2-(2\omega_c)^2}-1\right) = 2S_{2,NL}^2. \tag{15.3.35c}$$

Employing the dispersion relation condition (15.3.2), shifting the nonlocal terms to the right hand side of the equation while keeping local terms on the left, and ignoring terms of higher order than $k_\parallel^2$ arising from $\left(S_{1,NL}^1 \cdot S_{2,NL}^2\right)$ and $\left(S_{2,NL}^2\right)^2$, one obtains

$$\left(\epsilon_m\Omega^2-\alpha c|k_\perp|\right)\left(\alpha+c|k_\perp|\right)-\omega_c^2\epsilon_m c|k_\perp| = RHS, \tag{15.3.36}$$

where

$$RHS \equiv \frac{\alpha^2c|k_\perp|\left(\frac{\sigma^s}{n^s}\right)\left(\frac{k_\parallel^2}{m}\right)}{\Omega^2-(2\omega_c)^2}\left(\frac{3c|k_\perp|}{\alpha}-\frac{\epsilon_m(\Omega^2+8\omega_c^2)}{\alpha c|k_\perp|}+\frac{4\Omega^2-\omega_c^2}{\Omega^2-\omega_c^2}\right). \tag{15.3.37}$$

If $RHS=0$, then (15.3.36) is just the local mode dispersion formula in (15.3.9). Since the nonlocal terms in (15.3.37) are assumed small, we can obtain first order corrections to the local mode solutions by using the local mode values Ω_L, $|k_\perp|_L$ in

(15.3.37); i.e., $RHS \rightarrow RHS(\Omega_L,\ |k_\perp|_L)$. One can then re-solve for $\Omega,\ k_\perp$ using (15.3.36) with $RHS(\Omega_L,\ |k_\perp|_L)$ to obtain the nonlocal corrections to the local mode values. (It might be noted that the small dimensionless expansion parameter in (15.3.35a–15.3.35c) is $\left(\frac{\sigma^s}{n^s}\cdot\frac{1}{mc^2}\right)\sim 10^{-2}$ for a magnetic field $H_m = 10$ Tesla.)

The exact mode values $\Omega_L,\ |k_\perp|_L$ for the local case were shown to be the solutions for the cubic equation in $|k_\perp|_L$, (15.3.12)–(15.3.15), with Ω_L determined by $\epsilon_m\Omega_L^2 = c^2k_\parallel^2 - c^2|k_\perp|_L^2$, (15.3.16). With the nonlocal term, $RHS(\Omega_L,\ |k_\perp|_L)$, present in (15.3.36), one can re-solve the cubic equation (15.3.12) but with $r \rightarrow -\alpha c^2k_\parallel^2 + RHS(\Omega_L,\ |k_\perp|_L)$ to obtain the corrections to the local mode values. As before, the only acceptable solutions are those which have $\Omega,\ |k_\perp|$ real and positive for evanescent waves and are consistent with a low wavenumber approximation; i.e., have nonlocal shifts that are small in comparison to the local values. It should be noted that for numerical evaluation of these equations, it is necessary to calculate σ^s, the 2D energy density in (15.2.62), which incorporates quantum effects. This requires that one determine the chemical potential ζ in the Fermi function f_0, which is obtained from the user specified density n^s, (15.2.61). While this is straightforward for the non-degenerate case where $n^s = (e^{\zeta\beta}m\hbar\omega_c)/(2\pi\hbar^2\tanh(\hbar\omega_c\beta/2))$ as per (15.2.41), it is not as simple in general since one must effectively determine the ζ roots of the transcendental equation (15.2.61). A possible numerical technique for doing this is to use optimization methodologies to find the ζ values that minimize the difference between n^s and the right hand side of (15.2.61), with this difference required to approach zero. Possible optimization procedures include Monte Carlo or Genetic Algorithm searches for the ζ roots followed by quasi-Newton gradient minimization, which algorithms are generally available in standard mathematical software libraries (e.g. IMSL).

One can also use (15.3.36), (15.3.37) to find the nonlocal corrections to the approximate local mode solutions shown here for the cases $\omega_c << ck_\parallel$ or α, $\omega_c >> ck_\parallel$ or α, and the electrostatic limit $c \rightarrow \infty$. For ω_c small, one can employ the development leading to (15.3.23), (15.3.24), but now adding the nonlocal contributions from $RHS(\Omega_0,\ c|k_\perp|_0)$, (15.3.37), evaluated at $\omega_c \rightarrow 0$. For the case $\frac{c|k_\parallel|}{\alpha} >> 1$ where $\Omega_0^2 \cong \frac{\alpha ck_\parallel}{\epsilon_m}$ and $|k_\perp|_0 \cong k_\parallel$

$$RHS(\Omega_0,\ |k_\perp|_0) \cong 3\epsilon_m ck_\parallel \left(\frac{\sigma^s}{n^s}\right)\left(\frac{k_\parallel^2}{m}\right). \tag{15.3.38a}$$

Following the development leading to (15.3.23), one obtains

$$\Omega^2 \cong \frac{\alpha ck_\parallel}{\epsilon_m} + \omega_c^2 + \left(\frac{3\sigma^s}{n^s}\cdot\frac{k_\parallel^2}{m}\right), \tag{15.3.38b}$$

where the last term on the right is the nonlocal shift (compare with (15.3.25)). For ω_c small but $\frac{ck_\parallel}{\alpha} << 1$, one has from (15.3.26) $\Omega_0^2 \cong \frac{ck_\parallel^2}{\epsilon_m}$, $|k_\perp|_0 \cong k_\parallel\left(\frac{ck_\parallel}{\alpha}\right)$ so that

$$RHS(\Omega_0,\ |k_\perp|_0) \cong 3\alpha\epsilon_m \left(\frac{\sigma^s}{n^s}\right)\left(\frac{k_\parallel^2}{m}\right), \tag{15.3.39a}$$

and

$$\Omega^2 \cong \frac{c^2k_\parallel^2}{\epsilon_m}\left(1-\frac{c^2k_\parallel^2}{\alpha^2}\right)+2\omega_c^2\left(\frac{ck_\parallel}{\alpha}\right)^4+\left(\frac{6\sigma^s}{n^s}\cdot\frac{k_\parallel^2}{m}\right)\left(\frac{ck_\parallel}{\alpha}\right)^2, \tag{15.3.39b}$$

where the last term on the right provides the nonlocal correction (compare with (15.3.26)).

Considering next the case $\omega_c >> ck_\parallel$ or α, as per the discussion leading to (15.3.28), if ω_c is large then $c|k_\perp|$ must be small for local terms to remain finite, and this must also hold true in the presence of small nonlocal corrections. One can therefore write (as in (15.3.28))

$$c|k_\perp| = \frac{1}{\omega_c^2\epsilon_m}(\epsilon_m\Omega^2-\alpha c|k_\perp|)(\alpha+c|k_\perp|)-\frac{RHS}{\omega_c^2\epsilon_m}, \tag{15.3.40}$$

where RHS is approximately evaluated from (15.3.37) using the large magnetic field values $|k_\perp| \to 0$, $\Omega^2 \to \frac{(ck_\parallel)^2}{\epsilon_m}$. This results in

$$RHS(\omega_c \text{ large}) \cong 2\alpha\epsilon_m\left(\frac{\sigma^s}{n^s}\right)\left(\frac{k_\parallel^2}{m}\right). \tag{15.3.41a}$$

Substituting into (15.3.40), (15.3.36) yields

$$c|k_\perp| \cong \frac{\alpha c^2k_\parallel^2}{\epsilon_m\omega_c^2}\left(1-\frac{2\epsilon_m}{mc^2}\cdot\frac{\sigma^s}{n^s}\right) \tag{15.3.41b}$$

and

$$\Omega^2 \cong \frac{c^2k_\parallel^2}{\epsilon_m}-\frac{\alpha^2c^4k_\parallel^4}{\epsilon_m^3\omega_c^2}\left(1-\frac{2\epsilon_m}{mc^2}\cdot\frac{\sigma^s}{n^s}\right)^2 \tag{15.3.41c}$$

(compare with (15.3.30)).

Finally, in the electrostatic limit $c \to \infty$ (but $\alpha c = \frac{2\pi e^2 n^s}{m}$, $\omega_c \to 0$), where $c|k_\perp| \cong ck_\parallel$, $\Omega^2 \cong \frac{\alpha ck_\parallel}{\epsilon_m}$, (15.3.37) yields

$$RHS(c \to \infty) \cong 3\epsilon_m ck_\parallel\left(\frac{\sigma^s}{n^s}\cdot\frac{k_\parallel^2}{m}\right) \tag{15.3.42a}$$

and substituting into (15.3.36)

$$\Omega^2 = \frac{\alpha c k_{\parallel}}{m} + \omega_c^2 + \frac{RHS}{\epsilon_m c k_{\parallel}} \tag{15.3.42b}$$

or

$$\Omega^2 = \frac{\alpha c k_{\parallel}}{m} + \omega_c^2 + \left(\frac{3\sigma^s}{n^s} \cdot \frac{k_{\parallel}^2}{m} \right) \tag{15.3.42c}$$

As expected these results agree with (15.3.38) since in both cases $\omega_c << ck_{\parallel}$ and $\alpha << ck_{\parallel}$.

In regard to polarization of the nonlocally corrected modes, we note that our examination of the role of nonlocality has addressed only the leading nonlocal corrections, which are small by definition, and therefore do not have a significant affect on the polarization of the local modes.

In considering nonlocal corrections it is particularly interesting to note that a high magnetic field tends to suppress nonlocal structure through wavenumber parameters such as $k_{\parallel}^2/m\omega_c^2\beta$ (nondegenerate), $k_{\parallel}^2\zeta/m\omega_c^2$ (degenerate), $\hbar k_{\parallel}^2/m\omega_c$ (quantum); and in the degenerate quantum strong field limit nonlocal structure involving σ^s/n^s vanishes as $\sigma^s \to 0$.

(c) **Nonlocal "Bernstein" Resonances**

The formulae for σ_j^i in (15.3.4) show that the magnetoconductivity tensor possesses a simple first-order pole at $\Omega = 2\omega_c$. The presence of this pole induces a nonlocal root of the plasma-wave dispersion relation, in the nature of a "Bernstein"-type mode [22, 27–29]. For low wave number, the Bernstein mode should be near $\Omega \sim 2\omega_c$ and in view of this we estimate it by putting $\Omega \to 2\omega_c$ in the local terms of σ_j^i and neglect nonlocal shifting of the local structure. The expressions for σ_j^i, (15.3.4), then take the approximate form

$$\sigma_1^1 \cong \sigma_2^2 \cong -\frac{2e^2n^s}{im\omega_c} \left(\frac{1}{3} + \frac{\sigma^s k_{\parallel}^2}{mn^s} \frac{1}{\Omega_+^2 - (2\omega_c)^2} + \dots \right), \tag{15.3.43a}$$

$$\sigma_2^1 = -\sigma_1^2 \cong -\frac{2e^2n^s}{m\omega_c} \left(\frac{1}{6} + \frac{\sigma^s k_{\parallel}^2}{mn^s} \frac{1}{\Omega_+^2 - (2\omega_c)^2} + \dots \right). \tag{15.3.43b}$$

Substituting (15.3.43) into the dispersion relation (15.3.2), and defining

$$\alpha = \frac{2\pi e^2 n^s}{mc}, \tag{15.3.44a}$$

$$Z = \frac{\sigma^s k_{\parallel}^2}{mn^s} \frac{1}{\Omega_+^2 - (2\omega_c)^2}, \tag{15.3.44b}$$

one obtains the approximate dispersion relation

$$0 = \det \left| \underset{\sim}{1} + \alpha c|k_\perp| \begin{pmatrix} \frac{\frac{4}{3}+4Z}{c^2|k_\perp|^2} & \frac{i(\frac{2}{3}+4Z)}{c^2|k_\perp|^2} \\ \frac{i(\frac{1}{6}+Z)}{\epsilon_m\omega_c^2} & \frac{-(\frac{1}{3}+Z)}{\epsilon_m\omega_c^2} \end{pmatrix} \right|, \tag{15.3.45}$$

which has the solution (we note that the terms quadratic in Z cancel out of (15.3.45) and, in consequence, there is only a single "Bernstein"-type mode at $\Omega \sim 2\omega_c$):

$$Z = -\frac{1}{3}\left[\frac{\alpha^2c^2|k_\perp|^2 + \alpha c|k_\perp|\left(\alpha^2 - 3\epsilon_m\omega_c^2\right) - 4\alpha^2\epsilon_m\omega_c^2}{\alpha^2c^2|k_\perp|^2 + \frac{4}{3}\alpha^3c|k_\perp| - 4\alpha^2\epsilon_m\omega_c^2}\right], \tag{15.3.46}$$

Since $Z \equiv \frac{\sigma^s k_\parallel^2}{mn^s}\left(\Omega^2 - (2\omega_c)^2\right)^{-1}$, one obtains,

$$\Omega^2 = (2\omega_c)^2 - \frac{3\sigma^s}{mn^s}k_\parallel^2\,[\ldots], \tag{15.3.47a}$$

where

$$[..] = \frac{\alpha^2c^2|k_\perp|^2 + \frac{4}{3}\alpha^3c|k_\perp| - 4\alpha^2\epsilon_m\omega_c^2}{\alpha^2c^2|k_\perp|^2 + \alpha c|k_\perp|\left(\alpha^2 - 3\epsilon_m\omega_c^2\right) - 4\alpha^2\epsilon_m\omega_c^2}. \tag{15.3.47b}$$

Therefore,

$$(c|k_\perp|)^2 = \left(c^2k_\parallel^2 - \epsilon_m\Omega^2\right) \cong c^2k_\parallel^2 - 4\epsilon_m\omega_c^2 + \frac{3\epsilon_m\sigma^s k_\parallel^2}{mn^s}\,[...], \tag{15.3.48}$$

with n^s, σ^s given in (15.3.5). Hence, (15.3.47a), (15.3.47b) provides the frequency solution for the $n = 2$ "Bernstein"-type mode in terms of the frequency parameters $ck_\parallel$, α, ω_c and the nonlocality parameter $(\sigma^s/n^s)\frac{k_\parallel^2}{m}$, (where σ^s/n^s is the average energy of a Landau-quantized electron constrained to move on a two-dimensional surface). Note that Landau-quantization effects are present in the structure of this mode through the average energy σ^s/n^s.

The polarization of the $n = 2$ "Bernstein" mode can be obtained from (15.3.3) and (15.3.43), (15.3.44), viz:

$$\frac{E_s^1}{E_s^2} = \frac{E_s^x}{E_s^y} = i\left[\frac{\epsilon_m\omega_c^2 - \alpha c|k_\perp|\left(\frac{1}{3} + Z\right)}{\alpha c|k_\perp|\left(\frac{1}{6} + Z\right)}\right], \tag{15.3.49}$$

where Z is given in (15.3.46) and $c|k_\perp|$ is determined from (15.3.48).

The dispersion relation in (15.3.47a), (15.3.47b) simplifies considerably for various limiting values of $ck_\parallel$, ω_c and α. In the electrostatic limit $c \to \infty$ (corresponding to $ck_\parallel \gg \omega_c$ and α; with α/ω_c arbitrary) one obtains

$$\Omega^2 \cong (2\omega_c)^2 - \frac{3\sigma^s/n^s}{m}k_\parallel^2. \tag{15.3.50}$$

From (15.3.46), $Z \to -1/3$ as $c \to \infty$, and as per (15.3.49) the "Bernstein" mode is essentially longitudinally polarized for this case. When $\omega_c \ll ck_\parallel$ or α, corresponding to a weak magnetic field limit, (15.3.47) reduces to

$$\Omega^2 \cong (2\omega_c)^2 - \frac{3\sigma^s/n^s}{m}k_\parallel^2 \left(\frac{1 + \frac{4}{3}\frac{\alpha}{ck_\parallel}}{1 + \frac{\alpha}{ck_\parallel}} \right). \tag{15.3.51}$$

If $\alpha \ll ck_\parallel$, then [27]

$$\Omega^2 = (2\omega_c)^2 - \frac{3\sigma^s/n^s}{m}k_\parallel^2. \tag{15.3.52}$$

Since $Z \to -1/3$ and $\omega_c \to 0$ in (15.3.49), the mode is longitudinally polarized. However, if $\alpha \gg ck_\parallel$, then

$$\Omega^2 = (2\omega_c)^2 - \frac{4\sigma^s/n^s}{m}k_\parallel^2, \tag{15.3.53}$$

and $Z \to -\frac{1}{4}$ in this limit and $\epsilon_m\omega_c^2 \ll \alpha ck_\parallel$ or $\alpha c|k_\perp|$. Equation (15.3.49) then shows $\frac{E_s^1}{E_s^2} \cong i$. In this case, the mode has both longitudinal and transverse components that are approximately equal in magnitude and 90° out of phase.

Landau quantization effects are present in the structure of the $n = 2$ "Bernstein" type mode through the energy parameter σ^s and quantum magnetic field effects will also be present in the higher order "Bernstein" type modes as a consequence of higher order wave number corrections to the plasma magnetoconductivity tensor. These and other nonlocal phenomena can significantly affect the 2D plasma wave spectrum. For example, as discussed in [29], nonlocal modifications (~10%) to the 2D plasma wave spectrum of $Si(100) - SiO_2$ interfaces can occur for wave numbers $k_\parallel$ of the order of $[100 \text{ Angstroms}]^{-1}$.

15.4 Appendix 1: Alternative Formulation of the 2D Magnetoplasma Dispersion Relation

We consider a 2D plasma in a uniform, static magnetic field perpendicular to the plane of the plasma. The electrons are assumed mobile while the positive charges are taken to be fixed. From Gauss' and Stokes' Laws, one has the boundary conditions

$$E_\perp(0) = \frac{2\pi n^s}{\epsilon_m} \quad \text{and} \quad \hat{z} \times \vec{H}(0) = \frac{2\pi \vec{J}^s}{c}, \tag{15.4.1}$$

where $\perp$, $\parallel$ refer to field components perpendicular, parallel to the plasma surface at $z = 0$ and n^s, $\vec{J}^s$ are the surface charge and current densities respectively. For a monochromatic wave with frequency Ω and wavenumber $\vec{k}$, Maxwell's curl equation yields

$$\vec{k} \times \vec{E}(0) = \frac{\Omega}{c}\vec{H}(0), \therefore \hat{z} \times (\vec{k} \times \vec{E}(0)) = \frac{2\pi\Omega}{c^2}\vec{J}^s, \tag{15.4.2}$$

and one has

$$E_{\perp}(0)\vec{k}_{\parallel} - \vec{E}_{\parallel}(0)k_{\perp} = \frac{2\pi\Omega}{c^2}\vec{J}^s. \tag{15.4.3}$$

Eliminating $E_{\perp}(0)$ in terms of n^s using (15.4.1) and employing current continuity $\Omega n^s - \vec{k}_{\parallel} \cdot \vec{J}^s = 0$, one obtains

$$\vec{E}_{\parallel}(0) = -\frac{2\pi\Omega}{c^2 k_{\perp}}\left(\underset{\sim}{1} - \frac{\vec{k}_{\parallel}\vec{k}_{\parallel}}{k^2}\right) \cdot \vec{J}^s, \tag{15.4.4}$$

where $k^2 = k_{\perp}^2 + k_{\parallel}^2 = \frac{\epsilon_m \Omega^2}{c^2}$. One can then use Ohm's Law $\vec{J}^s = \underset{\sim}{\sigma}_2 \cdot \vec{E}_{\parallel}(0)$ to obtain

$$\left[\underset{\sim}{1} + \frac{2\pi\Omega}{c^2 k_{\perp}}\left(\underset{\sim}{1} - \frac{\vec{k}_{\parallel}\vec{k}_{\parallel}}{k^2}\right) \cdot \underset{\sim}{\sigma}_2\right] \cdot \vec{E}_{\parallel}(0) = 0. \tag{15.4.5}$$

The dispersion relation for 2D plasma waves is obtained by setting the determinant of the bracketed expression in (15.4.5) to zero (compare with (15.1.16) in the text). One can also define a 2D dielectric tensor $\underset{\sim}{\epsilon}_2$:

$$\underset{\sim}{\epsilon}_2 \equiv \epsilon_m \underset{\sim}{1} + \frac{2\pi k_{\perp}}{\Omega}\underset{\sim}{\sigma}_2, \tag{15.4.6}$$

where ϵ_m is the surrounding bulk medium permittivity. From (15.4.5), the 2D plasma dispersion relations are

$$\det\left|\underset{\sim}{1} + \frac{k^2}{k_{\perp}^2}\left(\underset{\sim}{1} - \frac{\vec{k}_{\parallel}\vec{k}_{\parallel}}{k^2}\right)\left(\frac{\underset{\sim}{\epsilon}_2}{\epsilon_m} - \underset{\sim}{1}\right)\right| = 0, \tag{15.4.7}$$

and with some simple manipulation, this can also be expressed as

$$\det\left|\left(\underset{\sim}{1} - \frac{\vec{k}_{\parallel}\vec{k}_{\parallel}}{k_{\parallel}^2}\right) - \frac{\Omega^2}{c^2 k_{\parallel}^2}\left(\underset{\sim}{1} - \frac{\vec{k}_{\parallel}\vec{k}_{\parallel}}{k^2}\right) \cdot \underset{\sim}{\epsilon}_2\right| = 0. \tag{15.4.8}$$

15.5 Appendix 2: The Time-Ordered Exponential Time Development Operator

An operator $U(t, t_0)$ defining the time development of a state $\Phi(t)$ from an initial state $\Phi(t_0)$,

$$\Phi(t) = U(t, t_0)\Phi(t_0),$$

subject to the initial condition $U(t_0, t_0) = 1$, satisfies the Schrödinger equation ($\hbar \to 1$)

$$i(\partial/\partial t)U(t, t_0) = H(t)U(t, t_0).$$

Integration of this with respect to t from t_0 to t and applying the initial condition leads to an integral equation,

$$U(t, t_0) = 1 - i\int_{t_0}^{t} dt H(t)U(t, t_0),$$

which can be solved by repeated iterations as

$$\begin{aligned}
U(t, t_0) &= 1 - i\int_{t_0}^{t} dt_1 H(t_1)\left[1 - i\int_{t_0}^{t_1} dt_2 H(t_2)U(t_2, t_0)\right] \\
&= 1 - i\int_{t_0}^{t} dt_1 H(t_1) + (-i)^2\int_{t_0}^{t} dt_1\int_{t_0}^{t_1} dt_2 H(t_1)H(t_2) + \ldots \\
&+ (-i)^n\int_{t_0}^{t} dt_1\int_{t_0}^{t_1} dt_2 \ldots\int_{t_0}^{t_{n-1}} dt_n H(t_1)H(t_2)\ldots H(t_n) + \ldots .
\end{aligned}$$

The n'th integrand involves products of the time dependent Hamiltonian $H(t)$ repeated n times with the highest time argument t, on the left, proceeding to the lower time arguments in succession with the lowest time argument t_n on the right. This is a defacto time ordering of the Hamiltonian products, denoted by the "+" subscript. $H(t_1)\ldots H(t_n) \equiv (H(t_1)\ldots(t_n))_+$. It is important to recognize and respect this time ordering because time dependent Hamiltonians generally fail to commute at different times (ie: $[H(t), H(t')] \neq 0$).

In the special case of a time-independent Hamiltonian, the iteration series for $U(t, t_0)$ is just the usual time development operator $U(t, t_0) = e^{-iH(t-t_0)}$. In view of this, it has become conventional to symbolize the iteration series above (for a time dependent $H(t)$) as $U(t, t_0) = (\exp[-i\int_{t_0}^{t} H(t)dt])_+$, (a "time-ordered exponential"), but this has meaning *only* as the iteration series above with the time ordered Hamiltonian products under multiple time integrations.

Having identified the correct time development operator for a time dependent Hamiltonian, it is important to recognize that the upper limit of the time integration involved in (15.2.13) is imaginary, $\tau = -i\beta = -i/\kappa_B T$, so the time displacement involved in $< \vec{J}^s >^{i\tau}$ is also imaginary. This imparts an imaginary time periodicity property to the time t-dependence of functions such as $< \vec{J}^s(t) >^{i\tau}$ of (15.2.13) and associated nonequilibrium Green's functions, with imaginary period $\tau = -i\beta$: This is discussed in detail in [21], sections 7.4 and 9.1. It should be borne in mind that these considerations are in the context of the many-body-problem, which involves a constant particle number such that $[N, H(t)] \equiv 0$ for all times t: This means that the appearance of N in the grand canonical ensemble averaging process is essentially "inert" with respect to the time development process in our present considerations. (If N were not a fixed constant in time, $[N, H(t)] \neq 0$, the situation would be more complicated.)

15.6 Appendix 3: Semiclassical Model

Although our principal interest is in quantum magnetic field effects due to Landau quantization in the 2D magnetopolariton spectrum, it is of interest to explore the precise meaning of the semiclassical model as defined by Chiu and Quinn [30]: That definition neglects the dHvA oscillatory terms embedded in the quantity δ given by (r = positive integers $\neq 0$)

$$\delta \equiv \frac{\pi}{\beta\zeta} \sum_{r\neq 0} (-1)^r \frac{\cos(2\pi r\zeta/\hbar\omega_c)}{\sinh(2\pi^2 r/\hbar\omega_c\beta)}. \tag{15.6.1}$$

These oscillatory terms have maximum amplitude at zero temperature ($\hbar\omega_c\beta >> 1$), taking the form

$$\delta_{T=0} = \frac{\hbar\omega_c}{2\pi\zeta} \sum_{r\neq 0} \frac{(-1)^r}{r} \cos(2\pi r\zeta/\hbar\omega_c), \tag{15.6.2}$$

so that its vanishing in the degenerate limit further requires that $\hbar\omega_c/\zeta \to 0$. In the nondegenerate statistical regime, $\hbar\omega_c\beta << 1$, (15.6.1) reduces to

$$\delta_{T\to\infty} = \frac{2\pi}{\beta\zeta} \sum_{r\neq 0} (-1)^r \cos(2\pi r\zeta/\hbar\omega_c) e^{-2\pi^2 r/\hbar\omega_c\beta}, \tag{15.6.3}$$

which vanishes exponentially with temperature. While this "semiclassical" model of the magnetopolariton spectrum is of some interest, it is antithetical to the arduous task of a fully quantum mechanical analysis of Landau quantization effects in a magnetic field, and it ignores valuable physical information embedded in the structure of dHvA oscillations.

References

1. N.J.M. Horing, D. Miessein, G. Gumbs, J. Optical Soc. Amer. **A32**, 1184 (2015)
2. N.J.M. Horing, T.Y. Bagaeva, V.V. Popov, J. Optical Soc. Amer. **B24**, 2428 (2007)
3. P.D. Toit, Spin current: the contribution of spin to the probability current for nonrelativistic particles. Honors Thesis, Brigham Young University, 2001
4. J.M. Wilkes, arXiv: 1908.03276v2 [quant-ph], 14 Oct. 2019, Eqs. (13, 14) and references 2,5,7,9 therein
5. M. Nowakowski, Amer. J. Phys. **67**(10), 916 (1999)
6. J.H. Garcia, M. Vila, A.W. Cummings, S. Roche, Chem. Soc. Rev., Royal Soc. Chem. (2018). https://doi.org/10.1039/c7cs00864c
7. J. Sinova et al., Phys. Rev. Lett. **92**, 126603 (2004)
8. Ma. Luo, Zhibing Li, Phys. Rev. B **96**, 165424 (2017)
9. M. Inglot et al., Appl. Phys. Lett. **112**, 231102 (2018)
10. A. Avsar, et al., Nature Communications: https://doi.org/10.1038/ncomms5875 (2014)
11. J. Ingla-Aynes, A.A. Kaverzin, B.K. Van Wees, arXiv: 2004.11043, v 1 [cond-mat. mes-hall], 23 Apr 2020
12. Y. Anugrah et al., AIP Adv. **8**, 015129 (2018)
13. N. Bolivar, E. Medina, B. Berche, arXiv: 1402.4103, v1[cond-mat. mes-hall], 17 Feb 2014
14. J. Linder, M.E. Bathen, Phys. Rev. **93**, 224509 (2016)
15. M. Weiler et al., Phys Rev. Lett. **108**, 106602 (2012)
16. V.P. Amin, J. Li, M.D. Stiles, P.M. Haney, Phys. Rev. **B99**, 220405 (R) (2019)
17. G. Tang, J. Thinga, J. Wang, Phys. Rev. B **97**, 155430 (2018)
18. D.C. Bardos, N.E. Frankel, Phys. Rev. B **49**, 4096 (1994)
19. P.C. Martin, J. Schwinger, Phys. Rev. **115**, 1342 (1959)
20. L.P. Kadanoff, G. Baym, *Quantum Statistical Mechanics* (W. A. Benjamin, Inc., New York, 1962)
21. N.J.M. Horing, *Quantum Statistical Field Theory*, Chapters 9 and 10 (Oxford University Press, 2017)
22. M. Orman, Electromagentic response of a quantized magnetoplasma in two and three dimensions. Ph.D. Dissertation, Stevens Institute of Technology, Hoboken, NJ (1974)
23. N.J. Horing, Ann. Phys. (NY) **31**, 1 (1965)
24. M. Abramowitz, I.A. Stegun (eds.), *Handbook of Mathematical Functions* (Dover Publications, New York, 1965)
25. A. Erdelyi, W. Magnus, F. Oberhettinger, F.G. Tricomi, *Higher Transcendental Functions*, vol. 2 (McGraw Hill Book Company, New York, 1953)
26. F. Stern, Phys. Rev. Lett. **18**(14), 546 (1967)
27. N.J.M. Horing, M. Yildiz, Ann. Phys. (NY) **97**, 216 (1976)
28. I.B. Bernstein, Phys. Rev. **109**, 10 (1958)
29. M. Orman, N.J.M. Horing, Solid State Commun. (Pergamon Press) **15**, 1381–1385 (1974)
30. K.W. Chiu, J.J. Quinn, Phys. Rev. **B9**, 4724 (1974)

Chapter 16
Quantum Dynamics in a 1D Dot/Antidot Lattice: Landau Minibands and Graphene Wave Packet Motion in a Magnetic Field

Norman J. M. Horing, R. A. W. Ayyubi, K. Sabeeh, and Sina Bahrami

Abstract This work is focused on the analysis of the quantum dynamics of a model one-dimensional lattice array of quantum dots on a two dimensional electron sheet/layer in a normal magnetic field. Our analysis is carried out with the derivation of the Green's function for the quantum dot lattice subject to Landau quantization using the corresponding "no-lattice" Green's function, for propagation along the axis of the 1D quantum dot lattice. The frequency/energy poles of this Green's function provide the dispersion relation for the Landau quantized energy spectrum, which exhibits Landau minibands, rather than discrete Landau levels. In the case of non-relativistic carriers, the dispersion relation is explicitly exhibited in a closed form in terms of the Jacobi Theta Function of the third kind, and an approximate solution displaying the Landau miniband formation is obtained in terms of Laguerre polynomials. We also examine the case of "relativistic" graphene carriers, employing the appropriate Landau quantized graphene Green's function in a study of the associated wave packet dynamics in a graphene antidot lattice in a normal magnetic field. In this, we analyze the effects of pseudospin on the wave packet dynamics in Landau quantized graphene, including the role of Zitterbewegung in circular orbit motion.

16.1 Introduction

The present work is concerned with a lattice of quantum dots/antidots in a normal magnetic field. Quantum dot/antidot systems have been under exploration as a mechanism for quantum transport for quite some time [1–3]. Quantum dot applica-

N. J. M. Horing (✉) · S. Bahrami
Department of Physics, Stevens Institute of Technology, Hoboken, NJ 07030, USA
e-mail: nhoring@stevens.edu

R. A. W. Ayyubi · K. Sabeeh
Department of Physics, Quaid-i-Azam University, Islamabad 45320, Pakistan

H. Ünlü and N. J. M. Horing (eds.), *Progress in Nanoscale and Low-Dimensional Materials and Devices*, Topics in Applied Physics 144,
https://doi.org/10.1007/978-3-030-93460-6_16

tions have also been discussed in graphene and have even reached into biology and medicine [4, 5]. In regard to the inclusion of a magnetic field, its role as a probe of the properties of matter has always been well appreciated [6], but its importance is further amplified by its splintering of the energy spectrum into a multitude of Landau eigenstates [7], which can influence electronic conduction in quantum dot/antidot transport (beyond the relatively poor semiclassical treatment of the magnetic field restricted by an approximation limited to the Peierls phase factor alone). Moreover, the role of a row of quantum dots in the form of a periodic lattice brings into view several important features in the presence of a magnetic field, which are addressed in this chapter:

- The formation of Landau Miniband Structure, [8] and
- The circular features of "relativistic" graphene carrier wave packet dynamics in a magnetic field and its behavior in an antidot lattice, including "Zitterbewegung": with an examination of the important role of pseudospin in the preparation of the incident wave packet.

Our study of a quantum dot/antidot lattice in a magnetic field begins with a "first principles" derivation of the associated nonrelativistic magnetic field Green's function, obtained explicitly in a closed form analytic representation in terms of known functions for propagation along the axis of the dot lattice [8]. This Green's function can serve as a basic element in facilitating further transport calculations. We have also extracted the desired eigen-energy information by an analysis of the magnetic field lattice Green's function's frequency poles, obtaining the desired spectral information—which shows that there is a proliferation of eigen-energy states in Landau minibands that must be taken into account in quantum dot/antidot lattice transport in a magnetic field, a fact often neglected. This concise analysis avoids potential calculational difficulties using a tractable model for quantum dots/antidots in a lattice array, subject to Landau quantization [9]. Similar phenomenology in a different system involving a superconductor with an Abrikosov lattice of vortices was discussed by Chen and Fal'ko [10].

In connection with the analysis of wave packet dynamics in a graphene quantum antidot lattice in a magnetic field, we determine the "relativistic" graphene antidot lattice Green's function subject to Landau quantization, again in a closed form analytic representation in terms of known functions for propagation along the axis of the antidot lattice. It is employed to trace the dynamical development of a wave packet and is seen to exhibit "Zitterbewegung" rapid oscillatory features in circular motion generated by the magnetic field. We find that the pseudospin polarization of the incident wave packet plays an important role in its dynamics, with both propagation of the packet and Zitterbewegung strongly dependent on it. In particular, the packet is found to propagate in the direction of initial pseudospin polarization . This study also suggests that a graphene antidot lattice can serve as a channel for electron transport that is tunable by pseudospin polarization, antidot potential and applied normal magnetic field strength.

16.2 Determination of the Quantum Dot Lattice Green's Function

Considering a two-dimensional (2D) sheet of nonrelativistic Schrödinger electrons in a lattice formed by a one dimensional periodic array of quantum dot potential wells, we examine the role of a quantizing magnetic field **B** perpendicular to the 2D sheet. In this analysis we explicitly construct the appropriate Green's function $G(x_1, x_2; y_1, y_2; \omega)$ describing the Landau-quantized electron dynamics for this 2D Krönig-Penney-type model of a 1D array of dots in a strong magnetic field [11]. We examine its frequency poles to establish the dispersion relation for the eigen-energy spectrum of this 2D system with an array of dots represented by a row of Dirac-delta functions in a high magnetic field.

In accordance with the Krönig-Penney model for a quantum dot lattice, we introduce an infinite periodic lattice array of identical dot potential wells on the x-axis at $x_n = nd$, $y \equiv 0$ as

$$U(\mathbf{r}) = U(x, y) = \alpha \sum_{n=-\infty}^{\infty} \delta(x - nd)\delta(y), \tag{16.1}$$

where $\alpha < 0$ is the product of the quantum dot potential well depth (U_0) of a typical potential well times its area (a^2), and d is the uniform spacing of the wells. The 2D Schrödinger Green's function in frequency representation is given by the Dyson-type integral equation

$$\begin{aligned} G(x_1, x_2; y_1, y_2; \omega) &= G_0(x_1, x_2; y_1, y_2; \omega) \\ &+ \int dx_3 \int dy_3 G_0(x_1, x_3, y_1, y_3; \omega) U(x_3, y_3) G(x_3, x_2, y_3, y_2; \omega); \end{aligned} \tag{16.2}$$

or (suppress ω)

$$\begin{aligned} G(x_1, x_2; y_1, y_2) &= G_0(x_1, x_2; y_1, y_2) \\ &+ \alpha \sum_{n=-\infty}^{\infty} G_0(x_1, nd; y_1, 0) G(nd; x_2; 0, y_2), \end{aligned} \tag{16.3}$$

where $G_0(\mathbf{r}_1, \mathbf{r}_2)$ is the infinite-space Green's function for 2D electrons in a perpendicular magnetic field in the complete absence of potential wells/dots. Note that the solution for $G(x_1, x_2; y_1, y_2)$ devolves upon the determination of $G(nd, x_2; 0, y_2)$ at a discrete set of values. Therefore we set $x_1 = md$, and $y_1 = 0$ in (16.3):

$$G(md, x_2; 0, y_2) = G_0(md, x_2; 0, y_2) + \alpha \sum_{n=-\infty}^{\infty} G_0(md; nd; 0, 0) G(nd, x_2; 0, y_2). \tag{16.4}$$

Addressing the presence of a perpendicular magnetic field, we limit our attention to electron propagation confined to the x-axis of the lattice, with $y \equiv y_1 \equiv y_2 \equiv 0$. In this case, the infinite-space magnetic field Green's function $G_0(md, nd; 0, 0)$ is spatially translationally invariant and

$$G_0(md, nd; 0, 0) = \dot{G}_0([m-n]d), \tag{16.5}$$

acts as a translationally invariant position-space matrix, indicated by an overhead dot on the right of (16.5). Suppressing x_2, ω for the moment, these equations may be solved using the periodicity of the lattice in a Fourier series defined by (r are integers here)

$$\tilde{G}(p) = \sum_{r=-\infty}^{\infty} e^{ipdr} G(rd) \tag{16.6}$$

with

$$G(md) = \frac{d}{2\pi} \int_{-\pi/d}^{\pi/d} dp\, e^{-ipdm} \tilde{G}(p), \tag{16.7}$$

and

$$\dot{G}_0([m-n]d) = \frac{d}{2\pi} \int_{-\pi/2}^{\pi/2} dp\, e^{-ipd[m-n]} \dot{\tilde{G}}_0(p). \tag{16.8}$$

Correspondingly, (16.4) becomes

$$\begin{aligned} \tilde{G}(p) &= \tilde{G}_0(p) + \alpha \dot{\tilde{G}}_0(p) \sum_{n=-\infty}^{\infty} e^{ipdn} G(nd) \\ &= \tilde{G}_0(p) + \frac{\alpha d}{2\pi} \dot{\tilde{G}}_0(p) \int_{-\pi/d}^{\pi/d} dq \left(\sum_{n=-\infty}^{\infty} e^{in[p-q]d} \right) \tilde{G}(q). \end{aligned} \tag{16.9}$$

Employing the Poisson Sum Formula as

$$\sum_{n=-\infty}^{\infty} e^{in[p-q]d} = \frac{2\pi}{d} \sum_{m=-\infty}^{\infty} \delta(p - q - 2\pi m/d),$$

we have

$$\tilde{G}(p) = \tilde{G}_0(p) + \alpha \sum_{m=-\infty}^{\infty} \int_{-\pi/d}^{\pi/d} dq\, \delta(p - q - 2\pi m/d) \dot{\tilde{G}}_0(p) \tilde{G}(q). \tag{16.10}$$

Since the q-integral is extended only over the first Brillouin zone, we have

$$\tilde{G}(p) = \tilde{G}_0(p) + \alpha\dot{\tilde{G}}_0(p)\tilde{G}(p), \tag{16.11}$$

with the solution given by (restore x_2, ω)

$$\tilde{G}(p; x_2; 0, 0; \omega) = \left[1 - \alpha\dot{\tilde{G}}_0(p; 0, 0; \omega)\right]^{-1} \tilde{G}_0(p; x_2; 0, 0; \omega). \tag{16.12}$$

Taken jointly with (16.7) and (16.3), the result of (16.12) completes the description of the Green's function for the 2D Krönig-Penney-like model for a 1D quantum dot lattice, for electron propagation confined to the axis of the lattice ($y \equiv y_1 \equiv y_2 \equiv 0$ and we suppress further reference to y):

$$G(x_1, x_2; \omega) = G_0(x_1, x_2; \omega) + \alpha \sum_{n=-\infty}^{\infty} G_0(x_1, nd; \omega) \times \frac{d}{2\pi} \int_{-\frac{\pi}{d}}^{\frac{\pi}{d}} dp\, e^{-ipdn} \left[1 - \alpha\dot{\tilde{G}}_0(p; 0, 0; \omega)\right]^{-1} \tilde{G}_0(p; x_2; \omega). \tag{16.13}$$

16.3 Role of the Magnetic Field

The 2D Schrödinger Green's function, G_0, in a perpendicular magnetic field in the absence of any potential barriers has been determined in position representation as [12]

$$G_0(\mathbf{r}, \mathbf{r}'; t, t') = C(\mathbf{r}, \mathbf{r}')G_0'(\mathbf{r} - \mathbf{r}'; t - t'), \tag{16.14}$$

where the Peierls phase factor $C(\mathbf{r}, \mathbf{r}')$ embodies all *non*-spatially-translationally invariant structure and gauge dependence ($\mathbf{B}$ is the magnetic field),

$$C(\mathbf{r}, \mathbf{r}') = \exp\left[\frac{ie}{2\hbar c}\mathbf{r} \cdot \mathbf{B} \times \mathbf{r}' - \phi(\mathbf{r}) + \phi(\mathbf{r}')\right], \tag{16.15}$$

and $\phi(\mathbf{r})$ is an arbitrary gauge function, which we discard. It is important to note that $C(\mathbf{r}, \mathbf{r}')$ enters $G_0(md; nd)$ on the right of (16.4) in the form $C(md\hat{x}, nd\hat{x}) = \exp\left[\frac{ie}{2\hbar c} md\hat{x} \cdot \mathbf{B} \times nd\hat{x}\right] \equiv 1$ since $\hat{x} \cdot \mathbf{B} \times \hat{x} = 0$ for our choice $y \equiv y_1 \equiv y_2 \equiv 0$. Therefore $C(\mathbf{r}, \mathbf{r}')$ does not enter the denominator factor of (16.13), although it may be present in the numerator factors of the final Green's function, except when eliminated by restricting considerations to $\mathbf{r} \parallel \mathbf{r}'$, as we have done by taking $y \equiv 0$ for propagation on the lattice axis. This means that the eigen-energy spectrum given by the vanishing of the denominator is unaffected by $C(\mathbf{r}, \mathbf{r}')$.

The solution for the translationally invariant part of the nonrelativistic retarded Green's function, $G_0'(\mathbf{R}, \omega)$, is given by ($\mathbf{R} = \mathbf{r} - \mathbf{r}'$; $X = x - x'$; $Y = y - y'$;

$\hbar \to 1$; frequency representation) [12]

$$G_0'(\mathbf{R};\omega) = \frac{-m\omega_c}{4\pi}\int_0^\infty d\tau \frac{e^{i\omega\tau}}{\sin(\omega_c\tau/2)} \exp\left\{\frac{im\omega_c[X^2+Y^2]}{4\tan(\omega_c\tau/2)}\right\}, \tag{16.16}$$

where m is the mass and ω_c is the cyclotron frequency. Expanding the $\tau-$integrand as a generator of Laguerre polynomials [13], L_n, we obtain another useful representation as

$$G_0'(\mathbf{R};\omega) = \frac{m\omega_c}{2\pi}\exp(-m\omega_c R^2/4)\sum_{n=0}^{\infty} L_n\left(\frac{m\omega_c R^2}{2}\right)\frac{1}{\omega-(n+\frac{1}{2})\omega_c}. \tag{16.17}$$

16.4 Energy Spectrum: Landau Minibands

The energy spectrum of the 2D Krönig-Penney-type model for a quantum dot lattice in a normal magnetic field is given by the frequency poles of the Green's function arising from the vanishing of the denominator on the right hand side of (16.12):

$$\left[1-\alpha\dot{\ddot{G}}_0(p;0,0;\omega)\right] = 0. \tag{16.18}$$

Employing (16.16) and forming the Fourier series of (16.6) yields

$$\dot{\ddot{G}}_0(p;0,0;\omega) = -\frac{m\omega_c}{4\pi}\sum_{r=-\infty}^{\infty} e^{ipdr}\int_0^\infty dT\frac{e^{i\omega T}}{\sin(\omega_c T/2)}\exp\left[\frac{im\omega_c d^2 r^2}{4\tan(\omega_c T/2)}\right]. \tag{16.19}$$

Noting that the T-integral is a half-time-axis transform of a periodic function, the semi-infinite range of integration may be divided into individual periods which are summed and translated to the fundamental interval. Defining $z=\omega_c T/2$, the result takes the form

$$\dot{\ddot{G}}_0(p;0,0;\omega) = -\frac{m}{2\pi}\left[\sum_{n=0}^{\infty}(-1)^n\exp\left(\frac{i2\pi\omega n}{\omega_c}\right)\right]\int_0^\pi dz\frac{\exp(i2\omega z/\omega_c)}{\sin(z)} \times\sum_{r=-\infty}^{\infty} e^{ipdr}\exp\left[\frac{im\omega_c d^2}{4\tan(z)}r^2\right]. \tag{16.20}$$

The n-sum is readily evaluated as

$$\sum_{n=0}^{\infty}(-1)^n\exp(i2\pi\omega n/\omega_c) = \frac{\exp(-i\pi\omega/\omega_c)}{2\cos(\pi\omega/\omega_c)}, \tag{16.21}$$

and the r-sum is just the Jacobi Theta Function of the third kind, $\Theta_3(\alpha, \beta)$: This yields the dispersion relation for the eigen-energies of the 2D Krönig-Penney-type model of a quantum dot lattice in a magnetic field as (restore $\hbar$)

$$1 = -\frac{\alpha m}{4\pi\hbar^2} \frac{\exp(-i\pi\omega/\omega_c)}{\cos(\pi\omega/\omega_c)} \int_0^{\pi} dz \frac{\exp(i2\omega z/\omega_c)}{\sin(z)} \Theta_3\left(\frac{pd}{2\pi}\middle| \frac{m\omega_c d^2}{4\pi\hbar\tan(z)}\right). \tag{16.22}$$

It should be noted that the limit $d \to 0$ results in divergence of the T-integral of (16.19). Consequently, the limit of the Jacobi Theta function diverges as $d \to 0$. This is an artifact of the δ-function single-point confinement potentials assumed for the dots. More realistically, a spatial spreading of the δ-function dot potentials gives rise to an integral equation which "smears" the Green's function solution over its area, a^2, rendering the result finite. This divergence may be circumvented by considering the restriction $d \geq a > 0$ in the results above.

Alternatively, one can employ (16.17) jointly with (16.6) to rewrite the dispersion relation in the form

$$1 = \frac{\alpha m\omega_c}{2\pi\hbar^2} \sum_{r=-\infty}^{\infty} e^{ipdr} \exp[-m\omega_c d^2 r^2/4\hbar] \sum_{n=0}^{\infty} \frac{L_n(m\omega_c d^2 r^2/2\hbar)}{\omega - (n+1/2)\omega_c}. \tag{16.23}$$

(The divergence discussed above is now manifested in the r-sum as $d \to 0$, and it can be circumvented by the same restriction.) For small dot radius ($\frac{|\alpha|m}{2\pi\hbar^2} \ll 1$), (16.23) requires that $\omega \to \omega_n$ closely approach the pole position $(n+1/2)\omega_c$, so that particular pole is the predominant influence in determining the energy root ω_n: Consequently, a reasonable first approximation may be undertaken dropping all other terms of the $n-$sum, leading to the result

$$\omega_n \cong (n+1/2)\omega_c + \frac{\alpha m\omega_c}{2\pi\hbar^2} \sum_{r=-\infty}^{\infty} e^{ipdr} \exp[-m\omega_c d^2 r^2/4\hbar] L_n(m\omega_c d^2 r^2/2\hbar). \tag{16.24}$$

When $m\omega_c d^2/4\hbar > 1$, it suffices to retain only $r = -1, 0, 1$ of the r-sum: Applying this to a GaAs-based antidot lattice having dot potential well depth about 100 meV and diameter 2nm, we find a proliferation of subband energies that are quantized by the magnetic field with frequencies $\omega_n (n = 0, 1, 2 \ldots \infty)$ approximately given by

$$\omega_n = (n+1/2)\omega_c + \frac{\alpha m\omega_c}{2\pi\hbar^2}(1 + 2\cos(pd)\exp[-m\omega_c d^2/4\hbar] L_n(m\omega_c d^2/2\hbar)), \tag{16.25}$$

for the case in which the lattice period d is larger than the orbit radius of the lowest Landau level. Equation (16.25) indicates that the quantum dot lattice broadens each Landau eigenstate into a subband (miniband) of width

$$\Delta\omega_n = \frac{2\alpha m\omega_c}{\pi\hbar^2}\exp[-m\omega_c d^2/4\hbar]L_n(m\omega_c d^2/2\hbar), \tag{16.26}$$

and generates an effective mass in the neighborhood of the nth subband minimum $p = \pi/d$ given by

$$\frac{1}{m_n^*} = \frac{d^2\Delta\omega_n}{2\hbar} = \frac{d^2\Delta E_n}{2\hbar^2}. \tag{16.27}$$

16.5 Landau Quatized Graphene: Wave Packet Dynamics in an Antidot Lattice

Addressing "relativistic" Dirac materials, we focus on graphene in a normal magnetic field (in the absence of any lattice) with Hamiltonian ($\hbar = c = 1$)

$$H = \gamma\boldsymbol{\sigma}_\nu \cdot (\mathbf{p} - \mathbf{eA}), \tag{16.28}$$

where $\sigma_\nu = [\sigma_x, 1_\nu\sigma_y]$ and σ_x, σ_y are Pauli spin matrices, which act on the sublattice/pseudospin space and represent the sublattice degree of freedom of graphene's honeycomb lattice structure, also $1_\nu = 1$ or -1 for K or K' valleys of the low energy graphene band structure, and $\gamma = \frac{3}{2}\alpha_h d \approx 10^6 ms^{-1}$ is the characteristic speed (α_h is the hopping amplitude originating from the tight binding approximation in which the lattice spacing is d). $\mathbf{A}$ is the vector potential of constant, uniform normal magnetic field $\mathbf{B}$. Furthermore, we employ an antidot lattice in the present case by reversing the sign of $\alpha > 0$ in (16.1) and in the antidot lattice Green's functions of (16.13). Again (16.14) and (16.15) are applicable and the 2×2 "relativistic" Landau quantized graphene matrix Green's function has a spatially translationally invariant part $\mathcal{G}'(\mathbf{r} - \mathbf{r}'; t - t')$ given by the equation

$$\left[i\frac{\partial}{\partial T} - \gamma\sigma_\nu \cdot \left(\frac{1}{i}\frac{\partial}{\partial \mathbf{R}} - \frac{e}{2}\mathbf{B} \times \mathbf{R}\right)\right]\mathcal{G}'(\mathbf{R}, T) = I_2\delta^2(\mathbf{R})\delta(T), \tag{16.29}$$

where $\mathbf{R} = \mathbf{r} - \mathbf{r}'$ such that $X = x - x'$, $Y = y - y'$ and $T = t - t'$ and I_2 is the 2×2 unit matrix.

The diagonal elements of the Green's function matrix (ω representation) for this system of monolayer graphene in a uniform, constant, perpendicular magnetic field are ($\mathcal{G}'_{\substack{11\\22}}$ collectively represents $\mathcal{G}'_{11}, \mathcal{G}'_{22}$) [9]

$$\mathcal{G}'_{\substack{11\\22}}(\mathbf{R}; \omega) = \frac{1}{4\pi\hbar\gamma^2}\ \omega e^{-\frac{\omega_g^2}{8\gamma^2}(X^2+Y^2)}\sum_{n=0}^{\infty}\frac{L_n\left[\frac{\omega_g^2}{4\gamma^2}\left(X^2+Y^2\right)\right]}{\frac{\omega^2}{\omega_g^2} - \left(n + \frac{1-1_\nu}{2}\right)}. \tag{16.30}$$

We introduced the notation

$$\omega_g = \gamma\sqrt{\frac{2eB}{\hbar}}, \tag{16.31}$$

which is the cyclotron frequency for Dirac fermions.

16.5.1 Landau Quantization

The poles of $\mathcal{G}'_{\substack{11\\22}}(\mathbf{R};\omega)$ show that the energy spectrum is Landau quantized, with Landau level index given by n. The frequency poles of (16.30) provide the energy spectrum of the infinite graphene sheet in perpendicular magnetic field (at the location of K-point) [14] as

$$\omega_K = \pm\omega_g\sqrt{n}. \tag{16.32}$$

Here the positive energy values in above equation correspond to conduction band (electrons), and the negative energy values correspond to valence band (holes). Furthermore, these levels are not equidistant as in the case of Landau quantization in conventional electronic systems. From (16.32), the maximum energy separation occurs between the zero'th and first Landau level. Quantum-Hall effect in graphene is the consequence of this large gap, even at normal (room) temperature [15].

And the energy pole positions at the location of K' point for an infinite graphene sheet in a perpendicular magnetic field are given by (16.30) as

$$\omega_{K'} = \pm\omega_g\sqrt{(n+1)}. \tag{16.33}$$

The residues obtained at the pole positions of equations (16.32) and (16.33) represent the relative strengths of the modes.

The K' point lacks the energy level with $E = 0$ as shown in Figs. 16.1a and b. One can expect some interesting features due to the presence of two sublattices in graphene, which are labeled sublattice A and sublattice B. Actually, occupation of these sublattices gives rise to many interesting properties. Specifically, the Landau levels corresponding to $n \neq 0$ have non-zero energies and consequently wave functions at these Landau levels carry non-zero amplitudes on sublattices A and B. But, the Landau level corresponding to $n = 0$ shows unique behaviour i.e the amplitude of the wave function is non-zero only at one sublattice (sublattice A for K' valley or sublattice B for K valley) because that state with energy $E = 0$ occurs only at one point of the Brillouin zone i.e K or K' point. This asymmetry property of the wave functions corresponding to the $n = 0$ Landau level makes the zero'th Landau level play a remarkable role in various magnetic applications of graphene. The reason behind the asymmetry of sublattices A and B is asymmetric positions of the nearest neighboring atoms in sublattices A and B.

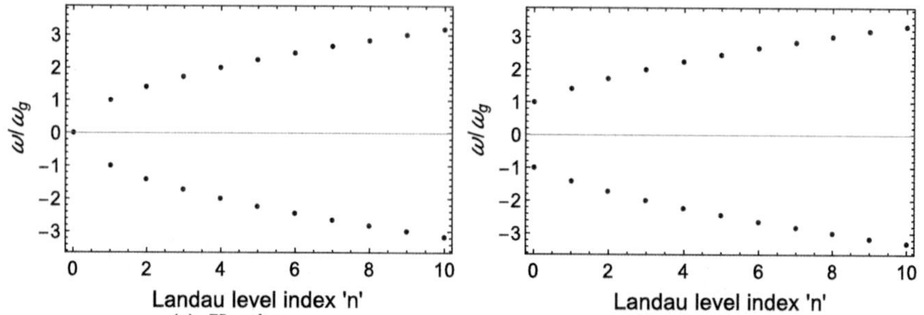

Fig. 16.1 Landau levels (n=0–10) and their energies for a Landau quantized monolayer graphene. **a** At K point in graphene. **b** At K' point in graphene

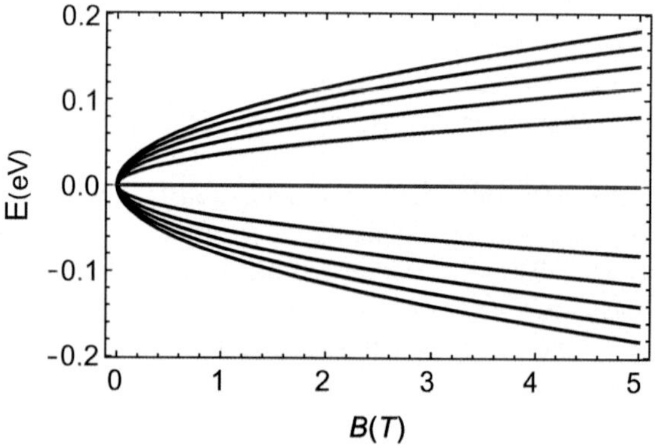

Fig. 16.2 Energy levels of graphene subjected to Landau quantization for K point as a function of magnetic field in units of Tesla

Also, note that the energy of a Landau level increases with $\sqrt{B}$ according to (16.31). Figure 16.2 clearly shows the increase in energy of the first five Landau levels at the location of K point with the variation in the magnetic field from zero to 5 $Tesla$, it also shows that the positive and negative energy Landau levels meet at the point with $E = 0$ (Dirac point) when there is no magnetic field.

16.5.2 Density of States

The Density of states (DOS) per unit area $\rho(\mathbf{r}; E)$ is the imaginary part of the diagonal matrix element of the Green's function, [W. Huaiyu, "Green's Function in Condensed Matter Physics", Science Press, 2012.]

$$\rho(\mathbf{r}; E) = \mp\frac{1}{\pi} Im\mathcal{G}^{\pm}(\mathbf{r}, \mathbf{r}; E), \tag{16.34}$$

and the total density of states is the imaginary part of the trace of the diagonal matrix elements of the Green's function:

$$\rho(E) = \int d\mathbf{r}\rho(\mathbf{r}; E) = \mp\frac{1}{\pi} Im\left[Tr\mathcal{G}^{\pm}(E)\right], \tag{16.35}$$

where (η is a small positive number introduced to facilitate computation)

$$\mathcal{G}^{\pm}(\mathbf{r}, \mathbf{r}'; E) = \lim_{\eta\to 0^+} \mathcal{G}(\mathbf{r}, \mathbf{r}'; E \pm i\eta) = \mathcal{G}(\mathbf{r}, \mathbf{r}'; E \pm i0^+). \tag{16.36}$$

The notation 0^+ represents a positive small number and its opposite number is 0^-. (i.e: $E - i0^+ = E + i0^-$).

The Green's function for the K-point given in (16.30) (using $R^2 = X^2 + Y^2$) can be written as

$$\mathcal{G}'_{11}(\mathbf{R}; \omega)_K = \frac{1}{4\pi\hbar\gamma^2}\omega e^{-\left(\frac{\omega_g^2}{8\gamma^2}R^2\right)} \sum_{n=0}^{\infty} \frac{L_n\left[\frac{\omega_g^2}{4\gamma^2}R^2\right]}{\frac{\omega^2}{\omega_g^2} - n}. \tag{16.37}$$

Putting $\mathbf{r} = \mathbf{r}', \mathbf{R} \to 0$, Peierls phase factor $C(\mathbf{r}, \mathbf{r}') = 1$, so the energy representation of the Green's function (using notation $\mathcal{G}_{11}(\mathbf{r}, \mathbf{r}; E)_K = \mathcal{G}(\mathbf{r}, \mathbf{r}; E)_K$) is given by ($E = \hbar\omega$; $E_g = \hbar\omega_g$)

$$\mathcal{G}(\mathbf{r}, \mathbf{r}; E)_K = \frac{E_g^2}{4\pi(\hbar\gamma)^2} E \sum_{n=0}^{\infty} \frac{1}{E^2 - nE_g^2} \tag{16.38}$$

Using (16.36), (16.38) may be written as

$$\mathcal{G}^{+}(\mathbf{r}, \mathbf{r}; E)_K = \frac{E_g^2}{4\pi(\hbar\gamma)^2} \lim_{\eta\to 0^+} \left[\sum_{n=0}^{\infty} \frac{(E + i\eta)}{(E + i\eta)^2 - nE_g^2}\right].$$

It is straightforward to separate the imaginary part of the above equation for use in (16.34), finally

$$\left(\frac{\rho(E)}{\rho_0}\right)_K = \lim_{\eta'\to 0^+} \left[\sum_{n=0}^{\infty} \frac{\left(E'^2 + \eta'^2 + n\right)\eta'}{\left(E'^2 - \eta'^2 - n\right)^2 + 4E'^2\eta'^2}\right], \tag{16.39}$$

where we have used

$$E_g = \hbar\omega_g, \qquad \rho_0 = \frac{E_g}{4\pi^2(\hbar\gamma)^2}, \qquad E' = \frac{E}{E_g}, \qquad and \qquad \eta' = \frac{\eta}{E_g}, \tag{16.40}$$

and for K'-point, n is to be replaced by $n + 1$ in (16.39).

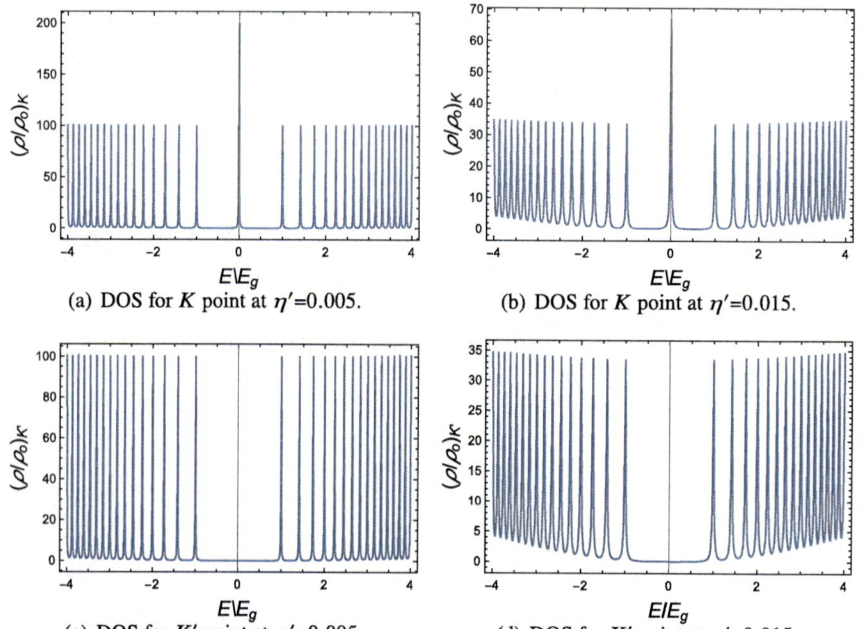

(a) DOS for K point at η'=0.005.

(b) DOS for K point at η'=0.015.

(c) DOS for K' point at η'=0.005.

(d) DOS for K' point at η'=0.015.

Fig. 16.3 Density of states (DOS) as a function of energy at the sites of both Dirac points in a Landau quantized monolayer graphene at two different values of η'. **a** DOS at $\eta' = 0.005$ for K point. **b** DOS at $\eta' = 0.015$ for K point. **c** DOS at $\eta' = 0.005$ for K' point. **d** DOS at $\eta' = 0.015$ for K' point

Calculated results for both (K and K') Dirac points are exhibited in Fig. 16.3a–d. These plotted results for the DOS are in good agreement with analytic results obtained in (16.32) and (16.33), i.e corresponding to the energy of each Landau level; the densities of states show sharp peaks, and for the energies between the Landau levels, the density of states is zero. Ideally, these lines are very sharp like a Dirac delta function of zero width. Actually each line corresponds to a unique Landau level (energy spectrum) with no continuum states, but we introduced a small artificial width "η" to facilitate the computation. Physically, the small artificial width corresponds to the continuum of energy states which are induced by interactions of electrons with surroundings (substrate and impurities, etc.).

Also note that, generally, the K and K' spectra (DOS) for graphene are closely aligned, except for the central mode of the K point, which has no K' counterpart as shown by the Fig. 16.3a, c and b, d.

16.5.3 Off-Diagonal Elements

The off-diagonal matrix elements can be calculated using the relation ($\gamma_\nu \equiv 1_\nu \gamma$) [9]

$$\omega \mathcal{G}'_{\substack{21\\12}} = [\gamma \Pi_{XY} \pm i\gamma_\nu \Pi_{YX}]\mathcal{G}'_{\substack{11\\22}}. \tag{16.41}$$

In the above equation, $\Pi_{XY} \equiv \frac{1}{i}\frac{\partial}{\partial X} + \frac{eB}{2}Y$ and $\Pi_{YX} \equiv \frac{1}{i}\frac{\partial}{\partial Y} - \frac{eB}{2}X$ are canonical momentum operators and $\mathcal{G}'_{\substack{21\\12}} = \mathcal{G}'_{21}$ or $\mathcal{G}'_{12}$ corresponds to the upper or lower $\pm$, $\mp$ signs elsewhere in the equations. Making use of (16.41), the off-diagonal elements of the Green's function matrix for K and K' can be separately expressed as [16, 17]

$$K(1_\nu = +1) : \mathcal{G}'_{\substack{21\\12}}(\mathbf{R}; \omega) = \frac{\omega_g^2}{8\pi\hbar\gamma^3} e^{-\frac{\omega_g^2}{8\gamma^2}(X^2+Y^2)} (iX \mp Y) \sum_{n=1}^{\infty} \frac{L_{n-1}^1\left[\frac{\omega_g^2}{4\gamma^2}(X^2+Y^2)\right]}{\frac{\omega^2}{\omega_g^2} - n}, \tag{16.42}$$

$$K'(1_\nu = -1) : \mathcal{G}'_{\substack{21\\12}}(\mathbf{R}; \omega) = \frac{\omega_g^2}{8\pi\hbar\gamma^3} e^{-\frac{\omega_g^2}{8\gamma^2}(X^2+Y^2)} (iX \pm Y) \sum_{n=1}^{\infty} \frac{L_{n-1}^1\left[\frac{\omega_g^2}{4\gamma^2}(X^2+Y^2)\right]}{\frac{\omega^2}{\omega_g^2} - (n+1)}. \tag{16.43}$$

Clearly, the off-diagonal elements interchange with the interchange of Dirac points (K and K').

The Green's function matrix in time representation can be obtained by Fourier transform of (16.30), (16.42), (16.43) as

$$\mathcal{G}'_{\mu\nu}(\mathbf{R}; t) = \int_{-\infty}^{\infty} \frac{d\omega}{2\pi} e^{-i\omega t} \mathcal{G}'_{\mu\nu}(\mathbf{R}; \omega), \tag{16.44}$$

where $\mu, \nu = 1, 2$ denote matrix indices.

Noting that $\mathcal{G}'_{\mu\nu}(\mathbf{R}; \omega)$ has real energy poles at $\epsilon_K = \pm\omega_g\sqrt{n}$; $\epsilon_{K'} = \pm\omega_g\sqrt{n+1}$ we employ contour integration with $\omega \to \omega + i0^+$ for the retarded Green's function with the contour closed in the lower half plane running clockwise from $+\infty$ to $-\infty$. For the K-point, we have (for the K'-point $\sqrt{n} \to \sqrt{n+1}$)

$$\mathcal{G}'_{\substack{11\\22}}\left(x, x'; y, y'; t\right) = -i\eta_+(t)\frac{\omega_g^2}{4\pi\gamma^2} e^{-\frac{1}{2}\zeta} \sum_{n=0}^{\infty} L_n[\zeta] \cos\left(\frac{\omega_g l}{\gamma}\frac{t}{\tau_o}\sqrt{n}\right), \tag{16.45}$$

$$\mathcal{G}'_{\substack{21\\12}}\left(x,x';y,y';t\right) = -\eta_+(t)\frac{\omega_g^3 l}{8\pi\gamma^3}e^{-\frac{1}{2}\zeta}\left[i\left(\frac{x-x'}{l}\right)\mp\left(\frac{y-y'}{l}\right)\right]\sum_{n=1}^{\infty}\frac{L_{n-1}^1[\zeta]}{\sqrt{n}} \times \sin\left(\frac{\omega_g l}{\gamma}\frac{t}{\tau_o}\sqrt{n}\right). \tag{16.46}$$

Here, we have introduced an arbitrary constant length, l, chosen for convenience to be the width of an impressed wave packet, and $\eta_+(t) = 0, 1$ for $t < 0, t > 1$, respectively, is the Heaviside unit step function; also, we have defined

$$\zeta = \frac{1}{4}\left(\frac{\omega_g l}{\gamma}\right)^2\left[\left(\frac{x-x'}{l}\right)^2 + \left(\frac{y-y'}{l}\right)^2\right],$$

and $\tau_o = l/\gamma$.

To study the wave packet dynamics, we take the initial wave function to be a Gaussian wave packet having nonvanishing average momentum $p_{0x} = \hbar k_{0x}$ and width l,

$$\psi(\mathbf{r},0) = \frac{f(\mathbf{r})}{\sqrt{|\,c_1\,|^2 + |\,c_2\,|^2}}\begin{pmatrix}c_1\\c_2\end{pmatrix}, \tag{16.47}$$

$$f(\mathbf{r}) = \frac{1}{l\sqrt{\pi}}exp\left(-\frac{x^2+y^2}{2l^2} + ik_{0x}x\right),$$

where c_1 and c_2 are the coefficients which set the initial pseudospin polarization. Also $\psi(\mathbf{r},0)$ can be taken as a smooth enveloping function considering that lattice period is much smaller than the width l of the initial wave packet. The Green's function matrix elements $\mathcal{G}_{\mu\nu}(\mathbf{r},\mathbf{r}',t)$ determine the time evolution of an arbitrary initial state $\psi(\mathbf{r},0)$; in Schrödinger representation it is given by

$$\psi_\mu(\mathbf{r},t) = \int d\mathbf{r}'\mathcal{G}_{\mu\nu}(\mathbf{r},\mathbf{r}',t)\psi_\nu(\mathbf{r}',0), \tag{16.48}$$

where $\mu, \nu = 1, 2$ denote the matrix indices, which correspond to upper component $\psi_1(\mathbf{r},t)$ and lower component $\psi_2(\mathbf{r},t)$ of state $\psi_\mu(\mathbf{r},t)$. The probability density is

$$\rho(\mathbf{r},t) = |\,\psi(\mathbf{r},t)\,|^2 = |\,\psi_1(\mathbf{r},t)\,|^2 + |\,\psi_2(\mathbf{r},t)\,|^2\,. \tag{16.49}$$

And to study ZB, the average value of coordinates can be represented as

$$\bar{x}_j = \int \psi_1^*(\mathbf{r},t)x_j\psi_1(\mathbf{r},t)d\mathbf{r} + \int \psi_2^*(\mathbf{r},t)x_j\psi_2(\mathbf{r},t)d\mathbf{r}, \tag{16.50}$$

where $j = 1, 2$ with $x_1 = x$ and $x_2 = y$. It is interesting to note that Zitterbewegung oscillations can also be observed in laboratories if they fulfill appropriate requirements: [18]

- Oscillations should have some considerable lifetime so that they can be observed easily.
- Frequency of the oscillations should not be greater than 10^{15} Hz i.e frequency should lie in currently detectable regime.
- Amplitude of the oscillations should not be less than a few angstroms.

16.6 Wave Packet Dynamics with Various Pseudospin Polarizations: Zitterbewegung

To obtain results for the temporal evolution of the initial Gaussian wave packet, ZB oscillations and the effect of initial-pseudospin polarization, we have performed numerical calculations. To facilitate it, we introduced the following dimensionless variables: [19–21]

- A dimensionless parameter which is suitable to replace wave vector k_{0x} is $a_0 = k_{0x}l$.
- Distance of propagation of the wave packet can be measured in units of initial width l of wave packet.
- Time can be measured in $\tau_o = l/\gamma$ (γ is Fermi velocity, 10^6 m/s) units.
- Some other variables can be combined to produce dimensionless variables, e.g $b = \frac{\omega_g l}{\gamma}$.

The Landau level summation is performed up to the $10th$ Landau level in all calculations since the results become convergent in this limit.

We consider four cases with different initial pseudospin polarizations $\{c_1, c_2\}$ for the Gaussian wave packet given in (16.47).

Case-1: $\{c_1, c_2\} = \{1, 0\}$ which corresponds to initial electron probability of one at the sites of sublattice A.

Case-2: $\{c_1, c_2\} = \{1, 1\}$ corresponds to the situation where the pseudospin is directed along the x-axis.

Case-3: $\{c_1, c_2\} = \{1, i\}$ corresponds to the pseudospin directed along y-axis.

Case-4: $\{c_1, c_2\} = \{1, e^{i\pi/4}\}$ implies that at time t=0, the pseudospin lies in x-y plane making an angle of 45° with x-axis.

The numerical results obtained from (16.48), (16.49) are plotted in Fig. 16.4.

In Fig. 16.4, the electronic probability density is plotted for parameters t=1, 5, $10\tau_o$, with momentum $k_{0x} = 0.6nm^{-1}$, width l=2 nm so that a_0=$k_{0x}l$=1.2 and τ_o=2 femtosecond and B=3.3 in units of $Tesla$, i.e. $\omega_g \simeq 1 \times 10^{14}$ Hz. B=3.3T is chosen to facilitate comparison with work in the literature and also because it facilitates numerical computation as ω_g is a round figure at this value [19]. Left panel shows the initial wave packet at a very small time t=1τ_o, and as we move from left to right the time evolution of the initial Gaussian packet can be seen for different initial pseudospin polarizations. The strength of the electron probability density $\rho(x, y, t)$ is given by the color bar on the right side.

As one can see in Fig. 16.4a, b, c, the wave packet spreads and propagates in the plane of the graphene sheet in the form of rings for the pseudospin c_1=1 and c_2=0. Initially in Fig. 16.4a, maximum probability of the electron is located at the origin, but as time increases, the wave packet propagates and the electron density can be found at a radius of $r \simeq 24$ nm in $20\,fs$. Similarly, for the other three cases, the wave packet propagates with its maximum probability density in the direction of the initial pseudospin polarizations, but in the shape of incomplete rings; this is because the probability density gradually decreases in the directions away from the direction of pseudospin polarization. It propagates in the x direction when the initial pseudospin polarization is along the x axis (see Figs. 16.4d, e, f), in the y direction when initial pseudospin polarization is along the y axis (see Figs. 16.4g, h, i) and it propagates in (x, y)=(1, 1) direction when initial pseudospin is polarized along (x, y)=(1, 1) direction (see Figs. 16.4j, k, l). This is because of the conservation of chirality, in which momentum gets aligned with pseudospin and $\boldsymbol{\sigma}_\nu.\mathbf{p}$ remains conserved. Also, the distance covered by the wave packet in these three cases is $r \simeq 24$ nm in $20\,fs$. Hence the direction of propagation of a wave packet in Landau quantized graphene can be controlled using pseudospin polarization. Also, the wave packet propagates without any splitting; splitting was observed in the case of monolayer graphene in the absence of a magnetic field [19]. We propose that, experimentally, this type of controlled propagation of a wave packet in any direction can easily be obtained using photonic graphene as test beds [22].

Also note that the electronic probability densities plotted in Fig. 16.4 are not symmetric with respect to both x and y axes: $\rho(x, y, t)_{\{c_1,c_2\}} \neq \rho(-x, y, t)_{\{c_1,c_2\}}$ and $\rho(x, y, t)_{\{c_1,c_2\}} \neq \rho(x, -y, t)_{\{c_1,c_2\}}$ for any pseudospin polarization (subscript $\{c_1, c_2\}$ defines the corresponding pseudospin polarization): This means that center of the wave packet is oscillating along both the x as well as the y directions; these oscillations can be readily recognized as Zitterbewegung oscillations.

To examine this trembling motion we use (16.50) and solve it numerically for expectation values of both x and y coordinates. Figures 16.5 a, c, e, g and b, d, f, h show the oscillations in the wave packet's center along x and y directions respectively. Results are plotted for two values of initial momentum $k_{0x} = 0.6$ and 0.8 i.e $a_0 = k_{0x}\, l = 1.2$ and 1.6, given by solid and dotted lines for four different values of pseudospin polarization. It is clear from Fig. 16.5 that an increase in momentum results in a decrease in amplitude of Zitterbewegung oscillations without any other change in the behaviour of ZB. In further discussion, ZB oscillations corresponding to different pseudospin polarizations will be referred to as $\bar{x}_{\{c_1,c_2\}}$ and $\bar{y}_{\{c_1,c_2\}}$.

The ZB oscillations shown in Fig. 16.5 are of the order of *nanometers* (easily detectable), except $x(t)_{\{1,0\}}$ and $x(t)_{\{1,i\}}$ which are of the order of $0.1\,pm$. On comparing the ZB oscillations $x(t)_{\{1,0\}}$, $x(t)_{\{1,1\}}$, $x(t)_{\{1,i\}}$, we have concluded that when the direction of initial momentum (x-axis) is perpendicular to initial pseudospin polarizations (z, y-axis) then the ZB oscillations have very small amplitude in the direction of initial momentum. Note that, ZB oscillations in Figs. 16.5g and h are

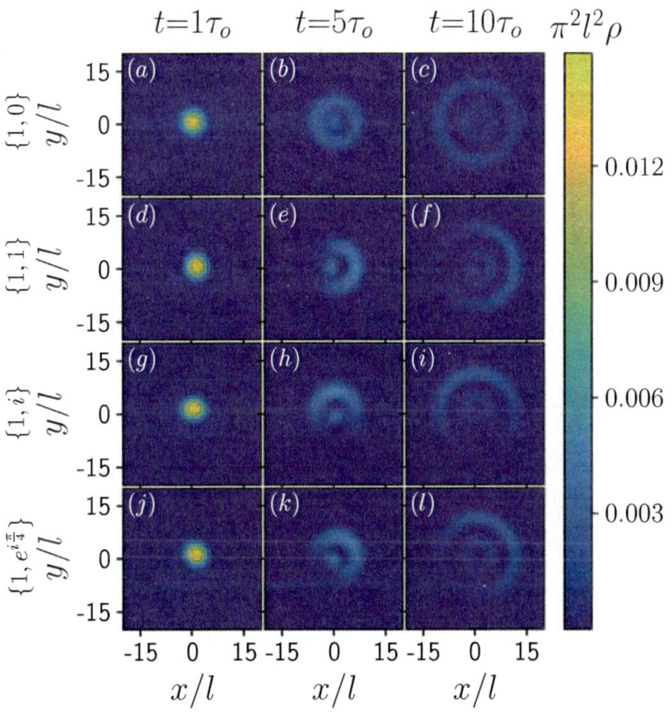

Fig. 16.4 (Color online) The electron probability density $\rho(x, y, t)$ for an initial Gaussian wave packet with $B = 3.3Tesla$, a_0=$k_{0x}l$=1.2 at times $1\tau_o$, $5\tau_o$ and $10\tau_o$. Time increases as we move from left to right. Color bar given at extreme right side shows the strength of probability density from its minimum to its maximum. Up to down: Four rows with initial pseudospin polarizations $\{1, 0\}$,$\{1, 1\}$,$\{1, i\}$ and $\{1, e^{i\frac{\pi}{4}}\}$ with pseudospin and propagation directed in the radial, x, y and ($x = 1$, $y = 1$) directions respectively. [Reproduced from "R. A. W. Ayyubi, N. J. M. Horing, and K. Sabeeh. Effect of pseudospin polarization on wave packet dynamics in graphene antidot lattices (GALs) in the presence of a normal magnetic field. Journal of Applied Physics, 129(7):074301, 2021", with the permission of AIP Publishing. https://aip.scitation.org/doi/10.1063/5.0038316]

very similar to those of Fig. 16.5c and f respectively. This is because pseudospin polarization $\{1, e^{i\frac{\pi}{4}}\}$ has projections on both x and y axes so ZB oscillations in this case have detectable amplitude in both directions.

In Figs. 16.5a–h we have seen that initially the amplitude of the Zitterbewegung oscillations increases, then these oscillations seems to die out but reappear for all pseudospin polarizations. For example in Fig. 16.5c these oscillations reappear at $t \simeq 18, 55, 84\tau_o$... . Hence, when there is a magnetic field applied to the system, Zitterbewegung oscillations are not transient; rather they are recurrent. Also, in the presence of the magnetic field, several ZB frequencies appear (see Figs. 16.5a–h). This is different from the ZB phenomenon observed in monolayer graphene without a magnetic field, in which ZB oscillations are transient having a single frequency [19].

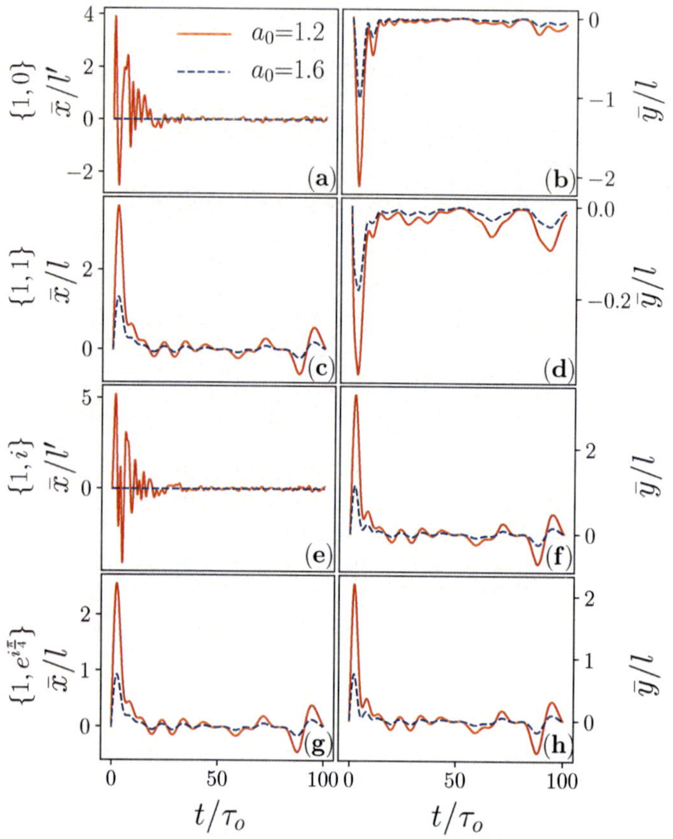

Fig. 16.5 Average coordinates $\bar{x}(t)$ (left column) and $\bar{y}(t)$ (right column) versus time ($\tau_o = l/\gamma$) corresponding to four different values of pseudospin polarization (changing from up to down), at two values of momentum a_0 with $B = 3.3Tesla$. **a**, **c**, **e**, **g**: $\bar{x}(t)$ versus time corresponding to pseudospin polarizations $\{1, 0\}$,$\{1, 1\}$, $\{1, i\}$ and $\{1, e^{i\frac{\pi}{4}}\}$ respectively. **b**, **d**, **f**, **h**: $\bar{y}(t)$ versus time corresponding to pseudospin polarizations $\{1, 0\}$,$\{1, 1\}$, $\{1, i\}$ and $\{1, e^{i\frac{\pi}{4}}\}$ respectively. Here $l'=l \times 10^{-5}$. [Reproduced from "R. A. W. Ayyubi, N. J. M. Horing, and K. Sabeeh. Effect of pseudospin polarization on wave packet dynamics in graphene antidot lattices (GALs) in the presence of a normal magnetic field. Journal of Applied Physics, 129(7):074301, 2021", with the permission of AIP Publishing. https://aip.scitation.org/doi/10.1063/5.0038316]

Hence, the results in hand are in agreement with the previous studies in both respects i.e., (1) in the presence of a magnetic field ZB oscillations are permanent and (2) ZB oscillations strongly depend on initial pseudospin polarization [23, 24]. Finally, in Fig. 16.6 we have plotted the average coordinates $\bar{x}(t)_{\{c_1,c_2\}}$ and $\bar{y}(t)_{\{c_1,c_2\}}$ against each other to study the ZB trajectory of the center of the wave packet corresponding to the Figs. 16.5c, d. Initially due to large ZB, the center of the packet sweeps a large area. For a better understanding, we have shown a zoomed

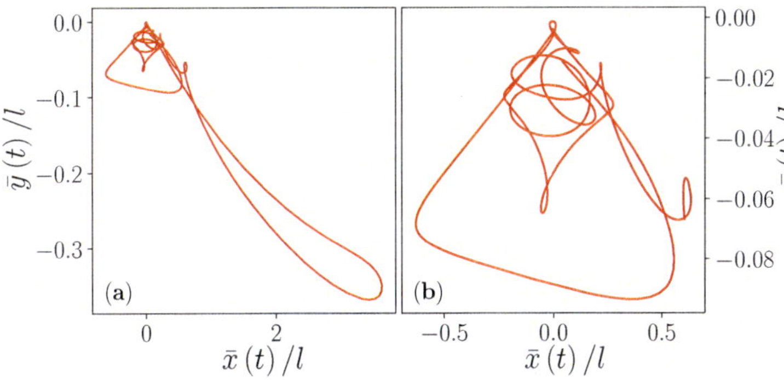

Fig. 16.6 Average coordinates $\bar{x}(t)$ versus $\bar{y}(t)$ corresponding to pseudospin polarization $\{c_1, c_2\}$=$\{1, 1\}$ for momentum $a_0 = 1.2$ with $B = 3.3Tesla$. **a** ZB Trajectory for $t = 0$ to $t = 100\tau_o(0.2ps)$. **b** Zoomed view of the same trajectory for $t = 7\tau_o$ to $t = 100\tau_o$. [Reproduced from "R. A. W. Ayyubi, N. J. M. Horing, and K. Sabeeh. Effect of pseudospin polarization on wave packet dynamics in graphene antidot lattices (GALs) in the presence of a normal magnetic field. Journal of Applied Physics, 129(7):074301, 2021", with the permission of AIP Publishing. https://aip.scitation.org/doi/10.1063/5.0038316]

view of the ZB trajectory in Fig. 16.6b. Due to the presence of recurrent ZB, these trajectories do not disappear with time (infinite trajectories) [21].

It is well known that the electronic band structure of graphene can be modified by introducing external ID potentials using nanopatterning [25]. These 1D potentials can not only modify the energy spectrum of the system but they also help us to control the charge transport properties of the system [26–30]. Potentials such as antidot lattices can be carved on graphene by various techniques and lattice parameters can be tuned [31, 32]. With the aim of controlling electron propagation in graphene along a 1D channel, we introduce a 1D antidot lattice in the following section.

16.7 Graphene Antidot Lattice in the Presence of a Magnetic Field

As discussed in Sect. 16.2 we consider a two-dimensional graphene sheet having a one dimensional lattice of quantum antidot potential barriers, with a quantizing magnetic field B, which is perpendicular to the plane of the graphene sheet. The antidot array is modeled by Dirac delta functions periodically spaced along the x-axis at $x_{n'} = n'd$, $y \equiv 0$ as

$$U(\mathbf{r}) = U(x, y) = \alpha \sum_{n'=-\infty}^{\infty} \delta(x - n'd)\delta(y), \tag{16.51}$$

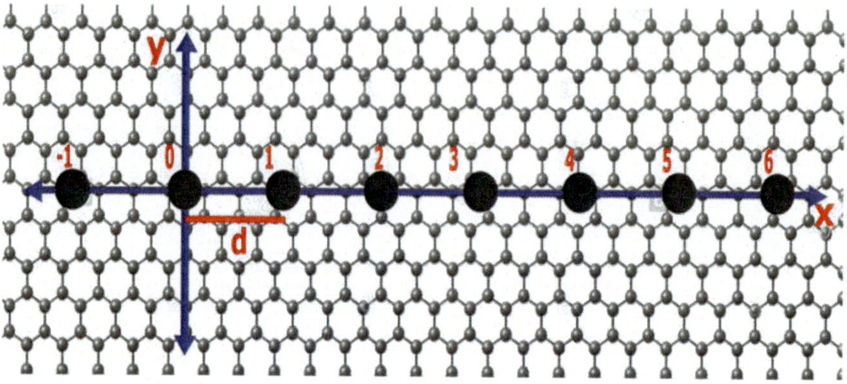

Fig. 16.7 Two dimensional graphene sheet having a one dimensional lattice of quantum antidot barriers at $y = 0$. Solid black circles represent periodically placed quantum antidots along the x-axis. a^2 is the dot area and d is the uniform separation of the periodically placed quantum antidot potential barriers. [Reproduced from "R. A. W. Ayyubi, N. J. M. Horing, and K. Sabeeh. Effect of pseudospin polarization on wave packet dynamics in graphene antidot lattices (GALs) in the presence of a normal magnetic field. Journal of Applied Physics, 129(7):074301, 2021", with the permission of AIP Publishing. https://aip.scitation.org/doi/10.1063/5.0038316]

where $\alpha > 0$ for antidot potential strength, with integers $n' = -\infty$ to $+\infty$ and

$$\alpha = U_0 \ a^2, \tag{16.52}$$

where U_0 is the barrier height, a^2 is area and d is the uniform separation of the periodic quantum antidot potential barriers as shown in Fig. 16.7. The associated Green's function $\mathcal{G}(x_1, x_2; 0, 0; \omega)$ for the two dimensional Krönig-Penney-like model having a one dimensional antidot lattice, for an electron propagating directly along the axis of the antidot lattice ($y \equiv 0 \equiv y_1 \equiv y_2$ and suppressing further reference to y) is completely analogous to that of the quantum dot lattice discussed in Sect. 16.2. This Green's function is given by [33]

$$\begin{aligned}\mathcal{G}(x_1; x_2; \omega) = \mathcal{G}^0(x_1; x_2; \omega) + \frac{\alpha d}{2\pi} \sum_{n'=-\infty}^{\infty} \mathcal{G}^0(x_1; n'd; \omega) \\ \times \int_{-\frac{\pi}{d}}^{\frac{\pi}{d}} dp \ e^{-ipn'd} \left[1 - \alpha \dot{\tilde{\mathcal{G}}}^0(p; 0, 0; \omega)\right]^{-1} \tilde{\mathcal{G}}^0(p; x_2; \omega).\end{aligned} \tag{16.53}$$

Here, $\mathcal{G}^0(\mathbf{r}_1, \mathbf{r}_1')$ is the Green's function for graphene in a perpendicular magnetic field in the complete absence of quantum antidot potential barriers, which is given by (16.30),(16.42) and (16.43), while the overhead dot represents the spatially translationally invariant Green's function i.e. $\mathcal{G}^0(m'd, n'd; \omega) = \dot{\mathcal{G}}^0([m' - n']d; \omega)$ where m' and n' are integers. Also, the Green's function $\tilde{\mathcal{G}}(p)$ can be expanded in a Fourier series due to the periodicity of the lattice, given by [33]

$$\tilde{\mathcal{G}}(p) = \sum_{r=-\infty}^{\infty} e^{iprd}\mathcal{G}(rd), \tag{16.54}$$

where $r = 0, 1, 2, 3...$ are integers, with

$$\mathcal{G}(m'd) = \frac{d}{2\pi}\int_{-\frac{\pi}{d}}^{\frac{\pi}{d}} dpe^{-ipdm'}\tilde{\mathcal{G}}(p), \tag{16.55}$$

and

$$\dot{\mathcal{G}}^0([m'-n']d) = \frac{d}{2\pi}\int_{-\frac{\pi}{d}}^{\frac{\pi}{d}} dpe^{-ip(m'-n')d}\dot{\tilde{\mathcal{G}}}^0(p). \tag{16.56}$$

It is important to note that the Peierls phase factor is $C(\mathbf{r}, \mathbf{r}')$=1 for the case involving propagation directly along the axis of the antidot lattice (our choice $y \equiv 0 \equiv y_1 \equiv y_2$ results in $\mathbf{r} \parallel \mathbf{r}'$). Therefore $C(\mathbf{r}, \mathbf{r}')$ does not appear in (16.53) and the eigen-energy spectrum given by poles of (16.53) is unaffected by $C(\mathbf{r}, \mathbf{r}')$.

16.7.1 Landau Minibands

The energy spectrum of this system can be obtained from the vanishing of the frequency poles of the Green's function of (16.53):

$$Det(I_2 - \alpha\dot{\tilde{\mathcal{G}}}^0(p; 0, 0; \omega)) = 0. \tag{16.57}$$

Equations (16.30) and (16.42) taken jointly with (16.54) yield ($Y = 0$, $X^2 + Y^2 = X^2 = (rd)^2$)

$$\dot{\tilde{\mathcal{G}}}^0_{\substack{11\\22}}(p; 0, 0; \omega)_K = \frac{1}{4\pi\hbar\gamma^2}\omega\sum_{r=-\infty}^{\infty} e^{iprd}e^{-\frac{\omega_g^2}{8\gamma^2}(rd)^2}\sum_{n=0}^{\infty}\frac{L_n\left[\frac{\omega_g^2}{4\gamma^2}(rd)^2\right]}{\frac{\omega^2}{\omega_g^2}-n}, \tag{16.58}$$

and

$$\dot{\tilde{\mathcal{G}}}^0_{\substack{21\\12}}(p; 0, 0; \omega)_K = \frac{i\omega_g^2}{8\pi\hbar\gamma^3}\sum_{r=-\infty}^{\infty}(rd)\, e^{iprd}e^{-\frac{\omega_g^2}{8\gamma^2}(rd)^2}\sum_{n=1}^{\infty}\frac{L_{n-1}^1\left[\frac{\omega_g^2}{4\gamma^2}(rd)^2\right]}{\frac{\omega^2}{\omega_g^2}-n}, \tag{16.59}$$

for the K point in graphene. It is clear from (16.58) and (16.59) that $\dot{\tilde{\mathcal{G}}}^0_{11}(p; 0, 0; \omega)_K$
$= \dot{\tilde{\mathcal{G}}}^0_{22}(p; 0, 0; \omega)_K$ and $\dot{\tilde{\mathcal{G}}}^0_{12}(p; 0, 0; \omega)_K = \dot{\tilde{\mathcal{G}}}^0_{21}(p; 0, 0; \omega)_K$. Hence (16.57) can be written as

$$1 - 2\alpha\dot{\tilde{\mathcal{G}}}^0_{11} + \alpha^2\left(\dot{\tilde{\mathcal{G}}}^{0\,2}_{11} - \dot{\tilde{\mathcal{G}}}^{0\,2}_{12}\right) = 0. \tag{16.60}$$

Note that since the antidot radius is very small (i.e $\frac{\alpha\omega_g}{4\pi\hbar\gamma^2} \ll 1$), a root of (16.60) ω approaches the pole position i.e $\omega \to \omega_n$, so that the n-th pole has the primary influence in determining the eigen-energy root ω_n. Therefore, we can make a reasonable first approximation by dropping all other terms of the n-sum. Also, for $\frac{\omega_g d}{8\gamma} > 1$, it suffices to keep only $r = -1, 0, 1$ terms of the r-sum in (16.58) and (16.59). This imposes the following condition on antidot spacing [21]

$$d > \frac{145}{\sqrt{B(Tesla)}}(nm). \tag{16.61}$$

Finally, (16.60) can be written as

$$1 - 2\frac{\omega_g\Omega\omega_n}{\omega_n^2 - n\omega_g^2} + \frac{\omega_g^2\Omega^2\omega_n^2}{\left(\omega_n^2 - n\omega_g^2\right)^2} - \frac{\omega_g^4\kappa^2}{\left(\omega_n^2 - n\omega_g^2\right)^2}, \tag{16.62}$$

where we have defined

$$\Omega = \frac{\alpha\omega_g}{4\pi\hbar\gamma^2}\left[1 + 2\cos pde^{-\frac{\omega_g^2}{8\gamma^2}d^2} L_n\left(\frac{\omega_g^2}{4\gamma^2}d^2\right)\right], \tag{16.63}$$

and

$$\kappa = \frac{\alpha\omega_g}{4\pi\hbar\gamma^2}\left(\frac{\omega_g d}{\gamma}\right)\sin pde^{-\frac{\omega_g^2}{8\gamma^2}d^2} L_n\left(\frac{\omega_g^2}{4\gamma^2}d^2\right). \tag{16.64}$$

The four roots of (16.62) describe the energy spectrum at the K point of graphene having a 1-D antidot lattice placed in a uniform normal magnetic field:

$$\omega_{n,K} = \frac{\Omega \pm \sqrt{\Omega^2 + 4(n \pm' \kappa)}}{2}\omega_g. \tag{16.65}$$

Similarly, for the K' point, n will be replaced by $n + 1$ on the right hand side of (16.65).

In this, we have the energy spectrum composed of broadened Landau levels (Landau minibands) for a graphene antidot lattice in a quantizing magnetic field. Each Landau level has split into two branches and each branch has broadened into a small continuous band (subband) of energy states instead of a single energy. Figure 16.8a and b show the Landau minibands at the location of K and K' points. The broadening is so small that it can not be observed with the naked eye, so we multiplied a broadening factor $\beta = 200$ with the oscillatory terms $\cos(pd)$ and $\sin(pd)$. The parameter β introduces an increase in the amplitude of the minibands to facilitate observation of the broadening of the Landau minibands; i.e we used

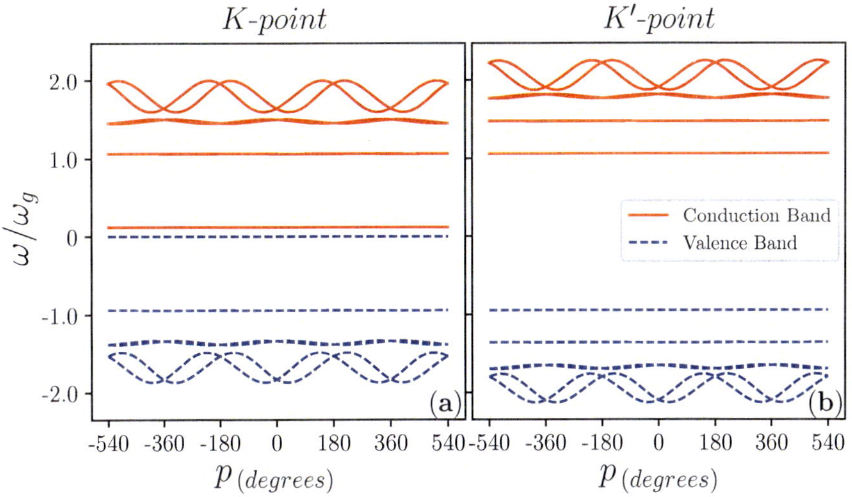

Fig. 16.8 Landau minibands ($n = 0 - 3$) in Landau quantized graphene having a one dimensional antidot lattice with antidot potential U_0=100 meV, antidot radius a=10 nm and spacing d=100 nm in the presence of magnetic field B=3.3$Tesla$. **a** K point. **b** K' point. [Reproduced from "R. A. W. Ayyubi, N. J. M. Horing, and K. Sabeeh. Effect of pseudospin polarization on wave packet dynamics in graphene antidot lattices (GALs) in the presence of a normal magnetic field. Journal of Applied Physics, 129(7):074301, 2021", with the permission of AIP Publishing. https://aip.scitation.org/doi/10.1063/5.0038316]

$$\Omega = \frac{\alpha\omega_g^2}{4\pi\hbar\gamma^2}\left(1 + 2\beta\cos(pd)e^{-\frac{\omega_g^2}{8\gamma^2}d^2}L_n\left[\frac{\omega_g^2}{4\gamma^2}d^2\right]\right)$$

and

$$\kappa = \frac{\alpha\omega_g}{4\pi\hbar\gamma^2}\left(\frac{\omega_g d}{\gamma}\right)\beta\sin(pd)e^{-\frac{\omega_g^2}{8\gamma^2}d^2}L_n\left(\frac{\omega_g^2}{4\gamma^2}d^2\right)$$

to exhibit the Landau minibands plotted in Fig. 16.8. The energy spectra of the K and K' points differ by a unit shift even in the presence of an antidot lattice. The presence of antidots induced a gap between the conduction and valence bands at both K and K' points. In the case of Fig. 16.8, if we consider $\beta = 1$ with all other parameters having same values, then the gap opened at the locations of the K and K' points is 7.96 meV and 132 meV, respectively. This gap between the conduction and valence bands increases with increase in strength of the magnetic field and reaches 36.25 and 283.4 meV for K and K' points respectively when B=15$Tesla$. Similarly, this gap shows an increase with increase in strength of the antidot lattice. Note that when the antidot strength α approach zero (which means no antidot lattice), Ω and κ also approach zero, i.e

$$\lim_{\alpha\to 0}\Omega = 0 = \lim_{\alpha\to 0}\kappa.$$

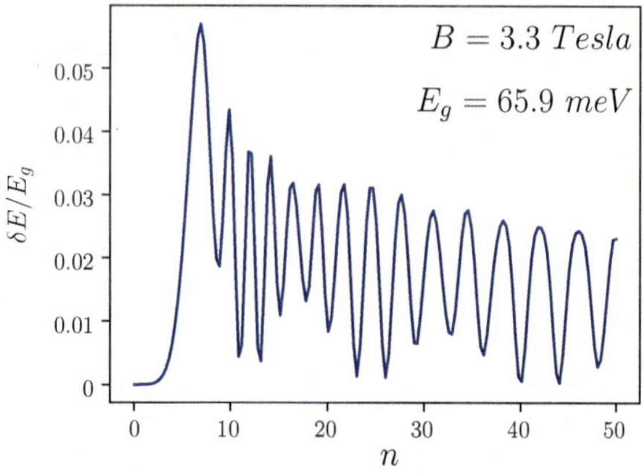

Fig. 16.9 Dimensionless width $\delta E/E_g$ of the Landau minibands as a function of Landau level index n for K point at U_0=100 meV, a=10 nm, d=100 nm and B=3.3 in units of $Tesla$ so that $E_g = \hbar\omega_g = 65.9meV$

In this case (16.65) reduces to the case of a discrete eigen-energy spectrum of graphene in a normal, uniform magnetic field. The width of Landau minibands for the K point of graphene is

$$\Delta\omega_{n,K} = [\omega_{n,K}]_{p=\frac{2\pi}{d}} - [\omega_{n,K}]_{p=\frac{\pi}{d}}.$$

Evaluation of this expression for the width of Landau minibands yields

$$\frac{\Delta\omega_{n,K}}{\omega_g} = \alpha'\varrho \pm \frac{1}{2}\sqrt{\alpha'^2(1+\varrho)^2 + (2\sqrt{n})^2} \mp \frac{1}{2}\sqrt{\alpha'^2(1-\varrho)^2 + (2\sqrt{n})^2}, \tag{16.66}$$

where we have defined

$$\alpha' = \frac{\alpha\omega_g}{4\pi\hbar\gamma^2} \qquad \varrho = 2e^{-\frac{\omega_g^2}{8\gamma^2}d^2} L_n\left[\frac{\omega_g^2}{4\gamma^2}d^2\right].$$

Upper and lower signs are used for conduction and valence bands respectively. For the width of the Landau minibands for the K' point, we have $(2\sqrt{n+1})^2$ in place of $(2\sqrt{n})^2$ in (16.66).

The width of Landau minibands as a function of Landau index n for the conduction band at the K point is plotted in Fig. 16.9. The parameters used are the same as the parameters used in Fig. 16.8.

To investigate electron propagation in the system under study, first we have to find the Green's function matrix of the system in time representation. This is evaluated in Sect. 16.8.

16.8 Green's Functions: Frequency and Time Representation

To evaluate the full Green's function matrix $\mathcal{G}(x_1; x_2; \omega)_K$ given in (16.53), we have to find the $\mathcal{G}^0(x_1; x_2; \omega)$, $\mathcal{G}^0(x_1; nd; \omega)$, $\dot{\tilde{\mathcal{G}}}^0(p; 0, 0; \omega)$ and $\tilde{\mathcal{G}}^0(p; x_2; \omega)$ matrices. These four matrix Green's functions can be easily determined using $\mathcal{G}^0(x_1; x_2; \omega)$, which is the Green's function in the presence of a perpendicular and uniform magnetic field in the absence of an antidot lattice given in (16.30) and (16.42).

For propagation along the antidot lattice i.e the x axis only ($y_1 = 0 = y_2$, $X = x_1 - x_2$ and $C(\mathbf{r}_1; \mathbf{r}_2) = 1$), (16.30) and (16.42) reduce to (for K-point)

$$\mathcal{G}^0_{\substack{11\\22}}(x_1, x_2; \omega)_K = \frac{1}{4\pi\hbar\gamma^2}\omega e^{-\frac{\omega_g^2}{8\gamma^2}(x_1-x_2)^2}\sum_{n=0}^{\infty}\frac{L_n\left[\frac{\omega_g^2}{4\gamma^2}(x_1-x_2)^2\right]}{\frac{\omega^2}{\omega_g^2}-n}, \tag{16.67}$$

$$\mathcal{G}^0_{\substack{21\\12}}(x_1, x_2; \omega)_K = \frac{i\omega_g^2}{8\pi\hbar\gamma^3}e^{-\frac{\omega_g^2}{8\gamma^2}(x_1-x_2)^2}(x_1-x_2)\sum_{n=1}^{\infty}\frac{L^1_{n-1}\left[\frac{\omega_g^2}{4\gamma^2}(x_1-x_2)^2\right]}{\frac{\omega^2}{\omega_g^2}-n}, \tag{16.68}$$

respectively. Matrix elements of $\mathcal{G}^0(x_1; n'd; \omega)_K$ can be obtained by taking x_2=$n'd$ in (16.67) and (16.68), while matrix elements of $\dot{\tilde{\mathcal{G}}}^0(p; 0, 0; \omega)_K$ are given in (16.58) and (16.59).

Similarly, the matrix $\tilde{\mathcal{G}}^0(p; x_2; \omega)_K$ can be obtained by using x_1=rd in (16.67) and (16.68) jointly with (16.54). Note that for $\frac{\omega_g d}{8\gamma} > 1$, it suffices to keep only $r = -1, 0, 1$ of the r-sum in the expressions for matrix elements of $\dot{\tilde{\mathcal{G}}}^0(p; 0, 0; \omega)_K$ and $\tilde{\mathcal{G}}^0(p; x_2; \omega)_K$, as discussed earlier. The above four matrix Green's functions completely determine the full Green's function $\mathcal{G}(x_1; x_2; \omega)_K$ given in (16.53).

As discussed earlier, for temporal evolution of the wave packet we require the time representation of the Green's function. Hence, to find the time representation of the full Green's function, we have to take the Fourier transform of $\mathcal{G}(x_1; x_2; \omega)_K$ matrix using (16.44) and (16.53):

$$\mathcal{G}_{\mu\nu}(x_1, x_2; t)_K = \int_{-\infty}^{\infty} d\omega e^{-i\omega t}\mathcal{G}^0(x_1; x_2; \omega) + \frac{\alpha d}{2\pi}\int_{-\infty}^{\infty} d\omega e^{-i\omega t} \times \sum_{n'=-\infty}^{\infty}\mathcal{G}^0(x_1; n'd; \omega)\int_{-\frac{\pi}{d}}^{\frac{\pi}{d}} dp e^{-ipn'd}\left[1 - \alpha\dot{\tilde{\mathcal{G}}}^0(p; 0, 0; \omega)\right]^{-1}\tilde{\mathcal{G}}^0(p; x_2; \omega). \tag{16.69}$$

There are two integrals in the above equation. The first integral was evaluated in (16.45) and (16.46) as

$$\mathcal{G}^0_{\substack{11\\22}}(x_1,x_2;t) = -i\eta_+(t)\frac{\omega_g^2}{4\pi\gamma^2}e^{-\frac{\omega_g^2}{8\gamma^2}(x_1-x_2)^2}\sum_{n=0}^{\infty}L_n\left[\frac{\omega_g^2}{4\gamma^2}(x_1-x_2)^2\right] \times \cos\left(\frac{\omega_g l}{\gamma}\frac{t}{\tau_o}\sqrt{n}\right), \quad (16.70)$$

and

$$\mathcal{G}^0_{\substack{21\\12}}(x_1,x_2;t) = -i\eta_+(t)\frac{\omega_g^3 l}{8\pi\gamma^3}e^{-\frac{\omega_g^2}{8\gamma^2}(x_1-x_2)^2}\left(\frac{x_1-x_2}{l}\right)\sum_{n=1}^{\infty}\frac{L^1_{n-1}\left[\frac{\omega_g^2}{4\gamma^2}(x_1-x_2)^2\right]}{\sqrt{n}} \times \sin\left(\frac{\omega_g l}{\gamma}\frac{t}{\tau_o}\sqrt{n}\right). \quad (16.71)$$

Evaluation of the second term of (16.69) is a lengthy process. The matrix elements $V_{i,j}(t)$ $(i, j{=}1, 2)$ of the second term in time representation are given by (see Appendix 2) [21]

$$V_{\substack{11\\22}} = -i\eta_+(t)\frac{\omega_g}{2\pi}\sum_{n'=-\infty}^{\infty}\left\{\gamma_1\left[c_{\substack{11\\22}}(\eta)\right]_{\eta=0} + \sum_{n=1}^{\infty}\cos\left(\frac{\omega_g l}{\gamma}\frac{t}{\tau_o}\sqrt{n}\right) \times \left[\gamma_1 L_n[\Upsilon]c_{\substack{11\\22}}(\eta) + \frac{\gamma_2}{\sqrt{n}}L^1_{n-1}[\Upsilon]c_{\substack{21\\12}}(\eta)\right]_{\eta=\sqrt{n}}\right\}, \quad (16.72)$$

and

$$V_{\substack{12\\21}} = -\eta_+(t)\frac{\omega_g}{2\pi}\sum_{n'=-\infty}^{\infty}\left\{i\gamma_1\left[c_{\substack{12\\21}}(\eta)\right]_{\eta=0} + \sum_{n=1}^{\infty}\sin\left(\frac{\omega_g l}{\gamma}\frac{t}{\tau_o}\sqrt{n}\right) \times \left[\gamma_1 L_n[\Upsilon]c_{\substack{12\\21}}(\eta) + \frac{\gamma_2}{\sqrt{n}}L^1_{n-1}[\Upsilon]c_{\substack{22\\11}}(\eta)\right]_{\eta=\sqrt{n}}\right\}, \quad (16.73)$$

where

$$\Upsilon = \frac{\omega_g^2 l^2}{4\gamma^2}\left(\frac{x_1-n'd}{l}\right)^2, \qquad \gamma_1 = \frac{\alpha\omega_g}{4\pi\hbar\gamma^2}e^{-\frac{\Upsilon}{2}}, \quad (16.74)$$

$$\gamma_2 = \frac{i\alpha\omega_g^2 l}{8\pi\hbar\gamma^3}\left(\frac{x_1-n'd}{l}\right)e^{-\frac{\Upsilon}{2}}, \quad (16.75)$$

and $c_{ij}(\eta)$ (where $\eta{=}\frac{\omega}{\omega_g}$) are the matrix elements of the following integral (see (16.91) and (16.94); notation $q \equiv pd$):

$$\int_{-\pi}^{\pi} dq e^{-iqn'} \left[I - \alpha \dot{\tilde{\mathcal{G}}}^0(p; 0, 0; \eta) \right]^{-1} \tilde{\mathcal{G}}^0(p; x_2; \eta).$$

These considerations yield the time representation of the full Green's function, i.e. $\mathcal{G}(x_1, x_2; t)_K$ for a Landau-quantized monolayer graphene having a one dimensional antidot lattice. In Sect. 16.9, this Green's function $\mathcal{G}(x_1, x_2; t)_K$ will be employed to study the temporal dynamics of a wave packet in the lattice system.

16.9 Wave Packet Dynamics Along a One-Dimensional Antidot Lattice

With the solution of (16.69) in hand, in the form of (16.70), (16.71), (16.72) and (16.73), the temporal study of an electron wave packet propagating along the axis of the antidot lattice due to an initial, incident wave packet given by

$$\psi(\mathbf{r}, 0) = \frac{f(\mathbf{r})}{\sqrt{\mid c_1 \mid^2 + \mid c_2 \mid^2}} \begin{pmatrix} c_1 \\ c_2 \end{pmatrix}, \tag{16.76}$$

where

$$f(\mathbf{r}) = \frac{1}{l\sqrt{\pi}} exp \left(-\frac{x^2 + y^2}{2l^2} + ik_{0x}x \right) \delta(y)$$

can be made using (16.48) and (16.49). In Fig. 16.10, results for the probability density $\rho(x, 0, t)$ of an electron along the x-axis of the antidot lattice ($y = 0$) are plotted for four different cases to examine the effect of the antidot lattice.

In Fig. 16.10, $\rho(x, 0, t)$ is represented at times $t = 1\tau_o$ and $t = 5\tau_o$ with solid and dashed lines, respectively. Moving from left to right, the three columns correspond to three different pseudospin polarizations $\{c_1, c_2\}$, while from up to down the four rows represent the increase in antidot strength $E_A = U_0$ from zero meV to 330 meV with constant magnetic field strength $E_B = 66$ meV ($E_B \equiv \hbar\omega_g$) at $B = 3.3 Tesla$. All three columns represent the propagation of the wave packet along the x axis ($y = 0$). One can see that with the increase in antidot strength, the probability density of finding the electron starts increasing along the axis of the antidot lattice. Row 1 corresponds to the situation when there is no antidot lattice and the wave packet propagates in graphene under the effect of the perpendicular magnetic field only. Similarly, rows 2, 3 and 4 correspond to $E_A = U_0 = 13.2$, 66 and 330 meV respectively. Clearly, the probability of finding an electron along the antidot direction increases with the introduction of the antidot lattice on a Landau quantized graphene sheet; this can be treated as the propagation of a wave packet through a quantum antidot wire, which is clearly supporting the propagation through it. Also, the probability density gets more confined with increase of antidot strength; this can be seen in rows 2, 3 and 4 of Fig. 16.10. Moving from up to down across the rows 2–4, the spread of the packet gets smaller with a clear increase in magnitude of probability density, which

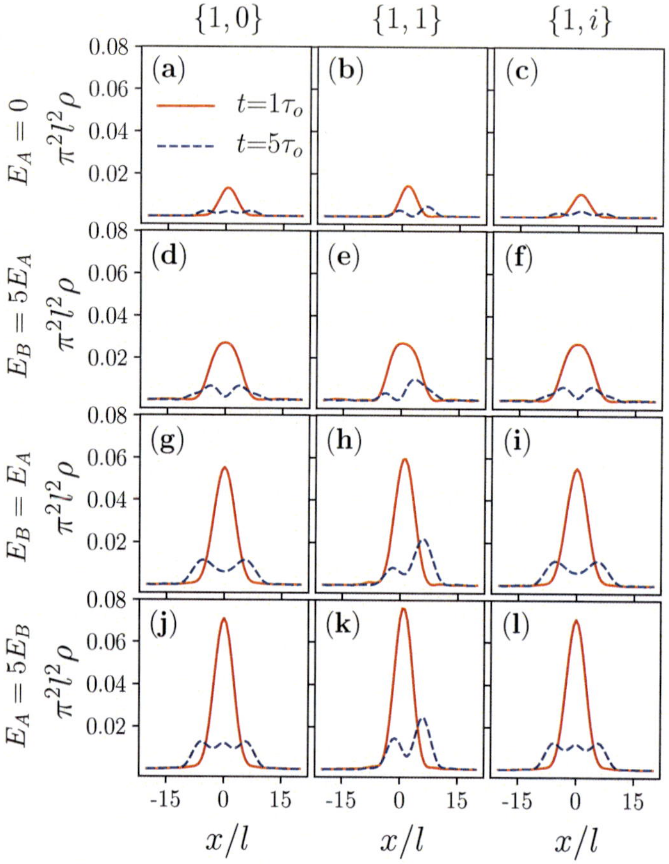

Fig. 16.10 Electron probability density $\rho(x, 0, t)$ for K point along the x axis in the presence of a uniform magnetic field $B = 3.3Tesla$ having strength $E_B = \hbar\omega_g = 66$ meV for $a_0 = k_{0x}l = 1.2$ with $l = 2$ nm; also antidot lattice parameters, spacing $d = 100$ nm and radius $a = 10$ nm. Left to right: Three columns with initial pseudospin polarization $\{1, 0\}$, $\{1, 1\}$ and $\{1, i\}$, respectively. Up to down: Four rows with antidot strength $E_A = U_0 = 0$, 13, 66 and 330 meV respectively. [Reproduced from "R. A. W. Ayyubi, N. J. M. Horing, and K. Sabeeh. Effect of pseudospin polarization on wave packet dynamics in graphene antidot lattices (GALs) in the presence of a normal magnetic field. Journal of Applied Physics, 129(7):074301, 2021", with the permission of AIP Publishing. https://aip.scitation.org/doi/10.1063/5.0038316]

means that the probability of finding an electronic current along the antidot lattice increases. This may be referred to as collimation of the electronic beam along the axis of antidot lattice.

Further, we explore the effect of initial pseudospin polarization on the wave packet dynamics in the presence of an antidot lattice, as shown in Fig. 16.10. In Fig. 16.10, it can be seen that propagation of the wave packet is strongly affected by the change of pseudospins polarization across the rows. Columns 1 and 3 correspond to pseudospins $\{1, 0\}$ and $\{1, i\}$, respectively, and the propagation corresponding to these two columns is similar because the pseudospins are both perpendicular to the axis of the antidot lattice, and the center of the wave packet does not propagate with an increase in time; it is always located at $x = 0$ without being affected by antidot strength, as shown in Figs. 16.10a, d, g, h and c, f, i, l. But, one can see in Figs. 16.10b, e, h, k (column 2) that the center of the wave packet propagates along the axis of the antidot lattice when the initial pseudospin polarization is along the axis of the antidot lattice i.e. x axis: At $t = 5\tau_o$ the dashed lines clearly show the propagation of the wave packet when compared with the solid lines at $t = 1\tau_o$. Hence, the wave packet only propagates along the one dimensional antidot lattice when the initial pseudospin polarization is parallel to the axis of antidot lattice [21].

16.9.1 Experimental Relevance

We now address the question of experimental relevance of this work. In this regard, we note that it is possible to initialize the pseudospin and study the dynamics of an electron wave packet in graphene by means of pump-probe laser spectroscopy. Experiments have been performed where pseudospin initialization and its subsequent relaxation have been probed on the femtosecond scale in graphene [34–36]. Recently it has been shown that pesudospin can be manipulated by a coupled waveguide-cavity configuration in graphene [37].

Another possibility is the creation of artificial graphene (honeycomb lattices) in cold atom systems. In these systems, it is possible to generate effective fields and study transport in graphene-like structures [38–41].

Recently it has become possible to treat pseudospin as a real measurable angular momentum at par with electron spin in photonic graphene (honeycomb array of evanescently coupled waveguides) [22, 39]. These photonic systems allow a great degree of control of initial conditions in the study of wave packet dynamics.

16.10 Discussion

In summary, this work first addressed the role of a normal quantizing magnetic field on two-dimensional Schrödinger electrons in an quantum dot lattice. The quantum dots are modeled by a row of uniformly spaced Dirac delta-function potential profiles on the x-axis, and the associated Green's function was formulated as an integral equation. It was seen to devolve upon a discrete matrix equation that was solved exactly for propagation confined to the x-axis of the lattice (with simplification of the Peierls phase factor $C(\mathbf{r}, \mathbf{r}') \rightarrow 1$ due to $y \equiv 0$). The frequency poles of this Green's function describe the eigenenergy dispersion relation, which has been exhibited in closed form in terms of the Jacobi Theta Function of the third kind. An alternative formulation of the dispersion relation is presented in terms of Laguerre polynomials, and was solved approximately, exhibiting the proliferation of broadened Landau quantized eigenstates into Landau minibands.

Furthermore, we studied the evolution of a two dimensional Gaussian wave packet in a graphene sheet placed in a uniform and perpendicular magnetic field. We have observed that the temporal dynamics of the wave packet strongly depends on the initial pseudospin polarization. We have shown that direction of propagation of the wave packet can be controlled through pseudospin polarization.

Further, we have studied the effect of pseudospin on trembling motion (ZB) of the Gaussian wave packet in a graphene sheet placed in a quantizing magnetic field for non zero values of the initial momentum k_{0x}. Initially, these ZB oscillations seemingly die out but reappear i.e. the amplitude of the ZB oscillations start to grow again without any fundamental frequency. So, for finite magnetic fields, ZB oscillations have a recurrent character and they do not die with time. The quantized (discrete) energy spectrum, which is a consequence of the magnetic field, is the main reason for the recurrent character of ZB oscillations. This property is completely different from the zero magnetic field case, in which the energy spectrum is not quantized (discrete) and ZB of the wave packets has a transient character [42]. Most importantly, we have found that in the presence of a magnetic field, ZB has strong dependence on initial pseudospin polarization, which can be regarded as pseudospinorial Zitterbewegung (PZB), a name given in a recent study [24].

In our study of wave packet dynamics in a one dimensional antidot lattice in graphene in the presence of the magnetic field, we determined the associated Green's function matrices of the system both in frequency and time representations. From the poles of the Green's function, we found that along the axis of the antidot lattice, the energy spectrum is composed of Landau minibands with a unit shift at K and K' points of graphene. In the time evolution of the wave packet, we found that the wave packet dynamics are highly dependent on the initial pseudospin polarization, as indicated above and, specifically, the center of the Gaussian wave packet can be made to propagate along the antidot lattice by tuning the pseudospin parallel to the axis of the antidot lattice. Also, when the strength of the antidot potential is greater than the strength of the magnetic field, the wave packet becomes more confined in space and the probability of finding it on the axis of the antidot lattice significantly increases. Hence, we propose that quantum antidot channels can be made such that propagation of the wave packet can be controlled using pseudospin polarization and the strength of the antidot potential. Finally, we would like to point out that this work may lead to new insights for controlling currents in both natural and artificial graphene systems by tuning initial pseudospin polarization, both in Landau quantized graphene and in a graphene antidot lattice in the presence of a magnetic field. This may lead to the preparation of graphene based nano gates in which two different pseudospin polarizations perpendicular to each other can be used to get "ON" and "OFF" states [21].

Acknowledgement This chapter replicates some work presented in our earlier publication "Landau Minibands in an Antidot Lattice" in AdvNanoBio M & D 1 (1), 24 (2017) [8]. The appearance of the work in this chapter is facilitated by the gracious consent of the journal editor, Prof. Dr. Anton Ficai of SciEdTech.eu.

Appendix A

16.11 Zitterbewegung Phenomenon

Zitterbewegung is the jittery motion of the Dirac electron. It occurs when one tries to confine the Dirac electrons. According to the Heisenberg uncertainty principle, localization of the electron wave packet leads to uncertainty in momentum. For particles with zero rest mass (massless Dirac particles), uncertainty in momentum translates into uncertainty in energy of the particle (This should be contrasted with the nonrelativistic case, where the position-momentum uncertainty relation is independent of the energy-time uncertainty relation) [27].

16.11.1 *Prediction and Interpretation of Zitterbewegung by Schrödinger*

Schrödinger discovered a highly oscillatory motion of the electron with velocity c during his work on the time evolution of the position operator, which he named Zitterbewegung [43]. Schrödinger attempted to explain this phenomenon in terms of microscopic dynamical variables i.e coordinate and momentum. To determine the time evolution of the position operator, Schrödinger used Dirac's Hamiltonian for the free electron-positron system, which is

$$H = c\boldsymbol{\alpha}.\mathbf{q} + mc^2\beta. \left[\beta = \begin{pmatrix} I & 0 \\ 0 & -I \end{pmatrix}\right] \tag{16.77}$$

In the above equation $\boldsymbol{\alpha}$ and β satisfy the following anti-commutation relations:

$$\{\boldsymbol{\alpha}_i, \boldsymbol{\alpha}_j\} = 2\delta_{ij} I; \qquad \{\boldsymbol{\alpha}_i, \beta\} = 0; \qquad \beta^2 = I,$$

where $i, j = 1, 2, 3$ and I is a unit matrix. Also the momentum operator $\mathbf{q}$ and coordinate operator $\mathbf{x}$ commute with $\boldsymbol{\alpha}$ and β, and satisfy the canonical anti-commutation relations for Fermions

$$\{\mathbf{x}_i, \mathbf{x}_j\} = \{\mathbf{q}_i, \mathbf{q}_j\} = 0 \qquad \textit{and} \qquad \{\mathbf{x}_i, \mathbf{q}_j\} = i\hbar\delta_{ij} I.$$

In the Heisenberg picture, the time derivative of any one of these operators, say A, which does not have explicit time dependence is given by

$$\frac{dA}{dt} = \frac{i}{\hbar}[H, A]. \tag{16.78}$$

As a result

$$\frac{d\mathbf{q}}{dt} = 0, \qquad \frac{dH}{dt} = 0, \qquad \frac{d\mathbf{x}}{dt} = c\boldsymbol{\alpha}, \tag{16.79}$$

and (16.77) and (16.78) for the operator $\boldsymbol{\alpha}$ give

$$-i\hbar\frac{d\boldsymbol{\alpha}}{dt} = [H, \boldsymbol{\alpha}] = 2H\alpha - \{H, \boldsymbol{\alpha}\} = 2H\boldsymbol{\alpha} - 2c\mathbf{q}. \tag{16.80}$$

Above equation can be written as

$$-i\hbar\frac{d\boldsymbol{\alpha}}{dt} = 2H\boldsymbol{\eta},$$

where we have defined

$$\boldsymbol{\eta} = \boldsymbol{\alpha} - cH^{-1}\mathbf{q}. \tag{16.81}$$

Schrödinger noted that

$$-i\hbar\frac{d\boldsymbol{\eta}}{dt} = -i\hbar\frac{d\boldsymbol{\alpha}}{dt} = 2H\boldsymbol{\eta},$$

which yields

$$\boldsymbol{\eta}(t) = e^{\frac{2iHt}{\hbar}}\boldsymbol{\eta}_0 \qquad with \qquad \boldsymbol{\eta}_0(0) = \boldsymbol{\alpha}(0) - cH^{-1}\mathbf{q}. \tag{16.82}$$

Using $\{H, \boldsymbol{\eta}\} = 0 = \{H, \boldsymbol{\eta}_0\}$, (16.82) can also be written as

$$\boldsymbol{\eta}(t) = \boldsymbol{\eta}_0 e^{\frac{-2iHt}{\hbar}}. \tag{16.83}$$

Combining (16.80), (16.81) and (16.83), Schrödinger obtained

$$\frac{d\mathbf{x}}{dt} = c\boldsymbol{\alpha} = c^2H^{-1}\mathbf{q} + c\boldsymbol{\eta}_0 e^{\frac{-2iHt}{\hbar}},$$

which on integration yields

$$\mathbf{x}(t) = \mathbf{a} + c^2H^{-1}\mathbf{q}t + \frac{i}{\hbar}c\boldsymbol{\eta}_0 H^{-1}e^{\frac{-2iHt}{\hbar}}, \tag{16.84}$$

where $\mathbf{a}$ is an operator which is a constant of integration with the definition

$$\mathbf{a} = \mathbf{x}(0) - \frac{i}{2}\hbar c^2 H^{-2}\mathbf{q}. \tag{16.85}$$

Defining

$$\mathbf{x}_A(t) = \mathbf{a} + c^2 H^{-1}\mathbf{q}t,$$

equation (16.84) will become

$$\mathbf{x}(t) = \mathbf{x}_A(t) + \boldsymbol{\xi}(t),$$

where we have defined

$$\boldsymbol{\xi}(t) = \frac{i}{2}\hbar c\eta_0 H^{-1} e^{\frac{-2iHt}{\hbar}} = \frac{i}{2}\hbar c\eta H^{-1}. \tag{16.86}$$

Here, the term $\boldsymbol{\xi}(t)$ corresponds to a microscopic coordinate which oscillates at high frequency known as Zitterbewegung (ZB). This motion is superimposed on a macroscopic type of motion associated with the coordinate $\mathbf{x}_A$. Characteristic amplitude associated with Zitterbewegung is $\frac{\hbar}{2mc}$ and this amplitude is equal to half of the Compton wavelength of an electron, while the characteristic angular frequency of the Zitterbewegung is $\frac{2mc^2}{\hbar}$.

Schrödinger went on to note, that if the orbital-angular momentum $\mathbf{L}$ and spin vector $\mathbf{S}$ are introduced as

$$\mathbf{S} = -\frac{i}{4}\hbar\boldsymbol{\alpha} \times \boldsymbol{\alpha} \qquad \textit{and} \qquad \mathbf{L} = \mathbf{x} \times \mathbf{q}$$

then both $\mathbf{L}$ and $\mathbf{S}$ alone are not constant but their sum $\mathbf{L} + \mathbf{S}$ is a constant of the motion. Schrödinger found

$$\mathbf{S}(t) = \mathbf{S}_A - \eta(t) \times \mathbf{q}$$

and

$$\mathbf{L}(t) = \mathbf{L}_A + \eta(t) \times \mathbf{q},$$

where $\mathbf{S}_A$ and $\mathbf{L}_A$ are constants and the term $\eta(t) \times \mathbf{q}$ is oscillatory. Finally he found that

$$\mathbf{L} + \mathbf{S} = \mathbf{L}_A + \mathbf{S}_A. \tag{16.87}$$

This constant of the motion allowed Schrödinger to explain Zitterbewegung in terms of microscopic dynamical variables spin and orbital angular momentum [43].

16.11.2 Zitterbewegung: Interpretation in Terms of Interference Between Positive and Negative Energy States

It can be easily verified that (16.84) along with (16.82) and (16.85) can be written as

$$\mathbf{x}(t) = \mathbf{x}(0) + \frac{c^2\mathbf{q}}{H}t + \frac{\hbar c}{2iH}\left(e^{\frac{2iHt}{\hbar}} - 1\right)\left(\boldsymbol{\alpha}(0) - \frac{c\mathbf{q}}{H}\right). \tag{16.88}$$

Also (16.80) can be brought to the form

$$H\boldsymbol{\alpha} + \boldsymbol{\alpha}H = 2c\mathbf{q},$$

and hence

$$H\left(\boldsymbol{\alpha} - \frac{c\mathbf{q}}{H}\right) - \left(\boldsymbol{\alpha} - \frac{c\mathbf{q}}{H}\right)H = 0. \tag{16.89}$$

Along with the initial term $\mathbf{x}(0)$ at $t = 0$, (16.88) also carries two more terms, one of them is linear in time t and corresponds to group velocity, while the second term is oscillatory in nature and is known as Zitterbewegung. $\boldsymbol{\alpha} - \frac{c\mathbf{q}}{H}$ is an operator whose matrix elements are required to evaluate the average value $\int \psi^{\dagger}(0, \mathbf{x})\mathbf{x}(t)\psi(0, \mathbf{x})d^3x$. Nonvanishing matrix elements of this operator only lie between states of the same momentum. Also from (16.89), the anticommutator only vanishes when the energies are of opposite sign. Hence it can be concluded that ZB is a consequence of interference between positive and negative energy states as predicted by relativistic quantum mechanics [44].

Appendix B

16.12 Contour Integration

To solve the second integral in (16.69), let us denote the integral by I [21]

$$I = \frac{\alpha d\omega_g}{2\pi}\sum_{n'=-\infty}^{\infty}\int_{-\infty}^{\infty} d\eta e^{-i\omega_g\eta t}\mathcal{G}^0(x_1; n'd; \eta)\int_{-\frac{\pi}{d}}^{\frac{\pi}{d}} dp e^{-ipn'd}$$
$$\times\left[I_2 - \alpha\dot{\tilde{\mathcal{G}}}^0(p; 0, 0; \eta)\right]^{-1}\tilde{\mathcal{G}}^0(p; x_2; \eta). \tag{16.90}$$

(I_2 is unit matrix of order 2) where

$$\eta = \frac{\omega}{\omega_g} \qquad and \qquad d\omega = \omega_g d\eta.$$

Note that in the above integral, each "no lattice" Green's function has real poles at η=$\pm\sqrt{n}$. These "no lattice" poles of the term

$$T_1(\eta) = \left[I_2 - \alpha\dot{\tilde{\mathcal{G}}}^0(p; 0, 0; \eta)\right]^{-1} \tilde{\mathcal{G}}^0(p; x_2; \eta) \tag{16.91}$$

can be seen to cancel by re-expressing the "no lattice" Green's functions $\dot{\tilde{\mathcal{G}}}^0(p; 0, 0; \eta)$ and $\tilde{\mathcal{G}}^0(p; x_2; \eta)$. For this purpose, by using (16.58) and (16.59), one can easily write $\alpha\dot{\tilde{\mathcal{G}}}^0(p; 0, 0; \eta)$ matrix as

$$\alpha\dot{\tilde{\mathcal{G}}}^0(p; 0, 0; \eta) = \frac{1}{\prod_{n=0}^{\infty}(\eta^2 - n)} \begin{pmatrix} a_{11} & a_{12} \\ a_{12} & a_{11} \end{pmatrix} \tag{16.92}$$

where we have defined (j corresponds to the index r in (16.58, 16.59))

$$a_{11}(\eta) = \alpha_1 \eta \sum_{m=0}^{\infty} L_m \left[\frac{\omega_g^2 l^2}{4\gamma^2}\left(j\frac{d}{l}\right)^2\right] \prod_{\substack{n=0 \\ n\neq m}}^{\infty}(\eta^2 - n),$$

and

$$a_{12}(\eta) = \alpha_2 \sum_{m=1}^{\infty} L_{m-1}^1 \left[\frac{\omega_g^2 l^2}{4\gamma^2}\left(j\frac{d}{l}\right)^2\right] \prod_{\substack{n=0 \\ n\neq m+1}}^{\infty}(\eta^2 - n),$$

with

$$\alpha_1 = \frac{\alpha\omega_g}{4\pi\hbar\gamma^2} \sum_{j=-\infty}^{\infty} e^{ipjd} e^{-\frac{\omega_g^2 l^2}{8\gamma^2}\left(j\frac{d}{l}\right)^2},$$

and

$$\alpha_2 = \frac{i\alpha\omega_g^2 l}{8\pi\hbar\gamma^3} \sum_{j=-\infty}^{\infty} \left(j\frac{d}{l}\right) e^{ipjd} e^{-\frac{\omega_g^2 l^2}{8\gamma^2}\left(j\frac{d}{l}\right)^2}.$$

Similarly, the $\tilde{\mathcal{G}}^0(p; x_2; \eta)$ matrix can be written as

$$\tilde{\mathcal{G}}^0(p; x_2; \eta) = \frac{1}{\prod_{n=0}^{\infty}(\eta^2 - n)} \begin{pmatrix} b_{11} & b_{12} \\ b_{12} & b_{11} \end{pmatrix}, \tag{16.93}$$

where we have defined

$$b_{11}(\eta) = \beta_1 \eta \sum_{m=0}^{\infty} L_m \left[\frac{\omega_g^2 l^2}{4\gamma^2} \left(j\frac{d}{l} \right)^2 \right] \prod_{\substack{n=0 \\ n \neq m}}^{\infty} (\eta^2 - n),$$

and

$$b_{12}(\eta) = \beta_2 \sum_{m=1}^{\infty} L_{m-1}^1 \left[\frac{\omega_g^2}{4\gamma^2 l^2} \left(j\frac{d}{l} \right)^2 \right] \prod_{\substack{n=0 \\ n \neq m+1}}^{\infty} (\eta^2 - n),$$

with

$$\beta_1 = \frac{\omega_g}{4\pi\hbar\gamma^2} \sum_{j=-\infty}^{\infty} e^{ipjd} e^{-\frac{\omega_g^2 l^2}{8\gamma^2}\left(\frac{jd-x_2}{l}\right)^2},$$

and

$$\beta_2 = \frac{i\omega_g^2 l}{8\pi\hbar\gamma^3} \sum_{j=-\infty}^{\infty} \left(\frac{jd - x_2}{l} \right) e^{ipjd} e^{-\frac{\omega_g^2 l^2}{8\gamma^2}\left(\frac{jd-x_2}{l}\right)^2}.$$

Note that in the above equations, n is a Landau index and m is a dummy index for the Landau levels; the maximum value of n and m is the same.

Finally substituting (16.92) and (16.93) in (16.91), the matrix T_1 becomes

$$T_1(\eta) = \left[\begin{pmatrix} 1 & 0 \\ 0 & 1 \end{pmatrix} - \frac{1}{\prod_{n=0}^{\infty}(\eta^2 - n)} \begin{pmatrix} a_{11}(\eta) & a_{12}(\eta) \\ a_{12}(\eta) & a_{11}(\eta) \end{pmatrix} \right]^{-1} \frac{1}{\prod_{n=0}^{\infty}(\eta^2 - n)} \begin{pmatrix} b_{11}(\eta) & b_{12}(\eta) \\ b_{12}(\eta) & b_{11}(\eta) \end{pmatrix}$$

$$= \left(\frac{1}{\prod_{n=0}^{\infty}(\eta^2 - n)} \right)^{-1} \left[\begin{pmatrix} \prod_{n=0}^{\infty}(\eta^2 - n) & 0 \\ 0 & \prod_{n=0}^{\infty}(\eta^2 - n) \end{pmatrix} - \begin{pmatrix} a_{11}(\eta) & a_{12}(\eta) \\ a_{12}(\eta) & a_{11}(\eta) \end{pmatrix} \right]^{-1} \times \frac{1}{\prod_{n=0}^{\infty}(\eta^2 - n)} \begin{pmatrix} b_{11}(\eta) & b_{12}(\eta) \\ b_{12}(\eta) & b_{11}(\eta) \end{pmatrix}$$

$$= \left[\begin{pmatrix} \prod_{n=0}^{\infty}(\eta^2 - n) & 0 \\ 0 & \prod_{n=0}^{\infty}(\eta^2 - n) \end{pmatrix} - \begin{pmatrix} a_{11}(\eta) & a_{12}(\eta) \\ a_{12}(\eta) & a_{11}(\eta) \end{pmatrix} \right]^{-1} \begin{pmatrix} b_{11}(\eta) & b_{12}(\eta) \\ b_{12}(\eta) & b_{11}(\eta) \end{pmatrix}.$$

In the above expression, the real "no lattice" poles ($\eta = \pm\sqrt{n}$) cancel. The above expression can be solved numerically. In this computation, it is also necessary to address the poles of the *actual* lattice Green's function, which are defined by the zeros of the denominator, $det \left[I_2 - \alpha \dot{\tilde{\mathcal{G}}}^0(p; 0, 0; \eta) \right] = 0$, and these roots depend on p (and are *not* spaced by integer multiples of ω_g): These poles are treated using the line broadening discussion of Sect. 16.5.2. The resultant 2×2 matrix with $c_{ij}(\eta)$ (where $i, j = 1, 2$) as matrix elements can be written as (substituting $q = pd$)

$$Q(\eta) = \int_{-\pi}^{\pi} dq e^{-iqn'} T_1(\eta) = \begin{pmatrix} c_{11}(\eta) & c_{12}(\eta) \\ c_{21}(\eta) & c_{22}(\eta) \end{pmatrix}. \tag{16.94}$$

Note that the above integral can be numerically calculated by applying the trapezoidal rule in the limits $-\pi$ to π, while keeping the trapezoidal step equal to $\frac{\pi}{10}$; this gives an accuracy up to five decimal points for each value of n'.

Putting (16.94) in (16.90), the integral I becomes

$$I = \frac{\alpha\omega_g}{2\pi} \sum_{n'=-\infty}^{\infty} \int_{-\infty}^{\infty} d\eta e^{-i\omega_g \eta t} \mathcal{G}^0(x_1, n'd; \eta).Q(\eta). \tag{16.95}$$

It can be seen that the "no lattice" poles of the second integral (p integral) in (16.90) have been removed; and the poles of the *actual* lattice Green's function have been dealt with using the line broadening of Sect. 16.5.2, so the only remaining poles that we have to deal with now are the poles of the matrix $\mathcal{G}^0(x_1, n'd; \eta)$.

Using (16.67) and (16.68), the matrix $\alpha\mathcal{G}^0(x_1, n'd; \eta)$ can be written as

$$\alpha\mathcal{G}^0(x_1, n'd; \eta) = \begin{pmatrix} \gamma_1 \frac{1}{\eta} & 0 \\ 0 & \gamma_1 \frac{1}{\eta} \end{pmatrix} + \sum_{n=1}^{\infty} \begin{pmatrix} \gamma_1 \eta \frac{L_n[\Upsilon]}{\eta^2 - n} & \gamma_2 \frac{L^1_{n-1}[\Upsilon]}{\eta^2 - n} \\ \gamma_2 \frac{L^1_{n-1}[\Upsilon]}{\eta^2 - n} & \gamma_1 \eta \frac{L_n[\Upsilon]}{\eta^2 - n} \end{pmatrix}, \tag{16.96}$$

where Υ, γ_1 and γ_2 are defined in (16.74) and (16.75). Using the matrices $Q(\eta)$ and $\alpha\mathcal{G}^0(x_1, n'd; \eta)$ in (16.95), and breaking the matrix into two matrices, we get

$$\begin{aligned} I = &\frac{\omega_g}{2\pi} \sum_{n'=-\infty}^{\infty} \int_{-\infty}^{\infty} d\eta e^{-i\omega_g \eta t} \begin{pmatrix} \gamma_1 \frac{1}{\eta} c_{11}(\eta) & \gamma_1 \frac{1}{\eta} c_{12}(\eta) \\ \gamma_1 \frac{1}{\eta} c_{21}(\eta) & \gamma_1 \frac{1}{\eta} c_{22}(\eta) \end{pmatrix} \\ &+ \frac{\omega_g}{2\pi} \sum_{n'=-\infty}^{\infty} \int_{-\infty}^{\infty} d\eta e^{-i\omega_g \eta t} \begin{pmatrix} M_{11}(\eta) & M_{12}(\eta) \\ M_{21}(\eta) & M_{22}(\eta) \end{pmatrix}. \end{aligned} \tag{16.97}$$

(The first and second matrices correspond to n=0 and $n > 0$ Landau minibands, respectively.) In the above expression

$$M_{\substack{11\\22}}(\eta) = \sum_{n=1}^{\infty} \left(\gamma_1 \eta \frac{L_n[\Upsilon]}{\eta^2 - n} c_{\substack{11\\22}}(\eta) + \gamma_2 \frac{L^1_{n-1}(\Upsilon)}{\eta^2 - n} c_{\substack{21\\12}}(\eta) \right),$$

and

$$M_{\substack{12\\21}}(\eta) = \sum_{n=1}^{\infty} \left(\gamma_1 \eta \frac{L_n[\Upsilon]}{\eta^2 - n} c_{\substack{12\\21}}(\eta) + \gamma_2 \frac{L^1_{n-1}[\Upsilon]}{\eta^2 - n} c_{\substack{22\\11}}(\eta) \right).$$

In (16.97), the first matrix has a pole at η=0, while the second matrix has poles at η=$\pm\sqrt{n}$. We now use contour integration with the Jordan lemma (closing the contour in the lower half plane for $t > 0$) to evaluate the integrals. Results for the two terms in (16.97) are

$$\int_{-\infty}^{\infty} d\eta e^{-i\omega_g \eta t} \frac{c_{ij}(\eta)}{\eta} = -i\pi \eta_+(t) \left[c_{ij}(\eta)\right]_{\eta=0}, \tag{16.98}$$

$$\int_{-\infty}^{\infty} d\eta e^{-i\omega_g \eta t} M_{\substack{11\\22}}(\eta) = -i\pi \eta_+(t) \sum_{n=1}^{\infty} \cos\left(\frac{\omega_g l}{\gamma}\frac{t}{\tau_o}\sqrt{n}\right)$$
$$\times \left[\gamma_1 L_n[\Upsilon] c_{\substack{11\\22}}(\eta) + \frac{\gamma_2}{\sqrt{n}} L_{n-1}^1[\Upsilon] c_{\substack{21\\12}}(\eta)\right]_{\eta=\sqrt{n}} \tag{16.99}$$

and

$$\int_{-\infty}^{\infty} d\eta e^{-i\omega_g \eta t} M_{\substack{12\\21}}(\eta) = -\pi \eta_+(t) \sum_{n=1}^{\infty} \sin\left(\frac{\omega_g l}{\gamma}\frac{t}{\tau_o}\sqrt{n}\right)$$
$$\times \left[\gamma_1 L_n[\Upsilon] c_{\substack{12\\21}}(\eta) + \frac{\gamma_2}{\sqrt{n}} L_{n-1}^1[\Upsilon] c_{\substack{22\\11}}(\eta)\right]_{\eta=\sqrt{n}}. \tag{16.100}$$

($\eta_+(t)$ is the Heaviside unit step function and i, j=1, 2). In the calculation of the expressions given by (16.98), (16.99) and (16.100), we have also used

$$\left[c_{\substack{11\\22}}(\eta)\right]_{\eta=-\sqrt{n}} = \left[c_{\substack{11\\22}}(\eta)\right]_{\eta=\sqrt{n}},$$

and

$$\left[c_{\substack{12\\21}}(\eta)\right]_{\eta=-\sqrt{n}} = -\left[c_{\substack{12\\21}}(\eta)\right]_{\eta=\sqrt{n}},$$

which we found during the calculations [21].

Hence, (16.98) along with (16.99) and (16.100) provide the complete solution for the integral I, which is the time representation of the second term of the full Green's function $\mathcal{G}(x_1, x_2; t)_K$ given in (16.69).

References

1. L.L. Sohn, L.P. Kouwenhoven, G. Schoen, *Mesoscopic Electron Transport* (Springer and Kluwer, Netherlands, 1997)
2. C.R. Kagan, C.B. Murray, Charge transport in strongly coupled quantum dot solids. Nat. Nanotechnol. **10**(12), 1013–1026 (2015)
3. V.P. Kunets, M. Rebello Sousa Dias, T. Rembert, M.E. Ware, Y.I. Mazur, V. Lopez-Richard, H.A. Mantooth, G.E. Marques, G.J. Salamo, Electron transport in quantum dot chains: dimensionality effects and hopping conductance. J. Appl. Phys. **113**(18), 183709 (2013)
4. T. Jamieson, R. Bakhshi, D. Petrova, R. Pocock, M. Imani, A.M. Seifalian, Biological applications of quantum dots. Biomaterials **28**(31), 4717–4732 (2007)

5. L. Qi, X. Gao, Emerging application of quantum dots for drug delivery and therapy. Expert Opin. Drug Deliv. **5**(3), 263–267 (2008). (PMID: 18318649)
6. National Research Council, *High Magnetic Field Science and Its Application in the United States: Current Status and Future Directions* (The National Academies Press, Washington, D.C., 2013)
7. D. Lai, Matter in strong magnetic fields. Rev. Mod. Phys. **73**, 629–662 (2001)
8. N.J.M. Horing, S. Bahrami, Landau minibands in an antidot lattice. AdvNanoBio M & D **1**(1), 24 (2017)
9. N.J. Morgenstern Horing, S.Y. Liu, Green's functions for a graphene sheet and quantum dot in a normal magnetic field. J. Phys. A: Math. Theor. **42**(22), 225301 (2009)
10. Xi Chen, V.I. Fal'ko, Hierarchy of gaps and magnetic minibands in graphene in the presence of the Abrikosov vortex lattice. Phys. Rev. B **93**, 035427 (2016)
11. C. Kittel. *Introduction to Solid State Physics*, 7th edn. (Wiley, 1991)
12. N.J. Morgenstern Horing, M.M. Yildiz, Quantum theory of longitudinal dielectric response properties of a two-dimensional plasma in a magnetic field. Ann. Phys. **97**(1), 216–241 (1976)
13. A. Erdelyi, H. Bateman, *Higher Transcendental Functions*, vol. 2 (McGraw-Hill, 1953)
14. T. Ando, Theory of electronic states and transport in carbon nanotubes. J. Phys. Soc. Jpn. **74**(3), 777–817 (2005)
15. Y. Zheng, T. Ando, Hall conductivity of a two-dimensional graphite system. Phys. Rev. B **65**, 245420 (2002)
16. N.J.M. Horing, Landau quantized dynamics and spectra for group-vi dichalcogenides, including a model quantum wire. AIP Adv. **7**(6), 065316 (2017)
17. N.J.M. Horing, Addendum: "Landau quantized dynamics and spectra for group-vi dichalcogenides, including a model quantum wire" [AIP Advances 7, 065316 (2017)]. AIP Adv. **8**(4), 049901 (2018)
18. Q. Wang, R. Shen, L. Sheng, B.G. Wang, D.Y. Xing, Transient zitterbewegung of graphene superlattices. Phys. Rev. A **89**, 022121 (2014)
19. G.M. Maksimova, V.Y. Demikhovskii, E.V. Frolova, Wave packet dynamics in a monolayer graphene. Phys. Rev. B **78**, 235321 (2008)
20. V.Y. Demikhovskii, G.M. Maksimova, E.V. Frolova, Wave packet dynamics in a two-dimensional electron gas with spin orbit coupling: Splitting and zitterbewegung. Phys. Rev. B **78**, 115401 (2008)
21. R.A.W. Ayyubi, N.J.M. Horing, K. Sabeeh, Effect of pseudospin polarization on wave packet dynamics in graphene antidot lattices (GALs) in the presence of a normal magnetic field. J. Appl. Phys. **129**(7), 074301 (2021)
22. D. Song, V. Paltoglou, S. Liu, Y. Zhu, D. Gallardo, L. Tang, J. Xu, M. Ablowitz, N.K. Efremidis, Z. Chen, Unveiling pseudospin and angular momentum in photonic graphene. Nat. Commun. **6**, 6272 (2015)
23. Tomasz M. Rusin, Wlodek Zawadzki, Zitterbewegung of electrons in graphene in a magnetic field. Phys. Rev. B **78**, 125419 (2008)
24. E. Serna, I. Rodríguez Vargas, R. Pérez-Álvarez, L. Diago-Cisneros, Pseudospin-dependent zitterbewegung in monolayer graphene. J. Appl. Phys. **125**(20), 203902 (2019)
25. L. Esaki, R. Tsu, Superlattice and negative differential conductivity in semiconductors. IBM J. Res. Dev. **14**(1), 61–65 (1970)
26. C. Heide, T. Higuchi, H.B. Weber, P. Hommelhoff, Coherent electron trajectory control in graphene. Phys. Rev. Lett. **121**, 207401 (2018)
27. A.H. Castro Neto, F. Guinea, N.M.R. Peres, K.S. Novoselov, A.K. Geim, The electronic properties of graphene. Rev. Mod. Phys. **81**, 109–162 (2009)
28. C.H. Park, Y.W. Son, L. Yang, M.L. Cohen, S.G. Louie, Electron beam supercollimation in graphene superlattices. Nano Lett. **8**(9), 2920 (2008)
29. S. Choi, C.-H. Park, S.G. Louie, Electron supercollimation in graphene and Dirac fermion materials using one-dimensional disorder potentials. Phys. Rev. Lett. **113**, 026802 (2014)
30. C.-H. Park, L. Yang, Y.-W. Son, M.L. Cohen, S.G. Louie, Anisotropic behaviours of massless Dirac fermions in graphene under periodic potentials (2008)

31. T.G. Pedersen, C. Flindt, J. Pedersen, N.A. Mortensen, A.-P. Jauho, K. Pedersen, Graphene antidot lattices: designed defects and spin qubits. Phys. Rev. Lett. **100**, 136804 (2008)
32. A.J.M. Giesbers, E.C. Peters, M. Burghard, K. Kern, Charge transport gap in graphene antidot lattices. Phys. Rev. B **86**, 045445 (2012)
33. N.J. Morgenstern Horing, Dichalcogenide Landau miniband dynamics and spectrum in an antidot superlattice. AIP Adv. **10**(3), 035203 (2020)
34. T. Danz, A. Neff, J.H. Gaida, R. Bormann, C. Ropers, S. Schäfer, Ultrafast sublattice pseudospin relaxation in graphene probed by polarization-resolved photoluminescence. Phys. Rev. B **95**, 241412 (2017)
35. S. Aeschlimann, R. Krause, M. Chávez-Cervantes, H. Bromberger, R. Jago, E. Malić, A. Al-Temimy, C. Coletti, A. Cavalleri, I. Gierz, Ultrafast momentum imaging of pseudospin-flip excitations in graphene. Phys. Rev. B **96**, 020301 (2017)
36. M. Trushin, A. Grupp, G. Soavi, A. Budweg, D. De Fazio, U. Sassi, A. Lombardo, A.C. Ferrari, W. Belzig, A. Leitenstorfer, D. Brida, Ultrafast pseudospin dynamics in graphene. Phys. Rev. B **92**, 165429 (2015)
37. Chen-Di Han, Hong-Ya Xu, Ying-Cheng Lai, Pseudospin modulation in coupled graphene systems. Phys. Rev. Research **2**, 033406 (2020)
38. M. Polini, F. Guinea, M. Lewenstein, H. Manoharan, V. Pellegrini, Artificial honeycomb lattices for electrons, atoms and photons. Nat. Nanotechnol. **8**, 625–633 (2013)
39. K. Gomes, W. Mar, W. Ko, F. Guinea, H. Manoharan, Designer Dirac fermions and topological phases in molecular graphene. Nature **483**, 306–310 (2012)
40. O. Bahat-Treidel, O. Peleg, M. Grobman, N. Shapira, M. Segev, T. Pereg-Barnea, Klein tunneling in deformed honeycomb lattices. Phys. Rev. Lett. **104**, 063901 (2010)
41. P. Soltan-Panahi, J. Struck, P. Hauke, A. Bick, W. Plenkers, G. Meineke, C. Becker, P. Windpassinger, M. Lewenstein, K. Sengstock, Multi-component quantum gases in spin-dependent hexagonal lattices. Nat. Phys. **7**, 05 (2010)
42. Tomasz M. Rusin, Wlodek Zawadzki, Transient zitterbewegung of charge carriers in mono- and bilayer graphene, and carbon nanotubes. Phys. Rev. B **76**, 195439 (2007)
43. A.O. Barut, A.J. Bracken, Zitterbewegung and the internal geometry of the electron. Phys. Rev. D **23**, 2454–2463 (1981)
44. F. Schwabl, *Advanced Quantum Mechanics*, 4th edn. (Springer, Berlin, 2008)

Chapter 17
Numerical Analysis of the Helmholtz Green's Function for Scalar Wave Propagation Through a Nano-hole on a Plasmonic Layer

Désiré Miessein, Norman J. M. Horing, Godfrey Gumbs, and Harry Lenzing

Abstract A detailed numerical study of the Helmholtz Green's function for the description of scalar wave propagation through a nano-hole on a plasmonic layer is presented in this chapter. In conjunction with this, we briefly review the analytic formulation taking the nano-hole radius as the smallest length parameter of the system. Figures exhibiting the numerical results for this Green's function in various ranges of the transmission region are presented.

17.1 Introduction

The transmission properties of a scalar field propagating through a nano-hole in a two-dimensional (2D) plasmonic layer have been analyzed using a Green's function technique in conjunction with an integral equation formulation [1–4]. The nano-hole is taken to lie on a plasmonic sheet (located on the plane $z = 0$ embedded in a three-dimensional (3D) bulk host medium with background dielectric constant $\varepsilon_b^{(3D)}$). In Sect. 17.2 of this paper, we briefly review in some detail the analytic determination of the scalar Helmholtz Green's function with the presence of the layer in which a two-dimensional plasma is embedded. Sect. 17.3 reviews the scalar Helmholtz Green's function solution for the 2D plasmonic layer embedded in a 3D host medium with the presence of a nano-hole aperture in the subwavelength regime. The results of our

D. Miessein
Department of Physics and Engineering Physics, Fordham University, Bronx, NY 10458, USA

D. Miessein · N. J. M. Horing (✉) · H. Lenzing
Department of Physics, Stevens Institute of Technology, Hoboken, NJ 07030, USA
e-mail: nhoring@stevens.edu

G. Gumbs
Department of Physics and Astronomy, Hunter College of the City University of New York, New York, NY 10065, USA

H. Ünlü and N. J. M. Horing (eds.), *Progress in Nanoscale and Low-Dimensional Materials and Devices*, Topics in Applied Physics 144,
https://doi.org/10.1007/978-3-030-93460-6_17

thorough numerical analysis of the perforated layer Helmholtz Green's function are discussed in Sect. 17.4 with illustrative figures showing results in the near, middle and far field zones of the transmission region. Finally, conclusions are summarized in Sect. 17.5.

17.2 Green's Function Solution for Full 2D Plasmonic Layer Embedded in a 3D Bulk Host Medium

17.2.1 Integral Equation for the Scalar Green's Function and Solution

We consider a two dimensional plasmonic layer S_1 to have a dynamic, nonlocal $2D$ polarizability $\alpha_{fs}^{(2D)}(\vec{k}_\parallel, \omega)$, located on the plane $z = 0$, embedded in a three dimensional bulk host medium with background dielectric constant $\varepsilon_b^{(3D)}$ (Fig. 17.1). The associated Helmholtz Green's function including the two-dimensional plasmonic sheet, G_{fs} *without* a nano-hole, satisfies the integro-differential equation (position/frequency representation) [1–4]

$$\left(-\vec{\nabla}^2 - \frac{\omega^2}{c^2}\varepsilon_b^{(3D)}\right) G_{fs}(\vec{r}, \vec{r}'; \omega) - \frac{\omega^2}{c^2} \int d^3\vec{r}'\alpha_{fs}^{(2D)}(\vec{r}, \vec{r}''; \omega) G_{fs}(\vec{r}'', \vec{r}'; \omega)$$
$$= \delta^{(3)}(\vec{r} - \vec{r}'). \tag{17.2.1}$$

The polarizability $\alpha_{fs}^{(2D)}$ of the full 2D plasmonic layer has the form

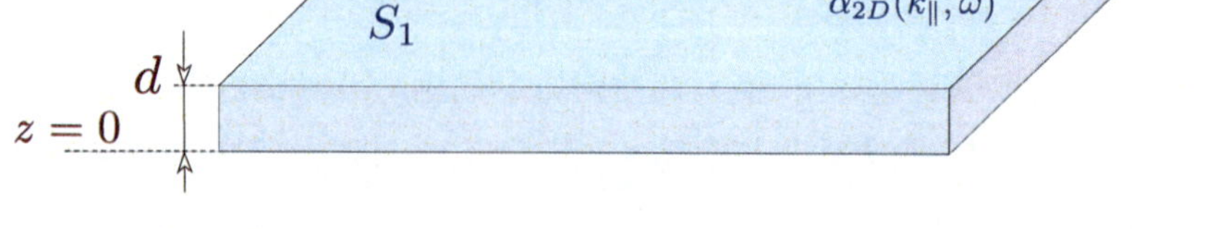

Fig. 17.1 (Color online) Schematic illustration of a two dimensional plasmonic layer S_1 of thickness d embedded at $z = 0$ in a three dimensional bulk medium with dielectric constant $\varepsilon_b^{(3D)}$

$$\alpha_{fs}^{(2D)}(\vec{r}, \vec{r}\,'; \omega) = \alpha_{fs}^{(2D)}(\vec{r}_\parallel, \vec{r}_\parallel'; \omega)\, d\, \delta(z)\delta(z'), \tag{17.2.2}$$

where d is the thickness of the plasmonic sheet, $\vec{r} = (\vec{r}_\parallel; z)$ and $\alpha_{fs}^{(2D)}(\vec{r}_\parallel, \vec{r}_\parallel'; \omega)$ is the 2D plasmonic polarizability of the 2D sheet; $\delta(z)$ is the Dirac delta function needed to confine the polarizability onto the plane of the 2D layer at $z = 0$.

To solve (17.2.1), we employ the bulk Helmholtz Green's function [5]

$$\left(-\vec{\nabla}^2 - \frac{\omega^2}{c^2}\varepsilon_b^{(3D)}\right) G_{3D}(\vec{r}, \vec{r}\,'; \omega) = \delta^{3D}(\vec{r} - \vec{r}\,'). \tag{17.2.3}$$

After performing the 2D spatial Fourier transform of G_{3D} in the plane of the translationally invariant 2D homogeneous plasmonic sheet ($\vec{\bar{r}}_\parallel = \vec{r}_\parallel - \vec{r}_\parallel'$)

$$G_{3D}(\vec{k}_\parallel; z, z'; \omega) = \int d^2\vec{\bar{r}}_\parallel\, G_{3D}(\vec{\bar{r}}_\parallel; z, z'; \omega)\, e^{-i\,\vec{k}_\parallel . \vec{\bar{r}}_\parallel}, \tag{17.2.4}$$

equation (17.2.3) becomes

$$\left[\frac{\partial^2}{\partial z^2} + k_\perp^2\right] G_{3D}(\vec{k}_\parallel; z, z'; \omega) = -\delta(z - z') \tag{17.2.5}$$

where $k_\perp^2 = q_\omega^2 - k_\parallel^2$ and $q_\omega = \frac{\omega}{c}\sqrt{\varepsilon_b^{(3D)}}$. This has the well-known solution [5]

$$G_{3D}(\vec{k}_\parallel; z, z'; \omega) = -\frac{e^{ik_\perp |z-z'|}}{2ik_\perp}. \tag{17.2.6}$$

Employing $G_{3D}(\vec{r}, \vec{r}\,'; \omega)$, (17.2.1) can be conveniently rewritten as an inhomogeneous integral equation as follows:

$$G_{fs}(\vec{r}, \vec{r}\,'; \omega) = G_{3D}(\vec{r}, \vec{r}\,'; \omega) + \frac{\omega^2}{c^2}\int d^3\vec{r}\,'' \int d^3\vec{r}\,''' G_{3D}(\vec{r}, \vec{r}\,''; \omega)\alpha_{fs}^{2D}(\vec{r}\,'', \vec{r}\,'''; \omega) G_{fs}(\vec{r}\,''', \vec{r}\,'; \omega). \tag{17.2.7}$$

Introducing (17.2.2) in (17.2.7) and Fourier transforming the resulting equation in the lateral plane of translational invariance ($\vec{r}_\parallel - \vec{r}_\parallel' \to \vec{k}_\parallel$), we obtain

$$G_{fs}(\vec{k}_\parallel; z, z'; \omega) = G_{3D}(\vec{k}_\parallel; z, z'; \omega) + \frac{\omega^2 d}{c^2} G_{3D}(\vec{k}_\parallel; z, 0; \omega)\alpha_{fs}^{(2D)}(\vec{k}_\parallel; \omega) G_{fs}(\vec{k}_\parallel; 0, z'; \omega). \tag{17.2.8}$$

Solving for $G_{fs}(\vec{k}_\parallel; 0, z'; \omega)$ algebraically, we obtain

$$G_{fs}(\vec{k}_\parallel; z, z'; \omega) = G_{3D}(\vec{k}_\parallel; z, z'; \omega) + \frac{\omega^2 d}{c^2} \frac{G_{3D}(\vec{k}_\parallel; z, 0; \omega)\alpha_{fs}^{(2D)}(\vec{k}_\parallel; \omega)G_{3D}(\vec{k}_\parallel; 0, z'; \omega)}{1 - \frac{\omega^2}{c^2} d\alpha_{fs}^{(2D)}(\vec{k}_\parallel; \omega)G_{3D}(\vec{k}_\parallel; 0, 0; \omega)}, \tag{17.2.9}$$

and using (17.2.6), this leads to the full sheet Green's function as

$$G_{fs}(\vec{k}_\parallel; z, z'; \omega) = -\frac{e^{ik_\perp|z-z'|}}{2ik_\perp} + \frac{\gamma e^{ik_\perp(|z|+|z'|)}}{2ik_\perp(2ik_\perp + \gamma)}, \tag{17.2.10}$$

where

$$\gamma = \omega^2 d\, \alpha_{fs}^{(2D)}(\vec{k}_\parallel; \omega)/c^2. \tag{17.2.11}$$

Our analysis (below) of the Green's function in the presence of an aperture will be seen to devolve upon the evaluation of $G_{fs}(\vec{r}, \vec{r}\,''; z, z''; \omega)$ at the aperture position $\vec{r}\,'' = 0$, $z'' = 0$ in position representation as given by

$$G_{fs}(\vec{r}_\parallel, 0; z, 0; \omega) = \frac{1}{(2\pi)^2} \int d^2\vec{k}_\parallel e^{i\vec{k}_\parallel.\vec{r}_\parallel} G_{fs}(\vec{k}_\parallel; z, 0; \omega) \tag{17.2.12}$$

where

$$G_{fs}(\vec{k}_\parallel; z, 0; \omega) = -\frac{e^{ik_\perp|z|}}{2ik_\perp + \gamma}, \tag{17.2.13}$$

so that

$$G_{fs}(\vec{r}_\parallel, 0; z, 0; \omega) = -\frac{1}{4i\pi} \int_0^\infty dk_\parallel \frac{k_\parallel J_0(k_\parallel r_\parallel) e^{ik_\perp|z|}}{k_\perp + \frac{\gamma}{2i}}. \tag{17.2.14}$$

Noting (17.2.11) and employing the $2D$ polarizability of the layer in the local limit as [6]

$$\alpha_{fs}^{(2D)}(\vec{k}_\parallel; \omega) = \left(\frac{2\pi i e^2 n_{2D}}{m^* \omega^2}\right) \sqrt{q_\omega^2 - k_\parallel^2} \tag{17.2.15}$$

Our calculations employ typical values for GaAs ($n_{2D} = 4 \times 10^{15}/\mathrm{cm}^2$ is the 2D equilibrium density on the sheet, $m^* = 0.065 m_0$ is the effective mass and $\mu \equiv \frac{\pi e^2 d\, n_{2D}}{m^* c^2}$), and we have

$$G_{fs}(\vec{r}_\parallel, 0; z, 0; \omega) = -\frac{1}{4i\pi[1+\mu]} \int_0^\infty \frac{dk_\parallel k_\parallel J_0(k_\parallel r_\parallel) e^{i|z|\sqrt{q_\omega^2 - k_\parallel^2}}}{\sqrt{q_\omega^2 - k_\parallel^2}}. \tag{17.2.16}$$

The $k_{\|}$-integral of (17.2.16) is readly evaluated as [7]

$$G_{fs}(\vec{r}_{\|}, 0; z, 0; \omega) = \frac{1}{4\pi[1+\mu]} \frac{e^{i\,q_{\omega}\sqrt{r_{\|}^2+|z|^2}}}{\sqrt{r_{\|}^2+|z|^2}}. \tag{17.2.17}$$

17.3 Green's Function Solution for a Perforated 2D Plasmonic Layer with a Nano-hole Embedded in a 3D Bulk Host Medium

17.3.1 Integral Equation: Scalar Helmholtz Green's Function for a Perforated Plasmonic Layer

We consider a 2D plasmonic layer S_1 which is perforated by a nano-scale aperture S_2 of radius R, as depicted in Fig. 17.2, lying in the $(x\text{–}y)$-plane. The presence of the nano-hole in the layer is represented by subtracting the part of the polarizability associated with the hole from the polarizability of the full layer, (Fig. 17.2) [1–4]

$$\alpha^{(2D)}(\vec{r}, \vec{r}'; \omega) = \alpha_{fs}^{(2D)}(\vec{r}, \vec{r}'; \omega) - \alpha_{h}^{(2D)}(\vec{r}, \vec{r}'; \omega) \tag{17.3.1}$$

where $\alpha_{h}^{(2D)}(\vec{r}, \vec{r}'; \omega)$ is the part of the layer polarizability removed by the nano-hole. The resulting Green's function for the perforated plasmonic layer with the hole satisfies the integral equation given by

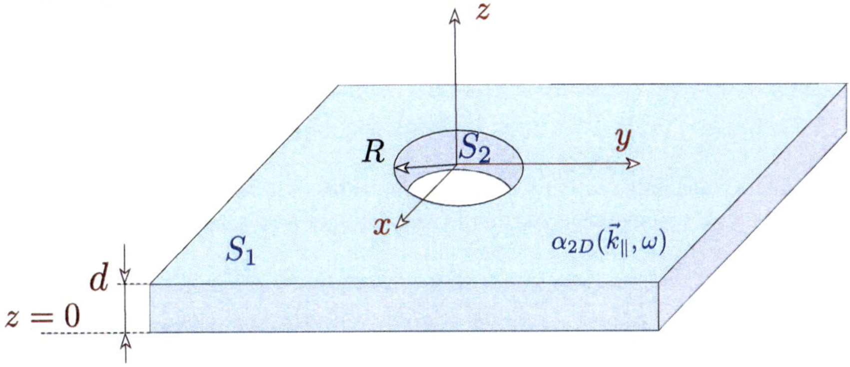

Fig. 17.2 (Color online) Schematic representation of a perforated 2D plasmonic layer (thickness d, embedded at $z = 0$ in a three dimensional bulk medium) with a nano-hole of radius R at the origin of the $(x\text{–}y)$-plane

$$G(\vec{r},\vec{r}';\omega) = G_{fs}(\vec{r},\vec{r}';\omega) - \frac{\omega^2}{c^2}\int d^3\vec{r}''\int d^3\vec{r}'''G_{fs}(\vec{r},\vec{r}'';\omega)\alpha_h^{2D}(\vec{r}'',\vec{r}''';\omega)G(\vec{r}''',\vec{r}';\omega). \tag{17.3.2}$$

Here, the polarizability of the nano-hole is defined as [8, 9]

$$\alpha_h^{(2D)}(\vec{r},\vec{r}';\omega) = \alpha_{fs}^{(2D)}(x,y,x',y';\omega)\eta_+(R-|x|)\eta_+(R-|y|)\eta_+(R-|x'|)\eta_+(R-|y|')d\,\delta(z)\delta(z'), \tag{17.3.3}$$

where $\eta_+(R-|x|)$ is the Heaviside unit step function representing a cut-off imposed to confine the integration range on the 2D sheet to the nano-hole dimensions; and the Dirac delta function $\delta(z)$ is needed to localize the polarizability onto the plane of the 2D plasmonic layer. A simple approximation of (17.3.3) for very small radius leads to (A represents the area of the aperture)

$$\alpha_h^{(2D)}(\vec{r},\vec{r}';\omega) \approx A^2\alpha_{fs}^{(2D)}(\vec{r}_\parallel,\vec{r}_\parallel{}';\omega)\delta^{2D}(\vec{r}_\parallel)\delta^{2D}(\vec{r}_\parallel{}')d\delta(z)\delta(z'), \tag{17.3.4}$$

and employing it in (17.3.2) to execute all positional integrations, we obtain

$$G(\vec{r}_\parallel,\vec{r}_\parallel{}';z,z';\omega) \approx G_{fs}(\vec{r}_\parallel,\vec{r}_\parallel{}';z,z';\omega) - \beta G_{fs}(\vec{r}_\parallel,0;z,0;\omega)G(0,\vec{r}_\parallel{}';0,z';\omega), \tag{17.3.5}$$

where $\beta = \gamma A^2$ with $\gamma = \frac{\omega^2 d}{c^2}\alpha_{fs}^{2D}([\vec{r}_\parallel = 0] - [\vec{r}_\parallel' = 0];\omega)$. To solve (17.3.5), we set $\vec{r}_\parallel = 0$ and $z = 0$, and determine $G(0,\vec{r}_\parallel{}';0,z';\omega)$ as

$$G(0,\vec{r}_\parallel{}';0,z';\omega) = \frac{G_{fs}(0,\vec{r}_\parallel{}';0,z';\omega)}{1+\beta G_{fs}(0,0;0,0;\omega)}. \tag{17.3.6}$$

Substituting (17.3.6) into (17.3.5) yields the algebraic closed form analytic solution:

$$G(\vec{r}_\parallel,\vec{r}_\parallel';z,z';\omega) \approx G_{fs}(\vec{r}_\parallel,\vec{r}_\parallel';z,z';\omega) - \frac{\beta G_{fs}(\vec{r}_\parallel,0;z,0;\omega)G_{fs}(0,\vec{r}_\parallel';0,z';\omega)}{1+\beta G_{fs}([\vec{r}_\parallel = 0]-[\vec{r}_\parallel' = 0];z=0,z'=0;\omega)}. \tag{17.3.7}$$

Noting that a transmitted scalar wave emerging from the aperture is controlled by $G(\vec{r}_\parallel,0;z,0;\omega)$, we have analyzed this quantity setting $\vec{r}_\parallel' = 0$ and $z' = 0$ in (17.3.7), leading to

$$G(\vec{r}_\parallel,0;z,0;\omega) \approx \frac{G_{fs}(\vec{r}_\parallel,0;z,0;\omega)}{1+\beta G_{fs}([\vec{r}_\parallel = 0]-[\vec{r}_\parallel' = 0];z=0,z'=0;\omega)}. \tag{17.3.8}$$

Equation (17.3.8) is an approximate analytical Green's function solution to (17.3.2), in which
$G_{fs}([\vec{r}_\parallel = 0]-[\vec{r}_\parallel' = 0];z=0,z'=0;\omega)$ is found to involve a divergent integral

when all its arguments vanish. This divergence may be seen in $\alpha^{(2D)}_{fs}$ and G_{fs} setting $\vec{r}_\parallel = 0$, $\vec{r}\,'_\parallel = 0$, $z = 0$ and $z' = 0$ as follows:

$$\alpha^{(2D)}_{fs}(0, 0; \omega) = \frac{1}{2\pi}\left(\frac{2\pi i e^2 n_{2D}}{m^*\omega^2}\right)\int_0^\infty dk_\parallel k_\parallel \sqrt{q_\omega^2 - k_\parallel^2}, \tag{17.3.9}$$

and (17.2.16) is given by

$$G_{fs}(0, 0; 0, 0; \omega) = \frac{i}{4\pi[1+\mu]}\int_0^\infty \frac{dk_\parallel k_\parallel}{\sqrt{q_\omega^2 - k_\parallel^2}}. \tag{17.3.10}$$

The divergence of $\alpha^{(2D)}_{fs}$ and G_{fs} is an artifact of limiting the radius of the aperture to be vanishingly small (zero) in the kernel of the integral equation, (17.3.2)–(17.3.4). A more realistic consideration involves a cut off at a small but finite radius R, alternatively an upper limit on the wavenumber integration, $k_\parallel \sim \frac{1}{R}$, yielding the convergent integrals

$$\alpha^{(2D)}_{fs}(0, 0; \omega) = \frac{q_\omega^3}{2\pi}\left(\frac{2\pi i e^2 n_{2D}}{m^*\omega^2}\right)\left\{\int_0^1 dy y\sqrt{1-y^2} + i\int_1^{\frac{1}{q_\omega R}} dy y\sqrt{y^2-1}\right\} \tag{17.3.11}$$

and

$$G_{fs}([\vec{r}_\parallel = 0] - [\vec{r}\,'_\parallel = 0]; z = 0, z' = 0; \omega) \approx \frac{i q_\omega}{4\pi[1+\mu]}\left\{\int_0^1 dy\, \frac{y}{\sqrt{1-y^2}} - i\int_1^{\frac{1}{q_\omega R}} dy\, \frac{y}{\sqrt{y^2-1}}\right\}, \tag{17.3.12}$$

where we have introduced the dimensionless notation $y = k_\parallel / q_\omega$, so that

$$k_\perp = \begin{cases} q_\omega\sqrt{1-y^2} & \text{for } 0 \le y < 1\,; \\ i\,q_\omega\sqrt{y^2-1} & \text{for } y > 1\,; \end{cases} \tag{17.3.13}$$

and the following results are obtained

$$\alpha^{(2D)}_{fs}([\vec{r}_\parallel = 0], [\vec{r}\,'_\parallel = 0]; \omega) \approx \frac{1}{6\pi R^3}\left(\frac{2\pi i e^2 n_{2D}}{m^*\omega^2}\right)\left[(q_\omega R)^3 + i\left[1 - (q_\omega R)^2\right]^{3/2}\right] \tag{17.3.14}$$

and

$$G_{fs}([\vec{r}_\parallel = 0] - [\vec{r}\,'_\parallel = 0]; z = 0, z' = 0; \omega) \approx \frac{1}{4\pi(1+\mu)R}\left[i\,q_\omega R + \sqrt{1-(q_\omega R)^2}\right], \tag{17.3.15}$$

for $q_\omega R < 1$. Furthermore, substituting (17.3.14) into the expression for γ, we have

$$\gamma = \frac{\mu}{3\pi R^3}\left[i\ (q_\omega R)^3 - \left[1 - (q_\omega R)^2\right]^{3/2}\right] \tag{17.3.16}$$

with $\mu = \pi e^2 d\, n_{2D}/(m^* c^2)$.

17.4 Numerical Results

Our numerical results for the real and imaginary parts of the Green's function in (17.3.8), $\mathrm{Re}[G(\vec{r}_\parallel, 0; z, 0; \omega)]$ and $\mathrm{Im}[G(\vec{r}_\parallel, 0; z, 0; \omega)]$, respectively, for frequency $f = 300$ THz are presented in 3D and density plots in Figs. 17.3, 17.4, 17.5, 17.6, 17.7, 17.8, 17.9,17.10, and 17.11, 17.12, 17.13, 17.14 as functions of x and y for several values of distance z away from the layer screen: We chose $z = 50\,R$ (near-field), $z = 300\,R$ (middle-field) and $z = 1000\,R$ (far-field). These figures reveal the structure of the Green's function for the perforated layer in terms of near-field ($z = 50\,R$), middle-field ($z = 300\,R$) and far-field ($z = 1000\,R$) radiation zones for $R = 5$ nm.

Further detail concerning $\mathrm{Re}[G(\vec{r}_\parallel, 0; z, 0; \omega)]$ and $\mathrm{Im}[G(\vec{r}_\parallel, 0; z, 0; \omega)]$ is provided in the Figs. 17.15, 17.16, 17.17, 17.18, 17.19, 17.20 below for the three z-radiation zones described by $z = 50\,R$ (near-field), $z = 300\,R$ (middle-field) and $z = 1000\,R$ (far-field):

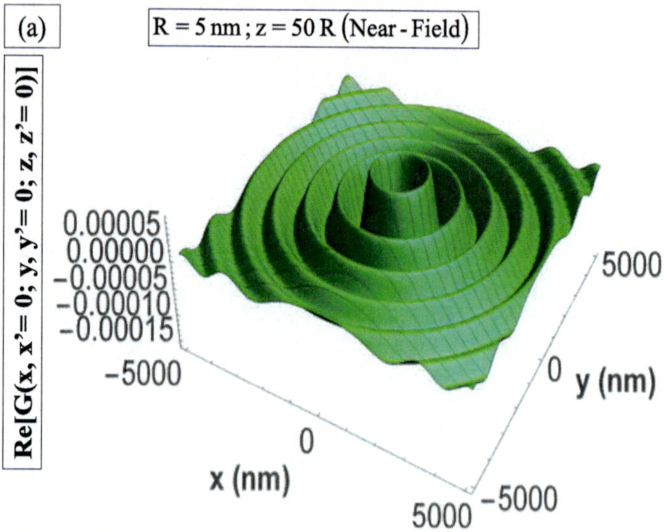

Fig. 17.3 (Color online) Figure **a** exhibits $\mathrm{Re}[G(\vec{r}_\parallel, 0; z, 0; \omega)]$ in terms of 3D plot for a perforated 2D plasmonic layer of GaAs in the presence of a nano-hole of radius $R = 5$ nm at $z = 50\,R$ (Near-Field) for $\varepsilon_b = 1$, $n_{2D} = 4 \times 10^{15}\,\mathrm{cm}^{-2}$, $d = 10$ nm and $m^* = 0.065 m_0$ where m_0 is the free-electron mass

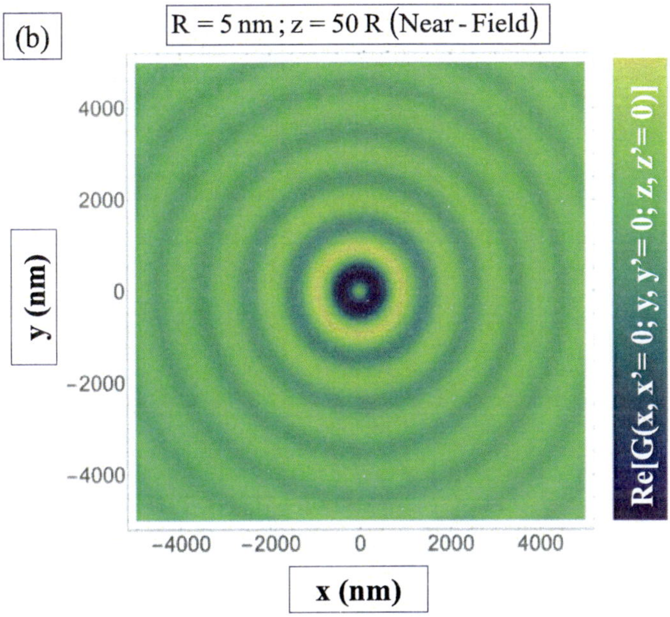

Fig. 17.4 (Color online) Figure **b** exhibits $\text{Re}[G(\vec{r}_{\parallel}, 0; z, 0; \omega)]$ in terms of density plot for a perforated 2D plasmonic layer of GaAs in the presence of a nano-hole of radius $R = 5\,\text{nm}$ at $z = 50\,R$ (Near-Field) for $\varepsilon_b = 1$, $n_{2D} = 4 \times 10^{15}\,\text{cm}^{-2}$, $d = 10\,\text{nm}$ and $m^* = 0.065m_0$ where m_0 is the free-electron mass

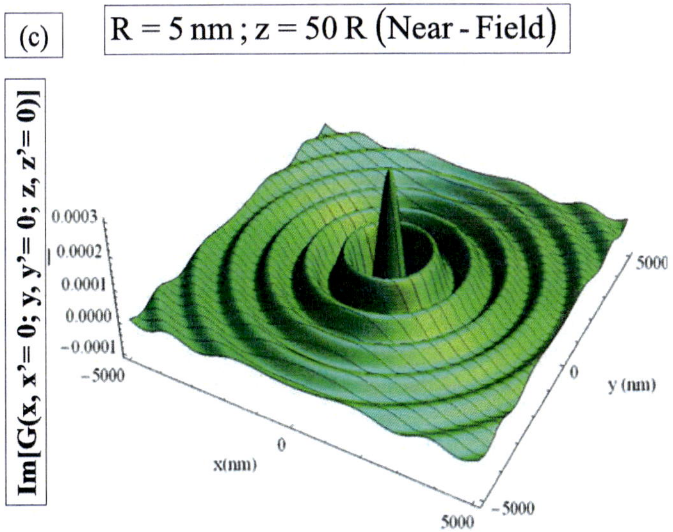

Fig. 17.5 (Color online) Figure **c** presents $\text{Im}[G(\vec{r}_{\parallel}, 0; z, 0; \omega)]$ in terms of 3D plot for a perforated 2D plasmonic layer of GaAs in the presence of a nano-hole of radius $R = 5\,\text{nm}$ at $z = 50\,R$ (Near-Field) for $\varepsilon_b = 1$, $n_{2D} = 4 \times 10^{15}\,\text{cm}^{-2}$, $d = 10\,\text{nm}$ and $m^* = 0.065m_0$ where m_0 is the free-electron mass

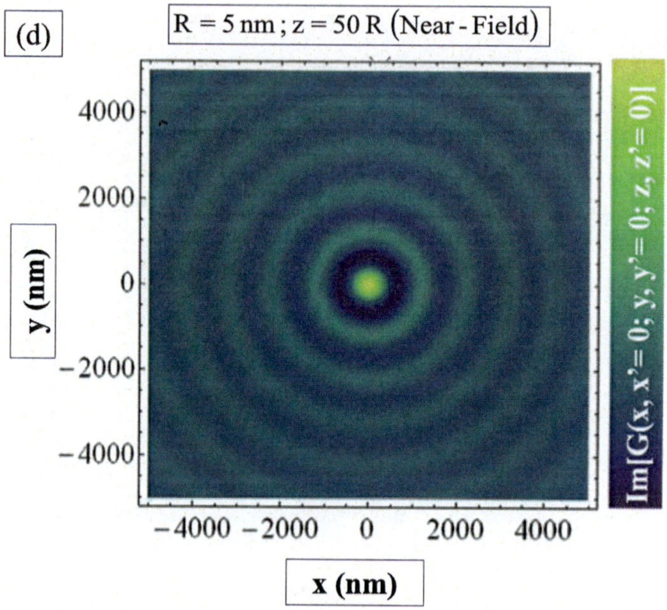

Fig. 17.6 (Color online) Figure **d** presents $\text{Im}[G(\vec{r}_{\parallel}, 0; z, 0; \omega)]$ in terms of density plot for a perforated 2D plasmonic layer of GaAs in the presence of a nano-hole of radius $R = 5\,\text{nm}$ at $z = 50\,R$ (Near-Field) for $\varepsilon_b = 1$, $n_{2D} = 4 \times 10^{15}\,\text{cm}^{-2}$, $d = 10\,\text{nm}$ and $m^* = 0.065m_0$ where m_0 is the free-electron mass

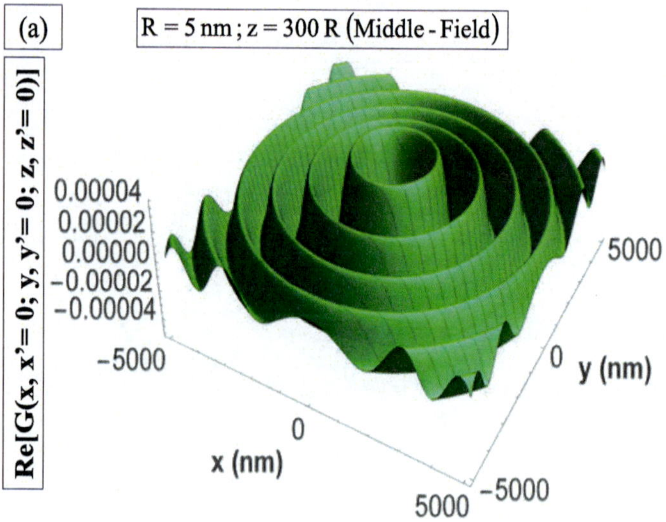

Fig. 17.7 (Color online) Figure **a** exhibits $\text{Re}[G(\vec{r}_{\parallel}, 0; z, 0; \omega)]$ in terms of 3D plot for a perforated 2D plasmonic layer of GaAs in the presence of a nano-hole of radius $R = 5\,\text{nm}$ at $z = 300\,R$ (Middle-Field) for $\varepsilon_b = 1$, $n_{2D} = 4 \times 10^{15}\,\text{cm}^{-2}$, $d = 10\,\text{nm}$ and $m^* = 0.065m_0$ where m_0 is the free-electron mass

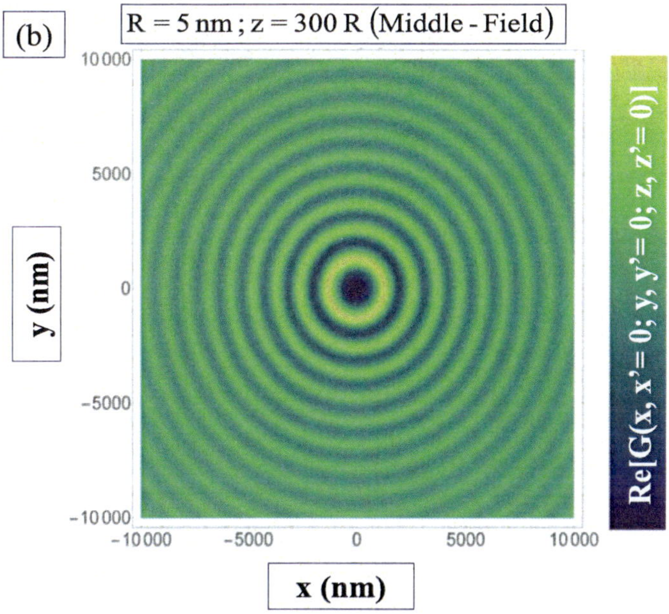

Fig. 17.8 (Color online) Figure **b** exhibits $\mathrm{Re}[G(\vec{r}_{\parallel}, 0; z, 0; \omega)]$ in terms of density plot for a perforated 2D plasmonic layer of GaAs in the presence of a nano-hole of radius $R = 5\,\mathrm{nm}$ at $z = 300\,R$ (Middle-Field) for $\varepsilon_b = 1, n_{2D} = 4 \times 10^{15}\,\mathrm{cm}^{-2}, d = 10\,\mathrm{nm}$ and $m^* = 0.065m_0$ where m_0 is the free-electron mass

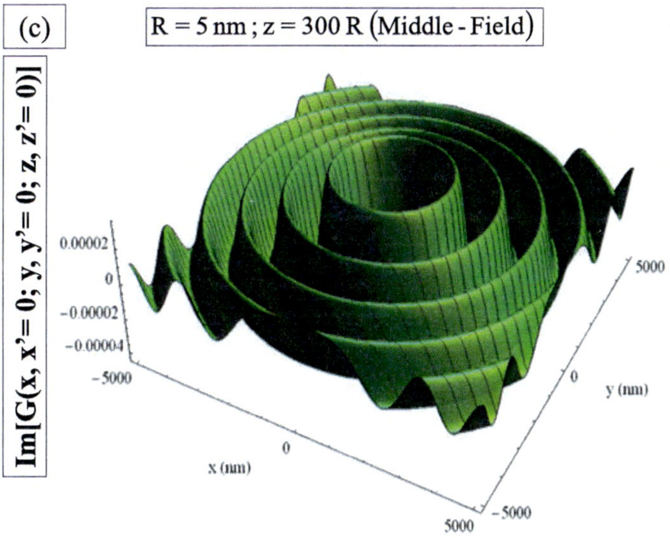

Fig. 17.9 (Color online) Figure **c** presents $\mathrm{Im}[G(\vec{r}_{\parallel}, 0; z, 0; \omega)]$ in terms of 3D plot for a perforated 2D plasmonic layer of GaAs in the presence of a nano-hole of radius $R = 5\,\mathrm{nm}$ at $z = 300\,R$ (Middle-Field) for $\varepsilon_b = 1, n_{2D} = 4 \times 10^{15}\,\mathrm{cm}^{-2}, d = 10\,\mathrm{nm}$ and $m^* = 0.065m_0$ where m_0 is the free-electron mass

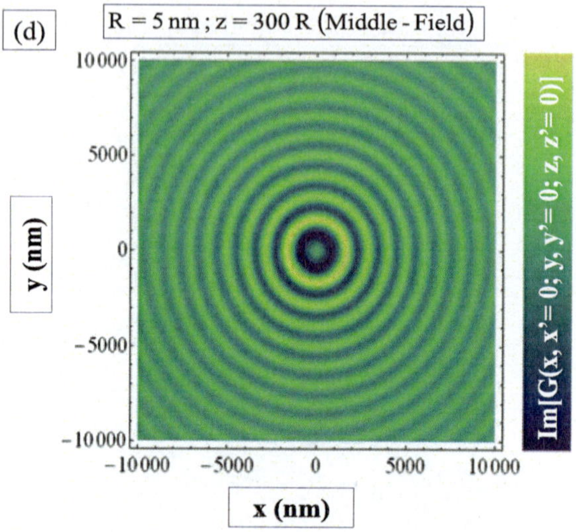

Fig. 17.10 (Color online) Figure **d** presents $\text{Im}[G(\vec{r}_{\|}, 0; z, 0; \omega)]$ in terms of density plot for a perforated 2D plasmonic layer of GaAs in the presence of a nano-hole of radius $R = 5\,\text{nm}$ at $z = 300\,R$ (Middle-Field) for $\varepsilon_b = 1, n_{2D} = 4 \times 10^{15}\,\text{cm}^{-2}, d = 10\,\text{nm}$ and $m^* = 0.065m_0$ where m_0 is the free-electron mass

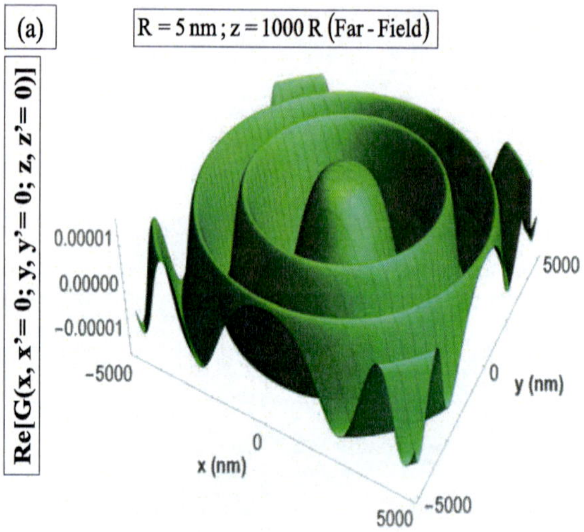

Fig. 17.11 (Color online) Figure **a** exhibits $\text{Re}[G(\vec{r}_{\|}, 0; z, 0; \omega)]$ in terms of 3D plot for a perforated 2D plasmonic layer of GaAs in the presence of a nano-hole of radius $R = 5\,\text{nm}$ at $z = 1000\,R$ (Far-Field) for $\varepsilon_b = 1, n_{2D} = 4 \times 10^{15}\,\text{cm}^{-2}, d = 10\,\text{nm}$ and $m^* = 0.065m_0$ where m_0 is the free-electron mass

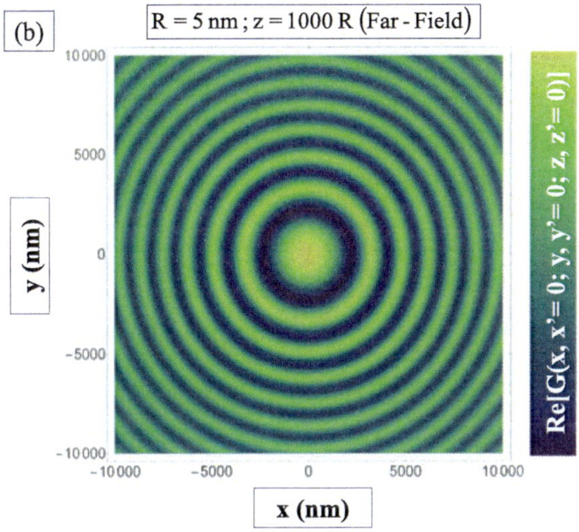

Fig. 17.12 (Color online) Figure **b** exhibits Re$[G(\vec{r}_{\|}, 0; z, 0; \omega)]$ in terms of density plot for a perforated 2D plasmonic layer of GaAs in the presence of a nano-hole of radius $R = 5\,\text{nm}$ at $z = 1000\,R$ (Far-Field) for $\varepsilon_b = 1$, $n_{2D} = 4 \times 10^{15}\,\text{cm}^{-2}$, $d = 10\,\text{nm}$ and $m^* = 0.065 m_0$ where m_0 is the free-electron mass

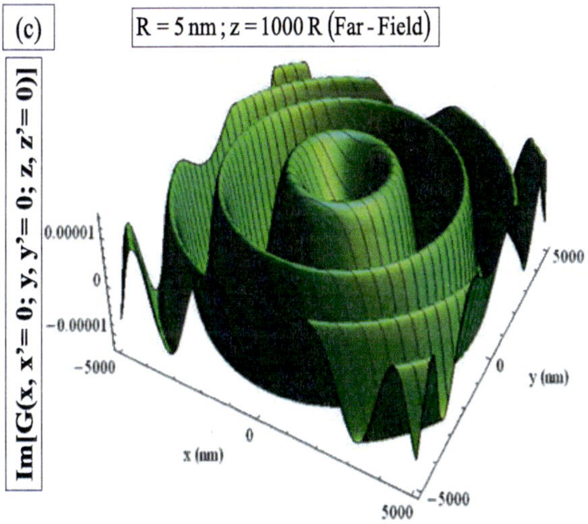

Fig. 17.13 (Color online) Figure **c** presents Im$[G(\vec{r}_{\|}, 0; z, 0; \omega)]$ in terms of 3D plot for a perforated 2D plasmonic layer of GaAs in the presence of a nano-hole of radius $R = 5\,\text{nm}$ at $z = 1000\,R$ (Far-Field) for $\varepsilon_b = 1, n_{2D} = 4 \times 10^{15}\,\text{cm}^{-2}, d = 10\,\text{nm}$ and $m^* = 0.065 m_0$ where m_0 is the free-electron mass

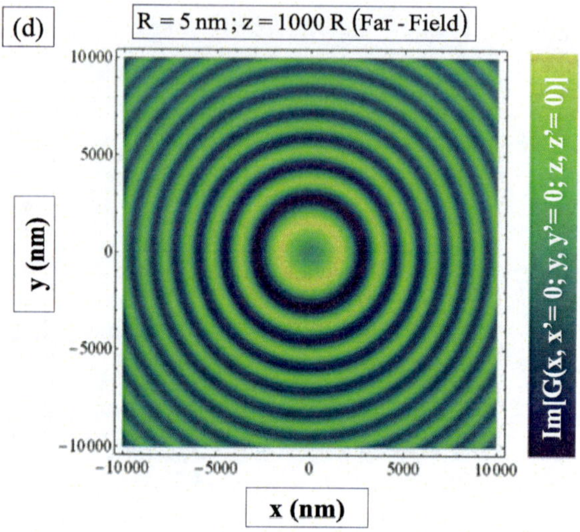

Fig. 17.14 (Color online) Figure **d** presents $\text{Im}[G(\vec{r}_{\parallel}, 0; z, 0; \omega)]$ in terms of density plot for a perforated 2D plasmonic layer of GaAs in the presence of a nano-hole of radius $R = 5\,\text{nm}$ at $z = 1000\,R$ (Far-Field) for $\varepsilon_b = 1$, $n_{2D} = 4 \times 10^{15}\,\text{cm}^{-2}$, $d = 10\,\text{nm}$ and $m^* = 0.065m_0$ where m_0 is the free-electron mass

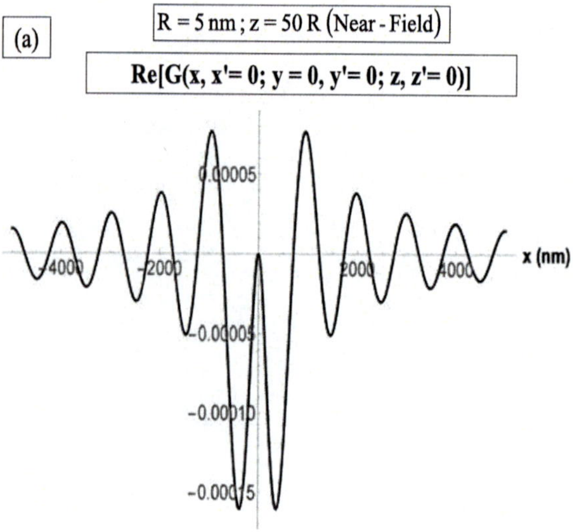

Fig. 17.15 **a** Exhibits $\text{Re}[G(x, x' = 0; y = 0, y' = 0; z = 50\,R, z' = 0; \omega)]$ as a function of x for a perforated 2D plasmonic layer of GaAs in the presence of a nano-hole of radius $R = 5\,\text{nm}$ at $z = 50\,R$ (Near-Field) for $\varepsilon_b = 1$, $n_{2D} = 4 \times 10^{15}\,\text{cm}^{-2}$, $d = 10\,\text{nm}$ and $m^* = 0.065m_0$ where m_0 is the free-electron mass

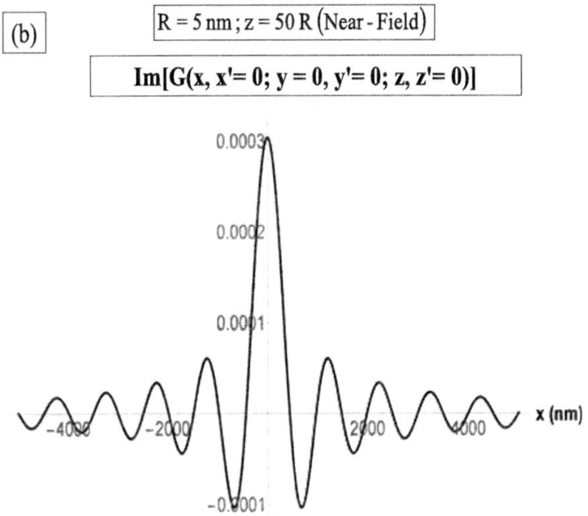

Fig. 17.16 **b** Presents Im $[G(x, x' = 0; y = 0, y' = 0; z = 50\,R, z' = 0; \omega)]$ as a function of x for a perforated 2D plasmonic layer of GaAs in the presence of a nano-hole of radius $R = 5\,\text{nm}$ at $z = 50\,R$ (Near-Field) for $\varepsilon_b = 1, n_{2D} = 4 \times 10^{15}\,\text{cm}^{-2}, d = 10\,\text{nm}$ and $m^* = 0.065m_0$ where m_0 is the free-electron mass

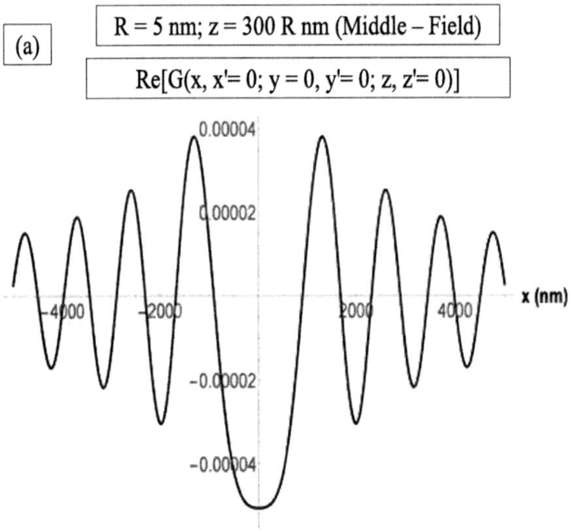

Fig. 17.17 **a** Exhibits Re$[G(x, x' = 0; y = 0, y' = 0; z = 300\,R, z' = 0; \omega)]$ as a function of x for a perforated 2D plasmonic layer of GaAs in the presence of a nano-hole of radius $R = 5\,\text{nm}$ at $z = 300\,R$ (Middle-Field) for $\varepsilon_b = 1, n_{2D} = 4 \times 10^{15}\,\text{cm}^{-2}, d = 10\,\text{nm}$ and $m^* = 0.065m_0$ where m_0 is the free-electron mass

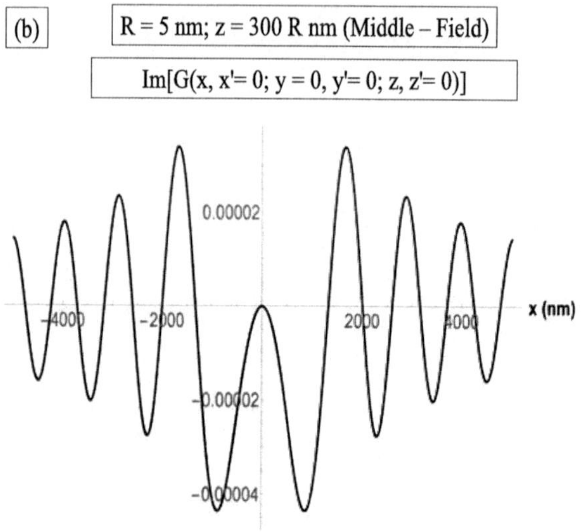

Fig. 17.18 **b** Presents Im $[G(x, x' = 0; y = 0, y' = 0; z = 300\,R, z' = 0; \omega)]$ as a function of x for a perforated 2D plasmonic layer of GaAs in the presence of a nano-hole of radius $R = 5\,\text{nm}$ at $z = 300\,R$ (Middle-Field) for $\varepsilon_b = 1, n_{2D} = 4 \times 10^{15}\,\text{cm}^{-2}, d = 10\,\text{nm}$ and $m^* = 0.065 m_0$ where m_0 is the free-electron mass

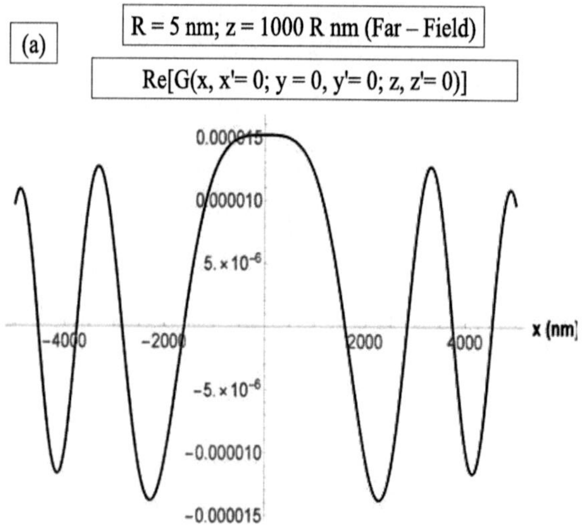

Fig. 17.19 **a** Exhibits Re$[G(x, x' = 0; y = 0, y' = 0; z = 1000\,R, z' = 0; \omega)]$ as a function of x for a perforated 2D plasmonic layer of GaAs in the presence of a nano-hole of radius $R = 5\,\text{nm}$ at $z = 1000\,R$ (Far-Field) for $\varepsilon_b = 1$, $n_{2D} = 4 \times 10^{15}\,\text{cm}^{-2}$, $d = 10\,\text{nm}$ and $m^* = 0.065 m_0$ where m_0 is the free-electron mass

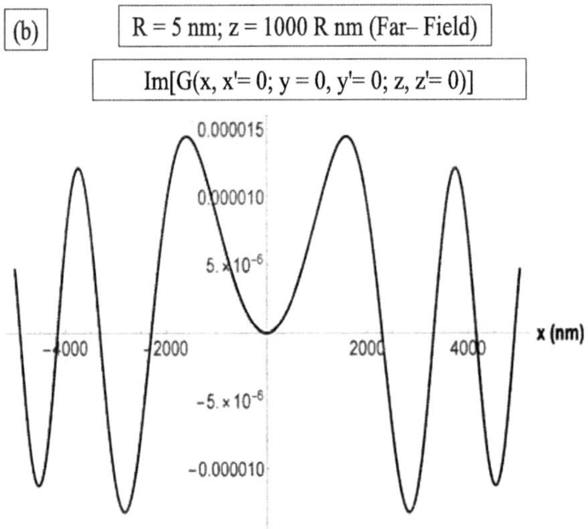

Fig. 17.20 **b** Presents Im $[G(x, x' = 0; y = 0, y' = 0; z = 1000\,R, z' = 0; \omega)]$ as a function of x for a perforated 2D plasmonic layer of GaAs in the presence of a nano-hole of radius $R = 5\,\text{nm}$ at $z = 1000\,R$ (Far-Field) for $\varepsilon_b = 1$, $n_{2D} = 4 \times 10^{15}\,\text{cm}^{-2}$, $d = 10\,\text{nm}$ and $m^* = 0.065m_0$ where m_0 is the free-electron mass

17.5 Concluding Remarks

In this chapter we have presented a thorough numerical analysis of the closed-form expression for the scalar Green's function of a perforated, thin 2D plasmonic layer embedded in a 3D host medium in the presence of a nano-hole.

Inspection of the resulting Green's function figures shows that for large $r_{\|} \mapsto x > 2500\,\text{nm}$ the spatial dependence of the Green's function becomes becomes oscillatory as a function of $r_{\|}(x)$ with peaks uniformly spaced. In this regard, it should be noted that our designation of near, middle and far radiation zones is defined in terms of z-values $(50R, 300R, 1000R)$ *alone*, to the exclusion of $r_{\|}$: In consequence of this exclusion, the figures actually carry useful information for $r_{\|}(x)$ in *all* radiation zones as conventionally defined in terms of the incident wavelength $\lambda \sim 2\pi/q_\omega$. Furthermore, this approach to oscillatory behavior as a function of $r_{\|}$ with uniformly spaced peaks is accompanied by a geometric $1/r_{\|}$ - diminution of the amplitude of the Green's function as $r_{\|}$ increases.

Acknowledgements Thanks are extended to Professors M. L. Glasser and Dr. Erik Lenzing for numerous helpful discussions as well as Liuba Zhemchuzhna and Dr. Andrii Iurov for helpful comments.

References

1. D. Miessein, Ph.D Thesis: Scalar and Electromagetic Wave Transmission Through a Subwavelength Nano-hole in a 2D Plasmonic Semiconductor Layer (Stevens Institute of Technology, Hoboken, 2013)
2. N.J.M. Horing, D. Miessein, Wave propagation and diffraction through a subwavelength nano-hole in a 2D plasmonic screen, in *Book Chapter "Low-Dimensional and Nanostructured Materials and Devices"*, H. Ünlü, N.J.M. Horing, J. Dabowski, Springer International Publishing, Switzerland (2016)
3. N.J.M. Horing, D. Miessein, G. Gumbs, Electromagnetics wave transmission through a subwavelength nano-hole in a two-dimensional plasmonic layer. J. Opt. Soc. Am. A **32**, 1184–1198 (2015)
4. D. Miessein, N.J.M. Horing, H. Lenzing, G. Gumbs, Incident-angle dependence of electromagnetic transmission through a plasmonic screen with a nano-aperture. Adv. Nano Bio. MD **1**(1), 54–70 (2017). ISSN: 2559-1118
5. P.M. Morse, H. Feshbach, *Methods of Theoretical Physics*, vol. 2 (McGraw-Hill, 1953), p. 234
6. M. Orman, Ph.D Thesis: Electromagnetic Response of a Quantized Magnetoplasma in 2 and 3 Dimensions, Stevens Institute of Technology (1974)
7. Bateman, *Higher Transcendental Functions*, vol. 2 (McGraw-Hill, 1953), p. 95, Eq. 52
8. W.C. Chew, *Waves and Fields In Inhomogeneous Media* (IEEE, Inc, 1995), p. 385
9. W.C. Chew, *Waves and Fields In Inhomogeneous Media* (IEEE, Inc, 1995), p. 377, Eq. (7.1.14)

Chapter 18
Near Zone Electromagnetic Wave Transmission Through a Nano-Hole in a Plasmonic Layer

Désiré Miessein, Norman J. M. Horing, Harry Lenzing, and Godfrey Gumbs

Abstract We examine the role of the angle of incidence in the transmission of an electromagnetic wave through a subwavelength nano-hole in a thin semiconductor plasmonic layer, focusing on the near zone. In this, we treat both p- and s-polarizations of the incident wave for a variety of incident angles using a dyadic Green's function formulation that includes both (1) the electromagnetic field transmitted directly through the 2D plasmonic layer and (2) the radiation emanating from the nano-hole, explicitly exhibiting interference fringes. Employing an integral equation formulation, this dyadic Green's function approach incorporates the role of the 2D plasmon of the semiconductor layer, which is smeared due to its lateral wave number dependence. The interference fringes, clustered near the nano-hole, flatten to a uniform level of transmission directly through the semiconductor sheet alone at large distances from the nano-hole. As the incident angle increases, the axis of the relatively large central transmission maximum through the nano-hole follows it, accompanied by a spatial compression of interference fringe maxima forward of the large central transmission maximum, and a spatial thinning of the fringe maxima behind it. For p- polarization, the transmission results show a strong increase as the incident angle θ_0 increases, mainly in the dominant E_z component (notwithstanding a concomitant decrease of the E_x component as θ_0 increases). In the case of s-polarization of the incident EM wave, the transmission decreases as θ_0 increases.

D. Miessein
Department of Physics and Engineering Physics, Fordham University,
Bronx 10458, NY, USA
e-mail: dmiessein@fordham.edu

D. Miessein · N. J. M. Horing (✉) · H. Lenzing
Department of Physics, Stevens Institute of Technology, Hoboken 07030, NJ, USA
e-mail: nhoring@stevens.edu

H. Lenzing
e-mail: harry07757@comcast.net

G. Gumbs
Department of Physics and Astronomy, Hunter College of the City University of New York,
New York 10065, USA
e-mail: ggumbs@hunter.cuny.edu

H. Ünlü and N. J. M. Horing (eds.), *Progress in Nanoscale and Low-Dimensional Materials and Devices*, Topics in Applied Physics 144,
https://doi.org/10.1007/978-3-030-93460-6_18

18.1 Introduction

This chapter is concerned with the role of the angle of incidence in the transmission/diffraction of an electromagnetic wave [1] through a subwavelength nano-hole [2–4] in a thin semiconductor plasmonic layer (Fig. 18.1), focusing on the near zone. We treat both $p-$ and $s-$ polarizations of the incident wave for a variety of incident angles using a dyadic Green's function formulation [5–8] that includes both (1) the electromagnetic field transmitted directly through the $2D$ plasmonic layer and (2) the radiation emanating from the nano-hole, explicitly exhibiting interference fringes. Employing an integral equation formulation, this dyadic Green's function approach incorporates the role of the $2D$ plasmon of the semiconductor layer, which is smeared due to its lateral wavenumber dependence.

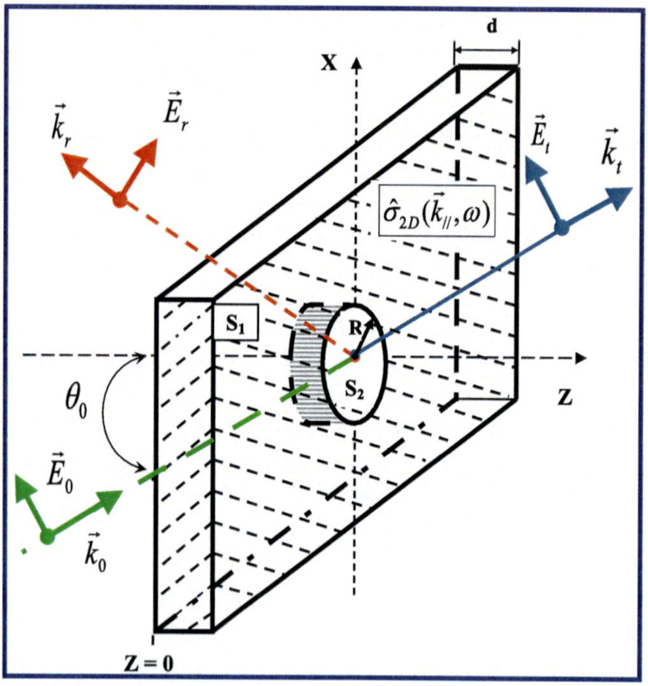

Fig. 18.1 2D plasmonic layer (S_1) (thickness d, embedded at $z = 0$ in a 3D bulk medium) with a nano-hole (S_2) of radius R at the origin of the $(x - y)$ plane, shown with incident, reflected and transmitted wave vectors $(\vec{k}_0, \vec{k}_r, \vec{k}_t)$ for corresponding waves $(\vec{E}_0(x, y, z; t), \vec{E}_r(x, y, z; t), \vec{E}_t(x, y, z; t))$

18.2 Analytical Formulation

The underlying Green's function is determined by a succession of integral equations, the first of which is given in dyadic notation by [9, 10] [z coordinate represents distance perpendicular to plane; suppress lateral wavevector $\vec{k}_{\parallel} = (k_x, k_y)$ (parallel to the $x - y$ plane of the layer) and frequency ω notation]

$$\widehat{G}_{fs}(z, z') = \widehat{G}_{3D}(z, z') - \frac{4\pi i\omega}{c^2} \int dz'' \int dz''' \widehat{G}_{3D}(z, z'') \widehat{\sigma}^{2D}_{fs}(z'', z''') \widehat{G}(z''', z'), \tag{18.1}$$

where, $\widehat{G}_{fs}(z, z')$ is the dyadic Green's function for the *non-perforated* $2D$ plasmonic layer, $\widehat{\sigma}^{(2D)}_{fs}(z, z')$ is its conductivity dyadic and $\widehat{G}_{3D}(z, z')$ refers to the bulk host medium dyadic Green's function [11–13]. Furthermore, the dyadic Green's function for the $2D$ layer perforated by a nano-hole obeys a second integral equation [14]

$$\widehat{G}(z, z') = \widehat{G}_{fs}(z, z') - \frac{4\pi i\omega}{c^2} \int dz'' \int dz''' \widehat{G}_{fs}(z, z'') \widehat{\sigma}^{2D}_{hole}(z'', z''') \widehat{G}(z''', z'), \tag{18.2}$$

where $\widehat{\sigma}^{(2D)}_{hole}(z, z')$ represents the conductivity dyadic for the material excised from the $2D$ layer to create the nano-hole. Equations (18.1) and (18.2) are solved exactly analytically for a nano-small layer thickness(d) and hole radius(R) [14–16].

The resulting 9 elements of $\widehat{G}(\vec{r}_{\parallel}, \vec{r}^{\,\prime}_{\parallel}; z, z'; \omega)$, which are obtained by solving (18.1) and (18.2) in succession, are given by ($\vec{r}_{\parallel} = (x, y)$ parallel to the plane of the layer; $\vec{k}_{\parallel} = (k_x, k_y)$ is the conjugate wavevector):

$$G^{xx}(\vec{r}_{\parallel}, 0; z, 0; \omega) = \frac{G^{xx}_{fs}(\vec{r}_{\parallel}, 0; z, 0; \omega)}{1 + \beta G^{xx}_{fs}(0, 0; 0, 0; \omega)}, \tag{18.3}$$

where (A represents the area of the nano-hole)

$$\beta = \left(\frac{4 i\pi\omega}{c^2} \sigma^{(2D)}_{fs}(\omega) \right) A^2, \tag{18.4}$$

and ($\varepsilon(\omega)$ is the dielectric function of the 2D plasmonic layer of thickness d embedded in a bulk medium of dielectric constant $\varepsilon^{(3D)}_b$)

$$\sigma^{(2D)}_{fs}(\omega) = \frac{i\omega}{4\pi} \left[\varepsilon^{(3D)}_b - \varepsilon(\omega) \right] d. \tag{18.5}$$

Here, we employ a model 2D polarizability given by $\alpha^{(2D)}(\omega) \simeq -\omega_{P_{3D}}^2 d/\omega^2 = -4\pi e^2 n_{2D}/m^*\omega^2$, where $n_{2D} = 4 \times 10^{15}/\mathrm{cm}^2$ and $m^* = 0.065\, m_0$ for a GaAs layer 10 nm thick. Greater accuracy can be achieved using the exact local 2D polarizability $\alpha_{fs}^{2D}(\overrightarrow{k_{||}};\omega) = (\frac{2\pi i e^2 n_{(2D)}}{m^*\omega^2})\sqrt{q_\omega^2 - k_{||}^2}$ ($k_{||}$ = lateral wavenumber representation; $q_\omega = (\omega/c)\sqrt{\varepsilon_b^{(3D)}}$).

Furthermore, ($J_0(x)$ is the Bessel function of order 0; $q_\omega = (\omega/c)\sqrt{\varepsilon_b^{(3D)}}$; $\Gamma = \frac{2\pi\omega\sigma_{fs}^{(2D)}(\omega)}{c^2}$; $\rho = q_\omega r_{||}$ and $|s| = q_\omega |z|$)

$$G_{fs}^{xx}(\vec{r}_{||}, 0; z, 0; \omega) = -\frac{1}{4\pi i}\frac{q_\omega^2}{\Gamma}\left\{\int_0^1 dy\, y\, J_0(\rho y)\, e^{i|s|\sqrt{1-y^2}} + \int_1^\infty dy\, y\, J_0(\rho y)\, e^{-|s|\sqrt{y^2-1}}\right\}$$
$$+\frac{q_\omega}{4\pi i}\frac{q_\omega^2}{\Gamma^2}\left\{\int_0^1 dy\, y\, J_0(\rho y)\left[\frac{e^{i|s|\sqrt{1-y^2}}}{\frac{q_\omega}{\Gamma}+\sqrt{1-y^2}}\right] - i\int_1^\infty dy\, y\, J_0(\rho y)\left[\frac{e^{-|s|\sqrt{y^2-1}}}{-i\frac{q_\omega}{\Gamma}+\sqrt{y^2-1}}\right]\right\}, \quad (18.6)$$

and (R is the radius of the nano-hole)

$$G_{fs}^{xx}(0, 0; 0, 0; \omega) = -\frac{1}{4\pi i R}\left\{\frac{1}{2\Gamma R} + i\frac{(q_\omega R)^2}{(\Gamma R)^2}\sqrt{1-(q_\omega R)^2} - \frac{(q_\omega R)^3}{(\Gamma R)^2}\right\}$$
$$+\frac{1}{4\pi i R}\left\{\frac{(q_\omega R)^4}{(\Gamma R)^3}\ln\left[\frac{(q_\omega R)^2 + i R\Gamma\sqrt{1-(q_\omega R)^2}}{(q_\omega R)^2 + R\Gamma(q_\omega R)}\right]\right\}. \quad (18.7)$$

Also,

$$G^{xy}(\vec{r}_{||}, 0; z, 0; \omega) = 0, \quad (18.8)$$

and

$$G^{xz}(\vec{r}_{||}, 0; z, 0; \omega) = \frac{G_{fs}^{xz}(\vec{r}_{||}, 0; z, 0; \omega)}{1 + \beta G_{fs}^{zz}(0, 0; 0, 0; \omega)}, \quad (18.9)$$

where

$$G_{fs}^{xz}(\vec{r}_{||}, 0; z, 0; \omega) = \frac{sgn(z)}{4\pi i}\frac{q_\omega^2}{\Gamma}\int_0^1 \frac{dy\, y^2\, J_0(\rho\, y)\sqrt{1-y^2}\, e^{i|s|\sqrt{1-y^2}}}{y^2 + \alpha_2\sqrt{1-y^2}}$$
$$+\frac{sgn(z)}{4\pi}\frac{q_\omega^2}{\Gamma}\int_1^\infty \frac{dy\, y^2\, J_0(\rho\, y)\sqrt{y^2-1}\, e^{-|s|\sqrt{y^2-1}}}{y^2 + i\,\alpha_2\sqrt{y^2-1}}. \quad (18.10)$$

and

$$G^{zz}_{fs}(0,0;0,0;\omega) = -\frac{q_\omega^2}{4\pi\, i\,\Gamma}\left\{\int_0^1 du\, u\left[\frac{u^2 - i\,\alpha_0\, u - 1}{u^2 - \alpha_2\, u - 1}\right]\right\} - \frac{q_\omega^2}{4\pi\, i\,\Gamma}\left\{\int_0^{\frac{1}{\alpha_3}\sqrt{1-\alpha_3^2}} dv\, v\left[\frac{v^2 - \alpha_0\, v + 1}{v^2 + i\,\alpha_2\, v + 1}\right]\right\}, \quad (18.11)$$

where we have defined $\alpha_0 = \frac{2}{d\, q_\omega}$ and $\alpha_1 = \frac{q_\omega}{\Gamma}$; $\alpha_2 = \alpha_1 + i\,\alpha_0$; $\alpha_3 = q_\omega\, R$ and $\Gamma = \frac{2\pi\omega\,\sigma^{(2D)}_{fs}(\omega)}{c^2}$ with $\sigma^{(2D)}_{fs}(\omega) = \frac{i\omega}{4\pi}\left[\varepsilon^{(3D)}_b - \varepsilon(\omega)\right] d$.

Moreover,

$$G^{yx}(\vec{r}_\parallel, 0; z, 0; \omega) = 0, \quad (18.12)$$

and

$$G^{yy}(\vec{r}_\parallel, 0; z, 0; \omega) = \frac{G^{yy}_{fs}(\vec{r}_\parallel, 0; z, 0; \omega)}{1 + \beta\, G^{yy}_{fs}(0,0;0,0;\omega)}, \quad (18.13)$$

where

$$G^{yy}_{fs}(\vec{r}_\parallel, 0; z, 0; \omega) = -\frac{q_\omega}{4\pi\, i}\left\{\int_0^1 \frac{dy\, y\, J_0(\rho y)\, e^{i\,|s|\sqrt{1-y^2}}}{\frac{\Gamma}{q_\omega} + \sqrt{1-y^2}}\right\} + \frac{q_\omega}{4\pi\, i}\left\{i\int_1^\infty \frac{dy\, y\, J_0(\rho y)\, e^{-|s|\sqrt{y^2-1}}}{-i\,\frac{\Gamma}{q_\omega} + \sqrt{y^2-1}}\right\}, \quad (18.14)$$

and

$$G^{yy}_{fs}(0,0;0,0;\omega) = -\frac{1}{4\pi\, i\, R}\left\{q_\omega R - i\sqrt{1-(q_\omega R)^2} - \Gamma R \ln\left[\frac{R\Gamma + q_\omega\, R}{R\Gamma}\right]\right\} - \frac{1}{4\pi\, i\, R}\left\{\Gamma R \ln\left[\frac{R\Gamma + i\sqrt{1-(q_\omega R)^2}}{R\Gamma}\right]\right\}; \quad (18.15)$$

$$G^{yz}(\vec{r}_\parallel, 0; z, 0; \omega) = 0. \quad (18.16)$$

Furthermore,

$$G^{zx}(\vec{r}_{\parallel}, 0; z, 0; \omega) = \frac{G^{zx}_{fs}(\vec{r}_{\parallel}, 0; z, 0; \omega)}{1 + \beta G^{xx}_{fs}(0, 0; 0, 0; \omega)}, \tag{18.17}$$

with

$$\begin{aligned} G^{zx}_{fs}(\vec{r}_{\parallel}, 0; z, 0; \omega) = {} & \frac{sgn(z)\ q^2_{\omega}}{4\pi i \quad \Gamma} \left\{ \int_0^1 \frac{dy\, y^2\, J_0(\rho\, y)\, e^{i\,|s|\,\sqrt{1-y^2}}}{\frac{q_\omega}{\Gamma} + \sqrt{1-y^2}} \right\} \\ & - \frac{sgn(z)\ q^2_{\omega}}{4\pi i \quad \Gamma} \left\{ i \int_1^{\infty} \frac{dy\, y^2\, J_0(\rho\, y)\, e^{-|s|\,\sqrt{y^2-1}}}{-i\,\frac{q_\omega}{\Gamma} + \sqrt{y^2-1}} \right\}, \end{aligned} \tag{18.18}$$

and $G^{xx}_{fs}(0, 0; 0, 0; \omega)$ is given by (18.7). Also,

$$G^{zy}(\vec{r}_{\parallel}, 0; z, 0; \omega) = 0; \tag{18.19}$$

$$G^{zz}(\vec{r}_{\parallel}, 0; z, 0; \omega) = \frac{G^{zz}_{fs}(\vec{r}_{\parallel}, 0; z, 0; \omega)}{1 + \beta G^{zz}_{fs}(0, 0; 0, 0; \omega)}, \tag{18.20}$$

$$\begin{aligned} G^{zz}_{fs}(\vec{r}_{\parallel}, 0; z, 0; \omega) = {} & -\frac{q_\omega}{4\pi i}\frac{q_\omega}{\Gamma} \int_0^1 dy\, y\, J_0(\rho y)\, e^{i\,|s|\sqrt{1-y^2}} \left\{ \frac{y^2}{y^2 + \alpha_2\sqrt{1-y^2}} \right\} \\ & - \frac{q_\omega}{4\pi i}\frac{q_\omega}{\Gamma} \int_1^{\infty} dy\, y\, J_0(\rho y)\, e^{-|s|\sqrt{y^2-1}} \left\{ \frac{y^2}{y^2 + i\,\alpha_2\sqrt{y^2-1}} \right\}, \end{aligned} \tag{18.21}$$

where $\alpha_0 = \frac{2}{d\,q_\omega}$, $\alpha_1 = \frac{q_\omega}{\Gamma}$, $\alpha_2 = \alpha_1 + i\,\alpha_0$ and $G^{zz}_{fs}(0, 0; 0, 0; \omega)$ is given by (18.11).

In this chapter, we present calculated results for the transmission of an incident plane electromagnetic wave $\vec{E}_0$ incident on the system for various angles of incidence $\theta_0 = 30°, 60°, 80°$, in particular for the "near zone" taken as $z \leq 50\, R$ (with nanohole radius R=5nm), although the lateral distance $\vec{r}_{\parallel}$ of the field point from the hole parallel to the plane can extend up to $1000\, R - 2000\, R$. The transmitted field $\vec{E}$ due to the incident field $\vec{E}_0$ is obtained using $\widehat{G}$ as (dyadic electromagnetic notation, and positional matrix product notation implies $z-$ integration) [14]

$$\vec{E} = \vec{E}_0 + \frac{4\pi\, i\, \omega}{c^2} \widehat{G}\, \left[\widehat{\sigma}^{2D}_{fs} - \widehat{\sigma}^{2D}_{hole}\right] \vec{E}_0\,. \tag{18.22}$$

Calculated results for the transmitted field $\vec{E}$ are exhibited below for both p-polarization of the incident field $\vec{E}_0 = E_0\,(\widehat{e}_x\, cos\, \theta_0 + \widehat{e}_z\, sin\, \theta_0)$ and for s-polarization $\vec{E}_0 = E_0\, \widehat{e}_y$.

18.3 Calculated Transmission Results: Near Zone

The results of our calculations are exhibited in terms of transmitted field-component powers $| E_{\|}(x, y, z; t)/E_0 |^2$ and $| E_z(x, y, z; t)/E_0 |^2$ for $p-$polarization (with the $x - z$ plane as the plane of incidence) in the line graphs of Figs. 18.2, 18.3, 18.4, 18.5, 18.6, 18.7, 18.8, 18.9, 18.10 for the special choice $y \equiv 0$; also for $s-$polarization, $| E_y(x, y, z; t)/E_0 |^2$ results are presented as functions of x for $y = 0$ in the line graphs of Fig. 18.5. Furthermore, the calculated results for the transmitted component powers are also exhibited as functions of both x and y for both polarizations in both $3D$ and density plots for the various angles of incidence in Figs. 18.11, 18.12, 18.13, 18.14, 18.15, 18.16, 18.17, 18.18, 18.19, 18.20, 18.21, 18.22, 18.23, 18.24, 18.25, 18.26, 18.27, 18.28. It should be noted that the interference fringes, clustered near the nano-hole, flatten to a uniform level of transmission directly through the semiconductor sheet alone at large lateral distances from the nano-hole. As the incident angle increases, the axis of the relatively large central transmission maximum through the nano-hole follows it, accompanied by a spatial compression of interference fringe maxima forward of the large central transmission maximum, and a spatial thinning of the fringe maxima behind it. For p-polarization, transmission results show a strong increase as the incident angle θ_0 increases, mainly in the dominant E_z component (notwithstanding a concomitant decrease of the E_x component as θ_0 increases). In the case of $s-$ polarization of the incident electromagnetic wave, the transmission decreases as θ_0 increases.

p**-polarization: Near-Field**, $z = 50\,R$; $\theta_0 = 30°$

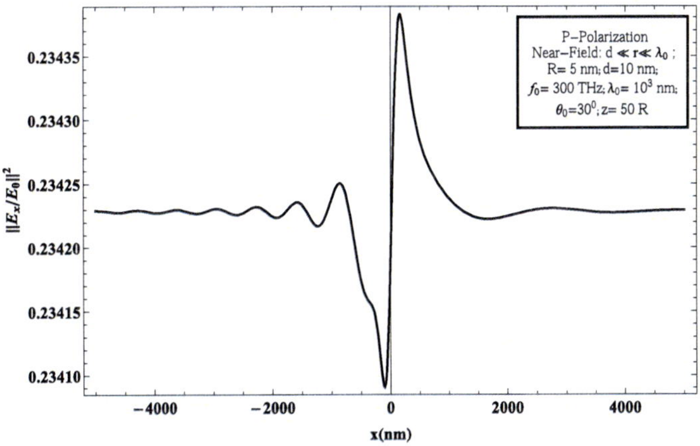

Fig. 18.2 p-polarization-Near-Field, $z = 50\,R; \theta_0 = 30°$: $| E_x(x, y, z; t)/E_0 |^2$ produced by a perforated 2D plasmonic layer of GaAs as a function of lateral distance $r_{\|} = x$ $(y = 0)$ from the aperture

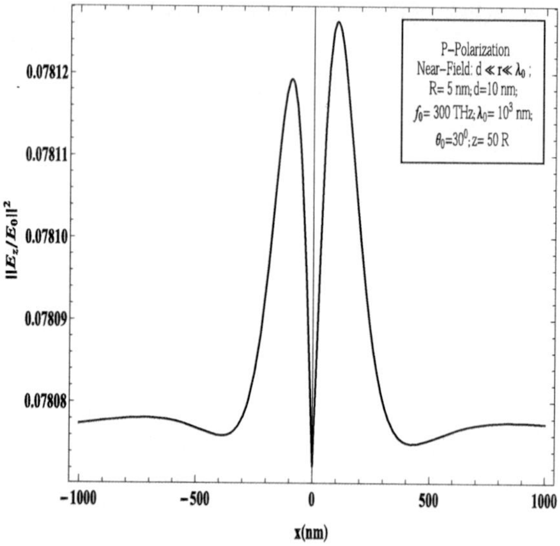

Fig. 18.3 p-polarization-Near-Field, $z = 50\,R$; $\theta_0 = 30°$: $\mid E_z(x, y, z; t)/E_0 \mid^2$ produced by a perforated 2D plasmonic layer of GaAs as a function of lateral distance $r_{\parallel} = x\ (y = 0)$ from the aperture

p-polarization: Near-Field, $z = 50\,R$; $\theta_0 = 60°$

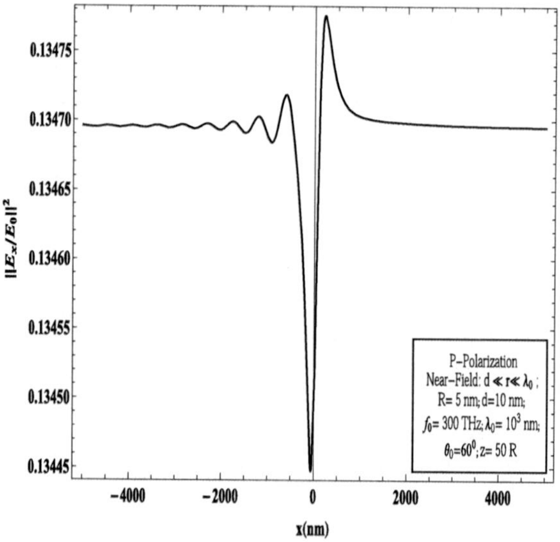

Fig. 18.4 p-polarization-Near-Field, $z = 50\,R$; $\theta_0 = 60°$: $\mid E_x(x, y, z; t)/E_0 \mid^2$ produced by a perforated 2D plasmonic layer of GaAs as a function of lateral distance $r_{\parallel} = x\ (y = 0)$ from the aperture

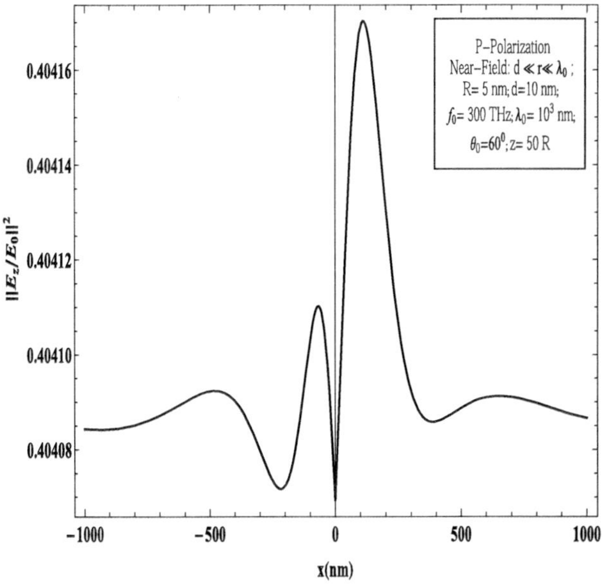

Fig. 18.5 p-polarization-Near-Field, $z = 50\,R$; $\theta_0 = 60°$: $\mid E_z(x, y, z; t)/E_0 \mid^2$ produced by a perforated 2D plasmonic layer of GaAs as a function of lateral distance $r_{\|} = x$ $(y = 0)$ from the aperture

p-polarization: Near-Field, $z = 50\,R$; $\theta_0 = 80°$

Fig. 18.6 p-polarization-Near-Field, $z = 50\,R$; $\theta_0 = 80°$: $\mid E_x(x, y, z; t)/E_0 \mid^2$ produced by a perforated 2D plasmonic layer of GaAs as a function of lateral distance $r_{\|} = x$ $(y = 0)$ from the aperture

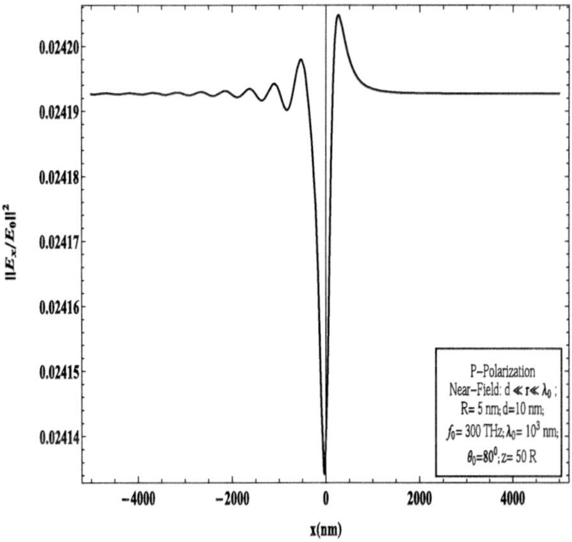

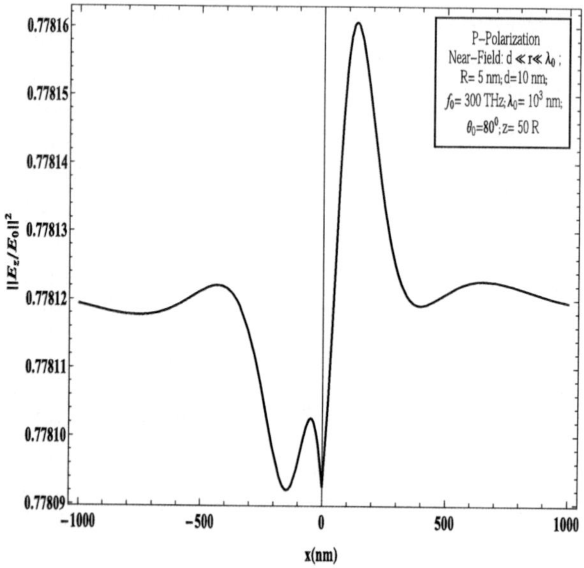

Fig. 18.7 p-polarization-Near-Field, $z = 50\,R; \theta_0 = 80°$: $|\ E_z(x, y, z; t)/E_0\ |^2$ produced by a perforated 2D plasmonic layer of GaAs as a function of lateral distance $r_{\parallel} = x\ (y = 0)$ from the aperture

s-polarization: Near-Field, $z = 50\ R$

Fig. 18.8 s-polarization-Near-Field, $z = 50\,R$: $|\ E_y(x, y, z; t)/E_0\ |^2$ $\theta_0 = 30°$, produced by a perforated 2D plasmonic layer of GaAs as a function of lateral distance $r_{\parallel} = x\ (y = 0)$ from the aperture

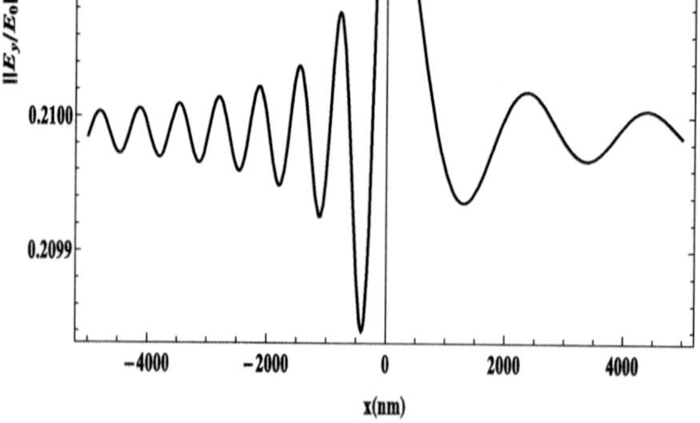

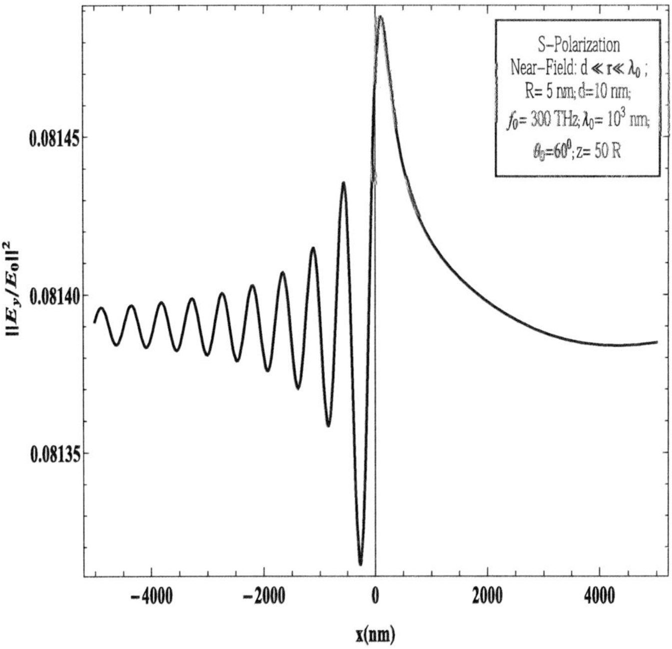

Fig. 18.9 s-polarization-Near-Field, $z = 50\,R$: $|\ E_y(x, y, z; t)/E_0\ |^2$, $\theta_0 = 60°$ produced by a perforated 2D plasmonic layer of GaAs as a function of lateral distance $r_{\parallel} = x$ ($y = 0$) from the aperture

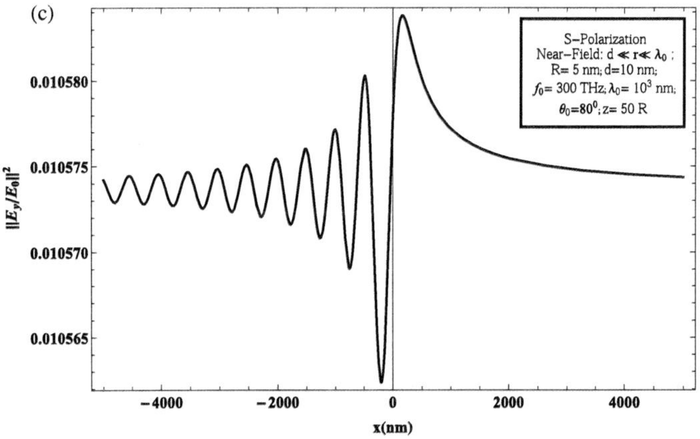

Fig. 18.10 s-polarization-Near-Field, $z = 50\,R$: $|\ E_y(x, y, z; t)/E_0\ |^2$, $\theta_0 = 80°$ produced by a perforated 2D plasmonic layer of GaAs as a function of lateral distance $r_{\parallel} = x$ ($y = 0$) from the aperture

p-polarization: Near-Field, $z = 50\,R$; $\theta_0 = 30°$

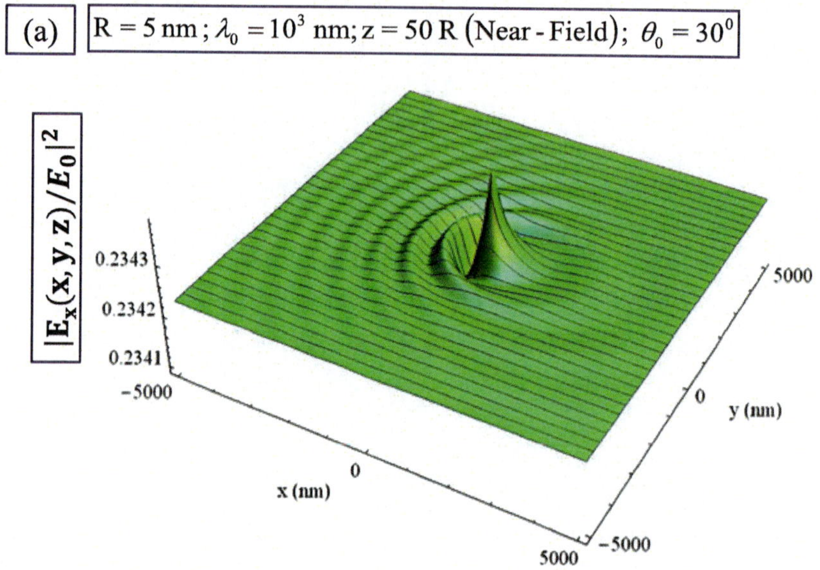

Fig. 18.11 p-polarization: Near-Field, $z = 50\,R$; $\theta_0 = 30°$—**a** Field-component distribution for GaAs layer in terms of 3D plot: $|\ E_x(x, y, z; t)/E_0\ |^2$ as functions of x and y for fixed z

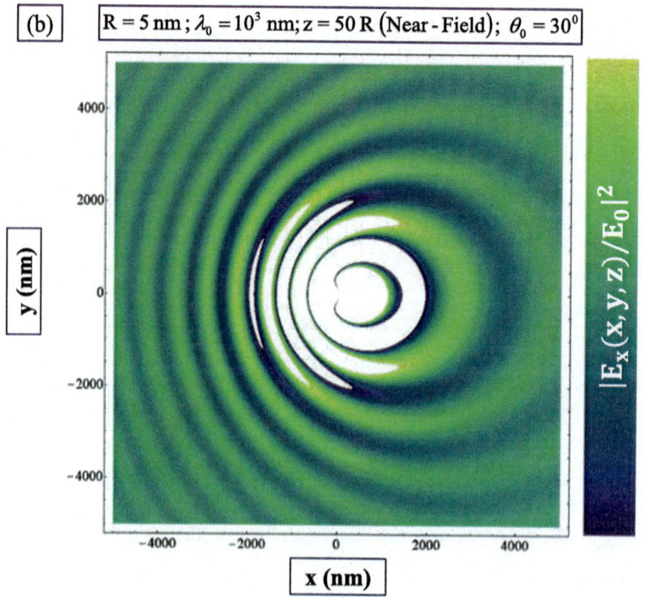

Fig. 18.12 p-polarization: Near-Field, $z = 50\,R$; $\theta_0 = 30°$—**b** Field-component distribution for GaAs layer in terms of density plot: $|\ E_x(x, y, z; t)/E_0\ |^2$ as functions of x and y for fixed z

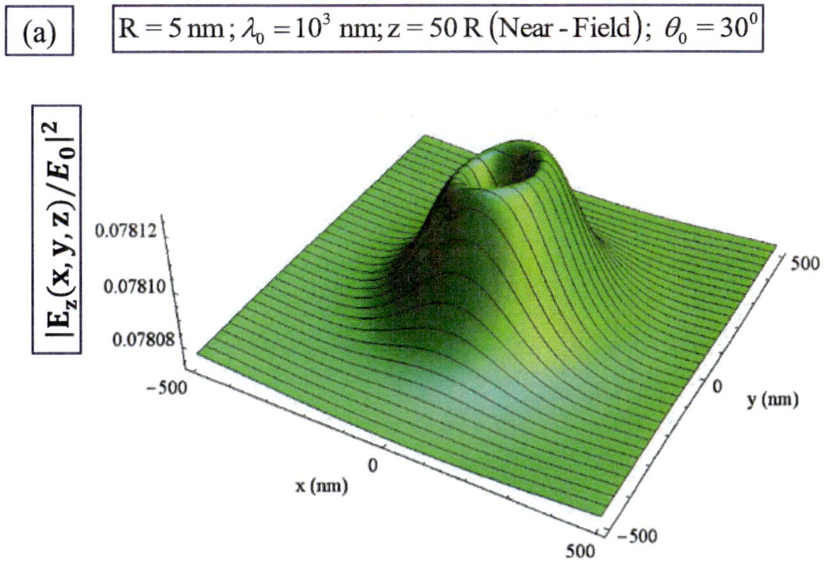

Fig. 18.13 p-polarization: Near-Field, $z = 50\,R$; $\theta_0 = 30°$—**a** Field-component distribution for GaAs layer in terms of 3D plot: $|\ E_z(x, y, z; t)/E_0\ |^2$ as functions of x and y for fixed z

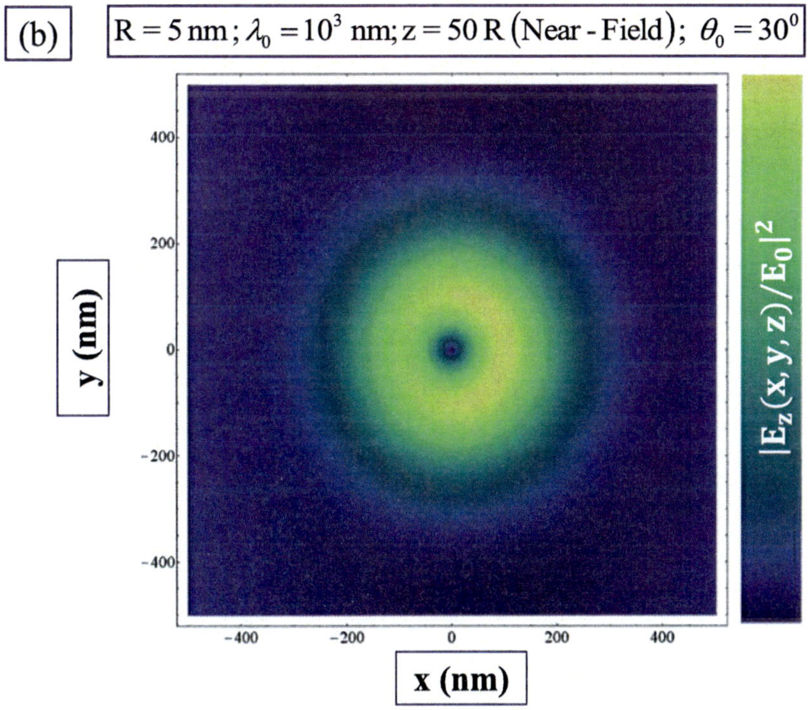

Fig. 18.14 p-polarization: Near-Field, $z = 50\,R$; $\theta_0 = 30°$—**b** Field-component distribution for GaAs layer in terms of density plots: $|\ E_z(x, y, z; t)/E_0\ |^2$ as functions of x and y for fixed z

p-polarization: Near-Field, $z = 50\,R$; $\theta_0 = 60°$

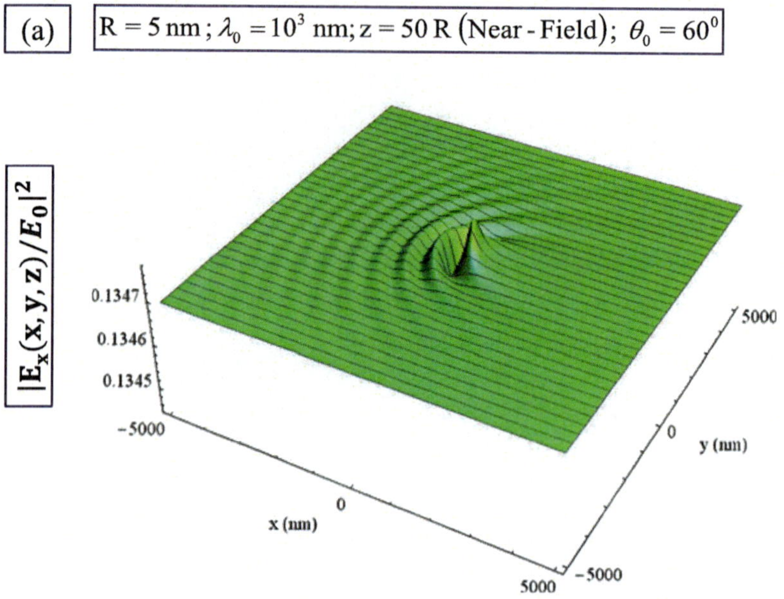

Fig. 18.15 p-polarization: Near-Field, $z = 50\,R$; $\theta_0 = 60°$—**a** Field-component distribution for GaAs layer in terms of 3D plot: $\mid E_x(x, y, z; t)/E_0 \mid^2$ as functions of x and y for fixed z

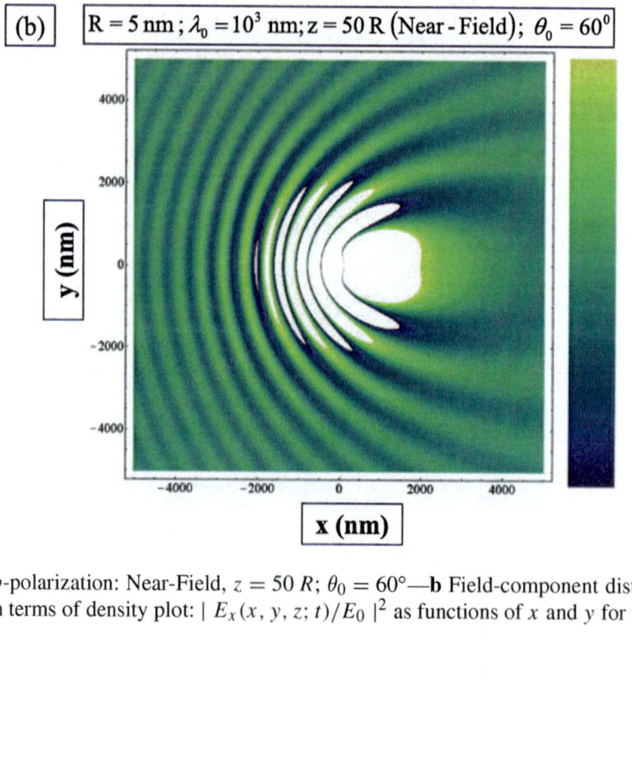

Fig. 18.16 p-polarization: Near-Field, $z = 50\,R$; $\theta_0 = 60°$—**b** Field-component distribution for GaAs layer in terms of density plot: $\mid E_x(x, y, z; t)/E_0 \mid^2$ as functions of x and y for fixed z

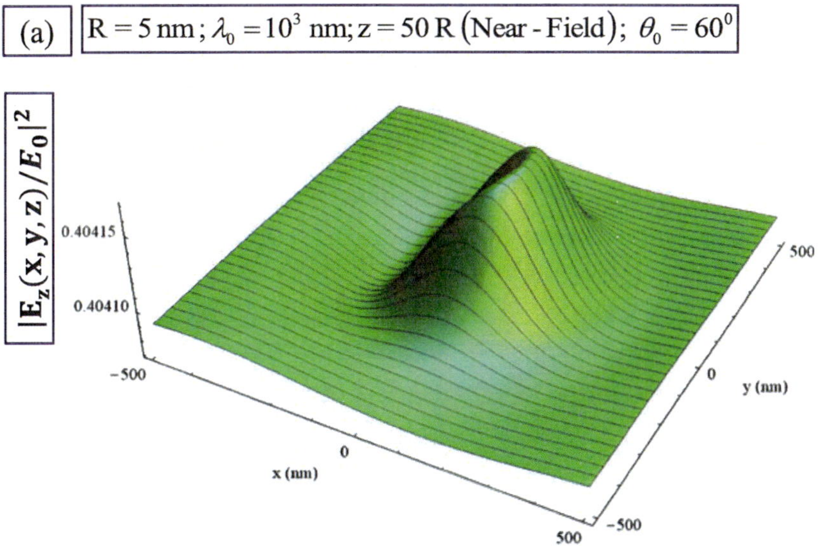

Fig. 18.17 p-polarization: Near-Field, $z = 50\,R$; $\theta_0 = 60°$—**a** Field-component distribution for GaAs layer in terms of 3D plot: $|\ E_z(x, y, z; t)/E_0\ |^2$ as functions of x and y for fixed z

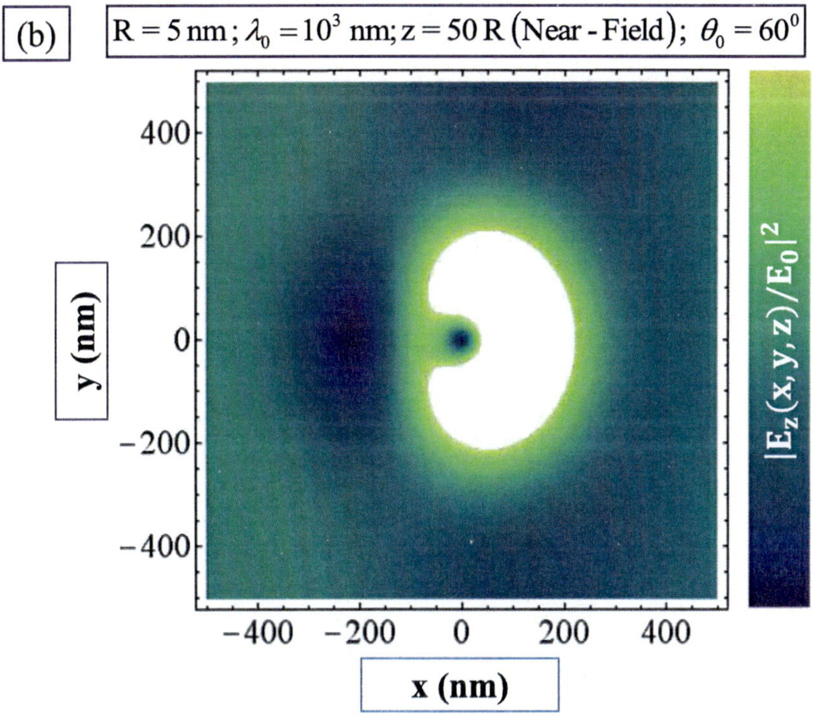

Fig. 18.18 p-polarization: Near-Field, $z = 50\,R$; $\theta_0 = 60°$—**b** Field-component distribution for GaAs layer in terms of density plot: $|\ E_z(x, y, z; t)/E_0\ |^2$ as functions of x and y for fixed z

p-**polarization: Near-Field,** $z = 50\ R$; $\theta_0 = 80°$

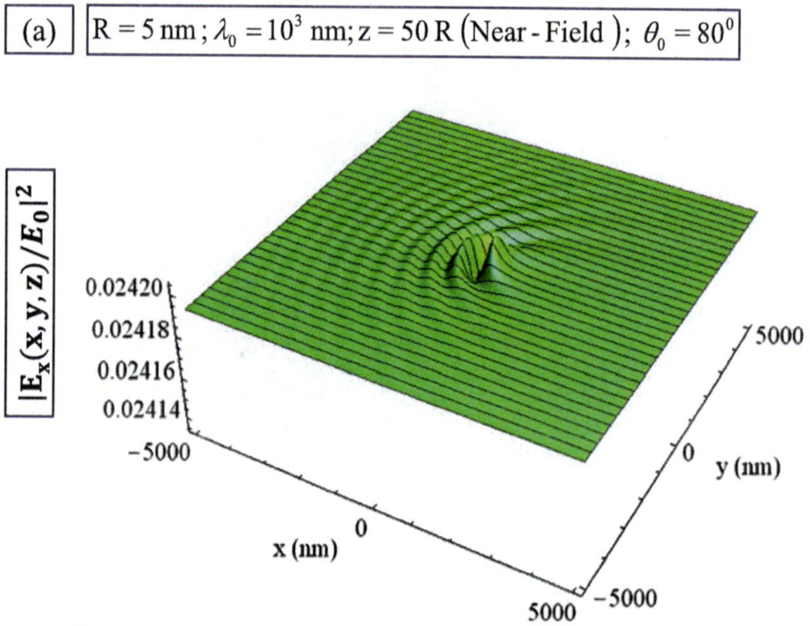

Fig. 18.19 *p*-polarization: Near-Field, $z = 50\ R$; $\theta_0 = 80°$—**a** Field-component distribution for GaAs layer in terms of 3D plot: $| E_x(x, y, z; t)/E_0 |^2$ as functions of x and y for fixed z

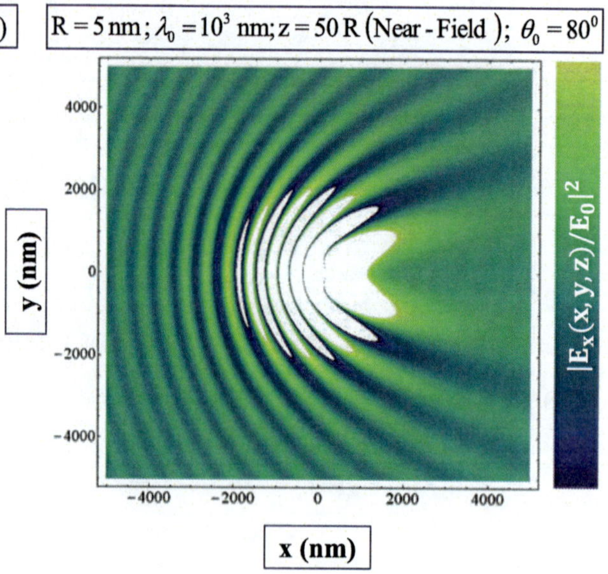

Fig. 18.20 *p*-polarization: Near-Field, $z = 50\ R$; $\theta_0 = 80°$—**b** Field-component distribution for GaAs layer in terms of density plot: $| E_x(x, y, z; t)/E_0 |^2$ as functions of x and y for fixed z

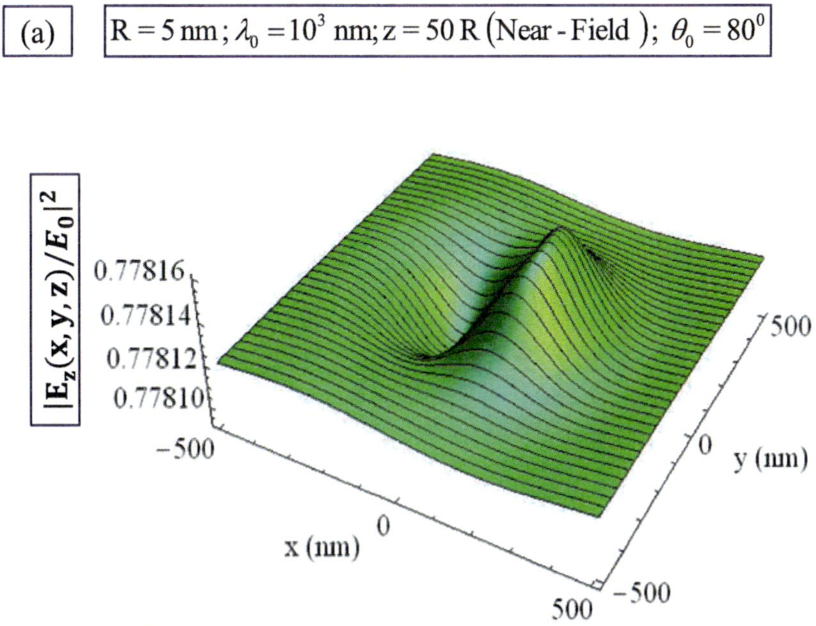

Fig. 18.21 p-polarization: Near-Field, $z = 50\,R$; $\theta_0 = 80°$—**a** Field-component distribution for GaAs layer in terms of 3D plot: $\mid E_z(x, y, z; t)/E_0 \mid^2$ as functions of x and y for fixed z

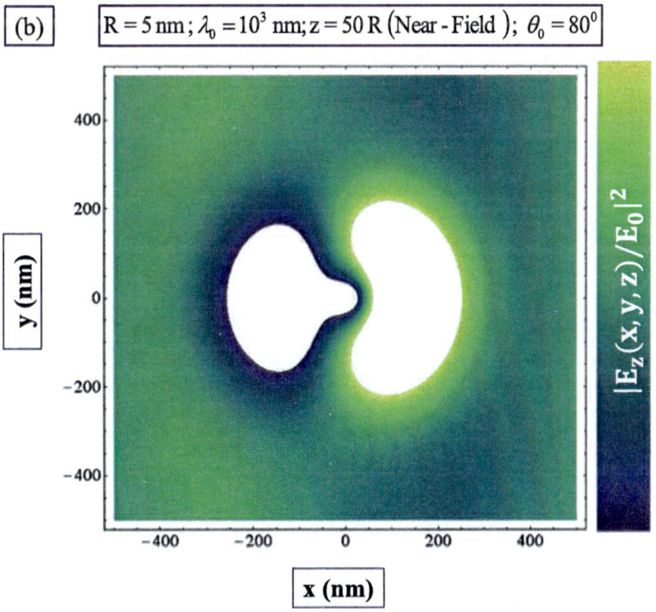

Fig. 18.22 p-polarization: Near-Field, $z = 50\,R$; $\theta_0 = 80°$—**b** Field-component distribution for GaAs layer in terms of density plot: $\mid E_z(x, y, z; t)/E_0 \mid^2$ as functions of x and y for fixed z

s-polarization: Near-Field, $z = 50\ R$; $\theta_0 = 30°$

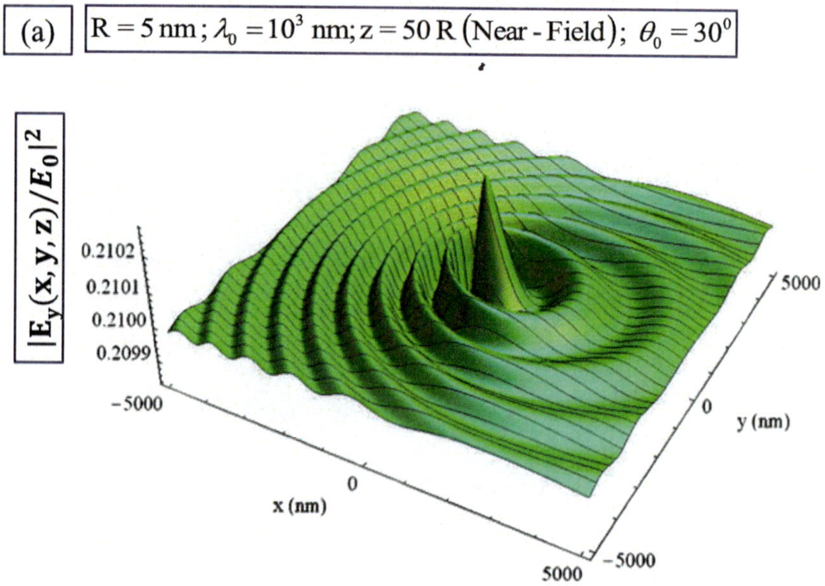

Fig. 18.23 s-polarization: Near-Field, $z = 50\ R$ —**a** Field distribution for GaAs layer in terms of 3D plot: $\mid E_y(x, y, z; t)/E_0 \mid^2$, $\theta_0 = 30°$ as functions of x and y for fixed z

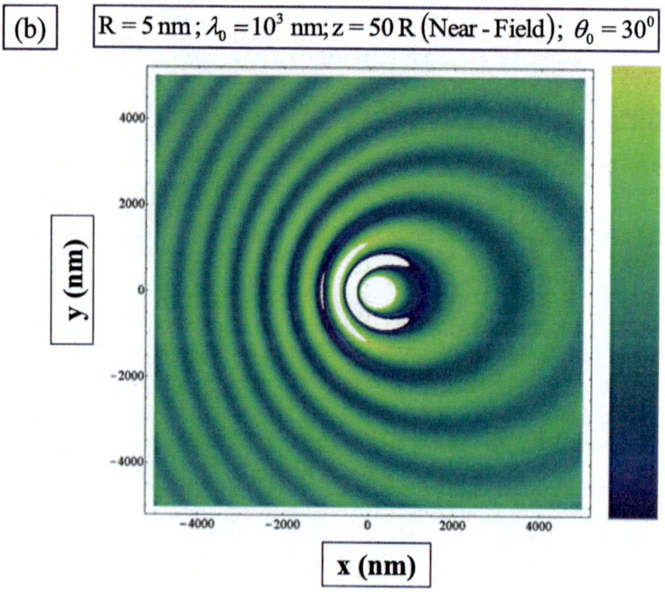

Fig. 18.24 s-polarization: Near-Field, $z = 50\ R$ —**b** Field distribution for GaAs layer in terms of density plot: $\mid E_y(x, y, z; t)/E_0 \mid^2$, $\theta_0 = 30°$ as functions of x and y for fixed z

s-polarization: Near-Field, $z = 50\ R$; $\theta_0 = 60°$

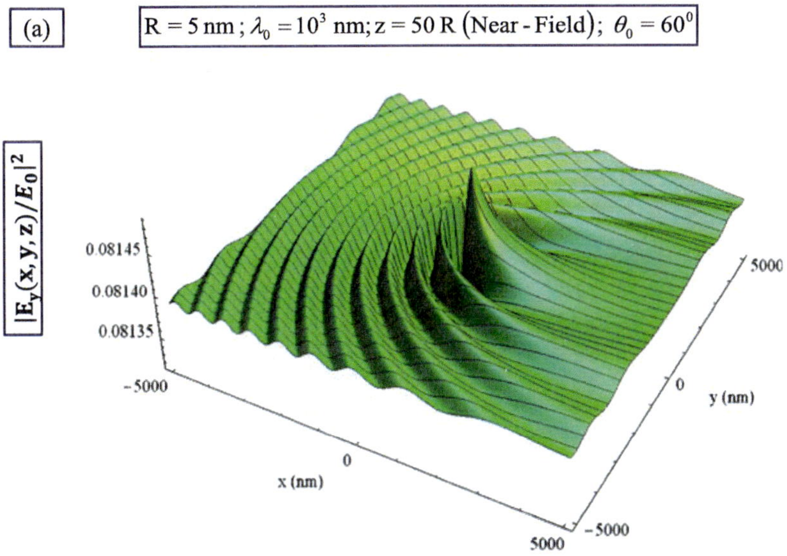

Fig. 18.25 s-polarization: Near-Field, $z = 50\ R$ —**a** Field distribution for GaAs layer in terms of 3D plot: $|\ E_y(x, y, z; t)/E_0\ |^2$, $\theta_0 = 60°$ as functions of x and y for fixed z

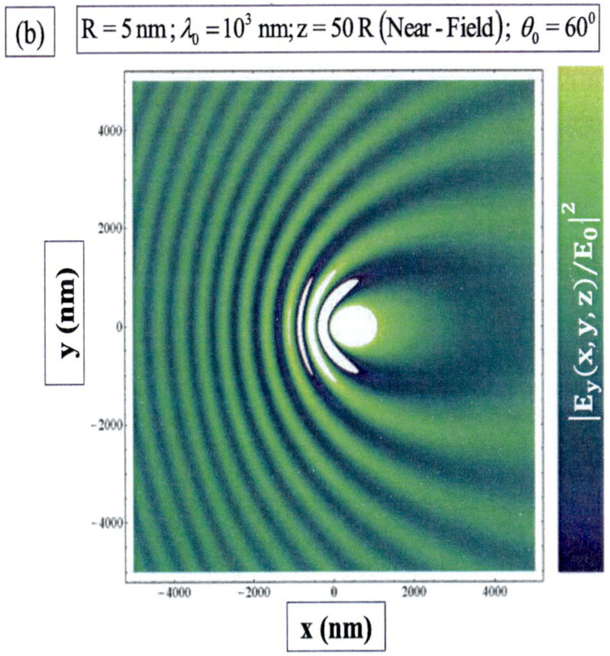

Fig. 18.26 s-polarization: Near-Field, $z = 50\ R$ —**b** Field distribution for GaAs layer in terms of density plot: $|\ E_y(x, y, z; t)/E_0\ |^2$, $\theta_0 = 60°$ as functions of x and y for fixed z

s-polarization: Near-Field, $z = 50\ R$; $\theta_0 = 80°$

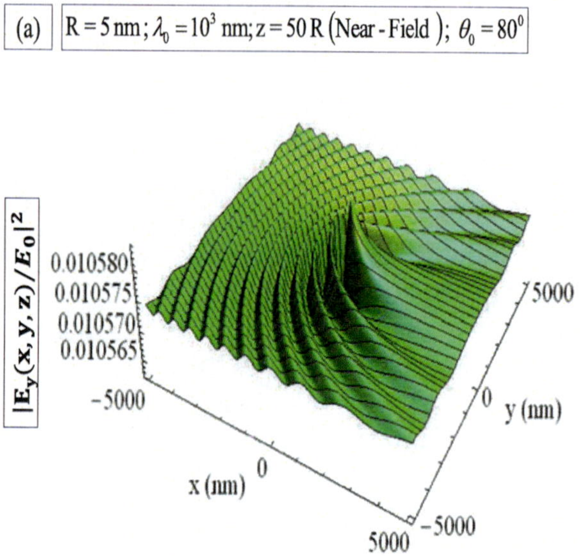

Fig. 18.27 s-polarization: Near-Field, $z = 50\ R$ —**a** Field distribution for GaAs layer in terms of 3D plot: $|\ E_y(x, y, z; t)/E_0\ |^2$, $\theta_0 = 80°$ as functions of x and y for fixed z

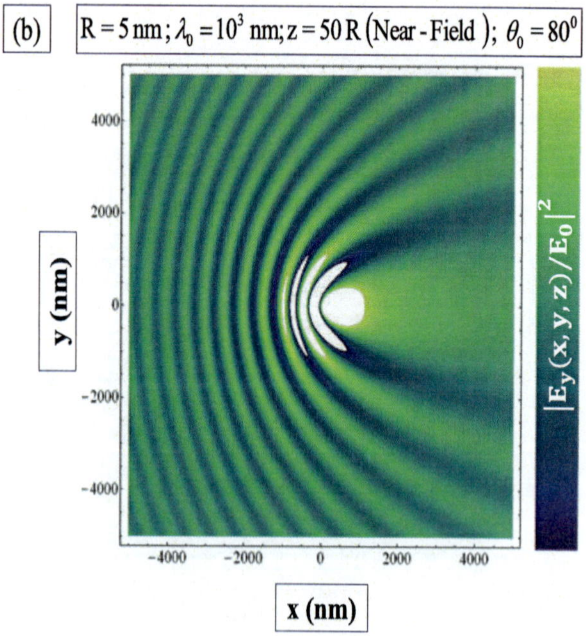

Fig. 18.28 s-polarization: Near-Field, $z = 50\ R$ —**b** Field distribution for GaAs layer in terms of density plot: $|\ E_y(x, y, z; t)/E_0\ |^2$, $\theta_0 = 80°$ as functions of x and y for fixed z

Acknowledgements It is a pleasure to acknowledge helpful discussions with Prof. M. L. Glasser, Dr. Andrii Iurov and Dr. Nan Chen. We thank Fordham University for use of computation facilities.

References

1. H.A. Bethe, Theory of diffraction by small holes. Phys. Rev. **66**, 163–182 (1944)
2. C. Genet, T. Ebbesen, Light in tiny holes. Nature **445**, 39–46 (2007)
3. S.V. Kukhlevsky, M. Mechler, O. Samek, K. Janssens, Analytical model of the enhanced light transmission through subwavelength metal slits: Green's function formalism versus Rayleigh's expansion. Appl. Phys. B: Lasers Opt. **N0**(84), 19–24 (2006)
4. S.V. Kukhlevsky, M. Mechler, L. Csapo, K. Janssens, O. Samek, Enhanced transmission versus localization of a light pulse by a subwavelength metal slit. Phys. Rev. B. **70**, 195428-1–195428-9 (2004)
5. C.-T. Tai, *Dyadic Green's Functions in Electromagnetic Theory, Intext Educational Publishers (1971); Reprinted as Dyadic Green's Functions in Electromagnetic Theory* (IEEE Press, Piscataway, NJ, 1994)
6. C.-T. Tai, *General Vector and Dyadic Analysis: Applied Mathematics in Field Theory*, 2nd edn. (Wiley-IEEE Press, New York, 1997)
7. R.E. Collin, *Field Theory of Guided Waves*, 2nd ed. (IEEE Press, 1991)
8. W.C. Chew, *Waves and Fields In Inhomogeneous Media* (IEEE, Inc, 1995)
9. N.J.M. Horing, T.Y. Bagaeva, V.V. Popov, Excitation of radiative polaritons in a two-dimensional excitonic layer by a light pulse. J. Opt. Soc. Am. B. **24**, 2428–2435 (2007)
10. N.J.M. Horing, Radiative plasmonic-polariton dispersion relation for a thin metallic foil with interband damping transitions. IEEE Sens. J. **8**(6), 771–774 (2008)
11. H. Levine, J. Schwinger, On the theory of diffraction by an aperture in an infinite plane screen I. Phys. Rev. **74**, 958–974 (1948)
12. H. Levine, J. Schwinger, On the theory of diffraction by an aperture in an infinite plane screen II. Phys. Rev. **75**, 1423–1432 (1949)
13. H. Levine, J. Schwinger, On the theory of electromagnetic wave diffraction by an aperture in an infinite plane conducting screen. Commun. Pure Appl. Math. **3**, 355–391 (1950)
14. N.J.M. Horing, Désiré Miessein and G. Gumbs. J. Opt. Soc. Am. A **32**, 1184–1198 (2015)
15. Désiré Miessein, N.J.M. Horing, H. Lenzing, G. Gumbs, Incident-angle dependence of electromagnetic transmission through a plasmonic screen with a nano-aperture. Adv. Nano Bio. MD **1**(1), 54–70 (2017). ISSN: 2559-1118
16. N.J.M. Horing, *Quantum Statistical Field Theory: Schwinger's Variational Method* (Oxford University Press, 2017)

Chapter 19
Spin Dependent Thermoelectric Currents of Tunnel Junctions, and Other Nanostructures: Onsager Response-Theory

K. H. Bennemann

Abstract Spin Currents in Tunnel Junctions, for example, those induced by thermoelectric forces due to temperature and magnetization gradients, etc., are discussed in this chapter. Using Onsager response theory, in particular for magnetic tunnel junctions, metallic rings and quantum dots, yields directly, spin dependently, all thermoelectric and thermomagnetic effects like the Seebeck and Peltier ones and Josephson–like Spin currents driven by the phase gradient of the magnetization. The results can be compared with recent experiments determining the spin dependent Seebeck effect and other thermoelectric effects. The Onsager theory directly yields coupled currents, like the heat current due to a spin current and other thermoelectric effects. The Onsager theory can be extended towards an electronic theory by expressing the Onsager coefficients in terms of current correlation functions, and then calculating these using Lagrange formalism, symmetry and scaling analysis. We discuss in particular the spin currents resulting from the force due to magnetic phase gradients at tunnel junctions of magnetic materials, both in normal and superconducting singlet and triplet states. Note, Onsager theory can also be applied to spin currents in molecules and in magnetic ionic liquids, and also to dynamics in cosmology problems.

19.1 Introduction

Recently, spin dependent currents in nanostructures and tunnel junctions have been discussed intensively [1–5]. In particular the spin dependent thermoelectric and thermomagnetic effects like Seebeck effect and the heat due to spin dependent currents in ferromagnets, spin Peltier effect receive special attention [4, 5]. The interdependence of the various currents is most interesting and is expected by Onsager theory. Onsager theory $j = LX$ for the currents j driven by the spin dependent generalized

K. H. Bennemann (✉)
Institute for Theoretical Physics, Freie Universität Berlin, Arnimallee 14, 14195 Berlin, Germany
e-mail: khb@zedat.fu-berlin.de

H. Ünlü and N. J. M. Horing (eds.), *Progress in Nanoscale and Low-Dimensional Materials and Devices*, Topics in Applied Physics 144,
https://doi.org/10.1007/978-3-030-93460-6_19

thermodynamical forces X like temperature gradient or magnetization gradient etc. yields the spin dependent thermoelectric effects [6]. In particular this holds for nanostructures like tunnel junctions and metallic rings [2, 7] and tunnel currents through molecules and spin currents in magnetic ionic liquids. Note, even if originally one has a homogeneous magnetization a temperature gradient ΔT will induce a difference ΔM in the magnetization and $\Delta M \propto \Delta T$.

Also, of course, even in a homogeneous ferromagnet one gets for itinerant electrons $j_\uparrow \neq j_\downarrow$ for the spin currents due to the spin dependent density of states $(N_\uparrow(\varepsilon) \neq N_\downarrow(\varepsilon))$ etc. Already the Boltzmann equation yields this in a qualitative correct way.

In analogy to the Josephson current in superconductors due to the phase difference of the order parameter, one expects also for ferromagnets, magnetic tunnel junctions, with a gradient in the magnetization its magnitude and phase ($M = |M| \exp(i\phi)$) a similar Josephson like spin current [2, 3, 6]. Such currents are also expected for metallic rings with inhomogeneous magnetization. Of course, in inhomogeneous ferromagnets, see Fig. 19.1, coupled currents involving spin and charge are expected. This is elegantly described by Onsager theory, see also Bennemann [3]. Note, the magnetization may result from local magnetic moments (for example, in rare–earth) or from spins of itinerant electrons in transition metals or rare–earth.

The Onsager theory may also be applied to describe thermoelectric and thermomagnetic effects in magnetic ionic liquids. Then spin dependent pressure effects are expected in case of pressure gradients, possibly interfering with other gradients.

Interesting are in particular inhomogeneous systems (nanostructures, tunnel junctions) like $(FM_1|N|FM_2)$, $(FM|SC|FM)$, etc., to study spin lifetimes of electron spins injected from a ferromagnet (FM) into a nonmagnetic metal (N) and to study spin currents in superconductors (SC) also analysis of singlet vs. triplet superconductivity, (FM|SC) interfaces [2, 7]. This may be used to test triplet superconductivity.

As indicated already by the giant Faraday effect in graphene and a few layers of graphene, one expects for graphene structures (due to the relatively long spin

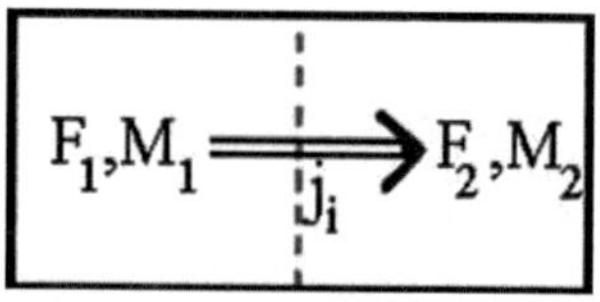

Fig. 19.1 Illustration of an inhomogeneous ferromagnetic tunnel junction with temperature gradient $\Delta T = T_1 - T_2$ and magnetization gradient $\Delta M(t) = M_1 - M_2$ and possibly other gradients. Coupled spin dependent currents j_i are expected in accordance with Onsager theory. Note that the currents may induce a temperature gradient and thus affect $\Delta M(t)$. In particular one gets Josephson like spin currents due to a phase difference of the magnetizations ($j_s^J \propto \frac{dM}{dt} \propto \overrightarrow{M_1} \times \overrightarrow{H_{eff.}} + ...$) and for example from the coupled currents a spin dependent Seebeck effect ($\Delta(\mu_\uparrow - \mu_\downarrow) \propto \Delta T$ and Peltier effect (heat$\propto$spin current). Clearly the various currents will depend on the magnetic configuration of M_1 and M_2, ferromagnetic vs. antiferromagnetic configuration of the two magnets and thus also the electric potential gradient $\Delta\varphi$ depends on $\overrightarrow{M_1}$ and $\overrightarrow{M_2}$), ($\Delta\varphi \propto \Delta T$)

mean free paths) interesting spin dependent thermoelectric effects (see MacDonald et al., Bennemann and others). For example, for tunnel junctions involving graphene between the two ferromagnets, the Josephson–like spin current driven by a gradient of the phase of the magnetizations on the left and right sides of the tunnel junction could be observed.

For tunneling involving triplet superconductivity and ferromagnetism, the interplay of the order parameters yields novel properties of tunnel junctions. For example, one gets Cooper pair tunneling even for no phase difference between the superconductors on both sides of the tunnel junction [7]. Note, regarding the Josephson like spin current driven by the phase gradient of the magnetization on the left side and right side of the tunnel junction [2, 6], see Fig. 19.2, this might require relatively long spin mean free paths. Thus, weak spin–orbit scattering and tunneling for example through graphene favors this spin current. Strong spin–orbit scattering is expected to suppress this Josephson spin current.

Spin currents in metallic rings, in particular persistent ones, are interesting. One expects that the Aharonov–Bohm effect, spin–orbit coupling and interferences of magnetism and superconductivity yield novel behavior [8].

For fluctuating spin currents (in z–direction) one has according to the Maxwell equations, also accompanying electromagnetic fields, see $4\pi j_{s,z} = -4\pi\mu_B\partial_t\langle S_z\rangle = \partial_x E_y - \partial_y E_x$, etc. Generally, the connection between spin currents and magnetization dynamics is given by $\partial_t M_i + \partial_\mu j_{i\mu,\sigma} = 0$ [2]. According to Kirchhoff, for example, the emissivity (e) of a tunnel junction (or thin film) is related to the magnetization dynamics and magnetic resistance ($\Delta e/e \simeq a(GMR)$, where Δe is the change in the emissivity due to changing the magnetic configuration ($\uparrow / \uparrow$) to ($\uparrow / \downarrow$) of neighboring thin films or tunnel junctions, GMR is the giant magnetoresistance).

In general, nonequilibrium thermodynamics describes the thermoelectric and thermomagnetic effects. The currents j_i, including spin currents, are driven by the spin dependent thermoelectric forces $X_i = -\frac{\partial \Delta S}{\partial x_i}$, where S is the entropy and $\dot{x_i} = j_i$ (x_i = fluctuations of usual thermodynamical variables, x_i and X_i are conjugate variables). Thus, the currents can be calculated from

$$j_i \propto (1/X_i)d\Delta F/dt, \tag{19.1}$$

where F is the free–energy determined for example by an electronic hamiltonian. Note, (19.1) is of basic significance, since it relates the currents to the free–energy, see F. Bloch, S. de Groot and others [8, 9]. Hence, the currents may be calculated from the free–energy and this obviously permits application of scaling theory regarding phase transition behavior.

In the case of itinerant electrons the spin dependent currents result from the gradients $\Delta\mu_\sigma$ of the spin σ dependent chemical potentials μ_σ. Note, $\mu_\uparrow - \mu_\downarrow \simeq 2\mu_0 H_{eff}$, where H_{eff} is the effective molecular field acting on the itinerant spins [9], $\mu_{\uparrow(\downarrow)} = \mu \mp \mu_0 H_{eff}$.

Above (19.1) follows from $d\Delta S = \Sigma_i X_i x_i$, $j_i = \dot{x}_i$ and then

$$d\Delta \dot{F} = -\sum_i (T j_i X_i) + ... \qquad (19.2)$$

This yields, in particular, $j_i \Delta\varphi = -d\dot{F}$ [8]. As discussed later and as quantum mechanically expected, of course, the phase of the driving force $X_i(t)$ plays an important general role, see for example Josephson currents in superconductors or spin currents in magnets, etc. Clearly, in general, (19.1) includes also contributions due to time dependencies of phases occurring in the free–energy and applies also to superconductors.

The following study may be useful to demonstrate how Onsager theory yields the interdependence of the various currents (in nanostructures). Onsager theory is most useful to describe directly all thermoelectric effects etc., spin dependently. Already, known results [1, 4, 5] and new results [2] are presented. Interesting behavior is expected by manipulating, for example, the phase gradient of the magnetic material in Fig. 19.2. This may help to apply studies by Bloch [8] and others to spintronics and to new problems.

19.2 Theory

19.2.1 Onsager Theory

As a general framework for deriving the coupled spin dependent currents in tunnel junctions (and nanostructures in general) driven by thermoelectric forces X_i, like magnetization–, temperature– or chemical potential–gradients one may use Onsager theory, see Kubo, Landau, de Groot et al. [9].

Generally for deriving the spin dependent thermoelectric and thermomagnetic currents one may use the Onsager theory. The coupled spin dependent currents j_i are given according to Onsager theory by (expanding $\dot{x}_i = f(x_l) = L_{ij} X_j$)

$$j_i(t) = L_{ij} X_j(t) + L_{ijl} X_j X_l + ..., \qquad (19.3)$$

with driving forces [9, 10] $X_i = -\partial\Delta S/\partial x_i$, and using for the entropy S the expression $d\Delta S = \Sigma_i X_i x_i$. Note, x_i denotes the extensive thermodynamical variables like E, V, e etc. Then, from thermodynamics one has

$$d\Delta S = \Delta(1/T)dE + \Delta(p/T)dV - \Sigma_\sigma \Delta(\mu_\sigma/T)dN_\sigma$$
$$+ \Delta(H'_{eff}/T)dM_L, \qquad (19.4)$$

where μ_σ is the spin dependent chemical potential of itinerant electrons, M_L the magnetization of local magnetic moments and H'_{eff} the effective molecular field acting on the local spins, $\mu_\sigma = \mu(\varphi) - \sigma\mu_0 H_{eff}$, $\mu(\varphi) = \mu(0) - e\varphi$, and φ is the potential acting on the electron charge, respectively. Note, the term $\Delta\mu_\sigma dN_\sigma$ can also be put into the form $(\Delta\mu dN - \Delta H_{eff})$, where H_{eff} is the molecular field acting on the itinerant electron spins with magnetization M. We may put $X_1 \equiv X_E = \Delta T/T^2$, $X_{2\sigma} = \Delta(\mu_\sigma/T)$, $X_3 = X_M = -\Delta(H'_{eff}/T)$, $X_4 = -\Delta(p/T)$, $X_{5\sigma} = -\Delta(p_\sigma/T)$, the partial pressure of the electrons with spin σ, etc.

Thus, one finds for the coupled currents $j_i = L_{ij}X_j + ...$ driven by the forces X_i (with $i = 1 = E, i = 2 = e, i = 3 = \uparrow$, etc.) the expressions [9] of a coupled set of Equations with Onsager coefficients as unknowns to be determined, for example, by experiments, various conductivities or they may be determined from the free-energy using for F an electronic theory. Special situations may describe, for example, decoupling of charge and spin currents. Near phase transitions scaling theory can be applied, in particular. Taking into account the spatial anisotropy induced by the molecular field H_{EFF} and by an external magnetic field H one has j_i

$$j_{M_L} = L_{41}\Delta T/T^2 + \Sigma L_{42}^{\sigma}\Delta(\mu_\sigma/T) - L_{43}\Delta(H'_{eff}/T) + ... \quad . \tag{19.5}$$

Note, the replacement $\uparrow \rightarrow \downarrow$ yields $j_\downarrow$. The spin currents $j_\uparrow$ and $j_\downarrow$ may be coupled by spin flip processes, in particular spin–orbit interaction. Then a term proportional to $\Delta\mu_\downarrow$ could also contribute to $j_\uparrow$. As usual, symmetries may reduce the number of different Onsager coefficients L_{ij}, for example $L_{ij}(H_{eff}) = L_{ji}(-H_{eff})$ may hold etc.

The most important and central property of the Onsager equations is the interdependence of the various currents driven by the forces X_i. In particular, the driving force

$$X_{2\sigma} = \Delta(\mu_\sigma/T) \propto -\Delta(H_{eff}/T) + ... \propto -\Delta(M/T) + ... \tag{19.6}$$

causes correlated currents due to gradients of the magnetization with respect to phase and magnetization magnitude,respectively. The phase gradient driven spin currents are of the Josephson type [2, 6]. The Onsager equations show that the spin Josephson current is accompanied by a contribution to j_e, j_E, for example, or, better, ΔM due to a phase gradient also induces a contribution to the other currents, j_e, etc. This is immediately obvious from Onsager theory and yields new behavior.

Note, the Onsager Equations apply also to superconductors and yield different behavior for the single particle currents regarding singlet and triplet superconductors, in particular for j_e and $j_\uparrow$. The current of the Cooper pairs may be added to above Onsager equations. In the case of triplet pairing, the spin or angular momentum current of the Cooper pairs is of particular interest.

Also note, the Onsager theory applies to ions (in liquids, gases), in particular, magnetic ones. A specially interesting application of Onsager theory may be to a lattice of atoms or molecules and of quantum dots.

The Onsager equations are very useful for deriving directly the thermoelectric and thermomagnetic effects. The Onsager coefficients may be determined experimentally, by various conductivities, and may be calculated from the free–energy using, for example, an electronic theory. Scaling theory may be applied to the coupled currents near phase transitions.

Special situations are easily described by the Onsager equations. For example, decoupling of charge and spin current is described by $j_e = 0$ and $j_\uparrow$ or $j_s = j_\uparrow - j_\downarrow$ not equal to zero. e giant magnetoresistance (GMR) or tunnel resistance (TMR) the tunnel currents depend on the relative orientation of the magnetizations. The Seebeck and Peltier effects will reflect this.

Regarding a gradient in the phase of the magnetization, one gets a spin current (as mentioned already) from the gradient of the phase of the magnetization for a tunnel junction (or for film multilayers). Using the continuity equation $\partial_t M_i + \partial_\mu j_{i\mu,\sigma} = 0$ under certain conditions, or using the Landau–Lifshitz equation $dM/dt = a\overrightarrow{M} \times \overrightarrow{H_{eff}} + ...$, where $\overrightarrow{H_{eff}}$ refers to the effective molecular field, one may derive a spin current including a Josephson like spin current j^J of the form

$$j_\sigma = j_\sigma^1(\varphi) + j^J . \tag{19.7}$$

Here, $j_\sigma^1(\varphi)$ is the spin current due to the electrical potential φ and may result from the spin dependent density of states. The Josephson like spin current driven by a phase gradient of the magnetization is given by

$$j^J \propto dM/dt \propto \overrightarrow{M_L} \times \overrightarrow{M_R} + ... \propto |M_L||M_R| \sin(\phi_L - \phi_R + ...), \tag{19.8}$$

where $(\phi_L - \phi_R)$ is the phase difference of the magnetization on the left and right side of a tunnel junction (or of two films). ϕ is the canonical conjugate of S^z. Note, damping of spin transport may approximately be taken into account, see the Landau–Lifshitz equation or Landau–Lifshitz–Gilbert equation, in the coefficient in front of the term $(\overrightarrow{M_L} \times \overrightarrow{M_R})$.

Note, the previous equation for J^J also follows from

$$j^J = \sum_\sigma \sigma j_\sigma = \mu_B(\dot{S}_z^L - \dot{S}_z^R) \tag{19.9}$$

and charge current $j_{charge} = \sum_{\sigma j_\sigma}$ due to electrons with spin σ. R and L refer to right and left side of tunnel junction.

Here, L and R refer to left and right sides of the tunnel junction. Then, in accordance with general thermodynamics theory (see Onsager), her treating $\|S_z\|$ and θ as canonical conjugate variables with $\dot{S}_z = -\frac{\partial H}{\partial \theta}$ and $\dot{\theta} = \frac{\partial H}{\partial S_z}$, just interchanging θ and $\|S_z\|$. Furthermore, assuming exchange coupling, $H = H_0 + H_{int}$, $H_{int} = -J \sum \overrightarrow{S_L} \cdot \overrightarrow{S_R}$ one gets $\overline{\|S_z\|} \exp i\theta$,

$$j^J = \mu_B J |S_L| \cdot |S_R| \sin(\theta_L - \theta_R) + \dots \tag{19.10}$$

Here, the spin current due to a phase gradient $\Delta\theta$ may be analyzed for special situations like $j_{charge} = 0$ and $4j^J \neq 0$.

Note, using for a tunnel junction the hamiltonian

$$H_{int.} = -E_J S^2 \cos(\theta_L - \theta_R) + \mu_B^2/2C_s(S_L^z - S_R^z)^2 \tag{19.11}$$

one gets see Nogueira, Bennemann

$$j^J = \frac{2E_J S^2}{\mu_B} \sin\Delta\theta, \tag{19.12}$$

and

$$\Delta\dot{\theta} = 2\mu_B V_S + (...)h. \tag{19.13}$$

Here, h refers to an external magnetic field and $V_S = \frac{\mu_B}{C_S}(S_L^z - S_R^{z)}$. C_S denotes the spin capacitance. Note, the current j^J induces electric fields $\nabla x E_i = 4\pi j^J$ for no external fields.

Note, in the spirit of Onsager theory j^J carries an energy current like the charge current j_{charge}. Both affect the gradient of the magnetization.

The Onsager equations for the coupled currents of the itinerant electrons may be rewritten by introducing the spin dependent Peltier (P) and Seebeck (S) coefficients, respectively, For a tunnel junction with a magnetic metal A on the left side and a metal B on the right side, see Fig., one gets heat generation

$$j_E^A - j_E^B = (\Pi_i^A - \Pi_i^B)j_i \tag{19.14}$$

where i refers to $i = e$ for electronic current $i = s$ for spin current j^J. Of course the current discontinuity depends on the relative direction of the megnetization of metals A and B.

More and detailed experimental studies are needed to determine the Onsager coefficients, to check on previous equations, and to determine different Onsager coefficients in external magnetic fields. The Onsager Equations show again that, in particular, the Josephson spin current due to ΔM, with respect to its phase gradient, is accompanied by corresponding contributions to j_e, j_E, etc. [8–15]. As discussed already, the Seebeck coefficient describes the generation of a spin voltage by a temperature gradient [10]. Note, this is a characteristic result of Onsager theory and which, of course, can also be derived using an electronic theory. Note again, one gets directly from $\mu_\sigma = \mu(0) - eV - \mu_0\sigma H_{eff}$ that the spin voltage and the Seebeck coefficient is controlled by $H_{eff}(T)$. Recent studies by others also derived this result, see MacDonald, et al. [1, 4, 5, 10]. As discussed, if one expands μ_σ in terms of n_σ, T and pressure, etc. one gets the Seebeck effect. Clearly, S_σ, etc., are given by the energy spectrum, density of states $N_\sigma(\varepsilon)$ and electron populations. Note, the form $\Delta(\mu_\uparrow - \mu_\downarrow) = eS_s\Delta T + ...$ is already most practical for an electronic calculation.

An external magnetic field $\overrightarrow{H}$ affects the spin voltage. Regarding the spatial dependence x of the spin dependent chemical potential $\mu_\sigma(x,t)$, note that for $\frac{dT}{dx} = const.$, the gradient of the spin voltage varies linearly for a (one–dim.) tunnel junction in x–direction.

The spin Seebeck effect means (as discussed before) that a spin current can be induced in a magnetic metal without an electric current ($\Delta\varphi = 0$), since ΔT causes a contribution to the spin voltage $\Delta(\mu_\uparrow - \mu_\downarrow) \neq 0$. The spin current j_s is expressed by L^σ_{ij} and, approximately, $j_s \propto (N_\uparrow(\varepsilon_F) - N_\downarrow(\varepsilon_F)) + \ldots$ As is clear, this spin current depends on the spin mean free path, but it might disappear due to spin–flip scattering far less than spin currents injected into metals. For example, in a tunnel junction involving tunneling through graphene (with spin dissipation length $\sim$ nm or more) one might get relatively large spin currents induced by a temperature gradient. This is also the case for the spin currents resulting from the gradient of the phase of the magnetization. Regarding dynamics, the time dependence of the gradient of the magnetization phase is of interest.

19.2.2 Tunnel Junctions Involving Superconductors

It is also of interest is also to analyze the phase gradient driven currents in superconductors and to use these as a spin filter, see the illustration in Fig. 19.3 [10]. As known, a singlet superconductor may block a spin current and affect the currents driven, for example, by the gradients ΔT, $\Delta M = M_1 - M_2$, etc. Depending on the energy gain due to j_e vs. loss of energy due to (singlet) Cooper pair breaking, one may have that the currents weaken the superconducting state. Note, $\Delta M(t)$ may cause Josephson like spin current ($j_s \propto \sin\Delta\phi + \ldots$) [12].

If the two ferromagnets are separated by a triplet superconductor, then the relative orientation of the angular momentum $\overrightarrow{d}$ of the triplet Cooper pairs with respect to the magnetizations $\overrightarrow{M_1}$ and $\overrightarrow{M_2}$ controls the tunnel currents [2, 7, 12].

Note, $\overrightarrow{d}$ may be oriented via an external magnetic field. It is of particular interest to study the effect of superconductivity, triplet superconductivity, on the (giant) magnetoresistance in the case of two antiferromagnetically (af) oriented ferromagnets, see Fig. 19.3. One expects for the parallel configuration of $\overrightarrow{d}$, $\overrightarrow{M_1}$ and $\overrightarrow{M_2}$ the lowest

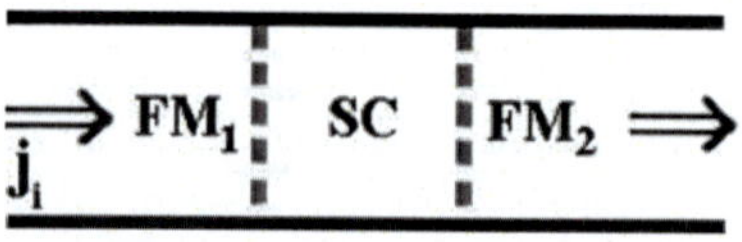

Fig. 19.2 Illustration of a magnetic tunnel junction consisting of two ferromagnetic metals $(FM)_1$ and $(FM)_2$ separated by a superconductor (SC). The electron current j_e, as well as $j_\uparrow$ and $j_\downarrow$ and $j_s = j_\uparrow - j_\downarrow$, depend on the relative orientation of the magnetisations $\overrightarrow{M_1}$ and $\overrightarrow{M_2}$ and on the superconducting state, singulet vs. triplet Cooper pairs. Note, $M_2 - M_1 = \Delta M$ may cause Josephson–like spin current, which is particularly affected in the case of a triplet superconductor by the phase of its order parameter. Of course, the spin current is destructively affected by spin flip scattering

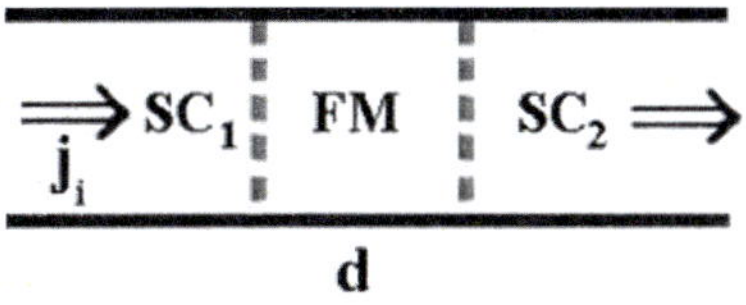

Fig. 19.3 Tunnel current j between two superconductors SC_1 and SC_2, which depends on the relative phase of the order parameter of the two superconductors and the phase of the magnetization. Of course, the thickness d of the ferromagnet controls the current, in particular spin polarized ones. The tunnel current may be manipulated optically via hot electrons in the ferromagnet

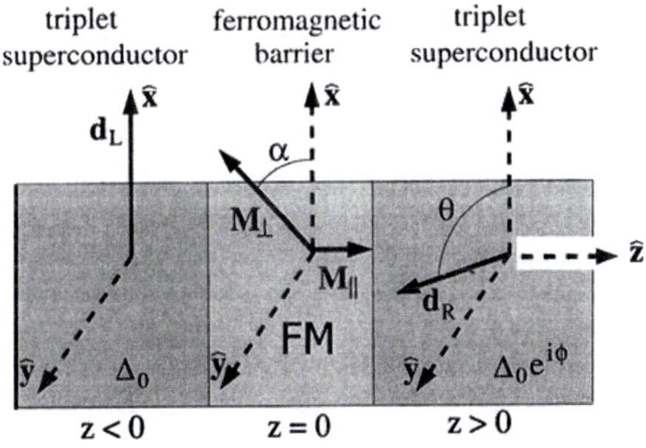

Fig. 19.4 Illustration of a tunnel junction ($TSC/FM/TSC$). The phases Θ, ϕ and α of the superconducting order parameter, Cooper pair condensate and of the magnetization $\overrightarrow{M}$, respectively, control the tunnel currents. The magnetization may be decomposed into components $M_\perp$ and $M_\parallel$. Due to the spin and angular momentum of the Cooper pairs one expects that the current through the FM depends sensitively on the relative direction of $\overrightarrow{M}$, spin relaxation, spin flip scattering resulting (for example) from spin–orbit coupling, population of the Andreev states and the thickness d of FM

resistance, while the largest one for af configuration of $\overrightarrow{d}$, and magnetizations. Of course, Onsager theory can be used to describe the system illustrated in Figs. 19.2 and 19.3. Note, $M_2 - M_1$ may act like a magnetization gradient.

Related properties are expected for the tunnel junction shown in Fig. 19.4. One may use this to distinguish singlet from triplet superconductivity. Onsager theory can be used to describe such a system phenomenologically. Josephson currents j_J sensitively characterize such tunnel junctions. The current j_J decreases for increasing thickness d of the ferromagnet and for decreasing Cooper pair binding energy (T_c). Also in the spirit of Onsager theory the difference $(\Delta_2 - \Delta_1)$ of the superconducting order parameters acts like a gradient inducing corresponding currents.

In the case of triplet superconductivity (TSC), the Josephson current j_J depends in an interesting way on T_c and the angle between the magnetization $\overrightarrow{M}$ and direction normal to $\overrightarrow{j_J}$. The current should depend on the triplet state and impurity scattering (in particular spin orbit scattering). Hot electrons in the ferromagnet FM modulate j_J. Generally the spin polarization of the currents may be manipulated by the gradient $\Delta M(t)$.

In view of the significance of the occurrence of triplet superconductivity in metals we sketch the situation in the following Fig. 19.4. The current carried by Andreev states is calculated using [7, 12]

$$j_J = -(e/\hbar) \sum \frac{\partial E_i}{\partial \phi} \tanh(E_i/2kT), \qquad (19.15)$$

where the sum is extended over all Andreev states with energy E_i and mediating the tunneling. Here, ϕ is the phase difference between the Cooper condensates on the left and right side of the tunnel junction. One expects j_J to depend characteristically on the phases of all order parameters, on the relative orientation of the Cooper pair vectors $\overrightarrow{d_L}$, $\overrightarrow{d_R}$ and magnetization $\overrightarrow{M}$, respectively. The triplet Cooper pairs are described by $\Delta(k) = \sum_l d_l(k)(\sigma_l i\sigma_2)$, $l = 1, 2, 3$ where σ_l are the Pauli spin matrices and d_l are the spin components of the superconducting order parameter, see Bennemann and Ketterson [7, 12]. Note, the triplet Cooper pairs have a spin and orbital momentum.

For the transport of angular momentum, obviously the phases of all three order parameters are of importance for tunneling. Even for no phase difference $\phi = \phi_L - \phi_R$ between the triplet Cooper pair condensates on both sides of the tunnel junction one gets for arbitrary phase of the magnetization of the ferromagnet a Josephson current. In the ferromagnet the Andreev states carry the current of the tunneling electrons and temperature controls its population. Also, of course, the magnitude of the magnetization and electron spin relaxation in the FM matter. As physically expected the Josephson current may change sensitively upon rotation of $M_\perp$, change of the direction of $\overrightarrow{M}$. Model calculations yield results shown in Fig.7(a) [7, 12]. This implies that the tunnel junction (TSC|FM|TSC) may act like a switch turning on and off the current j_J. This behavior suggests a sensitive dependence of the current j_J on an external magnetic field.

In Fig. 19.5b model calculation results, strongly simplifying the influence of the FM metal, are given for the temperature dependence of the Josephson current [7, 12]. These should reflect the temperature controlled occupation of the Andreev states. The change in sign of $j_J(T)$ as a function of T occurs only if Andreev states are non degenerate and in the case of two Andreev states 1 and 2 with derivatives $\frac{\partial E_1}{\partial \phi}$ and $\frac{\partial E_2}{\partial \phi}$ that have opposite signs. Also note, the sign change of j_J for increasing temperature may be suppressed by electron scattering in the FM [7, 12].

Clearly, in view of the importance of studying triplet superconductivity improved calculations of the current j_J are needed. The FM tunnel junction metal must be taken into account in a more realistic way. In case of paramagnetism $M = \chi H$, the direction of the external magnetic field can be used to manipulate the current. In view of the rich behaviour of $(TSC/FM/TSC)$ tunnel junctions one also expects

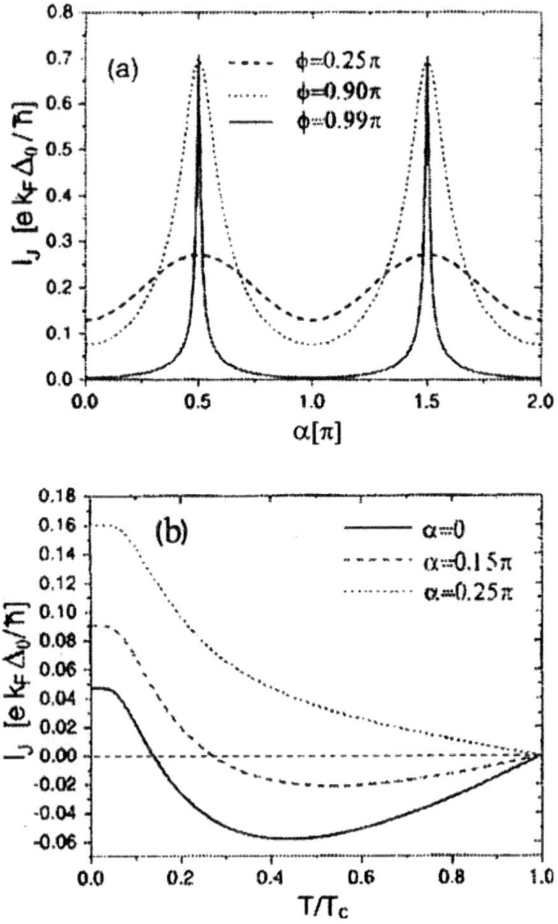

Fig. 19.5 Results for a tunnel junction (TSC/FM/TSC) sketched in Fig. 19.4, see Morr et al. [7, 12]. Dependence of the Josephson current on **a** phase α of the magnetization $\overrightarrow{M}$ and **b** temperature T for various values of α. Results refer to model calculations simplifying the coupling of the Cooper pairs to the FM. For increasing electron scattering at the ferromagnetic barrier, j_J does not change sign for increasing temperature any more. The Andreev states carrying the current are determined using the Bogoliubov–de Gennes method. T_c is the superconducting transition temperature

interesting behavior for the currents. Again, the phases of the three order parameters control the currents. One expects

$$j_s^J \simeq A(\Theta)\sin(\Delta\phi + \eta), \tag{19.16}$$

where Θ refers to the relative phase of the triplet Cooper pairs and $\Delta\phi$ to the phase difference of the magnetization on left and right side of the tunnel junction. Of course,the Josephson spin current j_s^J can also be manipulated optically by changing the population of the electronic states, by an external magnetic field and by applying a temperature(or pressure) gradient. A spin current may result due to $N_\sigma L(\varepsilon) \neq N_\sigma R\varepsilon$. Then, $j = j_s + j_s^J$.

Interesting spin-dependent effects are also expected for superconductors due to geometrically induced transitions of BCS and BEC transitions, $\rightleftharpoons$, in magneto optics of nanostructures, see Hübner, Bennemann studies, and for systems of quantum dots, see Garcia, Jensen, Bennemann [21, 22].

19.3 Summary

Various experiments can be used to determine the generally spin dependent Onsager coefficients L^σ. The spin dependent forces X^σ can generally be manipulated by light creating hot electrons and thus changing the various gradients ΔT, ΔM, etc. For the tunnel system shown in Fig. 19.2 one might expect interesting behavior if, for example, the metal 3 is replaced by Ce which electronic properties, valency changes upon photon induced and controlled population of the s, d and f–states. Similarly, currents change dramatically if, in Fig. 19.2, the material 3 consists of semiconductors like Si, Ge etc or magnetic semiconductors which conductivity is strongly affected by hot electrons [13].

Note, the Onsager coefficients are given by the current–current correlation function [9, 14, 15]

$$L_{il}(t) = \langle j_i(t) j_l(0) \rangle, \quad (19.17)$$

with $j_i(t)$ calculated using response theory, Heisenberg or v.Neumann equation of motion ($\dot{\rho} \propto [\rho, H] + ...$), or as a functional of the free–energy F [9, 14, 15], see also Bloch using $jV = dF/dt$ with V being the potential associated with the force $X_{i\sigma}(t)$. Then, (see response theory)

$$j_i(t) = \int_\infty^t dt' L_{il}(t - t^{'}) X_l(t'), \quad (19.18)$$

where the $L_{il}(t)$ are now calculated within an electronic theory. The analysis is simplified if $L_{ij} \propto \delta(t - t^{'})$ (Markov processes, see Kubo [9]).

An electronic theory may be useful in order to apply field theoretical arguments and symmetry considerations, see for example Nogueira [14]. The currents are also directly calculated as derivatives of the electronic free–energy [8]. One expects generally the formulae $jV_{th} = dF/dt$, where V_{th} denotes the "potential" associated with the driving force $X_{i\sigma(t)}$ of thermodynamics [9]. Note, for large gradients X_σ nonlinear contributions to the currents may play a role.

In the case of magnetic multilayer structures, for example a ferromagnet A on a ferromagnet B, one may induce spin currents by shining light on the surface of the thin magnetic film. This creates hot electrons and a temperature gradient and thus induces a spin current etc. Note, according to Maxwell equations $\Delta\overrightarrow{M(t)}$ will generate electric fields E_i, $i = x, y, z$, which should be reflected in the observed currents.

Onsager theory also applies to currents in (gases) liquids of magnetic ions and in the presence of a polarizing external magnetic field. Separating, for example, such a system by a wall (with appropriate holes) into two compartments A and B, one may induce charge and spin currents driven by $X_{i\sigma}$, voltage gradient, thermal gradients, magnetic field gradient, etc. . One gets (approximately) for the currents through the wall holes, see for example Kubo [9],

$$j_\sigma^\alpha \propto p_\sigma^\alpha / \sqrt{T^\alpha} + ..., \alpha = A, B, \tag{19.19}$$

where p_σ is the partial pressure due to ions with spin σ. Obviously, while for the stationary state $j_i^A = j_i^B$, gradients $X_{i\sigma}$ cause corresponding currents. For example, in case of no pressure difference charge (and spin) may flow from A to B due to a temperature gradient and $T_A < T_B$.

Onsager theory has many applications in thermodynamics. As an example, note that in case of magnetostriction thermodynamics yields [9]

$$\Delta M = - \left.\frac{\partial V}{\partial H_{eff}}\right)_{p,T} \Delta p + ... \tag{19.20}$$

Hence, a pressure gradient changes the magnetization and may cause spin current. In particular, a time dependent pressure p(t) drives a magnetization dynamics.

Onsager theory could also be applied to spin currents in topological insulators and at the interface of semiconductors in the presence of strong magnetic fields. Interplay of spin–orbit coupling and magnetic field and magnetism of a substrate of semiconductor should yield interesting results [7, 18].

Regarding (magnetic) atoms on lattices, including tunnel junctions (structures), spin dependent gradient forces X_i acting on the atoms may cause currents and novel behavior. Photon assistance of atom or molecule tunneling might be of particular importance.

Coupled currents under the influence of a magnetic field and radiation fields also likely play an important role in interstellar and galactic interactions and could be treated using same theory as above, using Onsager theory for magnetic, ionic gases.

Finally, interesting and novel behavior may occur if (diffusion) currents are accompanied by coupled chemical reactions, see de Groot [9]. Then in magnetic systems, spin and magnetization may play a role and cause magnetic effects [13–23].

This discussion demonstrates the many options for inducing spin currents and the powerful general analysis Onsager theory offers.

Acknowledgements I thank C. Bennemann for help and many useful and critical discussions. This study is dedicated to Prof. J.B. Ketterson (USA) for lifelong help, suggestions and assistance. Similarly, I thank also Prof V. Bortolani (Italy) and Prof. B. Alascio (Argentina) for help during many years. Last but not least I thank in particular F. Nogueira and M. Garcia for ideas and interesting discussions. Also, I thank E.M. Steinebach and H.W. Dube (Wasbüttel, Germany) for acting as "spiritus rector" throughout this study.

References

1. G.E. Bauer, A.M. MacDonald, S. Maekawa, Spin electronics. Solid State Commun. **150**, 459 (2010)
2. K.H. Bennemann, Magnetic Nanostructures. J. Phys. Condens. Matter **22**, 243201 (2010)
3. K.H. Bennemann, Photoinduced phase transitions. J. Phys Condens. Matter **23**, 073202 (2011)
4. K. Uchida, S. Takahashi et al., Observation of the spin seebeck effect. Nature **455**, 778 (2008); F.L. Bakker, A. Slachter, J.P. Adam, B.J. van Wees, Phys. Rev. Lett. (2010); on spin caloritronics: M. Johnson, R.H. Silsbee, Phys. Rev. B **35**, 4959 (1987), M. Johnson, Sol. State Commun. **150**, 543 (2010); L. Gravier, S. Serrano-Giusan, F. Reuse, and J.P. Ansermet, Phys. Rev. B **73**, 024419 (2006)
5. A. Slachter et al., Thermally driven injection from a ferromagnet into a non–magnetic metal. Nat. Phys. **6**, 879 (2010)
6. F. Nogueira, K.H. Bennemann, Europhys. Lett. **67**, 620 (2004). Note, the gradient of both the magnitude and phase of the magnetization (assuming to be canonical conjugate variables and for which one may use then the classical Hamilton equations) yield for magnetic tunnel junctions with hamiltonian $H = -E_J S^2 \cos(\theta_L - \theta_R) + \mu_B^2/2C_s(S^{z_L} - S^{z_R})^2$ the spin current $J = 2\frac{E_J}{S^2}\mu_B \sin\Delta\theta$, and $\Delta\theta \doteq 2\mu_B V_S + (...)h$. Here, h refers to an external field and $V_s = \frac{\mu_B}{C_S}(S_L^z - S_R^z)$ denotes the spin capacitance, see Nogueira, Bennemann. Note , J^J induces electric fields and $\nabla X E_i = 4\pi J_i^J$
7. B. Kastening, D. Morr, D. Manske, K.H. Bennemann, Phys. Rev. Lett. **96**, 047009 (2006); Phys. Rev. **79**, 144508 (2009)
8. F. Bloch, Phys. Rev. B **2**, 109 (1970)
9. R. Kubo et al., *Statistical Mechanics* (North–Holland Publishing Co., Amsterdam, 1965); L.D. Landau, E.M. Lifshitz, Statistical Mechanics (revised by E.M. Lifshitz and L.P. Pitaevski), Pergamon Press 1980; S.R. de Groot, Thermodynamik irreversibler Prozesse, Hochschultaschenbücher, Bd. 18, 1960, and Thermodynamics of Irreversible Processes (North–Holland, 1952)
10. S. Takahashi, H. Imamura, S. Maekawa, Phys. Rev. Lett. **82**, 3911 (1999)
11. Note, taking into account magnitude and phase of the spin magnetization one gets generally $j_M \propto \Delta|M| + b_1|M|\nabla \exp i\psi$ and similarly for j_σ Josephson like currents. Of course, spin flip scattering plays a role and should be considered
12. P.M. Brydon, B. Kastening, D. Morr, D. Manske, Interplay of ferromagnetism and triplet superconductivity in a Josephson junction, arXiv 0709.2918v1, 19 Sept 2007; K.H. Bennemann, J. Ketterson, Superconductivity, vol. 2 (Springer, 2008)
13. K.H. Bennemann, Ultrafast dynamics in solids. Ann. Phys. (Berlin) **18**, 480 (2009)
14. K.H. Bennemann, F. Nogueira, see Lecture–Notes, FU-Berlin. Of course, expressing the Onsager coefficients L_{ij} by current–current correlation functions and calculating these within Lagrangian theory and electronic Hamiltonian, it is in principle straightforward to get instead of phenomenological Onsager theory results which refer to an electronic Hamiltonian, for example Hubbard hamiltonian. This might be particularly useful for understanding many–body effects, see F. Nogueira, Introduction to the field theory of classical and quantum phase transitions (Lecture Notes, FU–Berlin, September 2010); see also L.D. Landau and E.M. Lifshitz, Kinetics, vol. 10 (Pergamon Press)
15. F. Nogueira, K.H. Bennemann, *Current Correlation Functions* (FU–Berlin, to be Published, 2010)
16. B. Tatievski, P. Stampfli, K.H. Bennemann, Ann. der Physik **4**, 202 (1995); Comp. Mat. Sci.**2**, 459 (1994); B. Tatievski, Diploma– thesis (FU–Berlin, 1993)
17. A.C. Bleszynski-Jayich, W.E. Shanks, B. Peaudecerf, E. Ginossar, F. von Oppen, L. Glazman, J.G. Harris, Science **326**, 272 (2009); P. Michetti, P. Recher, Bound States and persistent currents in topological insulator rings (2010). arXiv: 1011.5166v1
18. E.I. Rashba, Phys. Rev. B **68**, 241315-1 (2003); B.A. Bernevig, Phys. Rev.B **71**, 073201 (2005); A.V. Balatsky, B.L. Altshuler, Phys. Rev. Lett. **70**, 1678 (1993). Spin–orbit coupling causes in a ring a persistent current due to the induced electron phase ϕ_{AC} (Aharonov–Casher effect),

see studies in above references. Using Faraday's law $\partial_t \phi_{AC} = -c \oint E_{AC} \bullet dl$ it is $j_{\varphi,\sigma_z} = -(c/2\pi R)\frac{\partial E}{\partial \phi_{AB}}$, $\sigma_z = \pm 1/2$, and electron energies $E_i = E_i(\phi_{AB} + \phi_{AC})$

19. Assuming particle–hole symmetry it is $j = -\Sigma_i \frac{dE_i}{d\phi} \tanh(\frac{E_i}{2kT})$. Here, i refers to the electron states with energy E_i and for example to the polygonal orbits of the ring, see Stampfli *et al.*, $E_{t,p} \propto \cos(SB)$ assuming that all orbitals t,p enclose same area S. Then approximately $j = -\int dEN(E)\frac{\partial E}{\partial \phi} \tanh(E/2kT) + \ldots$ and thus at $T = 0$ it is $j = -\Sigma_i \frac{dE_i}{d\phi}$
20. Note, the Landau–Lifshitz equation, $\frac{d\vec{M}}{dt} = \gamma \vec{M} \times \overrightarrow{H_{eff}} + \alpha \vec{M} \times \frac{d\vec{M}}{dt} + \beta \vec{j} \times \partial_t \vec{M}$, includes damping. A time dependent spin current yields in accordance with the Maxwell equations an electric field
21. A.A. Shanenko, M.D. Critoru, A. Vagov, F.M. Peeters, Phys. Rev. B **78**, 024505 (2008), and More Recent Publ. by F.M. Peeters et al
22. M.E. Garcia, *Habilitation thesis, Physik* (FU Berlin, 1999)
23. To include chemical reactions occurring in (open) magnetic systems with diffusion currents etc. one may extend Onsager theory taking into account spin and magnetization. For simplicity one may first neglect nonlinear behavior. (However, note frequently this may not be valid.). Then, $j_i = L_{ij} X_j + \ldots$ and chemical reactions are taken into account by the forces $X_j = A_j/T$ with spin dependent chemical affinities $A_j = -\sum_k \nu_{k,j} \mu_k$. Here, the substance k with chemical potential μ_k is also characterized by its spin. The stoichometrical coefficients $\nu_{k,j}$ characterize the chemical reaction which couples to the currents. It is straightforward to work out details of the theory, see de Groot, Landau. Thus one gets spin dependent coupling of diffusion currents and chemical reactions, for example effects due to concentration gradients depending on spin and magnetization, etc.

Chapter 20
Bulk to Low Dimensional 2D Thermoelectric Materials: Latest Theoretical Research and Future View

T. Seddik and M. Batouche

Abstract The thermoelectric (TE) materials convert heat directly into electricity or transport thermal energy by the application of an electric current. This class of materials has been known for over a century, but the limited efficiency has slowed down their development. However, due to the discovery of new materials and the control of dimensionality, thermoelectric has regained its interest in recent years. Inspired by the fascinating properties of 2D graphite many new 2D materials are designed. The prediction of low dimensional materials and the exploration of their electronic and thermoelectric properties have demonstrated that the low dimensional approach is effective for the enhancement of the figure of merit *ZT*. In this review, we discuss the thermoelectricity phenomena and the latest theoretical investigations performed in developing and designing new 2D thermoelectric materials. Besides, the different strategies to improve the thermoelectric performance that could be used in the new 2D thermoelectric materials will be highlighted.

20.1 Introduction

In recent years, thermoelectricity has become very attractive as a new application that uses a clean energy source. So, many projects are being studied to apply thermo-generators in different fields such as automotive, microelectronics, medical…etc. These thermo-generators are typically made of materials called thermoelectric materials which have the ability to convert heat directly into electricity or vise versa. This class of materials has been known for over a century [1], but the limited efficiency has slowed down their development.

The thermoelectric performance of a material is essentially characterized by a factor stand for figure of merit, $ZT = \frac{S^2\sigma T}{\kappa}$, where S, σ, and κ denote the Seebeck coefficient, electrical and thermal conductivity, respectively. Thus the major target is

T. Seddik (✉) · M. Batouche
Laboratory of Quantum Physics of Matter and Mathematical Modeling, Mustapha Stambouli University, 29000 Mascara, Algeria
e-mail: seddik.taib@univ-mascara.dz

H. Ünlü and N. J. M. Horing (eds.), *Progress in Nanoscale and Low-Dimensional Materials and Devices*, Topics in Applied Physics 144,
https://doi.org/10.1007/978-3-030-93460-6_20

the improvement of ZT, however search for the materials with low κ value, the high value of S and σ had been a challenge to all the researchers in recent past. It is because of the reason that the dependence of S, is inversely proportional to the concentration of charge carriers. Hence, for the materials if the Seebeck coefficient decreases, the electrical conductivity is increased with increasing carriers concentration. Likewise, such increase in electrical conductivity results in increasing thermal conductivity, which shows an adverse effect on the improvement of the ZT factor.

One powerful strategy to increase ZT value is the control of dimensionality of the bulk TE materials. This strategy was initially introduced by Hicks and Dresselhaus where they examined the thermoelectric of 2D quantum well structure and reported that the quantum-confinement can considerably improve the power factor PF [2, 3]. The mechanisms behind this are: (i) dramatic change of density of states (DOS) caused by the increasing size; (ii) Scattering of phonons more effectively than electrons by increasing boundary and interfacial scattering [2–4]. Consequently, the PF of 2D materials is significantly increased and the lattice thermal conductivity is decreased compared to their bulk counterparts. Besides, inspired by the fascinating properties of 2D graphene [5–8] many new 2D materials are designed such as Graphene family, Penta-graphene family, Dichalogenide family, Trihalide family, Mxene family, Phosphorus family, …etc. [5–18]. Owing to their remarkable properties compared to the bulk counterpart these 2D materials have attracted broad attention in order to design a new class of thermoelectric materials with high performance [19–27].

In this review, we discuss the thermoelectricity phenomena and the latest theoretical investigations performed in developing and designing new 2D thermoelectric materials. Besides, the different strategies to improve the thermoelectric performance that could be used in the new 2D thermoelectric materials will be highlighted.

20.2 Principles of Thermoelectric Energy Conversion

A thermoelectric phenomenon is the generation of electricity (occurrence of electrical potential difference ΔV) when the material is exposed to a temperature difference ΔT via the Seebeck effect [28]. This phenomenon is mainly explained by the displacement of charge carriers from the hot side to the cold side of a material to return to an equilibrium state under a temperature gradient, creating a diffusion current. For material in an open circuit, this current is zero and a voltage ΔV appears which is depending on the used material. This thermoelectric voltage difference per unit temperature is termed the Seebeck coefficient. Conversely to this process i.e., when this material is subject to electrical potential difference ΔV the thermal energy is transporting, it's the Peltier effect [29]. These effects are usually used for power generation and cooling or heating applications through thermo-generator Fig. 20.1, which consist of two legs one of them is an n-type semiconductor and the other is a p-type semiconductor. Typically many thermo-generators are electrically connected to form a thermoelectric device (module) Fig. 20.1.

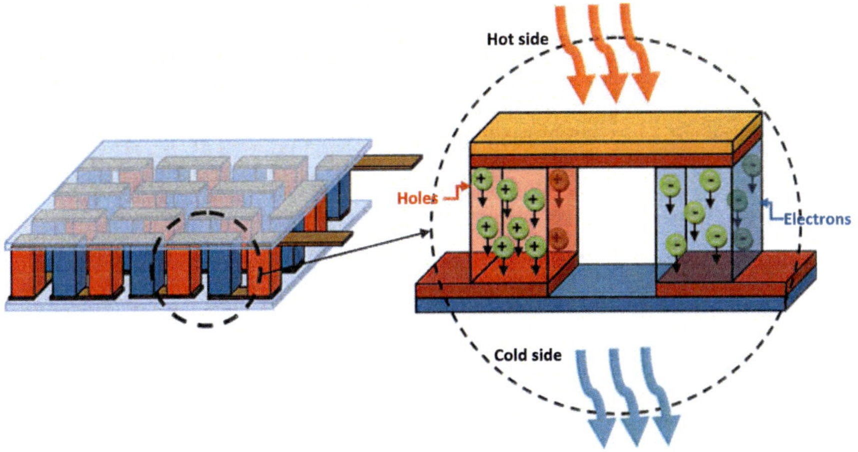

Fig. 20.1 Schematic diagram of thermo-generator which consists of two legs one of them is an n-type semiconductor and the other is a p-type semiconductor (right) and module (left)

20.2.1 Thermoelectric Figure of Merit

The conversion efficiency η of module for power generation applications can be defined as follow:

$$\eta = \frac{T_H - T_C}{T_H} \frac{\sqrt{1 + \overline{ZT}} - 1}{\sqrt{1 + \overline{ZT}} + \left({T_C}/{T_H}\right)} \tag{20.1}$$

on the other hand, the coefficient of performance (COP) of module for cooling applications is given by the following equation:

$$COP = \frac{T_H}{T_H - T_C} \frac{\sqrt{1 + \overline{ZT}} - \left({T_H}/{T_C}\right)}{\sqrt{1 + \overline{ZT}} + 1} \tag{20.2}$$

where T_H and T_C, are the hot-side and cold-side temperature, respectively, and $\overline{ZT}$ is the average figure of merit which is the average of the dimensionless figure of merit ZT of the materials included in the module. This figure of merit is given by:

$$ZT = \frac{S^2\sigma}{\kappa} T \tag{20.3}$$

Here, S denotes Seebeck coefficient, σ the electrical conductivity, and κ the thermal conductivity. The Seebeck coefficient is expressed by the following equation:

$$S = \frac{8\pi^2 k_B^2}{3eh^2} m^* T \left(\frac{\pi}{3n}\right)^{\frac{2}{3}} \tag{20.4}$$

where e is the electronic charge, k_B is the Boltzmann's constant, T is the temperature, h is the Planck's constant, n is the carrier concentration, and $m*$ is the effective mass of the carrier. Besides, the electrical conductivity σ which represents the ability of the material to conduct electric current is given by:

$$\sigma = ne\mu \tag{20.5}$$

where μ is the mobility of the charge carriers. Moreover, the thermal conductivity κ which represents the material's capability to transfer heat subject to temperature gradient across its points is given by:

$$\kappa = \kappa_e + \kappa_l \tag{20.6}$$

here, κ_e and κ_l are the electronic and the lattice thermal conductivity, respectively.

The first one κ_e is determined by the Wiedemann–Franz law $\kappa_e = L\sigma T$, where L denotes Lorenz number. On the other hand, κ_l is given by $\kappa_l = 1/3\, cv_s l_p$, where v_s is the sound velocity, c the specific heat, and l_p is the phonon mean free path.

So for higher η and *COP* we desire a material with *ZT* as high as possible, hence improving *ZT* factor is a major target. However, search for materials with lower κ value and higher S and σ value is a very big challenge for researchers. It is because of the reason that the dependence of S, is inversely proportional to the concentration of charge carriers Fig. 20.2. Hence, for the materials, if the Seebeck coefficient decreases, the electrical conductivity is increased with increasing carriers concentration. Likewise, such an increase in electrical conductivity results in increasing thermal conductivity, which shows an adverse effect on the improvement of the *ZT* factor.

20.3 Theoretical Investigations of 2D Thermoelectric Materials

20.3.1 Theoretical Background

From Sect. 20.2.1 we have observed that higher *ZT* requires high Seebeck coefficient S and electric conductivity σ, and low thermal conductivity κ, which is very difficult to realize since these three thermoelectric quantities are inter-related for the 3D crystalline system. However, reducing dimensionality is considered as an effective way to enhance *ZT* value of bulk materials. The mechanisms behind this are: (i) dramatic change of density of states (DOS) (Fig. 20.3) caused by the increasing size.

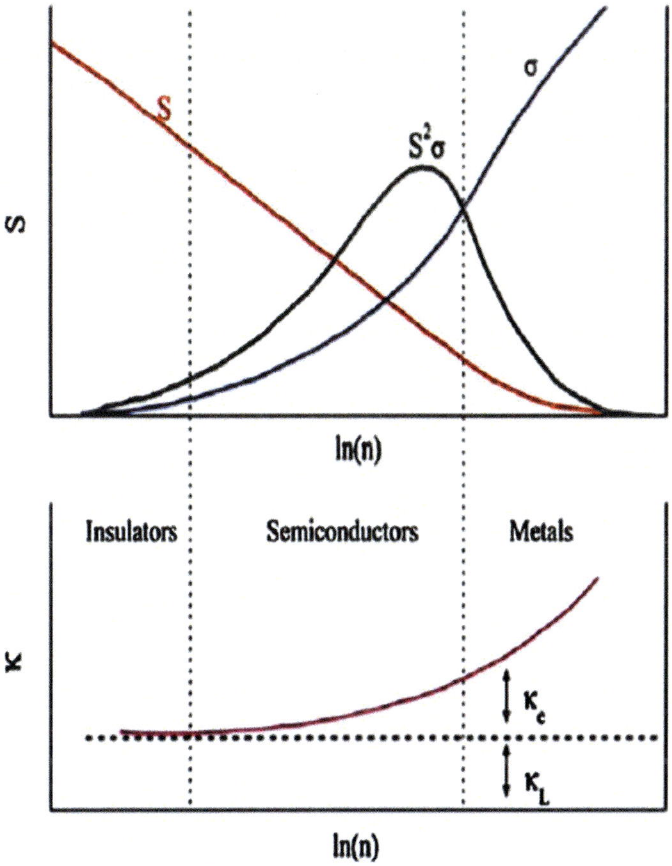

Fig. 20.2 Seebeck coefficient S, electrical conductivity σ, power factor (PF = $S^2\sigma$), electronic thermal conductivity κ_e, and lattice thermal conductivity κ_L as function of charge carrier concentration *n*.. Reprinted from [30], Copyright (2012), with permission from Elsevier

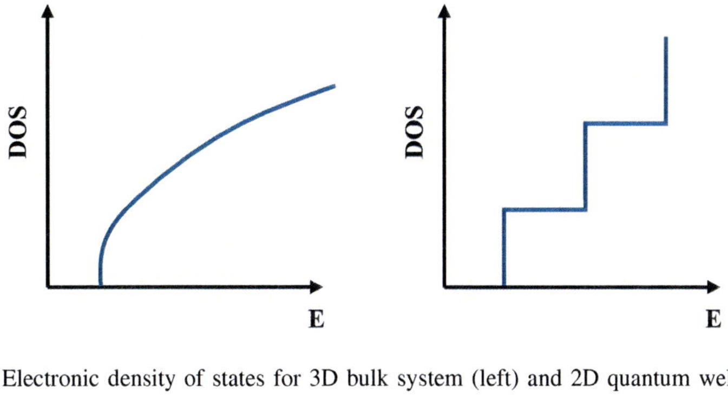

Fig. 20.3 Electronic density of states for 3D bulk system (left) and 2D quantum well structure (right)

Compared to the 3D bulk materials the Fermi level in 2D materials is located near a sharp change in DOS, which lead to significant enhancement in *S*. (ii) Scattering of phonons more effectively than electrons by increasing boundary and interfacial scattering, leading to a decrease in the lattice thermal conductivity [2–4].

This dimensionality effect was the subject of Hicks and Dresselhaus works where they examined the thermoelectric of 2D quantum well structure [2, 3]. They assumed a quantum confinement in the *z*-direction and in the *x*–*y* plane the electrons are in parabolic bands. The electronic dispersion relation in *x*–*y* plane is parabolic and given by [31]:

$$E_{2D}\left(k_x, k_y\right) = \frac{\hbar^2 k_x^2}{2m_x} + \frac{\hbar^2 k_y^2}{2m_y} + \frac{\hbar^2 \pi^2}{2m_z d_w^2} \tag{20.7}$$

where m_x, m_y, and m_z are the effective mass along *x*, *y*, and *z*-direction, and d_W is the thickness of the quantum well.

Compared to the expression of *ZT* for 3D bulk system given by [31]:

$$ZT_{3D} = \frac{\left[\frac{5F_{3/2}}{3F_{1/2}} - \zeta^*\right]^2 \frac{3}{2} F_{1/2}}{\frac{1}{B_{3D}} + \frac{7}{2} F_{5/2} - \frac{25F_{3/2}^2}{6F_{1/2}}} \tag{20.8}$$

where F_i denotes Fermi–Dirac function given by:

$$F_i = F_i\left(\zeta^*\right) = \int_0^\infty \frac{x^i dx}{exp(x - \zeta^*) + 1} \tag{20.9}$$

here $x = E/k_B T$ is a dimensionless energy and $\zeta^* = \zeta/k_B T$ is a dimensionless electrochemical potential, and B_{3D} is a dimensionless material parameter described by:

$$B_{3D} = \frac{1}{3\pi^2}\left(\frac{2k_B T}{2}\right)^{3/2}\left(m_x m_y m_z\right)^{1/2} \frac{k_B^2 T \mu_x}{e\kappa_{ph}} \tag{20.10}$$

here μ_x is the mobility in *x* direction.

Using 2D electronic dispersion relation the *ZT* expression for 2D system is given by [31]:

$$ZT_{2D} = \frac{\left[\frac{2F_1}{F_0} - \zeta^*\right]^2 F_0}{\frac{1}{B_{2D}} + 3F_2 - \frac{4F_1^2}{F_0}} \tag{20.11}$$

where the material parameter B_{2D} is expressed by the following equation:

$$B_{2D} = \frac{1}{\pi d_W}\left(\frac{k_B T}{2}\right)\left(m_x m_y\right)^{1/2}\frac{k_B^2 T \mu_x}{e\kappa_{ph}} \tag{20.12}$$

From ZT expression (20.8) and (20.11) one can observe that the figure of merit value depends on reduced chemical potential $\zeta *$ and material parameter B. Thus for 3D bulk materials, varying doping level to optimize $\zeta *$ or increasing B by reducing lattice thermal conductivity and/or maximize mobility, the ZT_{3D} can be enhanced. On the other hand for 2D system, in addition of these parameters one more variables of length scales d_W can be varied to maximize ZT_{2D} value. Apparently minimizing the d_W increase the material parameter B_{2D} and hence the ZT_{2D} can be largely enhanced [32].

20.3.2 Recent Theoretical Results

Since the discovery of graphene in 2004 [5], there has been a surge of interest in 2D materials for various applications. These include Graphene family, Penta-graphene family, Dichalogenide family, Trihalide family, Mxene family, Oxide family, Phosphorus family, and various others [5–18]. Owing to their remarkable properties compared to the bulk counterpart these 2D materials have attracted broad attention for thermoelectric applications. Using ab initio method based on density functional theory [33] within the semi-classical Boltzmann transport theory [34] many researchers group have explored 2D materials in order to designing new class of thermoelectric materials with high performance.

20.3.2.1 Graphene Family

The fact that the graphene have poor thermoelectric performance [35] leads to design and development of new 2D carbon structure [36–40] and new graphene-like 2D materials such as graphene-like BN, ZnO, AlN, BeN_2, …etc. [41–50].

Unlike graphene the graphyne exhibit an opening band gap which make it favorable candidate for thermoelectric application [19, 51]. Jiang et al. [52] have theoretically investigated the electronic and thermoelectric of γ-graphyne and they reported a high Seebeck coefficient and low thermal conductivity of γ-graphene compared to doses of graphene. Besides, they estimated a ZT_{max} of about 1.5 for the p-type doping along the x-direction and 1.0 for the n-type doping along the y-direction Fig. 20.4b. This interesting thermoelectric performance is interpreting by the existence of the carbon–carbon triple bonds in the γ-graphene structure and the opening of the band gap [52].

Moreover, many other work on 2D graphene-lick materials are achieved such as AlX (X = S, Se, and Te) [41], InN [42], Cd*Ch* (*Ch* = S, Se, and Te) [43], B*Pn* (*Pn* = P, As, and Sb) [44], SiC_x [45], SrS [46], XBi (X = Si, Ge, and Sn) [47], ZnO [48], BN [49] and BaX (X = O, S, Se, and Te) [50]. These materials show promising thermoelectric properties however theirs *ZT* are still far from the desired value for thermoelectric application except for BX (X = As and Sb).

In the case of the BX (X = As, Sb) monolayer Zhou et al. [44] predict a higher *ZT* value along the zigzag direction compared to those along armchair direction for both *p*- and *n*-type carriers, with ZT_{max} achieved at 1300 K and 700 K for BAs and BSb monolayer, respectively. At these temperature and for hole concentration of about 5.7×10^{19} cm^{-3} and 2.7×10^{19} cm^{-3} the expected maximum *ZT* values are 3.7 and 3.3 for 2D BAs and BSb materials, respectively, along the zigzag direction. They also found that along the armchair direction the BSb monolayer has almost same thermoelectric performance for *p* and *n*-type doping [44].

20.3.2.2 Penta-Graphene Family

The 2D carbon allotrope material named penta-graphene (PG) has proposed for the first time by Zhang et al. [16]. This PG is completely composed of carbon pentagons and shows a good dynamical and mechanical stability [16]. It is estimated that PG posses a band gap of about 3.25 eV, making it a potential candidate for optoelectronic device [53–57]. Furthermore, it is showed that PG has a κ_l of about 197.85 Wm^{-1} K^{-1}[58] at room temperature, which is much lower compared to the graphene one. C. P. Chen et al. reported that even though the PG posses low κ_l its *ZT* value is still lower (0.053) at room temperature, however this *ZT* can reached a value about 0.481 under strain effect for p-type PG [20]. Besides, a new two-dimensional materials with pentagonal structure investigated such as penta-SiC_2, penta-silicene, penta-SiN_2, penta-CB_2, penta- MAs_2 (M = Ni, Pd, Pt), penta-PdX_2, and penta-PtX_2 (X = S, Se, and Te) [59–67]. Y. S. Lan et al. investigated theoretically the electronic and thermoelectric properties of penta-PdCh_2 (*Ch* = S, Se et Te) [21]. They show that 2D penta-PdCh_2 materials are semiconductors with indirect band gap of about 1.14 eV, 1.34 eV and 1.24 eV, respectively. They also reported a lower lattice thermal conductivity along the *x*-direction compared to the *y*-direction with κ_l ($PdTe_2$) < κ_l ($PdSe_2$) < κ_l (PdS_2). This trend leads to higher *ZT* value of about 2.42 along *x*-direction for the p-type $PdTe_2$ material at 3.78×10^{12} cm^{-2} carriers' concentration and room temperature [21].

Later on, another first principle study has carried on other anisotropic 2D-material named penta-silicene [68]. This 2D-materials exhibit semiconductor character with a nearly direct band gap of about 0.68 eV using HSE06 functional. As a result the penta-silicene shows a reducing Seebeck coefficient at low concentration due to the bipolar effect and thus a lower power factor PF. At 300 K, Zhibin Gao et al. estimated a maximum PF value of about 156.40 and 40.23 mW/mK^2 at 5.09×10^{13} cm^{-2} hole carriers concentration along *x*- and *y*-direction, respectively. However, the PF_{max} is 61.75 and 39.22 mW/mK^2 at 1.55×10^{13} cm^{-2} electron carriers concentration along *x*- and *y*-direction, respectively. Such higher PF value of penta-silicene for p-type doping is explained by the presence of four hole pockets and relatively flat bands in the valence band (see Fig. 20.5) [68]. Furthermore, the calculated *ZT* display an anisotropic behavior with a higher value equal to 3.43 for p-type doping along the *x*-direction and 3.04 for n-type doping along the *y*-direction at room temperature Fig. 20.6. This makes penta-silicene a promising candidate for thermoelectric application at near room temperature.

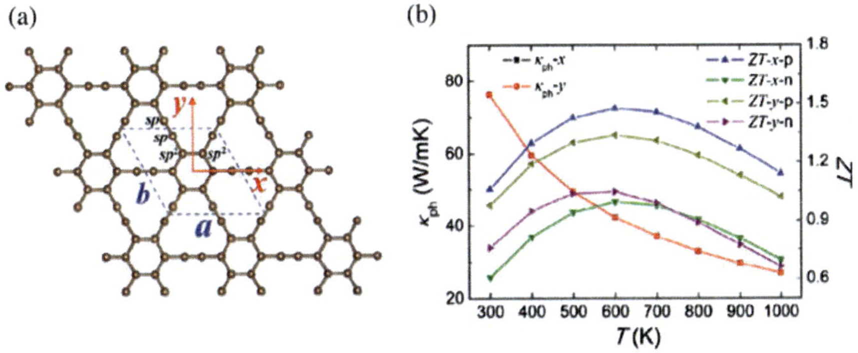

Fig. 20.4 **a** γ-graphyne structure, **b** the temperature dependence of phonon thermal conductivity of γ-graphyne along the *x*- and *y*-directions (left), and the corresponding ZT values (right). Reprinted from [52], Copyright (2017), with permission from Elsevier

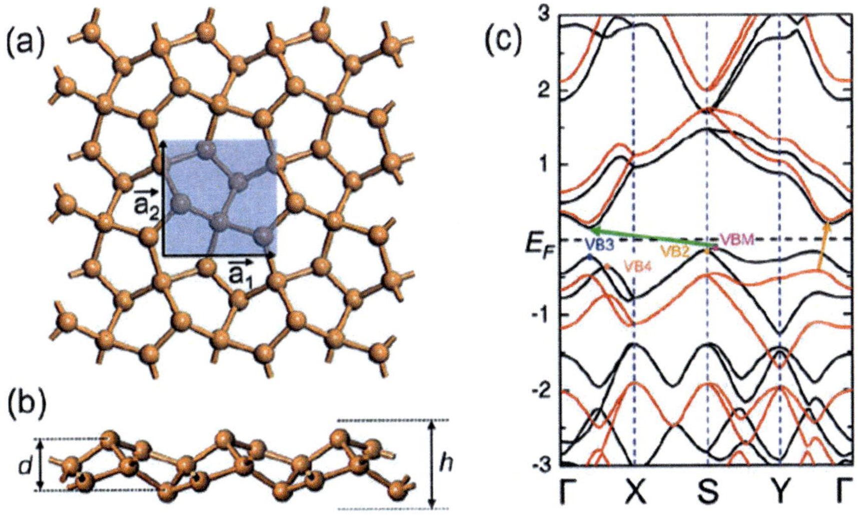

Fig. 20.5 The penta-silicene structure in a 3 × 3 supercell **a** top view and **b** side view, **c** Electronic band structure of penta-silicene using the DFT-PBE (black) and DFT-HSE06 (red) approximation. Reprinted and adapted with permission from [68]. Copyright (2020). American Chemical Society

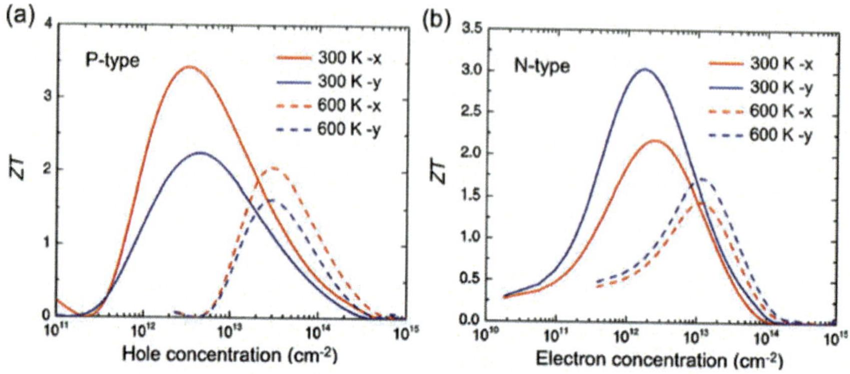

Fig. 20.6 Calculated figure of merit *ZT* of **a** p-type and **b** n-type penta-silicene as a function of the carrier concentration along x- and y-axes at 300 and 600 K. Reprinted with permission from [68]. Copyright (2020). American Chemical Society

Very recently, Tao et al. have theoretically studied a new 2D pentagonal materials based on platinum $PtCh_2$ (Ch = S, Se, and Te) [69]. Similarly to the penta-$PdCh_2$ the penta-$PtCh_2$ presents anisotropic thermoelectric properties with higher *ZT* values along the *x*-direction. Besides, the *ZT* values of this 2D materials are enhanced with the increasing temperature to attain a maximum at 600 K equal to 1.46 (0.71), 3.42 (1.33), and 5.03 (4.85) for p-type (n-type) PtS_2, $PtSe_2$, and $PtTe_2$, respectively along *x*-direction. This result suggests high thermoelectric properties of penta-$PtCh_2$ at medium temperature.

20.3.2.3 Phosphorus Family

After the exfoliation of the black-phosphoren from the bulk BP [70] the 2D phosphoren materials became more and more attractive for nanoelectronics applications [71–75]. Previously, both black-P and Blue-P are estimated to have a band gap value of about 1.59 eV and 2.73 eV, respectively by Cem Sevik and Hâldun Sevinçli using HSE06 functional [22]. These authors also pointed motivating thermoelectric performance of black and blue-P. At room temperature they reported a *ZT* value to be 0.4/0.13 in the armchair/zigzag direction for black-P and 1.25 for blue-P. Very recently, Cui et al. have studied the thermoelectric performance of black-P with a variety of thickness [23]. They estimated a maximum *ZT* value of 0.9 for bi-layer black-P (p-type) in the armchair direction, and point out the strong dependence of thermoelectric performance of black-p on layer numbers Fig. 20.7a. According to these authors these results is principally owing to the high and degenerate valence band near the Fermi level of bi-layer black-P. Besides, they show that the formation of the black-P/hBN heterostructure can improve the ZT_{max} of the black-P to be 0.65 and 1.2 at 300 K and 500 K, respectively Figs. 20.7b, 20.8.

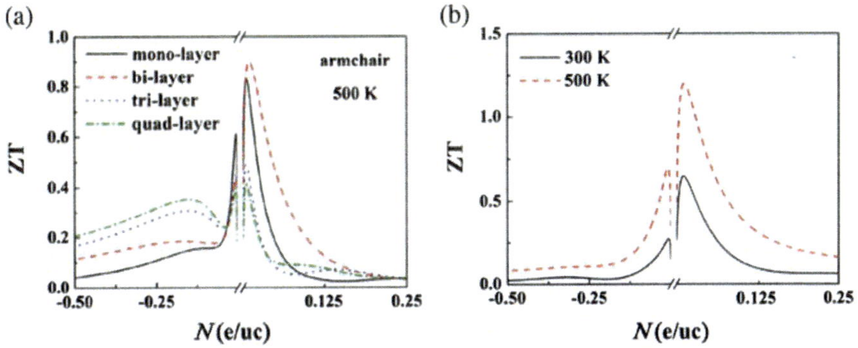

Fig. 20.7 The calculated ZT values as a function of carrier concentration along armchair direction of 2D black-P at 500 K (left panel) and black-P/h-BN heterostructure at 300 and 500 K (right panel). Reprinted from [23], Copyright (2021), with permission from Elsevier

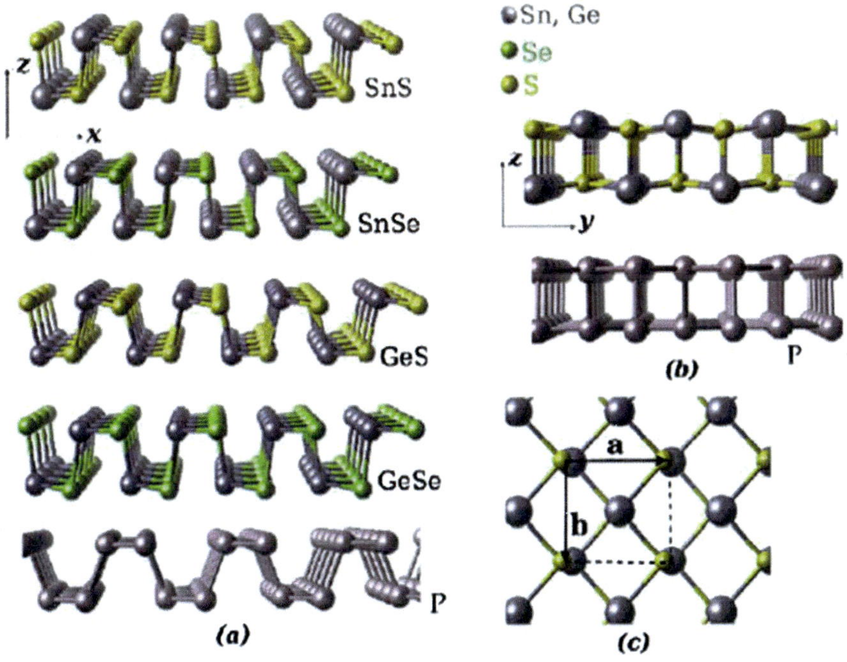

Fig. 20.8 Phosphorene-like structure of monolayers of group-IV monochalcogenides (**a**) Side views of the x–z plane for the four compounds and for phosphorene. **b** Side view of the y–z plane of SnS and phosphorene. **c** Top view of the structures. Reprinted figure with permission from [76]. Copyright (2015) by the American Physical Society

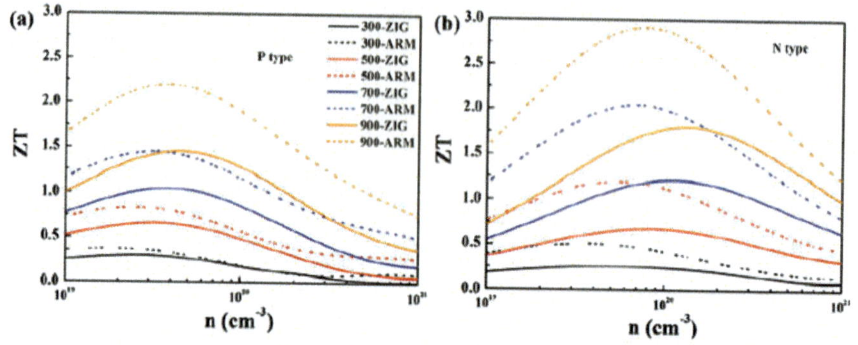

Fig. 20.9 *ZT* value of **a** p-type and **b** n-type 2D SnTe as function of carrier concentration at different temperature in armchair (dashed line) and zigzag (solid line) directions. Reprinted from [77], with the permission of AIP Publishing

Moreover, other 2D black-P analog materials with buckled structure (Fig. 20.9) have been investigated for thermoelectric applications, such as IV*Ch* (IV = Ge, Sn and *Ch* = S, Se, Te).

Using density functional theory combined with Boltzmann transport theory A. Shafique and Y. H. Shin show high thermoelectric performance of these 2D IV*Ch* (IV = Ge, Sn and *Ch* = S, Se) materials with ZT_{max} reached 1.75/1.88, 2.63/2.46, 1.85/1.29, and 1.99/1.73 for SnS, SnSe, GeS, and GeSe along armchair/zigzag direction, respectively at 700 K [24]. Later on, Li et al. investigated the thermoelectric and thermal transport of the monolayer SnTe [77]. They show that this 2D SnTe has interesting electronic transport properties owing to their specific Fermi surface and band structure. Thus, this 2D SnTe can achieved a *ZT* value of about 2.2 and 2.9 in the armchair direction for p-type and n-type doping, respectively, at 900 K Fig. 20.9.

20.3.2.4 Transition Metal Dichalcogenides and Oxides Family

The 2D transition metal dichalcogenides and oxides (TMDs and TMOs) have attracted much attention owing to theirs interesting electronic and optical properties as compared to their corresponding Bulk counterparts [78–85]. These 2D TMDs and TMOs have hexagonal 1H phase, however some of them possess another phase noted 1 T-CdI_2 [25]. Recently many researchers highlighted their fascinating thermoelectric properties [25, 86–91]. Based on DFT calculation Özbal et al. [25] have investigated on thermoelectric of 2D TMDs and TMOs materials. They have reported an exciting figure of merit value in the 1H phase of about 1.41 (1.42), 1.38 (1.17), and 1.57 (1.28) for p-type (n-type) $ZrSe_2$, HfS_2, and $HfSe_2$, respectively, at room temperature Fig. 20.10. At 800 K these values attain a maximum of about 2.96 (3.61), 3.03 (2.92), and 3.30 (3.04) for p-type (n-type) $ZrSe_2$, HfS_2, and $HfSe_2$, respectively [25]. On the other hand the 1 T phase of Zr/Hf oxides (both p-type and n-type) and n-type Zr/HfCh_2 (*Ch* = S and Se) shows considerable *ZT* value at 800 K Fig. 20.10.

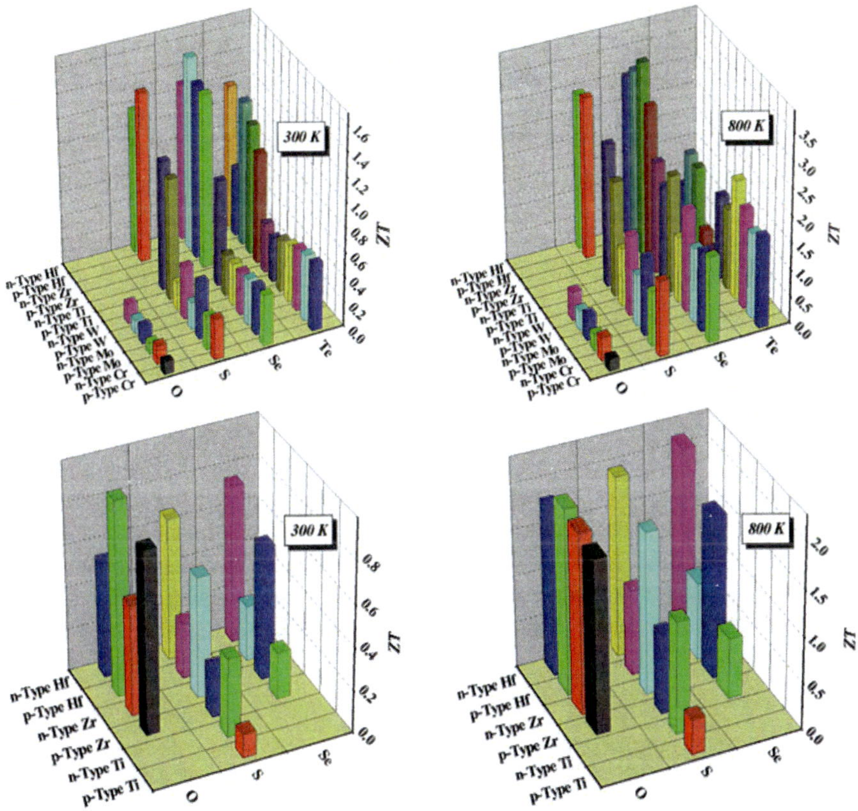

Fig. 20.10 *ZT* value of 2D transition metal dichalcogenides and oxides at 300 K and 800 K for 1H phase (top panels) and 1 T phase (bottom panels). All data are taken from [25]

Accordingly, compared to the bulk counterpart [92] the 2D TMDs and TMOs materials display fascinating thermoelectric properties makes them promising candidates for thermo-generator device.

20.3.2.5 MXenes Family

The MXenes 2D materials are generally formed by removing the A layers in MAX phases based nitride and carbides. These 2D materials display fascinating properties making them attractive for electromagnetic interference shielding [93], water purification [94], energy storage [95–97], biosensing [98], antibacterial activity [99], and ion sieving [100].

Among them, semiconducting MXenes based transition metal (M = Sc, Ti, Zr and Hf) shows promising thermoelectric properties, e.g., a $ZT_{\max}$ value of 0.5 and 0.45 is reported for p-type $Sc_2C(OH)_2$ and n-type Sc_2CF_2 and

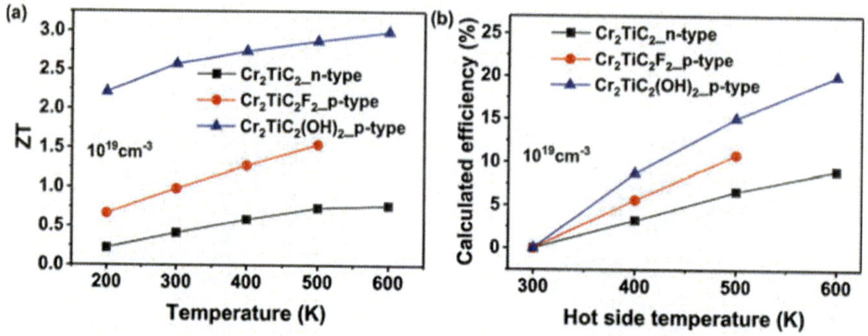

Fig. 20.11 The calculated ZT **a** and TE conversion efficiency **b** for Cr_2TiC_2 and $Cr_2TiC_2T_2$ (T= − F and − OH). Reprinted with permission from [101]. Copyright (2019). American Chemical Society

Ti_2CO_2, respectively [26, 27]. Moreover, Ziang Jing et al. have theoretically studied the thermoelectric performance of double transition metal 2D MXenes $Cr_2TiC_2T_2$ (T = − OH or − F) [101]. They estimated a *ZT* value of 2.59 and 3 for p-type $Cr_2TiC_2(OH)_2$ MXene at 300 K and 600 K, respectively Fig. 20.11. This material displays a thermoelectric conversion efficiency of 20%, showing interesting TE performance [101].

20.4 Strategies for Enhancement of Thermoelectric Figure of Merit

From above we have seen that the dimensionality play a major role to boost thermoelectric performance due to the enhanced power factor (PF = $S^2*\sigma$) caused by a sharper density of states near the Fermi level, or the reduced lattice thermal conductivity caused by the increased phonon scattering, however it's not always the case. Therefore, it is necessary to overcome this problem by using other strategy to improving *ZT* value of 2D materials. The band convergence is one of the promising solutions which can be realized by alloying, doping or using the strain method. For example, theoretically, Basant Roondhe et al. show that oxygen doped h-BN leads to two times larger PF than pure 2D h-BN [102]. This is due to the increase in the electrical conductivity resulting from the reduction of the band gap value from 4.36 eV for h-BN to 0.7 eV for BNO. Further, Lv et al. demonstrate that the thermoelectric performance of 2D phosphorene can be tuned by the applied strain [103]. They reported that the uniaxial strain improves simultaneously the Seebeck coefficient and electrical conductivity leading to *ZT* value of 1.65 for 5% zigzag-direction strain and 2.12 for 8% armchair direction strain at room temperature. Building 2D heterostructure is also an effective solution for improving thermoelectric performance due to the interfaces phenomena which leads to some new properties. Ding et al. show that making 2D $ZrSe_2$/$HfSe_2$ heterostructure can improve the ZT value to be 5.3 for n-type doping and 3.2 for p-type doping which is four times higher than those of $ZrSe_2$ and $HfSe_2$ monolayer [105]. In addition, Cui et al. show that the formation of the black-P/hBN heterostructure can improve the ZT value of the black-P to be 0.65 and 1.2 at 300 K and 500 K, respectively [23].

20.5 Conclusion

The thermoelectric materials have the ability to convert heat directly into electricity or vise versa. This class of materials has been known for over a century, but the limited efficiency has slowed down their development. In this review, we have seen that to overcome this problem a powerful solution is adopted which is the reduction of dimensionality. This strategy plays a major role to boost thermoelectric performance due to the enhanced power factor PF caused by a sharper density of states near the Fermi level, or the reduced lattice thermal conductivity caused by the increased phonon scattering. Consequently, new 2D materials such as graphene family, phosphorus family, penta-graphene family, MXenes family, and transition metal dichalcogenides and oxides family have been developed and designed to be the next-generation high-performance thermoelectric materials. However, more efforts are needed to understand the electrical and phonon transport mechanism in 2D materials. Although the power factor is enhanced, it is still difficult to decouple the different transport coefficients. Besides, it is still challenging to manufacture 2D thermoelectric materials for future practical applications.

References

1. H.J. Goldsmid, R.W. Douglas, British. J. Appl. Phys. **5**(11), 386–390 (1954)
2. L.D. Hicks, T.C. Harman, M.S. Dresselhaus, Appl. Phys. Lett. **63**, 3230 (1993)
3. L.D. Hicks, M.S. Dresselhaus, Phys. Rev. B **47**, 12727 (1993)
4. M.S. Dresselhaus, G. Chen, M.Y. Tang, R. Yang, H. Lee, D. Wang, Z. Ren, J.P. Fleurial, P. Gogna, Adv. Mater. **19**, 1043–1053 (2007)
5. K.S. Novoselov, A.K. Geim, S.V. Morozov, D. Jiang, Y. Zhang, S.V. Dubonos, I.V. Grigorieva, A.A. Firsov, Science **306**(5696), 666–669 (2004)
6. A.K. Geim, K.S. Novoselov, Nat. Mater. **6**, 183 (2007)
7. A.C. Neto, F. Guinea, N.M. Peres, K.S. Novoselov, A.K. Geim, Rev. Mod. Phys. **81**, 109 (2009)
8. A.K. Geim, Science **324**, 1530 (2009)
9. F.Q. Wang, S. Zhang, J. Yu, Q. Wang, Nanoscale **7**, 15962 (2015)
10. T. Zhao, S. Zhang, Y. Guo, Q. Wang, Nanoscale **8**, 233 (2016)
11. S. Zhang, J. Zhou, Q. Wang, P.J. Jena, Phys. Chem. C **120**, 3993 (2016)
12. C. Zhang, J. Sun, Phys. Chem. Lett. 2664 (2016)
13. J. Liu, Q. Sun, Y. Kawazoe, P. Jena, Phys. Chem. Chem. Phys. **18**, 8777 (2016)
14. X. Li, S. Zhang, F.Q. Wang, Y. Guo, J. Liu, Q. Wang, Phys. Chem. Chem. Phys. **18**, 14191 (2016)

15. Y. Guo, S. Zhang, T. Zhao, Q. Wang, Nanoscale **8**, 10598 (2016)
16. S. Zhang, J. Zhou, Q. Wang, X. Chen, Y. Kawazoe, P. Jena, Proc. Natl. Acad. Sci. **112**, 2372 (2015)
17. S. Zhang, Y. Li, T. Zhao, Q. Wang, Sci. Rep. 4 (2014)
18. D. Li, Y. Gong, Y. Chen, J. Lin, Q. Khan, Y. Zhang, Y. Li, H. Zhang, H. Xie, Nano-Micro Lett. **12**, 36 (2020)
19. X. Tan, H. Shao, T. Hu, G. Liu, J. Jiang, H. Jiang, Phys. Chem. Chem. Phys. **17**, 22872–22881 (2015)
20. C. P. Chen, C. Liu, L. L. Liu, L. S. Zhao and X. C. Wang, Mater. Res. Express, **4**, 105031 (2017)
21. Y. Lan, X. Chen, C. Hu, Y. Cheng, Q. Chen, J. Mater. Chem. A. **7**, 11134–11142 (2019)
22. C. Sevik, H. Sevinçli, Nanotechnology **27**, 355705 (2016)
23. Y. F. Cui, S. Duan, X. Chen, M. M. Yang, B. C. Yang, W. C. Yi, X. B. Liu, Vacuum 183, 109790 (2021)
24. A. Shafique, Y.H. Shin, Sci. Rep. **7**, 506 (2017)
25. G. Özbal, R.T. Senger, C. Sevik, H. Sevinçli, Phys. Rev. B **100**, 85415 (2019)
26. S. Kumar, U. Schwingenschlögl, Phys. Rev. B: Condens. Matter Mater. Phys. **94**(3), 035405 (2016)
27. A.N. Gandi, H.N. Alshareef, U. Schwingenschlögl, Chem. Mater. **28**(6), 1647–1652 (2016)
28. T.J. Seebeck, Abhandlungen der Deutschen Akademie der Wissenschafren zu Berlin **265**, 1822–1823 (1821)
29. J.C. Peltier, Ann. Chim. LVl, 371 (1834)
30. Z.G. Chena, G. Hana, L. Yanga, L. Chenga, J. Zou, Progress in Natural Science: Materials International **22**(6), 535–549 (2012)
31. M. S. Dresselhaus, Y.-M. Lin, S. B. Cronin, O. Rabin, M. R. Black, G. Dresselhaus, in *Recent Trends in Thermoelectric Materials Research III SEMICONDUCTORS AND SEMIMETALS*, vol. 71, ed. By Terry M. Tritt (Academic Press, San Diego, 2001), pp. 1–8
32. L.D. Hicks, T.C. Harman, X. Sun, M.S. Dresselhaus, Phys. Rev. B **53**, R10493–R10496 (1996)
33. P. Hohenberg, W. Kohn, Phys. Rev. B **136**, 864 (1964)
34. P. B. Allen, in *Quantum Theory of Real Materials,* ed. By J. R. Chelikowsky, S. G. Louie (Kluwer, Boston, 1996), pp. 219–250
35. J. Yu, Q. Sun, P. Jena, Proceedings SPIE 10174, International Symposium on Clusters and Nanomaterials, 101740C, 21 December 2016. https://doi.org/10.1117/12.2267739
36. H. Lu, S.D. Li, J. Mater. Chem. C **1**(23), 3677–3680 (2013)
37. Y. Li, L. Xu, H. Liu, Y. Li, Chem. Soc. Rev. **43**(8), 2572–2586 (2014)
38. Q. Song, B. Wang, K. Deng, X. Feng, M. Wagner, J.D. Gale, K. Müllen, L. Zhi, J. Mater. Chem. C **1**(1), 38–41 (2013)
39. X. L. Sheng, H. J. Cui, F. Ye, Q. B. Yan, Q. R. Zheng, G. Su, J. Appl. Phys. **112**(7), 074315 (2012)
40. B.R. Sharma, A. Manjanath, A.K. Singh, Sci. Rep. **4**(1), 1–6 (2014)
41. X. Chen, Y. Huang, J. Liu, H. Yuan, H. Chen, ACS Omega **4**(18), 17773–17781 (2019)
42. M. Yeganeh, F. Kafi, A. Boochani, Superlattices Microstruct. **138**, 106367 (2020)
43. M. Naseri, D.M. Hoat, J.F. Rivas-Silva, G. H. Cocoletzi, Optik – Int. J. Light Electron Optics **210**, 164567 (2020)
44. Z. Z. Zhou, H. J. Liu, D. D. Fan, G. H. Cao, J. Phys.: Condens. Matter **31**, 385701 (2019)
45. N. R. Abdullah, G. A. Mohammed, H. O. Rashid, V. Gudmundsson, Phys. Lett. A **384**, 126578 (2020)
46. A. Yari, A. Boochani, S. Rezaee, Phil. Mag. **100**(24), 3108–3124 (2020)
47. A. Bafekry, M. Yagmurcukardes, B. Akgenc, M. Ghergherehchi, B. Mortazav, Phys. Chem. Chem. Phys. (2021). https://doi.org/10.1039/D1CP01183A
48. B. Ul Haq, S. AlFaify, T. Alshahrani, R. Ahmed, Q. Mahmood, D. M. Hoat, S.A. Tahir, Phys. E Low-dimensional Syst. Nanostruct. **126**, 114444 (2021)
49. X. Jiang, C. Ban, L. Li, C. Wang, W. Chen, X. Liu, AIP Advances **11**, 055120 (2021)
50. P. Kumar, K. Rajput, D.R. Roy, Phys. E **127**, 114523 (2021)

51. N. Narita, S. Nagai, S. Suzuki, K. Nakao, Phys. Rev. B **58**(16), 11009–11014 (1998)
52. P.H. Jiang, H.J. Liu, L. Cheng, D.D. Fan, J. Zhang, J. Wei, J.H. Liang, J. Shi, Carbon **113**, 108–113 (2017)
53. J. Wang, Z. Wang, R.J. Zhang, Y.X. Zheng, L.Y. Chen, S.Y. Wang, C.C. Tsoo, H.J. Huangd, W.S. Su, Phys. Chem. Chem. Phys. **20**, 18110–18116 (2018)
54. T. Stauber, J.I. Beltran, J. Schliemann, Sci. Rep. **6**, 22672 (2016)
55. B. Rajbanshi, S. Sarkar, B. Mandal, P. Sarkar, Carbon **100**, 118 (2016)
56. H. Einollahzadeh, R.S. Dariani, S.M. Fazelib, Solid State Commun. **229**, 1 (2016)
57. Z. Wang, F. Dong, B. Shen, R.J. Zhang, Y.X. Zheng, L.Y. Chen, S.Y. Wang, C.Z. Wang, K.M. Ho, Y.-J. Fan, B.-Y. Jin, W.-S. Su, Carbon **101**, 77 (2016)
58. H. Liu, G. Qin, Y. Lin, M. Hu, Nano. Lett. **16**, 3831–3842 (2016)
59. A. Lopez-Bezanilla, P.B. Littlewood, The. J. Phys. Chem. C **119**, 19469–19474 (2015)
60. F. Li, K. Tu, H. Zhang, Z. Chen, Phys. Chem. Chem. Phys. **17**, 24151–24156 (2015)
61. S. Deng, L. Li, Y. Zhang, A.C.S. Appl, Nano Mater. **1**, 1932–1939 (2018)
62. J. Lin, S. Zuluaga, P. Yu, Z. Liu, S.T. Pantelides, K. Suenaga, Phys. Rev. Lett. **119**, 016101 (2017)
63. E. Li, D. Wang, P. Fan, R. Zhang, Y.-Y. Zhang, G. Li, J. Mao, Y. Wang, X. Lin, S. Du, H.-J. Gao, Nano Res. **11**, 5858–5865 (2018)
64. A.A. Puretzky, A.D. Oyedele, K. Xiao, A.V. Haglund, B.G. Sumpter, D. Mandrus, D.B. Geohegan, L. Liang, 2D Mater. **5**, 035016 (2018)
65. D. Qin, P. Yan, G. Ding, X. Ge, H. Song, G. Gao, Sci. Rep. **8**, 2764 (2018)
66. M. Sun, J.-P. Chou, L. Shi, J. Gao, A. Hu, W. Tang, G. Zhang, ACS Omega **3**, 5971–5979 (2018)
67. Q.L. Wei, H.Y. Yang, Y.Y. Wu, Y.B. Liu, Y.H. Li, Nanomaterials **10**, 2043 (2020)
68. Z. Gao, J.S. Wang, A.C.S. Appl, Mater. Interfaces **12**, 14298–14307 (2020)
69. W.L. Tao, Y.Q. Zhao, Z.Y. Zeng, X.R. Chen, H.Y. Geng, A.C.S. Appl, Mater. Interfaces **13**(7), 8700–8709 (2021)
70. J. Qiao, X. Kong, Z.X. Hu, F. Yang, W. Ji, Nat. Commun. **5**, 4475 (2014)
71. X. Wang, A.M. Jones, K.L. Seyler, V. Tran, Y. Jia, H. Zhao, H. Wang, L. Yang, X. Xu, F. Xia, Nat. Nano. **10**, 517–521 (2015)
72. V. Tran, R. Soklaski, Y. Liang, L. Yang, Phys. Rev. B **89**, 235319 (2014)
73. J. Xiao, M. Long, X. Zhang, J. Ouyang, H. Xu, Y. Gao, Sci. Rep. **5**, 9961 (2015)
74. X.B. Li, P. Guo, T.F. Cao, H. Liu, W.M. Lau, L.M. Liu, Sci. Rep. **5**, 10848 (2015)
75. J.S. Kim, Y. Liu, W. Zhu, S. Kim, D. Wu, L. Tao, A. Dodabalapur, K. Lai, D. Akinwande, Sci. Rep. **5**, 8989 (2015)
76. Lídia C. Gomes, A. Carvalho, Phys. Rev. B **92**, 085406 (2015)
77. Y. Li, M.N. Wu, T. Ding, K. Ma, F.S. Liu, W.Q. Ao, J.Q. Li, Appl. Phys. Lett. **114**, 083901 (2019)
78. Q.H. Wang, K. Kalantar-Zadeh, A. Kis, J.N. Coleman, M.S. Strano, Nat. Nanotechnol. **7**, 699 (2012)
79. K.F. Mak, J. Shan, Nat. Photonics **10**, 216 (2016)
80. F.K. Perkins et al., Nano. Lett. **13**, 668–673 (2013)
81. D. Voiry, J. Yang, M. Chhowalla, Adv. Mater. **28**, 6197–6206 (2016)
82. J. Lee et al., Adv. Mater. **29**, 1606667 (2017)
83. A. Arab, Q. Li, Sci. Rep. **5**, 13706 (2015)
84. J.Y. Oh et al., Energy Environ. Sci. **9**, 1696–1705 (2016)
85. M.J. Mleczko et al., Sci. Adv. **3**, e1700481 (2017)
86. D. Fan, H. Liu, L. Cheng, P. Jiang, J. Shi, X. Tang, Appl. Phys. Lett. **105**(13), 133113 (2014)
87. K.X. Chen, Z.Y. Luo, D.C. Mo, S.S. Lyu, Phys. Chem. Chem. Phys. **18**(24), 16337–16344 (2016)
88. C. Wan, X. Gu, F. Dang, T. Itoh, Y. Wang, H. Sasaki, M. Kondo, K. Koga, K. Yabuki, G.J. Snyder, R. Yang, K. Koumoto, Nat. Mater. **14**, 622–627 (2015)
89. J. Wang, Sci. Rep. **7**, 41418 (2017)
90. D. Singh, R. Ahuja, A.C.S. Appl, Energy Mater. **2**, 6891–6903 (2019)

91. J. Tseng, X. Luo, J. Phys. Chem. Solids **139**, 109322n (2020)
92. M. Abdulsalam, E. Rugut, D.P. Joubert, Mater. Today Commun. **25**, 101434 (2020)
93. F. Shahzad, M. Alhabeb, C.B. Hatter, B. Anasori, S.M. Hong, C.M. Koo, Y. Gogotsi, Science **353**, 1137–1140 (2016)
94. Q.R. Zhang, J. Teng, G.D. Zou, Q.M. Peng, Q. Du, T.F. Jiao, J.Y. Xiang, Nanoscale **8**, 7085–7093 (2016)
95. M.R. Lukatskaya, O. Mashtalir, C.E. Ren, Y. DallAgnese, P. Rozier, P.L. Taberna, M. Naguib, P. Simon, M.W. Barsoum, Y. Gogotsi, Science **341**, 1502–1505 (2013)
96. O. Mashtalir, M.R. Lukatskaya, M.Q. Zhao, M.W. Barsoum, Y. Gogotsi, Adv. Mater. **27**, 3501–3506 (2015)
97. Z. Ling, C.E. Ren, M.Q. Zhao, J. Yang, J.M. Giammarco, J.S. Qiu, M.W. Barsoum, Y. Gogotsi, Proc. Natl. Acad. Sci. U.S.A. **111**, 16676–16681 (2014)
98. H. Liu, C.Y. Duan, C.H. Yang, W.Q. Shen, F. Wang, Z.F. Zhu, Sens. Actuators B **218**, 60–66 (2015)
99. K. Rasool, M. Helal, A. Ali, C.E. Ren, Y. Gogotsi, K.A. Mahmoud, ACS Nano **10**, 3674–3684 (2016)
100. C.E. Ren, K.B. Hatzell, M. Alhabeb, Z. Ling, K.A. Mahmoud, Y. Gogotsi, J. Phys. Chem. Lett. **6**, 4026–4031 (2015)
101. Z. Jing, H. Wang, X. Feng, B. Xiao, Y. Ding, K. Wu, Y. Cheng, J. Phys. Chem. Lett. **10**, 5721–5728 (2019)
102. B. Roondhe, V. Sharma, H.L. Kagdada, D.K. Singh, T.S. Dasgupta, R. Ahuja, Appl. Surface Sci. **533**, 147513 (2020)
103. H.Y. Lv, W.J. Lu, D.F. Shao, Y.P. Sun, Phys. Rev. B **90**, 085433 (2014)
104. G. Ding, C. Wang, G. Gao, K. Yao, C. Dun, C. Feng, D. Li, G. Zhang, Nanoscale **10**, 7077–7084 (2018)

Chapter 21
Reversible DC Electric Field Modification of Optical Properties of CdTe Nanocrystals

Rabia Ince, Melda Patan Alper, and Mehmet Hikmet Yukselici

Abstract Modifiable nonlinear optical properties have the potential for innovative device applications. Quantum dot systems have an energy level system that is modified under external electric fields, modifying their optical properties. The matrix surrounding the quantum dot systems influences this optical modification. In this work absorption spectra of cadmium telluride quantum dot systems in solution and solid glass matrixes were studied under increasing applied D.C electric fields between 0.25 and 5 kV/cm. Quantum dot systems in both solid glass and liquid solution matrixes gave rise to electro-optic effects, which modified their optical properties (absorption spectra, bandgap energy and refractive index). The effect is reversible, so that removal of the electric field returns the original electro absorption spectra. This work showed all nanoparticle samples except one displayed a linear decrease of band gap energy with electric field; this amounted to equivalent changes in quantum dot radii of 0.02 nm per kV in glass matrixes and about 0.03 nm per kV for solution matrixes. Evidence of improved mono-dispersity was also observed.

21.1 Introduction

Optical absorption is a nonlinear optical property whose perturbation causes refractive index modifications. This has the potential for device applications in far-infrared

R. Ince (✉)
Diamond Light Source, Harwell Science and Innovation Campus, Didcot OX11 ODE, United Kingdom
e-mail: rabia.ince@diamond.ac.uk

R. Ince · M. P. Alper
Yeditepe University, Istanbul, Turkey
e-mail: mpatan@yeditepe.edu.tr

M. H. Yukselici
Yıldız Technical University, Istanbul, Turkey
e-mail: yukseli@yildiz.edu.tr

H. Ünlü and N. J. M. Horing (eds.), *Progress in Nanoscale and Low-Dimensional Materials and Devices*, Topics in Applied Physics 144,
https://doi.org/10.1007/978-3-030-93460-6_21

laser amplifiers, photodetectors, and high speed electro-optical modulators. Variable refractive index, energy bandgap and optical absorption are examples of nonlinear optical properties utilized in innovative optical devices such as distributed feedback lasers.

Quantum dots offer an energy level system that can be modified by external conditions, such as electric field and surrounding matrix, leading to changes of their optical properties. Electric field modification of nanoparticle nonlinear optical properties is unique and offers a simple, low-cost method of controlling their optical properties. The literature discusses the theoretical effect of electric fields on CdTe quantum dot systems [1–3] for the case of a spherical quantum-dot subjected to an external perpendicular electric field in the presence of an impurity.

In this work an experimental investigation of the effects of high voltage electric fields (0.25–5 kV/cm) on the absorption coefficients and bandgap energies of a range of prepared samples of cadmium telluride nanoparticles in both solid glass matrixes and solution matrixes has been carried out.

21.1.1 Electro-Optic Modification of Spectra

High voltage DC and AC electric fields perturb the energy levels of isolated atoms, leading to a shift in the absorption wavelength of the atom. Semiconductor nanoparticles exhibiting atom-like quantized energy states are perturbed in the presence of electric fields. The application of an electric field to quantum dot systems leads to electro-optic modulation of absorption spectra attributed to the quantum confined Stark effect, (QCSE) [4–10]. Figure 21.1 shows the QCSE mechanism and outcome of this effect. The electrons in the valence band can be excited by less energy than the band gap energy according to the red-shifted band gap energy, $E' = E_g - qxF$ where q is the charge per carrier, x position and F the electric field applied.

QCSE is caused when electrons and holes are forced in opposite directions by the applied field, decreasing wave function overlap and hence absorption peaks; a red shift of the energy band gap also occurs. In this study red shifts were observed alongside a decrease in absorption magnitude, which are characteristic of the QCSE.

21.2 Theory

When the radius of a semiconductor sphere approaches a few nanometres, Bohr radii of their excitons becomes larger than the sphere, efficiently confining them. This quantum size effect causes electron–hole pairs to have discrete transition energies greater than their band gap energy, E_g, by a size-dependent energy shift, $\Delta E_{n,l}$, which radically changes the chemical, physical, optical and electric properties of the whole quantum system.

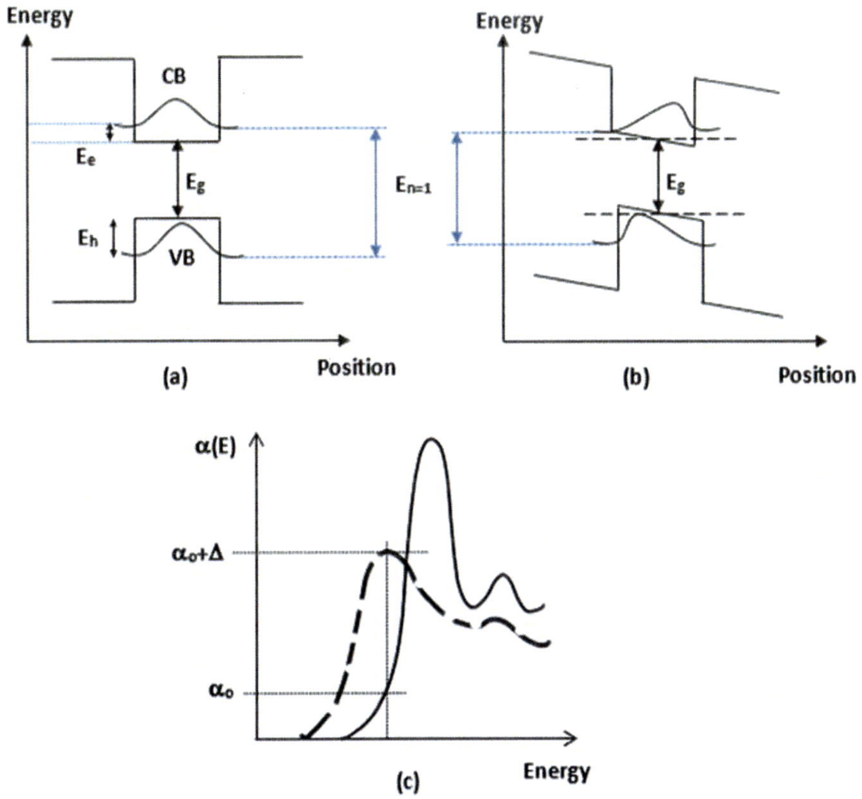

Fig. 21.1 E_g = bulk band gap, E_e = electron energy, E_h = hole energy, absorbed energy = E_g + E_e + E_h (**a**) wave function for n = 1 level of a quantum confined system at zero applied electric field (**b**) wave function of n = 1 level of the system is perturbated by applied electric field causing red shift (decrease in transition energy $E_{n=1}$), reduction in confinement energy, and reduction in wavefunction overlap so peaks don't coincide causing intensity decrease (**c**) Change in absorption spectrum after application of electric field (dashed line) [11]

The effective mass approximation [12–15] of (21.1) relates bandgap in eV to the quantum system properties. It is useful to compare energy bandgap modification to equivalent nanoparticle size modification with applied DC electric fields across nanoparticle samples.

$$E_{n,l}{}^{*} = \frac{c^{*}h}{\lambda_{onset}} = E_g^{bulk} + \frac{h^2}{8r^2}\left(\frac{1}{m_e{}^{*}} + \frac{1}{m_h{}^{*}}\right) - \frac{1.786e^2}{4\pi\varepsilon\varepsilon_o r} \tag{21.1}$$

where, r is the particle radius, $m_e{}^{*}$ is the effective mass of excited electron, $m_h{}^{*}$ is the effective mass of excited hole, ε is the relative permittivity, ε_0 is the permittivity of free space, and e is the electronic charge. The first term represents bulk band gap and the second confinement energy. The third term represents Coulomb interaction, polarization is negligible here, so a term to represent it is omitted. More terms should be added to the right side such as shift in energy if the temperature or pressure on the sample change during the measurement as well as shift due to electric field applied, as in this study. The electric field induced shift ΔE in the band edge is given by the equation [16, 17]

$$\Delta\mathrm{E(meV)} = -24 \times 10^{-7}\ \mathrm{M}\ \ [\mathrm{F(kV/cm)}]^2[\mathrm{R_{ave}(nm)}]^4 \tag{21.2}$$

where M is the translational mass of the electron–hole pair in units of electron rest mass and F is the electric field.

21.2.1 Size Distribution of the Crystals n(R)

Absorption spectra were recorded of all nanoparticle systems, allowing bandgap energies, $E_{n,l}$ to be determined from extrapolation of linearized Tauc plots [18–20]. This allowed an estimation of equivalent nanoparticle radius r (from (21.1)) for samples with and without H.V. fields, these could then be compared with their band gap modifications.

Absorption edges of the samples in this work were broader than bulk crystals due to the size distribution of the nanocrystals $n(R)$. This is an important quantity because monodispersed samples are desirable. In the dilute concentration limit, $n(R)$ can be obtained by assuming absorbance A at any wavelength is related to the total volume, V, of spherical particles with radius greater than or equal to the size corresponding to the absorption onset [21]:

$$A(r) \propto \int_r^{\infty} \frac{4}{3}\pi R^3 n(R) dr \tag{21.3}$$

Since $n(\mathrm{R}) = 0$ when $R \to \infty$

$$n(R) \propto -\frac{dA/dR}{V} \tag{21.4}$$

Particle size distributions were obtained from the absorption spectrum using (21.1) and (21.4). The derivative of the absorption curve was obtained by converting the wavelength axis to energy units, allowing radii to be extracted via (21.1). The gradients at each radius were substituted into (21.4) to determine size distributions from the full width at half maximum (FWHM) for the sample spectra.

21.3 Sample Preparation

21.3.1 Semiconductor Nanoparticles Grown in Glass Matrixes

CdTe colloidally coloured Schott optical filter glass samples (RG830) were heat treated over three temperature and time cycles to enhance diffusion of the Cd^{2+} and Te^{2-} ions as shown in Table 21.1. This caused CdTe quantum dots to form which had temperature–time dependent dimensions. Three samples were obtained from the same glass matrix which were melted at 1100 °C for 15 min to dissolve the colloids, these were quenched at room temperature. The samples subsequently underwent a two-stage heat treatment: at a temperature below the glass transition temperature to initiate nucleation, followed by annealing at 660 °C using a furnace of well-known temperature profile to allow the nanoparticles to grow.

21.3.2 Synthesis of Semiconductor Nanoparticles and Their Absorption Spectra

Nanoparticles in liquid matrixes were obtained through a synthesis route developed by trial-and-error modification of literature methods [22, 23] into a two-stage process using a NaHTe solution precursor. Sodium borohydride (1.5312 g) and Te (0.3828 g) powder were mixed in a 1:4 ratio with HPLC water (doubly distilled) which was left to cool for 24 h until a clear purple colour change occurred. This was then used for the second stage of the synthesis.

Cadmium chloride (0.2 g) was dissolved in 150 ml of HPLC water in a round bottomed flask, and the PH adjusted to 7–8 by the addition of sodium hydroxide with a syringe. The resulting mixture was then heated using a controller (hot plate with magnetic stirrer) until it reached a steady temperature of 98 °C. Mercaptropropionic acid (0.2 l) was quickly added into the mixture using a syringe, followed by the injection of 0.3 mmol of the clear purple precursor. This changed its colour to yellow on maintaining a constant temperature for 20 min.

The first quantum dot sample was then extracted using a syringe and deposited in a test tube. As the solution gradually changed colour through red, orange, brown over several hours, six further samples were extracted every hour. Each was centrifuged at

Table 21.1 Solid nanoparticle samples and annealing times

Sample matrix	Sample no	Second stage heat treatment time (hr)
Glass	1	3
Glass	2	8
Glass	3	12

6000 rpm for 15 min to remove excess surfactants; their peak absorption wavelengths were immediately determined with a uv–vis spectrometer through transmission measurement. The spectra are shown in Fig. 21.2.

The blue shift was observed in the onset of the asymptotic optical absorption for both sets due to the quantum size effect in Fig. 21.2a and b.

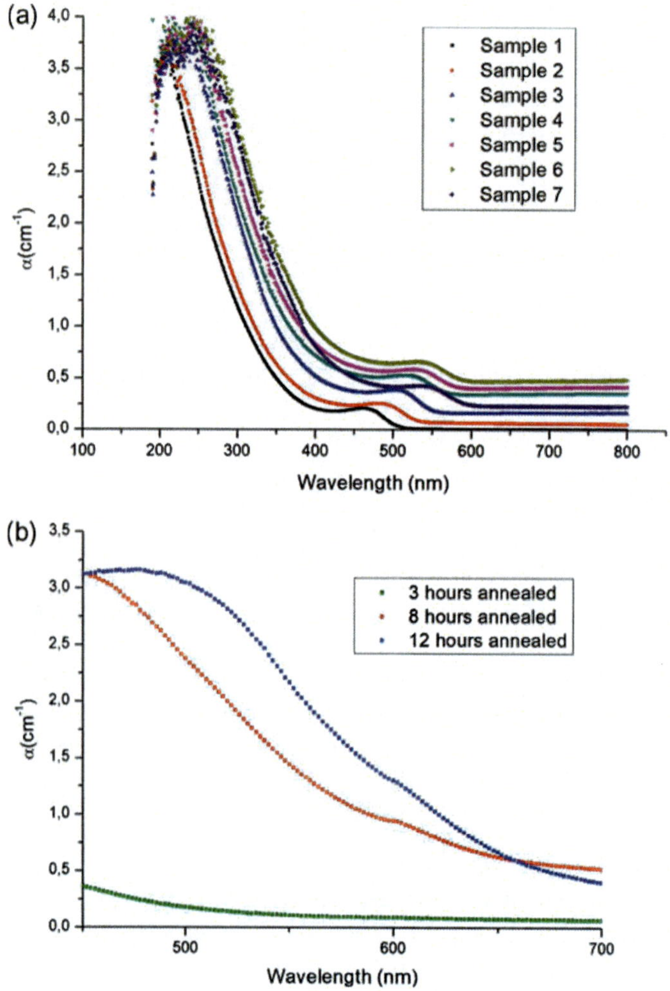

Fig. 21.2 Absorption. spectra of nanoparticles in (**a**) solution samples and (**b**) glass matrixes

21.4 Measurement of Spectra

The spectra of the solid and liquid samples were measured at zero field and several non-zero applied fields to determine their modification of band gap energies $E_{n,l}$, using Tauc plots [18–20] and equivalent nanoparticle radii (21.1). A secondary goal was to determine the effect of matrix type (solid and liquid) on nanoparticle properties. These observations provide information on controlling the non-linear properties of nanoparticles.

21.4.1 Sample Rigs for Electro Optical Measurements

Solid samples of varying thickness (0.9–0.14 mm) containing nanoparticles of varying size were sandwiched between two indium-tin-oxide (ITO) covered glass slides. Tiny spots of epoxy were carefully applied to their edges, as shown in Fig. 21.3 and the structures allowed to dry overnight. Optical transmission spectra of the sample were recorded with and without high electric DC fields (0.25–5 kV/cm).

To perform the electro-optical measurements on solution samples, the quantum dot solutions were placed inside a 1 cm quartz cell containing copper electrodes as shown in Fig. 21.4. Optical transmission spectra of the samples were recorded with and without high electric DC fields (0.25–5 kV/cm).

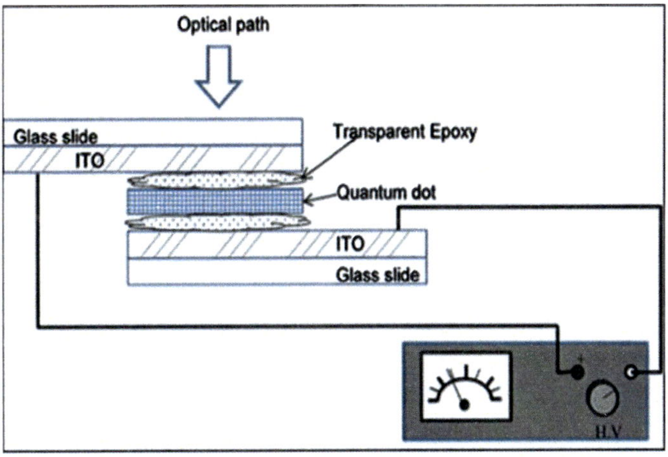

Fig. 21.3 Sample geometry and electrical/optical configurations for solid glass samples

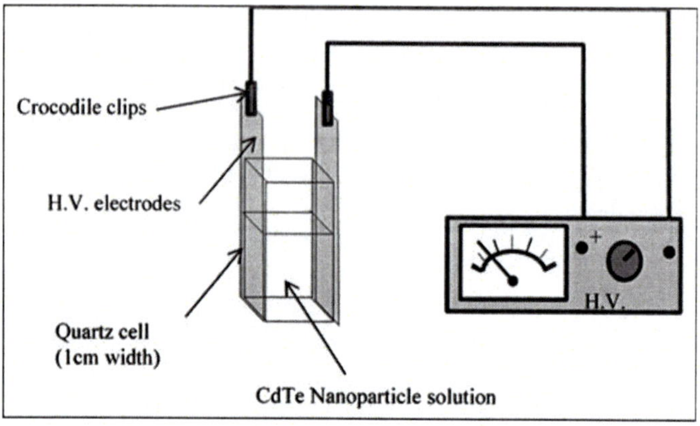

Fig. 21.4 Electrical/optical configurations for solution samples

21.4.2 Determination of Optical Band Gap from Tauc Plots Under H.V. Fields

Figure 21.5 shows the home-made absorption measurement system, constructed to allow samples to be simultaneously present in both the optical path and the H.V. field. Using the test rigs, it was adapted for measuring both quantum dots in glass and quantum dots in solution using a monochromator [Oriel, Cornerstone 130 motorized

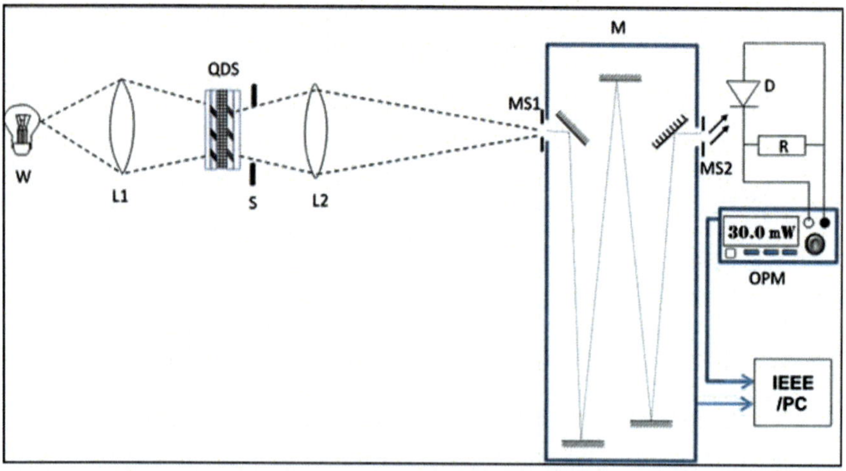

Fig. 21.5 Optical spectrometer set-up for spectral measurements. W: 50 W Tungsten lamp, L1, L2 lenses, QDS (quantum dot samples), S: slit, M: monochromator, MS1: monochromator entrance slit, MS2: monochromator exit slit, D: silicon detector, R: load, OPM: optical power meter, PC: personal computer and IEEE

monochromator 1/8 m, 1200 lines/mm grating]. Electric fields were applied along axes perpendicular to the optical path in the range 0.25–5.5 kV/cm for solid samples and 0.25–2.5 kV/cm for solution samples.

Samples under electric field might be modelled as a resistor and capacitor connected in parallel. As the voltage is applied, the capacitor is quickly charged, with some current passing through the resistor. Since the conductivity of water is ~6 orders of magnitude greater than that of glass, the electrical breakdown occurs at a lower potential in water than glass. This means the maximum voltage that can be applied is 2.5 kV for water samples and 5.5 kV for glass samples.

Light from a 50 W tungsten lamp was focused onto a 2 mm diameter area on the sample. Transmitted light was collected and re-focused at the entrance slit of the monochromator. During source wavelength scanning from 450 to 900 nm, transmitted light intensity was transferred to a computer via an optical power meter.

A reference measurement was recorded (I_o) without any sample, followed by transmission spectra with each sample. Absorption coefficients (α) were determined from Lambert–Beer's law [24] as shown in (21.5), where I is transmitted light intensity and, d is nanoparticle sample thickness.

$$\alpha d = \ln(I_o/I) \tag{21.5}$$

Wavelengths from the spectrometer were converted into energy units during the scan, allowing absorption coefficient to be plotted against source energy. The optical band gap of semiconductor samples was then determined using the Tauc equation [18–20], shown in (21.6), linearized to (21.7), so that Tauc plots could be used to extract band gap energy, E_{g}, as shown in Fig. 21.6.

$$\alpha E = A(E - E_g)^n \tag{21.6}$$

$$(\alpha h\upsilon)^2 = A^2 h\upsilon - A^2 E_g \tag{21.7}$$

where A is a constant, υ is photon frequency, and E_g is optical band gap with $n = ½$ for CdTe.

Equivalent radii (r) were determined by substitution of E_g into (21.1). Relative permittivity, ε, in (21.1) were extracted from Hilbert transforms of absorption spectra of each sample [25–27].

21.5 Results

All samples displayed absorption peaks red shifts and post absorption edge oscillations, characteristic of QCSE (as shown in Fig. 21.1).

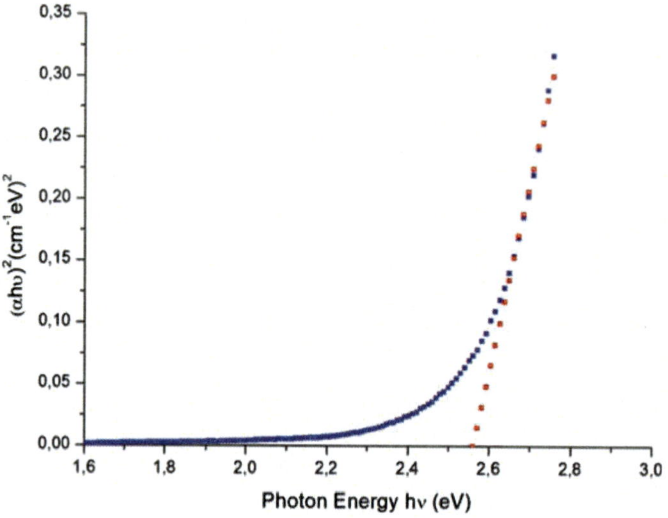

Fig. 21.6 Tauc plot of $(\alpha h\upsilon)^2$ versus $h\upsilon$ for a representative CdTe Solid Sample without field (dashed line extrapolates for a linear fit to obtain optical bandgap)

21.5.1 Effect of Annealing Time on Quantum Dot Growth

Bandgaps (E_g) versus annealing times in solution and glass matrixes are plotted in Fig. 21.7, showing nanoparticle growth as logarithmic functions of time (t). Nanoparticles in solution samples appear earlier, and grow faster, than those in glass matrixes [28]. Nanoparticles in glass matrixes were never observed until at least 2 h after annealing.

Figure 21.8 show equivalent radii for these nanoparticle bandgaps (from (21.1)) at no field (E = 0) and 1 kV (E = 1 kV) applied field conditions.

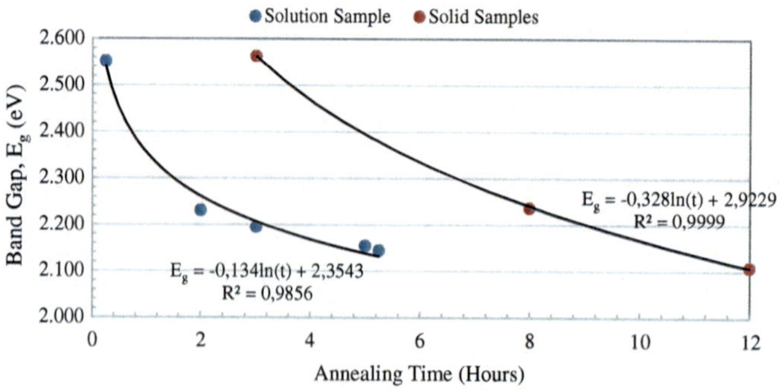

Fig. 21.7 Typical best fits of band gap energy versus annealing time for a solid and solution sample

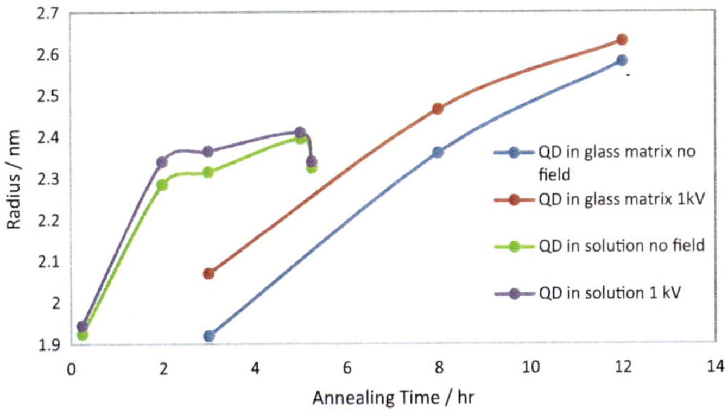

Fig. 21.8 Nanoparticle equivalent radius before and after application of a 1 kV DC electric field for annealing times of CdTe nanoparticles

21.5.2 *Effect of Step H.V. Field on Quantum Dot Samples in Glass and Solution Matrixes*

Spectral absorption peak width was affected by electric field, it is an indication of nanoparticle size distribution. Derivatives of sample Gaussian absorption curves [21] allowed determination of FWHM in Table 21.2. Figure 21.9 shows the decrease in FWHM with field on and off for a given sample (solid sample 2). Similar plots for the other samples yield Fig. 21.10; the decrease in FWHM per 1 kV field for all samples.

Nanoparticles synthesized in solution have a narrower size distribution compared to those in glass. Table 21.2, the FWHM tends to be narrower with applied field. To determine the measurement uncertainty of the particle size distribution, best Gauss curve fits were found and FWHM values extracted for 0 V and 1 kV measurement conditions. The standard uncertainty was calculated from the difference between curve 1 FWHM and curve 2 FWHM values.

Table 21.2 Electric field induced change in FWHM of quantum dot samples

Matrix type	Sample no	Equivalent radius (nm)		FWHM change (nm) per kV	Uncertainty (nm) per kV
		Zero field	1 kV		
Glass	1	0.25	0.23	−0.02	±0.028
Glass	2	0.50	0.43	−0.07	±0.014
Glass	3	0.39	0.36	−0.03	±0.014
Solution	1	0.16	0.15	−0.01	±0.028
Solution	3	0.18	0.15	−0.03	±0.028
Solution	4	0.23	0.20	−0.03	±0.028
Solution	6	0.29	0.24	−0.05	±0.014
Solution	7	0.20	0.17	−0.03	±0.028

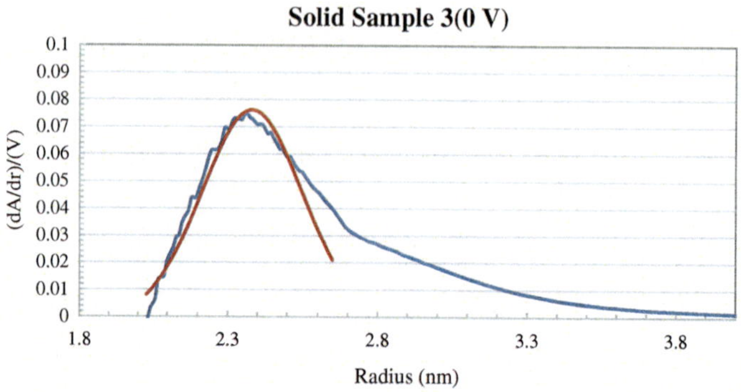

Fig. 21.9 Decrease in FWHM on application of an electric field on solid sample 2 (blue: field off, red: field on)

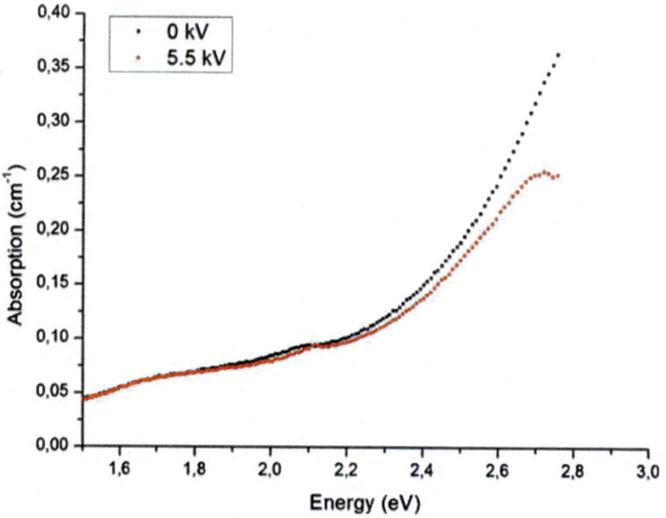

Fig. 21.10 Modification of absorption spectrum of CdTe nanoparticles in glass matrix

In Fig. 21.10, the absorption spectra of CdTe nanoparticles in solid glass is plotted against photon energy with zero field and maximum field. As the electric field increases, the absorption decreases. Application of DC electric field also caused blue shifts of the band edges of the nanoparticles (Figs. 21.10 and 21.11); although Tauc's law showed the band gap itself to be red shifted.

First band gap energies change from 2.55 to 2.52 eV for solution sample 1 and the greatest decrease in the band gap was recorded at 2.5 kV shown in Fig. 21.11. High voltage values could not be achieved in liquid samples, the maximum applied voltage possible in liquid samples was 2.5 kV.

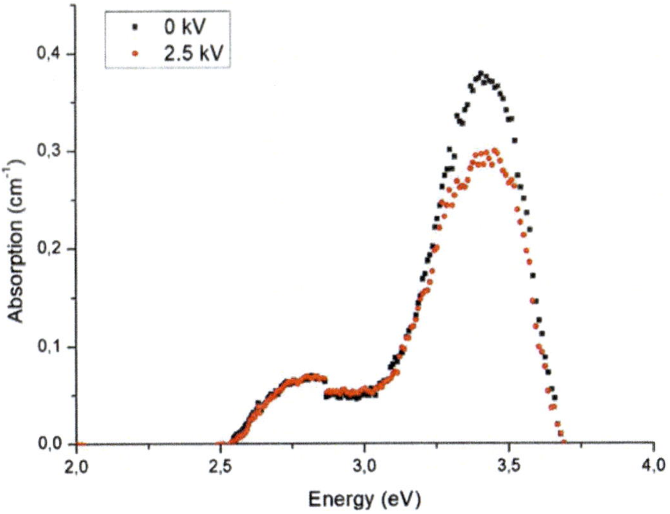

Fig. 21.11 Electric field modification of absorption spectrum of CdTe nanoparticles in solution

The bandgap energy was also modified by applied H.V. electric field The in situ optical transmission measurements were conducted for both solid and aqueous samples and the onset of the asymptotic optical absorption edge (band edge: Eedge) were determined under increasing high voltages applied to the samples. A typical plot for band gap energy versus applied high voltage is presented in Fig. 21.12 for solid samples and in Fig. 21.13 for aqueous samples, respectively. In the same plots the average radii calculated for each sample are also shown. As the applied high voltage increases the band edge decreases and the estimated average particle radius increases for both sets. The band edge is red-shifted by 164 meV under a maximum

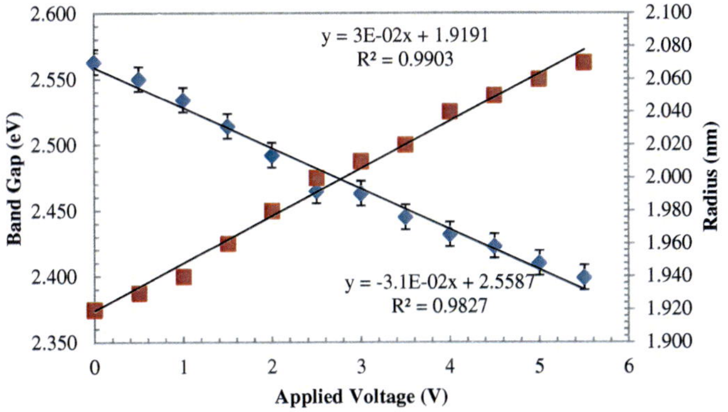

Fig. 21.12 Solid sample 1: Eg • (diamond) and radius (r) ■ variation with applied voltage (R^2 is the coefficient of fit)

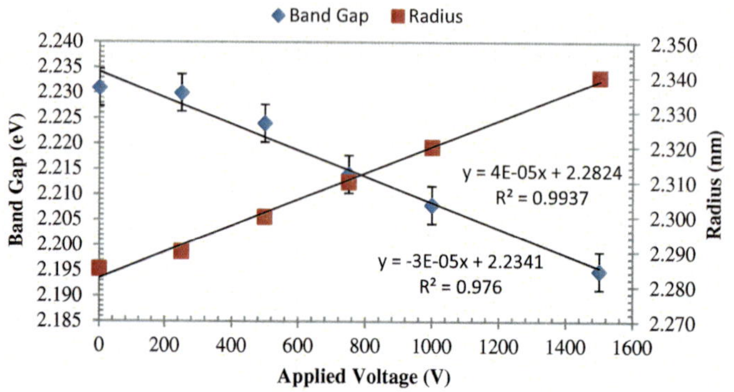

Fig. 21.13 Solution Sample 3: Bandgap (Eg) ◆ and radius (r) ■ variation with applied voltage

electric field of 5.5 kV/cm for the glass sample heat-treated at 660 °C for 3 h, while the same value is just 30 meV for the aqueous sample. The bandgap returns to its zero-field value within ±0.2% after removal of the applied field.

For the solid samples, the band gap energy and radius versus applied electric field graphs give information about the general trend of nanoparticles. The trend demonstrated that CdTe nanoparticles suffer a red shift, decreasing their band gap and leading to a higher calculated radius with applied electric field; quantum dot radius varies from 1.92 to 2.40 nm. As a result of quantum confinement, the band gap of the semiconductor nanocrystal QDs became smaller, causing their computed radii to increase. This demonstrates that the band gap of semiconductor nanocrystal QDs can be tuned using an applied H.V. electric field. Bandgaps of samples under increasing applied H.V. are shown below, together with their equivalent nanoparticle radii for each solution sample. Their fitting equations and correlations (R^2) are given inside the figures.

The band gap energies of the liquid samples grown for ~6 h range from 2.55 to 2.14 eV before applying the electric field. With no applied electric field, CdTe nanoparticles in solution were found to have a narrower size distribution compared to those in glass. The electro optical measurement results showed clearly that increasing electric field causes reduction in band gap energy.

21.6 Discussion

21.6.1 *Modification of Nanoparticle Bandgap in Both Matrixes*

The electric field induced red shift of nanoparticle bandgap in solid matrixes (Fig. 21.12) is linear except for one sample with rates of change in radius 0.02 to 0.07 nm per kV.

The electric field induced red shift of nanoparticle bandgap in solution matrixes (Fig. 21.13); are all linear, with rates of change in radius 0.01 nm/kV to just under 0.05 nm/kV.

The decrease in bandgap is expected and typical of QCSE dominated spectra, since the electron and hole wavefunctions are separated by the applied electric field. This effect has also been observed in the literature [1] where the absorption edge of a QCSE device is modulated with a bias voltage, resulting in a red shift. This decreases the absorption peak height (Fig. 21.1) since electrons and holes move closer together in energy, increasing the wavelength of the absorption [29, 30]. The literature shows that applied electric fields influence excited-state dynamics of CdTe quantum systems [31–35] and can influence electron–ion dynamics causing a decrease in nanoparticle size [36], due to Rayleigh instability.

The work presented in this chapter is the first time the modification in energy band gaps have been quantified against electric field magnitudes, and compared with changes in equivalent nanoparticle radius. The equivalent nanoparticle sizes increase with applied electric field strength. This could be explained by the electric field repelling electrons within the nanoparticles, causing a decreased electrical attraction within the nanoparticle, which increases their size due to Rayleigh instability.

The literature [1] shows that strain caused by bias voltage on quantum wells increases the lattice constant of quantum systems. In this work, atoms forming the nanoparticles are constrained within either a glass or solution matrix, which may cause strain due to increased lattice constant as the electron and holes are separated by the H.V. field. The QCSE effect is flexible, since removal of the electric field returns the original characteristic graphs.

21.6.2 *Decrease in Sample Bandwidth with Electric Field*

Widths of absorption peaks decreased with the 1 kV electric field application, indicating a narrower size distribution. One solid sample showed a larger decrease in size distribution compared to the other samples. This correlates with its comparative large change in bandgap under electric fields compared to the other samples. This, again may be explained by its radius and matrix combination allowing excited-state dynamics [31–35] to have a stronger influence upon it, leading to more monodispersity.

Nanoparticle samples in glass matrixes must experience a higher electric field for the same applied voltage ($E = V/d$), since they are only (0.9–1.4 mm) thick compared to the solution samples, which occupy 1 cm in their sample cells. There is also a higher relative electric permittivity (K) in solution samples (mainly water of approximately K = 80) compared to glass (approximately K = 5–10). Since, E = $V/(Kd)$ it should cause the electric fields experienced by solutions to be a factor ~80 times weaker than that of the glass matrix samples. However, since the voltage is continuously connected electric permittivity differences may only have a secondary effect on the local electric field experienced by nanoparticles.

21.7 Conclusion

The different sized CdTe quantum dots in aqueous and solid samples were characterized by uv–vis spectra and demonstrated that absorption wavelength gradually increases with increasing growth time. This work shows that a H.V. electric field modifies the spectra of quantum systems in both glass and solution matrixes. Samples generally displayed a linear decrease in their band gap energy with electric field increase; this amounted to equivalent maximum changes in nanoparticle radii of 0.07 nm for glass matrixes and 0.05 nm for solution matrixes, as summarised in Table 21.2. The former are 2–3 times stronger than quantum systems in the solutions as they are tightly constrained to their matrixes, as well as having a local environment with a higher dielectric constant. Deformation of nanoparticles due to an electric field reduces the energy required to break it down, which increases the disintegration probability due to thermal fluctuations. Monodispersity improves during the time an electric field is applied, but the changes were completely reversible once the electric field was removed, and the quantum systems did not disintegrate; most probably due to the strength of the matrix. This type of temporary modification of a quantum system has important applications in cases where optical properties need to be modulated temporarily, such as for localized surface plasmon resonances or refractive index variations.

Electric field dependent spatial separation of electron and hole wavefunctions may cause the linear bandgap modifications observed here through increasing strain on the matrix.

If nanoparticle size is modified it can be explained by the electric field repelling electrons from the nanoparticles, the decreased charge interaction of the nanoparticles then increases their size due to Rayleigh instability. This possibility can only be elucidated through X-ray microdiffraction studies.

References

1. F. Henneberger, J. Puls, C. Spiegelberg, A. Schulzgen, H. Rossman, V. Jungnickel, A.I. Ekimov, Semicond. Sci. Technol. **16**, A41 (1991)
2. H. Rossmann, A. Schulzgen, F. Henneberger, M. Muller, Phys. Status Solidi (b) **159**, 287 (1990)
3. M.K. Varsha, J. El Hamdaoui, L.M. Pérez, V. Prasad, M. El-Yadri, D. Laroze, E.M. Feddi, Quantum confined stark effect on the linear and nonlinear optical properties of SiGe/Si semi oblate and prolate quantum dots grown in Si wetting layer. Nanomaterials **11**, 1513–1536 (2021)
4. K. Li, S. Zhu, S. Dai, Z. Li, H. Yin, Z. Chen, Shape effect on the electronic state and nonlinear optical properties in the regulable Y-shaped quantum dots under applied electric field. Opt Exp **29**(4), 5848–5855 (2021)
5. M. Choubani, H. Maaref, F. Saidi, Nonlinear optical properties of lens-shaped core/shell quantum dots coupled with a wetting layer: effects of transverse electric field, pressure, and temperature. J. Phys. Chem. Solids **138**, 109226–109230 (2020)
6. J.S. Weiner, D.A.B. Miller, D.S. Chemla et al., Strong polarization-sensitive electro-absorption in GaAs/AlGaAs quantum well waveguides. Appl. Phys. Lett. **47**, 1148–1150 (1985)
7. L.V. Keldysh, Behaviour of non-metallic crystals in strong electric fields. J. Exptl. Theor. Phys. **33**, 994–1003 (1958)
8. L.V. Keldysh, Ionization in the field of a strong electromagnetic wave. J. Exptl. Theor. Phys. **47**, 1945–1957 (1964)
9. R. Williams, Electric field induced light absorption in CdS. Phys. Rev. **117**, 1487–1490 (1960)
10. J.I. Pankove, *Optical Processes in Semiconductors* (Dover Publications Inc., New York, 1971)
11. H. Haug, S.W. Koch, *Quantum Theory of the Optical and Electronic Properties of Semiconductors*, 3rd edn. (World Scientific, 1994), pp. 343–345
12. C. Kittel, *Introduction to Solid State Physics*, 7th edn. (J. Wiley and Sons Ltd., New York, 1996), pp. 364–366
13. L.E. Brus, Electronic wave functions in semiconductor clusters: experiment and theory. J. Phys. Chem. **90**, 2555–2560 (1986)
14. L.E. Brus, Electron-electron and electron-hole interactions in small semiconductor crystallites: the size dependence of the lowest excited electronic state. J. Chem Phys **80**, 4403–4409 (1984)
15. E.O. Chukwuocha, M.C. Onyeaju, Simulation of quantum dots in the confinement regime. Int. J. Appl. Sci. Eng. Res. **1**(6), 784–792 (2012)
16. L.E. Brus, A simple model for the ionization potential, electron affinity, and aqueous redox potentials of small semiconductor crystallites. J. Chem. Phys. **79**, 5566–5571 (1983)
17. HC van de Hulst, *Light Scattering by Small Particles* (Dover, 1981), pp. 114–130
18. J. Tauc, Optical properties and electronic structure of amorphous Ge and Si. Mat Res Bull **3**, 37–46 (1968)
19. R. López, R. Gómez, Band-gap energy estimation from diffuse reflectance measurements on sol–gel and commercial TiO_2: a comparative study. J. Sol-Gel Sci. Technol. **61**, 1–7 (2012)
20. B. Ghosh, S. Hussain, D. Ghosh et al., Studies of CdTe films deposited by pulsed laser deposition technique. Physica B **407**, 4214–4220 (2012)
21. N.S. Pesika, K.J. Stebe, P.C. Searson, Determination of the particle size distribution of quantum nanocrystals from absorbance spectra. Adv. Mater. **15**, 1289–1291 (2003)
22. K.T. Yong, W.C. Law, I. Roy et al., Aqueous phase synthesis of CdTe quantum dots for biophotonic. J. Biophoton. **4**, 9–20 (2011)
23. Z.J. Li et al., A robust artificial catalyst in situ formed from CdTe QDs and inorganic cobalt salts for photocatalytic hydrogen evolution. Energy Environ. Sci. **6**, 465–469 (2013)
24. L.H. Lajunen, P. Perämäki, Spectrochemical analysis by atomic absorption and emission. J. Chem. Edu. **86**, 78–199 (2009)
25. F. Castro, B. Nabet, Numerical computation of the complex dielectric permittivity using Hilbert transform and FFT techniques. J Franklin Inst. **336B**, 53–64 (1999)
26. K.A. Whittaker, J. Keaveney, I. Hughes, C. Adams, The Hilbert transform: applications to atomic spectra. Phys. Rev. A **91**, 1–5 (2014)

27. M.G. Feeney, R. Ince, M.H. Yukselici, C. Allahverdi, Interferometric investigation and simulation of refractive index in glass matrixes containing nanoparticles of varying sizes. Appl. Opt. **50**, 3259–3267 (2011)
28. M.S. Abd El-Sadek, S. Moorthy Babu, Growth and optical characterization of colloidal CdTe nanoparticles capped by a bifunctional molecule. Physica B **405**, 3279–3283 (2010)
29. P. Guyot-Sionnest, E. Lhuillier, H. Liu, A mirage study of CdSe colloidal quantum dot films, Urbach tail, and surface states. J. Chem. Phys. **137**, 154704–154705 (2012)
30. D.A.B. Miller, D.S. Chemla, T.C. Damen et al., Band edge electro absorption in quantum well structures: the quantum-confined Stark effect. Phys. Rev. Lett. **53**, 2173–2176 (1984)
31. M.S. Mehata, Enhancement of charge transfer and quenching of photoluminescence of capped Cds quantum dots. Nat. Sci. Rep. **5**(12056), 1–11 (2015)
32. Y. Zhang, W. Cheng, T. Zhang, T. Cui, Y. Wang, W.W. Yu, Size- and temperature-dependent quantum confined dielectric effect in colloidal PbSe and CdSe nanocrystals. J. Nanosci. Nanotechnol. **12**, 6224–6230 (2012)
33. R. Ohshima, T. Nakabayashi, Y. Kobayashi, N. Tamai, N. Ohta, External electric field effects on state energy and photoexcitation dynamics of water-soluble CdTe nanoparticles. J. Phys. Chem. C **115**, 15274–15281 (2011)
34. N. Zamani, A. Keshavarz, H. Nadgaran, Quadratic Electro-Optic effect and Electro Absorption process of multi-layer spherical quantum dot enhanced by metal nanoparticles. Plasmonics **12**, 383–391 (2017)
35. N.V. Tepliakov, M.Y. Leonov, A.V. Baranov, A.V. Fedorov, I.D. Rukhlenko, Quantum theory of electroabsorption in semiconductor nanocrystals. Opt. Exp. **24**, A52–A57 (2016)
36. D. Sapkota, Y. Li, O.R. Musaev, J.M. Wrobel, M.B. Kruger, Effect of electric fields on tin nanoparticles prepared by laser ablation in water. J. Laser App. **29**, 012002, 1–4 (2017)

Chapter 22
Perpendicular Andreev Reflection: Solid State Signature of Black Hole Horizon

Z. Faraei and S. A. Jafari

Abstract In this chapter, we introduce the Dirac and Weyl equation in solids. We classify the boundary conditions in Weyl materials and show how Fermi arcs naturally follow from the boundary conditions. Then the Green's function is used to study superconducting proximity in Dirac and Weyl materials. In the case of Dirac materials we find that a novel form of parity breaking superconductivity, Δ_5, namely the pseudoscalar superconductivity can be induced. We discuss its transport signatures. We further discuss the induction of superconductivity in Fermi arc states. We discuss how the simultaneous presence of pseudoscalar superconductivity, Δ_5, combined with the conventional superconducting parameter Δ_s can give rise to chiral supercurrent.

22.1 Introduction

In 1937 physicist Conyers Herring, while studying the properties of the electronic energy bands of solids, asked himself: "Under what conditions do electronic bands in solids have the same energy?" Such degeneracies can arise in solids with a high crystal symmetry, but he argued that accidental band touchings can occur in solid crystals that lack certain symmetries [1]. Near these band touching points, the description of the low-energy features of the band structures requires 4×4 matrices: a two dimensional space (spanned by Pauli matrices τ^μ [2]) for the two bands in question and another two dimensional space (spanned by σ^μ) to account for the spins. Taylor expanding the band touching in leading order gives a linear momentum dependence for the energy eigenvalues $\varepsilon_\pm = \pm\hbar v_F|\boldsymbol{k}|$, where the $\pm$ labels the states above/below

Z. Faraei · S. A. Jafari
Department of Physics, Sharif University of Technology, Tehran 11155-9161, Iran
e-mail: jafari@sharif.edu

Z. Faraei (✉)
Department of Physics, Institute for Advanced Studies in Basic Sciences (IASBS), Zanjan 45137-66731, Iran
e-mail: z.faraei@iasbs.ac.ir

H. Ünlü and N. J. M. Horing (eds.), *Progress in Nanoscale and Low-Dimensional Materials and Devices*, Topics in Applied Physics 144,
https://doi.org/10.1007/978-3-030-93460-6_22

the touching point and Fermi velocity v_F is the group velocity of the band structure at the touching point. This state of affairs can be formally encoded into the "Dirac equation" [3]. Although the symmetries of this *emergent* Dirac equation is the same as the Dirac equation that describes the electrons in CERN experiments, the important difference is that the speed of light, c, in the original Dirac equation is replaced by v_F, which is typically 2–3 orders of magnitude smaller than c. The emergent Lorentz symmetry of the solid-state Dirac equation enjoys the parity (P) and time reversal (T) symmetries of the original Dirac equation. When the chemical potential is right at the crossing point, it further enjoys the charge conjugation (C) symmetry which in the solid state context is nothing but the particle-hole symmetry.

The Dirac Hamiltonian being composed of 4×4 matrices can be gapped out by adding a Dirac mass term. When the mass (gap) vanishes, the Dirac Hamiltonian can be brought into block diagonal format composed of two 2×2 equations as $\chi\boldsymbol{\sigma}.\boldsymbol{k}$ [4] where $\chi = \pm$ labels the two sectors and is called the chirality. In the solid state, the so called Weyl semi-metals are materials in which either the P or T is broken. Therefore the band touching points will be grouped into Weyl points of opposite chirality. The electronic excitations around such points are the Weyl fermions in condensed matter [5].

The candidates for the Weyl fermions in the standard model of particle physics are neutrinos. However, the claim of small mass, has ruled out this possibility [6]. However, quantum condensed matter continues to offer massless Weyl fermions in solid-state systems. Examples of materials that exhibit strong spin-orbit coupling and offer massless Weyl fermions are tantalum or niobium arsenide (TaAs and NbAs) [7–10]. Such a band-crossing is not limited to electronic bands. One can also have such band crossings in photonic bands [11] which is not going to be our focus.

There is a topological significance associated with the Weyl points, namely the $\chi = \pm$ nodes act as source/sink of Berry monopole in the momentum space [5, 12–17]. The simplest manifestation of these fictitious magnetic monopoles is exotic surface states [12]. The surface of Weyl semimetals has a Fermi surface that forms pieces of constant energy surface which do not close on themselves referred to as "Fermi arcs" [1, 5]. These Fermi arcs terminate at the location of the bulk Weyl points ensuring their topological nature [5]. Angle-resolved photoemission spectroscopy (ARPES) is an appropriate method to observe the Fermi arc shapes [18–20]. The ARPES technique involves illuminating the sample with high-energy photons on a material and measuring the energy, momentum and spin of the ejected electrons both from the surface and the bulk. This allows for the explicit determination of both bulk Weyl nodes and the Fermi-arc surface states (Fig. 22.1).

Electrons in every material are characterized by the momentum $\boldsymbol{k}$ and spin $s_z = \uparrow, \downarrow$. In Weyl semimetals, in addition to the above attributes, one further needs to specify the chirality $\tau_z = \pm 1$ that determines to which Weyl node the fermion belongs. In the absence of chirailty the problem of forming Cooper pair involves the specification of the spin (s_z) and orbital ($\boldsymbol{k}$) for the pair. Now in the Weyl semimetals, one has to further specify the total chirality τ of the Cooper pair. Therefore the superconducting pairing in Weyl semimetals can be both even and odd with respect to exchanging the chirality attribute of the electrons forming the Cooper pair. The

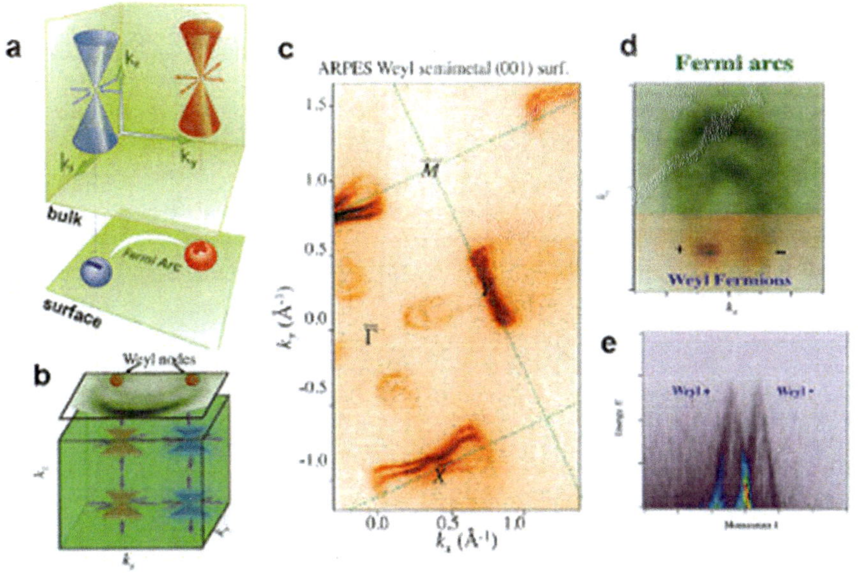

Fig. 22.1 Weyl fermion and Fermi arcs **a** Schematic of the band structure of a Weyl fermion semimetal. **b** Correspondence of the bulk Weyl fermions to surface Fermi arc states. **c** ARPES mapping of the TaAs Fermi surface. **d** The Fermi arc surface states and Weyl nodes on the (001) surface of TaAs. **e** Linear dispersion of Weyl quasi-particles in TaAs. (Adapted from [7])

conventional superconductivity does not know about the chirality, and hence can be considered to be even under the chirality flip. Therefore a striking possibility for the formation of the Cooper pairs in Weyl semimetals is that they can form chirality-odd pairings. We denote the superconducting order parameter that is even (odd) under the chirality flip by Δ_s (Δ_5). We will show that this promotes the superconducting order as $e^{i\phi}\Delta_s \to e^{i\phi}(\Delta_s + i\gamma^5\Delta_5)$, where a γ^5 is a 4×4 matrix whose eigenvalues denote the chirality.

22.2 Dirac and Weyl Hamiltonian in Solids

Let us start with the Hamiltonian for the electron in a periodic potential with spin-orbit (SO) interaction

$$H = \frac{p^2}{2m_e} + V(\boldsymbol{r}) + \frac{\hbar}{4m_e^2c^2}\boldsymbol{\sigma}.\nabla V \times \boldsymbol{p}, \tag{22.1}$$

where m_e is the bare electron mass, c the velocity of light, $\boldsymbol{\sigma}$ the Pauli spin matrix vector, $V(\boldsymbol{r})$ the periodic crystal potential, and $\boldsymbol{p} = -i\hbar\nabla$ the momentum operator.

The above Hamiltonian can be viewed as substitution $\boldsymbol{p} \to \boldsymbol{p} + \left(\frac{\hbar}{4mc^2}\right) \boldsymbol{\sigma} \times \nabla V$ and retaining the leading power of $\hbar/(m_e c)^2$.

The band structure of any solid around the band extremum can be expanded within the so called $\boldsymbol{k}.\boldsymbol{p}$ theory. This can also be done for solids with strong spin-orbit interactions and hyperbolic bands which results in the Dirac theory. Perhaps the oldest known example of a Dirac solid is Bismuth for which Wolf derived an effective Hamiltonian that is an anisotropic extension of the well-known Dirac theory. In the case of Bismuth, it has been shown that only two bands are enough for understanding experimental results. So, the matrix elements of the electrons' equation of motion reduces to a 4×4 matrix which describes conduction and valence bands with up and down spins. Cohen and Blount represented these matrix elements as:

$$\begin{pmatrix} \varepsilon_g - \varepsilon & 0 & \hbar\boldsymbol{k} \cdot \boldsymbol{v}_{13} & \hbar\boldsymbol{k} \cdot \boldsymbol{v}_{14} \\ 0 & \varepsilon_g - \varepsilon & \hbar\boldsymbol{k} \cdot \boldsymbol{v}_{23} & \hbar\boldsymbol{k} \cdot \boldsymbol{v}_{24} \\ \hbar\boldsymbol{k} \cdot \boldsymbol{v}_{31} & \hbar\boldsymbol{k} \cdot \boldsymbol{v}_{32} & -\varepsilon_g - \varepsilon & 0 \\ \hbar\boldsymbol{k} \cdot \boldsymbol{v}_{41} & \hbar\boldsymbol{k} \cdot \boldsymbol{v}_{42} & 0 & -\varepsilon_g - \varepsilon \end{pmatrix} \begin{pmatrix} c_1 \\ c_2 \\ c_3 \\ c_4 \end{pmatrix} = 0, \tag{22.2}$$

where ε_g is the band extremum and $c_1 \ldots c_4$ are the expansion coefficients of the Hamiltonian eigenfunction in the basis $(a\uparrow, a\downarrow, b\uparrow, b\downarrow)$ where a and b label the conduction and valence bands, respectively. We further assume that the a and b bands are very close to each other. The matrix elements of the velocity operator $\partial H/\partial \boldsymbol{p}$ are given by $v_{ij} = \langle\psi_i|\,|\boldsymbol{v}\psi_j\rangle$ and are influenced by the Peierls-like substitution in (22.1). In the limit of vanishing SO coupling, there is no coupling between the opposite spins in the a and b bands. But when SO coupling dominates over the $p^2/2m_e + V(r)$ term, there will be a full-fledged coupling between the spin states of the a and b bands in a manner represented in (22.2).

The Dirac equation in particle physics enjoys symmetry under parity (P) and time-reversal (Θ). The first thing parity does is to give zero velocity matrix elements for aa and bb, meaning that $\boldsymbol{v}_{12} = \boldsymbol{v}_{34} = 0$. This is responsible for the fact that the wave vector $\boldsymbol{k}$ appears only block-off-diagonally in the band space in (22.2). Further, imposing the $P\Theta$ symmetry on the Hamiltonian (22.2) gives, $\boldsymbol{v}_{13} = \boldsymbol{v}_{42}$ and $\boldsymbol{v}_{14} = -\boldsymbol{v}_{32}$. Take for example, $\boldsymbol{v}_{13}$ which is $\boldsymbol{v}_{13} = \langle\psi_{+\uparrow}|\,|\boldsymbol{v}\psi_{-\uparrow}\rangle = \langle PT\boldsymbol{v}\psi_{-\uparrow}|\,|PT\psi_{+\uparrow}\rangle = \langle \boldsymbol{v}PT\psi_{-\uparrow}|\,|\psi_{+\downarrow}\rangle = \langle\psi_{-\downarrow}|\,|\boldsymbol{v}\psi_{+\downarrow}\rangle = \boldsymbol{v}_{42}$ [21].

The difference between v_{13} and v_{14} magnitudes depends on the strength of the spin-orbit interaction. When the SO coupling is the largest energy scale, the spin $\uparrow$ state in band a will be equally coupled to both $\uparrow$ and $\downarrow$ states in band b, and therefore $|\boldsymbol{v}_{13}| = |\boldsymbol{v}_{14}|$. In general, we need to specify four vectors $Re(\boldsymbol{v}_{13})$, $Im(\boldsymbol{v}_{13})$, $Re(\boldsymbol{v}_{14})$ and $Im(\boldsymbol{v}_{14})$ in order to fix the two band model. Wolf used the freedom in choosing the basis functions to enforce $Re(\boldsymbol{v}_{13}) = 0$ which gives,

$$H = \frac{\varepsilon_g}{2}\beta + i\hbar\boldsymbol{k} \cdot \left[\sum_{\mu=1}^{3} \boldsymbol{W}(\mu)\beta\alpha_\mu\right], \tag{22.3}$$

where the vectors $\boldsymbol{W}(\mu)$ are given by $\boldsymbol{W}(1) = Im(\boldsymbol{v}_{14})$, $\boldsymbol{W}(2) = Re(\boldsymbol{v}_{14})$, $\boldsymbol{W}(3) = Im(\boldsymbol{v}_{13})$, and the 4×4 matrices β and α_μ are,

$$\beta = \begin{pmatrix} I & 0 \\ 0 & -I \end{pmatrix} \quad , \quad \alpha_\mu = \begin{pmatrix} 0 & \sigma_\mu \\ \sigma_\mu & 0 \end{pmatrix} . \tag{22.4}$$

If we assume isotropic velocity vectors as $\boldsymbol{W}(1) = (v_F, 0, 0)$, $\boldsymbol{W}(2) = (0, v_F, 0)$ and $\boldsymbol{W}(3) = (0, 0, v_F)$, leads to the isotropic Wolff Hamiltonian for 3DDMs

$$H_{0D}(\boldsymbol{k}) = \begin{bmatrix} \varepsilon_g & i\hbar v_F \boldsymbol{k}.\boldsymbol{\sigma} \\ -i\hbar v_F \boldsymbol{k}.\boldsymbol{\sigma} & -\varepsilon_g \end{bmatrix}, \tag{22.5}$$

where v_F is the Fermi velocity, $\pm\varepsilon_g$ sets band edges for conduction and valence bands. From this point we set $\hbar$ and v_F equal to 1 and will restore the constants whenever required.

Let us briefly remind ourselves of the properties of the Dirac Hamiltonian. An $n \times n$ Hamiltonian matrix of the form

$$H_{0D}(\boldsymbol{k}) = \boldsymbol{k}.\vec{\alpha} + m\beta, \tag{22.6}$$

where $\vec{\alpha}$ and $\vec{\beta}$ are $n \times n$ matrices satisfying $\vec{\alpha}\beta = -\beta\vec{\alpha}$, $\alpha^i\alpha^j = -\alpha^j\alpha^i$ when $i \neq j$ and $(\alpha^i)^2 = \beta^2 = \mathbb{1}$ represents a Dirac Hamiltonian. The spectrum $\varepsilon(\boldsymbol{k})$ of this Hamiltonian is trivially obtained by squaring the above matrix. The anti-commutation of dislike matrices eliminates cross terms, and the fact that every matrix squares to $\mathbb{1}$ immediately gives $H^2(\boldsymbol{k}) = \mathbb{1}(k^2 + m^2)$, and therefore $\varepsilon(\boldsymbol{k}) = \pm\sqrt{k^2 + m^2}$. This dispersion reminds us of the energy-momentum relation for relativistic particles. In the context of the standard model of particles, the $\pm$ branches correspond to fermion/anti-fermion. In our present solid-state setting, the $\pm$ signs correspond to the conduction/valence bands. Furthermore, in the context of high-energy physics, the parameter m is the "mass" of the particles, while in our solid-state setting it corresponds to the energy gap between the valence and conduction bands.

It is customary to use the α^i and β matrices to define a new set of matrices by

$$\gamma^0 = \beta, \quad \gamma^i = \beta\alpha^i, \tag{22.7}$$

where i runs over the space indices. For example in 2D materials we have two γ^x, γ^y etc. The matrices γ^μ where the "Lorentz" index $\mu = 0, 1, \ldots$ includes both "time" (0) and "space" directions $(1, 2, \ldots)$. These matrices satisfy the so-called Clifford Algebra,

$$\{\gamma^\mu, \gamma^\nu\} = \eta^{\mu\nu}, \tag{22.8}$$

where $\eta^{\mu\nu} = \text{diag}(1, -\mathbf{1})$, is the Minkowski metric, where $\mathbf{1}$ represents a unit matrix of dimension equal to the space dimension. It turns out that the representations of the above algebra is only possible for even values of n. For one space dimension the $n = 2$ is sufficient and one can choose e.g., $\alpha_x = \sigma_x$. Then for the mass term one has two choices $\beta = \sigma_y$ (corresponding to the Peierls distortion) or $\beta = \sigma_z$ corresponding to "sublattice" symmetry breaking. For two space dimensions still $n = 2$ is enough and one possible choice is $\alpha_x = \sigma_1$ and $\alpha_y = \sigma_2$. Then for β, the only possible choice will be $\beta = \sigma_3$. For three space dimensions, the 2×2 Pauli matrices are not enough, and one has to think of $n = 4$-dimensional matrices which can be constructed from the direct product of Pauli matrices τ and σ.

With this reminder, we are now ready to return to the effective theory of Bismuth. We make the following choice for the γ^μ matrices,

$$\gamma^0 = \tau_3 \otimes \sigma_0, \quad \vec{\gamma} = \tau_1 \otimes i\vec{\sigma}. \tag{22.9}$$

The Pauli matrices $\vec{\tau}$ act on the space of conduction and valence bands, while $\vec{\sigma}$ act on the spin space. In terms of what we construct in (22.9), $\beta = \gamma^0$ and $\vec{\alpha} = \gamma^0\vec{\gamma}$. The Clifford algebra for γ^μ matrices implies $\vec{\alpha}\beta = -\beta\vec{\alpha}$ and $-\vec{\gamma}^2 = \vec{\alpha}^2 = \beta^2 = 1$.

In this way, after a minimal substitution $\boldsymbol{k} \to \boldsymbol{k} - e\boldsymbol{A}$, the Dirac equation for a charge $-e$ electron in a background field $\boldsymbol{A}$ is given by,

$$[i\gamma^0\gamma^j(\partial_j - i.e.A_j) + m\gamma^0]\psi_e = \varepsilon_e\psi_e, \tag{22.10}$$

where ψ_e is the wave function of an electron with momentum $\boldsymbol{k}$ and energy $\varepsilon_e = \sqrt{k^2 + m^2}$ close to the Dirac point.

The Dirac Hamiltonian (22.6) has three important symmetries: charge conjugation (C), parity (P) and time reversal (Θ). The charge conjugation is nothing but the particle-hole symmetry when the chemical potential is precisely at the middle of the conduction and valence band. When the Dirac mass is zero, the 4×4 Dirac equation decouples into two 2×2 sectors. These two sectors are labeled by the eigenvalues $\pm$ of the matrix $\gamma^5 = i\gamma^0\gamma^1\gamma^2\gamma^3$. Using the Clifford anticommutations algebra one can immediately verify that $(\gamma^5)^2 = \mathbb{1}$ and that γ^5 anticommutes with all γ^μ matrices. The eigenstates with $+1$ (-1) eigenvalue are called right-handed (left-handed) fermions [22]. In a real "Weyl" semimetal, breaking either the P or Θ gives rise to the separation of the two 2×2 block of right/left handed fermions in the momentum space [14].

22.3 Chirality-Odd Superconductivity

In addition to its momentum $\boldsymbol{k}$ and spin s, an electron in a Dirac/Weyl solid requires another attribute, namely its chirality $\tau = \pm$ (the eigenvalues of γ^5) that specifies whether it is right-handed or left-handed. In the absence of the chirality attribute, the Cooper pair can be either odd or even with respect to the exchange of the spin of the two electrons forming the Cooper pair. These are spin singlet/triplet pairing. With the new chirality attribute available in the Dirac/Weyl materials, the Cooper pairing becomes richer, and can be even/odd with respect to the exchange of the chirality of the two fermions forming the Cooper pairs.

Let us formalize the above concept as follows. Assuming the spin part of the pairing is singlet, the chirality-even pairing is

$$e^{i\phi_s}\Delta_s\left[\psi^*_{-\uparrow}\psi^*_{+\downarrow}-\psi^*_{-\downarrow}\psi^*_{+\uparrow}+\psi^*_{+\uparrow}\psi^*_{-\downarrow}-\psi^*_{+\downarrow}\psi^*_{-\uparrow}\right]. \tag{22.11}$$

The above expression is clearly even under $+ \leftrightarrow -$. One can also construct a pairing that is odd under the above chirality flip operation as follows,

$$e^{i\phi_5}\Delta_5\left[\psi^*_{-\uparrow}\psi^*_{+\downarrow}-\psi^*_{-\downarrow}\psi^*_{+\uparrow}-\psi^*_{+\uparrow}\psi^*_{-\downarrow}+\psi^*_{+\downarrow}\psi^*_{-\uparrow}\right], \tag{22.12}$$

where $\tau = \pm$ and $\uparrow\downarrow$ denote the chirality and spin of the electrons, respectively. The name Δ_5 suggests a close connection with the matrix γ^5. If the above two forms coexist with each other, the requirement of a single transition temperature T_c restricts $\phi_5 - \phi_s$ to be $0, \pi$ [23]. The remaining parameter Δ_s is real (and can be chosen to be positive). The significance of the odd-chirality pairing is that, being odd with respect to the change in the chirality attribute $\pm \to \mp$, this form of superconductivity, in addition to the usual $U(1)$ symmetry, spontaneously breaks additional parity symmetry too. The above discussion is for the superconductivity formed by attractive interactions in the Dirac/Weyl material. In the following we will present the details of a proximity induced superconductivity for Dirac/Weyl materials and will show how the above form of superconductivity (in addition to other interesting forms of superconductivity) are induced. For simplicity, in the following we will assume that we are dealing with a single Dirac cone material [24].

Assume that a conventional s-wave superconductor, characterized with a scalar superconducting gap $\Delta_{\rm sc}$, is placed next to the three dimensional Dirac material (3DDM). We furthermore assume that the Dirac mass $m = 0$. The present tunneling calculation can be done for non-zero m as well [25]. We consider a planar interface perpendicular to the z-axis, located at $z = 0$. The subscript S denotes the superconductor region for $z < 0$ while D stands for the 3DDM region for $z > 0$. Combining the Nambu space of both the superconductor and the 3DDM, the Green's function in the absence of tunneling is given by

$$\mathbf{G}_0 = \begin{bmatrix} G_{0S} & 0 \\ 0 & G_{0D} \end{bmatrix}, \tag{22.13}$$

where $G_{0S(D)} = [i\omega_n - \mathcal{H}_{S(D)}]^{-1}$ is a 4×4 (8×8) Green's function matrix in the Nambu space of the conventional superconductor (Dirac material) mentioned above. ω_n are Matsubara frequencies, $\mathcal{H}_S$ is the standard BCS Hamiltonian and $\mathcal{H}_D$ is the *Dirac*-Bogoliubov-de Gennes (DBdG) Hamiltonian,

$$\begin{bmatrix} H^e_D - \mu & \Delta \\ \Delta^\dagger & \mu - H^h_D \end{bmatrix}\begin{bmatrix} \psi_e \\ \psi_h \end{bmatrix} = \varepsilon \begin{bmatrix} \psi_e \\ \psi_h \end{bmatrix},$$

The Dirac Hamiltonian H^e is given in (22.10) by setting $A_j = 0$. The Dirac Hamiltonian H^h for holes can be obtained from H^e by charge conjugation operation [25]. The end result is the intuitive expression $H^h = H^e$,

$$\mathcal{H}_{\rm D} = \begin{bmatrix} m\gamma^0 + k_\mu\gamma^0\gamma^\mu - \mu & \Delta \\ \Delta^\dagger & \mu - m\gamma^0 - k_\mu\gamma^0\gamma^\mu \end{bmatrix}. \tag{22.14}$$

The Nambu spinor is given by $\psi^T = [\psi_e{}^T(\boldsymbol{k})\ ,\ \psi_h{}^T(\boldsymbol{k})]$ where for a 3DDM it becomes,

$$\psi_e^T(\boldsymbol{k}) = \left[c_{\boldsymbol{k},+,\uparrow}\ c_{\boldsymbol{k},+,\downarrow}\ c_{\boldsymbol{k},-,\uparrow}\ c_{\boldsymbol{k},-,\downarrow}\right], \tag{22.15}$$

and its charge conjugate is

$$\psi_h^T(\boldsymbol{k}) = \left[c^\dagger_{-\boldsymbol{k},-,\downarrow}\ -c^\dagger_{-\boldsymbol{k},-,\uparrow}\ -c^\dagger_{-\boldsymbol{k},+,\downarrow}\ c^\dagger_{-\boldsymbol{k},+,\uparrow}\right], \tag{22.16}$$

in which, the subscripts $\pm$ are chirality indices and refer to the upper and lower bands with dispersion $\varepsilon = \pm|\boldsymbol{k}|$.

When the superconductor and 3DDM are brought together, the coupling between the two in the combined Nambu space can be described by a 4×8 tunneling matrix $\mathbf{t}$. This matrix has two blocks, one for the electron tunneling (τ_e) and the other for the holes (τ_h). The elements of τ_e (τ_h) connect an electron (hole) annihilation operator from the Dirac material to an electron (hole) creation operator in the superconductor. The tunneling matrix is given by,

$$\mathbf{t} = \sum_{\langle \boldsymbol{k},\boldsymbol{k}'\rangle} e^{-i(\boldsymbol{k}'-\boldsymbol{k}).\vec{r}} \begin{bmatrix} \tau_e & 0 \\ 0 & \tau_h \end{bmatrix}. \tag{22.17}$$

Here $\sum_{\langle \boldsymbol{k},\boldsymbol{k}'\rangle}$ denotes the summation over $\boldsymbol{k}$ and $\boldsymbol{k}'$ with the limitation $\boldsymbol{k}_{||} = \boldsymbol{k}'_{||}$, where $||$ means parallel to the interface. Indeed, we consider a system with an interface parallel to the xy-plane, so $p_x = \hbar k_x$ and $p_y = \hbar k_y$ are good quantum numbers and remain unchanged through the tunneling process. $\tau_e = (t_+\ \ t_-) \otimes \mathbb{1}$, as mentioned before, describes the electron transfer from the 3DDM side to the superconductor, and $\tau_h = (t_-\ \ t_+) \otimes \mathbb{1}$ represents the hole tunneling matrix, provided that the spin direction remains unchanged. Here t_+ and t_- are the spin-independent tunneling amplitudes of the superconductor to positive and negative energy states of the 3DDM.

The Green's function of 3DDM gets dressed at each order of tunneling and acquires off-diagonal matrix elements in the Nambu space which are anomalous Green's functions and correspond to induced superconducting correlations.

The leading correction to the Green's function that generates superconducting correlations is $G = G_0 + G_0 T G_0 T G_0$. Reading the Nambu off-diagonal component from the 8×8 Dirac sector in the above correction gives,

$$\mathbf{F}_2/g = A + B(\boldsymbol{k}.\vec{\alpha}) + (\boldsymbol{k}.\vec{\alpha})\tau^{+\dagger}\tau^-(\boldsymbol{k}.\vec{\alpha}), \tag{22.18}$$

where $A = (i\omega_n + \mu)\tau_e^\dagger \tau_h(i\omega_n - \mu)$, B are 4×4 matrices independent of $\boldsymbol{k}$, and

$$g = \frac{m^*\pi}{\sqrt{\omega_n^2 + \Delta_{sc}^2}} \left[\frac{e^{-\kappa_+ z}}{\kappa_+} - \frac{e^{-\kappa_- z}}{\kappa_-} \right] (\omega_n^2 + k^2)^{-2},$$

with $\kappa_\pm = \sqrt{k_{||}{}^2 \pm 2im^*(\omega_n{}^2 + \Delta_{sc}{}^2)^{1/2}}$ resulting from integration over the k'_z on the superconductor side. The m^* is the effective mass in the superconductor defining the dispersion of the underlying band structure by $(k'^2_x + k'^2_y + k'^2_z)/(2m^*)$ and $k_{||}{}^2 = k_x^2 + k_y^2$. The above factor is an even function of $\boldsymbol{k}$ and ω_n and will not affect our symmetry considerations regarding the even/odd behavior under the space ($\boldsymbol{k} \to -\boldsymbol{k}$) and time reversals.

22.4 Classification of Superconducting Order in 3DDM

The 4×4 Nambu-off-diagonal superconducting correlations are induced by the tunneling of Cooper pairs from the superconductor into 3DDM. Denoting the $\mathbf{F}_2/g$ by the superconducting matrix Δ whose matrix elements are defined as $\Delta_{\alpha\sigma,\alpha'\sigma'} = \langle \psi_{e\alpha\sigma} \bar{\psi}_{h\alpha'\sigma'} \rangle$ [26] with $\alpha, \alpha' = \pm$ and $\sigma, \sigma' = \uparrow, \downarrow$, or more explicitly [25]

$$\Delta = \begin{bmatrix} \Delta_{+\uparrow-\downarrow} & -\Delta_{+\uparrow-\uparrow} & \Delta_{+\uparrow+\downarrow} & -\Delta_{+\uparrow+\uparrow} \\ \Delta_{+\downarrow-\downarrow} & -\Delta_{+\downarrow-\uparrow} & \Delta_{+\downarrow+\downarrow} & -\Delta_{+\downarrow+\uparrow} \\ \Delta_{-\uparrow-\downarrow} & -\Delta_{-\uparrow-\uparrow} & \Delta_{-\uparrow+\downarrow} & -\Delta_{-\uparrow+\uparrow} \\ \Delta_{-\downarrow-\downarrow} & -\Delta_{-\downarrow-\uparrow} & \Delta_{-\downarrow+\downarrow} & -\Delta_{-\downarrow+\uparrow} \end{bmatrix}. \tag{22.19}$$

Fierz decomposing the above matrix

$$\Delta = \Delta^s \mathbb{1} + \Delta_\mu \gamma^\mu + \Delta_{\mu\nu}\sigma^{\mu\nu} + \Delta_{5\mu}\gamma^5\gamma^\mu + \Delta_5\gamma^5, \tag{22.20}$$

where $\sigma^{\mu\nu} = i[\gamma^\mu, \gamma^\nu]/2$, one can read off the superconducting in various "channels". Significance of the expansion (22.20) is that, assuming that the above superconducting matrix respects the Lorentz symmetry of the underlying 3DDM, the above expression must be scalar with respect to the Lorentz transformation. Therefore, with respect to the Lorentz transformation, the Δ_s is scalar, Δ_μ is a four-vector, Δ_5 is pseudo-scalar, $\Delta_{5\mu}$ is pseudo-vector and $\Delta_{\mu\nu}$ is a tensor. At the second order, the actual values of the above superconducting orders are summarized in Table 22.1.

As can be seen the pseudo-scalar pairing Δ_5 is driven by the different tunneling rates for right and left-handed fermions and vanishes once $t_+ = t_-$. Such a difference in a material setting is quite feasible. Because fine tuning to the $t_+ = t_-$ point requires the energy levels of the superconductor to be very precisely and evenly placed between the positive and negative energy states of the 3DDM. Any deviation from such a perfect symmetry produces a difference between t_+ and t_-, and therefore, the induction of the pseudo-scalar superconductivity is generic. Another important

Table 22.1 Fierz decomposition of the gap matrix for 3DDM . The indices i, j, l correspond to three spatial directions 1, 2, 3 and ϵ_{ijl} is the totally antisymmetric tensor

$\Delta^s : -t_+t_-(\omega_n^2 + \mu^2 + k^2)$
$\Delta_5 : \frac{i}{2}(t_+^2 - t_-^2)(\omega_n^2 + \mu^2 + k^2)$
$\Delta_\mu : 0$
$\Delta_{50} : \frac{i}{2}(t_+^2 + t_-^2)(\omega_n^2 + \mu^2 - k^2)$
$\Delta_{5j} : [\omega_n(t_+^2 + t_-^2)]k_j$
$\Delta_{0j} : 2\mu t_+t_-k_j$
$\Delta_{ij} : -i\mu(t_+^2 - t_-^2)\epsilon_{ijl}k_l$

feature of Δ_5 is that $\pi/2$ is out of phase with respect to Δ_s. This extra factor of "i" has important implications. Comparing a purely Δ_s superconductor with a purely Δ_5 one, every time an electron undergoes the Andreev reflection, a $\pi/2$ phase is accumulated. Therefore, in one complete round involving two reflections from the right and left superconductor in a Josephson setup, a total phase of π will be accumulated. Therefore, one needs two complete rounds to return the wave function of the fermion to its original value. This is the simple explanation of 4π periodic Josephson effect observed in the Josephson setup of the Dirac superconductors.

When forming a Cooper pair, the pair has to be odd with respect to exchanging *all* the attributes of the first electron with the second one. In the present case, the attributes are: spin (↑ or ↓), chirality (±), momentum $\boldsymbol{k}$ and the frequency ω_n. Exchanging the coordinates of the electron amounts to parity transformation (P), $\boldsymbol{k} \to -\boldsymbol{k}$. Table 22.2 summarizes the symmetry aspects of the induced superconducting correlations.

The first column is the superconducting amplitude in various channels (scalar, pseudo-scalar, vector, pseudo-vector, and tensor). The second column indicates the explicit expression for the Cooper pairs which is obtained by the Fierz decomposition of (22.19) such that it satisfies (22.20). The third column (S) indicates the sign arising from the exchange of the spins of the electrons in a Cooper pair. The fourth column indicates the sign that arises from the exchange in the chirality attributes + and − of the electrons in the Cooper pair. The fifth column (P) indicates the sign that arises from $\boldsymbol{k} \to -\boldsymbol{k}$ in Table 22.1. Although in the present second order perturbation result, summarized in Table 22.1, there are no Δ_j contributions, but since the only vector in the problem is k_j, the only acceptable functional dependence of Δ_j on k_j can have odd parity. That is why we have used quotation marks to indicate the putative parity (perhaps at higher orders of perturbation theory) of Δ_j. For the non-zero Dirac mass m, one can obtain non-zero Δ_0 [25] which turns out to have even parity. The last column follows from total antisymmetry under exchange of all attributes which agrees with Table 22.1. For any deviation of m from 0, frequencies other than those indicated in this column can mix. At $m = 0$, there would be no Δ_0 in the leading order perturbation result of Table 22.1, but if anything appears in higher orders it must have an odd order of frequency. Any $m \neq 0$ results in the mixing of a little bit of the opposite frequency symmetry (i.e., even frequency).

Table 22.2 Pairing symmetries of chiral fermions in three dimensional Dirac semimetals under the *exchange* spin (S), chirality (τ), space coordinate (P), and the time or frequency (ω) of the fermions forming the Cooper pair. The first (second) row is even (odd) under the chirality flip $+ \leftrightarrow -$.

Δ	Cooper pairing	S	τ	P	ω
Δ^s	$\Delta_{+\uparrow-\downarrow} - \Delta_{+\downarrow-\uparrow} + \Delta_{-\uparrow+\downarrow} - \Delta_{-\downarrow+\uparrow}$	$-$	$+$	$+$	$+$
Δ_5	$\Delta_{+\uparrow+\downarrow} - \Delta_{+\downarrow+\uparrow} - \Delta_{-\uparrow-\downarrow} + \Delta_{-\downarrow-\uparrow}$	$-$	$+$	$+$	$+$
Δ_μ	$\Delta_{+\uparrow-\downarrow} - \Delta_{+\downarrow-\uparrow} - \Delta_{-\uparrow+\downarrow} + \Delta_{-\downarrow+\uparrow}$	$-$	$-$	$+$	"$-$"
Δ_1	$\Delta_{+\uparrow+\uparrow} - \Delta_{+\downarrow+\downarrow} + \Delta_{-\uparrow-\uparrow} - \Delta_{-\downarrow-\downarrow}$	$+$	$+$	"$-$"	$+$
Δ_2	$\Delta_{+\uparrow+\uparrow} + \Delta_{+\downarrow+\downarrow} + \Delta_{-\uparrow-\uparrow} + \Delta_{-\downarrow-\downarrow}$	$+$	$+$	"$-$"	$+$
Δ_3	$-\Delta_{+\uparrow+\downarrow} - \Delta_{+\downarrow+\uparrow} - \Delta_{-\uparrow-\downarrow} - \Delta_{-\downarrow-\uparrow}$	$+$	$+$	"$-$"	$+$
Δ_{50}	$\Delta_{+\uparrow+\downarrow} - \Delta_{+\downarrow+\uparrow} + \Delta_{-\uparrow-\downarrow} - \Delta_{-\downarrow-\uparrow}$	$-$	$+$	$+$	$+$
Δ_{51}	$\Delta_{+\uparrow-\uparrow} - \Delta_{+\downarrow-\downarrow} - \Delta_{-\uparrow+\uparrow} + \Delta_{-\downarrow+\downarrow}$	$+$	$-$	$-$	$-$
Δ_{52}	$-\Delta_{+\uparrow-\uparrow} - \Delta_{+\downarrow-\downarrow} + \Delta_{-\uparrow+\uparrow} + \Delta_{-\downarrow+\downarrow}$	$+$	$-$	$-$	$-$
Δ_{53}	$-\Delta_{+\uparrow-\downarrow} - \Delta_{+\downarrow-\uparrow} + \Delta_{-\uparrow+\downarrow} + \Delta_{-\downarrow+\uparrow}$	$+$	$-$	$-$	$-$
Δ_{01}	$-\Delta_{+\uparrow+\uparrow} + \Delta_{+\downarrow+\downarrow} + \Delta_{-\uparrow-\uparrow} - \Delta_{-\downarrow-\downarrow}$	$+$	$+$	$-$	$+$
Δ_{02}	$\Delta_{+\uparrow+\uparrow} + \Delta_{+\downarrow+\downarrow} - \Delta_{-\uparrow-\uparrow} - \Delta_{-\downarrow-\downarrow}$	$+$	$+$	$-$	$+$
Δ_{03}	$\Delta_{+\uparrow+\downarrow} + \Delta_{+\downarrow+\uparrow} - \Delta_{-\uparrow-\downarrow} - \Delta_{-\downarrow-\uparrow}$	$+$	$+$	$-$	$+$
Δ_{12}	$\Delta_{+\uparrow-\downarrow} + \Delta_{+\downarrow-\uparrow} + \Delta_{-\uparrow+\downarrow} + \Delta_{-\downarrow+\uparrow}$	$+$	$+$	$-$	$+$
Δ_{23}	$-\Delta_{+\uparrow-\uparrow} + \Delta_{+\downarrow-\downarrow} - \Delta_{-\uparrow+\uparrow} + \Delta_{-\downarrow+\downarrow}$	$+$	$+$	$-$	$+$
Δ_{13}	$-\Delta_{+\uparrow-\uparrow} - \Delta_{+\downarrow-\downarrow} - \Delta_{-\uparrow+\uparrow} - \Delta_{-\downarrow+\downarrow}$	$+$	$+$	$-$	$+$

22.5 Implications of Pseudo-Scalar Superconductivity

So far we have shown that in a proximity induced setup, both Δ_s and Δ_5 can be *induced* into a Dirac material by a simple BCS superconductor. What can happen when attractive interactions in a Dirac matter are introduced to form Cooper pair? Assuming that the pairing is a spin-singlet, obviously in addition to the even chirality pairing Δ_s, there is always a second option of forming odd-chirality pairing Δ_5. Any arbitrary attractive interaction can be decomposed into scalar and pseudo-scalar channels. The strength of the resulting attraction in the above two channels is the same. Therefore it is likely that when a three dimensional Dirac/Weyl semimetal is allowed to form its own Cooper pairs, the Δ_s and Δ_5 might co-exist. In such a case, insisting that a superconduting matrix embodying $\Delta_s \mathbb{1}$ and $\Delta_5\gamma^5$ be simultaneously present in the Nambu-off-diagonal part of a Dirac-BdG equation, the hermiticity requirement for the Dirac-BdG equation forces them into a combination of the following form

$$e^{i\phi}\left(\Delta_s \mathbb{1} \pm i|\Delta_5|\gamma^5\right). \tag{22.21}$$

This generic form is consistent with the result listed in Table 22.1, according to which the order parameter multiplying γ^5 involves a real number and an additional i with respect to the scalar (BCS) order parameter. Indeed the pair of order parameters (Δ_s, Δ_5) behave like the real and the imaginary parts of a "complex" number. Note that this complex algebraic structure is apart from the common $U(1)$ phase above. As such, the chiral angle χ can be defined as $\tan\chi = \pm|\Delta_5|/\Delta_s$. Therefore the most generic form of a superconducting order in chiral fermions involves a new phase variable, χ.

The first non-trivial result following from (22.21) is that once a small value of Δ_5 is introduced (either by proximity, or by formation of odd-chirality Cooper pairs), there will be another chiral Josephson current I_5, in addition to the usual Josephson current I_s, and is controlled by the difference in the chiral angle χ of the left-side and right-side superconductors as $I_5 \propto \sin(\chi_R - \chi_L)$. The continuum limit of this current will be $I_5 \propto \nabla\chi$. Therefore if the chiral angle χ can vary in space, in addition to the current I arising from the spatial gradient of the superconducting phase $\nabla\phi$, there will be another chiral current driven by $\nabla\chi$ [23].

22.6 Boundary Conditions

So far we have been dealing with the Dirac/Weyl equation in infinite space. However, every real material is finite and bounded by a surface. Therefore an important question is "What are the appropriate boundary conditions (BCs) for the Dirac/Weyl equation?". An important effect of the BC in Weyl semimetals is that there will be peculiar surface states known as Fermi arcs. Unlike the familiar *closed* Fermi surfaces, the Fermi arcs are not closed. The arc connects the projections of the two Weyl nodes. Therefore, all the perturbations that do not gap out the Weyl nodes will not be able to destroy the Fermi arcs. In this section we will focus on Weyl semimetals and will discuss and classify the BCs for these materials.

The relevance of the Fermi arcs to our supeconducting hetereostructure is as follows. When the interface which is separating the supercondutor from the non-superconduting side of the junction, is perpendicular to the vector (say $\boldsymbol{b}$) that separates the Weyl nodes in the momentum space, then both Weyl nodes will have the same projection, and therefore there will be no Fermi arcs. Otherwise, there will be a Fermi arc. To appreciate the special role of the Fermi arc states, we first need to discuss the interesting phenomenon of *chirality blockade* suggested by [27] according to which, the bulk cannot contribute to the Andreev transport. The reason is that bulk fermions are described by the Hamiltonian $h = \tau\boldsymbol{\sigma}.\boldsymbol{p}$. An incident electron from one Weyl node with momentum $\boldsymbol{p}$, spin $\boldsymbol{\sigma}$ and chirality τ will be reflected with momentum $-\boldsymbol{p}$. If the supercondutor is a spin singlet, the net reflected momentum of the whole must be $-\boldsymbol{\sigma}$ such that the helicity $\boldsymbol{\sigma}.\boldsymbol{p}$ (which for massless Weyl fermions is the same as chirality) does not change upon Andreev reflection of the bulk degrees of freedom. But if we insist on a superconducting condensate whose total center of mass momentum is zero, such a process is impossible, and hence the name "chirality blockade".

Since the chirality blockade eliminates the bulk states from the transport in heterostructures involving the superconductors, when the interface is such that it supports the Fermi arcs, the Fermi arc states will be the only players in such transport phenomena. As such, the setups involving spin-singlet superconductors, is a natural framework to obtain a form of transport that is the sole effect of the Fermi arcs, and the low-energy excitations around the Weyl node have no contributions.

Therefore for our purposes it is of utmost importance to understand the formation of the Fermi arc states. Some researchers might find it convenient to "write down" a separate effective Hamiltonian for the two-dimensional surface Fermi arc states. However, we must emphasize that the only Hamiltonian governing the dynamics of the fermions in Weyl semimetals is the Weyl Hamiltonian $\tau\boldsymbol{\sigma}.\boldsymbol{p}$ of the bulk. It is normal practice in quantum theory to model surfaces using BCs. Therefore one expects to obtain the Fermi arcs by starting from the bulk Weyl Hamiltonian, and imposing the appropriate BCs on them. Indeed this point of view can lead to a distinct collective behavior in the two-dimensional electron gas formed by the Fermi arc states. The collective density (plasmon) oscillations in normal two dimensional electron gases can be shown from generic hydrodynamic arguments to behave as $\omega_{\rm pl}(q) \propto \sqrt{q}$. However, the peculiar Fermi arc states that have their "roots in the bulk" (see the rest of this section for derivations) will lead to $\omega_{\rm pl}^{\rm arc}(q) \propto q$, namely a sound-like density oscillation [28].

22.6.1 Classification of Boundary Conditions in WSMs

In a time reversal symmetry breaking Weyl semimetal, there will be two Weyl nodes separated by a vector $2\boldsymbol{b}$. The low energy effective theory around these nodes located at $\pm\boldsymbol{b}$ that correspond to chirality $\tau = \pm$ is given by

$$\hat{H}_\tau = \tau\boldsymbol{\sigma} \cdot (-i\nabla - \boldsymbol{b}). \tag{22.22}$$

By inversion symmetry, the band touching points come in pairs, at $-\boldsymbol{b}$ and $\boldsymbol{b}$, and these have opposite chiralities.

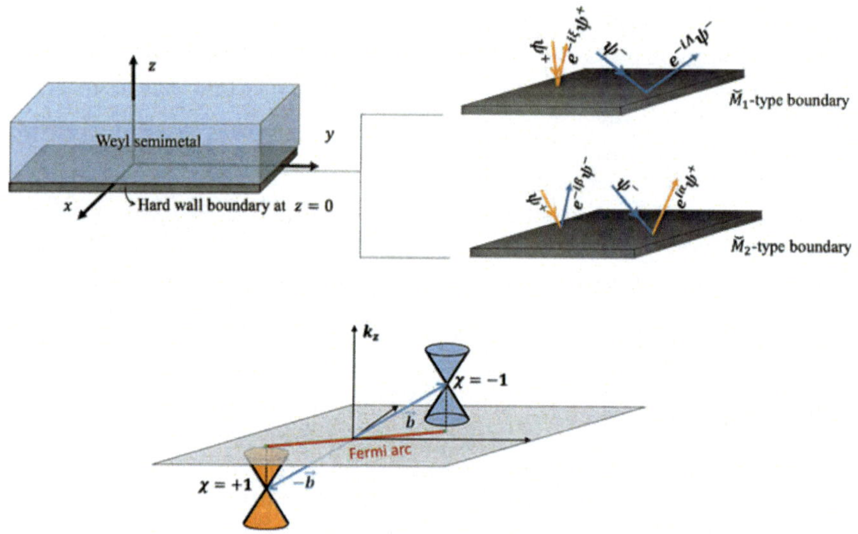

Fig. 22.2 (Top left): Schematic representation of a semi-infinite Weyl semimetal with a hard wall boundary at $z = 0$. (Top right): Reflection processes at the surface with $\check{M}_1$ and $\check{M}_2$ as the boundary matrices. The blue (orange) arrows are the electrons that come from the $\chi = -1$ ($\chi = +1$) Weyl node corresponding to the left (right)-handed Weyl fermions. In both cases, the reflected electron picks up an additional phase which comes from the boundary matrix angles. These angles turn out to be related to the slope of the Fermi arc. In the case of the $\check{M}_1$-type BC, the reflected electron involves a spin flip to make the spin parallel to the plane of the boundary. In the $\check{M}_2$-type BC, the reflected electron has the same spin as the incident electron, but involves a chirality flip such that the reflected electron is a mixture of both chiralities. Both types of BC are compatible with the Fermi arcs. (Bottom): Schematic illustration of the dispersion relation of the bulk states (Weyl cones) and the surface states, the Fermi arc (red line) connecting the projections of the two Weyl nodes with opposite chiralities (Adapted from [29].)

Within the geometry depicted in Fig. 22.2, we have a surface at position $z = 0$. We are dealing with a quantum mechanical problem, namely a Hamiltonian supplemented with the "appropriate" [30] boundary conditions. In fact a "good" boundary condition is the one that ensures the Weyl equation (22.22) is Hermitian. Such a "good" BC for Weyl fermions of single chirality has been discussed by Witten which immediately reproduces "Fermi rays" ending in the projection of the Weyl node at the boundary surface [30, 31]. Incorporating the band bending near the boundary of the Weyl semi-metal turns the "Fermi rays" emanating from the projection of the Weyl node into a "Fermi spiral" [32]. Accounting for simultaneous presence of two chiralities leads to an additional "good" BC that will be discussed below.

Here is the essential line of thought of the argument. Following [33], a physically sensible BC is the one that prohibits the current transmission through the boundary [34]. This is known as hard wall BC that can be effectively incorporated into the Hamiltonian with an additional confinement potential at the boundary as,

$$\left[i\hat{\tau}_z \otimes (\boldsymbol{\sigma} \cdot \nabla) + \hat{\tau}_0 \otimes (\boldsymbol{\sigma} \cdot \boldsymbol{b}) + \check{M}\delta(z)\right]\Psi = E\Psi. \tag{22.23}$$

Note that we have combined the space of chiralities acted upon by the Pauli matrices τ_μ and the spin space corresponding to the Paul matrices σ_μ to generate a 4×4 matrix. The vedge on the matrix M that incorporates the effect of the boundary, emphasizes its 4×4 nature. The matrix $\check{M}$ is a hermitian, unitary matrix with $\check{M}^2 = \check{1}$. The hermiticity follows from the fact that it is part of the Hamiltonian. Unitarity is imposed by requiring that it does not change the total amplitude of the wave function at the surface. Finally, integrating the above differential equation across an infinitesimal region surrounding the boundary gives,

$$\left[i\hat{\tau}_z \otimes (\vec{\sigma} \cdot \hat{z})\right]\Psi|_{z=0} = \check{M}\Psi|_{z=0}. \tag{22.24}$$

Multiplying this equation by its own conjugate transpose, given that M is a Hermitian matrix, we find the requirements that $M^2 = \check{1}$.

To satisfy the above BC, it is necessary and sufficient to choose $\check{M}$ such that $\{\hat{\tau}_z \otimes \hat{\sigma}_z, \check{M}\} = 0$. There are eight possible matrices satisfying these constraints on $\check{M}$, which can be parameterized with eight parameters as follows,

$$\begin{aligned}\check{M} = \hat{\tau}_0 \otimes (a_1\hat{\sigma}_x + a_2\hat{\sigma}_y) + \hat{\tau}_z \otimes (a_3\hat{\sigma}_x + a_4\hat{\sigma}_y)\\ +\hat{\tau}_x \otimes (b_1\hat{\sigma}_0 + b_2\hat{\sigma}_z) + \hat{\tau}_y \otimes (b_3\hat{\sigma}_0 + b_4\hat{\sigma}_z).\end{aligned} \tag{22.25}$$

The constraint $\check{M}\check{M}^\dagger = \check{1}$ forces $\check{M}$ to have either of the following basic forms,

$$\check{M}_1 = \tfrac{\hat{\tau}_0+\hat{\tau}_z}{2} \otimes (\cos\Lambda\ \hat{\sigma}_x + \sin\Lambda\ \hat{\sigma}_y) + \tfrac{\hat{\tau}_0-\hat{\tau}_z}{2} \otimes (\cos\xi\ \hat{\sigma}_x + \sin\xi\ \hat{\sigma}_y) \tag{22.26}$$

$$\check{M}_2 = (\cos\alpha\ \hat{\tau}_x + \sin\alpha\ \hat{\tau}_y) \otimes \tfrac{\hat{\sigma}_0+\hat{\sigma}_z}{2} + (\cos\beta\ \hat{\tau}_x + \sin\beta\ \hat{\tau}_y) \otimes \tfrac{\hat{\sigma}_0-\hat{\sigma}_z}{2} \tag{22.27}$$

where as far as the hard wall BC is concerned, the angles Λ, ξ, α and β are some arbitrary parameters. They describe the amount of mixing between spins ($\check{M}_1$) or chiralities ($\check{M}_2$). Any other linear combination of the form

$$\check{M} = \check{M}_3 = (\cos\gamma)\check{M}_1 + (\sin\gamma)\check{M}_2, \tag{22.28}$$

is also an eligible BC. Further constraints on these parameters can be obtained from additional physical requirements. For example, if we consider the charge conjugation operator in the Weyl representation $C = \hat{\tau}_z \otimes \hat{\sigma}_y K$ [35] (where K is the complex conjugation operator) and demand $C^{-1}\check{M}_1 C = \check{M}_1$, we obtain $\Lambda = \xi = \pi/2$. Similarly if we require $C^{-1}\check{M}_2 C = \check{M}_2$ then the boundary condition parameters will be constrained as $\beta + \alpha = \pi$. Alternatively, they can be obtained from explicit solutions which have already encoded all the appropriate symmetries.

The main difference between these two basic BCs is that $\check{M}_1$ is block diagonal in the chirality space while $\check{M}_2$ is block-off-diagonal, such that with $\check{M}_2$ type BC, scattering at the surface can change the chirality,

$$\check{M}_1 = \begin{pmatrix} 0 & e^{-i\Lambda} & 0 & 0 \\ e^{i\Lambda} & 0 & 0 & 0 \\ 0 & 0 & 0 & e^{-i\xi} \\ 0 & 0 & e^{i\xi} & 0 \end{pmatrix}, \quad \check{M}_2 = \begin{pmatrix} 0 & 0 & e^{-i\alpha} & 0 \\ 0 & 0 & 0 & e^{-i\beta} \\ e^{i\alpha} & 0 & 0 & 0 \\ 0 & e^{i\beta} & 0 & 0 \end{pmatrix}. \tag{22.29}$$

Asymptotic solutions for a Weyl system with $\check{M}_2$ as the boundary matrix, correspond to incoming waves with one chirality and are reflected as a mixture of both chiralities. But $\check{M}_1$ preserves the chirality index. So, the $\check{M}$-matrix specifies the behavior of the Weyl fermions in a solid when they hit the boundary surface. In the following, we show that either of $\check{M}_1$ or $\check{M}_2$ can consistently reproduce the Fermi arc. In the experimental observation of the Landau quantization of Fermi arcs [36, 37] one can infer that the relevant boundary conditions are modeled by $\check{M}_2$. The BC modeled by $\check{M}_1$ in a background B field (the Landau problem) gives a trivial $\psi = 0$ solution only. Since the two BCs cannot simultaneously hold, the BC in the above experiments can only be modeled with $\check{M}_2$ matrix.

22.7 Green's Function of Semi-infinite Weyl Semimetals

The Green's function is defined by the equation,

$$[\varepsilon - \check{H}(\vec{r})]\check{G}(\vec{r}, \vec{r}\,') = \delta(\vec{r} - \vec{r}\,'), \tag{22.30}$$

where the most general form of $\check{G}(\vec{r}, \vec{r}\,')$ is,

$$\check{G}(\vec{r}, \vec{r}\,') = \begin{pmatrix} \hat{G}_{--} & \hat{G}_{-+} \\ \hat{G}_{+-} & \hat{G}_{++} \end{pmatrix}. \tag{22.31}$$

In the above equation $\hat{G}_{\chi\chi'}$ itself is a 2×2 matrix in the spin space and has the following form,

$$\hat{G}_{\chi\chi'} = \begin{pmatrix} G^{\uparrow\uparrow}_{\chi\chi'}(z, z') & G^{\uparrow\downarrow}_{\chi\chi'}(z, z') \\ G^{\downarrow\uparrow}_{\chi\chi'}(z, z') & G^{\downarrow\downarrow}_{\chi\chi'}(z, z') \end{pmatrix} e^{[ik_x(x-x')+ik_y(y-y')]}. \tag{22.32}$$

Imposing the BC on $\check{G}(\vec{r}, \vec{r}\,')$ gives $\check{M}\check{G}(\vec{r}, \vec{r}\,')|_{z=0} = \check{G}(\vec{r}, \vec{r}\,')|_{z=0}$. Note that since we are considering a system that is infinite along the x and y directions such that the momentum along the x and y axes are good quantum numbers, a plane wave part has been factorized in (22.32).

Equation (22.30) implies two coupled equations for $G^{\sigma\sigma'}_{\chi\chi'}$ where σ and σ' denote the spin projections $\uparrow$ and $\downarrow$, and $\sigma =\uparrow$ corresponds to $\bar{\sigma} =\downarrow$ and vice versa. The above four equations decouple into two equations of the following form,

$$(\varepsilon - i\chi\sigma\partial_z + \sigma b_z)G^{\bar{\sigma}\sigma}_{\chi\chi'} - \chi(k^\chi_x + i\sigma k^\chi_y)G^{\sigma\sigma}_{\chi\chi'} = 0 \tag{22.33}$$

$$(\varepsilon + i\chi\sigma\partial_z - \sigma b_z)G^{\sigma\sigma}_{\chi\chi'} - \chi(k^\chi_x - i\sigma k^\chi_y)G^{\bar{\sigma}\sigma}_{\chi\chi'} = \frac{\delta_{\chi\chi'}}{4\pi^2}\delta(z - z'),$$

where $\sigma =\uparrow, \downarrow$ and $k^\chi_{x(y)} = k_{x(y)} - \chi b_{x(y)}$. Elimination of $G^{\sigma\sigma}_{\chi\chi'}$ between the coupled equations in (22.33) gives,

$$[q^2_\chi + (i\chi\partial_z - b_z)^2]G^{\bar{\sigma}\sigma}_{\chi\chi'}(z, z') = -\frac{k^\chi_x + i\sigma k^\chi_y}{4\pi^2\chi}\delta(z - z')\delta_{\chi\chi'},$$

where $q^2_\chi = (k^\chi_x)^2 + (k^\chi_y)^2 - \varepsilon^2$. We seek a solution of the form,

$$G^{\bar{\sigma}\sigma}_{\chi\chi'}(z, z') = -\frac{\chi(k^\chi_x + i\sigma k^\chi_y)}{8\pi^2(q_\chi + i\chi b_z)}e^{-(q_\chi + i\chi b_z)|z-z'|}\delta_{\chi\chi'} + C^{\bar{\sigma}\sigma}_{\chi\chi'}(z')e^{-(q_\chi + i\chi b_z)z}, \tag{22.34}$$

where the first term is the Green's function of the infinite system (a homogeneous differential equation) and is chirality diagonal (non-zero only for $\chi = \chi'$). The poles of the first term $q_\chi + i\chi b_z = 0$ give the bulk dispersion relation. The second term is a solution of the inhomogeneous differential equation and therefore incorporates the effects of the boundary. The coefficients $C^{\sigma\bar{\sigma}}_{\chi\chi'}(z')$ are determined in such a way as to satisfy the BC.

Once the spin-flip components of the Green's function in (22.34) are known, the spin-diagonal components are immediately obtained from the first line of (22.33) as,

$$G^{\sigma\sigma}_{\chi\chi'}(z, z') = \frac{\varepsilon - i\chi\sigma\partial_z + \sigma b_z}{\chi(k^\chi_x + i\sigma k^\chi_y)}G^{\bar{\sigma}\sigma}_{\chi\chi'}(z, z'). \tag{22.35}$$

Therefore we only need to obtain the coefficients $C^{\sigma\sigma'}_{\chi\chi'}$ for the two BCs corresponding to $\check{M}_{1,2}$ and demonstrate that in the absence of a magnetic field, both BCs give rise to Fermi arcs.

22.7.1 $\check{M}_1$-Type BC

If we choose $\check{M}_1$ as the boundary matrix, then the boundary does not mix the chiralities so we can treat them separately and solve the problem for just one chirality. In this situation the elements of the full Green function that mix the chiralities are zero. This leaves only $\hat{G}_{\chi\chi}$ to be computed which in the spin-space is,

$$\hat{G}_{\chi\chi}(\vec{r},\vec{r}\,') = \begin{pmatrix} 1 & \frac{\varepsilon - i\chi\partial_z + b_z}{\chi(k_x^\chi + ik_y^\chi)} \\ \frac{\varepsilon + i\chi\partial_z - b_z}{\chi(k_x^\chi - ik_y^\chi)} & 1 \end{pmatrix} \tag{22.36}$$

$$\times \begin{pmatrix} 0 & G_{\chi\chi}^{\uparrow\downarrow}(z,z') \\ G_{\chi\chi}^{\downarrow\uparrow}(z,z') & 0 \end{pmatrix} e^{[ik_x(x-x')+ik_y(y-y')]}.$$

Operating with the $\check{M}_1$-type matrix on the Green's function we obtain,

$$e^{i\sigma\theta_\chi} G_{\chi\chi'}^{\sigma\sigma'}(z,z')|_{z=0} = G_{\chi\chi'}^{\bar{\sigma}\sigma'}(z,z')|_{z=0} \tag{22.37}$$

where $\theta_- = \Lambda$, $\theta_+ = \xi$ and $\sigma\theta_\chi = \pm\theta_\chi$ for $\sigma = \uparrow / \downarrow$. With this BC, the coefficients of (22.34) are obtained as,

$$C_{\chi\chi'}^{\bar{\sigma}\sigma} = \frac{\varepsilon - i\chi\sigma q_\chi + 2\sigma b_z - \chi e^{-i\sigma\theta_\chi}(k_x^\chi + i\sigma k_y^\chi)}{\varepsilon + i\chi\sigma q_\chi - \chi e^{-i\sigma\theta_\chi}(k_x^\chi + i\sigma k_y^\chi)} \times \frac{\chi(k_x^\chi + i\sigma k_y^\chi)}{8\pi^2(q_\chi + i\chi b_z)} e^{-(q_\chi + i\chi b_z)z'} \delta_{\chi\chi'}. \tag{22.38}$$

The dispersion relation of the Fermi arc states corresponds to the poles of the above Green's function arising from the first line. The denominator vanishes when [38]

$$\varepsilon + i\chi\sigma q_\chi - \chi e^{-i\sigma\theta_\chi}(k_x^\chi + i\sigma k_y^\chi) = 0. \tag{22.39}$$

This equation with the definition of q_χ as $q_\chi^2 + \varepsilon^2 = (k_x^\chi)^2 + (k_y^\chi)^2$ yields the dispersion energy of the surface states as:

$$\varepsilon = \chi|k^\chi|\cos(\phi_\chi - \theta_\chi), \tag{22.40}$$

$$q_\chi = |k^\chi|\sin(\phi_\chi - \theta_\chi), \tag{22.41}$$

where $(|k^\chi|, \phi_\chi)$ are the polar coordinates corresponding to (k_x^χ, k_y^χ) and as defined above θ_χ are the angles arising from BC. So we have found surface localized states that are supported on the *line segment* $(k_x - \chi b_x)\cos\theta_\chi + (k_y - \chi b_y)\sin\theta_\chi - \varepsilon\chi = 0$ in the $k_x - k_y$ plane. At $\varepsilon = 0$ (which corresponds to the Fermi energy at bulk Weyl nodes), for $\chi = -1$, this is a straight line with a slope of $-\cot\Lambda$ that ends at $(k_x, k_y) = (b_x, b_y)$, and for $\chi = 1$, this is a line with a slope of $-\cot\xi$ that ends at $(k_x, k_y) = (-b_x, -b_y)$; see Fig. 22.3. At the endpoints where $q_\chi \to 0$, the localization length of the surface states set by q_χ^{-1} (see the second term in (22.34)) diverges, and therefore the Fermi arc states penetrate infinitely deep into the bulk. This way, tunneling from one surface to the opposite surface becomes possible. This effect has been discussed in the presence of a background B field by [40]. Note that, although the Fermi arc states close to the arc ends have extremely long localization lengths, they are distinct from the bulk states (those arising from the homogeneous part of (22.34)). To summarize, (22.40) is a Fermi arc that has its two end associated with

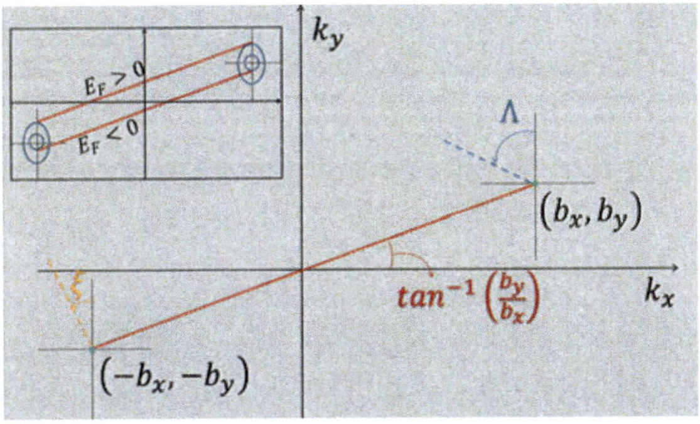

Fig. 22.3 Constant energy curves obtained from the poles of Green's function, (22.38), of a confined Weyl semimetal. For every chirality we obtain an independent energy dispersion that terminates at the projections of the Weyl nodes on the surface of the Brillouin zone. The slope of the Fermi rays denoted as dashed lines emanating from the projection of the nodes are determined from the boundary matrix parameters (Λ and ξ). In the absence of a mechanism for bending the Fermi rays, they orient themselves along the line segment connecting the projections of the two Weyl nodes. Inset: When the Fermi energy is $E_F \neq 0$, the Fermi arc has a non-zero intercept with the k_y and k_x axes, consistent with the formation of the bulk Fermi surfaces surrounding each Weyl point [39] (Adapted from [29])

the band crossing points $\vec{b}$ and $-\vec{b}$ and connects the projection of the Weyl points on the surface of the Brillouin zone, which in the real space are localized in the physical surface of the sample.

By Nielsen-Ninomiya theorem [13], the arc starting from the projection of one Weyl node has to end in the projection of another node with the opposite chirality. Therefore the above two slopes must be equal, i.e., $-\cot(\Lambda) = -\cot(\xi)$ which constrains the BC angles by $\Lambda - \xi = m\pi$. Furthermore the actual slope of the Fermi arcs is $\frac{b_y}{b_x}$ which fixes the BC angle as $\Lambda = -\cot^{-1}(\frac{b_y}{b_x}) + m'\pi$. In the above equations m and m' are integers.

For $\varepsilon \neq 0$ which corresponds to the situation where the Fermi level is shifted above or below the Weyl points, the line equation acquires an intercept with respect to (k_x, k_y) axis denoted in Fig. 22.3. This result is in agreement with the Haldane's prediction [39] according to which the end points of the Fermi arc are two points on the projected bulk Fermi surfaces (the inset in Fig. 22.3). The sign of ε, determines the sign of the intercept. For $\varepsilon > 0$, the shift of the end points of the Fermi arcs is positive. This is satisfied if $-\frac{\varepsilon}{\sin\Lambda} > 0$ and $\frac{\varepsilon}{\sin\xi} > 0$. Therefore the signs of $\sin\Lambda$ and $\sin\xi$ are negative and positive respectively, which confines m to odd integers and m' to even integers.

Since we have considered only the linear corrections around the band touching points, the Fermi arc is obtained as a straight line-segment. But in realistic materials the Fermi arcs must be bent at least for two reasons: (i) higher order corrections taking the band curvature into account leads to bending in the Fermi arc, which are captured in lattice models of Weyl semi-metals [41]; (ii) when the boundary is not atomically sharp, the boundary or interface can be modeled with an effective Hamiltonian which in turn gives rise to a curvature on the Fermi arc [32, 42].

22.7.2 $\check{M}_2$ as the Boundary Matrix

Now if we use the $\check{M}_2$-type BC, all the elements of Green's function in the chirality space are nonzero and the BC relates the wave functions with opposite chiralities:

$$e^{-i\chi\theta_\sigma} G^{\sigma\sigma'}_{\chi\chi'}(z,z')|_{z=0} = G^{\sigma\sigma'}_{\bar{\chi}\chi'}(z,z')|_{z=0}, \tag{22.42}$$

where $\theta_\uparrow = \alpha, \theta_\downarrow = \beta$ and $\bar{\chi} = -\chi$. Imposing the $\check{M}_2$-type BC gives the coefficients $C^{\bar{\sigma}\sigma}_{\chi\chi}$ and $C^{\bar{\sigma}\sigma}_{\bar{\chi}\chi}$ as,

$$C^{\bar{\sigma}\sigma}_{\chi\chi} = \frac{i\chi\sigma(k^{\chi}_x + i\sigma k^{\chi}_y)(k^{\bar{\chi}}_x + i\sigma k^{\bar{\chi}}_y)}{8\pi^2 D^{\bar{\sigma}\sigma}_{\chi\chi}} e^{-(q_\chi + i\chi b_z)z'}, \tag{22.43}$$

$$C^{\bar{\sigma}\sigma}_{\bar{\chi}\chi} = \frac{\chi(k^{\chi}_x + i\sigma k^{\chi}_y)}{8\pi^2(q_\chi + i\chi b_z)}\left(\frac{N^{\bar{\sigma}\sigma}_{\chi\chi}}{D^{\bar{\sigma}\sigma}_{\chi\chi}}\right) e^{-(q_\chi + i\chi b_z)z'}, \tag{22.44}$$

where

$$D^{\bar{\sigma}\sigma}_{\chi\chi} = \bar{\chi} e^{i\chi\theta_{\bar{\sigma}}}(\varepsilon + i\chi\sigma q_\chi)(k^{\bar{\chi}}_x + i\sigma k^{\bar{\chi}}_y) - \chi e^{i\chi\theta_\sigma}(\varepsilon + i\bar{\chi}\sigma q_{\bar{\chi}})(k^{\chi}_x + i\sigma k^{\chi}_y),$$

and

$$N^{\bar{\sigma}\sigma}_{\chi\chi} = D^{\bar{\sigma}\sigma}_{\chi\chi} + 2i\sigma e^{i\chi\theta_{\bar{\sigma}}}(q_\chi + i\chi b_z)(k^{\bar{\chi}}_x + i\sigma k^{\bar{\chi}}_y). \tag{22.45}$$

The dispersion relation is obtained from $D^{\bar{\sigma}\sigma}_{\chi\chi} = 0$. This equation gives the same Fermi arc as (22.40). To see this, in the above equation substitute for $\varepsilon + i\chi\sigma q_\chi$ and $\varepsilon + i\bar{\chi}\sigma q_{\bar{\chi}}$ from (22.39), which then gives $\alpha - \beta = \Lambda - \xi \mod 2\pi$. Therefore this equation gives the Fermi arcs in the same way that (22.39) does. Consequently, both $\check{M}_1$ and $\check{M}_2$ reproduce the Fermi arcs on the surface. Therefore as long as the experimentally observed Fermi arcs are concerned, both types of BC are compatible with the Fermi arcs. However, when we turn on a background magnetic field to study the Landau problem of the surface states, the $\check{M}_1$-type BC can only produce the trivial solution $\check{\psi} = 0$, while the $\check{M}_2$-type BC is compatible with non-trivial Landau orbitals [29]. Note that the BC proposed by Witten [30] only includes the $\check{M}_1$-type BC which does not give rise to Landau quantization of the surface states in Weyl materials.

22.8 Superconducting Proximity in Fermi Arc States

As pointed out, the electron system at the surface of Weyl semimetals is composed of Fermi arc states that differ from an ordinary two-dimensional electron gas in that the Fermi arc states leak more and more into the bulk as one approaches the end of the Fermi arcs. How will this difference manifest itself when we consider the superconducting proximity of the Fermi arc states? In a standard 2D electron system brought into proximity with a conventional BCS superconductor, the only possibility is to induce conventional superconductivity with a clean BCS gap. However, as we will show in the following section, fascinating forms of topologically non-trivial superconductivity can be induced into the "2D" Fermi arc states.

In (22.14), for a given electron Hamiltonian, we constructed the hole Hamiltonian required in the Nambu representation by operating on the electron part by a charge conjugation operation. Alternatively one can operate with the time-reversal operation [43] (22.23) describes the electrons in a Weyl semimetal. Let us call H_e the Hamiltonian of the electrons away from the boundaries (where the $\check{M}$ term has no effect). Then the hole Hamiltonian $H_h(\boldsymbol{k}) = \sigma_y H_e^*(\boldsymbol{k})\sigma_y$ is

$$H_h(\boldsymbol{k}) = \tau_z(\vec{\sigma}.\boldsymbol{k}) - \tau_0(\vec{\sigma}.\vec{b}), \tag{22.46}$$

where $\boldsymbol{k} = -i\nabla$ is understood. This can be combined with the electronic part to give the Bogoliubov-De Gennes Hamiltonian,

$$H_W = \begin{pmatrix} H_e & 0 \\ 0 & -H_h \end{pmatrix}. \tag{22.47}$$

Let us now return to the question of the boundary. The crucial point in constructing Green's function for holes is that, the particle-hole transformation should also operate the on matrix $\check{M}$ in (22.23) that encodes the BC information. Starting with the BC matrix $\check{M}_1$ for the electrons in (22.26) and operating with time-reversal, the BC for the holes will be $\sigma_y \check{M}_1^* \sigma_y = -\check{M}_1$. This is eventually equivalent to the substitution $\Lambda \to \pi + \Lambda$ and $\xi \to \pi + \xi$. This is quite intuitive, since the reflection of an electron with its in-plane spin rotated by an angle Λ after the TR operation can be equivalently viewed as the rotation of the spin of a hole by an angle $\pi + \Lambda$. Similarly for the $\check{M}_2$-type BC in (22.27) we have

$$\sigma_y \check{M}_2^* \sigma_y = (\cos\alpha\, \hat{\tau}_x - \sin\alpha\, \hat{\tau}_y) \otimes \frac{\hat{\sigma}_0 - \hat{\sigma}_z}{2} + (\cos\beta\, \hat{\tau}_x - \sin\beta\, \hat{\tau}_y) \otimes \frac{\hat{\sigma}_0 + \hat{\sigma}_z}{2}.$$

Therefore the $\check{M}_2$ BC matrix for the holes is obtained from the corresponding $\check{M}_2$ of the electrons by the replacement $\alpha \leftrightarrow -\beta$.

Now, we are ready to set up Green's function for the holes that satisfies

$$[\varepsilon + H_h + \check{M}_h \delta(z)] G_h = \delta(\vec{r} - \vec{r}'),$$

where the matrix $\check{M}_h$ can be any of the matrices discussed above. For the electron wave function $\psi_e = \left[\psi_{-\uparrow}\ ,\ \psi_{-\downarrow}\ ,\ \psi_{+\uparrow}\ ,\ \psi_{+\downarrow}\right]^T$ the hole wave function obtained from time-reversal will be $\psi_h = \left[-\psi^*_{+\downarrow}\ ,\ \psi^*_{+\uparrow}\ ,\ -\psi^*_{-\downarrow}\ ,\ \psi^*_{-\uparrow}\right]^T$. Intuitively, the time-reversal flips the chirality $+ \leftrightarrow -$ and up to a phase factor i it replaces $\uparrow \rightarrow -\downarrow$ and $\downarrow \rightarrow \uparrow$.

So that the $\check{G}_h(\vec{r}, \vec{r}\,')$ will be arranged into the following matrix,

$$\check{G}_h(\vec{r}, \vec{r}\,') = \begin{pmatrix} [\hat{G}_{++}(\vec{r}, \vec{r}\,')]_h & [\hat{G}_{+-}(\vec{r}, \vec{r}\,')]_h \\ [\hat{G}_{-+}(\vec{r}, \vec{r}\,')]_h & [\hat{G}_{--}(\vec{r}, \vec{r}\,')]_h \end{pmatrix}. \tag{22.48}$$

In the above equation $[\hat{G}_{\chi\chi'}(\vec{r}, \vec{r}\,')]_h$ has the following form in the spin space

$$[\hat{G}_{\chi\chi'}]_h = \begin{pmatrix} G^{\downarrow\downarrow}_{\chi\chi'}(z, z') & G^{\downarrow\uparrow}_{\chi\chi'}(z, z') \\ G^{\uparrow\downarrow}_{\chi\chi'}(z, z') & G^{\uparrow\uparrow}_{\chi\chi'}(z, z') \end{pmatrix} e^{[ik_x(x-x')+ik_y(y-y')]}, \tag{22.49}$$

where every element in the above equation is obtained from the corresponding element of Green's function for the electron by appropriate replacements of the angular parameters α, β, Λ and ξ as discussed above. After this replacement (and of course changing the sign of the energy) the spin-off-diagonal elements of Green's functions for the hole become,

$$G^{\bar{\sigma}\sigma}_{\chi\chi'}(z, z') = C^{\bar{\sigma}\sigma}_{\chi\chi'}(z')e^{-(q_\chi+i\chi b_z)z} - \frac{\chi(k^\chi_x - i\sigma k^\chi_y)}{8\pi^2(q_\chi + i\chi b_z)}e^{-(q_\chi+i\chi b_z)|z-z'|}\delta_{\chi\chi'}, \tag{22.50}$$

whereas the spin-diagonal components are,

$$G^{\sigma\sigma}_{\chi\chi'}(z, z') = \frac{\varepsilon + i\chi\sigma\partial_z - \sigma b_z}{\chi(k^\chi_x - i\sigma k^\chi_y)} G^{\bar{\sigma}\sigma}_{\chi\chi'}(z, z'), \tag{22.51}$$

where as before, $k^\chi_{x(y)} = k_{x(y)} + \chi b_{x(y)}$. The value of these matrix elements is the same as those for the electrons, except for the replacement $\sigma \rightarrow -\sigma$.

Up to this point the above expressions are valid for any BC. For the $\check{M}_1$-type BC we have:

$$\begin{aligned} C^{\bar{\sigma}\sigma}_{\chi\chi'} = & \frac{\varepsilon + i\chi\sigma q_\chi - 2\sigma b_z - \chi e^{i\sigma\theta_\chi}(k^\chi_x - i\sigma k^\chi_y)}{\varepsilon - i\chi\sigma q_\chi - \chi e^{i\sigma\theta_\chi}(k^\chi_x - i\sigma k^\chi_y)} \\ & \times \frac{\chi(k^\chi_x - i\sigma k^\chi_y)}{8\pi^2(q_\chi + i\chi b_z)} e^{-(q_\chi+i\chi b_z)z'}\delta_{\chi\chi'}. \end{aligned} \tag{22.52}$$

where $\theta_- = \Lambda + \pi$ and $\theta_+ = \xi + \pi$, and for the $\check{M}_2$-type BC, the spin-off-diagonal components are,

$$C_{\bar{\chi}\chi}^{\bar{\sigma}\sigma} = \frac{-i\chi\sigma(k_x^{\chi} - i\sigma k_y^{\chi})(k_x^{\bar{\chi}} - i\sigma k_y^{\bar{\chi}})}{8\pi^2 D_{\chi\chi}^{\bar{\sigma}\sigma}} e^{-(q_\chi + i\chi b_z)z'} \tag{22.53}$$

and

$$C_{\chi\chi}^{\bar{\sigma}\sigma} = \frac{\chi(k_x^{\chi} - i\sigma k_y^{\chi})}{8\pi^2(q_\chi + i\chi b_z)} \left(\frac{N_{\chi\chi}^{\bar{\sigma}\sigma}}{D_{\chi\chi}^{\bar{\sigma}\sigma}}\right) e^{-(q_\chi + i\chi b_z)z'}, \tag{22.54}$$

where

$$D_{\chi\chi}^{\bar{\sigma}\sigma} = \bar{\chi}e^{i\chi\theta_{\bar{\sigma}}}(\varepsilon - i\chi\sigma q_\chi)(k_x^{\bar{\chi}} - i\sigma k_y^{\bar{\chi}}) - \chi e^{i\chi\theta_\sigma}(\varepsilon - i\bar{\chi}\sigma q_{\bar{\chi}})(k_x^{\chi} - i\sigma k_y^{\chi}) \tag{22.55}$$

and

$$N_{\chi\chi}^{\bar{\sigma}\sigma} = D_{\chi\chi}^{\bar{\sigma}\sigma} - 2i\sigma e^{i\chi\theta_{\bar{\sigma}}}(q_\chi + i\chi b_z)(k_x^{\bar{\chi}} - i\sigma k_y^{\bar{\chi}}), \tag{22.56}$$

with $\theta_\uparrow = -\beta$ and $\theta_\downarrow = -\alpha$.

For practical calculations one has to specialize to a specific coordinate system. The coordinate system can be chosen in such a way that the Fermi arc lies along the k_x axis. This does not harm the generality of the approach, since always, by appropriate rotation along the k_z axis, a new coordinate system can be chosen in such a way that the new k_x is along the Fermi arc [29].

22.9 Inducing Superconductivity in Fermi Arcs

Now, we bring a conventional s-wave superconductor (SC) near the WSM. The SC occupies the $z > 0$ part of the space and WSM occupies the $z < 0$ part with their interface being at $z = 0$. The bulk Hamiltonian of the SC is:

$$H_s = [|\boldsymbol{k}_s|^2/(2m)\hat{\kappa}_3 + \Delta_s\hat{\kappa}_1] \otimes \hat{\sigma}_0, \tag{22.57}$$

where $\boldsymbol{k}_s$ denotes the momentum in the SC, m is the electron mass in the SC, Δ_s is the superconducting gap and $\hat{\kappa}_{(i=0\ldots3)}$ are the Pauli matrices acting in the particle-hole space. As before, the coupling between WSM and SC can be incorporated by:

$$\mathcal{T} = \begin{pmatrix} 0 & \check{t}^\dagger \\ \check{t} & 0 \end{pmatrix}, \tag{22.58}$$

where, assuming that the tunneling amplitude t is the same for right handed and left handed electrons, and the 4×8 matrix $\check{t}$ is constructed as $\check{t} = t/2(\check{t}_+ \ \check{t}_-)$ from 4×4 matrices $\check{t}_\alpha = (\hat{\tau}_z + \alpha\hat{\tau}_0 + \hat{\tau}_1 + i\alpha\hat{\tau}_y) \otimes \hat{\sigma}_0$, with $\alpha = \pm$. Based on the Dyson equation, Green's function of the WSM becomes:

$$\mathcal{G}_W = \mathcal{G}_W^0 + \sum_{k_s} \mathcal{G}_W^0 . \breve{t}^\dagger . \breve{g}_s . \breve{t} . \mathcal{G}_W, \tag{22.59}$$

where we use the symbols $\breve{g}_s$ to denote 4×4 matrices and $\mathcal{G}$ for 8×8 matrices. The superscript 0 in $\mathcal{G}^0$ denotes the Green's function in the Nambu-space in the absence of tunneling.

Assuming that the superconductivity at the surface of the SC is of the same form as its bulk, and that $\breve{t}$ and $\breve{t}^\dagger$ in (22.59) are independent of $\boldsymbol{k}_s$, we can perform the sum over $\boldsymbol{k}_s$ to obtain the self energy [44]:

$$\sum_{k_s} \breve{t}^\dagger . \breve{g}_s . \breve{t} = \frac{s}{\sqrt{\Delta^2 - \epsilon^2}} (\epsilon \hat{\kappa}_0 - \Delta \hat{\kappa}_1) \otimes (\hat{\tau}_0 + \hat{\tau}_x) \otimes \hat{\sigma}_0. \tag{22.60}$$

where $s = \pi \rho_0 t$ and ρ_0 is the density of states of the superconductor at its Fermi level before becoming a superconductor. Substituting this result is (22.59), we can derive Green's function for the surface of the WSM in presence of a SC.

22.9.1 $\check{M}_1$-Type BC

Once Green's functions obtained in (22.50) and (22.51) are corrected with the Dyson (22.59), the poles of the new Green's function give us the dispersion relation of the excitations on the surface. For an $\check{M}_1$-type BC, we obtain the following secular equation for the poles of Green's function:

$$[\mathcal{F}(\epsilon, \boldsymbol{k}) + 4\epsilon b k_y s^2]^2 - 16 s^4 \epsilon^4 (k_x^2 + k_y^2) = 0, \tag{22.61}$$

where $\mathcal{F}(\epsilon, \boldsymbol{k}) = \sqrt{\Delta^2 - \epsilon^2}[4 s^4(-b^2 + k_x^2 + k_y^2) - (\epsilon^2 - k_y^2)]$ and the tunneling strength s quantifies the ability of the electrons in the superconductor to tunnel into WSM. The states at the Fermi level correspond to $\epsilon = 0$ which will be equivalent to $\mathcal{F}^2(0, \boldsymbol{k}) = 0$. Therefore the solutions of $\mathcal{F}(0, \boldsymbol{k}) = 0$ will be twofold degenerate. These solutions are given by the following ellipse in the $k_x - k_y$ plane (see Fig. 22.4):

$$k_x^2 + (\frac{1 + 4s^4}{4s^4}) k_y^2 = 1. \tag{22.62}$$

The major axis of this ellipse is horizontal with magnitude 1 (note that in our units a momentum of size 1 actually means b) and coincides with the Fermi arc of a pristine Weyl semimetal before bringing the superconductor to its proximity. This is similar to the zero-energy surfaces due to the Fermi arcs of *doped* WSMs [45]. Furthermore, the magnitude of the minor axis, $\tilde{b} = \frac{2s^2}{\sqrt{1+4s^4}}$ is determined by the combination s of the tunneling amplitude t and the density of states ρ_0 of the superconductor in its normal phase. As such, when the superconducting agent is an undoped Dirac

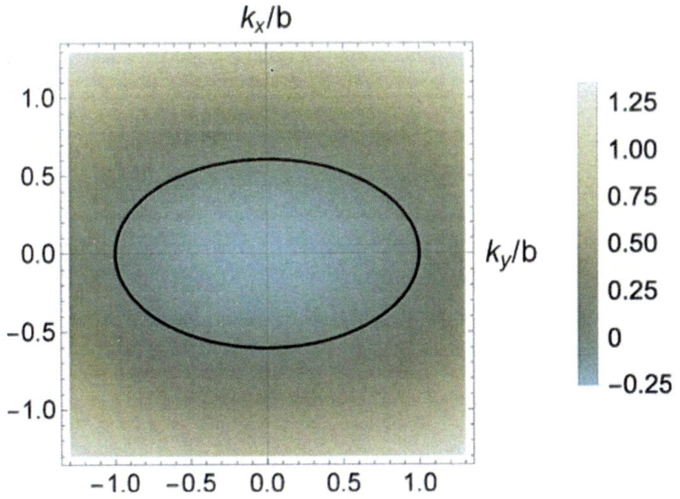

Fig. 22.4 Bogoliubov Fermi contour for the first type boundary conditions. The major axis of the ellipse coincides with the Fermi arc of the WSM before bringing the SC to contact with it. The Pfaffian (see the text) changes sign across this contour and excitations around the elliptic contour are linearly dispersing. The minor axis of the ellipse depends on the tunneling strength as in (22.62). By turning off the tunneling the minor axis becomes zero, and the ellipse reduces to the Fermi arc (Adapted from [46])

superconductor [25], due to $\rho_0 = 0$, the minor axis will be of zero length, and the ellipse will collapse into the Fermi arc. It is curious that although the very existence of the ellipse depends on the superconducting gap Δ of the s-wave superconductor that proximitizes the WSM, the minor axis does not depend on the superconducting gap Δ and is only controlled by the tunneling strength s. Therefore, the ability of the electrons to tunnel into Fermi arcs, gives rise to a superconducting state with an ellipse shaped Bogoluibov Fermi contour (BFC) that originates from the arc.

Let us discuss the topological significance of the BFCs. At $\epsilon = 0$, the denominator of Green's function $(\epsilon - H)^{-1}$ will become the determinant of the Hamiltonian, i.e., $\mathcal{F}^2(0, \boldsymbol{k}) = \det H(\boldsymbol{k})$, where $H(\boldsymbol{k})$ is the Hamiltonian of the entire system. The above relations mean that $\mathcal{F}(0, \boldsymbol{k})$ is actually the Pfaffian of the Hamiltonian. Following [47, 48] we use $\mathcal{F}(0, \boldsymbol{k})$ to construct the Z_2 topological index ν that protects the zero-energy ellipse of Bogoliubov quasi-particles as $(-1)^\nu = \mathrm{sgn}\big[\mathcal{F}(\boldsymbol{k}_-)\mathcal{F}(\boldsymbol{k}_+)\big]$ where $\boldsymbol{k}_+(\boldsymbol{k}_-)$ refers to the momenta inside (outside) of the BFC [47, 48]. As can be seen $\mathcal{F}(0, \vec{k})$ changes its sign across the elliptic zero energy contour and therefore we are dealing with the $\nu = -1$ situation which is Z_2-non-trivial. In our two-dimensional case, the Z_2 index is only consistent with the DIII class which belongs to the BdG family [49, 50]. In this class, particle-hole and sublattice symmetry must be present; which is the case by construction. The TR must be broken, which is again the case, as the parent WSM is characterized by the TR breaking parameter $\vec{b}$. The meaning of $\nu = -1$ is that the weak perturbations within the DIII class are not able to destroy the elliptic Fermi contour of the Bogoliubov quasiparticles. A simple consequence of this robustness is that by changing the tunneling parameter s, only the minor axis of the ellipse changes, but it cannot be cut into pieces or destroyed. As we will see in the following, with the $\check{M}_2$-type BC, we will have a totally different situation.

In terms of the Altland-Zirnbauer [51] classification, the induced superconductivity on Fermi arc states belongs to the DIII class. The interpretation of its Z_2 index is connected with the existence of the (elliptic) BFC. Once the Fermi contour is formed, the Fermi contour *itself* can be further classified as a singularity of Green's function in the momentum space by a winding number [15]. This is defined as

$$n_1 = \mathrm{tr}\frac{1}{2\pi i}\oint_C G\partial_\ell G^{-1} d\ell, \tag{22.63}$$

where the closed path C is any contour enclosing the Fermi contour (ellipse in our case) and ℓ parameterizes this path. For the Fermi contour of two-dimensional metals, as long as it has the Fermi liquid structure $G(i\omega, p) \propto (i\omega - p)^{-1}$, where p is the momentum deviation from the Fermi contour, the above winding number will be ± 1. However, an essential difference between the elliptic Fermi contour of Bogoliubov quasiparticles compared to the Fermi contour of Fermi liquids is that, due to two-fold degeneracy, the pole structure near the Fermi contour is given by $G(i\omega, p) \propto (i\omega - p)^{-2}$. This form of Fermi contour will give $n_1 = \pm 2$. This means that in principle there can be perturbations outside the DIII class which can break the $n_1 = 2$ topological charge into two $n_1 = 1$ (Fermi liquid-like) Fermi contours.

To gain further insight into the physical nature of this BFC, let us study the excitations around this elliptic Fermi contour. In the radial direction, a little away from the ellipse we can use a small parameter η to parameterize the momenta at $\epsilon = 0$ as $k_x = (1+\eta)\cos\phi$ and $k_y = (\tilde{b}+\eta)\sin\phi$. Let us assume that by approaching the ellipse, the energy vanishes as $\alpha\eta^\gamma$. With this choice, the lowest order terms of (22.61) are:

$$\begin{aligned} &\frac{4\tilde{b}^4}{1-\tilde{b}^2}\alpha^2\eta^{2\gamma}\sin^2\phi \\ &+\left(\frac{2\Delta\tilde{b}^2}{1-\tilde{b}^2}\right)^2\eta^2(\cos^2\phi+\frac{1}{\tilde{b}}\sin^2\phi)^2 \\ &+\left[\frac{8\Delta\tilde{b}^4}{(1-\tilde{b}^2)^{3/2}}\right]\alpha\eta^{\gamma+1}(\cos^2\phi+\frac{1}{\tilde{b}}\sin^2\phi)\sin\phi \\ &=\frac{4\tilde{b}^2}{1-\tilde{b}^2}\alpha^4\eta^{4\gamma}(\cos^2\phi+\frac{1}{\tilde{b}}\sin^2\phi) \end{aligned} \tag{22.64}$$

If $\gamma > 1$, then only the second term on the left-hand side is the leading order term and should be zero but this is generically impossible. On the other hand if $\gamma < 1$, then the first term in (22.64) is the leading order term and leads to $\alpha = 0$. We thus conclude that $\gamma = 1$ and that around the BFC, the energy disperses linearly. There are only two exceptions to $\gamma = 1$: at $\phi = 0$ and $\phi = \pi$ (related to the former by symmetry) which correspond to dispersion along k_x axis. These two peculiar points correspond to the projection of the Weyl nodes on the $k_x - k_y$ surface. In this case, $\sin\phi = 0$ and (22.64) reduces to,

$$\left(\frac{\Delta\tilde{b}^2}{1-\tilde{b}^2}\right)\eta^2 = \alpha^4\eta^{4\gamma}. \tag{22.65}$$

from which we obtain $\gamma = 1/2$. Therefore the singular behavior at $\phi = 0$ means that by departing from the projection of the Weyl nodes in the k_x direction inward to the ellipse, we obtain a peculiar $\varepsilon(p_x, p_y = 0) \sim \sqrt{p_x}$ where p_x and p_y measure the momenta from the two ends of the major axis of the ellipse.

22.9.2 $\check{M}_2$-Type BC

Unlike the $\check{M}_1$-type BC where a robust BFC is obtained which can be distorted but not destroyed by changing the parameters of the Hamiltonian (in our case the combination $s = \pi\rho_0 t$), for $\check{M}_2$-type BC, instead of BFC we will have a set of Bogoliubov-Weyl (BW) nodes. To see this, let us look into the zeros of the determinant appearing in the denominator of Green's function, which at $\epsilon = 0$ becomes,

$$\begin{aligned}&\left\{\left[3(b^2 - k_x^2 + ky^2)^2 + 4k_x^2(b^2 - 4k_y + 3k_y^2)\right]s^4 - b^2k_y^2\right\}^2 \\ &+ 16b^2k_x^2k_y^2(2k_y - b)^2\,s^4 = 0.\end{aligned} \tag{22.66}$$

This expression being the sum of two complete squares appearing on the first and second lines, respectively, can only vanish when each term vanishes separately. From the second line there are three possibilities, namely $k_x = 0$, $k_y = b/2$, or $k_y = 0$. The third case does not give any zeros for the first line. The first two cases, however, give two pairs of solutions as follows (note that we are working in units of $b = 1$). On the $k_x = 0$ line there are two values of k_y^2 as long as tunneling is less than $s_{\max} = \left[(4 - \sqrt{15})/6\right]^{1/4} \approx 0.38$. As can be seen in Fig. 22.5a, the two solutions move toward each other and meet at $s_{\max}$. Beyond $s_{\max}$ there is no zero energy solution on the k_y axis, meaning that the two BW nodes annihilate each other upon collision. This indicates that they are carrying opposite topological charges. Their partner in negative k_y axis also behaves similarly. This has been schematically shown in the second row of this figure. Figure 22.5b shows that on the $k_y = 1/2$ line, the blue pair of BW nodes start at $k_x^2 \approx 0$ for very small $s \approx 0$. As can be seen k_x^2 increases linearly as we increase s. Beyond $s_{\min} = (4/75)^{1/4} \approx 0.48$, a second pair of (red) BW nodes appear on the $k_y = 1/2$ line and start their journey from the $k_x^2 = 0$ point. By further increasing s, the blue and red BW nodes further depart from each other.

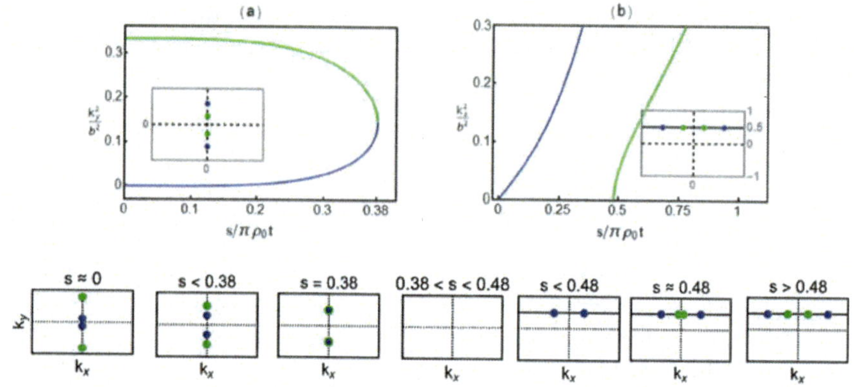

Fig. 22.5 Bogoliubov-Weyl nodes with second type boundary conditions. **a** k_y^2 coordinate of the position of BW nodes as a function of dimensionless tunneling strength s. **b** The k_x^2 coordinate of the nodes lying on $k_y = b/2$ as a function of s. Inset in both (**a**) and (**b**) indicates location of BW nodes. Panels in the second row show the schematic evolution of BW nodes upon varying s (Adapted from [46])

22.9.3 Linear Combination of $\check{M}_1$-Type and $\check{M}_2$-Type BCs

Let us now see what happens if the boundary condition is neither $\check{M}_1$ nor $\check{M}_2$ type, but a linear combination of the form $M = \frac{1}{1+\lambda^2}(M_1 + \lambda M_2) \approx M_1 + \lambda M_2$. Assuming that Green's functions of a Weyl semimetal with type-1 and type-2 BCs are g_1 and g_2, respectively. Green's function g for the mixed BC can be perturbatively obtained for small λ as follows:

$$g = \frac{g_1}{1 + \lambda g_1 M_2} \approx g_1 - \lambda g_1 M_2 g_1, \tag{22.67}$$

where higher powers of λ are ignored, and because of the $\delta(z)$ in (22.23) it is understood that the self-energy corrections arising from mixing of the $\check{M}_2$-type boundary condition of strength λ are non-zero only at the interface. When combined with (22.59) the above equation gives,

$$\det g_1 \to \det g_1 \left[1 - \lambda \mathrm{tr} \check{M}_2 g_1\right]. \tag{22.68}$$

As perturbative thinking is valid the poles of $\det g_1$ as long determine the pole of the above renormalized Green's function. Therefore, as long as λ remains within the reach of perturbation theory, the above self-energy arising from the chirality flip ($\check{M}_2$) at the interface will only produce a renormalization of spectral features of g_1. So the picture will be as follows. At $\lambda = 0$ (i.e., pure $\check{M}_1$-type BC) we have a robust Bogoliubov Fermi contour which is protected by topology. At $\lambda = \infty$ (i.e., pure $\check{M}_2$-type BC) we have a superconducting phase with point nodes in the spectrum of

Bogoliubov quasiparticles of the proximitized surface. These nodes are also protected by the topological charge of the BW nodes. Therefore it is likely that a phase transition at a finite λ separates the physics of $\check{M}_1$ from $\check{M}_2$ BC.

22.10 Pairing Symmetry of Fermi Arc States: Majorana Fermi Contour

So far we have shown that the $\check{M}_1$-type BC gives a topologically protected BFC. Now we are going to discuss its consequences. As pointed out earlier, the Cooper pairs can be either chirality-even or chirality-odd. In the following we discuss these two cases separately.

22.10.1 Chirality-Even Pairing

It is useful to form combinations of the pairing amplitudes which are even or odd under the exchange of the chirality index. Each of these Δs is a 2×2 matrix in the spin space and can be written as a sum of singlet and triplet components,

$$\hat{\Delta} = i\sigma_y(d_0 + \vec{d}.\vec{\sigma}). \tag{22.69}$$

The even-chirality part of anomalous Green's function is

$$\hat{F}_+ = h\begin{bmatrix} -ik_x + k_y & -b \\ -b & ik_x + k_y \end{bmatrix}, \tag{22.70}$$

$$h = \frac{4\Delta_s k_y s^2[\mathcal{F} - 4b\epsilon s^2(\epsilon - k_y)]}{(\mathcal{F} + 4\epsilon b k_y s^2)^2 - 16\epsilon^2 s^4 k^2} \tag{22.71}$$

which gives, $d_0 = 0$ and $\vec{d} = (ik_x, -ik_y, -b)h$. The spin-singlet pairing is absent, and therefore the spin angular momentum of the Cooper pairs is even with respect to the exchange of the spin attribute of the electrons forming the Cooper pair. Since chirality is already assumed to be even, the orbital part will be necessarily odd. It is evident from the $\vec{d}$ vector that in this channel a substantial $p + ip$ pairing exists. However it has been multiplied by a factor h which needs to be integrated over ϵ to give the induced pairing. In the weak tunneling regime where s is small, working to order s^2 which amounts to ignoring s^4 in comparison with s^2 in the numerator, and ignoring s altogether in the denominator, allows us to analytically calculate this function which gives the following strength for the pairing that after restoration of $\hbar$ and v_F becomes,

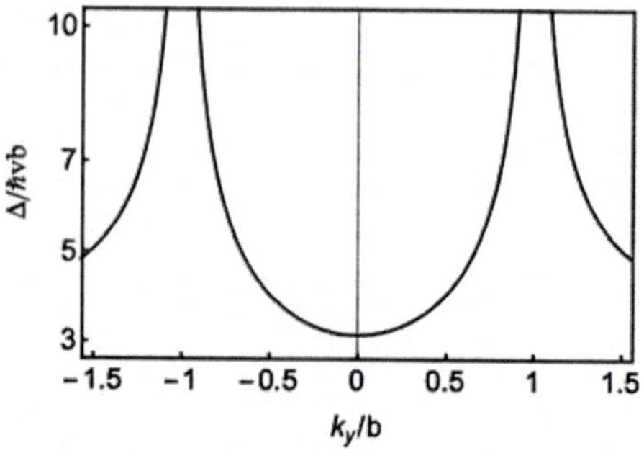

Fig. 22.6 The k_y dependence of (22.72) for $k_x = 0$ (Adapted from [46])

$$\Delta \sim \frac{\pi \Delta_s s^2 \hbar v_F \sqrt{k_x^2 + k_y^2 + b^2}}{\sqrt{|k_y^2 - \Delta_s^2|}}. \tag{22.72}$$

This function has been plotted in Fig. 22.6. The induced pairing in the middle, $k_x = 0$ of the Fermi arc $k_y = 0$ is simply $\pi \hbar v_F b s^2$ which conforms to the Golden rule intuition. Even on the BFC, according to (22.62), the minor axis is controlled by s^2 and hence even on the BFC, k_y remains small. As can be seen in Fig. 22.6, the k_y dependence near $k_y \approx 0$ is very weak, and therefore this factor will not introduce higher angular momenta and the orbital (angular momentum) part will be entirely given by the $p + ip$ form. The Majorana Fermi sea can be turned into a set of localized Majorana zero modes when an external B field is applied. Since every vortex can then bind a Majorana zero mode [52], the applied magnetic field perpendicular to the surface, can induce a delocalized-localization phase transition.

22.10.2 Odd-Chirality Pairing

The amplitude for odd-chirality pairing is

$$\hat{\Delta}_- = \frac{4\Delta k_y s^2[\{-4b\epsilon s^2(\epsilon - k_y)]}{(\mathcal{F} + 4\epsilon bkys^2)^2 - 16\epsilon^2 s^4 k^2} \begin{bmatrix} -ib & -k_x + ik_y \\ -k_x - ik_y & ib \end{bmatrix},$$

and so $d_0 = ik_y h$ and $\vec{d} = (-ib\ ,\ 0\ ,\ k_x)h$. This can be considered as a manifestation of pseudo-scalar pairing in the Fermi arc states. The integration over energy in weak tunneling regime gives the same formula (22.72). Although the singlet pairing amplitude d_0 is zero on the Fermi arc ($k_y = 0$), nevertheless on the BFC it becomes non-zero. From (22.62), this is proportional to the minor axis $\tilde{b} \propto s^2$. Therefore the singlet component of the pairing on the BFC will be controlled by the tunneling strength. On the contrary, the triplet component $\vec{d}$ of the induced pairing depends on

b and k_x. The z-component of this pairing changes from $+b$ to $-b$ by spanning the BFC, while its x-component remains constant, $-ib$.

22.10.3 Majorana Nature of BFC

The elliptic BFC in our problem is distinct from the underlying Fermi arc. Outside the BFC the Bogoliubov quasi particles are more electron-like, while inside the elliptic BFC the excitations are more hole-like. Right on the BFC the excitations will be equally electron-like and hole-like, so that the average charge of the excitations is zero. Therefore the BFC is actually a Majorana Fermi contour. The fact that it is protected by a Z_2 topological index already manifests as the simple fact that changing the tunneling strength s, does not destroy the elliptic BFC. It can only modify the aspect ratio, but maintains the elliptic shape of the BFC. Now the question is, what is the experimental signature of such a Majorana Fermi contour. In a transport setting the portion of the current which passes through the BFC surface states will appear as a zero-bias feature. At zero temperature, the strength of such a zero-bias peak which is determined by the number of channels is proportional to the perimeter of the Majorana Fermi contour,

$$\frac{dI}{dV} \propto 4bE\left(\frac{1}{\sqrt{1+4s^4}}\right), \tag{22.73}$$

where E is the elliptic function of the second kind, and we have restored the length $2b$ of the Fermi arc which determines the major axis of the ellipse. For low temperatures, the peak will acquire thermal broadening, but still remains proportional to the above value. According to [53], the effective length of the Fermi arc can be controlled by coupling to radiation. To that extent, the linear dependence of the above formula to the length $2b$ of the Fermi arc can be checked in transport measurements. However, a more exciting verification of the above formula would be to apply a magnetic field to form vortices in the superconductor.

The BFC will also have a clear thermodynamic signature in the specific heat. Since the two-dimensional BFC supports linearly dispersing excitations around it (except for the two nodal points which are of measure zero), the resulting density of states will be linear in energy. Therefore the contribution of these excitations to the specific heat will be $\sim T^2$. This situation is similar to graphene [54]. This contribution can fortunately be separated from other degrees of freedom that contribute to the absorption of heat. First of all, the bulk degrees of freedom of the superconductor have no sub-gap excitations. Secondly the bulk degrees of freedom of the WSM disperse linearly but in *three* space dimensions. By power counting, they will contribute a T^3 term. Therefore the T^2 term due to excitations around the BFC will take over at low temperatures and can be separated from the bulk of the WSM and SC.

References

1. C. Herring, Phys. Rev. **52**, 365 (1937). https://doi.org/10.1103/PhysRev.52.365
2. Where $\mu = 0$ denotes the unit 2×2 matrix and $\mu = 1, 2, 3$ stand for three Pauli matrices
3. P.A.M. Dirac, Proc. R. Soc. **117**, 610 (1928). https://doi.org/10.1098/rspa.1928.0023
4. We quite often work in units $\hbar = 1$
5. A. Vishwanath, Physics **8**, 84 (2015). https://doi.org/10.1103/Physics.8.84
6. K. Hirata, T. Kajita, M. Koshiba, M. Nakahata, S. Ohara, Y. Oyama, N. Sato, A. Suzuki, M. Takita, Y. Totsuka, T. Kifune, T. Suda, K. Nakamura, K. Takahashi, T. Tanimori, K. Miyano, M. Yamada, E. Beier, L. Feldscher, E. Frank, W. Frati, S. Kim, A. Mann, F. Newcomer, R. Van Berg, W. Zhang, B. Cortez, Phys. Lett. B **205**, 416 (1988). https://doi.org/10.1016/0370-2693(88)91690-5
7. S.-Y. Xu, I. Belopolski, N. Alidoust, M. Neupane, G. Bian, C. Zhang, R. Sankar, G. Chang, Z. Yuan, C.-C. Lee, S.-M. Huang, H. Zheng, J. Ma, D.S. Sanchez, B. Wang, A. Bansil, F. Chou, P.P. Shibayev, H. Lin, S. Jia, M.Z. Hasan, Science **349**, 613 (2015). https://doi.org/10.1126/science.aaa9297
8. N. B. I. Y. Z. B. G. C. T. R. Z. H. S. V. N. S. D. S. C. G. Z. C. M. D. W. Y. H. L. L. C. C. H. S. M. W. B. B. A. J. H. T. N. T. Xu, S. Y. Alidoust, A. Kaminski, H. Lin, S. Jia, and author M.Z. Hasan, Nat. Phys. **11**, 748 (2015). https://doi.org/10.1038/nphys3437
9. B.Q. Lv, H.M. Weng, B.B. Fu, X.P. Wang, H. Miao, J. Ma, P. Richard, X.C. Huang, L.X. Zhao, G.F. Chen, Z. Fang, X. Dai, T. Qian, H. Ding, Phys. Rev. X **5**, 031013 (2015). https://doi.org/10.1103/PhysRevX.5.031013
10. B. Lv, N. Xu, H.M. Weng, J. Ma, P. Richard, X. Huang, L. Zhao, C. Chen, G.F. Matt, F. Bisti, J. Strocov, V.N. Mesot, Z. Fang, X. Dai, T. Qian, M. Shi, H. Ding, Phys. Rev. X **5**, 031013 (2015). https://doi.org/10.1103/PhysRevX.5.031013
11. L. Lu, Z. Wang, D. Ye, L. Ran, L. Fu, J.D. Joannopoulos, M. Soljačić, Science **349**, 622 (2015). https://doi.org/10.1126/science.aaa9273
12. X. Wan, A.M. Turner, A. Vishwanath, S.Y. Savrasov, Phys. Rev. B **83**, 205101 (2011). https://doi.org/10.1103/PhysRevB.83.205101
13. H. Nielsen, M. Ninomiya, Phys. Lett. B **130**, 389 (1983). https://doi.org/10.1016/0370-2693(83)91529-0
14. S. Murakami, New J. Phys. **9**, 356 (2007). https://doi.org/10.1088/1367-2630/9/9/356
15. G.E. Volovik, *The Universe in a Helium Droplet* (Publisher Oxford University Press, 2003). https://global.oup.com/academic/product/the-universe-in-a-helium-droplet-9780198507826?cc=de&lang=en&
16. A.A. Burkov, L. Balents, Phys. Rev. Lett. **107**, 127205 (2011). https://doi.org/10.1103/PhysRevLett.107.127205
17. A.A. Burkov, M.D. Hook, L. Balents, Phys. Rev. B **84**, 235126 (2011). https://doi.org/10.1103/PhysRevB.84.235126
18. S.-M. Huang, S.-Y. Xu, I. Belopolski, C.-C. Lee, G. Chang, B. Wang, N. Alidoust, G. Bian, G. Bian, M. Neupane, C. Zhang, S. Jia, A. Bansil, H. Lin, M.Z. Hasan, Nat. Commun. **6**, 7373 (2015). https://doi.org/10.1038/ncomms8373
19. S.-Y. Xu, C. Liu, S.K. Kushwaha, R. Sankar, J.W. Krizan, I. Belopolski, M. Neupane, G. Bian, N. Alidoust, T.-R. Chang, H.-T. Jeng, C.-Y. Huang, W.-F. Tsai, H. Lin, P.P. Shibayev, F.-C. Chou, R.J. Cava, M.Z. Hasan, Science **347**, 294 (2015). https://doi.org/10.1126/science.1256742
20. S.-Y. Xu, I. Belopolski, N. Alidoust, M. Neupane, G. Bian, C. Zhang, R. Sankar, G. Chang, Z. Yuan, C.-C. Lee, S.-M. Huang, H. Zheng, J. Ma, D.S. Sanchez, B. Wang, A. Bansil, F. Chou, P.P. Shibayev, H. Lin, S. Jia, M.Z. Hasan, Science **349**, 613 (2015). https://doi.org/10.1126/science.aaa9297
21. Note that since Θ is anti-unitary, one has $\langle\phi|\psi\rangle = \langle P\Theta\psi|P\Theta\phi\rangle$
22. For one dimensional space, with $\beta = 2$ one has $\gamma^0 = 2$ and $\gamma^x = -i3$ and therefore $\gamma^5 = \gamma^0\gamma^1 = 2(-i3) = 1$. This matrix squares to 1 and anticommutes with all other γ's. But in

two space dimensions each of the γ^0, γ^1, γ^2 will be proportional to one of the Pauli matrices. Hence their product will be 1 that squares to 1, but does *not* anticommute with all the other γ's. Therefore in even space dimensions, there is no concept of chirality
23. Z. Faraei, A. Jafari, Unpublished (2020)
24. B.-J. Yang, N. Nagaosa, Nat. Commun. **5**, 4898 (2014). https://doi.org/10.1038/ncomms5898
25. Z. Faraei, S.A. Jafari, Phys. Rev. B **96**, 134516 (2017). https://doi.org/10.1103/PhysRevB.96.134516
26. Note that in order to form a Lorentz scalar from spinor ψ, one defines $\bar{\psi} = \psi^\dagger\gamma^0$
27. N. Bovenzi, M. Breitkreiz, P. Baireuther, T.E. O'Brien, J. Tworzydło, I.D.I. Adagideli, C.W.J. Beenakker, Phys. Rev. B **96**, 035437 (2017). https://doi.org/10.1103/PhysRevB.96.035437
28. F. Adinehvand, Z. Faraei, T. Farajollahpour, S.A. Jafari, Phys. Rev. B **100**, 195408 (2019). https://doi.org/10.1103/PhysRevB.100.195408
29. Z. Faraei, T. Farajollahpour, S.A. Jafari, Phys. Rev. B **98**, 195402 (2018). https://doi.org/10.1103/PhysRevB.98.195402
30. E. Witten, Riv. Nuovo Cim. **39**, 313 (2016). https://doi.org/10.1393/ncr/i2016-10125-3
31. K. Hashimoto, T. Kimura, X. Wu, Prog. Theor. Exp. Phys. **2017** (2017). https://doi.org/10.1093/ptep/ptx053
32. S. Li, A.V. Andreev, Phys. Rev. B **92**, 201107 (2015). https://doi.org/10.1103/PhysRevB.92.201107
33. E. McCann, V.I.F. Ko, J. Phys. Condens. Matter **16**, 2371 (2004). https://doi.org/10.1088/0953-8984/16/13/016
34. A.R. Akhmerov, C.W.J. Beenakker, Phys. Rev. B **77**, 085423 (2008). https://doi.org/10.1103/PhysRevB.77.085423
35. A. Zee, *Quantum Field Theory in a Nutshell* (Publisher Princeton University Press, Princeton, 2010)
36. X. Yuan, Z. Yan, C. Song, M. Zhang, Z. Li, C. Zhang, Y. Liu, W. Wang, M. Zhao, Z. Lin, T. Xie, J. Ludwig, Y. Jiang, X. Zhang, C. Shang, Z. Ye, J. Wang, F. Chen, Z. Xia, D. Smirnov, X. Chen, Z. Wang, H. Yan, F. Xiu, Nat. Commun. **9**, 1854 (2018). https://doi.org/10.1038/s41467-018-04080-4
37. S. Jeon, A. Zhou, Brian B. Gyenis, I. Feldman, Benjamin E. Kimchi, Q.D.C.R.J.V.A. Potter, Andrew C. Gibson, A. Yazdani, Nat. Mater. **13**, 851 (2014). https://doi.org/10.1038/nmat4023
38. Note that the poles of the second line in the above equation give the bulk excitations. It is interesting to note that the coefficient $C^{\bar{\sigma}\sigma}_{\chi\chi'}$ that formalizes the effect of the boundary, does contain information about the bulk excitations as well. This effect becomes important when convoluting the above Green's functions to compute various susceptibilities [28]
39. F.D.M. Haldane (2014). arXiv:1401.0529
40. A.C. Potter, I. Kimchi, A. Vishwanath, Nat. Commun. **5**, 5161 (2014). https://doi.org/10.1038/ncomms6161
41. N.P. Armitage, E.J. Mele, A. Vishwanath, Rev. Mod. Phys. **90**, 015001 (2018). https://doi.org/10.1103/RevModPhys.90.015001
42. S. Tchoumakov, M. Civelli, M.O. Goerbig, Phys. Rev. B **95**, 125306 (2017). https://doi.org/10.1103/PhysRevB.95.125306
43. As Feynman puts it, the electrons and holes are time-reversals of each other. See also the works of Carlo Beenakker [55] to construct the hole part
44. U. Khanna, S. Rao, A. Kundu, Phys. Rev. B **95**, 201115 (2017). https://doi.org/10.1103/PhysRevB.95.201115
45. Y. Li, F.D.M. Haldane, Phys. Rev. Lett. **120**, 067003 (2018). https://doi.org/10.1103/PhysRevLett.120.067003
46. Z. Faraei, S.A. Jafari, Phys. Rev. B **100**, 035447 (2019). https://doi.org/10.1103/PhysRevB.100.035447
47. D.F. Agterberg, P.M.R. Brydon, C. Timm, Phys. Rev. Lett. **118**, 127001 (2017). https://doi.org/10.1103/PhysRevLett.118.127001
48. P.M.R. Brydon, D.F. Agterberg, H. Menke, C. Timm, Phys. Rev. B **98**, 224509 (2018). https://doi.org/10.1103/PhysRevB.98.224509

49. A.P. Schnyder, S. Ryu, A. Furusaki, A.W.W. Ludwig, Phys. Rev. B **78**, 195125 (2008). https://doi.org/10.1103/PhysRevB.78.195125
50. C.-K. Chiu, J.C.Y. Teo, A.P. Schnyder, S. Ryu, Rev. Mod. Phys. **88**, 035005 (2016). https://doi.org/10.1103/RevModPhys.88.035005
51. A. Altland, M.R. Zirnbauer, Phys. Rev. B **55**, 1142 (1997). https://doi.org/10.1103/PhysRevB.55.1142
52. M. Stone, R. Roy, Phys. Rev. B **69**, 184511 (2004). https://doi.org/10.1103/PhysRevB.69.184511
53. U. Khanna, S. Rao, A. Kundu, Phys. Rev. B **95**, 201115 (2017). https://doi.org/10.1103/PhysRevB.95.201115
54. E. Pop, V. Varshney, A.K. Roy, MRS Bull. **37**, 1273 (2012). https://doi.org/10.1557/mrs.2012.203
55. C.W.J. Beenakker, Phys. Rev. Lett. **97**, 067007 (2006). https://doi.org/10.1103/PhysRevLett.97.067007

Chapter 23
Atomistic Tight-Binding Study of Core/Shell Nanocrystals

Worasak Sukkabot

Abstract Progressive technologies in the synthetic chemistry of semiconductor nanostructures have made it possible to access high quality semiconductor nanostructures with precise size, shape and composition. Colloidal core/shell nanocrystals are composed of a core made from one material terminated by a shell of another material. Because of the improved photoluminescence quantum yields, high photostability and size-tunable emission properties, core/shell nanocrystals are tremendously attractive for the active applications. The purpose of this chapter is to present the atomistic tight-binding theory to study the electronic structures and optical properties of core/shell nanocrystals with the purpose to evidently understand the significance of core and growth shell. Owing to the heterostructure of core/shell nanocrystal, the valence force field method is utilized to optimize the structural geometry. To analyze the electronic structures and optical properties of the core/shell nanocrystals with the corresponding structural parameters, some of the calculations are demonstrated. Finally, all-inclusive information based on atomistic tight-binding theory successfully conveys the natural behaviors of core/shell nanocrystals and carries a guideline for the design of their electronic and optical properties before applying to the novel electronic nanodevices.

23.1 Introduction

Advanced technologies in the synthetic chemistry of semiconductor nanostructures have made it possible to access high quality semiconductor nanostructures with controlled size, shape and composition. Nanostructures containing more than one material can easily be synthesized nowadays. Core/shell nanocrystal is one of the nanostructures composed of a core made from one material terminated by a shell of another material. Therefore, the structural, electronic, magnetic and optical properties of the core/shell nanocrystals can be manipulated not only by the core but also by the

W. Sukkabot (✉)
Faculty of Science, Department of Physics, Ubon Ratchathani University, 85 Sathollmark Rd., Warinchamrab, Ubon Ratchathani 34190, Thailand

H. Ünlü and N. J. M. Horing (eds.), *Progress in Nanoscale and Low-Dimensional Materials and Devices*, Topics in Applied Physics 144,
https://doi.org/10.1007/978-3-030-93460-6_23

growth shell. The advantages in the improved photoluminescence quantum yields, high photostability and size-tunable emission properties make core/shell nanocrystals extremely gorgeous for widespread applications such as light-emitting diodes [1, 2], solar cells [3–6], lasers [7], biological imaging [8–11] and quantum information [12–16]. Before the authentic manufacture, the theoretical description of semiconductor nanostructures is of crucial importance because of its allowance to analyze and predict the fundamental physics. Here, it is the aim of this chapter to provide the reader with a comprehensive overview of the band structure calculations of core/shell semiconductor nanocrystals. We take a special emphasis on the empirical tight-binding description and valence force field method in order to deliver some of their calculations.

To study the electronic and optical properties of semiconductor nanostructures, the band structures are required to be understood. At the moment, there are several theoretical calculations of the band structures such as k.p method [17–25], tight-binding model [26–35], pseudopotential model [36–42] and density functional theory [43–46]. Each method has their own technique to calculate the band structures of semiconductors. Conventionally, the nanostructures are investigated by k.p method in the framework of the envelope function approximation (EFA) [47–49]. This approach mainly lacks the atomic detail. Because of a reasonable compromise between the computational resource and the reliability of the results, k.p method still remains to be implemented. Advanced technologies of the synthesis make it possible to fabricate the high quality semiconductor nanostructures with complexity where k.p method hardly deals with. Hence, the envelope function concept is replaced by density functional theory. However, the realistic nanostructures contain an amount of atoms where density functional theory encounters the dimensionality of the problem. Presently, advanced density functional calculations under large parallel supercomputers can be applied to study the structures with several thousand atoms. With the aim to mainly preserve both of the computational efficiency and atomistic detail, the empirical methods are the suitable candidates for modelling the nanostructure devices. Two empirical methods are proposed for atomistic nanostructure description, namely the empirical tight-binding approach (ETB) and the empirical pseudopotential technique (EPM). ETB method yields the simple and appealing calculations using bonding properties in the framework of orbital occupations and orbital overlap. ETB approach overcomes k.p method due to the consideration of atomic details, while the computationally consuming time of ETB model is comparable to that of k.p method. Another strength is that the scalability of ETB approach can be investigated up to the level of the density function theory. However, the disadvantage of ETB method is the large number of parameters involved to accurately reproduce the band structures. The EPM represents the precise band structures with a few parameters, thus reducing the restrictions of ETB approach. For the demonstration of the accurate band structures, EPM method must be based on the wave function expansion produced from the large number of plane waves. Within the tight-binding model, the wave function is spanned with a small basis set depending on the included orbitals, number of atoms and neighboring interactions. Hence, ETB model is computationally less expensive than EPM model. According to the argument, ETB model is a good applicant for the study of relatively big and complicated systems in which both of the computational efficiency and atomistic description are preserved.

The main objective of this chapter is to give the comprehensive description of computational tool for the nanostructure devices. The computational tool consists of the valence force field method and empirical tight-binding theory. Valence force field method is implemented to optimize the atomic positions. After obtaining the relaxed structures, the electronic and optical properties are determined in the framework of empirical tight-binding model. This chapter is also to deliver primarily our own work and a rudimentary attempt is made to cover the wide literature on this subject. The chapter is organized as follow. Section 23.2 provides the description of the valence force method. In Sect. 23.3, the principle of the empirical tight-binding method is applied to the bulk semiconductors with different tight-binding hybridizations. The implementation of the sp^3s^* empirical tight-binding method into the core/shell semiconductor nanocrystals is demonstrated in Sect. 23.4. Finally, the summary is provided in Sect. 23.5.

23.2 Valence Force Field

Due to the lattice mismatch between core and shell material, the atomic positions inside and around core/shell nanocrystal are distorted. To optimize the structural geometry, there are two major methods, continuum elasticity [50–53] and atomistic elasticity approach [50, 54–61]. In this work, the valence force field method (VFF), one of the atomistic elasticity approaches, is implemented to relax the atomic positions of the core/shell nanocrystal. The advantage of the valence force field is that it includes atomic scale information such as inter-atomic potential and correct point group symmetry. For the demonstration, the atoms in core/shell nanocrystal are considered as point particles and the bonds are termed as springs. Therefore, the equation of the spring deformation is theoretically used to explain the behavior of the stretching and bending bonds. The objective of this approach is to minimize the energy associated with a given atomic structure. The total energy of this model is the summation of the energy of the stretching and bending of the bonds. The expression of total energy is described in term of the fitting parameters which explain the behaviors of different kinds of the atoms in core/shell nanocrystal. After relaxing the atomic positions, the strain distribution and strain tensors are attained.

23.2.1 Valence Force Field Method (VFF)

In the realistic structures, core and shell are chemically synthesized from different materials with their own lattice constants, leading to the lattice-mismatch-induced strain in such structure. For example, the mismatch for InAs core passivated by GaAs shell is 6%, while for InAs core terminated by InP shell it is about 3%. To calculate the optimized atomic positions of core/shell nanocrystal, strain relaxation has been provided. In this work, the valence force field method is implemented for atomistic optimization. The elastic energy of all atoms is expressed as a function of the atomic positions R_i as:

$$E = \sum_{ij} V_2(R_i - R_j) + \sum_{ij} V_3(\theta_{ijk})$$

where V_2 is the two-body term called stretching function, V_3 is the three body term of the bond angle called bending function and θ_{ijk} is the angle subtended at the atom i by atoms j and k. The actual formula for the total elastic energy is given by:

$$\begin{aligned} \mathrm{E} = & \sum_{i}\sum_{j=1}^{4} \frac{3\alpha_{ij}}{16(d_{ij}^0)^2}\left[(R_j - R_i)^2 - (d_{ij}^0)^2\right]^2 \\ & + \sum_{i}\sum_{j,k>1}^{4} \frac{3\beta_{ijk}}{8d_{ij}^0 d_{ik}^0}\left[(R_j - R_i)\cdot(R_k - R_i) - \cos\theta d_{ij}^0 d_{ik}^0\right]^2 \end{aligned}$$

Here, d_{ij}^0 denotes the bulk equilibrium bond length between the nearest-neighboring atom i and j in the corresponding binary compound and $\theta = \arccos(1/3)$ is the ideal bond angle of zinc-blende structure. The first term is a sum over all atom i and its nearest neighbours j. The second term is a sum over all atoms i and distinct pairs of its nearest neighbours j and k. The fitting dependent parameters α and β are the bond-stretching and bond-bending force constants, respectively. For the bending term at the interface where the species j and k are different, the average of the corresponding values of these pure semiconductors is utilized. Table 23.1 lists the parameters α and β of the the important semiconductors (mainly in III–V and II–VI group) [62].

Table 23.1 The parameters α and β of the the important semiconductors (mainly in III–V and II–VI group) [62]

Semiconductor	α(N/m)	β(N/m)
InAs	35.04	6.14
InP	38.56	6.75
InSb	28.68	5.02
GaAs	41.70	7.30
GaP	48.18	8.44
GaSb	31.68	5.55
CdS	33.43	3.54
CdSe	29.64	3.14
ZnO	69.67	7.38
ZnS	42.16	4.46

To calculate the relaxed atomic positions of the core/shell nanocrystals, the total elastic energy is minimized with respect to the atomic positions R_i. Here, the conjugate gradient method (CGDESCENT) [63–65] developed by William W. Hager is implemented. Once the relaxed atomic positions of all atoms are known, the strain distribution is realized through the strain tensors. The cation site is formed a tetrahedron bonding to its four nearest neighboring anions. The strain tensors ε are calculated from the correlation between the distorted and ideal tetrahedron edges. The distorted tetrahedron edges, R_{12}, R_{23} and R_{34}, are connected to the ideal tetrahedron edges, R^0_{12}, R^0_{23} and R^0_{34}, from this relation: [50]

$$\begin{bmatrix} R_{12,x} & R_{23,x} & R_{34,x} \\ R_{12,y} & R_{23,y} & R_{34,y} \\ R_{12,z} & R_{23,z} & R_{34,z} \end{bmatrix} = \begin{bmatrix} 1+\varepsilon_{xx} & \varepsilon_{yx} & \varepsilon_{zx} \\ \varepsilon_{xy} & 1+\varepsilon_{yy} & \varepsilon_{zy} \\ \varepsilon_{xz} & \varepsilon_{yz} & 1+\varepsilon_{zz} \end{bmatrix} \begin{bmatrix} R^0_{12,x} & R^0_{23,x} & R^0_{34,x} \\ R^0_{12,y} & R^0_{23,y} & R^0_{34,y} \\ R^0_{12,z} & R^0_{23,z} & R^0_{34,z} \end{bmatrix}$$

The strain tensors are then calculated by the matrix inversion as:

$$\begin{bmatrix} \varepsilon_{xx} & \varepsilon_{yx} & \varepsilon_{zx} \\ \varepsilon_{xy} & \varepsilon_{yy} & \varepsilon_{zy} \\ \varepsilon_{xz} & \varepsilon_{yz} & \varepsilon_{zz} \end{bmatrix} = \begin{bmatrix} R_{12,x} & R_{23,x} & R_{34,x} \\ R_{12,y} & R_{23,y} & R_{34,y} \\ R_{12,z} & R_{23,z} & R_{34,z} \end{bmatrix} \begin{bmatrix} R^0_{12,x} & R^0_{23,x} & R^0_{34,x} \\ R^0_{12,y} & R^0_{23,y} & R^0_{34,y} \\ R^0_{12,z} & R^0_{23,z} & R^0_{34,z} \end{bmatrix}^{-1} - I$$

where I is the 3×3 identity matrix.

23.2.2 *Examples of the Calculations*

Here, the examples in the implementation of the valence force field method on the core/shell nanostructures are demonstrated. The nanocrystals and nanorods of CdSe and CdS are excellent candidates for the novel optical and electronic applications. Luo and Wang [66] optimized the atomic positions of CdSe/CdS core/shell nanorod. The spherical CdSe core with a diameter of 3.44 nm at the right-hand side of the nanorod was surrounded by CdS rod with diameter of 4.30 nm and the length of 15.48 nm. Due to the lattice mismatch between core and shell material, the compressive strain was found in CdSe core, while the tensile strain was observed nearby CdSe core and in CdSe shell. In CdS shell far away from CdSe core, there was no strain influence. CdSe/ZnSe core/shell nanocrystal is one of the outstanding candidates to provide the high luminescence quantum yields. Due to the lattice mismatch between core and shell, the induced strain in CdSe/ZnSe core/shell nanocrystals with experimentally synthesized sizes was studied by Sukkabot [67]. The ZnSe shell thicknesses (t_s) of 0.9, 1.8 and 2.7 monolayer (ML) were consecutively passivated on CdSe core with diameter of 3.6 nm. Figure 23.1 displayed the hydrostatic strains ($e_{xx} + e_{yy} + e_{zz}$) in the z direction through the middle of the core/shell nanocrystal under different shell thicknesses. The hydrostatic strains were mainly sensitive with the growth shell

Fig. 23.1 Hydrostatic strains ($e_{xx} + e_{zz} + e_{zz}$) of CdSe/ZnSe core/shell nanocrystals as a function of ZnSe shell thicknesses (ML) [67]

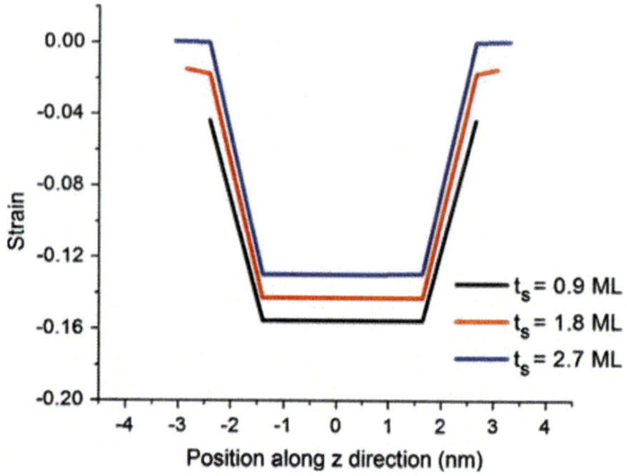

thickness. The compressive strains were probed in the core region within all shell dimensions, whereas there was no tensile strain in core and shell.

23.3 Empirical Tight-Binding Method

The empirical tight-binding theory was initially developed by Slater and Koster [68]. Within the empirical tight-binding theory, Hamiltonian matrix elements are elucidated by fitting parameters. These parameterizations can reproduce the reference quantities such as the band structure, effective mass or band gap derived by first-principles calculations or extracted from experimental data. The empirical tight-binding method is often used to study electronic and optical properties of nanostructure devices. Before developing the nanostructure description, it is essential to describe bulk semiconductors and matrix element parameters. Here, sp^3s* empirical tight-binding method is used to determine the bulk semiconductors. The historical parameterization is described with the increasing including orbitals starting from the minimal sp^3 basis.

23.3.1 The Bulk Hamiltonian

In the term of the bulk semiconductor, the one-electron wavefunctions |nk> can be written as Bloch functions due to the symmetry of the system as given by:

$$\left|\psi_{n,k}\right\rangle = \sum_{\alpha,\beta} C_{\alpha\beta}(n,k)\left\{\sum_{R} e^{ik\cdot R_\beta}\left|\alpha, R_\beta\right\rangle\right\}$$

For the convenience, $R_\beta = R + d_\beta$ where R is the lattice vector and d_β is the atomic position. β and α refer to the index of the atom and to the atomic orbital index, respectively. $C_{\alpha\beta}(n, k)$ are the coefficients of the linear combination which depend on the band index n and the k vector in Brillouin zone. The expansion of the one-electron wavefunction utilizes a complete set of basis functions. In the tight-binding method, the basis used in the linear combination contains the atomic states of the outermost occupied valence electron states. The coefficients $C_{\alpha\beta}(n, k)$ of the linear expansion are carried out by solving the well-known Schrodinger equation:

$$H\left|\psi_{n,k}\right\rangle = E_n(k)\left|\psi_{n,k}\right\rangle$$

where $E_n(k)$ symbolizes the energy band dispersion. To obtain the energies and one-electron wavefunctions, quantum matrix is implemented with assistance of the hopping matrix elements. Then, the one-electron tight-binding Hamiltonian is written in term of the localized basis as:

$$H = \sum_{\alpha, R_\beta} \left|\alpha, R_\beta\right\rangle \varepsilon_{\alpha,\beta} \left\langle\alpha, R_\beta\right| + \sum_{\alpha,\alpha', R_{\beta'} \neq R_\beta} \left|\alpha', R_{\beta'}\right\rangle t_{\alpha\alpha'} \left\langle\alpha, R_\beta\right|$$

The first term stands for $R_{\alpha'} = R_\alpha$ and $\alpha' = \alpha$, namely the on-site matrix elements:

$$\varepsilon_{\alpha,\beta} = \left\langle\alpha, R_\beta\right|H\left|\alpha, R_\beta\right\rangle$$

The second term stands for $R_{\beta'} \neq R_\beta$, called the off-site matrix elements:

$$t_{\alpha'\alpha}(R_{\beta'} - R_\beta) = \left\langle\alpha', R_{\beta'}\right|H\left|\alpha, R_\beta\right\rangle$$

For the demonstration of the bulk band calculations, the tight-binding method with the combination of minimal sp^3 basis (one s orbital and three p orbitals) plus the excited s state with s-like symmetry (s*), called sp^3s* empirical tight-binding approach. This empirical tight-binding model has been widely implemented for various nanostructures. In this chapter, sp^3s* empirical tight-binding method with the nearest-neighboring interaction is applied to the zinc-blende structures. The independent tight-binding parameterizations consisting of on-site energies ($\varepsilon_{s,a(c)}$, $\varepsilon_{p,a(c)}$ and $\varepsilon_{s*,a(c)}$) and off-site matrix elements (V) are demonstrated as:

$$V_{ss} = 4t_{ss}(R_a - R_c)$$

$$V_{xx} = 4t_{p_x p_x}(R_a - R_c)$$

$$V_{xy} = 4t_{p_x p_y}(R_a - R_c)$$

$$V_{sp_x} = 4t_{sp_x}(R_a - R_c)$$

$$V_{p_x s} = 4t_{p_x s}(R_a - R_c)$$

$$V_{s*p_x} = 4t_{s*p_x}(R_a - R_c)$$

$$V_{p_x s*} = 4t_{p_x s*}(R_a - R_c)$$

Here, a and c present the anion and cation atoms. The interaction between s * orbitals is neglected. Finally, the 10 × 10 Hamiltonian matrix is given by:

$$\begin{bmatrix} \varepsilon_{s,a} & 0 & 0 & 0 & 0 & V_{ss}g_0 & V_{sp}g_1 & V_{sp}g_2 & V_{sp}g_3 & 0 \\ 0 & \varepsilon_{p,a} & 0 & 0 & 0 & -V_{ps}g_1 & V_{xx}g_0 & V_{xy}g_3 & V_{xy}g_2 & -V_{ps*}g_1 \\ 0 & 0 & \varepsilon_{p,a} & 0 & 0 & -V_{ps}g_2 & V_{xy}g_3 & V_{xx}g_0 & V_{xy}g_1 & -V_{ps*}g_2 \\ 0 & 0 & 0 & \varepsilon_{p,a} & 0 & -V_{ps}g_3 & V_{xy}g_2 & V_{xy}g_1 & V_{xx}g_0 & -V_{ps*}g_3 \\ 0 & 0 & 0 & 0 & \varepsilon_{s*,a} & 0 & V_{s*p}g_1 & V_{s*p}g_2 & V_{s*p}g_3 & 0 \\ V_{ss}g_0^* & -V_{ps}g_1^* & -V_{ps}g_2^* & -V_{ps}g_3^* & 0 & \varepsilon_{s.c} & 0 & 0 & 0 & 0 \\ V_{sp}g_1^* & V_{xx}g_0^* & V_{xy}g_3^* & V_{xy}g_2^* & V_{s*p}g_1^* & 0 & \varepsilon_{p.c} & 0 & 0 & 0 \\ V_{sp}g_2^* & V_{xy}g_3^* & V_{xx}g_0^* & V_{xy}g_1^* & V_{s*p}g_2^* & 0 & 0 & \varepsilon_{p.c} & 0 & 0 \\ V_{sp}g_3^* & V_{xy}g_2^* & V_{xy}g_1^* & V_{xx}g_0^* & V_{s*p}g_3^* & 0 & 0 & 0 & \varepsilon_{p.c} & 0 \\ 0 & V_{ps*}g_1^* & V_{ps*}g_2^* & V_{ps*}g_3^* & 0 & 0 & 0 & 0 & 0 & \varepsilon_{s*.c} \end{bmatrix}$$

In this matrix, the g_i coefficients depend on the wave vectors k. These values are defined as:

$$g_0(k) = \frac{1}{4}\left[e^{ik\cdot\tau_1} + e^{ik\cdot\tau_2} + e^{ik\cdot\tau_3} + e^{ik\cdot\tau_4}\right]$$

$$g_1(k) = \frac{1}{4}\left[e^{ik\cdot\tau_1} - e^{ik\cdot\tau_2} + e^{ik\cdot\tau_3} - e^{ik\cdot\tau_4}\right]$$

$$g_2(k) = \frac{1}{4}\left[e^{ik\cdot\tau_1} - e^{ik\cdot\tau_2} - e^{ik\cdot\tau_3} + e^{ik\cdot\tau_4}\right]$$

$$g_3(k) = \frac{1}{4}\left[e^{ik\cdot\tau_1} + e^{ik\cdot\tau_2} - e^{ik\cdot\tau_3} - e^{ik\cdot\tau_4}\right]$$

where the vectors τ_i are the distances between the nearest-neighboring atoms in the zinc-blende structure given by:

$$\tau_1 = \frac{A_0}{4}\left(\hat{i} + \hat{j} + \hat{k}\right)$$

$$\tau_2 = \frac{A_0}{4}\left(-\hat{i} - \hat{j} + \hat{k}\right)$$

$$\tau_3 = \frac{A_0}{4}\left(\hat{i} - \hat{j} - \hat{k}\right)$$

$$\tau_4 = \frac{A_0}{4}\left(-\hat{i} + \hat{j} - \hat{k}\right)$$

Here, A_0 is lattice constant of the studied zinc-blende semiconductor.

For the realistic situation, the spin–orbit interaction is included in sp^3s* empirical tight-binding Hamiltonian as described by Chadi [69]. Considering a spherical symmetric potential, the spin–orbit interaction is defined as:

$$H_{so} = \frac{\hbar}{4m^2c^2}\frac{1}{r}\frac{dV_C(r)}{dr}L \cdot \sigma$$

Here, σ are the Pauli matrices, V_C is the crystal potential and L is the angular momentum. Then, the spin–orbit matrix elements with spin components s (s = ↑, ↓) are given by:

$$\left\langle \alpha, R_\beta, s \middle| H_{SO} \middle| \alpha', R_{\beta'}, s' \right\rangle = \left\langle \alpha, R_\beta \middle| \frac{\hbar}{4m^2c^2}\frac{1}{r}\frac{dV_C(r)}{dr} L \middle| \alpha', R_{\beta'} \right\rangle \cdot \langle s|\sigma|s\rangle$$

For the tight-binding approximation, contributions on the same atom are only considered. Then, the non-zero matrix elements of spin–orbit interaction are demonstrated:

$$\left\langle p_x, R_\beta, \uparrow \middle| H_{SO} \middle| p_y, R_\beta, \uparrow \right\rangle = -i\lambda_\beta$$

$$\left\langle p_x, R_\beta, \downarrow \middle| H_{SO} \middle| p_y, R_\beta, \downarrow \right\rangle = i\lambda_\beta$$

$$\left\langle p_z, R_\beta, \uparrow \middle| H_{SO} \middle| p_x, R_\beta, \downarrow \right\rangle = -\lambda_\beta$$

$$\left\langle p_z, R_\beta, \downarrow \middle| H_{SO} \middle| p_x, R_\beta, \uparrow \right\rangle = \lambda_\beta$$

$$\left\langle p_y, R_\beta, \uparrow \middle| H_{SO} \middle| p_z, R_\beta, \downarrow \right\rangle = -i\lambda_\beta$$

$$\left\langle p_y, R_\beta, \downarrow \middle| H_{SO} \middle| p_z, R_\beta, \uparrow \right\rangle = -i\lambda_\beta$$

$\lambda_\beta(\lambda_a, \lambda_c)$ are the spin–orbit splittings of the anion and cation of p orbitals. In empirical tight-binding method, these values are the fitting parameters. Finally, the introduction of the spin–orbit interaction into sp^3s* empirical tight-binding Hamiltonian of a zinc-blende structure doubles to the 20×20 Hamiltonian matrix.

23.3.2 Empirical Tight-Binding Parameterization

To obtain the reliable results of the empirical tight-binding method, the parameterization is a main point which depends on the interesting applications. In the band structure analysis, the data such as effective masses and band gaps need to be reproduced with the density functional calculations and experiments. These parameterizations can be implemented to study the structural, electronic and optical properties of nanostructure devices. There are numerous tight-binding parameterizations of semiconductors introduced by several authors. J. C. Slater and G. F. Koster initially studied the linear combination of atomic orbitals (LCAO) or tight binding, approximation with orthogonalized plane-wave methods to calculate the band structures of simple cubic, face-centered cubic, body-centered cubic and diamond structures. Using the sp^3 tight-binding method with the nearest neighboring interaction [70], the valence bands are described in a proper feature in the comparison with the pseudopotential method. However, sp^3 empirical tight-binding method fails to generate the accurate conduction bands. This problem is revised by including more excited energy orbitals or the orders of the neighboring interactions. Niquet et al. [71] proposed the sp^3 tight-binding model with up to third-nearest neighboring interactions to Si band structure. The tight-binding band structure was excellent to fit with the GW model.

The modification of the parameterizations has been reported by Vogl et al. [72] by introducing an additional s* orbital with higher energy, called sp^3s^* empirical tight-binding method. Using this basis, both valence and conduction bands of various IV, III–V and II–VI semiconductors with direct and indirect band gap are accurately reported. The band structure of Si semiconductor plotted along the high symmetry of the Brillouin zone is illustrated in Fig. 23.2. This sp^3s^* empirical tight-binding model has been successfully applied to semiconductor nanostructures. However, the sp^3s^* tight-binding model cannot properly fit the X point in the Brillouin zone. Therefore, its applicability to model the nanostructures is limited in the semiconductor nanostructures where the natural properties are described in X point of the Brillouin zone.

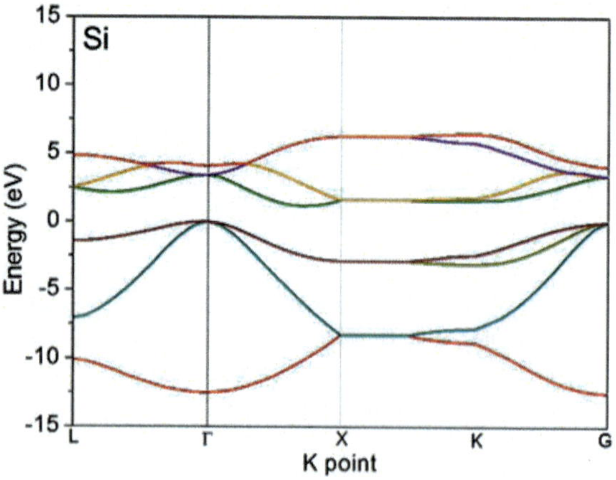

Fig. 23.2 The sp^3s^* tight-binding band structure of Si semiconductor along the high symmetry of the Brillouin zone

Table 23.2 Empirical tight-binding models with their references of parameterizations

Semiconductor	sp^3 TB	sp^3s^* TB	$sp^3d^5s^*$ TB
IV			
Si	[76–78]	[72, 75]	[73]
Ge	[76–78]	[72]	[73]
SiC	[79]	[72]	[73]
III–V			
GaAs	[85]	[72, 75, 80–84]	[73]
AlAs	[70]	[72, 75, 80–83]	[73]
InAs	[85, 87]	[72, 75, 80, 83, 86]	[73]
GaP	[85]	[72, 75, 83]	[73]
AlP		[72, 75]	[73]
InP	[85]	[72, 75, 80, 83]	[73]
GaSb		[72, 75]	[73]
II–VI			
ZnSe		[72, 75, 84, 88–92]	
CdSe	[93]	[88, 94, 95]	
ZnTe		[72, 75, 88–90]	
CdTe	[98, 99]	[88, 96, 97]	
HgTe	[98, 99]	[88, 96]	
ZnS	[93]	[84, 88, 89]	
CdS	[93, 100]	[88]	
ZnO	[93, 101]		

In order to overcome the restrictions, the sp^3s^* tight-binding model is improved by increasing the number of neighboring interactions or the number of orbitals.

Jancu et al. [73, 74] included the d orbital into sp^3s^* tight-binding model with the combination of the nearest-neighboring interaction and spin-orbit coupling, called $sp^3d^5s^*$ empirical tight-binding method. The additional d orbitals play the crucial role for the corrections of lowest conduction bands at X point. The band structures for many IV, III–V and nitride semiconductors such as C, Si, Ge, AlP, GaP, InP, AlAs, GaAs, InAs, AlSb, GaSb, InSb, GaN, AlN and InN are in an excellent agreement with pseudopotential method and experiments. Table 23.2 itemizes the references of the tight-binding parameterizations for the important semiconductors (mainly in IV, III–V and II–VI group).

23.4 Empirical Tight-Binding Theory of Core/Shell Nanocrystals

23.4.1 The sp^3s^* Empirical Tight-Binding Description

For the demonstration of the studied nanostructures, sp^3s^* empirical tight-binding theory with the combination of the nearest neighboring interaction and spin–orbit coupling is utilized to investigate the core/shell semiconductor nanocrystals. The sp^3s^* empirical tight-binding method was initially introduced by Vogl [72] to obtain more accurate band structures than sp^3 TB and consume the computationally requirement less than $sp^3d^5s^*$ TB. The computational process starts with a definition of the atomic positions of the core/shell nanocrystals with the zinc-blende or wurtzite crystal structure depending on the available experiments. The single wave function is written as a linear combination of atomic orbitals localized on each atom given by:

$$|\psi\rangle = \sum_{R=1}^{N_{at}} \sum_{\alpha=1}^{10} C_{R,\alpha} |R, \alpha\rangle$$

where α stands for the localized atomic orbitals on atom R with N_{at} being the total number of atoms inside the system. The index of α from 1 to 5 stands for $s\uparrow$, $p_x\uparrow$, $p_y\uparrow$, $p_z\uparrow$ and $s*\uparrow$, while the index of α from 6 to 10 is defined as $s\downarrow$, $p_x\downarrow$, $p_y\downarrow$, $p_z\downarrow$ and $s*\downarrow$. The coefficients $C_{R,\alpha}$ determining the ith single-particle state and the corresponding single-particle energies are found by diagonalizing the empirical tight-binding Hamiltonian. The sp^3s^* empirical tight-binding Hamiltonian is given by:

$$H_{TB} = \sum_{R=1}^{N_{at}} \sum_{\alpha=1}^{10} \varepsilon_{R\alpha} c_{R\alpha}^{\dagger} c_{R\alpha} + \sum_{R=1}^{N_{at}} \sum_{\alpha=1}^{10} \sum_{\alpha'=1}^{10} \lambda_{R\alpha\alpha'} c_{R\alpha}^{\dagger} c_{R\alpha'} + \sum_{R=1}^{N_{at}} \sum_{R'=1}^{N_{at}} \sum_{\alpha=1}^{10} \sum_{\alpha'=1}^{10} t_{R\alpha,R'\alpha'} c_{R\alpha}^{\dagger} c_{R'\alpha'}$$

where the operator $c_{R\alpha}^{\dagger}$ ($c_{R\alpha}$) creates (annihilates) the particle on the orbital α of atom R. The on-site orbital energies $\varepsilon_{R\alpha}$, spin–orbit coupling constant $\lambda_{R\alpha\alpha'}$ and hopping matrix elements $t_{R\alpha,R'\alpha'}$ connecting different orbitals situated on neighboring atoms are described in this Hamiltonian. For the heterostructure, the valence-band offset (E_v) between core and shell is included in the tight-binding Hamiltonian. Therefore, all diagonal matrix elements of the core are shifted by E_v as compared to the shell diagonal matrix elements. The strain effect in the heterostructure is encompassed to provide a realistic description of the electronic states. The changes due to the strain are treated only by the scaling of inter-site matrix elements defined as:

$$t_{R\alpha,R'\alpha'} = t^0_{R\alpha,R'\alpha'} \left(\frac{d^0_{R'_J R_J}}{d_{R'_J R_J}} \right)^{n_{kl}}$$

where $t^0_{R\alpha,R'\alpha'}$ and $t_{R\alpha,R'\alpha'}$ are the ideal and distorted hopping matrix elements, respectively. $d^0_{R'_J R_J}$ and $d_{R'_J R_J}$ are the bond lengths of unstrained and strained binary materials, respectively. From Harrison's d^{-2} rule [102], $n_{kl} = 2.0$ is employed. Besides, the core/shell nanocrystals are passivated with the hydrogen atoms at the surface to avoid the formation of dangling bonds because these bonds may produce gap states and impede the calculations.

23.4.2 Oscillation Strength

To study the optical properties of the nanostructures, the oscillator strengths between the electron and hole states are demonstrated and analyzed. The oscillator strengths are calculated to determine optically allowed states of the dynamic polarization. The oscillation strengths f_{ij} between electron (i) and hole (j) states are defined as:

$$f_{ij} = \frac{2m_0}{\hbar^2} \left| \hat{e} \cdot \underset{ij}{\vec{D}} \right|^2 \times \left(E_i - E_j \right)$$

where m_0 is the free-electron mass. E_i and E_j are the energies of electron (i) and hole (j) levels, respectively. $\hat{e}$ are the polarized vectors (xy plane [110] and z axis [001]). $\underset{ij}{\vec{D}}$ are the interacted dipole moments between transition states of electron (i) and hole (j) levels.

23.4.3 Optical Spectra

To understand the optical properties of nanostructure devices, the optical spectra are calculated with the combination of single-particle spectra obtained from empirical tight-binding model and Fermi's Golden rule. The formula of the optical intensity $(I(E))$ is given by:

$$I(E) = \frac{2\pi}{\hbar} \sum_{n,m} \left| \left\langle \psi^v_{n,k=0} \middle| E\, \vec{e} \cdot \vec{r} \middle| \psi^c_{m,k=0} \right\rangle \right|^2 \delta\left(E^v_{n,k=0} - E^c_{m,k=0} - E \right)$$

Here, $k = 0$ is the Gamma point hin Brillouin zone. The indexes of m and n are symbolized for mth electron and nth hole levels. $\psi^c_{m,k=0}$ and $\psi^v_{n,k=0}$ are the electron and hole wave functions, respectively. $E^c_{m,k=0}$ and $E^v_{n,k=0}$ are the electron and hole energies, correspondingly. $\vec{e}$ are polarized vectors.

23.4.4 Radiative Lifetime

To understand the recombination between electron and hole, the calculations of radiative lifetime (τ_{ij}) between electron (i) and hole (j) states are determined. The equation is given by:

$$\tau_{ij} = \frac{3h^2 m_0 c^3}{8\pi^2 n e^2 \left(E^e_i - E^h_j\right)^2} \frac{1}{f_{ij}}$$

For the demonstration, m_0 is the free-electron mass, n is the refractive index and f_{ij} are the oscillator strengths between electron (i) and hole (j) states. E^e_i and E^h_j are the single-particle energies of electron (i) and hole (j) states, respectively.

23.4.5 Many-Body Hamiltonian

Once the single-particle states $\psi(r)$ and energies E_n are obtained by diagonalizing the empirical tight-binding Hamiltonian, the many-body Hamiltonian for the interacting electrons and holes is written in second quantization as: [103–105]

$$\begin{aligned} H_{many-body} = & \sum_i E^e_i c^\dagger_i c_i + \sum_i E^h_i h^\dagger_i h_i + \frac{1}{2} \sum_{ijkl} V^{ee}_{ijkl} c^\dagger_i c^\dagger_j c_k c_l \\ & + \frac{1}{2} \sum_{ijkl} V^{hh}_{ijkl} h^\dagger_i h^\dagger_j h_k h_l - \sum_{ijkl} V^{eh,dir}_{ijkl} c^\dagger_i h^\dagger_j h_k c_l + \sum_{ijkl} V^{eh,exchg}_{ijkl} c^\dagger_i h^\dagger_j c_k h_l \end{aligned}$$

Here, the operators $c^\dagger_i (c_i)$ and $h^\dagger_i (h_i)$ create (annihilate) the electron or hole in the state with energies $E^e_i (E^h_i)$. The two-body coulomb matrix elements are V^{ee}_{ijkl} for electron–electron interaction, V^{hh}_{ijkl} for hole-hole interaction, $V^{eh,dir}_{ijkl}$ for electron–hole direct coulomb interaction and $V^{eh,exchg}_{ijkl}$ for electron–hole exchange interaction. In the single excitonic case (contain one electron and one hole), the third and fourth term are therefore absent. Then, the single excitonic Hamiltonian is re-written as:

$$\begin{aligned} H_{\text{single}-exciton} = & \sum_i E^e_i c^\dagger_i c_i + \sum_i E^h_i h^\dagger_i h_i \\ & - \sum_{ijkl} V^{eh,dir}_{ijkl} c^\dagger_i h^\dagger_j h_k c_l + \sum_{ijkl} V^{eh,exchg}_{ijkl} c^\dagger_i h^\dagger_j c_k h_l \end{aligned}$$

The wave functions of the excitonic states are defined as the product of electron and hole wave functions:

$$\Phi_{ex} = \sum_{i,j} C_{i,j} \Psi_i^e(r_e) \Psi_j^h(r_h)$$

Ψ_i^e and Ψ_j^h are the single-particle *i*th electron and *j*th hole wave functions, respectively. Finally, the energies and the coefficients $C_{i,j}$ of the excitonic states can be obtained by diagonalizing the single excitonic Hamiltonian.

23.4.6 Examples of the Tight-Binding Calculations

Here, the examples in the implementation of the empirical tight-binding method with different orbital hybridizations on the different nanostructures are presented. In addition, the results corresponding to the electronic structures and optical properties of the examples are mainly presented in this chapter. Here, the example begins with the simple sp^3 empirical tight-binding model. Niquet et al. [71] reported the electronic calculations of Si nanostructures using the sp^3 empirical tight-binding model with third-nearest neighboring interactions and spin–orbit coupling. Using this parameterization, the band gap, Luttinger parameters and effective masses were properly fitted to experiment. The comparison of the confinement energies of Si nanocrystals was realized as a function of diameters. The empirical tight-binding data was compared against pseudopotential (PP) and local density approximations (LDA). The confinement energies of tight-binding model were not only in a good agreement with PP model in large sizes but also in an excellent agreement with LDA within the small clusters. Sukkabot [106, 107] reported the theoretical investigation of electronic structures and optical properties of InN and InSb nanocrystals via the sp^3s^* empirical tight-binding model with the nearest neighboring interactions and spin–orbit coupling. The structural and optical properties of InN and InSb nanocrystals are mainly dependent on their sizes, compositions and crystal structures. The example of the calculated band gaps in these nanocrystals was demonstrated. Figure 23.3 displayed the excitonic gaps of InSb nanocrystals as a function of the diameters. The optical band gaps of InSb nanocrystals were reduced with the increasing diameters due to the quantum confinement. The excitonic energies of tight-binding model were more consistent with the experimental data than those originally calculated by Efros and Rosen.

In case of the core/shell nanocrystals, there are several scientific works. This chapter mainly focuses on some important results of these nanostructures via sp^3s^* empirical tight-binding theory. Sukkabot [67] studied the electronic structures and optical properties of CdSe/ZnSe core/shell nanocrystals with experimentally synthesized sizes by sp^3s^* empirical tight-binding theory as a function of the growth shell thicknesses. The structural parameters of the CdSe/ZnSe core/shell nanocrystals were

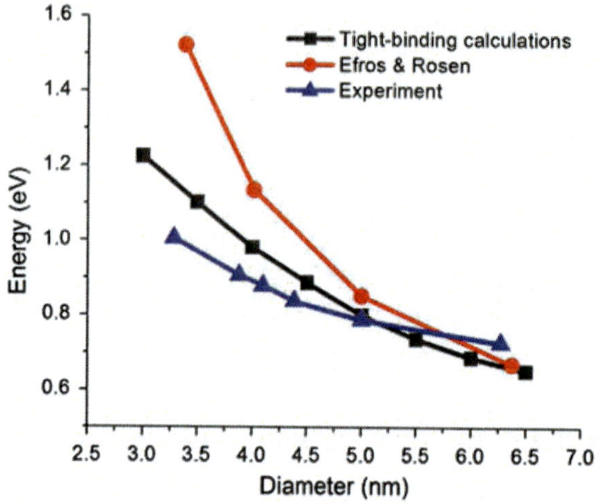

Fig. 23.3 The excitonic gaps of InSb zinc-blende nanocrystals as a function of diameters [106]

employed from the experimental data of Reiss et al. [108]. The spherical shape of CdSe/ZnSe core/shell nanocrystals was consisted of CdSe core with diameter of 3.6 nm and ZnSe terminated shell thicknesses (t_s) of 0.9, 1.8 and 2.7 monolayer (ML). For the demonstration of the electronic properties, the energies of electron and hole states under different shell thickness were displayed in Figs. 23.4 and 23.5, respectively. In the presence of the ZnSe shell, the electron and hole levels were improved in the comparison with those of single CdSe nanocrystals because of the

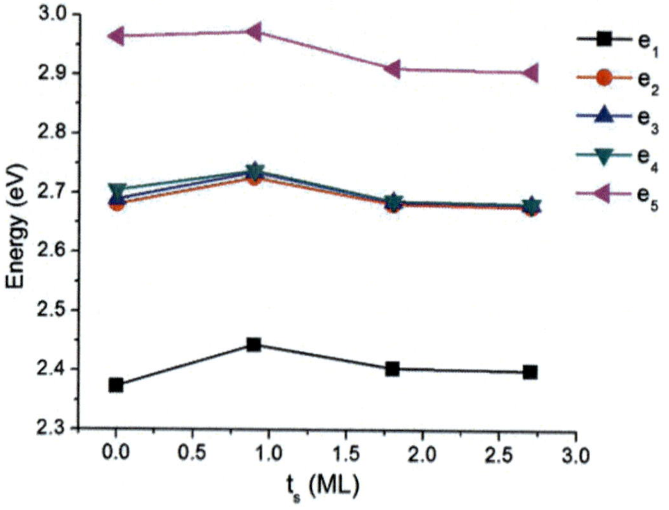

Fig. 23.4 Electron energies as a function of ZnSe shell thicknesses (ML) [67]

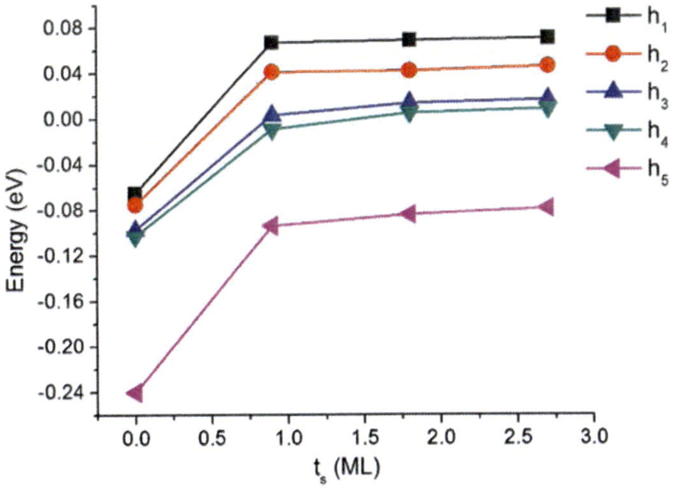

Fig. 23.5 Hole energies as a function of ZnSe shell thicknesses (ML) [67]

valence band offset between core and shell. Figure 23.6 illustrated the single-particle gaps, excitonic gaps and the size-tuneable absorption spectra of CdSe/ZnSe core/shell nanocrystals as a function of the ZnSe shell thicknesses. The single-particle gap was computed from the difference between the maximum hole and minimum electron state. The excitonic enegies were achieved by diagonalizing the single excitonic Hamiltonian using the configuration interaction method. The results underlined that the reduction of the single-particle and excitonic gaps with increasing shell thickness reflected the quantum confinement. As the comparison, the sp^3s^* empirical

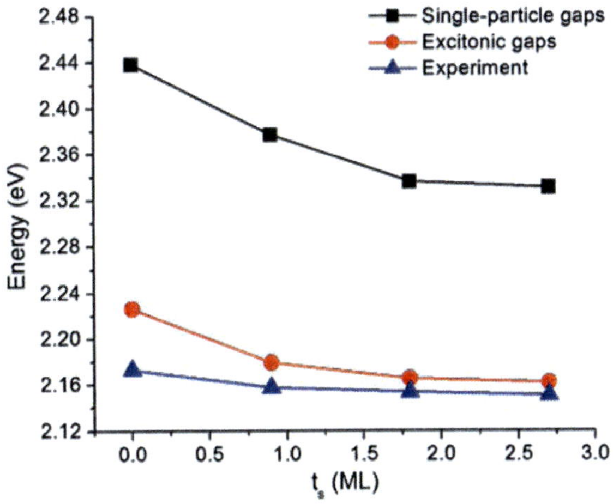

Fig. 23.6 The single-particle and excitonic gaps as a function of ZnSe shell thickness (ML) [67]

tight-binding calculations coincided well with optical band gaps from the experiment. In order to understand the behavior of orbital localized in the CdSe/ZnSe core/shell nanocrystals, the theoretical analysis of electron and hole character was also discussed. The ground electron states were characterized the s-like behavior in Fig. 23.7. In Figs. 23.8 and 23.9, the admixture of the p orbital contributed to the first-two hole states. In the presence of the ZnSe shell, the heavy-hole-like state in h_1 and light-hole-like state in h_2 were significantly improved in the comparison with the single CdSe nanocrystal. For the demonstration of the optical properties, the oscillator strength spectra (f_{ij}^{xy} and f_{ij}^{z}) of dipole moments determined by the spatial symmetrical xy plane and z component were displayed in Figs. 23.10 and 23.11 as a function of the inter-band transition states between the electron (e_i) and hole levels (h_j) under

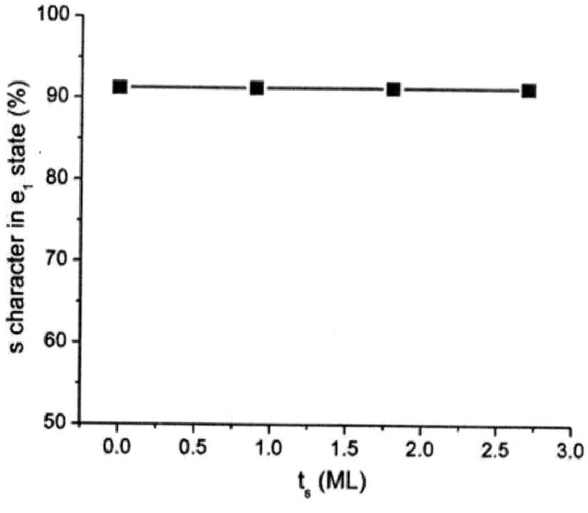

Fig. 23.7 The s character in the ground electron state (e_1) as a function of ZnSe shell thickness (ML) [67]

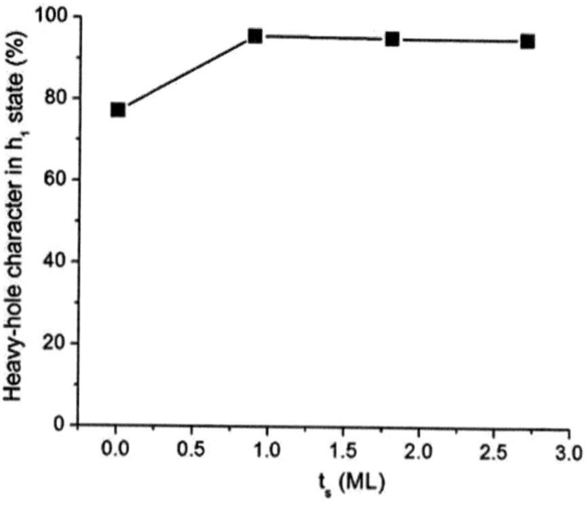

Fig. 23.8 The heavy-hole character in the first hole state (h_1) as a function of ZnSe shell thickness (ML) [67]

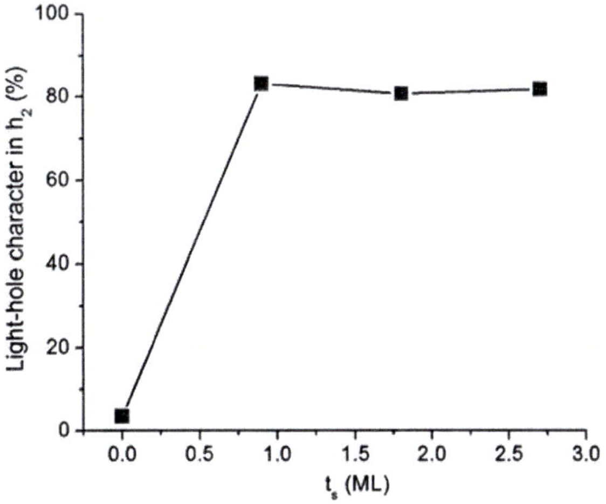

Fig. 23.9 The light-hole character in the second hole state (h_2) as a function of ZnSe shell thickness (ML) [67]

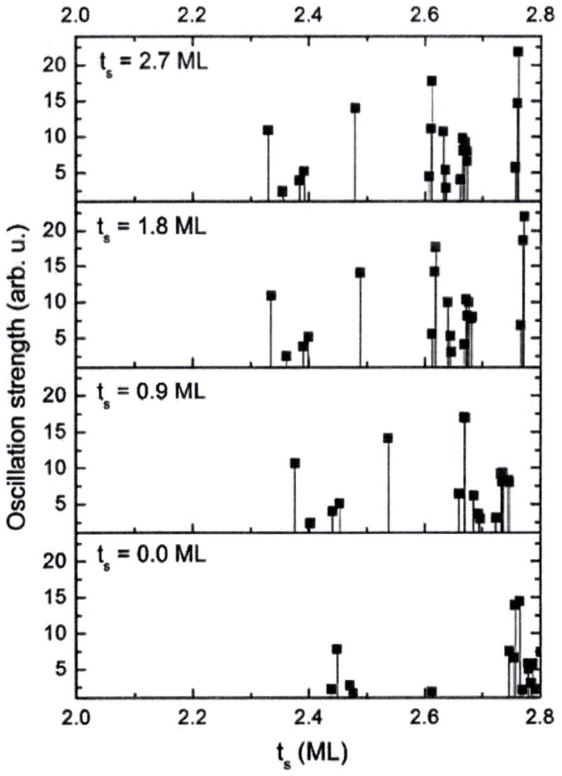

Fig. 23.10 Oscillation strengths in xy plane (f_{ij}^{xy}) as a function of ZnSe shell thickness (ML) [67]

Fig. 23.11 Oscillation strengths along z direction (f_{ij}^z) as a function of ZnSe shell thickness (ML) [67]

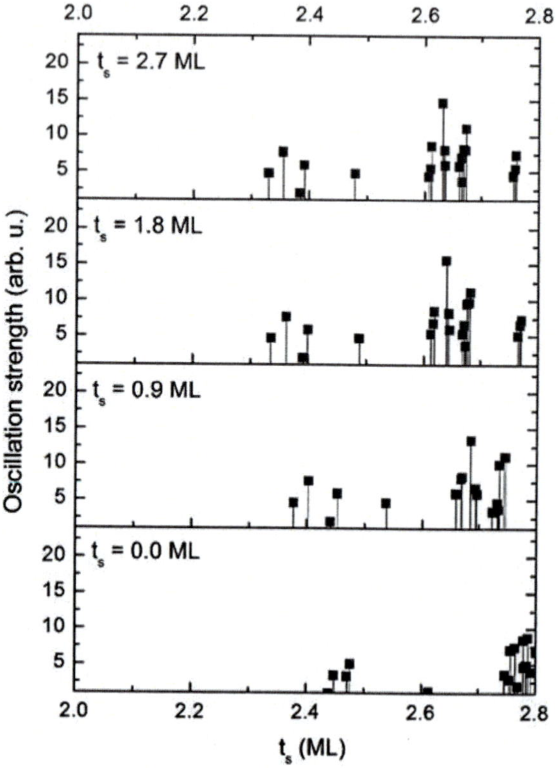

various shell thicknesses, respectively. The oscillation strengths were sensitive with the shell thickness and polarized vectors (xy and z direction). The minor red shifts in the oscillation strength spectra were reported with the increasing shell sizes. The oscillation strengths were enhanced when passivating the ZnSe growth shell on CdSe core, showing the improved optical properties. To analyze the optical properties, the first-two inter-band transition was determined. By considering the first peak of the spectra contributed from $e_1 - h_1$ transition, in the presence of the ZnSe shell the magnitudes of f_{ij}^{xy} were greater than those of f_{ij}^z. This was due to the fact that the dipole moment interactions between s-like (e_1) and heavy-hole-like states (h_1) along [109] plane were mainly more promoted than those along the [001] direction as described by the orbital characters in the previous Fig. 23.8. Therefore, the first peaks of the emission spectra were mostly donated by f_{ij}^{xy} terms. For the second peaks from $e_1 - h_2$ transition, when coating ZnSe shell the magnitudes of f_{ij}^z were higher than those of f_{ij}^{xy} because the dipole moment interactions between s-like (e_1) and light-hole-like states (h_2) along the z direction were augmented. Hence, the majorities of the second peaks were from f_{ij}^z. As can be seen, the electronic structures and optical properties were mainly manipulated by the growth shell thickness with the implementation to novel optoelectronic applications.

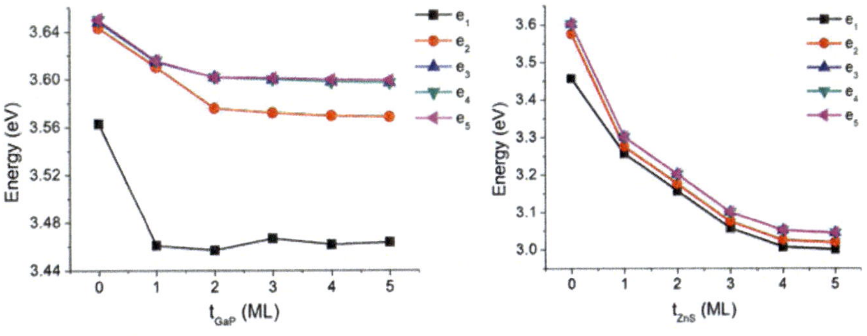

Fig. 23.12 Electron energies of InP/GaP core/shell and InP/GaP/ZnS core/shell/shell nanocrystals as a function of internal shell (t_{GaP}) and external shell (t_{ZnS}) thicknesses, respectively [109]

In addition, the core/shell nanocrystals could be passivated by another growth shell of different material, called core/shell/shell nanocrystal. Thus, the electronic structures and optical properties were widely controlled within more structural parameters (external shell). Sukkabot [109] reported the improvement of the luminescence efficiency for InP nanocrystals by coating a GaP/ZnS multi shell in zinc-blende phase using the atomistic tight-binding theory and configuration interaction method. InP/GaP/ZnS core/shell/shell nanocrystals were considered to be a non-toxic material for customer applications. For the demonstration of the impact for the interior shell on the natural behaviors, InP core with diameter of 3.10 nm was passivated by layers of GaP internal shell (t_{GaP}) from 0 to 5 monolayers (ML). To study the external shell dependence, thick layers of the ZnS external shell (t_{ZnS}) from 0 to 5 ML were terminated on InP/GaP core/shell nanocrystal with InP core diameter of 3.10 nm and GaP internal shell thickness of 2 ML. Figure 23.12 demonstrated the electron energies of InP/GaP core/shell and InP/GaP/ZnS core/shell/shell nanocrystals as a function of the internal and external growth shell thicknesses. The electron energies were reduced with the increasing internal and external coated shell thicknesses. The hole energies of InP/GaP core/shell and InP/GaP/ZnS core/shell/shell nanocrystals were illustrated in Fig. 23.13 under different internal and external growth shell thicknesses. With the increasing growth shell thicknesses, the hole energies of InP/GaP core/shell and InP/GaP/ZnS core/shell/shell nanocrystals were improved. For the active applications in the optoelectronic devices, the optical band gaps of InP/GaP core/shell and InP/GaP/ZnS core/shell/shell nanocrystals as a functions of the internal and external growth shell thicknesses were presented in Fig. 23.14. The reduction of optical band gaps was obtained with the increasing internal and external growth shell thicknesses due to the quantum confinement effect. The optical band gaps changing across the visible wave lengths were carried out by varying the internal and external growth shell thicknesses. The optical properties of InP/GaP core/shell and InP/GaP/ZnS core/shell/shell nanocrystals were analyzed in the framework of ground-state oscillation strengths in Fig. 23.15. The oscillation strengths were improved when termi-

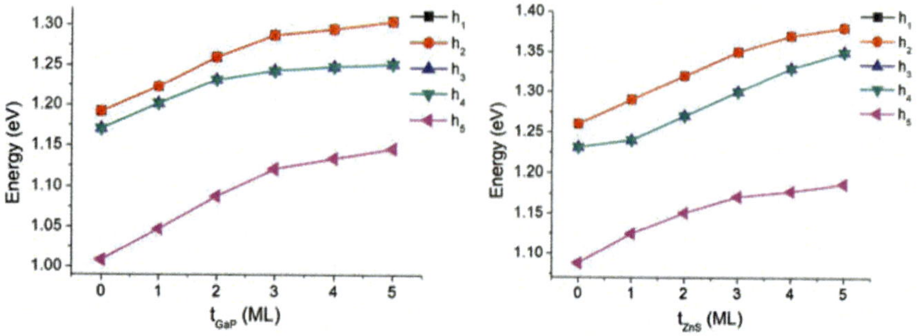

Fig. 23.13 Hole energies of InP/GaP core/shell and InP/GaP/ZnS core/shell/shell nanocrystals as a function of internal shell (t_{GaP}) and external shell (t_{ZnS}) thicknesses, respectively [109]

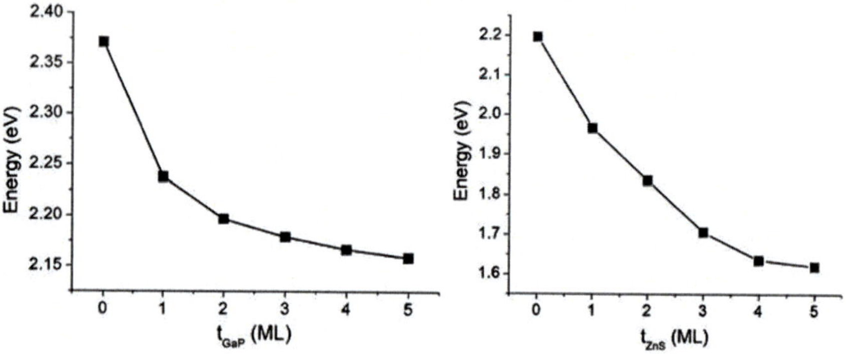

Fig. 23.14 Optical band gaps of InP/GaP core/shell and InP/GaP/ZnS core/shell/shell nanocrystals as a function of internal shell (t_{GaP}) and external shell (t_{ZnS}) thicknesses, respectively [109]

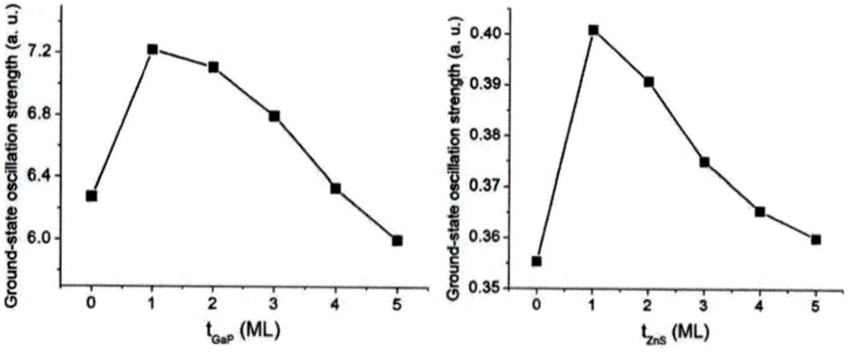

Fig. 23.15 Oscillation strengths of InP/GaP core/shell and InP/GaP/ZnS core/shell/shell nanocrystals as a function of internal shell (t_{GaP}) and external shell (t_{ZnS}) thicknesses, respectively [109]

Table 23.3 The references of theoretical studies for the important core/shell nanostructures (mainly in IV, III–V and II–VI group)

Systems	References		
	DFT	EPM	ETB
Si/Ge	[110, 111]		[112, 113]
InAs/InP			[114]
InAs/GaAs		[115]	[116, 117]
CdSe/CdS	[118, 119]	[66, 120]	[121, 122]
CdSe/ZnS	[123]		[124]
CdSe/ZnSe			[67]
CdTe/CdSe	[125]		[126]
ZnSe/ZnS	[127]		[128, 129]
ZnTe/ZnS			[130]
ZnTe/ZnSe			[131]

nating the internal and external shell, implying the enhancement of the optical properties. The oscillation strengths were progressively reduced as the internal and external shell thicknesses were increased. Passivation of GaP shell on InP core obviously improved the optical properties in the comparison with InP single nanocrystal. In addition, the optical properties in InP/GaP/ZnS core/shell/shell nanocrystals were enhanced when comparing to InP/GaP core/shell nanocrystals. By changing the sizes of the internal external coated shell, the guideline for designing the electronic structures and optical performance of these systems was finally achieved.

At the moment, there are several semiconductor core/shell nanocrystals, especially in IV, III–V and II–VI group. These systems have been widely determined by various experimental and theoretical methods. The theoretical description of semiconductor nanostructures is of essential prominence to examine fundamental physics before the actual manufacture. Table 23.3 itemizes the references of theoretical studies for the important core/shell nanostructures (mainly in IV, III–V and II–VI group). Here, theoretical investigations mainly focus on density functional theory (DFT), empirical pseudopotential method (EPM) and empirical tight-binding method (ETB).

23.5 Conclusion

Semiconductor core/shell nanocrystals are commonly applied in modern electronic and optoelectronic devices. Before actual fabrication, the theoretical description of semiconductor core/shell nanocrystals is of crucial importance. It is well known that atomistic approaches are essential to model the structural, electronic and optical properties of the approaching nanometric sizes. Here, the empirical tight-binding approach is successfully utilized because both of the computational efficiency and atomistic description are preserved. For the demonstration, the procedure of the computations is structured as follow. The first step is to generate the atomic positions of core/shell nanocrystals with desired shape, core diameter and growth shell thickness. Due to the lattice mismatch induced by the difference in core and shell material, the valence force field method (VFF) is implemented to optimize the atomic positions. The total energy of this approach is described as the summation of the energies of the stretching and bending bonds with the integration of the fitting parameters which explain the behaviors of each binary. After minimizing the total energy by conjugate gradient method, the relaxed atomic positions and the strain distribution are achieved. Using the Schrodinger equation, the single-particle spectra are numerically obtained in the framework of atomistic tight-binding theory. Due to the optimized computational consume with the satisfied bulk band structure calculations, sp^3s^* empirical tight-binding theory with the combination of the nearest neighboring interaction and spin–orbit coupling is utilized to investigate the core/shell semiconductor nanocrystals. After obtaining the single particle states and energies by diagonalizing the empirical tight-binding Hamiltonian, the natural properties of the studied core/shell nanocrystals are carried out. Implementing the configuration interaction description, the many-body Hamiltonian is computationally solved via the exact diagonalization method in the combination of the single-particle spectra. Using the tight-binding model, the electronic structures and optical properties of several semiconductor core/shell nanocrystals are evaluated and present a good agreement with the other theoretical and experimental data. Therefore, the empirical tight-binding model is very promising for the simulation tool to support the project of nanostructure devices.

References

1. M.J. Bowers, J.R. McBride, S.J. Rosenthal, J. AM. Chem. Soc. **127**, 15378 (2005)
2. P.O. Anikeeva, C.F. Madigan, S.A. Coe-Sullivan, J.S. Steckel, M.G. Bawnendi, V. Bulovic, Chem. Phys. Lett. **424**, 120 (2006)
3. A.J. Nozik, Phys. E **14**, 115 (2002)
4. I. Gur, N.A. Fromer, M.L. Geier, A.P. Alivisatos, Science **310**, 462 (2005)
5. M.C. Hanna, A.J. Nozik, J. Appl. Phys. **100**, 074510 (2006)
6. P.V. Kamat, J. Phys. Chem. C **112**, 18737 (2008)
7. V.I. Klimov, A.A. Mikhailovsky, S. Xu, A. Malko, J.A. Hollingsworth, C.A. Leatherdale, H.J. Eisler, M.G. Bawendi, Science **290**, 314 (2000)
8. S. Achilefu, Technol. Cancer Res. Treat **3**, 393 (2004)
9. M. Bruchez, M. Moronne, P. Gin, S. Weiss, A.P. Alivisatos, Science **281**, 2013 (1998)
10. A. Mukherjee, S. Ghosh, J. Phys. D Appl. Phys. **45**, 195103 (2012)
11. M. Dahan, S. Levi, C. Luccardini et al., Science **302**, 442 (2003)
12. D. Loss, D.P. DiVincenzo, Phys. Rev. A **57**, 120 (1998)
13. C. Filgueiras, O. Rojas, M. Rojas, Annalen Der Physik, 2000207 (2020)
14. N.W. Hendrickx, W.I.L. Lawrie, L. Petit, A. Sammak, G. Scappucci, M. Veldhorst, Nat. Commun. **11**, 3478 (2020)
15. H. Qiao, Y.P. Kandel, K. Deng, S. Fallahi, G.C. Gardner, M.J. Manfra, E. Barnes, J.M. Nichol, Phys. Rev. X **10**, 031006 (2020)
16. B. Lassen, M. Willatzen, R. Melnik, L.C.L.Y. Voon, J. Mater. Res. **21**, 2927 (2006)

17. W. Kohn, J. Luttinger, Phys. Rev. **98**, 915 (1955)
18. J. Luttinger, Phys. Rev. **102**, 1030 (1956)
19. D.S. Citrin, Y.-C. Chang, Phys. Rev. B **40**, 5507 (1989)
20. J.-B. Xia, Phys. Rev. B **43**, 9856 (1991)
21. V.V.R. Kishore, B. Partoens, F.M. Peeters, Phys. Rev. B **86**, 165439 (2012)
22. V.R. Kishore, N. Čukarić, B. Partoens, M. Tadić, F. Peeters, J. Phys. Condens. Matt. **24**, 135302 (2012)
23. B. Lassen, L. Lew Yan Voon, M. Willatzen, R. Melnik, Solid State Comm. **132**, 141 (2004)
24. P. Redliński, F. Peeters, Phys. Rev. B 77, 075329 (2008)
25. V.R. Kishore, B. Partoens, F. Peeters, Phys. Rev. B **82**, 235425 (2010)
26. M. Persson, A. Di Carlo, J. Appl. Phys. **104**, 073718 (2008)
27. M. Persson, H.Q. Xu, Nano Lett. **4**, 2409 (2004)
28. M. Persson, H.Q. Xu, Phys. Rev. B **73**, 125346 (2006)
29. M. Persson, H.Q. Xu, Phys. Rev. B **73**, 035328 (2006)
30. Y. Niquet, Nano Lett. **7**, 1105 (2007)
31. Y. Niquet, Phys. Rev. B 74 (2006) 155304.
32. M. Luisier, A. Schenk, W. Fichtner, G. Klimeck, Phys. Rev. B **74**, 205323 (2006)
33. G. Liao, N. Luo, Z. Yang, K. Chen, H.Q. Xu, J. Appl. Phys. **118**, 094308 (2015)
34. G. Liao, N. Luo, K.-Q. Chen, H.Q. Xu, J. Phys. Condens. Matter **28**, 135303 (2016)
35. G. Liao, N. Luo, K.-Q. Chen, H.Q. Xu, Sci. Rep. **6**, 28240 (2016)
36. P.Y. Yu, M. Cordona, Fundamentals of Semiconductors (2001)
37. A. Di Carlo, Semicond. Sci. Technol. **18**, R1 (2003)
38. P. Harrison, Quantum Wells, Wires and Dots. Wiley (2006)
39. H. Haug, S.W. Koch, Quantum Theory of the Optical and Electronic Properties of Semiconductors. World Scientific (2005)
40. J.R. Chelikowsky, M.L. Cohen, Phys. Rev. B **14**(2), 556 (1976)
41. M.L. Cohen, V. Heine, The Fitting of Pseudopotentials to Experimental Data and Their Subsequent Application, volume 24 of Solid State Physics. Academic Press, New York (1970)
42. M.L. Cohen, T.K. Bergstresser, Phys. Rev. **141**(2), 789 (1966)
43. F. Ning, L.-M. Tang, Y. Zhang, K.-Q. Chen, J. Appl. Phys. **114**, 224304 (2013)
44. S. Cahangirov, S. Ciraci, Phys. Rev. B **79**, 165118 (2009)
45. A. Srivastava, N. Tyagi, R. Ahuja, Solid State Sci. **23**, 35 (2013)
46. C.L. Dos Santos, P. Piquini, Phys. Rev. B **81**, 075408 (2010)
47. S. Schulz, Electronic and Optical Properties of Quantum Dots: A Tight-Binding Approach (2007)
48. W. Trzeciakowski, Phys. Rev. B **38**, 12493 (1988)
49. M.G. Burt, J. Phys. Condens. Matter **4**, 6651 (1992)
50. C. Pryor, J. Kim, L.W. Wang, A.J. Williamson, A. Zunger, J. Appl. Phys. **83**, 2548 (1998)
51. V.G. Malyshkin, I.P. Ipatova, V.A. Shchukin, J. Appl. Phys. **74**, 7198 (1993)
52. B. Jogai, J. Appl. Phys. **88**, 5050 (2000)
53. B. Jogai, J. Appl. Phys. **90**, 699 (2001)
54. O. Stier, Electronic and Optical Properties of Quantum Dots and Wires (2000)
55. R. Maranganti, P. Sharma, A Handbook of Theoretical and Computational Nanotechnology. American Scientific Publishers (2005)
56. T.S. Marshall, T.M. Wilson, Phys. Rev. B **50**, 15034 (1994)
57. H. Jiang, J. Singh, Phys. Rev. B **56**, 4696 (1997)
58. P.N. Keating, Phys. Rev. **145**, 637645 (1966)
59. R.B. Capaz, P. Kratzer, Q.K.K. Liu, R. Santoprete, B. Koiller, M. Scheffler, Phys. Rev. B **68**, 235311 (2003)
60. J.H. Seok, J.Y. Kim, Appl. Phys. Lett. **78**, 3124 (2001)
61. L.-W. Kim, A. Zunger, Phys. Rev. B **57**, R9408 (1997)
62. D.S. Yadav, C. Kumar, Int. J. Phys. Sci. **8**, 1174 (2013)
63. W.W. Hager, H. Zhang, SIAM J. Optim. **16**, 170 (2005)
64. W.W. Hager, H. Zhang, ACM Trans. Math. Softw. **32**, 113 (2006)

65. W.W. Hager, H. Zhang, Pac. J. Optim. **2**, 35 (2006)
66. Y. Luo, L.-W. Wang, ACS Nano **4**, 91 (2010)
67. W. Sukkabot, Comput. Mater. Sci. **96**, 336 (2015)
68. J.C. Slater, G.F. Koster, Phys. Rev. **94**, 1498 (1954)
69. D.J. Chadi, Phys. Rev. B **16**, 790 (1977)
70. D.J. Chadi, M.L. Cohen, Phys. Stat. Sol. (b) **68**, 405 (1975)
71. Y.M. Niquet et al., Phys. Rev. B **62**, 5109 (2000)
72. P. Vogl, H.P. Hjalmarson, J.D. Dow, J. Phys. Chem. Solids **44**, 365 (1983)
73. J.-M. Jancu, R. Scholz, F. Beltram, F. Bassani, Phys. Rev. B **57**, 6493 (1998)
74. J.-M. Jancu, F. Bassani, F. Della Sala, R. Scholz, Appl. Phys. Lett. **81**(25), 4838 (2003)
75. G. Klimeck, R.C. Bowen, T.B. Boykin, C. Salazar-Lazaro, T.A. Cwik, A. Stoica, Superlattices Microstruct. **27**, 77 (2000)
76. C. Tserbak, H.M. Polatoglou, G. Theodorou, Phys. Rev. B **47**, 7104 (1993)
77. Q.M. Ma, K.L. Wang, J.N. Schulman, Phys. Rev. B **47**, 1936 (1993)
78. G. Grosso, C. Piermarocchi, Phys. Rev. B **51**, 16772 (1995)
79. D.N. Talwar, Z.C. Feng, Phys. Rev. B **44**, 3191 (1991)
80. J.N. Schulman, Y.-C. Chang, Phys. Rev. B **31**, 2056 (1985)
81. T.B. Boykin, J.P.A. van der Wagt, J.S.Jr. Harris, Phys. Rev. B **43**, 4777 (1991)
82. A. Di Carlo, P. Lugli, Semicond. Sci. Technol. **10**, 1673 (1995)
83. A. Di Carlo, A. Reale, L. Tocca, P. Lugli, IEEE J. Quantum Electron. **34**, 1730 (1998)
84. H. Dierks, G.Z. Czycholl, Phys. Rev. B **99**, 207 (1996)
85. K. Shim, H. Rabitz, Phys. Rev. B **57**, 12874 (1998)
86. T.B. Boykin, Phys. Rev. B **51**, 4289 (1995)
87. G. Theodorou, G. Tsegas, Phys. Rev. B **61**, 10782 (2000)
88. D. Olguin, R. Baquero, R. de Coss, Rev. Mex. Fis. **47**(1), 43 (2001)
89. D. Bertho, D. Boiron, A. Simon, C. Jouanin, C. Proester, Phys. Rev. B **44**, 6118 (1991)
90. Z.Q. Li, Q. Pötz, Phys. Rev. B **46**, 2109 (1992)
91. D. Bertho, J.M. Jancu, C. Jouanin, Phys. Rev. B **48**, 2452 (1993)
92. E.G. Wang, C.-F. Chen, C.S. Ting, J. Appl. Phys. **78**, 1832 (1995)
93. A. Kobayashi, O.F. Sankey, S.M. Volz, J.D. Down, Phys. Rev. B **28**, 935 (1983)
94. P.E. Lippens, M. Lannoo, Phys. Rev. B **41**, 6079 (1990)
95. J. Perez-Conde, A.K. Bhattacharjee, Phys. Rev. B **63**, 245318 (2001)
96. M. Fornari, H. H. Chen, L. Fu, R. D. Graft, D. J. Lohrmann, S. Moroni, G. Pastori Parravicini, L. Resca, M.A. Stroscio, Phys. Rev. B **55**, 16339 (1997)
97. J. Perez-Conde, A.K. Bhattacharjee, M. Chamarro, P. Lavallard, V.D. Petrikov, A.A. Lipovskii, Phys. Rev. B **64**, 113303 (2001)
98. A. Kobayashi, O.F. Sankey, J.D. Down, Phys. Rev. B **25**, 6367 (1983)
99. K.C. Hass, H. Eherenreich, B. Velicky, Phys. Rev. B **27**, 1088 (1983)
100. M. Dib, M. Chamarro, V. Voliotis, J.L. Fave, C. Guenaud, P. Roussignol, T. Gacoin, J.P. Boilot, C. Delerue, G. Allan, M. Lannoo, Phys. Status Solidi B **212**, 293 (1999)
101. J.P. LaFemina, C.B. Duke, J. Vac. Sci. Technol. A **9**, 1847 (1991)
102. W.A. Harrison, Elementary Electronic Structure. World Scientific Publishing Company (1999)
103. M. Zielinski, M. Korkusinski, P. Hawrylak, Phys. Rev. B **81**, 085301 (2010)
104. M. Korkusinski, M.E. Reimer, R.L. Williams, and P. Hawrylak, Phys. Rev. B **79**, 035309 (2009)
105. K. Leung, K.B. Whaley, Phys. Rev. B **56**, 7455 (1997)
106. W. Sukkabot, Mater. Sci. Semicond. Process. **27**, 51 (2014)
107. W. Sukkabot, Mater. Sci. Semicond. Process. **38**, 142 (2015)
108. P. Reiss, J. Bleuse, A. Pron, Nano Lett. **2**(7), 781 (2002)
109. W. Sukkabot, Comput. Mater. Sci. **161**, 46 (2019)
110. N. Scott Bobbitt, J.R. Chelikowsky, J. Chem. Phys. **144**(12), 124110 (2016)
111. E.L. de Oliveira, E.L. Albuquerque, J.S. de Sousa, G.A. Farias, F.M. Peeters, J. Phys. Chem. C **116**, 7 (2012)
112. M.O. Nestoklon, A.N. Poddubny, P. Voisin, K. Dohnalova, J. Phys. Chem. C **120**, 33 (2016)

113. I.D. Avdeev, A.V. Belolipetsky, N.N. Ha, M.O. Nestoklon, I.N. Yassievich, J. Appl. Phys. **127**, 114301 (2020)
114. W. Sukkabot, Phys. E Low-Dimens. Syst. Nanostruct. **63**, 235 (2014)
115. H.Y.S. Al-Zahrani, J. Pal, M.A. Migliorato, G. Tse, D. Yu, Nano Energy **14**, 382 (2015)
116. Y.M. Niquet, Phys. Rev. B **74**, 155304 (2006)
117. Y.M. Niquet, Nano Lett. **7**(4), 1105 (2007)
118. V. Kocevski, J. Rusz, O. Eriksson, D.D. Sarma, Sci. Rep. **5**, 10865 (2015)
119. L. Zhu, et al., IOP Conf. Ser. Mater. Sci. Eng. **490**, 022021 (2019)
120. H. Eshet, M. Grünwald, E. Rabani, Nano Lett. **13**(12), 5880 (2013)
121. W. Sukkabot, Mater. Sci. Semicond. Process. **34**, 14 (2015)
122. A. Jain, O. Voznyy, S. Hoogland, M. Korkusinski, P. Hawrylak, E.H. Sargent, Nano Lett. **16**(10), 6491 (2016)
123. X. Zhai, R. Zhang, J. Lin, Y. Gong, Y. Tian, W. Yang, X. Zhang, Cryst. Growth Des. **15**(3), 1344 (2015)
124. W. Sukkabot, Mater. Sci. Semicond. Process. **27**, 1020 (2014)
125. L. Zhang, Z. Lin, J.-W. Luo, A. Franceschetti, ACS Nano **6**(9), 8325 (2012)
126. W. Sukkabot, Comput. Mater. Sci. **111**, 23 (2106)
127. S.C. Pandey, J. Wang, T.J. Mountziaris, D. Maroudasa, J. Appl. Phys. **111**, 113526 (2012)
128. W. Sukkabot, Superlattices Microstruct. **75**, 739 (2014)
129. W. Sukkabot, Mater. Sci. Semicond. Process. **41**, 252 (2016)
130. W. Sukkabot, Comput. Mater. Sci. **101**, 275 (2015)
131. W. Sukkabot, Physica E **74**, 457 (2015)

Chapter 24
Tight Binding and Density Functional Theory of Tailoring Electronic Properties in $Al_{1-x}In_xN$/AlN/GaN High Electron Mobility Transistors (HEMTs)

Meziani Abdelhakim, Abdul-Rahman Allouche, Telia Azzedine, and Hilmi Unlu

Abstract Group III-nitrides have acquired an important role in science and technology of compound semiconductors for fabricating the novel electronic and optical devices. AlN, InN, GaN and to a lesser extent BN and their alloys InGaN, AlGaN, and AlInN are of great interest for the high temperature electronics and optoelectronics applications. Recent growth technologies leading to high quality films renewed interest especially when the ternary $Al_{1-x}In_xN$ alloy presents a band gap that covers the widest energy range (0.69–6.25 eV) compared to other nitride compounds. This material provides a unique range of composition coverage making it an ideal candidate to fabricate electronic and optoelectronic devices such as high-power high-frequency field-effect transistors, blue and ultraviolet light-emitting and laser diodes, resonant- cavity light-emitting diodes, surface-emitting lasers, and solar blind ultraviolet photodetectors. In this chapter, we present a semiempirical tight binding theory and density functional theory analysis of the lattice mismatch and thermal strain effects on the tailoring of the electronic, optical and elastic properties of the zincblende and wurtzite phases of the $Al_{1-x}In_xN$/GaN ternary/binary heterostructure. Results are presented and compared with available data for the wurtzite phase, yielding a useful database for the modeling of AlInN high electron mobility field effect transistors (HEMTs).

M. Abdelhakim (✉)
Physics Department, Frères Mentouri University Constantine, Constantine, Algeria
e-mail: meziani-hakim@umc.edu.dz

A.-R. Allouche
Institute of Light and Matter (ILM), UMR5306 University of Lyon1, CNRS, University of Lyon, Villeurbanne Cedex, France

T. Azzedine
Electronic Department, Frères Mentouri University Constantine, Constantine, Algeria

H. Unlu
Department of Physics Engineering, İstanbul Technical University, Istanbul, Turkey
e-mail: hunlu@itu.edu.tr

H. Ünlü and N. J. M. Horing (eds.), *Progress in Nanoscale and Low-Dimensional Materials and Devices*, Topics in Applied Physics 144,
https://doi.org/10.1007/978-3-030-93460-6_24

24.1 Introduction

Since the development of a commercial blue light emitting diode based on GaN [1], group III nitride binaries (InN, GaN, AlN) and their ternary and quaternary alloys (InGaN, AlGaN, AlInN and AlInGaN) have revolutionized solid state lighting and continue to attract substantial research interest due to their unique properties and importance for micro and nanoscale optoelectronics and electronics. Consequently, the theoretical and experimental investigation of structural, electronic and optical properties of Group III-nitrides GaN, InN, AlN binaries and their ternary alloys InGaN, AlGaN and AlInN, and the quaternary alloy AlInGaN have been the subject of intensive research for the device scientists and engineers in recent years due to their potential application for the fabrication of novel electronic and optoelectronic devices fabrication [2]. Among these semiconductor structures ternary $Al_{1-x}In_xN$ alloy has a band gap that covers a very wide energy range (0.69–6.25 eV), which makes it an ideal candidate to fabricate electronic and optoelectronic devices such as high-power high-frequency field-effect transistors, blue and ultraviolet light emitting and laser diodes, resonant-cavity light emitting diodes, surface-emitting lasers, and solar blind ultraviolet photodetectors [3, 4] (Fig. 24.1).

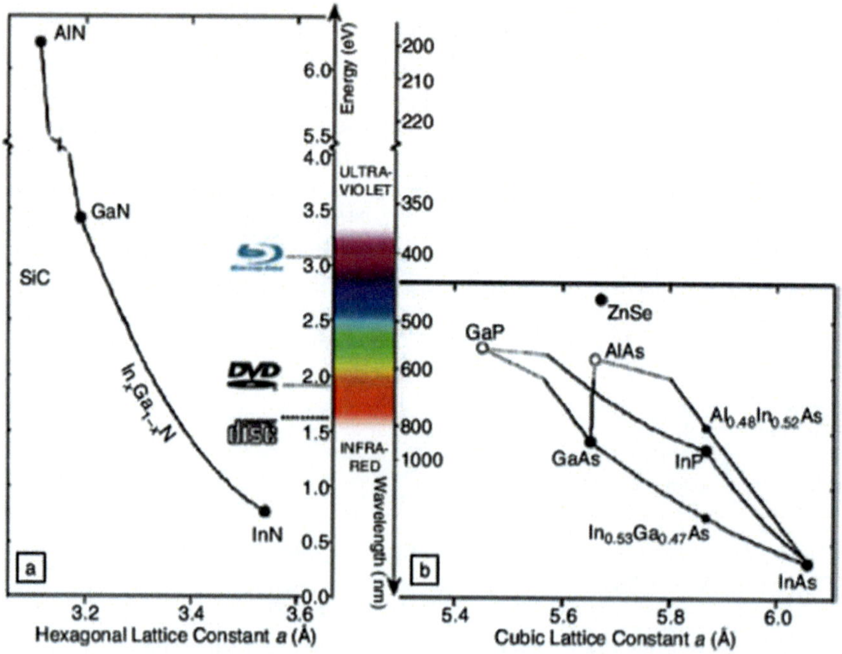

Fig. 24.1 Band gaps versus structural parameters of III–V wurtzitic nitrides (left). Color range of emitted light (middle). Relationship between band gap and structure parameters on some material used in LED (right) [5]

Under normal growth conditions, AlN and InN crystallize in hexagonal wurtzite structure (B4 phase), but growth of zincblende structure (B3 phase) on cubic substrates can also been achieved [6], whereas the rocksalt structure (B1 phase) is only obtained at high pressure [7, 8]. Over the last few years, the theoretical and experimental studies of these semiconductors under high pressure have become an extremely important subject exhibiting high growth. The effects of pressure are more easily incorporated into first-principles simulations and varying gaps by means of pressure variation are another way of tailoring gaps and providing a wider spectrum for optoelectronic devices.

The architecture of electronic or optical devices requires heterostructure stacking of different compounds with various lattice parameter constants thus creating strains on the thin films. This has an immediate effect on the structural properties on the strained alloy and therefore on its electronic and optical properties. In the case of AlInN based devices, GaN is often chosen as a buffer layer resulting in a biaxial strain on the semiconductor alloy with an immediate modification of the lattice parameters hence its electronic and optical properties. Many of the charge transport parameters of AlInN/AlN/GaN high electron mobility transistors (HEMTs) depend on the density of two-dimensional electron gas (2DEG), which is greatly influenced by the interface strain.

In this chapter we will review theoretical studies about the pressure effects on the variation of the electronic properties of AlInN ternary alloy zincblende and wurtzite phases. We will discuss use of the first nearest neighbor semi-empirical sp^3s^* tight binding theory and the density functional theory (DFT), with the modified Becke-Johnson (mBJLDA) potential, to investigate the composition effects on the lattice and energy band structure of $Al_{1-x}In_xN$ ternary alloy in the zincblende and wurtzite phases. In these studies, one first optimizes the lattice constant and then calculates the composition effect on the energy band gaps at high symmetry points of the first Brillouin zone, the electron effective mass and the density of states. We shall compare the theoretical predictions of some DFT studies with mBJLDA functional and first nearest neighbor (NN) semi-empirical sp^3s^* Tight binding method for the lattice and band structure properties of ternary $Al_{1-x}In_xN$ as a function of alloy composition against the available experimental data.

24.2 Semi-empirical Tight Binding Theory

In the semi-empirical sp^3s^* tight binding model (TB) every atom is described by the valence s orbital and the outer p orbital and a fictitious s^* state to take into account the effects of d orbitals in the energy band calculations [9]. We used the semi-empirical sp^3s^* TB with first nearest neighbor (NN) interactions taken into account to calculate the electronic properties of ternary $Al_{1-x}In_xN$ alloy. The spin orbit effect is neglected because of the type of atoms used in this study. The Schrodinger equation in matrix form is

$$\sum_{\beta} \left[H_{\alpha\beta}(k) - S_{\alpha\beta}(k)E \right] u_{\beta} = 0, \tag{24.1}$$

where E is the energy eigenvalue of the Hamiltonian matrix:

$$H_{\alpha\beta}(k) = < \chi_{\alpha}(k)|H|\chi_{\beta}(k) > \tag{24.2}$$

with the overlap integral between the atomic-like orbitals, with α and β that correspond to cation (c) and anion (a) s, p atomic orbitals, respectively.

$$S_{\alpha\beta}(k) = < \chi_{\alpha}(k)|\chi_{\beta}(k) > \tag{24.3}$$

where χ(k) is the basis function formed by the linear combination of s, p atomic orbitals of cation and anion of binary compound. The sp^3s^* Hamiltonian $H_{\alpha\beta}$ has six diagonal elements (on-site atomic energies:E_{sa}, E_{sc}, E_{pa},E_{pc}, E_{s^*a} and E_{s^*c}) and seven off-diagonal elements (interaction integrals, known as hopping terms:E_{ss}, E_{xx}, $E_{s_a p_c}$, $E_{s_c p_a}$,E_{xy}, E_{s^*p} and E_{ps^*}). The general form of sp^3s^* TB Hamiltonian matrix for semiconductors with zincblende crystal structure without spin–orbit interaction is given by [10]:

$$H = \begin{pmatrix} E_{sa} & V_{ss}g_1 & 0 & 0 & 0 & V_{sapc}g_2 & V_{sapc}g_3 & V_{sapc}g_4 & 0 & 0 \\ V_{ss}g_1' & E_{sc} & -V_{pasc}g_2' & -V_{pasc}g_3' & -V_{pasc}g_4' & 0 & 0 & 0 & 0 & 0 \\ 0 & -V_{pasc}g_2 & E_{pa} & 0 & 0 & V_{xx}g_1 & V_{xy}g_4 & V_{xy}g_3 & 0 & -V_{pasc*}g_2 \\ 0 & -V_{pasc}g_3 & 0 & E_{pa} & 0 & V_{xy}g_4 & V_{xx}g_1 & V_{xy}g_2 & 0 & -V_{pasc*}g_3 \\ 0 & V_{pasc}g_4 & 0 & 0 & E_{pa} & V_{xy}g_3 & V_{xy}g_2 & V_{xx}g_1 & 0 & -V_{pasc*}g_4 \\ V_{sapc}g_2' & 0 & V_{xx}g_1' & V_{xy}g_4' & V_{xy}g_3' & E_{pc} & 0 & 0 & V_{s*apc}g_2 & 0 \\ V_{sapc}g_3' & 0 & V_{xy}g_4' & V_{xx}g_1' & V_{xy}g_2' & 0 & E_{pc} & 0 & V_{s*apc}g_3 & 0 \\ V_{sapc}g_4' & 0 & V_{xy}g_3' & V_{xy}g_2' & V_{xx}g_1' & 0 & 0 & E_{pc} & V_{s*apc}g_4 & 0 \\ 0 & 0 & 0 & 0 & 0 & V_{s*apc}g_2' & V_{s*apc}g_3' & V_{s*apc}g_4' & E_{s*a} & 0 \\ 0 & 0 & -V_{pasc*}g_2' & -V_{pasc*}g_3' & -V_{pasc*}g_4' & 0 & 0 & 0 & 0 & E_{s*c} \end{pmatrix} \tag{24.4}$$

$$\begin{aligned} g_1 &= \frac{1}{4}(\exp(id_1k_0) + \exp(id_2k_0) + \exp(id_3k_0) + \exp(id_4k_0)), \\ g_2 &= \frac{1}{4}(\exp(id_1k_0) + \exp(id_2k_0) - \exp(id_3k_0) - \exp(id_4k_0)), \\ g_3 &= \frac{1}{4}(\exp(id_1k_0) - \exp(id_2k_0) + \exp(id_3k_0) - \exp(id_4k_0)), \\ g_4 &= \frac{1}{4}(\exp(id_1k_0) - \exp(id_2k_0) - \exp(id_3k_0) + \exp(id_4k_0)) \end{aligned} \tag{24.5}$$

where we have assumed that atom 1 is located at the origin and d_i (i = 1,4) are the positions of its four nearest neighbors and a is the lattice parameter.

$$d_1 = \frac{a}{4}(1, 1, 1)\,,\ d_2 = \tfrac{a}{4}(1, -1, -1)\ ,\ d_3 = \tfrac{a}{4}(-1, 1, -1)\ ,\ d_4 = \tfrac{a}{4}(-1, -1, 1) \tag{24.6}$$

The general form of sp^3s^* TB Hamiltonian matrix for semiconductors with wurtzite crystal structure without spin–orbit interaction, can be expressed in a block form as below [11].

$$H = \begin{pmatrix} E_a & H_0 & H_{13} & H_{14} \\ H_0 & E_a & H_{14} & H_{24} \\ H'_{13} & H'_{14} & E_c & H_0 \\ H'_{14} & H'_{24} & H_0 & E_c \end{pmatrix} \tag{24.7}$$

where the different matrix elements are 4 × 4 matrix blocks. The diagonal matrices E_a and E_c contain the orbital energies. The matrices H_{ij} represent the interaction up to nearest neighbors.

$$E_a = \begin{pmatrix} E_{sa} & 0 & 0 & 0 \\ 0 & E_{pza} & 0 & 0 \\ 0 & 0 & E_{pxa} & 0 \\ 0 & 0 & 0 & E_{pxa} \end{pmatrix},\quad E_c = \begin{pmatrix} E_{sc} & 0 & 0 & 0 \\ 0 & E_{pc} & 0 & 0 \\ 0 & 0 & E_{pc} & 0 \\ 0 & 0 & 0 & E_{pc} \end{pmatrix}$$

$$H_{13} = g_1 M_{13} = g_1 \begin{pmatrix} f_0 U_{1ss} & f_0 U_{1sz} & f_1 U_{1sx} & \frac{\sqrt{3}}{2} f_3 U_{1sx} \\ f_0 U_{1zs} & f_0 U_{1zz} & f_1 U_{1zx} & \frac{\sqrt{3}}{2} f_3 U_{1zx} \\ f_1 U_{1xs} & f_1 U_{1xz} & f_1 U_{1xx} + \frac{3}{4} f_2 \left(U_{1xx} + U_{1yy}\right) & \frac{\sqrt{3}}{4} f_3 \left(U_{1xx} - U_{1yy}\right) \\ \frac{\sqrt{3}}{2} f_3 U_{1xs} & \frac{\sqrt{3}}{2} f_3 U_{1xz} & -\frac{\sqrt{3}}{4} f_3 \left(U_{1xx} - U_{1yy}\right) & f_1 U_{1yy} + \frac{3}{4} f_2 \left(U_{1xx} + U_{1yy}\right) \end{pmatrix}$$

$$H_{14} = g_1 M_{14} = g_1 \begin{pmatrix} U_{ss} & U_{sz} & 0 & 0 \\ U_{zs} & U_{zz} & 0 & 0 \\ 0 & 0 & U_{xx} & 0 \\ 0 & 0 & 0 & U_{xx} \end{pmatrix}$$

$$H_{24} = g_2 M_{24} = g_2 \begin{pmatrix} f'_0 U_{1ss} & f'_0 U_{1sz} & -f'_1 U_{1sx} & -\frac{\sqrt{3}}{2} f'_3 U_{1sx} \\ f'_0 U_{1zs} & f'_0 U_{1zz} & -f'_1 U_{1zx} & -\frac{\sqrt{3}}{2} f'_3 U_{1zx} \\ -f'_1 U_{1xs} & -f'_1 U_{1xz} & f'_1 U_{1xx} + \frac{3}{4} f'_2 \left(U_{1xx} + U_{1yy}\right) & -\frac{\sqrt{3}}{4} f'_3 \left(U_{1xx} - U_{1yy}\right) \\ -\frac{\sqrt{3}}{2} f'_3 U_{1xs} & -\frac{\sqrt{3}}{2} f'_3 U_{1xz} & -\frac{\sqrt{3}}{4} f'_3 \left(U_{1xx} - U_{1yy}\right) & f'_1 U_{1yy} + \frac{3}{4} f'_2 \left(U_{1xx} + U_{1yy}\right) \end{pmatrix}$$

$$U_{ss} = V_{ss}/4 = -\sqrt{\frac{3}{4}} V_{sapc}\,,\ U_{zs} = -U_{sz}\,,$$

$$U_{zz} = \left((V_{xx}/4) + \left(V_{xy}/2\right)\right),\ U_{xx} = \left((V_{xx}/4) - \left(V_{xy}/4\right)\right),$$

$$U_{1ss} = U_{ss}\ \ U_{1sz} = -U_{sz}/3;\ U_{1zs} = -U_{zs}/3;\ U_{1sx} = V_{sapc}/\sqrt{6},$$

$$U_{1xs} = -V_{scpa}/\sqrt{6};\ U_{1xz} = V_{xy}/\left(3\sqrt{2}\right);\ U_{1zx} = U_{1xz};$$

$$U_{1xx} = ((8U_{zz}/9) + (U_{xx}/9)),\ U_{1zz} = ((U_{zz}/9) + (8U_{xx}/9)),\ U_{1yy} = U_{xx}$$

Table 24.1 AlN and InN onsite and hopping parameters

		AlN	InN
B3 (eV)	Onsite	$E_{sa} = -11.5047$; $E_{pa} = 4.3815$; $E_{sc} = 0.5047$; $E_{pc} = 10.2184$; $E_{s*a} = 12.04$; $E_{s*c} = 13.74$;	$E_{sa} = -12.8605$; $E_{pa} = 1.98$; $E_{sc} = -0.3994$; $E_{pc} = 8.02$; $E_{s*a} = 10.63$; $E_{s*c} = 13.00$;
	Hopping	$V_{ss} = -9.8077$; $V_{xx} = 6.031$; $V_{xy} = 8.6191$; $V_{sapc} = 9.4$; $V_{pasc} = 8.5$; $V_{s*apc} = 8.03$; $V_{pas*c} = 2.47$;	$V_{ss} = 4.2285$; $V_{xx} = 3.65$; $V_{xy} = 6.405$; $V_{sapc} = 3.81$; $V_{pasc} = 5.75$; $V_{s*apc} = 6.88$; $V_{pas*c} = 3.36$;
B4 (eV)	Onsite	$E_{sa} = -12.104$ $E_{pxa} = 3.581$; $E_{pza} = 3.725$; $E_{sc} = -0.096$; $E_{pc} = 8.95$	$E_{sa} = -6.791$; $E_{pxa} = 0.000$; $E_{pza} = 0.000$; $E_{sc} = -3.015$; $E_{pc} = 8.822$;
	Hopping	$V_{ss} = -10.735$; $V_{xx} = 5.808$; $V_{xy} = 7.486$; $V_{scpa} = 9.755$; $V_{sapc} = 10.092$;	$V_{ss} = -5.371$; $V_{xx} = 0.022$; $V_{xy} = 6.373$; $V_{scpa} = 0.370$; $V_{sapc} = 18.0$

The tight binding AlN, InN and GaN self-energies parameters have been extracted from [12, 13] and modified so as to fit with high symmetry points. AlN and InN onsite and hopping parameters are listed in Table 24.1.

Within this scheme one can reproduce the conduction and valence band energies of zincblende and wurtzite III-N semiconductors with direct and indirect bandgaps. The energies of the on-site and off-site elements of the TB Hamiltonian matrix are obtained from the fitting of the experimental data of bandgap energies to the TB calculations at high symmetry points. Using the derived parameters, the energy band structure is obtained by diagonalizing the Hamiltonian matrix at each point of the Brillouin zone.

In order to study the band structure of ternary/binary system the first step is to reliably and precisely determine the composition effects on the band structure of ternary semiconductor. In this respect, the empirical tight binding scheme can be used to incorporate the composition effects for a realistic description of the composition dependent band structure. The common practice is to employ so called the virtual crystal approximation (VCA) to the tight binding Hamiltonian matrix elements and bond length with and without the compositional disorder of the alloy semiconductor. The calculation of atomic energies for the $Al_{1-x}In_xN$ ternary is carried out by using the virtual crystal approximation. The tight binding parameters representing the diagonal terms in the Hamiltonian matrix for $A_x B_{1-x} C$ ternary semiconductor are taken as

$$E_{\alpha\beta}(x) = xE_{\alpha\beta}(AC) + (1-x)E_{\alpha\beta}(BC) \tag{24.8}$$

where $E_{\alpha\beta} = E_{ss}, E_{xx}, E_{s_a p_c}, E_{s_c p_a}, E_{xy}, E_{s^* p}$ and E_{ps^*} are the atomic energies of s and p states of the anion and cation atoms forming the AC and BC compounds binary such as AlN and InN, respectively.

24.3 Density Functional Theory Calculations

Density functional theory (DFT) is a powerful *ab-initio* method for computing the physical properties of solids such as the structural, electronic magnetic properties of semiconductors to name a few from the first principles without any fitting parameters. The conventional density functional theory (DFT) enables us to have a parameter-free description of structural and electronic properties of semiconductors and has its foundations in the work of Kohn and his collaborators [14, 15]. The variation method is used to calculate the ground state properties of a many-body system with the charge density, expressed in terms of single-particle electronic orbitals

$$n(\vec{r}) = \sum_{i=1}^{N} \sum_{k} f_{i,k} |\varphi_{i,k}(r) > |^2 \tag{24.9}$$

which plays the central role in the calculation of the structural and electronic properties of semiconductors. The major consequence of the Kohn–Hohenberg–Sham studies [14, 15] is that the single particle states $\varphi_{i,k}(r)$ are determined by the solution of a set of equations similar to the Schrodinger equation:

$$H|\varphi_{i,k} >= (-\frac{\hbar^2}{2m}\nabla_i^2 + V_{ext} + V_{eff})|\varphi_{i,k} >= E_i|\varphi_{i,k} > \tag{24.10}$$

where $\varphi_{i,k}(r)$ are the one-electron wave functions, V_{ext} is the external potential of the nuclei, and $V_{eff}(r)$ the effective potential defined as

$$V_{eff}(r) = V_H(r) + V_{xc}(r) = V_H(r) + \frac{\delta E_{xc}(n(r)}{\delta n(r)} \tag{24.11}$$

where $V_H(r)$ is the Hartree Coulomb term and $E_{xc}(n(r)$ is the exchange–correlation functional. Since the electron density n(r) is involved in the definition of effective potential, (24.10) must be solved consistently with (24.9) and (24.11). Although the set of (24.9)–(24.11) yields, in principle, an exact solution to the quantum–mechanical problem, there are approximations involved in the exchange–correlation potential $V_{xc}(r)$. In this respect, Local Density Approximation (LDA) [16] and Generalized Gradient Approximations (GGA) [17] have proven to be effective for a large number of semiconductors.

Existing codes differ by generally the basis set used to expand the Kohn–Sham orbitals. Among the variety of basis three main sorts of sets are used: local, nonlocal and augmented. Our DFT-based ab-initio calculations were performed by means of Wien2k [18] and VASP [19] codes. Wien2k uses a mixed set the full potential linearized augmented plane wave (FP-LAPW) method. In this method, the unit cell of the system is partitioned into atom-centered spheres, called Muffin–Tin (MT), of selected radii designed by RMT and an interstitial region. Inside each atomic sphere the wave function is approximated by a linear combination of radial functions times spherical harmonics, while in the interstitial region a plane wave expansion is used. The VASP, an acronym for The Vienna Ab initio Simulation, on the other hand, uses a plane wave basis for the valence band electrons whereas the all-electron wavefunction and core potential beyond a certain cutoff distance from the ionic core are replaced by a pseudo-wavefunction and pseudopotential. This assumption is justified by the fact that the core electrons do not participate in chemical bonds and are therefore can be frozen and that the wave functions near the nucleus present strong oscillations and therefore require a large number of plane waves.

24.3.1 Wien2k Code Tuning

The energy separation between valence and core states was taken to a value of 6 Ry, furthermore the azimutal quantum number expansion was carried out up to $l_{max} = 10$. The plane waves were expanded up to a cut-off parameter, K_{max}, so that $R_{MT}K_{max} = 7$ the RMT being the radius of muffin tin spheres. The convergence of total energy was set to an accuracy of 0.1 mRy. The ternary calculation were carried out using a supercell of 16 atoms ($2 \times 2 \times 1$) in an ordered form for the wurtzite structure, and a 8-atom supercell for the zincblende phase (Fig. 24.2). The sampling density of the reciprocal space defined by the number of k-points, whose value needed a convergence test with respect to the energy, was found to be 1500 k for both binary compounds. A proportionally smaller number of k-points were used for the ternary calculation.

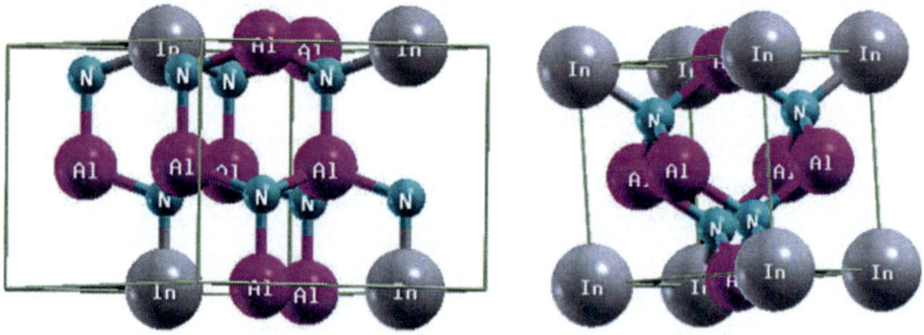

Fig. 24.2 $Al_{75}In_{25}N$ wurtzite cell (left) zincblende cell (right)

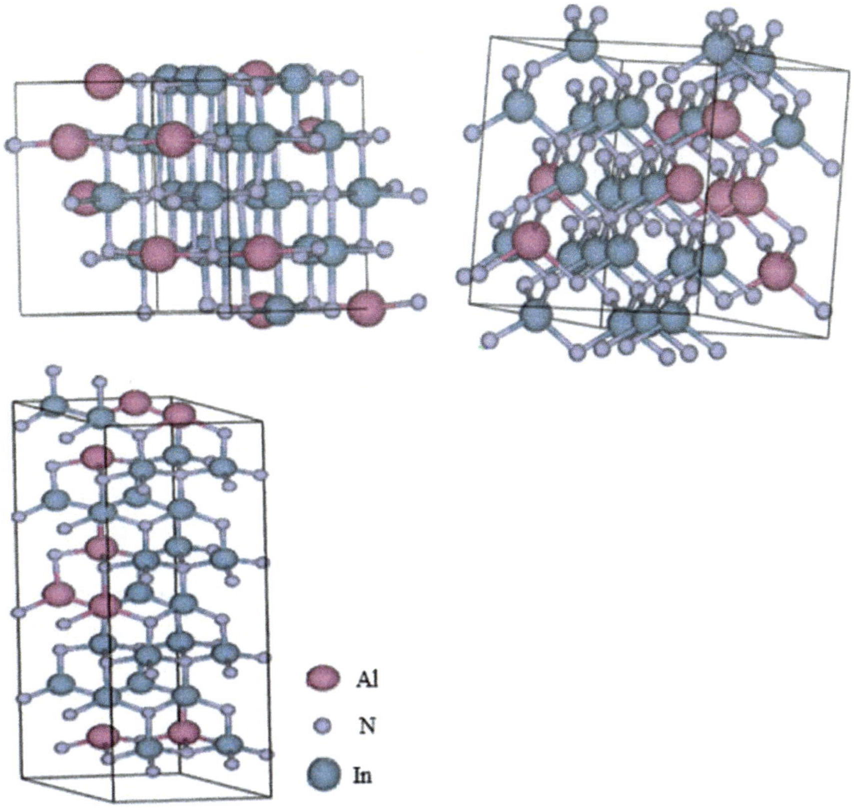

Fig. 24.3 Special quasi-random 64-atom supercell $Al_{0.75}In_{0.25}N$ derived with ATAT. Top left: rocksalt structure top right: zincblende structure, bottom: wurtzite structure

24.3.2 Vasp Code Tuning

In our computational setup, calculations for all phases were performed with a plane wave basis set with an energy cut-off of 600 eV and an energy convergence criteria fixed at 0.01 meV. The interaction between the valence electrons and the ionic cores were treated by the projector-augmented-wave (PAW) method. The Brillouin integration for binaries was performed using a Gamma centered grid of $8 \times 8 \times 8$ for both B1 and B3 and $8 \times 8 \times 6$ for B4. For ternary $Al_{1-x}In_xN$, a $4 \times 4 \times 4$ k-points sampling grid was selected for B1 and B3 phases and a $6 \times 6 \times 2$ for B4.

The ternary $Al_{1-x}In_xN$ (x varying from 0 to 1) was obtained by replacing, in a supercell of 64 atoms, the Al with In atoms in a proportion step variation x of 0.125. Alloy disorder was taken into account for all phases by using a special quasi-random structure (SQS) [20, 21] in the 64-atom supercell and was implemented by means of the Alloy Theoretic Automated Toolkit (ATAT) [22]. The SQS approach proved to be an efficient method for calculating random alloy physical properties [23, 24], it is designed in a manner that In or Al atomic sites of the supercell are occupied in such a way that the pair correlation function reproduces that of an infinite random alloy.

For both codes, the Perdew, Burke and Ernezerhof (PBE)sol [25] exchange correlation functional, a modified version for solids of PBE, was utilized for total energy and lattice parameter calculations while, since PBE greatly underestimates band gaps, the Tran-Blaha potential [26] a modified version of the Becke and Johnson [27] (mBJ) potential was employed in the energy gap calculations. With the mBJ potential a remarkable improvement on gaps accuracy is obtained nearly as precise as those obtained by much more time consuming approach like HSE functionals or G_0W_0 method [28].

24.4 Group III-Nitrides Basic Structures

III-nitride semiconductors AlN, GaN, InN and their alloys can crystallize in three structures (Fig. 24.4): the hexagonal wurtzite phase (B4), the cubic zincblende phase (B3) and the rocksalt phase (B1). The thermodynamically stable phase at room temperature is the wurtzite phase, but also nitride epitaxial growth with zincblende structure can be achieved on (001) oriented cubic substrates. Finally, the rocksalt structure can be induced in III-nitrides at high pressures.

The structural properties calculations were carried out starting from a sequence of different volumes, the corresponding energies were computed and the energy variation with the volume (V) plot illustrated in Fig. 24.5. Results clearly confirmed the B4 phase as the most favorable for the entire In range with B3 is slightly higher in ground state energy.

In addition, the equilibrium structural parameters of bulk AlN and InN binaries and their ternary $Al_{1-x}In_xN$ are determined by calculating the total energy and pressure at various volumes (DFT calculation) and fitting it to the pressure versus volume Birch–Murnaghan equation of state [29] in its third-order expansion of volume as expressed in (24.12).

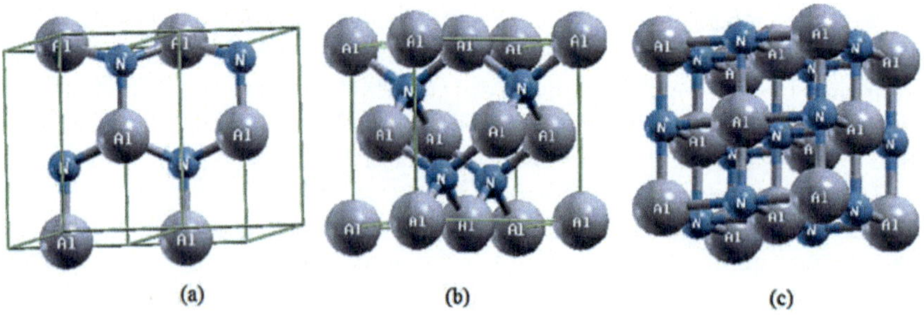

Fig. 24.4 AlN structure: B4 (**a**), B3 (**b**) and B1 (**c**)

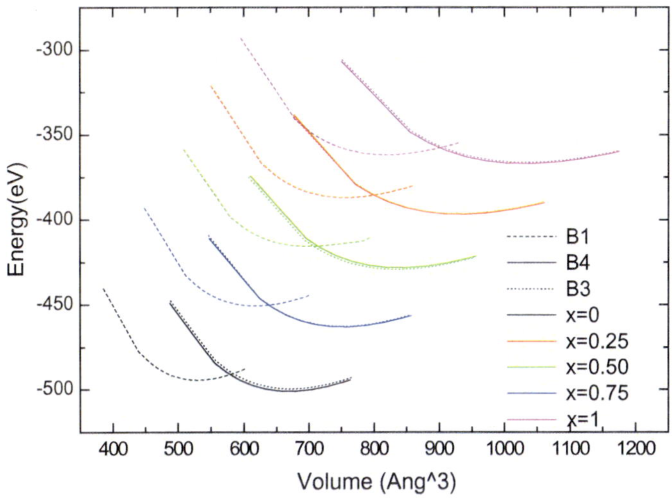

Fig. 24.5 $Al_{1-x}In_xN$ total energy versus cell volume (DFT calculation)

$$P(V) = \frac{3}{2}B_0 + \left[(V_0/V)^{\frac{7}{3}} - (V_0/V)^{\frac{5}{3}}\right]\left\{1 + \frac{3}{4}(B_0^{'} - 4)\left[(V_0/V)^{\frac{2}{3}} - 1\right]\right\} \quad (24.12)$$

where B_0 is the bulk modulus and $B_0^{'}$ its first derivative.

The lattice constants (Fig. 24.6) are increasing with In contents. The phases B3 is presenting the largest lattice constant ranging from 4.37 Å (x = 0) to 5.05 Å (x = 1) compared to B1 varying from 4.04 to 4.67 Å, whereas the parameter a and c of the B4 phase vary respectively from 3.11 to 3.58 Å and from 4.98 to 5.77 Å.

The lattice constants of the $Al_{1-x}In_xN$ ternary alloys may be expressed in terms of the binaries parents by:

$$a(x) = x.a_{\mathrm{InN}} + (1 - x).a_{\mathrm{AlN}} + b.x.(1 - x) \quad (24.13)$$

with the term *b.x.(1−x)* representing the linearity correction term due to the lattice distortion.

We observe very little deviation from linearity for all phases, with the bowing parameter b calculated values of 0.020; −0.253; 0.006 for B3, B1 and B4 phases respectively.

The response to pressure or strain in the process of growth or in the making of heterostructures needs the knowledge of the bulk modulus (BM). The B1, B3 and B4 BM computed values are represented in Table 24.2 and drawn in Fig. 24.7. One may note that BM decreases as the In content is increased which can be a consequence of a longer lattice parameter leading to a weaker bond. Furthermore, the bulk modulus presents the largest value for B1 phase and close values for B3 and B4 phases. This

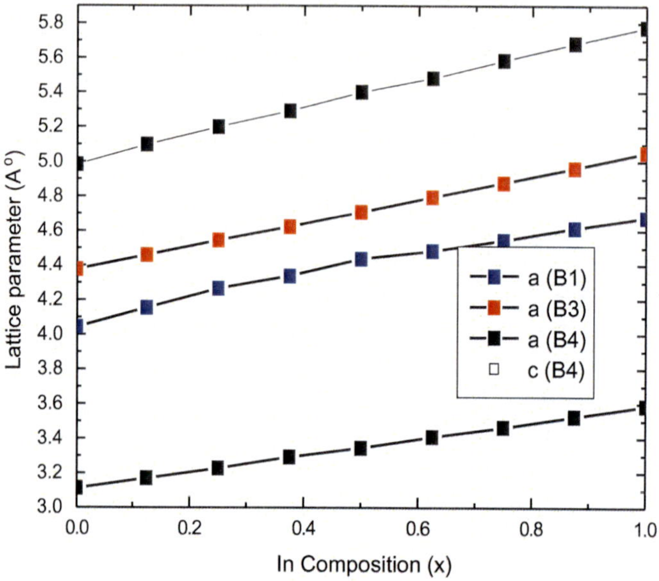

Fig. 24.6 Lattice parameters of $Al_{1-x}In_xN$ versus In composition (B1, B3 and B4 phases)

Table 24.2 Bulk modulus B and its fist derivative B'_0 for $Al_{1-x}In_xN$ (B1, B3 and B4)

Composition x	$ssB(GPa)$			$B'(GPa)$		
	B_1	B_3	B_4	B_1	B_3	B_4
0 TW Exp Others	264.32 221[a] 272[b]	202.43 – 211.78[c]	203.33 207.9[a] 209[b]	3.935 4.8[b] 3.8[b]	3.87 – 3.90[c], 3.7[b]	3.866 5.7–6.3[b] 3.7[a]
0.25 TW	223.92	183.51	184.66	4.546	3.964	3.648
0.5 TW	167.15	167.69	166.90	5.955	3.991	3.909
0.75 TW	203.33	152.65	149.04	4.762	4.278	4.414
1 TW Exp Others	193.39 170[a] 186[e]	142.56 137[a] 144[c]	141.82 125[a] 143[e]	4.931 5.0[b] 4.6[b]	4.443 – 4.56[c]	4.477 12.7[d] 4.6[b]

[a][30], [b] [31], [c][32], [d][33], [e][34].

may be traced back to the fact that the B1 phase has a shorter lattice constant and B3 and B4 phases have similar bond lengths and coordinate numbers. There also seems to be a discontinuity in the BM B1 curve at a composition of 0.5 this anomaly may be due to the fact that AlN and InN have different phase transition pressures and subsequently is the evolution of the $Al_{1-x}In_xN$ and $In_{1-x}Al_xN$ curves.

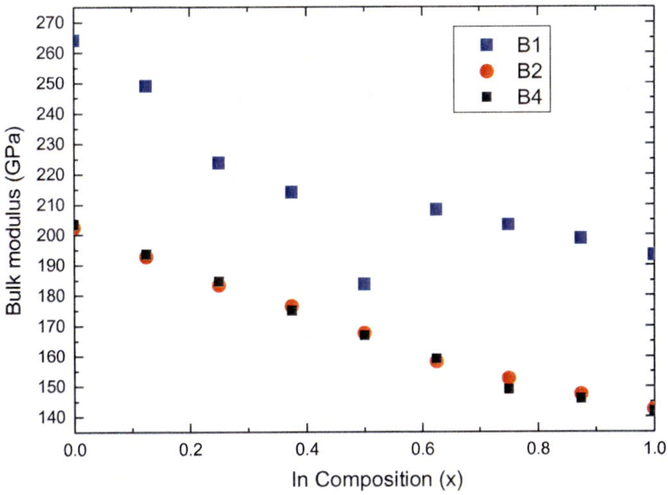

Fig. 24.7 Bulk modulus of $Al_{1-x}In_xN$ versus In composition (B1, B3 and B4 phases)

24.5 Tailoring Bandgap by Alloying

The energy band diagram of the wurtzite and zincblende binaries determined by tight binding are respectively shown in Figs. 24.8 and 24.9. The graphs clearly indicate that, apart from zincblende AlN which is an indirect semiconductor, wurtzite AlN and InN in both phases have direct band gaps.

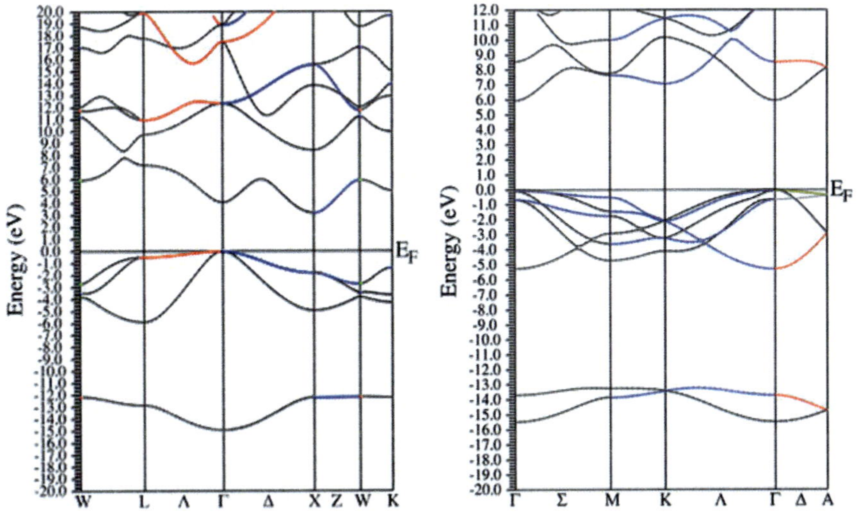

Fig. 24.8 TB AlN band diagram: zincblende (left) wurtzite (right)

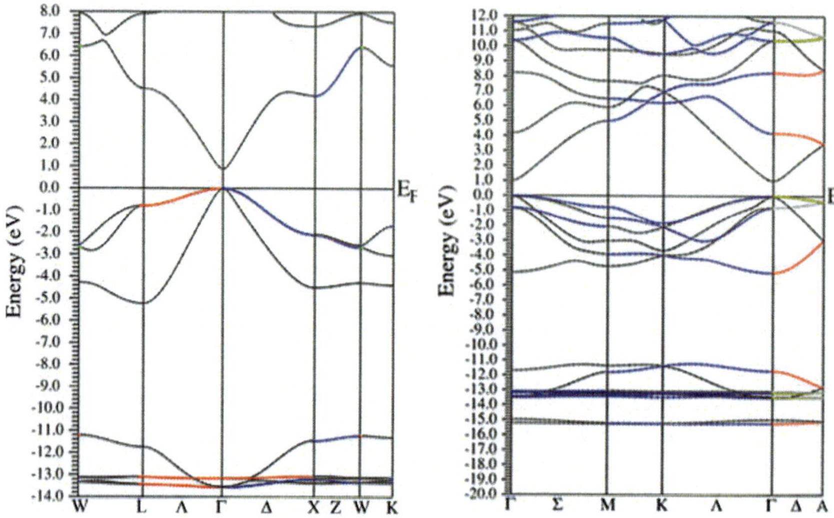

Fig. 24.9 TB InN band diagram: zincblende (left) wurtzite (right)

Zincblende AlN is found to be of indirect gap, the calculated gap values at gamma and X high symmetry points are respectively 5.60 eV and 4.81 eV, while, on the other hand, zincblende structure InN is a direct semiconductor with a gap value of 0.59 eV. When alloying AlN and InN the band gap is found to be indirect up to a crossover whose concentration value was found at $x = 0.17$ with a corresponding gap of 4.95 eV.

The wurtzite phase, in contrast, shows direct gap for both AlN and InN, with respective values of 5.34 and 0.78 eV, comparatively the rocksalt phase shows the largest band gap range varying from 5.74 (AlN) to 0.80 eV (InN). One has to note, at that point, that the indium band gap (B4 phase) has long been overestimated at 1.7 eV it was not until 2001 that a new value was reported [35].

The computed $\Gamma-\Gamma$ energy band gap values, of $Al_{1-x}In_xN$ for all three phases B1, B3 and B4 as a function of indium proportion are presented in Table 24.3 and plotted in Fig. 24.10. The results show that the band gaps are decreasing with In composition.

The results confirm that AlInN has the largest spectrum compared to the other III–V semiconductors where the gap spectra for the phosphides (AlInP) the antimonides (AlInSb) and the arsenides (AlInAs) varies respectively in the range [3.63, 1.42], [2.38, 0.23] and [3.09, 0.41] eV [36].

The energy band gap variation with In contents can be expressed as follows:

$$E_g(x) = xE_{g,AlN} + (1-x)E_{g,InN} + bx(1-x) \tag{24.14}$$

where E_{gAlN} and E_{gInN} are respectively the band gap energy of InN and AlN and b the band gap bowing parameter for $Al_{1-x}In_xN$ ternary alloy. The obtained values of b are presented in Table 24.4. The band-gap bowing parameter yielded by our

Table 24.3 Band gap of $Al_{1-x}In_xN$ in rocksalt (B1), zincblende (B3) and wurtzite (B4) phases

Composition x	Eg (eV)		
	B_1	B_3	B_4
0 TW Exp. Others	5.743 – 5.42[c], 5.40[d]	4.817 5.34[a] 4.09[c], 4.36[e]	5.347 6.23[b] 4.26[c], 4.22[f],
0.25 TW	3.794	2.961	2.924
0.5 TW	2.037	2.044	1.794
0.75 TW	1.346	1.256	1.203
1 TW Exp. Others	0.800 1.0[a]	0.595 0.7[g], 0.6[h] 0.53[a], 0.78[b],	0.782 0.9[a], 0.78[b] 0.69[a], 0.90[l]

[a][37], [b][38], [c] [39], [d][40], [e][41], [f][42], [g][43]; [h][44], [l][45].

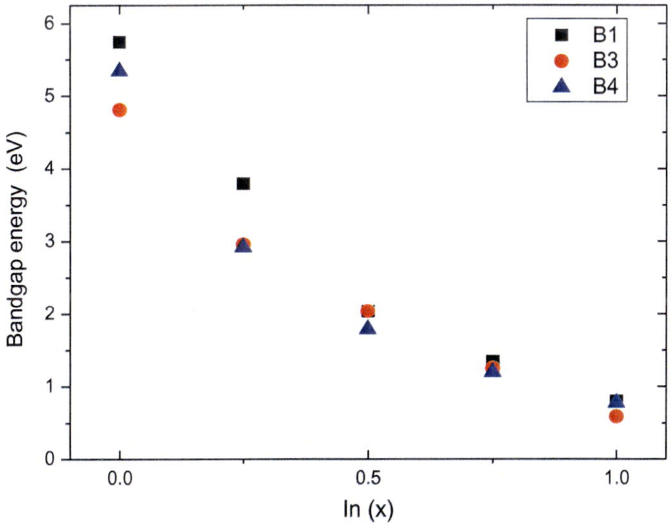

Fig. 24.10 Band gap variation of $Al_{1-x}In_xN$ rocksalt B1, zincblende B3and wurtzite B4

calculations has a constant value for both B1 and B3 phases. Yet, it depends on the alloy composition x for the B4 phase. A similar band-gap bowing parameter composition dependent has been reported in numerous works for the same material system of interest [46–48].

The strong dependence of bowing parameter on In-composition is a general trend in all the In-containing nitride alloys. Physical explanation seem still in debate and there a number of theories regarding the cause among which: a large size mismatch between cation atoms [48], charge transfer contributions due to a large electronegativity differences between aluminum and indium atoms [46], localized electronic states [47], indium clustering [49]. As far as the B1 and B3 phases are concerned in the present contribution, the physical reason for the non-dependence of the band-gap bowing parameter on the alloy composition x is not clear.

Table 24.4 AlInN Band gap bowing parameter of zincblende, wurtzite and rocksalt

Phase	Bowing parameter this work	Bowing parameter others
B3	2.83	2.5[a]
B1	4.37	–
B4:	11.82	11.74 (x = 0.13)[b]
x = 0.125	6.80	8.12
x = 0.25	5.07	5.15
x = 0.50	3.85	4.24
x = 0.75	3.61	3.87
x = 0.875	–	–

[a][38], [b][47]

24.6 Tailoring Effective Mass by Alloying

In addition to the band gap, the effective mass of electrons stands to be another significant physical parameters giving more insights as regards to the electronic transport electronic devices engineering especially the newly HEMT's using AlInN active layers [4]. The obtained results are presented in Table 24.5 where the $m_{\parallel}$ and $m_{\perp}$ represent respectively the masses in the direction parallel and perpendicular to the c-axis. Plots of the calculated electron effective mass of $Al_{1-x}In_xN$ alloy, as a function of indium composition, for B1, B3 and B4 phases are illustrated in Fig. 24.11 where the B4 mass is taken as the mean mass value defined by $m^* = \sqrt[3]{m_{\parallel} m_{\perp}^2}$. The variation of the electron effective mass of $Al_{1-x}In_xN$ alloy, with indium composition show a decrease with In composition for all phases and an overall a lighter relative electron mass of the B4 structure and since the electron effective mass is in inverse proportion to the electron mobility, one may expect, if we exclude the alloy scattering

Table 24.5 Electron effective masses (in free electron mass units) of $Al_{1-x}In_xN$ (B1, B3 and B4)

Composition x	B_1	B_3	B_4	
	m	m	$m_{\perp}$	$m_{\parallel}$
0 TW Exp. Others	0.583 – –	0.523 – 0.32[g]	0.306 0.29–0.45[a] 0.32[g],	0.289 0.29–0.45 0.284[m]
0.25 TW	0.438	0.260	0.293	0.253
0.5 TW	0.202	0.174	0.169	0.151
0.75 TW	0.134	0.093	0.077	0.069
1 TW Exp. Others	0.096 – –	0.077 0.04[j] –	0.049 0.07[a] 0.129[m]	0.046 0.07[a] 0.089[m]

[a][36], [g][42], [j][50], [m][51]

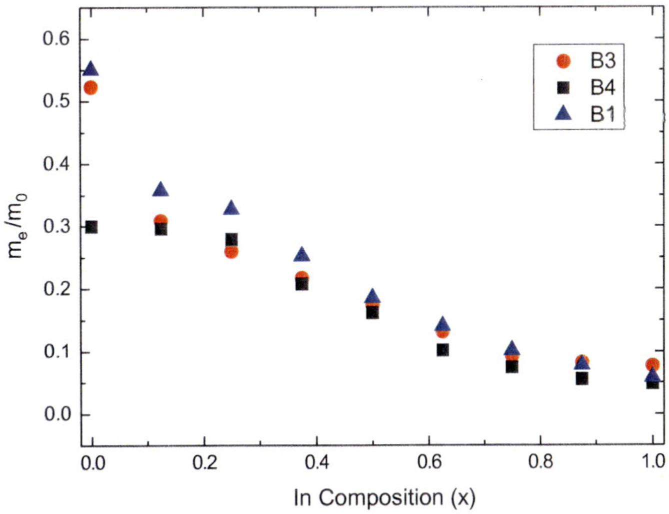

Fig. 24.11 Electron effective mass of $Al_{1-x}In_xN$ zincblende and wurtzite

and the effects of disorder and eventual clustering, an increase of the mobility richer In ternary compounds.

The obtained masses of the stable B4 varies in the range [0.29, 0.05] which, to some extent, are heavier compared to AlInP ([0.22, 0.08]), AlInSb ([0.14, 0.014]) and AlInAs ([0.15, 0.026]) [36] the other In based III–V's.

Assuming that the electron effective mass versus alloy content x curve is quadratic $b_m(x)$, similarly to the band-gap energy, one defines an electron effective mass bowing parameter by the relation,

$$m_n^*(x) = (1 - x)m_n^*(AC) + xm_n^*(BC) + b_m(x)x(1 - x) \quad (24.15)$$

where $m_n^*(AC)$ ad $m_n^*(BC)$ are the effective electron mass of AC and BC binaries such as AlN and InN group III-nitrides and $b_m(x)$ is a coefficient representing the composition dependent bowing effect on ternary effective mass. The computed bowing parameters of the three phases for various In concentrations (Table 24.6) show small values.

Table 24.6 Electron effective mass bowing factor

x	0.25	0.5	0.75
b (B1)	−0.124	−0.55	−0.447
b (B3)	−0.808	−0.504	−0.507
b (B4)	0.273	0.094	0.0386

24.7 Density of States

The number of states available per unit energy (density of states (DOS)) is an important factor in studying the electronic properties. The calculated total and partial density of states of zincblende and wurtzite are shown respectively in Figs. 24.12 and 24.13. The vertical dashed line represents the Fermi level which is set to zero. It is clear that the density of states is higher in wurtzite phase than in zincblende phase. The upper energy valence bands range from −5 to 0 eV. The major contribution comes from the nitrogen atom for all compositions in the two phases. The p-orbital seems to bring all the contribution of nitrogen to the binaries. The lower conduction band of wurtzite phase extends continuously from the value of the gap up to about 15 eV with a similar contribution of all atoms either in wurtzite or zincblende phase. In the case of AlN, the p-orbital is prevalent while for InN, N and In, respectively, contributes through the p-and s-orbitals. The ternary compounds present in the case of wurtzite phase a continuous lower conduction band. However, there is a discontinuity for zincblende phase that seems to shift towards higher energies with increasing indium content. In the case of the valence band we notice that the DOS peak increases with indium.

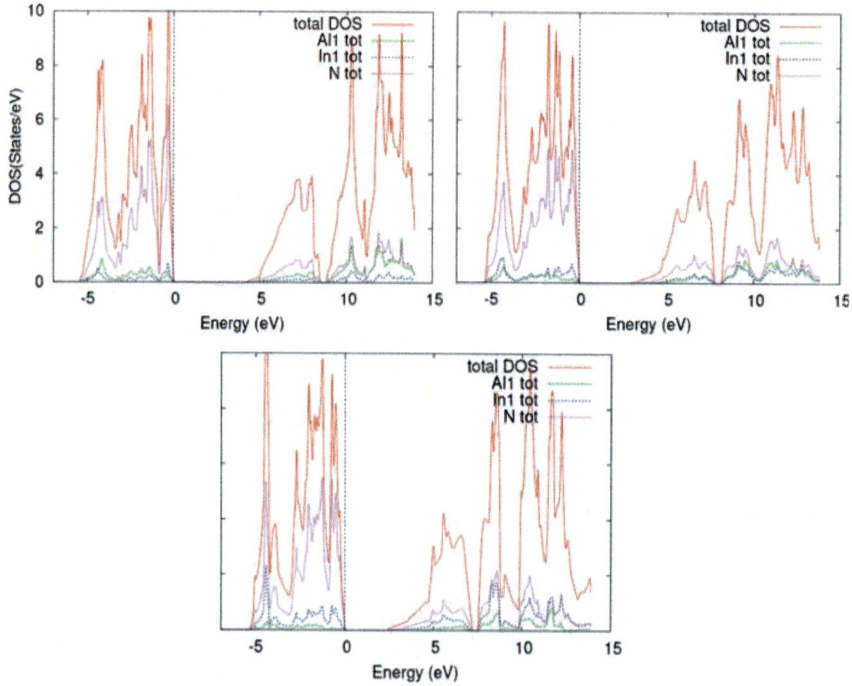

Fig. 24.12 Total and partial DOS of $Al_{1-x}In_xN$ (B3) for x = 0.25 (top left), 0.50 (top right) and 0.75 (bottom)

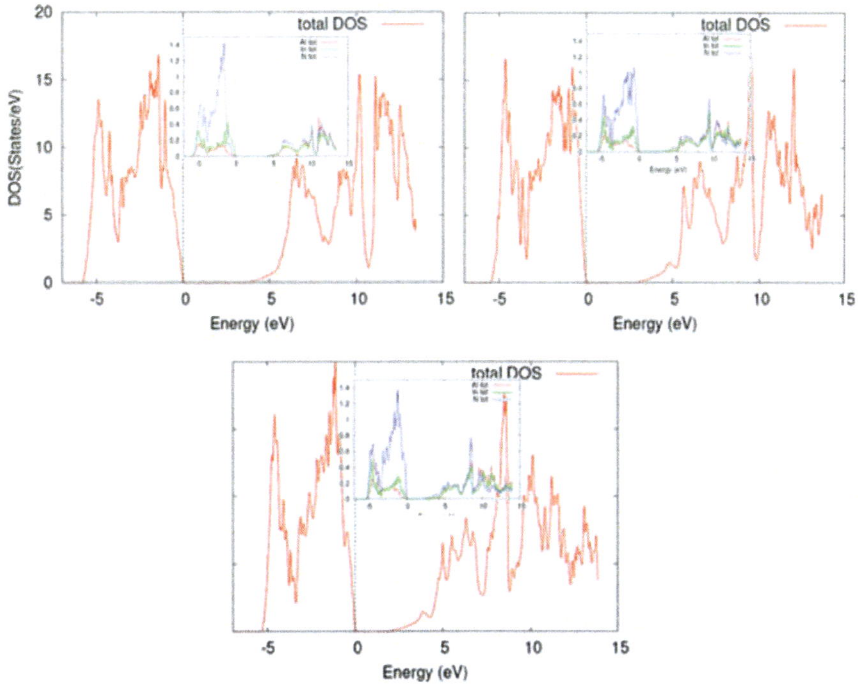

Fig. 24.13 Total and partial DOS of $Al_{1-x}In_xN$ (B3) for x = 0.25 (top left), 0.50 (top right) and 0.75 (down)

24.8 High Pressure Effects on Bandgap Energy

24.8.1 Phase Transition Pressures

High pressure studies of semiconductors have given valuable insights to their room pressure properties and have also been used to generate new phases. Accordingly, the changes in optical or electronic properties can be interpreted in a straightforward manner. Nitrides are most commonly found in the wurtzite crystalline structure and to a lesser extent in zincbende phase rearranging into the rocksalt structure under extreme high pressure.

Before investigating the high pressure effects on the bandgap, the phase transition pressure of respectively the zincblende and wurtzite to the rocksalt phase should first be determined.

The phase stability is determined by the minimum of the Gibbs free energy G:

$$G = E + P.V - T.S \tag{24.16}$$

where E, P, V, T, S stand respectively for internal energy, pressure, volume, temperature and entropy. The Gibbs free energy reduces to enthalpy H ($H = E + P.V$)

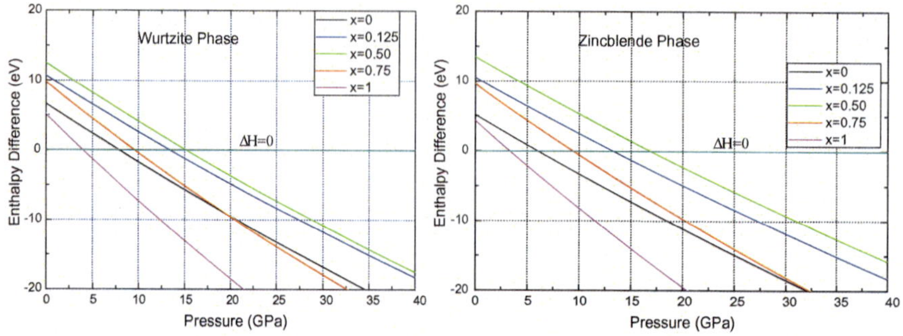

Fig. 24.14 $Al_{1-x}In_xN$ enthalpy difference of B4–B1 (left) and B3–B1(right)

Table 24.7 B4–B1 and B3–B1 phase transition pressures

Composition x	Transition pressure of $Al_{1-x}In_xN$ (GPa)	
	B4–B1	B3–B1
0 TW	7.84	6.02
Others	12.5[a], 9.2[b]	7.1[b]
Exp.	14,20[c]	
0.25 TW	15.01	14.69
0.5 TW	15.19	16.94
0.75 TW	9.61	9.46
1 TW	4.02	3.37
Others	4.93[d],11.1[b]	10.5[b]
Exp.	11.6[b]	

[a][52], [b] [30], [c][33], [d] [53]

given that calculations are made at 0°K. By using the obtained values for total energy and corresponding volume one can calculate the variation of enthalpy with pressure for each phase. The variation of the relative enthalpy difference of B4 and B3, for various In proportions, are plotted in Fig. 24.14 with the phase transition pressures determined at the x-axis intersection of the various enthalpy curves ($\Delta H = 0$).

The computed phase transition pressures are summarized in Table 24.7. The results show a structural phase transition varying, with In concentration, from 4.02 ($x = 0$) to 7.84 GPa ($x = 1$) for B4–B1 transition and from 3.37 ($x = 0$) to 6.02 GPa ($x = 1$) for B3–B1 transition with a maximum value around 16 GPa for $x = 0.5$.

24.8.2 Bandgap Energy Variation with Pressure

In order to study of the effects on the energy band gap of $Al_{1-x}In_xN$ in its B3 and B4 phases the pressure is varied from tensile to compressive up to the transition phase pressure. Results show a linear increase of the gap for either phases (Fig. 24.15),

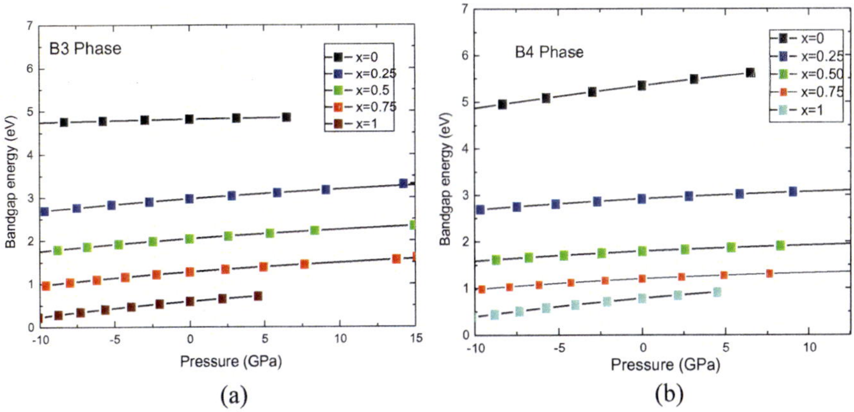

Fig. 24.15 Band gap of $Al_{1-x}In_xN$ versus pressure B3 phase (**a**) and B4 phase (**b**)

with the slopes varying with In concentration from 36.8 meV/GPa (x = 0) to 24.5 (x = 1) for B4 and from 41.5 (x = 0) to 29.5 meV/GPa (x = 1) for B3. The magnitude of these slopes is typical of III-group nitrides which, due to their large ionicity, and independently of their phase, exhibit much lower band gap pressure coefficients than other III-V group compounds such as GaAs (117 meV/GPa) [54].

24.9 Strain Effects on Heterostructure Bandgap

If we consider the deposition of a thin layer of material (f) on a substrate of material (S) the atoms of the film f align with those of S causing a deformation of the thin layer cell. The film takes the form of the substrate in the growth plane inducing a change of the cell parameters in the normal direction. The lattice mismatch causes therefore a biaxial compression (dilation) where the new parameters of the epilayer correspond to those of the substrate in the in-plane (as, b_s) (Fig. 24.16) while the parameter out of plane is dilated (compressed).

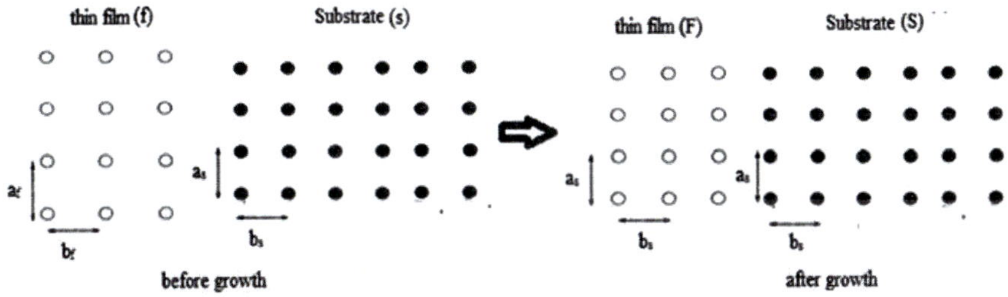

Fig. 24.16 In-plane lattice modification under strain

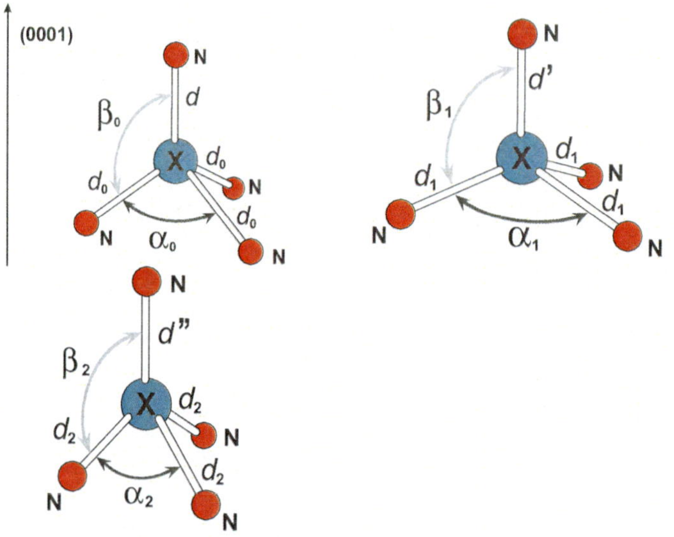

Fig. 24.17 XN (X = Al,In) unstrained (top left), biaxially compressed (top right), biaxially dilated (bottom)

The effect of depositing a AlInN film on a GaN buffer, and due to the difference in the lattice structure parameters between the film and the buffer, leads to strain effects. This results in changes in the structural parameters and therefore affects the physical properties of the alloy. If we consider a thin AlInN film, the first layers will be affected by the buffer. The first direct effect will be the change in the bond length and the angles hence a deformation of the wurtzite cell (Fig. 24.17).

In order to calculate electronic structure, one has first to calculate or measure the position dependent strain tensor $\varepsilon_{\alpha\beta}$. The basic approach to calculate the variation on the geometry of the unit cell under such strains is to use classical harmonic elasticity. The growth of both the zincblende or wurtzite AlInN structures on a GaN substrate yields a tetragonal structure whose basal parameter corresponding to the substrate (GaN) lattice parameter while the perpendicular c parameter depends on the layer structure. For a film grown on a buffer with parallel lattice constant a_s, the strain components are:

$$\varepsilon_{\|} = \varepsilon_{xx} = \varepsilon_{yy} = \frac{a_s - a_{eq}}{a_{eq}} \qquad \varepsilon_z = \frac{c - c_{eq}}{c_{eq}} \tag{24.17}$$

where a_{eq} and c_{eq} are the equilibrium lattice constant of the unstrained material and c is the perpendicular lattice constant of the strained film.

For a cubic system the value of the c lattice parameter is determined from deriving the strain energy with respect to parallel strain yielding [55]:

Table 24.8 Perpendicular lattice parameter of the tetragonal strained structure

	AlN	$Al_{0.75}In_{0.25}$ N	$Al_{0.50}In_{0.50}$ N	$Al_{0.25}In_{0.75}$ N	InN
C_{zb} (Ang)	2.062	1.856	1.618	1.333	0.976
C_{wz} (Ang)	4.913	5.204	5.543	5.964	6.535

$$c = \big((2 - 3q)\varepsilon_{\|}(a_s) + 1\big)a_{eq} \tag{24.18}$$

q is the 'epitaxial strain reduction factor' for orientation G of the c axis is:

$$q = 1 - \frac{B}{C_{11} + \gamma\Delta} \tag{24.19}$$

where Δ is the elastic anisotropy:

$$\Delta = C_{44} - \frac{1}{2}(C_{11} - C_{12}) \tag{24.20}$$

γ is a directional geometric factor.

B is the bulk modulus:

$$B = \frac{2}{3}(C_{11} + 2C_{12}) \tag{24.21}$$

For an hexagonal structure the c lattice parameter is given by [56]:

$$c = c_{eq}(1 - 2\frac{C_{13}}{C_{33}}\varepsilon_{xx}) \tag{24.22}$$

The computed lattice parameters of the resulting tetragonal structure are listed in Table 24.8. The basal parameter a_s is that of the substrate i.e. $a_s = a_{GaN} = 3.189$ Ang, while c_{wz} and c_{zb}, correspond respectively to the perpendicular direction parameters of the formerly wurtzite and zincblende structures.

24.9.1 Tight Binding Prediction of Bandgap

Under the strain TB matrix elements $H_{lR,mR}$ are modified since the lattice parameter and the bonding length differ from those of the unstrained material. Considering only scaling of the inter-site matrix elements, the matrix elements of the strained $H_{lR,mR}$ and unstrained $H^0{}_{lR,mR}$ structure are related by:

$$H_{lR',mR} = H^0_{lR',mR} f(d^0_{R'-R}, d_{R'-R}) \tag{24.23}$$

where the function $f(d^0_{R'-R}, d_{R'-R})$ describes the influence of the bond length and angle on the hopping matrix elements, ($d^0_{R'-R}$ and $d_{R'-R}$) being, respectively, the bond vectors between the atomic positions of the unstrained and strained material. A very accurate band structure description TB model can be achieved by scaling only the inter-site matrix elements without the need to adjust the diagonal matrix elements [57]. By using Harrison's d^{-2} rule [58] which is a reasonable approximation we get the following relation:

$$H_{lR',mR} = H^0_{lR',mR}\left[\frac{d^0_{R'-R}}{d_{R'-R}}\right]^2 \tag{24.24}$$

All these facts taken into account result in a change of the lattice parameters of AlInN which induces a whole new energy band diagram and therefore a modified band gap and electronic effective mass. An example (x = 0.75) of energy band diagram of a unstrained and strained AlInN film is presented in Fig. 24.18.

The band gap is extracted from the band energy diagram for various In concentration x and plotted in Fig. 24.19. Below the matching concentration the strained mass is lower than the unstrained one whereas the trend reverses for higher In contents.

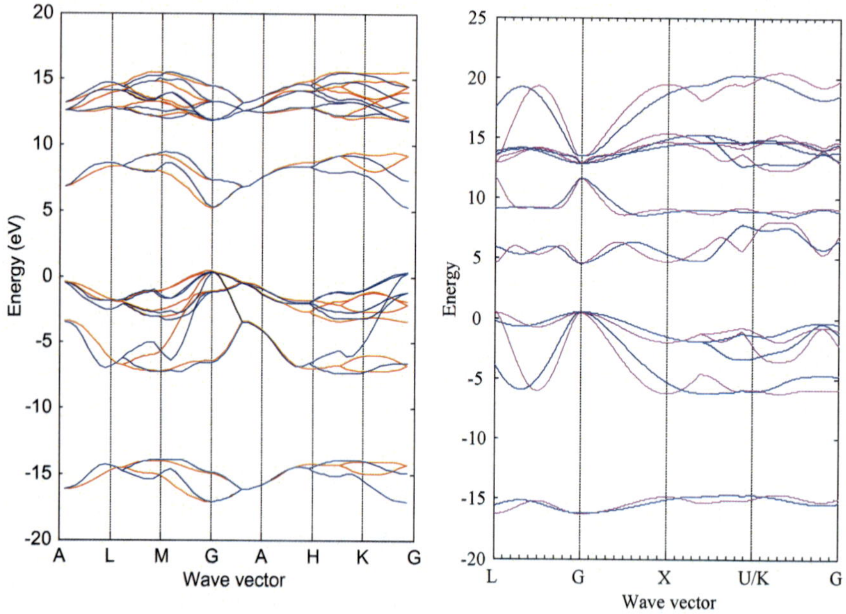

Fig. 24.18 Unstrained (blue) and strained (red) $Al_{0.75}In_{0.25}$ N wurtzite (right) zincblende (left) band diagram

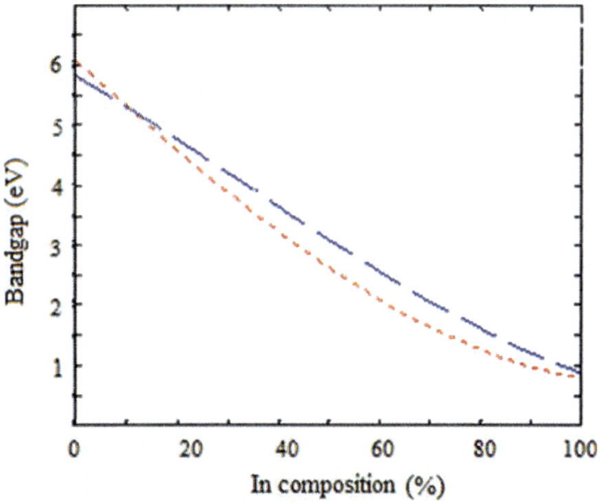

Fig. 24.19 Band gap variation of strained (red) and unstrained (blue) $Al_x In_{1-x}N$ (TB)

24.9.2 Density Functional Theory Predicton of Bandgap

The strain results in a deformation of the wurtzite cell, leading to a tetragonal structure. All calculation have been made using 16 atom supercell with the new cell parameters determined by (24.16) and (24.21). Figures 24.20 and 24.21 represent respectively the energy band diagram of unstrained and strained $Al_{0.75}$ $In_{0.25}$ N in wurtzite and zincblende form. We notice for the wurtzite diagram a widening in the

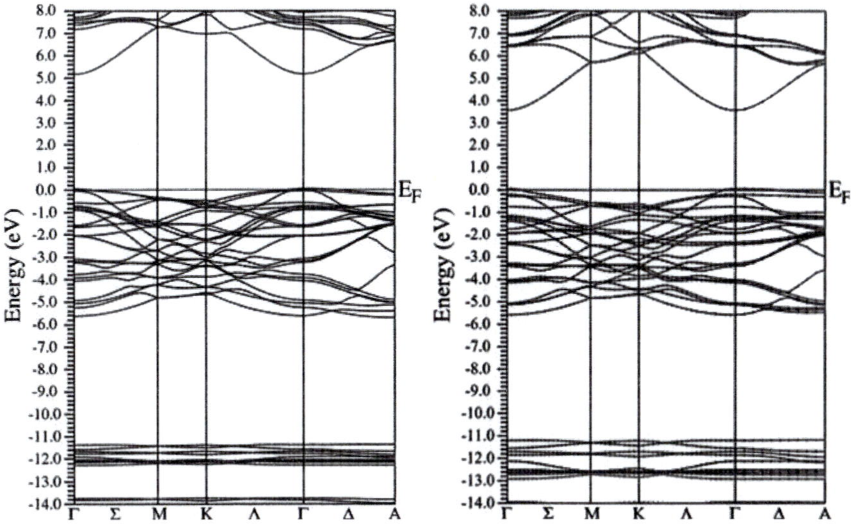

Fig. 24.20 Energy band diagram (DFT) $Al_{0.75}$ $In_{0.25}$ N (wz). left: unstrained right: strained

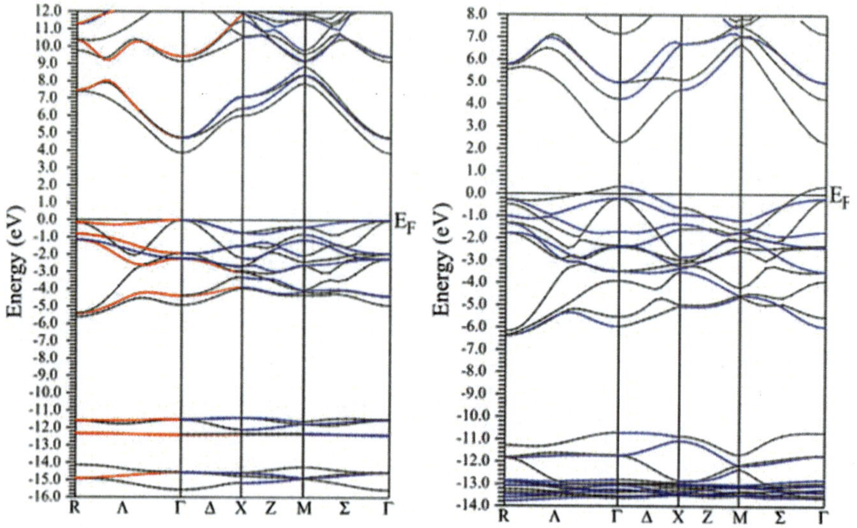

Fig. 24.21 Energy band diagram (DFT) $Al_{0.75}$ $In_{0.25}$ N (zb). left: unstrained, right: strained

Table 24.9 Band gap for strained and unstrained $Al_{1-x}In_xN$ in the wurtzite phase

	AlN	$Al_{0.75}In_{0.25}$ N	$Al_{0.50}In_{0.50}$ N	$Al_{0.25}In_{0.75}$ N	InN
TB unstrained	6.231	4.303	2.656	1.448	0.790
TB strained	5.825	4.492	3.097	1.833	0.879
DFT unstrained	5.950	3.769	2.802	1.786	0.998
DFT strained	5.238	3.368	3.175	2.145	1.523

shape of the lower conduction band and a narrowing of the band gap. Moreover the degeneracy is lifted for the higher valence bands. The same observation is valid for zincblende where the lowering on the lowest CB is more severe than in the other CB bands.

The energy band gaps extracted from these band diagrams are reported in Table 24.9 with the corresponding plots in Fig. 24.22.

24.10 Strain Effects on Electron Effective Mass

Computed electron mass for the strained and unstrained configurations is presented in Table 24.10 and plotted in Figs. 24.23 and 24.24 respectively for the TB and the DFT methods.

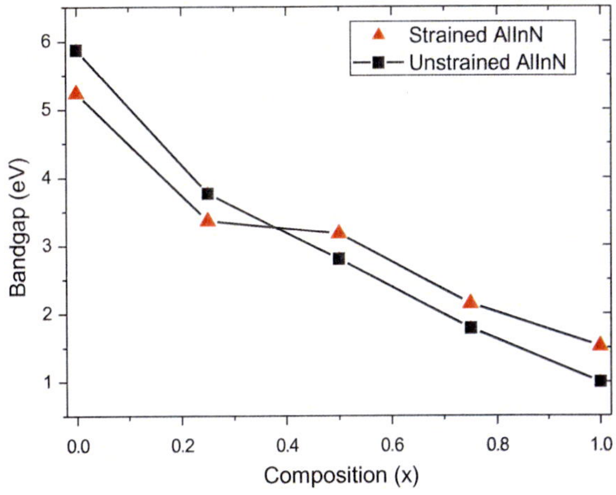

Fig. 24.22 Band gap of unstrained and strained wurtzite AlInN versus In composition

Table 24.10 Strained and unstrained $Al_xIn_{1-x}N$ electron effective masses in wurtzite phase

		AlN	$Al_{0.75}In_{0.25}$ N	$Al_{0.50}In_{0.50}$ N	$Al_{0.25}In_{0.75}$ N	InN
TB unstrained	m_e/m_0	0.321	0.259	0.194	0.127	0.054
TB strained		0.294	0.257	0.215	0.166	0.113
DFT unstrained	m_e/m_0	0.306	0.293	0.169	0.077	0.049
DFT strained		0.300	0.209	0.196	0.185	0.141

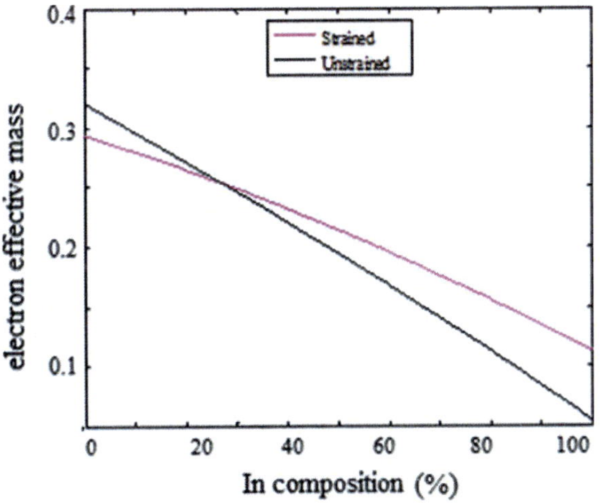

Fig. 24.23 Electron effective mass of unstrained and strained AlInN versus In composition (TB)

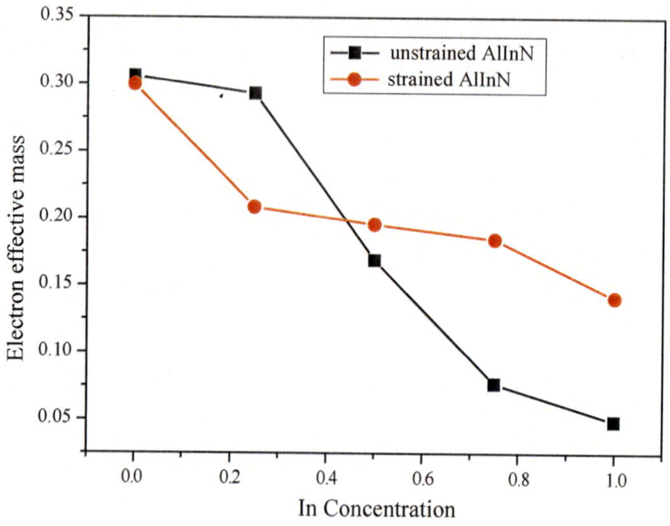

Fig. 24.24 Electron effective mass of unstrained and strained AlInN versus In composition (DFT)

Both methods show decreasing mass with In proportion. We notice two different sections: above around 25% the strained mass is larger than the unstrained while below this mark which coincides with the matching proportion between AlInN and GaN (18%) the strained mass is lighter.

24.11 Polarization Effects

Piezoelectricity is a charge that builds up in certain crystals arising from an applied mechanical stress. A piezoelectric potential can be created in crystals lacking a center of symmetry, due to polarization of ions under applied stress and strain. This property is common to both the zincblende and wurtzite crystal structures.

The polarization is related to stress by the relation:

$$P_{pz} = e_{ij}\varepsilon_j \tag{24.25}$$

where $\overline{\overline{e}}$ and $\overline{\overline{\varepsilon}}$ are respectively the piezoelectric and the strain tensors.

For zincblende materials, the relation between the induced polarization and strain is given by:

$$\begin{bmatrix} P_x \\ P_y \\ P_z \end{bmatrix} = \begin{bmatrix} 0\,0\,0\,e_{14} & 0 & 0 \\ 0\,0\,0 & 0\ e_{14} & 0 \\ 0\,0\,0 & 0 & 0\ e_{14} \end{bmatrix} \begin{bmatrix} \varepsilon_{xx} \\ \varepsilon_{yy} \\ \varepsilon_{zz} \\ \varepsilon_{yz} \\ \varepsilon_{zx} \\ \varepsilon_{xy} \end{bmatrix} \tag{24.26}$$

where e_{14} is the non zero the piezoelectric tensor coefficient, and ε_{xx},ε_{yy} are the strain tensor in-plane components and ε_{zz} the normal one. As a result of the special form of the piezoelectric tensor, in zincblende semiconductors a biaxial strain does not generate piezoelectricity on (001) direction.

Crystals with a cubic structure have four equivalent bonds and due to this symmetry the center of electronic charges coincides with the nucleus. On the other hand, for materials with a hexagonal structure, the [0001] direction carries a bond that is longer than the others, thus creating an electric dipole.

In strained wurtzite structures along or perpendicular to c-axis, the relation is:

$$\begin{bmatrix} P_x \\ P_y \\ P_z \end{bmatrix} = \begin{bmatrix} 0 & 0 & 0 & 0 & e_{15} & 0 \\ 0 & 0 & 0 & e_{15} & 0 & 0 \\ e_{31} & e_{31} & e_{33} & 0 & 0 & 0 \end{bmatrix} \begin{bmatrix} \varepsilon_{xx} \\ \varepsilon_{yy} \\ \varepsilon_{zz} \\ \varepsilon_{yz} \\ \varepsilon_{zx} \\ \varepsilon_{xy} \end{bmatrix} \tag{24.27}$$

e_{31}, e_{33} and e_{15} are the piezoelectric tensor elements.

Thus for layers grown in the [0001] orientation, a piezoelectric polarization will be aligned along the c-axis direction and given by:

$$P_{pz} = e_{31}(\varepsilon_{xx} + \varepsilon_{yy}) + e_{33}\varepsilon_{zz} \tag{24.28}$$

Using (24.17) and (24.22) the piezoelectric polarization, of an AlInN wurtzite layer grown on a GaN buffer, takes the following expression:

$$P_{piezo}^{Al_{1-x}In_xN}(x) = 2(e_{31}(x) - e_{33}(x)\frac{C_{13}(x)}{C_{33}(x)})\varepsilon_{xx}(x) \tag{24.29}$$

where $e_{ij}(x)$ are the piezo-coefficients and $C_{ij}(x)$ the elastic parameters. They are related to the binaries e_{ij}^{AlN}, e_{ij}^{InN}, C_{ij}^{AlN} and C_{ij}^{InN} by the following relations:

$$e_{ij}^{Al_{1-x}In_xN} = (1-x)e_{ij}^{AlN} + xe_{ij}^{InN} \tag{24.30}$$

$$C_{ij}^{Al_{1-x}In_xN} = (1-x)C_{ij}^{AlN} + xC_{ij}^{InN} \tag{24.31}$$

Table 24.11 Piezoelectric and elastic constants and spontaneous polarization of nitrides binaries [59, 60]

Parameters	AlN	InN	GaN
$e_{31}(cm^{-2})$	−0.58	−0.412	−0.55
$e_{33}(cm^{-2})$	1.55	0.815	1.12
$C_{13}(GPa)$	100	92	110
$C_{33}(GPa)$	390	224	387
$P_{sp}(cm^{-2})$	−0.09	−0.042	−0.034

Table 24.12 Piezoelectric and spontaneous polarization of $Al_{1-x}In_xN$ versus In composition

x	0.00	0.25	0.50	0.75	1.00
P_{sp}	−0.0900	−0.0647	−0.0483	−0.0407	−0.0420
P_{pz}	−0.0477	0.0217	0.0808	0.1278	0.1657

III–V nitrides show also in addition to piezoelectric polarization a spontaneous polarization. Spontaneous polarization (Psp) is an intrinsic property related to the bonding nature of the material, whose origin can be attributed to the fact that the center of the negative charges (electrons) in the solid does not coincide with the center for the positive charges (nuclei).

The spontaneous polarization of the ternary AlInN takes the form:

$$P_{sp}^{Al_{1-x}In_xN}(x) = (1-x)P_{sp}^{AlN} + xP_{sp}^{InN} + bx(1-x) \tag{24.32}$$

where P_{sp}^{AlN} and P_{sp}^{InN} are respectively the spontaneous polarizations of AlN and InN and b the bowing factor ($b_{PspAlInN} = 0.071\ cm^{-2}$) [59].

The values of the binaries AlN and InN spontaneous polarization and piezoelectric coefficients e_{31} and e_{33} used in this work, are presented in Table 24.11.

The resulting values of piezoelectric and spontaneous polarization are listed in Table 24.12, and their variation with In concentration plotted in Fig. 24.25.

The plot shows that the piezoelectric polarization is always negative for the In composition corresponding to a layer of $Al_{1-x}In_xN$ in tension. An In concentration close to 18% gives a lattice matching with GaN layer and therefore a zero piezoelectric polarization then with the composition rising, the layer will be in compression and the positive piezoelectric polarization will oppose spontaneous polarization. At high In concentration the piezoelectric becomes the predominant polarization.

24.12 Hemt Device

The High Electron mobility Transistor (HEMT) is a field effect transistor consisting of a heterostructure of two semiconductors of different gaps. A thin-layer of a ternary alloy Al(Ga, In)N, called the barrier layer is grown on an unintentionally doped GaN

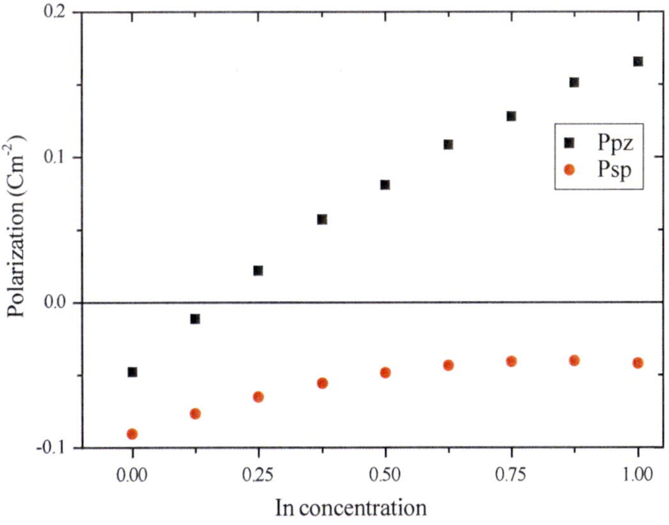

Fig. 24.25 $Al_{1-x}In_xn$ Spontaneous and Piezoelectric polarization versus In concentration

buffer layer acting as a channel resulting in a quantum well at the interface between the two materials (Fig. 24.26). A large density of high electron mobility electrons is then created in the undoped zone resulting in a High Electron Mobility Transistor. Over the last few years, GaN HEMTs have attracted attention owing to their high-power performance and the ability to operate at high frequencies. Nitride HEMTs generally use a heterostructure AlGaN/GaN, however the difference in lattice constants of two layers induces a stress that limits their reliability [61, 62]). As an alternative, Kuzmik [63] proposed using AlInN/GaN HEMTs that have the advantages that at 18% of In $Al_{0.82}$ $In_{0.18}$ N and GaN are lattice matched removing strains in the heterostructures.

Fig. 24.26 Band diagram for AlInN / GaN heterojunctions HEMT

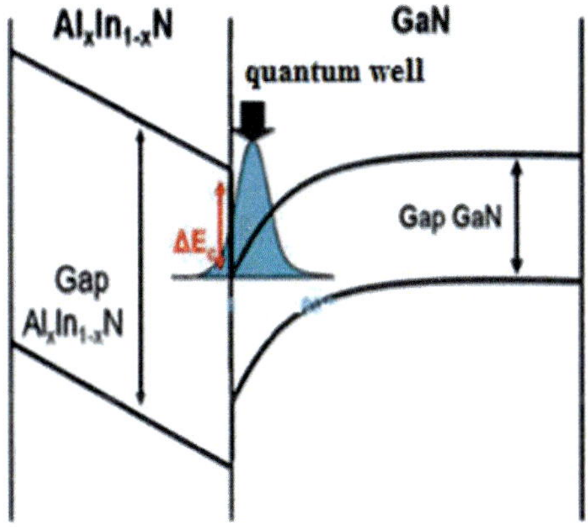

Moreover, charges induced by the spontaneous polarization are almost three times higher than in the traditional AlGaN/GaN HEMT [63].

24.12.1 Hemt Physical Processes

In the AlInN HEMTs a heterostructure AlInN/GaN is used with a AlInN doped wide band gap layer and an smaller gap of undoped GaN channel layer (Fig. 24.27). As a result of the difference of band gaps between the two layers, a quantum well is created within the structure. The high concentration of electrons of the doped layer of AlInN fall completely into the GaN layer. Owing to the GaN undoped layer, a thin channel of highly mobility electrons at very high concentration is created in the GaN-side conduction band forming what is called a two-dimensional electron gas (2DEG). In a HEMT the conduction is hence provided by a large density n_s of high mobility electrons supplied by the 2DEG layer whose origin are the spontaneous polarization of the polar material and the piezoelectric polarization. In addition to that in order to provide a better carrier confinement and to reduce the alloy scattering 2DEG one generally adds a spacer layer of AlN at the AlInN/GaN interface [64, 65].

The schematic band diagram of the above heterostructure is shown in Fig. 24.28, where:

e is the electron charge, Φ_b the Schottky contact barrier height.

d^{AlInN} and d^{AlN} are respectively the barrier layer AlInN and cap layer AlN thickness.

$\Delta E_{AlInN/GaN}(x)$ and $\Delta E_{AlN/GaN}$ are the conduction band discontinuity between AlInN/GaN and AlN/GaN.

E_0 is the triangular quantum well ground state level considered usually the only state occupied and E_F is Fermi energy level given respectively by [66]:

$$E_0 = \left(\frac{9\pi \hbar e^2}{8\varepsilon_0\sqrt{8m^*(x)}} \frac{n_s(x)}{\varepsilon(x)} \right)^{\frac{2}{3}} \tag{24.33}$$

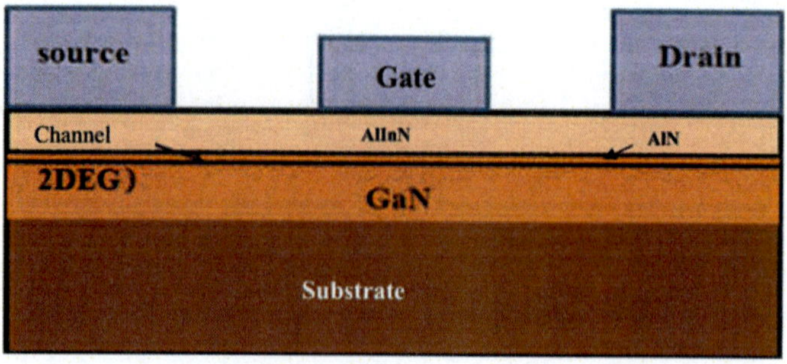

Fig. 24.27 Structure of an AlInN HEMT transistor

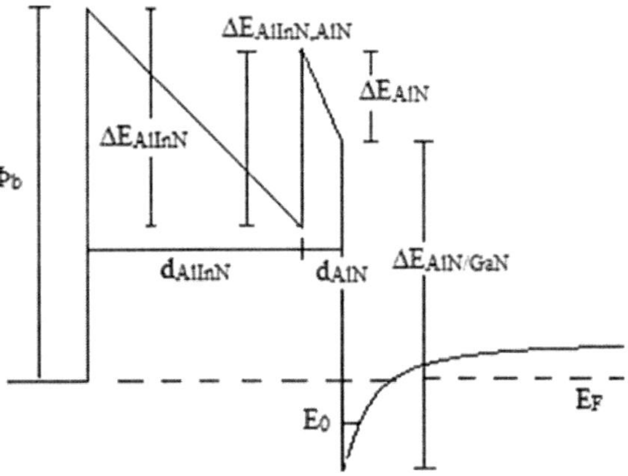

Fig. 24.28 Schematic band diagram of an AlInN/AlN/GaN

$$E_F = E_0(x) + \frac{\pi \hbar^2}{m^*(x)} n_s(x) \tag{24.34}$$

The corresponding balance equation can be written as follow:

$$e\Phi_b(x) - E_{c,AlInN/AlN} d_{AlInN} + \Delta E_{c,AlInN/AlN} - E_{c,AlN} d_{AlN} - \Delta E_{c,AlN/GaN} + E_0 + (E_F - E_0) = 0 \tag{24.35}$$

where:

$\varepsilon_{\text{AlInN}}$ and ε_{AlN} are respectively the alloy relative permittivity of respectively AlInN and AlN.

$$\varepsilon^{Al_{1-x}In_xN}(x) = (1-x)\varepsilon^{AlN} + x\varepsilon^{InN} \tag{24.36}$$

$E_{c,\text{AlN}}$ and $E_{c,\text{AlInN}}$ are the fields across respectively the AlN and the AlInN barrier and have the following expressions:

$$E_{c,AlN} = e(\sigma^{AlN/GaN} - en_s)/\varepsilon_0 \varepsilon_{AlN} \tag{24.37}$$

$$E_{c,AlInN} = e(\sigma^{AlInN/AlN}(x) - en_s)/\varepsilon_0 \varepsilon_{AlInN}(x) \tag{24.38}$$

σ is the charge density induced by the spontaneous and piezoelectric polarizations given by:

$$\sigma^{Al_{1-x}In_xN} = -P_{sp}^{Al_{1-x}In_xN} - P_{piezo}^{Al_{1-x}In_xN} \tag{24.39}$$

$$\sigma^{AlN} = -P_{sp}^{AlN} - P_{piezo}^{AlN} \tag{24.40}$$

Considering that that ground state energy value E_0 nearly corresponds to the bottom of conduction band one then gets the carrier density n expression of the 2DEG:

$$n_s = \frac{1}{e}\left(\frac{d^{AlInN}}{\varepsilon^{AlInN}} + \frac{d^{AlN}}{\varepsilon^{AlN}}\right)^{-1}\left[\frac{d^{AlInN}}{\varepsilon^{AlInN}}\sigma^{AlInN} + \frac{d^{AlN}}{\varepsilon^{AlN}}\sigma^{AlN} - \frac{\varepsilon_0}{e}(e\Phi_b(x) - \Delta E_c^{AlN/GaN} + \Delta E_c^{AlInN/AlN}(x))\right] \quad (24.41)$$

Assuming a Ni/Au gate, the Schottky contact barrier height Φ_b is [63]:

$$\Phi_b(Ni/Al_{1-x}In_xN) = 3.05 - 3.7x \quad (24.42)$$

The conduction band discontinuities $\Delta E_c^{AlInN/GaN}(x)$ and $\Delta E_c^{AlInN/AlN}(x)$ take the following value [66]:

$$\Delta E_c^{AlInN/GaN(AlN)}(x) = 0.7(E_g^{Al_{1-x}In_xN}(x) - E_g^{GaN(AlN)}) \quad (24.43)$$

Replacing in (24.39) the various terms by their values and taking a buffer (AlInN) thickness of 11 nm and a spacer thickness of 1 nm, one gets the variation of the carrier density with In concentration as shown in Fig. 24.29.

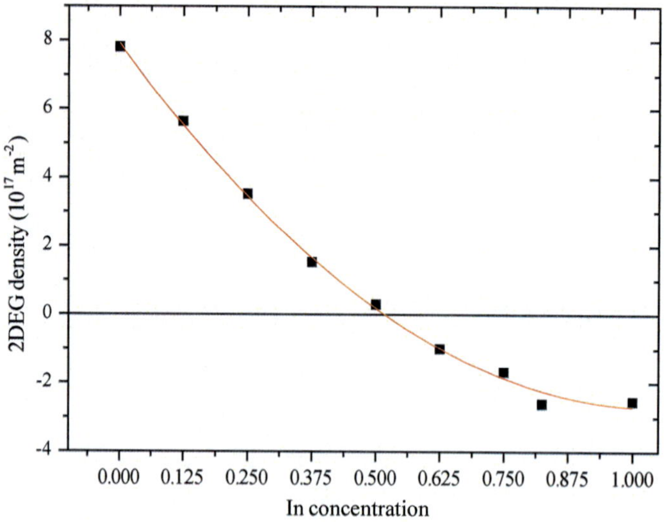

Fig. 24.29 Variation of the 2DEG density with In concentration

One is mainly interested in the region around the matching point (MP) x = 0.18, at that concentration the electron gas may, theoretically, reach a density of 4.50 10^{17} m^{-2} for a AlInN/AlN/GaN configuration. Another noticeable fact is that at above approximately 40% of In content the negative values obtained for the gas density correspond to a gas of holes carriers (2DHG).

Since the stress is composition dependent, to provide estimation of the factors influencing the change the channel resistance with concentration is investigated. The channel resistivity is inversely related to the 2DEG sheet carrier density n_s and electron mobility μ_e and given by [60]:

$$\rho_{ch} = \frac{1}{q n_s \mu_e} \quad (24.44)$$

In the presence of a variable stress, the relative variation in channel resistivity can be written as:

$$\frac{\Delta\rho_{ch}}{\rho_{ch}} = \frac{\Delta n_s}{n_s} + \frac{\Delta\mu_e}{\mu_e} \quad (24.45)$$

To evaluate the effect of stress on the channel resistivity, the effect of stress on both the 2DEG sheet carrier density and mobility needs to be considered. If we consider that "The mobility variation is dominated by a change in the effective mass through band warping" [67] we get then the following expression:

$$\frac{\Delta\rho_{ch}}{\rho_{ch}} = \frac{\Delta n_s}{n_s} + \frac{\Delta m_e}{m_e} \quad (24.46)$$

If we take the zero stress concentration x = 0.18 as reference, the fluctuation of the 2DEG density and the electron effective mass around the matching point as respectively:

$\Delta n_s = n_s(x) - n_s(0.18)$, $\Delta m_e = m_e(x) - m_e(0.18)$.

where m_e (x) is the electron effective mass of the strained structure.

The effect of a ±3% of indium fluctuation, around the matching concentration (x = 0.18), on the variation of the channel resistance can reach 14% mainly due to 2DEG variation (Fig. 24.30).

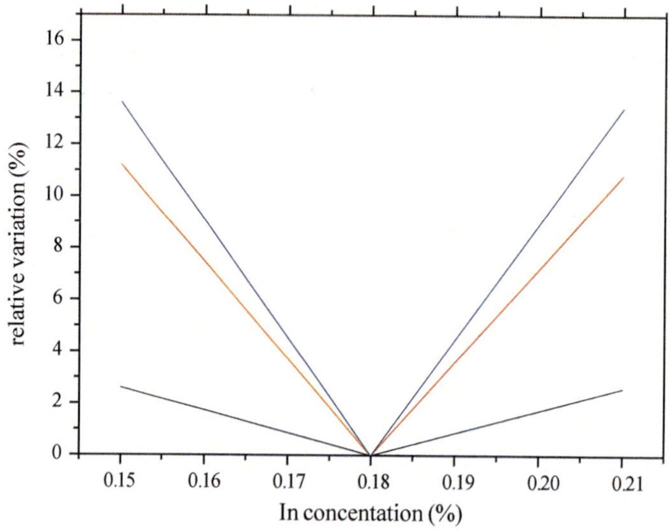

Fig. 24.30 Channel resistance variation $\Delta R/R$ (blue), 2DEG relative variation $\Delta n_s/n_s$ (red) electron effective mass variation $\Delta m_e/m_e$ (black)

References

1. S. Nakamura, M. Senoh, T. Muka, p-GaN/n-InGaN/n-GaN double heterostructure blue-light-emitting diodes. Jpn. J. Appl. Phys., Part 2, **32**, L8–L11 (1993)
2. J. Wu, When group-III nitrides go infrared: new properties and perspectives. J. Appl. Phys. **106**(1), 011101 (2009)
3. J. Kuzmik, Power electronics on InAlN/(In)GaN: prospect for a record performance. IEEE Electron Device Lett. **22**, 510 (2001)
4. R. Butté, J.F. Carlin, E. Feltin, M. Gonschorek, S. Nicolay, G. Christmann, D. Simeonov, A. Castiglia, G. Dorsaz, H.J. Buehlmann, Current status of AlInN layers lattice-matched to GaN for photonics and electronics. J. Phys. D: Appl. Phys. **40**, 6328–6344 (2007)
5. C.J. Humphreys, Solid state lighting. MRS Bull. **33**, 459–470 (2008)
6. D.J. As, D. Schikora, K. Lischka, Molecular beam epitaxy of cubic III-nitrides on GaAs substrates. Phys. Stat. Sol. (c), **0**(6), 1607–1626 (2003)
7. H. Xia, Q. Xia, A.L. Ruoff, High-pressure structure of gallium nitride: Wurtzite-to-rocksalt phase transition. Phys. Rev. B **47**, 12925 (1993)
8. C. Pinquier, F. Demangeot, J. Frandon, J. C. Chervin, A. Polian, B. Couzinet, P. Munsch, O. Briot, S. Ruffenach, B. Gil, B. Maleyre, Raman scattering study of wurtzite and rocksalt InN under high pressure. Phys. Rev. B73, 115211 (2006)
9. P. Vogl, H.P. Hjalmarson, J.D. Dow, A Semi-empirical tight-binding theory of the electronic structure of semiconductors. J. Phys. Chem. Solids **44**, 365 (1983)
10. M. Cardona, P.Y. Yu, *Fundamentals of Semiconductors* (Springer-Verlag, 1996)
11. A. Kobayashi, O.F. Sankey, S.M. Volz, J.D. Dow, Semiempirical tight-binding band structures of wurtzite semiconductors AlN, CdS, CdSe, ZnS, and ZnO. Phys. Rev. B **28**, 935 (1983)
12. W. Paul, Band structure of the intermetallic semiconductors from pressure experiments. J. Appl. Phys. **32** Supplement, 2082 (1961)
13. H. Ünlü, H. H. Gürel, Ö. Akinci, M.R. Karim, Modeling of low dimensional semiconductors: characterisation, modeling and applications, in H. Ünlü, N.J.M. Horing (eds.) (Springer, 2013)
14. P. Hohenberg, W. Kohn, Inhomogeneous electron gas. Phys. Rev. **136**, B864 (1964)

15. W. Kohn, L.J. Sham, Self-consistent equations including exchange and correlation. Phys. Rev. **140**, A1133 (1965)
16. J.P. Perdew, Y. Wang, Accurate and simple analytic representation of the electron-gas correlation energy. Phys. Rev. B **45**(13), 13244–13249 (1992)
17. J.P. Perdew, K. Burke, M. Ernzerhof, Generalized gradient approximation made simple. Phys. Rev. Lett. **77**, 3865 (1996)
18. P. Blaha, K. Schwarz, G.K.H. Madsen, D. Kvasnicka, J. Luitz, *WIEN2k, An Augmented Plane Wave + Local Orbital Program for Calculating Crystal Properties*. Karlheinz Schwartz, Techn. Universität Wien, Austria (2001)
19. G. Kresse, J. Furthmuller, Efficient iterative schemes for ab initio total-energy calculations using a plane-wave basis set. Comput. Mater. Sci. **6**, 15 (1996)
20. A. Zunger, S.-H. Wei, L.G. Ferreira, J.E. Bernard, Special quasirandom structures. Phys. Rev. Lett **65**, 354 (1990)
21. S.-H. Wei, L.G. Ferreira, J.E. Bernard, A. Zunger, Electronic properties of random alloys: special quasirandom structures. Phys. Rev. B **42**, 9622 (1990)
22. A. Walle, Multicomponent multisublattice alloys, non configurational entropy and other additions to the alloy theoretic automated toolkit. Calphad **33**, 266 (2009)
23. F. Bernardini, V. Fiorentini, Nonlinear behavior of spontaneous and piezoelectric polarization in III–V nitride alloys. Phys. Status Solidi A **190**, 65–73 (2002)
24. D. Shin, A.van de Walle, Y. Wang, Z. Liu, First-principles study of ternary fcc solution phases from special quasirandom structures. Phys. Rev. B **76**, 144204 (2007)
25. P. Perdew, A. Ruzsinszky, G.I. Csonka, O.A. Vydrov, G.E. Scuseria, L.A. Constantin, X. Zhou, K. Burke, Restoring the density-gradient expansion for exchange in solids and surfaces. Phys. Rev. Lett. **100**, 136406 (2008)
26. F. Tran, P. Blaha, Accurate band gaps of semiconductors and insulators with a semilocal exchange-correlation potential. Phys. Rev. Lett. **102**, 226401 (2009)
27. A.D. Becke, E.R. Johnson, A simple effective potential for exchange. J. Chem. Phys. **124**, 221101 (2006)
28. J. Heyd, G.E. Scuseria, Hybrid functionals based on a screened Coulomb potential. J. Chem. Phys. **118**, 8207 (2003)
29. F.D. Murnaghan, The compressibility of media under extreme pressures. Proc. Natl. Acad. Sci. USA **30**, 244–247 (1944)
30. J. Serrano, A. Rubio, E. Hernandez, A. Munoz, A. Mujica, Theoretical study of the relative stability of structural phases in group-III nitrides at high pressures. Phys. Rev. B **62**, 16612 (2000)
31. S. Goumri-Said, M.B. Kanoun, A.E. Merad, G. Merad, H. Aourag, Zinblende AlN and GaN under pressure: structural, electronic, elastic and piezoelectric properties. Chem. Phys. **302**, 135 (2004)
32. S. Adachi, *Properties of Group-IV, III-V and II-VI Semiconductors*, Wiley Series in Materials for Electronic & Optoelectronic Applications (2005)
33. M. Ueno, M. Yoshida, A. Onodera, O. Shimomura, K. Takemura, Stability of the wurtzite-type structure under high pressure: GaN and InN. Phys. Rev. B **49**, 14 (1994)
34. R. Oliva, A. Segura, J. Ibáñez, T. Yamaguchi, Y. Nanishi, L. Artús, Pressure dependence of the refractive index in wurtzite and rocksalt indium nitride. Appl. Phys. Lett. **105**, 232111 (2014)
35. P. Rinke, M. Scheffler, A. Qteish, M. Winkelnkemper, D. Bimberg, J. Neugebauer, Band gap and band parameters of InN and GaN from quasiparticle energy calculations based on exact-exchange density-functional theory. Appl. Phys. Lett. **89**, 161919 (2006)
36. I. Vurgaftman, J.R. Meyer, L.R. Ram-Mohan, Band parameters for III–V compound semiconductors and their alloys. J. Appl. Phys. **89**, 5815 (2001)
37. P. Rinke, M. Winkelnkemper, A. Qteish, D. Bimberg, J. Neugebauer, M. Scheffler, Consistent set of band parameters for the group-III nitrides AlN, GaN and InN. Phys. Rev. B **77**, 075202 (2008)
38. I. Vurgaftman, J.R. Meyer, Band parameters for nitrogen containing semiconductors. J. Appl. Phys. **36**, 3675 (2003)

39. M. Rezek, S. Katircioglu, A comparative study of the electronic properties of aluminum nitride compounds. Turk. J. Phys. **40**, 219–230 (2016)
40. B.-T. Liou, C.-W. Liu, Electronic and structural properties of zinc blende $Al_xIn_{1-x}N$. Optics Commun. **274**, 361–365 (2007)
41. C. Persson, A. Ferreira da Silva, R. Ahuya, B. Johansson, Effective electronic masses in wurtzite and zinc-blende GaN and AlN. J. Cryst. Growth **231**, 397–406 (2001)
42. R. Ahmed, S.J. Hashemifar, H. Akbarzadeh, M. Ahmed, F. E-Aleem, Ab-initio study of structural and electronic properties of III-arsenide binary compounds. Comput. Mater. Sci. **39**, 580–586 (2007)
43. F. Wang, S.F. Li, Q. Sun, Y. Jia, First-principles study of structural and electronic properties of zincblende $Al_xIn_{1-x}N$. Solid State Sci. **12**, 1641–1644 (2010)
44. J. Schörmann, D.J. As, K. Lischka, P. Schley, R. Goldhahn, S.F. Li, W. Löffler, Hetterich, H. Kalt, Molecular beam epitaxy of phase pure cubic InN. Appl. Phys. Lett. **89**, 261903 (2006)
45. B. UlHaq, R. Ahmed, A. Shaari, F. ElHaj Hassan, M.B. Kanoun, S. Goumri-Said, Study of wurtzite and zincblende GaN/InN based solar cells alloys: first-principles investigation within the improved modified Becke-Johnson potential. Sol. Energy **107**, 543–552 (2014)
46. E. Iliopoulos, A. Adikimenakis, C. Giesen et al., Energy bandgap bowing of InAlN alloys studied by spectroscopic ellipsometry. Appl. Phys. Lett. **92**, 191907 (2008)
47. S. Schulz, M.A. Caro, L. Tan, P.J. Parbrook, R.W. Martin, E.P. O'Reilly, Composition-dependent band gap and band-edge bowing in AlInN: a combined theoretical and experimental study. Appl. Phys. Express **6**, 121001 (2013)
48. C.G. Van de Walle, M.D. McCluskey, C.P. Master et al., Large and composition dependent band gap InGaN alloys. Mat. Sci. Eng. B. **59**, 274–278 (2009)
49. I. Gorczyca, S. Lepkowski, T. Suski et al., Influence of Indium clustering on the band structure of semiconducting ternary and quaternary alloys. Phys. Rev. B. **80**, 75202 (2006)
50. P. Schley, C. Naprerala, R. Goldhahn et al., Band gap and effective electron mass of cubic InN. Phys. Status Solidi C. **5**, 2342–2335 (2008)
51. E. Lopez-Apreza, J. Arriaga, D. Olguin, Ab initio calculation of structural and electronic properties of $Al_xGa_{1-x}N$ and $In_xGa_{1-x}N$ alloys. Rev. Mex. Fis. **56**(3), 183–194 (2010)
52. P.E. Van Camp, V.F. Van Doren, J.T. Devreese, High-pressure properties of wurtzite and rocksalt-type aluminum nitride. Phys. Rev. B. **44**, 9056–9059 (1991)
53. A. Munoz, K. Kunc, Structure and static properties of indium nitride at low and moderate pressures. Phys. J. Condens. Matter, **5**, 6015–6022 (1993)
54. A.R. Goñi, K. Syassen, in W. Paul (ed.) *Semiconductor Semimetal* (Elsevier, 1998), pp. 247–425
55. A. Breidi, A.V. Postnikov, F. El Haj Hassan, Cinnabar and SC16 high-pressure phases of ZnSe: An ab initio study. Phys. Rev. B **81**(20), 205213 (2010)
56. Z. Yang, Z. Xu, Electronic and optical properties of unstrained and strained wurtzite GaN. Phys. Rev. B **15**, 54 (1996)
57. C. Priester, D. Bertho, C. Jouanin, Band-offset determination at strained II-VI heterojunctions within a self-consistent tight-binding model. Phys. B **191**, 1–15 (1993)
58. A. Harrison, *Electronic Structure and the Properties of Solids* (Dover, 1989)
59. C. Wood, D. Jena, *Polarization Effects in Semiconductors From Ab-Initio Theory to Device Applications* (Springer, 2008)
60. J. Piprek, *Nitride Semiconductor Devices Principles and Simulation* (Wiley-Vch, Verlag, 2007)
61. Y. Park, C. Floresca, U. Chowdhury, J.L. Jimenez, C. Lee, Physical degradation of GaN HEMT devices under high drain bias reliability testing. Microelectron. Reliab. **49**, 478–483 (2009)
62. A. Chini, V. Di Lecce, M. Esposto, G. Meneghesso, E. Zanon, Evaluation and numerical simulations of GaN HEMTs electrical degradation. IEEE Electron. Device Lett. **30**(10)
63. J. Kuzmık, InAlN/(In)GaN high electron mobility transistors: some aspects of the quantum well heterostructure proposal. Semicond. Sci. Technol. **17**, 540–544 (2002)
64. M.W. Wang, J.O. McCaldin, J.F. Swenberg, T.C. McGill, R.J. Hauenstein, Schottky-based band lineups for refractory semiconductors. Appl. Phys. Lett. **66** (1974)
65. I.P. Smorchkova, L. Chen, T. Mates, L. Shen, S. Heikman, B. Moran, S. Keller, S.P. DenBaars, J.S. Speck, U.K. Mishra, AlN/GaN and (Al, Ga) N/AlN/GaN two-dimensional electron gas structures grown by plasma-assisted molecular-beam epitaxy. J. Appl. Phys. **90**, 5196 (2001)

66. O. Ambacher, J. Smart, J.R. Shealy, N.G. Weimann, K. Chu, M. Murphy, W.J. Schaff, L.F. Eastman, Two-dimensional electron gases induced by spontaneous and piezoelectric polarization charges in N- and Ga-face AlGaN/GaN heterostructures. J. Appl. Phys. **85**, 15 (1999)
67. A.D. Koehler et al., Extraction of AlGaN/GaN HEMT gauge factor in the presence of traps. IEEE Electron Device Lett. **31**(7) (2010)

Chapter 25
Nonlinear Optical Properties of Low Dimensional Quantum Systems

V. Ustoglu Ünal, M. Tomak, E. Akşahin, and O. Zorlu

Abstract Low dimensional semiconductor structures such as quantum wells, wires and dots demonstrate interesting nonlinear optical properties. The low dimensionality modulates the electronic structure leading to distinct nonlinear optical properties as compared to their bulk counterparts. In this review, we concentrate more on quantum wells and dots although we also consider some works on quantum wires. The nonlinear optical parameters such as the second and the third order susceptibilities are specifically discussed.

25.1 Introduction

The optical properties of semiconductor nanostructures are studied by a number of references [1–20]. These low-dimensional systems can be grown as *quantum wells* (QWs), *quantum wires, and quantum dots* (QDs) which are effectively two, one and zero dimensional electronic systems. When sufficiently thin layers of different band gap materials are sandwiched, *quantum properties* can be observed due to the quantized energy levels. Therefore, these nanostructures, where electrons or holes are confined are referred to as artificial atoms if the confinement extends to three dimensions. The electron wave functions behave like the standing waves of a particle in a square well potential in two-dimensional confinement.

These nanostructure materials show interesting properties which are completely different from their solid-state bulk counterparts. They have very rich physics and applications in fields such as photonics, quantum information technology, microelectronic and optoelectronic devices such as infrared photo detectors, quantum dot

V. U. Ünal (✉) · E. Akşahin · O. Zorlu
Department of Physics, Yeditepe University, Kayisdagi, 34755 Atasehir, Istanbul, Turkey
e-mail: vunal@yeditepe.edu.tr

M. Tomak
Department of Physics, Middle East Technical University, 06531 Ankara, Turkey

H. Ünlü and N. J. M. Horing (eds.), *Progress in Nanoscale and Low-Dimensional Materials and Devices*, Topics in Applied Physics 144,
https://doi.org/10.1007/978-3-030-93460-6_25

lasers, data transmission based on quantum cryptography, electro-optical modulators, light emitting diodes, laser amplifiers, electron transistors, optical memory and communication technology.

The polarization of the medium, especially the nonlinearity, is the main interest in the related scientific works. The nonlinear effects can be enhanced more strongly in these low-dimensional asymmetrical quantum systems than in bulk materials. The nonlinear optical properties together with the second and third order susceptibilities and the effects of electric field have been studied in many works [21–23]. Researchers are very interested in the study of optical properties of these low dimensional structures and excitonic effects, the attractive part is that, their applications in technology, interesting contents and demanding physics [1–3].

If we list the important points that influence the linear and nonlinear optical properties; the first is the shape of the confining potential, the second is the Coulomb interaction, the third is the external electric and magnetic fields and the fourth one is the incident field intensity. The confining potentials could be used in the optimization of these systems. The Coulomb interaction in the exciton effects considerably on the nonlinear optical coefficients. In some studies [24–26], the Coulomb interaction which forms the exciton is not included in the strong confinement limit or they handled this term as a perturbation. But its affect is not negligible and should be taken carefully in the calculations of nonlinear optical properties of the low dimensional systems. The nonlinear optical coefficients are considerably larger if one takes the excitonic effects into account.

The externally applied electric and magnetic fields create changes in the results obtained for the elements of the dipole matrix. The nonlinear optical response increases strongly with the asymmetry in confinement potential profile and this asymmetry may be obtained by the application of an external electric field. The electronic structure and related optical properties are influenced by the external fields.

There are also some other factors such as contamination, temperature, noise, etc. A very large dipole strength and a narrow bandwith have been observed in the experiments for GaAs QDs which suggest that the intersubband optical transitions in QDs may have huge nonlinearities [27].

25.2 Theoretical Framework and Discussions

25.2.1 Quantum Wells (QWs)

Quantum wells can be imagined as a sandwich consisting of a layer of material with small bandgap embedded between two layers of material with a larger bandgap. Such a double heterostructure confines carriers into the small bandgap material, in the growth direction. QWs are very thin quasi-3D planes, and just one of the three main parts of quantum devices. If we slice a narrow strip from one of the planes it results into 1D-quantum wire. Then dicing up a 1D-wire yields a 0D-quantum dot.

Reducing the number of dimensions in this manner forces electrons to behave more like in a super-atom.

The shape of the confining potential plays important role on the nonlinear optical response. Studies on different confining potentials are based on the better control of the atomic layers and effective design of QWs. The results of studies with confining potentials, such as Woods–Saxon, Pöschl-Teller (PT) and Modified Pöschl-Teller (MPT) in [29–31, 36] have shown that it is possible to optimize nonlinear optical properties of these low dimensional systems.

$AlGaAs - GaAs$ QWs are studied in detail [28–31], the linear and nonlinear changes in the refractive index, absorption coefficients, second and third harmonic generations together with electric field effects are analysed.

The one dimensional effective-mass equation, defining the quantization of motion along the growth direction, is solved using the effective-mass approximation and the transverse energy of the nth-subband E_n and the envelope wave functions $\varphi_n(z)$ are obtained.

In quantum wells, the Schrödinger equation is written using one dimensional confinement potential $V(z)$,(PT/MPT):

$$-\frac{\hbar^2}{2}\frac{d}{dz}\left(\frac{1}{m^*}\frac{d\varphi_n(z)}{dz}\right)+V(z)\varphi_n(z)=E_n\varphi_n(z), \tag{25.1}$$

Two parameters are changing the well width and depth and help us to control the MPT potential, including the wave functions and the energies. In PT potential, three parameters tune the degree of the asymmetry.

The generalization is usually done that the Hamiltonian of the system can be written in two parts,

$$H = H_0 + e\left|\vec{F}\right|z \tag{25.2}$$

where H is the total Hamiltonian, H_0 is the unperturbed part, $\vec{F}$ is the external electric field and e is the absolute value of the electron charge. Electric field is taken as a perturbation [5, 6]. In the dipole approximation, the interaction of an electromagnetic wave with the two-level electron system is defined by the second term in (25.2).

$$e\left|\vec{F}\right|z\langle\langle\left|E_0^{(0)}-E_1^{(0)}\right| \tag{25.3}$$

is considered. The wavefunctions and the corresponding energy levels are determined using the time-independent, non-degenerate perturbation theory;

$$\Delta_n = E_n - E_n^{(0)} = e\left|\vec{F}\right|z_{nn} + e^2\left|\vec{F}\right|^2\sum\nolimits_{j\neq n}\frac{\left|z_{nj}\right|^2}{E_n^{(0)}-E_j^{(0)}} + \ldots, \tag{25.4}$$

and

$$\varphi_n(z) = \varphi_n^{(0)}(z) + e\left|\vec{F}\right|\sum\nolimits_{j\neq n}\varphi_j^{(0)}(z)\frac{|z_{nj}|}{E_n^{(0)} - E_j^{(0)}} + \ldots, \qquad (25.5)$$

with no degeneracy, $\varphi_0^{(0)}$ is expected to be a parity eigenstate. The unperturbed wave functions are used to determine the z_{nj} matrix.

Incident optical radiation of angular frequency w has the polarization along the growth direction z.

The incident radiation over all frequencies can be written as

$$E(t) = \sum\nolimits_j E(w_j)exp(-iw_jt). \qquad (25.6)$$

In the linear case, the polarization and the strength are linearly proportional, but the nonlinear optical response of a material can be described more generally as,

$$P(t) = \chi_\omega^{(1)}E(t) + \chi_0^{(2)}|E|^2 + \chi_{2\omega}^{(2)}E^2(t) + \chi_{3\omega}^{(3)}E^3(t) + \cdots = \sum_j P(\omega_j)e^{-i\omega_jt}, \qquad (25.7)$$

where $\chi_\omega^{(1)}$, $\chi_0^{(2)}$, and $\chi_{3\omega}^{(3)}$ are the linear, optical rectification and third-order nonlinear optical susceptibilities, respectively.

The polarization can also be written in terms of the expectation value of electric dipole moment $\mu(t)$, which can be calculated using the density matrix $\widehat{\rho}$;

$$P(t) = N\langle\mu(t)\rangle = N\mathrm{tr}(\widehat{\rho}\widehat{\mu}) = N\sum_{ij}\rho_{ij}\mu_{ij}, \qquad (25.8)$$

where N is the free electron density and $\langle\mu(t)\rangle = \sum_j\langle\mu(\omega_j)\rangle e^{-i\omega_jt}$. Indices i and j run over all of the energy eigenstates of the system and ρ_{ij} represents the elements of the density matrix.

By direct time differentiation of (25.8) and using Schrödinger's equation for the time evolution of the probability amplitudes, the one-electron density matrix equation with intra-band relaxation is obtained including the damping terms;

$$\frac{\partial\rho_{ij}}{\partial t} = \frac{1}{i\hbar}[H_o - qzE(t), \rho]_{ij} - \Gamma_{ij}\left(\rho - \rho^{(o)}\right)_{ij} \qquad (25.9)$$

assuming ρ_{ij} relaxes to its equilibrium value $\rho_{ij}^{(0)}$ at a single relaxation rate Γ_o only. The Γ_{ij} elements are taken to be equal to one value Γ_o only. Equation (25.9) is solved via the iterative method [5].

The iterative solution of (25.9) gives the density matrix steady state solution $\rho_{ij}^{(0)}$ and the higher order corrections $\rho_{ij}^{(1)}$, $\rho_{ij}^{(2)}$ etc.

The linear susceptibility can be obtained using

$$\langle\mu(t)\rangle = \sum_{ij} \rho_{ij}^{(1)} \mu_{ij}, \tag{25.10}$$

and

$$P^{(1)}(\omega_j) = N\langle\mu(\omega_j)\rangle = N\sum_{ij} \rho_{ij}\mu_{ij} = \chi^{(1)}(\omega_j)E(\omega_j). \tag{25.11}$$

The calculations in (25.10, 25.11) can be carried to higher orders. Using the third-order correction $\rho_{ij}^{(3)}$, the expression for the third-order nonlinear susceptibility can be obtained. But these expressions can be simplified using the two-level approximation.

In most studies, the one-electron density matrix formalism is considered for a two-level approximation of the system of QWs, QDs, where the two energy states, 0 for the ground state and 1 for the first excited state. More commonly, the effective mass approximation is used for the derivation of electronic structure of these low dimensional systems.

The susceptibilities become particularly large when one of the frequencies of the incident field (or sum or differences) becomes equal to a transition frequency of the system, resonance response. Density matrix formulation allows one to describe these near resonant situations, relaxation processes, thus it can show explicitly how accurately one needs to set the incident field frequency to that of the system resonance and how strongly the system response at the resonance. This formalism provides more generally valid results. Thus the first order and the third order susceptibilities can be written as [5, 32];

$$\chi^{(1)}(\omega) = \frac{\rho_s|\mu_{10}|^2}{E_{10} - \hbar\omega - i\hbar\Gamma_o}, \tag{25.12}$$

$$\chi^{(3)}(\omega, I) = \frac{2\pi I\rho_s|\mu_{10}|^4}{n_r c(E_{10} - \hbar\omega - i\hbar\Gamma_o)} \times \left[\frac{4}{(E_{10} - \hbar\omega)^2 + (\hbar\Gamma_o)^2} - \frac{|\mu_{11} - \mu_{00}|^2}{|\mu_{10}|^2} \frac{1}{(E_{10} - \hbar\omega - i\hbar\Gamma_o)(E_{10} - i\hbar\Gamma_o)}\right], \tag{25.13}$$

Here, I is the intensity of the incident field, ρ_s is the electron density, energy difference $E_{10} = E_1 - E_0$ and the second-order nonlinear optical rectification coefficient is given by [33–35],

$$\chi_0^{(2)} = \frac{4\rho_s\left(\mu_{01}^{'}\right)^2\left|\mu_{11}^{'} - \mu_{00}^{'}\right|\left\{\left(E_{10}^{'}\right)^2\left(1 + \frac{T_1}{T_2}\right) + \left[(\hbar\omega)^2 + \left(\frac{\hbar}{T_2}\right)^2\right]\left(\frac{T_1}{T_2} - 1\right)\right\}}{\varepsilon_o\left[\left(E_{10}^{'} - \hbar\omega\right)^2 + \left(\frac{\hbar}{T_2}\right)^2\right]\left[\left(E_{10}^{'} + \hbar\omega\right)^2 + \left(\frac{\hbar}{T_2}\right)^2\right]}. \tag{25.14}$$

where μ_{ij}, $\mu_{ij}^{'}$ define the unperturbed, perturbed matrix element of the electric dipole operator, obtained by unperturbed, perturbed wave functions, μ_{ij} (i, j = 0,1).

$$\mu_{ij} =< \varphi_i|qz|\varphi_j > \delta_{k_n k_n^{'}}. \tag{25.15}$$

The frequency-dependent refractive index is

$$n(\omega) = Re\left[\sqrt{\varepsilon + 4\pi\chi(w)}\right], \tag{25.16}$$

for the well material with the static dielectric constant ε.

If electron density is small, the refractive index is expressed as

$$n(\omega) = \sqrt{\varepsilon}\left\{1 + Re\left[\frac{4\pi\chi(w)}{2\varepsilon}\right]\right\}. \tag{25.17}$$

The linear change in the refractive index due to the incident field can be written as

$$\Delta n^{(1)}(\omega) = Re\left[\frac{2\pi\chi^{(1)}(w)}{\sqrt{\varepsilon}}\right] = Re\left[\frac{2\pi\chi^{(1)}(w)}{n_r}\right]. \tag{25.18}$$

where n_r is the refractive index.

We can write the third order nonlinear change using the (25.18) written for the third order nonlinear susceptibility;

$$\Delta n^{(3)}(\omega) = \frac{2\pi Re\left[\chi^{(3)}(w)\right]}{n_r}. \tag{25.19}$$

The total optical absorption with its linear and third order nonlinear parts are combined as;

$$\alpha(\omega, I) = \alpha^{(1)}(\omega) + \alpha^{(3)}(\omega, I), \tag{25.20}$$

where the linear absorption coefficient is

$$\alpha^{(1)}(\omega) = \frac{\omega}{n_r}\sqrt{\frac{\mu}{\varepsilon_o}}\mathrm{Im}\left[\chi^{(1)}(\omega)\right] \tag{25.21}$$

and the nonlinear part is,

$$\alpha^{(3)}(\omega, I) = \frac{\omega}{n_r}\sqrt{\frac{\mu}{\varepsilon_o}}Im\left[\chi^{(3)}(\omega, I)\right], \tag{25.22}$$

where μ is the permeability of the system [31, 32, 36–38].

Let us give some examples. The one-dimensional confining potential is represented by a Pöschl-Teller (PT) potential by Tomak and others [28–30] and by Modified Pöschl-Teller (MPT) potential by Ustoglu Unal, Aksahin and Aytekin [31]. They used PT and MPT in getting the optical properties of QWs.

These studies have shown that the shape; depth and the width of the potential, are important parameters on the optical properties and should be handled sensitively, Figs. 25.1 and 25.2. The optical intensity of the incident photons is another control parameter on the results. These parameters can give the opportunity for the optimization of nonlinear optical properties: that is the applied electric and magnetic fields, dot size and the intensity of the incoming photons as well as the shape of the confinement.

The mentioned works compared the results with different confinements; the infinite PT potential and the finite MPT potential with well depth −0.34eV. The main differences in their results related to the difference in the depths are presented in Table 25.1. In the MPT model, the negative third order nonlinear conribution is quite small, the peak values are smaller than those obtained in the PT well. Therefore total change in the refractive indices in MPT are found to be bigger than those in PT. This

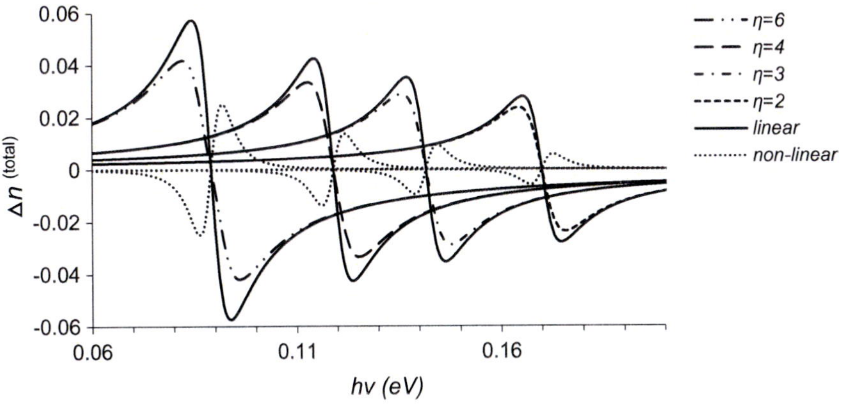

Fig. 25.1 The change in the refractive indexe as a function of photon energy for various QW width-(η) values with $I = 1.0\ \mathrm{MW/cm^2}$, $V_o = -0.34\mathrm{eV}$

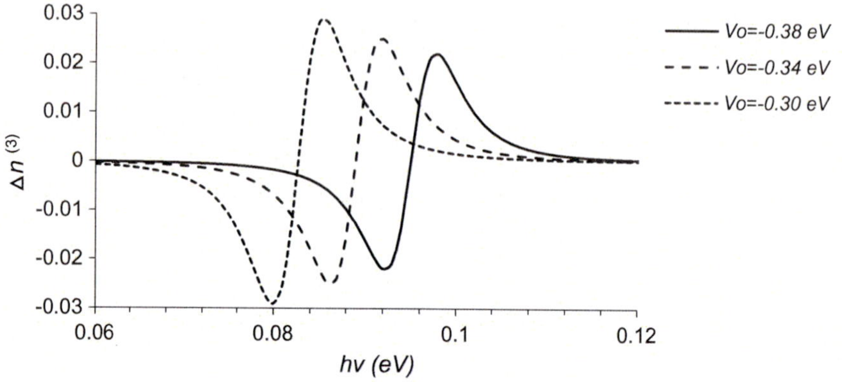

Fig. 25.2 The non-linear change in the refractive index as a function of photon energy for various well depths

Table 25.1 Comparison of the infinite PT potential results in Yıldırım and Tomak [29] with those in the finite MPT potential in Ustoglu Unal, Aksahin and Aytekin [31]

Potential type	Well width (Å)	$\Delta n^{(1)}$ Peak	$\Delta n^{(3)}$ Peak	$\Delta n^{(total)}$ Peak
PT	126.5	0.03 (0.171 eV)	−0.006 (0.172 eV)	0.025 (0.17 eV)
MPT	98	0.028 (0.165 eV)	−0.006 (0.167 eV)	0.024 (0.164 eV)
MPT	126.5	0.057 (0.084 eV)	−0.025 (0.086 eV)	0.042 (0.082 eV)

is mainly due to the difference in the calculated values of the dipole matrix elements related to the different well depths.

The external electric field mainly causes an asymmetry in the QW and as the strength of the field increases, the effects on the optical coefficients become more clear, the nonlinear optical properties increase. Various applied fields are studied as in Fig. 25.3, their resultant effects are shown for the second term in (25.13). Strong electric fields increases the contribution of the second term in the third-order susceptibility. The change in the total refractive index can be increased if the external electric fields and the optical intensities reduce.

25.2.2 *Quantum Dots (QDs)*

The nanostructures with three-dimensional confinement of electrons and holes are called Quantum Dots (QDs). When a thin layer of a semiconductor is grown on top of a substrate, which has a different lattice constant, the thin layers order, or *self-assemble* into QDs. QD is a region of space (1–100 nm at low temperatures) in a crystalline semiconductor matrix, with sizes comparable to the exciton Bohr

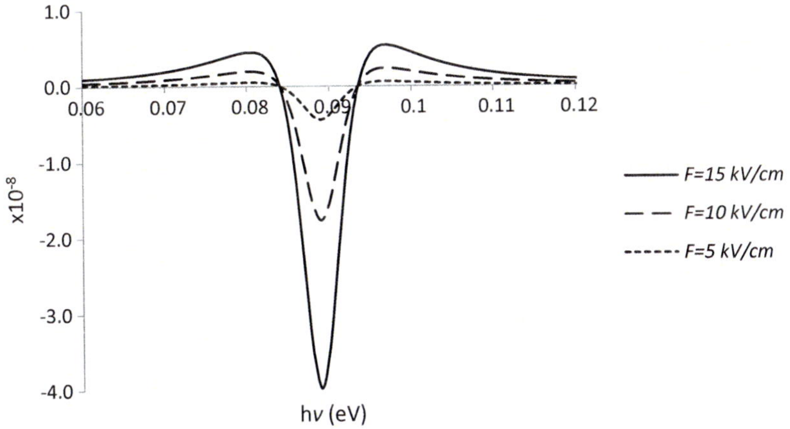

Fig. 25.3 Contribution of the nonlinear term as a function of photon energy, various applied electric fields F

radius a. QDs in *AlGaAs* based material have been obtained with radii typically in the range of 50 nm or larger. For such sizes the spacing between energy levels can be larger than the thermal energy. In a quantum dot, the electron is confined in all three-dimensions, thus reducing the degrees of freedom to zero. [11, 23]. The quantization of the electronic levels appears as a result of confinement of the exciton. Thus, these nanostructures show unique electrical and optical behaviors which make them very useful as device elements in number of applications such as batteries, transistors, sensors, photovoltaics, etc.

It is clearly shown excitonic effects are essential in increment of the nonlinear optical coefficients. Approximations about the shape of the dot are done; one-dimensional [40–42], two-dimensional [23, 36], disk-like [7, 12, 13, 39], or semi-spherical [43] shapes are analyzed by researchers. The electronic structure of the QD may be calculated more simply in this way.

Another simplification in the calculations can be done by considering only the strong confinement limit.

To investigate the optical properties of QDs, the system may be off-resonantly excited by an optical pulse. In [36–38], it is shown that the Woods–Saxon Potential and Parabolic Potential can be used to model the confinement in quantum dots with considerable success. One could engineer the structure of materials by means of external electric field and confinement strengths and tailor the energy spectrum to produce desirable nonlinear optical effects.

Excitonic effects

In some studies, analitical solutions are obtained for the exciton in a spherical QDs, based on the effective mass approximation. The linear and nonlinear optical properties are calculated within the density matrix formalism. The effects of external electric field are searched. Sometimes, the strength of the confinement is not defined specificially as strong or intermediate. On the optical absorption and the refractive index calculations, the challenging effective mechanisms are the Coulomb interaction, the confinement potential and the applied static electric field.

Some references considers just the strong/intermediate confinement limit where they take the interaction of the electron and hole via Coulomb force as a perturbation or simply neglect it [12, 23–26, 40, 41]. The computational method followed by the references varies from a variational approach [8, 13, 44] to numerical matrix diagonalization [7, 9]. Usually, a small electric field is considered and the Coulomb term is treated as a perturbation [37]. In [37], to get the electronic structure of three dimensional spherical QD, an analytical solution is set out. This is especially vital to compare various contributions those take part in the QDs electronic structure. Considering a parabolic confinement potential for the exciton moving in a spherical QD, the Hamiltonian operator within the effective mass approximation may be expressed as

$$H = -\frac{\hbar^2}{2m_e}\vec{\nabla}_e^2 + \frac{1}{2}k_e r_e^2 - \frac{\hbar^2}{2m_h}\vec{\nabla}_h^2 + \frac{1}{2}k_h r_h^2 - \frac{\gamma}{|\vec{r}_h - \vec{r}_e|} + |q|\vec{F}\cdot(\vec{r}_e - \vec{r}_h), \tag{25.23}$$

where m_e, m_h denote the effective masses, and $\vec{r}_e, \vec{r}_h$ are the position vectors of the electron e and hole h. $k_e(k_h) = m_e(m_h)\omega_o^2$, where ω_o is the harmonic frequency which controls the confinement strength. q is the electron charge, $\vec{F}$ represents the external electric field and γ is a positive constant.

The Hamiltonian operator can be decomposed in terms of the relative coordinate $|\vec{r}| = |\vec{r}_h - \vec{r}_e|$ and the center-of-mass coordinate

$$\vec{R} = \left(m_e\vec{r}_e - m_h\vec{r}_h\right)/(m_e + m_h). \tag{25.24}$$

and written as

$$H = H_{C.M} + H_{rel} \tag{25.25}$$

For the center of mass, the Hamiltonian operator is shown as

$$H_{C.M} = -\frac{\hbar^2}{2M}\vec{\nabla}_{C.M}^2 + \frac{1}{2}M\omega_o^2 R^2 \tag{25.26}$$

where the total mass is $M = m_e + m_h$. The Hamiltonian in relative coordinates is shown as,

$$H_{rel} = H_o + H' \tag{25.27}$$

Unperturbed H_o is solved including the electric field, without the Coulomb interaction, and the unperturbed eigenenergies and eigenfunctions are obtained. Coulomb interaction is taken as a perturbation. Using the perturbation theory, the new perturbed energy, $E_n^{'}$, and the wave functions, $\psi_n^{'}$, are calculated. These wavefunctions and energies related to the parabolic confinement are the main part of the work and simple solution of the problem is presented in [37, 38].

The results which present the changes in the calculated optical properties due to the Coulomb interaction are clearly shown in Figs. 25.4, 25.5, 25.6 and 25.7. It

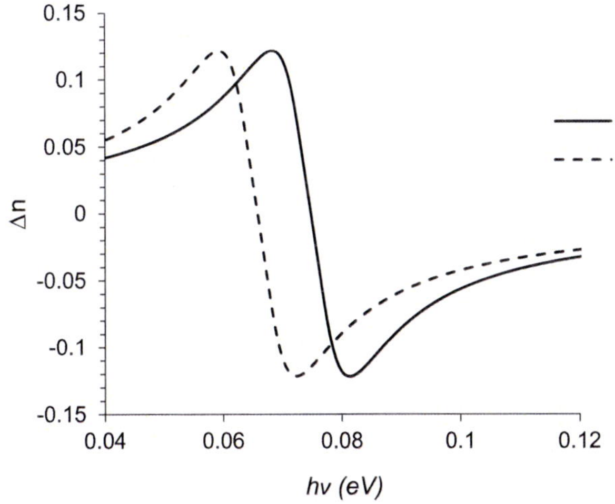

Fig. 25.4 Effect of Coulomb interaction on the refractive index calculations, $F = 25\text{kV/cm}$ and $I = 0.5 \times 10^{10}\text{W/m}^2$

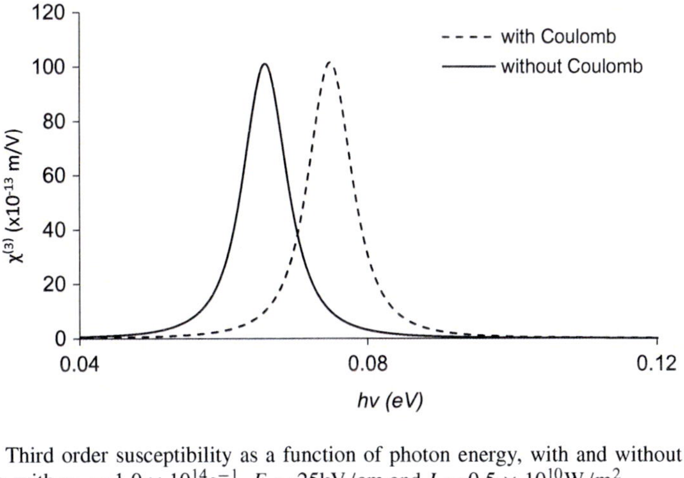

Fig. 25.5 Third order susceptibility as a function of photon energy, with and without Coulomb interaction, with $\omega_o = 1.0 \times 10^{14}\text{s}^{-1}$, $F = 25\text{kV/cm}$ and $I = 0.5 \times 10^{10}\text{W/m}^2$

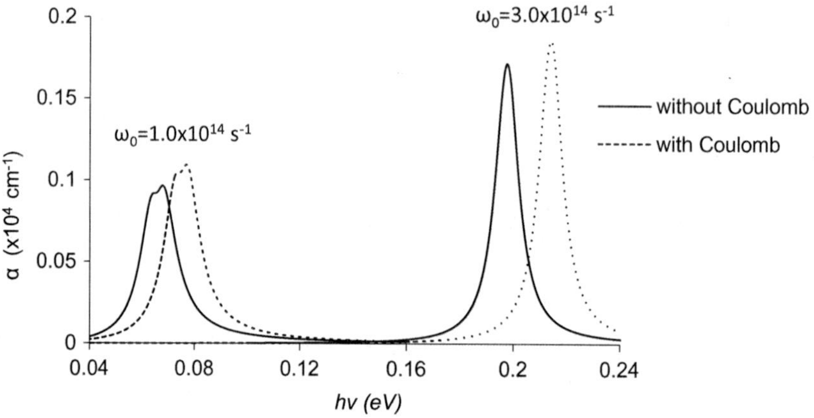

Fig. 25.6 Effect of the Coulomb interaction on the total absorption coefficient, $F = 25\text{kV/cm}$ and $I = 0.5 \times 10^{10}\text{W/m}^2$

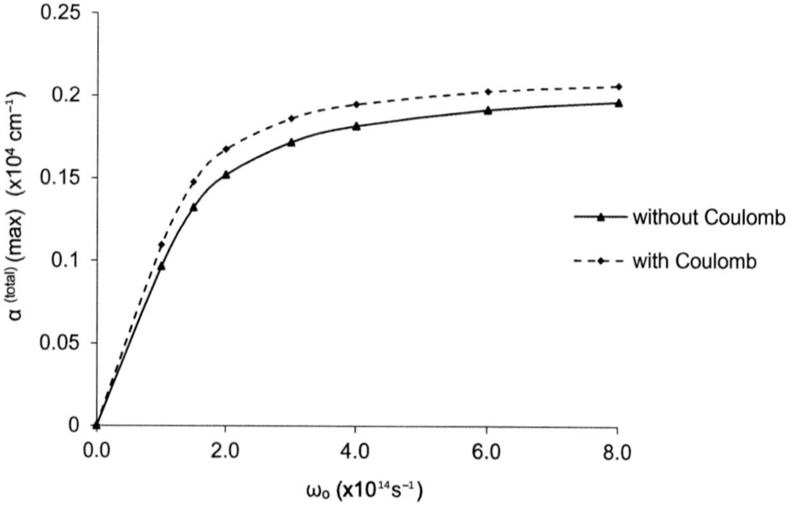

Fig. 25.7 Effect of the Coulomb interaction on the maximum absorption coefficient as a function of harmonic frequency ω_o

can be seen that the linear absorption coefficient $\alpha^{(1)}$ and the nonlinear absorption coefficient $\alpha^{(3)}$ act in opposite ways; linear part is large and positive, unlike the smaller and negatively acting nonlinear part. Another critical characteristics is that the blue shifted peaks as a result of increasing harmonic frequency, that is decreasing QD size.

The change in harmonic frequency has different effects on the linear and nonlinear absorption coefficients. As ω_o increases, there is no curicial change in linear absorption coefficient but the nonlinear one decreases in negative height and therefore becomes more positive. As harmonic frequency increases, the energies are getting smaller than those calculated without the Coulomb interaction. This could be due to the stronger localization created by the Coulomb interaction and thus the total absorption coefficient increases. The Coulomb interaction increases both the linear and nonlinear coefficients. It causes the blue shift of the peaks. This result is equivalent to the results of stronger confinements. One can conclude that the presence of the Coulomb interaction increases the absorption, acting as a confining factor. It is effectively increasing the confinement, blue-shifting the peaks but not changing their peak value. The Coulomb interaction has a considerable effect on the position of the peaks.

The calculation of third order susceptibility may not be successful if the Coulomb interaction is ignored. These three parameters, the confinement, the Coulomb interaction and the incident photon energies, have appreciable affect on the electronic structure, matrix elements and the optical coefficients.

Applied electric field has an important effect on all calculations. The effect of the electric field on absorption coefficients, susceptibilities and optical rectification are clear in Figs. 25.8, 25.9 and 25.10. Electric field decreases the absorption peak value and obvious red shift is observed. The electric field is effectively increasing the distance between the particles, thus the resultant effect looks like in that of a reducing confinement. The third-order susceptibility peaks are always blue shifted, along with the values of the refractive index change, Δn, with increasing field values but change in the total refractive index peak values is hardly noticable.

The electric field also increases the second-order nonlinear optical rectification coefficient $\chi_0^{(2)}$ values for both narrow and wider dots. External electric field creates

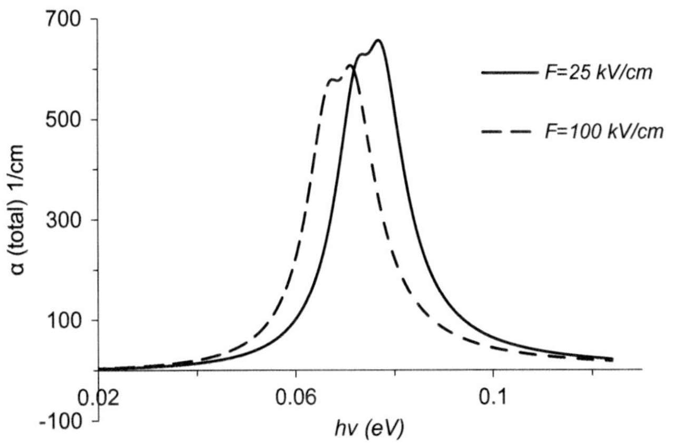

Fig. 25.8 Effect of applied electric field F on the total absorption coefficient, $\omega_o = 2.0 \times 10^{14}\text{s}^{-1}$ and $I = 0.5 \times 10^{10}\text{W/m}^2$

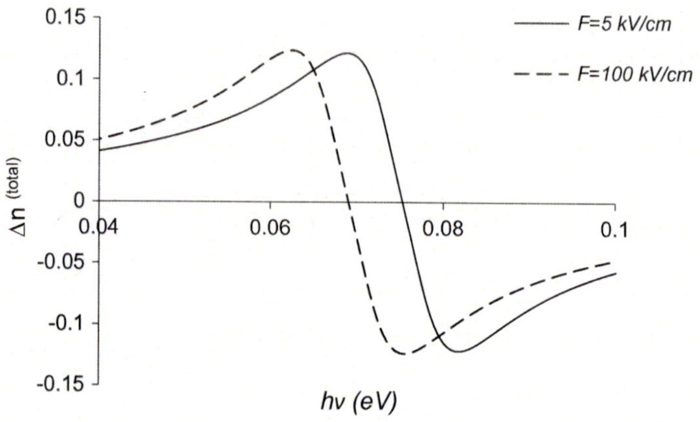

Fig. 25.9 Effect of applied electric field F on the total refractive index change, with$\omega_o = 1.0 \times 10^{14}\mathrm{s}^{-1}$,$I = 0.5 \times 10^{10}\mathrm{W/m^2}$

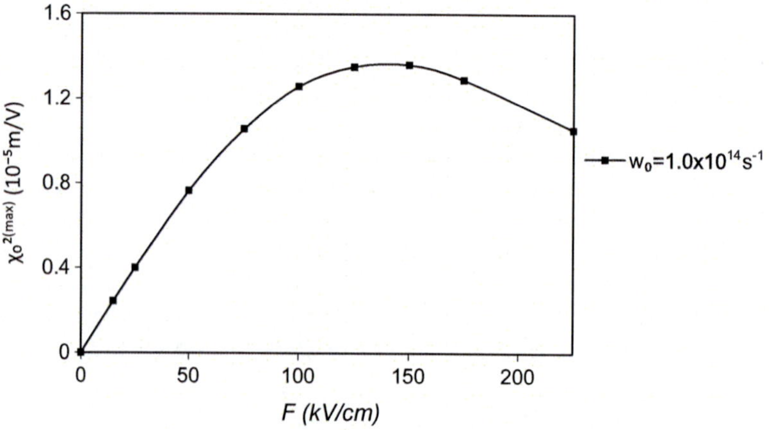

Fig. 25.10 Second order non-linear optical rectification coefficient as a function of electric field F, with harmonic frequency ω_o

asymmetry in QDs and when the QD has some asymmetry, the second order nonlinear optical property becomes more powerful than higher ones. A red shift is observed when electric field strength is increased. For smaller harmonic frequency values, the decrease is observed clearly and forms up a peak at weak electric field strengths. The $\chi_0^{(2)}$ peaks are blue shifted in the strong confinement as shown in Fig. 25.11, [37, 38].

The intensity of incoming photons is another important parameter as presented in Fig. 25.12. Increase in the intensity causes a reduction in the absorption coefficient. Absorption saturates towards $I = 1.0 \times 10^{10} W/m^2$. The peak may turn into two peaks at large intensities due to the perturbational nature of the approach [42]. The refractive index changes considerably if the field intensity increases. Especially at

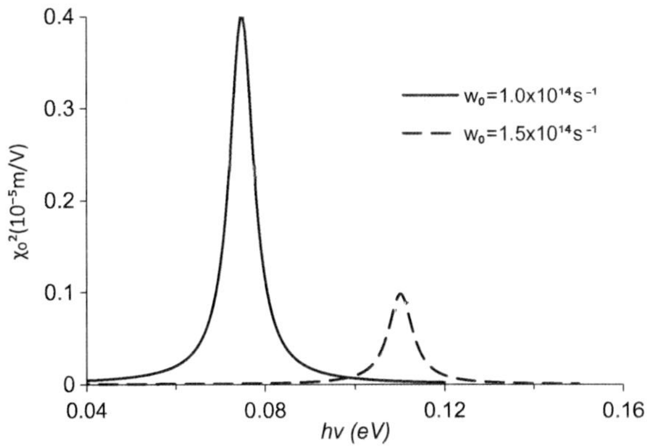

Fig. 25.11 Effect of harmonic frequency ω_o on the second order non-linear optical rectification coefficient, with applied electric field $F = 25\mathrm{kV/cm}$

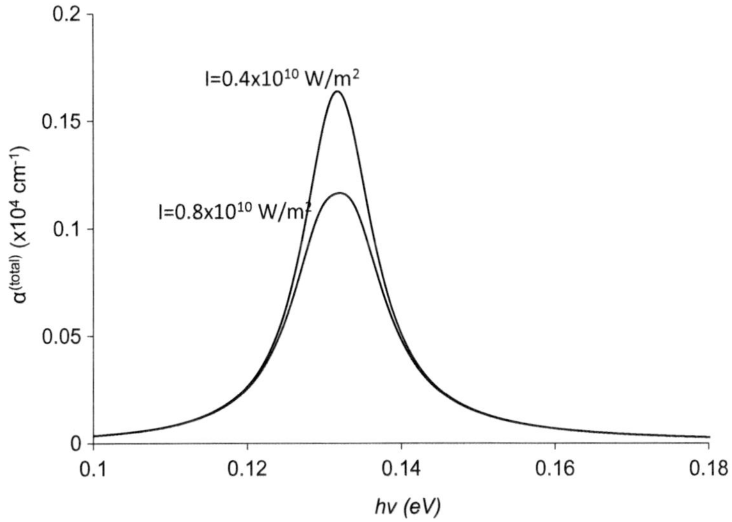

Fig. 25.12 Variation of the absorption coefficient with the intensity of the incident field, with $\omega_o = 2.0 \times 10^{14} s^{-1}$ and $F = 25\mathrm{kV/cm}$

the high intensities, the intensity dependent refractive index should be considered carefully.

The shape of QDs is another effective factor on the optical properties. As an example, in [39], disc-like and spherical QDs are discussed by calculating their optical response. When we take a disc-like QD, the peaks are red-shifted and have lower values compared to those in the spherical case. The elements $|\mu_{10}|^2$ of dipole matrix for the disc-like case is also lower than that for the spherical QD, Fig. 25.13.

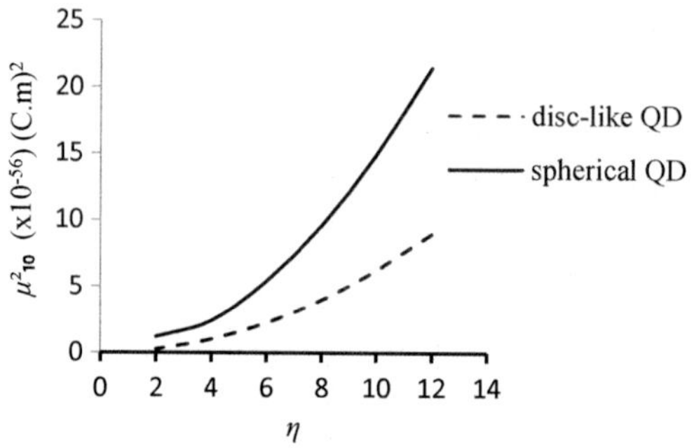

Fig. 25.13 Effect of dot-shape on the dipole matrix element $|\mu_{10}|^2$, as a function of QD size (η)

The peaks in both refractive index and absorption graphs depend on $|\mu_{10}|^2$, this is the reason why larger peaks appear in the spherical QDs.

The peak values of the coefficients studied are always red shifted with increasing dot size. The effect of harmonic frequency ω_o, which changes the confinement, on the linear refractive index calculations, $\Delta n^{(1)}$ can be followed in Fig. 25.14. The linear change is plotted as a function of incident photon energy $h\nu$. The larger ω_o corresponds to smaller dot size. For smaller dot sizes, that is for stronger confinement, the energy difference between states becomes larger, the blue shift of the peaks occurs.

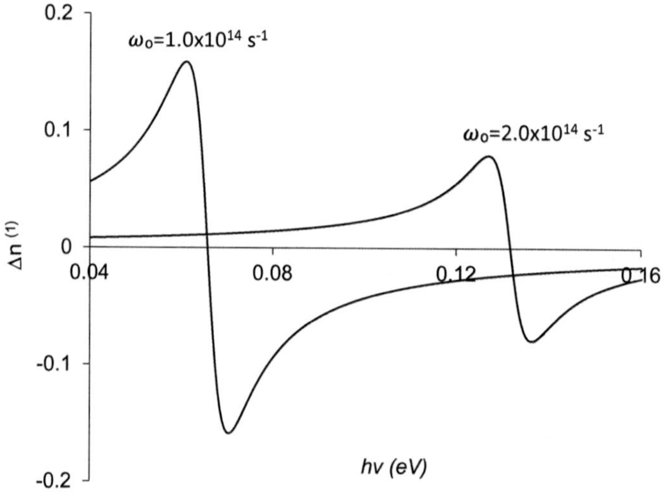

Fig. 25.14 Effect of dot-size (ω^o) on the linear refractive index change, as a function of photon energy, $F = 25\text{kV/cm}$, $I = 0.5 \times 10^{10}\text{W/m}^2$

25.2.3 *Quantum Wires*

Quantum wires are very narrow structures where electron may only transport in a very few transverse modes (with energies less than the Fermi energy). Semiconductor quantum wires can be used to make switchable high-speed lasers as the length and width of a quantum wire can be controlled. By altering the width, the number of transverse propagating modes is changed and as a result the conductance is variable.

In many calculations an exciton in a cylindrical quantum wire with a parabolic potential is studied. The center of mass problem is exactly solvable since H_{CM} correspond to a two-dimensional harmonic oscillator plus the kinetic energy of the free motion along the QW. In terms of the relative coordinate $|\vec{r}| = |\vec{r}_h - \vec{r}_e|$ and the center-of-mass coordinate $\vec{R} = (m_e \vec{r}_e - m_h \vec{r}_h)/(m_e + m_h)$.

Just in studies for QDs, the Hamiltonian operator may be written as $H = H_{C.M} + H_{rel}$, where $H_{C.M} = -\frac{\hbar^2}{2M}\vec{\nabla}^2_{C.M} + \frac{1}{2}M\omega_o^2 R^2$ and $M = m_e + m_h$.

The Hamiltonian is written in relative coordinates, $H_{rel} = H_o + H'$.

The Hamiltonian operator within the effective mass approximation may be written as;

$$H = -\frac{\hbar^2}{2M}\vec{\nabla}^2_R + \frac{\omega_0^2}{2}M\vec{R}^2 - \frac{\hbar^2}{2\mu}\vec{\nabla}^2_r + \frac{\omega_0^2}{2}\mu\vec{r}^2 - \gamma\frac{e^2}{|\vec{r}|} + |e|\vec{F}.(\vec{r}) \tag{25.28}$$

where ω_o is the frequency of the parabolic confining potential, $\vec{F}$ is the external electric field applied in r direction.

Using the Schrödinger equation, the eigenfunctions and eigenvalues related to the states of exciton wave functions, the total refractive index change and linear absorption coefficient may be plotted as in Figs. 25.15 and 25.16.

25.3 Conclusion

Semiconductor heterostructures, quantum wells, quantum-well wires, quantum dots, and superlattices, represent some of the most important objects in modern semiconductor physics.

These structures exhibit very different confining potentials. Quantum confinement effects appear when the geometric dimensions of the nanostructures reach the value of the bulk excitonic Bohr radius.

The mechanism of the nonlinearity in quantum dots is of a basically different kind. With decreasing size, the optical nonlinearity is strongly influenced by quantum confinement. In the case of quantum dots, the zero-dimensional structures can be considered as experimental realization of discrete-level systems in semiconductors.

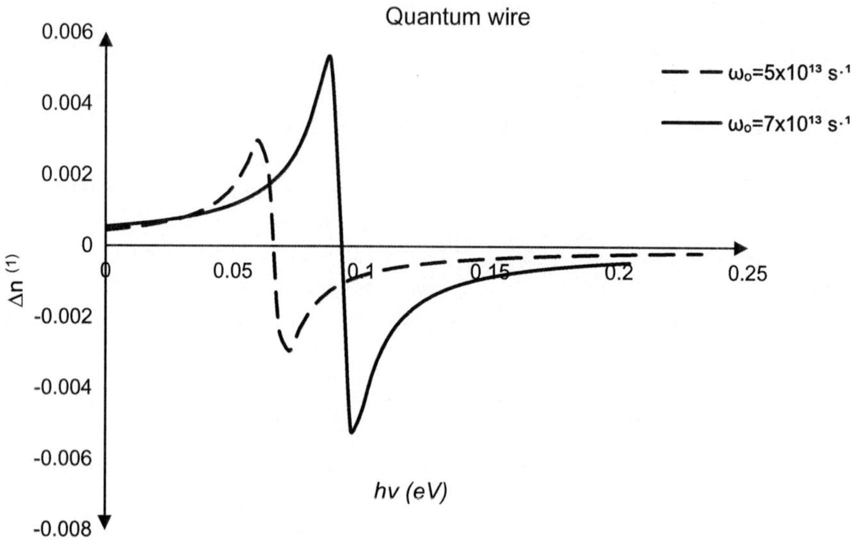

Fig. 25.15 Linear refractive index change as a function of photon energy, with different wire sizes (ω_o)

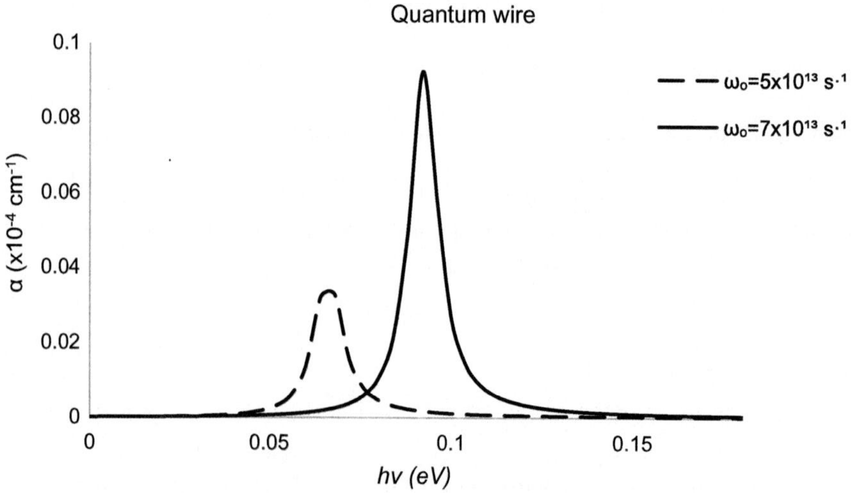

Fig. 25.16 Linear absorption coefficient as a function of photon energy, with different wire sizes (ω_o)

The motivation behind introducing these structures is an attempt to tailor the electronic and optical properties of these materials for exploitation in devices. The quantum confinement of carriers in the low-dimensional system leads to the formation of discrete energy levels and the drastic changes of physical and chemical properties such as the novel nonlinear optical effects. Practical use of these low dimensional systems; fast optical switching, phase conjugation and wave guiding, efficient emission of radiation and lasing with tunability over a wide spectral range, microelectrodes and photocatalysts in photochemical reactions, a promising basis for binary logic operation; applications in waveguiding devices, optical switching technology, optical computing and quantum cryptography, light-emitting devices, laser characteristics, etc.

The aim generally is to test how far a correlation between shape, size, external electric field, intensity of the incident field, confinement potential and electronic states can be proven in the optical properties as a sign of three-dimensional quantum confinement.

As noted above, the shape has a significant effect on the QD physical properties. The shape of a quantum dot affects not only the energy of charges in QDs, but also their electronic density of states. We have shown that refractive index and absorption peaks are larger in the spherical QDs compared to those in the disc-like QD shape. It is related to the changes in the dipole matrix elements as a result of shape changes. The peak values of the coefficients studied are always red shifted with increasing dot size.

In many studies, different confinements are investigated, the exact solutions to the problems with or without electric field is interesting. Calculations have been made in the effective mass approximation and using perturbation theory to include the effects of applied electric field. We have given the examples of the Pöschl-Teller and Modified Pöschl-Teller potentials that can have tunable asymmetry and the corresponding Schrödinger equation is analytically solvable. It yields promising nonlinear optical properties. MPT gives larger changes in refractive index values compared to those in PT, mainly due to the difference in well depths. The Woods–Saxon and the Parabolic potential studies have also shown that one could control the confinement strengths and tailor the energy spectrum to produce desirable nonlinear optical effects.

The excitonic properties are investigated in a three-dimensional spherical quantum dot under various confinement potentials plus a static electric field using the effective mass approximation. The novelty of some calculations is the use of rather simple analytical expressions for the exciton wavefunction, in the calculation of corresponding binding energy and, the first and third order susceptibilities (under the density matrix formalism). Therefore, the competing role of different mechanisms (confinement, Coulomb interaction, etc.) can be easily disentangled when determining the optical absorption properties of the system.

The main trends are very well known in the Literature, even under more realistic descriptions of the quantum dot itself. However, by the careful use of the analytical work, it is shown that, whereas the customary infinite-barrier confinement in spherical quantum dots leads to infinite series when evaluating the first-order correction to the unperturbed exciton wavefunction, the parabolic confinement leads to a great simplification of such corrections. In other words, the unique properties of the harmonic-oscillator wavefunctions are the very basis of the possibility of the relatively simple analytical treatment of the problem.

The considerable contribution of the Coulomb interaction on the calculations of nonlinear optical properties of QDs in all confinement limits are presented, no assumptions are needed. One must underline the fact that the Coulomb effect should be considered in the Hamiltonian even in the strong confinement limit. The third order nonlinear susceptibility should not be calculated by ignoring the Coulomb interaction. This interaction increases the absorption. This is natural since the Coulomb interaction results in more localization. Its effect is equivalent to the effect of stronger confinement.

The second order nonlinear optical property is stronger than higher order optical nonlinearities especially when there is an asymmetry in QDs. The asymmetry can be realized by the applied electric field. The results show that the optical rectification susceptibility obtained in the spherical QD can reach the magnitude of 10^{-4} m/V. The peak values illustrate the effects of both the harmonic frequency and the external electric field on the results.

It is observed that linear optical coefficient peaks obtained for quantum wire are smaller than those for quantum wells and dots and seems to be more sensitive to the changes in harmonic frequency ω_o which controls the size. In quantum wires, the peaks have an obvious blue shift.

The dynamic effects of all important physical quantities related to the low dimensional qunantum systems must be studied precisely to understand their optical responses clearly. The neglect of any of them may lead to an incomplete conclusions. The very interesting part is that it is possible to optimize the linear and nonlinear optical properties. By controlling those parameters mentioned so far, one can get rid of any unwanted effects in these systems, which is not an available option for bulk materials.

References

1. L. Banyai, S.W. Koch, *Semiconductor Quantum Dots*. World Scientific (1993)
2. T. Chakraborty, *Quantum Dots: A Survey of the Properties of Artificial Atoms*. Elsevier Science (1999)
3. S. Glutsch, *Excitons in Low-Dimensional Semiconductors: Theory Numerical Methods Applications* (Springer Series in Solid-State Sciences) (2004)
4. P. Atkins, R. Friedman, *Molecular Quantum Mechanics*. Oxford University Press, Fourth ed. (2005)
5. R.W. Boyd, *Nonlinear Optics*, 3rd edn. (Academic Press, San Diego, Elsevier, 2008)
6. J.J. Sakurai, *Modern Quantum Mechanics*, Rev. (Addison-Wesley, MA, 1994)
7. W. Xie, Physica E **43**, 1704 (2011)
8. N.S. Minimala, A.J. Peter, C.W. Lee, Physica E **48**, 133 (2013)
9. R. Khordad, *Superlattices Microstruct*, 54–7 (2013)

10. X. Li, C. Zhang, Superlatt. Microstruct. **60**, 40 (2013)
11. M.G. Barseghyan, A. Hakimyfard, S.Y. Lopez et al., Physica E-low-dimensional systems & nanostructures **43**, 529–533 (2010)
12. W. Xie, J. of Lumin. **131**, 943 (2011)
13. R.E. Acosta et al., Physica E **44**, 1936 (2012)
14. W. Fanghui, W. Shanshan, S. Zhaonan et al., Fullerenes nanotubes and carbon Nanostruct. **23**, 769–776 (2015)
15. K. Wen-Jian, L. Qing, P. Jiang-Yong et al., J. Display Technol. **11**, 461–465 (2015)
16. L. Dallali, Jaziri, Sihem, Martinez-Pastor. Solid State Commun. **209**, 33–37 (2015)
17. T.T.K. Chi, U.T.D. Thuy, T.T.T. Huyen et al., J. of Electronic Mater. **45**, 2449–2454 (2016)
18. S. Ortakaya, M. Kirak, A. Guldeste, J. Nonlinear Opt. Phys. Mat. **26**, 1750035 (2017)
19. L. Shi, Z.W. Yan, Superlattices Microstruct. **109**, 382–393 (2017)
20. M.V. Stephanie, A.A. Iskandar. M.-O. Tjia, J. Nonlinear Optics. Phys. Mat. **27**(2), 1850018 (2018)
21. B. Çakır, Y. Yakar, A. Özmen, Progress Electromag. Res. **21**, 77–92 (2011)
22. I. Karabulut, H. Safak, M. Tomak, J. Phys. D 155104 (2008)
23. İ. Karabulut, S. Baskoutas, AIP Publising of Physics, 103 (2008)
24. Y.Z. Hu, S.W. Koch, M. Lindberg, N. Peyghambarian, E.L. Pollock, F.F. Abraham, Phys. Rev. Lett. **64**, 1805 (1990)
25. G.W. Bryant, Phys. Rev. B **41**, 1243 (1990)
26. Y.Z. Hu, S.W. Koch, D.B.T. Thoai, Mod. Phys. Lett. B **4**, 1009 (1990)
27. P. Harrison, Quantum Wells, Wires and Dots, Theoretical and Computational Physics of Semiconductor Nanostructures, Second ed., Wiley-Interscience (2005)
28. H. Yıldırım, M. Tomak, Phys. Rev. B (72) 115340(1–6) (2005)
29. H. Yıldırım, M. Tomak, J. Appl. Phys. **99**, 093103 (2006)
30. H. Yıldırım, M. Tomak, Phys. Stat. Sol. (b) **243**, 15 4057–4063 (2006)
31. V. Ustoglu Unal, E. Aksahin, O. Aytekin, Physics E, **47**, 103–108 (2013)
32. D. Ahn and S.L. Chuang, IEEE J. Quantum Electron. QE-23 2196–2204 (1987)
33. S. Baskoutas, E. Paspalakis, A.F. Terzis, Phys. Rev. B **74**, 153306 (2006)
34. B. Chen, K. XGuo, R.Z. Wang, Y.B. Zheng, B. Li, Eur. Phys. J. B **66**, 227 (2008)
35. E. Rosencher, Ph. Bois, Phys. Rev. B **44**, 11315 (1991)
36. O. Aytekin, S. Turgut, V. Üstoğlu Ünal, E. Akşahin, M. Tomak, Physica E, **54**, 257–261 (2013)
37. E. Aksahin, V. Ustoglu Unal, M. Tomak, Eur. Phys. J. B. **87**, 265 (2014)
38. Ü.V. Üstoğlu, E. Akşahin, M. Tomak, E. Physica, Low-dimensional systems and nanostructures **74**, 258–263 (2015)
39. V.Ü. Ünal, E. Birsey, E. Aksahin, J. Nonlinear Opt. Phys. Mater. **27**(3), 1850034 (2018)
40. Y.-B. Yu, S.-N. Zhu, K.-X. Guo, Phys.Letts. A **335**, 175 (2005)
41. M.J. Karimi, G. Rezai, Physica B **406**, 4428 (2011)
42. E. Paspalakis, J. Boviatsis, S. Baskoutas, J. Appl. Phys. **114**, 15310 (2013)
43. J. Florez, A. Camacho, Nano. Res. Letts. **6**, 268 (2011)
44. A.J. Peter, C.W. Lee, Current. Appl. Phys. **13**, 390 (2013)

Chapter 26
One-Dimensional Silicon Nano-/microstructures Based Opto-Electronic Devices

H. Karaağaç, E. Peksu, B. Alhalaili, and M. Saif Islam

Abstract One-dimensional (1D) nanostructures, including nanorods, nanowiskers, nanowires, nanotubes and nanobelts, have been receiving a great deal of research attention from industry and academia in recent years. Due to their special and outstanding many characteristics, such as effective light–trapping ability, bandgap tunability, efficient charge carrier collection, high carrier mobility, large surface-to-volume ratio and excellent thermal conductivity, such nanostructures play a very important role in the manufacture of high-performance devices with novel functionalities. To date, a number of materials, such as TiO_2 (titanium-oxide), ZnO_2 (zinc-oxide), Si (silicon), C (carbon), Ga_2O_3 (gallium oxide) and SnO_2 (tin-oxide), have been employed in the production of one-dimensional 1D structures for the fabrication of high-performance electronic and opto-electronic devices. Among them, silicon (Si) is particularly attractive material for a wide range of opto-electronic device application owing to its highly developed technology and outstanding features such as high thermal conductivity, facile doping control, hardness and excellent optical and electrical properties. In this chapter, although we discuss the recent advances in optoelectronic applications of 1D Si nano-/microstructures, it focuses mainly on our own recent studies based on the synthesis of ordered and disordered Si-nanowires/micropillars and their applications in photodetection and harvesting solar

H. Karaağaç (✉) · E. Peksu
Department of Physics Engineering, Istanbul Technical University, Maslak, Istanbul 34469, Turkey
e-mail: karaagacha@itu.edu.tr

E. Peksu · B. Alhalaili · M. S. Islam
Department of Electrical and Computer Engineering, University of California at Davis, Davis, CA 95616, USA
e-mail: bhalaili@kisr.edu.kw

M. S. Islam
e-mail: sislam@ucdavis.edu

B. Alhalaili
Nanotechnology and Advanced Materials Program, Kuwait Institute for Scientific Research, Safat 13109, Kuwait

H. Ünlü and N. J. M. Horing (eds.), *Progress in Nanoscale and Low-Dimensional Materials and Devices*, Topics in Applied Physics 144,
https://doi.org/10.1007/978-3-030-93460-6_26

energy. In particular, a special focus will be given on the fabrication of Si nano-/microstructures based solar cells with transferred 1D nano-/microstructures from Si-wafer to glass substrates via using a fracture-transfer printing technique, which have demonstrated the possibility of the fabrication of low-cost, transparent, flexible and high-efficient next generation 1D Si nano-/microstructures based next generation opto-electronic devices.

26.1 Introduction

Due to their capacity for interpreting basic physical concepts and building a wide range of functional optoelectronic devices, including light-emitting diodes (LEDs), solar cells, photodetectors and laser diodes at the nanoscale, one-dimensional (1D) inorganic nanostructures such as nanowires, nanopillars and nanotubes have attracted a great deal of research interest in recent years [1]. Their function as interconnectors and the key components in these devices is expected to play an important role in realizing high-performance next generation systems requiring some special characteristics such as high-speed, great flexibility, low-weight, low-cost, excellent light absorption capability, high storage-capacity, good thermal stability, high transparency and good thermal and electric conductivity [1, 2]. Photodetection and solar energy harvesting, in particular, have recently been among the most popular opto-electronic device applications of 1D nanostructured materials.

The finite supplies of conventional energy resources based on oil, natural gas and coal have motivated researchers to explore renewable energy resources like solar, biomass, geothermal and wind, as a substitute for fossil fuels. Among renewable energy options, solar energy has an immense potential to become a significant source of safe, abundant, clean, sustainable and cost-effective energy that can meet global energy demand in the near future. Today, one of the most powerful ways of harvesting energy from the sun is the use of solar cells, devices that capture light from the sun and convert it directly into electricity. Silicon (Si), the second-most abundant element in the earth's crust, is the most widely known material employed in solar cells, mainly due to its low cost, long lifetime, high efficiency and mature technology. In addition to traditional Si-based solar cells, the manufacture of highly efficient low-cost inorganic solar cells with a wide variety of device architectures and material combinations has also been explored in recent years. In particular, considerable research efforts have been devoted to implement new photovoltaic absorber materials that can embody earth-abundant and environmentally friendly constitute elements for the construction of low-cost and high-efficiency solar cells. It is a well-known fact that today's thin film solar cells are primarily based on materials such as $CuInSe_2$ (CIS), $CuIn_xGa_{1-x}Se_2$ (CIGS) and CdTe, consisting of rare, costly and toxic elements. This is considered to be as one of the main obstacles for the expansion, mass production and commercialization of thin-film solar cells. Therefore, the introduction of new photovoltaic absorber materials that can encapsulate earth-abundant, inexpensive and environment friendly constituent elements for the realization of low-cost, large-scale, and highly efficient thin film solar cells has been a major research endeavor in recent years. Kesterite Cu_2ZnSnS_4 (CZTS), a quaternary (I_2–II–IV–VI_4) compound having p-type conductivity with a high absorption coefficient (10^{-4} cm^{-1}) and ideal direct optic-bandgap energy (~1.45 eV) for the solar spectrum, has been recently proposed as a promising photovoltaic semiconductor that addresses almost all the issues related to the limitations of all the above materials. While a significant amount of work has been conducted so far on the development and characterization of CZTS thin films, relatively little attention has been paid to the realization of next-generation CZTS solar cells based on one-dimensional nanostructures. Therefore, we will primarily concentrate on our own studies on the realization of one-dimensional Si-nano-/microstructures based CZTS solar cells such as Si (wafer)/Si-nanowires (NWs)/CZTS, glass/Si-NWs/CZTS, and glass/Si-micropillars (MPs)/CZTS structured devices, in the first part of the present chapter. The manufactured solar cells are expected to combine the performance of the 1st generation solar cells with the advantages of the solar cells of the 3rd generation, allowing the realization of low-cost, environmentally friendly and high-efficiency solar cells."

Recently, several groups have demonstrated periodic photon trapping structures on silicon to convert normally incident beams of light by almost ninety degrees into laterally propagating modes of light along the plane of a silicon substrate [3–6]. Such surface structures are mostly1D nano/micro holes (these can be thought of as the inverse of a nanopillar or nanowire structures) and can not only increase the propagation length of light, contributing to more than an order of magnitude improvement in light absorption efficiency but also offer an opportunity to design solar cells and photodetectors with a very thin absorption layer. An added advantage is the enhanced speed of operation in photodetectors because of thin absorption layer. These devices can be designed to be flexible and offer opportunity for using commercially viable CMOS compatible processes.

New methods, architectures and materials, known to be an important tool for a wide variety of opto-electronic device applications such as photodetectors and solar cells, have been continuously proposed in photoelectric research to enhance the sensing performance of the devices [7, 8]. Today, in the application of environmental, military, and even biological fields such as flame sensing, emissions control, light-wave communications, and biological agent detection, photodetectors are considered to be very crucial instruments [9]. Over the last two decades, the use of one-dimensional nanostructures, such as nanowires, nanotubes and nanopillars, in devices has received much attention as an effort to boost the performance of photodetectors [10–16]. Incorporation of these nanostructures is expected to provide more favorable features, such as the decoupling of absorption and carrier collection direction, bandgap tunability, unique surface chemistry, effective light-trapping, efficient charge collection and separation, large surface-to-volume ratio, compared to their planar counterparts [1]. In the second part of this chapter, we will comprehensively review recent progress on photodetectors based on 1D nanostructures, with particular emphasis on our own recent studies based on photodetectors manufactured with

1D Si nano-/microstructures as well as the inverse of 1D nanopillar or nanowire structures or holes.

This chapter is divided into three sections. The first section gives a brief overview of the techniques used to fabricate arrays of one-dimensional Si-nano-/microstructures. This is followed by an overview of the recent studies on the opto-electronic device application of 1D ordered and disordered Si nano-/microstructures such as photodetectors and solar cells. In third section, future research opportunities and challenges for 1D Si nano-/macrostructures based solar cells and photodetectors are discussed.

26.2 Fabrication of One-Dimensional (1D)Silicon Nano-/microstructures

To date, a number of top down or bottom up synthesis routes, including wet-chemical etching, gas-phase growth, and dry etching, have been reported for the fabrication of arrays of 1D well-ordered and disordered Si nano-/microstructures with well-controlled geometries that can be employed for a wide range of electronic and opto-electronic device applications [17–21]. In this section of the chapter, recent advances in the fabrication of the arrays of these structures via wet-chemical etching and dry etching techniques are provided.

26.2.1 Dry Etching

Dry etching methods, like inductively coupled plasma reactive ion etching (ICP RIE), plasma etching and deep reactive ion etching (DRIE), instead of other methodologies are often embraced to synthesize 1D Si nano-/microstructures with a high degree of reliability, stability, homogeneity and controllability in terms of many parameters such as the crystallographic orientation, diameter, interspace, spatial arrangement and length [22]. In general, for this approach, an etch-mask transferred on Si-wafer with a desired pattern is required for the fabrication of regular and homogenous arrays of 1D Si nano-/microstructures with different aspect-ratios. Over the years, a number of routes have been suggested for the preparation of etch masks on Si-wafers to control the spacing, position and diameter of the resultant Si nano-/microstructures, such as patterning a photoresist, silicon nitrate of silicon and alumina (hard mask) via lithography, using anodized aluminum oxide (AAO), block copolymers (BCL) and self-assembled close-packed monolayer of nanospheres [23, 24]. Nanosphere lithography (NSL) is one of the most widely preferred techniques for the preparation of a large-scale etch mask [25, 26]. A typical process of patterning a layer via NSL to fabricate 1D Si nano-/microstructures is depicted in Fig. 26.1 [27]. Intense plasma is used to adjust the size of the nanospheres in a controlled manner. In this way, the

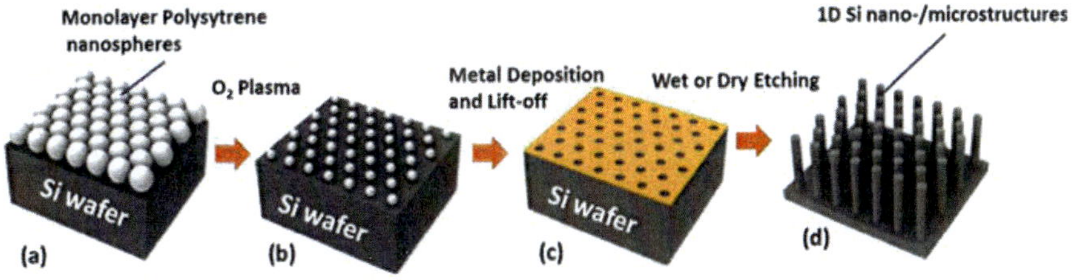

Fig. 26.1 The process flow for the fabrication of 1D Si nano-/microstructures using nanosphere lithography combined with a dry or wet etching technique: **a** Deposition of a monolayer polystyrene nanospheres onto the Si wafer, **b** size reduction of nanosphere mask via O_2 plasma, **c** a noble metal layer deposition and removal of nanosphere, and **d** nano-/microstructures formation via a dry or wet chemical etching process [27]

mask layer can then be modified for various configurations, which is subsequently carved into Si wafer in order to fabricate 1D nano-/microstructures with different length, geometry, diameter, density and spacing.

The ability to prepare the masks in the required direction will have a direct effect on determining the diameter, interspace and spatial distribution of the 1D nano-/microstructures to be synthesized. To transfer a pattern to a silicon wafer, the sample configured with the mask is exposed to the etching plasma consisted of accelerated electrons and ions, which subsequently removes the Si atoms from the wafer's surface either by chemical reaction with the reactive gases (e.g., SF_6 and C_4F_8) or momentum transfer of ions and yields 1D Si nano-/microstructures with specific diameter, geometry, length, density, interspace and aspect ratio.

By adjusting the applied RF (radio frequency) power and flow rate of the gases supplied into the chamber, the etching speed can be controlled, respective role of each is accelerating ions towards the material being etched and triggering radical-material chemical reactions.

To date, a considerable amount of literature has been published on the synthesis of ordered and disordered 1D Si nano-/microstructures using different forms of dry etching techniques [28–30]. Among them, deep reactive ion etching (DRIE), a process that combine physical and chemical effects, has attracted a special attention due to its repeatability, versatility, wide parameter tunability, process controllability, minimized undercut and high mask selectivity [31, 32]. It has been considered one of the most widely preferred dry etching technique in semiconductor industry, such as silicon based Micro-Electro-Mechanical Systems (MEMS), complex microfluidics devices and micro-optics [33, 34]. There are two major processes for the DRIE to achieve deep etches: cryogenic and Bosch. Figure 26.2 illustrates one cycle of a standard DRIE process flow for the BOSCH recipe for the production of Si nano/microstructures [35].

The Bosch process, as seen in the Fig. 26.2, is based on switching between depositing a protective layer on the sidewalls with a particular material like C_4F_8 (Teflon-like material) and etching the bottom part of the feature with a reactive gas such as SF_6, which enables the realization of deep and high aspect features needed

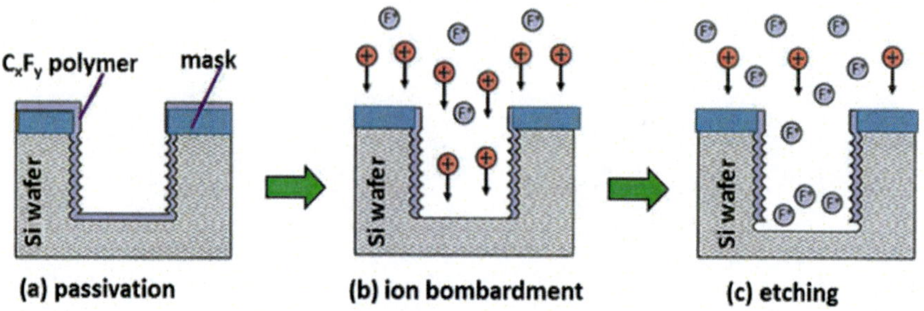

Fig. 26.2 A schematic one cycle of the BOSCH process for the production of Si nano/microstructures: **a** Passivation via a Teflon like polymer, **b** vertically directed ions to remove the passivation layer, and **c** etching phase [35]

for the manufacture of 1D Si nano-/microstructures based high performance electronic and optoelectronic devices. The full etch process requires multiple etching and passivation cycles with specific switching times to achieve 1D nano/microstructures with certain length, diameter and roughness. With the Bosch process, aspect ratios of 30:1 and etch rates of 10–20 nm/min are achievable [36, 37].

To date, DRIE has been extensively used to manufacture 1D Si nano-/microstructures for a wide range of applications [17–19]. For example, Cutarelli et al. fabricated 3D vertically-aligned Si micropillar arrays that could be employed as culture systems for human iPSC (Induced pluripotent stem cells)-derived cortical progenitors [38]. Recently, Mikulik et al. reported the synthesis of Si micropillars (MPs) arrays via combining photolithographic patterning with DRIE technique to study the effect of Al_2O_3 passivation on interface trap density of MPs using electrochemical impedance spectroscopy [39]. The findings revealed that Al_2O_3 passivated Si micropillars with optimum parameters could be used for high-performance solar cells and photodetectors. In another study, MP array was produced as an effective anti-reflection coating to reduce the visible light reflection from a radial p–n junction solar cell [40]. The results demonstrated that the constructed solar cell with Si MPs arrays exhibits a better short current density and open-circuit voltage compared to its planer counterpart, which was attributed to the reduced reflection in the wavelength range of 400–1000 nm. In a different study, the quality of the manufactured Si nanopillars synthesized by metal-assisted chemical etching (MACE) process was compared with those by DRIE process [41]. Optical lithography and DRIE were used by Oates et al. [42] fabricated arrays of vertical Si pillars for the manufacture of a model system for radial junction solar cells. The built solar cell based on MPs showed superior optical properties to its planer counterpart.

In our very recent study [20], ordered Si micropillars (MPs) were successfully fabricated using the DRIE technique. The study's main goal was to transfer MPs from a mother substrate (Si wafer) via a fracture-transfer printing technique to a number of carrier substrates that could be amorphous, transparent, or flexible while preserving the original hierarchy on the mother substrate. Figure 26.3a shows SEM pictures of the fabricated Si-MPs captured from various angles. The periodicity diameter and

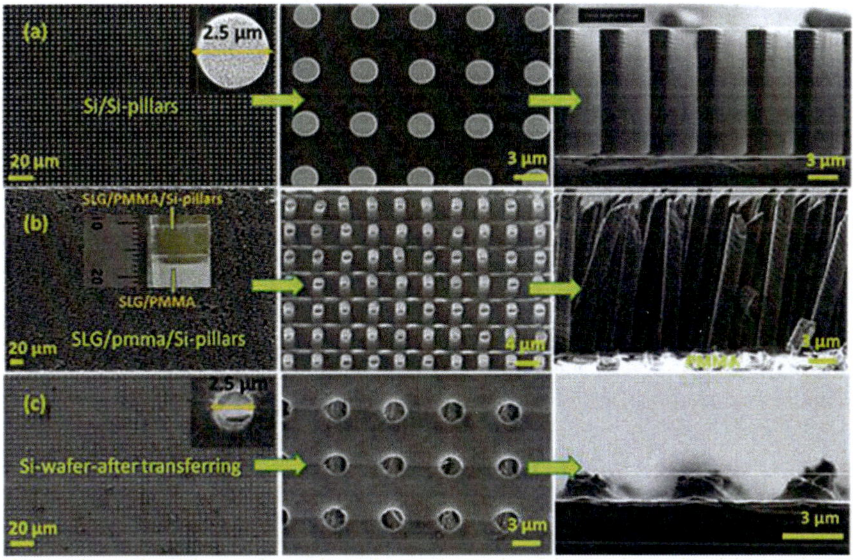

Fig. 26.3 The captured SEM images of ordered Si micropillars (MPs) fabricated using the DRIE technique. **a** The top and cross-sectional SEM images of the manufactured Si MPs on the Si-wafer. **b** The SEM images showing the transferred MPs on the SLG substrate. Inset image shows the photograph of the transferred MPs on the SLG substrate. **c** The Si-wafer following the transfer process [20]

length of the ordered Si-MPs were found to be 3 µm, 2 µm, and 16 µm, respectively, which can be seen in the recorded micrographs. Following the production of the pillars, they were detached from the Si-wafer using the transfer printing technique and successfully transferred onto an SLG substrate, as seen in Fig. 26.3b.The carrier substrates (soda-lime-glass (SLG)) were first cleaned using the standard cleaning procedures before initiating the transfer process. Following the cleaning process, PMMA (Polymethyl methacrylate) thin film with layers (~2 µm thick) was deposited onto the carrier substrates by spin coating technique.

The quality of transferred Si-NWs onto carrier substrates is determined by the thickness of the PMMA film, which plays a significant role in maintaining the local orientation, order, and alignment of MPs. If PMMA's thickness is less than a critical value, Si NWs may not retain its original hierarchy because of the poor adhesion between Si MPs and PMMA material. Therefore, the thickness of PMMA on the carrier substrate was adjusted accordingly based on the length and density of MPs on the mother substrate in order to preserve their vertical orientation and hierarchy throughout the transfer process. Si MPs on the mother substrate were then placed onto the carrier substrate without any vertical force being embossed. The combined mother-carrier substrates were subsequently heated above the glass transition temperature of the PMMA, which is around 160 °C [43]. Embossing forces ranging from 1 to 100 N were applied to the mother substrate in the vertical direction to penetrate the nanowires (NWs) into the PMMA polymer, which were determined based on the density of MPs on the mother substrate. The unified mother-carrier substrates were then cooled to room temperature. A lateral force was applied to the edge of the Si-wafer to separate Si MPs from the mother substrate, as detailed in our previous study [44]. The transfer of MPs from the Si-substrate has been successfully accomplished by maintaining their original order and orientation on the mother substrate, as can be seen from the recorded top and cross-sectional SEM photographs shown in Fig. 26.3b. Furthermore, it was revealed that the area transferred from the mother substrate was approximately 11 mm × 9 mm, as seen in the inset image in Fig. 26.3b. The SEM micrographs of the mother substrate following the transfer process were also obtained (see Fig. 26.3c). They revealed that the Si-MPs were perfectly broken from the roots during the transfer process, emphasizing the success of the conducted study. Furthermore, as seen in the SEM images, the transfer process can be carried out without producing too many defects, such as semi-broken pillars and non-transferred pillars on the mother substrate, which is a significant step forward in the manufacturing of next-generation core–shell structured semi-transparent and flexible solar cells on a variety of carrier substrates that can be flexible, amorphous, and transparent.

26.2.2 Wet Etching

Wet etching, unlike dry etching, is not a vacuum based approach; instead, it takes place in a liquid under atmospheric conditions, allowing for the fabrication of large scale Si nano-/microstructures in a cost effective way. The process is based on the conversion of solid substances into liquid compounds when immersed in a liquid solution by means of special chemical etchants, such as alkaline and acidic solutions. It is usually an isotropic and a highly selective process. Different combinations of HF (hydrofluoric acid)-HNO_3 (nitric acid) and HF-NH_4F (ammonium fluoride) are the most extensively utilized etchants for Si and SiO_2, respectively. In general, for the manufacturing of 1D Si nano-/microstructures, two kinds of wet-etching techniques are widely employed: electrochemical wet etching and metal-assisted chemical etching (MACE) [45]. Among them, MACE technique has recently drawn particular attention because of its simplicity, low-cost and versatility [46]. It can be employed not only for the realization of 1D Si nano-/microstructures, but also the fabrication of more complicated structures, which are crucial for the fabrication of numerous applications such as 3D (three dimensional) accelerometers, square diaphragms and comb-drive resonators [47].

At the beginning of MACE process, a patterned metal layer, like Ag, Au, Pt and Cu, is deposited onto a planar Si wafer via a number of etch masks such as a monolayer polystyrene nanospheres with hexagonal close-packed arrays, heat-deformed colloidal monolayers and the periodic nanoparticle patterns. After that, the Si wafer decorated with a patterned noble metal layer is immersed into an aqueous solution containing HF and an oxidant. The patterned metal layer on the Si wafer acts as an effective catalysis for the MACE process, which induce local oxidation and reduction reactions [48]. In other words, metals act as local cathode to catalyze the

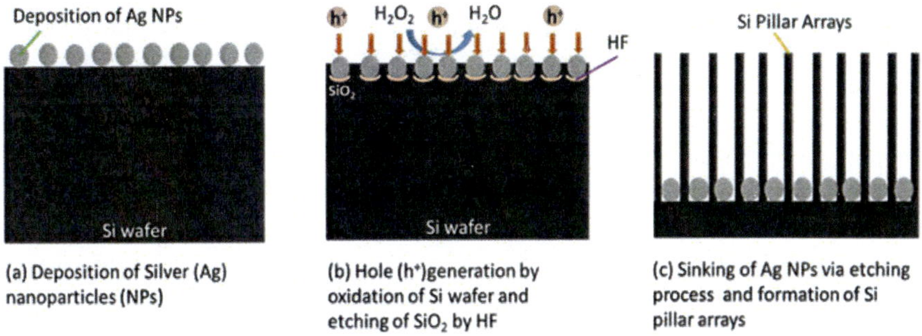

Fig. 26.4 Overall metal assisted chemical etching process for the fabrication of Si pillar arrays [49]

reduction process of oxidants such as H_2O_2 and generating holes. The generated holes are then injected into the Si semiconductor, oxidizing it before being etched away by the HF in the solution. The Si material beneath the metal is etched much faster than the uncoated bare Si regions because of the catalytic activity of the metal. In the end, the metal sinks down the Si wafer as a result of the selective etching, and the transferred metal pattern is engraved into the wafer [48]. By changing the etching mask pattern and growth parameters, such as reaction temperature, precursor concentration, and etching time, 1D Si nano-/microstructures of various lengths, geometry, diameters, and interspaces can be manufactured. A typical MACE process for the fabrication of Si pillar arrays is shown in Fig. 26.4, which is derived from [49].

The MACE technique has been extensively used in recent years to fabricate one-dimensional Si-nano/microstructures arrays. For instance, in an effort to build stretchable Si-organic hybrid solar cells, S.-S. Yoon and D.-Y. Khang successfully produced ordered vertical Si MPs arrays with diameters and lengths of 5–25 μm and 30–40 μm, respectively [50]. The findings showed that many process parameters, such as the etching time and the pattern dimension transferred on the Si wafer, could be used to easily control the length, interspacing, and diameter of the MPs. The study by H. Lin et al. [51] reported the fabrication of Si nano-/microstructure arrays using the MACE technique to construct radial junction solar cells. The effect of different catalysts, like Au, Ag and Ag/Au, used in the MACE process on the quality of the grown Si nano-/microstructures was examined in detail. The results revealed the advantages of using an Ag/Au bilayer as catalyst in the MACE process to produce high-performance solar cells with shape-controlled Si nano/microstructures. In another study, p-type 3D Si micropillars (MPs) with lengths ranging from 20 to 25 μm were produced via the MACE process to investigate the effect of porous Si/SiO_x shell on the characteristics of Si MPs based optoelectronic device applications [52]. In a different study, densely-packed Si MPs were successfully fabricated via the MACE technique to examine the effect of Au contamination on the performance of 3D radial junction solar cells [53]. The study concluded that cost-effective and high-performance solar cells could be manufactured with low-quality materials if they were constructed with a 3D radial junction architecture. In another study, the same group used the MACE process to fabricate smooth micropillar arrays with high aspect ratios for the manufacture of radial junction solar cells [54]. The effect of EtOH (ethanol) and HF concentration on the quality of the developed Si MPs was investigated in depth in that analysis, and it was revealed that the presence of EtOH and a high concentration of HF were essential in order to achieve Si MPs with smooth sidewalls.

Recently, we synthesized disordered Si NWs via electroless etching (EE) technique for the fabrication of a third generation solar cell device structure with n-Si-NWs/p-CZTS material combination [21]. The EE technique was particularly preferred for the production of nanowires as it is a simple and low-cost wet etching process that enables wafer-scale production of vertically well-aligned Si NWs. The EE technique offers a direct control over a number of physical parameters of the synthesized Si nanowires, like length and density, by adjusting growth parameters such as precursor concentration, reaction temperature, and growth time. Top-view and tilted-view SEM images of the fabricated Si nanowires (NWs) using the EE technique are seen in Fig. 26.5. The captured micrographs show an approximately homogeneous etching profile with a depth of 3.5 μm. In addition, it can be deduced from the images that the average diameter of the produced nanowires (NWs) varies from 20 to 80 nm.

In 2020, we published a study aiming at fabricating disordered and well-ordered one-dimensional Si nanostructures (nanowires and pillars) using a method that combined nanosphere lithography with the metal assisted chemical etching (MACE) technique in order to construct CZTS (Cu_2ZnSnS_4)-based semi-transparent, flexible, and lightweight solar cells [20]. The electroless etching (EE) technique was chosen for the fabrication of disordered Si nanowires (NWs). As mentioned earlier, this technique is advantageous because it allows Si-NWs to retain the bulk crystalline properties of the Si from which they are developed and the fabrication of nanowire (NW) arrays at large scale. The disordered Si-NWs were produced using an n-type Si wafer (100) with an electrical resistivity of 10Ω.cm as the starting material. The n-type Si wafer was subjected to a mixture solution of $AgNO_3$/HF for a specific time to fabricate n-type Si NWs. The fabrication of Si NW arrays was achieved by

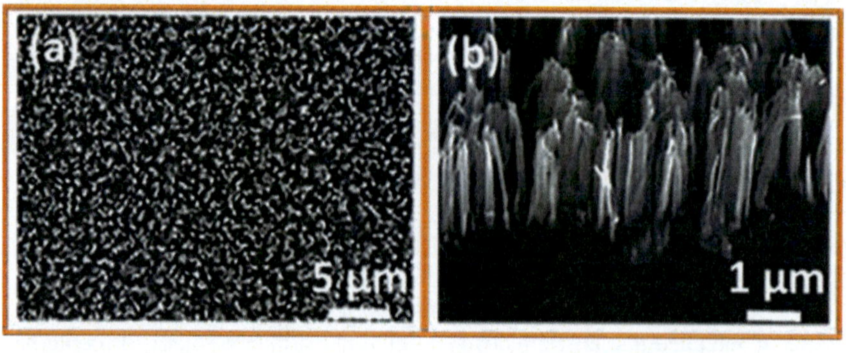

Fig. 26.5 **a** Top-view and **b** tilted-view SEM images of the fabricated Si NWs using the electroless etching (EE) technique [21]

local oxidations by Ag ions adsorbed onto the wafer and a subsequent HF etching process for the oxidized Si, as described elsewhere [21, 55–57]. Following nanowire synthesis, they were printed on Ag-paste coated glass and copper-foil substrates using the fracture-transfer printing technique, the details of which are given elsewhere [20].

Figure 26.6a shows the SEM micrographs of disordered Si NWs synthesized using the EE technique. It can be deduced from the images that the length and density of the synthesized Si NWs are around 8.0 μm and 5×10^7 NWs/mm^2, respectively. The Si-NWs were then successfully transferred from the Si-wafer to the 1st carrier substrate (Soda-lime-glass (SLG)) while preserving the hierarchy of the NWs on the wafer, as shown in Fig. 26.6b. The SEM image of the mother substrate was analyzed again following the transfer process (see Fig. 26.6c), which clearly revealed that the vast majority of NWs was successfully fractured from the roots of the NWs located on the mother substrate. Figure 26.6d shows the SEM images captured for the Si NWs transferred from the 1st carrier substrate to an Ag (silver)-paste (5 μm thick) pre-coated SLG substrate. In terms of the characteristics of the Si NW packs, Fig. 26.6a, b reveal that the bottom and top portions of the Si NWs do not have the same appearance. However, as seen in Fig. 26.6d, the top–bottom-top switching processes result in the emergence of the original Si NWs synthesized on the mother substrate at the final phase of the process, highlighting the success of the transfer operation.

The constructed SLG/Ag-paste/n-Si NWs structure was employed for the realization of core–shell structured CZTS based solar cell on an amorphous substrate (SLG). In addition to an amorphous substrate, the Si NWs were also successfully transferred to flexible copper-foil substrates using the same stages, as seen in the SEM images provided in Fig. 26.7. It can be deduced from the images that the transfer process to the Cu foil substrate was completed in acceptable proportions by preserving the original NW order on the mother substrate. The electrical resistance between the Cu foil and the tips of the transferred Si NW arrays was determined to be around 300 Ω, suggesting the existence of a physical connection between the bottom part of the

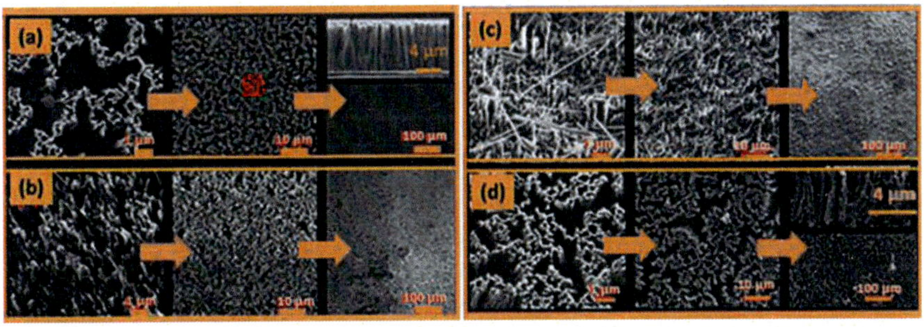

Fig. 26.6 The SEM micrographs captured for **a** the fabricated Si-NWs on the Si-wafer using the electroless etching technique, **b** the transferred Si NWs on 1st carrier substrate (SLG), **c** the Si-wafer following the transferring operation, and **d** the transferred Si-NWs on the Ag-paste pre-coated SLG substrate [20]

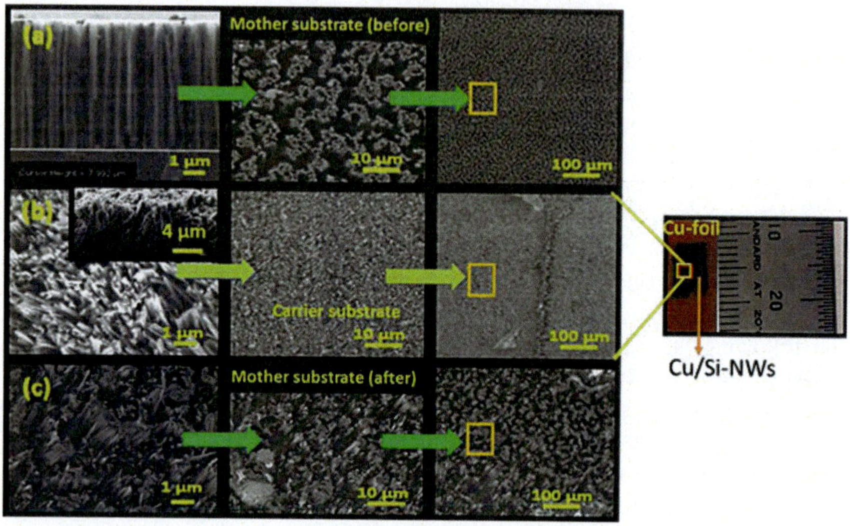

Fig. 26.7 The top and cross-sectional SEM micrographs captured for **a** the fabricated disordered Si-NWs on the Si-wafer, **b** the transferred Si NWs on Cu foil carrier substrate, and **c** Si-wafer after the transferring operation. Inset in **b** shows the photograph of the transferred NWs on the Cu-foil [20]

nanowire (NW) and the Cu-foil. Due to the high flexibility of the Cu-foil, which prevents the formation of homogeneous lateral force during the transferring process, some nanowires (NWs) were not fractured from the bottom section, as can be seen in Fig. 26.7c.

For the realization of high-aspect ratio ordered one-dimensional (1D) Si-nano-/microstructures, nanosphere lithography technique (NSL), a simple and cost-effective method, was combined with the MACE method, stages of which are presented in Fig. 26.8. Figure 26.8a shows a monolayer of polystyrene spheres with a diameter of 460 nm deposited on an n-type Si wafer via spin coating process. As shown in Fig. 26.8b–e, the polystyrene nanospheres were then exposed to O_2 (oxygen) plasma, formed at various powers for specific time intervals, so as to reduce their diameters. Figure 26.8d illustrates how a 70 W power and a 60 s time interval were found to be sufficient to reduce nanosphere diameters from 460 to 300 nm. Furthermore, as can be seen in Fig. 26.8e, spheres with smaller diameters could be achieved by either increasing plasma power or etching time.

Thermal evaporation was used to deposit a thin film of Au (20 nm) on the nanosphere-layer following the size reduction (see Fig. 26.8f). Sonication in toluene was used to remove the Au-decorated nanospheres, resulting in a patterned Au layer on the Si wafer, as seen in Fig. 26.8g. The MACE process was eventually conducted by dipping the Si wafer with the patterned metal layer in a solution containing (H_2O_2: HF: H_2O; V:V:V; 1:4:5) for a few minutes at room temperature. As seen in Fig. 26.8h, aforementioned process steps can be used to fabricate well-aligned Si-nanopillar (NPs) of the predefined diameters and lengths. By changing certain

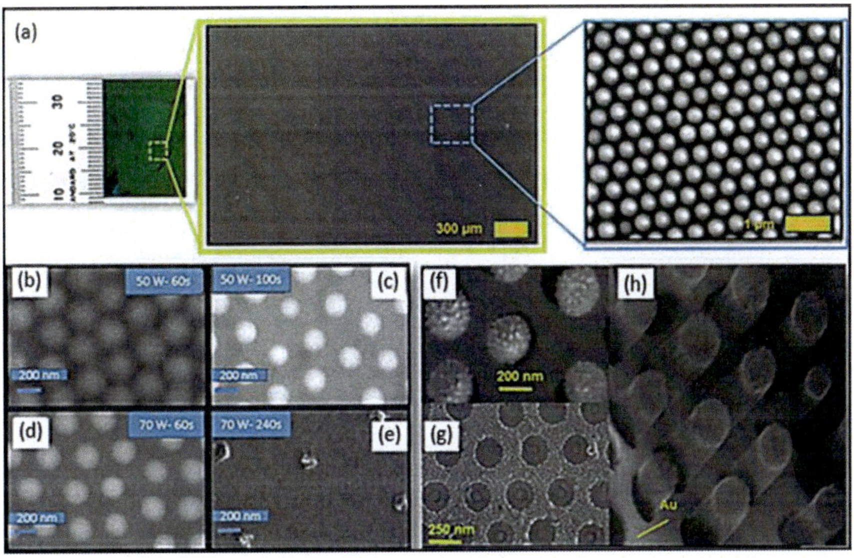

Fig. 26.8 Si nanopillar (NPs) fabrication steps by NSL technique. **a** Photograph and top-view SEM images illustrating hexagonally close packed monolayer of polystyrene nanospheres (etching mask) with diameter of 400 nm on an n-type Si-wafer surface. **b–e** SEM images of monolayer nanospheres subjected to O_2 plasma treatment for different powers and periods of time. **f** A thin film of Au (20 nm) deposited on nanospheres. **g** The recorded SEM image following the extraction of etching mask (monolayer nanospheres) from the Si-wafer. **h** Tilted-view SEM image of ordered arrays of Si NPs fabricated by the MACE process [20]

growth parameters during the growth stage, it is possible to fabricate ordered arrays of Si nanopillar (NPs) in a variety of diameters and lengths. The etching time is one of these growing parameters that can be used to adjust the length of NPs. For example, to investigate the effect of etching time on nanopillar length, the MACE process was carried out at room temperature over time intervals of 30 to 90 s for the Si-wafer decorated with a monolayer polystyrene nanospheres of diameter 300 nm. The length of the NPs increases with increasing etching time, reaching a length of ~1 μm for 90 s etching duration, as can be deduced from the recorded SEM micrographs presented in Fig. 26.9.

Using the same process steps, Si micropillars (MPs) with dimensions ~8 μm in diameter and 300 nm in length were also successfully obtained, as shown in Fig. 26.10a. Following the transferring steps outlined above, the grown ordered Si micropillars (MPs) were successfully moved onto the ITO (indium-tin-oxide) and Ag-paste coated glass substrates (see Fig. 26.10b–e), while holding their initial pattern on the Si wafer. This achievement will undoubtedly pave the way for the manufacture of third-generation ordered arrays of Si MPs-based solar cells on wide range of substrates, including the amorphous, transparent and flexible ones, in the near future.

As can be seen in Fig. 26.11, the synthesis of ordered arrays of Si pillars micropil-

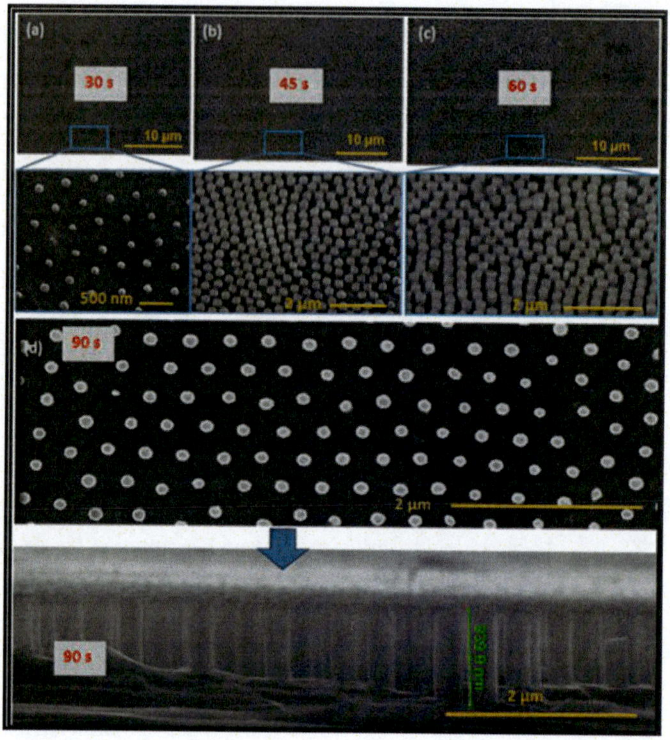

Fig. 26.9 The SEM images captured for the ordered arrays of Si nanopillar (NPs)fabricated by the MACE process conducted at different etching times: **a** 30 s, **b** 45 s, **c** 60 and **d** 90 s. **e** Cross-sectional SEM image of the Si NPs obtained following 90 s etching time at room temperature [20]

lars (MPs) with extremely large diameters was also accomplished within the context of the study by following the same stages of the route of MACE process coupled with the NSL. The recorded SEM micrographs provided in Fig. 26.11a, b revealed that the diameter and length of the micropillars (MPs) were around 2 µm and 7.5 µm, respectively. The initial diameter of the polystyrene spheres used to prepare the etch mask for the MACE process was 3 µm. However, following an oxygen plasma treatment at 70 W for 60 s, their diameters were reduced to 2 µm. It is also apparent from the top-view SEM images given in Fig. 26.11b that the KOH (potassium hydroxide, silicon etchant) etching solution can be employed to reduce the diameters of the Si MPs even more. The Si micropillars (MPs) were subsequently transferred onto the Ag-paste pre-coated glass substrates using the above-mentioned transfer process by preserving the initial hierarchy that existed on the Si wafer, as seen in the top-view and tilted-view SEM pictures provided in Fig. 26.11b. The transfer of MPs from the Si wafer to the Ag paste pre-coated glass substrate was a huge success and achieved for the first time in this work.

Fig. 26.10 Transfer of Si nanopillar (NPs) onto the ITO and Ag-paste coated SLG carrier substrates. **a** Top-view SEM image of the synthesized NPs with diameter of 300 nm and length of ~8 μm in length on the Si-wafer using the MACE process combined with the NSL. Transfer of NPs via the transferring material (PMMA) onto **b**, **c** the ITO and **d**, **e** Ag-paste coated SLG substrates [20]

26.3 Optoelectronic Device Applications

One-dimensional (1D) Si nano-/microstructures arrays can be employed for a wide range of opto-electronic device application owing to their highly developed technology and outstanding features such as high thermal conductivity, facile doping control, hardness and excellent optical and electrical properties. In this section, however, the recent studies on 1D Si nano-/microstructures based solar cells and photodetectors will be discussed.

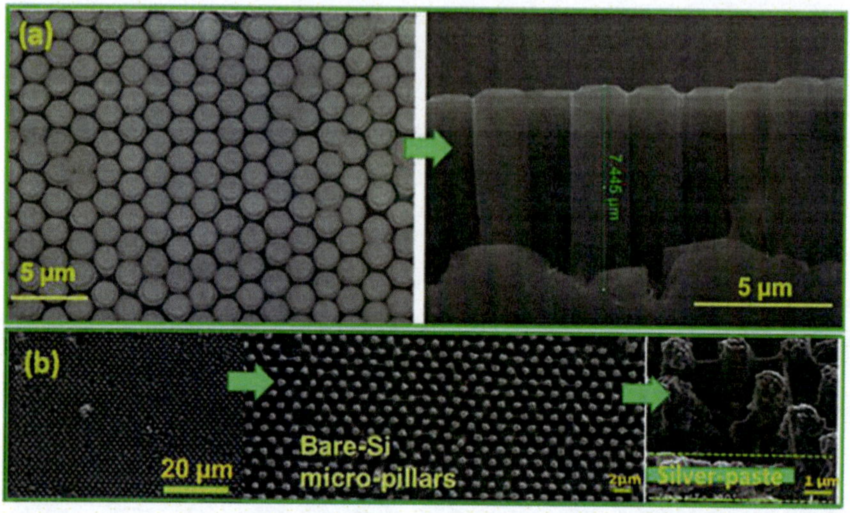

Fig. 26.11 The SEM micrographs obtained for the ordered arrays of Si micropillars (MPs) fabricated via an etch mask consisted of polystyrene microspheres (~3 μm in diameter). **a** The top-view and cross-sectional SEM images of the ordered arrays of Si MPs on the Si wafer produced by using the MACE process coupled with the NSL. **b** Top and tilted-view SEM images captured for the ordered arrays of Si MPs transferred onto Ag-paste pre-coated glass substrate [20]

26.3.1 Solar Cells

If designed with the right architectures, like radial p–n junction, and optimized structural parameters such as diameter and array periodicity, one-dimensional Si nano-/microstructured solar cells will combine the benefits of first and second-generation solar cells to realize both high power conversion efficiency and low-cost goals in solar cells. Therefore, there has been an increase of interest in the fabrication of these structure-based solar cells in recent years, which offer lower optical reflection, less material consumption, better absorption, and superior charge collection compared to the traditional planer bulk Si systems such as crystalline-Si wafer and polycrystalline/multicrystalline/amorphous-film based ones, which are dominating the photovoltaic market [58].

Recently, a Si/InP structured Si nanowires based heterojunction solar cell was reported in [59]. In order to construct the structure, p-type Silicon NWs arrays on Si (100) were decorated with an n-type InP layer using atmospheric pressure metal organic chemical vapor deposition. The open-circuit voltage (V_{oc}), short-circuit current density (J_{sc}) and power conversion efficiency (η) were found to be 0.56 V, 14.26 mA/cm^2 and 4.39%, respectively. The findings revealed that the core–shell radial heterojunction could provide a significant improvement in efficiency as compared to the traditional nanowire-based solar cells, which was attributed to the effective light trapping and better absorption ability of the cell in core–shell arrays. Singh et al. [40] constructed Si micropillars (MPs) based radial-junction solar cells as an attempt to improve the charge carrier collection and light management. The periodicity, length and diameter of the ordered Si MPs arrays with vertical sidewalls, produced via the DRIE technique, were 1.5–3.3 μm, 8 μm and 4 μm, respectively. The constructed device exhibited a 27% increase in J_{sc} as compared to its planar counterpart, which was ascribed to the improvement in the extraction of photo-generated carriers through the formed radial p_n junction using Si MPs. In addition, the recombination losses were minimized in the device by planarizing Si/metal interface to improve the V_{oc} parameter. The resulting cell exhibited a V_{oc} of 0.62 V, the highest voltage reported so far for the MPs based solar cells. In another study, as a proof-of-concept demonstration, stretchable bifacial Si-organic hybrid solar cells based on Si MPs were fabricated by embedding Si MPs into polydimethylsiloxane (PDMS) [50]. The cell made up of MPs with a diameter of 5 μm exhibited the highest efficiency, 3.3%, and did not display any performance degradation even when stretched by 40%. The preliminary results suggest that the same approach can be employed for the realization of large scale wearable energy-harvesting and attachable personal health tracking devices. In 2016, a radial junction solar cell based on Si micro-/nano hybrid structures, using the MACE process, with an efficiency of 17.6% was successfully fabricated [60]. Owing to the peculiar design of the produced micro-nano hybrid structure, the cell had a very high short-circuit current density (39.5 mA/cm^2), which resulted in a substantial increase in efficiency, 61.5%, over a planer Si solar cell owing to its excellent light absorption capacity (97%). The results indicated that the manufactured hybrid structure through growing nanostructures on top of microwire arrays could be used for the production of high-performance radial junction solar cells. Baytemir et al. [54] demonstrated a 3D radial junction solar cell based on ordered Si MPs with high aspect ratios and low sidewall-roughness for the manufacture of high-performance solar cells. A power conversion efficiency of 17.26% has been recorded, which is one of the highest values reported to date for Si-MPs based radial junction solar cells. It was realized that solar cells with higher efficiency could be accomplished by lowering the surface recombination at the interface and series resistance of the cells with a set of optimization process, including the passivation of the MPs with SiO_2/SiN_x and improving the quality of metal contacts. In a different study, highly ordered and well-aligned Si nanopillars (NPs) were employed for the realization of axial p–n junction solar cells [61]. An approach combining the use of colloidal particle self-assembly with cryogenic silicon plasma etching process was chosen for the production of Si NPs. The V_{oc}, J_{sc} and FF of the p^+n junction solar cell constructed with an axial architecture were determined to be 0.37 V, 18.9 mA/cm^2 and 0.59, respectively. The Methyl-termination and H-termination processes were also found to be very effective in preventing leakage current by providing good and stable electrical passivation for Si NPs. In 2018, Ioannis Leontis et al. [62] investigated the effect of Si nanowires (NWs), grown on the front solar cell surface via the MACE technique, as light-trapping material on the performance of crystalline Si solar cells. The cells were constructed with nanowires (NWs) having different lengths, such as 0.5, 1, and 1.5 μm. The highest efficiency, 13.4%, was recorded for the case of 1 μm without any passivation layer, which was attributed to the lower light reflection and decreased density of states compared to the other two NW lengths. In addition, the performance of the cell decorated with 1 μm NWs was compared to that of a reference planar Si solar cell, revealing that the use of NWs increases power conversion efficiency by 45%. In 2020, Myunghae Seo et al. [63] successfully synthesized hourglass-shaped silicon nanowires (HG-Si NWs) using inductively coupled plasma reactive ion etching for the realization of solar cells that can achieve light absorption over the entire solar spectrum and overcome the limits of the conventional bulk Si solar cells. The results revealed that only adjusting the diameter of the HG-Si NWs would increase the J_{SC} and result in a 1.85 times higher power conversion efficiency as compared to the reference cylindrical Si NW-based solar cells due to their excellent absorption capability. In 2017, Pankaj Yadav et al. [64] conducted a detailed study on the manufacture of large-scale ordered Si micropillars based solar cells. The fabricated solar cell had a power conversion efficiency of 16.90%, which was the highest ever recorded for Si solar cells based on nano-/microstructures. The recorded high efficiency was mainly ascribed to the improvement in photocurrent and voltage following the interfacial engineering of the cell. The findings indicated that the microstructured Si solar cell could have better advantages compared to its planer counterpart in terms of production costs and scalability, both of which are critical for the widespread adoption of these devices. In another study, Hsin-Ping Wang et al. [65] synthesized wafer-scale Si micro-pyramids and nanowire arrays using a time-effective, simple and low-cost fabrication technique (electroless etching) for the manufacture of cost-effective and high-performance nano-/microstructured Si solar cells. Without any additional process such as the inclusion of passivation or anti-reflective layers, the solar-cell with optimized structure exhibited excellent light-harvesting ability with negligible electrical losses, resulting in a 10.48% power conversion efficiency. In 2017, Hong-Sik Kim et al. [66] reported a record high-efficiency of 16.92% for the Si microwires (MWs) based solar cells. The effect of NW parameters including diameter and periodicity on the operation and electrical properties was studied in detail. The performance of solar cells was compared with that of conventional planar-Si solar cells. The findings showed that incorporating Si MWs into the device structure boosts the V_{oc} of the cells by increasing the shunt resistance as compared to the standard flat-Si solar cell. In a recent paper by Laxmikanta Karmakar and Debajyoti Das [67], a solar cell based on n-type Si nanopillars, grown via PTNCC (p-to-n type conductivity conversation) process, with a proof-of-concept power conversion efficiency of 4.22% was achieved. In the constructed solar cell structure, the ITO (indium-tin-oxide) and Ag paste were employed as front and back electrodes, respectively. Melvin David Kumar et al. [68]synthesized ordered Si pyramids via wet etching process as an attempt to manufacture high-performance Si solar cells. Melvin David Kumar et al. [68]used a wet etching route to synthesize ordered Si pyramids in an effort to manufacture high-performance Si solar cells. The ITO decorated ordered Si pyramids, derived from Si pillars via wet etching, was found to be very effective in the absorption of photons with long-wavelengths. The J_{sc} and power conversion efficiency of the pyramid patterned solar cell were calculated to be 36.2 mA/cm^2 and 16.3%, respectively.

Very recently, we reported a study on an n-Si NWs/p-Cu_2ZnSnS_4 (CZTS) core–shell solar cell by using disordered Si nanowires (NWs) synthesized via electroless etching (EE) technique [21]. The aim of study was to fill a gap in the literature by examining the photovoltaic performance of a sol–gel derived CZTS thin film in a core–shell solar cell device configuration for the first time. The constructed hybrid structure was expected to combine the performance of first-generation solar cells, like high efficiency and mature technology of silicon, with the advantages of third-generation solar cells such as an excellent light trapping, enhanced charge collection and large surface-to-volume ratio, which allows for the manufacture of low-cost and high efficiency Si nano-/microstructures based solar cells. A sol–gel derived thin layer (600 nm) of CZTS thin film was used to cap the synthesized Si NWs, 20–80 nm in diameter and ~3.5 μm in height, as the shell component of the core–shell device architecture. A thermally evaporated 60-nm thick silver dot contacts and 150-nm thick silver thin film layer were assigned as the ohmic top-contact and a back contact of the solar cell, respectively, schematic representation of which is provided in Fig. 26.12.

Figure 26.13 demonstrates cross-sectional and tilted-view SEM images of the synthesized bare Fig. 26.13a, b and CZTS-coated Fig. 26.13c, d Si NWs. The SEM images reveal that vast majority of the NWs are completely decorated with the CZTS layer both longitudinally and radially, suggesting a high contact interface between Si-NWs and the CZTS absorber layer.

The current (I)-voltage (V) characteristic of the cell under the AM 1.5 G (AM1.5G stands for the standard spectrum at the Earth's surface) illumination is given in Fig. 26.14, from which the open circuit voltage (V_{oc}), short circuit current (I_{sc}), fill factor (FF) and power conversion efficiency (η) were found to be 0.40 ± 0.01 V, 9.60

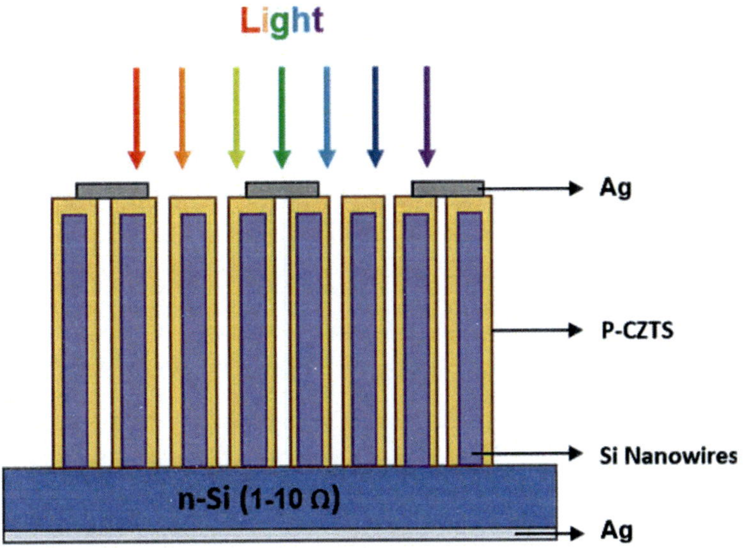

Fig. 26.12 Schematic representation of the manufactured Ag/n-Si/n-Si NWs/p-CZTS/Ag core–shell solar cell [21]

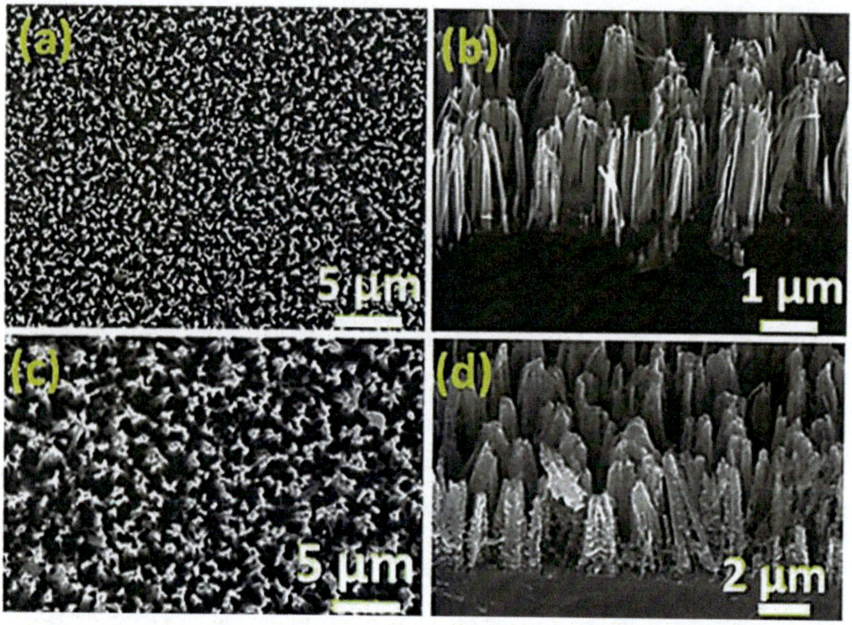

Fig. 26.13 Cross-sectional and tilted-view SEM images of the synthesized (a-b) bare and (c-d) CZTS-coated Si NWs so as to construct the n-Si NWs/p-CZTS core–shell solar cell [21]

± 0.20 mA/cm^2, 26 ± 1% and 1.0 ± 0.2%, respectively. Since these are the first findings reported for an n-Si-NWs/p-CZTS core–shell solar cell, it is impossible to compare them with those published in the literature. Nonetheless, as compared to traditional planer CZTS and CIGS thin film solar cells, the achieved power conversion efficiency (1%) is clearly low. A solar cell with a Mo/p-CZTS/n-Si/Al planer heterojunction configuration [69] has exhibited the highest power conversion efficiency reported so far. The V_{oc}, J_{sc}, and η parameters of the cell were calculated to be 0.52 V, 3.28 mA/cm^2 and 1.13%, respectively. Despite the fact that the cell's efficiency is slightly higher than that accomplished with current study, the J_{sc} is significantly lower than that obtained with the NW-based device configuration. The observed enhancement in the J_{sc} for the cell with Si nanowires (NWs) configuration compared to its planar counterpart can be attributed to unique features of Si NWs such as excellent light trapping, enhanced charge collection and large surface-to-volume ratio.

The reflectance spectra of CZTS thin films deposited on Si NWs and planer-Si wafers are shown in Fig. 26.14b. As can be deduced from the inset figure, the average reflectance measured for the Si-NWs coated with CZTS is between 0.01% and 0.3%. The reflectance measurement for the CZTS thin film deposited on the planer Si wafer was also carried out as reference, which showed higher values for the same wavelength range, varying from 20 to 40%. The recorded extremely low

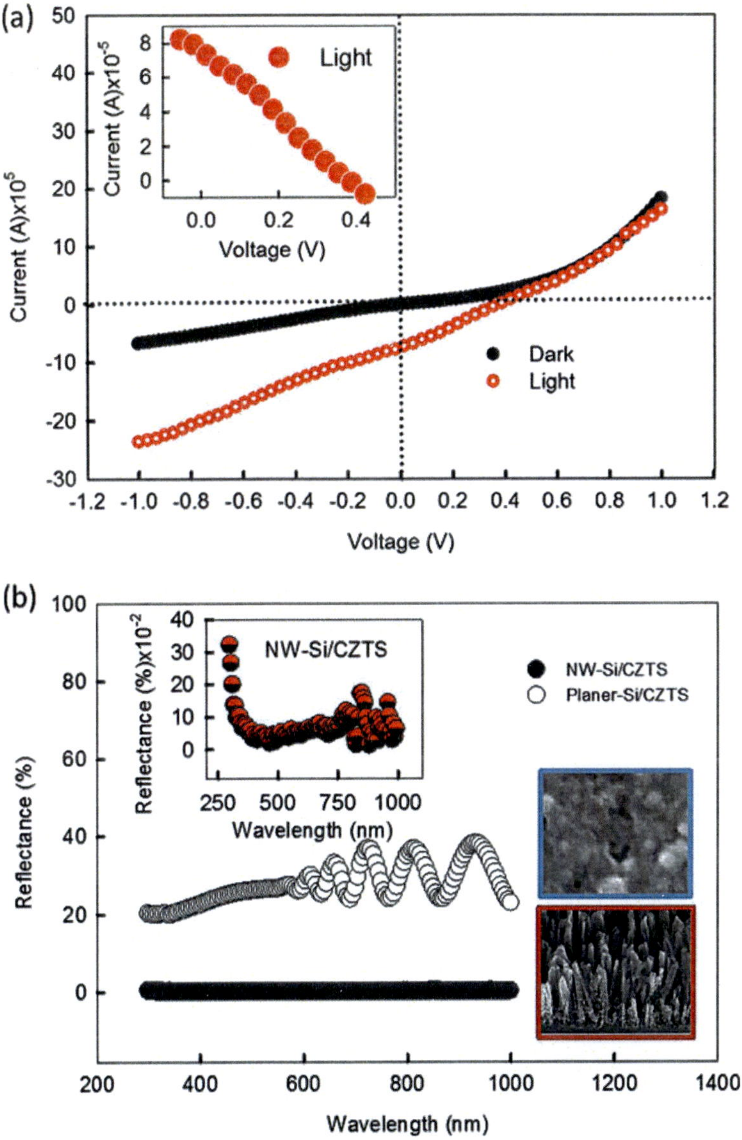

Fig. 26.14 **a** The current–voltage (I-V) characteristic of the n-Si-NWs/p-CZTS core–shell solar cell recorded under dark and an illumination of 100 mW/cm^2. The inset plot shows the I-V curve under light in the first quadrant. **b** The spectral reflectance measured for the disordered arrays of Si nanowires (NWs) based solar cell and its planer equivalent. The inset plot presents the zoomed-in spectral reflectance of the device constructed with the Si NWs. The inset pictures display the SEM micrographs captured for both device configurations [21]

reflectance for the solar cell with Si NWs configuration can thus be related to the measured higher J_{sc} in the present study as compared to its planer equivalents.

In 2020, we conducted a study to manufacture Si nano-/microstructures based core solar cells on low-cost, lightweight and amorphous surfaces by transferring them from Si wafer onto a number of substrates through a transfer printing method, as stated earlier [20, 70]. Following the synthesis of 1D ordered and disordered Si nano-/microstructures using the MACE process in conjunction with the NSL, the fracture-transfer printing technique was used to transfer them onto Ag-paste pre-coated glass substrates. The disordered Si nanowires (NWs) and ordered arrays of Si micropillars (MPs) transferred onto the Ag-paste pre-coated glass substrates were then decorated with 600 nm thick e-beam (electron-beam) evaporated p-type Cu_2ZnSnS_4 (CZTS) thin films for the realization of the SLG/Ag-paste/n-Si-nano/microstructures/p-CZTS structured core–shell solar cells. The structural, electrical, and optical properties of CZTS thin films obtained in a single step via the e-beam technique have been described in detail elsewhere [71]. To complete the solar cell device structure constructed on the Ag-paste coated glass substrate, a thin film of ITO (indium-tin-oxide, 100 nm) and Al (120 nm) as top contacts were deposited on the CZTS decorated Si nano-/microstructures using the DC-sputtering technique.

The first prototype solar cell was constructed via the 1D Si micropillars fabricated using the MACE technique in conjunction with the NLS, the diameters of which were further reduced using a KOH etching solution, as described in the wet etching part. Figure 26.15a shows top and tilted-view SEM pictures captured for the CZTS coated Si MPs transferred onto the Ag-paste coated glass substrate. From the images, it can be observed that the Si MPs are continuously covered with the CZTS layer. The I–V characteristics of the built SLG/Ag-paste/n-Si MPs/p-CZTS/ITO/Al solar cell recorded in the dark and under light illumination are seen in Fig. 26.15b. The formation of a p-n hetero-junction between the n-Si MPs and CZTS film was revealed by the observed bias-polarity dependent rectification of the current in the plots. The V_{oc}, J_{sc}, FF and η of the solar cell were found to be 0.42 V, 7.93 mA/cm^2, 26% and 0.87%, respectively. This study is the first to report on the performance of the solar cell in the SLG/Ag-paste/n-Si MPs/p-CZTS/ITO/Al device structure. The efficiency measured is substantially higher than those reported in the literature for solar cells with similar configurations, which range from 0.1% to 0.46% [72, 73]. Even higher efficiencies can be achieved with this device configuration by optimizing certain manufacturing processes such as improving the quality of the CZTS layer and applying surface passivation processes to the Si MPs.

By coating the Si nanowires (NWs) transferred onto the Ag-paste pre-coated glass substrate with a 600 nm thick electron-beam evaporated CZTS thin layer, the SLG/Ag-paste/n-Si-NWs/p-CZTS/ITO/Al structured core–shell solar cell was also manufactured as an application of the transferred disordered Si NWs. As stated earlier, the Al dot-contacts and Ag-paste layer were assigned as the top and bottom contacts of the constructed solar cell, respectively. The SEM images of the synthesized disordered Si NWs on the Si wafer and those transferred onto the Ag-paste pre-coated SLG glass substrate, coated with the CZTS layer, are shown in Fig. 26.16. The captured SEM micrographs clearly demonstrate that the Si-NWs were not only

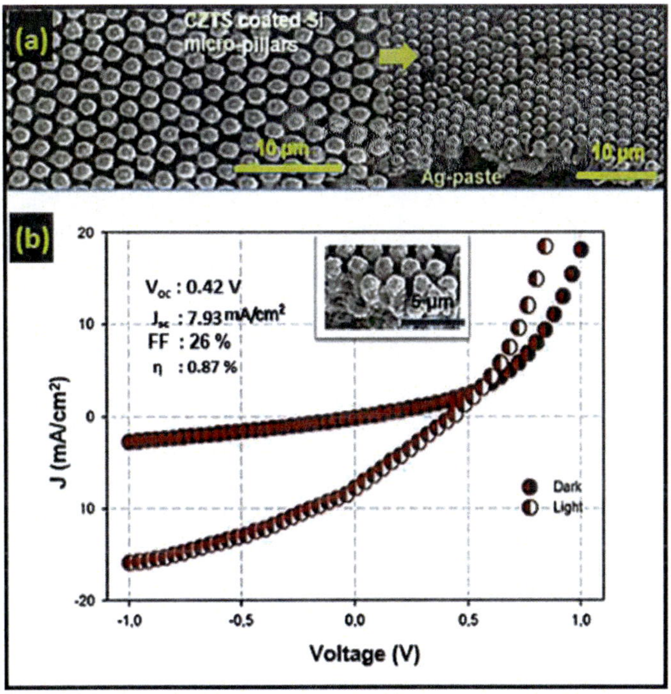

Fig. 26.15 **a** The top and tilted-view SEM images captured for the transferred Si-MPs on Ag-paste pre-coated glass substrate (SLG) decorated with a thin layer of e-beam evaporated p-type CZTS. **b** The I-V characteristics of the SLG/Ag-paste/n-Si NWs/p-CZTS/ITO/Al core–shell solar cell measured in dark and under light illumination. The tilted view SEM image of the fabricated solar cell is seen in the inset micrograph [20]

perfectly printed on the Ag-paste pre-coated SLG substrate, but that they are also fully coated with the CZTS layer. The I-V characteristics of the manufactured solar cell recorded in the dark and under light illumination are presented in Fig. 26.17a, b.

The semilog plot of the I-V characteristic under light illumination is also plotted in Fig. 26.17b to clearly show the observed photovoltaic effect. The V_{oc}, J_{sc}, FF and η of the SLG/Ag-paste/n-Si NWs/p-CZTS/ITO/Al solar cell were determined to be 0.40 V, 19, 20 mA/cm^2, 17% and 1.31%, respectively. These are the first published results for an SLG/Ag/n-Si-NW/p-CZTS/ITO/Al structured solar cell constructed on the SLG substrate with transferred Si NWs, which exhibits the highest energy conversion efficiency ever achieved in a solar cell based on a Si and CZTS material combination built on an SLG surface [21, 69, 74, 75]. The higher efficiency achieved in this study over planar counterparts may be attributed to the unique properties of the Si NWs, including the excellent light trapping, efficient charge extraction and large surface to volume ratio. The results demonstrate the promise of next generation solar cells that can be manufactured on a wide range of substrates, including transparent, amorphous, and flexible ones. As highlighted earlier, even though the

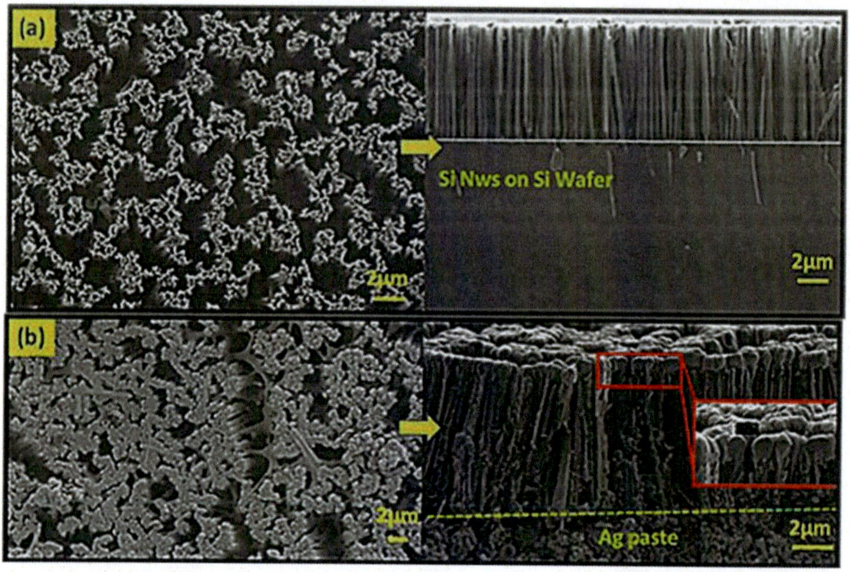

Fig. 26.16 The top and cross-sectional SEM images of **a** the Si NWs on mother substrate with ~ 8.0 μm lengths, **b** after the transferred onto the Ag-paste covered 2nd carrier (SLG) substrate and decorated with a p-type CZTS absorber layer to fabricate a SLG/Ag-paste/n-Si NWs/p-CZTS/ITO/Al structured solar cell. The inset is an enlarged view of the Si NWs coated with a 600 nm thick CZTS layer [20]

current solar cell's power energy conversion efficiency is still lower than that of its traditional planar equivalents, higher-efficiency NW-based solar cells can be manufactured by adjusting several production processes, including improving the quality of the absorber layer and applying surface passivation processes to the Si NW to eliminate potential surface defects, such as surface recombination centers.

26.3.2 Photodetectors

Photodetectors with nano-/microwires

Nanowire (NW)-based photodetector (PDs) can be used in a wide range of technologies. However, for the nanowire (NW) integration process to be economically viable and attractive for CMOS applications, it must offer the possibility of direct growth and integration of non-Si-based NW devices onto large Si wafers. Currently, issues such as lattice constant, material and thermal mismatch, lack of good control over atomic structures, assembly into functional devices, difficulty in forming ternary and quaternary NW alloys, surface states, persistent photocurrent, contact resistance and noise, controlled doping for sharp homo- and heterojunctions, nano-to-micro impedance matching, catalyst-induced contamination, etc., make it difficult to grow

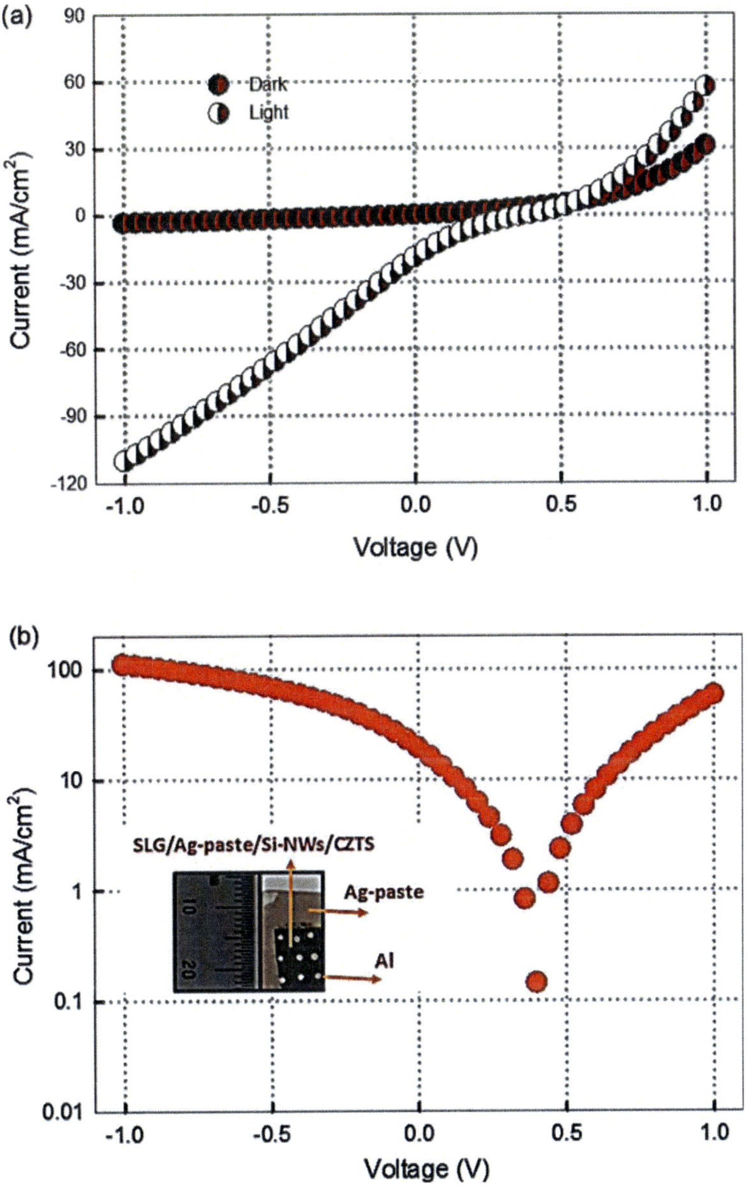

Fig. 26.17 **a** The current (I)-voltage (V) characteristics of the fabricated solar cell measured in dark and under AM 1.5G illumination. **b** The semi log plot of I-V characteristics that clearly reveals the photovoltaic effect. Inset Figure displays the photograph of the SLG/Ag-paste/n-Si NWs/p-CZTS/ITO/Al solar cell [20]

and design nanowire (NW) devices on Si or any other substrates, which is a setback in the application of NW PDs at the industrial level.

A successful nanowire (NW) detection and sensing device would require a meticulous control of surface states, defects, traps, orientation, polarization and light coupling as well as good control on generating sharp axial or radial junctions in the nanowires (NWs). There are obvious challenges in implementing the nanowire (NW) PDs. Most conventional semiconductor PDs operate at high quantum efficiency, whereas NW PDs' efficiencies are low in values. Development of feasible techniques is now crucial to match the efficiency of NW PDs to that of classical counterparts and a combination of plasmonic techniques and NW heteroepitaxy may offer the potential to realize substrate independent NW PDs with high efficiency and bandwidth.

The basic devices and laboratory demonstrations of the key elements of technology exist for NW PDs and no physical breakthrough is required to implement the device in most systems. Substantial technological work remains. Controlling the physical properties such as diameter, doping uniformity, orientations, identical surface and contacts are required to meet the strict requirements for practical systems with future generations of silicon photonics and CMOS. In addition, work will be required in developing impedance matching techniques between NW PDs and conventional high-speed circuits. The technology for integrating III-V and II-VI NWs with silicon integrated circuits is still at an early stage, though there have been considerable key demonstrations in last few years. Several groups introduced NW PDs using hybrid approaches, such as transfer printing or pick-and-place, such hybrid approaches require no modifications to the current process for fabricating silicon integrated circuits except to add pro- cesses to fabricated NW PDs [70, 76, 77]. Commercially viable approaches will focus on monolithic integration via nanoheteroepitaxy. In the next section, we will present a recent approach based on inverse of nanowires or negative nano-/microwires which is potentially mass manufacturable and CMOS compatible.

Photodiodes with Photon-Trapping Inverse Nano/microwires (Holes)

A periodic array of micro-nano scale holes were recently used to bend a vertically oriented beam of light by 90 degrees (Fig. 26.18a) and make it propagate horizontally within a thin Si film on an SOI substrate [4, 78]. Such light-bending and lateral slow propagating modes enhance photon-material interactions and dramatically increase the light absorption capabilities of semiconductors that are weak in absorption characteristics. Using this technology, researchers developed extremely fast Si and Ge-based photodetectors with high quantum efficiency (~10×) [79], which opens opportunities for integrating amplifiers, equalizers, and other CMOS circuit elements with photodiodes, overcoming the challenges of cost, the complexity of integration, and mass production of receivers for rapidly growing data centers. Figure 26.18b–d shows a top-illuminated all-silicon PD designed and fabricated using IC compatible processes to facilitate very-large-scale integration (VLSI) and leverage cost reduction enabled by the CMOS industry. The PD structure (Fig. 26.18b) was epitaxially grown on a silicon-on-insulator (SOI) substrate that has 0.25 μm device layer (*p*-Si). The thin

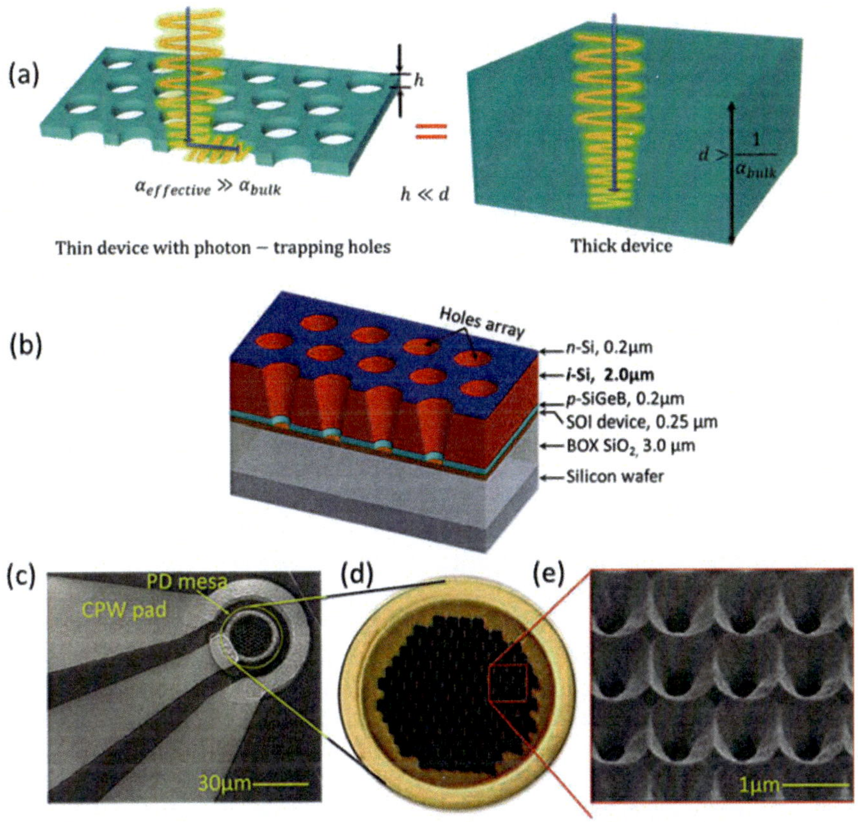

Fig. 26.18 **a** Schematics comparing light propagation in a thin slab with hole arrays and a thick semiconductor without holes (adapted from reference [80]. **b** Schematics of cross-section of the device on an SOI substrate showing individual layer thicknesses and an array of tapered holes etched into the active PD region. **c–e** Micrographs of a Si PD with a high-speed coplanar waveguide (CPW) and integrated micro and nanoscale holes

absorption region consists of 2 μm thick *i*-Si to minimize the transit time for electrons and holes. A lattice-matched 0.2 μm *p*++ $Si_{0.988}Ge_{0.01}B_{0.002}$ was used as the bottom *p*-contact layer, as well as an etch-stop layer for the *n*-mesa isolation process. A 0.3 μm phosphorus doped $n{+}{+}$ thin layer served as the top *n*-contact. The devices have mesa structures with *nip* on a SOI wafer with a 3 μm SiO_2 layer.

In order to address the weak absorption in indirect bandgap semiconductor Si, an array of 2D periodic micro or nanoscale holes were integrated in the active region of the PDs. The diameters of these holes range from 600 to 1700 nm and with periods ranging from 900 to 2000 nm. The details of the fabrication processes are discussed in reference [4]. Holes were fabricated with both uniform cylindrical shapes as well as gradual funnel shapes with square or hexagonal lattices in the active regions of the PDs with ~ 65° tapered sidewalls as shown in Fig. 26.18b–e.

The nanohole array supports a set of modes with wave vectors in the perpendicular direction, as well as in-plane (x–y) directions. Depending on the parameters of the 2D hole array, collective modes are formed similar to the modes in photonic crystals. Under certain parameters, slow lateral modes can appear to contribute to considerably higher photon absorption [3]. The insets in Fig. 26.19a show FDTD numerical simulations that depict the formation of lateral modes around holes when a vertically oriented beam illuminates them. As time increases from left to right, lateral components of the electric field spread into Si as cylindrical waves. Based on the simulations, for a design to operate around 850 nm, the hole periods of 900 nm or bigger values were found to excite modes that propagate in the lateral x–y plane. Thus, a successful conversion of an initial *incident vertical plane waves* to an ensemble of *lateral collective modes* was realized in a 2D periodic array of holes. Strong lateral modes propagating deep into *i*-region enable the absorption of a wide range of the wavelength and contribute to above 80% absorption, as shown for three different side-wall angles of holes in Fig. 26.19a. Since a PD has a much larger lateral size than its vertical thickness, lateral waves are almost completely absorbed, and as a result, electron–hole pairs are efficiently generated. These generated carriers drift quickly under the strong vertical electric field in the device and create a high-speed current output.

Through extensive simulations and experimental implementations of photon-trapping structures in silicon photodetectors, a direct correlation between the enhancement of absorption and physical parameters of the photon-trapping structures was also divulged. By employing cylindrical nanoholes, inverted pyramids, and funnel-shaped surface formations, close to 1000% enhancement in the quantum efficiency was achieved compared to the control devices [3, 79, 81]. This was made possible by bending the incident beam of light and enabling lateral propagation of modes to prolong the light-matter interactions and suppress back reflection using the inverse nanowires or nanoholes. The enhancement in absorption also comes with a considerable reduction in device capacitance and thereby an improvement in the time response. The combined effect collectively can help to overcome the trade-off between the efficiency and speed of operation of the PDs. State-of-the-art CMOS fabrication processes could enable near-perfect EQE and above 50% capacitance reduction, by increasing the number of photon-trapping nanoholes integrated into the devices, especially in photodiodes with small surface area, required to operate at high speed. This open opportunities for the development of complete CMOS integrated receivers operating with high sensitivity and high speed and can be expanded to other semiconductors such as germanium (Ge), gallium arsenide (GaAs), and indium phosphide (InP) based ternary and quaternary materials [82].

26.4 Summary and Future Outlook

Due to their outstanding properties such as high charge carrier collection, unique surface chemistry, enhanced charge-carrier generation, large surface-to-volume ratio,

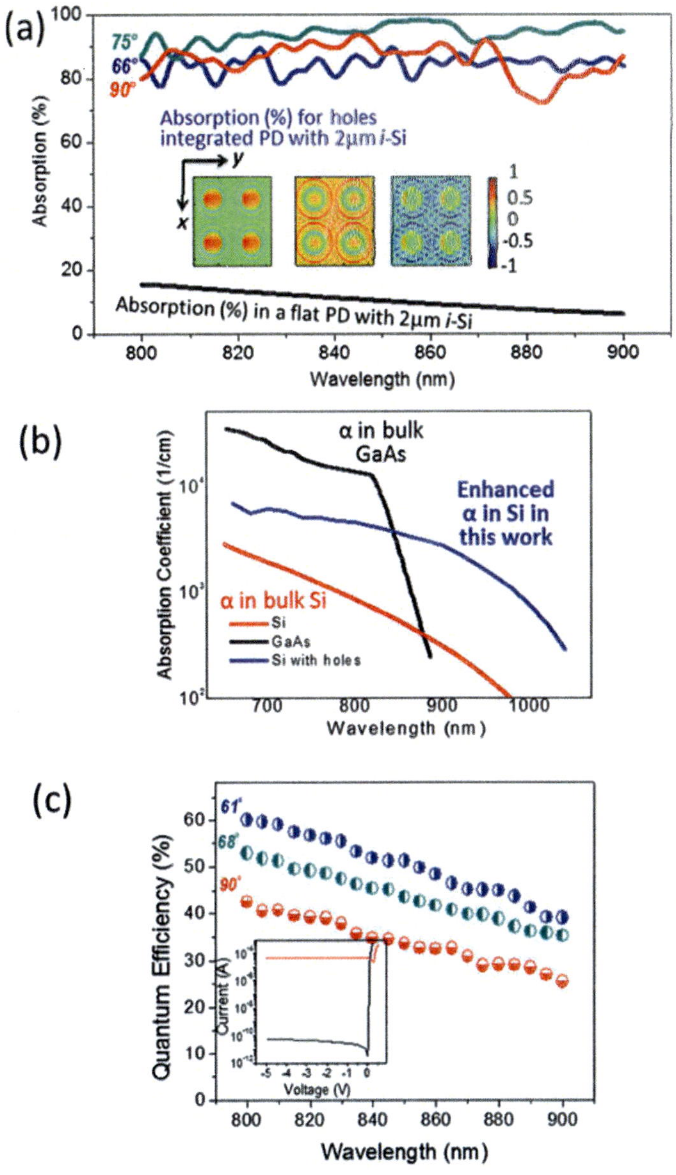

Fig. 26.19 **a** Absorption in a *pin* PD with integrated holes of ~2 μm depth etched in funnel shapes (diameter/period (d/p):1300/2000 nm) with angles of 75° and 66° and cylindrical shape with 90° slopes. The bottom curve (black) shows the absorption in a regular Si PD with 2 μm *i*-Si. **b** Measured enhanced EQE versus wavelengths in these PDs enabled by integrated holes (d/p: 700/1000 **nm**) with angles of 61° and 66° and cylindrical shape with 90° slopes. The inset shows I-V curves for dark and optical illumination. **c** Plot of $\alpha_{effective}$ between 600–1050 nm showing how its value in our PDs increased by more than an order of magnitude to the level of α_{GaAs}

the decoupled charge transport and light-incidence direction, and excellent light absorption/trapping, 1D Si nano-/microstructures have been shown to be potential building blocks for the realization of cost-effective and high-performance photodetectors and solar cells. Therefore, in this chapter, a number of top-down approaches, like wet and dry etching methods, for the synthesis of ordered and disordered arrays of 1D Si nano-/microstructures are first introduced. The advantages and disadvantages of the wet and dry etching processes were also briefly discussed. A special attention has been paid to the fabrication of ordered arrays of 1D Si nano-/microstructures by coupling the MACE or DRIE process with the NSL technique. The recent studies on these structure-based solar cells and photodetectors are then discussed, with a main focus on our own recent studies.

The results demonstrate that large-scale arrays of 1D Si nano/microstructures with different diameters, interspace and lengths ranging from a few micrometers to tens of nanometers can be achieved in a cost-effective way by controlling several parameters such as the dimension and feature size of etching mask (or template) prepared via the NSL, the etching time, the reaction temperature and precursor concentration. The synthesized arrays of Si nano-/microstructures have been shown to have a wide range of applications, including light sensing and solar energy harvesting. The recent studies have demonstrated that both ordered and disordered arrays of Si nano-/microstructures can be successfully transferred to a wide range of substrates, including Ag-pre-coated glasses, transparent-conductive-oxides, and metal foils, which is crucial for scalable production of low cost, lightweight and flexible electronics and opto-electronic components. As this approach offers a way to grow low-cost device manufacturing on a diverse range of substrate surfaces with highly crystalline semiconductor materials, it can be taken as a giant step towards the realization of next-generation opto-electronic devices. The performance of the aforementioned opto-electronic devices can be improved even further in future studies since the introduced fabrication routes have a full control over the size and morphology of the manufactured 1D arrays of Si nano-/microstructures.

Despite the amazing progress in nano-/microstructure production and applications, there are still some problems in the application of 1D ordered and disordered Si nano-/microstructures that need to be resolved in future study for the realization of high-performance solar cells that can be alternative to the conventional planer solar cells. One of these issues is the presence of high-density defects at the interface/surface resulting from the 1D structures' large surface-to-volume ratio. Coating Si nano-/microstructures with a suitable surface passivation material, such as porous silicon, a thin transparent conductive oxide layer, SiN_x and Al_2O_3, is one of the most effective ways to address this issue. Other critical factors that need to be optimized for the realization of high-performance core–shell solar cells, for instance, include the alignment and dimensions of Si nano-/microstructures such as diameter, density, pitch and aspect ratio as the core component, the quality of the absorption layer in terms of its thickness, series resistance, electrical and optical properties as the shell component, the nature of interfacial junction in hetero-structures and its depth, the doping profile uniformity along the nano-/microstructures' length, and the electrical quality of the metal contacts required for the extraction of the photo-generated current. The compatibility of the contacts and the density of the impurities at the junction, in particular, should be optimized for the realization of core–shell solar cells with high open-circuit voltages, as they play a critical role in determining the level of shunt resistance. Another major issue is the high series resistance of solar cells, which is accompanied by a high bulk resistivity of the absorber layer. The series resistance has a direct effect on the short-circuit current density of the radial p–n junction based solar cells and can be expected to be dramatically lowered by optimizing the quality of the shell layer in terms of uniformity, thickness and crystallinity, and eliminating bulk defects causing a recombination problem.

Si-based high-speed photodiodes with light-trapping mechanism enabled by inverse-nanowires or nanoholes can possess a new category of high speed and highly efficient photodiodes that can be implemented in an all-Si platform. Monolithic integration of surface illuminated Si-based high-speed photodiodes with CMOS circuits will open new opportunities for optical communication industry to design low-cost on-board optical modules with less complexity and low-parasitic [3]. The nature of inverse or negative structures such as holes, inverted pyramids or cones can keep the surface material unfragmented and offer solution to the first challenge by designing contacts that can be fabricated with CMOS-compatible conventional processes. On the other hand, the challenges additional surface area and surface damage along with higher surface resistance are added issues of this approach. Therefore, CMOS-compatible surface passivation techniques need to be developed to fully tap into the potential of micro-/nanostructures in Si PDs [6]. Inherently, having less material compared to bulk counterparts, micro-/nanostructures cause increase in resistance, which can limit the speed of a photodiode. However, this drawback can be suppressed with smart designs of micro-/nanostructures to reduce the capacitance at the same time. Reduction in capacitance can compensate for the increase in resistance and eliminate RC time limited response. Even though there are challenges that need to be addressed to benefit the full potential of Si-based PDs with micro-/nanostructures, the advantages of monolithic integration and reduced cost of the transceivers that utilized such PDs, promise real solutions to meet the high demand of connectivity in modern data centers that are expected to connect numerous computer systems serving artificial intelligence, internet of things, autonomous vehicles, and social infrastructures [83, 84].

References

1. T. Zhai, L. Li, Y. Ma, M. Liao, X. Wang, X. Fang, J. Yao, Y. Bando, D. Golberg, One-dimensional inorganic nanostructures: synthesis, field-emission and photodetection. Chem. Soc. Rev. **40**(5), 2986–3004 (2011). https://doi.org/10.1039/C0CS00126K
2. X. Xu, X. Fang, T. Zhai, H. Zeng, B. Liu, X. Hu, Y. Bando, D. Golberg, Tube-in-Tube TiO2 nanotubes with porous walls: fabrication, formation mechanism, and photocatalytic properties. Small **7**(4), 445–449 (2011). https://doi.org/10.1002/smll.201001849

3. C. Bartolo-Perez, W. Qarony, S. Ghandiparsi, A.S. Mayet, A. Ahamed, H. Cansizoglu, Y. Gao, E. Ponizovskaya Devine, T. Yamada, A.F. Elrefaie, S.-Y. Wang, M.S. Islam, Maximizing absorption in photon-trapping ultrafast silicon photodetectors. Adv. Photon. Res. **2**(6), 2000190 (2021). https://doi.org/10.1002/adpr.202000190
4. Y. Gao, H. Cansizoglu, K.G. Polat, S. Ghandiparsi, A. Kaya, H. Mamtaz, A. Mayet, Y. Wang, X. Zhang, T. Yamada, E.P. Devine, A. Elrefaie, S.-Y. Wang, M. Islam, Photon-trapping microstructures enable high-speed high-efficiency silicon photodiodes. Nat. Photonics **11**, 301–308 (2017)
5. S. Ghandiparsi, A.F. Elrefaie, A.S. Mayet, T. Landolsi, C. Bartolo-Perez, H. Cansizoglu, Y. Gao, H.H. Mamtaz, H.R. Golgir, E.P. Devine, T. Yamada, S.Y. Wang, M.S. Islam, High-speed high-efficiency photon-trapping broadband silicon pin photodiodes for short-reach optical interconnects in data centers. J. Lightwave Technol. **37**(23), 5748–5755 (2019). https://doi.org/10.1109/JLT.2019.2937906
6. A.S. Mayet, H. Cansizoglu, Y. Gao, S. Ghandiparsi, A. Kaya, C. Bartolo-Perez, B. AlHalaili, T. Yamada, E. Ponizovskaya Devine, A.F. Elrefaie, S.-Y. Wang, M.S. Islam, Surface passivation of silicon photonic devices with high surface-to-volume-ratio nanostructures. J. Opt. Soc. Am. B **35**(5), 1059–1065 (2018). https://doi.org/10.1364/JOSAB.35.001059
7. M. Liu, Y. Lu, Z.B. Xie, G.M. Chow, Enhancing near-infrared solar cell response using upconverting transparentceramics. Sol. Energy Mater. Sol. Cells **95**(2), 800–803 (2011). https://doi.org/10.1016/j.solmat.2010.09.018
8. Y. Luo, Y. Huang, G. Zang, X. Ren, X. Duan, S. Cai, Q. Wang, X. Zhang, J. Wang, Zero bias PIN photodetector based on gradient band distribution and doping gradient profile. Infrared Phys. Technol. **67**, 391–396 (2014). https://doi.org/10.1016/j.infrared.2014.09.008
9. O. Guller, E. Peksu, H. Karaagac, Synthesis of TiO2 nanorods for Schottky-Type UV-photodetectors and third-generation solar cells. physica status solidi (a) **215**(4), 1700404 (2018). https://doi.org/10.1002/pssa.201700404
10. T. Gao, Q.H. Li, T.H. Wang, CdS nanobelts as photoconductors. Appl. Phys. Lett. **86**(17):173105 (2005).https://doi.org/10.1063/1.1915514
11. J.S. Jie, W.J. Zhang, Y. Jiang, X.M. Meng, Y.Q. Li, S.T. Lee, Photoconductive characteristics of single-crystal CdS nanoribbons. Nano Lett. **6**(9), 1887–1892 (2006). https://doi.org/10.1021/nl060867g
12. H. Kind, H. Yan, B. Messer, M. Law, P. Yang, Nanowire ultraviolet photodetectors and optical switches. Adv. Mater. **14**(2), 158–160 (2002). https://doi.org/10.1002/1521-4095(20020116)14:2%3c158::AID-ADMA158%3e3.0.CO;2-W
13. X. Liu, C. Li, S. Han, J. Han, C. Zhou, Synthesis and electronic transport studies of CdO nanoneedles. Appl. Phys. Lett. **82**, 1950–1952 (2003). https://doi.org/10.1063/1.1562331
14. M. Salvato, M. Scagliotti, M. De Crescenzi, M. Boscardin, C. Attanasio, G. Avallone, C. Cirillo, P. Prosposito, F. De Matteis, R. Messi, P. Castrucci, Time response in carbon nanotube/Si based photodetectors. Sens. Actuators, A **292**, 71–76 (2019). https://doi.org/10.1016/j.sna.2019.04.004
15. G. Shen, D. Chen, One-dimensional nanostructures for photodetectors. Recent Pat. Nanotechnol. **4**, 20–31 (2010). https://doi.org/10.2174/187221010790712101
16. H. Wu, Y. Sun, D. Lin, R. Zhang, C. Zhang, W. Pan, GaN nanofibers based on electrospinning: facile synthesis, controlled assembly, precise doping, and application as high performance UV photodetector. Adv. Mater. **21**(2), 227–231 (2009). https://doi.org/10.1002/adma.200800529
17. Y. Cui, Z. Zhong, D. Wang, W.U. Wang, C.M. Lieber, High performance silicon nanowire field effect transistors. Nano Lett. **3**(2), 149–152 (2003). https://doi.org/10.1021/nl0258751
18. K. Das, S. Mukherjee, S. Manna, S. Ray, A. Raychaudhuri, Single Si nanowire (diameter $\leq$ 100nm) based polarization sensitive near-infrared photodetector with ultra-high responsivity. Nanoscale 6 (2014).https://doi.org/10.1039/C4NR03170A
19. H. Karaagac, M.S. Islam, Enhanced field ionization enabled by metal induced surface states on semiconductor nanotips. Adv. Func. Mater. **24**(15), 2224–2232 (2014). https://doi.org/10.1002/adfm.201303308

20. E. Peksu, O. Guller, M. Parlak, M.S. Islam, H. Karaagac, Towards the fabrication of third generation solar cells on amorphous, flexible and transparent substrates with well-ordered and disordered Si-nanowires/pillars. Physica E: Low-Dimen. Syst. Nanostructures **124**, 114382 (2020).https://doi.org/10.1016/j.physe.2020.114382
21. E. Peksu, H. Karaagac, A third generation solar cell based on wet-chemically etched Si nanowires and sol-gel derived Cu_2ZnSnS_4 thin films. J. Alloy. Compd. **774**, 1117–1122 (2019). https://doi.org/10.1016/j.jallcom.2018.10.012
22. Z. Zhang, R. Zou, L. Yu, J. Hu, One-dimensional silicon-based semiconductor nanomaterials: synthesis, structures, properties and applications. Critical Reviews in Solid State and Material Sciences **36**, 148–173 (2011)
23. M.D. Henry, S. Walavalkar, A. Homyk, A. Scherer, Alumina etch masks for fabrication of high-aspect-ratio silicon micropillars and nanopillars. Nanotechnology **20**(25), 255305 (2009).https://doi.org/10.1088/0957-4484/20/25/255305
24. H.P. Yoon, Y.A. Yuwen, C.E. Kendrick, G.D. Barber, N.J. Podraza, J.M. Redwing, T.E. Mallouk, C.R. Wronski, T.S. Mayer, Enhanced conversion efficiencies for pillar array solar cells fabricated from crystalline silicon with short minority carrier diffusion lengths. Appl. Phys. Lett. **96**(21), 213503 (2010).https://doi.org/10.1063/1.3432449
25. J. Kim, H. Han, Y.H. Kim, S.-H. Choi, J.-C. Kim, W. Lee, Au–Ag Bilayered metal mesh as a Si etching catalyst for controlled fabrication of Si nanowires. ACS Nano **5**(4), 3222–3229 (2011). https://doi.org/10.1021/nn2003458
26. K. Peng, M. Zhang, A. Lu, N.-B. Wong, R. Zhang, S.-T. Lee, Ordered silicon nanowire arrays via nanosphere lithography and metal-induced etching. Appl. Phys. Lett. **90**(16), 163123 (2007).https://doi.org/10.1063/1.2724897
27. H.J. In, C.R. Field, P.E. Pehrsson, Periodically porous top electrodes on vertical nanowire arrays for highly sensitive gas detection. Nanotechnology **22**(35), 355501 (2011).https://doi.org/10.1088/0957-4484/22/35/355501
28. X.D. Bai, Z. Xu, S. Liu, E.G. Wang, Aligned 1D silicon nanostructure arrays by plasma etching. Sci. Technol. Adv. Mater. **6**(7), 804–808 (2005). https://doi.org/10.1016/j.stam.2005.05.015
29. C.-H. Hsu, H.-C. Lo, C.-F. Chen, C.T. Wu, J.-S. Hwang, D. Das, J. Tsai, L.-C. Chen, K.-H. Chen, Generally applicable self-masked dry etching technique for nanotip array fabrication. Nano Lett. **4**(3), 471–475 (2004). https://doi.org/10.1021/nl049925t
30. Y. Hung, S. Lee, B.J. Thibeault, L.A. Coldren, Fabrication of highly ordered silicon nanowire arrays with controllable sidewall profiles for achieving low-surface reflection. IEEE J. Sel. Top. Quantum Electron. **17**(4), 869–877 (2011). https://doi.org/10.1109/JSTQE.2010.2068540
31. D. Baklykov, M. Andronic, O. Sorokina, S. Avdeev, K. Buzaverov, I. Ryzhikov, I. Rodionov, Self-controlled cleaving method for silicon DRIE process cross-section characterization. Micromachines **12**(5) (2021). https://doi.org/10.3390/mi12050534
32. E. Baquedano, R.V. Martinez, J.M. Llorens, P.A. Postigo, Fabrication of silicon nanobelts and nanopillars by soft lithography for hydrophobic and hydrophilic photonic surfaces. Nanomaterials (Basel) **7**(5), 109 (2017). https://doi.org/10.3390/nano7050109
33. H. Berthet, J. Jundt, J. Durivault, B. Mercier, D. Angelescu, Time-of-flight thermal flowrate sensor for lab-on-chip applications. Lab. Chip **11**(2), 215–223 (2011). https://doi.org/10.1039/C0LC00229A
34. B. Saadany, M. Malak, M. Kubota, F. Marty, Y. Mita, D. Khalil, T. Bourouina, Free-space tunable and drop optical filters using vertical bragg mirrors on silicon. IEEE J. Sel. Top. Quantum Electron. **12**(6), 1480–1488 (2006). https://doi.org/10.1109/JSTQE.2006.884082
35. M. Sasaki, Silicon etching for multiple-height microstructures, in J. Yan (ed.) *Micro and Nano Fabrication Technology*. Springer Singapore, Singapore, pp 1–22 (2018). https://doi.org/10.1007/978-981-10-6588-0_29-1
36. R. Abdolvand, F. Ayazi, An advanced reactive ion etching process for very high aspect-ratio sub-micron wide trenches in silicon. Sens. Actuators, A **144**(1), 109–116 (2008). https://doi.org/10.1016/j.sna.2007.12.026
37. W.J. Park, J.H. Kim, S.M. Cho, S.G. Yoon, S.J. Suh, D.H. Yoon, High aspect ratio via etching conditions for deep trench of silicon. Surf. Coat. Technol. **171**(1), 290–295 (2003). https://doi.org/10.1016/S0257-8972(03)00288-3

38. A. Cutarelli, S. Ghio, J. Zasso, A. Speccher, G. Scarduelli, M. Roccuzzo, M. Crivellari, N. Maria Pugno, S. Casarosa, M. Boscardin, L. Conti, Vertically-aligned functionalized silicon micropillars for 3D culture of human pluripotent stem cell-derived cortical progenitors. Cells **9**(1), 88 (2020)
39. D. Mikulik, A.C. Meng, R. Berrazouane, J. Stückelberger, P. Romero-Gomez, K. Tang, F.-J. Haug, Fontcuberta i Morral A, McIntyre PC, Surface defect passivation of silicon micropillars. Adv. Mater. Interfaces **5**(20), 1800865 (2018). https://doi.org/10.1002/admi.201800865
40. V.K. Singh, J. Nagaraju, S. Avasthi, Radial junction silicon solar cells with micro-pillar array and planar electrode interface for improved photon management and carrier extraction. Curr. Appl. Phys. **19**(3), 341–346 (2019). https://doi.org/10.1016/j.cap.2018.12.016
41. R. Chaudhary, Comparison between silicon nanopillars prepared by Bosch process and metal assisted chemical etching. Protocols Reports Paper **52**, 1–8 (2018)
42. A. Oates, F.J. Cabrera-España, A. Agrawal, H.S. Reehal, Fabrication and characterisation of Si micropillar PV structures. Mater. Res. Innov. **18**(7), 500–504 (2014). https://doi.org/10.1179/1433075X14Y.0000000244
43. M.Y. Efremov, E.A. Olson, M. Zhang, L.H. Allen, Glass transition of thin films of poly (2-vinyl pyridine) and poly (methyl methacrylate): nanocalorimetry measurements. Thermochim. Acta **403**, 37–41 (2003)
44. V.J. Logeeswaran, J. Oh, A.P. Nayak, A.M. Katzenmeyer, K.H. Gilchrist, S. Grego, N.P. Kobayashi, S.Y. Wang, A.A. Talin, N.K. Dhar, M.S. Islam, A Perspective on nanowire photodetectors: current status, future challenges, and opportunities. IEEE J. Sel. Top. Quantum Electron. **17**(4), 1002–1032 (2011). https://doi.org/10.1109/JSTQE.2010.2093508
45. R. Elbersen, W. Vijselaar, R.M. Tiggelaar, H. Gardeniers, J. Huskens, Fabrication and doping methods for silicon nano- and micropillar arrays for solar-cell applications: a review. Adv. Mater. **27**(43), 6781–6796 (2015). https://doi.org/10.1002/adma.201502632
46. H. Han, Z. Huang, W. Lee, Metal-assisted chemical etching of silicon and nanotechnology applications. Nano Today **9**(3), 271–304 (2014). https://doi.org/10.1016/j.nantod.2014.04.013
47. K.E. Bean, Anisotropic etching of silicon. IEEE Trans. Electron Dev. **25**(10), 1185–1193 (1978). https://doi.org/10.1109/T-ED.1978.19250
48. X. Li, Metal assisted chemical etching for high aspect ratio nanostructures: a review of characteristics and applications in photovoltaics. Curr. Opin. Solid State Mater. Sci. **16**(2), 71–81 (2012). https://doi.org/10.1016/j.cossms.2011.11.002
49. R. Bhujel, U. Rizal, A. Agarwal, B.S. Swain, B.P. Swain, Synthesis and characterization of silicon nanowires by electroless etching. J. Mater. Eng. Perform. **27**(6), 2655–2660 (2018). https://doi.org/10.1007/s11665-018-3179-z
50. S.-S. Yoon, D.-Y. Khang, Stretchable, Bifacial Si-organic hybrid solar cells by vertical array of SI micropillars embedded into elastomeric substrates. ACS Appl. Mater. Interfaces. **11**(3), 3290–3298 (2019). https://doi.org/10.1021/acsami.8b17826
51. H. Lin, F. Wu, P. Gao, W. Shen, Shape-controlled silicon microwire arrays from Au–Ag-catalyzed metal-assisted chemical etching for radial junction solar cells. ACS Appl. Energy Mater. **2**(8), 5871–5876 (2019). https://doi.org/10.1021/acsaem.9b01006
52. Y. Qian, D.J. Magginetti, S. Jeon, Y. Yoon, T.L. Olsen, M. Wang, J.M. Gerton, H.P. Yoon, Heterogeneous optoelectronic characteristics of Si micropillar arrays fabricated by metal-assisted chemical etching. Sci. Rep. **10**(1), 16349 (2020). https://doi.org/10.1038/s41598-020-73445-x
53. G. Baytemir, F. Es, R. Turan, Comparison of influence of gold contamination on the performances of planar and three dimensional c-Si solar cells. Renew. Energy **142**, 393–399 (2019). https://doi.org/10.1016/j.renene.2019.04.081
54. G. Baytemir, E.H. Ciftpinar, R. Turan, Enhanced metal assisted etching method for high aspect ratio microstructures: applications in silicon micropillar array solar cells. Sol. Energy **194**, 148–155 (2019). https://doi.org/10.1016/j.solener.2019.10.033
55. H, Karaagac, A hybrid solar cell based on silicon nanowire and organic thin film. physica status solidi (a) **211**(11):2503–2508 (2014). https://doi.org/10.1002/pssa.201431320

56. H. Karaagac, M. Parlak, M.S. Islam, Synthesis of Si nanowires by electroless etching technique and their integration into I–III–VI_2 thin films for solar cells. MRS Proc. **1408**, 1–6 (2012). https://doi.org/10.1557/opl.2012.40
57. H. Karaagac, M. Parlak, E. Yengel, M.S. Islam, Heterojunction solar cells with integrated Si and ZnO nanowires and a chalcopyrite thin film. Mater. Chem. Phys. **140**(1), 382–390 (2013). https://doi.org/10.1016/j.matchemphys.2013.03.053
58. Y. Li, Q. Chen, D. He, J. Li, Radial junction Si micro/nano-wire array photovoltaics: Recent progress from theoretical investigation to experimental realization. Nano Energy **7**, 10–24 (2014). https://doi.org/10.1016/j.nanoen.2014.04.015
59. B. Pal, K.J. Sarkar, P. Banerji, Fabrication and studies on Si/InP core-shell nanowire based solar cell using etched Si nanowire arrays. Solar Energy Mater. Solar Cells **204**, 110217 (2020).https://doi.org/10.1016/j.solmat.2019.110217
60. K. Lee, I. Hwang, N. Kim, D. Choi, H.-D. Um, S. Kim, K. Seo, 17.6%-Efficient radial junction solar cells using silicon nano/micro hybrid structures. Nanoscale **8**(30), 14473–14479 (2016). https://doi.org/10.1039/C6NR04611H
61. A. Smyrnakis, P. Dimitrakis, P. Normand, E. Gogolides, Fabrication of axial p-n junction silicon nanopillar devices and application in photovoltaics. Microelectron. Eng. **174**, 74–79 (2017). https://doi.org/10.1016/j.mee.2017.02.011
62. I. Leontis, M.A. Botzakaki, S.N. Georga, A.G. Nassiopoulou, Study of Si nanowires produced by metal-assisted chemical etching as a light-trapping material in n-type c-Si solar cells. ACS Omega **3**(9), 10898–10906 (2018). https://doi.org/10.1021/acsomega.8b01049
63. M. Seo, S. Yoon, H. Cho, S. Lee, K. Kim, B.D. Kong, M. Meyyappan, C. Baek, Solar cell using hourglass-shaped silicon nanowires for increased light-trapping path. IEEE J. Photovoltaics **10**(2), 475–479 (2020). https://doi.org/10.1109/JPHOTOV.2020.2964329
64. P. Yadav, M. Patel, H. Kim, Y. Cho, H. Kim, J. Kim, J. Yi, D.-W. Kim, Routes for realizing high-performing Si solar cells by using periodic structures. Mater. Res. Bull. **94**, 92–99 (2017). https://doi.org/10.1016/j.materresbull.2017.05.027
65. H.-P. Wang, D. Periyanagounder, A.-C. Li, J.-H. He, Fabrication of silicon hierarchical structures for solar cell applications. IEEE Access, 1–1 (2018). https://doi.org/10.1109/ACCESS.2018.2885169
66. H.-S. Kim, D.B. Patel, H. Kim, M. Patel, K.R. Chauhan, W. Park, J. Kim, Electrical and optical properties of Si microwire solar cells. Sol. Energy Mater. Sol. Cells **164**, 7–12 (2017). https://doi.org/10.1016/j.solmat.2017.01.046
67. L. Karmakar, D. Das, Single-step fabrication of single-junction c–Si nano-structured solar cells by optimization of plasma etching parameters. J. Alloys Compounds **847**, 155352 (2020).https://doi.org/10.1016/j.jallcom.2020.155352
68. M.D. Kumar, H. Kim, J. Kim, Periodically patterned Si pyramids for realizing high efficient solar cells by wet etching process. Sol. Energy **117**, 180–186 (2015). https://doi.org/10.1016/j.solener.2015.04.034
69. M.Y. Yeh, P.H. Lei, S.H. Lin, C.D. Yang, Copper-Zinc-Tin-Sulphur thin film using spin-coating technology. Materials **9**(7), 526 (2016). https://doi.org/10.3390/ma9070526
70. L. Vj, A.M. Katzenmeyer, M.S. Islam, Harvesting and transferring vertical pillar arrays of single-crystal semiconductor devices to arbitrary substrates. IEEE T Electron Dev **57**(8), 1856–1864 (2010). https://doi.org/10.1109/TED.2010.2051195
71. E. Peksu, M. Terlemezoglu, M. Parlak, H. Karaagac, Characterization of one-step deposited Cu_2ZnSnS_4 thin films derived from a single crystalline powder. Renew. Energy **143**, 1133–1142 (2019). https://doi.org/10.1016/j.renene.2019.05.076
72. E.C. Garnett, P. Yang, Silicon nanowire radial p−n junction solar cells. J. Am. Chem. Soc. **130**(29), 9224–9225 (2008). https://doi.org/10.1021/ja8032907
73. L. Tsakalakos, J. Balch, J. Fronheiser, B.A. Korevaar, O. Sulima, J. Rand, Silicon nanowire solar cells. Appl. Phys. Lett. **91**(23), 233117 (2007).https://doi.org/10.1063/1.2821113
74. F. Jiang, H. Shen, W. Wang, L. Zhang, Preparation and properties of Cu2ZnSnS4 absorber and Cu2ZnSnS4/amorphous silicon thin-film solar cell. Appl. Phys. Express **4**, 4101 (2011). https://doi.org/10.1143/APEX.4.074101

75. N. Song, M. Young, F. Liu, P. Erslev, S. Wilson, S.P. Harvey, G. Teeter, Y. Huang, X. Hao, M.A. Green, Epitaxial Cu2ZnSnS4 thin film on Si (111) 4° substrate. Appl. Phys. Lett. **106**(25), 252102 (2015).https://doi.org/10.1063/1.4922992
76. L. Vj, J. Oh, A.P. Nayak, A.M. Katzenmeyer, K.H. Gilchrist, S. Grego, N.P. Kobayashi, S.Y. Wang, A.A. Talin, N.K. Dhar, M.S. Islam, A perspective on nanowire photodetectors: current status, future challenges, and opportunities. IEEE J. Sel. Top. Quantum Electron. **17**(4), 1002–1032 (2011). https://doi.org/10.1109/JSTQE.2010.2093508
77. V.J. Logeeswaran, A. Sarkar, M.S. Islam, N.P. Kobayashi, J. Straznicky, X. Li, W. Wu, S. Mathai, M.R.T. Tan, S.-Y. Wang, R.S. Williams, A 14-ps full width at half maximum high-speed photoconductor fabricated with intersecting InP nanowires on an amorphous surface. Appl. Phys. A **91**(1), 1–5 (2008). https://doi.org/10.1007/s00339-007-4394-x
78. K. Zang, X. Jiang, Y. Huo, X. Ding, M. Morea, X. Chen, C.-Y. Lu, J. Ma, M. Zhou, Z. Xia, Z. Yu, T.I. Kamins, Q. Zhang, J.S. Harris, Silicon single-photon avalanche diodes with nano-structured light trapping. Nat. Commun. **8**(1), 628 (2017). https://doi.org/10.1038/s41467-017-00733-y
79. H. Cansizoglu, C. Bartolo-Perez, Y. Gao, E. Ponizovskaya Devine, S. Ghandiparsi, K.G. Polat, H.H. Mamtaz, T. Yamada, A.F. Elrefaie, S.-Y. Wang, M.S. Islam, Surface-illuminated photon-trapping high-speed Ge-on-Si photodiodes with improved efficiency up to 1700 nm. Photon Res. **6**(7), 734–742 (2018). https://doi.org/10.1364/PRJ.6.000734
80. C. Bartolo-Perez, S. Chandiparsi, A.S. Mayet, H. Cansizoglu, Y. Gao, W. Qarony, A. AhAmed, S.-Y. Wang, S.R. Cherry, M. Saif Islam, G. Ariño-Estrada, Avalanche photodetectors with photon trapping structures for biomedical imaging applications. Opt. Express **29**(12), 19024–19033 (2021). https://doi.org/10.1364/OE.421857
81. H. Cansizoglu, A.S. Mayet, S. Ghandiparsi, Y. Gao, C. Bartolo-Perez, H.H. Mamtaz, E.P. Devine, T. Yamada, A.F. Elrefaie, S.Y. Wang, M.S. Islam, Dramatically enhanced efficiency in ultra-fast silicon MSM photodiodes via light trapping structures. IEEE Photonics Technol. Lett. **31**(20), 1619–1622 (2019). https://doi.org/10.1109/LPT.2019.2939541
82. S. Ghandiparsi, A.F. Elrefaie, H. Cansizoglu, Y. Gao, C. Bartolo-Perez, H.H. Mamtaz, A. Mayet, T. Yamada, E.P. Devine, S.-Y. Wang, M.S. Islam, High-speed high-efficiency broad-band silicon photodiodes for short-reach optical interconnects in data centers, in *Optical Fiber Communication Conference, San Diego, California*, 2018/03/11 2018. OSA Technical Digest (online). Optical Society of America, p W1I.7. https://doi.org/10.1364/OFC.2018.W1I.7
83. H. Cansizoglu, E.P. Devine, Y. Gao, S. Ghandiparsi, T. Yamada, A.F. Elrefaie, S. Wang, M.S. Islam, A New Paradigm in high-speed and high-efficiency silicon photodiodes for communication—part i: enhancing photon-material interactions via low-dimensional structures. IEEE T Electron Dev. **65**(2), 372–381 (2018). https://doi.org/10.1109/TED.2017.2779145
84. H. Cansizoglu, A.F. Elrefaie, C. Bartolo-Perez, T. Yamada, Y. Gao, A.S. Mayet, M.F. Cansizoglu, E.P. Devine, S.Y. Wang, M.S. Islam, A New paradigm in high-speed and high-efficiency silicon photodiodes for communication—part II: device and vlsi integration challenges for low-dimensional structures. IEEE T Electron Dev. **65**(2), 382–391 (2018). https://doi.org/10.1109/TED.2017.2779500

Chapter 27
Two-Dimensional Nanomaterials Based Biosensors

Bahadır Salmankurt and Hikmet Hakan Gürel

Abstract Graphene became the first 2D (two-dimensional) nanostructure which was discovered in 2004. After the synthesis of graphene revealed its unique properties, researchers set out to discover new 2D nanomaterials: Phosphorene is one of the new 2D nanomaterials. It can be described as a counterpart of graphene. Like graphene, it has excellent biocompatibility and unique properties making phosphorene very suitable for biosensing applications. Two forms of phosphorene, which are called as BP (Black Phosphorene) and BuP (Blue Phosphorene), have been demonstrated by both experimental and theoretical studies. BuP possesses a buckled honeycomb lattice, whereas BP exhibits a puckered non-planar structure. There is now increasing interest in the unique biological and medical properties of these 2D materials. Our main focus is on the interaction between DNA/RNA nucleobases (NB) and monolayer graphene/phosphorene. Better understanding of the interaction between DNA/RNA nucleobases with these 2D surfaces will provide a better understanding of the same interaction mechanisms for amino acids, peptides and proteins. According to both experimental and theoretical studies, the interactions of biomolecules and 2D materials are long-ranged and very weak. Considering the nature of this interaction, it is very important to focus on vdW (Van der Waals) interactions. The application of some external mechanisms, such as charging, can modify the strength of binding. In this work, the binding mechanism of DNA/RNA nucleobases on 2D monolayer graphene/phosphorene has been studied using the DFT (Density Functional Theory) formalism including vdW-DF2 scheme. In this chapter, we report on the trends of the binding energies and on the effects of the charging on the structural and electronic properties of the graphene/phosphorene nucleobases systems. The results presented in this study will be useful for advances in biosensing applications.

B. Salmankurt
Remote Education Center, Sakarya University of Applied Sciences, 54187 Sakarya, Turkey

H. H. Gürel (✉)
Information Systems Engineering Department, Technology Faculty, Kocaeli University, 41001 Kocaeli, Turkey
e-mail: hhakan.gurel@kocaeli.edu.tr

H. Ünlü and N. J. M. Horing (eds.), *Progress in Nanoscale and Low-Dimensional Materials and Devices*, Topics in Applied Physics 144,
https://doi.org/10.1007/978-3-030-93460-6_27

27.1 Introduction

After the discovery of graphene, two-dimensional materials became popular and scientists predicted many two-dimensional materials, theoretically [1]. Many of these materials have also been experimentally synthesized over time [2]. The interactions of two-dimensional materials and biological molecules and the interfaces they form are still an active research area [3]. It is possible to develop various biomedical applications such as biosensors using these interfaces. When the surface of the two-dimensional material interacts with the biomolecules, the electronic and optical properties of the surface are modified [4]. The fixation of organic molecules on the surface, defects, adatoms and external effects such as electric field and charging can be used to ensure to bind organic molecules to the 2D surface [5]. Studies on DNA/RNA nucleobases and amino acids continue with increasing interest [6].

Because of the large surface-to-volume ratio, 2D materials are good candidates as molecular diagnostics, DNA sequencing and biosensors [6–10]. Recognition, detecting and distinguishing of DNA/RNA nucleobases (NB) are very important especially for genomics, forensic sciences, diagnosis and treatment of some of the diseases [10].

BP (Black Phosphorene) and BuP (Blue Phosphorene) differ from graphene in terms of both structural and electronic properties. Unlike graphene, BuP possesses silicene like structure (a buckled honeycomb lattice), whereas BP exhibits a puckered non-planar structure, as shown in Fig. 27.1. Furthermore, BP has a direct band gap with 0.90 eV. Similar to BP, BuP is an indirect band gap semiconductor with 1.92 eV, as shown in Fig. 27.2 [11, 12].

In this work, the interaction mechanism of NB with Graphene, BuP and BP were performed by using Density Functional Theory (DFT). Hence, obtained results in

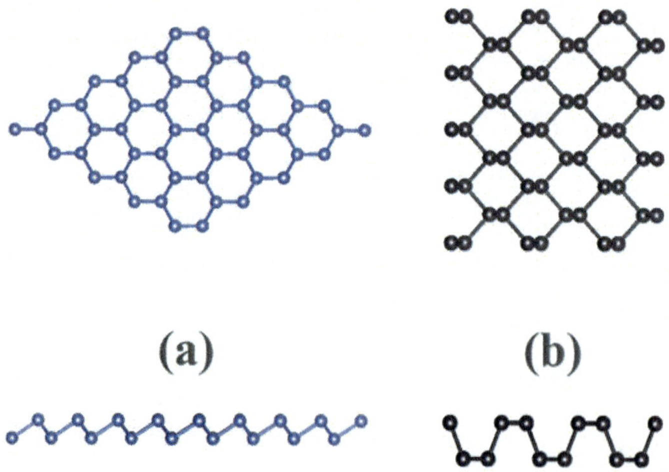

Fig. 27.1 Top and side view of **a** BlueP and **b** BlackP

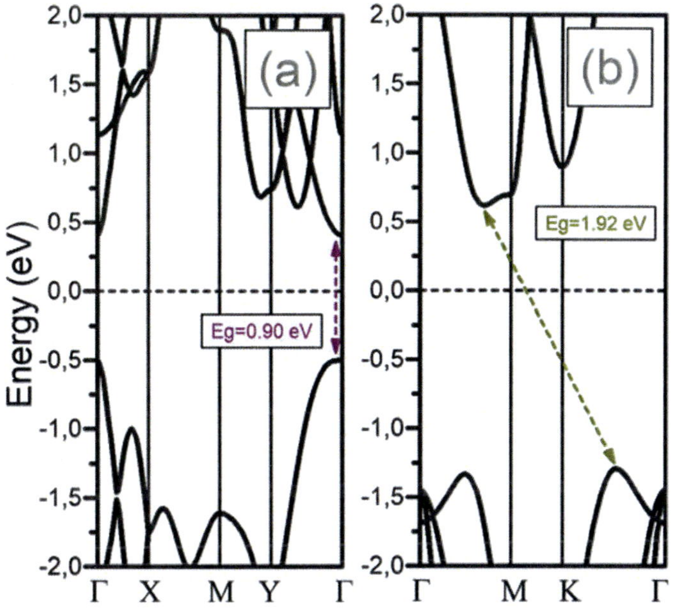

Fig. 27.2 Band structure of 1 × 1 **a** BP and **b** BuP

this study are expected to increase the significance of 2D monolayers for biosensing applications. Moreover, recent progress in charging molecule-interface interactions has shown us a new way for modifying the binding mechanism [13–15]. Adding or removing charge to/from the systems not only allow us to identify a NB but also to change the binding energies. So, in this work, the charging process is applied to all NB-2D systems. Obtained results have provided a good way for understanding of binding mechanism for each system.

27.2 Computational Details

We have performed Density Functional Methods (DFT) for exploring the adsorption geometries, energies and dynamics of NB/2D interfaces. Our calculations are made by using the PW-PP approach within DFT as implemented in the Quantum Espresso code [16]. The DFT has been realized in the GGA, using the PBE method with DFT-D2 to define vdW interactions [17, 18]. The electron-ion interaction is described by using ultrasoft PP [19]. The Brodyden -Fletcher-Goldfarb-Shanno (BFGS) minimization scheme was used in geometry optimization [20]. In order to simulate the two-dimensional monolayer system, we used a 5x5 supercell geometry (Fig. 27.1) with a vacuum space of about 19 A in the z-direction. It allows us the interaction between two adjacent unit cells in the periodic cell is negligible. To explore the effects of larger supercell geometry, we also calculated the interaction energy between two periodic molecules along the xy plane for 5x5 supercell which is about 10 meV. Thus, it is computationally effective strategy to continue to calculations with 5x5 supercell geometry.

Also, we have realized that it is suitable to use 3 × 4 and and 4 × 4 supercell geometries (Figure 27.1) for BlackP and BlueP, respectively. For the Brillouin zone integration, we used a 3 × 3 × 1 k-points mesh [21]. The maximum value of PW cutoff energy is 35 Ry, while the electronic charge density is expanded in a basis cutoff up to 380 Ry. Integration up to the Fermi surface is done by a smearing technique using a Methfessel-Paxton (MP) smearing of 0.005 Ry. The relaxation procedure started in the middle of the supercell on xy plane. To calculated the band structure, it has been used the special 60 k points along the high symmetry directions. The geometric structures are drawn by using Vesta software [22].

27.3 Results and Discussions

27.3.1 *Relaxed Adsorption Structures and Adsorption Energies*

It is well known that binding energy E_b, can be expressed as follows.

$$E_b = E_{2D} + E_{NB} - E_{2D+NB} \quad (27.1)$$

E_{2D} is the total energies of the bare Graphene, BP or BuP, E_{NB} is total energies of bare the NB (NB= Adenine (A), Guanine (G), Cytosine (C), Thymine(T) and Uracil (U)) and E_{2D+NB} is the total energies of NB adsorbed to each 2D supercell. The binding energy can be taken as an indicator of strength of interaction between biomolecule and nanolayer. As an example the optimized geometries of Guanine physiosorbed on a graphene, BuP and BP sheet are presented in Fig 27.3. respectively.

The adsorption energies and average vertical distances are shown in Table 27.1 for graphene and in Table 27.2 for BP and BuP, respectively.

All NB are placing parallel to graphene, BuP and BP layer. It is possible that being parallel is the energetically more favourable due to larger contact area. It is clear that the nature of the interaction between DNA/RNA nucleobases on the 2D monolayer surface comes from vdW interaction. Hence, it is included Grimme-vdW-D2 correction to mimic vdW interaction. Guanine has the largest adsorption energy among all the nucleobases for all 2D monolayers. The larger binding energy is come from its relative high polarizability compared to other nucleobases [26].

GR-NB interactions show that the interacting energies are in the order of G>A>T>C>U [9, 10, 13, 23, 27, 28]. Furthermore, the adsorption of NB on WS_2, MoS_2 and h-BN, are similar to that of GR. The order of binding energies of the

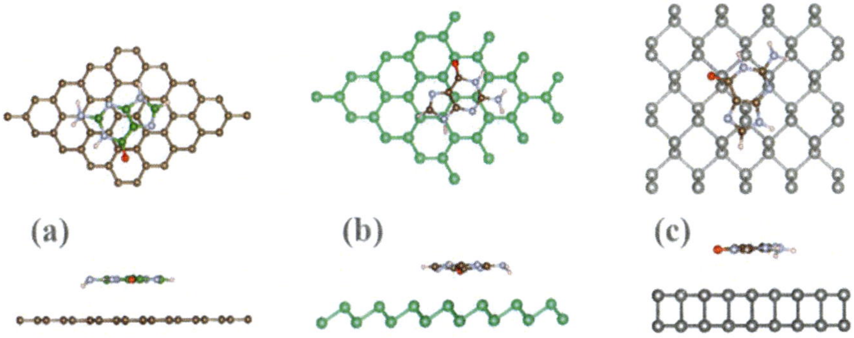

Fig. 27.3 Top and side views of optimized binding geometry of individual Guanine (G) on **a** Graphene, **b** BuP and **c** BP, respectively. Part **a** is reprinted from, Hikmet Hakan Gürel and Bahadır Salmankurt. Binding mechanisms of DNA/RNA nucleobases adsorbed on graphene under charging: first-principles van der Waals study. Mater. Res. Express 4 065,401 [13]. https://doi.org/10.1088/2053-1591/aa6e67.

Table 27.1 Calculated binding energies and average distance of Nucleobases on Phosphorene (in eV) with previous studies. It is reprinted from, Hikmet Hakan Gürel and Bahadır Salmankurt. Binding mechanisms of DNA/RNA nucleobases adsorbed on graphene under charging: first-principles van der Waals study. Mater. Res. Express 4 065,401 [13]. https://doi.org/10.1088/2053-1591/aa6e67.

NB	Graphene				
	DFT-D2 [13]	PBE+vdW [23]	vdW-DF [23]	DFT-D2 [23]	d
G	−0.70	−0.99	−0.74	−0.77	3.15
A	−0.59	−0.85	−0.63	−0.64	3.20
C	−0.52	−0.76	−0.58	−0.57	3.17
T	−0.57	−0.76	−0.60	−0.58	3.19
U	−0.50	–	–	−0.50	3.14

Table 27.2 Calculated binding energies (in eV) and average distance (Å) of Nucleobases on BP and BuP with previous studies

NB	Phosphorene						
	BP	BuP	BuP [24]	BP [25]	BP [26]	d_{BP}	d_{BuP}
G	−0.65	−0.64	−0.62	−0.84	−0.97	3.21	3.20
A	−0.51	−0.46	−0.48	−0.70	−0.89	3.33	3.26
C	−0.49	−0.46	–	−0.65	−0.75	3.16	3.26
T	−0.47	−0.40	−0.46	−0.65	−0.79	3.32	3.44
U	−0.44	−0.37	–	–	−0.69	3.30	3.33

NB with BP is G>A>C>T>U and with BuP is G>C>A>T>U. Our order of binding energies of BP is excellent agreement with the previous reported data [24, 29]. The binding energy order of Bp and BuP is significantly different from that of other 2D materials such as graphene, MoS_2 and WS_2 (G>A>T>C>U) calculated by DFT-D2 method [9]. Main difference possibly comes from the structure of BlueP and BlackP. Because of the same buckled structure, both silicene, germanene and BlueP have some binding energy order [30]. The presented binding energies are in an agreement with previous obtained data [31–33].

27.3.2 *Effects of Charging on Electronic Energy and Band Structure*

In recent years, it is possible to see several computational investigations on charged nanostructures in the literature [34–40]. In this part, we have focused electronic properties and band structure of the system how can be affected when they are charged. The charging mechanism can be calculated via increasing/decreasing the number of electrons in the unit cell.

In this section, it is explained the interaction mechanisms of NB and graphene, BP and BuP can be changed by applied charging.

The charging can be defined as follows, $Q > 0$ shows a missing electron, while $Q < 0$ shows an extra electron for per unit cell (ucell). Also, $Q = 0$ shows the neutral case. It is shown that band structure of U on graphene surface for the neutral $Q = 0$ e/ucell, $Q = -1$ e/ucell and $Q = +1$ e/ucell cases in Fig. 27.4. It can be easily seen that in Fig. 27.4, Fermi level moves down for $Q = +1$ e/ucell and Fermi level moves up for $Q = -1$ e/ucell. The movement of Fermi level can be observed for other nucleobases.

The graphene band structure has a Dirac semimetal character. It keeps Dirac semimetal character even when the NB adsorbed on the graphene. On the other hand, this Dirac semimetal character has been changed under charging. According to Fig. 27.4, changing the number of electrons of NB and graphene system leads to NB on graphene surface system becomes metal. Hence, it is possible to modify the electronic properties and band structure of DNA/RNA nucleobases and graphene system by changing the number of electrons of the system.

In Fig. 27.5, it is shown that the band structure of U on BuP as an example for neutral and charged cases ($Q = \pm 1$ e/ucell). After adding an extra electron ($Q = -1$ e/ucell) to the U on BlueP system, Fermi level moves up with respect to position of neutral case. On the other side, after removing an electron ($Q = +1$ e/ucell) from the U on BuP system, Fermi level moves down with respect to position of neutral case. Adsorption process under charging with one added/removed electron breaks the semiconductor character of both BuP and BP. After adsorption of all DNA/RNA NB on BP/BuP system, under charging with one added/removed electron ($Q = \pm 1$

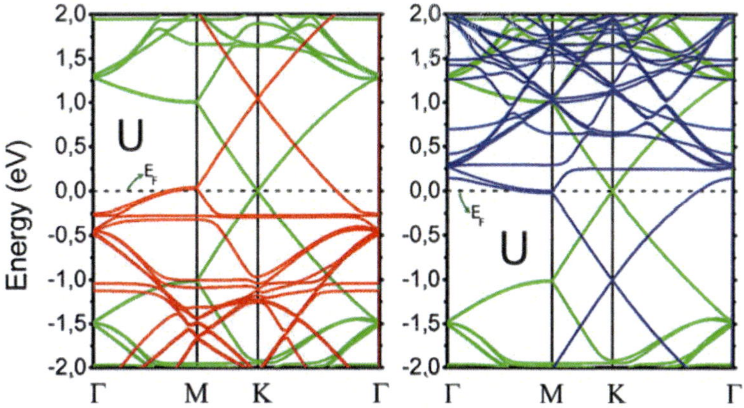

Fig. 27.4 (Color Online) Band structure of Uracil on GR. Green line shows the Q = 0 neutral case; blue line shows Q = −1 e/ucell case; red line shows Q = +1 e/ucell case. It is reprinted from, Hikmet Hakan Gürel and Bahadır Salmankurt. Binding mechanisms of DNA/RNA nucleobases adsorbed on graphene under charging: first-principles van der Waals study. Mater. Res. Express 4 065,401 [13]. https://doi.org/10.1088/2053-1591/aa6e67. © IOP Publishing. Reproduced with permission. All rights reserved

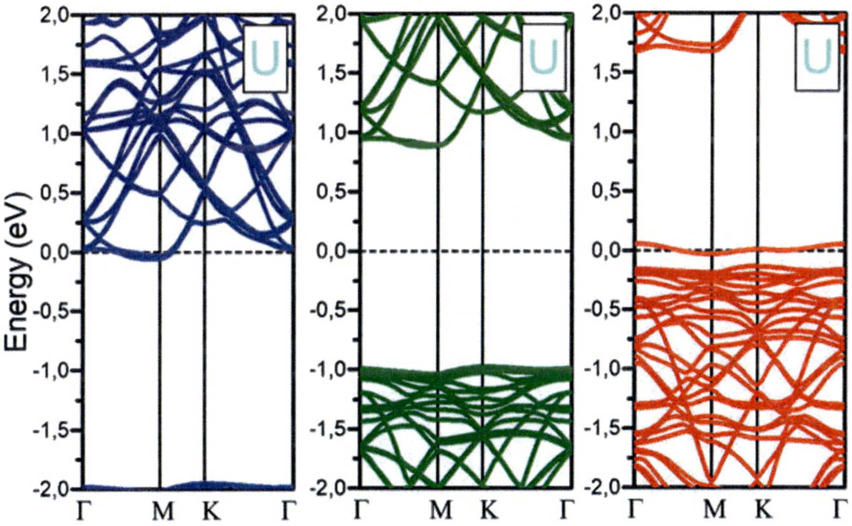

Fig. 27.5 Band structure of U on BlueP. The band structure in the middle shows the neutral Q = 0; on the left shows Q = −1 e/ucell; on the right shows Q = +1 e/ucell. Dash line shows that the Fermi level and it is set to zero

e/ucell), becomes metallic such as U on BuP as shown in Fig. 27.5. It is possible say that semiconductor–metal transition occurs while charging for NB on BP and BuP.

27.3.3 Simulating Adsorbent Behavior Under Charging via Pulling Mechanism

The pulling mechanism can be utilized to show the binding character of NB on 2D systems under charging. It has been calculated total energies of the optimized geometries of NB and graphene, BP and BuP system while a NB is pulled out along z direction perpendicular the monolayer plane. For avoiding movement of 2D monolayer while pulling process, it has been fixed the carbon/phosphor atoms at the corner of the monolayer. The pulling energy can be expressed as follows; $Ep = E_T [NB + 2D; Q; d] - E_T [NB + 2D;Q; d = 0]$. NB indicates each of nucleobases. Ep gets its maximum value by E*pmax*. Obtained pulling energy results for graphene are presented in Fig. 27.6. The obtained results suggest that E*pmax* can be considered as a strength of the binding between NB and 2D monolayer system. We can simply say that E*pmax* can be considered as the energy barrier to pull out the adsorbed NB from 2D monolayer surface. In case of NB and graphene system, there is a strong correlation between E*pmax* and the charging. When we allow to be $Q < 0$, E*pmax* increases. On the other hand, for $Q > 0$ decreasing of the E*pmax* is observed. For example, in adenine and graphene system, E*pmax* is 0.33 eV for $Q = +1$ e/ucell. However, this pulling energy up to 0.72 eV for $Q = -1$ e/ucell. It suggests 118% more energy to pull out A from graphene surface with respect to $Q = +1$ e/ucell. Hence, in uracil and graphene system, the E*pmax* energies are 0.47 eV and 0.53 eV for $Q = +1$ e/ucell and $Q = -1$ e/ucell, respectively. The amount of changing of E*pmax* is different among NB. The difference depends on the amount of charging on NB and graphene. Considering Bader charge calculations, when it is added an extra electron to the system charge distribution between graphene and adenine is much lower than uracil and graphene system. Because of the charge redistribution, the electrostatic repulsion/attraction is much bigger for uracil and graphene system than adenine and graphene system. Because of this reason, when it is added an extra electron, it is harder to pull out adenine from graphene plain than uracil. When it is removed an electron from the system just direct opposite, it becomes much harder to pull out uracil from graphene plain than adenine.

Pulling behaviour of NB on BP and BuP are shown in Fig. 27.7 and Fig. 27.8, respectively. According to our latest calculations, For BP, when one more electron added to system ($Q = -1$ e/ucell), Ep_{max} slightly increases with respect to its neutral case except for U. For BuP, when one more electron added to system ($Q = -1$ e/ucell), increasing of Ep_{max} with respect to its neutral case is much higher than BP. For example, we get maximum rising in C+BuP system for $Q = +1$ e/ucell. It means that we need to 100% more energy to pull out C from BuP with respect to neutral state $Q = 0$. It is also needed 55% more energy to pull out U from BuP surface for

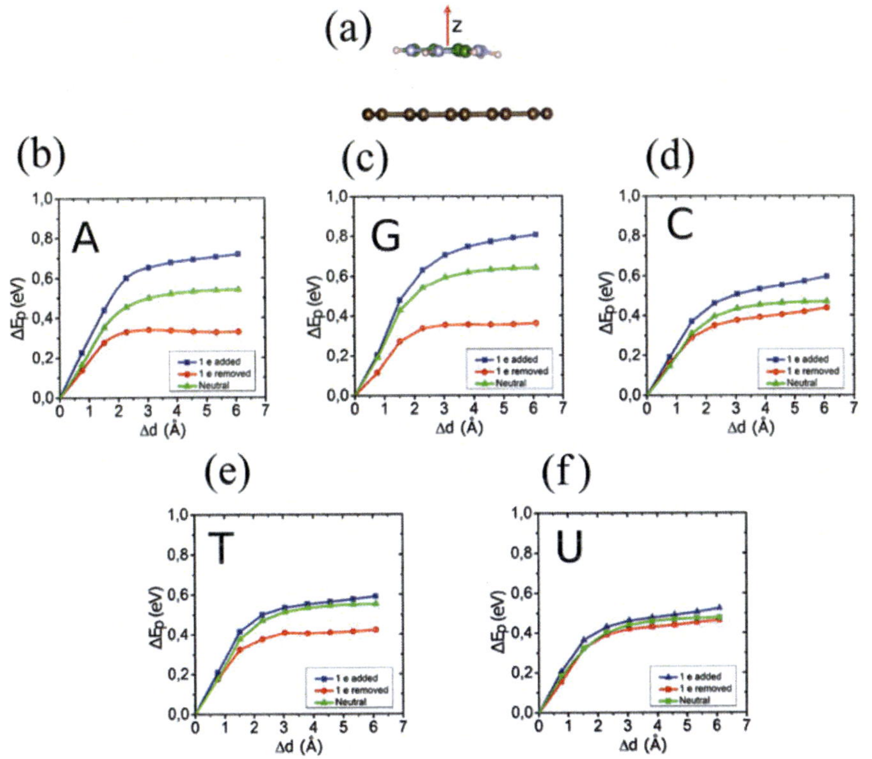

Fig. 27.6 (Color Online) Pulling technique of a NB from graphene surface (**a**). Pulling energies of Adenine (**b**), Guanine (**c**), Cytosine (**d**), Thymine (**e**) and Uracil (**f**) from graphene surface. Green line shows the neutral Q = 0; blue line ishows Q = −1 e/ucell; red line shows Q = +1 e/ucell. It is reprinted from, Hikmet Hakan Gürel and Bahadır Salmankurt. Binding mechanisms of DNA/RNA nucleobases adsorbed on graphene under charging: first-principles van der Waals study. Mater. Res. Express 4 065,401. https://doi.org/10.1088/2053-1591/aa6e67. © IOP Publishing. Reproduced with permission. All rights reserved

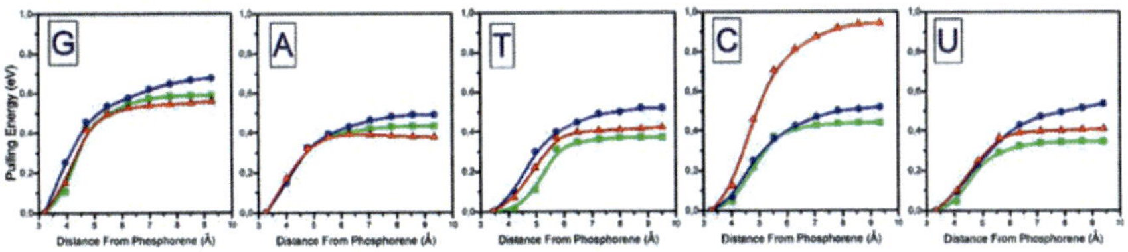

Fig. 27.7 (Color Online) Pulling energies of a NB from BuP layer. Green line shows the neutral Q = 0; blue line shows Q = −1 e/ucell; red line shows Q = +1 e/ucell

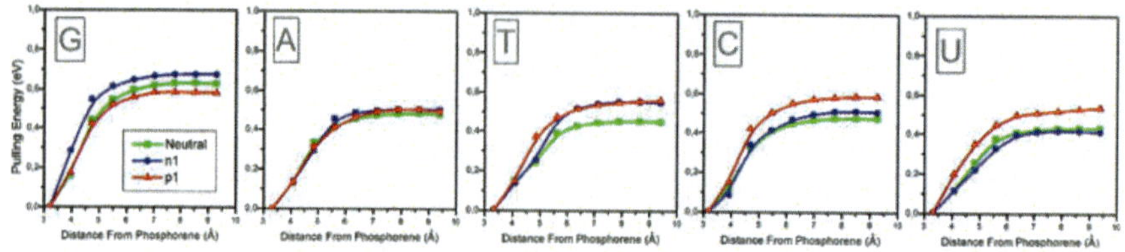

Fig. 27.8 (Color Online) Pulling energies of a NB from BP. Green line shows the neutral Q = 0; blue line shows Q = −1 e/ucell; red line shows Q = +1 e/ucell

under Q = −1 e/ucell. It is possible to increase these examples. There are several factors to affect the pulling mechanism under charging. The main reason behind these differences is the amount of charge redistribution on system. Considering the obtained Bader charge analysis, the charging can change the charge redistribution so electrostatic repulsion/attraction can be different between BP/BuP and NB. Another reason is that structural properties of the monolayers and NB can play a key role on this different behaviour. Binding mechanism is also affected by structural geometry of BP and BuP. In this context, we also need to consider the structural geometry of NB such as ring geometry, number of O atoms, etc. It is observed that O atom responds charging much more than other atoms. It may come from the electronegativity of O. We simply conclude that NB act mainly electron acceptor from BP/BuP monolayer.

It is also calculated magnetic moments of the NB 2D monolayer system under for neutral and charged cases. It is applicable method to tune the magnetic moment of the 2D monolayer via adsorbed with NB or changing the number of the electrons of the systems. However, the spin polarized simulations show that it is not observed any magnetic moment ($\mu = 0$).

27.4 Conclusion

In this study, binding mechanism of DNA/RNA NB on 2D graphene, BP and BuP monolayer have been simulated using DFT method including vdW interactions. It is aimed to investigate effects of charging on binding mechanism. Our calculations show that the nature of adsorption is mainly due to van der Waals interactions. All the NB are physisorbed onto graphene, BP, and BuP surface and the strength of the interaction of the DNA/RNA NB with BlackP is G>A>C>T>U with BlueP is G>C>A>T>U. The order of the binding energy of graphene, MoS_2 and WS_2 is G>A>T>C>U and it is significantly different from BP and BuP.

In neutral cases, adsorption of NB on BP has no effect on its direct bandgap character. On the other side, adsorption of NB on BuP changes its indirect band gap and also band gap value.

After adsorption of G and A, BuP becomes direct band gap semiconductor with lower band gap value with respect to pristine BuP. According to our band structure calculations Fermi level moves down for Q = +1 e/ucell and moves up for Q = −1 e/ucell. In case of Q = ±1 e/ucell, semiconductor character of both NB adsorbed on BuP and BP are broken, and they become metal. There is a semiconductor–metal transition for both BP and BuP for different level of charging. In this study, we showed that binding energies could be modifying via adding or removing to/from NB and graphene, BP and BuP system. Our recent analysis is clearly showed that binding mechanism, pulling energy, electronic properties are strongly depended to the total number of charges of the system. Graphene, BP and BuP are good candidates for the applications in fast sequencing devices for DNA/RNA NB. Furthermore, our results have the potential to contribute to experimental studies and may be valuable for advances in the fields of biosensors and nano-biotechnology.

Acknowledgements H.H. Gurel acknowledges The Scientific and Technological Research Council of Turkey (TÜBİTAK, Grant Number: MFAG-114F453), for financial support.

References

1. M. Xu, T. Liang, M. Shi, H. Chen, Chem. Rev. **113**, 3766 (2013)
2. M. Naguib, Y. Gogotsi, Acc. Chem. Res. **48**, 128 (2015)
3. A. Rochefort, J.D. Wuest, Langmuir **25**, 210 (2009)
4. Y.H. Lu, W. Chen, Y.P. Feng, P.M. He, J. Phys. Chem. B **113**, 2 (2009)
5. T. Zhang, Q. Xue, M. Shan, Z. Jiao, X. Zhou, C. Ling, Z. Yan, The Journal of Physical Chemistry C **116**, 19918 (2012)
6. Y. Ding, X. Wang, L. Xie, X. Yao, W. Xu, Chem. Commun. **54**, 9259 (2018)
7. Y. Yang, A.M. Asiri, Z. Tang, D. Du, Y. Lin, Mater. Today **16**, 365 (2013)
8. L. Kou, T. Frauenheim, C. Chen, The Journal of Physical Chemistry Letters **5**, 2675 (2014)
9. H. Vovusha, B. Sanyal, RSC Adv. **5**, 67427 (2015)
10. Y. Yin, J. Cervenka, N.V. Medhekar, The Journal of Physical Chemistry Letters **8**, 3087 (2017)
11. H. Liu, A.T. Neal, Z. Zhu, Z. Luo, X. Xu, D. Tománek, P.D. Ye, ACS Nano **8**, 4033 (2014)
12. J. Xie, M.S. Si, D.Z. Yang, Z.Y. Zhang, D.S. Xue, J. Appl. Phys. **116**, 073704 (2014)
13. H.H. Gürel, B. Salmankurt, Mater. Res. Express **4**, 065401 (2017)
14. X. Li, Y. Yin, X. Chang, Y. Xiong, L. Zhu, W. Xing, Q. Xue, Chem. Eng. J. **387**, 123403 (2020)
15. X. Tan, L. Kou, S.C. Smith, Chemsuschem **8**, 2987 (2015)
16. P. Giannozzi, S. Baroni, N. Bonini, M. Calandra, R. Car, et al., J. Phys. Condens. Matter **21**, 395502 (2009)
17. J.P. Perdew, K. Burke, M. Ernzerhof, Phys. Rev. Lett. **77**, 3865 (1996)
18. S. Grimme, Semiempirical GGA-type density functional constructed with a long-range dispersion correction. J. Comput. Chem. **27**, 1787 (2006)
19. D. Vanderbilt, Phys. Rev. B **41**, 7892 (1990)
20. T.H. Fischer, J. Almlof, J. Phys. Chem. **96**, 9768 (1992)
21. H.J. Monkhorst, J.D. Pack, Phys. Rev. B **13**, 5188 (1976)
22. K. Momma, F. Izumi, J. Appl. Crystallogr. **41**, 653 (2008)
23. J.H. Lee, Y.K. Choi, H.J. Kim, R.H. Scheicher, J.H. Cho, J. Phys. Chem. C **117**, 13435 (2013)
24. T. Gorkan, Y. Kadioglu, E. Akturk, S. Ciraci, Phys. Chem. Chem. Phys. **22**, 26552 (2020)
25. Y. Kadioglu et al., J. Phys. Chem. C **123**, 23691 (2019)
26. D. Cortés-Arriagada, J. Phys. Chem. C **122**, 4870 (2018)
27. D. Le, A. Kara, E. Schröder, P. Hyldgaard, T.S. Rahman, J. Phy. Condens. Matter **24**, 424210 (2012)
28. S. Gowtham, R.H. Scheicher, R. Ahuja, R. Pandey, S.P. Karna, Phys. Rev. B **76**, 033401 (2007)

29. R.L. Kumawat, B. Pathak, J. Phys. Chem. C **123**, 22377 (2019)
30. T. Hussain, H. Vovusha, T. Kaewmaraya, V. Amornkitbamrung, R. Ahuja, Sens. Actuators, B Chem. **255**, 2713 (2018)
31. S. Cui, H. Pu, S.A. Wells, Z. Wen, S. Mao, J. Chang et al., Nat. Commun. **6**, 1 (2015)
32. Y. Jing, et al, Nanotechnology **26**, 095201 (2015)
33. R.L. Kumawat, P. Garg, S. Kumar, B. Pathak, ACS Appl. Mater. Interfaces **11**, 219 (2018)
34. M. Topsakal, H.H. Gürel, S. Ciraci, J. Phys. Chem. C **117**, 5943 (2013)
35. H.H. Gürel, S. Ciraci, J. Phys. Condens. Matter **25**, 435304 (2013)
36. R. Poloni, A.S. Miguel, M. Fernandez-Serra, J. Phys. Condens. Matter **24**, 095501 (2012)
37. C. Attaccalite, L.Wirtz, M. Lazzeri, F. Mauri, A. Rubio, Nano Lett. **10**, 1172 (2010)
38. H.H. Gürel, V.O. Özçelik, S. Ciraci, J. Phys. Condens. Matter **25**, 305007 (2013)
39. H.H. Gürel, S. Ciraci, J. Phys. Condens. Matter **25**, 275302 (2013)
40. H.H. Gürel, M. Topsakal, S. Ciraci Low-Dimensional and Nanostructured Materials and Devices (pp. 261–90). Springer, Berlin (2016)

Chapter 28
Recent Applications of Microfluidics in Bionanotechnology

B. Hacısalihoğlu and Z. P. Çakar

Abstract Microfluidic systems provide many advantages for biological studies, such as rapid and precise control of the cellular microenvironment, multiplexed analysis and integration of multiple experimental steps on a single platform. This chapter reviews major recent microfluidic applications in bionanotechnology, including microfluidic bioreactors, microbial strain development, microbial behavior and single-cell studies. Additionally, recent applications of microfluidics in common molecular biology methods such as nucleic acid amplification and DNA microarrays are discussed.

28.1 Introduction

Microfluidics has been established as a major distinct technological area and microfluidic platforms with novel functions have been continuously introduced in recent years. Besides, studies over the past 30 years resulted in microdevices that enable efficient analysis of systems which were elusive to be investigated by conventional technologies [1]. Microfluidic devices allow miniaturization and parallelization of multiple experiments with the ability of precise and automated control of liquids at minute volumes. Hence, microfluidics offers several advantages in biological experiments such as reduced manual labor and errors as well as the lower consumption of reagents and costs. Developments in the fabrication of microfluidic devices have led to the production of devices at a precision of single micron or even lower [2].

B. Hacısalihoğlu
Independent Researcher, Erzurum, Turkey

Z. P. Çakar (✉)
Department of Molecular Biology & Genetics, Faculty of Science and Letters, Istanbul Technical University, 34469 Maslak, Istanbul, Turkey
e-mail: cakarp@itu.edu.tr

Dr. Orhan Öcalgiray Molecular Biology, Biotechnology and Genetics Research Center (ITU-MOBGAM), Istanbul Technical University, 34469 Istanbul, Maslak, Turkey

H. Ünlü and N. J. M. Horing (eds.), *Progress in Nanoscale and Low-Dimensional Materials and Devices*, Topics in Applied Physics 144,
https://doi.org/10.1007/978-3-030-93460-6_28

Microfluidics has contributed to the biological sciences by, for example, in vitro diagnostics [3], bio-based production processes [4] and microbial ecology [5]. In vitro testing of biological samples has a major role in clinical diagnosis. Point-of-care (POC) tests with a rapid and accurate response and multiplexing capacity could open up new directions in health care delivery in low-resource settings [6]. Microfluidics-based POC tests partially or fully integrate chemical lysis, isolation and amplification of nucleic acids and optical detection on a platform [7]. Microfluidics has also potential to be used in the development of POC tools for rapid and low-cost identification of bacterial strains with antimicrobial resistance [8].

Microfluidic cultivation devices enable to study the effect of large-scale bioreactor conditions on cellular growth dynamics and product formation, providing clues to the strain development for the improved performance in the industrial bioprocesses [9]. Microfluidics is also compatible with conventional cultivation systems and can be used to separate dead cells and debris [10] and for the rapid detection of bacterial contamination in mammalian cell cultures [11].

Microfluidics can enable spatial control and constant tracking of single- cells [2]. Additionally, microfluidic devices provide rapid molecular transport within the channels, hence, allow dynamic control of the chemical environment of cells through fluid manipulation. Such devices are promising tools to investigate stem cell differentiation at single-cell resolution and the formation of organoids [12].

We had previously reviewed the microfluidics applications in bionanotechnology extensively [13]. However, since that review (published in early 2016), there has been an increasing interest in this field, resulting in many recent publications. In this respect, this chapter reviews recent literature on a range of current and potential applications of microfluidics within bionanotechnology. Firstly, it focuses on microfluidic bioreactors and then discusses microfluidics applications in microbial behavior studies, strain development and single-cell studies. Finally, the recent examples of the implementation of microfluidics to commonly used biological techniques such as nucleic acid amplification and DNA microarrays are discussed.

28.2 Microfluidic Bioreactors

Microfluidic bioreactors are promising bioprocessing systems with improved parameter control, the low consumption of reagents, parallel operation, and automation. Microfluidics bioreactor platforms are mainly in the form of droplets and chambers that support a stream of fluids. The discrete droplets are generated in the micropores that shear two immiscible phases in one another [14]. The droplets allow compartmentalization of different cells and experiment conditions in a volume down to even femtoliter scale [15]. As the droplets form a mixed population, encoding and decoding of the droplet content are critical for the tracking of multiple conditions. Spectroscopic encoding based on fluorescent dyes is frequently used for this purpose. Recently, a method based on colored polystyrene beads has been suggested as a potent droplet encoding system. The colored beads were encapsulated to gather

Table 28.1 Selected reports on microfluidic bioreactors

Improvement area	Reference
Encoding/ decoding of the droplet content	[16]
Mimicking biological membranes	[20]
Biomass trapping for perfusion cultures	[23]
Precise control of pH and dilution rate	[24]
Screening and optimization of bioproducts	[25, 26]
Multilayered vessels for the functional organization of different cell types in 3D cell culture	[28]

with biological samples and the content of the droplets in-flow was quantified via automated analysis of images obtained by bright-field microscopy [16]. That report and some other selected reports on microfluidic bioreactors are listed in Table 28.1.

Droplet microfluidics has been recently used to evaluate microbial producer strains for valuable chemicals [17], to study enzyme activity [18] and for the screening of enzyme evolution libraries [19]. By droplet-based microfluidic devices, it is possible to form giant liposomes or giant unilamellar vesicles (GUVs) that contain a lipid bilayer resembling a biological membrane. GUVs have potential as a microbial culture system owing to its natural permeability that allows modification of aqueous conditions inside or outside of the compartment. Immobilization of GUVs provides a direct observation system and improves stability under culture conditions [20]. Liquid droplets coated with hydrophobic particles, which are called liquid marbles, expand the applicability of droplet fluidics as an in vitro 3D model of cell–cell interactions [21]. Recently, real-time differentiation of the embryonic body has been monitored by using a transparent silica nanoparticle coating on the liquid marbles [22].

Microfluidic devices in chamber format have been described for cultivations. Pico-liter sized chambers enable the restriction of cellular growth to monolayers and provide a proper arrangement for imaging and image analysis [23]. Microfluidic chamber systems that allow the flow of fresh media while trapping the biomass in special zones in the bioreactor are suitable for perfusion cultures, which have potential applicability in the optimization of macroscale fermentations with immobilized cells. Such a microfluidic perfusion culture was assessed for the continuous invertase production by the yeast *Saccharomyces cerevisiae*. Precise control over the process parameters such as dilution rate and pH supported stable cell growth and enzyme activity for an extended duration [24].

Microfluidic reactors are also efficiently used in strain screening and optimization of microbial production of biochemical compounds. A microfluidic device containing 12 individual bioreactors has been demonstrated as an efficient system to reduce the time required to collect meaningful data during the screening of yeast strains for lactic acid production [25] Down-scaling of complex processes also allows throughput for production optimization. The optimized enzymatic production of levodopa and dopamine in a microfluidic system maintained the increased yields when up-scaled to a milliliter system [26]. Besides, such systems can be implemented in algae culturing to accelerate the tedious work of screening competent strains and fertilizers [27]. Microfluidic bioreactors provide novel platforms for 3D cell culture and tissue engineering, as well. Multilayered vessels imitate the natural structure of vascular walls and enable the functional organization of different cell types [28].

28.3 Microbial Behaviour Studies by Microfluidics

Microfluidic technologies have been recognized for providing a novel in situ system to study microbial behavior. Some of the recent reports on microfluidics-based monitoring of microbial behaviour are listed in Table 28.2. Different geometrical structures at the sub-micron scale have been fabricated; the motility characteristics of bacteria in such structures could be relevant in the context of host-microbe interactions [29]. Recently, microfluidic approaches have been adapted for the interrogation of microbial interactions in a variety of ecosystems. Microbial habitats are often defined by their heterogeneity, from which certain microorganisms benefit by their chemotactic ability. A chemotaxis assay has been demonstrated to analyze aquatic microbial behavior at the microscale, which is difficult to achieve by conventional oceanographic techniques. The platform was designed as an array of wells that support the chemotactic accumulation of microorganisms, where each well was connected to the exterior seawater environment by microfluidic channels [30]. More recently, Täuber et al. designed a microfluidic platform that facilitates analysis of cellular behavior under dynamic environmental conditions at the single-cell level. A model organism, *Corynebacterium glutamicum*, was cultivated for the first time, using nutrient profiles mimicking the environmental conditions present in large-scale bioreactors [31].

Microfluidics overcomes the limitations in real-time microscopic studies with opaque backgrounds such as soil. A microfluidic device, Tracking Root Interactions System (TRIS), allowed the growth of roots of *Arabidopsis* seedlings into nine independent microfluidic chambers. The platform allowed live-imaging of chemotactic colonization of fluorescently labeled bacteria on the root sections, which could give clues to the complex root microenvironment [32]. Microfluidics also opens up the opportunity to study chemotaxis in the context of variations among individuals of a microbial population. In a recent study, the differences in chemotactic sensitivity among a clonal bacterial population have been established by using a microfluidic

Table 28.2 Recent examples of microfluidics-based monitoring of microbial behaviour

Microbial behaviour	Reference
Bacterial motility in different geometrical structures	[29]
Aquatic microbial behavior at the microscale	[30]
Cellular behaviour under dynamic environmental conditions at the single-cell level	[31]
Chemotactic colonization of bacteria on the plant root	[32]
Response to antimicrobial agents	[34]

device in the form of an iterative T-maze which had a different range of chemical gradient at each junction, so that bacteria were compelled to decision-making [33].

Microfluidics has been applied to study microbial response to drugs and antibiotics for over a decade. Advanced microscopy techniques implemented within microfluidic systems allow the analysis of single cells quantitatively which has enabled the comparison of responses of wild-type and mutant bacteria in response to antimicrobial agents [34]. Conventional antibiotic susceptibility tests employ either disc diffusion or broth dilution methods to assess the inhibition of bacterial growth in the presence of various concentrations of antibiotics. Microfluidic systems reduce the occurrence of false results by eliminating manual intervention. However, such systems necessitate a reliable concentration gradient. To this end, centrifugal microfluidic platforms that exploit centrifugal flow for mixing the fluids have been presented [35]. Alternatively, Shi et al. developed a microfluidic chip with nine improved micro-channels for mixing the liquids under laminar flow [36].

28.4 Strain Development

The advancement of sustainable bioindustries requires new tools that expedite the strain construction process, which typically consists of iterative design-built-and-test cycles. Microfluidics has the potential to be incorporated within the synthetic biology workflow including gene assembly, DNA delivery into host strains and phenotypic evaluation of the engineered strains [37]. Biological applications often require efficient merging and mixing of reagents. Digital microfluidic (DMF) devices based on the electro-wetting phenomenon address droplets discretely, therefore they are advantageous for such fluidic operations. DMF has been applied for the automation of bacterial transformation by electroporation [38] and heat-shock protocols [39], which are important techniques in the rational construction of mutant libraries.

After construction, the genetic libraries—typically of high diversity- must be screened to detect the rather rare mutants with phenotypes of interest. Microfluidics offers compartmentalized cell culture settings suitable for high-throughput screening. Moreover, microfluidic devices that have the feature of targeted cell isolation from a heterogeneous population have been developed. A recent example was based on a self-assembled nanoporous membrane that enabled nutrient transport into micro-compartments and supported chemostat-like growth conditions. The compartments were connected to another H-shaped channel made up of a UV-curable material which enabled selective blocking of the micro-chambers that contained non-target cells [40]. Syntrophic Co-culture Amplification of Production phenotype (SnoCAP) has also been introduced recently to facilitate the phenotype evaluation in strain development. It employed a metabolic cross-feeding between producer and sensor strains to translate the concentration of target molecules into distinct co-culture growth features [41].

Microfluidic screening systems have also been used in the context of non-directed strain improvement. Non-GMO approaches such as random mutagenesis and evolutionary engineering are preferred especially in the development process of strains used in the food industry [42, 43]. The application of screening systems using microfluidics in strain improvement of lactic acid bacteria has been recently reviewed [44]. In one of the studies, drop-based microfluidics enabled the detection of the rare *Lactococcus lactis* variants with improved riboflavin production from a mutant library obtained by mild chemical mutagenesis [45]. Microdroplets have also been employed in the evolution-based selection of microalgae variants with overproduction of lipids [46]. Most recently, a microbial microdroplet culture system (MMC) provided an integrated platform for automated, high-throughput cultivation and adaptive evolution of microorganisms. In a proof-of-concept study, adaptive evolution of methanol-essential *Escherichia coli* strain for 18 days in MMC resulted in strains with higher growth rates [47].

28.5 Single-Cell Studies

The role of non-genetic heterogeneity of cells in the biological processes such as tumor development and stem cell differentiation have been already recognized for a few decades. However, considering the sizes and concentration of cellular components, the conventional analysis tools are limited in terms of their sensitivity and detection limits. In this respect, several methods have been developed for microfluidic single-cell studies. A successful captivation of the single cells is the basis of single-cell studies. Effective emulsification of single cells is critical and can be a challenge due to cell sedimentation and aggregation. The efficiency of single-cell encapsulation of THP-1, a human monocytic leukemia cell line, in microdroplets was increased by the use of neutral buoyancy [48]. There are also several single-cell trapping strategies based on hydrodynamic methods, electrophoresis, and colloids [49]. Of these, hydrodynamic methods such as droplet microfluidics can achieve high throughput but may require integration to other methods to improve the accuracy of the analysis [50]. Microfluidic platforms that integrate isolation and downstream analysis are promising for the detection of rare circulating tumor cells, which have been considered as an alternative to biopsy [51].

Zhou et al. integrated a microfluidic device to impedance spectroscopy. The differentiation of single mouse embryonic stem cells captured by hydrodynamic trapping was analysed by tracking the variation of the cellular electrical parameters at different time points [52]. Electrical impedance spectroscopy was also proposed as a label-free tool for the dynamic single-cell studies of microorganisms. Zhu et al. described the use of impedance spectroscopy in the microfluidic analysis of the growth of single *Schizosaccharomyces pombe* cells [53].

In contrast to the analysis of total RNA obtained by multicell lysate, single-cell transcriptome analysis provides information about cell-specific regulations. A "digital hydrolysate" approach has been introduced to determine the number of single-cell RNA sequencing analysis required to represent a cell population. Multiple single-cell transcriptomes were computationally averaged and compared to conventional multicell transcriptome analysis. It was concluded that 15 single-cell transcriptomes were sufficient to characterize the fluorescence ubiquitination-based cell cycle indicator (FUCCI)-expressing-cells [54].

Single-cell gene expression has the potential to provide clues about cancer treatment selections. In this respect, microfluidics has been used for multiplexed gene expression profiling of human cancer cells in response to drug treatments. The designed microchip integrated the cell lysis and reverse transcription of the released mRNA, and it enabled the detection of the genotoxic effect of the tested drugs [55]. Cancer stem cells are a small portion of the heterogeneous cell population of cancer and are responsible for the metastases. Microfluidics has been effectively applied to identify and quantify cancer stem cells based on the detection of single-cell derived spheres [56]. Additionally, single-cell RNA sequencing has been used to investigate the gene expression dynamics during the early differentiation of human pluripotent stem cells, which enabled the identification of the key transcription factors and signalling pathways involved in the process [57].

28.6 Nucleic Acid Amplification

Nucleic acid amplification techniques enable the detection of the low-copy number of DNA and rare variants. Thus, they have been increasingly used in medical diagnosis [58], forensic sciences [59], and environmental sampling [60].

Polymerase chain reaction (PCR) was the first nucleic acid amplification method developed on microfluidics systems. Selected reports on the improvement of nucleic acid amplification using microfluidic systems are listed in Table 28.3. Recent advances in microfluidic PCR include improvements in the chamber design to prevent bubble formation and reagent evaporation [61]. Another study focused on the design of the microchannel and heat transfer system to enhance the accuracy of the process [62]. Besides, PCR devices that amplify DNA extremely fast (<2 s/cycle) with subsequent detection [63] and melting analysis have been developed [64]. Elimination of post-PCR analysis makes these systems suitable for POC testing [65].

Loop-mediated isothermal amplification (LAMP) is a relatively new nucleic acid amplification method. LAMP is conducted at constant temperatures ranging from 37 to 65 °C. Because of its simpler heating system, LAMP can be an alternative

Table 28.3 Selected reports on improvements of nucleic acid amplification

Improvement area	Reference
Prevention of bubble formation and evaporation	[61]
Heat transfer system for PCR	[62]
Integration to subsequent detection step	[63]
Low-copy nucleic acid detection	[67]
Integration to preceding nucleic acid isolation	[69]
Real-time analysis and quantification of nucleic acid amplification	[71]

to PCR for the implementation in POC testing devices. While the specificity of the PCR depends on the optimized annealing temperatures, LAMP achieves specificity of amplification by more complex primer designs that allow auto-cycling strand displacement after primer annealing [66]. Implementation of microfluidics to reverse transcription LAMP has been useful to detect disease-causing viruses. A wax-printed paper microfluidic chip that conducts reverse transcription LAMP has been developed for rapid detection of Zika virus as fast as 15 min at a limit of detection as low as 1 copy/μL [67]. In another study, the prior sample preparation step has been integrated into isothermal nucleic acid amplification of H1N1 influenza virus on a microfluidic chip utilizing capillary forces through a polydimethylsiloxane (PDMS) surface treatment [68]. Recently, automation of magnetic-bead extraction of nucleic acids has been demonstrated with a direct transfer to nucleic acid amplification [69].

Microfluidic LAMP devices have also the potential to be used in the detection of mutations associated with diseases. Cao et al. designed a centrifugal microfluidic chip integrated with visual detection of LAMP amplification of the molecular markers of myeloproliferative neoplasms. The reaction mixture was centrifuged into the microchambers, in which the target primers were immobilized, and the amplification was detected by color-change [70]. Hardinge et al. combined LAMP with the bioluminescent assay in real-time (LAMP-BART). They encapsulated droplets containing LAMB-BART reagents and DNA stably within hydrogel shells. DNA amplification could be determined from the light output and quantified in real-time [71].

28.7 DNA Microarrays

DNA microarrays have been frequently applied in profiling of gene expression, detection of single nucleotide polymorphisms and nucleic acid-based diagnostics. Microfluidic systems offer advantages over conventional DNA microarray formats in terms of shortened assay time and consistency of the hybridization quality due to the elimination of manual operation [72]. The hybridization signal strength and occurrence of false-positive results in the microfluidic DNA microarray is an important issue, especially for point mutation detection during the screening of genetically inherited diseases. An automated microfluidic DNA microarray device has been developed to achieve precise control over the hybridization parameters, including microfluidic flow and temperature. In this platform, the specificity of the point mutation detection was improved by the application of a graphene oxide treatment step, which quenched the dyes bound to non-hybridized DNA [72].

One of the time-consuming steps of the commercial DNA microarrays is the hybridization of nucleic acids to probes, which is usually performed overnight. A microfluidic DNA microarray platform can enable a significantly reduced hybridization time. A system based on the active electrophoresis that transports the biotin- and fluorophore-labeled DNA samples through the streptavidin-functionalized hydrogel could discriminate the genotypes of five different clinical targets as rapid as 3 h [73].

The morphology of the oligonucleotide spots is another important factor during the signal detection on the hybridization arrays. A uniform deposition and immobilization of the biomolecules are necessary for the proper spot formation, thus for the accuracy of the miniaturized microarrays. The surface characteristics of the microarray substrates can be modified by coating with polymers to control the size and the spreading of the droplets in which the biomolecules are suspended. Sola et al. introduced an *N*, *N*-dimethylacrylamide (DMA) derived polymer coating to increase the hydrophobicity of the surface, which helped prevent non-specific interactions with the surface [74]. This approach was more recently extended to accommodate probes with different chemistries on multi-chemistry DNA arrays [75]. Additionally, integration of cell-free expression systems to DNA microarray on microfluidic platforms enables the production of protein microarrays in a cost-effective way. Such systems lay the foundation for the analysis of complex protein networks [76].

28.8 Conclusions

Microfluidics has provided researchers several advantages such as rapid and precise control over experimental parameters, reduced manual labor, parallelization of experiments and increased throughput. Novel microfluidic tools can enable to study biological phenomena at a single-cell level, which are elusive by macro-scale techniques. Miniaturization of common molecular biology methods such as DNA microarray has improved the assay performances. Future advances in microfluidics can make the large-scale single-cell analysis a practical reality.

Acknowledgements This chapter is dedicated to the memory of Prof. Dr. Fatma Neşe Kök, our beloved Head of the Department of Molecular Biology and Genetics, Istanbul Technical University (ITU), and Director of ITU-MOBGAM Research Center, who passed away on May 28th, 2022.

References

1. N. Convery, N. Gadegaard, 30 years of microfluidics. Micro Nano Eng **2**, 76–91 (2019)
2. O. Scheler, W. Postek, P. Garstecki, Recent developments of microfluidics as a tool for biotechnology and microbiology. Curr Opin Biotechnol **55**, 60–67 (2019)
3. X. Sun, J.J. Wan, K. Qian, Designed microdevices for *in vitro* diagnostics. Small Methods **1**, 1700196 (2017)
4. R.C. Prado, E.R. Borges, Microbioreactors as engineering tools for bioprocess development. Braz J Chem Eng **35**, 1163–1182 (2018)
5. J. Kehe, A. Kulesa, A. Ortiz, C.M. Ackerman, S.G. Thakku, D. Sellers, S. Kuehn, J. Gore, J. Friedman, P.C. Blainey, Massively parallel screening of synthetic microbial communities. Proc Natl Acad Sci USA **116**, 12804–12809 (2019)
6. E. Primiceri, M.S. Chiriacò, F.M. Notarangelo, A. Crocamo, D. Ardissino, M. Cereda, A.P. Bramanti, M.A. Bianchessi, G. Giannelli, G. Maruccio, Key enabling technologies for point-of-care diagnostics. Sensors **18**, 3607 (2018)

7. M. Mauk, J. Song, C. Liu, H. Bau, Simple approaches to minimally-instrumented, microfluidic-based point-of-care nucleic acid amplification tests. Biosensors **8**, 17 (2018)
8. A. Aroonnual, T. Janvilisri, P. Ounjai, S. Chankhamhaengdecha, Microfluidics: innovative approaches for rapid diagnosis of antibiotic-resistant bacteria. Essays Biochem **61**, 91–101 (2017)
9. P. Ho, C. Westerwalbesloh, E. Kaganovitch, A. Grünberger, P. Neubauer, D. Kohlheyer, E.V. Lieres, Reproduction of large-scale bioreactor conditions on microfluidic chips. Microorganisms **7**, 105 (2019)
10. T. Kwon, R. Yao, J.P. Hamel, J. Han, Continuous removal of small nonviable suspended mammalian cells and debris from bioreactors using inertial microfluidics. Lab Chip **18**, 2826–2837 (2018)
11. C. Surrette, B. Scherer, A. Corwin, G. Grossmann, A.M. Kaushik, K. Hsieh, P. Zhang, J.C. Liao, P.K. Wong, T.H. Wang, C.M. Puleo, Rapid microbiology screening in pharmaceutical workflows SLAS Technol. 23, 387–394 (2018)
12. S. Sart, S.N. Agathos, Towards three-dimensional dynamic regulation and *in situ* characterization of single stem cell phenotype using microfluidics. Mol Biotechnol **60**, 843–861 (2018)
13. Z.P. Çakar, B. Sönmez, Microfluidics and its applications in bionanotechnology', in *Low-dimensional and nanostructured materials and devices properties, synthesis, characterization, modelling and application* eds. H. Ünlü, NJM. Horing, J. Dabowski (Springer International Publishing Switzerland, 2016), pp. 589–608
14. L. Shang, Y. Cheng, Y. Zhao, Emerging droplet microfluidics. Chem Rev **117**, 7964–8040 (2017)
15. C. Martino, A.J. deMello, Droplet-based microfluidics for artificial cell generation: a brief review. Interface Focus **6**, 20160011 (2016)
16. C.M. Svensson, O. Shvydkiv, S. Dietrich, L. Mahler, T. Weber, M. Choudhary, M. Tovar, M.T. Figge, M. Roth, Coding of experimental conditions in microfluidic droplet assays using colored beads and machine learning supported image analysis. Small **15**, 1802384 (2019)
17. S. Siedler, N.K. Khatri, A. Zsohár, I. Kjærbølling, M. Vogt, P. Hammar, C.F. Nielsen, J. Marienhagen, M.O.A. Sommer, H.N. Joensson, Development of a bacterial biosensor for rapid screening of yeast p-coumaric acid production. ACS Synthetic Biol. **6**, 1860–1869 (2017)
18. M. Girault, T. Beneyton, D. Pekin, L. Buisson, S. Bichon, C. Charbonnier, Y. Del Amo, J.C. Baret, High-content screening of plankton alkaline phosphatase activity in microfluidics. Anal Chem **90**, 4174–4181 (2018)
19. X.W. Diefenbach, I. Farasat, E.D. Guetschow, C.J. Welch, R.T. Kennedy, S. Sun, J.C. Moore, Enabling biocatalysis by high-throughput protein engineering using droplet microfluidics coupled to mass spectrometry. ACS Omega **3**, 1498–1508 (2018)
20. M. Morita, K. Katoh, N. Noda, Direct observation of bacterial growth in giant unilamellar vesicles: a novel tool for bacterial cultures. ChemistryOpen **7**, 845–849 (2018)
21. R.K. Vadivelu, H. Kamble, A. Munaz, N.T. Nguyen, Liquid marbles as bioreactors for the study of three-dimensional cell interactions. Biomed Microdevices **19**, 31 (2017)
22. K. Lin, R. Chen, L. Zhang, D. Zang, X. Geng, W. Shen, Transparent bioreactors based on nanoparticle-coated liquid marbles for *in situ* observation of suspending embryonic body formation and differentiation. ACS Appl Mater Interfaces **11**, 8789–8796 (2018)
23. E. Kaganovitch, X. Steurer, D. Dogan, C. Probst, W. Wiechert, D. Kohlheyer, Microbial single-cell analysis in picoliter-sized batch cultivation chambers. N Biotechnol **47**, 50–59 (2018)
24. E.J. Brás, V. Chu, M.R. Aires-Barros, J.P. Conde, P. Fernandes, A microfluidic platform for physical entrapment of yeast cells with continuous production of invertase. J Chem Technol Biotechnol **92**, 334–341 (2017)
25. D. Totaro, B. Radoman, B. Schmelzer, M. Rothbauer, M.G. Steiger, T. Mayr, M. Sauer, P. Ertl, D. Mattanovich, Microscale perfusion-based cultivation for *Pichia pastoris* clone screening enables accelerated and optimized recombinant protein production processes. Biotechnol. J. **15**, e2000215 (2020)

26. E.J.S. Brás, C. Domingues, V. Chu, P. Fernandes, J.P. Conde, Microfluidic bioreactors for enzymatic synthesis in packed-bed reactors-Multi-step reactions and upscaling. J Biotechnol **323**, 24–32 (2020)
27. Y. Wang, H. Zhao, X. Liu, W. Lin, Y. Jiang, J. Li, Q. Zhang, G. Zheng, An integrated digital microfluidic bioreactor for fully automatic screening of microalgal growth and stress-induced lipid accumulation. Biotechnol Bioeng **118**, 294–304 (2021)
28. J. Liu, H. Zheng, X. Dai, P.S.P. Poh, H.G. Machens, A.F. Schilling, Transparent pdms bioreactors for the fabrication and analysis of multi-layer pre-vascularized hydrogels under continuous perfusion. Front. Bioeng. Biotechnol. **8**, 568934 (2020)
29. M. Nayak, A.S. Perumal, D.V. Nicolau, F.C. van Delft, Bacterial motility behaviour in sub-ten micron wide geometries, in *16th IEEE International New Circuits and Systems Conference (NEWCAS)*, (2018), pp 382–384
30. B.S. Lambert, J.B. Raina, V.I. Fernandez, C. Rinke, N. Siboni, F. Rubino, P. Hugenholtz, G.W. Tyson, J.R. Seymour, R. Stocker, A microfluidics-based in situ chemotaxis assay to study the behaviour of aquatic microbial communities. **2**, 1344–1349 (2017)
31. S. Täuber, C. Golze, P. Ho, E. von Lieres, A. Grünberger, dMSCC: a microfluidic platform for microbial single-cell cultivation of *Corynebacterium glutamicum* under dynamic environmental medium conditions. Lab Chip **20**, 4442–4455 (2020)
32. H. Massalha, E. Korenblum, S. Malitsky, O.H. Shapiro, A. Aharoni, Live imaging of root-bacteria interactions in a microfluidics setup. Proc Natl Acad Sci USA **114**, 4549–4554 (2017)
33. M.M. Salek, F. Carrara, V. Fernandez, J.S. Guasto, R. Stocker, Bacterial chemotaxis in a microfluidic T-maze reveals strong phenotypic heterogeneity in chemotactic sensitivity. Nat Commun **10**, 1877 (2019)
34. M. Elitas, On-chip isoniazid exposure of mycobacterium smegmatis penicillin-binding protein (PBP) mutant using time-lapse fluorescent microscopy. Micromachines **9**, 561 (2018)
35. M. Tang, X. Huang, Q. Chu, X. Ning, Y. Wang, S.K. Kong, X. Zhang, G. Wang, H.P. Ho, A linear concentration gradient generator based on multi-layered centrifugal microfluidics and its application in antimicrobial susceptibility testing. Lab Chip **18**, 1452–1460 (2018)
36. H. Shi, Z. Hou, Y. Zhao, K. Nie, B. Dong, L. Chao, S. Shang, M. Long, Z. Liu, Rapid and steady concentration gradient generation platform for an antimicrobial susceptibility test. Chem Eng J **359**, 1327–1338 (2019)
37. P.C. Gach, K. Iwai, P.W. Kim, N.J. Hillson, A.K. Singh, Droplet microfluidics for synthetic biology. Lab Chip **17**, 3388–3400 (2017)
38. A.C. Madison, M.W. Royal, F. Vigneault, L. Chen, P.B. Griffin, M. Horowitz, G.M. Church, R.B. Fair, Scalable device for automated microbial electroporation in a digital microfluidic platform. ACS Synth Biol **6**, 1701–1709 (2017)
39. E. Moazami, J.M. Perry, G. Soffer, M.C. Husser, S.C.C. Shih, Integration of world-to-chip interfaces with digital microfluidics for bacterial transformation and enzymatic assays. Anal Chem **91**, 5159–5168 (2019)
40. J. Lee, J. Park, T. Kim, Dynamic culture and selective extraction of target microbial cells in self-assembled particle membrane-integrated microfluidic bioreactor array. Anal Chem **91**, 6162–6171 (2019)
41. T.E. Saleski, A.R. Kerner, M.T. Chung, C.M. Jackman, A. Khasbaatar, K. Kurabayashi, X.N. Lin, Syntrophic co-culture amplification of production phenotype for high-throughput screening of microbial strain libraries. Metab Eng **54**, 232–243 (2019)
42. Z.P. Çakar, U.O.S. Seker, C. Tamerler, M. Sonderegger, U. Sauer, Evolutionary engineering of multiple-stress resistant *Saccharomyces cerevisiae*. FEMS Yeast Res **5**, 569–578 (2005)
43. Z.P. Çakar, B. Turanlı-Yıldız, C. Alkım, Ü. Yılmaz, Evolutionary engineering of *Saccharomyces cerevisiae* for improved industrially important properties. FEMS Yeast Res **12**, 171–182 (2012)
44. J. Chen, M. Vestergaard, J. Shen, C. Solem, M. Dufva, P.R. Jensen, Droplet-based microfluidics as a future tool for strain improvement in lactic acid bacteria. FEMS Microbiol. Lett. 365, fny258 (2018)

45. J. Chen, M. Vestergaard, T.G. Jensen, J. Shen, M. Dufva, C. Solem, P.R. Jensen, Finding the needle in the haystack—the use of microfluidic droplet technology to identify vitamin-secreting lactic acid bacteria. MBio **8**: e00526-17 (2017)
46. M. Li, M. van Zee, C.T. Riche, B. Tofig, S.D. Gallaher, S.S. Merchant, R. Damoiseaux, K. Goda, D. Di Carlo, A gelatin microdroplet platform for high-throughput sorting of hyperproducing single-cell-derived microalgal clones. Small **14**, 1803315 (2018)
47. X. Jian, X. Guo, J. Wang, Z.L. Tan, X.H. Xing, L. Wang, C. Zhang, Microbial microdroplet culture system (MMC): an integrated platform for automated, high-throughput microbial cultivation and adaptive evolution. Biotechnol Bioeng **117**, 1724–1737 (2020)
48. H. Liu, M. Li, Y. Wang, J. Piper, L. Jiang, Improving single-cell encapsulation efficiency and reliability through neutral buoyancy of suspension. Micromachines (Basel) **11**, 94 (2020)
49. B. Deng, H. Wang, Z. Tan, Y. Quan, Microfluidic cell trapping for single-cell analysis. Micromachines **10**, 409 (2019)
50. T. Luo, L. Fan, R. Zhu, D. Sun, Microfluidic single-cell manipulation and analysis: methods and applications. Micromachines **10**, 104 (2019)
51. M. Xu, H. Zhao, J. Chen, W. Liu, E. Li, Q. Wang, L. Zhang, An integrated microfluidic chip and its clinical application for circulating tumor cell isolation and single-cell analysis. Cytometry A **97**, 46–53 (2020)
52. Y. Zhou, S. Basu, E. Laue, A.A. Seshia, Single cell studies of mouse embryonic stem cell (mESC) differentiation by electrical impedance measurements in a microfluidic device. Biosens Bioelectron **81**, 249–258 (2016)
53. Z. Zhu, O. Frey, A. Hierlemann, Wide-band electrical impedance spectroscopy (EIS) measures *S. pombe* cell growth in vivo, in *Schizosaccharomyces pombe: Methods in Molecular Biology*. ed. by T. Singleton (Humana Press, New York, 2018), pp. 135–153
54. Y. Chen, J. Millstein, Y. Liu, G.Y. Chen, X. Chen, A. Stucky, C. Qu, J.B. Fan, X. Chang, A. Soleimany, K. Wang, J. Zhong, J. Liu, F.D. Gilliland, Z. Li, X. Zhang, J.F. Zhong, Single-cell digital lysates generated by phase-switch microfluidic device reveal transcriptome perturbation of cell cycle. ACS Nano **12**, 4687–4694 (2018)
55. H. Dong, H. Sun, A microchip for integrated single-cell gene expression profiling and genotoxicity detection. Sensors **16**, 1489 (2016)
56. D. Belgorosky, T. Fernández-Cabada, A.B. Peñaherrera-Pazmiño, Y. Langle, R. Booth, S. Bhansali, M.S. Pérez, A.M. Eiján, B. Lerner, Analysis of tumoral spheres growing in a multichamber microfluidic device. J Cell Physiol **233**, 6327–6336 (2018)
57. X. Han, H. Chen, D. Huang, H. Chen, L. Fei, C. Cheng, H. Huang, G.C. Yuan, G. Guo, Mapping human pluripotent stem cell differentiation pathways using high throughput single-cell RNA-sequencing. Genome Biol **19**, 47 (2018)
58. L. Detemmerman, S. Olivier, V. Bours, F. Boemer, Innovative PCR without DNA extraction for African sickle cell disease diagnosis. Hematology **2**, 181–186 (2018)
59. B. Bruijns, A. van Asten, R. Tiggelaar, H. Gardeniers, Microfluidic devices for forensic DNA analysis: a review. Biosensors **6**, 41 (2016)
60. R. Martzy, C. Kolm, K. Brunner, R.L. Mach, R. Krska, H. Šinkovec, R. Sommer, A.H. Farnleitner, G.H. Reischer, A loop-mediated isothermal amplification (LAMP) assay for the rapid detection of *Enterococcus spp.* in water. Water Res **122**, 62–69 (2017)
61. S.H. Lee, J. Song, B. Cho, S. Hong, O. Hoxha, T. Kang, D. Kim, L.P. Lee, Bubble-free rapid microfluidic PCR. Biosens Bioelectron **126**, 725–733 (2019)
62. M. Mollajan, S.R. Bazaz, A.A. Mehrizi, A thoroughgoing design of a rapid-cycle microfluidic droplet-based PCR device to amplify rare DNA strands. J. Appl. Fluid Mech. **11(1)**, 21–29 (2018)
63. A.R. Jafek, S. Harbertson, H. Brady, R. Samuel, B.K. Gale, Instrumentation for xPCR incorporating qPCR and HRMA. Anal Chem **90**, 7190–7196 (2018)
64. J.T. Myrick, R.J. Pryor, R.A. Palais, S.J. Ison, L. Sanford, Z.L. Dwight, J.J. Huuskonen, S.O. Sundberg, C.T. Wittwer, Integrated extreme real-time PCR and high-speed melting analysis in 52 to 87 seconds. Clin Chem **65**, 263–271 (2019)

65. M. Mauk, J. Song, H.H. Bau, R. Gross, F.D. Bushman, R.G. Collman, C. Liu, Miniaturized devices for point of care molecular detection of HIV. Lab Chip **17**, 382–394 (2017)
66. H. Zhang, Y. Xu, Z. Fohlerova, H. Chang, C. Iliescu, P. Neuzil, LAMP-on-a-chip: revising microfluidic platforms for loop-mediated DNA amplification. Trends Anal Chem **113**, 44–53 (2019)
67. K. Kaarj, P. Akarapipad, J.Y. Yoon, Simpler, faster, and sensitive Zika virus assay using smartphone detection of loop-mediated isothermal amplification on paper microfluidic chips. Sci Rep **8**, 12438 (2018)
68. Y.D. Ma, Y.S. Chen, G.B. Lee, An integrated self-driven microfluidic device for rapid detection of the influenza A (H1N1) virus by reverse transcription loop-mediated isothermal amplification. Sensor Actuat B- Chem 126647 (2019)
69. D.E. Gaddes, P.W. Lee, A.Y. Trick, P. Athamanolap, C.M. O'Keefe, C. Puleo, K. Hsieh, T.H. Wang, Facile coupling of droplet magnetofluidic-enabled automated sample preparation for digital nucleic acid amplification testing and analysis. Anal Chem **92**, 13254–13261 (2020)
70. G. Cao, J. Kong, Z. Xing, Y. Tang, X. Zhang, X. Xu, Z. Kang, X. Fang, M. Guan, Rapid detection of CALR type 1 and type 2 mutations using PNA-LNA clamping loop-mediated isothermal amplification on a CD-like microfluidic chip. Anal. Chim. Acta **1024**, 123–135 (2018)
71. P. Hardinge, D.K. Baxani, T. McCloy, J.A.H. Murray, O.K. Castell, Bioluminescent detection of isothermal DNA amplification in microfluidic generated droplets and artificial cells. Sci Rep **10**, 21886 (2020)
72. S.H. Huang, Y.S. Chang, J.M.J. Juang, K.W. Chang, M.H. Tsai, T.P. Lu, L.C. Lai, E.Y. Chuang, N.T. Huang, An automated microfluidic DNA microarray platform for genetic variant detection in inherited arrhythmic diseases. Analyst **143**, 1367–1377 (2018)
73. Y.K. Jung, J. Kim, R.A. Mathies, Microfluidic hydrogel arrays for direct genotyping of clinical samples. Biosens Bioelectron **79**, 371–378 (2016)
74. L. Sola, F. Damin, M. Cretich, M. Chiari, Novel polymeric coatings with tailored hydrophobicity to control spot size and morphology in DNA microarray. Sensor Actuat B-Chemical **231**, 412–422 (2016)
75. L. Sola, F. Damin, M. Chiari, Array of multifunctional polymers for localized immobilization of biomolecules on microarray substrates. Anal Chim Acta **1047**, 188–196 (2019)
76. N. Kilb, T. Herz, J. Burger, J. Woehrle, P.A. Meyer, G. Roth, Protein microarray copying: easy on-demand protein microarray generation compatible with fluorescence and label-free real-time analysis. ChemBioChem **20**, 1554–1562 (2019)

Chapter 29
Synthesis and Biological Use of Nanomaterials

Manolya Kukut Hatipoglu and Pinar Akkus Sut

Abstract Nanomaterials are the most evolving and developing area of the nanotechnology. They are considered as unique materials regarding their size-dependent properties. Hence, they became essential in human life in particular they are a very valuable tool in modern medicine. Nanomaterials are used for both diagnostic and treatment purposes including design of fluorescent biological markers/labels, synthesis of molecules for diagnosis, drug and gene delivery systems, bio-detection of antibodies and proteins, genetic and tissue engineering, detection and treatment of tumors and contrast agent enhancement for magnetic resonance imaging. The type of the nanomaterial (or nanodevices) varies with the need and the purpose of the applications. Liposomes and micelles, dendrimers, quantum dots, magnetic nanoparticles, gold nanoparticles and silver nanoparticles are some of the most important nanoparticles to be used in biology and medicine. The aim of this chapter is firstly to explain the rationale of utilization of nanomaterials to biology and medicine, secondly to overview the synthesis methods and biological applications of nanomaterials.

29.1 Introduction

Nanomaterials are the advanced products of nanotechnology that are interpreted in assorted fields such as medicine, cosmetics, electronics, coatings and many more. The application of nanoscale technology to medicine is an important contribution to molecular imaging, diagnostics and drug delivery systems. Nanomaterials offers solutions in molecular levels by enabling detection of single molecules in complex

---

M. Kukut Hatipoglu (✉)
Materials Science and Nanotechnology Engineering Department, Yeditepe University, Faculty of Engineering and Architecture, Kayisdagi, Istanbul, Turkey
e-mail: sevim.hatipoglu@fda.hhs.gov

The Food and Drug Administration, 645 S Newstead Ave, St. Louis, MO 63110, USA

P. Akkus Sut
Genetics and Bioengineering Department, Yeditepe University, Faculty of Engineering and Architecture, Kayisdagi, Istanbul, Turkey

H. Ünlü and N. J. M. Horing (eds.), *Progress in Nanoscale and Low-Dimensional Materials and Devices*, Topics in Applied Physics 144,
https://doi.org/10.1007/978-3-030-93460-6_29

biological environments and targeting specific tissues of related diseases for the treatment. Nanomaterials are providing better treatment options compared to conventional methods that are used in physiology, pathology and other specialized medical sciences. Targeted delivery with more precision, lower toxicity and less adverse effects of drugs are just a few advantages that the use nanomaterials brings. The synthesis/preparation methods of nanomaterials varies depending on the need. There is a vast number of techniques starting from a great collection of precursors. These techniques include both dry/physical and wet methods based on the type of nanomaterial [1]. Ball milling, vapor condensation, chemical vapor deposition and laser ablation are considered as dry methods whereas sol-gel, co-precipitation, microemulsion and solvothermal methods are known as wet techniques [1]. There are also green synthesis approaches where microorganisms and plants are used to produce nanomaterials. This chapter contains synthesis/preparation methods and biological use of most common nanomaterials including liposomes, micelles, metal nanoparticles, magnetic nanoparticles, dendrimers and carbon nanotubes. Quantum dots, one of the common nanomaterials that are employed in medicine, are not subject of this chapter. The synthesis and biological applications of quantum dots was reviewed in detail elsewhere [2].

29.2 Liposomes

Liposomes are micrometer or nanometer sized artificial carriers [3]. They can be synthesized from cholesterol and natural non-toxic phospholipids as spherical vesicles [3]. Liposomes have become promising drug delivery systems due to their size, amphiphilic character and biocompatibility since their discovery in the 1970s [4–6]. Therefore, they have been used as one of the earliest nanoscale platform for therapeutic delivery to the central nervous system (CNS) [7]. Properties of liposomes varies to a great extend based on lipid composition, surface charge, size and method of preparation [3]. Liposomes are bilayer particles that give the opportunity to adjust the rigidity or fluidity and charge by the choice of its components. As an example, the use of phosphatidylcholine from egg or soybean creates more permeable and less stable bilayers than long acyl chains which form rigid and relatively impermeable bilayer structure [8–10]. The phospholipid nature of forming closed structures such as spherical particles in aqueous solutions enables transportation of lipid and liquid drugs. Since the liposomes behave amphiphilic in aqueous environment, self-assembly and thermodynamic phase properties entropically affect the enclosure of their hydrophobic sections into spherical bilayers (see Fig. 29.1) [3, 11].

Liposomes can be fabricated with particle size ranging from 10 nm to several micrometers as spherical vehicles. Generally, the polar head groups of the lipid bilayer located in the pathway of interior and exterior aqueous phases. A representative phospholipid structure is shown in Fig. 29.2 [11].

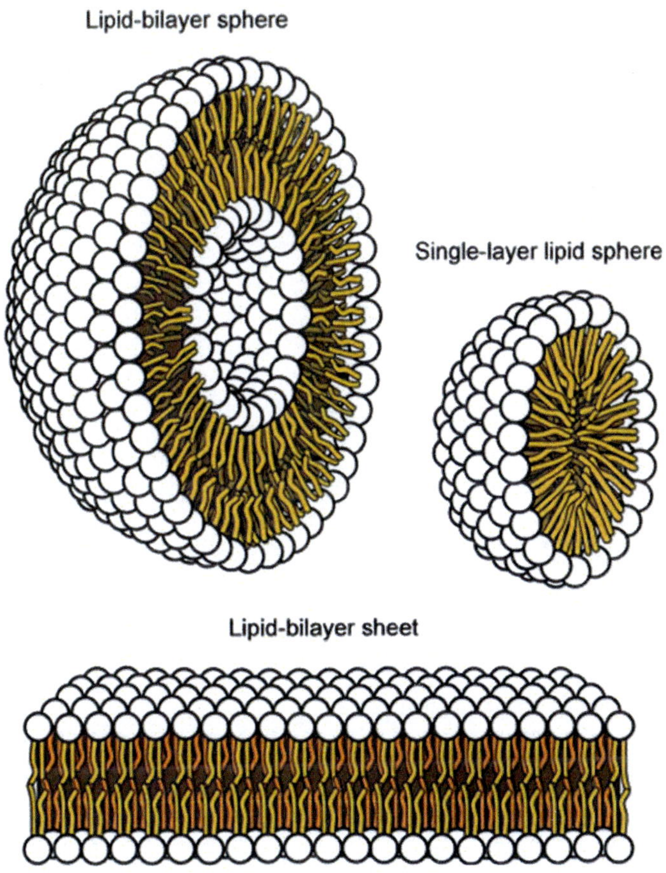

Fig. 29.1 Illustration of a liposome structure. Reprinted from [11] with permission from creative commons

29.2.1 Classification of Liposomes

Liposomes are divided into five categories in terms of work and mechanism of intracellular delivery including; conventional liposomes, pH sensitive liposomes, cationic liposomes, immune liposomes and long circulating liposomes. A list of various types of liposomes are collected in Table 29.1 [12].

Conventional liposomes; were first synthesized by Immordino et al. in 2006 for pharmaceutical applications [13]. This type of liposomes are usually fabricated from phospholipids or lipids such as 1,2-distearoryl-sn-glycero-3-phosphatidyl choline (DSPC), sphingomyelin, egg phosphatidylcholine and monosialoganglioside [14]. ***pH-sensitive liposomes***; were first used in 1987 by Wang and Huang for in vitro gene transfection [15]. There are vast compositions of pH-sensitive liposomes which genuinely bind to cell surface followed by internalization into

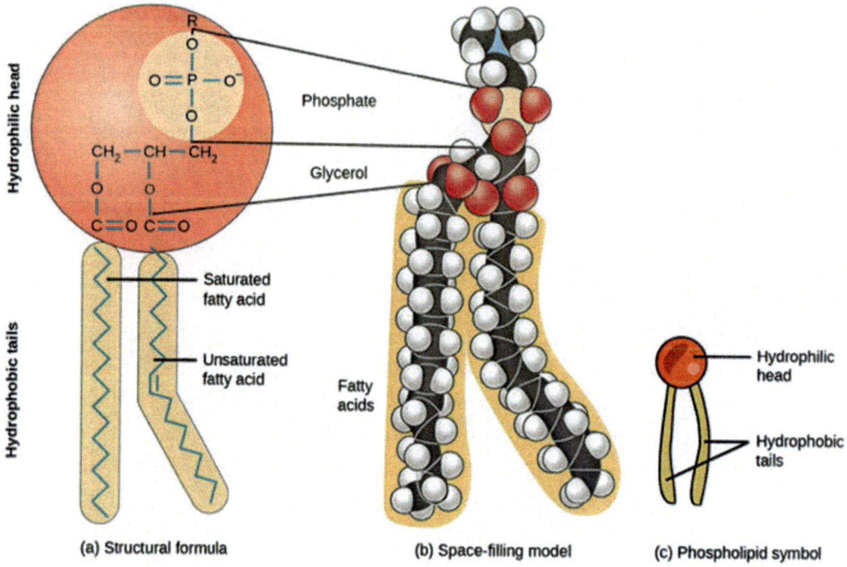

Fig. 29.2 A representative phospholipid structure. Reprinted from [11] with permission from creative commons

Table 29.1 List of different types of liposomes [12]

Vesicle type	Main components
Emulsomes	A mixture of fats and triglycerides stabilized by high proportion of lecithin
Enzymosomes	Complexes of lipids and enzymatic proteins
Sphyngosomes	Sphingolipids containing amide and ether bonds
Transfersomes	A mixture of single chain surfactant, phospholipids and ethanol (10%)
Ethosomes	Phospholipids and ethanol (20–40%)
Pharmacosomes	Conjugate of drug and phospholipid
Virosomes	Viral glycoproteins
Aquasomes	Tin oxide, diamonds or brushite core covered with oligomeric film
Bilosomes	Bile salts and acids (deoxycholic acid)
Niosomes	Non-ionic surfactants (span and tween)

endosomes. The internal environment of endosomes have more acidic pH than the external intermediate. Since the inner pH of endosomes around 6.50 [16], the content of conventional liposomes were delivered to lysosomes and stained. To overcome the accumulation of liposomes pH-sensitive liposomes were proposed which was inspired by viruses that would deliver their cargo to the cytosol after penetration to endosomal membrane before reaching the lysosomes [17]. ***Cationic liposomes***, synthesis of these particles is relatively simple compared to other types of liposomes. The procedure only requires the mixing of cationic lipids such as

1,2-di-O-octadecenyl-3-trimethylammonium propane (DOTMA), 1,2-dilauroyl-sn-glycero-3-ethylphosphocholine (EPC), dimethyldioctadecylammonium (DDAB), 1,2-dioleyloxy-3-dimethylaminopropane (DODMA) with their cargo and the it is followed by the administration of the particles for the delivery. ***Immune liposomes***; in addition to delivery of therapeutics, liposomes can be used in generation of immune response by acting as immunological adjuvant. Encapsulation or incorporation of antigens can enhance the immune response such as macrophage activation, antibody production [18], effective induction of cytotoxic cell [19] and subsequent antitumor activity [20]. Biodegradability, non-toxic nature, low antigenicity and targeting potential of liposomes makes them favorable among other particles. There is great number of data showing that liposomes are outstanding adjuvants for generation of immune response [21, 22, 23]. Immune liposomes are also valuable tools considering the given data for immunoassays and diagnostic tests. ***Long circulating liposomes***; disposition of liposomes generally occurs through mononuclear phagocytic system with intrahepatic uptake. One of the first examples of extended time in the systemic circulation of liposomes was the administration of ganglioside GM1 or phosphatidylinositol at 5–10% into the bilayer [24, 25]. Maximum circulation time was reached after modification of lipids with polyethylene glycol. It was also shown that, both size and number of bilayers were two major factors on determination systemic circulation time which limits their uptake by immunoglobulins, macrophage uptake or fibronectins.

Depending on their size and number of bilayers liposomes can be divided in two groups as multilamellar vesicles (MLV) and unilamellar vesicles. Unilamellar vesicles can be divided in two sub-groups as large unilamellar vesicles (LUV) and small unilamellar vesicles (SUV) [26]. Unilamellar liposomes consist of single phospholipid bilayer encapsulating the aqueous solution. Multilamellar liposomes bears several unilamellar vesicles one inside in another creating an onion like structure. The phospholipid layers in multilamellar liposomes are separated by the layer of water.

29.2.2 Synthesis Methods of Liposomes

The synthesis of liposomes composed of four stages including; (a) drying down the lipids from organic solvent, (b) dispersing the lipid in aqueous media, (c) purifying the resultant liposome and (d) characterization of the final product. Methods of liposome loading is divided in to two as passive loading where liposomes are formed simultaneously with drug loading, and active loading which requires generation of liposomes with transmembrane gradient prior to loading of the active agents. Passive loading techniques of liposomes is composed of three methods referred as mechanical dispersion, solvent dispersion and detergent removal method. ***Mechanical dispersion method,*** contains different types of techniques such as sonication, French pressure cell, freeze-thawing, lipid film hydration, micro emulsification and membrane extrusion. *Sonication* is the most common route of SUV synthesis. Probe sonication and sonication baths can be used in preparation of MLVs but there are major drawbacks of this method which can specified as low internal volume or encapsulation efficiency, degradation of phospholipids and cargo of the liposomes, metal pollution caused by the sonicator probe tip and fabrication of MVLs and SUVs at the same time. *French pressure cell*, is an extrusion technique for preparation of MLVs. This method offers gentle handling of delicate materials such as proteins. Working with small volumes and difficulty in controlling the temperature are main disadvantages of this method. *Freeze-thawing*, method can be used to prepare unilamellar vesicles by instant freezing and slow thawing cycles of SUVs. The fusion of SUVs during the process results in unilamellar vesicles where the encapsulation efficiency ranges from 20 to 30% [27]. ***Solvent dispersion method***, involves injection of a solution of lipids containing diethyl ether or ether methanol mixture into an aqueous solution of the cargo under reduced pressure. Liposomes can be obtained after the removal of the solvent from the system. Another solvent evaporation set up is injection of lipid solution in ethanol to an excess buffer where MLVs can be formed. One of the handicaps of this process is controlling the uniformity of the particle size [28]. Reverse phase evaporation method is another technique that is used in synthesis of liposomes. It offers entrapment of large amounts of aqueous material and a high aqueous space to lipid ratio. Formation of inverted micelles by sonication of an aqueous phase in the presence of an organic solvent containing amphiphilic lipids. The evaporation of organic solvent results in viscous gel form of inverted micelles. The presence of excess amount of phospholipids creates liposomes out of these inverted micelles. The combination of different lipids or volume to lipid ratios can be used to produce multilamellar liposomes [29, 30]. ***Detergent removal method***; this method includes a few different techniques such as dialysis and gel-permeation chromatography. Detergents are used to solubilize lipids during liposome synthesis. The micelles in the mixture behave better after removal of detergent from the environment. For this purpose detergents can removed from the solution by dialysis bags. Another method for the removal of the detergent is gel-permeation chromatography which would eliminate the detergent by size special chromatography.

29.2.3 Biological Applications of Liposomes

Liposomes are not only experimental tool for delivery purposes, they are also commercially available products for clinical and veterinary use. They have utilized to several fields of life science. Liposomes have shown great ability on delivery of drugs, gene, vaccine and diagnostic products. They are also used as cosmetic ingredients and encapsulation of food [31, 32].

For the drug delivery purposes, liposomes are used to modify pharmacokinetics of drug to obtain higher therapeutic efficacy with minimum toxicity [3]. In addition to their well-established biocompatibility and safety profiles; they offer enhanced encapsulation efficiency for both hydrophilic and hydrophobic drugs, improved bioavailability of drugs and controlled release of their cargo [33, 34]. They are used in parenteral, topical, oral, ophthalmic and pulmonary administrations for vast number of diseases including but not limited to CNS diseases, inflammatory diseases, bacterial and fungal infections, skin conditions and administration of local anesthetics [35]. Chemotherapeutic agents are known by their poor pharmacokinetic properties. This is where liposomes come to play to enhance the physical and chemical stability of the chemotherapeutics. There is a great number of examples of liposomes that are used as chemotherapeutic agent carriers for breast cancer, multi-drug resistant breast cancer, ovarian carcinoma, non-small cell lung carcinoma, multidrug resistant ovarian carcinoma, gastric cancer, brain cancer, head and neck cancer [36, 37, 38, 39]. The first introduction of liposome based therapies was on 1990s for Kaposi's sarcoma, ovarian cancer, multiple myeloma, and metastatic breast cancer with the approval of FDA. The stealth liposomes and an uncoated liposome formulations were used in Doxil/Caelyx (Janssen Pharmaceutica NV, Beerse, Belgium) and Myocet (GP Pharm SA, Barcelona, Spain/Teva Pharmaceutical Industries Ltd, Krakow, Poland), respectively. The efficacy of liposomal formulations was found to be higher than conventional therapy on patients suffering from breast, advanced or recurrent ovarian, or prostate cancer, as well as other types of tumors [40–49]. In a recent study, docetaxel (DTX) loaded nanoliposomes have shown superior biodistribution than free iodinated DTX due to enhanced permeability and retention (EPR) effect. The size and encapsulation efficiency of DTX-liposomes were determined as 115 nm and 34–67%, respectively. As indicated in Fig. 29.3, in vivo experiments showed that DTX-liposomes significantly delayed the tumor growth and prolonged the survival time in comparison with control group on BALB/c mice bearing $_{4}T_{1}$ or TUBO breast cancer carcinoma tumors [50].

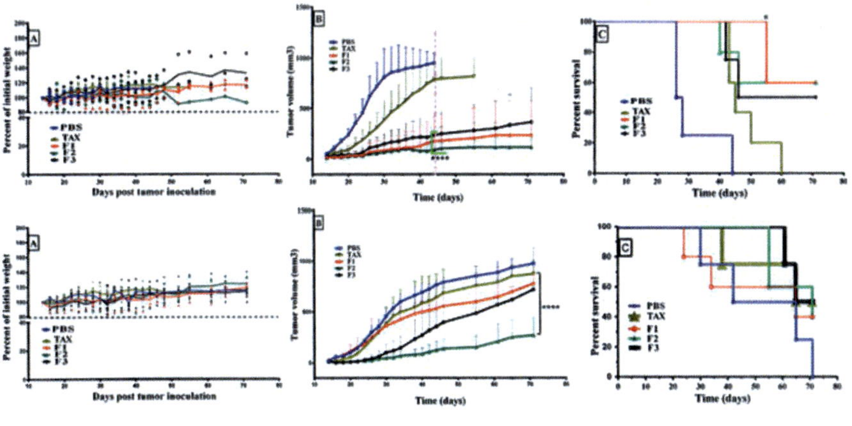

Fig. 29.3 Therapeutic efficacy of DTX loaded liposomes in comparison with control groups (PBS, and Taxotere) in BABL/c models of 4T1 (upper panel) or TUBO (lower panel) mammary carcinoma. **a**: Body weight, **b**: Average tumor volume (mm^3) in all treated groups and **c**: Survival of all groups was monitored (n = 4 or 5, mean ± S.D). Reprinted from [50] with permission from the Creative Commons Attribution (CC BY)

Recently, another approach was applied to be used in cancer by designing aptamer-gold nanoparticle loaded pH-sensitive liposomes (Apt-Au@MSL) for encapsulation morin for treatment of cancer [51]. In this study, Ding et al. proposed the synthesis of a pH-sensitive liposome that also had targeting properties. Since the environment that surround the tumors are lower than the normal tissue, a pH responsive carrier can release their cargo in acidic environment via dialysis. It has been shown that, Apt-Au@MSL were monodisperse and can effectively suppress proliferation on SGC-7911 model cancer cell line in vitro [51]. Additionally, in vivo study on xenograft mouse model of BALB/c male nude mice confirmed that Apt- Au@MSL could inhibit tumor growth and also it can promote tumor apoptosis. Figure 29.4 shows the extracted tumors from mice 24 days after the first administration of 2 mg/kg dose of liposomes intravenously. Ding et al. concluded as Apt- Au@MSL can be used in targeting and delivery of chemotherapeutics in the future.

In another example of liposomes were employed to carry dacarbazine which is used in the treatment of malignant melanoma, Hodgkin's disease, and soft tissue sarcomas. The short half-life and poor hydrophilicity limit the its applications. The

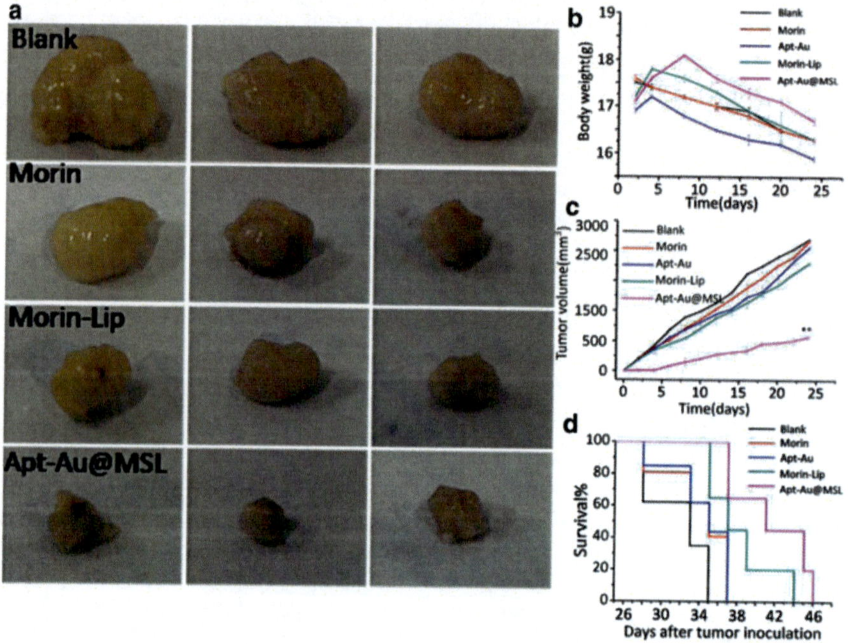

Fig. 29.4 **a** In vivo applications of Apt-Au@MSL and photographs of the mice tumor taken 24 days. A dosage of 2 mg/kg was administrated intravenously for all mice (n = 6–8). **b** Tumor weight of mice in different groups after 24 days. **c** Tumor volume index for the different treatment groups. The tumor sizes were measured at the indicated time points. **d** Survival rate of the mice in different group after tumor inoculation. Data are means ± SD (n = 6–8). In vivo therapeutic effects of Apt-Au@MSL in SGC-7901-bearing mice. Data are means ± SD, *P < 0.05, **P < 0.01. Reprinted from [51] with permission from Copyright (2004) National Academy of Sciences, U.S.A

encapsulation of dacarbazine into nanoliposomes resulted in releasing 50% of drug within the first 2 h where the rest of it released up to 30 h [52]. It was speculated that this carriage system can be beneficial for the drugs similar to dacarbazine after the adjustment of the required dose [53]. Liposomes can also be used in immunotherapy for cancer treatment. For instance, immune checkpoint blockades (ICBs) can block important inhibitory pathways which enables tumor cells to evade immune attack [54]. In this manner, ICBs can be promising for the treatment of melanoma. Hei et al. used programmed death ligand 1 monoclonal antibodies (aPDL1s) as an ICB by attaching on the surface of a liposome bearing catalase (CAT) as an encapsulated cargo. The rationale behind the employment of liposomes was to overcome the limitations which caused by the low selectivity in vivo and immunosuppressed tumor microenvironment including hypoxia [54]. The study showed that immunoliposomes (CAT@aPDL1-SSLs), enhanced the cellular uptake both in vivo and in vitro, provided a good biodistribution as a result of proper targeting and inhibited the growth of tumor [54]. A different application of liposome was encapsulation of raloxifene (RLX) which is a selective estrogen receptor modulator. It is commonly used to lower the risk of breast cancer in post-menopausal women [55]. RLX suffer from the similar problems as many anticancer drugs does; low bioavailability due to poor hydrophilicity and substantial first pass mechanism [56]. Liposomal formulation of RLX with glyceryl caprylate and glyceryl monostearate as lipids, enhanced the pharmacokinetic parameters as well as the oral bioavailability [57]. In addition to that, the degree of absorption was 3.75-fold higher which points out that the liposomes were by-passing the first pass metabolism [57]. Curcumin is another promising anticancer agent lacking bioavailability [58]. It has shown a cytotoxic effect on many human cancer cell lines solo or accompanied to chemotherapy [33, 59, 60]. In a recent study, curcumin loaded liposomes showed better pharmacokinetic parameters with the twofold increase in the half-life compared to control group for a potential brain tumor treatment [61]. More to this, in vivo studies indicated that liposomal formulation raised the accumulation of curcumin in the brain which resulted in an increase in the inhibitory effect of curcumin approximately fourfold [61]. Liposomes are also frequently used in delivery of chemotherapeutics to the CNS. Cytarabine is an FDA approved antimetabolic cancer agent that is also available as DepoCyt®. It is used in many cancer applications. Liposomal formulation of cytarabine produces a foam-like material providing a sustained drug release, an improved pharmacokinetic profile, low cytotoxicity up to 14 days and a longer half-life of 43 h [62–66]. Dong et al. employed liposomes to obviate a common problem of multiple drug resistance that usually arise due to long-term administration of chemotherapeutic agents [67]. In this approach, more than one drug were encapsulated into same particle, aiming to both control the release of the chemotherapeutic agents and enhancing the anticancer effect through the synergistic effect of accompanying two drugs (Doxorubicin (DOX) and vincristine (VCR)). The results showed a prolonged release and a sustained antitumor activity effect on lymph cancer [67]. Wang et al. flowed a similar approach to benefit from the synergistic effect by incorporating paclitaxel (PTX) and DOX for the treatment of lung cancer [68]. In vivo studies showed that liposomal formulation of PTX and DOX had improved tumor targeting and enhanced inhibitory effect [68].

Cancer therapy may appear to be dominating the implement of liposomal formulations due to poor bioavailability of anticancer drugs. Nevertheless, liposomes are good candidates for many other diseases including but not limited to gene delivery, vaccine delivery and several infectious diseases [31]. Liposomes made of cationic lipids can achieve encapsulation and the delivery of genetic materials via electrostatic forces with negatively charged phosphate backbones of nucleic acids. In addition to that, positive surface charge of liposome promotes the interaction between the particle and cell membranes which leads to higher internalization [69, 70]. This type of liposomes referred as lipoplexes. Besides having good advantages as easy and safe production, cost effectiveness, biodegradability, biocompatibility and lack of dangerous immunogenicity the limited endosomal escaping ability causes decreases transfection efficiency. This condition leads to exposure of genetic materials enzymatic and acid degradation in lysosomes [70]. The incorporation of 1,2-dioleoyl-*sn*-glycero-3-phosphoethanolamine (DOPE) to the liposomal formulation helps to solve the problem by helping lipids particles to escape by membrane fusion with endosomes resulting in early cytoplasmic release of the cargo which are referred as fusogenic liposomes. There are several reported studies in this perspective on delivering genes encoding for tumor suppression proteins [71]. In a study performed on 2016 by He et al. suggested that lipoplexes were found to be promising on therapeutic gene expression regulated by an hTERT promoter [72]. The combination of the lipids in the preparation stage of liposomes determines the application area of the particles. For instance the incorporation of phosphatidylserine, N-[1-(2,3-Dioleoyloxy)propyl]-N,N,N-trimethylammonium methyl-sulfate (DOTAP), fatty acids and monophosphoryl lipids in liposome production render the particles for immune-stimulating purposes. These liposomes can generate an immune response on macrophage uptake followed by endolysomal degradation [73]. Generation of an immune response also depended on the lamellarity, size and the surface charge of the particles [74].

Army liposome formulation (ALF) was first synthesized and tested by US Army in 1980s [75]. It contains saturated phospholipids, cholesterol and monophosphoryl lipid to act as immunostimulant. ALF-based adjuvants were found to be promising for the development of vaccines for paracitic, bacterial or viral targets including P. falciparum malaria and human immunodeficiency virus (HIV-1) [75]. Furthermore, ALF-based adjuvants pave the way of development of vaccine which was effective on group B Neisseria meningitidis caused meningitis [76], various number of immunotherapeutic cancer vaccines [77–79], and an Alzheimer disease vaccine which is currently under phase 1/2a testing [80]. van Dissel et al. showed efficient delivery of TB vaccine by combination of various lipids including a glycolipid, trehalose 6,6'-dibehenate, and a cationic lipid DDAB [81]. There is a number commercial vaccines of liposomal formulations such as; Epaxal® for hepatitis A, Inflexal®V for influenza virus and Mosquirix® for malaria [31]. In spite of successful applications, some liposomal formulations were not able to implemented. Major histocompability targeting class I complex (Stimuvax®) vaccine for the treatment lung carcinoma failed as a vaccine application.

Liposomes have also been utilized in biomedical applications. Magnetoliposomes, superparamagnetic particles bearing liposomes, are used in magnetic resonance imaging (MRI). Modification of liposomes with radioisotopes enables them to be employed in nuclear imaging. ^{64}Cu, ^{18}F, ^{89}Zr or ^{52}Mn and ^{99m}TC, ^{111}In or ^{67}Ga labeled liposomes are used in positron emission tomography and single photon emission computed tomography, respectively. Liposomes functionalized with quantum dots (QDs) or fluorescent dyes are implemented to diagnostic platform. Analytical areas have used the leverage of liposomes by incorporating them in immunoassays, biosensor analysis and liquid chromatography. Direct enzyme-linked immunosorbent assay (ELISA) is a common use of fluorophore conjugated liposomes that carry secondary antibody to be bound to antigen.

Liposomes have drawn attention due their demonstrated clinical success as carriers of diverse chemical and macromolecular species. A better adaptation of liposomes in therapeutic applications would enhance availability and accessibility of liposomal products would lead to more effective treatments.

29.3 Micelles

Micelles are colloidal spherical lipids ranging from 5–100 nm [82, 83]. The type of head groups and length of the alkyl chains determines the size of micelles [84, 85]. The hydrophobic ends of phospholipids (tail) arranges facing inward while hydrophilic ends (head) point outside [86, 87]. They can be prepared from a fatty acid, a salt of a fatty acid (soap), phospholipids or any other molecules with a similar head and tail orientation. The structure of a micelle is shown in Scheme 29.1.

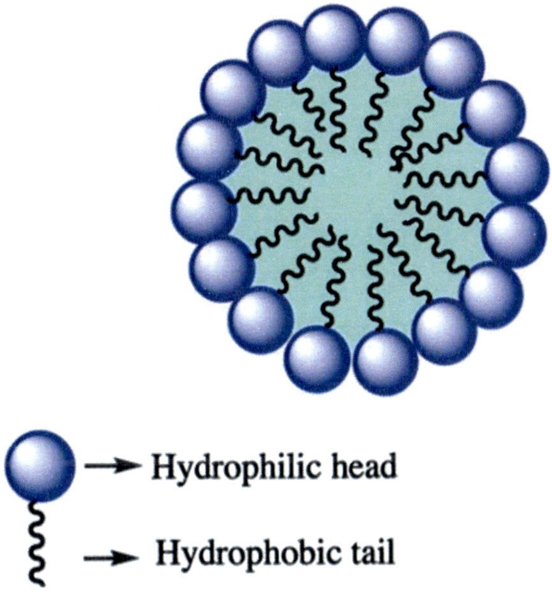

Scheme 29.1 Structure of a micelle

Micelles formed from a lipid, might have lower critical micelle concentration (CMC) [88]. CMC can be described as measure of the surface tension as a function of surfactant concentration. Micelle formation starts as the surfactant concentration reaches CMC. At this point, hydrophobic units of the micelle forming materials becomes important. Amphiphilic copolymers were developed as alternatives due to limitations in the synthesis procedure caused by CMC [89]. For instance, distearoylphosphatidyl ethanolamine (DSPE) was employed as hydrophobic unit in the synthesis of a di-block copolymer of polyethylene oxide (PEO) to prepare 22 nm micelles [90]. The interference of surrounding water with polar head groups leads to separation of hydrophobic and hydrophilic segments during micelle preparation. In this case, porous micelles form which are favorable to be used in biological applications and drug delivery systems [91, 92]. This depicted micelles can enhance drug solubility, prolong circulation time and reduce toxicity. In addition, micelles are more capable of penetrating into tumor cells owing to EPR effect than liposomes. On the other hand due to their smaller size the circulation time in the body is relatively short compared to liposomes [93].

29.3.1 Synthesis of Micelles

Micelles are synthesized from amphiphilic materials in aqueous solution by self-assembly as shown in Scheme 29.2 [94, 95]. In general, small units referred as monomers forms in the solution than it is followed by aggregation and self- assembly resulting in micelle formation [83]. As mentioned before CMC is required to form

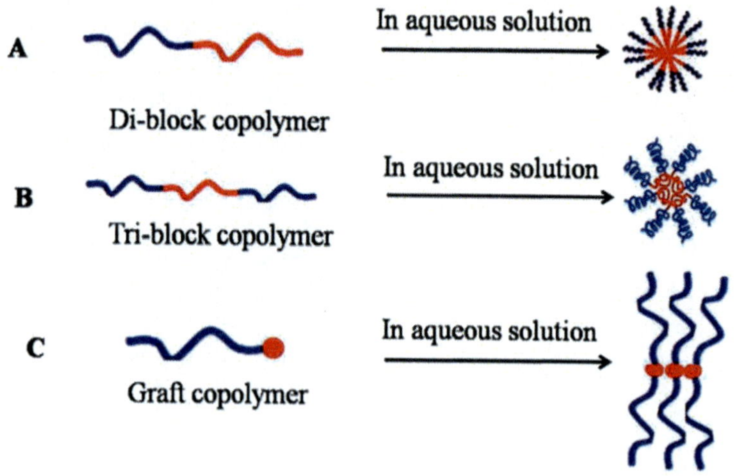

Scheme 29.2 Polymeric micelle structures. Reprinted from [101] with permission of the Creative Commons Attribution (CC BY)

micelles. Dehydration of the hydrophobic tails forms a state entropy as the core of micelles are established by Van der Waals bonds [83]. The obtained structure contains crosslinked hydrogen bonds between hydrophilic shell and surrounding water in the final step of micelle formation [96]. Micelles can be in various forms including spheres, rods, tubules and lamellae which can be altered by the solvent, length of blocker chain, structure of the materials to be used in the synthesis of the micelles and temperature [97, 98, 99]. Polymeric micelles are more durable in the case of dilution due to low CMC, that gives them an enhanced circulation time in the body compared to surfactant micelles [100]. The structure of polymers used in micelle fabrication is shown in Scheme 29.2.

A great number of polymers with biocompatibility and biodegradability can be benefited in order to produce micelles. Some of the polymers used to manufacture micelles are listed as follows; poly(ethylene glycol) (PEG), poly(*N*-vinyl pyrrolidone) (PVP), poly(*N*-isopropylacrylamide) pNIPAM, poly(propylene oxide) (PPO), poly(D,L-lactic acid) (PDLLA), poly(ε-caprolactone) (PCL), poly(L-aspartate) and poloxamers [83, 102–104]. The right selection of the polymeric material can make the micelles by-pass the reticuloendothelial system (RES) resulting as an increase in circulating time in the body [100]. Micelles fabricated from amphiphilic polymers form a core–shell structure with a hydrophobic core and hydrophilic shell [105]. The structure bring the features of encapsulating poorly water-soluble drugs, long circulation half-life, sustained release of the cargo and option of functionalization the surface for targeting [106]. Conjugation of PEG with diacyl-lipids, oleic acid (OA) modified chitosan and PEG-phosphatidylethanolamine (PEG-PE) are a few examples of the hybrid polymers that widely took place in micelle synthesis [107–109].

29.3.2 Biological Applications of Micelles

The most common route for the drug carriers to deliver their cargo is intravenous administration [110]. Since low molecular weight drugs are eliminated rapidly by liver and/or kidneys, encapsulation of these in micelles lead an increment in their molecular weight and results in higher bioavailability [110]. Furthermore, EPR effect of the micelles offers passive targeting of tumor cells which caused by the toxicity of solubilizing agents that are generally accompanied with hydrophobic drugs, becomes eliminated [110]. Delivery of anticancer drugs via micelles have drawn a lot of attention. Paclitaxel (Ptx) is an anticancer agent used in treatments against breast, ovarian, colon and neck cancers. As a common problem of most anticancer agents, intravenous administration of Ptx is challenging as the formulation in solution is not stable, catheter use can cause infection, there is also a possible risk of extravasation of thrombosis and on top of these the chance of occurrence of several adverse reactions is high. After taking into consideration of these drawbacks oral administration of Ptx seems to be a better choice than intravenous administration [111]. Even though oral administration is preferred over intravenous injection, low oral bioavailability is still an important issue to deal with. Moreover, Ptx is a substrate to Gp-p and has high affinity to intestinal enzymes [112, 113]. In a study conducted by Bromberg et al. has found promising results of oral administration of Ptx via polymeric micelles. A formulation of Pluronic P85 and polyacrylic acid (PAA) was used in comparison between oral and intravenous delivery routes of encapsulated Ptx in polymeric micelles. The results were found encouraging due to the area under the curve of oral administration which was similar to intravenous administration. Also, drug bioavailability had improved by prolonged retention time and suppression of efflux through the membrane [113]. In recent studies, Ptx bearing polymeric micelles have shown a great potential of high drug capacity and good efficiency in metastatic breast cancer patients [114] and non-small cell lung carcinoma patients [115]. Besides Ptx, there is a number of examples for micelles as carriers of poor-water soluble drugs including but not limited to tamoxifen, porphyrin, camptothecin, vitamin K and curcumin [116–118]. PEG conjugated curcumin was found to be effective on many cancer cell lines such as breast [119], colon [120], prostate [121], kidney [122], liver [123], lymphoid and myeloid tissues [124] and melanoma [125].

Micellar nanocarriers which have the ability to from a linkage via electrostatic interactions with nucleic acids, are referred as micelleplexes. One of the successful utilization of micelleplexes was performed by Lee et al. [126]. They developed a theranostic micelleplex that mimics the combined therapy of chemotherapeutic drug SN38 and siRNA for the treatment of colorectal cancer. This delivery system consists of poly(2-(dimethylamino) ethyl methacrylate (PDMA)-block-poly(ε-caprolate) (PDMA-b-PCL) and an ultra-small super- paramagnetic iron oxide (USPIO) and targets human vascular endothelial growth factor [126]. The study concluded that the delivery system had shown a significant reduction in tumor growth and these micelleplexes could be used as a diagnostic tool for various types of diseases [126]. A study years before Lee et al. another group designed a new nanocarrier for siRNA delivery [127]. The aim of the study was to get through the endosomal barrier with high efficiency to silence the target gene via specific siRNA. To achieve this goal, pH responsive micelles (PDMA-b-poly(2-(diisopropylamino) ethyl methacrylate (PDPA)) were used to bind and co-deliver the anti-fungal drug (amphotericin B) simultaneously by the synergic effect. They have used the self-assemble ability of PDPA which is also an amphiphilic polymer, to encapsulate amphotericin B while cationic polymer bound to siRNA molecules. The results indicated that these micelleplexes were able to overcome the endosomal barrier for an efficient delivery of their cargo to endoplasm [127]. Micelleplexes were also employed in the treatment of glioblastoma multiforme (GBM) known as a very aggressive and fatal brain tumor. The goal of Zhang et al. was to deliver anti-miRNA in the cerebrospinal fluid (CSF) for a gene-based therapy. A hydrophilic arginine-rich cell-penetrating peptide (R_{15}) containing micelleplex system made of methoxy-poly (ethylene glycol-b-lactide) (PEG-b-PLA) block copolymer was developed. In this system, high levels of glutathione was exploited to incorporate with a reducible disulfide bond between the PEG-b-PLA and R_{15} which promoted the release of anti-miRNA release. It was reported that this delivery system of reducible micelleplexes were found promising for gene therapy to be used in the future GBM treatment [128]. Prior to this work, a similar approach was followed where mPEG-PLA-b-polyarginine (R_{15}) was employed as cationic polymer to encapsulate and deliver siRNAs to cancer cells. This nanosystem was tested on breast cancer cell line in vitro and in vivo. Zhao et al. have shown that these micelleplexes efficiently protect the genetic material and deliver them to the target cells. They have also shown a good bioavailability, an effective uptake by the cells and an increased transfection efficiency demonstrated by the significant gene silencing in vitro and the significant reduction of the tumor growth in vivo [129]. In a recent study, an acid activatable cationic micelleplexes were developed to be utilized in cancer immunotherapy mediated by photodynamic therapy. The aim was to inhibit the function of the programmed cell death receptor 1 (PD-1) and its ligand programmed cell death ligand (PD-L1) by the delivery of siRNA to block the PD-1/PD-L1 pathway. The micelleplexes were made from a photosensitizer pheophorbide A (PPa) modified pH-responsive poly(ethylene glycol)-block-poly(diiso- propanol amino ethyl methacrylate-co-hydroxyethyl methacrylate) (PEG-b-P (DPA-co-HEA) by co-assembly of an amphiphilic polycation 1,2-epoxytetradecane alkylated oligoethyleneimine (OEI-C14). The results indicated that micelleplexes were able to escape from immune resistance which was demonstrated by an efficient tumor growth. They also concluded that PD-1 pathway was blocked, tumor-specific ROS production and stimulation of immune responses were induced [130]. The final example for micelleplexes is smart particles that were utilized to deliver siRNAs to a particular cell type. Lu et al. developed micelleplexes of PEI grafted PCL-b-PEG polymer with a conjugation of folate (PEI-PCL-PEG-Fol). This structure was specific to target cells owing to folate in the structure and having core–shell type of particles resulted in good transfection in in vitro and in vivo. It can be said that this nanocarrier system could be used in delivery of siRNAs into ovarian cancer cells [131].

Micelles are also engaged in enhancing the solubility of drugs. The lipophilic segment of the block copolymers, the core of micelles, entraps the water insoluble drug molecules. This process boosts their aqueous solubility and bioavailability of the drugs [132]. In general, the drugs metabolized through gastrointestinal uptake are affected by the particle size. It has been reported as, the particles with 100 nm size are uptaken in GI tract 15–250 fold higher than micrometer size drug particles [133]. The effect of nano-sized micelle systems for these drugs is more significant than the others. The chain length of lipophilic segment of the block copolymers is the major factor on solubilization capacity of micelles besides concentration of polymers, and the temperature of the process parameters [134, 135]. In this aspect, Bemabeau et al. solubilized PTX in a mixed micellar system by combination of two biocompatible copolymers; polyvinyl caprolactam–polyvinyl acetate–polyethylene glycol (Soluplus®) and D-α-tocopheryl polyethylene-glycol 1000 succinate (TPGS). They have shown that the solubility of PTX was increased 60,000 and 38,000 fold when it was formulated with only Soluplus® micelles and combination of Soluplus® and TPGS, respectively. Mixed micelles were tested on human cancer cell lines including ovarian (SKOV-3), breast (MCF7) and triple negative breast (MDA-MB-231) cancer cells in vitro. The results showed that micelles provided a significant increase in cellular uptake compared to free drug shown in Fig. 29.5 [136].

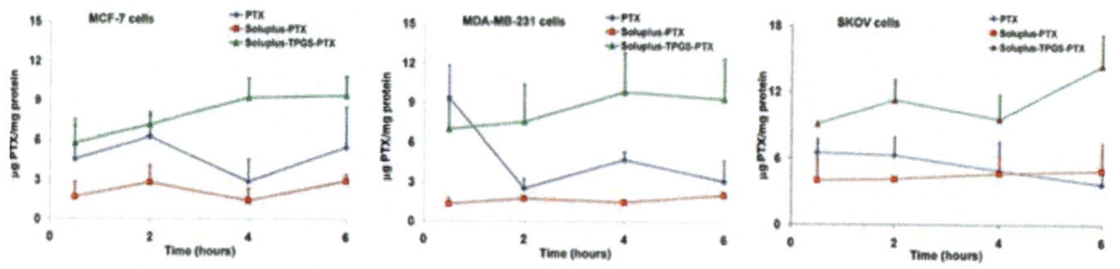

Fig. 29.5 Time-dependent intracellular/cell PTX levels in MCF-7, MDA-MB-231 and SKOV cancer cell lines for drug-loaded single and mixed micelles in comparison with PTX solution. Drug amount was normalized by protein concentrations of the cell lysates. Results are expressed as mean ± S.D. (n = 3). Reprinted from [136] with permission

Micellar nanocarriers are versatile systems for the delivery of biologics and drugs. As summarized in this part of the chapter micellar formulations employed in solubilization of poorly water-soluble drugs, encapsulation of the nucleic acids due to electrostatic interactions, targeted delivery, cancer treatment and many more fields. The possibility of utilization of these particles depending on the need make them unique transporters. Therefore, micelles will continue to be popular among carriers.

29.4 Magnetic Nanoparticles

Iron, cobalt, nickel or metal oxides are such magnetic materials that are found in different fields of technology, have a great impact on advancement of modern world. Devices as motors, generators, sensors, videotapes and hard discs were evolved by the use of magnetic nanoparticles due to high interest on miniaturization efforts. Particularly, magnetic nanoparticles (MNPs) exhibit superparamagnetic features (iron oxides) at room temperature which is described by being composed of one single domain where ferromagnetic or ferrimagnetic particles lose their magnetism below their Curie temperature. The superparamagnetic particles own an important place in data analysis and medicine. MNPs in each field have diverse characteristics such as geometry, stability and physical properties.

Magnetic property of a material associated with spin motions of the electrons in the orbital that angular momentum and magnetic moment are correlated with it [137]. Magnetic moment interactions of atoms influence magnetic order below a certain critical temperature. This results in a change on the materials behavior under magnetic field at different temperatures [138]. Bulk magnetic materials are formed by regions that are referred as magnetic domains. Magnetic properties differ with the size of the material meaning that once the volume reaches nano dimension, the magnetic properties are no longer alike with the bulk material. In general, MNPs show superparamagnetic behavior which means magnetization of each MNP can be manipulated by the change in thermal energy. A change in the temperature results in random rotations in the magnetic moment of each individual MNP. Due to this influence of temperature, in the absence of magnetic field (see Fig. 29.6) the net magnetic

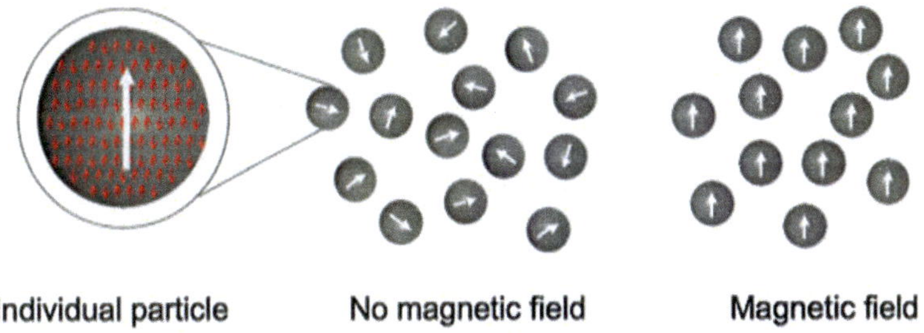

Fig. 29.6 A schematic representation of superparamagnetic particles under a magnetic field. Reproduced from [138] with permission from the Centre National de la Recherche Scientifique (CNRS) and The Royal Society of Chemistry

moment will be zero whereas a net statistical alignment of magnetic moments will take place under a magnetic field [138]. Aside from having a magnetic moment of a single atom as in paramagnetic materials in the case of MNPs they contain various atoms which show similar behavior. After removal of the magnetic field, remanent magnetization provides colloidal stability and helps to avoid agglomeration.

29.4.1 Synthesis of Magnetic Nanoparticles

The properties of the MNPs such as size, shape, surface coating and colloidal stability can be tailored for a given biomedical applications. Despite having the ability to tune MNP according to a specific need, the choice of magnetic material is limited to iron based magnetic oxides. Note that, the use of magnetite (mixed Fe^{2+} and Fe^{3+} ions), and maghemite have already been approved by FDA for medical purposes. Due to larger magnetization and ferrimagnetism of both materials, they are often preferred for the synthesis of MNPs. Even though iron oxides possess the first place in biomedical applications, particularly magnetic ferrites of the general formula of MFe_2O_4 are being intensively investigated. These ferrites gives the leverage of fine-tuning of magnetic properties where M can be Zn^{2+}, Mn^{2+}, Co^{2+}, Ni^{2+}, Mg^{2+} etc. It is possible to incorporate divalent cation into ferrite structure that can change both the saturation of magnetisation and magnetic anisotropy of the material. The synthesis methods of MNPs can be summed up in two categories as hydrolytic and non-hydrolytic approaches. Magnetic ferrites are generally fabricated by coprecipitation which is a subset of hydrolytic synthesis for biomedical applications. In general, Massart method is employed that involves alkaline coprecipitation of stoichiometric amounts of ferrous and ferric salts [139]. The average MNP size differs from 3 to 20 nm that mostly depends on temperature, pH, concentration of the cations and the nature of base. It should be noted that, the pH has the major influence on size of MNPs during synthesis. Stable ionic ferrofluids can be prepared in a wide range of pH owing to electrostatic repulsion on the surface. There is a number of pluses of the coprecipitation approach including the use of cheap chemicals, mild reaction conditions, direct synthesis in water, easy scale-up procedures, high concentration of ferrofluid products. The ease of altering the core and surface properties can be considered as the most important advantage of this approach. Availability of diverse reactive sites offers an easy surface modification by direct incorporation of additives [140]. The challenging part of hydrolytic approach is having to deal with large number of parameters in order to control properties of final product. Since the major handicap is limited control over MNP size, reverse micelles are used to carry out the synthesis of MNPs [141]. Although, reverse micelle synthesis method can improve the particle size in the 4–12 nm range; low yields, poor crystallinity, reproducibility and scaling up are disadvantages of this technique. Another method for the synthesis of MNPs is a non-hydrolytic technique where decomposition of iron (III) acetylacetonate ($Fe(acac)_3$) takes place in the presence of mixture of surfactants and polyols acting as reducing agent. The control of size is achieved through a seeded growth mechanism that leads to particles with 20 nm of size [142]. Regarding to advantages of employing cheap and non-toxic starting materials, polyol-based synthesis methods are referred as valuable methods for large scale production.

29.4.2 Biomedical Applications of Magnetic Nanoparticles

MNPs have attracted a great interest regarding their success in extensive use as magnetic fluids, magnetic energy storage, catalysis, environmental benefits, magnetic inks, and magnetic resonance imaging (MRI) [143–146]. The utilization potential of inorganic nanomaterials with metal based configurations enables manipulation of them in numerous applications including biomedicine [147]. Precision in the synthesis, morphology, surface modification of magnetic nanoparticles for specific applications owes the success to the progress in nanoparticle research field. For instance, dextran-coated- gadolinium nanoparticles were employed in imaging kidneys, liver, spleen, tumor and tumor angiogenesis with no side effects or toxicity [146].

MNPs are not only used in imaging but also in hyperthermia, protein and DNA separation, biosensing, bacteria detection, contrast agent for MR, in vivo enzyme activity, stem cells and drug delivery systems.

29.4.2.1 Hyperthermia

Hyperthermia is one the methods used in cancer treatment which is performed by increasing the temperature of a specific region up to 41–45 °C in the whole body with ultrasounds, microwaves or destroying the tissue of diseased area with radiofrequency [148]. This method built on the thermosensitive character of the cancer cells over healthy ones. The high glycolytic activity and relatively lower pH results in thermos sensitivity on cancer cells [149]. Even though hyperthermia seems like a local treatment, it suffers from low penetration depths and limited targeting addition to being very aggressive. To overcome these drawbacks, MNP-mediated hyperthermia (MH) is being used. MNPs transfers the electromagnetic energy into heat owing to their magnetic properties [150, 151]. The size of MNPs employed in MH differs from 1 to 200 nm. The type of nanoparticles can be oxides, metallic, or metallic shells and are stabilized by an organic or inorganic coatings [152, 153]. The small size of the MNPs bring in a major advantage on diffusion leading to their distribution in the target tissue. For instance, Jordan et al. conducted a study to investigate the silane and dextrane magnetite nanoparticles ranging from 3–13 nm for stimulation of hyperthermia in vitro. The study indicated that the nanoparticles were able to induce hyperthermia, differential endocytosis and tissue necrosis [154]. Another group investigated the effects of nanoparticle characteristics including shapes, aspect ratios and hydrodynamic volume on human breast tissue (ex vivo) and in mouse tumors in vivo. It was shown that magnetic nanoparticles with different features created a significant increase in tissue temperatures in both in vivo and ex vivo experiments [155]. Maier-Hauff et al. carried out a study on GBM with surface modified iron oxide nanoparticles. The magnetic nanoparticles were coated with silane and the sizes were determined approximately 15 nm. It was shown that the all patients were able to endure the treatment with no major side effects. However minor side effects were observed at an intratumoral temperature of 44.6 °C with the evidence of tumor control [156]. Another example of surface modified iron oxide nanoparticles was an antibody linked nanoparticles with size of 20 nm. The surface of the nanoparticles were covered with dextran and PEG followed by conjugation of Chimeric L6 antibody. These particles showed a good targeting ability on human breast cancer xenografts in mice. The surface modified particles were also found to be efficient on the control of the tumor growth by MH therapy [157]. The effect of polyaniline coating of superparamagnetic iron oxide nanoparticles (SPIONs) was reported indicating that the coated SPOINs (65 nm) had higher specific absorption rate (SAR) compared to bare particles. It was speculated that the higher SAR was the result of better stability and also particles were able to induce magnetic hyperthermia in tumor cells [158]. Liu et al. have proven that a ferrimagnetic vortex-domain nanoring (FVIO) was effective against MCF-7 breast cancer cells and it had high SAR [159]. In a recent study, $Mg_{0.13}$-γFe_2O_3 nanospheres were developed as MH agent by doping γFe_2O_3 with magnesium. These particles displayed an efficient heating and boosted antitumor activity in vitro and in vivo [160].

In spite of the efforts put in MH therapy, it still remains in its early development stage. The challenges of cancer treatment varies with the type of the cancer. MH therapy needs more efficient particles in terms of targeting, biocompatibility, and a better temperature management.

29.4.2.2 Drug Release

The common problems and major drawbacks of controlled drug delivery systems such as poor body distribution, limited effectiveness and low selectivity can be eliminated by the use of MNP [161]. Smart and stimuli responsive systems of MNPs are under investigation to deliver therapeutics with site specific manner at the right concentrations within the body under externally applied magnetic field [162]. The magnetically drug delivery approach seems more abler way than conventional drug delivery systems since it meets the two major requirements of a delivery system; targeting and organ controlled drug release [161, 163]. All types of magnetic particles such as ferrimagnetic, paramagnetic, super paramagnetic, anti-ferromagnetic, ferromagnetic and diamagnetic are employed in drug delivery systems. Among the types of magnetic particles super paramagnetic ones are the far most interesting ones regarding drug delivery purposes owing to their ability to develop magnetization under an external magnetic field. In addition magnetic nanoparticles offer a variety of surface modifications with various polymers or surfactants which makes them tunable depending on the aim [164]. Regarding to surface modifications of magnetic nanoparticles the most common and promising approach is introducing polymers as an external stimuli-sensitive element to the system. In this context, delivery of ibuprofen (IBU) from a poly(l-lactic acid) (PLLA) membrane with inclusions of a zeolite (Faujasite) and magnetostrictive Terfenol-D particles was monitored. The synthesis, characterization and drug release profile of the particles were investigated. The effect of external magnetic field on drug release rate was observed as more than 30%. In addition, an increase in the magnetic field intensity was seen, it also accelerated the release of IBU. Furthermore, in a different approach of employing magnetic nanocomposites with temperature responsive hydrogels was carried out. The magnetic field was used as a trigger where it generated the heat that led to release of larger amounts of drug [165].

A total new methodology enlighten the way for the advancement of smart delivery materials and systems [166]. It was reported that, a delivery system made of hybrid polymers bearing γ-Fe_2O_3 super-paramagnetic iron oxide NPs with encapsulated DOX was able to increase cytotoxicity to 18% and set off intracellular drug release when exposed to a high frequency magnetic field. Even so; in clinical tests, field of intensity and the frequency was lowered due to the patients discomfort [167, 168]. A multifunctional mesoporous Fe_3O_4/silica nanoparticle had the ability to be manipulated by an external magnetic field to increase the concentration close vicinity to the tumor site [169].

The nanoparticles had also provided the option to switch on the drug delivery upon demand. These nanoparticles could integrate themselves into tumor cells by taking the advantage of their surface coatings of (poly(ethyleneglycol))-3-aminophenylboronic acid-mediated endocytosis. The integration of the particles into tumor cells caused disorientation as a result of high concentration glutathione in cytoplasm which triggered the release of the drug followed by cell apoptosis.

Apart from the above-mentioned studies magnetic nanoparticles are used in MRI assisted drug delivery by iron oxide nanoparticles (IONs). Xie et al. modified the surface of oleate coated IONs with dopamine-plus-human serum albumin (HSA) in order to explore their potential on U87MG xenograft model [170]. The results revealed that the surface coating led to a prolonged circulation half-life and accumulation in tumor site due to EPR effect. Same group proceeded the with the same particles to encapsulate DOX into HAS matrices [171] for tumor targeting. The particles had the hydrodynamic diameter of 51 nm where they displayed better uptake by breast cancer cells (4T1) compared to free DOX. This study was followed by an in vivo xenograft experiment on 4T1 murine breast cancer model and demonstrated promising results as having higher tumor accumulation and suppression than free DOX. A similar approach was applied for the treatment and diagnosis of breast cancer by using folic acid conjugated IONs [172]. The efficacy of folic acid containing nanocarriers was tested on nude mice with xenograft MCF-7 breast cancer tumor and the accumulation of the particles in the tumor was demonstrated.

Another platform where these nanoparticles employed is magnetically guided drug delivery. The targeted delivery of drugs to the specific sites in the body is possible under an external field by taking advantage of magnetic properties of these particles. The efficiency of this method was proved by detection of the accumulation of nanoparticles in pre-determined sites such as tumors or inflamed areas [173]. Manipulation of MNPs was investigated by many groups such as Wagstaff et al. where they developed a delivery system made of gold coated IONs for the release of cisplatin [174]. Cisplatin release was evaluated in vitro on human ovarian carcinoma cell lines (A2780) and the cell growth inhibition when orienting nanoparticles with a bare magnet was monitored. Another delivery system for cisplatin was developed by Unterweger et al. [175]. The group synthesized cisplatin containing IONs with dextran/hyaluronic acid. The rationale behind hyaluronic acid use was to increase incorporation of cisplatin and targeting of overexpressed CD44 receptors in cancer cells [176]. The drug release kinetics found to be promising and showed that the presence of hyaluronidases increases drug release rate. PVA coated IONs loaded with DOX had a similar approach for the utilization of magnetic particles in drug delivery [177]. The results revealed that the external magnetic field had a control over drug delivery properties of these NPs. In a different study, lauric acid (LA) and HSA coated IONs bearing mitoxantrone demonstrated a linear drug release profile with an enhanced stability for more than 72 h. The same research group has also shown a significant site specific targeting and therapeutic effect via magneto-guided assay in another study. Natesan et al. also studied magnetically driven targeted delivery for cancer treatment. They demonstrated that chitosan and artemisinin coated IONs could accumulate at the tumor region in in vivo on 4T1- breast tumor bearing BALB/c mice model [178].

29.4.2.3 Molecular Detection

The signaling processes need reciprocal recognition done by molecular pairs through specific interactions. Labeling is one these biomolecular entities with MNPs results in specifically binding to the biomolecular counterpart. This makes it possible to control the localization of specific targets by an externally applied magnetic field and under proper conditions isolating them. This methodology is used in protein and DNA separation. In this technique, MNP bearing immobilized affinity tag is mixed with the desired molecules. The affinity species tightly bind to the ligands bearing MNPs after a pre-determined time and the isolated can be collected by magnetic decantation. In the final step, purified target molecules can be MNPs free by a proper elution procedure. Figure 29.7 illustrates the separation procedure [138] (Fig. 29.7).

The grey rods on the surface of the spherical MNP in the figure represent the immobilized tag which will tightly bind the DNA strand or protein of interest. The separation process follows the steps: (1) mixing MNPs and the solution with different components, (2) particular proteins or DNA (green rings) bind to the MNPs and (3) the magnetic field is applied to trigger magnetic decantation, followed by further washing steps and collecting the molecule of interest [138].

DNA and RNA can be isolated with this method where selected oligonucleotides grafted on MNPs that would lead to capture of complementary strands [179]. MNPs can also be used in biosensors for the determination of concentration of various analytes including glucose and calcium ion [180, 181]. For instance, CLIO-glucose and concavalin-A combination was able to provide clinically meaningful data for measuring glucose concentration over semi-permeable membrane. The same methodology can serve in simultaneous detection of metabolites by MRI in vivo [182]. Magnetic separation is also used for the detection of the bacteria. As the organism selectively incorporates with the MNPs, it would be isolated from the biological samples and identification can be performed through conventional methods. For this purpose, El-Boubbou et al. produced the silica-coated magnetic glycol-NPs to detect *E. coli* strains in a very short period of time such as 5 min where 88% of the bacteria was removed from the sample [183]. Another application that serves this purpose was single domain Ab functionalized MNPs which showed

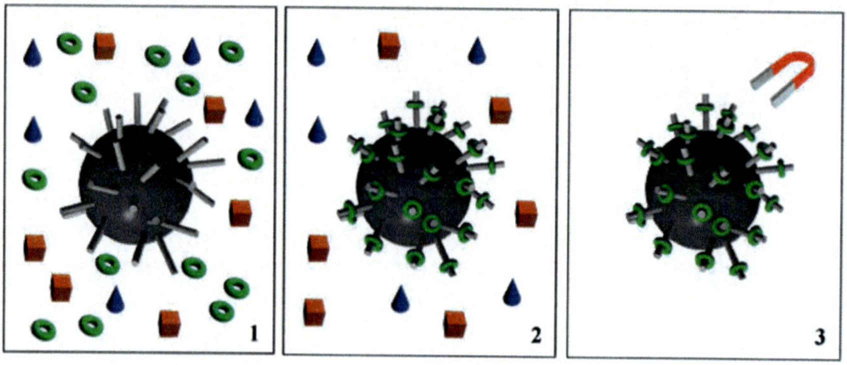

Fig. 29.7 Illustration of magnetic separation of DNA or proteins in solution. Reproduced from [138] with permission from the Centre National de la Recherche Scientifique (CNRS) and The Royal Society of Chemistry

high selectivity and efficiency on *Staphylococcus aureus* to target and capture from a mixed cell population [184].

29.4.2.4 Contrast Agent

In vivo molecular imaging is classified as a non-invasive route to investigate diseases by the National Cancer Institute of the Unites of America [185]. The purpose of monitoring the molecular characteristics of physiological and pathological changes in living organisms is to be couple of steps ahead of the diseases before a required invasive procedure. Generation of 3D images, excellent spatial resolution within soft tissues, no required irradiation and having a good signal-to-noise ratio are among the advantages of MRI. Paramagnetic materials such as gadolinium (Gd), europium (Eu), neodymium (Nd) and manganese (Mn) and IONs as superparamagnetic materials can be used as MRI contrast agents. The most important criteria for the contrast agents to meet as to be used in targeting is being able to identify cells and/or diseases and/or function-specific biomarkers. The desired biomarker should only be expressed on the specific cell types, ideally. Moreover, disease-specific biomarkers should be different from healthy ones. In general, MNPs which are intended to be used in targeting are composed of at least two segments that are targeting and affinity components. MNPs without targeting components are cleared by the monocytes or macrophages [138]. In a target-specific detection study of MPS, two different breast cancer types were visualized.

MNPs can also be utilized as multi modal magnetic imaging probes where they can function as magneto-optical and magneto-radioactive probes. Most of the MNP probes bear organic fluorophores due to their optical properties. Hence, these fluorophore groups bring the advantage of detection in broad range for in vivo and in vitro applications. For instance, fluorophore containing MNP probes are favored in NIR fluorescence applications which provides an excellent contrast between the target and background tissues [138]. Gold MNPs were used in imaging of small animals, protein purification systems and sensitive biosensing applications [186, 187]. In a study, gold MNPs were employed in colorimetric detection of human α-thrombin which was produced by matrix metalloproteinases expressed in tumors [188–191]. Another application of MNPs in MRI was detection of enzyme activity. Enzymes can be used as biomarkers for various processes and they are essential molecules in physiological and pathological events [138]. Detection of in vivo enzyme activity is carried out by MR contrast agents that benefits from latter principle. The detection process starts with the functionalization of MNPs with affinity to the enzyme modified substrates, followed by enzyme activation which leads to induced molecular and MR signal changes. This detection technique has been used to visualize cell surface ADP-ribosyltransferse 2 upon lymphoma cells [190]. The sensors for the enzyme activity demonstration are generally made of Gd-complexes. There are only a few examples where iron oxide nanosensors used for this purpose and designed to measure enzyme activity of proteases, methylases and restriction endonucleases [189, 191, 192].

Dendritic cell (DC) tracking is one of many areas where MNPs are employed for imaging. DCs derive from bone marrow hematopoietic cells and they enhance the antitumor immune responses such as anti-cancer reactions that puts them under focus of interest [193]. DC-based vaccines were studied for several types of conditions for treatment including skin, prostate, breast and neuronal cancers [194, 195, 196]. Although these therapies were accomplished successfully, the methods used to monitor DC migration is invasive [197]. Cellular MRI can be a useful tool to visualize DC migration in a non-invasive manner. There are several labels for complex MNPs including fluorophores, multifunctional polymers containing NPs [198–200]. In general, mice and other small animals were used in DC tracking studies by MRI [191, 200]. Addition to DC tracking, MNPs has great potential to be used in monitoring stem cell migration where stem cell transplants were utilized for the treatment of degenerative diseases. The capability of multiple cell cycle divisions and differentiation efficiency of stem cells make them great candidates in this area [201, 202]. Unlike DC tracking, several clinical stem cell trials are in progress. Surface functionalized MNPs with PEG, silica, dextran and polystyrene were used in order to increase stability of the particles and avoid the toxicity cause by agglomeration [203–205]. An example of MNP labelled stem cells for cardiovascular disease showed that this techniques was not only offers a potential regeneration of the heart tissue but also gave opportunity to follow long term migration of the cells [206, 207, 208]. Even though many clinical trials approved by FDA, there are still a vast number of issues to be addressed.

29.5 Dendrimers

Dendrimers are branched synthetic polymers and layered architectures. The molecular weight and chemical composition of these molecules can be determined based on their synthesis [209–212]. Dendrimers, molecules with high branching and symmetry, were first reported in the late 1970s and early 80s with the work of the Vögtle, Denkewalter, Tomalia, and Newkome groups, and afterwards synthesis, analysis and applications of dendrimers were reported by many researchers [213, 214]. A dendrimer is a polymeric molecule consisting of a large number of branched monomer layers emerging from the core central structure [215]. The structure of a dendrimer is shown in Fig. 29.8. When the core structure is removed from the dendrimer, each of the remaining identical structures are called dendrons, and the number of dendrons depends on the multiplicity of the core. Dendrons consist of three parts: core, interior (branches) and periphery [215]. The most important feature of dendrimers is their multivalency. Contrary to linear polymers, as the weight and generation increase, the terminal ends become more tightly packed and this feature allows the researchers to load higher rates of drug or spectroscopic markers for treatment and imaging studies. The presence of a large number of end groups is important in determining the solubility of the dendrimer.

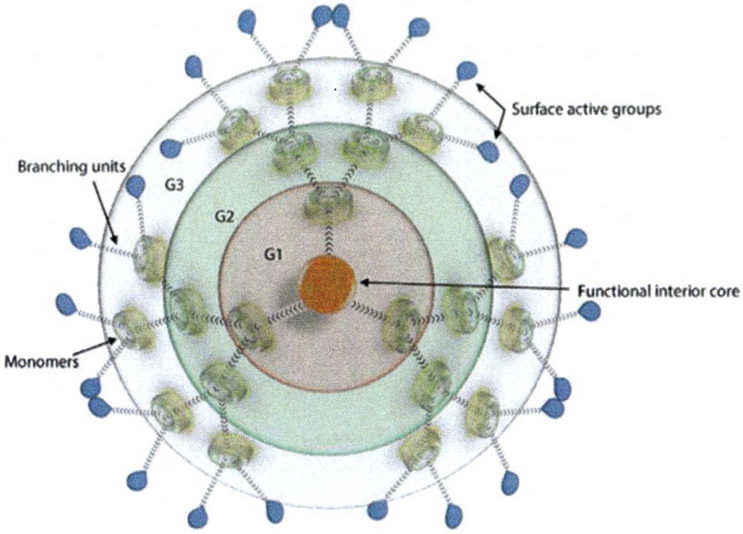

Fig. 29.8 Structure of a dendrimer. Reprinted with permission from [216]

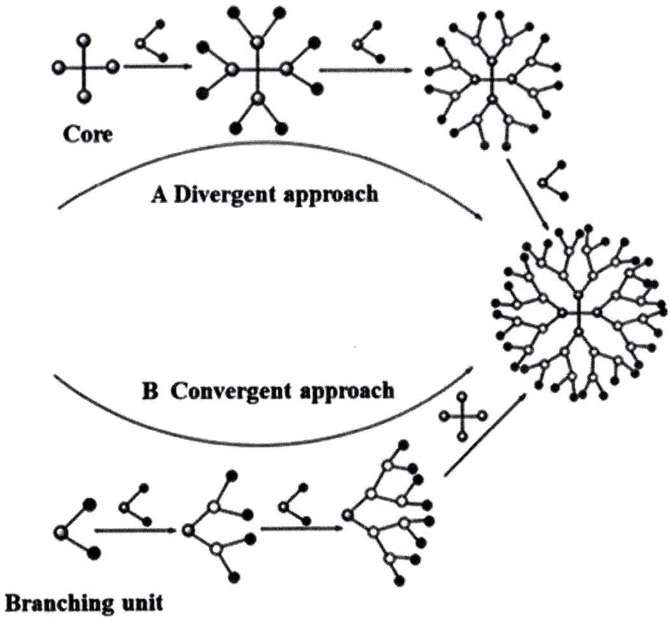

Fig. 29.9 Synthesis of a dendron structure. Convergent and divergent synthesis methods. Reprinted with permission from [217, 218]

29.5.1 Synthesis of Dendrimers

Synthesis of dendrimers proceeds by repeating a 2-step reaction sequence; generation growth step and activation step. The smooth synthesis of dendrimer structures depends on the pureness and high reaction yields of the reactions carried out without side reactions. There are two synthetic methods for dendrimer synthesis referred as divergent approach and convergent approach, which differ in direction of dendrimer growth. In the divergent approach developed by Tomali et al. and Newkome et al., the growth of the dendrimer occurs by the gradual addition of sequential layers, which are building blocks, starting from the core (see Fig. 29.9) [213, 214]. In the convergent approach developed by Hawker and Frechet [219], in contrast to the divergent approach, dendrimer growth occurs inward, starting from the end chains (see Fig. 29.9). Since divergent approach enables the control of final dendritic structure, it is found as a better option for large-scale production. High costs associated with stepwise synthesis are important factors to consider for the dendrimer production. However, time-consuming and challenging steps of these approaches requires new techniques for the preparation of dendrimers.

29.5.2 Biological Applications of Dendrimers

Potential applications of dendrimer and dendrimer based structures are studied by numerous research groups. Based on their behavior in in vivo applications, these branched polymers use to be components of structures such as tissue treatment scaffolds, targeted carriers of chemotherapeutic agents, molecular imaging systems, and optical oxygen sensors [215, 220–223]. Functional end groups of dendrimers allow conjugations with various biologically active molecules. The use of dendrimers as a carrier or a scaffold is an advantage for diagnosis and therapy applications. Amphiphilic characters and interior spaces make it possible to use hydrophilic or hydrophobic drugs in encapsulation. Unlike conventional amphiphilic polymers, they are effective in maintaining the stability of drug formulations. Medium-sized dendrimers are used as MRI agents in the diagnosis of lymphatic systems. Dendrimers with multivalent character and high branching mimicking natural extracellular matrices are preferred in tissue engineering applications. The common dendrimers used in biological applications are polyamidoamines [224], polyamines [225], polyamides (polypeptides) [226], poly (aryl ethers) [219], polyesters [227, 228], carbohydrates [229] and DNA [230, 231]. Polyamidoamine (PAMAM) dendrimers, which are commercially available with various generations and peripheral functionalities, are the most studied dendrimers [215] (Fig. 29.10).

The controlled functionalization of the surface and their compact special structures determines the use of dendrimers in drug delivery systems [233]. As they employed for the drug delivery purposes, the interaction between drug molecules

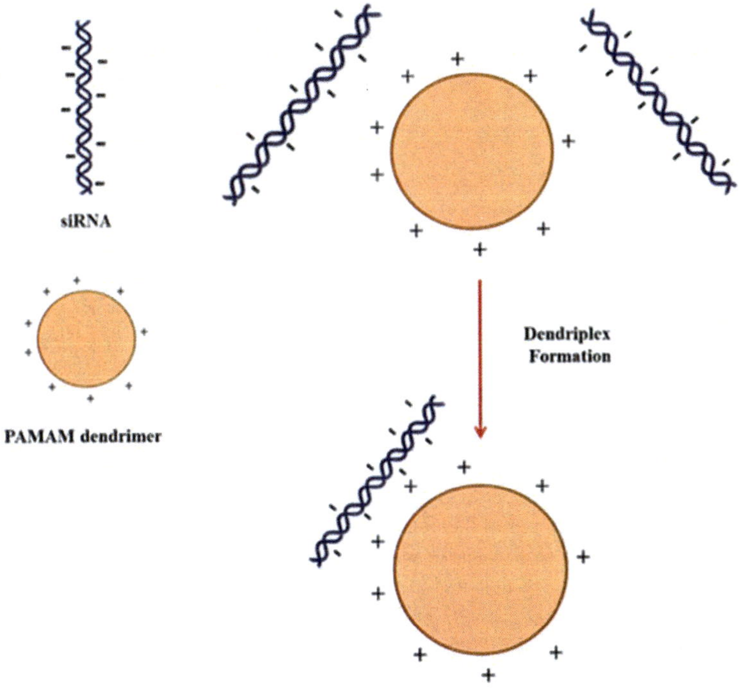

Fig. 29.10 siRNA delivery with PAMAM dendrimer

and dendrimers occurs through covalent and non-covalent conjugation. While covalent conjugation with drug molecules is preferred for the delivery large amounts of drug molecules, non-covalent conjugation method is favorable for poor water soluble drugs [234, 235]. Figure 29.11 shows the covalent and non-covalent interactions of dendrimers with drug molecules. Drug molecules that are covalently bound to dendrimers are released into the cell environment by the breakdown of the bonds by intracellular enzymes. Covalent drug loading provides an advantage with more drug molecules to be loaded into dendrimers, and also provides a controlled release of drugs. Ibuprofen [30], propanolol [236], methotrexate [237], dexamethasone [238] paclitaxel [239], cisplatin [240], penicillin [241], venlafaxine [242], doxorubicin [243], and paclitaxel [244] are among the drug molecules covalently loaded on dendrimers.

Micelle-like structures of dendrimers are important properties that allow non-covalent interactions for encapsulation of drug molecules. The hydrophilic groups located in the terminal part of dendrimers help to overcome the solubility limitation of hydrophobic drugs [245]. Dendrimer structures can increase the oral bioavailability of poor water soluble drugs such as propanolol, camptothecin analogues, silbylin, DOX, and naproxen by increasing their solubilities, drug penetrations and avoiding the drug efflux pump [236, 246].

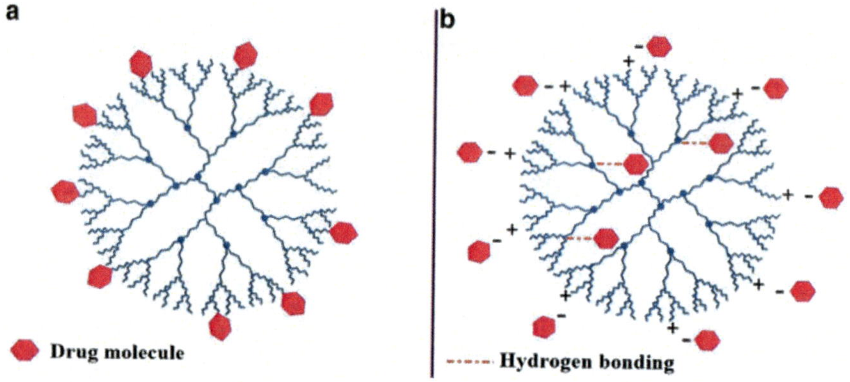

Fig. 29.11 Interactions of drug molecules with dendrimer structure, **a** Covalent interactions and **b** Non-covalent interactions. Reproduced from [232]

It has been reported in studies performed with dendrimers that many drugs such as indomethacin [247], methotrexate [235], doxorubicin [248], ibuprofen [249], 5-fluorouracil [250], flurbiprofen [251], piroxicam [252], 5-amino salicylic acid [253], rifampicin [254], lamivudine [255], betamethasone [256], and PTX [257] have been encapsulated and solubilized. An important anticancer agent, camptothecin which has limited use in clinical application caused its poor water solubility. It has been reported that the solubility of camptothecin can be altered by the use of PAMAM dendrimers. In the study a polyester dendrimer structure consisting of natural metabolites such as glycerol and succinic acid was used as the carrier for camptothecin, and it was reported that there was a 16-fold increase in camptothecin uptake for MCF-7 cells. In addition to that, an increase in the retention of the drug in the cell has also been observed [258].

Active targeting is also possible by integrating targeting agents such as antibody molecules, folic acid, etc. in dendrimer based drug delivery systems. Folic acid is one of the molecules commonly used among the targeting agents mentioned. In a study where folic acid used as a targeting agent after conjugation with PAMAM; the efficacy of the drug, methotrexate, was found 10 times higher compared to the free drug in vivo [259]. Another article stated that the dendrimer structure conjugated to J591 anti-prostate specific membrane antibody can specifically bind to prostate cancer cells such as LNCaP containing prostate-specific antigen [260].

While many studies have been reported in the literature regarding in vitro administration of entirely dendrimer-based drug carriers, few studies on in vivo therapeutic applications are noted. The carrier system obtained by complexing the surface groups of PAMAM dendrimers terminated with G-4 carboxylate with cisplatin is one of the first examples of anti-tumor drug transport with dendrimers [240]. The addition of cisplatin to the dendrimer structure, it led to increase in the solubility approximately tenfold while the drug induced the cross-linking between dendrimer molecules. This structure results in formation of aggregates of 30–40 nm in diameter. This made it possible to follow the aggregates when given to the mice intravenously rather than the subcutaneous administration to tumors with the passive targeting mechanism through EPR effect. When the free drug and drug-dendrimer aggregates given in the same dose were compared, it was reported that platinum level in tumor tissue was 5 times higher with the use of dendrimer-drug aggregates.

Dendrimers such as PAMAM and poly(propyleneimine) [PPI], which have cationic primary amine end groups, efficiently transport genetic materials into the cell by forming nanoscale complexes with DNA and siRNA called dendriplex [261]. The first use of dendrimers as gene delivery agents is the work of Haensler and Szoka in 1993, which includes gene transfection with PAMAM dendrimers [261]. In that study, it was reported that high efficacy DNA transfection was depended on the dendrimer-DNA ratio and the diameter of the dendrimer. It has also been shown that the PAMAM dendrimer-DNA complex exhibits less toxicity than polylysine and has higher transfection efficiency.

Genetic material transfection with dendrimers on different cell lines was reported by Baker in 1996. It was shown that the protonated dendrimer structures form complexes were stable under physiological conditions with the DNA molecule having negative charge, and under specific conditions the transfection efficiency was found as 10 to 100 times higher than the commercial cationic lipid molecules.

The success of PAMAM dendrimers in DNA transfection has led to the development of commercial kits [262]. The successful results of PAMAM dendrimers in DNA-based gene delivery systems has drawn the attention for the their potential use in siRNA transport studies and PAMAM-based dendrimers were the most studied type of dendrimers for siRNA studies [262].

The extensive use of dendrimers in the field of imaging owes the popularity to their large number of reactive chain ends. Controllable chain ends gives opportunity to modify and attach contrast agents that produces images with high sensitivity. Dendrimer-based MRI contrast agents are used to monitor heart, blood vessels and many other organs. Dendrimers are also used as chelating group carriers for MRI agents [263]. Medium-sized dendrimers of around 5 nm are employed for MRI contrast agents in the diagnosis of lymphatic systems. Dendrimers also enables active targeting by utilization of antibodies to their structure. The number of the dendrimer generations are affect the application areas such as G2 dendrimers are suitable for renal imaging whereas G7 for intratumoral vasculature imaging [264].

Photonic oxygen sensing is another imaging applications of dendrimer structures. Since the oxygen level in the tumor tissues gives an idea in determining the response to the treatment, it is important to develop methods that allows the determination of the oxygen level in the tumor tissue [265]. Phosphorescence lifetimes of dendrimers are inversely proportional to the oxygen levels in the environment and can be determined in vitro and in vivo [265]. Vinogradov et al. prepared different sized dendrimer structures in which the hydrophobic metalloporphyrins were encapsulated in their cores. The phosphorescence of these water soluble dendrimer oxygen sensors was quenching when they collide with dissolved oxygen [266].

Dendrimers are excellent artificial macromolecules that can mimic the protein structure thanks to their high molecular weights and branched structures. The use of dendrimers that mimic the surface structures of proteins involved in inhibition of

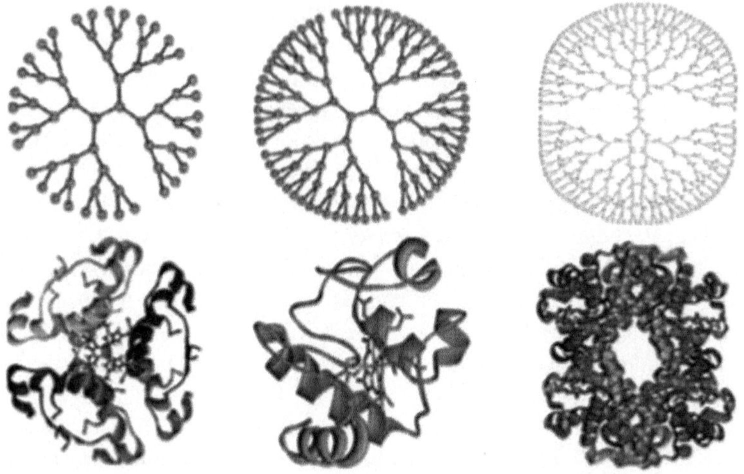

Fig. 29.12 Comparison of protein and PAMAM dendrimers: **a** insulin, **b** cytochrome C, **c** hemoglobin, and PAMAM dendrimers generations **d** 3, **e** 4, and **f** 5. Reprinted with permission from [268]

angiogenesis in antitumor systems has been reported [267]. Recent studies suggested that arginine-based dendrimers suppress and prevent the growth and metastasis of solid tumors [268].

The use of dendrimer structures has also been reported in the synthetic production of collagen which is an important extracellular matrice structure component [269]. With the development of such synthetic collagen mimetic molecules, it is considered to overcome allergic responses or disease transitions caused by the collagen sources in use. Kinberger et al. carried out the synthesis of a collagen mimicking dendrimer structure, TMA[TRIS[(Gly-Pro-Nleu)$_6$-OMe]$_3$]$_3$, which exhibited a triple-helical structure [270].The distribution of ligands on the dendrimer surface had an invaluable binding inhibitory effect between cells and microorganisms (Fig. 29.12).

29.6 Metal Nanoparticles

29.6.1 Silver Nanoparticles (AgNPs)

Their unique chemical, biological and physical properties have made silver nanoparticles (AgNPs) the most studied nanomaterials in recent years [271]. AgNPs, whose optical, thermal and catalytic properties vary in size and shape, have applications in many areas such as biosensors, electronic compounds, textile products, food industry, cosmetics, medical products etc. [272]. The biggest factor of their use in the aforementioned applications is their wide antimicrobial spectra that they exhibit against microorganisms (See Fig. 29.13). They are widely used in medical applications due to their antimicrobial properties against microorganisms such as bacteria, viruses, fungi. They are also used in medical imaging applications owing to their light absorption properties compared to other nanoparticles [273].

29.6.1.1 Synthesis of AgNPs

There are many synthesis methods for AgNP including chemical, physical, biological, microwave-assisted, electrochemical, photochemical methods. Considering the terms such as repeatability, low cost, easy and fast synthesis, the most used method for the preparation of AgNPs is the chemical reduction of silver salts in the presence of reducing agents [274]. The molecules selected as starting material, reducing agent and stabilizer for the synthesis of AgNPs are the factors that needs to be adjusted carefully. The reducing capacity of the reducing agent is an important factor influencing the size and dispersibility of the particles in AgNP synthesis. Synthesis of well-dispersed nano sized particles is possible when using a strong reducing agent in the synthesis procedure.

Sodium borohydride and citrate are the most used reducing among others. However, many different chemicals such as hydroxylamine, hydrazine, aniline, formaldehyde, and ascorbic acid are used as reducers in AgNP synthesis.

The early studies on synthesis of AgNPs have shown the ability of citrate has on reducing the metal cation and stabilize the formed nanoparticles. It was followed by Turkevitch where the synthesis process involved boiling of the gold salt with sodium citrate. This method has shown that it was possible to synthesize gold nanoparticles (AuNPs) with the same size and shape. As the same method was used for the synthesis of AgNPs, it was observed that silver crystallites with different sizes and shapes ranging from 60–200 nm are obtained.

In addition to chemical reduction, there are also physical and biological synthesis approaches for AgNP synthesis. In general, biological synthesis involves the use of organisms such as plants, bacteria and fungi. Biosynthesis methods generally carried out by employing microorganisms and plants which are environmentally friendly biomolecules without toxicity and they have the role of natural stabilizer for nanoparticles. Biosynthesis is under investigation as an alternative to chemical methods regarding the ability to block nanoparticle aggregation and provide easy synthesis opportunities of biomolecules [274, 275]. Synthesis of AgNPs can be performed intracellularly or extracellularly using strains of bacteria such as *Bacillus subtillus* [276], *Staphylococcus aureus* [277], *Escherichia coli* [278], and *Lactobacillus acidophilus* [278]. Even though the time need to synthesize AgNPs with plant extracts, producing monodispersed nanoparticles is a huge issue to overcome [279, 280]. The type material including plant roots, leaves, seeds and fruits in the synthesis determines the shapes and sizes of the nanoparticles.

UV-initiated photoreduction method also used to produce AgNPs in the presence of citrate, polyvinylpyrrolidone, collagen, etc. [281]. The size of the particles obtained in this method, which is quite simple and effective, varies depending on the irradiation time. As the process is terminated, particles with a fixed size can be obtained. It is also possible to determine the shape of AgNPs formed by the sono-electrochemistry technique method. Nitrilotriacetate can be employed in order to prevent agglomeration of nanoparticles [281]. Electrochemical synthesis of AgNPs is also among the current synthesis methods. Adjusments of the electrolysis parameters and components of the electrolytic solution, it is possible to obtain different sizes of AgNP [281].

The most commonly used methods among the physical synthesis are evaporation–condensation and laser ablation. The use of atmospheric pressure and tube furnace brings in the advantage of no contamination which is generally caused by the solvents in chemical synthesis methods. However the size of the furnace and the high energy consumptions are the factor to be considered for the physical synthesis of AgNPs [275].

Laser ablation is another technique to synthesize AgNPs which allows the manipulation of properties by adjusting the applied laser wavelength, laser pulse duration [275].

29.6.1.2 Biological Applications of AgNPs

AgNPs are widely used in the food preservation (storage), health surveys, detection and diagnosis platforms, data storage, textile and medical device coatings, and in many environmental applications owing to their specific features including broad antibacterial activity, high resistance to oxidation, and high thermal conductivity [272]. The most important feature that make AgNPs favorable in most of these applications is their wide range of antimicrobial effects. Studies based on the antimicrobial properties of AgNPs expanded over time including development and preparation of anti-cancer agents, drug delivery systems and different materials. Reproducibility and stability in their production are the crucial parameters that play a major role on utilization of AgNPs and AgNP based systems for applications in the medical field [273].

Factors such as size, shape, concentration, surface charge, and colloidal state are elements on determination of the antimicrobial properties of AgNPs [282, 283]. Some studies indicates that antibacterial effects depends on these parameters, while others suggest the release of silver ions from AgNPs play an important role on it. The mechanism of antimicrobial effect of AgNPs starts with binding to the cell surface and causing a change in permeability on the membrane followed by the disruption in the cellular respiratory mechanism and subsequent release of intracellular metallic silver ions [284].

AgNPs are also effective on biofilm-forming microorganisms [280]. AgNPs combat these microorganisms by destabilizing their exopolymeric components in the extracellular matrix or by interacting with bacterial communication molecules. AgNPs are also employed in the prevention of the development of such life-threatening infections by providing antimicrobial properties to medical devices and clinical materials in the high risk category to be exposed to microbial contamination caused by hospital-acquired infections [285, 286]. The studies performed with the materials such as central venous catheters coated with AgNPs showed that the biofilm formation on catheters by Gram (+) or (−) bacteria have been inhibited [287, 288].

In orthopedics, the use of AgNPs in various applications such as bone cements, tumor prostheses and trauma implants is becoming increasingly common due to its antimicrobial properties [289]. The use of AgNPs in hydroxy apatite coatings to block biofilm formation plays an essential role in order to prevent the occurrence of disease after orthopedic surgery [290].

Research about the effects of AgNPs on wound healing revealed that AgNPs lowers matrix metalloproteinase activity and increasing neutrophil apoptosis in the wound area. In 1995, Dr. Robert Burrell developed the first nanosilver-containing wound dressing, anticoat, which sped up the healing process with no scar. Huang et al. reported that AgNPs accelerated the healing process compared to conventional silver dressings [290]. The adaptation of AgNPs for wound healing in diabetic patients which takes long periods of time to heal and mostly get secondary infections before complete recovery is a vital goal [291]. The presence of AgNPs in adsorbent wound dressings leads to exudation of bacteria resulting in their complete destruction [292, 293].

It has been demonstrated that AgNPs can be used to assist regeneration of nerve cells in post-traumatic injuries or degenerative neural diseases. AgNPs have been reported among nanoparticles, which are known to increase neuron growth and affect electrical activity, with their effects to increase differentiation of some nerve cell types [294].

AgNP-based nanosystems are evaluated and intensively studied as suitable special carriers for the transport of different therapeutic molecules such as anti-inflammatories [295, 296], antioxidants [297], antimicrobials [298], and anti-cancer agents [299]. The reason for the great interest in AgNPs in these studies is the wide range of organic molecule binding capacity, adjustable absorption capacity, and their low toxicity. There is a number of studies of AgNPs as active or passive carriers in cancer treatment serving as an effective anti-tumor drug delivery system [300, 301]. The interactions of AgNPs with cancer cells are shown in Fig. 29.14. Ramar et al. demonstrated that AgNPs exhibited dose-dependent cytotoxicity by inducing apoptosis on MCF-7 breast cancer cells [302]. A study carried out by Gurunathan et al. has shown that AgNP induced cell death on MDA-MB-231 cells through caspase 3 activation, reactive oxygen species production and DNA fragmentation [303].

In a study of AgNPs on multi-drug resistant cancer cells, it was reported that AgNPs modified with TAT peptide could kill tumor cells, prevent their growth and show similar efficacy with DOX, which has serious toxic effects in clinical use [306].

AgNPs synthesized using Cleome viscosa plant extract are considered as alternatives to improve cancer treatment. It has been reported that the particles tested for human ovarian teratocarcinoma (PA1) and human lung adenocarcinoma (A549) cell lines could hinder the growth of cancer cells and could have a great potential for cancer treatment [307]. In a study with gefitinib, AgNPs used to transport gefitinib. This delivery system was found to be reducing the side effects from gefitinib and improving the effectiveness of the drug [305].

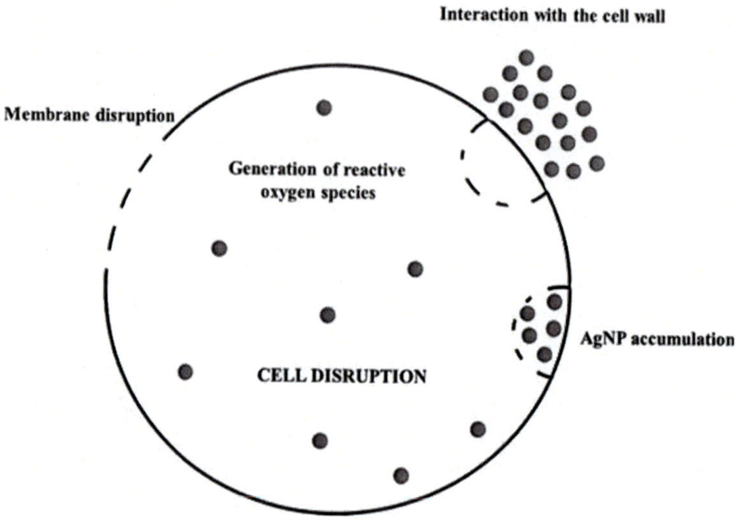

Fig. 29.13 Mechanism of antimicrobial action. Reproduced with permission from [304]

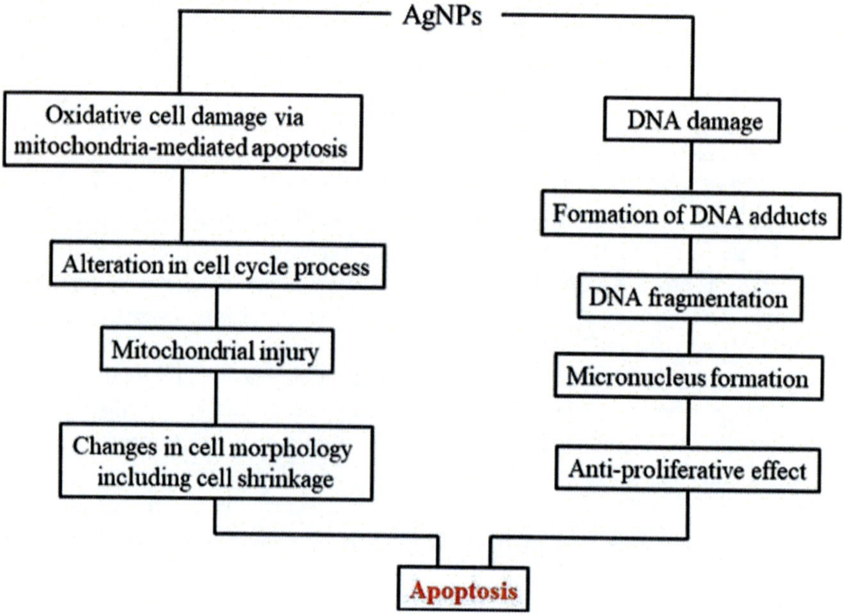

Fig. 29.14 Interactions of AgNP with cancer cells. Reproduced from [305]

Antiviral effect of AgNPs were examined on HIV-1 resulted in proofing acitivity of the particles on the inhibition of virus binding to carrier cells [308]. The effect of different sizes of AgNPs on Hepatitis B was investigated by using the HepAD38 cell line as an infection model, showed that the particles prevented the production of hepatitis B virus RNA and extracellular virions in vitro [309]. A similar study to that on Herpes simplex virus 2, AgNPs were reported to be active in the early phases of viral replication [310].

The main applications of AgNPs in the food industry are food packaging and the determination of contaminants in foods. AgNPs are used in antibacterial packaging to increase the shelf life and quality of foods [311, 312]. It has been reported that polymer composites containing AgNP can protect foods against bacteria, fungi, etc. which cause contamination in foods. In addition, it is stated that long term AgNP release prevents the microbial growth for a long time. It is also noted if the inner surfaces of food containers such as canned food covered with AgNPs, biofilm formations originating from microorganisms can be prevented. Studies on the insecticidal activities of AgNPs are also present in the literature. AgNPs synthesized using Euphorbia prostrate aqueous leaf extract have been reported to show insecticidal activity against the Sitophilus oryzae pest that affects rice, wheat and maize grains [313]. AgNPs have shown an effective insecticidal activity at every growing stage of fruit fly Drosophila melanogaster [314].

AgNPs are used in various analytical applications due to their responsiveness to stimuli, having many sorption zones on their surfaces and high molar absorptivity. Given the plasma membrane permeability, small AgNPs tend to accumulate in the internal parts of the cells.

One of the important application areas of AgNPs is their utilization to biosensors and medical diagnostics regards to their optic features. As the surface electrons of AgNPs are excited by the light, a collective resonant oscillation occurs which is also known as surface plasmon resonance. The resonant oscillation increases by the intensity of the applied light and this process leads to diffusion of the light or converts to heat. This enhanced scattering is useful particularly in diagnostic purposes. The plasmonic properties of AgNPs has given a rise to their use in surface enhanced Raman spectroscopy applications in diagnostics [315, 316]. While it is possible to detect various single biological molecules using AgNPs in this way, they constitute a potential for early detection and diagnostic applications of diseases such as cancer [317]. AgNPs enables to diagnose diseases by determining specific proteins, peptides and other biological molecules in the blood through determination of the changes in their Raman signals [315].

29.6.2 Gold Nanoparticles (AuNPs)

Gold nanoparticles are nanotechnology products that are widely studied in biotechnology due to their unique optical and physical features. AuNPs continues to offer innovations and advantages in biosensors, diagnosis of microorganisms, development of targeted drug delivery systems, immunoassays, cancer treatment and optical imaging processes. Due to their biocompatibility, easy and variety of functionalization options and non-toxicity AuNPs play a role in the development of more reliable and biocompatible systems, especially in the diagnosis and treatment of diseases.

29.6.2.1 Synthesis of AuNPs

A great number of different methods exists in the literature for synthesis of AuNPs such as chemical reduction, physical, photochemical, biological reductions and microwave irradiation. The applied synthesis method give opportunity to modify the properties of AuNPs such as size and shape as desired.

The recent studies on developing new strategies on the synthesis of AuNPs employed different stabilizer molecules such as thiol compounds, polymers, organic molecules etc. The common AuNP synthesis method in which sodium citrate is used as a reducing agent is developed by Turkevich [318]. This method involves the reduction of gold salts such as hydrogen tetrachloroaurate with the use of citrate as reducing agent. In the Turkevich method [318], AuNP production is carried out by adding sodium citrate solution while mixing to the boiling $HAuCl_4$ solution. The initially colorless $HAuCl_4$ solution turns wine red, showing the formation of AuNP, after the addition of sodium citrate. With this method, it is possible to obtain AuNPs with monodisperse and sizes ranging between 10 to 20 nm. The synthesis of larger AuNPs with high polydispersity and low efficiency can be listed as disadvantages of this method. It is possible to synthesize monodisperse AuNPs with different methods as shown by Brown and Natan [319]. Another AuNP synthesis methods known as the seeding approach, the surface of AuNPs was used as a catalyst in the reduction of Au^{3+} by hydroxylamine. Later on, Murphy and his colleagues took advantage of this seeding approach in determining the shape and size of AuNPs [320].

It is also possible to synthesize spherical AuNP that can be dispersed in organic solvents, and one of the best known method for this synthesis is the Brust-Schiffrin method [321]. The method allows to obtain AuNPs less than 10 nm.

29.6.2.2 Biological Applications of AuNPs

Biomedical applications of AuNPs go back to the middle ages with the use of potable gold applications in the treatment of several diseases [322]. Colloidal gold applications, which cause significant side effects to be observed mostly due to the use in oxidation state, started to be used more safely in the biomedical field with the understanding of the nature of AuNPs and the development of the nanotechnology concept.

The size, shape, and functionalized surface structures, which allow specific or non-specific interactions with the cellular lipid layer, are important factors affecting the utilization of AuNPs in biological systems. AuNPs applications are widely examined in tumor diagnosis and treatment due to their physical and chemical properties. The ability to form chemical bonds with groups containing -S and -N allows AuNPs to be modified with many molecules in order to be employed in effective targeting and drug delivery systems [323]. An illustration of the biomedical application of AuNPs is shown in Fig. 29.15.

Photothermal therapy (PTT) applications, which are benefited from the heat generated via nanoparticles applied in the tumor areas, in response to laser light, are among the most important applications of AuNPs [325]. Noble metal nanoparticles are promising agents for photothermal therapy applications due to their surface plasmon resonance (SPR) properties to produce heat by absorbing laser. Gold nanoparticles have maximum absorption in the visible area or near IR region and generate heat when stimulated with light [325]. As AuNPs are directed into the tumor tissue, the particles heat up rapidly and kill the tumor cells after being exposed to the light. Global AuNPs with high NIR absorption greater than 50 nm are widely used in photothermal therapy. As a PTT mediator, AuNPs offer advantages such as high biocompatibility, high efficiency controllable surface synthesis possibilities and conjugation with drug molecules [325]. Larger AuNPs are preferred for high resolution and sensitive cancer imaging-guided PTT, and smaller AuNPs having higher absorption efficiency that is preferred for higher photothermal efficiency.

The high surface-to-volume ratio surfaces of gold nanoparticles allow multiple binding of hundreds of molecules, such as ligands, antibodies, therapeutic, diagnostic

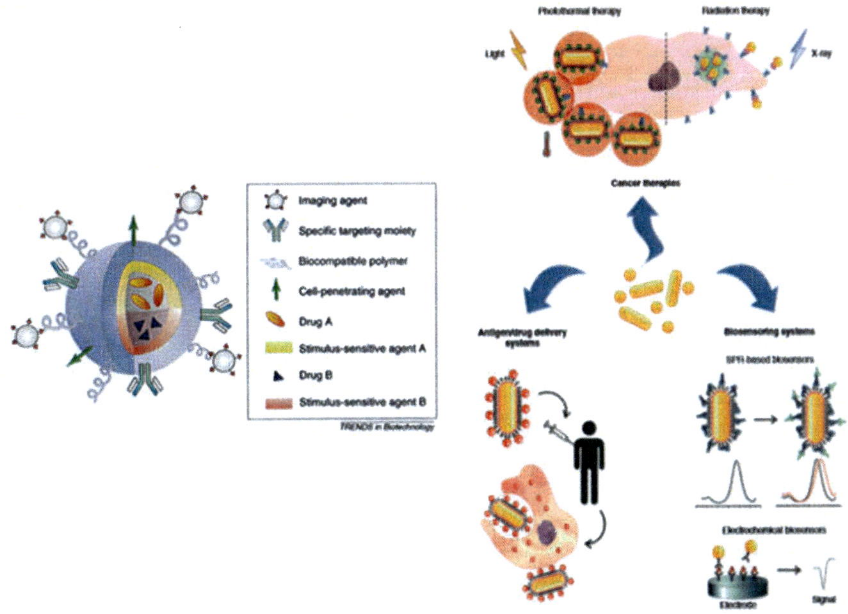

Fig. 29.15 Biomedical applications for different types of gold nanoparticles. Reprinted with permission from [324]

and targeting agents. PTT application using AuNPs can be applied together with chemotherapy, immunotherapy, and gene therapy [325]. Commonly used iodinated molecules for imaging are low molecular weight vascular contrast agents, have short retention times and are eliminated through the kidneys. AuNPs with high X-ray absorption coefficient offer advantages over conventional contrast agents, in terms of the length of vascular retention times, targeting, non-toxicity, and ease of surface functionalization. It has been demonstrated by Hainfeld that AuNPs can be used to increase vascular contrast in vivo when administered in CT imaging [326].

By taking advantage the basic features they have, AuNPs can be used as effective agents for the detection of different biological analytes such as nucleotides, proteins, toxins, and saccharides [327, 328]. As a result of utilization of the AuNP in the sensing applications; surface plasmon resonance, colorimetric, fluorescence-based, electrical and electrochemical, surface-enhanced Raman scattering-based, quartz crystal microbalance-based and Bio-Barcode assay sensors are available [329]. Among the different AuNP types, gold nanorods attract much attention for bio-sensing due to their important and unique optical properties [330] (Fig. 29.16).

Colorimetric sensing applications are based on visible color changes as a result of aggregation of AuNPs. The surface plasmon resonance properties of AuNPs cause a shift at the maximum characteristic absorbance wavelength in UV–Vis region as a result of a change in the particle size, morphology and particle distance [331–333]. A change in the color of AuNPs is observed with the appereance of the shift in UV–Vis spectrum [334, 335]. This behavior of AuNPs is used to create optical sensing technologies by making use of nanoparticle aggregation induced by interaction with the target substance in covalent or non-covalent manner. A general illustration for various applications of AuNPs is shown in Fig. 29.17.

The most promising and continuously developing applications of AuNPs in the medical field is targeted drug delivery systems. Conjugation of AuNPs with antitumor agents such as DOX [337], tamoxifen [338], methotrexate [339], dodecylcysteine

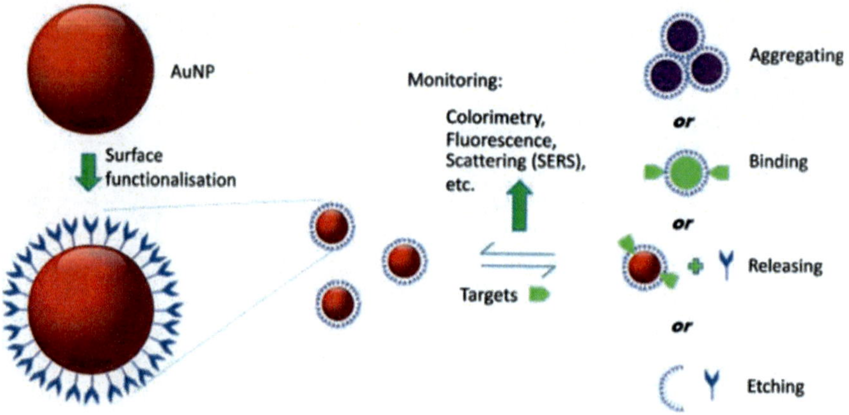

Fig. 29.16 Gold nanoparticle-based sensing in analytical science. Reprinted with permission from [331]

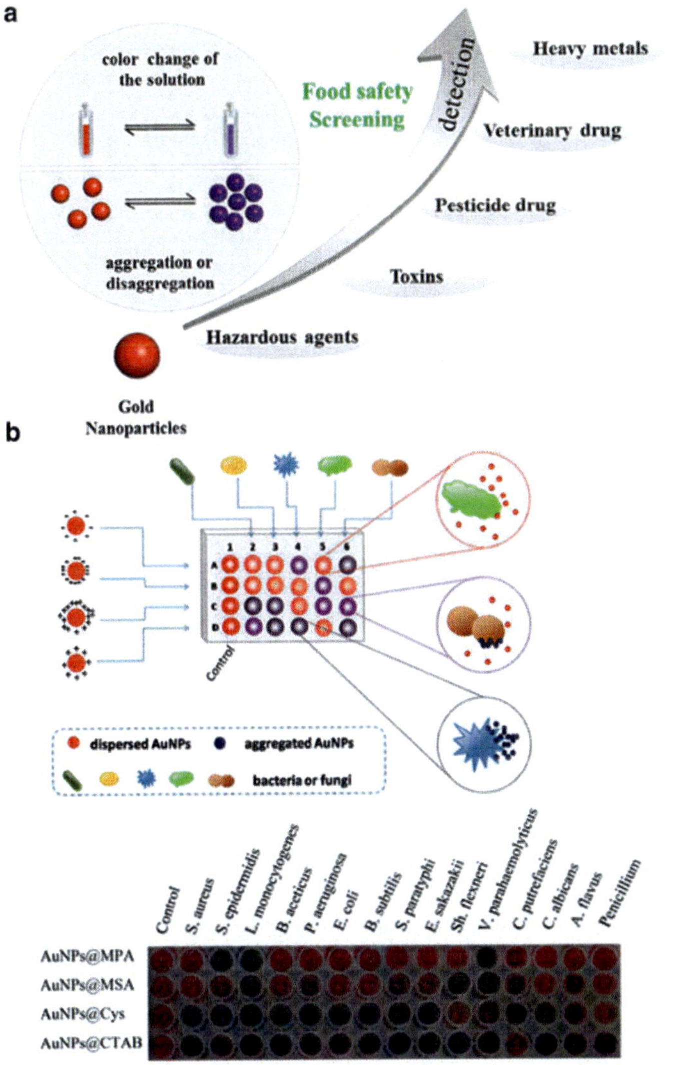

Fig. 29.17 Applications of gold nanoparticle colorimetric sensors. Reprinted with permission from **a** [232], **b** [336]

[340], docetaxel [341], platinum complexes [342], and Herceptin [343] via adsorption or alkanethiol linker molecules. The effectiveness of the carrier system can be altered by including targeting molecules to the carrier system. The vast majority of studies reported that drug molecules are more effective when conjugated with AuNPs.

The use of platinum-based anticancer drugs such as cisplatin, carboplatin, and oxaliplatin, which are the main therapeutics of chemotherapy applications, have limited use in clinical applications due of their undesirable effects. AuNPs appear as an alternative carrier in this context. AuNP-platinum based drug delivery system has been reported to show superior results than oxaloplatin, in terms of cytotoxicity [344, 345]. In another study it was noted that when methotrexate used in the treatment of various types of cancer is given with AuNP, it suppressed tumor growth [339, 346]. Surface functionalizations of AuNPs also increases the circulation time of drug-AuNP conjugates and delivers the drug to the target cells or tissues with reduced toxicity. An application of AuNPs for pancreatic cancer treatment as a carrier of gemcitabine demonstrated the inhibition of the growth of tumor cells in an advanced stage of the disease [347].

29.7 Carbon Nanotubes (CNTs)

Carbon nanotubes (CNTs) are nanostructures that consist of cylindrical graphene layers and exhibit unique physical, mechanical and chemical properties. They are grouped as single (SWCNT) and multi-walled (MWCNT) carbon nanotubes according to the number of graphene layers they contain [348] (see Fig. 29.18). Due to their unique structural [349], dimensional [349], electrical [349], thermal [350], mechanical [350] and optical properties [351], CNTs are widely studied in different areas such as composite materials [352], nanoelectronic applications [353], and hydrogen storage [354]. Recently, applications in the biomedical field have taken a new turn [355, 356].

SWCNTs are 1D nanomaterial with sizes ranging from 50 nm to 1 cm in length and 1 to 2 nm in diameter. SWCNTs offer new opportunities in biological environments depending on their unique behavior than other kinds of nanoparticles. The fact that a flexible 1D nanotubes have multiple binding sites which allows nanotubes to be

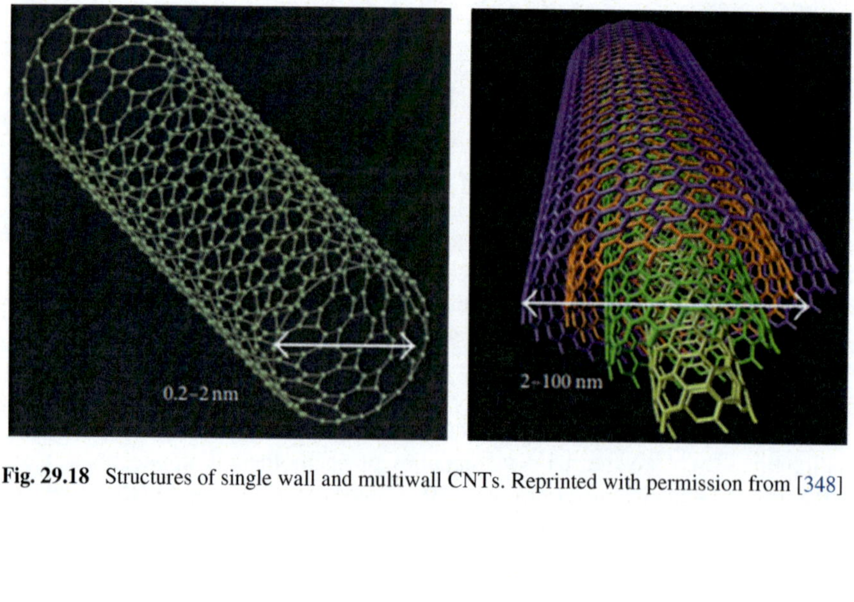

Fig. 29.18 Structures of single wall and multiwall CNTs. Reprinted with permission from [348]

conjugated with targeting agents for and increased affinity to target structures. Their high surface area allows many molecules to be attached along with the nanotube surface [357].

SWCNTs are highly absorbing materials. They exhibit strong optical absorption in the near infrared (NIR) range due to their E11 optical transitions and this situation makes them useful for photothermal therapy and photoacoustic imaging [358, 359]. Semiconductor SWCNTs with small band gaps in the 1 eV level exhibit photoluminescence in NIR range. Since SWCNTs have emission between 800–2000 nm [88, 360, 361] which covers the biological tissue permeability window, they are feasible for biological imaging. At the same time the enhanced resonance allow Raman detection and imaging possible [362, 363].

MWCNTs, unlike SWCNTs, consist of a large number of graphene layers and have a diameter of 10–100 nm. Although they are less interesting in terms of optical properties than SWCNTs, it was found out that they can be used in biological systems and they are found to be advantageous in order to transfer of large structures as plasmids to the cells [364, 365]. In general CNTs are promising structures to be employed as carriers for biological molecules to cells [366]. On the down side, CNTs have been shown to exhibit toxicity on cells when they are not functionalized [367, 368].

29.7.1 Synthesis of CNTs

Several synthesis methods are available to control diameter of the nanotubes, chirality and wall numbers in a controlled manner. Methods such as arc-discharge [369], laser ablation [370], flame synthesis [371], spray pyrolysis [372], and chemical vapor deposition [373] (CVD) are the five main techniques used in CNT synthesis [374]. Among these methods, the CVD method is widely preferred one due to simplicity of the process, low cost and easy scale up.

Although most of CNT walls are not reactive structures, fullerene-like types them allow functionalizations at some degree. The main factor that determines the reactivity of CNTs is π orbitals. The end parts of CNTs are more reactive as they are curved relative to their side surfaces. Surface modifications of CNTs can be carried out by covalent bonding through chemical reactions and non-covalent bonding by hydrophobic and hydrophilic interaction is also possible.

Oxidation of CNT surface via oxidizing agents such as nitric acid is one of the methods used in the functionalization of these particles [375]. In a study, it was shown that amino acids were able to covalently attach to the surface of SWCNT after oxidation of nanotubes [376]. Different modification procedures and bioactive molecules that can be attached on the CNT sidewalls are shown in Figs. 29.19 and 29.20 respectively. The high salt contents in most biological solutions hinder CNTs from being used directly due to the a potential of aggregation in these environments [377].

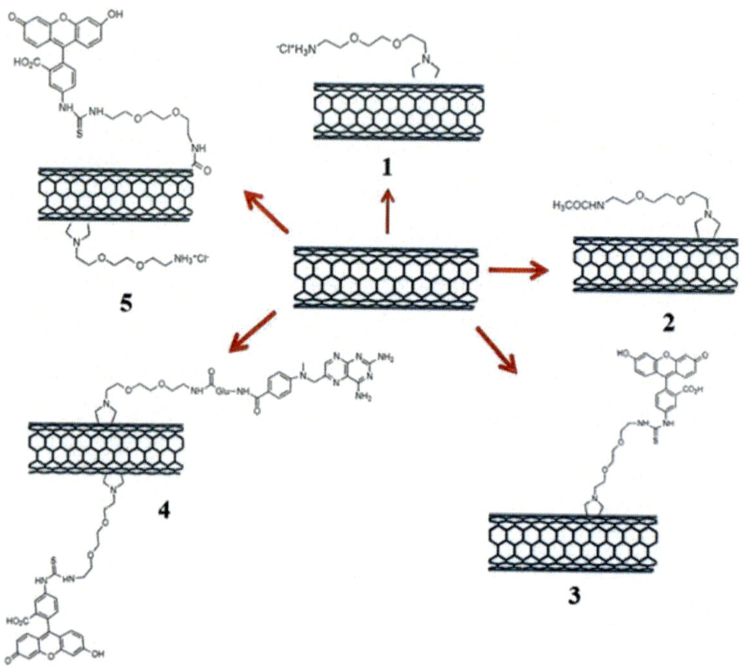

Fig. 29.19 Molecular structures of CNT covalently functionalized with different types of small molecules. 1, CNT-Ammonium; 2, CNT-Acetamido; 3, CNT-Fuorescein isothiocyanate (FITC); 4, CNT-Ammonium-FITC; 5, CNT-Methotrexate (MTX)-FITC; 6, shortened CNT-amphotericin B (AmB)-FITC; 7, shortened CNT-FITC. Reproduced from [348]

Non-covalent sidewall functionalizations are based on weak interactions such as hydrogen bonding, π-π stacking, electrostatic forces, van der Waals forces, and hydrophobic interactions [33]. These interactions are generally used to wrap small molecules, biochemically active molecules or molecules such as polymers onto CNT side walls.

Modification of oxidized CNTs with polymers such as hydrophilic PEG results in stable CNT-polymer conjugates in biological environments that can be utilized in in vivo and in vitro studies. Another way of functionalization is provided by cyclo additions performed on the aromatic side wall. [2 + 1] cycloadditions can be done by the reaction of CNTs with azide molecules through photochemistry or Bingel reaction with carbene-forming compounds [378, 379].

Another frequently used modification reaction for CNTs is the 1,3-dipolar cyclo addition reaction developed by Prato [380, 381]. It was accomplished by azomethine-ylide mediated condensation of α-amino acids and aldehydes to the graphitic surface with the coupled pyrrolidine ring formation on the CNT surface. Even though the surface modifications make CNTs stable and allow them to be employed in biological applications, they have negative effect on the photoluminescence and Raman scattering properties of the particle due to the damage caused to the structure [377].

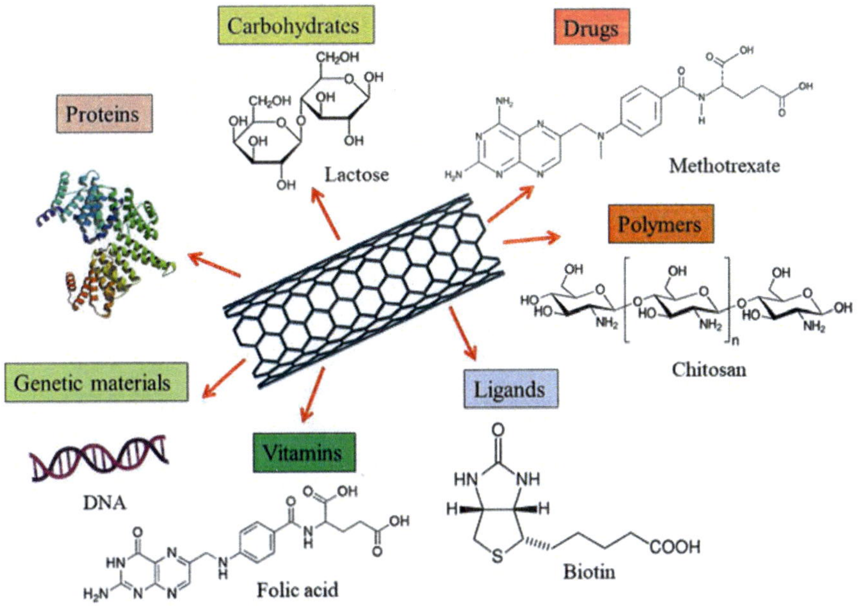

Fig. 29.20 Bioactive molecules that can be conjugated with CNTs. Reproduced from [348]

29.7.2 *Biological Applications of CNTs*

Until the beginning of the twenty-first century, carbon nanotubes, which came to the fore with their applications in the industrial field, have started to be reported in the biological field, and a great increase has been observed in the biomedicine field. Nanotubes that are stable in water and serum are biocompatible, non-toxic and have potential for biomedical applications. Although the in vivo distribution varies depending on their functionalization and size, they tend to accumulate in reticuloendothelial system (RES), which mostly includes liver and spleen. It is known that noncovalently functionalized CNTs mostly accumulate in the liver and spleen while covalently functionalized CNTs tend to be excreted through the urine. An illustration of CNTs use in biomedical applications is shown in Fig. 29.21.

The precise determination of biological materials using CNTs is possible with surface passivation via preventing the non-specific interactions of biological molecules with the hydrophobic nanotube surface.

In vivo and in vitro studies show that carbon nanotube-based drug delivery systems appear to be promising carrier systems for cancer therapy due to the decrease seen in side effects of anticancer drugs for the organism and increased the possibility of localization of the particles in the desired region. Another reason to prefer CNTs as a drug delivery system is their high cellular internalization, efficient drug loading capacity and exhibiting the selective targeting via surface modifications. Studies performed by functionalizing of CNTs with active or passive targeting agents are

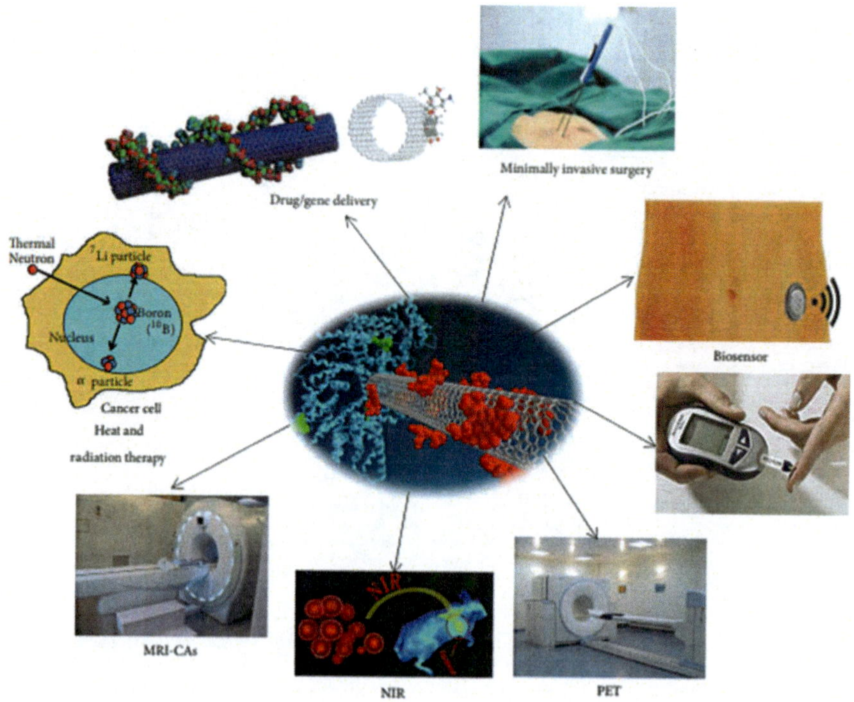

Fig. 29.21 CNTs in biomedical applications. Reprinted with permission from [348]

reported. Modifications with hydrophilic polymers such as chitosan, PEG and poly (acrylic acid) were the strategies used in the preparation of carrier systems for tumor tissues.

Meng et al. showed that CNTs modified with folic acid was wrapped with chitosan could penetrate into cancer cells through an energy-dependent endocytotic mechanism [382] (see Fig. 29.22).

Studies on the conjugation of CNTs with therapeutics such as DOX, cisplatin, paclitaxel have been extensively investigated [383]. For instance, DOX can be efficiently loaded into CNTs based on the π-π interactions between them. Most of the studies conducted on cancer cells in order to compare with free DOX and CNT transported DOX, concluded that transported DOX higher cancer cell killing capacity. In addition to this, transportation of DOX with folic acid modified MWCNTs prevented the tumor growth and decreased the side effects of DOX in vivo.

SWCNTs have been used in chemotherapy and NIR PTT for the delivery of DOX [384]. The results revealed that SWCNTs caused high amount of DOX accumulation in the nucleus of cancer cells and exhibited low drug release profile. As the NIR radiation included to the treatment, cell death was observed due to mitochondrial damage and increased reactive oxygen species kin the environment. A similar approach was followed for the treatment of breast cancer. MWCNTs were combined

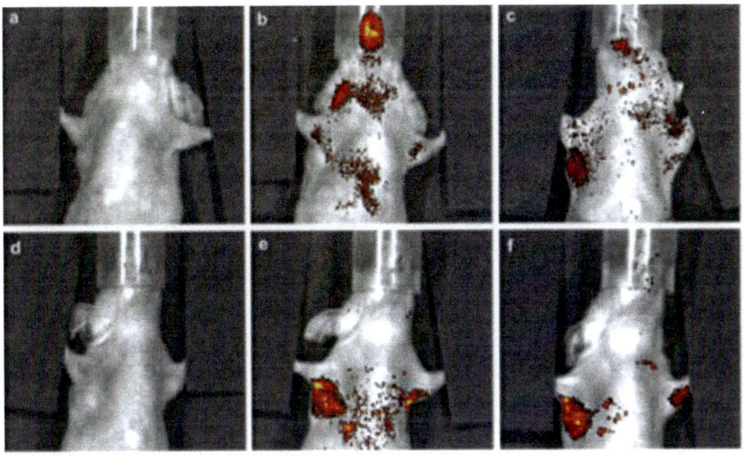

Fig. 29.22 In vivo fluorescent images of tumor-bearing BALB/c mice after intravenous injection of fluorescent Chitosan-SWNTs for **a** 0 h, **b** 2 h, **c** 20 **h**, and Folic acid/Chitosan-SWNTs for **d** 0 **h** **e** 2 h, **f** 20 h. Reprinted with permission from [382]

with methotrexate and delivered to the tumor site followed by thermal ablation. It was reported that the higher rate on the death of breast cancer cells were observed compared to free drug [385].

Biosensing studies are among main applications of CNTs. One of the first studies in this area was reported by Kirschner which involved modification of the surface of a biosensor for the detection of proteins by interacting them with fullerene and CNTs. Later on, various high-sensitivity DNA-biosensors using functionalized MWCNT have been reported by researchers. Studies have shown that specific DNA sequences and pathogenic microorganism DNAs such as *mycobacterium* tuberculosis and Hepatitis C DNAs can be detected with high sensitivity [386, 387]. Apart from DNA detection, CNT-based biosensors have also been reported to be used in detection of molecules such as pesticides [388], glucose [389], reactive oxygen species [390], and identification of diseases including malaria [391] and tuberculosis [392].

Modified CNTs have been prepared to be employed in the combat with infections from different microorganisms. In this aspect, antibiotic conjugated CNTs were produced against the resistance developed by bacteria to antibiotics [393]. CNTs modified with vancomycin hydrochloride have been reported to have high potential in wound healing processes due to their distinguished antimicrobial properties. Resistance to ciprofloxacin, which is used in the treatment of infections caused by Gram-positive and Gram-negative bacteria, is a problem all over the world [394]. Assali et al. conducted a study to overcome this problem and reported that ciprofloxacin loaded SWCNTs demonstrated a significant increase in the antibacterial activity compared to free drug against 3 bacterial strains [395]. It has been shown in several studies that the combination of CNTs with noble metals such as AgNPs is an effective option in the combat with pathogenic microorganisms, including antibiotic resistant microorganisms [396, 397]. Al Aani et al. demonstrated an increased antibacterial activity of AgNP coated MWCNT hybrid nanostructure against two important microorganisms such as *Staphylococcus aureus* and *Escherichia coli* [398]. CNT-copper hydroxide acetate hybrid nanostructure has shown to exhibit antimicrobial activity against the same bacteria [399].

CNTs are also utilized in neural tissue applications including the development of neuro implants for neuron growth and repair, as they can provide optimum conditions as high binding affinity and electrical conductivity required for neural regeneration [400]. It has been reported that hybrid CNT scaffold structures exhibit high electrical conductivity, elasticity and cellular compatibility were developed to make use of them in neural applications.

29.8 Conclusions

As it is reviewed in this chapter, the most of the effort put on the nanomaterials regardless of the type is about adopting them to medicine and commercialization. Nanoparticles are replacing contrast agents in terms of imaging, changing administration routes of the drugs into non-invasive manner, offering better treatment options with high efficacy and low toxicity particularly for cancers and several types of other diseases, providing precision in detection of various materials and biologics. These benefits and many more which are not listed here makes the utilization of nanomaterials inevitable. Since the nature and the behavior of nanoparticles in diversified environments have not been well understood yet, their potential to exhibit treat against human health is still unclear. The next step in the adaptation of nanomaterials to our daily life and medicine is to develop them as multifunctional and controllable nanodevices while keeping the potential of risk of cytotoxicity in mind.

References

1. M.I. Majeed, H.N. Bhatti, H. Nawaz, M. Kashif, Nanobiotechnology: applications of nanomaterials in biological research. in *Integrating Green Chemistry and Sustainable Engineering*, ed. by S. ul-Islam (2019), p. 581–615
2. M. Hatipoğlu Kukut, S. Keleştemur, M. Çulha, Synthesis and biological applications of quantum dots, in *Low-Dimensional and Nanostructured Materials and Devices*, ed. by H.N. Ünlü, J. Dabrowski (Cham, Springer, 2016)
3. A. Akbarzadeh, R. Rezaei-Sadabady, S. Davaran, S.W. Joo, N. Zarghami, Y. Hanifehpour et al., Liposome: classification, preparation, and applications. Nanoscale Res. Lett. **8**, 102 (2013)
4. D.C. Drummond, O. Meyer, K. Hong, D.B. Kirpotin, D. Papahadjopoulos, Optimizing liposomes for delivery of chemotherapeutic agents to solid tumors. Pharmacol. Rev. **51**, 691–743 (1999)

5. S. Kim, Liposomes as carriers of cancer chemotherapy. Current status and future prospects. Drugs **46**, 618–638 (1993)
6. V.P. Torchilin, Recent advances with liposomes as pharmaceutical carriers. Nat. Rev. Drug Discov. **4**, 145–160 (2005)
7. Micelles and liposomes, Lipid nanovehicles for intracerebral drug delivery, in *The Textbook of Nanoneuroscience and Nanoneurosurgery*. ed. by J.D.H. Babak Kateb (CRC Press, Boca Raton, 2014), p. 602
8. T.M. Allen, Liposomes. opportunities in drug delivery. Drugs **54**(Suppl 4), 8–14 (1997)
9. A. Gabizon, D. Goren, R. Cohen, Y. Barenholz, Development of liposomal anthracyclines: from basics to clinical applications. J. Control Release **53**, 275–279 (1998)
10. S.K. Sahoo, V. Labhasetwar, Nanotech approaches to drug delivery and imaging. Drug Discov. Today **8**, 1112–1120 (2003)
11. M.D. Mary Ann Clark, J. Choi, *Biology 2e*, 2nd edn. (OpenStax, March 18, 2018), p. 127–155
12. C.I. Nkanga, A.M. Bapolisi, N.I. Okafor, R.W.M. Krause (eds.), *General Perception of Liposomes: Formation, Manufacturing and Applications* (2019)
13. M.L. Immordino, F. Dosio, L. Cattel, Stealth liposomes: review of the basic science, rationale, and clinical applications, existing and potential. Int. J. Nanomedicine **1**, 297–315 (2006)
14. H. Daraee, A. Etemadi, M. Kouhi, S. Alimirzalu, A. Akbarzadeh, Application of liposomes in medicine and drug delivery. Artif. Cells Nanomed. Biotechnol. **44**, 381–391 (2016)
15. C.Y. Wang, L. Huang, pH-sensitive immunoliposomes mediate target-cell-specific delivery and controlled expression of a foreign gene in mouse. Proc. Natl. Acad. Sci. USA **84**, 7851–7855 (1987)
16. I. Mellman, R. Fuchs, A. Helenius, Acidification of the endocytic and exocytic pathways. Annu. Rev. Biochem. **55**, 663–700 (1986)
17. C.J. Chu, J. Dijkstra, M.Z. Lai, K. Hong, F.C. Szoka, Efficiency of cytoplasmic delivery by pH-sensitive liposomes to cells in culture. Pharm. Res. **7**, 824–834 (1990)
18. C.R. Alving, Liposomes as carriers of antigens and adjuvants. J. Immunol. Methods **140**, 1–13 (1991)
19. L. Raphael, B.H. Tom, Liposome facilitated xenogeneic approach for studying human colon cancer immunity: carrier and adjuvant effect of liposomes. Clin. Exp. Immunol. **55**, 1–13 (1984)
20. S.J. LeGrue, Carrier and adjuvant properties of liposome-borne tumor-specific antigens. Cancer Immunol. Immunother **17**, 135–141 (1984)
21. V.H. Engelhard, G.A. Powers, L.C. Moore, M.J. Holterman, M.C. Correa-Freire, Cytotoxic T lymphocyte recognition of HLA-A/B antigens introduced into EL4 cells by cell-liposome fusion. J. Immunol. **132**, 76–80 (1984)
22. G. Gregoriadis, D. Davis, A. Davies, Liposomes as immunological adjuvants: antigen incorporation studies. Vaccine **5**, 145–151 (1987)
23. G. Steele Jr., T. Ravikumar, D. Ross, V. King, R.E. Wilson, T. Dodson, Specific active immunotherapy with butanol-extracted, tumor-associated antigens incorporated into liposomes. Surgery **96**, 352–359 (1984)
24. T.M. Allen, A. Chonn, Large unilamellar liposomes with low uptake into the reticuloendothelial system. FEBS Lett. **223**, 42–46 (1987)
25. A. Gabizon, D. Papahadjopoulos, Liposome formulations with prolonged circulation time in blood and enhanced uptake by tumors. Proc. Natl. Acad. Sci. USA **85**, 6949–6953 (1988)
26. A. Sharma, U.S. Sharma, Liposomes in drug delivery: progress and limitations. Int. J. Pharm. **154**, 123–140 (1997)
27. U. Pick, Liposomes with a large trapping capacity prepared by freezing and thawing of sonicated phospholipid mixtures. Arch. Biochem. Biophys. **212**, 186–194 (1981)
28. S. Batzri, E.D. Korn, Single bilayer liposomes prepared without sonication. Biochim. Biophys. Acta **298**, 1015–1019 (1973)
29. K. Sahil, S. Premjeet, B. Ajay, A. Middha, K. Bhawna K (eds.), *Stealth Liposomes: A Review* (2011)

30. L.D. Mayer, M.B. Bally, M.J. Hope, P.R. Cullis, Techniques for encapsulating bioactive agents into liposomes. Chem. Phys. Lipids **40**, 333–345 (1986)
31. A.S. Abu Lila, T. Ishida, Liposomal delivery systems: design optimization and current applications. Biol. Pharm. Bull. **40**, 1–10 (2017)
32. G. Bozzuto, A. Molinari, Liposomes as nanomedical devices. Int. J. Nanomedicine **10**, 975–999 (2015)
33. S.M. El-Bahr, Effect of curcumin on hepatic antioxidant enzymes activities and gene expressions in rats intoxicated with aflatoxin B1. Phytother Res. **29**, 134–140 (2015)
34. S. Talegaonkar, A. Bhattacharyya, Potential of lipid nanoparticles (SLNs and NLCs) in enhancing oral bioavailability of drugs with poor intestinal permeability. AAPS PharmSciTech **20**, 121 (2019)
35. P. Jansook, Z. Fülöp, G.C. Ritthidej, Amphotericin B loaded solid lipid nanoparticles (SLNs) and nanostructured lipid carrier (NLCs): physicochemical and solid-solution state characterizations. Drug Dev. Ind. Pharm. **45**, 560–567 (2019)
36. X.G. Zhang, J. Miao, Y.Q. Dai, Y.Z. Du, H. Yuan, F.Q. Hu, Reversal activity of nanostructured lipid carriers loading cytotoxic drug in multi-drug resistant cancer cells. Int. J. Pharm. **361**, 239–244 (2008)
37. Y. Wang, H. Zhang, J. Hao, B. Li, M. Li, W. Xiuwen, Lung cancer combination therapy: co-delivery of paclitaxel and doxorubicin by nanostructured lipid carriers for synergistic effect. Drug Deliv. **23**, 1398–1403 (2016)
38. J. Yang, Z. Ju, S. Dong, Cisplatin and paclitaxel co-delivered by folate-decorated lipid carriers for the treatment of head and neck cancer. Drug Deliv. **24**, 792–799 (2016)
39. C.W. How, A. Rasedee, S. Manickam, R. Rosli, Tamoxifen-loaded nanostructured lipid carrier as a drug delivery system: characterization, stability assessment and cytotoxicity. Colloids Surf B Biointerfaces **112**, 393–399 (2013)
40. A. Gabizon, H. Shmeeda, Y. Barenholz, Pharmacokinetics of pegylated liposomal doxorubicin. Clin. Pharmacokinet. **42**, 419–436 (2003)
41. A.K. Singla, A. Garg, D. Aggarwal, Paclitaxel and its formulations. Int. J. Pharm. **235**, 179–192 (2002)
42. T.M. Allen, P.R. Cullis, Liposomal drug delivery systems: from concept to clinical applications. Adv. Drug Deliv. Rev. **65**, 36–48 (2013)
43. G. Batist, G. Ramakrishnan, C.S. Rao, A. Chandrasekharan, J. Gutheil, T. Guthrie et al., Reduced cardiotoxicity and preserved antitumor efficacy of liposome-encapsulated doxorubicin and cyclophosphamide compared with conventional doxorubicin and cyclophosphamide in a randomized, multicenter trial of metastatic breast cancer. J. Clin. Oncol. **19**, 1444–1454 (2001)
44. F.M. Muggia, J.D. Hainsworth, S. Jeffers, P. Miller, S. Groshen, M. Tan et al., Phase II study of liposomal doxorubicin in refractory ovarian cancer: Antitumor activity and toxicity modification by liposomal encapsulation. J. Clin. Oncol. **15**, 987–993 (1997)
45. L. Harris, G. Batist, R. Belt, D. Rovira, R. Navari, N. Azarnia et al., Liposome-encapsulated doxorubicin compared with conventional doxorubicin in a randomized multicenter trial as first-line therapy of metastatic breast carcinoma. Cancer **94**, 25–36 (2002)
46. O. Lyass, B. Uziely, R. Ben-Yosef, D. Tzemach, N.I. Heshing, M. Lotem et al., Correlation of toxicity with pharmacokinetics of pegylated liposomal doxorubicin (doxil) in metastatic breast carcinoma. Cancer **89**, 1037–1047 (2000)
47. M.R. Ranson, J. Carmichael, K. O'Byrne, S. Stewart, D. Smith, A. Howell, Treatment of advanced breast cancer with sterically stabilized liposomal doxorubicin: results of a multicenter phase II trial. J. Clin. Oncol. **15**, 3185–3191 (1997)
48. T. Tejada-Berges, C.O. Granai, M. Gordinier, W. Gajewski, Caelyx/doxil for the treatment of metastatic ovarian and breast cancer. Expert Rev. Anticancer Ther. **2**, 143–150 (2002)
49. G. Zuccari, A. Milelli, F. Pastorino, M. Loi, A. Petretto, A. Parise et al., Tumor vascular targeted liposomal-bortezomib minimizes side effects and increases therapeutic activity in human neuroblastoma. J. Control. Release **211**, 44–52 (2015)

50. R. Vakili-Ghartavol, S.M. Rezayat, R. Faridi-Majidi, K. Sadri, M.R. Jaafari, Optimization of docetaxel loading conditions in liposomes: proposing potential products for metastatic breast carcinoma chemotherapy. Sci. Rep. **10**, 5569 (2020)
51. X. Ding, C. Yin, W. Zhang, Y. Sun, Z. Zhang, E. Yang et al., Designing aptamer-gold nanoparticle-loaded pH-sensitive liposomes encapsulate morin for treating cancer. Nanoscale Res. Lett. **15**, 68 (2020)
52. M. Almousallam, C. Moia, H. Zhu, Development of nanostructured lipid carrier for dacarbazine delivery. Int. Nano Lett. **5**, 241–248 (2015)
53. H. Breithaupt, A. Dammann, K. Aigner, Pharmacokinetics of dacarbazine (DTIC) and its metabolite 5-aminoimidazole-4-carboxamide (AIC) following different dose schedules. Cancer Chemother. Pharmacol. **9**, 103–109 (1982)
54. Y. Hei, B. Teng, Z. Zeng, S. Zhang, Q. Li, J. Pan et al., Multifunctional immunoliposomes combining catalase and PD-L1 antibodies overcome tumor hypoxia and enhance immunotherapeutic effects against melanoma. Int. J. Nanomedicine **15**, 1677–1691 (2020)
55. B.L. Riggs, L.C. Hartmann, Selective estrogen-receptor modulators—mechanisms of action and application to clinical practice. N. Engl. J. Med. **348**, 618–629 (2003)
56. P.H. Patil, V.S. Belgamwar, P.R. Patil, S.J. Surana, Solubility enhancement of raloxifene using inclusion complexes and cogrinding method. J. Pharm. (Cairo) 527380 (2013)
57. N.V. Shah, A.K. Seth, R. Balaraman, C.J. Aundhia, R.A. Maheshwari, G.R. Parmar, Nanostructured lipid carriers for oral bioavailability enhancement of raloxifene: design and in vivo study. J. Adv. Res. **7**, 423–434 (2016)
58. W. Park, A.R. Amin, Z.G. Chen, D.M. Shin, New perspectives of curcumin in cancer prevention. Cancer Prev. Res. (Phila) **6**, 387–400 (2013)
59. T. Jia, L. Zhang, Y. Duan, M. Zhang, G. Wang, J. Zhang et al., The differential susceptibilities of MCF-7 and MDA-MB-231 cells to the cytotoxic effects of curcumin are associated with the PI3K/Akt-SKP2-Cip/Kips pathway. Cancer Cell Int. **14**, 126 (2014)
60. L. Vecchi Brumatti, A. Marcuzzi, P.M. Tricarico, V. Zanin, M. Girardelli, A.M. Bianco, Curcumin and inflammatory bowel disease: potential and limits of innovative treatments. Molecules **19**, 21127–21153 (2014)
61. Y. Chen, L. Pan, M. Jiang, D. Li, L. Jin, Nanostructured lipid carriers enhance the bioavailability and brain cancer inhibitory efficacy of curcumin both in vitro and in vivo. Drug Deliv. **23**, 1383–1392 (2016)
62. D. Bhojwani, C.-h Pui, Intrathecal liposomal cytarabine: more friend than foe? Leuk. Lymphoma **49**, 1427–1430 (2008)
63. L. Bomgaars, J.R. Geyer, J. Franklin, G. Dahl, J. Park, N.J. Winick et al., Phase I trial of intrathecal liposomal cytarabine in children with neoplastic meningitis. J. Clin. Oncol. **22**, 3916–3921 (2004)
64. S. Phuphanich, B. Maria, R. Braeckman, M. Chamberlain, A pharmacokinetic study of intra-CSF administered encapsulated cytarabine (DepoCyt®) for the treatment of neoplastic meningitis in patients with leukemia, lymphoma, or solid tumors as part of a phase III study. J. Neurooncol. **81**, 201–208 (2007)
65. A. Peyrl, R. Sauermann, F. Traunmueller, A.A. Azizi, M. Gruber-Olipitz, A. Gupper et al., Pharmacokinetics and safety of intrathecal liposomal cytarabine in children aged <3 years. Clin. Pharmacokinet. **48**, 265–271 (2009)
66. A. Peyrl, R. Sauermann, M. Chocholous, A.A. Azizi, W. Jäger, M. Höferl et al., Pharmacokinetics and toxicity of intrathecal liposomal cytarabine in children and adolescents following age-adapted dosing. Clin. Pharmacokinet. **53**, 165–173 (2014)
67. X. Dong, W. Wang, H. Qu, D. Han, J. Zheng, G. Sun, Targeted delivery of doxorubicin and vincristine to lymph cancer: evaluation of novel nanostructured lipid carriers in vitro and in vivo. Drug Deliv. **23**, 1374–1378 (2016)
68. Y. Wang, H. Zhang, J. Hao, B. Li, M. Li, W. Xiuwen, Lung cancer combination therapy: co-delivery of paclitaxel and doxorubicin by nanostructured lipid carriers for synergistic effect. Drug Deliv. **23**(4), 1398–1403 (2016)

69. O. Paecharoenchai, N. Niyomtham, A. Apirakaramwong, T. Ngawhirunpat, T. Rojanarata, B.E. Yingyongnarongkul et al., Structure relationship of cationic lipids on gene transfection mediated by cationic liposomes. AAPS PharmSciTech **13**, 1302–1308 (2012)
70. D.A. Balazs, W.T. Godbey, Liposomes for use in gene delivery. J. Drug Deliv. 326497 (2011)
71. S.M. Dizaj, S. Jafari, A.Y. Khosroushahi, A sight on the current nanoparticle-based gene delivery vectors. Nanoscale Res. Lett. **9**, 252 (2014)
72. Z.Y. He, F. Deng, X.W. Wei, C.C. Ma, M. Luo, P. Zhang et al., Ovarian cancer treatment with a tumor-targeting and gene expression-controllable lipoplex. Sci. Rep. **6**, 23764 (2016)
73. B.S. Pattni, V.V. Chupin, V.P. Torchilin, New developments in liposomal drug delivery. Chem. Rev. **115**, 10938–10966 (2015)
74. M.J. Hussain, A. Wilkinson, V.W. Bramwell, D. Christensen, Y. Perrie, Th1 immune responses can be modulated by varying dimethyldioctadecylammonium and distearoyl-sn-glycero-3-phosphocholine content in liposomal adjuvants. J. Pharm. Pharmacol. **66**, 358–366 (2014)
75. C.R. Alving, K.K. Peachman, G.R. Matyas, M. Rao, Z. Beck, Army liposome formulation (ALF) family of vaccine adjuvants. Expert Rev. Vaccines **19**, 279–292 (2020)
76. W.D. Zollinger, J.G. Babcock, E.E. Moran, B.L. Brandt, G.R. Matyas, N.M. Wassef et al., Phase I study of a Neisseria meningitidis liposomal vaccine containing purified outer membrane proteins and detoxified lipooligosaccharide. Vaccine **30**, 712–721 (2012)
77. D.T. Harris, G.R. Matyas, L.G. Gomella, E. Talor, M.D. Winship, L.E. Spitler et al., Immunologic approaches to the treatment of prostate cancer. Semin. Oncol **26**, 439–447 (1999)
78. R. Sangha, C. Butts, L-BLP25: a peptide vaccine strategy in non small cell lung cancer. Clin. Cancer Res. **13**, s4652-4654 (2007)
79. A. Wrona, Role of immunotherapy in stage III nonsmall cell lung cancer. Curr. Opin. Oncol **31**, 18–23 (2019)
80. C. Theunis, N. Crespo-Biel, V. Gafner, M. Pihlgren, M.P. López-Deber, P. Reis, et al., Efficacy and safety of a liposome-based vaccine against protein Tau, assessed in tau.P301L mice that model tauopathy. PLoS One **8**, e72301 (2013)
81. J.T. van Dissel, S.A. Joosten, S.T. Hoff, D. Soonawala, C. Prins, D.A. Hokey et al., A novel liposomal adjuvant system, CAF01, promotes long-lived Mycobacterium tuberculosis-specific T-cell responses in human. Vaccine **32**, 7098–7107 (2014)
82. A.V. Kabanov, E.V. Batrakova, N.S. Melik-Nubarov, N.A. Fedoseev, T.Y. Dorodnich, V.Y. Alakhov et al., A new class of drug carriers: micelles of poly(oxyethylene)-poly(oxypropylene) block copolymers as microcontainers for drug targeting from blood in brain. J. Control Release **22**, 141–157 (1992)
83. V.P. Torchilin, Micellar nanocarriers: pharmaceutical perspectives. Pharm. Res. **24**, 1–16 (2007)
84. L.L. Schramm, E.N. Stasiuk, D.G. Marangoni, 2 Surfactants and their applications. Annu. Rep. Sect. C (Phys. Chem.) **99**, 3–48 (2003)
85. M. Kellermann, W. Bauer, A. Hirsch, B. Schade, K. Ludwig, C. Böttcher, The first account of a structurally persistent micelle. Angew. Chem. Int. Ed. Engl. **43**, 2959–2962 (2004)
86. L.G. Chen, S.H. Strassburg, H. Bermudez, Micelle co-assembly in surfactant/ionic liquid mixtures. J. Colloid Interface Sci. **477**, 40–45 (2016)
87. K. Jia, J. Hu, J. Dong, X. Li, Light-responsive multillamellar vesicles in coumaric acid/alkyldimethylamine oxide binary systems: Effects of surfactant and hydrotrope structures. J. Colloid Interface Sci. **477**, 156–165 (2016)
88. V.K. Patil, I.T. Gawali, G.A. Usmani, Synthesis and properties of novel cationic triazolium gemini surfactants. J. Dispersion Sci. Technol. **37**, 1630–1637 (2016)
89. N. Nishiyama, K. Kataoka, Current state, achievements, and future prospects of polymeric micelles as nanocarriers for drug and gene delivery. Pharmacol. Ther. **112**, 630–648 (2006)
90. N.N. Gaber, Y. Darwis, K.K. Peh, Y.T. Tan, Characterization of polymeric micelles for pulmonary delivery of beclomethasone dipropionate. J. Nanosci. Nanotechnol. **6**, 3095–3101 (2006)

91. L.J. Magid, Z. Li, P.D. Butler, Flexibility of elongated sodium dodecyl sulfate micelles in aqueous sodium chloride: a small-angle neutron scattering study. Langmuir **16**, 10028–10036 (2000)
92. B. Aoun, V.K. Sharma, E. Pellegrini, S. Mitra, M. Johnson, R. Mukhopadhyay, Structure and dynamics of ionic micelles: MD simulation and neutron scattering study. J. Phys. Chem. B **119**, 5079–5086 (2015)
93. S.R. Croy, G.S. Kwon, Polymeric micelles for drug delivery. Curr. Pharm. Des. **12**, 4669–4684 (2006)
94. A.V. Kabanov, I.R. Nazarova, I.V. Astafieva, E.V. Batrakova, V.Y. Alakhov, A.A. Yaroslavov et al., Micelle formation and solubilization of fluorescent probes in poly(oxyethylene-b-oxypropylene-b-oxyethylene) solutions. Macromolecules **28**, 2303–2314 (1995)
95. A.I. Papaioannou, S. Papiris, G. Papadaki, E.D. Manali, A. Roussou, A. Spathis et al., Surfactant proteins in smoking-related lung disease. Curr. Top Med. Chem. **16**, 1574–1581 (2016)
96. A.R.V. Ferreira, C.A.V. Torres, F. Freitas, C. Sevrin, C. Grandfils, M.A.M. Reis et al., Development and characterization of bilayer films of FucoPol and chitosan. Carbohydr. Polym. **147**, 8–15 (2016)
97. M.-C. Jones, J.-C. Leroux, Polymeric micelles – a new generation of colloidal drug carriers. Eur. J. Pharm. Biopharm. **48**, 101–111 (1999)
98. G. Giorgio, G. Colafemmina, F. Mavelli, S. Murgia, G. Palazzo, The impact of alkanes on the structure of Triton X100 micelles. RSC Adv. **6**, 825–836 (2016)
99. M.J. Pottage, T.L. Greaves, C.J. Garvey, R.F. Tabor, The effects of alkylammonium counterions on the aggregation of fluorinated surfactants and surfactant ionic liquids. J. Colloid Interface Sci. **475**, 72–81 (2016)
100. M.L. Adams, A. Lavasanifar, G.S. Kwon, Amphiphilic block copolymers for drug delivery. J. Pharm. Sci. **92**, 1343–1355 (2003)
101. N.A.N. Hanafy, M. El-Kemary, S. Leporatti, Micelles structure development as a strategy to improve smart cancer therapy. Cancers **10**, 238 (2018)
102. D. Sutton, N. Nasongkla, E. Blanco, J. Gao, Functionalized micellar systems for cancer targeted drug delivery. Pharm Res **24**, 1029–1046 (2007)
103. J.E. Chung, M. Yokoyama, T. Okano, Inner core segment design for drug delivery control of thermo-responsive polymeric micelles. J. Control Release **65**, 93–103 (2000)
104. E. Blanco, C.W. Kessinger, B.D. Sumer, J. Gao, Multifunctional micellar nanomedicine for cancer therapy. Exp. Biol. Med. (Maywood) **234**, 123–131 (2009)
105. R. Tong, J. Cheng, Anticancer polymeric nanomedicines. Polym. Rev. **47**, 345–381 (2007)
106. W.S. Cheow, K. Hadinoto, Factors affecting drug encapsulation and stability of lipid-polymer hybrid nanoparticles. Colloids Surf B Biointerfaces **85**, 214–220 (2011)
107. J. Wang, D. Mongayt, V.P. Torchilin, Polymeric micelles for delivery of poorly soluble drugs: preparation and anticancer activity in vitro of paclitaxel incorporated into mixed micelles based on poly(ethylene glycol)-lipid conjugate and positively charged lipids. J. Drug Target **13**, 73–80 (2005)
108. J. Zhang, X.G. Chen, L. Huang, J.T. Han, X.F. Zhang, Self-assembled polymeric nanoparticles based on oleic acid-grafted chitosan oligosaccharide: biocompatibility, protein adsorption and cellular uptake. J. Mater. Sci. Mater. Med. **23**, 1775–1783 (2012)
109. C. Esquenet, P. Terech, F. Boué, E. Buhler, Structural and rheological properties of hydrophobically modified polysaccharide associative networks. Langmuir **20**, 3583–3592 (2004)
110. C. Oerlemans, W. Bult, M. Bos, G. Storm, J.F.W. Nijsen, W.E. Hennink, Polymeric micelles in anticancer therapy: targeting, imaging and triggered release. Pharm. Res. **27**, 2569–2589 (2010)
111. R. Mo, X. Jin, N. Li, C. Ju, M. Sun, C. Zhang et al., The mechanism of enhancement on oral absorption of paclitaxel by N-octyl-O-sulfate chitosan micelles. Biomaterials **32**, 4609–4620 (2011)

112. X. Dong, C.A. Mattingly, M.T. Tseng, M.J. Cho, Y. Liu, V.R. Adams et al., Doxorubicin and paclitaxel-loaded lipid-based nanoparticles overcome multidrug resistance by inhibiting P-glycoprotein and depleting ATP. Can. Res. **69**, 3918–3926 (2009)
113. L. Bromberg, Polymeric micelles in oral chemotherapy. J. Control Release **128**, 99–112 (2008)
114. T.Y. Kim, D.W. Kim, J.Y. Chung, S.G. Shin, S.C. Kim, D.S. Heo et al., Phase I and pharmacokinetic study of Genexol-PM, a cremophor-free, polymeric micelle-formulated paclitaxel, in patients with advanced malignancies. Clin. Cancer Res. **10**, 3708–3716 (2004)
115. K.S. Lee, H.C. Chung, S.A. Im, Y.H. Park, C.S. Kim, S.B. Kim et al., Multicenter phase II trial of Genexol-PM, a Cremophor-free, polymeric micelle formulation of paclitaxel, in patients with metastatic breast cancer. Breast Cancer Res. Treat **108**, 241–250 (2008)
116. Y. Matsumura, T. Hamaguchi, T. Ura, K. Muro, Y. Yamada, Y. Shimada et al., Phase I clinical trial and pharmacokinetic evaluation of NK911, a micelle-encapsulated doxorubicin. Br. J. Cancer **91**, 1775–1781 (2004)
117. Z. Gao, A.N. Lukyanov, A. Singhal, V.P. Torchilin, Diacyllipid-polymer micelles as nanocarriers for poorly soluble anticancer drugs. Nano Lett. **2**, 979–982 (2002)
118. L. Mu, A. Chrastina, T. Levchenko, V. Torchilin, Micelles from poly(ethylene glycol)–phosphatidyl ethanolamine conjugates (Peg-Pe) as pharmaceutical nanocarriers for poorly soluble drug camptothecin. J. Biomed. Nanotechnol. **1**, 190–195 (2005)
119. M. Yokoyama, T. Okano, Y. Sakurai, H. Ekimoto, C. Shibazaki, K. Kataoka, Toxicity and antitumor activity against solid tumors of micelle-forming polymeric anticancer drug and its extremely long circulation in blood. Cancer Res. **51**, 3229–3236 (1991)
120. D. Deeb, H. Jiang, X. Gao, S. Al-Holou, A.L. Danyluk, S.A. Dulchavsky et al., Curcumin [1,7-bis(4-hydroxy-3-methoxyphenyl)-1-6-heptadine-3,5-dione; $C_21H_20O_6$] sensitizes human prostate cancer cells to tumor necrosis factor-related apoptosis-inducing ligand/Apo2L-induced apoptosis by suppressing nuclear factor-kappaB via inhibition of the prosurvival Akt signaling pathway. J. Pharmacol. Exp. Ther. **321**, 616–625 (2007)
121. M.C. Jiang, H.F. Yang-Yen, J.J. Yen, J.K. Lin, Curcumin induces apoptosis in immortalized NIH 3T3 and malignant cancer cell lines. Nutr. Cancer **26**, 111–120 (1996)
122. J. Holy, Curcumin inhibits cell motility and alters microfilament organization and function in prostate cancer cells. Cell Motil Cytoskelet. **58**, 253–268 (2004)
123. J.H. Woo, Y.H. Kim, Y.J. Choi, D.G. Kim, K.S. Lee, J.H. Bae et al., Molecular mechanisms of curcumin-induced cytotoxicity: induction of apoptosis through generation of reactive oxygen species, down-regulation of Bcl-XL and IAP, the release of cytochrome c and inhibition of Akt. Carcinogenesis **24**, 1199–1208 (2003)
124. P. Yoysungnoen, P. Wirachwong, P. Bhattarakosol, H. Niimi, S. Patumraj, Effects of curcumin on tumor angiogenesis and biomarkers, COX-2 and VEGF, in hepatocellular carcinoma cell-implanted nude mice. Clin. Hemorheol. Microcirc. **34**, 109–115 (2006)
125. S. Anuchapreeda, P. Limtrakul, P. Thanarattanakorn, S. Sittipreechacharn, P. Chanarat, Inhibitory effect of curcumin on WT1 gene expression in patient leukemic cells. Arch. Pharm. Res. **29**, 80–87 (2006)
126. S.-Y. Lee, C.-Y. Yang, C.-L. Peng, M.-F. Wei, K.-C. Chen, C.-J. Yao et al., A theranostic micelleplex co-delivering SN-38 and VEGF siRNA for colorectal cancer therapy. Biomaterials **86**, 92–105 (2016)
127. H. Yu, Y. Zou, Y. Wang, X. Huang, G. Huang, B.D. Sumer et al., Overcoming endosomal barrier by amphotericin B-loaded dual pH-responsive PDMA-b-PDPA micelleplexes for siRNA delivery. ACS Nano **5**, 9246–9255 (2011)
128. Y. Zhang, J.S. Buhrman, Y. Liu, J.E. Rayahin, R.A. Gemeinhart, Reducible micelleplexes are stable systems for anti-miRNA delivery in cerebrospinal fluid. Mol. Pharm. **13**, 1791–1799 (2016)
129. Z.-X. Zhao, S.-Y. Gao, J.-C. Wang, C.-J. Chen, E.-Y. Zhao, W.-J. Hou et al., Self-assembly nanomicelles based on cationic mPEG-PLA-b-Polyarginine(R15) triblock copolymer for siRNA delivery. Biomaterials **33**, 6793–6807 (2012)
130. M. Wang, B. Wu, J.D. Tucker, P. Lu, Q. Lu, Poly(ester amine) constructed from polyethylenimine and pluronic for gene delivery in vitro and in vivo. Drug Deliv. **23**, 3224–3233 (2016)

131. L. Liu, M. Zheng, D. Librizzi, T. Renette, O.M. Merkel, T. Kissel, Efficient and tumor targeted siRNA delivery by polyethylenimine-graft-polycaprolactone-block-poly(ethylene glycol)-folate (PEI–PCL–PEG–Fol). Mol. Pharm. **13**, 134–143 (2016)
132. C. Sang-Cheol, Y. Dae-Ii, K. Sung-Chul, P. Eun-Seok, A polymeric micellar carrier for the solubilization of biphenyl dimethyl dicarboxylate. Arch. Pharmacal. Res. **26**, 173–181 (2003)
133. M.F. Francis, M. Cristea, Y. Yang, F.M. Winnik, Engineering polysaccharide-based polymeric micelles to enhance permeability of cyclosporin A across caco-2 cells. Pharm. Res. **22**, 209–219 (2005)
134. H. Lee, F. Zeng, M. Dunne, C. Allen, Methoxy poly(ethylene glycol)-block-Poly(δ-valerolactone) copolymer micelles for formulation of hydrophobic drugs. Biomacromol **6**, 3119–3128 (2005)
135. Z. Zhou, C. Chaibundit, A. D'Emanuele, K. Lennon, D. Attwood, C. Booth, Solubilisation of drugs in worm-like micelles of block copolymers of ethylene oxide and 1,2-butylene oxide in aqueous solution. Int. J. Pharm. **354**, 82–87 (2008)
136. E. Bernabeu, L. Gonzalez, M. Cagel, E.P. Gergic, M.A. Moretton, D.A. Chiappetta, Novel Soluplus®—TPGS mixed micelles for encapsulation of paclitaxel with enhanced in vitro cytotoxicity on breast and ovarian cancer cell lines. Colloids Surf. B **140**, 403–411 (2016)
137. Engelsberg S, Properties of materials: magnetism in solids. DH Martin, MIT Press, Cambridge, Mass., 1967. 462 pp., illus. $19.95. Science **158**(3805), 1168–1169 (1967)
138. M. Colombo, S. Carregal-Romero, M.F. Casula, L. Gutiérrez, M.P. Morales, I.B. Böhm et al., Biological applications of magnetic nanoparticles. Chem. Soc. Rev. **41**, 4306–4334 (2012)
139. R. Massart, Preparation of aqueous magnetic liquids in alkaline and acidic media. IEEE Trans. Magn. **17**, 1247–1248 (1981)
140. J.M. El Khoury, D. Caruntu, C.J. O'Connor, K.-U. Jeong, S.Z.D. Cheng, J. Hu, Poly(allylamine) stabilized iron oxide magnetic nanoparticles. J. Nanopart. Res. **9**, 959–964 (2007)
141. N. Feltin, M.P. Pileni, New technique for synthesizing iron ferrite magnetic nanosized particles. Langmuir **13**, 3927–3933 (1997)
142. S. Sun, H. Zeng, D.B. Robinson, S. Raoux, P.M. Rice, S.X. Wang et al., Monodisperse MFe_2O_4 (M = Fe Co, Mn) nanoparticles. J. Am. Chem. Soc. **126**, 273–279 (2004)
143. A.H. Lu, E.L. Salabas, F. Schüth, Magnetic nanoparticles: synthesis, protection, functionalization, and application. Angew. Chem. Int. Ed. Engl. **46**, 1222–1244 (2007)
144. Y. Wang, Q. Liu, Y. Sun, R. Wang, Magnetic field modulated SERS enhancement of CoPt hollow nanoparticles with sizes below 10 nm. Nanoscale **10**, 12650–12656 (2018)
145. P. Martins, M. Silva, S. Lanceros-Mendez, Determination of the magnetostrictive response of nanoparticles via magnetoelectric measurements. Nanoscale **7**, 9457–9461 (2015)
146. B. Zhang, W. Yang, J. Yu, W. Guo, J. Wang, S. Liu, et al., Green synthesis of sub-10 nm gadolinium-based nanoparticles for sparkling kidneys, tumor, and angiogenesis of tumor-bearing mice in magnetic resonance imaging. Adv. Healthc. Mater. 6 (2017)
147. D.A. Volgina, E.A. Stepanidenko, T.K. Kormilina, S.A. Cherevkov, A. Dubavik, M.A. Baranov et al., Study of the optical properties of CdZnSe/ZnS-quantum dot–Au-nanoparticle complexes. Opt. Spectrosc. **124**, 494–500 (2018)
148. R.W. Habash, R. Bansal, D. Krewski, H.T. Alhafid, Thermal therapy, part 2: hyperthermia techniques. Crit. Rev. Biomed. Eng. **34**, 491–542 (2006)
149. T. Koltai, Cancer: fundamentals behind pH targeting and the double-edged approach. Onco Targets Ther. **9**, 6343–6360 (2016)
150. N.E.Y. Suriyanto, S.D. Kumar, Physical mechanism and modeling of heat generation and transfer in magnetic fluid hyperthermia through Néelian and Brownian relaxation: a review. Biomed. Eng. Online **16**, 36 (2017)
151. Q.A. Pankhurst, J. Connolly, S.K. Jones, J. Dobson, Applications of magnetic nanoparticles in biomedicine. J. Phys. D Appl. Phys. **36**, R167–R181 (2003)
152. L.X. Mu, W.S. Shi, T.P. Zhang, H.Y. Zhang, Y. Wang, G.W. She, et al., ZnO nanowire-based all-optical switch with reset-set flip-flop function. Appl. Phys. Lett. 98 (2011)

153. E.W. Miller, S.X. Bian, C.J. Chang, A fluorescent sensor for imaging reversible redox cycles in living cells. J. Am. Chem. Soc. **129**, 3458–3459 (2007)
154. A. Jordan, R. Scholz, P. Wust, H. Schirra, S. Thomas, H. Schmidt et al., Endocytosis of dextran and silan-coated magnetite nanoparticles and the effect of intracellular hyperthermia on human mammary carcinoma cells in vitro. J. Magn. Magn. Mater. **194**, 185–196 (1999)
155. I. Hilger, W. Andrä, R. Hergt, R. Hiergeist, H. Schubert, W.A. Kaiser, Electromagnetic heating of breast tumors in interventional radiology: in vitro and in vivo studies in human cadavers and mice. Radiology **218**, 570–575 (2001)
156. K. Maier-Hauff, R. Rothe, R. Scholz, U. Gneveckow, P. Wust, B. Thiesen et al., Intracranial thermotherapy using magnetic nanoparticles combined with external beam radiotherapy: results of a feasibility study on patients with glioblastoma multiforme. J. Neurooncol. **81**, 53–60 (2007)
157. S.J. DeNardo, G.L. DeNardo, A. Natarajan, L.A. Miers, A.R. Foreman, C. Gruettner et al., Thermal dosimetry predictive of efficacy of 111In-ChL6 nanoparticle AMF–induced thermoablative therapy for human breast cancer in mice. J. Nucl. Med. **48**, 437–444 (2007)
158. S. Rana, N.V. Jadhav, K.C. Barick, B.N. Pandey, P.A. Hassan, Polyaniline shell cross-linked Fe_3O_4 magnetic nanoparticles for heat activated killing of cancer cells. Dalton Trans. **43**, 12263–12271 (2014)
159. X.L. Liu, Y. Yang, C.T. Ng, L.Y. Zhao, Y. Zhang, B.H. Bay et al., Magnetic vortex nanorings: a new class of hyperthermia agent for highly efficient in vivo regression of tumors. Adv. Mater. **27**, 1939–1944 (2015)
160. J.-t Jang, J. Lee, J. Seon, E. Ju, M. Kim, Y.I. Kim et al., Giant magnetic heat induction of magnesium-doped γ-Fe_2O_3 superparamagnetic nanoparticles for completely killing tumors. Adv. Mater. **30**, 1704362 (2018)
161. F. Assa, H. Jafarizadeh-Malmiri, H. Ajamein, H. Vaghari, N. Anarjan, O. Ahmadi et al., Chitosan magnetic nanoparticles for drug delivery systems. Crit. Rev. Biotechnol. **37**, 492–509 (2017)
162. J. Barbosa, D.M. Correia, R. Gonçalves, C. Ribeiro, G. Botelho, P. Martins et al., Magnetically controlled drug release system through magnetomechanical actuation. Adv. Healthc. Mater. **5**, 3027–3034 (2016)
163. P.J. Cregg, K. Murphy, A. Mardinoglu, Inclusion of interactions in mathematical modelling of implant assisted magnetic drug targeting. Appl. Math. Model. **36**, 1–34 (2012)
164. K. Hola, Z. Markova, G. Zoppellaro, J. Tucek, R. Zboril, Tailored functionalization of iron oxide nanoparticles for MRI, drug delivery, magnetic separation and immobilization of biosubstances. Biotechnol. Adv. **33**, 1162–1176 (2015)
165. N.S. Satarkar, J.Z. Hilt, Magnetic hydrogel nanocomposites for remote controlled pulsatile drug release. J. Control Release **130**, 246–251 (2008)
166. H. Oliveira, E. Pérez-Andrés, J. Thevenot, O. Sandre, E. Berra, S. Lecommandoux, Magnetic field triggered drug release from polymersomes for cancer therapeutics. J. Control Release **169**, 165–170 (2013)
167. W.H. Gmeiner, S. Ghosh, Nanotechnology for cancer treatment. Nanotechnol. Rev. **3**, 111–122 (2015)
168. X. Hua, Q. Yang, Z. Dong, J. Zhang, W. Zhang, Q. Wang et al., Magnetically triggered drug release from nanoparticles and its applications in anti-tumor treatment. Drug Deliv. **24**, 511–518 (2017)
169. Z. Tang, Y. Gao, D. Li, S. Zhou, Controllably switched drug release from successively dual-targeted nanoreservoirs. Adv. Healthc. Mater. **6**, 1600919 (2017)
170. J. Xie, K. Chen, J. Huang, S. Lee, J. Wang, J. Gao et al., PET/NIRF/MRI triple functional iron oxide nanoparticles. Biomaterials **31**, 3016–3022 (2010)
171. Q. Quan, J. Xie, H. Gao, M. Yang, F. Zhang, G. Liu et al., HSA coated iron oxide nanoparticles as drug delivery vehicles for cancer therapy. Mol. Pharm. **8**, 1669–1676 (2011)
172. Y. Huang, K. Mao, B. Zhang, Y. Zhao, Superparamagnetic iron oxide nanoparticles conjugated with folic acid for dual target-specific drug delivery and MRI in cancer theranostics. Mater. Sci. Eng. C Mater. Biol. Appl. **70**, 763–771 (2017)

173. J. Estelrich, E. Escribano, J. Queralt, M.A. Busquets, Iron oxide nanoparticles for magnetically-guided and magnetically-responsive drug delivery. Int. J. Mol. Sci. **16**, 8070–8101 (2015)
174. A.J. Wagstaff, S.D. Brown, M.R. Holden, G.E. Craig, J.A. Plumb, R.E. Brown et al., Cisplatin drug delivery using gold-coated iron oxide nanoparticles for enhanced tumour targeting with external magnetic fields. Inorg. Chim. Acta **393**, 328–333 (2012)
175. H. Unterweger, R. Tietze, C. Janko, J. Zaloga, S. Lyer, S. Dürr et al., Development and characterization of magnetic iron oxide nanoparticles with a cisplatin-bearing polymer coating for targeted drug delivery. Int. J. Nanomedicine **9**, 3659–3676 (2014)
176. G. Mattheolabakis, L. Milane, A. Singh, M.M. Amiji, Hyaluronic acid targeting of CD44 for cancer therapy: from receptor biology to nanomedicine. J. Drug Target **23**, 605–618 (2015)
177. M. Nadeem, M. Ahmad, M.S. Akhtar, A. Shaari, S. Riaz, S. Naseem, et al., Magnetic properties of polyvinyl alcohol and doxorubicine loaded iron oxide nanoparticles for anticancer drug delivery applications. PLoS One **11**, e0158084 (2016)
178. D. Thapa, V.R. Palkar, M.B. Kurup, S.K. Malik, Properties of magnetite nanoparticles synthesized through a novel chemical route. Mater. Lett. **58**, 2692–2694 (2004)
179. M. Uhlen, Magnetic separation of DNA. Nature **340**, 733–734 (1989)
180. J.B. Haun, T.J. Yoon, H. Lee, R. Weissleder, Magnetic nanoparticle biosensors. Wiley Interdiscip. Rev. Nanomed. Nanobiotechnol. **2**, 291–304 (2010)
181. T. Atanasijevic, M. Shusteff, P. Fam, A. Jasanoff, Calcium-sensitive MRI contrast agents based on superparamagnetic iron oxide nanoparticles and calmodulin. Proc. Natl. Acad. Sci. USA **103**, 14707–14712 (2006)
182. E.Y. Sun, L. Josephson, K.A. Kelly, R. Weissleder, Development of nanoparticle libraries for biosensing. Bioconjug. Chem. **17**, 109–113 (2006)
183. K. El-Boubbou, C. Gruden, X. Huang, Magnetic glyco-nanoparticles: a unique tool for rapid pathogen detection, decontamination, and strain differentiation. J. Am. Chem. Soc. **129**, 13392–13393 (2007)
184. S. Ryan, A.J. Kell, H. van Faassen, L.-L. Tay, B. Simard, R. MacKenzie et al., Single-domain antibody-nanoparticles: promising architectures for increased staphylococcus aureus detection specificity and sensitivity. Bioconjug. Chem. **20**, 1966–1974 (2009)
185. R. Weissleder, U. Mahmood, Molecular imaging. Radiology **219**, 316–333 (2001)
186. M. Swierczewska, S. Lee, X. Chen, Inorganic nanoparticles for multimodal molecular imaging. Mol. Imaging **10**, 3–16 (2011)
187. J. Xie, F. Zhang, M. Aronova, L. Zhu, X. Lin, Q. Quan et al., Manipulating the power of an additional phase: a flower-like Au-Fe_3O_4 optical nanosensor for imaging protease expressions in vivo. ACS Nano **5**, 3043–3051 (2011)
188. L. Tong, M. Zhao, S. Zhu, J. Chen, Synthesis and application of superparamagnetic iron oxide nanoparticles in targeted therapy and imaging of cancer. Front. Med. **5**, 379–387 (2011)
189. J.M. Perez, T. O'Loughin, F.J. Simeone, R. Weissleder, L. Josephson, DNA-based magnetic nanoparticle assembly acts as a magnetic relaxation nanoswitch allowing screening of DNA-cleaving agents. J. Am. Chem. Soc. **124**, 2856–2857 (2002)
190. P. Bannas, O. Graumann, P. Balcerak, K. Peldschus, M.G. Kaul, H. Hohenberg et al., Quantitative magnetic resonance imaging of enzyme activity on the cell surface: in vitro and in vivo monitoring of ADP-ribosyltransferase 2 on T cells. Mol. Imaging **9**, 211–222 (2010)
191. L. Josephson, J.M. Perez, R. Weissleder, Magnetic nanosensors for the detection of oligonucleotide sequences. Angew. Chem. Int. Ed. Engl. **40**, 3204–3206 (2001)
192. A. MartIn-Fontecha, S. Sebastiani, U.E. Höpken, M. Uguccioni, M. Lipp, A. Lanzavecchia et al., Regulation of dendritic cell migration to the draining lymph node: impact on T lymphocyte traffic and priming. J. Exp. Med. **198**, 615–621 (2003)
193. P.L. Triozzi, R. Khurram, W.A. Aldrich, M.J. Walker, J.A. Kim, S. Jaynes, Intratumoral injection of dendritic cells derived in vitro in patients with metastatic cancer. Cancer **89**, 2646–2654 (2000)
194. S. Tuyaerts, J.L. Aerts, J. Corthals, B. Neyns, C. Heirman, K. Breckpot et al., Current approaches in dendritic cell generation and future implications for cancer immunotherapy. Cancer Immunol. Immunother. CII **56**, 1513–1537 (2007)

195. G.J. Adema, I.J. de Vries, C.J. Punt, C.G. Figdor, Migration of dendritic cell based cancer vaccines: in vivo veritas? Curr. Opin. Immunol. **17**, 170–174 (2005)
196. J. Schlom, P.M. Arlen, J.L. Gulley, Cancer vaccines: moving beyond current paradigms. Clin. Cancer Res. Off. J. Am. Assoc. Cancer Res. **13**, 3776–3782 (2007)
197. R. Rohani, S.N. de Chickera, C. Willert, Y. Chen, G.A. Dekaban, P.J. Foster, In vivo cellular MRI of dendritic cell migration using micrometer-sized iron oxide (MPIO) particles. Mol. Imag. Biol. **13**, 679–694 (2011)
198. L.J. Cruz, I. Que, M. Aswendt, A. Chan, M. Hoehn, C. Löwik, Targeted nanoparticles for the non-invasive detection of traumatic brain injury by optical imaging and fluorine magnetic resonance imaging. Nano Res. **9**, 1276–1289 (2016)
199. Y.-W. Noh, Y.-S. Jang, K.-J. Ahn, Y.T. Lim, B.H. Chung, Simultaneous in vivo tracking of dendritic cells and priming of an antigen-specific immune response. Biomaterials **32**, 6254–6263 (2011)
200. W. Reichardt, C. Dürr, D. von Elverfeldt, E. Jüttner, U.V. Gerlach, M. Yamada et al., Impact of mammalian target of rapamycin inhibition on lymphoid homing and tolerogenic function of nanoparticle-labeled dendritic cells following allogeneic hematopoietic cell transplantation. J. Immunol. **181**, 4770 (2008)
201. I.L. Weissman, Translating stem and progenitor cell biology to the clinic: barriers and opportunities. Science **287**, 1442–1446 (2000)
202. M. Morigi, A. Benigni, G. Remuzzi, B. Imberti, The regenerative potential of stem cells in acute renal failure. Cell Transpl. **15**(Suppl 1), S111-117 (2006)
203. M.O. Oyewumi, R.A. Yokel, M. Jay, T. Coakley, R.J. Mumper, Comparison of cell uptake, biodistribution and tumor retention of folate-coated and PEG-coated gadolinium nanoparticles in tumor-bearing mice. J. Control Release **95**, 613–626 (2004)
204. F.H. Wang, I.H. Lee, N. Holmström, T. Yoshitake, D.K. Kim, M. Muhammed et al., Magnetic resonance tracking of nanoparticle labelled neural stem cells in a rat's spinal cord. Nanotechnology **17**, 1911–1915 (2006)
205. L. Seymour, E. Schacht, R. Duncan, The effect of size of polystyrene particles on their retention within the rat peritoneal compartment, and on their interaction with rat peritoneal macrophages in vitro. Cell Biol. Int. Rep. **15**, 277–286 (1991)
206. D.L. Kraitchman, A.W. Heldman, E. Atalar, L.C. Amado, B.J. Martin, M.F. Pittenger et al., In vivo magnetic resonance imaging of mesenchymal stem cells in myocardial infarction. Circulation **107**, 2290–2293 (2003)
207. L.C. Amado, A.P. Saliaris, K.H. Schuleri, M. St John, J.S. Xie, S. Cattaneo et al., Cardiac repair with intramyocardial injection of allogeneic mesenchymal stem cells after myocardial infarction. Proc. Natl. Acad. Sci. USA **102**, 11474–11479 (2005)
208. K.W. Au, S.Y. Liao, Y.K. Lee, W.H. Lai, K.M. Ng, Y.C. Chan et al., Effects of iron oxide nanoparticles on cardiac differentiation of embryonic stem cells. Biochem. Biophys. Res. Commun. **379**, 898–903 (2009)
209. S. Thakur, P. Kesharwani, R.K. Tekade, N.K. Jain, Impact of pegylation on biopharmaceutical properties of dendrimers. Polymer **59**, 67–92 (2015)
210. P. Kesharwani, R.K. Tekade, N.K. Jain, Generation dependent cancer targeting potential of poly(propyleneimine) dendrimer. Biomaterials **35**, 5539–5548 (2014)
211. B. Birdhariya, P. Kesharwani, N.K. Jain, Effect of surface capping on targeting potential of folate decorated poly (propylene imine) dendrimers. Drug Dev. Ind. Pharm. **41**, 1393–1399 (2015)
212. P. Kesharwani, R.K. Tekade, N.K. Jain, Formulation development and in vitro-in vivo assessment of the fourth-generation PPI dendrimer as a cancer-targeting vector. Nanomedicine (Lond) **9**, 2291–2308 (2014)
213. D.A. Tomalia, H. Baker, J. Dewald, M. Hall, G. Kallos, S. Martin et al., A new class of polymers: starburst-dendritic macromolecules. Polym. J. **17**, 117–132 (1985)
214. G.R. Newkome, Z. Yao, G.R. Baker, V.K. Gupta, Micelles. Part 1. Cascade molecules: a new approach to micelles. A [27]-arborol. J. Org. Chem. **50**, 2003–2004 (1985)

215. C.C. Lee, J.A. MacKay, J.M.J. Fréchet, F.C. Szoka, Designing dendrimers for biological applications. Nat. Biotechnol. **23**, 1517–1526 (2005)
216. M. Yousefi, A. Narmani, S.M. Jafari, Dendrimers as efficient nanocarriers for the protection and delivery of bioactive phytochemicals. Adv. Colloid Interface Sci. **278**, 102125 (2020)
217. U. Gupta, O. Perumal, Chapter 15—dendrimers and its biomedical applications, in *Natural and Synthetic Biomedical Polymers*. ed. by S.G. Kumbar, C.T. Laurencin, M. Deng (Elsevier, Oxford, 2014), pp. 243–257
218. E.R. Gillies, J.M.J. Fréchet, Designing macromolecules for therapeutic applications: polyester DendrimerPoly(ethylene oxide) "Bow-Tie" hybrids with tunable molecular weight and architecture. J. Am. Chem. Soc. **124**, 14137–14146 (2002)
219. C.J. Hawker, J.M.J. Frechet, Preparation of polymers with controlled molecular architecture. a new convergent approach to dendritic macromolecules. J. Am. Chem. Soc. **112**, 7638–7647 (1990)
220. E.R. Gillies, J.M. Fréchet, Dendrimers and dendritic polymers in drug delivery. Drug Discov. Today **10**, 35–43 (2005)
221. J.B. Wolinsky, M.W. Grinstaff, Therapeutic and diagnostic applications of dendrimers for cancer treatment. Adv. Drug Deliv. Rev. **60**, 1037–1055 (2008)
222. H. Xu, C.A. Regino, M. Bernardo, Y. Koyama, H. Kobayashi, P.L. Choyke et al., Toward improved syntheses of dendrimer-based magnetic resonance imaging contrast agents: new bifunctional diethylenetriaminepentaacetic acid ligands and nonaqueous conjugation chemistry. J. Med. Chem. **50**, 3185–3193 (2007)
223. S. Odai, H. Ito, T. Kamachi, Dendrimer porphyrins as the oxygen sensor for intracellular imaging to suppress interaction towards biological molecules. J. Clin. Biochem. Nutr. **65**, 178–184 (2019)
224. D.A. Tomalia, A.M. Naylor, W.A. Goddard Iii, Starburst dendrimers: molecular-level control of size, shape, surface chemistry, topology, and flexibility from atoms to macroscopic matter. Angew. Chem. Int. Ed. Engl. **29**, 138–175 (1990)
225. de Brabander-van den Berg EMM, Meijer EW, Poly(propylene imine) dendrimers: large-scale synthesis by hetereogeneously catalyzed hydrogenations. Angew. Chem. Int. Ed. Engl. **32**, 1308–1311 (1993)
226. K. Sadler, J.P. Tam, Peptide dendrimers: applications and synthesis. Rev. Mol. Biotechnol. **90**, 195–229 (2002)
227. H. Ihre, A. Hult, E. Söderlind, Synthesis, characterization, and 1H NMR self-diffusion studies of dendritic aliphatic polyesters based on 2,2-bis(hydroxymethyl)propionic acid and 1,1,1-tris(hydroxyphenyl)ethane. J. Am. Chem. Soc. **118**, 6388–6395 (1996)
228. M.W. Grinstaff, Biodendrimers: new polymeric biomaterials for tissue engineering. Chem—Eur. J. **8**, 2838–2846 (2002)
229. W.B. Turnbull, J.F. Stoddart, Design and synthesis of glycodendrimers. Rev. Mol. Biotechnol. **90**, 231–255 (2002)
230. T.W. Nilsen, J. Grayzel, W. Prensky, Dendritic nucleic acid structures. J. Theor. Biol. **187**, 273–284 (1997)
231. Y. Li, Y.D. Tseng, S.Y. Kwon, L. d'Espaux, J.S. Bunch, P.L. McEuen et al., Controlled assembly of dendrimer-like DNA. Nat. Mater. **3**, 38–42 (2004)
232. G. Liu, M. Lu, X. Huang, T. Li, D. Xu, Application of gold-nanoparticle colorimetric sensing to rapid food safety screening. Sensors (Basel). **18**(12), 4166 (2018). https://doi.org/10.3390/s18124166
233. S. Stevelmans, J.C.M. van Hest, J.F.G.A. Jansen, D.A.F.J van Boxtel, E.M.M. de Brabander-van den Berg, E.W. Meijer, Synthesis, characterization, and guest−host properties of inverted unimolecular dendritic micelles. J. Am. Chem. Soc. **118**, 7398–7399 (1996)
234. P. Kolhe, J. Khandare, O. Pillai, S. Kannan, M. Lieh-Lai, R.M. Kannan, Preparation, cellular transport, and activity of polyamidoamine-based dendritic nanodevices with a high drug payload. Biomaterials **27**, 660–669 (2006)
235. A.K. Patri, J.F. Kukowska-Latallo, J.R. Baker, Targeted drug delivery with dendrimers: comparison of the release kinetics of covalently conjugated drug and non-covalent drug inclusion complex. Adv. Drug Deliv. Rev. **57**, 2203–2214 (2005)

236. A. D'Emanuele, R. Jevprasesphant, J. Penny, D. Attwood, The use of a dendrimer-propranolol prodrug to bypass efflux transporters and enhance oral bioavailability. J. Control. Release Off. J. Control. Release Soc. **95**, 447–453 (2004)
237. R.B. Kolhatkar, P. Swaan, H. Ghandehari, Potential oral delivery of 7-ethyl-10-hydroxy-camptothecin (SN-38) using poly(amidoamine) dendrimers. Pharm. Res. **25**, 1723–1729 (2008)
238. A. Choksi, K.V.L. Sarojini, P. Vadnal, C. Dias, P.K. Suresh, J. Khandare, Comparative anti-inflammatory activity of poly(amidoamine) (PAMAM) dendrimer-dexamethasone conjugates with dexamethasone-liposomes. Int. J. Pharm. **449**, 28–36 (2013)
239. J.J. Khandare, S. Jayant, A. Singh, P. Chandna, Y. Wang, N. Vorsa et al., Dendrimer versus linear conjugate: influence of polymeric architecture on the delivery and anticancer effect of paclitaxel. Bioconjug. Chem. **17**, 1464–1472 (2006)
240. N.S. Malik, E.G. Evagorou, R.M. Duncan, Dendrimer-platinate: a novel approach to cancer chemotherapy. Anticancer Drugs **10**(8), 767–776 (1999)
241. H. Yang, S.T. Lopina, Penicillin V-conjugated PEG-PAMAM star polymers. J. Biomater. Sci. Polym. Ed. **14**, 1043–1056 (2003)
242. H. Yang, S.T. Lopina, Extended release of a novel antidepressant, venlafaxine, based on anionic polyamidoamine dendrimers and poly(ethylene glycol)-containing semi-interpenetrating networks. J. Biomed. Mater. Res. Part A **72A**, 107–114 (2005)
243. O.L. Padilla De Jesús, H.R. Ihre, L. Gagne, J.M.J. Fréchet, F.C. Szoka, Polyester dendritic systems for drug delivery applications. in vitro and in vivo evaluation. Bioconjug. Chem. **13**, 453–461 (2002)
244. S.-T. Lo, S. Stern, J.D. Clogston, J. Zheng, P.P. Adiseshaiah, M. Dobrovolskaia et al., Biological assessment of triazine dendrimer: toxicological profiles, solution behavior, biodistribution, drug release and efficacy in a pegylated, paclitaxel construct. Mol. Pharm. **7**, 993–1006 (2010)
245. A. D'Emanuele, D. Attwood, Dendrimer–drug interactions. Adv. Drug Deliv. Rev. **57**, 2147–2162 (2005)
246. S. Sadekar, H. Ghandehari, Transepithelial transport and toxicity of PAMAM dendrimers: implications for oral drug delivery. Adv. Drug Deliv. Rev. **64**, 571–588 (2012)
247. S. Chauhan Abhay, V. Diwan Prakash, K. Jain Narendra, V. Raghavan Kondapuram, Inventors; COUNCIL OF SCIENTIFIC AND INDUSTRIAL RESEARCH, assignee. Compositions and complexes containing a macromolecular compound as potential anti-inflammatory agents (2006). (2006/09/22/Application date)
248. W. Ke, Y. Zhao, R. Huang, C. Jiang, Y. Pei, Enhanced oral bioavailability of doxorubicin in a dendrimer drug delivery system. J. Pharm. Sci. **97**, 2208–2216 (2008)
249. O.M. Milhem, C. Myles, N.B. McKeown, D. Attwood, A. D'Emanuele, Polyamidoamine Starburst® dendrimers as solubility enhancers. Int. J. Pharm. **197**, 239–241 (2000)
250. D. Bhadra, S. Bhadra, S. Jain, N.K. Jain, A PEGylated dendritic nanoparticulate carrier of fluorouracil. Int. J. Pharm. **257**, 111–124 (2003)
251. A. Asthana, A.S. Chauhan, P.V. Diwan, N.K. Jain, Poly(amidoamine) (PAMAM) dendritic nanostructures for controlled sitespecific delivery of acidic anti-inflammatory active ingredient. AAPS PharmSciTech **6**, E536–E542 (2005)
252. R.N. Prajapati, R.K. Tekade, U. Gupta, V. Gajbhiye, N.K. Jain, Dendimer-mediated solubilization, formulation development and in vitro−in vivo assessment of piroxicam. Mol. Pharm. **6**, 940–950 (2009)
253. H. Namazi, M. Adeli, Dendrimers of citric acid and poly (ethylene glycol) as the new drug-delivery agents. Biomaterials **26**, 1175–1183 (2005)
254. P.V. Kumar, H. Agashe, T. Dutta, N.K. Jain, PEGylated dendritic architecture for development of a prolonged drug delivery system for an antitubercular drug. Curr. Drug Deliv. **4**, 11–19 (2007)
255. T. Dutta, M. Garg, N.K. Jain, Targeting of efavirenz loaded tuftsin conjugated poly(propyleneimine) dendrimers to HIV infected macrophages in vitro. Eur. J. Pharm. Sci. **34**, 181–189 (2008)

256. Z. Sideratou, D. Tsiourvas, C.M. Paleos, Solubilization and release properties of pegylated diaminobutane poly(propylene imine) dendrimers. J. Colloid Interface Sci. **242**, 272–276 (2001)
257. T. Ooya, J. Lee, K. Park, Effects of ethylene glycol-based graft, star-shaped, and dendritic polymers on solubilization and controlled release of paclitaxel. J. Control. Release Off. J. Control. Release Soc. **93**, 121–127 (2003)
258. M.T. Morgan, Y. Nakanishi, D.J. Kroll, A.P. Griset, M.A. Carnahan, M. Wathier et al., Dendrimer-encapsulated camptothecins: increased solubility, cellular uptake, and cellular retention affords enhanced anticancer activity in vitro. Cancer Res. **66**, 11913–11921 (2006)
259. S. Hong, P.R. Leroueil, I.J. Majoros, B.G. Orr, J.R. Baker, M.M. Banaszak Holl, The binding avidity of a nanoparticle-based multivalent targeted drug delivery platform. Chem. Biol. **14**, 107–115 (2007)
260. A.K. Patri, A. Myc, J. Beals, T.P. Thomas, N.H. Bander, J.R. Baker, Synthesis and in vitro testing of J591 antibody–dendrimer conjugates for targeted prostate cancer therapy. Bioconjug. Chem. **15**, 1174–1181 (2004)
261. J. Haensler, F.C. Szoka, Polyamidoamine cascade polymers mediate efficient transfection of cells in culture. Bioconjug. Chem. **4**, 372–379 (1993)
262. J.F. Kukowska-Latallo, A.U. Bielinska, J. Johnson, R. Spindler, D.A. Tomalia, J.R. Baker, Efficient transfer of genetic material into mammalian cells using starburst polyamidoamine dendrimers. Proc. Natl. Acad. Sci. **93**, 4897–4902 (1996)
263. V.V.K. Venuganti, O.P. Perumal, Poly(amidoamine) dendrimers as skin penetration enhancers: influence of charge, generation, and concentration. J. Pharm. Sci. **98**, 2345–2356 (2009)
264. P. Kesharwani, S. Banerjee, U. Gupta, M.C.I. Mohd Amin, S. Padhye, F.H. Sarkar et al., PAMAM dendrimers as promising nanocarriers for RNAi therapeutics. Mater. Today **18**, 565–572 (2015)
265. H. Kobayashi, M.W. Brechbiel, Dendrimer-based macromolecular MRI contrast agents: characteristics and application. Mol. Imaging **2**, 1–10 (2003)
266. L.S. Ziemer, W.M.F. Lee, S.A. Vinogradov, C. Sehgal, D.F. Wilson, Oxygen distribution in murine tumors: characterization using oxygen-dependent quenching of phosphorescence. J. Appl. Physiol. **98**, 1503–1510 (2005)
267. V. Rozhkov, D. Wilson, S. Vinogradov, Phosphorescent Pd porphyrin–dendrimers: tuning core accessibility by varying the hydrophobicity of the dendritic matrix. Macromolecules **35**, 1991–1993 (2002)
268. B. Noriega-Luna, L.A. Godínez, F.J. Rodríguez, A. Rodríguez, G. Zaldívar-Lelo de Larrea, C.F. Sosa-Ferreyra, et al., Applications of dendrimers in drug delivery agents, diagnosis, therapy, and detection. J. Nanomater. **2014**, 507273 (2014)
269. B. Noriega-Luna, L.A. Godínez, F.J. Rodríguez, A. Rodríguez, G. Zaldívar-Lelo de Larrea, C.F. Sosa-Ferreyra et al., Corrigendum to applications of dendrimers in drug delivery agents, diagnosis, therapy, and detection. J. Nanomater. **2020**, 3020287 (2020)
270. G.A. Kinberger, W. Cai, M. Goodman, Collagen mimetic dendrimers. J. Am. Chem. Soc. **124**, 15162–15163 (2002)
271. R.D. Kent, P.J. Vikesland, Controlled evaluation of silver nanoparticle dissolution using atomic force microscopy. Environ. Sci. Technol. **46**, 6977–6984 (2012)
272. E. Abbasi, M. Milani, S. Fekri Aval, M. Kouhi, A. Akbarzadeh, H. Tayefi Nasrabadi et al., Silver nanoparticles: synthesis methods, bio-applications and properties. Crit. Rev. Microbiol. **42**, 173–180 (2016)
273. I.A. Nedelcu, A. Ficai, M. Sonmez, D. Ficai, O. Oprea, E. Andronescu, Silver based materials for biomedical applications. Curr. Org. Chem. **18**, 173–184 (2014)
274. R. Behra, L. Sigg, M.J. Clift, F. Herzog, M. Minghetti, B. Johnston et al., Bioavailability of silver nanoparticles and ions: from a chemical and biochemical perspective. J. R. Soc. Interface **10**, 20130396 (2013)
275. S. Iravani, H. Korbekandi, S.V. Mirmohammadi, B. Zolfaghari, Synthesis of silver nanoparticles: chemical, physical and biological methods. Res. Pharm. Sci. **9**, 385–406 (2014)

276. N. Kannan, S. Subbalaxmi (eds.), *Green Synthesis of Silver Nanoparticles using Bacillus subtillus IA751 and Its Antimicrobial Activity* (2011)
277. A. Nanda, M. Saravanan, Biosynthesis of silver nanoparticles from Staphylococcus aureus and its antimicrobial activity against MRSA and MRSE. Nanomedicine **5**, 452–456 (2009)
278. A.R. Shahverdi, S. Minaeian, H.R. Shahverdi, H. Jamalifar, A.-A. Nohi, Rapid synthesis of silver nanoparticles using culture supernatants of Enterobacteria: a novel biological approach. Process Biochem. **42**, 919–923 (2007)
279. S. Ghosh, S. Patil, M. Ahire, R. Kitture, S. Kale, K. Pardesi et al., Synthesis of silver nanoparticles using Dioscorea bulbifera tuber extract and evaluation of its synergistic potential in combination with antimicrobial agents. Int. J. Nanomedicine **7**, 483–496 (2012)
280. G.R. Salunke, S. Ghosh, R.J. Santosh Kumar, S. Khade, P. Vashisth, T. Kale et al., Rapid efficient synthesis and characterization of silver, gold, and bimetallic nanoparticles from the medicinal plant Plumbago zeylanica and their application in biofilm control. Int. J. Nanomedicine **9**, 2635–2653 (2014)
281. H. Huang, Y. Yang, Preparation of silver nanoparticles in inorganic clay suspensions. Compos. Sci. Technol. **68**, 2948–2953 (2008)
282. J.-H. Lee, J.-M. Lim, P. Velmurugan, Y.-J. Park, Y.-J. Park, K.-S. Bang et al., Photobiologic-mediated fabrication of silver nanoparticles with antibacterial activity. J. Photochem. Photobiol., B **162**, 93–99 (2016)
283. I. Ghiuță, D. Cristea, C. Croitoru, J. Kost, R. Wenkert, I. Vyrides et al., Characterization and antimicrobial activity of silver nanoparticles, biosynthesized using Bacillus species. Appl. Surf. Sci. **438**, 66–73 (2018)
284. X. Yan, B. He, L. Liu, G. Qu, J. Shi, L. Hu et al., Antibacterial mechanism of silver nanoparticles in Pseudomonas aeruginosa: proteomics approach. Metallomics **10**, 557–564 (2018)
285. A.C. Burdușel, O. Gherasim, A.M. Grumezescu, L. Mogoantă, A. Ficai, E. Andronescu, Biomedical applications of silver nanoparticles: an up-to-date overview. Nanomater. (Basel) 8 (2018)
286. S. Rtimi, R. Sanjines, C. Pulgarin, J. Kiwi, Microstructure of Cu–Ag uniform nanoparticulate films on polyurethane 3D catheters: surface properties. ACS Appl. Mater. Interfaces. **8**, 56–63 (2016)
287. K.N. Stevens, S. Croes, R.S. Boersma, E.E. Stobberingh, C. van der Marel, F.H. van der Veen et al., Hydrophilic surface coatings with embedded biocidal silver nanoparticles and sodium heparin for central venous catheters. Biomaterials **32**, 1264–1269 (2011)
288. M. Antonelli, G. De Pascale, V.M. Ranieri, P. Pelaia, R. Tufano, O. Piazza et al., Comparison of triple-lumen central venous catheters impregnated with silver nanoparticles (AgTive®) vs conventional catheters in intensive care unit patients. J. Hosp. Infect. **82**, 101–107 (2012)
289. S. Castiglioni, A. Cazzaniga, L. Locatelli, J.A.M Maier, Silver nanoparticles in orthopedic applications: new insights on their effects on osteogenic cells. Nanomater. (Basel) 7 (2017)
290. Y. Huang, X. Li, Z. Liao, G. Zhang, Q. Liu, J. Tang et al., A randomized comparative trial between acticoat and SD-Ag in the treatment of residual burn wounds, including safety analysis. Burns **33**, 161–166 (2007)
291. M. Mishra, K. Tripathi, Diabetic delayed wound healing and the role of silver nanoparticles. Dig. J. Nanomater. Biostruct. **3**, 49–54 (2008)
292. V. Vijayakumar, S.K. Samal, S. Mohanty, S.K. Nayak, Recent advancements in biopolymer and metal nanoparticle-based materials in diabetic wound healing management. Int. J. Biol. Macromol. **122**, 137–148 (2019)
293. X. Dai, Q. Guo, Y. Zhao, P. Zhang, T. Zhang, X. Zhang et al., Functional silver nanoparticle as a benign antimicrobial agent that eradicates antibiotic-resistant bacteria and promotes wound healing. ACS Appl. Mater. Interfaces. **8**, 25798–25807 (2016)
294. P. Polak, O. Shefi, Nanometric agents in the service of neuroscience: manipulation of neuronal growth and activity using nanoparticles. Nanomedicine **11**, 1467–1479 (2015)
295. Q. Jiang, S. Yu, X. Li, C. Ma, A. Li, Evaluation of local anesthetic effects of Lidocaine-Ibuprofen ionic liquid stabilized silver nanoparticles in Male Swiss mice. J. Photochem. Photobiol. B **178**, 367–370 (2018)

296. C.S. Karthik, H.M. Manukumar, A.P. Ananda, S. Nagashree, K.P. Rakesh, L. Mallesha et al., Synthesis of novel benzodioxane midst piperazine moiety decorated chitosan silver nanoparticle against biohazard pathogens and as potential anti-inflammatory candidate: a molecular docking studies. Int. J. Biol. Macromol. **108**, 489–502 (2018)
297. N. Soni, R.C. Dhiman, Phytochemical, anti-oxidant, larvicidal, and antimicrobial activities of castor (Ricinus communis) synthesized silver nanoparticles. Chin. Herb. Med. **9**, 289–294 (2017)
298. H. Al-Obaidi, R. Kalgudi, M.G. Zariwala, Fabrication of inhaled hybrid silver/ciprofloxacin nanoparticles with synergetic effect against Pseudomonas aeruginosa. Eur. J. Pharm. Biopharm. **128**, 27–35 (2018)
299. P.D. Petrov, K. Yoncheva, V. Gancheva, S. Konstantinov, B. Trzebicka, Multifunctional block copolymer nanocarriers for co-delivery of silver nanoparticles and curcumin: synthesis and enhanced efficacy against tumor cells. Eur. Polym. J. **81**, 24–33 (2016)
300. M.E. Barbinta-Patrascu, N. Badea, C. Pirvu, M. Bacalum, C. Ungureanu, P.L. Nadejde et al., Multifunctional soft hybrid bio-platforms based on nano-silver and natural compounds. Mater. Sci. Eng. C **69**, 922–932 (2016)
301. S. Patra, S. Mukherjee, A.K. Barui, A. Ganguly, B. Sreedhar, C.R. Patra, Green synthesis, characterization of gold and silver nanoparticles and their potential application for cancer therapeutics. Mater. Sci. Eng. C Mater. Biol. Appl. **53**, 298–309 (2015)
302. M. Ramar, B. Manikandan, P.N. Marimuthu, T. Raman, A. Mahalingam, P. Subramanian et al., Synthesis of silver nanoparticles using solanum trilobatum fruits extract and its antibacterial, cytotoxic activity against human breast cancer cell line MCF 7. Spectrochim. Acta A Mol. Biomol. Spectrosc. **140**, 223–228 (2015)
303. S. Gurunathan, J.W. Han, V. Eppakayala, M. Jeyaraj, J.H. Kim, Cytotoxicity of biologically synthesized silver nanoparticles in MDA-MB-231 human breast cancer cells. Biomed. Res. Int. 535796 (2013)
304. S. Prabhu, E.K. Poulose, Silver nanoparticles: mechanism of antimicrobial action, synthesis, medical applications, and toxicity effects. Int. Nano Lett. **2**, 32 (2012)
305. S. Khalid, *Green Biosynthesis of Silver Nanoparticles Conjugated to Gefitinib as Delivery Vehicle* (2017)
306. J. Liu, Y. Zhao, Q. Guo, Z. Wang, H. Wang, Y. Yang et al., TAT-modified nanosilver for combating multidrug-resistant cancer. Biomaterials **33**, 6155–6161 (2012)
307. G. Lakshmanan, A. Sathiyaseelan, P.T. Kalaichelvan, K. Murugesan, Plant-mediated synthesis of silver nanoparticles using fruit extract of Cleome viscosa L.: assessment of their antibacterial and anticancer activity. Karbala Int. J. Mod. Sci. **4**, 61–68 (2018)
308. J.L. Elechiguerra, J.L. Burt, J.R. Morones, A. Camacho-Bragado, X. Gao, H.H. Lara et al., Interaction of silver nanoparticles with HIV-1. J. Nanobiotechnology **3**, 6 (2005)
309. L. Lu, R.W. Sun, R. Chen, C.K. Hui, C.M. Ho, J.M. Luk et al., Silver nanoparticles inhibit hepatitis B virus replication. Antivir. Ther. **13**, 253–262 (2008)
310. R.L. Hu, S.R. Li, F.J. Kong, R.J. Hou, X.L. Guan, F. Guo, Inhibition effect of silver nanoparticles on herpes simplex virus 2. Genet. Mol. Res. **13**, 7022–7028 (2014)
311. P.K. Singh, G. Jairath, S.S. Ahlawat, Nanotechnology: a future tool to improve quality and safety in meat industry. J. Food Sci. Technol. **53**, 1739–1749 (2016)
312. S. Azlin-Hasim, M.C. Cruz-Romero, E. Cummins, J.P. Kerry, M.A. Morris, The potential use of a layer-by-layer strategy to develop LDPE antimicrobial films coated with silver nanoparticles for packaging applications. J. Colloid Interface Sci. **461**, 239–248 (2016)
313. A.A. Zahir, A. Bagavan, C. Kamaraj, G.A. Elango, A.A. Rahuman (eds.), *Efficacy of Plant-Mediated Synthesized Silver Nanoparticles Against Sitophilus Oryzae* (2002)
314. A. Awwad, Toxicity of nanoparticles against drosophila melanogaster (Diptera: Drosophilidae). J. Nanomater. 9 (2015)
315. M.L. Coluccio, M. Francardi, F. Gentile, P. Candeloro, L. Ferrara, G. Perozziello et al., Plasmonic 3D-structures based on silver decorated nanotips for biological sensing. Opt. Lasers Eng. **76**, 45–51 (2016)

316. M. Larguinho, P.V. Baptista, Gold and silver nanoparticles for clinical diagnostics—from genomics to proteomics. J. Proteomics **75**, 2811–2823 (2012)
317. M. Larguinho, P.V. Baptista, Gold and silver nanoparticles for clinical diagnostics—from genomics to proteomics. J. Proteomics **75**(10), 2811–2823 (2012)
318. J. Turkevich, P.C. Stevenson, J. Hillier, A study of the nucleation and growth processes in the synthesis of colloidal gold. Discuss. Faraday Soc. **11**, 55–75 (1951)
319. K.R. Brown, M.J. Natan, Hydroxylamine seeding of colloidal Au nanoparticles in solution and on surfaces. Langmuir **14**, 726–728 (1998)
320. N.R. Jana, L. Gearheart, C.J. Murphy, Seeding growth for size control of 5–40 nm diameter gold nanoparticles. Langmuir **17**, 6782–6786 (2001)
321. M. Brust, M. Walker, D. Bethell, D.J. Schiffrin, R. Whyman, Synthesis of thiol-derivatised gold nanoparticles in a two-phase Liquid–Liquid system. J. Chem. Soc. Chem. Commun. 801–802 (1994)
322. H. Daraee, A. Eatemadi, E. Abbasi, S. Fekri Aval, M. Kouhi, A. Akbarzadeh, Application of gold nanoparticles in biomedical and drug delivery. Artif. Cells Nanomedicine Biotechnol. **44**, 410–422 (2016)
323. X. Bai, Y. Wang, Z. Song, Y. Feng, Y. Chen, D. Zhang et al., The basic properties of gold nanoparticles and their applications in tumor diagnosis and treatment. Int. J. Mol. Sci. **21**, 2480 (2020)
324. S. Her, D.A. Jaffray, C. Allen, Gold nanoparticles for applications in cancer radiotherapy: mechanisms and recent advancements. Adv. Drug Deliv. Rev. **109**, 84–101 (2017)
325. R.S. Riley, E.S. Day, Gold nanoparticle-mediated photothermal therapy: applications and opportunities for multimodal cancer treatment. WIREs Nanomedicine and Nanobiotechnology **9**, e1449 (2017)
326. J.F. Hainfeld, D.N. Slatkin, T.M. Focella, H.M. Smilowitz, Gold nanoparticles: a new X-ray contrast agent. Br. J. Radiol. **79**, 248–253 (2006)
327. C.S. Thaxton, D.G. Georganopoulou, C.A. Mirkin, Gold nanoparticle probes for the detection of nucleic acid targets. Clin. Chim. Acta **363**, 120–126 (2006)
328. K. Sato, K. Hosokawa, M. Maeda, Colorimetric biosensors based on DNA-nanoparticle conjugates. Anal. Sci. **23**, 17–20 (2007)
329. K. Saha, S.S. Agasti, C. Kim, X. Li, V.M. Rotello, Gold nanoparticles in chemical and biological sensing. Chem. Rev. **112**, 2739–2779 (2012)
330. Y. Lin, M. Zhao, Y. Guo, X. Ma, F. Luo, L. Guo et al., Multicolor colormetric biosensor for the determination of glucose based on the etching of gold nanorods. Sci. Rep. **6**, 37879 (2016)
331. G. De Luca, P. Bonaccorsi, V. Trovato, A. Mancuso, T. Papalia, A. Pistone et al., Tripodal tris-disulfides as capping agents for a controlled mixed functionalization of gold nanoparticles. New J. Chem. **42**, 16436–16440 (2018)
332. S.-H. Lo, M.-C. Wu, P. Venkatesan, S.-P. Wu, Colorimetric detection of chromium(III) using O-phospho-l-serine dithiocarbamic acid functionalized gold nanoparticles. Sens. Actuators B Chem. **220**, 772–778 (2015)
333. G. Yue, S. Su, N. Li, M. Shuai, X. Lai, D. Astruc et al., Gold nanoparticles as sensors in the colorimetric and fluorescence detection of chemical warfare agents. Coord. Chem. Rev. **311**, 75–84 (2016)
334. A.M. Baetsen-Young, M. Vasher, L.L. Matta, P. Colgan, E.C. Alocilja, B. Day, Direct colorimetric detection of unamplified pathogen DNA by dextrin-capped gold nanoparticles. Biosens. Bioelectron. **101**, 29–36 (2018)
335. K.L. Kelly, E. Coronado, L.L. Zhao, G.C. Schatz, The optical properties of metal nanoparticles: the influence of size, shape, and dielectric environment. J. Phys. Chem. B **107**, 668–677 (2003)
336. B. Li, X. Li, Y. Dong, B. Wang, D. Li, Y. Shi et al., Colorimetric sensor array based on gold nanoparticles with diverse surface charges for microorganisms identification. Anal. Chem. **89**, 10639–10643 (2017)
337. F. Wang, Y.-C. Wang, S. Dou, M.-H. Xiong, T.-M. Sun, J. Wang, Doxorubicin-tethered responsive gold nanoparticles facilitate intracellular drug delivery for overcoming multidrug resistance in cancer cells. ACS Nano **5**, 3679–3692 (2011)

338. E.C. Dreaden, S.C. Mwakwari, Q.H. Sodji, A.K. Oyelere, M.A. El-Sayed, Tamoxifen−poly(ethylene glycol)−thiol gold nanoparticle conjugates: enhanced potency and selective delivery for breast cancer treatment. Bioconjug. Chem. **20**, 2247–2253 (2009)
339. Y.-H. Chen, C.-Y. Tsai, P.-Y. Huang, M.-Y. Chang, P.-C. Cheng, C.-H. Chou et al., Methotrexate conjugated to gold nanoparticles inhibits tumor growth in a syngeneic lung tumor model. Mol. Pharm. **4**, 713–722 (2007)
340. E. Azzam, S. Morsy, Enhancement of the antitumour activity for the synthesised dodecylcysteine surfactant using gold nanoparticles. J. Surfactants Deterg. **11**, 195–199 (2008)
341. A. François, A. Laroche, N. Pinaud, L. Salmon, J. Ruiz, J. Robert et al., Encapsulation of docetaxel into PEGylated gold nanoparticles for vectorization to cancer cells. ChemMedChem **6**, 2003–2008 (2011)
342. S. Dhar, W.L. Daniel, D.A. Giljohann, C.A. Mirkin, S.J. Lippard, Polyvalent oligonucleotide gold nanoparticle conjugates as delivery vehicles for platinum(IV) warheads. J. Am. Chem. Soc. **131**, 14652–14653 (2009)
343. M. Eghtedari, A.V. Liopo, J.A. Copland, A.A. Oraevsky, M. Motamedi, Engineering of heterofunctional gold nanorods for the in vivo molecular targeting of breast cancer cells. Nano Lett. **9**, 287–291 (2009)
344. S.D. Brown, P. Nativo, J.-A. Smith, D. Stirling, P.R. Edwards, B. Venugopal et al., Gold nanoparticles for the improved anticancer drug delivery of the active component of oxaliplatin. J. Am. Chem. Soc. **132**, 4678–4684 (2010)
345. X. Zhao, J. Pan, W. Li, W. Yang, L. Qin, Y. Pan, Gold nanoparticles enhance cisplatin delivery and potentiate chemotherapy by decompressing colorectal cancer vessels. Int. J. Nanomedicine **13**, 6207–6221 (2018)
346. S. Pandey, Development and characterization of Methotrexate loaded PEGylated gold nanoparticles. J. Pharm. Appl. Sci. **3**, 34–41 (2016)
347. R.A. Kudgus, A. Szabolcs, J.A. Khan, C.A. Walden, J.M. Reid, J.D. Robertson, et al., Inhibiting the growth of pancreatic adenocarcinoma in vitro and in vivo through targeted treatment with designer gold nanotherapeutics. PLoS One **8**, e57522 (2013)
348. Z. Chen, A. Zhang, X. Wang, J. Zhu, Y. Fan, H. Yu et al., The advances of carbon nanotubes in cancer diagnostics and therapeutics. J. Nanomater. **2017**, 3418932 (2017)
349. N. Gulati, H. Gupta, Two faces of carbon nanotube: toxicities and pharmaceutical applications. Crit. Rev. Ther. Drug Carr. Syst. **29**, 65–88 (2012)
350. W. Lu, M. Zu, J.-H. Byun, B.-S. Kim, T.-W. Chou, State of the art of carbon nanotube fibers: opportunities and challenges. Adv. Mater. **24**, 1805–1833 (2012)
351. J.K. Young, E.R. Figueroa, R.A. Drezek, Tunable nanostructures as photothermal theranostic agents. Ann. Biomed. Eng. **40**, 438–459 (2012)
352. H. Ago, K. Petritsch, M.S.P. Shaffer, A.H. Windle, R.H. Friend, Composites of carbon nanotubes and conjugated polymers for photovoltaic devices. Adv. Mater. **11**, 1281–1285 (1999)
353. Q. Cao, J.A. Rogers, Random networks and aligned arrays of single-walled carbon nanotubes for electronic device applications. Nano Res. **1**, 259–272 (2008)
354. A.C. Dillon, K.M. Jones, T.A. Bekkedahl, C.H. Kiang, D.S. Bethune, M.J. Heben, Storage of hydrogen in single-walled carbon nanotubes. Nature **386**, 377–379 (1997)
355. R.J. Chen, S. Bangsaruntip, K.A. Drouvalakis, N. Wong Shi Kam, M. Shim, Y. Li et al., Noncovalent functionalization of carbon nanotubes for highly specific electronic biosensors. Proc. Natl. Acad. Sci. **100**, 4984–4989 (2003)
356. P. Cherukuri, S.M. Bachilo, S.H. Litovsky, R.B. Weisman, Near-infrared fluorescence microscopy of single-walled carbon nanotubes in phagocytic cells. J. Am. Chem. Soc. **126**, 15638–15639 (2004)
357. Z. Liu, S. Tabakman, K. Welsher, H. Dai, Carbon nanotubes in biology and medicine: in vitro and in vivo detection, imaging and drug delivery. Nano Res. **2**, 85–120 (2009)
358. N.W. Kam, M. O'Connell, J.A. Wisdom, H. Dai, Carbon nanotubes as multifunctional biological transporters and near-infrared agents for selective cancer cell destruction. Proc. Natl. Acad. Sci. USA **102**, 11600–11605 (2005)

359. P. Chakravarty, R. Marches, N.S. Zimmerman, A.D. Swafford, P. Bajaj, I.H. Musselman et al., Thermal ablation of tumor cells with antibody-functionalized single-walled carbon nanotubes. Proc. Natl. Acad. Sci. USA **105**, 8697–8702 (2008)
360. K. Welsher, Z. Liu, D. Daranciang, H. Dai, Selective probing and imaging of cells with single walled carbon nanotubes as near-infrared fluorescent molecules. Nano Lett. **8**, 586–590 (2008)
361. M.J. O'Connell, S.M. Bachilo, C.B. Huffman, V.C. Moore, M.S. Strano, E.H. Haroz et al., Band gap fluorescence from individual single-walled carbon nanotubes. Science **297**, 593–596 (2002)
362. A.M. Rao, E. Richter, S. Bandow, B. Chase, P.C. Eklund, K.A. Williams et al., Diameter-selective raman scattering from vibrational modes in carbon nanotubes. Science **275**, 187–191 (1997)
363. D.A. Heller, S. Baik, T.E. Eurell, M.S. Strano, Single-walled carbon nanotube spectroscopy in live cells: towards long-term labels and optical sensors. Adv. Mater. **17**, 2793–2799 (2005)
364. D. Pantarotto, R. Singh, D. McCarthy, M. Erhardt, J.-P. Briand, M. Prato et al., Functionalized carbon nanotubes for plasmid DNA gene delivery. Angew. Chem. Int. Ed. **43**, 5242–5246 (2004)
365. Y. Liu, D.-C. Wu, W.-D. Zhang, X. Jiang, C.-B. He, T.S. Chung et al., Polyethylenimine-grafted multiwalled carbon nanotubes for secure noncovalent immobilization and efficient delivery of DNA. Angew. Chem. Int. Ed. **44**, 4782–4785 (2005)
366. N.W.S. Kam, Z. Liu, H. Dai, Functionalization of carbon nanotubes via cleavable disulfide bonds for efficient intracellular delivery of siRNA and potent gene silencing. J. Am. Chem. Soc. **127**, 12492–12493 (2005)
367. L. Ding, J. Stilwell, T. Zhang, O. Elboudwarej, H. Jiang, J.P. Selegue et al., Molecular characterization of the cytotoxic mechanism of multiwall carbon nanotubes and nano-onions on human skin fibroblast. Nano Lett. **5**, 2448–2464 (2005)
368. M. Bottini, S. Bruckner, K. Nika, N. Bottini, S. Bellucci, A. Magrini et al., Multi-walled carbon nanotubes induce T lymphocyte apoptosis. Toxicol. Lett. **160**, 121–126 (2006)
369. S. Iijima, Helical microtubules of graphitic carbon. Nature **354**, 56–58 (1991)
370. W.K. Maser, E. Muñoz, A.M. Benito, M.T. Martínez, G.F. De la Fuente, E. Anglaret, et al., Single-wall carbon nanotubes: study of production parameters using cw CO_2-laser ablation technique. AIP Conf. Proc. 544, 213–216 (2000)
371. W. Merchan-Merchan, A.V. Saveliev, L. Kennedy, W.C. Jimenez, Combustion synthesis of carbon nanotubes and related nanostructures. Prog. Energy Combust. Sci. **36**, 696–727 (2010)
372. B. Parasuram, S. Sundaram, C. Sathiskumar, S. Karthikeyan, Synthesis of multi-walled carbon nanotubes using tire pyrolysis oil as a carbon precursor by spray pyrolysis method. Inorg. Nano-Met. Chem. **48**, 103–106 (2018)
373. B.D. Fahlman, Chemical vapor deposition of carbon nanotubes: an experiment in materials chemistry. J. Chem. Educ. **79**, 203 (2002)
374. Y. Yan, M.B. Chan-Park, Q. Zhang, Advances in carbon-nanotube assembly. Small **3**, 24–42 (2007)
375. S. Niyogi, M.A. Hamon, H. Hu, B. Zhao, P. Bhowmik, R. Sen et al., Chemistry of single-walled carbon nanotubes. Acc. Chem. Res. **35**, 1105–1113 (2002)
376. L. Zeng, L.B. Alemany, C.L. Edwards, A.R. Barron, Demonstration of covalent sidewall functionalization of single wall carbon nanotubes by NMR spectroscopy: Side chain length dependence on the observation of the sidewall sp3 carbons. Nano Res. **1**, 72–88 (2008)
377. Z. Liu, S. Tabakman, K. Welsher, H. Dai, Carbon nanotubes in biology and nedicine: in vitro and in vivo detection, imaging and drug delivery. Nano Res. **2**, 85–120 (2009)
378. K.M. Lee, L. Li, L. Dai, Asymmetric end-functionalization of multi-walled carbon nanotubes. J. Am. Chem. Soc. **127**, 4122–4123 (2005)
379. K.S. Coleman, S.R. Bailey, S. Fogden, M.L.H. Green, Functionalization of single-walled carbon nanotubes via the Bingel reaction. J. Am. Chem. Soc. **125**, 8722–8723 (2003)
380. V. Georgakilas, K. Kordatos, M. Prato, D.M. Guldi, M. Holzinger, A. Hirsch, Organic functionalization of carbon nanotubes. J. Am. Chem. Soc. **124**, 760–761 (2002)

381. N. Tagmatarchis, M. Prato, Functionalization of carbon nanotubes via 1,3-dipolar cycloadditions. J. Mater. Chem. **14**, 437–439 (2004)
382. L. Meng, X. Zhang, Q. Lu, Z. Fei, P.J. Dyson, Single walled carbon nanotubes as drug delivery vehicles: targeting doxorubicin to tumors. Biomaterials **33**, 1689–1698 (2012)
383. V.R. Raphey, T.K. Henna, K.P. Nivitha, P. Mufeedha, C. Sabu, K. Pramod, Advanced biomedical applications of carbon nanotube. Mater. Sci. Eng. C Mater. Biol. Appl. **100**, 616–630 (2019)
384. Y. Oh, J.-H. Jin, J. Oh, Photothermal-triggered control of sub-cellular drug accumulation using doxorubicin-loaded single-walled carbon nanotubes for the effective killing of human breast cancer cells. Nanotechnology **28**, 125101 (2017)
385. A. Karimi, M. Erfan, S.A. Mortazavi, F. Ghorbani-Bidkorbeh, F. Kobarfard, F.H. Shirazi, Functionalisation of carbon nanotubes by methotrexate and study of synchronous photothermal effect of carbon nanotube and anticancer drug on cancer cell death. IET Nanobiotechnol. **13**, 52–57 (2019)
386. S. Shahrokhian, R. Salimian, H.R. Kalhor, A simple label-free electrochemical DNA biosensor based on carbon nanotube–DNA interaction. RSC Adv. **6**, 15592–15598 (2016)
387. B. Zribi, E. Roy, A. Pallandre, S. Chebil, M. Koubaa, N. Mejri, et al., A microfluidic electrochemical biosensor based on multiwall carbon nanotube/ferrocene for genomic DNA detection of Mycobacterium tuberculosis in clinical isolates. Biomicrofluidics **10**, 014115 (2016)
388. R.P. Deo, J. Wang, I. Block, A. Mulchandani, K.A. Joshi, M. Trojanowicz et al., Determination of organophosphate pesticides at a carbon nanotube/organophosphorus hydrolase electrochemical biosensor. Anal. Chim. Acta **530**, 185–189 (2005)
389. S. Same, G. Samee, Carbon Nanotube (CNT) Biosensor for Diabetes Disease (2018)
390. F.J. Rawson, J. Hicks, N. Dodd, W. Abate, D.J. Garrett, N. Yip et al., Fast, Ultrasensitive Detection of Reactive Oxygen Species Using a Carbon Nanotube Based-Electrocatalytic Intracellular Sensor. ACS Appl. Mater. Interfaces. **7**, 23527–23537 (2015)
391. K.B. Paul, A.K. Panigrahi, V. Singh, S.G. Singh, A multi-walled carbon nanotube–zinc oxide nanofiber based flexible chemiresistive biosensor for malaria biomarker detection. Analyst **142**, 2128–2135 (2017)
392. Y. Chen, S. Guo, M. Zhao, P. Zhang, Z. Xin, J. Tao et al., Amperometric DNA biosensor for Mycobacterium tuberculosis detection using flower-like carbon nanotubes-polyaniline nanohybrid and enzyme-assisted signal amplification strategy. Biosens. Bioelectron. **119**, 215–220 (2018)
393. A. Khazi-Syed, M.T. Hasan, E. Campbell, R. Gonzalez-Rodriguez, A.V. Naumov, Single-walled carbon nanotube-assisted antibiotic delivery and imaging in S. epidermidis strains addressing antibiotic resistance. Nanomater. (Basel) 9 (2019)
394. A. Rehman, W.M. Patrick, I.L. Lamont, Mechanisms of ciprofloxacin resistance in Pseudomonas aeruginosa: new approaches to an old problem. J. Med. Microbiol. **68**, 1–10 (2019)
395. M. Assali, A.N. Zaid, F. Abdallah, M. Almasri, R. Khayyat, Single-walled carbon nanotubes-ciprofloxacin nanoantibiotic: strategy to improve ciprofloxacin antibacterial activity. Int. J. Nanomedicine **12**, 6647–6659 (2017)
396. R. Subbiah, M. Veerapandian, S. Sadhasivam, K. Yun, Triad CNT-NPs/polymer nanocomposites: fabrication, characterization, and preliminary antimicrobial study. Synth. React. Inorg. Met. Org. Nano-Met. Chem. **41**, 345–355 (2011)
397. R.P. Subbiah, H. Lee, M. Veerapandian, S. Sadhasivam, S.-w Seo, K. Yun, Structural and biological evaluation of a multifunctional SWCNT-AgNPs-DNA/PVA bio-nanofilm. Anal. Bioanal. Chem. **400**, 547–560 (2011)
398. S. Al Aani, V. Gomez, C.J. Wright, N. Hilal, Fabrication of antibacterial mixed matrix nanocomposite membranes using hybrid nanostructure of silver coated multi-walled carbon nanotubes. Chem. Eng. J. **326**, 721–736 (2017)
399. H. Palza, N. Saldias, P. Arriagada, P. Palma, J. Sanchez, Antibacterial carbon nanotubes by impregnation with copper nanostructures. JOM **69**, 1319–1324 (2017)

400. N. Singh, J. Chen, K.K. Koziol, K.R. Hallam, D. Janas, A.J. Patil et al., Chitin and carbon nanotube composites as biocompatible scaffolds for neuron growth. Nanoscale **8**, 8288–8299 (2016)
401. S. Kasai, H. Nagasawa, M. Shimamura, Y. Uto, H. Hori, Design and synthesis of antiangiogenic/heparin-binding arginine dendrimer mimicking the surface of endostatin. Bioorg. Med. Chem. Lett. **12**, 951–954 (2002)
402. H. Chen, K. Zhou, G. Zhao, Gold nanoparticles: from synthesis, properties to their potential application as colorimetric sensors in food safety screening. Trends Food Sci. Technol. **78**, 83–94 (2018)

Chapter 30
Recent Advances in Textile Wastewater Treatment Using Nanoporous Zeolites

Abdullah Zahid Turan and Mustafa Turan

Abstract This review study deals with low-cost nanoporous zeolite for color removal from textile wastewaters. Wastewater from textile industries contains colorants originating from printing or dyeing processes. Basic and reactive dyes are extensively used in the textile industry due to their favorable characteristics of bright color, being easily water soluble, simple application technique and low-energy consumption. Yet, they can not be easily removed via conventional treatment methods since they strongly resist to biodegradation in an aerobic environment. Oxidation and adsorption are two major technologies that are used for wastewater treatment in the textile industry while adsorption-based removal of colorants along with other organic pollutants from industrial wastewater is considered an important application. Zeolites are becoming widely used as alternative materials in areas where sorptive applications are required. Having said that, several studies have reported that zeolites are not suitable for the treatment of anionic contaminants and reactive dyes. Thus, in order to enhance the removal of anionic contaminants, modification of zeolites with quaternary amine cationic surfactants have been proposed. This review investigates the equilibrium and the dynamics of the adsorption of dye on zeolites. The dynamics of color/dye removal using bed depth service time (BDST) was modeled and design parameters of the fixed-bed system were determined. Textile and other synthetic dyes were discussed along with their removal via non-conventional low-cost adsorbents. In a nutshell, this chapter provides an updated literature on the application of nanoporous zeolites in the treatment of textile dyeing wastewaters. An extensive list of natural and modified zeolites in batch and fixed-bed systems was compiled. Dye adsorption capacities and other parameters for natural and modified zeolites were evaluated. Color removal efficiency for textile wastewater treatment systems coupled with zeolites were presented. Adsorption capacities and cost of zeolites were also reported in comparison with other adsorbents.

A. Z. Turan (✉)
TUBITAK Marmara Research Center, 41470 Kocaeli, Turkey
e-mail: abdullah.turan@tubitak.gov.tr

M. Turan
Department of Environmental Engineering, Istanbul Technical University, 34469 Istanbul, Turkey
e-mail: turanm@itu.edu.tr

H. Ünlü and N. J. M. Horing (eds.), *Progress in Nanoscale and Low-Dimensional Materials and Devices*, Topics in Applied Physics 144,
https://doi.org/10.1007/978-3-030-93460-6_30

30.1 Introduction

Many industries, such as dyestuffs, textile, paper and plastics, use dyes in order to color their products and also consume substantial volumes of water. Wastewater from textile industries contains colorants originating from printing or dyeing processes. Dyes used in the textile industry are classified into three classes: (i) anionic (direct, acid and reactive dyes), (ii) cationic (all basic dyes) and (iii) nonionic (dispersed dyes). Basic and reactive dyes are extensively used in the textile industry due to their favorable characteristics of bright color, being easily water soluble, simple application technique and low-energy consumption [1, 2].

The use of organic dyes has increased in these said industries i.e., for an annual world production of 30 million tonnes of textiles, 700,000 tonnes of different dyes are required [3]. Such voluminous quantities pose considerable environmental problems since reactive dyes are water-soluble and cannot be easily removed by conventional methods such as chemical coagulation/flocculation and biological methods [4]. Moreover, these dyes are hardly biodegraded in an aerobic environment [5].

The adsorption process provides an attractive alternative for the decolorization of dye wastewaters and low-cost and readily available adsorbents have so far been utilized for this purpose in various applications. Many non-conventional low-cost adsorbents, including natural materials, biosorbents and waste materials from industry and agriculture, have been proposed in several studies. Some of the reported sorbents include clay materials (sepiolite, bentonite, kaolinite), zeolites, siliceous materials (silica beads, alunite, perlite), biosorbents (chitosan, peat, biomass), agricultural wastes and industrial waste products.

Nanoporous zeolites have negative charges that arise due to isomorphous substitution of Al^{3+} for Si^{4+} and its typical unit cell formula is given either as $Na_6[(AlO_2)_6(SiO_2)_{30}]_{0.2}4H_2O$ or $(Na_2, K_2, Ca, Mg)_3[(AlO_2)_6(SiO_2)_{30}]_{0.2}4H_2O$ [6]. The framework structure may contain linked cages, cavities or channels which are of the right size to allow small molecules to enter (Fig. 30.1). The three-dimensional crystal structure of zeolite contains two-dimensional channels [7, 8] which embody some ion exchangeable cations such as Na^+, K^+, Ca^{2+} and Mg^{2+}. These cations can be exchanged with organic and inorganic cations [9, 10]. Such sorptive properties have been utilized for a variety of purposes such as removal of ammonia [11–13], heavy metals [14, 15] and dyes [16, 17].

This review study presents a thorough investigation of studies on zeolite-assisted treatment of textile wastewater and in this respect covers; (i) batch adsorption studies, where adsorption equilibrium, kinetic models and thermodynamic parameters are considered; (ii) column adsorption studies with adsorption column design parameters and bed depth service time (BDST) model; (iii) examination of dyes (textile and other synthetic) along with their removal via non-conventional low-cost adsorbents; (iv) description of dye/color removal efficiency and other parameters for zeolites in the batch and fixed-bed column systems; (v) investigation of dye/color removal efficiency for wastewater treatment systems coupled with zeolites; (vi) analysis of the cost of zeolite and other low-cost materials.

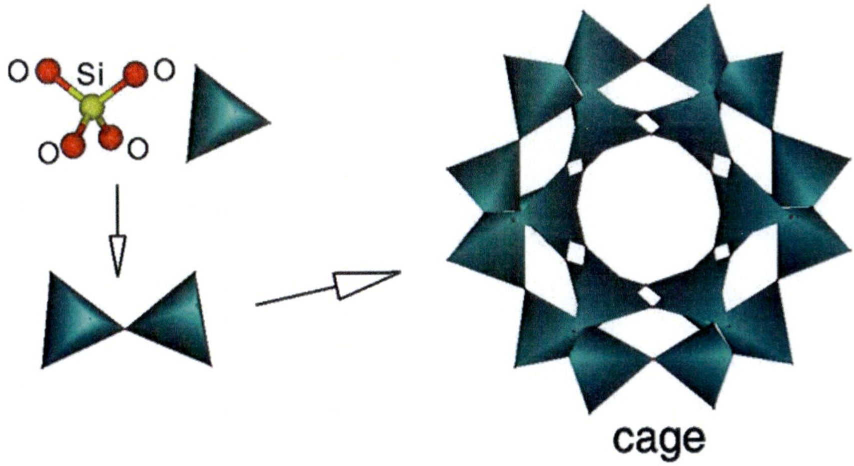

Fig. 30.1 Zeolite structure

30.2 Methodology

30.2.1 Batch Adsorption Studies

30.2.1.1 Equilibrium Isotherms

Sorption studies are performed by batch technique to obtain rate and equilibrium data. The removal efficiency (%) of adsorbent and the adsorption capacity of solution q_e (mg/g) is calculated, respectively, as follows;

$$\text{Removal efficiency}(\%) = (C_i - C_e)/C_i \times 100 \tag{30.1}$$

$$q_e = (C_i - C_e)V/m \tag{30.2}$$

where C_i and C_e are the initial and equilibrium solution concentrations (mg/L), respectively, V the volume of the solutions (L) and m the weight of adsorbent used (g). The equilibrium adsorption isotherm is fundamental in describing the interactive behavior between solutes and adsorbent and is important in the design of adsorption system. The widely used Langmuir isotherm [18] has found successful application in many real sorption processes and the lineer form of Langmuir model is expressed as;

$$C_e/q_e = 1/kb + C_e/b \tag{30.3}$$

where the values of b (mg/g) and k (L/mg) are maximum adsorption capacity of adsorbent and the adsorption energy coefficient, respectively. The well-known Freundlich isotherm [19] used for isothermal adsorption is a special case for heterogeneous surface energy, where the energy term in the Langmuir equation varies being dependent on surface coverage varying with sorption degree. The lineer form of Freundlich equation is given as;

$$\log q_e = \log K_f + (1/n) \log C_e \tag{30.4}$$

where K_f is a practical indicator of the adsorption capacity and n is the adsorption intensity. The magnitude of 1/n reveals the favorability of adsorption. Values, $n > 1$ indicate favorable adsorption condition [20]. The three-parameter Koble-Corrigan model is the combination of Langmuir and Freundlich models and is given as;

$$q_c = AC_e^m/(1 + C_e^m) \tag{30.5}$$

This model is valid when $m > 1$ [21]. Redlich-Peterson model has three constants and has been proposed to improve the fit by Langmuir and Freundlich models as follows;

$$q_e = K_{RP}C_e/(1 + a_{RP}C_e^b) \tag{30.6}$$

where K_{RP}, α_{RP} and β are the Redlich-Peterson parameters. Value of β is between 0 and 1. Equation (30.6) reduces to a linear isotherm at low surface coverage, to the Freundlich isotherm at high adsorbate concentration and to the Langmuir isotherm when $\beta = 1$ [22]. The derivation of the Temkin isotherm assumes that the fall in the heat of adsorption is linear rather than logarithmic, as implied in the Freundlich equation. The Temkin isotherm [23] is given as;

$$q_e = A + B \ln C_e \tag{30.7}$$

where A and B are isotherm constants.

30.2.1.2 Kinetics and Thermodynamics

In order to investigate the mechanism of adsorption, the pseudo-first-order adsorption, the pseudo-second-order adsorption and the intraparticle diffusion model are used to test dynamical experimental data. The first-order rate expression of Lagergren [24] is given as;

$$\log (q_e - q) = \log q_e - k_1t/2.303 \tag{30.8}$$

where q_e and q are the amounts of dye adsorbed on adsorbent (mg/g) at equilibrium and at time t, respectively and k_1 is the rate constant of first-order adsorption (1/min). The slopes and intercepts of plots of log $(q_e{-}q)$ versus t are used to determine the first-order rate constant k_1. In many cases, the first-order equation of Lagergren does not fit well to the whole range of contact time and is generally applicable over the initial stage of the adsorption processes [25]. The second-order kinetic model [26] is expressed as;

$$t/q = 1/\left(k_2 q_e^2\right) - t/q_e \tag{30.9}$$

and

$$h = k_2 q_e^2 \tag{30.10}$$

where k_2 (g/(mg min)) is the rate constant of second-order adsorption and h is the initial adsorption rate (mg/(g min)). The slopes and intercepts of plots of t/q versus t are used to calculate the second-order rate constant k_2 and q_e. This model assumes that adsorption is the rate-controlling step and it is more convenient for the prediction of adsorption behaviour over a wider range [25, 26]. The intraparticle diffusion equation is described as;

$$q = k_i t^{0.5} \tag{30.11}$$

where k_i is intraparticle diffusion rate constant (mg/(g $min^{0.5}$)). k_i is the slope of straight-line portions of plot of q versus $t^{0.5}$. The pseudo-second-order rate constant of solute adsorption is expressed as a function of temperature by the Arrhenius equation:

$$\ln k = \ln A - E_a/RT \tag{30.12}$$

where E_a is the Arrhenius activation energy of sorption, representing the minimum energy that reactants must have for the reaction to proceed, A is the Arrhenius factor, R is the gas constant (8.314 J/(mol K)) and T is the solution temperature. The thermodynamic parameters such as change in free energy (ΔG°), enthalpy (ΔH°) and entropy (ΔS°) are determined using the following equations [27, 28]:

$$K_C = C_{Ae}/C_e \tag{30.13}$$

$$DG^0 = -RT\ \ln K_C \tag{30.14}$$

$$\log K_C = DS^0/(2.303R) - DH^0/(2.303RT) \tag{30.15}$$

where K_C is the equilibrium constant, C_{Ae} is the amount of adsorbate (mg) adsorbed on the adsorbent per L of the solution at equilibrium, C_e is the equilibrium concentration (mg/L) of the dye in the solution, T is the solution temperature (K) and R is the gas constant. $\Delta H°$ and $\Delta S°$ are calculated from the slope and intercept of $\log K_C$ versus 1/T plot.

The magnitude of activation energy gives an idea about the nature of adsorption, which is mainly either physical or chemical. Generally, the absolute magnitude of the change in free energy for physisorption is between −20 and 0 kJ/mol, while chemisorption has a range of –80 to –400 kJ/mol [29]. The negative values of the standard enthalpy change ($\Delta H°$) indicate that the interaction of the adsorbate with adsorbent has an exothermic nature.

30.2.2 Column Adsorption Studies

30.2.2.1 Adsorption Column Design Parameters

Adsorption performance of fixed-bed process can be evaluated while bed volumes (BV) at breakthrough point ($C/C_0 = 0.1$) is maximum. The breakthrough curves are constructed by plotting the normalized effluent concentration (C/C_0) versus time (t) and/or bed volumes (BV). The bed volumes, BV and the empty bed contact time, EBCT, for a fixed-bed column process are defined, respectively, as follows;

$$BV = V_F/V_R = Q_t/V_R \tag{30.16}$$

$$EBCT = V_R/Q \tag{30.17}$$

where V_F is the total volume of wastewater treated during adsorption process at time t (L), V_R is the volume of zeolite bed (L), C_0 is the influent concentration (mg/L), C is the effluent concentration at time t (mg/L), Q is the fixed-bed feed flow rate (L/min) and t is the service time (min).

The formation and the movement of the adsorption zone can be evaluated numerically [30, 31]. The time required for the adsorption zone to become established and move completely out of the bed at exhaust time is;

$$t_{Exh} = V_{Exh}/Q = BV_{Exh}EBCT \tag{30.18}$$

The rate at which the adsorption zone (U_z) is moving up or down through the bed is;

$$U_z = h_z/t_z = h/(t_{Exh} - t_f) \tag{30.19}$$

From (30.19), the height of the adsorption zone (h_z) is obtained as;

$$h_z = h(t_z)/(t_{Exh} - t_f) \tag{30.20}$$

where V_{Exh} is the total volume of wastewater treated in the zeolite column at exhaust time (L), h_z is the height of adsorption zone (cm), h is the total bed height (cm) and t_f is the time required for the adsorption zone to initially form (min). t_f value can be found as follows;

$$t_f = (1 - F)t_z \tag{30.21}$$

At breakthrough point, the fraction (F) of adsorbent present in the adsorption zone still possessing the ability to remove solute is;

$$F = S_z/S_{max} = \int_{Vb}^{VExh} (C_0 - C)dV/C_0(V_{Exh} - V_b) \tag{30.22}$$

where V_b is the total volume of the wastewater treated to the breakthrough point (L), S_z is the amount of solute that has been removed by the adsorption zone from breakthrough to exhaustion and S_{max} is the amount of solute removed by the adsorption zone when completely exhausted. The percentage of column saturation at breakthrough point is;

$$\%\text{saturation}(h + (F - 1)h_z) \times 100/h \tag{30.23}$$

30.2.2.2 Bed Depth Service Time (BDST) Model

The objective of fixed-bed process is to reduce the effluent concentration to such a degree that it does not exceed a specific breakthrough concentration (C_b). The original study on the BDST model was carried out by Bohart and Adams [32] who proposed a relationship between bed depth, Z, and the time taken for breakthrough to occur. The service time, t, is related to the process conditions and operating parameters;

$$\ln(C_0/C_b - 1) = \ln\left(e^{kadN0Z/V} - 1\right) - k_{ad}C_0t \tag{30.24}$$

Hutchins [33] proposed a linear relationship between the bed depth and service time;

$$t = N_0Z/(C_0v) - (1/k_{ad}C_0)\ln(C_0/C_b - 1) \tag{30.25}$$

where C_b is the solute concentration at breakthrough point (mg/L), k_{ad} is the adsorption rate constant (L/(mg h)), N_0 is the adsorption capacity (mg of solute/L of adsorbent), Z is the bed depth of column (cm), v is the linear flow velocity of feed stream (cm/h), and t is the service time of column (h). The critical bed depth, Z_0, is the minimum column height necessary to produce an effluent concentration C_b. By letting $t = 0$, Z_0 is obtained from (30.25);

$$Z_0 = (v/k_{ad}N_0)\ \ln(C_0/C_b - 1) \tag{30.26}$$

Equation (30.25) enables the service time, t, of an adsorption bed to be determined by a specified bed depth, Z, of adsorbent. t and Z, which are are correlated with the process parameters such as the influent concentration of solute, solution flow-rate and the adsorption capacity [34] 30.23, can be expressed in the following form:

$$t = mZ + n \tag{30.27}$$

where

$$m = \text{slope} = N_0/C_0 v \tag{30.28}$$

$$n = \text{intercept} = -(1/k_{ad}C_0)\ \ln(C_0/C_b - 1) \tag{30.29}$$

30.2.3 *Dyes and Low-Cost Adsorbents*

30.2.3.1 Textile and Other Synthetic Dyes

Today, the annual worldwide production of more than 100,000 commercial dyes is estimated to be around 7×10^5–1×10^6 tons [35–38]. Having said that, it is also reported that 10–15% of the used dyes enter the environment through wastes [39]. The main consumers of dyes are textile, dyeing, paper and pulp, tannery and paint industries. The classification of synthetic dyes and some basic features of them are explained below [40]:

(i) Reactive dyes: They are usually used in cotton and other cellulosics. Their attachment is over a covalent bond with the fiber. Their structure is relatively simple, which is composed of chromophoric groups such as azo, anthraquinone, triarylmethane etc.

(ii) Acid dyes: These are mainly used in nylon, wool, silk and modified acrylics. Most of them are water-soluble and generaly in chemical forms like azo, anthraqui-none, triphenylmethane, azine, xanthene, nitro and nitroso.

(iii) Basic (cationic) dyes: These dyes generate coloured cations upon dissolved in water. They are often used for paper, polyacrylonitrile, modified nylons, modified polyesters and cation dyeable polyethylene terephthalate. These water-soluble dyes yield coloured cations in solution and that's why are called as cationic dyes. They are generally in diazahemicyanine, triarylmethane, cyanine, hemicyanine, thiazine, oxazine or acridine form.

(iv) Vat dyes: These water-insoluble dyes usually consist of anthraquinone (including polycyclic quinones) and indigoid structures and they are mainly used in cotton.

(v) Disperse dyes: Their highly water-insoluble character makes them good candidates for use in hydrophobic polyester fibers. They are often composed of azo, anthraquinone, styryl, nitro and benzodifuranone groups.

(vi) Direct dyes: These water-soluble anionic dyes show have high affinity for cellulosic fibers and are mainly utilized in the dyeing of cotton, rayon, paper and leather. Poly azo compounds, along with some stilbenes, phthalocyanines and oxazines are typical chemical structures.

(vii) Solvent dyes: These dyes are used for plastics, gasoline, lubricants, oils and waxes. They are water-insoluble and generally exhibit nonpolarity or weak polarity, that's why they are deprived of polar solubilizing groups such as sulfonic acid, carboxylic acid or quaternary ammonium. Their chemical structures are generally azo and anthraquinone, though phthalocyanine and triarylmethane are also possible.

(viii) Sulfur dyes: These dyes are mainly used for cotton and rayon and utilized in other sort of fibers only at a small extent.

30.2.3.2 Non-Conventional Low-Cost Adsorbents

The removal of colorants and other organic pollutants from industrial wastewater using suitable adsorbents is considered an important application of the adsorption process. Activated carbon and polymer resins appear to be the best adsorbents for removing chemicals from relatively concentrated wastewater, yet they are expensive and require regeneration [16, 41–43]. That's why several researchers have focused studies on low-cost adsorbent systems. A sorbent can be regarded as low-cost if it requires only minor processing, is wide-spread in nature or is a by-product or waste of a different process [44]. Materials to be used as adsorbents, either being natural or wastes from industrial and agricultural processes, can be classified as follows [45]:

(i) *Activated carbons from solid wastes*: Activated carbonss (AC) can be practically produced from almost any carbon-containing source, yet commercial production is generally from natural materials such as wood, coconut shell, lignite or coal.

(ii) *Agricultural solid wastes*: Potential adsorbents in this group are agricultural solid wastes and waste materials from forest industries such as sawdust and bark.

(iii) *Industrial by-products*: Industrial solid wastes such as metal hydroxide sludge, fly ash and red mud can be classified as low-cost materials.

(iv) *Clays*: Clays emerge as an important class of adsorbents thanks to their low cost, world-wide abundance and high sorption properties. Clay materials possess a layered structure and are classified accorfing to the differences in these structures. There exist several forms of clays such as smectites (montmorillonite, saponite), mica (illite), kaolinite, serpentine, pylophyllite (talc), vermiculite and sepiolite.

(v) *Siliceous materials*: The use of natural siliceous sorbents such as silica beads, glasses, alunite, perlite and dolomite for wastewater is increasing due to their abundance, availability and low cost. Within inorganic materials, silica beads are of particular insterest.

(vi) *Zeolites*: Zeolites are highly porous aluminosilicates with different cavity structures. Their structures consist of a three dimensional framework, having a negatively charged lattice. The negative charge is balanced by cations which are exchangeable with certain cations in solutions. Zeolites correspond to a wide spectrum i.e., there exist more than 40 natural species such as clinoptilolite, heulandite, phillipsite etc. The characteristics and applications of zeolites have been reviewed elsewhere [46]. Zeolites are attractive adsorbents thanks to their high surface area and low cost. Another advantage of zeolites over resins is their ion selectivities due to their rigid porous structure. Zeolites are becoming widely used as alternative materials in sorption-based processes. They have been extensively studied recently because of their capability in removing trace pollutants (e.g. heavy metals and phenols) thanks to their cage-like structures suitable for ion exchange [47–49].

(vii) *Biosorbents:* Sorption-based separation of species via biological materials is defined as biosorption. Chitin, chitosan, peat, yeasts, fungi and bacterial biomass can be regarded as common biosorbents, which can be used as chelating and complexing sorbents in order to concentrate and to remove dyes from solutions.

30.3 Dye/Color Removal from Textile Dyeing Effluents Using Nanoporous Zeolites

30.3.1 Dye Adsorption onto Natural and Synthetic Zeolites in the Batch Systems

An overview of nanoporous zeolites for the dye/color removal from textile wastewaters is presented in this article. Several studies have been conducted on the sorbent behavior of natural zeolites [16, 46–50]. Recently, natural zeolites have been used effectively for the removal of basic dyes [51, 52]. However, they were not suitable for the removal of reactive dyes due to extremely low sorption capacities. So, a technical challenge have arisen for the enhancement of these low-cost materials to be used as effective adsorbents. In this respect, within the scope of the recent studies of the authors, a quaternary amine, hexadecyltrimethylammonium bromide (HTAB, C19H42BrN) as seen in Figs. 30.2 and 30.3, was used to modify the surface of zeolite for the sorption of some anionic azo dyes [17, 53–56]. The adsorption mechanism of three reactive dyes by zeolite has been examined with the aim of identifying the ability of zeolite (clinoptilolite) to remove textile dyes from aqueous solutions [17]. The reactive dyes CI Reactive black 5, red 239 and yellow 176 are typical azo dyes extensively used in textile dyeing. The adsorption results indicate that the natural zeolite has a limited adsorption capacity for reactive dyes, yet is substantially improved upon modifying its surfaces with quaternary amines. The results are also supported by electrokinetic measurements. The adsorption data were fitted to the Langmuir

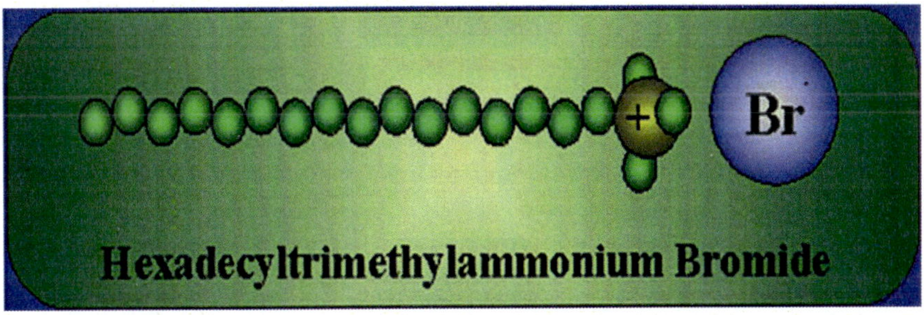

Fig. 30.2 Hexadeciltrimetilammonium Bromide (HTAB) [56]

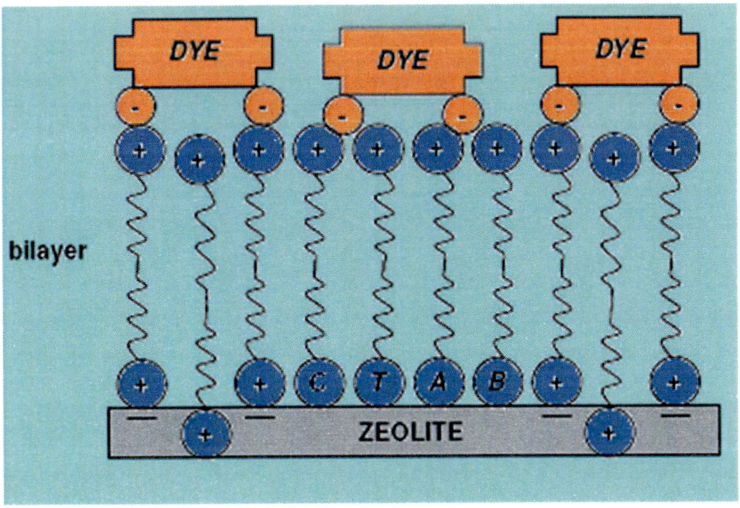

Fig. 30.3 A schematic illustration of the interaction of dye anions with quaternary amine on zeolite surface [17]

isotherm and it was found that the modified sepiolite yields adsorption capacities (q_e) of 111, 89 and 61 mg/g for red, yellow and black, respectively (Table 30.1).

Karadag [55] investigated the adsorption of Acid orange 8 on the surfactant-modified clinoptilolite. The maximum adsorption capacity was obtained as 18.98 mg/g at pH 8. Error analysis showed that the Koble-Corrigan model best fits the equilibrium adsorption data. Adsorption process was also found to be controlled by both surface and particle diffusions. Adsorption of cationic methylene blue (MB) and anionic Orange II (OII) onto unmodified and surfactant-modified zeolites was studied using a batch equilibration method [57]. For unmodified and 2% SDBS-modified zeolites, the optimum conditions for MB adsorption are at 25 °C, pH 6.85, and 1 h equilibrium time with adsorption capacities of 8.67 and 15.68 mg/g, respectively. For OII adsorption on unmodified and modified zeolites, the best condition is at 25 °C and pH 1 for 20 min equilibrium time with adsorption capacities of 0.63 and 3.38 mg/g, respectively (Table 30.1).

Alpat et al. [49] investigated the adsorption of Toluidine blue O (TBO) using clinoptilolite as a natural zeolite. It was found that adsorption of TBO on clinoptilolite was fitted to the second-order kinetic model with two-step diffusion process and that the adsorption isotherm follows both the Langmuir and Freundlich models. The maximum adsorption capacity of clinoptilolite for TBO was 2.1×10^{-4} mol/g at solution pH of 11.0. In 2006, the adsorption of methylene blue and rhodamine B on an Australian natural zeolite was studied [50]. The maximal adsorption capacity of zeolite for both basic dyes was found to be 2.8×10^{-5} and 7.9×10^{-5} mol/g at 50 oC, respectively (Table 30.1). The adsorption isotherms fitted both the Langmuir and Freundlich models and the adsorption was endothermic process. It was also found that only 60% capacity could be recovered by the two regeneration techniques. Wang et al. [58] researched an adsorption study on MCM-22, which is a novel nanoporous zeolite synthesized [59] for the removal of basic dyes including methylene blue (MB), crystal violet (CV), and rhodamine B (RB) from aqueous solution. The adsorption capacities for MB, CV, and RB can reach 1.8×10^{-4}, 1.2×10^{-4}, and 1.1×10^{-4} mol/g, respectively. The adsorption kinetics follows the pseudo-second-order model and the external diffusion is the controlling process. The adsorption isotherms can be well fitted with Langmuir and Freundlich models. Thermodynamic calculations indicate that the adsorption of basic dyes on MCM-22 is spontaneous and an endothermic reaction. The $\Delta H°$ for MB, CV, and RB is 5.4, 1.2, and 5.0 kJ/mol, respectively.

Within another study, the removal performance of clinoptilolite for Amido black 10B and Safranine T was compared [60]. The results indicate that clinoptilolite has a limited adsorption capacity for Amido black 10B (removal efficiency of 81.2%) and has a good adsorption capacity for Safranine T (removal efficiency of 16.3%). the adsorption isotherms for both dyes follow the Langmuir model. In a similar study, adsorption equilibrium and kinetic of methylene blue (MB) onto natural zeolite was studied [61]. The increase in temperature resulted in a higher MB adsorption. The adsorption isotherm and kinetic follow the Redlich–Peterson model and the pseudo-second-order model, respectively. The thermodynamics parameters indicated that the MB adsorption onto zeolite was spontaneous and endothermic process. Meshko et al. [16] evaluated the adsorption of basic dyes from aqueous

Table 30.1 Dye adsorption capacities and other parameters for natural, synthetic and modified zeolites

Adsorbent	Dye	Surface area	Adsorption capacity/Dye removal	Concentration range	Contact time	pH	Source
Zeolite (Turkish)	Cationic dye (Toluidine Blue O	40.79 m^2/g	2.1×10^{-4} mol/ g	1.0×10^{-3}–3.0×10^{-3} M	30 min	11	[51]
Zeolite (Avustralian)	Basic dye (Rhodamine B)	16 m^2/g	2.8×10^{-5} mol/g	10^{-6}–10^{-5} M	50–200 h	7	[52]
Zeolite (Avustralian)	Basic dye (Methylene Blue)	16 m^2/g	7.9×10^{-5} mol/g	10^{-6}–10^{5} M	50–200 h	7	[52]
Modified zeolite (Turkish)	Acid Orange 8	–	18.98 mg/g	25–200 mg/L	60 min	3	[55]
Zeolit (ZSM-5)	Cationic dye (Methylene Blue)	400–800 m^2/g	8.67 mg/g	10–60 mg/L	60 min	6.85	[57]
Zeolite (ZSM-5)	Anionic dye (Orange II)	400–800 m^2/g	0.63 mg/g	10–60 mg/L	60 min	6.85	[57]
Modified zeolite (ZSM-5)	Cationic dye (Methylene Blue)	400–800 m^2/g	15.68 mg/g	10–60 mg/L	60 min	6.85	[57]
Modified zeolite (ZSM-5)	Anionic dye (Orange II)	400–800 m^2/g	3.38 mg/g	10–60 mg/L	60 min	6.85	[57]
Zeolite (MCM-22)	Basic dye (Methylene Blue)	490 m^2/g	1.8×10^{-4} mol/g	10–6–10^{-5} M	10 days	–	[58]
Zeolite (MCM-22)	Basic dye (Crystal violet)	490 m^2/g	1.2×10^{-4} mol/g	10–6–10^{-5} M	10 days	–	[58]
Zeolite (MCM-22)	Basic dye (Rhodamine B)	490 m^2/g	1.1×10^{-4} mol/g	10–6–10^{-5} M	10 days	–	[58]
Zeolite (Chinese)	Acid Black 1(Amido Black 10B)	–	0.05513 mg/g	10–40 mg/L	600 min	7	[60]

(continued)

Table 30.1 (continued)

Adsorbent	Dye	Surface area	Adsorption capacity/Dye removal	Concentration range	Contact time	pH	Source
Zeolite (Chinese)	Basic dye (Safranine T)	–	0.0112 mg/g	10–40 mg/L	600 min	7	[60]
Zeolite (Chinese)	Basic blue 9 (Methylene Blue)	–	5.8 mg/g	16–230 mg/L	14 h	7.5	[61]
Zeolite (Macedonian)	Basic dye (MS-300)	–	14.91 mg/g	20–200 mg/L	4–10 days	–	[16]
Zeolite (Macedonian)	Basic dye (MG-400)	–	55.86 mg/g	50–500 mg/L	4–10 days	–	[16]
Modified zeolite (Mexican)	Reactive red 40	18.2–20.3 m^2/g	1.714 mg/g	2.5–20 mg/L	12 h	–	[62]
Modified zeolite (Mexican)	Reactive yellow 5	18.2–20.3 m^2/g	1.051 mg/g	2.5–20 mg/L	12 h	–	[62]
Zeolite (MCM-41)	Basic dye (Rhodamine B)	1003.5 m^2/g	9.4×10^{-4} mol/g	1.0×10^{-3}–1.2×10^{-2} M and	3 h	4	[64]
Zeolite (MCM-41)	Basic dye (Crystal violet)	1003.5 m^2/g	5.8×10^{-4} mol/g	1.0×10^{-3}–1.2×10^{-2} M and	3 h	4	[64]
Zeolite (MCM-41)	Basic dye (Methylene green)	1003.5 m^2/g	3.1×10^{-4} mol/g	1.0×10^{-3}–1.2×10^{-2} M and	3 h	4	[64]
Zeolite (MCM-41)	Acid dye (Acid red 1)	1003.5 m^2/g	0.5×10^{-4} mol/g	1.0×10^{-3}–1.2×10^{-2} M and	3 h	4	[64]
Zeolite (MCM-41)	Acid dye (Erioglaucine)	1003.5 m^2/g	0.5×10^{-4} mol/g	1.0×10^{-3}–1.2×10^{-2} M and	3 h	4	[64]
M-RZ (based on kaolin)	Basic blue 9 (Methylene Blue)	–	0.07 mmol/g	8×10^{-6}–4×10^{-5} M	48 h	–	[65]
M-JZ (based on Jordanian zeolite)	Basic blue 9 (Methylene blue)	–	0.08 mmol/g	8×10^{-6}–4×10^{-5} M	48 h	–	[65]

(continued)

Table 30.1 (continued)

Adsorbent	Dye	Surface area	Adsorption capacity/Dye removal	Concentration range	Contact time	pH	Source
Modified zeolite (Turkish)	Reactive black 5	11.8 m^2/g	12.93 mg/g	60–150 mg/L	30 min	5	[66]
Modified zeolite (Turkish)	Reactive red 239	11.8 m^2/g	15.94 mg/g	60–150 mg/L	30 min	5	[66]
Zeolite (Turkish)	Basic Red 46	11.8 m^2/g	8.56 mg/g	50–200 mg/L	60 min	7	[67]
HDTMA-zeolite (Turkish)	Reactive Yellow 176	11.8 m^2/g	13.15 mg/g	50–200 mg/L	60 min	7	[67]
CTAB-zeolite (Turkish)	Reactive Yellow 176	11.8 m^2/g	5.54 mg/g	50–200 mg/L	60 min	7	[67]
Modified zeolite (Turkish)	Reactive red 239	11.8 m^2/g	111 mg/g	1–1000 mg/L	4 h	7	[17, 53]
Modified zeolite (Turkish)	Reactive yellow 176	11.8 m^2/g	89 mg/g	1–1000 mg/L	4 h	7	[17, 53]
Modified zeolite (Turkish)	Reactive black 5	11.8 m^2/g	61 mg/g	1–1000 mg/L	4 h	7	[17, 53]
Zeolite (MCM-41)	Anionic dye (Methyl orange)	403 m^2/g	1.12 mmol/g	–	24 h	5.6	[68]
Zeolite (MCM-41)	Anionic dye (Orange IV)	403 m^2/g	1.09 mmol/g	–	24 h	5.6	[68]
Zeolite (MCM-41)	Anionic dye (Reactive Brilliant red)	403 m^2/g	0.43 mmol/g	–	24 h	5.6	[68]
Zeolite (MCM-41)	Anionic dye (Acid fuchsine)	403 m^2/g	0.34 mmol/g	–	24 h	5.6	[68]

(continued)

Table 30.1 (continued)

Adsorbent	Dye	Surface area	Adsorption capacity/Dye removal	Concentration range	Contact time	pH	Source
Modified NaY zeolite	Cibacron blue 3GA	900 m^2/g	20%	500 mg/L	2 h		[69]
Zeolite (Australian)	Basic dye (Methylene blue)	16 m^2/g	5×10^{-5} mol/g	2.7×10^{-5} M	24 h	8.5	[70]
Zeolite (MCM-41)	Basic dye (Methylene blue)	488 m^2/g	1.7×10^{-4} mol/g	2.7×10^{-5} M	24 h	8.5	[70]
Zeolite (MCM-41)	Basic green	458.4 m^2/g	36 mg/g	500 mg/L	24 h	4	[71]
Zeolite (MCM-41)	Basic violet	458.4 m^2/g	500 mg/g	500 mg/L	24 h	4	[71]
Zeolite (ZM 1–5)	Indigo carmine dye	567 m^2/g	100%	100 mg/L	60 min	2	[72]
Fly ash/zeolite (FA-ZA)	Basic dye (Methylene blue)	–	25.98 mg/g (at 30 °C)	20–70 mg/L	48 h	6.58	[73]
Fly ash/zeolite (FA-ZX)	Basic dye (Methylene blue)	–	35.42 mg/g (at 30 °C)	20–70 mg/L	48 h	6.58	[73]
Fly ash/zeolite	Basic dye (Methylene blue)	53.4–90.7 m^2/g	–	1.6–32 mg/L	120 min	–	[74]
Synthesized zeolite	Cationic dye (Methylene blue)	400 m^2/g	10.63–16.86 mg/g	10–50 mg/L	2 h	7	[75]
Zeolite (Australian)	Basic dye (Malachite green)	16 m^2/g	5×10^{-5} mol/g (at 30 °C)	6.1×10^{-5} M	200 h	6	[76]
Mesoporous carbon/zeolite X (C130-H)	Basic dye (Methylene blue)	764 m^2/g	223 mg/g (at 20 °C)	105–330 mg/L	24 h	–	[77]

(continued)

Table 30.1 (continued)

Adsorbent	Dye	Surface area	Adsorption capacity/Dye removal	Concentration range	Contact time	pH	Source
Mesoporous carbon/zeolite X (C130-OH)	Basic dye (Methylene blue)	764 m^2/g	380 mg/g (at 20 °C)	105–330 mg/L	24 h	–	[77]
Modified NaY zeolite	Cibacron blue 3GA	900 m^2/g	20%	500 mg/L	2 h	–	[78]
Modified zeolite (Turkish)	Reactive blue 4		81.4%	50 mg/L	60 min	2	[79]
Modified zeolite (Turkish)	Reactive yellow	–	97.7%	50 mg/L	40 min	8	[79]
Zeolite A	Basic dye (Methylene blue)	1.9 m^2/g	2.8×10^{-6} mol/g	0.5×10^{-6}–17.5×10^{-6} M	5 min	8	[80]
Modified zeolite (Mexican)	Reactive dye (Remazol yellow)	18.2–20.3 m^2/g	4.15 mg/g	20–100 mg/L	360 min	3–10	[82]
Modified zeolite (Turkish)	Reactive dye (Remazol Brillant blue R)	11.80 m^2/g	13.9 mg/g	100–1000 mg/L	24 h	–	[83]
Modified zeolite (Turkish)	Reactive dye (Remazol Yellow)	11.80 m^2/g	38.31 mg/g	100–1000 mg/L	24 h	–	[83]
Synthetic zeolite	Basic dye (Amido Black)	–	1.0 mg/g	4.9–14.5 mg/L	6 h	–	[84]
Synthetic zeolite	Basic dye (Thionine)	–	2.3 mg/g	4.9–14.5 mg/L	6 h	–	[84]
Modified zeolite (Turkish)	Reactive red 239	–	28.57 mg/g at 293 K	25–500 mg/L	4 h	2–12	[85]
Modified zeolite (Turkish)	Reactive blue 250	–	17.63 mg/g at 293 K	25–500 mg/L	4 h	2–12	[85]

(continued)

Table 30.1 (continued)

Adsorbent	Dye	Surface area	Adsorption capacity/Dye removal	Concentration range	Contact time	pH	Source
Zeolite (Algerian)	Cationic dye, methylene blue	24.59 m^2/g	22 mg/g	4–37 mg/L	120 min	2–12	[86]
Modified zeolite (DAAO)	Anionic dye, Congo red	–	69,44 mg/g	20–100 mg/L	24 h	5–9	[87]
Synthetic zeolite (ZIF-67)	Cationic dye, Malachite green	–	2430 mg/g	0–1000 mg/L	140 min	3–6	[88]
Zeolite (chabazite, Turkish)	Basic dye (cationic) Methylene blue	–	28.9 mmol/g	40 mg/L	3 h	3–9	[89]
Zeolite	Basic dye, Methylene blue						[90]
Mordenite zeolite nanoparticles	Basic dye, Methylene blue	28.26 m^2/g	2.12 mg/g	10–50 mg/L	120 min		[91]
Zeolite tuff (Romanian)	Basic Blue 41	–	192.31 mg/g	–	–	–	[92]
Zeolite (Jordan)	Basic dye, Bromocresol purple (BCP),	1918.6 m^2/g	175.5 mg/g	10–80 mg/L	205 min	2–8	[93]
Synthetic zeolite (Biopolymer/ZSM–5)	Basic dye (Crystal violet)	–	1217.3 mg/g	0–200 mg/L	120 min	1–11	[94]
Synthetic zeolite (Biopolymer/ZSM–5)	Basic dye (Methylene blue)	–	548.2 mg/g	0–200 mg/L	120 min	1–11	[94]
Synthetic zeolite (Biopolymer/ZSM–5)	Basic dye (Basic fuchsin)	–	237.5 mg/g	0–200 mg/L	120 min	1–11	[94]
Synthetic zeolite	Cationic dye, Malachite green	32.8 m^2/g	108.26 mg/g	25–200 mg/L	40 min	7	[95]

(continued)

Table 30.1 (continued)

Adsorbent	Dye	Surface area	Adsorption capacity/Dye removal	Concentration range	Contact time	pH	Source
Synthetic zeolite (ZnO@Ze)	Congo red (CR)	–	161.3	25–500 mg/L	60 min	3–12	[96]
Zeolite (Phillipsite,)	Basic dye, Safranin-T	–	65.35 mg/g	25–200 mg/L	1600 min	2–10	[97]
Zeolite (Clinoptilolite)	Basic dye, Safranin-T	–	42.9 mg/g	25–200 mg/L	1600 min	2–10	[97]
Zeolite (Heulandite)	Basic dye, Safranin-T	–	41.66 mg/g	25–200 mg/L	1600 min	2–10	[97]
Sodalite zeolite nanoparticle (SZN)	Direct red 23 (DR23)	–	2415 mg/g	50–125 mg/L	60 min	2.1–10	[98]
Sodalite zeolite nanoparticle (SZN)	Direct red 80 (DR80)	–	4842 mg/g	50–125 mg/L	60 min	2.1–10	[98]
Synthetic zeolite (Fe-ZSM-5)	Basic dye, Basic Fuchsin	399 m^2/g	251.87 mg/g	0–350 mg/L	250 min		[99]

CTAB: cetyltrimethylammonium bromide and HDTMA: hexadecyltrimethylammonium bromide solution onto granular activated carbon and natural zeolite using an agitated batch adsorber. Two basic dyes were used as commercial salts without purification. The dye Maxilon Schwarz FBL-01 300% (MS-300) contains 30–40% component A and 20–30% component B. The dye Maxilon Goldgelb GL EC 400% (MG-400) contains more than 90% component A.

The selectivity of a Mexican surfactant-modified clinoptilolite-rich tuff to retain azo dyes (Red-40, Yellow-5 and a mixture of them) from aqueous solutions has been investigated by Torres-Perez et al. [62]. The zeolitic material was modified with hexadecyltrimethylammonium bromide and then left in contact with azo dyes. It was found that in single and binary solutions, under the experimental conditions, the modified zeolitic material retain preferentially Red-40 in comparison to Yellow-5. The best models to describe the reaction rate and the sorption of azo dyes Red-40 and Yellow-5 by Mexican surfactant modified clinoptilolite-rich tuff were pseudo-second order and Langmuir models, respectively. Recently, the adsorption of three basic dyes as rhodamine B (RB), crystal violet (CV), and methylene green (MG) and two acid dyes as Acid red 1 (AR1) and Erioglaucine (EG) onto MCM-41 was studied to examine the potential of MCM-41 [63] for the removal of dyes from water solution by Lee et al. [64]. The adsorption capacity of MCM-41 for the five dyes followed a decreasing order of RB > CV > MG > EG ~ AR1. The fitness of both Langmuir and Freundlich adsorption model on describing the equilibrium isotherms of three basic dyes was examined. Thermodynamic calculations indicate that the adsorption of basic dyes on MCM-41 is an exothermic process and the change in enthalpy ($\Delta H°$) for RB, MG, Q and CV are −32.1, −7.5, and −26.3 kJ/mol, respectively.

Yousef et al. [65] investigated the effect of using Jordanian zeolitic tuff as filler on the mechanical performance and on the adsorption capacity of the geopolymers products. Geopolymers consist of an amorphous, three-dimensional structure resulting from the polymerization of aluminosilicate monomers. For methylene blue and copper (II) ions, geopolymers M-RZ and M-JZ (based on kaolin and Jordanian zeolite) have the highest adsorption capacities of $q_m = 0.07$ and 0.08 mmol/g, respectively. The adsorption of Reactive black 5 and Reactive red 239 in aqueous solution on cetyltrimethylammonium bromide (CTAB)-zeolite was studied in a batch system by Karadag et al. [66]. CTAB modification covered the zeolite surface with positive charges, and the adsorption capacity of zeolite increased. Chemical structure of Reactive black 5 and Reactive red 239 was shown in Fig. 30.4. The Freundlich isotherm agrees very well with the experimental data. Thermodynamic parameters such as change in free energy ($\Delta G°$), enthalpy ($\Delta H°$), and entropy ($\Delta S°$) were also determined. The maximum adsorption capacities changed from 12.93 to 12.18 mg/g for RR239 and from 15.94 to 11.02 mg/g for RB5 with increasing temperature from 30 to 40 °C.

Similarly, the adsorption of the dyes Basic red 46 (BR46) and Reactive yellow 176 (RY176) from aqueous solution onto natural and modified zeolites has been investigated [67]. The surfactants cetyltrimethylammonium bromide (CTAB) and hexadecyltrimethylammonium bromide (HDTMA) were used to modify the zeolite surface. BR46 adsorption onto natural zeolite increased to a minor extent with increasing pH, whereas pH did not significantly affect RY176 adsorption. The increase in ionic

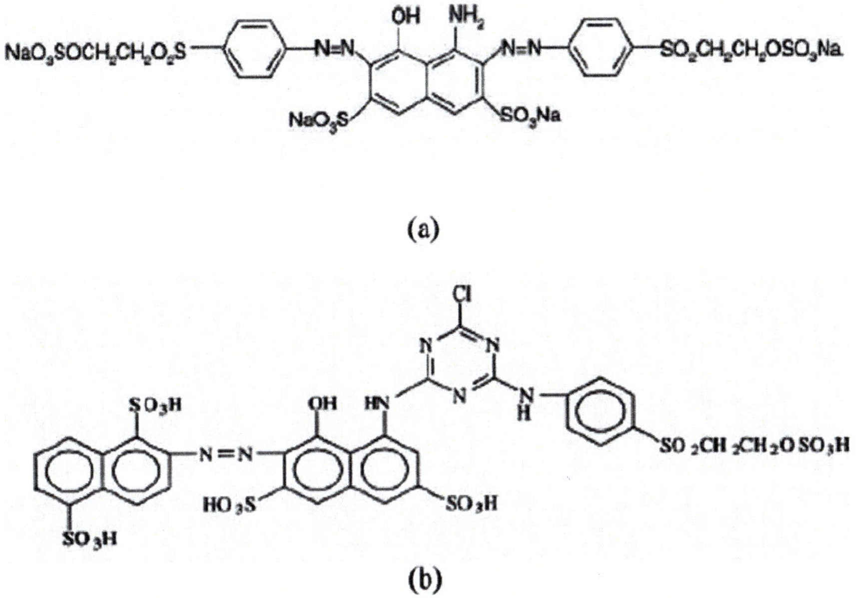

Fig. 30.4 Chemical structure of reactive dyes: **a** Reactive black 5, and **b** Reactive red 239 [66]

strength caused a decrease in adsorption of BR46 and an increase in adsorption of RY176. Adsorption of BR46 on natural zeolite is more favorable than RY176 on the CTAB and HDTMA modified zeolites. In 2009, the adsorption behavior of four anionic dyes as methyl orange (MO), Orange IV (OIV), Reactive brilliant red X-3B (X-3B), and Acid fuchsine (AF) on ammonium-functionalized MCM-41 (NH^{3+}-MCM-41) was investigated [68]. The intraparticle diffusion model was found to be the best in describing the adsorption kinetics for the four anionic dyes on NH^{3+}-MCM-41. The adsorption isotherm follows the Langmuir model. It was observed that the anion of weak acid inhibited the adsorption capacity significantly. Recently, the surfaces of NaY zeolite particles were modified by the alkylsilylation of *n*-octadecyltrichlorosilane (OTS) by Nur et al. [69]. Two kinds of modified NaY zeolites were prepared; one with its external surface partially and the other fully covered with alkylsilyl groups. Since the size of OTS is larger than the pore diameter of NaY, it is attached on the external surface, leaving the internal pore accessible to adsorbate molecules. As a result of alkylsilylation, the adsorption properties of these sorbents were improved. The adsorption properties of these materials were tested by their reaction in a mixture of Paraquat and Blue dye.

Natural zeolite and synthetic zeolite, MCM-22, were used for removal of a basic dye, methylene blue, from wastewater [70]. Fenton oxidation and high temperature combustion were also used for the regeneration of used materials. It is found that MCM-22 exhibits equilibrium adsorption at 1.7×10^{-4} mol/g, which is much higher than the adsorption of natural zeolite (5×10^{-5} mol/g) at initial dye concentration of 2.7×10^{-5} M and 30 °C. Higher solution pH results in higher adsorption capacity.

Physical regeneration under high temperature calcination will recover more than 88% adsorption capacity while chemical regeneration by Fenton oxidation will only restore 60% adsorption capacity. In a similar study [71], the adsorption of two basic dyes, Basic green 5 (BG5) and Basic violet 10 (BV10), onto MCM-41 was studied. It was found that the effect of interactions between large dyes (such as BV10) and MCM-41 on the pore structure stability of MCM-41 was insignificant. It was also found that the adsorption capacity for the two dyes was very distinct, though their molecular sizes were similar. Hydrothermally synthesized ZSM-5 zeolite was modified by manganese (Mn/ZSM-5) or lanthanum (La/ZSM-5) or mixture of both (Mn–La/ZSM-5) using impregnation technique [72]. These materials were tested for discoloration (adsorption) and mineralization (in the presence of UV irradiation) of Indigo Carmine (IC) dye. Mn–La/ZSM-5 showed high catalytic activity during IC adsorption (100% removal).

Wang et al. [73] investigated the removal of methylene blue (MB) by adsorption process on single-phase and high-crystalline zeolite A (FA-ZA) and X (FA-ZX). Both zeolite adsorbents, FA-ZA and FA-ZX, were synthesized from aluminosilicate gel, which is prepared from fly ash and subsequently subjected to a hydrothermal treatment at 100 °C. The adsorption capacity followed this: FA-ZX > FA-ZA. Another similar study [74] was also carried out to remove methylene blue using zeolites synthesized from fly ashes. Nearly 90 min of contact time were found to be sufficient for the adsorption of dye to reach equilibrium. The adsorption isotherm follows the Freundlich model and the adsorption kinetic follows the pseudo-second-order model. A novel zeolite was synthesized, characterized and employed for the adsorption of Methylene blue (cationic dye) and bisphenol-A in aqueous solution [75]. The pore properties of the synthesized zeolite have been determined using N_2 adsorption–desorption isotherms, indicating that it is a supermicroporous adsorbent with BET surface area of over 400 m^2/g (Table 30.1). It was found that the synthesized zeolite exhibited significantly higher adsorption capacity for methylene blue than that for bisphenol-A due to the difference in molecular properties.

In 2007, the removal of malachite green (MG) and Pb^{2+} ions onto natural zeolite was studied in a batch system [76]. In the single system, MG adsorption isotherm follows the Langmuir model while Pb adsorption follows the Freundlich isotherm. The adsorption capacities of MG and Pb at 30 °C, pH 6 are 5×10^{-5} and 10×10^{-5} mol/g, respectively. In the binary system, MG and Pb^{2+} exhibit competitive adsorption on the natural zeolite. The dynamic adsorption of malachite green and Pb^{2+} follows the first-order kinetics. Yan et al. [77] determined adsorption isotherms and kinetics of methylene blue (MB) onto mesoporous carbons. It was found that the adsorption capacities of MB at 20 °C onto the carbons prepared using acid- and alkaline-treated zeolite X as the template were 223 and 380 mg/g, respectively. Adsorption isotherm follows the Langmuir model and the adsorption kinetics of MB on mesoporous carbons also followed pseudo-second-order kinetic model. Zeolites, polymeric resins, ion exchangers and granulated ferric hydroxide (GEH) were investigated with different reactive dyes (M 600–1500 g/moL) [78]. Zeolites and microporous resins were not suitable due to extremely low sorption capacities. The macroporous resins without functional groups showed moderate maximum loadings (100–400 μmoL/g or 100–400 mg/g) but low affinity and were not suitable for large dyes (>1000 g/moL).

Karadag et al. [79] used cetyltrimethylammonium bromide (CTAB) modified zeolite for the removal of Reactive blue 4 (RB4) and Reactive yellow (RY) dyes. According to individual dye removal efficiencies, coagulation is the main treatment process of the combined process and RY has better performance than RB4. The combination of chemical coagulation followed by adsorption achieved greater than 99.5% color removal and 97.25% COD removal efficiencies for both dyes. In another study, type A zeolite, an insoluble crystalline sodium aluminosilicate, has been developed as a water-hardness controlling agent for laundry detergent products [80]. The isotherm of methylene blue on Na-A was Langmuirian over the concentration range studied, reaching constancy for solution concentrations above 2.0×10^{-6} M and indicating coverage in the plateau region of 2.8 × mol of MB/g of zeolite. Combining this coverage with the measured surface area gives a calculated area of zeolite per molecule of MB of 113 A^{o2}. This is similar to the calculated surface area for a flat orientation of MB, viz., 120 A^{o2} [81]. Recently, the sorption behaviour of dye remazol yellow using surfactant modified zeolitic rock and a carbonaceous material obtained by pyrolysis of sewage sludge and treated with HCl (1.0%) solution was determined [82]. Remazol yellow could be removed from the saturated modified zeolitic material by chemical and thermal treatment, but, the surfactant was removed together with the dye. However, thermal treatment was the best method found to remove the dye from the carbonaceous material.

Kuleyin and Aydin [83] also used surfactant-modified zeolite (SMZ) to remove Remazol Brillant blue R and Remazol yellow reactive dyes from aqueous solutions in the batch system. The Langmuir monolayer adsorption capacities of Remazol Brillant blue R and Remazol yellow were estimated as 13.9 and 38.31 mg/g, respectively (Table 30.1). Thermodynamic parameters such as $\Delta H°$, $\Delta S°$, and $\Delta G°$, at 20 °C, were found to be 5.2126 kJ/mol, 0.0273 kJ/mol K, and 22.7969 kJ/mol (Remazol Brillant blue R), and 29.9747 kJ/mol, 0.10875 kJ/mol K, and 21.8900 kJ/mol (Remazol yellow), respectively. Sayal et al. [84] studied the adsorption of two basic dyes (Amido black and Thionine) onto granular activated carbon as well as zeolite from fly ash and bentonite. The adsorption isotherm could be described by the Langmuir and Freundlich isotherm equations. The parameters in the adsorption isotherms are estimated from the experimental equilibrium data using non-linear regression analysis. Recently, the adsorption of reactive dyes, namely Reactive red 239 (RR-239) and Reactive blue 250 (RB-250) on modified zeolite (HMDA-Z) was investigated in batch system [85]. The most appropriate model was the Freundlich model. The kinetic studies indicated that the adsorption of reactive dyes followed the pseudo-second-order kinetic. Thermodynamic calculations showed that the adsorption of both investigated dyes was a spontaneous and endothermic process for RB-250 and an exothermic process for RR-239.

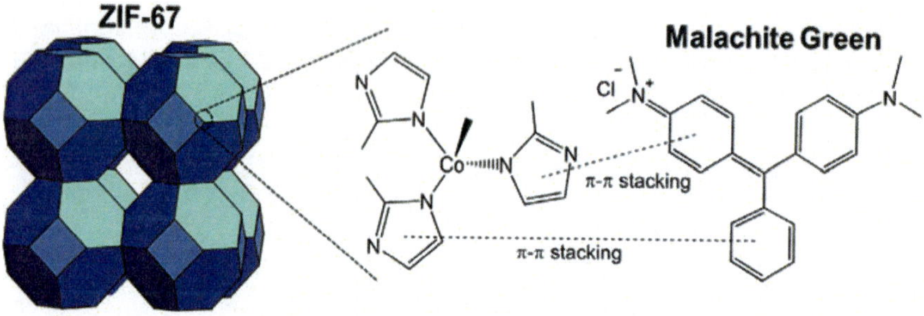

Fig. 30.5 Synthetic zeolite (ZIF-67) and cationic dye Malachite Green (MG) [88]

In 2013, kaolin and zeolite have been utilized as adsorbents for the removal of a cationic dye, methylene blue (MB), from aqueous solution within a batch process [86]. The results indicate that kaolin and zeolite could be interesting alternative materials with respect to more costly adsorbents used for dye removal. A novel N,N-dimethyl dehydroabietylamine oxide (DAAO) surfactant was synthesized via alkylation and oxidation of dehydroabietylamine [87]. The adsorptive removal of an anionic dye, Congo red (CR), from aqueous solution was then investigated. The Langmuir isotherm model fits well with the equilibrium adsorption data and adsorption capacity was found 69.94 mg/g. The results showed that the CR adsorption was best described by pseudo-second-order kinetics. In 2015, zeolitic imidazole frameworks (ZIFs) (Fig. 30.5), a new class of adsorbents, are proposed to adsorb malachite green (MG) in water [88]. The as-synthesized ZIF-67 was characterized and used to adsorb MG from water. The adsorption capacity of ZIF-67 for MG could be as high as 2430 mg/g at 20 °C, which could be improved at the higher temperatures. Such an ultra-high adsorption capacity of ZIF-67 was almost 10-times of those of conventional adsorbents, including activated carbons and biopolymers.

In 2016, chabazite, one of the common types of zeolite, was used to remove methylene blue (MB) dye from aqueous solutions [89]. The experimental data were well fitted to the Langmuir isotherm model. The adsorption kinetics of MB dye on chabazite could be described by a pseudo second-order model. Hor et al. [90] evaluated different facile and cost-effective physicochemical methods in enhancing the adsorption capacity of natural zeolite and subsequently demonstrated the removal of methylene blue (MB). Four different physicochemical methods, namely: acid treatment (AT), base treatment (BT), combined acid-thermal (ATT) and base-thermal treatments (BTT) were investigated. Batch adsorption studies were carried out using the modified zeolite adsorbents, in order to evaluate the highest removal efficiency of MB under varying adsorption conditions. Results showed that both the physicochemical methods of AT (using 0.4 M HCl) and BT (using 4.0 M NaOH) are able to enhance the adsorption capacity of natural zeolite from 41% to 98.8% and 52.2%, respectively. Nassar et al. [91] reported on the preparation of mordenite zeolite nanostructures using a low-cost hydrothermal treatment of silica gel, aluminumnitrate and sodium hydroxide. The influence of organic templates such as ethylene glycol, glycerol and polyethylene glycol 200 (PEG 200) on the zeolite products was studied. The PEG 200 organic template generated a mordenite product with 57.51 nm crystallite size and 28.26 m^2/g BET surface area (Table 30.1). The mordenite product also showed good adsorption properties toward the removal of methylene blue (MB) dye from wastewater.

In 2017, the adsorption of Basic blue 41 textile dye from aqueous solutions on a zeolite tuff (Romanian) was studied in batch mode [92]. The maximum adsorption capacity for the investigated basic textile dye (~40% pure dye) was 192.31 mg/g at 50 °C (at pH 5) for the average diameter particles tuff of 0.085 mm. The positive enthalpy change for the adsorption process confirms the endothermic nature ofadsorption and a free energy change confirms the spontaneity of the process. Aljerf [93] researched the importance of the adsorption properties of a modified clinoptilolite zeolite (CL) for the removal of BCP dye and some heavy metals as total chromium (tCr) in the ammoniac phase. The batch removal showed the optimal conditions for BCP adsorption: pH (6.5), time (t) (60 min), temperature (T) (303.15 K), sorbent dosage (m) (60.4 mg) and initial concentration (C_o) (11.7 mg/L). The maximum adsorption capacities of BCP and tCr onto the CL-SW were 175.5 mg/g and 37 mg/g, respectively. In 2018, a highly efficient and reusable zeolite was prepared using the biopolymer chitin as mesoporosity agent and applied to adsorb three cationic dyes [94]. The zeolite was synthetized by the hydrothermal method and was characterized. The potential of zeolite to adsorb crystal violet (CV), methylene blue (MB) and basic fuchsin (BF) was evaluated conducting kinetic, equilibrium, thermodynamic and regeneration studies. The treatment of simulated effluents was also performed (Table 30.2). The modification with chitin biopolymer provided attractive adsorptive characteristics for the zeolite structure. Maximum adsorption capacities predicted by the isotherm models were 1217.3, 548.2 and 237.5 mg/g for CV, MB and BF, respectively. The modified zeolite can be used for fifteen adsorption cycles maintaining the same adsorption capacity. Also, the adsorbent was able to treat a simulated textile effluent providing 85% of color removal.

Adsorption of the hazardous cationic dye malachite green (MG) on the synthesized zeolite was investigated [95]. The maximum equilibrium adsorption capacity was found to be 108.26 mg/g at 30 °C. The MG adsorption data fit better to the Freundlich model (R2 > 0.99) than the Langmuir model. Another study [96] focused

Table 30.2 Chemical composition of the simulated textile effluent [94]

Cationic dyes	λmax (nm)	Concentration (mg/L)
Basic Fuchsin	540	15
Methylene Blue	664	15
Crystal Violet	590	15
Inorganic compounds		
NaCl	–	100
$NaHCO_3$	–	100

on the adsorption properties of zinc oxide (ZnO) in the form of nanoflakes on the surface of zeolite for the removal of Congo red (CR) dye. The adsorption data follows pseudo second order kinetics and Langmuir isotherm model with a maximum adsorption capacity of 161.3 mg/g. The synthesized composite is easily regenerable and reusable and is capable of removing CR dye from the polluted water up to 90% (for 5 regeneration cycles). The adsorption properties of three types of natural zeolite (heulandite, clinoptilolite, and phillipsite) was also investigated [97]. The selected zeolite minerals were applied as adsorbents in the decolorization of safranin-T dye from water. The estimated uptake capacities for safranin revealed the higher capacity of heulandite than clinoptilolite and Phillipsite (heulandite > clinoptilolite > phillipsite). Monolayer adsorption form was suggested for the uptake by heulandite and clinoptilolite (Langmuir) and multilayer form was predicted for phillipsite (Freundlich).

In 2018, Sodalite zeolite nanoparticle (SZN) was synthesized and its surface was modified by different amounts of (3-aminopropyl) triethoxy silane (APTES) and denoted as SZN(0.5), SZN(1) and SZN (1.5) by Mahmoodi and Saffar-Dastgerdi [98]. The surface modified zeolite nanoparticles were used for adsorbing Direct red 23 (DR23) and Direct red 80 (DR80) from wastewater. The investigated data for the effect of APTES amount on the zeolite n terms of DR23 and DR80 removal was shown in Fig. 30.6. Adsorption capacity of SZN(0.5) was 2415 and 4842 mg/g for DR80 and DR23, respectively. The Gibbs free energy at 298, 313, 323 and 333 K were −12.53, −14.68, −16.12 and −17.55 kJ/mol for DR23 and −9.27, −11.73, −13.37 and −15.01 kJ/mol for DR80. In 2020, another study [99] reported the first use of Fe-ZSM-5 as a potential adsorbent for textile effluent purification. The Fe-ZSM-5 zeolite was synthesized through a hydrothermal method using tetrapropylammonium bromide as an organic structure-directing template for application as a binder for Basic fuchsin (BF) dye. N_2 adsorption–desorption measurement showed the microporous structure (pore size around 1.93 nm) of Fe-ZSM-5 with a

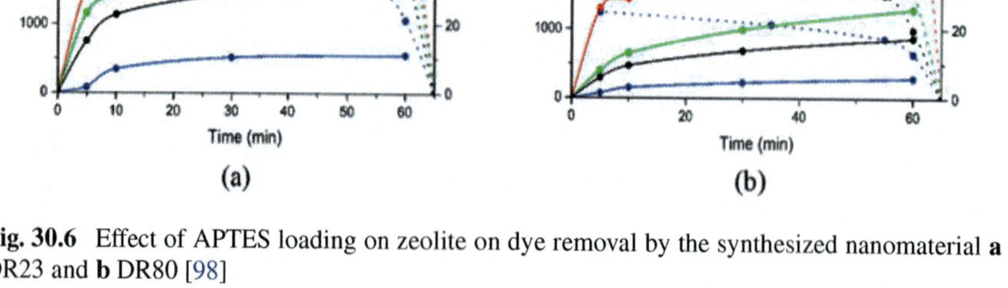

Fig. 30.6 Effect of APTES loading on zeolite on dye removal by the synthesized nanomaterial **a** DR23 and **b** DR80 [98]

BET surface area of 399 m^2/g. The experimental adsorption data were in good agreement with pseudo-second-order kinetics and Langmuir isotherm with a maximum adsorbed amount of 251.87 mg/g. BF dye adsorption onto Fe-ZSM-5 was a spontaneous and endothermic phenomenon. The highest BF dye removal performance was found to be 99.6% under optimum conditions of pH 5, Fe-ZSM-5 dose 0.75 g/L, initial BF concentration 10 mg/L for 120 min at 20 °C.

30.3.2 *Dye/Color Removal from Textile Wastewaters in the Fixed Bed Columns*

In spite of the promising results demonstrated in several studies, the real applicability of these natural materials to purify dye wastewaters is still not clear. Another problem of zeolites is their low permeability and this requires an artificial support when used in column operations. The sorption mechanism on zeolite particles is complex because of their porous structure, inner and outer charged surfaces, mineralogical heterogeneity and other imperfections on the surface [47]. Adsorption is advantageous to other techniques in respect of initial cost, flexibility and simplicity of design, ease of operation and insensitivity to toxic pollutants. In addition, adsorption is one of the most important unit processes in a wastewater treatment plant and the design of the adsorption column usually requires data generated in pilot-scale studies [31]. Several investigators reported studies in the fixed-bed reactors using zeolite as a cost-effective adsorbent for dye/color removal (Table 30.3). In the recent studies of the authors, the modification of zeolite (clinoptilolite) surface with a cationic surfactant to improve the removal efficiency of reactive azo dyes in the fixed-bed reactor was investigated (57,104–107).

Adsorption tests were conducted to find the uptake of three types of reactive dyes (Reactive black 5, red 239 and yellow 176) [54]. Each run consisted of modifying zeolite with HTAB in the column followed by removal of color from the modified zeolite bed. The breakthrough curves for modification process were obtained as the normalized effluent concentration (C/C_o) versus time or bed volumes (BV). Optimization studies show that 3 g/L of HTAB dosage at a flowrate of 0.025 L/min showed the best performance. Similarly, Taguchi method was applied as an experimental design to determine optimum conditions for color removal from textile dyebath house effluents in a zeolite fixed bed reactor [100]. The optimum conditions were found as: HTAB concentration of 1 g/L, HTAB feeding flowrate of 0.015 L/min, textile wastewater flowrate of 0.025 L/min and bed height of 50 cm. The treated wastewater volume reached a maximum at 217 BV (Table 30.3). Recently, the ability of surfactant-modified zeolite (SMZ) to remove color from real textile wastewater was investigated [101]. Reactive dyes and auxiliary chemicals used in dyeing process that produce textile wastewater are given in Table 30.4. Tests were performed in a fixed-bed column reactor and natural zeolite was modified with HTAB. Wastewater

Table 30.3 Dye/Color removal efficiency and other parameters for zeolites in fixed-bed systems

Adsorbent	Dye/textile wastewater	Influent concentration	Flowrate	Dye/color removal	Breakthrough time/bed volumes	Source
Modified zeolite (Turkish)	Reactive red 239	50 mg/L	25–75 mL/min	90%	70 BV	[54]
Modified zeolite (Turkish)	Reactive yellow 176	50 mg/L	25–75 mL/min	90%	120 BV	[54]
Modified zeolite (Turkish)	Reactive black 5	50 mg/L	25–75 mL/min	90%	137 BV	[54]
Modified zeolite (Turkish)	Real textile wastewater	Abs. wavelength 448 nm	0.025 L/min	90%	217 BV	[100]
Modified zeolite (Turkish)	Real textile wastewater	Diluted ratios of 25%	0.025 L/min	90%	107 BV	[101]
Modified zeolite (Turkish)	Real textile wastewater	Diluted ratios of 50%	0.050 L/min	90%	60 BV	[101]
Modified zeolite (Turkish)	Real textile wastewater	Diluted ratios of 75%	0.075 L/min	90%	39 BV	[101]
Modified zeolite (Turkish)	Reactive yellow 176	50 mg/L	50 mL/min	90%	480 min	[102]
Modified zeolite (Turkish)	Simulated textile wastewater	50 mg/L	50 mL/min	90%	360 min	[102]
Modified zeolite (Turkish)	Real textile wastewater	50 mg/L	50 mL/min	90%	180 min	[102]
Modified zeolite (Turkish)	Reactive black 5	50 mg/L	70 mL/min	90%	72 BV	[103]
Modified zeolite (Turkish)	Reactive yellow 176	50 mg/L	70 mL/min	90%	60 BV	[103]
Modified zeolite (Turkish)	Reactive red 239,	50 mg/L	70 mL/min	90%	48 BV	[103]
Zeolite (Chinese)	Basic blue 9 (Methylene blue)	30–72 mg/L	2.2–8.2 mL/min	80%	100 min (H_{bed} = 15 cm)	[104]
Zeolite (Macedonian)	Basic dye Maxilon Goldgelb GL EC	50–200 mg/L	2–40 cm^3/min	40%	15–100 min	[105]

(continued)

Table 30.3 (continued)

Adsorbent	Dye/textile wastewater	Influent concentration	Flowrate	Dye/color removal	Breakthrough time/bed volumes	Source
Zeolite (Macedonian)	Basic dye Schwarz FBL01	50–200 mg/L	25–50 cm^3/min	40%	15–100 min	[105]
Zeolite	Basic dye, Malachite green	50 mg/L	9 mL/min	–	–	[106]
Zeolite	RB5	22–32 mg/m^3.d	–	20–57%	11–17 h	[107]
Zeolite	RB19	22–32 mg/m^3.d	–	46–80%	11–17 h	[107]
Zeolite	RR22	22–32 mg/m^3.d	–	16–57%	11–17 h	[107]
Zeolite-GAC arrange	True color	680–750 PtCo	90, 270 and 510 mL/min	58.4%	7–123 s	[108]
Cu/zeolite Y	Quinoline	3.87 mM	0.97–3.89 cm^3/min	98%	4 h	[109]
Synthetic zeolite(CMC/ZSM-5/ZIF-8)	Basic blue 9 (Methylene blue)	4–8 mg/L	0.74 ml/min	–	0–350 min	[110]

Table 30.4 Reactive dyes and auxiliary chemicals used in dyeing process that produce textile wastewater [101]

Chemicals	Amount (kg)
Dyes	
Everzol orange 3R	11
Everzol red F2B	15
Everzol black GR	166
Everzol black HC	77
Remazol gelb 3R5	8
Remazol rot 3B5	8
Remazol black N150	82
Procion yellow HEXL	3
Procion yellow HEGG	3
Procion red HEGXL	1
Procion crimson HEXL	1
Procion blue HERD	2
Procion navy HEXL	3
Evercion yellow HE4R	15
Evercion red HE7B	22
Evercion blau HEGN	6
Evercion navy HER	26
Evercion navy ESL	32
Auxiliary chemicals	
Na2SO4	650
NaCl	850
Na2CO3	1500
NaOH	250
CH3·COOH	350
H2O2	220
Na2·S2O4	45
Ion holder	90
Detergent	80
Wetting	125

was diluted in the ratios of 25%-75% in order to assess the influence of wastewater strength. The theoretical service times evaluated from bed depth service time (BDST) approach for different column variables. The regeneration of the SMZ was also evaluated using a solution consisting of 30 g/L NaCl and 1.5 g/L NaOH at pH 12 and temperature 30 °C. The point of 0.1 C_{eff}/C_{inf} for 1st, 2nd and 3rd regeneration cycles were obtained at 360, 330 and 300 min (or 51, 46.7 and 42.5 BV), respectively.

Faki et al. [102] also studied adsorption of Reactive yellow 176 onto HTAB modified zeolite in a fixed-bed column system. Adsorption of Reactive yellow 176 dye by zeolite modified at different HTAB concentrations against service time are seen in Fig. 30.7. The bed depth service time (BDST) model proved to be effective in the comparison of column variables. The minimum bed height, the adsorption rate constant, and the adsorption capacity of the HTAB modified zeolite for yellow dye removal were found to be 12.02 cm, 6.432 × 10^{-3} L/(mg h), and 12.05 g/L, respectively. The breakthrough times (corresponding to $C_b/C_o = 0.1$) were found to be 8, 6, and 3 h for the aqueous yellow dye solution, the simulated textile wastewater, and the real textile wastewater, respectively. The zeolite column was regenerated at different temperatures and the higher temperature (60 °C) with a desorption efficiency of 90.6% was found to be more effective than that at lower temperature (30 °C) with a desorption efficiency of 23% by means of the desorption of yellow dye. In another study [103], adsorption of reactive dyes (Everzol black B, Everzol red 3BS, Everzol yellow 3RS H/C) on surfactant-modified clinoptilolite (zeolite) was investigated by batch and column systems. For black, yellow and red dyes, the breakthrough curves of zeolite exhibited as $C/C_o = 0$ value from the startup to 72, 60 and 48 BV, respectively (Table 30.3).

Han et al. [104] evaluated comparison analysis of linear least square method and nonlinear least square method for estimating the kinetic parameters using the experimental column data of methylene blue (MB) adsorption onto zeolite at different flow rates and initial concentration. The data were fitted to Thomas model equations

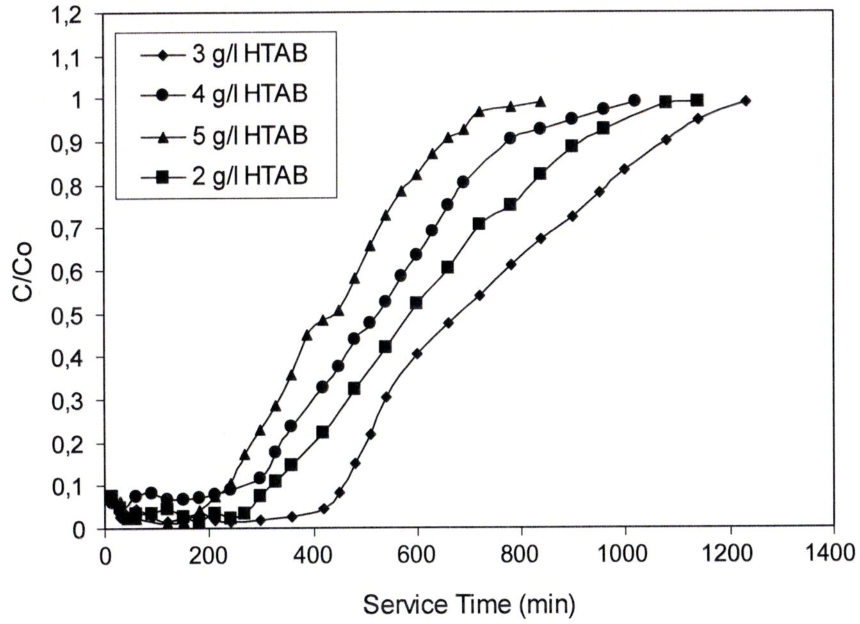

Fig. 30.7 Adsorption of Reactive yellow 176 dye by zeolite modified at different HTAB concentrations against service time (influent dye conc.) 50 mg/L, dye flow rate) 0.050 L/min, bed height) 25 cm) [102]

using linear and nonlinear regressive analysis, respectively. Markovska et al. [105] researched adsorption of basic dyes (Maxilon Goldgelb GL EC, Maxilon Shwarz FBL-01) from aqueous solution onto granular activated carbon and natural zeolite in a fixed bed system. The experimental results showed that both adsorbents removed the dyes effectively, and the effect of changing the column operating variables, such as flow rate and initial concentration, can be predicted by the BDST model. Wu et al. [106] used natural zeolite as an adsorbent for adsorption of malachite green (MG) from an aqueous solution in fixed-bed reactor. Adsorption reached saturation faster with increasing the flow rate and influent MG concentration. When the flow rate was 9 mL/min and the influent concentration of MG was 50 mg/L, the maximum adsorption quantity reached 23.55 mg/g according to the Thomas model.

Ojstrsek and Fakin [107] proposed to examine the potential of natural zeolite for colour and TOC reduction in a fixed-bed system. Assessment of the biofilter's performance was verified by monitoring absorbance and total organic carbon in initial and treated wastewaters. The results showed that when increasing the hydraulic retention time from 11 up to 17 h, the decolouration increased from 20 to 57% (for RB5), from 46 to 80% (for RB19) and from 16 to 57% (for RR22) while TOC removal efficiency rose from 30 to 75%. The removal of COD, ammonia and color in dye wastewater using granular activated carbon (GAC) and zeolite in the column studies was evaluated by Syafalni et al. [108]. The maximum removal efficiency was found at the surface loading rate of 2.84 mL/cm^2.min and bed height of 10 cm. The best removal of the contaminants among the all adsorption treatment was found using GAC (bottom layer) and zeolite (upper layer) in 6.35 cm diameter column with 59.46% removal of COD, 60.82% removal of ammonia and 58.4% removal of color. The adsorption data fitted well for both Langmuir and Freundlich isotherms. In another study [109] several Cu/zeolite Y samples were synthesized by different routes and examined for the catalytic wet peroxide oxidation of quinoline aqueous solution in continuous fixed-bed reactor. Aqueous ion-exchange (CuYAIE) and wet-impregnation (CuYIMP) methods promoted isolated $Cu^{1+/2+}$ species; however, large crystallites of CuO were present on the external surface of precipitation-impregnation (CuYPI) catalyst. CuYPI showed hierarchical porosity and increase of surface area from 567 to 909 m^2/g.

Zeolitic imidazolate frameworks are a class of metal–organic frameworks that are topologically isomorphic with zeolites. Bionanocomposites of carboxymethyl cellulose-based ZSM-5/zeolitic imidazolate framework (CMC/ZSM-5/ZIF-8) hollow beads with different compositions were synthesized and employed as an adsorbent for methylene blue removal from aqueous solution in batch and continuous fixed bed systems [110]. The equilibrium batch adsorption capacities for CMC, CMC/ZIF-8, CMC/ZSM-5, and CMC/ZSM-5/ZIF-8 adsorbents were 12.01, 13.06, 11.53, and 10.49 mg/g, respectively. All four adsorbents are consistent with all models but the pseudo-first-order model showed more consistency. The equilibrium continuous adsorption capacities for CMC, CMC/ZIF-8, CMC/ZSM-5, and CMC/ZSM-5/ZIF-8 adsorbents were 10.56, 11.87, 9.29, and 8.15 mg/g, respectively. The adsorbents showed more consistency with models of Thomas, Yoon-Nelson and Modified Dose Response. Besides, the generation process was successfully assessed in five steps. In another research [111], the efficiency of an integrated ultra-violet (UV) photocatalysis and anaerobic digestion (AD) system in the degradation of methylene blue (MB) dye was studied in up-flow fixed bed reactors. Zeolite was applied as support material for micro-organism and photocatalyst in the bioreactor and photoreactor, respectively. The integration of the two processes led to high colour reduction efficiencies of above 75%.

30.3.3 Dye Wastewater Treatment Systems Combined with Zeolites

Coloured wastewaters from dye production and application industries cause a serious threat for the environment due to high resistance to decomposition present in their chemical nature. Many conventional wastewater treatment methods are often inappropriate for such pollutants due to their inability to effectively remove and degrade persistent synthetic dyes. Wastewater treatment methods based on the advanced oxidation processes (AOPs), especially Fenton type processes, seem to offer promise due to their high efficiency for dye degradation, relatively low cost and the ease of process operation and maintenance. Through the combination of AOPs with zeolite adsorption, the overall process efficiency could be enhanced [112–114]. Catalytic wet hydrogen peroxide oxidation of an anionic dye has been explored. Copper (II) complex of NN'–ethylene bis(salicylidene-aminato) (salenH_2) has been encapsulated in super cages of zeolite-Y by flexible ligand method [112]. The results indicate that complete removal of color has been obtained after a period of less than 1 h at 60 °C, 0.175 M H_2O_2 and 0.3 g/L catalyst. More than 95% dye removal has been achieved using this catalyst for commercial effluent. Neamtu et al. [113] reported the degradation of a reactive azo dye, Procion Marine H-EXL, by catalytic wet hydrogen peroxide oxidation (CWHPO). The catalyst was prepared by ion-exchange, starting from a commercially available ultrastable Y zeolite. The results indicate that after only 10 min at 50 °C, 20 mmol/L H_2O_2 and 1 $_g$/L $FeY_{11.5}$ color removal was as high as 97% at pH = 3 and 53% at pH = 5. In another research [114], the addition of solid particles, namely synthetic zeolites NH_4ZSM5 and HY, was investigated in order to enhance the efficiency of AOPs for dye wastewater treatment. Degradation studies were carried out with Reactive blue 49 and Reactive blue 137. It was found that AOPs in combination with zeolites were shown to be very successful in colour removal (>98%) (Table 30.5).

Zeolites can be applied for the treatment of coloured wastewater through biological methods. Chang et al. [115] developed a biological aerated filter (BAF) to treat textile wastewater and tested its performance using natural zeolite and sand as media. Higher nitrogen removal in the biofilter with natural zeolite was attributable to its ion exchange capacity with NH_4^+. The results indicate that natural zeolite is a superior media for textile wastewater treatment while reductions for COD (86–92%), BOD (99%) and colour (77–79%) were not significantly dependent of hydraulic load.

Table 30.5 Dye/Color removal efficiency for wastewater treatment systems combined with zeolites

Adsorbent	Dye/textile wastewater	Surface area	Initial concen-tration	Dye/color removal	Contact time	pH	Source
AOP/Cu-Y zeolite	Acid dye	–	50 mg/L	–	3 h	6	[112]
CWHPO/Fe-Y zeolite	Reactive dye Procion Marine H_EXL	–	–	96%	30 min	5	[113]
AOPs/NH4ZSM5 and HY zeolites	Reactive Blue 49	–	–	>98%	2 h	3	[114]
BAF/zeolite	Real textile wastewater	–	–	78.9%	–	–	[115]
SAPO-Zeolite/membrane	Synthetic textile wastewater	–	0.01–0.09 g/L	20–100%	–	–	[116]
AOPs/Fe-ZSM5 zeolite	Acid Blue 74	187 m^2/g	8.56×10^{-5} mol/L	99%	120 min	5	[117]
CWPO/ Fe-Y zeolite	Azo dye (Congo red)	423 m^2/g		100%	4 h	7	[118]
CWHPO/Fe-Y zeolite	Reactive yellow 84	660–780 m^2/g	100 mg/L	78.87–99.96%	120 min	5	[119]
AOPs/FeY and ZMS-5 zeolites	Reactive Brilliant Blue KN-R	–	250–500 mg/L	95%	90 min	2.5	[120]
TiO_2/NaY/HY zeolite/ Photocat	Basic dye (Methylene blue)	661–698 m^2/g	50 ppm	57–82%	4 h	4.9–6.7	[121]
TiO_2/NaY zeo./silver/ Photocat	Basic dye (Malachite green)	620–668 m^2/g	50 ppm	100%	4 h	4.9–6.7	[122]

(continued)

Table 30.5 (continued)

Adsorbent	Dye/textile wastewater	Surface area	Initial concen-tration	Dye/color removal	Contact time	pH	Source
Pt-TiO_2/zeolite	Azo dye (Methyl orange)	118.7 m^2/g	–	86.2%	30 min	6.2	[123]
AOPs/Fe(III)/NaA zeolite	Direct blue-1	410 m^2/g	100 mg/L	80%	1 h	10.8	[124]
AOPs/Fe zeolite	Reactive orange 16	–	50 mg/L	90%	60 min	5.2	[125]
AOPs/TiO_2/NaY zeolite	Basic violet 10	900 m^2/g	50 mg/L	67%	4 h	10	[126]
CWPO /Mn-Y zeolite	Acid brown dye	319 m^2/g	10–100 mg/L	90%	20 min	2	[127]
TiO_2/Na-HZSM-5 nano-comp. Photocat	Azo dye (Methyl orange)	363.5 m^2/g		12–100%	4 h		[128]
TiO_2/NaA zeolite/Photocat	Basic dye (Methylene blue)	–	10 mg/L	92%	8 h		[129]
AOPs/Zeolite	Reactive Red 45	–	20 mg/L	100%	2 h		[130]
TiO_2/Clinop-zeo./Photocat	Azo dye (Disperse yellow 23)	50 m^2/g	20–80 ppm	30–97.4%	5 h	9–11	[131]
TiO_2/Clinop–zeo./Photocat	azo dye Acid Red 114	–	–	–	–	–	[132]
O_3/zeolite	Methylene blue (MB)	205 m^2/g	30 mg/L	25%	24 h	2–8	[133]
Macrocomposite (MC) (Zeolite/AC.4:1)	azo dye Acid Orange 7 (AO7)	84,73 m^2/g	100–400 mg/L	99%	30 h	2–7	[134]
Magnetic NaY zeolite	Methylene blue	–	10 mg/L	99%	45 min	10.3	[135]

(continued)

Table 30.5 (continued)

Adsorbent	Dye/textile wastewater	Surface area	Initial concen-tration	Dye/color removal	Contact time	pH	Source
Magnetic NaY zeolite	Crystal violet	–	10 mg/L	98.6%	45 min	10.3	[135]
Magnetic NaY zeolite	Fuchsine	–	10 mg/L	98.4%	45 min	10.3	[135]
Magnetic zeolite (MZ)	Crystal violet (CV)	–	100–1000 mg/L	95%	10 min	2–10	[136]
TiO2–zeolite nanocomposites	Reactive Black 5	324 m^2/g	1–10 ppm	–	95–120 min	5	[137]
Synthetic zeolite (Z-AC/C) composite	Methylene blue and Acid blue 29	–	50–400 mg/L	–	1800 min	3–13	[138]
PVA/CMC/ZSM-5 zeolite membranes	Methylene blue (MB)	–	10 ppm	97%	10 h	–	[139]
Nanocrystalline zeolite X (nZX)	Crystal violet (CV)	651.34 m^2/g	100 mg/L	99.52%	60 min	9	[140]
Nanozeolite X (nFAZX)	Methylene blue (MB)	–	12.50–62.50 mg/L	99.30%	2.50–12.50 min	3–11	[141]
Nanoporous Zeolite-X	RB 19 and AB 225	–	100–1000 mg/L	100%	105 min	3–7	[142]
PVA/STH/ZSM-5 Zeolite Membranes	Eriochrome black T (EBT)	–	10–50 mg/L	–	220 min	2–11	[143]
Zeolite/CeO_2 nanocomposite	Methylene blue (MB)	–	2–10 mg/L	93.9%	180 min	9	[145]
Zeolite (ZSM-12) based AS-Sorbents	Procion Blue MX-7RX (Reactive blue 161)	61.58 m^2/g	7.3–27.8 mg/L	–	1400 min	4–9	[146]

(continued)

Table 30.5 (continued)

Adsorbent	Dye/textile wastewater	Surface area	Initial concen-tration	Dye/color removal	Contact time	pH	Source
Hydrogel composite (ZHC) of GK and zeolite-Y	Cationic dye, Brilliant green (BG)	–	100–800 mg/L	98.8%	30 min	2–11	[147]
zwitterion composite (CHS-ECH/ZL)	Methylene blue and Reactive red 120	–	30–400 mg/L	–	0–600 min	4–10	[148]
PA-6/Modified zeolite nanofibrous composite	Acid Blue 74	–	20–30 mg/L	91.88%	20–100 min	3.5–10.5	[149]

AOPs = Advanced oxidation processes; CWHPO = catalytic wet hydrogen peroxide oxidation; CWPO = catalytic wet peroxide oxidation; BAF = Biological aerated filter

Recently, Ciobanu et al. [116] also used the combination of zeolite and membrane for wastewater treatment. The SAPO-5 zeolite-filled polyurethane membranes with perselectivity to the gases or liquids were prepared, characterized and utilized for the ultra- and microfiltration (UF + MF) of synthetic wastewater containing colophony. The study demonstrates the potential of the unfilled and zeolite-filled PU membranes for colophony removal from solutions with below 1 g/L concentrations. In 2008, decolourization and mineralization of an indigoid dye, Acid blue 74, was conducted using Fe-ZSM5 zeolite as a catalyst in the presence of UV-C light and H_2O_2 [117]. While optimal conditions were pH 5, 21.4 mmol/L of H_2O_2 and 0.5 g/L of the catalyst, 57% TOC of a solution containing 8.56×10^{-5} mol/L Acid blue 74 could be removed after 120 min in a 2 L batch photo-reactor.

In another study [118], the degradation of azo dye (Congo red) by catalytic wet peroxide oxidation was investigated using Fe exchanged commercial Y zeolite as a catalyst. The percent removals of dye, color and COD at optimum pH = 7, 90 °C using 0.6 mL H_2O_2 /350 mL solution and 1 g/L catalyst was 97% (in 4 h), 100% (in 45 min) and 58% (in 4 h), respectively. Dye removal in 45 min and at above conditions was 82%, whereas the color removal was 100% (Table 30.5). The catalytic effectiveness of three iron exchanged ultra-stable Y zeolites (USY) for the degradation of a reactive azo dye, Reactive yellow 84 (RY84) using hydrogen peroxide was evaluated [119]. The results show that it is possible to remove 96.90% of color, 70.70% of COD and 34.52% of TOC using Fe-Y11.5 catalyst after the 60 min of oxidation at the above-mentioned conditions. Using Fe-Y5 catalyst, 99.93% of color removal, more than 74.14% of COD and 64.21% of TOC removal were obtained. Chen et al. [120] examined the decolorization of an anthraquinone dye, Reactive brilliant Blue KN-R by hydrogen peroxide using Fe-containing Y and ZSM-5 zeolites as heterogeneous catalysts. Fe-containing Y and ZSM-5 zeolites generally exhibited similar or better catalytic efficiency compared with homogeneous Fenton reagent, with Fe-containing ZSM-5 being more efficient. Under the conditions of initial pH 2.5, 30.0 mmol/L H_2O_2 and 4.0 g/L catalyst, 250 mg/L KN-R could be decolorized to a degree over 90% within 20 min.

Zeolites have demonstrated their potential as unique and versatile host materials for variety of chemical transformations. Semiconductor-loaded zeolite and mesoporous materials such as natural zeolite, ZSM5, NaY zeolite, MCM-41 etc., have drawn attention as potential photocatalysts due to their unique pore structure and adsorption properties [120–123]. Titanium dioxide has proved to be a potential photocatalytic semiconductor as it allows for the complete degradation of pollutants under ultraviolet irradiation. The advantages using zeolite support for titania photocatalysis include formation of ultrafine titania particles during sol–gel deposition, increased adsorption in the pores and surface acidity which enhances electron-abstraction. A method of supporting TiO_2 on zeolite, without losing the photosensitization of TiO_2 and the adsorption properties of zeolite, is an important aspect while preparing zeolite-based photocatalysts. Tayade et al. [121] prepared TiO_2-coated zeolite photocatalysts by dispersing zeolite powders in dilute titanium tetraisopropoxide solution. The photocatalytic activity of TiO_2-coated NaY and HY zeolite was investigated by degradation of aqueous solution of methylene blue dye. The highest photocatalytic was obtained with 1% TiO_2-coated zeolite catalysts. The photocatalytic activity of TiO_2-coated catalyst was found to be higher than that of bare TiO_2 at a low amount of TiO_2 coating on the surface. Similarly, zeolite-based photocatalyst have been prepared using TiO_2-coated NaY zeolite by post-synthesis modification with silver metal ion exchange [122]. The photocatalytic activity of the catalysts was studied by degradation of aqueous nitrobenzene, acetophenone, methylene blue, and malachite green in the presence of UV light and the highest photocatalytic activity was obtained for silver metal ion exchanged photocatalysts coated with TiO_2 (2%, w/w). Huang et al. [123] also prepared Pt modified TiO_2 loaded on natural zeolites (Pt-TiO_2/zeolites) by solegel technique and photoreductive deposition method. Their photocatalytic activities were examined by the photocatalytic decolorization of methyl orange solution under UV light irradiation. The results show that Pt doping induced the enhancement of photocatalytic decolorization and optimal Pt doping is about 1.5 wt% with 86.2% of decolorization rate under 30 min irradiation time. The repeatability of photocatalytic activity was also tested and the decolorization rate was 81.9% of initial decolorization rate after five cycles (Table 30.5).

In 2009, the complex of Fe^{III}2[2-(hydroxy-benzeylidine)-amino] benzoic acid has been encapsulated in pores of NaA zeolite by two different synthesis procedures, namely a zeolite synthesis method in the presence of preformed iron complex ($Fe^{III}L/A_{en}$) and a solid–solid-type reaction ($Fe^{III}L/A_{ss}$) [124]. The resulting host/guest compound including H_2O_2/UV was active system for the oxidative color removal of Direct blue-1 in aqueous heterogeneous suspensions with very low catalyst/substrate ratio. Recently, the heterogeneous photo-Fenton-type oxidation of reactive azo dye solutions was also studied in a quartz batch reactor using artificial UVA as a light source [125]. Fe-exchanged zeolite has been used as a heterogeneous catalyst in the process. The optimal operation parameters were found as: 35 °C, pH as solution pH 5.2, 15 mmol H_2O_2 dosage, 1 g/L catalyst loading. Mineralization and comparison with homogenous photo-Fenton process were evaluated by analyzing color removal and total organic carbon (TOC) values. Wang et al. [126] evaluated the physical and chemical states of zeolite-supported TiO_2. The photocatalytic reaction followed first-order kinetics for all catalysts; optimum photodegradation efficiency was found to result from the use of a high TiO_2 concentration (20% TiO_2), a calcination temperature of 600 °C, alkaline pH (9–10) and a catalyst dosage of 5333 ppm. Aravindhan et al. [127] explored catalytic wet hydrogen peroxide oxidation of acid dye. Manganese (III) complex of N,N'-ethylene bis(salicylidene-aminato) (salenH2) has been encapsulated in super cages of zeolite-Y by flexible ligand method. The results indicate that after 20 min at 30 °C, 0.175 MH_2O_2 and 3 g/L catalyst, about 90% dye removal was obtained.

Guo et al. [128] prepared TiO_2 /Na-HZSM-5 nano-composite photocatalysts by dispersing TiO_2 onto the external surface of Na^+-modified nano-ZSM-5 zeolite using a sol–gel process and investigated the adsorption and photodegradation of methyle orange (MO) in aqueous solution. Results showed that the modification of support by Na does not affect the loading, dispersion and structure of loaded TiO_2, but reduces the acidity of the nano-zeolitic support by preferentially eliminating stronger acid sites. The MO photodegradation activity of TiO_2/Na-HZSM-5 nano-composites is better than that of bare TiO_2. Petkowicz et al. [129] estimated the degradation of methylene blue dye for a series of in situ generated titania resulting from the impregnation of $TiCl_4$ into a series of NaA zeolites. The highest catalyst activity was reached with samples having 10 wt.% Ti. Catalyst activity was shown to be comparable to that exhibited by commercial P-25 after 1 h of UV light exposition and experiments carried out in natural sunlight reached 92% decomposition of methylene blue. Koprivanac et al. [130] also used Reactive red 45 (RR45) in the advanced oxidation processes (AOPs) for the treatment of model wastewater containing organic dye. Mineralization rates of 40% were achieved with Fe(II)/H_2O_2 and Fe(II)/H_2O_2/zeolite processes. In the case of O_3/H_2O_2 and O_3/H_2O_2/zeolite processes, 60% of mineralization was obtained.

Recently, Nikazar et al. [131, 132] investigated photocatalytic degradation of some azo dyes. Photocatalytic degradation of Disperse yellow 23 in water was studied [131]. Titanium (IV) oxide was supported on clinoptilolite using the solid-state dispersion (SSD) method. The results show that the TiO_2 /Clinoptilolite (SSD) is an active photocatalyst. The maximum effect of photo degradation was observed at 10 wt % TiO_2, 90 wt % clinoptilolite zeolite and a first order reaction with $k = 0.0119$ 1/min was observed. Similarly, photocatalytic degradation of azo dye Acid Red 114 in water with TiO_2 supported on clinoptilolite was also studied [132]. A photocatalyst containing 10% TiO_2 and 90% clinoptilolite has the maximum efficiency on photo degradation of AR 114. The photodegradation conversion of AR 114 decreases with an increase in the initial con centration of AR 114. Valdes et al. [133] studied methylene blue (MB) removal systems based ozone oxidation, zeolite adsorption, and simultaneous adsorption-oxidation using ozone in presence of natural zeolite. It was found that ozone oxidation combined with zeolite increases the overall MB oxidation rate with respect to ozonation process and zeolite adsorption. In presence of free radical scavenger, only a 25% of reduction on MB removal rate are in the simultaneous treatment (Table 30.5), as compared with 70% when ozonation treatment is used.

Lim et al. [134] analyzed the adsorption behavior of azo dye Acid orange 7 (AO7) from aqueous solution onto macrocomposite (MC) under various experimental conditions. The adsorbent, MC, which consists of a mixture of zeolite and activated carbon, was found to be effective in removing AO7. The Langmuir isotherm model fitted the equilibrium data and the adsorption kinetic was found to be well described by the pseudo-second-order model. Thermodynamic analysis indicated that the adsorption process is a spontaneous and endothermic process. For the phytotoxicity test, treated AO7 was found to be less toxic. Recently, the adsorbent, magnetic NaY zeolite was synthesized for simultaneous removal of three toxic cationic dyes as methylene blue, crystal violet, and fuchsine, from aqueous solutions [135]. The percentage of dye removal was mathematically described as a function of experimental parameters and was modeled through central composite design (CCD) and optimum conditions of 10.3, 50 °C, 45 min, 10 mg/L, and 46.2 mg, for pH, temperature, time, initial dyes concentration, and adsorbent mass were resulted, respectively. It was obtained the percent of model dye removal prediction of 99.0, 98.6, and 98.4 for methylene blue, crystal violet, and fuchsine, respectively. Amodu et al. [136] reported the adsorption of crystal violet (CV) dye onto magnetic zeolite (MZ) nanoparticles, synthesized by direct fusion of fly ash (FA) and magnetite particles. The synthesised MZ showed high capacity for CV dye adsorption, removing 95% of the dye at an equilibrium adsorption time of 10 min and 25°C. Chong et al. [137] studied to synthesize a functional-form of titanium dioxide (TiO_2)–zeolite nanocomposite through the modified two-step sol–gel method for enhanced application and separation after advanced industrial dye wastewater treatment. The photoactivity of synthesized TiO_2–zeolite nanocomposite was measured and compared against the commercial TiO_2 particles. It was found that the TiO_2–zeolite nanocomposite shows a high apparent pseudo-first order reaction rate constant of 0.0419 L/min at lower dye concentration.

In 2017, cross-linked beads of activated oil palm ash zeolite/chitosan (Z-AC/C) composite were prepared through the hydrothermal treatment of NaOH activated oil palm ash followed by beading with chitosan [138]. The effects of initial dye concentration (50–400 mg/L), temperature (30–50 °C) and pH (3–13) on batch adsorption of methylene blue (MB) and acid blue 29 (AB29) were studied. Adsorption of both dyes was better described by Pseudo-second-order kinetics and Freundlich isotherm model. The maximum adsorption capacities of Z-AC/C were 151.51, 169.49, and 199.20 mg/g for MB and 212.76, 238.09, and 270.27 mg/g for AB29 at 30 °C, 40 °C, and 50 °C, respectively. In another study [139], an efficient biodegradable membranes were fabricated using polyvinyl alcohol (PVA), carboxymethyl cellulose (CMC) and ZSM-5 zeolite, and their utility for the removal of methylene blue (MB), was investigated [139]. A high dye removal (97%) and high adsorption capacity (7.83) were observed for 5 wt% of zeolite loaded sample for an initial dye concentration of 10 ppm with a contact time of 10 h at 30 °C. Similarly, a novel nanocrystalline zeolite X (nZX) was synthesized using coal fly ash of eastern India for adsorptive removal of crystal violet (CV) from aqueous solution. The nZX has a surface area, an average crystal size, an average particle size, crystallinity, and product yield of 651.34 m^2/g, 16.28 nm, 27.28 nm, 82% and 91% respectively. A maximum of 99.52% CV was removed at optimized condition of pH 9, 60 min contact time, 50 °C temperature, 100 mg/L initial dye concentration and 1.0 g/L adsorbent mass. The maximum Langmuir adsorption capacity of 234.57 mg/g was obtained. The adsorption of CV followed a pseudo-second order kinetic model.

Sono-assisted adsorptive elimination of dyes from industrial effluent is proved to be a fast, efficient and clean technique. Highly pure nanozeolite X (nFAZX) was synthesized from waste coal post-combustion fly ash (CFA) by pre-fusion hydrothermal method. Rapid sono-assisted adsorptive elimination of methylene blue (MB) dye from water solution was studied in presence nFAZX, CFA and commercial zeolite X (CZX) [141]. Optimization study was explored by response surface methodology (RSM). Maximum MB removal (64.52% by CFA, 99.30% by nFAZX, and 96.57% by CZX) was obtained at pH 8.25, 0.164 g/100 mL of adsorbent, 9.92 min sonication time, and 25 mg/L MB concentration. Also, The maximum Langmuir adsorption capacity of CFA, nFAZX and CZX for MB were 78.22 mg/g, 345.36 mg/g, and 250.41 mg/g respectively for 0.160 g/100 ml adsorbents. In another research [142], laccase enzyme from Polyporus durus ATCC 26,726 was immobilized on nanoporous Zeolite-X (ZX), with 83% immobilization yield. The immobilization process protected laccase against pH and temperature changes. The activation energy was lowered after immobilization. The thermodynamic analysis for substrate oxidation indicated that the enthalpy (ΔH°), entropy (ΔS°), and free energy of transition state (ΔG^o_{E-T}) were lower for immobilized laccase. Furthermore, there was an increase in turnover number (k_{cat}) after immobilization. The immobilized laccase could decolorize two dyes and two synthetic wastewater solutions. It retained 100% activity against AB 225 dye after 7 successive decolorization cycles and it could be reused for 11 times with 48% activity loss.

In 2020, the application of a novel polyvinyl alcohol/starch/ZSM-5 zeolite membrane as an adsorbent for the removal of Eriochrome black T (EBT) from aqueous medium was investigated [143]. Swelling measurement showed that with increase in zeolite content the hydrophobicity of the membrane also increases. The maximum adsorption capacity was estimated to be 2.17 mg/g for 5 wt% zeolite loaded membrane at an initial concentration of 10 mg/L and pH 3. A similar study [144] is mainly focused on the explanation of nanoparticles aggregation by deposition on natural zeolite, and utilization of this natural zeolite as supported material to nano zerovalent iron (NZ-nZVI) in the form of liquid slurry with sodium percarbonate acting as an oxidant in a Fenton like system for the removal of synthetic CI Acid orange 52 (AO52) azo dye, in textile effluent. The removal of CI Acid orange 52 dye was obtained as 94.86%. Furthermore, the actual dyeing effluent including green, magenta, and the blended colour was successfully decolourized by natural zeolite-supported nZVI/SPC Fenton process. Nyankson et al. [145] studied zeolite/Cerium oxide nanocomposite (Z/CeO_2–NC) hydrothermally synthesize and its adsorption capacity on methylene blue organic dye (MB). The thermodynamic Gibbs free energy (ΔG) parameter was determined to be negative, indicating a spontaneous adsorption process. The synthesized Z/CeO_2–NCs showed strong adsorption for methylene blue dye with increasing amount of CeO_2. The maximum adsorption efficiency was calculated as 93.9% and the maximum regeneration efficiency of the adsorbent was found to be 78.8% higher after it had been washed by alkaline solution of pH 13. Tony [146] showed that highly efficient low-cost alum sludge-based sorbents (AS-Sorbents), derived from alum sludge by-product waste is a significant potential for treating wastewaters contaminated by textile dyes. The experimental data revealed that AS-Sorbents exhibited a reasonable textile dye removal from real jeans dying effluent. The adsorption matrix was well fitted by Langmuir isotherm and followed second-order kinetic model while the maximum monolayer adsorption capacity was 6.5 mg/g. Thermodynamic parameters show that the process follows the physical adsorption matrix with a spontaneous and exothermic nature.

Mittal et al. [147] reported the feasibility of utilizing zeolite-Y incorporated hydrogel composite (ZHC) of gum karaya (GK) as a potential adsorbent to treat cationic dyes contaminated wastewater. For dye adsorption studies, Brilliant green dye (BG) was taken as the target pollutant. Adsorption isotherm followed Langmuir model with 1461.35 mg/g adsorption capacity and the adsorption kinetics was most appropriately explained using pseudo-second-order model. Jawad et al. [148]

attempted to develop zwitterion composite adsorbent is conducted by modifying chitosan (CHS) with a covalent cross-linker (epichlorohydrin, ECH) and an aluminosilicatemineral (zeolite, ZL). The zwitterion composite adsorbent of chitosan-epichlorohydrin/zeolite (CHS-ECH/ZL) was performed multifunctional tasks by removing a cationic dye as methylene blue (MB) and anionic dye Reactive red 120 (RR120) from aqueous solutions. The research findings revealed that the adsorption isotherm at equilibrium well explained in according to the Freundlich isotherm model, and the recorded adsorption capacities of CHS-ECH/ZL are 156.1 and 284.2 mg/g for MB and RR120 respectively at 30 °C (Table 30.5). Recently, Ghanavati et al. [149] studied to evaluate adsorption of Acid Blue 74 by electrospun PA-6/Modified zeolite nanofibrous composite. Design of the Response Surface Method (RSM) of Box-Behnken model (BBD) was used to determine the optimal conditions for the removal of Acid Blue 74. The investigation of kinetic and isotherm model showed that the reaction kinetics follows the pseudo-second-order model and the PA-6/Modified zeolite nanofibrous composite adsorption process follows the Freundlich model.

30.4 Cost Analysis

Zeolites correspond to a wide family of chemical structures, while there are 40 naturally occurring zeolite frameworks like clinoptilolite, which are readily available and abundant in nature. High ion-exchange capacity, relatively high specific surface areas and more significantly being low-cost materials, make zeolites attractive adsorbents for the textile wastewater treatment. Currently, their price varies between US$ 0.03–0.12/kg, depending on the quality of the mineral [150]. On the other hand, the cost of one of the most inexpensive commercially available carbon is about US$2/kg. An average natural zeolite costs about US$70/ton, including the cost of its purchase, transport and processing (chemicals, electrical energy, and labor required in the process), whereas modified zeolite costs about US$420/ton. Having said that, modified zeolite can be considered as a good alternative to commercially available activated carbon considering their enhanced adsorption performance [103, 151, 152]. Among low-cost materials, bagasse fly ash, peat, sphagnum moss peat, fullers earth, BF slag, bentonite, manganese oxide, fullers earth, carbonaceous adsorbent (fertilizer industry waste) can be considered as potential candidates with costs lower than 0.1 US$/kg as compared to commercial ACs [40]. Montmorillonite clay, for example, has a very large surface area and high cation exchange capacity and with its current market price (about US$ 0.04–0.12/kg), it is considered to be quite inexpensive compared to activated carbon [129]. The removal performances of Fullers earth and CAC for basic blue 9 were compared by Atun et al. [153]. They showed that the adsorption capacity is greater for Fullers earth than for CAC. Moreover, Fullers earth is an promising sorbent thanks to its low price of ca. US$ 0.04/kg.

Chitosan is produced by alkaline N-deacetylation of chitin, which is widely found in the exoskeleton of shellfish and crustaceans. It was estimated that chitosan could be produced from fish and crustaceans at a market price of US$ 15.43/kg [150]. Among their current commercial prices, zeolite is undoubtedly the most inexpensive alternative adsorbent compared to others and is 15 times cheaper than chitosan. Peat moss, a complex soil material containing lignin and cellulose as major constituents, is a natural substance widely available and abundant. Peat moss has a large surface area (>200 m^2/g) and is highly porous so that it can be used to bind heavy metals. Peat moss is a relatively inexpensive material and commercially sold at US$ 0.023/kg in the US [150]. A further study was also conducted to investigate the removal of Pb^{2+} and Cr^{6+} using blast-furnace slag sold at US$ 38/ton [150]. It was found that the maximum metals uptake observed under identical conditions is 40 and 7.5 mg/g of Pb^{2+} and Cr^{6+}, respectively. It can be concluded that its removal performance is comparable to the most inexpensive commercial activated carbon, as 1 g of activated carbon can adsorb 32.4 mg of Pb^{2+}. It was found that the performance of insoluble starch xanthate is better in terms of adsorption capacity and ease of operation. It was also reported that soluble starch xanthates cost only US$ 1.0/kg.

30.5 Conclusions

In this review, the recent development of nanoporous zeolites in the textile wastewaters treatment has been evaluated. Briefly, it can be claimed that natural zeolite is an inexpensive, locally available material and could be used effectively in place of commercial activated carbon for color removal from textile effluents. Synthetic and natural zeolites are important alternatives as adsorbents due to their high ion-exchange and adsorption capacities as well as thermal and mechanical stabilities.

Among 40 natural zeolite species, clinoptilolite is regarded as the most abundant form present in nature. Zeolites are considered as low-cost materials, where the prices vary between ca. US$ 0.03–0.12/kg, depending on the quality of the zeolite. The reactive dyes have negative sulfonate groups, which are repelled by the negatively charged zeolite surface. This induces a relatively low adsorption capacity for natural zeolite. For this reason, in order to increase the adsorption capacity, the surface of natural zeolite is modified with a cationic surfactant. A large number of researches on the batch and fixed-bed zeolite systems have been performed in recent years. The influence of coupling conventional systems with zeolites for textile wastewater treatment was investigated by various researchers. Moreover, zeolites, thanks to their large surface areas, internal pore volume, unique uniform pores and channel size, are regarded as promising hosts to disperse semiconductor photocatalyst on their surfaces. Extensive literature information on dye adsorption by a wide range of zeolites has been reviewed in this paper. The adsorption of Basic blue 41 textile dye from aqueous solutions on a zeolite tuff (Romanian) was studied in batch mode. The monolayer adsorption capacity of adsorbent was found to be 192.31 mg/g at 50 °C and pH 5 for the tuff particles of 0.085 mm. Besides, zeolitic imidazole frameworks (ZIFs), a new class of adsorbents, were proposed to adsorb malachite green (MG) in water. ZIF-67 exhibited 2430 mg/g MG adsorption at 20 °C and $\sim$3000 mg/g at

40 °C. However, despite a large number of papers published on zeolite as a low-cost adsorbent, there is still little information on the application of a full-scale zeolite treatment system. Once zeolite performance in removing dyes at low cost is proven at higher scales, they can be adopted and widely used in industries not only to minimize cost inefficiency, but also to improve profitability. Undoubtedly zeolite as a low-cost adsorbent offers a lot of promising benefits for commercial purpose in the future.

References

1. A.R. William, T.F. Leonard, Text. Chem. Color. **29**, 10 (1997)
2. D. Mohan, K.P. Singh, G. Singh, K. Kundan, Ind. Eng. Chem. Res. **41**, 3688 (2002)
3. R. Abraham, *Dyes Environmental Chemistry in Encyclopaedia of Chemical Technology*, (Wiley, NY, 8, 1993) pp. 672–783
4. T. Panswad, W. Luangdilok, Water Res. **34**, 4177 (2000)
5. U. Pagga, D. Brown, Chemosphere **15**, 479 (1986)
6. D.W. Breck, *Zeolite Molecular Sieves* (Wiley, NY, 1974)
7. W.J. Mortier, J.R. Pearce, Am. Min. **66**, 309 (1981)
8. M.W. Ackley, R.T. Yang, AIChE **37**, 1645 (1991)
9. L.L. Ames, Am. Min. **45**, 689 (1960)
10. G. Blanchard, M. Maunaye, G. Martin, Water Res. **18**, 1501 (1984)
11. M. Gaspard, A. Neveu, G. Matin, Water Res **17**, 279 (1983)
12. M. Turan, Application of nanoporous zeolites for the removal of ammonium from wastewaters: a review, in H. Ünlü, N.J.M. Horing, J. Dabowski (Eds.), Low-Dimensional and Nanostructured Materials and Devices Part of the series NanoScience and Technology. (Springer, Berlin Heidelberg, 2016) pp. 477–504
13. M. Turan, M.S. Çelik, J. Water SRT-Aqua **52**, 59 (2003)
14. M. Vaca-Mier, R.L. Callegas, R. Gehr, C.B. Jimenez, P.J. Alvarez, Water Res. **35**, 373 (2001)
15. M. Turan, U. Mart, B. Yüksel, M.S. Çelik, Chemosphere **60**, 1487 (2005)
16. V. Meshko, L. Markovska, M. Mincheva, A.E. Rodrigues, Water Res. **35**, 3357 (2001)
17. B. Armağan, O. Özdemir, M. Turan, M.S. Çelik, J. Chem. Technol. Biotechnol. **78**, 725 (2003)
18. I. Langmuir, J. Am. Chem. Soc. **40**, 1361 (1918)
19. H. Freundlich, Z. Phys, Chemei. **57**, 384 (1906)
20. G. McKay, H.S. Blair, J.R. Gardner, J. Appl. Polym. Sci. **27**, 3043 (1982)
21. Z. Aksu, I.A. Isoglu, Process Biochem. **40**, 3031 (2005)
22. J.A. Stephen, G. Quan, M. Ronan, A.J. Pauline, Bioresource Technol. **88**, 143 (2003)
23. X.C. Fu, W.X. Shen, T.Y. Yao, *Physical Chemistry*, 4th edn. (Higher Education Press, China, 1994), pp. 303–321
24. S. Lagergren, Kungliga Svenska Vetenskapsaka-demiens. Handlingar **24**, 1 (1898)
25. G. McKay, Y.S. Ho, Water Res. **33**, 578 (1999)
26. G. McKay, Y.S. Ho, Process Biochem. **34**, 451 (1999)
27. G.C. Catena, F.V. Bright, Anal. Chem. **61**, 905 (1989)
28. C. Namasivayam, R.T. Yamuna, Environ. Pollut. **89**, 1 (1995)
29. H. Nollet, M. Roels, P. Lutgen, P. Van der Meeren, W. Verstraete, Chemosphere **53**, 655 (2003)
30. L.D. Benefield, J.F. Judkins Jr., B.L. Weand, *Process Chemistry for Water and Wastewater Treatment* (Prentice Hall, Englewood Cliffs, NJ, 1982)
31. S. Kundu, A.K. Gupta, J. Colloid Interface Sci. **290**, 52 (2005)
32. G.S. Bohart, E.Q. Adams, J. Am. Chem. Soc. **42**, 523 (1920)
33. R.A. Hutchins, Chem. Eng. **80**, 133 (1973)
34. G.M. Walker, L.R. Weatherley, Water Res. **31**, 2093 (1997)

35. U. Meyer, Biodegradation of synthetic organic colorants, in *Microbial Degradation of Xenobiotic and Recalcitrant Compounds*. ed. by T. Leisinger, A.M. Cook, R. Hunter, J. Nuesch (Academic Press, London, 1981), pp. 371–385
36. H. Zollinger, *Colour Chemistry—Synthesis, Properties of Organic Dyes and Pigments* (VCH Publishers, New York, 1987)
37. K. Hunger, *Industrial Dyes: Chemistry, Properties, Applications* (Wiley-VCH, Weinheim, Cambridge, 2003)
38. R.M. Christie, *Environmental Aspects of Textile Dyeing*. (Woodhead, Boca Raton, Cambridge, 2007)
39. Q. Husain, Crit. Rev. Biotechnol. **26**, 201 (2006)
40. V.K. Gupta, Suhas. J. Environ. Manage. **90**, 2313 (2009)
41. G. McKay, J. Chem. Technol. Biotechnol. **31**, 717 (1981)
42. G. McKay, J. Chem. Technol. Biotechnol. **32**, 759 (1982)
43. D.J.W. Blum, I.H. Suffet, J.P. Duguet, Crit. Rev. Envir. Sci. Technol. **23**, 121 (1993)
44. S.E. Bailey, T.J. Olin, M. Bricka, D.D. Adrian, Water Res. **33**, 2469 (1999)
45. G. Crini, Biores. Technol. **97**, 1061 (2006)
46. H. Ghobarkar, O. Schaf, U. Guth, Prog. Solid State Chem. **27**, 29 (1999)
47. B. Armagan, M. Turan, M.S. Çelik, Desalination **170**, 33 (2004)
48. O. Ozdemir, B. Armagan, M. Turan, M.S. Çelik, Dyes Pigments **62**, 49 (2004)
49. M.T. Yagub, T.K. Sen, S. Afroze, H.M. Ang, Advan. Colloid. Interface Sci. **209**, 172 (2014)
50. G. Calzaferri, D. Bruhwiler, S. Megelski, M. Pfenniger, M. Pauchard, B. Hennessy, H. Maas, A. Devaux, A. Graf, Solid States Sci. **2**, 421 (2000)
51. S.K. Alpat, Ö. Özbayrak, Ş Alpat, H. Akçay, J. Hazard. Mater. **151**, 213 (2008)
52. S. Wang, Z.H. Zhu, J. Hazard. Mater. **B136**, 946 (2006)
53. B. Armagan, O. Ozdemir, M. Turan, M.S. Celik, J. Environ. Sci. Heal. **A38**, 2251 (2003)
54. Y.E. Benkli, M.F. Can, M. Turan, M.S. Çelik, Water Res. **39**, 487 (2005)
55. D. Karadag, Dyes Pigm. **74**, 659 (2007)
56. B. Ersoy, *The adsorption mechanisms of various cationic surface active agents onto clinoptilolite and capture of non-ionic organic contaminants in liquids by modified clinoptilolite.* Ph.D. thesis, Istanbul Technical University, Istanbul, Turkey (2000)
57. X. Jin, M.-qin Jiang, X.-quan Shan, Z.-guo Pei, Z. Chen, J. Colloid Interface Sci. **328**, 243 (2008).
58. S. Wang, H. Li, L. Xu, J. Colloid Interface Sci. **295**, 71 (2006)
59. M.K. Robin, P. Chu, US patent 4954325, 1990
60. M. Qiu, C. Qian, J. Xu, J. Wu, G. Wang, Desalination **243**, 286 (2009)
61. R. Han, J. Zhang, P. Han, Y. Wang, Z. Zhao, M. Tang, Chem. Eng. J. **145**, 496 (2009)
62. J. Torres-Perez, M. Solache-Rios, M.T. Olguin, Sep. Sci. Technol. **42**, 299 (2007)
63. J.S. Beck, C. Vartuli, W.J. Roth, M.E. Leonowicz, C.T. Kresge, K.D. Scmmitt, C.T-W. Chu, D.H. Olson, E.W. Sheppard, S.B. McCullen, J.B. Higgins, J.L. Schlenker, J. Am. Chem. Soc. **114**, 10834 (1992)
64. C.-K. Lee, S.-S. Liu, L.-C. Juang, C.-C. Wang, K.-S. Lin, M.-D. Lyu, J. Hazard. Mater. **147**, 997 (2007)
65. R.I. Yousef, B. El-Eswed, M. Alshaaer, F. Khalili, H. Khoury, J. Hazard. Mater. **165**, 379 (2009)
66. D. Karadag, M. Turan, E. Akgul, S. Tok, A. Faki, J. Chem. Eng. Data **52**, 1615 (2007)
67. D. Karadag, E. Akgul, S. Tok, F. Erturk, M.A. Kaya, M. Turan, J. Chem. Eng. Data **52**, 2436 (2007)
68. Q. Qin, J. Ma, K. Liu, J. Hazard. Mater. **162**, 133 (2009)
69. H. Nur, A. Faizal N. A. Manan, L. K. Wei, M. N. M. Muhid, H. Hamdan, J. Hazard. Mater. **B117**, 35 (2005)
70. S. Wang, H. Li, S. Xie, S. Liu, L. Xu, Chemosphere **65**, 82 (2006)
71. L.-C. Juang, C.-C. Wang, C.-K. Lee, Chemosphere **64**, 1920 (2006)
72. I. Othman, R.M. Mohamed, I.A. Ibrahim, M.M. Mohamed, Appl. Catal. A **299**, 95 (2006)
73. C. Wang, J. Li, L. Wang, X. Sun, J. Huang, Chinese J. Chem. Eng. **17**, 513 (2009)

74. D.A. Fungaro, M. Bruno, Quim. Nova **32**, 955 (2009)
75. W.-T. Tsai, K.-J. Hsien, H.-C. Hsu, J. Hazard. Mater. **166**, 635 (2009)
76. S. Wang, E. Ariyanto, J. Colloid Interface Sci. **314**, 25 (2007)
77. C. Yan, C. Wang, J. Yao, L. Zhang, X. Liu, Colloids Surf., A **333**, 115 (2009)
78. S. Karcher, A. Kornmüller, M. Jekel, Dyes Pigm. **51**, 111 (2001)
79. D. Karadag, S. Tok, E. Akgul, K. Ulucan, H. Evden, M.A. Kaya, Ind. Eng. Chem. Res. **45**, 3969 (2006)
80. A.C. Savitsky, B.H. Wiers, R.H. Wendt, Environ. Sci. Technol. **15**, 1191 (1981)
81. M.M. Allingham, J.M. Cullen, C.H. Giles, S.K. Jain, J.S.J. Woods, Appl. Chem. **8**, 108 (1958)
82. J. Torres-Pérez, M. Solache-Ríos, A. Colín-Cruz, Water Air Soil Pollut. **187**, 303 (2008)
83. A. Kuleyin, F. Aydin, AIChe. Environ. Prog. **30**, 141 (2011)
84. A. Sayal, V.K. Bulasara, S. Barman, Chem. Proc. Engng. Res. **2**, 54 (2012)
85. E. Alver, A.U. Metin, Chem. Engng. J. **200–202**, 59 (2012)
86. K. Rida, S. Bouraoui, S. Hadnine, Applied Clay Sci. **83–84**, 99 (2013)
87. S. Liu, Y. Ding, P. Li, K. Diao, X. Tan, F. Lei, Y. Zhan, Q. Li, B. Huang, Z. Huang, Chem. Engng J. **248**, 99 (2013)
88. K.-Y.A. Lin, H.-A. Chang, Chemosphere **139**, 624 (2015)
89. H. Aysan, S. Edebali, C. Ozdemir, M.C. Karakaya, N. Karakaya, Micropor. Mesopor. Mater. **235**, 78 (2016)
90. K.Y. Hor, J.M.C. Chee, M.N. Chong, B. Jin, C. Saint, P.E. Poh, R. Aryal, J. Clean. Product. **118**, 197 (2016)
91. M.Y. Nassar, E.A. Abdelrahman, A.A. Aly, T.Y. Mohamed, J. Molecul. Liquid. **248**, 302 (2017)
92. I. Humelnicu, A. Baiceanu, M.-E. Ignat, V. Dulman, Proc. Safe. Environ. Protect. **105**, 274 (2017)
93. L. Aljerf, J. Environ. Manage. **225**, 12 (2018)
94. G.V. Brião, S.L. Jahn, E.L. Foletto, G.L. Dotto, Colloid. Surf. A. **556**, 43 (2018)
95. A.G. Olaremu, A.O. Adeola, Int. J. Modern Res. Engng. Manage. **1**, 7 (2018)
96. S. Madan, R. Shaw, S. Tiwari, S.K. Tiwari, Applied. Surf. Sci. **487**, 907 (2019)
97. M.R. Abukhadra, A.S. Mohamed, SILICON **11**, 1635 (2019)
98. N.M. Mahmoodi, M.H. Saffar-Dastgerdi, Microchem. J. **145**, 74 (2019)
99. B.B. Mohammed, A. Hsini, Y. Abdellaoui, H.A. Oualid, M. Laab, M. El Ouardi, A.A. Addi, K. Yamni, N. Tijani, J. Environ. Chem. Engng. **8**, 104419 (2020)
100. A.B. Engin, Ö. Özdemir, M. Turan, A.Z. Turan, J. Hazard. Materials **159**, 348 (2008)
101. Ö. Özdemir, M. Turan, A.Z. Turan, A. Faki, A.B. Engin, J. Hazard. Mater. **166**, 647 (2009)
102. A. Faki, M. Turan, Ö. Özdemir, A.Z. Turan, Ind. Eng. Chem. Res. **47**, 6999 (2008)
103. B. Armağan, M. Turan, Ö. Özdemir, M.S. Çelik, J. Environ. Sci. Health A **39**, 1247 (2004)
104. R. Han, Y. Wang, W. Zou, Y. Wang, J. Shi, J. Hazard. Mater. **145**, 331 (2007)
105. L. Markovska, V. Meshko, V. Noveski, Korean J. Chem. Eng. **18**, 190 (2001)
106. X. Wu, Y. Wang, J. Liu, J. Ma, R. Han, Desal. Water Treat. **20**, 228 (2010)
107. A. Ojstrsek, D. Fakin, Desal. Water Treat. **33**, 147 (2011)
108. S. Syafalni, I. Abustan, I. Dahlan, C.K. Wah, G. Umar, Modern. Appl. Sci. **6**, 37 (2012)
109. L. Singh, P. Rekha, S. Chand, J. Environ. Manage. **215**, 1 (2018)
110. N. Marsiezade, V. Javanbakht, Int. J. Biologic. Macromol. **162**, 1140 (2020)
111. S. Apollo, M.S. Onyango, A. Ochieng, Chemical Engng J. **245**, 241 (2014)
112. N.N. Fathima, R. Aravindhan, J.R. Rao, B.U. Nair, Chemosphere **70**, 1146 (2008)
113. M. Neamtu, C. Zaharia, C. Catrinescu, A. Yediler, M. Macoveanu, A. Kettrup, Appl. Catal. B **48**, 287 (2004)
114. H. Kusic, A. L. Bozic, N. Koprivanac, S. papic, Dyes and Pigments **74**, 388 (2007)
115. W.-S. Chang, S.-W. Hong, J. Park, Proc. Biochem. **37**, 693 (2002)
116. G. Ciobanu, G. Carja, O. Ciobanu, Desalination **222**, 197 (2008)
117. M.B. Kasiri, H. Aleboyeh, A. Aleboyeh, Appl. Catal. B **8**, 49 (2008)
118. A.K. Kondru, P. Kumar, S. Chand, J. Hazard. Mater. **166**, 342 (2009)
119. M. Neamtu, C. Catrinescu, A. Kettrup, Appl. Catal. B **51**, 149 (2004)

120. A. Chen, X. Ma, H. Sun, J. Hazard. Mater. **156**, 568 (2008)
121. R.J. Tayade, R.G. Kulkarni, R.V. Jasra, Ind. Eng. Chem. Res. **46**, 369 (2007)
122. R.J. Tayade, P.K. Surolia, M.A. Lazar, R.V. Jasra, Ind. Eng. Chem. Res. **47**, 7545 (2008)
123. M. Huang, C. Xu, Z. Wu, Y. Huang, J. Lin, J. Wu, Dyes Pigm. **77**, 327 (2008)
124. T M. Salama, I. O. Ali, A. I. Hanafy, W. M. Al-Meligy, Mater. Chem. Physics **113**, 159 (2009)
125. M. Tekbaş, H.C. Yatmaz, N. Bektaş, Micropor. Mesopor. Mater. **115**, 594 (2008)
126. C.-C. Wang, C.-K. Lee, M.-Du Lyu, L.-C. Juang, Dyes and Pigments **76**, 817 (2008)
127. R. Aravindhan, N.N. Fathima, J.R. Rao, B.U. Nair, J. Hazard. Mater. **B138**, 152 (2006)
128. P. Guo, X. Wang, H. Guo, Appl. Catal. B **90**, 677 (2009)
129. D.I. Petkowicz, R. Brambilla, C. Radtke, C.D.S. da Silva, Z.N. da Rocha, S.B.C. Pergher, J.H.Z. dos Santos, Appl. Catal. A **357**, 125 (2009)
130. N. Koprivanac, B.R. Locke, S. Papić, A.L. Božić, D. Vujević, Z. Lazarević, Studies in Surface Sci. Catalysis **154**, 2548 (2004)
131. M. Nikazar, K. Gholivand, K. Mahanpoor, Kinet. Catal. **48**, 214 (2007)
132. M. Nikazar, K. Gholivand, K. Mahanpoor, Desalination **219**, 293 (2008)
133. H. Valdes, R.F. Tardon, C.A. Zaror, Water Sci. Technol. **60**, 1419 (2009)
134. C.K. Lim, H.H. Bay, C.H. Neoh, A. Aris, Z.A. Majid, Z. Ibrahim, Environ. Sci. Pollut. Res. **20**, 7243 (2013)
135. M. Shirani, A. Semnani, H. Haddadi, S. Habibollahi, Water Air Soil Pollut. **225**, 2054 (2014)
136. O.S. Amodu, T.V. Ojumu, S.K. Ntwampe, O.S. Ayanda, J. Encapsul. Adsorp. Sci. **5**, 191 (2015)
137. M.N. Chong, Z.Y. Tneu, P.E. Poh, B. Jin, R. Aryal, J. the Taiwan Inst. Chem. Eng. **50**, 288 (2015)
138. W.A. Khanday, M. Asif, B.H. Hameed, Int. J. Biologic. Macromol. **95**, 895 (2017)
139. R. Sabarish, G. Unnikrishnan, Carbohyd. Polym. **199**, 129 (2018)
140. R. Sabarish, G. Unnikrishnan, J. Taiwan Inst. Chem. Eng. **96**, 305 (2019)
141. S. Sivalingam, T. Kella, M. Maharana, S. Sen, J. Clean. Product. **208**, 1241 (2019)
142. H.R. Wehaidya, M.A. Abdel-Naby, H.M. El-Hennawi, H.F. Youssef, Biocatal. Agricultur. Biotechnol. **19**, 101135 (2019)
143. S. Radoor, J. Karayil, J. Parameswaranpillai, S. Siengchin, J Polymer Environ. **28**, 2631 (2020)
144. T. Rashid, D. Iqbal, A. Hazafa, S. Hussain, F. Sher, F. Sher, J. Environ. Chem. Engng. **8**, 104023 (2020)
145. E. Nyankson, J. Adjasoo, J.K. Efavi, A. Yaya, G. Manu, A. Kingsford, R.Y. Abrokwah, Scientific African **7**, e00257 (2020)
146. M.A. Tony, Int. J. Environ. Sci. Technol. **17**, 2485 (2020)
147. H. Mittal, R. Babu, A. A. Dabbawala, S. Stephen, S. M. Alhassan, Colloid. Surf. A **586**, 124161 (2020)
148. A.H. Jawad, A.S. Abdulhameed, A. Reghioua, Z.M. Yaseen, Int. J. Biologic. Macromol. **163**, 756 (2020)
149. L. Ghanavati, A.H. Hekmati, A. Rashidi, A. Shafiekhani, J. Textile Instit. (2020). https://doi.org/10.1080/00405000.2020.1840691
150. S. Babel, T.A. Kurniawan, J. Hazard. Mater. **97**, 219 (2003)
151. Ö. Özdemir, *Investigation of Color Removal from Textile Industry Dyehouse Wastewater in Modified Zeolite Fixed-Bed Reactor*. Ph.D. thesis, Istanbul Technical University, Istanbul, Turkey (2007)
152. R.S. Bowman, *Surfactant-Modified zeolite (SMZ)-A versatile, inexpensive sorbent for removing contaminants from water*, Research and Development Program Report (2006)
153. G. Atun, G. Hisarli, W.S. Sheldrick, M. Muhler, J. Colloid Int. Sci. **261**, 32 (2003)

Chapter 31
Removal of Heavy Metals and Dyes from Wastewaters by Raw and Activated Carbon Hazelnut Shells

Abdullah Zahid Turan and Mustafa Turan

Abstract In this article, the technical feasibility of raw and activated carbon hazelnut shells for the removal of heavy metals and dyes from wastewaters has been reviewed. Adsorption has been proved to be an excellent way to treat industrial waste effluents, offering significant advantages like the low-cost, availability, profitability, convenience in operation and efficiency. There is a great potential of hazelnut shells to be used in adsorption applications. Nanotechnology is an advanced field of science having the ability to solve the variety of environmental challenges by controlling the size and shape of the materials at a nanoscale. Carbon nanomaterials are unique because of their nontoxic nature, high surface area, easier biodegradation and particularly useful environmental remediation. Textile industry effluents and heavy metal contamination in water are major problems and pose a great risk to human health. Carbon nanomaterials namely carbon nanotubes, fullerenes, graphene, graphene oxide and activated carbon have great potential for removal of heavy metals and dyes from water because of their large surface area, nanoscale size and availability of different functionalities and they are easier to be chemically modified and recycled. Activated carbon was also prepared using agricultural by-products such as palm-tree cobs, grape seeds, several nutshells (almond shell, hazelnut shell, walnut shell and apricot stone), olive-waste cakes and corn cob due to the fact that activated carbon made from conventional raw materials are expensive. This review reports the removal of heavy metals and dyes from wastewaters using raw and activated carbon hazelnut shells in order to provide useful information on various aspects of utilization of the agricultural waste materials and carbon nanomaterials. The adsorption capacities of raw and activated carbon hazelnut shells under different experimental conditions are also reported and compared with other agro-based adsorbents.

A. Z. Turan
TUBITAK Marmara Research Center Energy Institute, 41470 Kocaeli, Turkey
e-mail: abdullah.turan@tubitak.gov.tr

M. Turan (✉)
Department of Environmental Engineering, Istanbul Technical University, 34469 Istanbul, Turkey
e-mail: turanm@itu.edu.tr

H. Ünlü and N. J. M. Horing (eds.), *Progress in Nanoscale and Low-Dimensional Materials and Devices*, Topics in Applied Physics 144,
https://doi.org/10.1007/978-3-030-93460-6_31

31.1 Introduction

The increasing contamination of urban and industrial wastewaters by toxic metal ions and colorants is a worrying environmental problem. These pollutants are of considerable concern because they are non-biodegradable and highly toxic [1]. The removal of these toxic pollutants from wastewater is difficult by conventional coagulation and the activated sludge process [2]. Therefore, the treatment methods, such as oxidation, chemical precipitation, membrane filtration, ion exchange and adsorption have been practiced for the removal of heavy metals and dyes [3]. One of the major methods for the removal of pollutants from aqueous effluent is adsorption by using porous solid adsorbents. Adsorption has demonstrated its efficiency and economic feasibility as a wastewater treatment process compared to the other purification and separation methods [4, 5], and has gained importance in industrial applications, such as removal of heavy metals cations [6, 7] and coloring materials [8, 9] from aqueous solution by adsorption onto activated carbon prepared from agricultural wastes. Many non-conventional low-cost adsorbents have been used for the removal of heavy metal and dye from wastewaters such as [10, 11]: (1) Clay materials (sepiolite, bentonite, kaolinite etc.) [6–8, 12, 13], (2) Zeolites [7–9, 14–18], (3) Siliceous materials (silica beads, alunite, perlite) [19–21], (4) Biosorbents (chitosan, peat, biomass) [22–26], (5) Agricultural wastes [27–29] and (6) Industrial waste products [30–34].

Nanotechnology is an advanced field of science having the ability to solve the variety of environmental challenges by controlling the size and shape of the materials at a nanoscale. Carbon nanomaterials are unique because of their nontoxic nature, high surface area, easier biodegradation, and particularly useful environmental remediation. Carbon nanomaterials namely carbon nanotubes, fullerenes, graphene, graphene oxide, and activated carbon have great potential for removal of heavy metals and dyes from water because of their large surface area, nanoscale size, and availability of different functionalities and they are easier to be chemically modified and recycled [35]. Activated carbon is the most used adsorbent owing to its high capacity, but needs a regeneration process being an expensive material. In general, the methods for production activated carbon consist of two steps: (1) The first step involves a chemical activation step where raw agricultural materials are impregnated with a solution of dehydrating agent (for example $ZnCl_2$, H_2SO_4) to retard the formation of tars during the carbonization process. (2) Furthermore, in physical activation, they are washed, dried and carbonized in an inert atmosphere to produce the final activated carbon [36–38].

Many efforts, however, have been made to investigate the use of various low cost organic adsorbents [39, 40]. They should be cheap, easily available and disposable without regeneration. Activated carbon was also prepared using agricultural by-products such as palm-tree cobs, grape seeds, several nutshells (almond shell, hazelnut shell, walnut shell and apricot stone), olive-waste cakes and corn cob due to the fact that activated carbon made from conventional raw materials is expensive [41–45]. Hazelnut shell is a biomass and especially important in Turkey, because Turkey is the biggest hazelnut producer country in the world [46]. North coasts of Turkey are so suitable to grow hazelnut that approximately 80% of the total hazelnut production in the world is supplied from this region [47]. Its major use today is as combustible owing to the considerable calorific value. Besides, there is a great potential of raw hazelnut shell to use in some industrial applications in that region. Figure 31.1 shows green and dried hazelnuts with husks. Various researches on hazelnut shell have been done in recent years.

In 2005, the possibility of the production of a methyl ester biodiesel from hazelnut waste/sunflower oil mixture using methanol, sulphuric acid and sodium hydroxide in a two-stage process was investigated [48]. The effects of particle size on the non-isothermal slow pyrolysis of hazelnut shell from ambient to 1173 K with a linear heating rate of 20 K/min under dynamic nitrogen atmosphere were also studied [46]. In another research, it was carried out to determine effects of composted hazelnut husk (CHH) on some chemical properties of soil and soil humic acid (HA) [49]. This material was also considered in some works as source for activated carbon mostly utilized in removal of dye and heavy metals. Hazelnut shells have the some polar functions such as alcoholic, carbonylic, carboxylic and phenolic groups, which are potentially involved in bonding with sorbed pollutants [50, 51]. This review reports the removal of heavy metals and dyes in the wastewater treatment using natural (raw) hazelnut shells and their activated carbon forms (activated carbon hazelnut shells) in order to provide useful information on various aspects of utilization of the agricultural waste materials and carbon nanomaterials. The adsorption capacities of raw and activated carbon hazelnut shells under different experimental conditions are also reported and compared with other agro-based adsorbents.

31.2 Pore Charateristics of Porous Solids

The shape of pores is preferably described in terms of cylinders (which may be the case for activated oxides like alumina or magnesia), prisms (some fibrous zeolites), cavities and windows (other zeolites), slits (possible in clays and activated carbons), or spheres (although, most often, the pores are on the contrary, the voids left between solid spheres in contact with each other, as it happens with gels: silica gel, zirconia gel etc.) [52]. Pore size has a precise meaning when the geometrical shape of the pores is well defined and known (e.g., cylindrical, slit-shaped etc.). Nevertheless, the limiting size of a pore, for most phenomena in which porous solids are involved, is that of its smallest dimension which, in the absence of any further precision, is referred to as the width of the pore.

Porous solids can be classified as follows: (i) Micropores (pore size up to 2 nm), (ii) Mesopores (pore size 2–50 nm) and (iii) Macropores (pore size larger than 50 nm) [52, 53]. Five types of hysteresis loops were identified and correlated with various pore shapes: Type A hysteresis corresponds to cylindrical pores; type B is related to slit-shaped pores; type C and D hysteresises are attributed to wedge-shaped pores, and type E hysteresis is produced by bottle neck pores (Fig. 31.2a). Based on pore connectivity, the types of pores in coal can be also

Fig. 31.1 **a** Green and **b** Dried Hazelnuts with husks

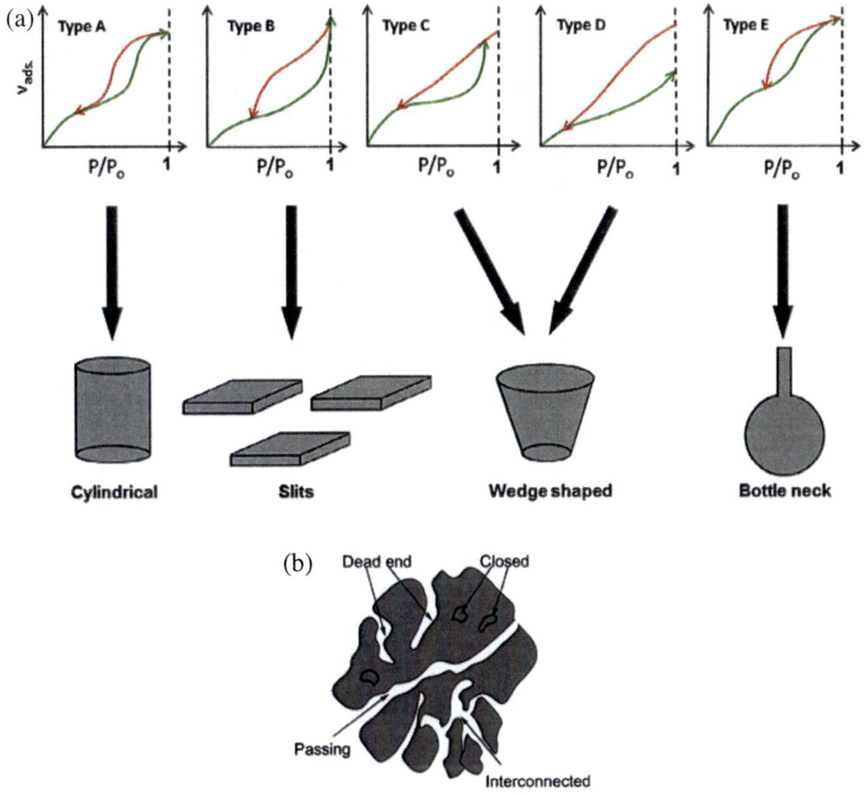

Fig. 31.2 **a** Hysteresis loops and their corresponding pore shapes; **b** Types of pores [55]

classified into passing, interconnected, dead end, and closed pores (Fig. 31.2b). The former three types are called open pores, and they have great influence on the gas adsorption, desorption and diffusion in coal [54, 55].

31.3 Activated Carbon Production from Hazelnut Shells

Carbons are widely used as industrial adsorbents because of their hydrophobic surface, high specific area and good thermal stability. With respect to other nanoporous materials like zeolites, characterized by different pore architecture and composition, porous carbons offer the advantage of a lower specific weight [56]. Carbons are prepared by synthesis and advanced separation systems and catalysis. The synthesis of ordered meso- and micro-porous carbons involves the use of mesoporous silica or zeolites as a template [57–62]. The preparation of carbon with nanopores, width less than 2 nm, is unattainable by this method, thus limiting its ultimate surface area, porosity, and performance. Stress graphitization of pure carbon has been reported at temperatures in excess of 2000 °C [63, 64].

Nanoporous carbon (NPC) is synthesized by pyrolysis of polymers, such as polyfurfuryl alcohol (PFA) [65]. Due to its strong resistance to transformation to graphite even when annealed at temperature above 2000 °C, it is classified as non-graphitizing, as defined by Franklin [66]. Commercially available activated carbon (CAC) is usually derived by thermal treatment from natural materials such as biomass, lignite or coal, but almost any carbonaceous materials may be used as precursor for the preparation of carbon adsorbents, while due to its availability and cheapness, coal is the most commonly used precursor for activated carbon production [67–71]. Biomass and other waste materials may also offer an inexpensive and renewable additional source of activated carbon [72]. A wide variety of carbons have been prepared from biomass and other wastes, such as olive-seed waste [45, 68], desert plant [69], biosorbent (Caulerpa lentillifera) [70], date pit [73], wheat bran [74], oil palm fiber [41, 75], almond shell, walnut shell, apricot stone [76], hazelnut shell [76, 77], peach stone [78], rice hull [79, 80], corn cob [36, 81], coir pith [25, 43, 82] and peanut hull [28, 83]. In recent years, growing research interest in the production of carbon based material has focused on agricultural residues. Figure 31.3 shows a process flow diagram of activated carbon preparation from agricultural residue [84]. Activated carbon can be defined as carbonaceous material having high porosity and internal surface area

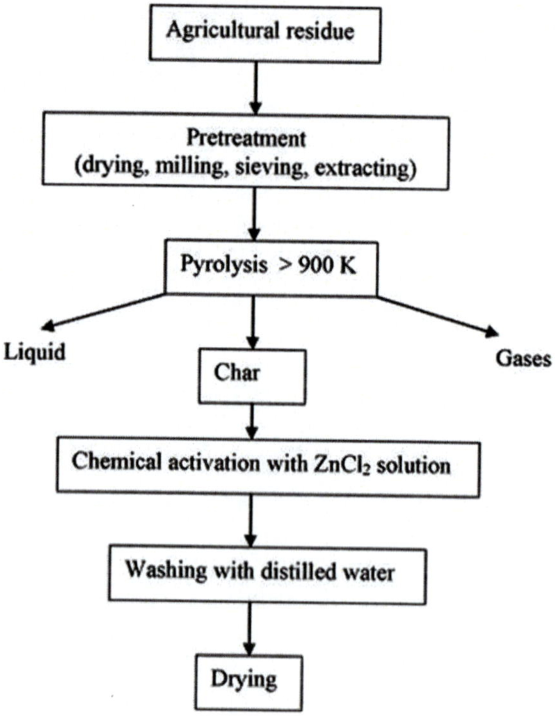

Fig. 31.3 Process flow diagram of activated carbon preparation from agricultural residue [84]

and cannot be characterized by any distinctive chemical formula. Depending on the usage, activated carbons are available as powder, granulated, pressed sticks in various size and shapes.

Hazelnut shell, as a promising biomass, has lower ash content and abundant hydrocarbon, which can be utilized along with municipal sewage sludge (MSS) by co-pyrolysis to decrease total content of pollution. In 2016, activated carbons were prepared from hazelnut shell taken from Eastern Blacksea region [85]. Activated carbons were prepared by chemical activation method with phosphoric acid and boric acid (phosphoric acid activation, phosphoric acid + boric acid activation and activated carbon + boric acid activation). In order to find the optimum activated carbon, different H_3PO_4 (35–65%) and H_3BO_3 (3–10%) concentration, activation temperature (400–800 °C) and carbonization time (30–120 min) was employed. As a result of studies, optimum activated carbon was obtained at 35% phosphoric acid concentration, 500 °C carbonization temperature and 60 min carbonization time. The obtained iodine number of activated carbon was 583.32 mg/I_2, while surface area was 1071.29 m^2/g. Also, hardness and adsorption tests produced results of high hardness of the active carbon (90%) and high adsorption capacity. Due to having low moisture and ash content with high surface area, hardness, hazelnut shell is considered to be a potential raw material for producing activated carbon.

Lewicka [86] researched producing activated carbons for CO_2 capture from hazelnut shells (HN), walnut shells (WN) and peanut shells (PN). Saturated solution of KOH was used as an activating agent in ratio 1:1. Samples were carbonized in the furnace in the range of temperatures 600–900 °C. Properties of carbons were tested by N_2 adsorption method, using BET equation, DFT method and volumetric CO_2 adsorption method. With the increase of carbonization temperature, specific surface area of studied samples increased. The largest surface area was calculated for samples carbonized at 900 °C and the highest values of CO_2 adsorption were obtained with the samples: PN900 at 0 °C (5.5 mmol/g) and WN900 at 25 °C (4.34 mmol/g). All of the samples had a well-developed microporous structure.

Recently, a low cost activated carbon was prepared from hazelnut shells by chemical activation with sodium hydroxide at 600 °C in a N_2 atmosphere and then through combining with magnetic $NiFe_2O_4$ nanoparticles by hydrothermal and co-precipitation methods [87]. Results indicated that the $NiFe_2O_4$ nanoparticles synthesized by the hydrothermal method had a higher saturation magnetization and smaller average particle size than those produced by the co-precipitation method. The specific surface area and total pore volume of the activated carbon decreased from 314 to 288 m^2/g and 0.3639 to 0.3338 cm^3/g, respectively by forming a hybrid with the magnetic $NiFe_2O_4$ nanoparticles synthesized by the hydrothermal method. $NiFe_2O_4$ nanoparticles were mainly distributed on the surface, although a few were inside the pores of the activated carbon. Their sizes were the same as those of the original ones. The saturation magnetization of the hybrids was lower than those of the original $NiFe_2O_4$ nanoparticles due to the existence of the activated carbon. They showed super paramagnetic behavior at room temperature and were easily separated from solutions by an external magnet.

The copyrolysis of MSS and hazelnut shell blend was analyzed by a method of multi-heating rates and different blend ratios with TG-DTG-MS under N_2 atmosphere by Zhao et al. [88]. The apparent activation energy of co-pyrolysis was calculated by three iso-conversional methods. Satava–Sestak method was used to determine mechanism function G(α) of copyrolysis, and Lorentzian function was used to simulate multi-peaks curves. The results showed that there were four thermal decomposition stages, where the biomass transformed at different temperature ranges. The apparent activation energy increased from 123.99 to 608.15 kJ/mol. The reaction mechanism of co-pyrolysis is random nucleation and nuclei growth. The apparent activation energy and mechanism function afford a theoretical groundwork for co-pyrolysis technology.

31.4 Removal of Heavy Metals from Wastewaters Using Hazelnut Shells

31.4.1 Removal of Heavy Metals from Wastewaters by Raw Hazelnut Shells

In recent years the most promising alternative method for removal of heavy metal ions uses the sorption by waste materials, both organic and inorganic, which are low cost and abundant. Cimino et al. [89] reported the use of hazelnut shell (HS) as a biosorbent to remove toxic ions such as cadmium, zinc, three- and hexavalent chromium, from aqueous solutions. General description of HS is given in Table 31.1. Batch tests showed that more Cr^{3+} ions were removed than both Cd^{2+} and Zn^{2+} ions. The Cr(VI) removal was pH dependent and fitted with the Langmuir isotherm model. In another investigation [90], the removal of Ni(II), Cd(II) and Pb(II) by shells of hazelnut (HS) and almond (AS) was studied. Figure 31.4 shows the sorption efficiency of heavy metal ions (HMI) on HS. The negative values of free change (ΔG^o) indicated the spontaneous nature of the adsorption of Ni(II), Cd(II) and Pb(II) onto shells of hazelnut and almond and the positive values of enthalpy change (ΔH^o) suggested the endothermic nature of the adsorption process. The selectivity order of the adsorbents was Pb(II) > Cd(II) > Ni(II), with adsorption capacities of 16.23, 5.42, and 3.83 mg/g, respectively (Table 31.2). The potential to remove Pb^{2+} ion from aqueous solutions using the shells of hazelnut (HS) and almond (AS) through biosorption was investigated in batch experiments [91]. The equilibrium sorption capacities of HS and AS were 28.18 and 8.08 mg/g for lead, respectively after equilibrium time of 2 h.

Demirbaş et al. [92] studied batch adsorption of Cu^{2+} ions onto hazelnut shells (HS). The capacity of the adsorption for the removal of copper ions from aqueous solution was investigated under different conditions. The sorption capacities increased with increasing pH and decreasing particle size values. The adsorption process becomes more favorable with increasing temperature. The enhancement of adsorption at higher temperatures may be attributed to the enlargement of pore

Table 31.1 Physical and chemical properties of HS raw material used in the experiments

Parameters	Aygün et al. [76]	Cimino et al. [89]	Bulut and Tez [90]	Pehlivan et al. [91]
Cellulose (%)	–	27.3	–	18.24
Hemicellulose (%)	–	24.6	–	28.90
Lignin (%)	–	40.7	–	48.57
Ash content (%)	0.49	1.40	2.10	1.46
C content (%)	51.4	46.8	42.67	48.92
Oxygen (%)	–	44.9	–	45.43
H content (%)	5.95	5.50	4.74	5.65
Nitrogen (%)	–	1.40	–	–
Moisture content (%)	7.7	–	5.49	–
Total loss of ignition (%)	–	–	81.10	
Water soluble components (%)	–	–	7.41	–
Insoluble components (%)	–	–	8.68	–
Particle size (mm)	1.0–1.25	0.3–0.8	0.8	–
Surface area (BET) (m^2/g)	–	–	4.31	–
Bulk density (g/mL)	–	–	0.46	–

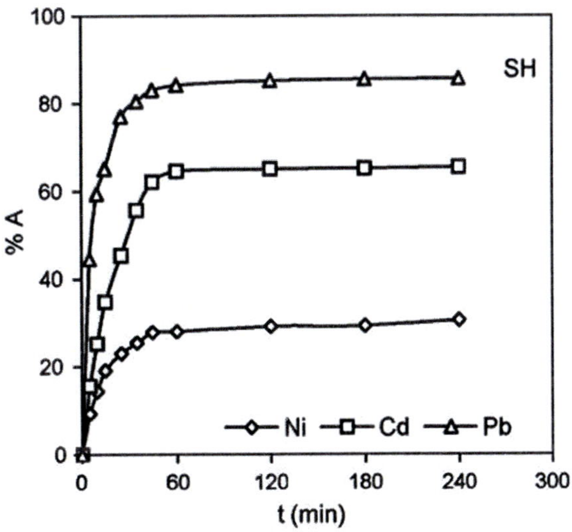

Fig. 31.4 Percent removal of HMI versus time on hazelnut shell (HS) (initial HMI concentration = 100 mg/L; temperature = 298 K; HS concentration = 10 g/L; agitation speed = 150 rpm) [90]

Table 31.2 Heavy metal adsorption capacities and other parameters for raw hazelnut shell (HS) and activated carbon hazelnut shell (ACHS)

Material	Metal ion	Surface area/particle size range	Adsorption capacity/ Removal efficiency	Concentration range	Contact time/flowrate	pH	Source
HS	Cd^{2+}	0.3–0.8 mm	5.42 mg/g	0.129–0.256 mmol/L	5 h	4	[89]
HS	Cr^{3+}	0.3–0.8 mm	3.08 mg/g	0.129–0.256 mmol/L	5 h	4	[89]
HS	Zn^{2+}	0.3–0.8 mm	1.78 mg/g	0.129–0.256 mmol/L	5 h	4	[89]
ACHS	Cr(VI)	0.3–0.8 mm	17.7 mg/g	0.135–0.2562 mmol/L	5 h	2	[89]
HS	Ni(II)	4.31 m^2/g/0.8 mm	3.83 mg/g	100 mg/L	240 min	4.99	[90]
HS	Cd(II)	4.31 m^2/g/0.8 mm	5.42 mg/g	100 mg/L	240 min	4.99	[90]
HS	Pb(II)	4.31 m^2/g/0.8 mm	16.23 mg/g	100 mg/L	240 min	4.99	[90]
HS	Pb^{2+}	75 μm	28.18 mg/g	0.1–1 mM	90 min	2–7	[91]
HS	Cu^{2+}	75–200 μm	0.104 mmol/g	1.575×10^{-5}–1.45×10^{-3} mol/L	24 h	5–7	[92]
HS	Cu(II)	75 μm	6.65 mg/g/75.6%	0.005–0.05 mmol/mL	240 min	6	[94]
HS	Cr (VI)	–	8.28 mg/g/88.46%	0.1–1 mM	240 min	2–3.5	[95]
AHS	Cu^{2+}	80 mesh	21.14 mg/g	0–1000 mg/L	250 min	2–10	[96]
AHS	Pb^{2+}	80 mesh	32.74 mg/g	0–1000 mg/L	250 min	2–10	[96]
H S	Pb(II)	0.5–0.63 mm	91.6–97.3%	20–80 mg/L	4–10 min	5	[97]
HS	Ni	–	–	2.5–125 mg/L	75 min	6	[98]
ACHS	Ni(II)	441 m^2/g/0.9–1.60 mm	11.64 mg/g	15–200 mg/L	180 min	3–8	[99]
HS	Pb^{2+}	5.92 m^2/g	41.9 mg/g	100 mg/L	5–120 min	1–6	[123]
ACHS	Pb^{2+}	736.49 m^2/g	151.9 mg/g	100 mg/L	5–120 min	1–6	[123]
ACHS	Pb(II)	1197.6 m^2/g/300–180 μm	123.8 mg/g	100 mg/L	60 min	4–6	[124]
ACHS	Co(II)	0.80–1.60 mm	13.88 mg/g	13.30–45.55 mg/L	24 h	2–8	[100]

(continued)

Table 31.2 (continued)

Material	Metal ion	Surface area/particle size range	Adsorption capacity/ Removal efficiency	Concentration range	Contact time/flowrate	pH	Source
ACHS	Zn^{2+}	786 m^2/g/0.18 mm	58.8%	–	3 mL/min	6–10	[101]
ACHS	Cu^{2+}	786 m^2/g/0.18 mm	92.9%	–	3 mL/min	6–10	[101]
ACHS	Pb^{2+}	786 m^2/g/0.18 mm	96.9%	–	3 mL/min	6–10	[101]
ACHS	Cd^{2+}	786 m^2/g/0.18 mm	90.5%	–	3 mL/min	6–10	[101]
ACHS	Cu^{2+}	10.1 m^2/g/0.83 mm	40 mg Cu^{2+}/gAc	–	6 h		[102]
ACHS	Cr(VI)	1.25–1.40 mm	170 mg/g	1000 mg/L	72 h	1.0	[110]
ACHS	Cu(II)	441m^2/g /1.0–1.2 mm	58.27 mg/g	25–200 mg/L	120 min	2–6	[111]
ACHS	Cu(II)	1651 m^2/g/1–1.6 mm	240 mg/g	0.013 and 0.260 mol/L	5 h		[109]
ACHS	Cr (VI)	445 m^2/g/0.5–1 mm	667.2 mg/g	–	20–120 min	3	[112]
ACHS	Cr (VI)	445 m^2/g/> 1 mm)	195.2 mg/g	–	20–120 min	3	[112]
ACHS	Cu(II)	646.80 m^2/g/1–1.14 mm	90.8 mg/g	25–1000 mg/L	5–480 min	2–5	[120]
ACHS	Cu(II)	1452 m^2/g	3.07 mmol/g (195 mg/g)	13–260 mmol/L	6 h	3.04–5.50	[121]
ACHS	Cd(II)	–	20.9 mg/g	10–110 mg/L	300 min	5–7	[122]
ACHH	Zn^{2+}	5 m^2/g/1.29 mm	26 mg/g	100–300 mg/L	6 h		[108]
ACHH	Pb^{2+}	980,9 m^2/g	109.9 mg/g	–	20 h	5	[125]
ACHH	Cu(II)	–	105.3 mg/g	–	240 min	5	[126]
ACHS	Cu(II)	1990.23 m^2/g/1 mm	42.28 mg/g	0–100 mg/L	24 h		[127]
ACHS	Cr (VI)	1990.23 m^2/g	89.5 mg/g	10–200 mg/L	10–240 min	1.5–7	[128]

ACHH: activated carbon hazelnut husk, AMPS: 2-acrylamide-2-methyl propane sulfonic acid. AHS: AMPS-modified hazelnut shell size and/or activation of the adsorbent surface [93]. The Langmuir isotherm model appears to fit the isotherm data. The capacity of the adsorption for the removal of copper ions was found 0.104 mmol/g at pH 5–7 (Table 31.2). In 2007, the batch removal of Cu(II) ions from aqueous solution using walnut (WS), hazelnut (HS), and almond (AS) shells was also investigated comparatively [94]. The sorption process conformed to the Langmuir isotherm with maximum Cu(II) ion sorption capacities of 6.74, 6.65, and 3.62 mg/g for WS, HS, and AS, respectively. The percentage removal Cu(II) ion was maximum at 10^{-3} mol/L solution concentration and initial pH of 6.0 (80.3, 75.6, and 75.0% by WS, HS, and AS, respectively). A further investigation [95] on the removal of Cr(VI) ion from aqueous solutions was performed through biosorption using WS, HS and AS in a batch system. The Langmuir isotherm was found suitable with maximum Cr(VI) ion sorption capacities of 8.01, 8.28, and 3.40 mg/g for WS, HS and AS, respectively. Percentage removal by WS, HS and AS was 85.32, 88.46, and 55.00%, respectively.

A high-efficiency, low-cost and environment-friendly 2-acrylamide-2-methyl propane sulfonic acid (AMPS)-modified hazelnut-shell-based adsorbent (AHS) was synthesized and used to adsorb Cu^{2+}, Pb^{2+}, methylene blue (MB) and malachite green (MG) from aqueous solutions by Lu et al. [96]. The maximum adsorption capacities were found to be 21.14 and 32.74 mg/g for Cu^{2+} and Pb^{2+}, respectively (Table 31.2). The adsorption capacities could be maintained above 90% even after ten adsorption–desorption cycles. Ahmed et al. [97] investigated the possibility to enhance Pb(II) removal from wastewater by ultrasound-assisted adsorption using hazelnut shell, a locally and readily available agricultural waste in batch system. Operation parameters were optimized using response surface methodology. The obtained results indicated high adsorption efficiency, between 91.6 and 97.3%, even for very short contact time (4–10 min) and for relatively large particles of grounded shell (particle size between 0.5 and 0.63 mm). Langmuir model fitted better to the isotherms of ultrasound-assisted adsorption. In another study [98], kinetics and isotherms of nickel uptake by activated carbon (AC), sawdust (SD), hazelnut (HS) and almond shells (AS) were compared. pH tests results showed that maximum absorption using AC, HS, SD and AS obtained at pH 6, 6, 6, and 7 respectively. Kinetic model fitting results showed that for sawdust and hazelnut shells, Lagergern model was a better representative. The results also revealed that rate of Nickel adsorption follows this order: AS < SD < HS < AC.

31.4.2 *Removal of Heavy Metals from Wastewaters by Activated Carbon Hazelnut Shells*

Hazelnut shell is an important agricultural residue, the amount produced annually in Turkey being estimated as ca. 3×10^5 tonne. At present, this agricultural waste material is used principally as a solid fuel. In recent years, metal ions such as Cr(VI), Cd(II), Zn(II) [89] and Ni(II) [99] have been removed efficiently from aqueous solutions by activated carbons generated from hazelnut shells (ACHS). Similarly, activated carbon prepared from hazelnut shells was also used to remove Co(II) from aqueous solution by adsorption in the batch mode [100]. The adsorption capacity calculated from the Langmuir isotherm was 13.88 mg Co(II)/g carbon at 303 K employing a pH value of 6 and a particle size of 1.00–1.20 mm (Table 31.2). Chemically-activated hazelnut shell (ACHS) was also prepared by adding 100 ml of double distilled sulphuric acid to 100 g of HS material. After heating in an air oven at 40 °C for 2 h, the product was washed by doubly distilled water until acid reaction was over. It was then suspended in 1% $NaHCO_3$ aqueous solution overnight to remove any residual acid. The material was washed with doubly distilled water and dried for 6 h in the air oven at 60 °C. Kazemipour et al. [101] studied the adsorption of Cu, Zn, Pb, and Cd onto the activated carbon produced from nutshells of walnut, hazelnut, pistachio, almond, and apricot stone. The aqueous adsorption tests indicate that the activated carbon has a notable adsorption capacity for Cu, Zn, Cd, and Pb. The maximum removal occurred at pH 6–10, flowrate of 3 mL/min, and 0.1 g of the adsorbent. The activated carbons produced from shells and stone have high surface areas between 635 and 1208 m^2/g.

Sayan [102] examined the combined chemical and physical activation process to prepare activated carbon from ultrasound-assisted KOH-impregnated hazelnut shell for removing heavy metals cations such as Cu^{2+}. For this purpose, hazelnut shells were impregnated with KOH solution under ultrasound irradiation. After filtration, hazelnut shells have been carbonized under inert N_2 atmosphere. It is now widely accepted that ultrasound power has great potential for uses, in addition to conventional applications in cleaning and plastic welding, in a wide variety of industrial fields such as electrochemistry, food technology, nanotechnology, chemical synthesis, dissolution and extraction, dispersion of solids, phase separation, water and sewage treatment [103]. Ultrasound produces its mechanical and chemical effects through the formation and collapse of "cavitations" bubbles [104]. A significant amount of research has been published concerning with this "sonochemical effect", and collected in various recent books [105, 106]. Ultrasound power (continuous mode) is adjusted using the relationship between the intensity setting of the generator, and ultrasound power absorbed by the reaction medium measured by the calorimetric method [107]. BET surface areas of the raw and activated carbon from hazelnut shell were calculated to be 0.188 and 10.1 m^2/g, respectively. Considering BET surface areas, the surface area of the activated carbon is 50 times greater than raw hazelnut shell surface area.

In a similar study [108], activated carbon adsorbent was prepared for removing heavy metal cations such as Zn^{2+} from aqueous solutions. The experiments were planned by statistical design methods. Finally, activated carbons were characterized by the evolution of their zinc adsorption capacity. BET surface areas of the raw and activated carbon from hazelnut shell were calculated to be 0.188 and 5 m^2/g, respectively. A further study [109] was also conducted on the removal of Cu(II) ions by ultrasound-assisted adsorption onto ACHS. The granular activated carbon was prepared from ground dried HS by simultaneous carbonization and activation by water steam at 950 °C for 2 h. The maximum adsorption capacity of the adsorbent for Cu(II), calculated from the Langmuir isotherms, in the presence of ultrasound

(3.77 mmol/g) is greater than that in the absence of ultrasound (3.14 mmol/g). The removal of Cu(II) ions was higher in the presence of ultrasound than in its absence, but ultrasound reduced the rate constant.

Kobya [110] studied the adsorption Cr(VI) from aqueous solutions onto ACHS. Adsorption of Cr(VI) is highly pH dependent and the best results are obtained in the pH range 1.0–2.0. At low pH a high percentage of Cr(VI) reduces to Cr(III) form. Chromium is adsorbed rapidly when lower concentrations are used. Adsorption capacity was calculated from the Langmuir isotherm as 170 mg/g at an initial pH of 1.0 for the 1000 mg/l Cr(VI) solution. Thermodynamic parameters were evaluated and the adsorption is endothermic showing monolayer adsorption of Cr(VI). Demirbaş et al. [111] utilized waste hazelnut shell as the raw material for the production of granular activated carbon by chemical activation and evaluated its adsorption capacity for Cu(II) ions from aqueous solutions. The adsorption capacity of Cu(II) increased with time and attained a maximum value at 60 min and thereafter, it reached a constant value indicating that no more Cu(II) ions were further removed from the solution (Fig. 31.5). Desorption efficiencies in four cycles were found to be in the range 74–79%. The monolayer adsorption capacity of Cu(II) ions was determined as 58.27 mg/g at 323 K (Table 31.2). Thermodynamic parameters were calculated for the Cu(II) ion–ACHS system and the positive value of ΔH (18.77 kJ/mol) showed that the adsorption was endothermic and physical in nature.

The possible use of activated hazelnut shell ash and activated bentonite as the adsorbents of Cr(VI) from synthetic solutions and the effect of operating parameters were investigated [112]. The activated hazelnut shell ash was prepared thermally in two sizes 0.5 and 1.0 mm. The maximum removal of Cr(VI) occurred at pH 3 by activated hazelnut shell ash and at pH 5 by activated bentonite. The Freundlich model agrees very well with experimental data. It is evident that after chemical modification or conversion by heating into activated carbon, the activated carbon hazelnut shells (ACHS) originating from raw hazelnut shells have demonstrated

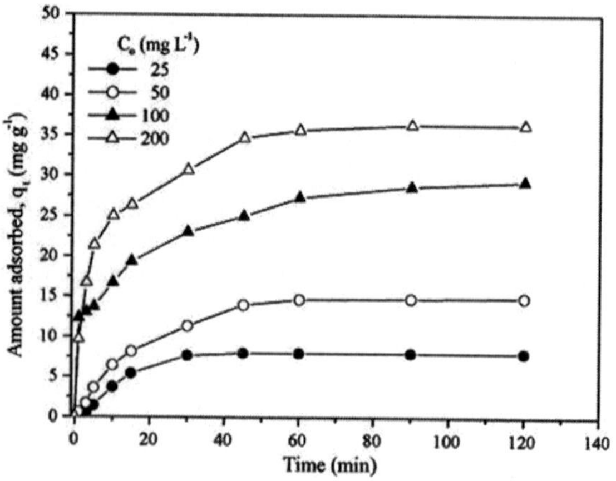

Fig. 31.5 Effect of initial concentration on the adsorption of Cu(II) by ACHS (Conditions: pH 6, 1.00–1.20 mm particle size, 3 g/L adsorbent dosage, 300 rpm, 120 min, agitation time and 293 K) [111]

outstanding metal removal capabilities. To justify their viability as effective adsorbents for heavy metal removal, the adsorption capacities of all low-cost adsorbents need to be compared. Adsorption capacities (mg/g) for the removal of heavy metal such as Cr(II), Cr(VI) and Ni (II) by activated hazelnut shell (ACHS) and other agricultural based activated carbons are reported in Table 31.3 [113–119]. Among these agricultural based activated carbons, activated carbon hazelnut shell-ash (Cu(VI): 667.2 mg/g), ACHS-ultrasound (Cu(VI): 240 mg/g), ACHS (Cr(VI): 170 mg/g), AC-coconut shell (Cr(VI): 107.1 mg/g) and AC-peanut hull (Cu(II): 65.57 mg/g) stand out for significantly higher metal adsorption capacities compared to the other adsorbents.

Recently, activated carbons prepared from hazelnut shells and apricot stones were used as adsorbents for the removal of copper(II) ions from aqueous solution [120]. The amount of Cu(II) ions adsorbed increased with increasing temperature, pH and initial Cu(II) ion concentration. The Freundlich model showed a better fit of the adsorption data. The physical and chemical properties of hazelnut shell and apricot stone activated carbons were determined for comparative purposes. Figure 31.6a shows that the hazelnut shell activated carbon had a porous structure, an observation supported by the BET surface area (646.80 m^2/g) and iodine number (276 mg/g) of this activated carbon. As can be seen from Fig. 31.6b, most of the adsorbed Cu(II) ions were located inside the pores. In 2013, the removal of copper(II) ion

Table 31.3 Recent reported adsorption capacities (mg/g) for heavy metal sorption on activated carbon hazelnut shell (ACHS) and other agricultural based activated carbons

Adsorbent	Metal ion	Surface area (m^2/g)	Adsorption capacity (mg/g)	Source
ACHS, Steam activated-ultrasound	Cu(II)	1651	240	[109]
ACHS	Cu(II)	441	58.27	[111]
AC-Rubber wood dust, H_3PO_4 activat	Cu(II)	1674	5.73	[113]
AC-Peanut hull	Cu(II)		65.57	[116]
AC-Olive stone	Cu(II)		9.21	[117]
AC-Almond Shell	Cu(II)		8.85	[117]
ACHS-ash	Cr(VI)	445	667.2	[112]
ACHS	Cr(VI)		170	[110]
AC-Coconut Shell	Cr(VI)		107.1	[114]
AC-Wood	Cr(VI)		87.6	[114]
P-Sawdust	Cr(VI)	320	37.8	[115]
ACHS	Ni(II)	441	11.64	[99]
AC-Coir pith	Ni(II)	592	62.5	[118]
AC-Almond husk	Ni(II)		37.17	[119]

AC: Activated carbon, P: Pyrolysis

(a)
(b)
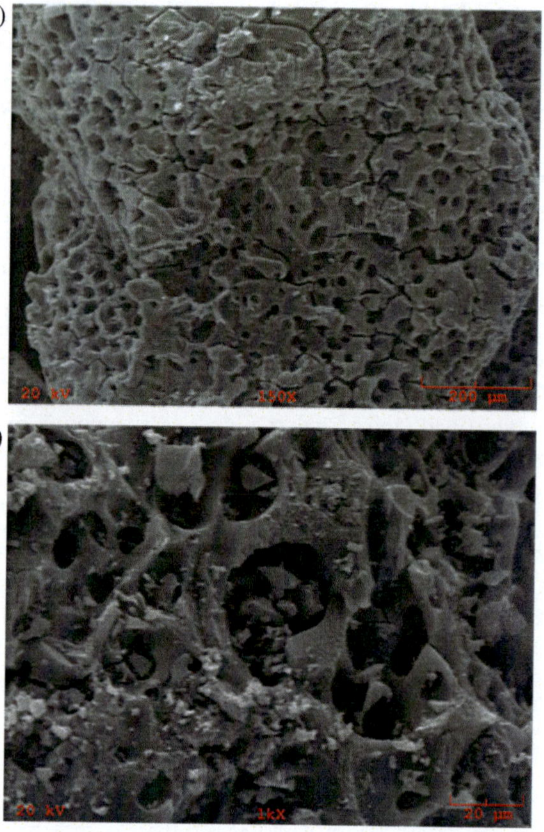

Fig. 31.6 Scanning electron micrographs of hazelnut shell activated carbon: **a** before and **b** after the adsorption of Cu(II) ions [120]

from aqueous solution by the granular activated carbon, obtained from hazelnut shells (ACHS), was investigated in the batch tests [121]. The ACHS was prepared from ground dried hazelnut shells by specific method carbonisation and water steam activation at 950 °C for 2 h. The granular activated carbon produced from hazelnut shells has a high specific surface area (1452 m^2/g) and highly developed microporous structure (micropore volume: 0.615 cm^3/g). The monolayer adsorption capacity of the ACHS-copper(II) ion, calculated from the Langmuir isotherms, was 3.07 mmol/g (195 mg/g).

Imamoglu [122] reported a batch adsorption of Cd(II) ions onto activated carbon (AC) produced from hazelnut husks. The AC showed a high affinity to Cd(II) ions at pH values between 5.0 and 7.0. Maximum Cd(II) adsorption capacity of AC was calculated to be 20.9 mg/g. Cd(II) adsorption was kinetics described well with the pseudo second order model. Şencan et al. [123] investigated the Pb^{+2} adsorption capacities of hazelnut shell and activated carbon obtained from hazelnut shell. The raw hazelnut shell's BET surface area is 5.92 m^2/g and the surface area of activated carbons which is pyrolyzed at 250 and 700 °C were determined 270.2 and 686.7 m^2/g, respectively. The surface area of hazelnut shell, which pyrolyzed at

700°C after being activated with $ZnCl_2$, was determined to be 736.49 m^2/g. Results show that physical adsorption process is dominant for the activated carbon pyrolysis at 700 °C but the chemical adsorption is dominant for the activated carbon pyrolysis at lower degrees and for raw hazelnut shell. Obtained results after the experimental studies are shown in Fig. 31.7. Fencan and Kılıç [124] used raw hazelnut shells to obtain charcoal by pyrolysis at 250 °C and the obtained material was subjected to physical, chemical, and physicochemical treatment methods to obtain activated carbons (ACs). The charcoal samples were examined for Pb(II) adsorption from aqueous solutions under different pH conditions of 4 to 6. Effect of microwave and ultrasound treatments was also examined during Pb(II) adsorption by the chemically treated AC. The results showed that chemical treatment with $Ca(OCl)_2$, microwave treatment for 5 min, ultrasound treatment for 20 min, and pyrolysis at 700 °C together were the most suitable combination enhancing the surface area of the adsorbent. This combination increased the surface area and the adsorption capacity of the adsorbent by 202 and 4.76 times, respectively, when compared to those of the raw hazelnut shell.

Hazelnut husk (HH) was also used to prepare an activated carbon (ACHH) by chemical activation using K_2CO_3 [125]. ACHH has a high surface area (980.9 m^2/g) and contains 2.60 mmol/g of total acidic functional groups. ACHH was used for the removal of Pb(II) from aqueous solutions, where a number of effective factors were investigated. For Pb(II) adsorption on ACHH, kinetics followed pseudo-second-order model and the Langmuir model showed a better fit of the adsorption data while maximum adsorption capacity was 109.9 mg/g. In a similar study [126], hazelnut husk (HH) was converted to carbonaceous material by chemical activation using potassium acetate. ACHH was used for the batch adsorption of Cu(II) ions from aqueous solutions and the adsorption capacity was found as 105.3 mg/g. Zhao

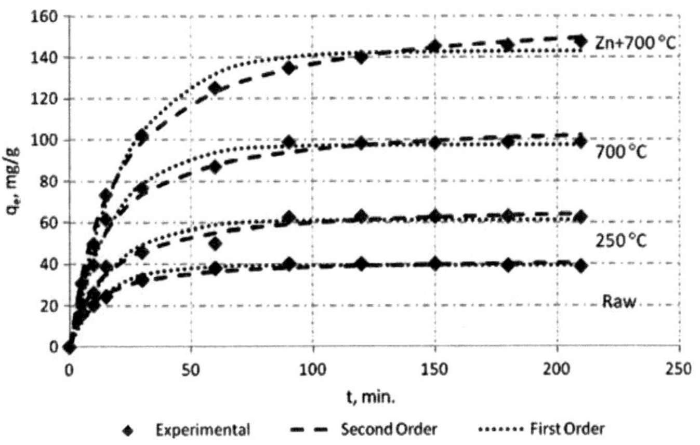

Fig. 31.7 Pseudo-first- and second-order kinetical graphics of the sorption process made with nutshell (conditions: 100 mg/L initial concentration, pH 5, 300–180 μm particle size, 1 g/L adsorbent dosage, 150 rpm, and 20 °C) [123].

et al. [127] applied the co-pyrolysis technology to municipal sewage sludge (MSS) and hazelnut shell with alkaline activating agent K_2CO_3 under N_2 atmosphere. The specific surface area reached 1990.23 m^2/g, and the iodine absorption number was 1068.22 mg/g after co-pyrolysis at 850 °C. The residual fractions of heavy metals in bio-char were above 92.95% after co-pyrolysis at 900 °C except Cd to prevent heavy metals digestion. The Cu(II) adsorption capacity of the bio-char was 42.28 mg/g after 24 h, and surface functional groups acted as active binding sites for Cu(II) adsorption. In 2021, in order to produce an adsorbent with high Cr(VI) adsorption capacity, two different composite adsorbents were synthesized by coating citric acid modified almond and hazelnut shells with chitosan [128]. Then, chitosan in the structure was cross-linked with glutaraldehyde. Cr(VI) adsorption performances of chitosan-citric acid modified almond shell composite (C-CA-AS) and chitosan-citric acid modified hazelnut shell composite (C-CA-HS) were investigated. In batch adsorption experiments, the optimum adsorbent dosages for the adsorption of Cr(VI) from the 55 mg/L Cr(VI) solution using C-CA-AS and C-CA-HS adsorbents were 1.8 and 1.0 g/L, respectively. The maximum adsorption capacities calculated by Langmuir isotherm were 100.3 mg/g for C-CA-AS and 89.5 mg/g for C-CA-HS. Adsorption was found to be compatible with pseudo-second-order kinetic model, and it was determined that adsorption was spontaneous and endothermic.

31.5 Removal of Dyes from Wastewaters Using Hazelnut Shells

31.5.1 Removal of Dyes from Wastewaters by Raw Hazelnut Shells

The colour removal from dyehouse effluents becomes day by day more than ever an important objective of wastewater treatment processes. The dye adsorption behaviour of ground hazelnut shells was compared with that of wood sawdust, a low cost adsorbent already experimented for dye removal by Ferrero [129]. Both batch adsorption of methylene blue and acid blue onto ground hazelnut shells (HS) were investigated in comparison with sawdust of various species of wood and fixed bed adsorption of methylene blue on HS columns. The equilibrium data were processed according to Langmuir's model and higher adsorption capacity values towards both dyes were shown by HS than wood sawdust. The adsorption capacity of hazelnut shells (HS) changed according to particle size of HS (125–500 μm) and dye concentration (50–1000 mg/L) (methylene blue: 5–76,9 mg/g, acid blue 25: −60.2 mg/g) (Table 31.4). In the subsequent research [129], fixed bed adsorption of methylene Blue was also performed on HS columns and the breakthrough curves were determined by varying bed depth, flow rate and influent concentration. The data were processed according to Bohart–Adams model and the column performances could be predicted by the bed

Table 31.4 Dye adsorption capacities and other parameters for raw hazelnut shell (HS) and activated carbon hazelnut shell (ACHS)

Material	Dye	Surface area/Particle size	Adsorption capacity/Removal efficiency	Concentration range	Contact time	pH	Source
HS	Methylene blue	21.6 m^2/g/125 μm	76.9 mg/g	500–1000 mg/L	360 min	3–4	[129]
HS	Acid blue 25	21.6 m^2/g/125 μm	60.2 mg/g	500–1000 mg/L	360 min	3–4	[129]
HS	Congo red		13.75 mg/g/43%	50–5000 mg/L	24 h		[136]
HS	Methylene blue	2.85 m^2/g/0–75 μm	11.95–27.51 × 10^{-5} mg/g	3 × 10^{-3} mol/L	24 h	3–9	[137]
HS	Methylene blue	2.85 m^2/g/0–200 μm	11–19.5 × 10^{-5} mg/g	1–3 × 10^{-4} mol/L	40 min	4.1–4.5	[138]
AMPS modified HS	Methylene blue	80 mesh	68.03 mg/g	0–1000 mg/L	250 min	2–10	[96]
AMPS modified HS	Malachite green	80 mesh	263.16 mg/g	0–1000 mg/L	250 min	2–10	[96]
ACHS	Methylene blue	793 m^2/g/1–1.25 mm	8.82 mg/g	–	24 h	4.8	[76]
ACHS	Acid dye	–	50.5 mg/g	–	–	–	[139]
ACHS	Direct yellow 50,	1.5 mm	11 mg/g/8%	300 mg/L	40 min	5.6	[140]
ACHS	Direct red 80,	1.5 mm	14 mg/g/12%	300 mg/L	40 min	6.8	[140]
ACHS	Direct blue 71	1.5 mm	26 mg/g/88%	300 mg/L	40 min	6.4	[140]
ACHS	Acid red 183	–	111 mg/g	–	–	–	[141]
ACHH	Methylene blue	770 m^2/g	204 mg/g	–	–	–	[142]
ACHS	Reactive red 2	75 μm	8.67 mg/g	20 mg/L	80 min	3–11	[151]
ACHH	Methylene blue	106–300 μm	476.2 mg/g	5–800 mg/L	120 min	3–12	[149]
ACHS	Alizarin yellowGG	124.3 m^2/g	3.1 mg/g	20–60 ppm	500 min	3–10	[150]

ACHH: Activated carbon hazelnut husk

depth service time (BDST) approach. Some theoretical models of fixed bed adsorption were developed for activated carbon and applied to other adsorbents [130–135]. Lu et al. [96] synthesized a high-efficiency, low-cost and environment-friendly 2-acrylamide-2-methyl propane sulfonic acid (AMPS)-modified hazelnut-shell-based adsorbent (AHS) and used to adsorb Cu^{2+}, Pb^{2+}, methylene blue (MB) and malachite green (MG) from aqueous solutions. Adsorption kinetics shows that the adsorption rate is well represented by the pseudo second-order rate model, and the Langmuir model gives the best fit adsorption isotherm. The Langmuir maximum adsorption capacities were found to be 68.03 mg/g for MB and 263.16 mg/g for MG, respectively (Table 31.4).

Recently, Carletto et al. [136] has investigated the adsorption of the direct azo dye Congo red on hazelnut shell within a concentration range of 50–5000 mg/L. Hazelnut shells were employed as organic support for Phanerochaete chrysosporium cultures to study the best cultural medium composition for the MnP production. Cultures of Phanerochaete chrysosporium were carried out with hazelnut shells coming from Congo red adsorption tests, showing that 43% of the adsorbed dye was degraded. In 2008, the batch adsorption of a basic dye, methylene blue (MB), was performed onto ground hazelnut shell [137]. The extent of the MB removal increased with the increase of solution pH, ionic strength and temperature but decreased with increase in the particle size. The maximum adsorption capacities for MB were 2.14–2.31 × 10^{-4} mol/g at temperature of 25, 35, 45, and 55°C, respectively. Adsorption heat revealed that the adsorption of MB is endothermic in nature. Dogan et al. [138] reported the adsorption kinetics of methylene blue (MB) on hazelnut shell. Adsorption activation energy was calculated to be 45.6 kJ/mol. The values of activation parameters such as free energy (ΔG°), enthalpy (ΔH°) and entropy (ΔS°) were also determined as 83.4 kJ/mol, 42.9 kJ/mol and −133.5 J/(mol K), respectively. Figure 31.8 shows the effects of contact time on the amount of MB adsorbed by hazelnut shell under different initial MB concentrations.

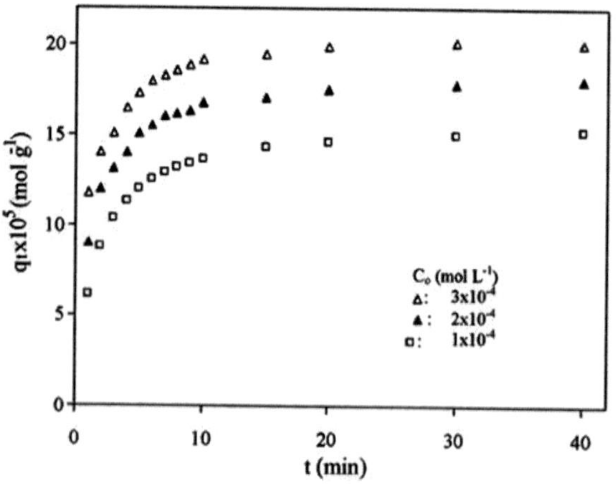

Fig. 31.8 The effect of initial dye concentration to the adsorption rate of MB on raw hazelnut Shell (HS) (*T*: 30 °C, pH: 4.1–4.5, SS: 200 rpm, PS: 0–75 μm) [138]

31.5.2 Removal of Dyes from Wastewaters by Activated Carbon Hazelnut Shells

An investigation of several agricultural wastes (almond shell, hazelnut shell, walnut shell and apricot stone) revealed that their suitability for granular activated carbon production is more governed by type-specific features rather than material characteristics (elemental composition). Granular activated carbons were evaluated for their physical (attrition, bulk density), chemical (elemental composition, % weight loss), surface (surface area, surface chemistry) and adsorption properties (iodine number, phenol and methylene blue adsorption) [76]. Adsorption isotherm data were fitted to both Langmuir and Freundlich models. Highest surface areas of activated carbons from ACAS, ACHS, and ACWS were found at 750 °C with 10 h activation time but for ACAST, it was obtained at 800 °C with 18 h activation time. ACHS has the highest surface area and iodine number. The iodine numbers of ACHS and ACWS are higher than that of other two activated carbons. All carbons have high bulk density and hardness. These properties can be explained by high lignin and low ash contents of the shells (Table 31.5).

The removal of dyestuff from aqueous solutions by different adsorbents, such as activated carbon, kaolinite and montmorillonite was studied [139]. The adsorption isotherm data were fitted to the Langmuir isotherm. For Acid dye, the adsorption capacities of RAC (a commercial activated carbon), ACHS, KC (raw kaolinite) and MC (montmorillonite) were obtained as 967.0, 50.5, 43.7, and 31.3 mg/g at 25 °C, respectively. Similarly, the removal of Direct dyes (Direct yellow 50 (DY50), Direct red 80 (DR80) and Direct blue 71 (DB71)) from an aqueous solution by different adsorbents such as RAC, ACHS, KC, and MC was also investigated [140]. Adsorption capacities of RAC, ACHS, KC, and MC increased in the following order: DY50 > DR80

Table 31.5 Physico-chemical characteristics of activated carbons [76]

Characteristics	ACAS	ACAST	ACHS	ACWS
Particle size (mm)	1.00–1.25	1.00–1.25	1.00–1.25	1.00–1.25
Burn off (wt.%)	67	70	63	70
Bulk density (kg/L^3)	0.80	0.82	0.79	0.81
BET surface area (m^2/g)	736	783	793	774
Iodine number (mg/g)	638	754	965	905
Attrition (%)	9.4	12	10	11
C (%)	79.2	76.0	72.0	75.3
H (%)	0.64	0.49	1.8	0.71
Acidic groups (mmol/g)	0.33	0.00	0.00	0.13
Basic groups (mmol/g)	0.66	0.31	0.39	0.36

ACAS: Activated carbon almond shell, ACAST: Activated carbon apricot fruit stone, ACHS: Activated carbon hazelnut shell, ACWS: Activated carbon walnut shell

> DB71, DB71 > DR80 > DY50, DR80 > DB71 > DY50, and DB71 > DR80, respectively. Recently, the removal of Acid red 183 from aqueous solution onto activated carbon, raw kaolinite and montmorillonite was also researched [141]. The adsorption capacities were 1495, 111, 29, and 19 mg/g for RAC, ACHS, KC and MC at 25 °C, respectively (Table 31.4). The activated carbons produced from hazelnut shells have high surface areas and highly developed micropore structure.

Özer et al. [142] prepared activated carbon from hazelnut husks (ACHH) using phosphoric acid activation and characterized by some physicochemical analyses. Surface area and zero point charge of pH (pH_{PZC}) were found to be 770 m^2/g and 4.5, respectively. Adsorption of methylene blue (MB) was elevated with increasing pH and temperature. The equilibrium data were analyzed by Langmuir and an adsorption capacity of 204 mg/g was found. Table 31.6 presents the dye adsorption capacities of the activated carbon hazelnut shell (ACHS) and other agricultural based activated carbons [69, 76, 140, 141, 143–148]. The removal capacities of AC-Bamboo (Methylene blue: 454.20 mg/g), AC-Vetiver roots (Methylene blue: 375 mg/g), and AC-Orange peel (Direct yellow 12: 75.76 mg/g) were found higher than that of activated carbon hazelnut shell (ACHS) (Methylene blue: 8.82 mg/g, and Direct yellow 50: 11 mg/g). For acid dye removal, the adsorption capacity of ACHS (Acid red 183: 111 mg/g), was significantly higher than that of AC-Guava seed (Acid red 1: 0.4 mg/g). Karaçetin et al. [149] analyzed activated carbon prepared from hazelnut husk (ACHH) using zinc chloride and investigated adsorption from aqueous solution of Methylene blue (MB) by the batch method. SEM images of HH

Table 31.6 Recent reported adsorption capacities (mg/g) for dye sorption on activated carbon hazelnut shell (ACHS) and other agricultural based activated carbons

Adsorbent	Dye	Surface area (m^2/g)	Adsorption capacity (mg/g)	Source
ACHS	Methylene blue	793	8.82	[76]
AC-Bamboo	Methylene blue	1896	454.20	[143]
AC-Vetiver roots	Methylene blue	1185	375	[145]
AC-Salsola vermiculata plant	Methylene blue	1178	130	[69]
AC-Coir pith	Methylene blue		5.87	[144]
AC-Apricot stone	Methylene blue	783	4.11	[76]
AC-Walnut shell	Methylene blue	774	3.53	[76]
AC-Almond shell	Methylene blue	736	1.33	[76]
ACHS	Direct yellow 50		14	[140]
AC-Orange peel	Direct yellow 12		75.76	[146]
ACHS	Direct blue 71		26	[140]
AC-Orange peel	Direct Blue 86		33.78	[148]
ACHS	Acid red 183		111	[141]
AC-Guava seed	Acid red1	12	0.4	[147]

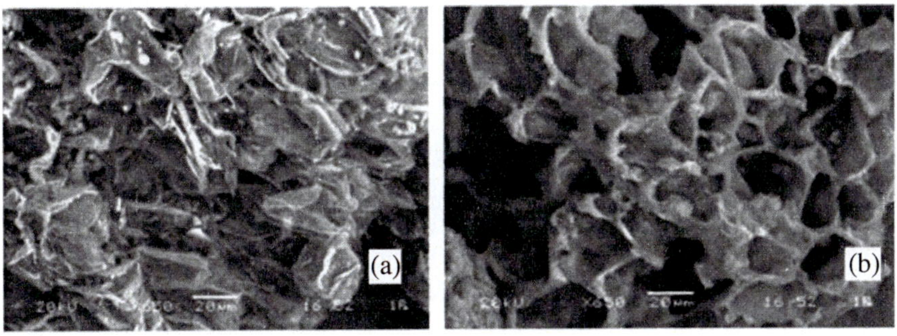

Fig. 31.9 **a** SEM micrographs of **a** HH and **b** ACHH [149]

and ACHH were obtained by magnifying 650 × and are shown in Fig. 31.9a and b, respectively. When surface patterns and porous form of both carbons are compared with each other, it is apparent that the ACHH has a porous structure. MB adsorption capacity of ACHH was found to be 476.2 mg/g and MB adsorption kinetics corresponds well with pseudo second order model. Thermodynamic studies revealed that MB adsorption is a spontaneous and endothermic process.

Low-cost biosorbents such as pine cone char, walnut shell char, and hazelnut shell char were prepared by pyrolysis process at different carbonization temperatures in the range of 400–700 °C [150]. Biochars with the highest surface area were used to remove Alizarin yellow GG from aqueous solution and the adsorption capacities of these materials were compared to commercially available activated carbon. The maximum removals of Alizarin yellow GG were about 89, 82, 78, and 52% for commercially available activated carbon (charcoal), pine cone char (259.740 m^2/g), walnut shell char (256.931 m^2/g), and hazelnut shell char (124.347 m^2/g), respectively (Table 31.4). Al-Sharify et al. [151] used the prepared activated carbon from hazelnut shells (HS) to remove Reactive red 2 dye (RR 2) from aqueous solutions. The adsorption behavior of dye onto the porous carbon was studied by varying parameters. The maximum adsorption capacity of dye was 8.67 mg/g at 45 °C and pH 3. The thermodynamic parameters (ΔG^o, ΔH^o, ΔS^o) were also determined, which revealed that the process of adsorption Reactive red 2 dye was spontaneous and endothermic while an increase in temperature favored the adsorption.

31.6 Conclusions

This review presents hazelnut shell as a low-cost material to be used for the removal of heavy metals and dyes from the wastewaters. Hazelnut shell is a biomass and Turkey has become the largest hazelnut producer per year worldwide, i.e. about 70% of total world hazelnut production, which corresponds to 675,000 tonnes of annual hazelnut production performed on ca. 700,000 hectares of land. Activated carbon has great potential for removal of heavy metals and dyes from water owing to its large surface area, nanoscale size, and availability of different functionalities, not to mention that they tend to be more easily modified and recycled. Generally, hazelnut shell has low removal efficiency and adsorption capacity for the removal of heavy metals and dyes. It is expected that activated carbons prepared from hazelnut shells using different activation processes have high surface areas and may show high removal performance. Extensive literature information on the removal of heavy metals and dyes from wastewaters using hazelnut shells has been reviewed in this paper. The raw hazelnut shell's BET surface area was found 5.92 m^2/g and the surface area of activated carbons which is pyrolyzed at 250 and 700 °C were determined 270.2 and 686.7 m^2/g, respectively. The surface area of raw hazelnut shell (HS), which pyrolyzed at 700 °C after being activated with $ZnCl_2$, was determined to be 736.49 m^2/g for activated carbon hazelnut shell (ACHS). The adsorption capacity of HS was 28.18 mg/g for lead. The maximum adsorption capacities of AMPS modified HS were found to be 21.14 mg/g for Cu^{2+}, 32.74 mg/g for Pb^{2+}, 68.03 mg/g for Methylene blue (MB) and 263.16 mg/g for Methylene green (MG), respectively. The maximum Pb^{+2} adsorption capacities were found to be 667.2 and 195.2 mg/g for ACHS (0.5 mm) and ACHS (1.0 mm), respectively. The adsorption capacity toward MB of HSs with $d_p = 125\ \mu m$ was found to be 76.9 mg/g. MB adsorption capacity of the activated carbon hazelnut husk (ACHH) was also found to be 476.2 mg/g. This study demonstrated that hazelnut shell and its activated carbon forms could be used as effective adsorbents for the treatment of wastewater containing heavy metals and dyes. Hazelnut shell as a low-cost adsorbent proposes a lot of promising benefits for commercial purposes in the future.

References

1. T. Chakrabarti, O.V.R. Subrahmanyan, B.B. Sundaresan, Biodegradation of recalcitrant industrial wastes, in *Biotreatment Systems*, vol. II (CRC Press, Inc. Boca Raton, FL, 1988)
2. T.J. Chern, S.N. Huang, Ind. Eng. Chem. Res. **37**, 253 (1998)
3. T.S. Anirudhan, C. Raji, K.P. Shubha, Water Res. **35**, 300 (2001)
4. G. McKay, *Use of Adsorbents for the Removal of Pollutants from Wastewater* (CRC Press, Inc., Tokyo, 1996)
5. B.S. Girgis, M.F. Ishak, Mater. Let. **39**, 107 (1999)
6. G. Rytwo, S. Nir, M. Crepsin, L. Margulies, J. Colloid Interface Sci. **222**, 12 (2000)
7. M. Turan, U. Mart, B. Yuksel, M.S. Celik, Chemosphere **60**, 1487 (2005)
8. O. Ozdemir, B. Armagan, M. Turan, M.S. Celik, Dyes Pigm. **62**, 51 (2004)
9. A.B. Engin, O. Ozdemir, M. Turan, A.Z. Turan, J. Hazard. Mater. **159**, 348 (2008)
10. G. Crini, Bioresource Technol. **97**, 1061 (2006)
11. S. Babel, T.A. Kurniawan, J. Hazard. Mater. **97**, 219 (2003)
12. C. Bilgic, Adsorption. Sci. Technol. **26**, 363 (2008)
13. A.R. Kul, N. Caliskan, Adsorption. Sci. Technol. **27**, 85 (2009)
14. V.J. Inglezakis, H. Grigoropoulou, J. Hazard. Mater. **112**, 37 (2004)
15. J. Peric, M. Trgo, N. Vukojevic Medvidovic, Water Res. **38**, 893 (2004)
16. Y.E. Benkli, M.F. Can, M. Turan, M.S. Celik, Water Res. **39**, 487 (2005)
17. D. Karadag, E. Akgul, S. Tok, F. Erturk, M.A. Kaya, M. Turan, J. Chem. Eng. Data **52**, 2436 (2007)

18. A. Faki, M. Turan, O. Ozdemir, A.Z. Turan, Ind. Eng. Chem. Res. **47**, 6999 (2008)
19. X. Huang, X. Liao, B. Shi, J. Hazard. Mater. **173**, 33 (2010)
20. S. Tunali, A.S. Özcan, A. Özcan, T. Gedikbey, J. Hazard. Mater. **135**, 141 (2006)
21. T. Mathialagan, T. Viraraghavan, J. Hazard. Mater. **94**, 291 (2002)
22. Y.S. Ho, G. McKay, Chem. Eng. J. **70**, 115 (1998)
23. F.C. Wu, R.L. Tseng, R.S. Juang, J. Hazard. Mater. **B73**, 63 (2000)
24. Z. Aksu, Biochem. Eng. J. **7**, 79 (2001)
25. C. Namasivayam, M.D. Kumar, R.A. Begum, Biomater. Bioenerg. **21**, 477 (2001)
26. M. Hasan, A.L. Ahmad, B.H. Hameed, Chem. Eng. J. **136**, 164 (2008)
27. M. Özacar, İA. Şengil, Process Biochem. **40**, 565 (2005)
28. H. Masitah, S.N. Azzahra, J. Engng Resear. Educat. **11**, 15 (2019)
29. F. Güzel, H. Yakut, G. Topal, J. Hazard. Mater. **153**, 1275 (2008)
30. Y.H. Magdy, A.A.M. Daifullah, Waste Manage. **18**, 219 (1998)
31. S. Netpradit, P. Thiravetyan, S. Towprayoon, J. Colloid Interf. Sci. **270**, 255 (2004)
32. F.L. Fu, Y. Xiong, B.P. Xie, R. Chen, Chemosphere **66**, 1 (2007)
33. M. Rao, A.V. Parawate, A.G. Bhole, Waste Manage. **22**, 821 (2002)
34. C. Namasivayam, R. Jeyakumar, R.T. Yamuna, Waste Manage. **14**, 643 (1994)
35. R. Baby, B. Saifullah, M.Z. Hussein, Nanoscale Res. Let. **14**, 341 (2019)
36. W.T. Tsai, C.Y. Chang, S.Y. Wang, C.F. Chang, S.F. Chien, H.F. Sun, Bioresource Technol. **78**, 203 (2001)
37. S.J. Park, W.Y. Jung, Carbon **40**, 2021 (2002)
38. J. Guo, A.C. Lua, Micropor. Mesopor. Mater. **32**, 111 (1999)
39. R. Sanghi, B. Bhattacharya, Color. Technol. **118**, 256 (2002)
40. S.J. Allen, B. Koumanova, J. Univ. Chem. Technol. Metall. **40**, 175 (2005)
41. J. Guo, A.C. Lua, Carbon **38**, 1985 (2000)
42. G.H. Oh, C.R. Park, Fuel **81**, 327 (2002)
43. C. Namasivayam, K. Kadirvelu, Bioresource Technol. **62**, 123 (1997)
44. K.J. Hu, J.L. Hu, K.P. Ho, K.W. Yeung, Carbohydrate Polym. **58**, 45 (2004)
45. A. Bacaoui, A. Yaacoubi, A. Dahbi, C. Bennouna, R.P.T. Luu, F.J. Maldonado-Hodar, J. Rivera-Utrilla, C. Moreno-Castilla, Carbon **39**, 425 (2001)
46. H. Haykiri-Acma, J. Anal. Appl. Pyrol. **75**, 211 (2006)
47. A. Topuz, M. Topakci, M. Canakci, I. Akinci, F. Ozdemir, J. Food Eng. **66**, 519 (2004)
48. N. Usta, E. Ozturk, O. Can, E.S. Conkur, S. Nas, A.H. Con, A.C. Can, M. Topcu, Energy Convers. Manage. **46**, 741 (2005)
49. F. Cimen, S.S. Ok, C. Kayran, S. Demirci, D.B. Ozenc, N. Ozenc, Biodegradation **18**, 295 (2007)
50. W. Heschel, E. Klose, Fuel **74**, 1786 (1995)
51. A. Demirbas, Energy **24**, 141 (1999)
52. J. Rouquerol, D. Avnir, C.W. Fairbridge, D.H. Everett, J.M. Haynes, N. Pernicone, J.D.F. Ramsay, K.S.W. Sing, K.K. Unger, Pure Appl Chem **66**, 1739 (1994)
53. J. Haber, Pure Appl. Chem. **63**, 1227 (1991)
54. J.H. De Boer, The structure and properties of porous materials. in *Proceedings of the tenth symposium of the colston research society*. (University of Bristol, Butterworths, London, 1958). pp. 68–94
55. B. Nie, X.F. Liu, L.L. Yang, J. Meng, X.C. Li, Fuel **158**, 908–917 (2015)
56. J. Weitkamp, M. Fritz, S. Ernst, Int. J. Hydrogen Energy **20**, 967 (1995)
57. P.M. Barata-Rodrigues, T.J. Mays, G.D. Moggridge, Carbon **41**, 2231 (2003)
58. J. Kim, J. Lee, T. Hyeon, Carbon **42**, 2711 (2004)
59. A.B. Fuertes, D.M. Nevskaia, Micropor. Mesopor. Mater. **62**, 177 (2003)
60. A.B. Fuertes, S. Alvarez, Carbon **42**, 3049 (2004)
61. T. Kyotani, Carbon **38**, 269 (2000)
62. A.B. Fuertes, T.A. Centeno, J. Mater. Chem. **15**, 1079 (2005)
63. R.E. Franklin, Nature **177**, 239 (1956)
64. H. Honda, K. Kobayash, S. Sugawara, Carbon **6**, 517 (1968)

65. C.L. Burket, R. Rajagopalan, H.C. Foley, Carbon **45**, 2307 (2007)
66. R. Franklin, Proc. Roy. Soc. **209**, 196 (1951)
67. M.J. Iqbal, M.N. Ashiq, J. Hazard. Mater. **B139**, 57 (2007)
68. G.G. Stavropoulos, A.A. Zabaniotou, Micropor. Mesopor. Mater. **82**, 79 (2005)
69. B. Bestani, N. Benderdouche, B. Benstaali, M. Belhakem, A. Addou, Bioresource Technol. **99**, 8441 (2008)
70. K. Marungrueng, P. Pavasant, Bioresource Technol. **98**, 1567 (2007)
71. M. Armandi, B. Bonelli, C.O. Arean, E. Garrone, Micropor. Mesopor. Mater. **112**, 411 (2008)
72. M. Rafatullah, O. Sulaiman, R. Hashim, A. Ahmad, J. Hazard. Mater. **177**, 70 (2010)
73. F. Banat, S. Al-Asheh, L. Al-Makhadmeh, Process Biochem. **39**, 193 (2003)
74. A. Özer, G. Dursun, J. Hazard. Mater. **146**, 262 (2007)
75. I.A.W. Tan, B.H. Hameed, A.L. Ahmad, Chem. Eng. J. **127**, 111 (2007)
76. A. Aygün, S. Yenisoy-Karakas, I. Duman, Micropor. Mesopor. Mater. **66**, 189 (2003)
77. S. Balci, T. Doğu, H. Yücel, J. Chem. Tech. Biotech. **60**, 419 (1994)
78. A.A. Attia, B.S. Girgis, N.A. Fathy, Dyes Pigm. **76**, 282 (2008)
79. K.S. Low, C.K. Lee, A.Y. Ng, Bioresource Technol. **68**, 205 (1999)
80. M.M. El-Halwany, Desalination **250**, 208 (2010)
81. T. Vaughan, C.W. Seo, W.E. Marshall, Bioresource Technol. **78**, 133 (2001)
82. C. Namasivayam, K. Kadirvelu, Ind. J. Chem. Technol. **5**, 334 (1998)
83. K. Periasamy, C. Namasivayam, Chemosphere **32**, 769 (1995)
84. A. Demirbas, J. Hazard. Mater. **167**, 1 (2009)
85. C. Kumaş, Z. Sayın, *16th International Symposium on Environmental Issues and Waste Management in Energy and Mineral Production*, (SWEMP2016–26, Istanbul, 2016)
86. K. Lewicka, Polish J. Chem. Technol. **19**, 38 (2017)
87. M.J. Livani, M. Ghorbani, H. Mehdipour, New Carbon Mater. **33**, 578 (2018)
88. B. Zhao, X. Xu, H. Li, X. Chen, F. Zeng, Bioresource Technol. **247**, 21 (2018)
89. G. Cimino, A. Passerini, G. Toscano, Water Res. **34**, 2955 (2000)
90. Y. Bulut, Z. Tez, J. Hazard. Mater. **149**, 35 (2007)
91. E. Pehlivan, T. Altun, S. Cetin, M.I. Bhanger, J. Hazard. Mater. **167**, 1203 (2009)
92. O. Demirbas, A. Karadag, M. Alkan, M. Dogan, J. Hazard. Mater. **153**, 677 (2008)
93. I. Gaballah, G. Kilbertus, J. Geochem. Explor. **62**, 241 (1998)
94. T. Altun, E. Pehlivan, Clean-Soil Air Water **35**, 601 (2007)
95. E. Pehlivan, T. Altun, J. Hazard. Mater. **155**, 378 (2008)
96. L. Lu, X. Jiang, L. Jia, T. Ai, H. Wu, Chem. Res. Chin. Univ. **33**, 112 (2017)
97. S.B. Ahmed, A. Stoica-Guzun, F.H. Kamar, T. Dobre, D. Gudovan, C. Busuioc, I.M. Jipa, Int. J. Environ. Sci. Technol. **16**, 1249 (2019)
98. M.M. Galehzan, S. Shamohammadi, J. Water Wastewater **24**, 71 (2013)
99. E. Demirbas, M. Kobya, S. Oncel, S. Sencan, Bioresource Technol. **84**, 291 (2002)
100. E. Demirbas, Adsorpt. Sci. Technol. **21**, 951 (2003)
101. M. Kazemipour, M. Ansari, S. Tajrobehkar, M. Majdzadeh, H.R. Kermani, J. Hazard. Mater. **150**, 322 (2008)
102. E. Sayan, Chem. Eng. J. **115**(213), 103104 (2006)
103. T.J. Mason, Ultrason. Sonochem. **10**, 175 (2003)
104. J.G. Price, *Current Trends in Sonochemistry* (The Royal Society of Chemistry, Cambridge, 1992)
105. L.A. Crum, T.J. Mason, J.L. Reisse, K.S. Suslick (eds.), *Sonochemistry and Sonoluminescence* (Kluwer Academic Publishers, Dordrecht, 1990)
106. R.V. Eldik, C.D. Hubbard, *Chemistry under Extreme or Non-classical Conditions* (John Wiley and Sons, New York, 1997)
107. T. Kimura, T. Sakamoto, J. Leveque, H. Sohmiya, M. Fujita, S. Ikeda, T. Ando, Ultrason. Sonochem. **3**, 157 (1996)
108. E. Şayan, J. Chem. Soc. Pak. **36**, 28 (2014)
109. D.D. Milenkovic, P.V. Dašic, V.B. Veljkovic, Ultrason. Sonochem. **16**, 557 (2009)
110. M. Kobya, Bioresource Technol. **91**, 317 (2004)

111. E. Demirbas, N. Dizge, M.T. Sulak, M. Kobya, Chem. Eng. J. **148**, 480 (2009)
112. Y. Bayrak, Y. Yesiloglu, U. Gecgel, Micropor. Mesopor. Mater. **91**, 107 (2006)
113. M.H. Kalavathy, T. Karthikeyan, S. Rajgopal, L.R. Miranda, J. Colloid. Interf. Sci. **292**, 354 (2005)
114. C. Selomulya, V. Meeyoo, R. Amal, J. Chem. Technol. Biotechnol. **74**, 111 (1999)
115. N.K. Hamadi, X.D. Chen, M.M. Farid, M.G.Q. Lu, Chem. Eng. J. **84**, 95 (2001)
116. K. Periasamy, C. Namasivayam, Chemosphere **32**, 769 (1996)
117. M.A. Ferro-Garcia, J. Rivera-Ultrilla, J. Rodriguez-Gordillo, I. Bautista-Toledo, Carbon **26**, 363 (1988)
118. K. Kadirvelu, K. Thamaraiselvi, C. Namasivayam, Sep. Purif. Technol. **24**, 497 (2001)
119. H. Hasar, J. Hazard. Mater. **B97**, 49 (2003)
120. D. Özçimen, A. Ersoy-Meriçboyu, Adsorpt. Sci. Technol. **28**, 327 (2010)
121. D.D. Milenković, M.M. Milosavljević, A.D. Marinković, V.R. Đokić, J.Z. Mitrović, A.L. Bojić, Water SA **39**, 515 (2013)
122. M. Imamoglu, J. Dispersion Sci. Technol. **34**, 1183 (2013)
123. A. Şencan, M. Karaboyacı, M. Kılıç, Environ. Sci. Pollut. Res. **22**, 3238 (2015)
124. A. Fencan, M. Kılıç, J. Chemistry. **651**, 1 (2015)
125. M. Imamoglu, H. Şahin, Ş Aydın, F. Tosunoğlu, H. Yılmaz, S.Z. Yıldız, Desal Water Treat. **57**, 4587 (2016)
126. M. Imamoglu, A. Ozturk, Ş Aydın, A. Manzak, A. Gündoğdu, C. Duran, J. Dispers. Sci. Technol. **39**, 1144 (2018)
127. B. Zhao, X. Xu, F. Zeng, H. Li, X. Chen, Environ. Sci. Pollut. Res. **25**, 19423 (2018)
128. T. Altun, H. Ecevit, B. Çiftçi, Arabian J. Geosci. **14**, 439 (2021)
129. F. Ferrero, J. Hazard. Mater. **142**, 144 (2007)
130. G.M. Walker, L.R. Weatherley, Water Res. **31**, 2093 (1997)
131. L. Markovska, V. Meshko, V. Noveski, Korean J Chem. Eng. **18**, 190 (2001)
132. F. Rozada, L.F. Calvo, A.I. Garcıa, J. Martın-Villacorta, M. Otero, Bioresource Technol. **87**, 221 (2003)
133. S. Netpradit, P. Thiravetyan, S. Towprayoon, Water Res. **38**, 71 (2004)
134. B.G.P. Kumar, L.R. Miranda, M. Velan, J. Hazard. Mater. **B126**, 63 (2005)
135. O. Ozdemir, M. Turan, A.Z. Turan, A. Faki, A.B. Engin, J. Hazard. Mater. **166**, 647 (2009)
136. R.A. Carletto, F. Chimirri, F. Bosco, F. Ferrero, BioResources **3**, 1146 (2008)
137. M. Dogan, H. Abak, M. Alkan, Water Air Soil Pollut. **192**, 141 (2008)
138. M. Dogan, H. Abak, M. Alkan, J. Hazard. Mater. **164**, 172 (2009)
139. O. Yavuz, A.H. Aydin, Fresenius Environ. Bull. **11**, 377 (2002)
140. O. Yavuz, A.H. Aydin, Polish J. Environ. Studies **15**, 155 (2006)
141. A.H. Aydin, O. Yavuz, Indian J Chem. Technol. **11**, 89 (2004)
142. C. Özer, M. Imamoglu, Y. Turhan, F. Boysan, Toxicolog. Environ. Chemist. **94**, 1283 (2012)
143. B.H. Hameed, A.T.M. Din, A.L. Ahmad, J. Hazard. Mater. **141**, 819 (2007)
144. D. Kavitha, C. Namasivayam, Bioresource Technol. **98**, 14 (2007)
145. S. Altenor, B. Carene, E. Emmanuel, J. Lambert, J.J. Ehrhardt, S. Gaspard, J. Hazard. Mater. **165**, 1029 (2009)
146. A. Khaled, A. El Nemr, A. El-Sikaily, O. Abdelwahab, Desalination **238**, 210 (2009)
147. M.P. Elizalde-González, V. Hernández- Montoya, Bioresource Technol. **100**, 2111 (2009)
148. A. El Nemr, O. Abdelwahab, A. El-Sikaily, A. Khaled, J. Hazard. Mater. **161**, 102 (2009)
149. G. Karaçetin, S. Sivrikaya, M. Imamoglu, J. Analytic. Appl. Pyrolysis **110**, 270 (2014)
150. N. Kaya, Z.Y. Uzun, Water Sci. Technol. **81**, 1191 (2020)
151. A.N. Al-Sharify, Z.H. Athab, A.F. Halbus, Iraqi Nation J. Chem. **51**, 273 (2013)

Index

H. Ünlü and N. J. M. Horing (eds.), *Progress in Nanoscale and Low-Dimensional Materials and Devices*, Topics in Applied Physics 144,
https://doi.org/10.1007/978-3-030-93460-6